T0172412

# Contemporary and Emerging Applications

## Volume II

# Chapman & Hall/CRC
# Computer and Information Science Series

*Series Editor: Sartaj Sahni*

**Computer-Aided Graphing and Simulation Tools for AutoCAD Users**
*P. A. Simionescu*

**Integration of Services into Workflow Applications**
*Paweł Czarnul*

**Handbook of Graph Theory, Combinatorial Optimization, and Algorithms**
*Krishnaiyan "KT" Thulasiraman, Subramanian Arumugam, Andreas Brandstädt, and Takao Nishizeki*

**From Action Systems to Distributed Systems**
The Refinement Approach
*Luigia Petre and Emil Sekerinski*

**Trustworthy Cyber-Physical Systems Engineering**
*Alexander Romanovsky and Fuyuki Ishikawa*

**X-Machines for Agent-Based Modeling**
FLAME Perspectives
*Mariam Kiran*

**From Internet of Things to Smart Cities**
Enabling Technologies
*Hongjian Sun, Chao Wang, and Bashar I. Ahmad*

**Evolutionary Multi-Objective System Design**
Theory and Applications
*Nadia Nedjah, Luiza De Macedo Mourelle, and Heitor Silverio Lopes*

**Networks of the Future**
Architectures, Technologies, and Implementations
*Mahmoud Elkhodr, Qusay F. Hassan, and Seyed Shahrestani*

**Computer Simulation**
A Foundational Approach Using Python
*Yahya E. Osais*

**Internet of Things**
Challenges, Advances, and Applications
*Qusay F. Hassan, Atta ur Rehman Khan, and Sajjad A. Madani*

**Handbook of Data Structures and Applications, Second Edition**
*Dinesh P. Mehta and Sartaj Sahni*

**Handbook of Approximation Algorithms and Metaheuristics, Second Edition**
Methodologies and Traditional Applications, Volume 1
*Teofilo F. Gonzalez*

**Handbook of Approximation Algorithms and Metaheuristics, Second Edition**
Contemporary and Emerging Applications, Volume 2
*Teofilo F. Gonzalez*

*For more information about this series please visit:*
*https://www.crcpress.com/Chapman--HallCRC-Computer-and-Information-Science-Series/book-series/CHCOMINFSCI*

# Handbook of Approximation Algorithms and Metaheuristics, Second Edition

## Contemporary and Emerging Applications

### Volume II

Edited by

## Teofilo F. Gonzalez

CRC Press

Taylor & Francis Group

Boca Raton  London  New York

CRC Press is an imprint of the
Taylor & Francis Group, an **informa** busines

A CHAPMAN & HALL BOOK

CRC Press
Taylor & Francis Group
6000 Broken Sound Parkway NW, Suite 300
Boca Raton, FL 33487-2742

© 2018 by Taylor & Francis Group, LLC
CRC Press is an imprint of Taylor & Francis Group, an Informa business

No claim to original U.S. Government works

Printed on acid-free paper

International Standard Book Number-13: 978-1-4987-6999-0 (Hardback)

This book contains information obtained from authentic and highly regarded sources. Reasonable efforts have been made to publish reliable data and information, but the author and publisher cannot assume responsibility for the validity of all materials or the consequences of their use. The authors and publishers have attempted to trace the copyright holders of all material reproduced in this publication and apologize to copyright holders if permission to publish in this form has not been obtained. If any copyright material has not been acknowledged please write and let us know so we may rectify in any future reprint.

Except as permitted under U.S. Copyright Law, no part of this book may be reprinted, reproduced, transmitted, or utilized in any form by any electronic, mechanical, or other means, now known or hereafter invented, including photocopying, microfilming, and recording, or in any information storage or retrieval system, without written permission from the publishers.

For permission to photocopy or use material electronically from this work, please access www.copyright.com (http://www.copyright.com/) or contact the Copyright Clearance Center, Inc. (CCC), 222 Rosewood Drive, Danvers, MA 01923, 978-750-8400. CCC is a not-for-profit organization that provides licenses and registration for a variety of users. For organizations that have been granted a photocopy license by the CCC, a separate system of payment has been arranged.

**Trademark Notice:** Product or corporate names may be trademarks or registered trademarks, and are used only for identification and explanation without intent to infringe.

**Visit the Taylor & Francis Web site at**
**http://www.taylorandfrancis.com**

**and the CRC Press Web site at**
**http://www.crcpress.com**

Printed and bound in the United States of America by Sheridan

*To my wife Dorothy, and our children:*
*Jeanmarie, Alexis, Julia, Teofilo, and Paolo.*

# Contents

## SECTION I  Computational Geometry and Graph Applications

## SECTION II  Large-Scale and Emerging Applications

# Preface

More than half a century ago the research community began analyzing formally the quality of the solutions generated by heuristics. The heuristics with guaranteed performance bounds eventually became known as approximation algorithms. The idea behind approximation algorithms was to develop procedures to generate provable near-optimal solutions to optimization problems that could not be solved efficiently by the computational techniques available at that time. With the advent of the theory of NP-completeness in the early 1970s, approximation algorithms became more prominent as the need to generate near optimal solutions for NP-hard optimization problems became the most important avenue for dealing with computational intractability. As it was established in the 1970s, for some problems one could generate near optimal solutions quickly, while for other problems it was established that generating provably good suboptimal solutions was as difficult as generating optimal ones. Other approaches based on probabilistic analysis and randomized algorithms became popular in the 1980s. The introduction of new techniques to solve linear programming problems started a new wave for developing approximation algorithms that matured and saw tremendous growth in the 1990s. To deal with the inapproximable problems, in a practical sense, there were a few techniques introduced in the 1980s and 1990s. These methodologies have been referred to as metaheuristics and may be viewed as problem independent methodologies that can be applied to sets of problems. There has been a tremendous amount of research in metaheuristics during the past three decades. During the last 25 years or so, approximation algorithms have attracted considerably more attention. This was a result of a stronger inapproximability methodology that could be applied to a wider range of problems and the development of new approximation algorithms. In the last decade there has been an explosion of new applications arising from most disciplines.

As we have witnessed, there has been tremendous growth in areas of approximation algorithms and metaheuristics. The second edition of this handbook includes new chapters, updated chapters and chapters with traditional content that did not warrant an update. For this second edition we have partitioned the handbook into two volumes. Volume 1 covers methodologies and traditional applications. Volume 2 covers contemporary and emerging applications. More specifically volume 1 discusses the different methodologies to design approximation algorithms and metaheuristics, as well as the application of these methodologies to traditional combinatorial optimization problems. Volume 2 discusses application of these methodologies to classical problems in computational geometry and graphs theory, as well as in large-scale and emerging application areas. Chapter 1 in both of these volumes presents an overview of approximation algorithms and metaheuristics as well as as an overview of both volumes of this handbook.

It has been a decade since the first edition and our authors have experienced all sorts of different transitions. Several authors expanded their families while writing chapters for the first edition. The babies born at that time are now more than ten years old! A few of the authors and the editor have now retired from their day to day obligations, but continue to be active in research. A couple of the authors became presidents of universities while others now hold prestigious chaired positions or high level positions at their institutions. But sadly, Rajeev Motwani and Ivan Stojmenovic, well-known researchers and authors of first edition chapters, passed away. Since their chapters did not changed significantly, they remain as co-authors of their chapters. Also, Imreh Csanád, a new author for the second edition, passed away in 2017. They are all missed greatly by their families, friends, and the entire research community.

We have collected in this volume a large amount of material with the goal of making it as complete as possible. We apologize in advance for any omissions and would like to invite all of you to propose new chapters for future editions of this handbook. Our research area will continue to grow and we are confident that the following words from an old song "The best is yet to come, you ain't seen nothing yet..." applies to our research area. We look forward to the next decade in which new challenges and opportunities await the new, as well as the established, researchers. We look forward to working on problems arising in new emerging applications and editing the third edition of this handbook.

I like to acknowledge the University of California, Santa Barbara (UCSB) for providing me the time and support needed to develop the first and second editions of this handbook for the past 12 years. I also want to thank my wife, Dorothy, our daughters, Jeanmarie, Alexis and Julia, and our sons, Teofilo and Paolo for their moral support, love, encouragement, understanding, and patience, throughout the project and my tenure at UCSB.

Teofilo F Gonzalez
Professor Emeritus of Computer Science
University of California, Santa Barbara

# Contributors

**Abdullah N. Arslan**
Department of Computer Science
Texas A & M University
Commerce, Texas

**Sudha Balla**
Department of Computer Science and
    Engineering
University of Connecticut
Storrs, Connecticut

**Roberto Battiti**
Department of Computer Science and
    Telecommunications
University of Trento
Trento, Italy

**Alan A. Bertossi**
Department of Computer Science
University of Bologna
Bologna, Italy

**Maria J. Blesa**
ALBCOM, Computer Science Department
Universitat Politècnica de Catalunya (UPC) -
    BarcelonaTech
Barcelona, Spain

**Christian Blum**
ALBCOM, Dept. Llenguatges i
    Sistemes Informátics
Universitat Politècnica de Catalunya (UPC) -
    BarcelonaTech
Barcelona, Spain

**Gruia Călinescu**
Department of Computer Science
Illinois Institute of Technology
Chicago, Illinois

**Kun-Mao Chao**
Department of Computer Science and
    Information Engineering
National Taiwan University
Taipei, Taiwan, Republic of China

**Ting Chen**
Department of Computer Science and
    Technology
Tsinghua University
Beijing, China

**Marco Chiarandini**
Department of Mathematics and
    Computer Science
University of Southern Denmark
Odense, Denmark

**Jason Cong**
Department of Computer Science
University of California
Los Angeles, California

**Artur Czumaj**
Department of Computer Science
University of Warwick
Coventry, United Kingdom

**Gianlorenzo D'Angelo**
Gran Sasso Science Institute
L'Aquila, Italy

**Jaime Davila**
Department of Computer Science and
    Engineering
University of Connecticut
Storrs, Connecticut

**Irina Dumitrescu**
School of Mathematics
University of New South Wales
Sydney, Australia

**Ömer Eğecioğlu**
Department of Computer Science
University of California
Santa Barbara, California

**Cristina G. Fernandes**
Department of Computer Science
University of Sao Paulo
Sao Paulo, Brazil

**Anurag Garg**
Department of Computer Science and
    Telecommunications
University of Trento
Trento, Italy

**Hossein Ghasemalizadeh**
Department of Computer Engineering
Shahid Bahonar University
Kerman, Iran

**Debasish Ghose**
Department of Aerospace Engineering
Indian Institute of Science
Bangalore, India

**Teofilo F. Gonzalez**
Department of Computer Science
University of California
Santa Barbara, California

**Arturo Gonzalez-Gutierrez**
Facultad de Ingeniería
Universidad Autónoma de Querétaro
Querétaro, México

**Joachim Gudmundsson**
University of Sydney
Sydney, Australia

**Sudipto Guha**
Department of Computer
    Information Sciences
University of Pennsylvania
Philadelphia, Pennsylvania

**Yao-Ting Huang**
Department of Computer Science and
    Information Engineering
National Chung Cheng University
Chiayi County, Taiwan, Republic of China

**Ragesh Jaiswal**
Department of Computer Science and
    Engineering
Indian Institute of Technology Delhi
New Delhi, India

**Konstantinos G. Kakoulis**
Department of Industrial Design Engineering
Western Macedonia
University of Applied Sciences
Florina, Greece

**Christian Knauer**
Universität Bayreuth
Bayreuth, Germany

**Stavros G. Kolliopoulos**
Department of Informatics and
    Telecommunications
National and Kapodistrian
University of Athens
Athens, Greece

**Andrzej Lingas**
Department of Computer Science
Lund University
Lund, Sweden

**Errol L. Lloyd**
Department of Computer and
    Information Sciences
University of Delaware
Newark, Delaware

**Ion Mandoiu**
Department of Computer Science
University of Connecticut
Storrs, Connecticut

**Alfredo Navarra**
Department of Computer Science and
 Mathematics
University of Perugia
Perugia, Italy

**Marius Nicolae**
Department of Computer Science and
 Engineering
University of Connecticut
Storrs, Connecticut

**Sotiris E. Nikoletseas**
Computer Engineering &
 Informatics Department
Patras University
Patras, Greece

**Zeev Nutov**
Department of Mathematics and
 Computer Science
The Open University of Israel
Raanana, Israel

**Stephan Olariu**
Department of Computer Science
Old Dominion University
Norfolk, Virginia

**Alex Olshevsky**
Department of Electrical and
 Computer Engineering
Boston University
Boston, Massachusetts

**Cristina M. Pinotti**
Department of Computer Science and
 Mathematics
University of Perugia
Perugia, Italy

**Sanguthevar Rajasekaran**
Department of Computer Science and
 Engineering
University of Connecticut
Storrs, Connecticut

**Christoforos L. Raptopoulos**
Patras University
Patras, Greece

**S. S. Ravi**
Biocomplexity Institute of Virginia Tech
Blacksbury, Virginia

**Mohammadreza Razzazi**
Department of Computer Engineering
Amirkabir University of Technology
Tehran, Iran

**Andréa W. Richa**
Department of Computer Science
Arizona State University
Tempe, Arizona

**Romeo Rizzi**
Department of Computer Science
University of Verona
Verona, Italy

**Daniel J. Rosenkrantz**
University at Albany – SUNY
Albany, New York

**Pedro M. Ruiz**
DIIC
University of Murcia
Murcia, Spain

**Christian Scheideler**
Department of Computer Science
University of Paderborn
Paderborn, Germany

**Stefan Schmid**
Department of Computer Science
Aalborg University
Aalborg, Denmark

**Sandeep Sen**
Department of Computer Science and
 Engineering
Indian Institute of Technology Delhi
New Delhi, India

**Joseph R. Shinnerl**
Mentor Graphics Corporation
Freemont, California

**Michiel Smid**
School of Computer Science
Carleton University
Ottawa, Canada

**Paul G. Spirakis**
Computer Engineering &
    Informatics Department
Patras University
Patras, Greece

**Ivan Stojmenovic**
University of Ottawa
Ottawa, Canada

**Thomas Stützle**
Institut de Recherches Interdisciplinaires
    et de Développements en Intelligence
    Artificielle (IRIDIA)
Université Libre de Bruxelles
Brussels, Belgium

**Chuan Yi Tang**
Department of Computer Science
National Tsing Hua University
Hsinchu, Taiwan, Republic of China

**Giri K. Tayi**
University at Albany – SUNY
Albany, New York

**Joseph Thomas**
Department of Aerospace Engineering
Indian Institute of Science
Bangalore, India

**Ioannis G. Tollis**
Department of Computer Science
University of Crete
Rethimno, Greece

**Sarat Chandra Varanasi**
Department of Computer Science
University of Texas at Dallas
Richardson, Texas

**Lan Wang**
Department of Computer Science
Old Dominion University
Norfolk, Virginia

**Xianping Wang**
Department of Computer Science
Old Dominion University
Norfolk, Virginia

**Bang Ye Wu**
Department of Computer Science and
    Information Engineering
National Chung Cheng University
Chiayi County, Taiwan,
Republic of China

**Weili Wu**
Department of Computer Science
University of Texas at Dallas
Richardson, Texas

**Zhigang Xiang**
Queens College of the City
    University of New York
Flushing, New York

**Wen Xu**
Department of Mathematics and
    Computer Science
Texas Woman's University
Denton, Texas

**Jing Yuan**
Department of Computer Science
University of Texas at Dallas
Richardson, Texas

**Alex Zelikovsky**
Department of Computer Science
Georgia State University
Atlanta, Georgia

**Kui Zhang**
Department of Mathematical Sciences
Michigan Technological University
Houghton, Michigan

**Si Qing Zheng**
Department of Computer Science
University of Texas at Dallas
Richardson, Texas

# 1

# Introduction, Overview, and Notation

Teofilo F. Gonzalez

## 1.1 Introduction

Approximation algorithms were formally introduced in the 1960s to generate near-optimal solutions to optimization problems that could not be solved efficiently by the computational techniques available at that time. With the advent of the theory of NP-completeness in the early 1970s, the area became more prominent as the need to generate near-optimal solutions for NP-hard optimization problems became the most important avenue for dealing with computational intractability. As it was established in the 1970s, for some problems it is possible to generate near-optimal solutions quickly, whereas for other problems generating provably good suboptimal solutions is as difficult as generating optimal ones. Computational approaches based on probabilistic analysis and randomized algorithms became popular in the 1980s. The introduction of new techniques to solve linear programming problems started a new wave of approximation algorithms that matured and saw tremendous growth in the 1990s. There were a few techniques introduced in the 1980s and 1990s to deal, in the practical sense, with inapproximable problems. These methodologies have been referred to as metaheuristics and include Simulated Annealing (SA), Ant Colony Optimization (ACO), Evolutionary Computation (EC), Tabu Search (TS), Memetic Algorithms (MAs), and so on. Other previously established methodologies such as local search, backtracking, and branch-and-bound were also explored at that time. There has been a tremendous amount of research in metaheuristics during the past three decades. These techniques have been evaluated experimentally and have demonstrated their usefulness for solving problems that are arising in practice. During the last 25 years or so, approximation algorithms have attracted considerably more attention. This was a result of a stronger inapproximability methodology that could be applied to a wider range of problems and the development of new approximation algorithms for problems arising in established and emerging application areas. Polynomial Time Approximation Schemes (PTASs) were introduced in the 1960s and the more powerful Fully Polynomial Time Approximation Schemes (FPTASs) were introduced in the 1970s. Asymptotic PTAS (APTAS) and Asymptotic FPTAS (AFPTAS), and Fully Polynomial Randomized Approximation Schemes (FPRASs) were introduced later on.

Today approximation algorithms enjoy a stature comparable to that of algorithms in general and the area of metaheuristics has established itself as an important research area. The new stature is a byproduct of a natural expansion of research into more practical areas where solutions to real-world problems are expected, as well as by the higher level of sophistication required to design and analyze these new procedures. The goal of approximation algorithms and metaheuristics is to provide the best possible solutions and to guarantee that such solutions satisfy certain criteria. This two-volume handbook houses these two approaches and thus covers all the aspects of approximations. We hope it will serve you as a valuable reference for approximation methodologies and applications.

Approximation algorithms and metaheuristics have been developed to solve a wide variety of problems. A good portion of these algorithms have only theoretical value due to the fact that their time complexity is a high-order polynomial or they have a huge constant associated with their time complexity bound. However, these results are important because they establish what is possible, and it may be that in the near future these algorithms will be transformed into practical ones. Other approximation algorithms do not suffer from this pitfall, but some were designed for problems with limited applicability. However, the remaining approximation algorithms have real-world applications. Given this, there is a huge number of important application areas, including new emerging ones, where approximation algorithms and metaheuristics have barely penetrated and we believe there is an enormous potential for their use. Our goal is to collect a wide portion of the approximation algorithms and metaheuristics in as many areas as possible, as well as to introduce and explain in detail the different methodologies used to design these algorithms.

## 1.2 Overview

Our overview in this section is devoted mainly to the earlier years. The individual chapters in the two volumes discuss in detail the recent research accomplishments in different subareas. This section will also serve as an overview of both volumes of this handbook. Volume 1, Chapter 2 discusses some of the basic methodologies and applies them to classical problems.

Even before the 1960s, researchers in applied mathematics and graph theory had established upper and lower bounds for certain properties of graphs. For example, bounds had been established for the chromatic number, achromatic number, chromatic index, maximum clique, maximum independent set, and so on. Some of these results could be seen as the precursors of approximation algorithms. By the 1960s it was understood that there were problems that could be solved efficiently, whereas for other problems all the known algorithms required exponential time in the worst case. Heuristics were being developed to find quick solutions to problems that appeared to be computationally difficult to solve. Researchers were experimenting with heuristics, branch-and-bound procedures, and iterative improvement frameworks and were evaluating their performance when solving actual problem instances. There were many claims being made, not all of which could be substantiated, about the performance of the procedures being developed to generate optimal and suboptimal solutions to combinatorial optimization problems.

Half a century ago (1966), Ronald L. Graham [1] formally introduced approximation algorithms. He analyzed the performance of list schedules for scheduling tasks on identical machines, a fundamental problem in scheduling theory.

*Problem*: Scheduling tasks on identical machines.

*Instance*: Set of $n$ tasks $(T_1, T_2, \ldots, T_n)$ with processing time requirements $t_1, t_2, \ldots, t_n$, partial order $C$ defined over the set of tasks to enforce task dependencies, and a set of $m$ identical machines.

*Objective*: Construct a schedule with minimum makespan. A *schedule* is an assignment of tasks to time intervals on the machines in such a way that (1) each task $T_i$ is processed continuously for $t_i$ units of time by one of the machines, (2) each machine processes at most one task at a time, and (3) the precedence constraints are satisfied (i.e., machines cannot commence the processing of a task until all of its predecessors have been completed). The *makespan* of a schedule is the time at which all the machines have completed processing the tasks.

The *list-scheduling* procedure is given an ordering of the tasks specified by a list $L$. Then the procedure finds the earliest time $t$ when a machine is idle, and an unassigned task is available (i.e., all its predecessor tasks have been completed). It assigns the leftmost available task in the list $L$ to an idle machine at time $t$ and this step is repeated until all the tasks have been scheduled.

The main result in Reference 1 was proving that for every problem instance $I$, the schedule generated by this policy has a makespan that is bounded above by $(2 - 1/m)$ times the optimal makespan for the instance. This is called the *approximation ratio* or *approximation factor* for the algorithm. We also say that the algorithm is a $(2 - 1/m)$-approximation algorithm. This criterion for measuring the quality of the solutions generated by an algorithm remains as one of the most important ones in use today. The second contribution in Reference 1 was showing that the approximation ratio $(2-1/m)$ is the best possible for list schedules, that is, the analysis of the approximation ratio for this algorithm cannot be improved. This was established by presenting problem instances (for all $m$ and $n \geq 2m - 1$) and lists for which the schedule generated by the procedure has a makespan equal to $2-1/m$ times the optimal makespan for the instance. A restricted version of the list-scheduling algorithm is analyzed in detail in Volume 1, Chapter 2.

The third important aspect of the results in Reference 1 was showing that list scheduling may have anomalies. To explain this we need to define some terms. The makespan of the list schedule for instance $I$ using list $L$ is denoted by $f_L(I)$. Suppose that instance $I'$ is a slightly modified version of instance $I$. The modification is such that we intuitively expect that $f_L(I') \leq f_L(I)$. But this is not always true, so there is an anomaly. For example, suppose that $I'$ is $I$, except that $I'$ has an additional machine. Intuitively $f_L(I') \leq f_L(I)$ because with one additional machine, tasks should be finished earlier or at worst at the same time as when there is one fewer machine. But this is not always the case for list schedules, there are problem instances and lists for which $f_L(I') > f_L(I)$. This is called an *anomaly*. Our expectation would be valid if list scheduling would generate minimum makespan schedules, but we have a procedure that generates suboptimal solutions. Such guarantees are not always possible in this type of environment. List schedules suffer from other anomalies, for example, relaxing the precedence constraints or decreasing the execution time of the tasks. In both cases, one would expect schedules with smaller or the same makespans. But, that is not always the case. Volume 1, Chapter 2 presents problem instances where anomalies occur. The main reason for discussing anomalies now is that even today, numerous papers are being published and systems are being deployed where "common sense"-based procedures are being introduced without any analytical justification and/or thorough experimental validation. Anomalies show that since we live for the most part in a "suboptimal world," the effect of our decisions is not always the intended one (unintended consequences). One can design approximation algorithms that do not suffer from certain types of anomalies but probably not for all possible ones.

Other classical problems with numerous applications are the traveling salesperson, Steiner tree, and spanning tree problems, which will be formally defined later on. Even before the 1960s there were several well-known polynomial time algorithms to construct minimum weight spanning trees for edge-weighted graphs [2]. These simple greedy algorithms have low-order polynomial time complexity bounds. It was well known at that time that the same type of procedures does not generate an optimal tour for the traveling salesperson problem (TSP) and does not construct optimal Steiner trees. However, in 1968 E. F. Moore (as discussed in Reference 3) showed that for any set of points $P$ in metric space $L_M \leq L_T \leq 2L_S$, where $L_M$, $L_T$, and $L_S$ are the total weight of a minimum weight spanning tree, a minimum weight tour (solution) for the TSP, and minimum weight Steiner tree for $P$, respectively. Since every spanning tree is a Steiner tree, the above-mentioned bounds show that when using a minimum weight spanning tree to approximate the Steiner tree we have a solution (Steiner tree) whose weight is at most twice the weight of an optimal Steiner tree. In other words, any algorithm that generates a minimum weight spanning tree is a 2-approximation algorithm for the Steiner tree problem. Furthermore, this approximation algorithm takes no more time than an algorithm that constructs a minimum weight spanning tree for edge weighted graphs [2], as such an algorithm can be used to construct an optimal spanning tree for a set of points in metric space. The above-mentioned bound is established by defining a transformation from any minimum weight Steiner tree into a TSP tour with weight at most $2L_S$. Therefore, $L_T \leq 2L_S$ [3]. Then by observing that the deletion of an edge in an optimum tour for the TSP problem results in a spanning tree, it follows that $L_M < L_T$. Volume 1, Chapter 3 discusses this approximation algorithm and its analysis in more detail. The Steiner ratio is defined as $L_S/L_M$. The earlier arguments show that the Steiner ratio is at least $\frac{1}{2}$. Gilbert and Pollak [3] conjectured that the Steiner ratio in the Euclidean plane equals $\frac{\sqrt{3}}{2}$ (the $0.86603\ldots$ conjecture). A proof of this conjecture and improved approximation algorithms for the Steiner tree problem are discussed in Volume 1, Chapter 36.

The above-mentioned constructive proof can be applied to a minimum weight spanning tree to generate a tour for the TSP problem. The construction takes polynomial time and results in a 2-approximation algorithm for the TSP problem. This approximation algorithm for the TSP is also referred to as the *double spanning tree algorithm* and is discussed in Volume 1, Chapters 3 and 27. Improved approximation algorithms for the TSP and algorithms for its generalizations are discussed in Volume 1, Chapters 3, 27, 34, 35, and Volume 2, Chapter 2. The approximation algorithm for the Steiner tree problem just discussed is explained in Volume 1, Chapter 3, and improved approximation algorithms and applications are discussed in Volume 1, Chapters 36, 37, and Volume 2, Chapter 2. Volume 2, Chapter 14, discusses approximation algorithms for computationally intractable variations of the spanning tree problem.

In 1969, Graham [4] studied the problem of scheduling tasks on identical machines but restricted to independent tasks, that is, the set of precedence constraints is empty. He analyzed the Largest Processing Time first (LPT) scheduling rule, which is list scheduling where the list of tasks $L$ is arranged in nonincreasing order of their processing requirements. His elegant proof established that the LPT procedure generates a schedule with makespan at most $\frac{4}{3} - \frac{1}{3m}$ times the makespan of an optimal schedule, that is, the LPT scheduling algorithm has a $\frac{4}{3} - \frac{1}{3m}$ approximation ratio. He also showed that the analysis is best possible for all $m$ and $n \geq 2m+1$. For $n \leq 2m$ tasks the approximation ratio is smaller and under some conditions, LPT generates an optimal makespan schedule. Graham [4], following a suggestion by D. Kleitman and D. Knuth, considered list schedules where the first portion of the list $L$ consists of $k$ tasks (without loss of generality, assume $k$ is a multiple of $m$) with the longest processing times arranged by their starting times in an optimal schedule for these $k$ tasks (only). Then the list $L$ has the remaining $n - k$ tasks in any order. The approximation ratio for this list schedule using list $L$ is $1 + \frac{m-1}{m+k}$. An optimal schedule for the longest $k$ tasks can be constructed in $O(n + m^k)$ time by a straight forward branch and bound algorithm. In other words, this algorithm has an approximation ratio $1 + \epsilon$ and time complexity $O(n + m^{(m-1-\epsilon m)/\epsilon})$. For any fixed constants $m$ and $\epsilon$, the algorithm constructs in polynomial (linear) time with respect to $n$, a schedule with makespan at most $1 + \epsilon$ times the optimal makespan. Note that for a fixed constant $m$ the time complexity is polynomial with respect to $n$, but it is not polynomial with respect to $1/\epsilon$. This was the first algorithm of its kind and later on, it was called *polynomial time approximation scheme* (PTAS). Volume 1, Chapter 8 discusses different PTASs. Additional PTAS appear in Volume 1, Chapter 36 and Volume 2, Chapters 2 and 5. The proof techniques presented in References 1, 4 are outlined in Volume 1, Chapter 2 and have been extended to apply to other problems. There is an extensive body of literature for approximation algorithms and metaheuristics for scheduling problems. Volume 1, Chapters 38, 39, and Volume 2, Chapter 25 discuss interesting approximation algorithms and heuristics for scheduling problems. The scheduling handbook [5] is an excellent source for scheduling algorithms, models, and performance analysis.

The development of NP-completeness theory in the early 1970s by Cook [6], Karp [7], and others formally introduced the notion that there is a large class of decision problems (the answer to these problems is a simple yes or no) that are computationally equivalent. This means that either every problem in this class has a polynomial time algorithm that solves it, or none of them do. Furthermore, this question is the same as the $P = NP$ question, a classical open problem in computational complexity. This question is to determine whether or not the set of languages recognized in polynomial time by deterministic Turing machines is the same as the set of languages recognized in polynomial time by nondeterministic Turing machines. The conjecture has been that $P \neq NP$, and thus the hardest problems in NP would not be solvable in polynomial time. The computationally equivalent decision problems in this class are called *NP-complete* problems. The scheduling on identical machines problem discussed earlier is an optimization problem. Its corresponding decision problem has its input augmented by an integer value $B$, and the yes–no question is to determine whether or not there is a schedule with makespan at most $B$. An optimization problem whose corresponding decision problem is NP-complete is called an *NP-hard* problem. Therefore, scheduling tasks on identical machines is an NP-hard problem. The TSP and the Steiner tree problem are also NP-hard problems. The minimum weight spanning tree problem can be solved in polynomial time and it is not an NP-hard problem under the assumption that $P \neq NP$. The next section discusses NP-completeness in more detail. There is a long list of practical problems arising in many different fields of study that are known to be NP-hard problems [8]. Because of this, the need to cope with these computationally intractable problems was recognized earlier on. Since then approximation algorithms became a central area of research activity. Approximation algorithms offered a way to circumvent computational intractability by paying a price when it comes to the quality of the solutions generated. But a solution can be generated quickly. In other words and another language, "no te fijes en lo bien, fijate en lo rápido." Words used to describe my golf playing ability when I was growing up.

In the early 1970s, Garey et al. [9] as well as Johnson [10,11] developed the first set of polynomial time approximation algorithms for the bin packing problem. The analysis of the approximation ratio for these algorithms is asymptotic, which is different from those for the scheduling problems discussed earlier. We will define this notion precisely in the next section, but the idea is that the ratio holds when the value of an optimal solution is greater than some constant. Research on the bin packing problem and its variants has attracted very talented investigators who have generated more than 1000 papers, most of which deal with approximations. This work has been driven by numerous applications in engineering and information sciences (Volume 1, Chapters 28, 29, 30, and 31).

Johnson [12] developed polynomial time algorithms for the sum of subsets, max satisfiability, set cover, graph coloring, and max clique problems. The algorithms for the first two problems have a constant ratio approximation, but for the other problems the approximation ratio is ln $n$ and $n^\epsilon$. Sahni [13,14] developed a PTAS for the knapsack problem. Rosenkrantz et al. [15] developed several constant ratio approximation algorithms for the TSP that satisfy the triangle inequality (or simply, defined over metric graphs). This version of the problem is defined over edge weighted complete graphs, rather than for points in metric space as in Reference 3. These algorithms have an approximation ratio of two.

Sahni and Gonzalez [16] showed that there were a few NP-hard optimization problems for which the existence of a constant ratio polynomial time approximation algorithm implies the existence of a polynomial time algorithm to generate an optimal solution. In other words, complexity of generating a constant ratio approximation and an optimal solution are computationally equivalent problems. For these problems, the approximation problem is NP-hard or simply inapproximable (under the assumption that $P \neq NP$). Later on, this notion was extended to mean that there is no polynomial time algorithm with approximation ratio $r$ for a problem under some complexity theoretic hypothesis. The approximation ratio $r$ is called the *inapproximability ratio* (see Chapter 17 in the first edition of this handbook).

The $k$-min-cluster problem is one of these inapproximable problems. Given an edge-weighted undirected graph, the $k$-min-cluster problem is to partition the set of vertices into $k$ sets so as to minimize the sum of the weight of the edges with endpoints in the same set. The $k$-maxcut problem is defined as the $k$-min-cluster problem, except that the objective is to maximize the sum of the weight of the edges with endpoints in different sets. Even though these two problems have exactly the same set of feasible and optimal solutions, there is a linear time algorithm for the $k$-maxcut problem that generates $k$-cuts with weight at least $\frac{k-1}{k}$ times the weight of an optimal $k$-cut [16], whereas approximating the $k$-min-cluster problem is a computationally intractable problem. The former problem has the property that a near-optimal solution may be obtained as long as partial decisions are made optimally, whereas for the $k$-min-cluster an optimal partial decision may turn out to force a terrible overall solution. For the $k$-min-cluster problem if one makes a mistake at some iteration, one will end up with a solution that is far from optimal. Whereas for the $k$-maxcut problem one can make many mistakes and still end up with a near-optimal solution. A similar situation arises when you make a mistake an exam where almost everyone receives a perfect score, versus a course where the average score is about 50% of the points.

Another interesting problem whose approximation problem is NP-hard is the TSP problem [16]. This is not exactly the same version of the TSP problem discussed earlier, which we said has several constant ratio polynomial time approximation algorithms. Given an edge-weighted undirected graph, the TSP is to find a least weight tour, that is, to find a least weight (simple) path that starts at vertex 1, visits each vertex in the graph *exactly* once, and ends at vertex 1. The weight of a path is the sum of the weight of its edges. The weights of the edges are unrelated and the approximation problem is NP-hard. The version of the TSP problem studied in Reference 15 is limited to metric graphs, that is, the graph is complete (all the edges are present) and the set of edge weights satisfies the triangle inequality (which means that the weight of the edge joining vertex $i$ and $j$ is less than or equal to the weight of any path from vertex $i$ to vertex $j$). This version of the TSP problem is equivalent to the one studied by E. F. Moore [3]. The approximation algorithms given in References 3, 15 can be easily adapted to provide a constant ratio approximation to the version of the TSP problem where the tour is defined as visiting each vertex in the graph *at least* once. Since Moore's approximation algorithms for the metric Steiner tree and metric TSP are based on the same idea, one would expect that the Steiner tree problem defined over arbitrarily weighted graphs is NP-hard to approximate. However, this is not the case. Moore's algorithm [3] can be modified to be a 2-approximation algorithm for this more general Steiner tree problem.

As pointed out in Reference 17, Levner and Gens [18] added a couple of problems to the list of problems that are NP-hard to approximate. Garey and Johnson [19] show that the max clique problem has the property that if for some constant $r$ there is a polynomial time $r$-approximation algorithm, then there is a polynomial time $r'$-approximation for any constant $r'$ such that $0 < r' < 1$. Since that time researchers have tried many different algorithms for the clique problem, none of which were constant ratio approximation algorithms, and it was conjectured that none existed under the assumption that $P \neq NP$. This conjecture has been proved.

A PTAS is said to be an FPTAS if its time complexity is polynomial with respect to $n$ (the problem size) and $1/\epsilon$. The first FPTAS was developed by Ibarra and Kim [20] for the knapsack problem. Sahni [21] developed three different techniques based on rounding, interval partitioning, and separation to construct FPTAS for sequencing and scheduling problems. These techniques have been extended to other problems and are discussed in Volume 1, Chapter 9. Horowitz and Sahni [22] developed FPTAS for scheduling on processors with different processing speeds. Reference 17 discusses a simple $O(n^3/\epsilon)$ FPTAS for the knapsack problem developed by Babat [23,24]. Lawler [25] developed techniques for maximum speed-up FPTAS for the knapsack and related problems. Volume 1, Chapter 9 presents different methodologies to design FPTAS. Garey and Johnson [26] showed that if any problem in a class of NP-hard optimization problems that satisfy certain properties has a FPTAS, then $P = NP$. The properties are that the objective function value of every feasible solution is a positive integer, and the problem is *strongly* NP-hard. A problem is strongly NP-hard if the problem is NP-hard even when the magnitude of the maximum number in the input is bounded by a polynomial on the input length. For example, the TSP problem is strongly NP-hard, whereas the knapsack problem is not, under the assumption that $P \neq NP$ (see Volume 1, Chapter 9).

Lin and Kernighan [27] developed elaborate heuristics that established experimentally that instances of the TSP with up to 110 cities can be solved to optimality with 95% confidence in $O(n^2)$ time. This was an iterative improvement procedure applied to a set of randomly selected feasible solutions. The process was to perform $k$ pairs of link (edge) interchanges that improved the length of the tour. However, Papadimitriou and Steiglitz [28] showed that for the TSP no local optimum of an efficiently searchable neighborhood can be within a constant factor of the optimal value unless $P = NP$. Since then there has been quite a bit of research activity in this area. Deterministic and stochastic local search in efficiently searchable, as well as in very large neighborhoods, are discussed in Volume 1, Chapters 16, 17, 18, and 19. Volume 1, Chapter 13 discusses issues relating to the empirical evaluation of approximation algorithms and metaheuristics.

Perhaps the best known approximation algorithm for the TSP defined over metric graphs is the one by Christofides [29]. The approximation ratio for this algorithm is $\frac{3}{2}$, which is smaller than the approximation ratio of 2 for the algorithms reported in References 3, 15. However, looking at the bigger picture that includes the time complexity of the approximation algorithms, Christofides algorithm is not of the same order as the ones given in References 3, 15. Therefore, neither approximation algorithm dominates the other as one has a smaller time complexity bound, whereas the other (Christofides algorithm) has a smaller worst case approximation ratio.

Ausiello et al. [30] introduced the differential ratio, which is another way of measuring the quality of the solutions generated by approximation algorithms. Differential ratio destroys the artificial dis-symmetry between "equivalent" minimization and maximization problems (e.g., the $k$-maxcut and the $k$-min cluster discussed earlier) when it comes to approximation. This ratio uses the difference between the worst possible solution minus the solution generated by the algorithm, divided by the difference between the worst solution minus the best solution. Cornuejols et al. [31] also discussed a variation of differential ratio approximations. They wanted the ratio to satisfy the following property: "A modification of the data that adds a constant to the objective function value should also leave the error measure unchanged." That is, the "error" by the approximation algorithm should be the same as before. Differential ratio and its extensions are discussed in Volume 1, Chapter 15, along with other similar notions [30]. Ausiello et al. [30] introduced *reductions that preserve approximability*. Since then there have been several new types of approximation preserving reductions. The main advantage of these reductions is that they enable us to define large classes of optimization problems that behave in the same way with respect to approximation. Informally, the class of NP Optimization (**NPO**) problems, is the set of all optimization problems Π which can be "recognized" in polynomial time (see Volume 1, Chapter 14 for a formal definition). An **NPO** problem Π is said to be in **APX**, if it has a constant approximation ratio polynomial time algorithm. The class **PTAS** consists of all **NPO** problems which have PTAS. The class **FPTAS** is defined similarly. Other classes, **Poly-APX**, **Log-APX**, and **Exp-APX**, have also been defined (see Volume 1, Chapter 14).

One of the main accomplishments at the end of the 1970s was the development of a polynomial time algorithm for Linear Programming (LP) problems by Khachiyan [32]. This result had a tremendous impact on approximation algorithm research and started a new wave of approximation algorithms. Two subsequent research accomplishments were at least as significant as Khachiyan's [32] result. The first one was a faster polynomial time algorithm for solving linear programming problems developed by Karmakar [33]. The other major accomplishment was the work of Grötschel et al. [34,35]. They showed that it is possible to solve a linear programming problem with an exponential number of constraints (with respect to the number of variables) in time which is polynomial in the number of variables and the number of bits used to describe the input, given a *separation oracle* plus a bounding ball and a lower bound on the volume of the feasible solution space. Given a solution, the separation oracle determines in polynomial time whether or not the solution is feasible, and if it is not it finds a constraint that is violated. Volume 1, Chapter 10 gives an example of the use of this approach. Important developments have taken place during the past 30 years. The books [35,36] are excellent references for linear programming theory, algorithms, and applications.

Because of the above-mentioned results, the approach of formulating the solution to an NP-hard problem as an integer linear programming problem and so solving the corresponding linear programming problem became very popular. This approach is discussed in Volume 1, Chapter 2. Once a fractional solution is obtained, one uses rounding to obtain a feasible solution to the original NP-hard problem. The rounding may be deterministic or randomized, and it may be very complex (meta-rounding). LP rounding is discussed in Volume 1, Chapters 2, 4, 7, 8, 10, 11, and Volume 2, Chapters 8 and 11.

Independently, Johnson [12] and Lovász [37] developed efficient algorithms for the set cover with approximation ratio of $1 + \ln d$, where $d$ is the maximum number of elements in each set. Chvátal [38] extended this result to the weighted set cover problem. Subsequently, Hochbaum [39] developed an algorithm with approximation ratio $f$, where $f$ is the maximum number of sets containing any of the elements in the set. This result is normally inferior to the one by Chvátal [38], but it is more attractive for the weighted vertex cover problem, which is a restricted version of the weighted set cover. For this subproblem it is a 2-approximation algorithm. A few months after Hochbaum's initial result,* Bar-Yehuda and Even [40] developed a primal-dual algorithm with the same approximation ratio as the one in Reference 39. The algorithm in Reference 40 does not require the solution of an LP problem, as in the case of the algorithm in Reference 39, and its time complexity is linear. But it uses linear programming theory to establish this result. This was the first primal-dual approximation algorithm, though some previous algorithms may also be viewed as falling into this category. An application of the primal-dual approach as well as related ones are discussed in Volume 1, Chapter 2. Volume 1, Chapters 4, 34, and Volume 2, Chapter 23 discuss several primal-dual approximation algorithms. Volume 1, Chapter 12 discusses "distributed" primal-dual algorithms. These algorithms make decisions by using only "local" information.

In the mid 1980s Bar-Yehuda and Even [41] developed a new framework parallel to the primal-dual methods. They called it *local ratio*; it is simple and requires no prior knowledge of linear programming. In Volume 1, Chapter 2 we explain the basics of this approach, and Volume 1, Chapter 6, and Reference 42 covers extensively this technique as well as its extensions.

Raghavan and Thompson [43] were the first to apply randomized rounding to relaxations of linear programming problems to generate solutions to the problem being approximated. This field has grown tremendously. LP randomized rounding is discussed in Volume 1, Chapters 2, 4, 7, 10, 11, and Volume 2, Chapter 11, and deterministic rounding is discussed in Volume 1, Chapters 2, 7, 8, 10, and Volume 2, Chapters 8 and 11. A disadvantage of LP-rounding is that a linear programming problem needs to be solved. This takes polynomial time with respect to the input length, but in this case it means the number of bits needed to represent the input. In contrast, algorithms based on the primal-dual approach are for the most part faster, since they take polynomial time with respect to the number of "objects" in the input. However, the LP-rounding approach can be applied to a much larger class of problems and it is more robust since the technique is more likely to be applicable after changing the objective function and/or constraints for a problem.

The first Asymptomatic PTAS (APTAS) was developed by Fernandez de la Vega and Lueker [44] for the bin packing problem. The first Asymptomatic FPTAS (AFPTAS) for the same problem was developed by Karmakar and Karp [45]. These approaches are discussed in Volume 1, Chapter 15. FPRASs are discussed in Volume 1, Chapter 11.

In the 1980s, new approximation algorithms were developed as well as PTAS and FPTAS based on different approaches. These results are reported throughout the handbook. One difference was the application of approximation algorithm to other areas of research activity (very large-scale integration (VLSI), bioinformatics, network problems) as well as other problems in established areas.

In the late 1980s Papadimitriou and Yannakakis [46] defined **MaxSNP** as a subclass of **NPO**. These problems can be approximated within a constant factor and have a nice logical characterization. They showed that if MAX3SAT, vertex cover, MAXCUT, and some other problems in the class could be approximated in polynomial time with an arbitrary precision, then all **MaxSNP** problems would. This fact was established by using *approximation preserving* reductions (see Volume 1, Chapter 14). In the 1990s Arora et al. [47], using complex arguments (see Chapter 17 in the 1st edition of this handbook), showed

---

* Here we are referring to the time when these results appeared as technical reports. Note that from the journal publication dates, the order is reversed. You will find throughout the chapters similar patterns. To add to the confusion, a large number of papers have also been published in conference proceedings. Since it would be very complex to include the dates when the initial technical report and conference proceedings were published, we only include the latest publication date. Please keep this in mind when you read the chapters and, in general, the computer science literature.

that MAX3SAT is hard to approximate within a factor of $1 + \epsilon$ for some $\epsilon > 0$ unless $P = NP$. Thus, all problems in **MaxSNP** do not admit a PTAS unless $P = NP$. This work led to major developments in the area of approximation algorithms, including inapproximability results for other problems, a bloom of approximation preserving reductions, discovery of new inapproximability classes, and construction of approximation algorithms achieving optimal or near-optimal ratios.

Feige et al. [48] showed that the clique problem could not be approximated to within some constant value. Applying the previous results in Reference 26 it showed that the clique problem is inapproximable to within any constant. Feige [49] showed that set cover is inapproximable within $\ln n$. Other inapproximable results appear in References 50, 51. Chapter 17 in the first edition of this handbook discusses all of this work in detail.

There are many other very interesting results that have been published in the past 25 years. Goemans and Williamson [52] developed improved approximation algorithms for the maxcut and satisfiability problems using *semidefinite programming* (*SDP*). This seminal work opened a new venue for the design of approximation algorithms. Chapter 8 of the first edition of this handbook discusses this work as well as developments in this area. Goemans and Williamson [53] also developed powerful techniques for designing approximation algorithms based on the primal-dual approach. The dual-fitting and factor revealing approach is used in Reference 54. Techniques and extensions of these approaches are discussed in Volume 1, Chapters 4, 12, 34, and Volume 2, Chapter 23.

This concludes our overview of Section I of Volume 1 of this handbook. Section 1.2.1 presents an overview of Section II of Volume 1 dealing with local search, artificial neural nets and metaheuristics. Section 1.2.2 presents an overview of multiobjective optimization, reoptimization, sensitivity analysis and stability all of which is in Section III of Volume 1. In the last couple of decades we have seen approximation algorithms being applied to traditional combinatorial optimization problems as well as problems arising in other areas of research activity. These areas include: VLSI design automation, networks (wired, sensor, and wireless), bioinformatics, game theory, computational geometry, and graph problems. In Sections 1.2.3 through 1.2.5 we elaborate further on these applications. Section 1.2.3 overviews traditional application covered in Section IV of Volume 1. Sections 1.2.4 and 1.2.5 overview contemporary and emerging application which are covered in this volume.

## 1.2.1 Local Search, Artificial Neural Networks, and Metaheuristics

Local search techniques have a long history; they range from simple constructive and iterative improvement algorithms to rather complex methodologies that require significant fine-tuning, such as evolutionary algorithms (EAs) or SA. Local search is perhaps one of the most natural ways to attempt to find an optimal or suboptimal solution to an optimization problem. The idea of local search is simple: Start from a solution and improve it by making local changes until no further progress is possible. Deterministic local search algorithms are discussed in Volume 1, Chapter 16. Volume 1, Chapter 17 covers stochastic local search algorithms. These are local search algorithms that make use of randomized decisions, for example, in the context of generating initial solutions or when determining search steps. When the neighborhood to search for the next solution is very large, finding the best neighbor to move to is many times an NP-hard problem. Therefore, an approximation solution is needed at this step. In Volume 1, Chapter 18 the issues related to very large-scale neighborhood search are discussed from the theoretical, algorithmic, and applications point of view.

Reactive Search advocates the use of simple subsymbolic machine learning to automate the parameter tuning process and make it an integral (and fully documented) part of the algorithm. Parameters are normally tuned through a feedback loop that many times depends on the user input. Reactive search attempts to mechanize this process. Volume 1, Chapter 19 discusses issues arising during this process.

Artificial neural networks have been proposed as a tool for machine learning and many results have been obtained regarding their application to practical problems in robotics control, vision, pattern recognition, grammatical inferences, and other areas. Recently, neural networks have found many applications in the forefront of Artificial Intelligence (AI) and Machine Learning (ML). For example, Google's open-source deep learning neural network tools as well as the Cloud, have been a catalyst for the development of these new applications.* Recently artificial neural networks have been used to improve energy utilization during certain periods in Google's massive data centers with impressive results. The whole process has been fully automated without the use of training data. Once trained (automatically or manually), the neural network will compute an input/output mapping which, if the training data was representative enough, will closely match the unknown rule which produced the original data. Neural networks are discussed in Volume 1, Chapter 20 and may be viewed as heuristics to solve a large class of problems.

The work of Lin and Kernighan [27] sparked the study of modern heuristics, which have evolved and are now called *metaheuristics*. The term *metaheuristics* was coined by Glover [55] in 1986 and in general means "to find beyond in an upper level." The most popular metaheuristics include: TS, SA, ACO, EC, iterated local search (ILC), MAs, plus many others that keep up popping-up every year. One of the motivations for the study of metaheuristics is that it was recognized early on that constant ratio polynomial time approximation algorithms are not likely to exist for a large class of practical problems [16]. Metaheuristics do not guarantee that near-optimal solutions will be found quickly for all problem instances. However, these complex programs do find near-optimal solutions for many problem instances that arise in practice. These procedures have wide range of applicability, which is their most appealing aspect.

There are many ways of viewing metaheuristics. Some are single point while others are population based. In the former case one solution is modified over and over again until the algorithm terminates. Whereas in the latter case a set of solutions is carried throughout the execution of the algorithm. Some metaheuristics have a fixed neighborhood where moves can be made, whereas others have variable neighborhoods throughout the execution of the procedure. Metaheuristic algorithms may use memory to influence future moves and some are memoryless. Metaheuristics may be nature-inspired or algorithmic based. But no matter how they work, they have all been used successfully to solve many practical problems. In what follows we discuss several metaheuristics.

The term *tabu search* was coined by Glover [55]. TS is based on *adaptive memory* and *responsive exploration*. The former allows for an effective and efficient search of the solution space. The latter is used to guide the search process by imposing restraints and inducements based on the information collected. Intensification and diversification are controlled by the information collected, rather than by a random process. Volume 1, Chapter 21 discusses many different aspects of TS as well as problems to which it has been applied. Most recently, applications in the field of quantum computing as well an open-source hybrid quantum solver for D-wave systems have emerged. These developments have placed TS at the forefront of the field of quantum computing.

In the early 1980s Kirkpatrick et al. [56] and independently Černý [57] introduced SA as a randomized local search algorithm to solve combinatorial optimization problems. SA is a local search algorithm, which means that it starts with an initial solution and then searches through the solution space by iteratively generating a new solution that is "near" to it. But, sometimes the moves are to a worse solution to escape local optimal solutions. This method is based on statistical mechanics (Metropolis algorithm). It was heavily inspired by an analogy between the physical annealing process of solids and the problem of solving large combinatorial optimization problems. Volume 1, Chapter 25 in the 1st edition of this handbook discusses this approach in detail.

EC is a metaphor for building, applying, and studying algorithms based on Darwinian principles of natural selection. Algorithms that are based on evolutionary principles are called EAs. They are inspired by nature's capability to evolve living beings well adapted to their environment. There have been a variety of slightly different EAs proposed over the years. Three different strands of EAs were developed independently of each other over time. These are *evolutionary programming* (*EP*) introduced by Fogel [58] and Fogel et al. [59], *evolutionary strategies* (*ESs*) proposed by Rechenberg [60], and *genetic algorithms* (*GAs*)

---

* https://www.tensorflow.org/ and https://aiexperiments.withgoogle.com

initiated by Holland [61]. GAs are mainly applied to solve discrete problems. *Genetic programming* (*GP*) and *scatter search* (*SS*) are more recent members of the EA family. EAs can be understood from a unified point of view with respect to their main components and the way they explore the search space. EC is discussed in Volume 1, Chapter 22.

Volume 1, Chapter 23 presents an overview of ACO—a metaheuristic inspired by the behavior of real ants. ACO was proposed by Dorigo et al. [62] in the early 1990s as a method for solving hard combinatorial optimization problems. ACO algorithms may be considered to be part of *swarm intelligence*, the research field that studies algorithms inspired by the observation of the behavior of *swarms*. Swarm intelligence algorithms are made up of simple individuals that cooperate through self-organization.

MAs were introduced by Moscato [63] in the late 1980s to denote a family of metaheuristics, which can be characterized as the hybridization of different algorithmic approaches for a given problem. It is a population-based approach in which a set of cooperating and competing agents are engaged in periods of individual improvement of the solutions while they sporadically interact. An important component is *problem and instance-dependent knowledge*, which is used to speed-up the search process. A complete description is given in Volume 1, Chapter 27 of the 1st edition of this handbook.

## 1.2.2 Multiobjective Optimization, Reoptimization, Sensitivity Analysis, and Stability

Volume 1, Chapter 24 discusses *multiobjective combinatorial optimization*. This is important in practice since quite often a decision is rarely made with only one criterion. There are many examples of such applications in the areas of transportation, communication, biology, finance, and also computer science. Volume 1, Chapter 24 covers stochastic local search algorithms for multiobjective optimization problems.

Volume 1, Chapter 25 discusses *reoptimization* which tries to address the question: Given an optimal or nearly optimal solution to some instance of an NP-hard optimization problem and a small local change is applied to the instance, can we use the knowledge of the old solution to facilitate computing a reasonable solution for the new locally modified instance? As pointed out in Volume 1, Chapter 25, we should not expect major results for optimal solutions to NP-hard problems, but there are some interesting results for approximations. Sensitivity analysis is the dual problem, meaning that given an optimal or suboptimal solution, find all the set of related instances for which the solution remains optimal or near-optimal.

Volume 1, Chapter 26 covers *sensitivity analysis*, which has been around for more than 40 years. The aim is to study how variations affect the optimal solution value. In particular, parametric analysis studies problems whose structure is fixed, but where cost coefficients vary continuously as a function of one or more parameters. This is important when selecting the model parameters in optimization problems. On the other hand, Volume 1, Chapter 27 considers a newer area which is called *stability*. By this we mean how the complexity of a problem depends on a parameter whose variation alters the space of allowable instances.

## 1.2.3 Traditional Applications

We have used the label "traditional applications" to refer to more established combinatorial optimization problems. Some of these application can be categorized differently and vice-versa. The problems studied in this part of the handbook fall into the following categories: bin packing, packing, traveling salesperson, Steiner tree, scheduling, planning, generalized assignment, linear ordering, and submodular functions maximization. Let us briefly discuss these categories.

One of the fundamental problems in approximations is the bin packing problem. Volume 1, Chapter 28 discusses online and offline algorithms for one-dimensional bin packing. Volume 1, Chapters 29 and 30 discuss variants of the bin packing problem. This include variations that fall into the following type of problems: the number of items packed is maximized while keeping the number of bins fixed; there is a bound on the number of items that can be packed in each bin; dynamic bin packing, where each item has an arrival and departure time; the item sizes are not known, but the ordering of the weights is known; items may be fragmented while packing them into fixed capacity bins, but certain items cannot be assigned to the same bin; bin stretching; variable sized bin packing problem; the bin covering problem; black and white bin packing; bin packing with rejection; batched bin packing; maximal resource bin packing; and bin packing with fragile items.

Volume 1, Chapter 31 discusses several ways to generalize the bin packing problem to more dimensions. Two- and three-dimensional strip packing, bin packing in dimensions two and higher, vector packing, and several other variations are discussed. Cutting and packing problems with important applications in the wood, glass, steel and leather industries, as well as in LSI and VLSI design, newspaper paging, and container and truck loading are discussed in Volume 1, Chapter 32. For several decades, cutting and packing problems have attracted the attention of researchers in various areas including operations research, computer science, and manufacturing. Volume 1, Chapter 33 survey heuristics, metaheuristics, and exact algorithms for two-dimensional packing of general shapes. These problems have many practical applications in various industries such as the garment, shoe, and shipbuilding industries and many variants have been considered in the literature.

Very interesting approximation algorithms for the prize collecting traveling salesperson problem is studied in Volume 1, Chapter 34. In this problem a salesperson has to collect a certain amount of prizes (the quota) by visiting cities. A known prize can be collected in every city. Volume 1, Chapter 35 discusses branch and bound algorithms for the TSP problem. These algorithms have been implemented to run in a multicomputer environment. A general software tool for running branch and bound algorithms in a distributed environment is discussed. This framework may be used for almost any divide-and-conquer computation. With minor adjustments this tool can take any algorithm defined as a computation over directed acyclic graph, where the nodes refer to computations and the edges specify a precedence relation between computations, and run in a distributed environment.

Approximation algorithms for the Steiner tree problem are discussed in Volume 1, Chapter 36. This problem has applications in several research areas. One of this area is VLSI physical design. In Volume 1, Chapter 37 practical approximations for a restricted Steiner tree problem are discussed.

Volume 1, Chapter 38 surveys problems at the intersection of two scientific fields: graph theory and scheduling. These problems can either be viewed as scheduling dependent jobs where jobs have resource requirements, or as graph coloring minimization involving different objective functions. Applications include: wire minimization in VLSI design, minimizing the distance traveled by a robot moving in a warehouse, session scheduling on a path, and resource constrained scheduling.

Automated planning consists of finding a sequence of actions that transforms an initial state into one of the goal states. Planning is widely applicable and has been used in such diverse application domains as spacecraft control, planetary rover operations, automated nursing aides, image processing, business process generation, computer security, and automated manufacturing. Volume 1, Chapter 39 discusses approximation algorithms and heuristics for problems falling into this category.

Volume 1, Chapter 40 presents heuristics and metaheuristics for the generalized assignment problem. This problem is a natural generalization of combinatorial optimization problems including bipartite matching, knapsack and bin packing problems, and has many important applications in flexible manufacturing systems, facility location, and vehicle routing problems. Computational evaluation of the different procedures is discussed.

The linear ordering problem is discussed in Volume 1, Chapter 41. Versions of this problem were initially studied back in 1938 and 1958 in Economics. Exact algorithms, constructive heuristics, local search algorithms, and metaheuristics as well as computational results are discussed extensively in this chapter.

Volume 1, Chapter 42 discusses approximation algorithms and metaheuristics for submodular function maximization. These problems play a major role in combinatorial optimization. A few examples of these functions include: cut functions of graphs and hypergraphs, rank functions of matroids, and covering functions.

## 1.2.4 Computational Geometry and Graph Applications

The problems falling into this category have applications in several fields of study, but can be viewed as computational geometry and graph problems. The problems studied in this part of the handbook fall into the following categories: connectivity problems, design and evaluation of geometric networks, pair decomposition, covering with unit balls, minimum edge length partitions, automatic placement of labels in maps and drawings, finding corridors, clustering, maximum planar subgraphs, disjoint path problems, $k$-connected subgraph problems, node connectivity in survivable network problems, optimum communication spanning trees, activation network design problems, graph coloring, algorithms for a special type of graphs, and facility dispersion.

Chapter 2 examines approximation schemes for various geometric minimum-cost $k$-connectivity problems and for geometric survivability problems, giving a detailed tutorial of the novel techniques developed for these algorithms.

Geometric networks arise in many applications. Road networks, railway networks, telecommunication, pattern matching, bio-informatics—any collection of objects in space that have some connections between them can be modeled as a geometric network. Chapter 3 considers the problem of designing a "good" network and the dual problem, that is, evaluating how "good" a given network is. Chapter 4 presents an overview of several proximity problems that can be solved efficiently using the well-separated pair decomposition (WSPD). A WSPD may be regarded as a "small" set edges that approximates the dense complete Euclidean graph.

Chapter 5 surveys approximation algorithms for covering problems with unit balls. This problem has many applications including: finding locations for emergency facilities, placing wireless sensors or antennas to cover targets, and image processing. Approximation algorithms for minimum edge length partitions of rectangles with interior points are discussed in Chapter 6. This problem has applications in the area of Computer-Aided Design (CAD) of integrated circuits and systems.

Automatic placement of labels in maps and drawings is discussed in Chapter 7. These problems have applications in information visualization, cartography, geographic information systems, graph drawing, and so on. The chapter discusses different methodologies that have been used to provide solutions to these important problems. Chapter 8 discusses approximation algorithms for finding corridors in rectangular partitions. These problems have applications in VLSI, finding corridors in floorplans, and so on. Approximation algorithms and heuristics are discussed in this chapter and results of empirical evaluations are presented.

Clustering is a very important problem that has a long list of applications. Classical algorithms for clustering are discussed and analyzed in Chapter 9. These problems include $k$-median, $k$-center, and $k$-means problems in metric and Euclidean space.

Chapter 10 discusses the problem of finding a planar subgraph of maximum weight in a given graph. Problems of this form have applications in circuit layout, facility layout, and graph drawing. Finding disjoint paths in graphs is a problem that has attracted considerable attention from at least three perspectives: graph theory, VLSI design, and network routing/flow. The corresponding literature is extensive. Chapter 11 explores offline approximation algorithms for problems on general graphs as influenced from the network flow perspective.

Chapter 12 discusses approximation algorithms and methodologies for the $k$-connected subgraph problem. The problems discussed include directed and undirected graphs, as well as general, metric, and special weight functions. A survey of approximation algorithms and the hardness of approximations for survivable networks problems are discussed in Chapter 13. These problems include the minimum cost spanning tree, traveling salesperson, Steiner tree, Steiner forest, and their directed variants.

Besides numerous network design applications, spanning trees also play important roles in several newly established research areas, such as biological sequence alignments and evolutionary tree construction. Chapter 14 explores the problem of designing approximation algorithms for spanning tree problems under different objective functions. It focuses on approximation algorithms for constructing efficient communication spanning trees.

Chapter 15 discusses the activation network design problem where the goal is to select a "cheap" graph that satisfies some property $G$, meaning that the graph belongs to a family $G$ of subgraphs of a given graph $G$. Many properties can be characterized by degree demands or pairwise connectivity demands.

Stochastic local search algorithms for the classical graph coloring problem are discussed in Chapter 16. This problem arises in many real-life applications such as register allocation, air traffic flow management, frequency assignment, light wavelengths assignment in optical networks, or timetabling. Chapter 17 discusses ACO for solving the maximum disjoint paths problems. This problem has many applications including the establishment of routes for connection requests between physically separated network endpoints.

Chapter 18 discusses efficient approximation algorithms for classical problems defined over random intersection graphs. These problems are inapproximable ones when defined over arbitrary graphs.

Facility dispersion problems are covered in Chapter 19. Dispersion problems arise in a number of applications, such as locating obnoxious facilities, choosing sites for business franchises, and selecting dissimilar solutions in multiobjective optimization. The facility location problem that model the placement of "desirable" facilities such as warehouses, hospitals, and fire stations, is discussed in Chapter 39 in the 1st edition of this handbook. This chapter covers approximation algorithms referred to as "dual fitting and factor revealing."

## 1.2.5 Large-Scale and Emerging Applications

The problems arise in the areas of wireless and sensor networks, multicasting, topology control, multimedia, data broadcast and aggregation, data analysis, computational biology, alignment problems, human genomics, VLSI placement, wavelets and streams, color quantization, digital reputation, influence maximization, and community detection. These may be referred to as "emerging" applications and normally involve large-scale problems instances. Some of these problems also fall in the other application areas.

Chapter 20 describes existing multicast routing protocols for ad hoc and sensor networks, and analyzes the issue of computing minimum cost multicast trees. The multicast routing problem, and approximation algorithms for mobile ad hoc networks (MANETs) and wireless sensor networks (WSNs) are presented. These algorithms offer better performance than Steiner trees.

Since flat networks do not scale, it is important to overlay a virtual infrastructure on a physical network. The design of the virtual infrastructure should be general enough so that it can be leveraged by a multitude of different protocols. Chapter 21 proposes a novel clustering scheme based on a number of properties of diameter-2 graphs. Extensive simulation results have shown the effectiveness of the clustering scheme when compared to other schemes proposed in the literature.

Ad hoc networks are formed by collections of nodes which communicate with each other through radio propagation. Topology control problems in such networks deal with the assignment of power values to the nodes so that the power assignment leads to a graph topology satisfying some specified properties. The problem is to minimize a specified function of the powers assigned to the nodes. Chapter 22 discusses some known approximation algorithms for this type of problems. The focus is on approximation algorithms with proven performance guarantees.

Recent progress in audio, video, and data storage technologies has given rise to a host of high-bandwidth real-time applications such as video conferencing. These applications require quality of service (QoS) guarantees from the underlying networks. Thus, multicast routing algorithms, which manage network resources efficiently and satisfy the QoS requirements, have come under increased scrutiny in recent years. Chapter 23 considers the problem of finding an optimal multicast tree with certain special characteristics. This problem is a generalization of the classical Steiner tree problem.

Scalability is especially critical for peer-to-peer systems. The basic idea of peer-to-peer systems is to have an open self-organizing system of peers that does not rely on any central server and where peers can join and leave at will. This has the benefit that individuals can cooperate without fees or an investment in additional high-performance hardware. Also, peer-to-peer systems can make use of the tremendous amount of resources (such as computation and storage) that otherwise sit idle on individual computers when they are not in use by their owners. Chapter 24 seeks ways of implementing join, leave, and route operations so that for any sequence of join, leave, and route requests can be executed quickly; the degree, diameter, and stretch factor of the resulting network are as small as possible; and the *expansion* of the resulting network is as large as possible. This is a multiobjective optimization problems for which they try to find good approximate solutions.

Scheduling problems modeling the broadcasting of data items over wireless channels are discussed in Chapter 25. The chapter covers exact and heuristic solutions for different versions of this problem.

Sensor networks are deployed to monitor a seemingly list of events in a wide rage of applications domains. By performing data analysis many patterns can be identified in sensor networks. Data analysis is based on data aggregation that must be performed at every node, but this is complicated by the fact that sensors deteriorate, often dramatically, over time. Chapter 26 discusses different strategies for aggregation time-discounted information in sensor networks. A related problem is studied in Chapter 27 where a limited number of storage nodes need to be placed in a wireless sensor network. In this chapter approximation and exact algorithm for this problem are discussed.

Chapter 28 considers two problems from computational biology, namely, primer selection and planted motif search. The closest string and the closest substring problems are closely related to the planted motif search problem. Representative approximation algorithms for these problems are discussed.

There are interesting algorithmic issues that arise when length constraints are taken into account in the formulation of a variety of problems on string similarity, particularly in the problems related to local alignment. Chapter 29 discusses these types of problems, which have their roots and most striking applications in computational biology. Chapter 30 discusses approximation algorithms for the selection of robust tag single nucleotide polymorphisms (SNPs). This is a problem in human genomics that arises in the current experimental environment.

VLSI has produced some of the largest combinatorial optimization problems ever considered. Placement is one of the most difficult of these. Placement problems with over 10 million variables and constraints are not unusual, and problem sizes continue to grow. Realistic objectives and constraints for placement incorporate complex models of signal timing, power consumption, wiring routability, manufacturability, noise, temperature, and so on. Chapter 31 considers VLSI placement algorithms.

Over the last decade the size of data seen by a computational problem have grown immensely. There appears to be more web pages than human beings, and web pages have been successfully indexed. Routers generate huge traffic logs, in the order of terabytes, in a short time. The same explosion of data is felt in observational sciences because our capabilities of measurement have grown significantly. Chapter 32 considers a processing mode where inputs item are not explicitly stored and the algorithm just passes over the data once.

Chapter 33 considers the problem of approximating "colors." Several algorithmic methodologies are presented and evaluated experimentally. These algorithms include some clustering approximation algorithms with different weights for the three dimensions.

Chapter 34 discusses a glowworm swarm optimization (GSO) algorithm for multimodal function optimization. This is a metaheuristic that is used in this chapter for odor source localization in an artificial olfaction system. This has applications to detect toxic gas leaks, fire origins of forest fires, leak point determination in pressurized systems, chemical discharge in water bodies, detection of mines and explosives, and so on.

A virtual community can be defined as a group of people sharing a common interest or goal who interact over a virtual medium, most commonly the Internet. Virtual communities are characterized by an absence of face-to-face interaction between participants which makes the task of measuring the trustworthiness of other participants harder than in non-virtual communities. This is due to the anonymity provided by the Internet, coupled with the loss of audiovisual cues that help in the establishment of trust. As a result, digital reputation management systems are an invaluable tool for measuring trust in virtual communities. Chapter 35 discusses various system which can be used to generate a good solution to this problem.

Chapter 36 continues with online social networks. This chapter discusses the use of social networks to detect the nodes that most influence the network. This has important applications, especially for advertising. Chapter 37 discusses community detection in online social networks. This problem is in some sense a clustering problem. Extracting the community structure and leveraging it to predict patterns in a dynamic network is an extremely important problem.

## 1.3 Definitions and Notation

One can use many different criteria to judge approximation algorithms and heuristics. For example, we could use the quality of the solution generated, and the time and space complexity needed to generate it. One may measure the criteria in different ways, for example, we could use the worst case, average case, median case, and so on. The evaluation could be analytical or experimental. Additional criteria include the following: characterization of data sets where the algorithm performs very well or very poorly; comparison with other algorithms using benchmarks or data sets arising in practice; tightness of bounds (quality of solution, time and space complexity); and the value of the constants associated with the time complexity bound including the ones for the lower order terms. For some researchers the most important aspect of an approximation algorithm is that it is complex to analyze, but for others it is more important that the algorithm be complex and involve the use of sophisticated data structures. For researchers working on problems directly applicable to the "real world," experimental evaluation, or evaluation on benchmarks is the more important criterion. Clearly, there is a wide variety of criteria one can use to evaluate approximation algorithms. The chapters in this handbook discuss different criteria to evaluate approximation algorithms.

For any given optimization problem $P$, let $A_1, A_2, \ldots$ be the set of current algorithms that generate a feasible solution for each instance of problem $P$. Suppose that we select a set of criteria $C$ and a way to measure it that we feel is the most important. How can we decide which algorithm is best for problem $P$ with respect to $C$? We may visualize every algorithm as a point in multidimensional space. Now the approach used to compare feasible solutions for multiobjective function problems (see Volume 1, Chapter 24) can also be used in this case to label some of the algorithms as current Pareto optimal with respect to $C$. Algorithm $A$ is said to be *dominated* by algorithm $B$ with respect to $C$, if for each criteria $c \in C$ algorithm $B$ is "not worse" than $A$, and for at least one criteria $c \in C$ algorithm $B$ is "better" than $A$. An algorithm is said to be a *current Pareto optimal* algorithm with respect to $C$ if none of the current algorithms dominates it.

In the next subsections we define time and space complexity, NP-completeness, and different ways to measure the quality of the solutions generated by the algorithms.

### 1.3.1 Time and Space Complexity

There are many different ways one can use to judge algorithms. The main ones we use are the time and space required to solve the problem. This is normally expressed in terms of $n$, the input size. It can be evaluated empirically or analytically. For the analytic evaluation we use the time and space complexity of the algorithm. Informally, this is a way to express the time the algorithm takes to solve a problem of size $n$ and the amount of space needed to run the algorithm.

It is clear that almost all algorithms take different time to execute with different data sets even when the input size is the same. If you code it and run it on a computer you will see more variation depending on the different hardware and software installed in the system. It is impossible to characterize exactly the time and space required by an algorithm. We need a short cut. The approach that has been taken is to count the number of "operations" performed by the algorithm in terms of the input size. "Operations" is not an exact term and refers to a set of "instructions" which is independent of the problem size being solved. Then we just need to count the total number of operations.

Counting the number of operations exactly, is very complex for a large number of algorithms. So we just take into consideration the "highest" order term. This is the $O$ notation.

> *Big oh Notation:* A (positive) function $f(n)$ is said to be $O(g(n))$ if there exist two constants $c \geq 1$ and $n_0 \geq 1$ such that $f(n) \leq c \cdot g(n)$ for all $n \geq n_0$.

The function $g(n)$ must include at least one term that is as large as the highest order term. For example, if $f(n) = n^3 + 20n^2$, then $g(n) = n^3$. Setting $n_0 = 1$ and $c = 21$ shows that $f(n)$ is $O(n^3)$. Note that $f(n)$ is also $O(n^4)$, but we like $g(n)$ to be the function with the smallest possible growth for $f(n)$. The function $f(n)$ cannot be $O(n^2)$ because it is impossible to find constants $c$ and $n_0$ such that $n^3 + 20n^2 \leq cn^2$ for all $n \geq n_0$.

Time and space complexity are normally expressed using the $O$ notation and describes the growth rate of the algorithm in terms of the problem size. Normally, the problem size is the number of vertices and edges in a graph, or the number of tasks and machines in a scheduling problem, and so on. But it can also be the number of bits used to represent the input.

When comparing two algorithms expressed in $O$ notation we have to be careful because the constants $c$ and $n_0$ are hidden. For large $n$, the algorithm with the smallest growth rate is the better one, but that might only hold for huge values of $n$. When two algorithms have similar constants $c$ and $n_0$, the algorithm with the smallest growth function has a smaller running time. The book [2] discusses in detail the $O$ notation as well as other notation.

## 1.3.2 NP-Completeness

Before the 1970s, researchers were aware that some problems could be computationally solved by algorithms of (low-order) polynomial time complexity ($O(n)$, $O(n^2)$, $O(n^3)$, etc.), whereas other problems had exponential time complexity, for example, $O(2^n)$ and $O(n!)$. It was clear that even for small values of $n$, exponential time complexity equates to computational intractability if the algorithm actually performs an exponential number of operations for some inputs. The convention of computational tractability being equated to polynomial time complexity does not really fit well in practice, as an algorithm with time complexity $O(n^{100})$ is not really tractable if it actually performs $n^{100}$ operations. But even under this relaxation of "tractability" there is a large class of problems that do not seem to have computational tractable algorithms for their solution.

We have been discussing optimization problems. But NP-completeness is defined for decision problems. A decision problem is simply one whose answer is "yes" or "no." The scheduling on identical machines problems discussed earlier is an optimization problem. Its corresponding decision problem has its input augmented by an integer value $B$ and the yes-no question is to determine whether or not there is a schedule with makespan at most $B$. Every optimization problem has a corresponding decision problem. Since the solution of an optimization problem can be used directly to solve the decision problem, we say that the optimization problem is at least as hard to solve as the decision problem. If we show that the decision problem is a computationally intractable problem, then the corresponding optimization problem is also intractable.

The development of NP-completeness theory in the early 1970s by Cook [6], Karp [7], and others formally introduced the notion that there is a large class of decision problems that are computationally equivalent. By this we mean that either every problem in this class has a polynomial time algorithm that solves it, or none of them do. Furthermore, this question is the same as the $P = NP$ question, an open problem in computational complexity. This question is to determine whether or not the set of languages recognized in polynomial time by deterministic Turing machines is the same as the set of languages recognized in polynomial time by nondeterministic Turing machines. The conjecture has been that $P \neq NP$, and thus these problems would not have polynomial time algorithms for their solution.

The decision problems in this class of problems are called *NP-complete* problems. Optimization problems whose corresponding decision problem is NP-complete are called *NP-hard* problems.

Scheduling tasks on identical machines is an NP-hard problem. The TSP and Steiner tree problem are also NP-hard problems. The minimum weight spanning tree problem can be solved in polynomial and it is not an NP-hard problem, under the assumption that $P \neq NP$. There is a long list of practical problems arising in many different fields of study that are known to be NP-hard problems. In fact almost all the optimization problems discussed in this handbook are NP-hard problems. The book [8] is an excellent source for information about NP-complete and NP-hard problems.

One establishes that a problem $Q$ is an NP-complete problem by showing that the problem is in NP and giving a polynomial time transformation from an NP-complete problem to the problem $Q$.

A problem is said to be in NP if one can show that a yes answer to it can be verified in polynomial time. For the scheduling problem defined earlier you may think of this as providing a procedure that given any instance of the problem and an assignment of tasks to machines, the algorithm verifies in polynomial time, with respect to the problem instance size, if the assignment is a schedule and its makespan has length at most $B$. This is equivalent to the task a grader or teaching assistant (TA) performs when grading a question of the form "Does the following instance of the scheduling problem have a schedule with makespan at most 300? If so, give a schedule." Just verifying that the "answer" is correct is a simple problem. But solving a problem instance with 10,000 tasks and 20 machines seems much harder than simply grading it. In our over simplification it seems that $P \neq NP$. Polynomial time verification of an yes answer does not seem to imply polynomial time solvability. It is interesting to note that undergraduate students pay to take courses and exams, but the graders or TAs are paid to grade the exams, which seems to be a computationally simpler task! Though, professors just write the exam questions and are paid significantly more than the TAs. Recently, I noticed that emeriti professors do not even do that and they are still paid!

A polynomial time transformation from decision problem $P_1$ to decision problem $P_2$ is an algorithm that takes as input any instance $I$ of problem $P_1$ and constructs an instance of $f(I)$ of $P_2$. The algorithm must take polynomial time with respect to the instance $I$. The transformation must be such that $f(I)$ is a yes-instance of $P_2$ if, and only if, $I$ is a yes-instance of $P_1$.

The implication of a polynomial transformation $P_1 \alpha P_2$ is that if $P_2$ can be solved in polynomial time, then so can $P_1$, and if $P_1$ cannot be solved in polynomial time, then $P_2$ cannot be solved in polynomial time.

Consider the partition problem. We are given $n$ items $1, 2, \ldots, n$. Item $j$ has size $s(j)$. The problem is to determine whether or not the set of items can be partitioned into two sets such that the sum of the size of the items in one set equals the sum of the size of the items in the other set. Now let us polynomially transform the partition problem to the problem of scheduling tasks on identical machines. Given an instance $I$ of partition, we define the instance $f(I)$ as follows. There are $n$ tasks and $m = 2$ machines. Task $i$ represents item $i$ and its processing time is $s(i)$. All the tasks are independent and $B = \sum_{i=1}^{i=n} s(i)/2$. Clearly, $f(I)$ has schedule with maskespan $B$ if, and only if, the instance $I$ has a partition.

A decision problem is said to be *strongly NP-complete* if the problem is NP-complete even when all the "numbers" in the problem instance are less than or equal to $p(n)$, where $p$ is a polynomial and $n$ is the "size" of the problem instance. Partition is not NP-complete in the strong sense (under the assumption that $P \neq NP$) because there is a polynomial time dynamic programming algorithm for its solution (see Volume 1, Chapter 9). The book by Garey and Johnson [8] is an excellent source for NP-Completeness information.

Resolving the $P = NP$? question is one of the most important ones in the field of computer science. The Clay Institute of Mathematics* has offered a $1,000,000 reward for answering this question. The reward will be given whether one proves that $P = NP$ or $P \neq NP$. No matter what, the person or people solving this problem for the first time will most likely be hired for life at the one of the most prestigious research

---

* http://www.claymath.org/

labs or universities. However, if someone proves that $P = NP$ and the algorithm(s) to solve NP-complete problems are really efficient, then programming will be simplified considerably. To solve an NP-complete problem you just need to write a procedure to check a yes-answer. All the theoretical machinery developed so far will then be used to produce a fast algorithm to solve any NP-complete problem. If this ever happens, the best strategy would be to buy a huge farm and fill it with computers that will run implementations of such algorithms solving practical instances of NP-hard problems. Such person would become the first "trillionaire" and will be able to hire Bill to clean his/her windows, Mark to clean his/her face and books, Jeff to do his/her shopping, and so on. But, to be honest, the probability of this event is very close to zero, and most likely zero.

### 1.3.3 Performance Evaluation of Algorithms

The main criterion used to compare approximation algorithms has been the quality of the solution generated. Let us consider the different ways to compare the quality of the solutions generated when measuring the worst case. That is the criterion discussed in Section 1.2.

For some problems it is very hard to judge the quality of the solution generated. For example, approximating colors can only be judged by viewing the resulting images and that is subjective (see Chapter 33). Chapter 35 covers digital reputation schemes. Here again it is difficult to judge the quality of the solution generated. Problems in the application areas of bioinformatics and VLSI CAD fall into this category because, in general, these are problems with multiobjective objective functions.

In what follows, we concentrate on problems where it is possible to judge the quality of the solution generated. At this point, we need to introduce additional notation. Let $P$ be an optimization problem and $A$ be an algorithm that generates a feasible solution for every instance $I$ of problem $P$. We use $\hat{f}_A(I)$ to denote the objective function value of the solution generated by algorithm $A$ for instance $I$. We drop $A$ and use $\hat{f}(I)$ when it is clear which algorithm is being used. Let $f^*(I)$ be the objective function value of an optimal solution for instance $I$. Note that normally we do not know the value of $f^*(I)$ exactly, but we have bounds which should be as tight as possible.

Let $G$ be an undirected graph that represents a set of cities (vertices) and roads (edges) between a pair of cities. Every edge has a positive number called the weight (or cost) and represents the cost of driving (gas plus tolls) between the pair of cities it joins. A *shortest path* from vertex $s$ to vertex $t$ in $G$ is a path from $s$ to $t$ ($st$-path) such that the sum of the weight of the edges in it is the "least" possible among all possible $st$-paths. There are well-known algorithms that solve this shortest path problem in polynomial time [2]. Let $A$ be an algorithm that generates a feasible solution ($st$-path) for every instance $I$ of problem $P$. If for every instance $I$ algorithm $A$ generates an $st$-path such that

$$\hat{f}(I) \leq f^*(I) + c,$$

where $c$ is some fixed constant, then $A$ is said to be an *absolute* approximation algorithm for problem $P$ with (additive) approximation bound $c$. Ideally, we would like to design a linear (or at least polynomial) time approximation algorithm with the smallest possible approximation bound. It is not difficult to see that this is not a good way of measuring the quality of a solution. Suppose that we have a graph $G$ and we are running an absolute approximation algorithm for the shortest path problem concurrently in two different countries with the edge weight expressed in the local currency. Furthermore, assume that there is a large exchange rate between the two currencies. Any approximation algorithm solving the weak currency instance will have a much harder time finding a solution within the bound of $c$, than when solving the strong currency instance. We can take this to the extreme. We now claim that the above-mentioned absolute approximation algorithm $A$ can be used to generate an optimal solution for any problem instance within the same time complexity bound.

The argument is simple. Given any instance $I$ of the shortest path problem, we construct an instance $I_{c+1}$ using the same graph, but every edge weight is multiplied by $c + 1$. Clearly, $f^*(I_{c+1}) = (c + 1)f^*(I)$.

The $st$-path for $I_{c+1}$ constructed by the algorithm is also an $st$-path in $I$ and its weight is $\hat{f}(I) = \hat{f}(I_{c+1})/(c+1)$. Since $\hat{f}(I_{c+1}) \leq f^*(I_{c+1}) + c$, then by substituting the above bounds we know that

$$\hat{f}(I) = \frac{\hat{f}(I_{c+1})}{c+1} \leq \frac{f^*(I_{c+1})}{c+1} + \frac{c}{c+1} = f^*(I) + \frac{c}{c+1}.$$

Since all the edges have integer weights and $c/(c+1)$ is less than one, it follows that the algorithm finds an optimal solution to the problem. In other words, for the shortest path problem any algorithm that generates a solution with (additive) approximation bound $c$ can be used to generate an optimal solution within the same time complexity bound. This same property can be established for almost all NP-hard optimization problems. Because of this the use of absolute approximation has never been given a serious consideration.

Sahni [14] defines as an $\epsilon$-approximation algorithm for problem $P$ an algorithm that generates a feasible solution for every problem instance $I$ of $P$ such that

$$\left| \frac{\hat{f}(I) - f^*(I)}{f^*(I)} \right| \leq \epsilon.$$

It is assumed that $f^*(I) > 0$. For a minimization problem $\epsilon > 0$ and for a maximization problem $0 < \epsilon < 1$. In both cases $\epsilon$ represents the percentage of error. The algorithm is called an $\epsilon$-approximation algorithm and the solution is said to be an $\epsilon$-approximate solution. Graham's list scheduling algorithm [1] is a $1 - 1/n$ approximation algorithm, and the Sahni and Gonzalez [16] algorithm for the $k$-maxcut problem is a $\frac{1}{k}$-approximation algorithm (see Section 1.2). Note that this notation is different from the one discussed in Section 1.2. The difference is 1 unit, that is, the $\epsilon$ in this notation corresponds to $1 + \epsilon$ in the other.

Johnson [12] used a slightly different, but equivalent notation. He uses the approximation ratio $\rho$ to mean that for every problem instance $I$ of $P$ the algorithm satisfies $\frac{\hat{f}(I)}{f^*(I)} \leq \rho$ for minimization problems, and $\frac{f^*(I)}{\hat{f}(I)} \leq \rho$ for maximization problems. The one for minimization problems is the same as the one given in Reference 1. The value for $\rho$ is always greater than one, and the closer to one, the better the solution generated by the algorithm. One refers to $\rho$ as the *approximation ratio* and the algorithm is a $\rho$-approximation algorithm. The list scheduling algorithm in the previous section is a $(2 - \frac{1}{m})$-approximation algorithm and the algorithm for the $k$-maxcut problem is a $(\frac{k}{k-1})$-approximation algorithm. Sometimes $1/\rho$ is used as the approximation ratio for maximization problems. Using this notation, the algorithm for the $k$-maxcut problem in the previous section is a $1 - \frac{1}{k}$-approximation algorithm.

All the above-mentioned forms are in use today. The most popular ones are $\rho$ for minimization and $1/\rho$ for maximization. These are referred to as approximation ratios or approximation factors. We refer to all these algorithms as $\epsilon$-approximation algorithms. The point to remember is that one needs to be aware of the differences and be alert when reading the literature. In the previous discussion we make $\epsilon$ and $\rho$ look as if they are fixed constants. But, they can be made dependent on the size of the problem instance $I$. For example, it may be $\ln n$, or $n^\epsilon$ for some problems, where $n$ is some parameter of the problem that depends on $I$, for example, the number of nodes in the input graph, and $\epsilon$ depends on the algorithm being used to generate the solutions. But it is most desirable that $\epsilon$ is a small constant.

Normally, one prefers an algorithm with a smaller approximation ratio. However, it is not always the case that an algorithm with a smaller approximation ratio always generates solutions closer to optimal than one with a larger approximation ratio. The main reason is that the notation is for the worst case ratio and the worst case does not always occur. But there are other reasons too. For example, the bound for the optimal solution value used in the analysis of two different algorithms may be different. Let $P$ be the shortest path minimization problem and let $A$ be an algorithm with approximation ratio 2. In this case we use $d$ as the lower bound for $f^*(I)$, where $d$ is some parameter of the problem instance. Algorithm $B$ is a 1.5-approximation algorithm, but $f^*(I)$ used to establish it is the exact optimal solution value. Suppose that for problem instance $I$ the value of $d$ is 5 and $f^*(I) = 8$. Algorithm $A$ will generate a path with weight at most 10, whereas algorithm $B$ will generate one with weight at most $1.5 \times 8 = 12$. So the solution generated by Algorithm $B$ may be worse than the one generated by $A$ even if both algorithms generate the worst values for the instance. One can argue that the average "error" makes more sense than worst case. The problem is how to define and establish bounds for average "error." There are many other pitfalls when using worst case ratios. It is important to keep in mind all of this when making comparisons between algorithms. In practice one may run several different approximation algorithms concurrently and output the best of the solutions. This has the disadvantage that the running time of this compound algorithm will be the one for the slowest algorithm.

There are a few problems for which the worst case approximation ratio applies only to problem instances where the value of the optimal solution is small. One such problem is the bin packing problem discussed in Section 1.2. Informally, $\rho_A^\infty$ is the smallest constant such that there exists a constant $K < \infty$ for which

$$\hat{f}(I) \leq \rho_A^\infty f^*(I) + K.$$

The *asymptotic approximation ratio* is the multiplicative constant and it hides the additive constant $K$. This is most useful when $K$ is small. Volume 1, Chapter 28 discusses this notation formally. The asymptotic notation is mainly used for bin packing and some of its variants.

Ausiello et al. [30] introduced the *differential ratio*. Informally, an algorithm is said to be a $\delta$ differential ratio approximation algorithm if for every instance $I$ of $P$

$$\frac{\omega(I) - \hat{f}(I)}{\omega(I) - f^*(I)} \leq \delta,$$

where $\omega(I)$ is the value of a worst solution for instance $I$. The worst case for the TSP over a complete graph is well defined, but for a scheduling problem, its value would be $\infty$ as one can keep on leaving machines idle. For problems when the worst solution is clear, differential ratio has some interesting properties for the complexity of the approximation problems. Volume 1, Chapter 15 discusses differential ratio approximation and its variations.

As said before there are many different criteria to compare algorithms. What if we use both the approximation ratio and time complexity? For example, the approximation algorithms in Reference 15 and the one in Reference 29 are current Pareto optimal with respect to this criteria for the TSP defined over metric graphs. Neither of the algorithms dominates the others in both time complexity and approximation ratio. The same can be said about the simple linear time (and very fast) approximation algorithm for the $k$-maxcut problem in Reference 16 and the complex one given in Reference 52 or the more recent ones that apply for all $k$.

The best algorithm to use also depends on the instance being solved. It makes a difference whether we are dealing with an instance of the TSP with optimal tour cost equal to one billion dollars and one with optimal cost equal to just a few pennies. Though, it also depends on the number of such instances being solved.

More elaborate approximation algorithms have been developed that generate a solution for any fixed constant $\epsilon$. Formally, a *PTAS* for problem $P$ is an algorithm $A$ that given any fixed constant $\epsilon > 0$ it constructs a solution to every instance $I$ problem $P$ such that $|\frac{\hat{f}(I) - f^*(I)}{f^*(I)}| \leq \epsilon$ in polynomial time with respect to the length of the instance $I$. Note that the time complexity may be exponential with respect to $1/\epsilon$. For example, the time complexity could be $O(n^{(1/\epsilon)})$ or $O(n + 4^{O(1/\epsilon)})$. Equivalent PTAS are also defined using different notation, for example based on $\frac{\hat{f}(I)}{f^*(I)} \leq 1 + \epsilon$ for minimization problems.

One would like to design PTAS for all problems, but that is not possible unless $P = NP$. Clearly, with respect to approximation ration the PTAS is better than the $\epsilon$-approximation algorithms for some $\epsilon$.

But their main draw back is that they are not practical because the time complexity is exponential on $1/\epsilon$. This does not preclude the existence of a practical PTAS for a "natural" occurring problem. However, a PTAS establishes that a problem can be approximated for all fixed constants. Different types of PTAS are discussed in Volume 1, Chapter 8. Additional PTAS are presented in Volume 1, Chapter 36 and Volume 2, Chapter 2.

A PTAS is said to be a FPTAS if its time complexity is polynomial with respect to $n$ (the problem size) and $1/\epsilon$. For reasonable values of $\epsilon$ most FPTASs have practical running times. Different methodologies for designing FPTAS are discussed in Volume 1, Chapter 9.

Approximation schemes based on asymptotic approximation and on randomized algorithms have been developed. Volume 1, Chapter 10 discusses asymptotic approximation schemes and Volume 1, Chapter 11 discusses randomized approximation schemes.

# References

1. Graham, R.L., Bounds for certain multiprocessing anomalies, *Bell Syst. Tech. J.*, 45, 1563, 1966.
2. Sahni, S., *Data Structures, Algorithms, and Applications in C++*, 2nd ed., Silicon Press, Summit, NJ, 2005.
3. Gilbert, E.N. and Pollak, H.O., Steiner minimal trees, *SIAM J. Appl. Math.*, 16(1), 1, 1968.
4. Graham, R.L., Bounds on multiprocessing timing anomalies, *SIAM J. Appl. Math.*, 17, 263, 1969.
5. Leung, J.Y.T., Ed., *Handbook of Scheduling: Algorithms, Models, and Performance Analysis*, Chapman & Hall/CRC, Boca Raton, FL, 2004.
6. Cook, S.A., The complexity of theorem-proving procedures, in *Proceedings of STOC'71*, 1971, p. 151.
7. Karp, R.M., Reducibility among combinatorial problems, in Miller, R.E. and Thatcher, J.W., (Eds.), *Complexity of Computer Computations*, Plenum Press, New York, 1972, p. 85.
8. Garey, M.R. and Johnson, D.S., *Computers and Intractability: A Guide to the Theory of NP-Completeness*, W. H. Freeman and Company, New York, 1979.
9. Garey, M.R., Graham, R.L., and Ullman, J.D., Worst-case analysis of memory allocation algorithms, in *Proceedings of STOC*, ACM, 1972, p. 143.
10. Johnson, D.S., *Near-Optimal Bin Packing Algorithms*, PhD Thesis, Massachusetts Institute of Technology, Department of Mathematics, Cambridge, MA, 1973.
11. Johnson, D.S., Fast algorithms for bin packing, *JCSS*, 8, 272, 1974.
12. Johnson, D.S., Approximation algorithms for combinatorial problems, *JCSS*, 9, 256, 1974.
13. Sahni, S., On the knapsack and other computationally related problems, PhD Thesis, Cornell University, 1973.
14. Sahni, S., Approximate algorithms for the 0/1 knapsack problem, *JACM*, 22(1), 115, 1975.
15. Rosenkrantz, R., Stearns, R., and Lewis, L., An analysis of several heuristics for the traveling salesman problem, *SIAM J. Comput.*, 6(3), 563, 1977.
16. Sahni, S. and Gonzalez, T., P-complete approximation problems, *JACM*, 23, 555, 1976.
17. Gens, G.V. and Levner, E., Complexity of approximation algorithms for combinatorial problems: A survey, *SIGACT News*, 12, 52, 1980.
18. Levner, E. and Gens, G.V., *Discrete Optimization Problems and Efficient Approximation Algorithms*, Central Economic and Mathematics Institute, Moscow, Russia, 1978 (in Russian).
19. Garey, M.R. and Johnson, D.S., The complexity of near-optimal graph coloring, *SIAM J. Comput.*, 4, 397, 1975.
20. Ibarra, O. and Kim, C., Fast approximation algorithms for the knapsack and sum of subset problems, *JACM*, 22(4), 463, 1975.
21. Sahni, S., Algorithms for scheduling independent tasks, *JACM*, 23(1), 116, 1976.
22. Horowitz, E. and Sahni, S., Exact and approximate algorithms for scheduling nonidentical processors, *JACM*, 23(2), 317, 1976.

23. Babat, L.G., Approximate computation of linear functions on vertices of the unit $N$-dimensional cube, in *Studies in Discrete Optimization,* Fridman, A.A., (Ed.), Nauka, Moscow, Russia, 1976 (in Russian).

24. Babat, L.G., A fixed-charge problem, *Izv. Akad. Nauk SSR, Techn, Kibernet.,* 3, 25, 1978 (in Russian).

25. Lawler, E., Fast approximation algorithms for knapsack problems, *Math. Oper. Res.,* 4, 339, 1979.

26. Garey, M.R. and Johnson, D.S., Strong NP-completeness results: Motivations, examples, and implications, *JACM,* 25, 499, 1978.

27. Lin, S. and Kernighan, B.W., An effective heuristic algorithm for the traveling salesman problem, *Oper. Res.,* 21(2), 498, 1973.

28. Papadimitriou, C.H. and Steiglitz, K., On the complexity of local search for the traveling salesman problem, *SIAM J. Comput.,* 6, 76, 1977.

29. Christofides, N., Worst-case analysis of a new heuristic for the traveling salesman problem. Technical Report 338 Grad School of Industrial Administration, CMU, 1976.

30. Ausiello, G., D'Atri, A., and Protasi, M., On the structure of combinatorial problems and structure preserving reductions, in *Proceedings of ICALP'77,* LNCS, 52, Springer-Verlag, Turku, Inland, 1977, p. 45.

31. Cornuejols, G., Fisher, M.L., and Nemhauser, G.L., Location of bank accounts to optimize float: An analytic study of exact and approximate algorithms, *Manag. Sci.,* 23(8), 789, 1977.

32. Khachiyan, L.G., A polynomial algorithms for the linear programming problem, *Dokl. Akad. Nauk SSSR,* 244(5), 1979 (in Russian).

33. Karmakar, N., A new polynomial-time algorithm for linear programming, *Combinatorica,* 4, 373, 1984.

34. Grötschel, M., Lovász, L., and Schrijver, A., The ellipsoid method and its consequences in combinatorial optimization, *Combinatorica,* 1, 169, 1981.

35. Schrijver, A., *Theory of Linear and Integer Programming,* Wiley-Interscience Series in Discrete Mathematics and Optimization, John Wiley, New York, 1998.

36. Vanderbei, R.J., *Linear Programming Foundations and Extensions,* Series: International Series in Operations Research & Management Science, Springer, Berlin, Germany, No. 196, 4th Edition, 2014.

37. Lovász, L., On the ratio of optimal integral and fractional covers, *Disc. Math.,* 13, 383, 1975.

38. Chvátal, V., A greedy heuristic for the set-covering problem, *Math. Oper. Res.,* 4(3), 233, 1979.

39. Hochbaum, D.S., Approximation algorithms for set covering and vertex covering problems, *SIAM J. Comput.,* 11, 555, 1982.

40. Bar-Yehuda, R. and Even, S., A linear time approximation algorithm for the weighted vertex cover problem, *J. Algorithms,* 2, 198, 1981.

41. Bar-Yehuda, R. and Even, S., A local-ratio theorem for approximating the weighted set cover problem, *Ann. Disc. Math.,* 25, 27, 1985.

42. Bar-Yehuda, R. and Bendel, K., Local ratio: A unified framework for approximation algorithms, *ACM Comput. Surv.,* 36(4), 422, 2004.

43. Raghavan, R. and Thompson, C., Randomized rounding: A technique for provably good algorithms and algorithmic proof, *Combinatorica,* 7, 365, 1987.

44. Fernandez de la Vega, W. and Lueker, G.S., Bin packing can be solved within $1 + \epsilon$ in linear time, *Combinatorica,* 1, 349, 1981.

45. Karmakar, N. and Karp, R.M., An efficient approximation scheme for the one-dimensional bin packing problem, in *Proceedings of FOCS,* 1982, p. 312.

46. Papadimitriou, C.H. and Yannakakis, M., Optimization, approximation and complexity classes, *J. Comput. Syst. Sci.,* 43, 425, 1991.

47. Arora, S., Lund, C., Motwani, R., Sudan, M., and Szegedy, M., Proof verification and hardness of approximation problems, in *Proceedings of FOCS,* 1992.

48. Feige, U., Goldwasser, S., Lovasz, L., Safra, S., and Szegedy, M., Interactive proofs and the hardness of approximating cliques, *JACM*, 43, 268, 1996.

49. Feige, U., A threshold of ln n for approximating set cover, *JACM*, 45(4), 634, 1998. (Prelim. version in STOC'96.)

50. Engebretsen, L. and Holmerin, J., Towards optimal lower bounds for clique and chromatic number, *TCS*, 299, 537, 2003.

51. Hastad, J., Some optimal inapproximability results, *JACM*, 48, 2001. (Prelim. version in STOC'97.)

52. Goemans, M.X. and Williamson, D.P., Improved approximation algorithms for maximum cut and satisfiability problems using semidefinite programming, *JACM*, 42(6), 1115, 1995.

53. Goemans, M.X. and Williamson, D.P., A general approximation technique for constrained forest problems, *SIAM J. Comput.*, 24(2), 296, 1995.

54. Jain, K., Mahdian, M., Markakis, E., Saberi, A., and Vazirani, V.V., Approximation algorithms for facility location via dual fitting with factor-revealing LP, *JACM*, 50, 795, 2003.

55. Glover, F., Future paths for integer programming and links to artificial intelligence, *Computers Oper. Res.*, 13, 533, 1986.

56. Kirkpatrick, S., Gelatt Jr., C.D., and Vecchi, M.P., Optimization by simulated annealing, *Science*, 220, 671, 1983.

57. Černý, V., Thermodynamical approach to the traveling salesman problem: An efficient simulation algorithm, *J. Optim. Theory Appl.*, 45, 41, 1985.

58. Fogel, L.J., Toward inductive inference automata, in *Proceedings of the International Federation for Information Processing Congress*, Munich, Germany, 1962, p. 395.

59. Fogel, L.J., Owens, A.J., and Walsh, M.J., *Artificial Intelligence through Simulated Evolution*, Wiley, New York, 1966.

60. Rechenberg, I., *Evolutionsstrategie: Optimierung technischer Systeme nach Prinzipien der biologischen Evolution*, Frommann-Holzboog, Stuttgart, Germany, 1973.

61. Holland, J.H., *Adaption in Natural and Artificial Systems*, The University of Michigan Press, Ann Harbor, MI, 1975.

62. Dorigo, M., Maniezzo, V., and Colorni, A., Positive feedback as a search strategy, Technical Report 91-016, Dipartimento di Elettronica, Politecnico di Milano, Italy, 1991.

63. Moscato, P., On genetic crossover operators for relative order preservation, C3P Report 778, California Institute of Technology, 1989.

# I

# Computational Geometry and Graph Applications

# 2

# Approximation Schemes for Minimum-Cost $k$-Connectivity Problems in Geometric Graphs[*]

[*] Research partially supported by the Centre for Discrete Mathematics and Its Applications, by the Engineering and Physical Sciences Research Council (EPSRC) awards EP/D063191/1 and EP/G064679/1, and by the National Science Foundation (NSF) Information Technology Research (ITR) grant CCR-0313219 and by Swedish Research Council (VR) grant 621-2002-4049.

Artur Czumaj

Andrzej Lingas

## 2.1 Introduction

### 2.1.1 Multiconnectivity Problems

We survey the area of the design of approximation schemes for geometric variants of the following classical optimization problem: For a given undirected weighted graph, find its minimum-cost subgraph that satisfies a priori given multiconnectivity requirements. We present the approximation schemes for various geometric minimum-cost $k$-connectivity problems and for geometric survivability problems, giving a detailed tutorial of the novel techniques developed for these algorithms.

A classical multiconnectivity graph problem is as follows: For a given undirected weighted graph, find its minimum-cost subgraph that satisfies a priori given connectivity requirements.

Multiconnectivity graph problems are central in algorithmic graph theory and have numerous applications in computer science and operation research, see, for example, [1–5]. They also play a very important role in the design of networks that arise in practical situations, see, for example, [1,3,6]. Typical application areas include telecommunication, computer, and road networks. Low-degree connectivity problems for geometrical graphs in the plane can often closely *approximate* such practical connectivity problems [3,5,7]. For instance, they can be used to model the design of low-cost telephone networks that can "survive" some types of edge and node failure. In such a model, the cost of the edge corresponds to the cost of lying a fiber-optic cable between the endpoints of the edge and the planned cost of the service of the cable. Furthermore, the minimum connectivity requirement for a pair of vertices corresponds to the minimum number of edge and/or node failures that must occur in the network before the pair is completely disconnected. In practice, the latter value tends to be quite low, usually no more than 2, as failures are assumed to be isolated accidents, such as fires at nodes [5,7]. Note that the cost of lying a fiber-optic cable between two points is roughly proportional to the length of the link [7].

In the currents work, we survey approximation results for these problems restricted to geometric graphs, briefly mentioning results for planar graphs in Section 2.7.4. Our main goal is not only to show the results but also to demonstrate a variety of new techniques developed to cope with these problems.

The most classical problem we study is the *(Euclidean) minimum-cost k-vertex-connected spanning subgraph (k-VCSS) problem*. We are given a set $S$ of $n$ points in the Euclidean space $\mathbb{R}^d$ and the aim is to find a minimum-cost $k$-vertex-connected Euclidean graph spanning points in $S$ (i.e., a subgraph of the complete Euclidean graph on $S$).

Throughout the paper, we shall assume that the cost of the graph is equal to the sum of the costs of the edges of the graph. Furthermore, in the geometric case, the cost of an edge connecting a pair of points $x, y \in \mathbb{R}^d$ is equal to the Euclidean distance between points $x$ and $y$, that is, $\sqrt{\sum_{i=1}^{d}(x_i - y_i)^2}$, where $x = (x_1, \ldots, x_d)$ and $y = (y_1, \ldots, y_d)$. More generally, the distance could be defined using other norms, such as $\ell_p$ norms for any $p > 1$; all results discussed in this survey can be extended from the Euclidean case to other $\ell_p$ norms.

By substituting the requirement of $k$-edge connectivity for that of $k$-vertex connectivity, we obtain the corresponding *(Euclidean) minimum-cost k-edge-connected spanning subgraph (k-ECSS) problem*.

We term the generalization of the latter problem that allows for parallel edges in the output graph spanning $S$ as the *(Euclidean) minimum-cost k-edge-connected spanning sub-multigraph (k-ECSSM) problem*.

The concept of minimum-cost $k$-connectivity naturally extends to include that of *Euclidean Steiner k-connectivity* by allowing the use of additional vertices, called *Steiner points*. For a given set $S$ of points in $\mathbb{R}^d$, we say that a geometric graph $G$ is a *Steiner k-VCSS (or, Steiner k-ECSS) for $S$* if the vertex set of $G$ is a *superset* of $S$ and for every pair of points from $S$, there are $k$ internally vertex-disjoint (edge-disjoint, respectively) paths connecting them in $G$. The problem of *(Euclidean) minimum-cost Steiner k-vertex- (or, k-edge-) connectivity* is to find a minimum-cost Steiner $k$-VCSS (or, Steiner $k$-ECSS) for $S$. For $k = 1$, it is simply the *Steiner minimal tree* (SMT) problem, which has been very extensively studied in the literature (see, e.g., [8,9] and Vol. 1, Chap. 36).

In a more general formulation of multiconnectivity graph problems, nonuniform connectivity constraints have to be satisfied. The *survivable network design problem* is defined as follows: For a given weighted undirected graph $G = (V, E)$ and a connectivity requirement function $r : V \times V \to \mathbb{N}$, find a minimum-cost subgraph of $G$ such that for any pair of vertices $x, y \in V$, the subgraph has $r_{x,y}$ internally vertex-disjoint (or edge-disjoint, depending on the context) paths between $x$ and $y$ [10]. Also in that case, the output may be allowed to be a multigraph [5]. The survivable network design problem arises in many aforementioned applications, for example, in telecommunication, communication network design, VLSI design, and so on.

In many applications of the survivable network design problem, often regarded as the most interesting ones [3,11], the connectivity requirement function is specified with the help of a single-argument function that assigns to each vertex $v$ its connectivity type $r_v \in \mathbb{N}$. Then, for any pair of vertices $v, u \in V$, the connectivity requirement $r_{u,v}$ is simply given as $\min\{r_u, r_v\}$ [3,5,7,12,13]. Following the literature, we assume this standard simplification of the connectivity requirements function in this paper. Note that, in particular, this includes the *Steiner tree problem* [14], in which $r_v \in \{0, 1\}$ for any vertex $v \in V$ (in fact, $r_v = 1$ for any terminal vertex $v$ and $r_v = 0$ for any Steiner possible point [nonterminal vertex] $v$). It also includes the most widely applied variant of the survivability problem in which $r_v \in \{0, 1, 2\}$ for any vertex $v \in V$ [3,5,7].

As all the aforementioned $k$-connectivity problems are known to be $\mathcal{NP}$-hard when restricted to even two-dimensions for $k \geq 2$ [15], we focus on efficient constructions of good approximations. We aim at developing a *polynomial-time approximation scheme*, a PTAS. This is a family of algorithms $\{\mathcal{A}_\varepsilon\}$ such that, for each fixed $\varepsilon > 0$, $\mathcal{A}_\varepsilon$ runs in time polynomial in the size of the input and produces a $(1 + \varepsilon)$-approximation (see Vol. 1, Chap. 9 and [16]).

## 2.1.2 History of the Multiconnectivity Problems

For a very extensive presentation of results concerning problems of finding minimum-cost $k$-vertex- and $k$-edge-connected spanning subgraphs, nonuniform connectivity, connectivity augmentation problems, and geometric problems, we refer the reader to References 1, 4, 17 and to various chapters of Reference 16, especially to References 2, 18. Here, we discuss mostly the work related to geometric graphs.

All the multiconnectivity problems discussed in this survey are known to be $\mathcal{NP}$-hard not only for general graphs, but also for several nontrivial classes of graphs. For general graphs, the multiconnectivity problems are even known to be APX-hard, that is, they do not have a PTAS unless $\mathcal{P} = \mathcal{NP}$ [19]. Despite the practical relevance of the multiconnectivity problems for geometrical graphs and the vast amount of practical heuristic results reported [3,5,7,13], very little theoretical research has been done towards developing efficient approximation algorithms for these problems until 1998. This contrasts with the very rich and successful theoretical investigations of the corresponding problems in general metric spaces and for general weighted graphs. And so, until 1998, even for the simplest and most fundamental multiconnectivity problem, that of finding a minimum-cost biconnected graph spanning a given set of points in the Euclidean plane, obtaining approximations achieving better than a $\frac{3}{2}$ ratio had been elusive (the ratio $\frac{3}{2}$ is the best polynomial-time approximation ratio known for general graphs the weights of which satisfy the triangle inequality [20]; for other results see [18,21]).

For many years, the algorithmic community has believed that traveling salesman problem (TSP) and the multiconnectivity problems discussed in this survey were also APX-hard for geometric and planar graphs, and hence they do not have a PTAS unless $\mathcal{P} = \mathcal{NP}$. However, the situation changed dramatically with the seminal works of Arora [14] and Mitchell [22], who showed that the TSP problem in geometric graph has a PTAS; soon after, a PTAS for TSP in planar graphs has been also developed by Arora et al. [23]. These results gave a hope that also the multiconnectivity problems in geometric and planar graphs can have PTASs. This has been proven affirmatively in a series of papers by Czumaj and Lingas [19,24,25] in the case of geometric graphs. The case of planar graphs seemed to be harder a decade ago, when only quasi-polynomial time approximation schemes (running time of the form $n^{\widetilde{\mathcal{O}}(\log n/\varepsilon)}$) were known for 2-connectivity problems [26,27]. More recently, fast polynomial-time approximations schemes have been developed for some variants of the aforementioned problems, see, for example, [28–31].

### 2.1.3 Overview of the Results

In this survey, we overview the PTASs for multiconnectivity problems in geometric graphs, as developed in a series of papers by Czumaj et al. [19,24,25,32] (see also [33,34] for full versions of the papers). Apart from presenting the specific results, we give a detailed tutorial of techniques developed during the design of PTASs for various $k$-connectivity problems in geometric graphs; we also emphasize the difference between these algorithms and the recent PTASs for TSP and related problems. In addition, we discuss lower bounds on approximability of these problems in higher dimensions [19] and extensions to include the survivability problems [32] and planar graphs [26,27,30,31].

## 2.2 Preliminaries

In this section, we introduce basic technical definitions and notions used in this survey. For simplicity of presentation, we shall assume that the quality of approximation $\varepsilon$ satisfies $n^{-1/4} < \varepsilon \leq 0.1$. Furthermore, we shall aim at achieving an approximation of $(1 + \mathcal{O}(\varepsilon))$ rather than $(1 + \varepsilon)$. Both these assumptions can be easily relaxed.

All algorithms for geometric graphs that we discuss in this survey are randomized. Even though the algorithms can be derandomized, and the final results are stated in deterministic versions as well, the randomized versions of the algorithms are more natural to present and are simpler.

For a given undirected graph $G$, a *traveling salesman tour* (TST) is any Hamiltonian cycle in $G$; a *traveling salesman path* is any Hamiltonian path in $G$. For a given set $S$ of points, a TST is any TST for the complete graph on $S$. For a given graph $G$ with cost on the edges, or for a set of points $S$ in a metric space, the *TSP* is to find a TST in $G$ or for $S$, respectively, that has a minimum total cost of the edges. For simplicity of our presentation, we define a TST for a set of *two* points to be the edge connecting these points.

We use term $L^d$-*cube* with $L \in \mathbb{R}$ to denote any axis-parallel $d$-dimensional cube in $\mathbb{R}^d$ of side-length $L$ in all $d$ dimensions. A *bounding box* of the input multiset of points in $\mathbb{R}^d$ is any $L^d$-cube in $\mathbb{R}^d$ enclosing these points (see Figure 2.1).

*The perturbation*: In our algorithms for multiconnectivity problems, we first *perturb* the input instance so that each node lies on the unit grid and every internode distance is at least 8. We begin with rescaling the input so that the smallest bounding box $L^d$ has $L = \mathcal{O}(n^3)$ being a power of 2. Next, we move every point to the nearest point on the unit grid, all coordinates of which are multipliers of 8 (what may merge some points). Then, it is easy to see that the perturbation ensures that any $k$-VCSS for the original input instance is now mapped into a $k$-VCSS whose cost (after rescaling) differs by at most an $\varepsilon$ fraction. Therefore, if we can find a $(1 + \mathcal{O}(\varepsilon))$-approximation for the $k$-VCSS problem for the perturbed instance, then we can directly obtain a $(1 + \mathcal{O}(\varepsilon))$-approximation for the $k$-VCSS problem for the original instance. Because of

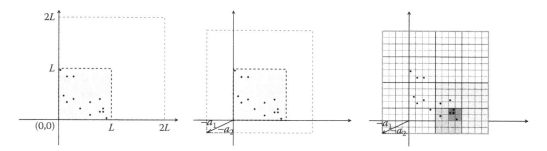

**FIGURE 2.1** Shifted dissection of a set of points in the bounding box $[0, L]^d$ in $\mathbb{R}^2$ (with $\mathbf{a} = (a_1, a_2)$).

that, from now on we assume that all input points have integer coordinates, lie in the cube $[0, L]^d$ with $L = \mathcal{O}(n^3)$ being a power of 2, and the distance between any two points is either 0 or is at least 8. (We chose $L = \mathcal{O}(n^3)$ for convenience only; a much smaller $L$ would be enough.)

*The dissection*: The concept of space partitioning via *dissections* (quadtrees) and *shifted dissections* plays the key role in all our algorithms. Following Reference 14, we define the geometric partitioning of a bounding box as follows: A *($2^d$-ary) dissection* of the bounding box $L^d$ of a set of points in $\mathbb{R}^d$ is its recursive partitioning into smaller subcubes, called *regions*. Each region $U^d$ of volume larger than 1 is recursively partitioned into $2^d$ regions $(U/2)^d$. A *$2^d$-tree* with respect to the ($2^d$-ary) dissection is a tree whose root corresponds to the bounding box, and other nonleaf nodes of which correspond to the regions containing at least two points from the input multiset. For a nonleaf node $v$ of the tree, the nodes corresponding to the $2^d$ regions partitioning the region corresponding to $v$ are the children of $v$ in the tree. Note that the dissection has $\Theta(L^d)$ regions and its recursion depth is logarithmic in $L$. Furthermore, if $L$ is a power of 2, the boundaries of all regions in the dissection have integer coordinates, and thus they are in the unit grid.

For any $d$-vector $\mathbf{a} = (a_1, \ldots, a_d)$, where all $a_i$ are $0 \leq a_i \leq L$, the *$\mathbf{a}$-shifted dissection* [14] of a set $X$ of points in the bounding box $[0, L]^d$ in $\mathbb{R}^d$ is the result of shifting all the regions in the dissection of $X$ in the bounding box $[0, 2L]^d$ by the vector $(-\mathbf{a})$. The *$\mathbf{a}$-shifted $2^d$-ary tree* with respect to the $\mathbf{a}$-shifted dissection is defined analogously. A *random-shifted dissection* of a set of points $X$ in a cube $L^d$ in $\mathbb{R}^d$ is an $\mathbf{a}$-shifted dissection of $X$ with $\mathbf{a} = (a_1, \ldots, a_d)$ and the elements $a_1, \ldots, a_d$ chosen independently and uniformly at random from $\{0, 1, \ldots, L\}$.

## 2.3 First Approach: Polynomial-Time *Pseudoapproximation* Schemes

After the development of PTASs for the TSP problem due to Arora [14] and Mitchell [22], it seemed to be almost a straightforward task to extend their schemes to obtain a PTAS for multiconnectivity problems, or at least for the most basic 2-VCSS and 2-ECSS problems. However, it turned out that the schemes, which work very well for TSP and for some number of related problems, including Minimum Steiner Tree, Min-Cost Perfect Matching, $k$-TSP, and $k$-Minimum Spanning Tree ($k$-MST), could not be extended in a simple way. The reason was that in all these approximation schemes, the key step was to find a low-cost solution that uses Steiner points. Although (by the triangle inequality) a Steiner point can be removed from a TST without any increase of its cost, such a transformation is impossible for $k$-VCSS and $k$-ECSS problems: For example, a minimum-cost 2-VCSS for a point set $S$ in $\mathbb{R}^2$ can have cost as much as $\frac{\sqrt{3}}{2}$ times larger than a Steiner 2-VCSS for $S$ [35] (see Figure 2.1).

Despite this difficulty, Czumaj and Lingas [24] showed that one can apply the approach developed by Arora [14] to design a "pseudoapproximation scheme": an algorithm that finds a Steiner $k$-VCSS for a point set $S$ whose cost is at most $(1 + \varepsilon)$ times larger than the cost of a minimum-cost $S$-VCSS for $S$. In other words, the algorithm finds a solution with Steiner points that has cost not much larger than an optimal solution that uses no Steiner points. Even though this is only a pseudoapproximation scheme and not a PTAS, in this section we shall present this algorithm in details, because the underlying ideas of this approach are used later in all other algorithms we discuss in this survey.

On a very high level, the approach of Arora [14,22,36,37] is a clever combination of the divide-and-conquer method with the dynamic programming approach, and as such, it follows a design of many classical PTASs. For the multiconnectivity problems, similarly as Arora, we hierarchically partition the cube containing the input points (via random-shifted dissection) into regions, and then prove the key technical result, that there is an approximate solution to the problem which can cross the boundaries of each region only in prespecified points a bounded number of times (*Structure Theorem*). This theorem states that for any problem instance, there is a $(1 + \mathcal{O}(\varepsilon))$-approximation that satisfies some basic local property: It is *m-portal respecting* and *r-light*, see the definitions in the following. This theorem is proven by taking an optimal solution to the problem and applying a sequence of transformations that increases the cost of the resulting graph and at the same time makes it *m*-portal respecting and *r*-light. Once the Structure Theorem is proved, a dynamic programming procedure finds in a polynomial-time an almost optimal solution that satisfies the basic local property. The dynamic programming procedure combines optimal partial solutions within regions into an optimal global solution under the crossing restrictions. To combine solutions efficiently, we derive a *k*-connectivity characteristic of a spanning subgraph within a region solely in terms of the set of prespecified points on the region boundary included in its vertex set. In our crucial theorem, we show that the connectivity characteristic of a union of two adjacent subgraphs can be computed from the connectivity characteristics of the subgraphs. This allows us to set up a dynamic programming procedure computing a $(1 + \mathcal{O}(\varepsilon))$-approximation of minimum-cost Euclidean graph that is *k*-vertex or *k*-edge connected and obeys the crossing restrictions.

In the following, we discuss this approach in more detail; we focus only on the *k*-VSCC problem and note that the extension to the *k*-ECSS problem is straightforward.

## 2.3.1 Special Forms of Geometric Graphs

### 2.3.1.1 *m*-Portal-Respecting Graphs

For every integer *m*, an *m-regular* set of *portals* in a $(d-1)$-dimensional region facet $U^{d-1}$ is an orthogonal lattice of *m* points in the facet in which the spacing between the portals is $(U + 1) \cdot m^{-1/(d-1)}$. A graph is *m-portal-respecting* (with respect to a shifted dissection) if whenever it crosses a facet in the dissection, it does so at a portal (see Figure 2.2). Observe that this restriction forces us to assume that an *m*-portal-respecting graph may have to *bend* some of its edges, that is, an edge may deviate from being a straight line connecting its endpoints and be rather a *straight-line path* between the endpoints. If we are allowed to bend the ends, then for any graph in a dissection, it is easy to make it *m*-portal-respecting by moving every crossing of every facet to its nearest portal. Arora [14] proved the following result that transforms a graph into an *m*-portal-respecting one at a small cost increase and without changing the connectivity.

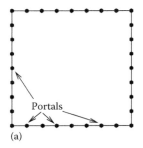 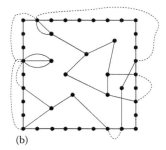

(a)                                                      (b)

**FIGURE 2.2** Portals (a) and a portal-respecting graph (b).

**Lemma 2.1** *Let G be a geometric graph in $\mathbb{R}^d$ for a set of (perturbed) points contained in a bounding box $L^d$. Pick a random-shifted dissection of $L^d$. Then, one can transform G into an m-portal-respecting graph by moving each crossing of each facet to its nearest portal so that the expected increase of the cost of the resulting graph is at most $\mathcal{O}(d \log L \, m^{-1/(d-1)}) \cdot \text{cost}(G)$.*

*Proof.* Pick any edge $(v, u)$. By the definition of the dissection, edge $(v, u)$ crosses the facets in the dissection at most $\mathcal{O}(\sqrt{d}) \cdot \mathfrak{c}(v, u)$ times, where $\mathfrak{c}(v, u)$ is the cost of the edge $(v, u)$.* To make this edge $m$-portal-respecting, we move each crossing of a facet to the nearest portal, which involves bending the edge that might increase its length. If the facet has side-length $L/2^i$, then this increases the distance by at most $\mathcal{O}(\sqrt{d} L/2^i) \, m^{-1/(d-1)}$, as the interportal distance is $\mathcal{O}(L/2^i) \, m^{-1/(d-1)}$. As we have chosen the dissection at random, the probability that a given facet has side-length $L/2^i$ is $\mathcal{O}(2^i/L)$. Hence, the expected increase of the cost of a given edge $(v, u)$ is

$$\sum_{i=0}^{\log L} \mathcal{O}(\sqrt{d}\, 2^i/L) \cdot \mathcal{O}(L/2^i) \, m^{-1/(d-1)} = \mathcal{O}(\sqrt{d} \log L \cdot m^{-1/(d-1)}).$$

The same arguments can be applied to all of at most $\mathcal{O}(\sqrt{d}) \cdot \mathfrak{c}(v, u)$ dissection crossings by any edge $(v, u)$. Therefore, the expected increase of the cost of the entire graph is at most

$$\sum_{(v,u)} \left( \mathcal{O}(\sqrt{d}) \cdot \mathfrak{c}(v, u) \right) \cdot \mathcal{O}(\sqrt{d} \log L \cdot m^{-1/(d-1)}) = \mathcal{O}(d \log L \cdot m^{-1/(d-1)}) \cdot \text{cost}(G). \qquad \blacksquare$$

Note that in our applications, we require this error term to be at most an $\mathcal{O}(\varepsilon)$ factor of the cost of the optimal solution, and therefore we set $m = (\mathcal{O}(d \log L/\varepsilon))^{d-1}$. By using the transformation from Lemma 2.1, from now on, we assume that we consider a geometric graph that is $m$-portal-respecting with $m = (\mathcal{O}(d \log L/\varepsilon))^{d-1}$.

*Special forms of geometric graphs: r-light graphs*: We say a geometric graph is *r-light* (with respect to a shifted dissection) if for each region in the dissection there are at most $r$ edges *crossing any of its facets*.

## 2.3.2 Dynamic Programming and Finding an Optimal *m*-Portal-Respecting *r*-Light Solution

In our presentation, we begin from the end and discuss first the goal of our analysis. In the following section, we show that for any set $S$ of $n$ (perturbed) points in $\mathbb{R}^d$ that are contained in a bounding box $L^d$, if we choose a random-shifted dissection of $L^d$, then with a good probability that there is an $m$-portal-respecting $r$-light (for the dissection chosen) Steiner $k$-VCSS for $S$ whose cost is at most $(1 + \mathcal{O}(\varepsilon))$ times the cost of the optimal $k$-VCSS for $S$, for appropriated values of $m$ and $r$. How can we use this existential result? The key observation is that if we restrict ourself to $m$-portal-respecting $r$-light graphs then we can use dynamic programming to actually find an almost optimal Steiner $k$-VCSS efficiently! In what follows, we briefly discuss main ideas of this result; see [24] for more details.

The key idea is that the subproblem (finding an optimal Steiner $k$-VCSS) inside a region in the dissection can be solved independently of the subproblems in other regions provided that we know which portals are used by the edges of the graph and the structure of the external $k$-connectivity properties outside that region. External $k$-connectivity properties outside a region are defined in terms of the portals used: If a portal is used by the graph outside the current region, we want to know which connections with other portals it supports. The concept of *connectivity characteristic* developed in Reference 24 aims at maintaining this structural properties of graphs contained in any region.

---

* The number of crossings of the facets in the dissection is upper bounded by a constant times the $\ell_1$ distance between $v$ and $u$, and this is upper bounded by $\mathcal{O}(\sqrt{d})$ times the $\ell_2$ distance between $v$ and $u$, that is, $\mathcal{O}(\sqrt{d}) \cdot \mathfrak{c}(v, u)$.

Let us define an *internal interface* of a region $Q$ to be any multiset of at most $m$ portals, such that for every facet of $Q$, the total multiplicity of all portals (in the multiset) is upper bounded by $r$. As our goal is to find an $m$-portal-respecting and $r$-light Steiner $k$-VCSS with low cost, we do not know its structure in advance (except that it is $m$-portal-respecting and $r$-light), and hence in our algorithm we have to consider all possible internal interfaces. Note that for $m$-portal-respecting and $r$-light graphs, the number of internal interfaces of a region is at most $m^{\mathcal{O}(d\,r)}$.

Next, for any region $Q$ and any given internal interface of $Q$, we define a *connectivity characteristic* to be a description of routing properties within the region and requirements on the routing properties in the complementary graph from the point of view of portals needed to preserve $k$-connectivity. Let $P$ be the multiset of points in the portals used in the internal interface. The connectivity characteristic consists of three parts corresponding to different aspects of $k$-vertex connectivity requirements for the graph as follows:

- Requirements for *internal connectivity*: What configurations of external disjoint paths ought to be outside the region to make any pair of vertices within region $k$-vertex connected; as for any pair of points in $Q$, all sets of disjoint paths leaving $Q$ must traverse through the portals, each such a set of vertex-disjoint paths can be encoded by a matching in the complete graph on $P$.
- Requirements for *internal/external connectivity*: What configurations of disjoint paths ought to be inside and outside $Q$ to ensure that any vertex inside region $Q$ has $k$ vertex-disjoint paths to any vertex outside the region; this can be encoded by a set of pairs consisting of a matching in the complete graph on $P$ and a subset of portals (that is used to encode the parts of the paths from a vertex inside $Q$ to the first portals, before they leave $Q$).
- Requirements for *external connectivity*: What configurations of internal disjoint paths ought to be inside $Q$ to ensure that any pair of vertices outside $Q$ are connected by $k$ vertex-disjoint paths; this can be encoded by matchings in the complete graph on $P$.

One can show that for a given region and its internal interface, there are at most $2^{(d\,r)^{\mathcal{O}(d\,r)}}$ connectivity characteristics.

The goal of the dynamic programming procedure is to determine for each region $Q$, for each possible internal interface of $Q$, and for each possible connectivity characteristic of $Q$, an (almost) optimal $m$-portal-respecting $r$-light graph within the region using given internal interface and having given connectivity characteristic. We maintain a lookup table that, for each region, each internal interface, and each connectivity characteristic, stores the optimal way to solve the subproblem inside the region. The lookup table is created bottom-up and the efficiency of this procedure relies on the efficiency of computing the connectivity characteristic for a region from its $2^d$ subregions one level down in the $2^d$-dissection tree. One can find a minimum-cost graph within region $Q$ having a given characteristic by combining minimum-cost graphs within subregions of $Q$, and this can be done in time $m^{d\,2^d\,r} \cdot 2^{(d\,r)^{\mathcal{O}(d\,r)}}$. This approach has to be refined for regions corresponding to the leaves in the $2^d$-dissection tree, in which we have to find an optimal graph directly. On the contrary, as we do not know the locations of Steiner points in an optimal solution, we can only find an approximate solution within every leaf region. Still, this is enough to conclude with the following result (see [24] for more details) as follows:

**Lemma 2.2** *Let $S$ be a (perturbed) point set in $\mathbb{R}^d$ contained in a bounding box $L^d$ and with minimum nonzero interdistance at least 8. Let $m$ and $r$ be integer parameters. Then, in time $n \cdot \log L \cdot m^{\mathcal{O}(d\,2^d\,r)} \cdot 2^{(d\,r)^{\mathcal{O}(d\,r)}}$ one can find a $(1 + \mathcal{O}(\varepsilon))$-approximation of a minimum-cost $m$-portal-respecting $r$-light Steiner $k$-VCSS for $S$.*

### 2.3.3 Patching Lemma: Reducing Number of Crossings Using Steiner Points

In this section, we discuss a *patching* procedure (initially used by Arora [14] for TSP) that is a key ingredient of our result that for any set of points and for a random-shifted dissection of its bounding box, there is always an $m$-portal-respecting $r$-light Steiner $k$-VCSS for $S$ whose cost is low. The patching procedure takes any facet crossed by more than $k$ edges and patches the crossings to reduce the number of crossings to at most $k$, by augmenting the original graph with new Steiner vertices and new edges (line segments).

**Lemma 2.3 (Patching Lemma)** *Let $\mathcal{F}$ be a $(d-1)$-dimensional facet of side-length $W$ and let $H$ be any Steiner $k$-VCSS (for some point set $S$) that crosses $\mathcal{F}$ exactly $\ell$ times, $\ell > k$. Then, one can break edges of $H$ in all but $k$ of the crossings and add to $H$ new Steiner vertices (that lie infinitesimally close to $\mathcal{F}$) and line segments of total cost at most $\mathcal{O}(k \cdot W \cdot \ell^{1-1/(d-1)})$ such that $H$ changes into a $k$-VCSS $H^*$ for $S$ that crosses $\mathcal{F}$ at most $k$ times.*

*Proof.* Let $x_1, \ldots, x_\ell$ be the points at which $H$ crosses the $(d-1)$-dimensional facet $W^{d-1}$-cube $\mathcal{F}$. For each $i$, $1 \le i \le \ell$, break the edge $(y_i, z_i)$ crossing $\mathcal{F}$ at $x_i$ into two parts, one on each side of $\mathcal{F}$; we assume that all vertices $y_1, \ldots, y_\ell$ are on the same side of $\mathcal{F}$. We consider $2k+4$ copies of each $x_i$, denoted by $x_{i,j}^+$, and $x_{i,j}^-$ with $0 \le j \le k+1$; $k+2$ copies for each side of $\mathcal{F}$. We assume that all copies are at distance zero from each other.

Now, we define $H^*$. $H^*$ is obtained from $H$ by removing all the edges crossing $\mathcal{F}$, and inserting the vertices $\{y_1, \ldots, y_\ell\} \cup \{z_1, \ldots, z_\ell\} \cup \bigcup_{1 \le i \le \ell \,\&\, 0 \le j \le k+1}\{x_{i,j}^+\} \cup \bigcup_{1 \le i \le \ell \,\&\, 0 \le j \le k+1}\{x_{i,j}^-\}$ and eight groups of edges:

1. Two halves of each edge crossing $\mathcal{F}$ in $H$ in the form of the edges $\{y_i, x_{i,0}^+\}$ and $\{x_{i,0}^-, z_i\}$, for all $1 \le i \le \ell$
2. Edges crossing $\mathcal{F}$ that connect $x_{i,k+1}^+$ with $x_{i,k+1}^-$, for all $1 \le i \le k$
3. $k$ Edges connecting $x_{i,0}^+$ with $x_{i,j}^+$, for all $1 \le i \le \ell$, $1 \le j \le k$
4. Edges connecting $x_{i,k+1}^+$ with $x_{i,j}^+$, for all $1 \le i \le \ell$, $1 \le j \le k$
5. Edges connecting $x_{i,0}^-$ with $x_{i,j}^-$, for all $1 \le i \le \ell$, $1 \le j \le k$
6. Edges connecting $x_{i,k+1}^-$ with $x_{i,j}^-$, for all $1 \le i \le \ell$, $1 \le j \le k$
7. Edges of a traveling salesman path for $\bigcup_{1 \le i \le \ell}\{x_{i,j}^+\}$, for all $1 \le j \le k$
8. Edges of a traveling salesman path for $\bigcup_{1 \le i \le \ell}\{x_{i,j}^-\}$, for all $1 \le j \le k$

(observe that all edges in groups (2)–(6) have cost zero [infinitesimally small], because we assumed that for every $i$ and $j$, $1 \le i < \ell$, $0 \le j \le k+1$, all nodes $x_{i,j}^+$, and $x_{i,j}^-$ are at distance zero from each other.) (see Figure 2.3).

It is easy to see that the cost of the nonzero length edges in $H^* \setminus H$ is bounded from above by the cost of the edges in $H$ plus the cost of $2k$ traveling salesman paths for the point sets $\bigcup_{1 \le i \le \ell}\{x_{i,j}^+\}$, $\bigcup_{1 \le i \le \ell}\{x_{i,j}^-\}$,

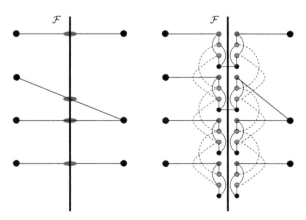

**FIGURE 2.3** Graph $H^*$ constructed in the Patching Lemma. Dotted lines correspond to the traveling salesman paths. In this example $d = 2$, $\ell = 4$, and $k = 2$.

$j = 1, \ldots, k$, respectively. Now, a well-known result about geometric TSP (see, e.g., Vol. 1, Chap. 6 in Reference 38) implies that for any set of $\ell$ points contained in a $(d-1)$-dimensional $W^{d-1}$ cube, there is a traveling salesman path of total length smaller than $\mathcal{O}(W \ell^{1-\frac{1}{d-1}})$. Therefore, we can conclude that the total additional cost is bounded by $\mathcal{O}(k\, W\, \ell^{1-\frac{1}{d-1}})$.

Finally, it is not hard to show that $H^*$ satisfies the vertex-connectivity requirements.   ∎

### 2.3.4 Structure Theorem: There Is Always a Good $r$-Light Steiner $k$-Vertex Connected Spanning Subgraph

Now, we are ready to present the first Structure Theorem for the $k$-VCSS problem. This theorem compares the cost of an $m$-portal-respecting $r$-light Steiner $k$-VCSS for a set of points with the cost of an optimal $k$-VCSS for this set of points, in which the optimal solution is not allowed to use Steiner points.

**Theorem 2.1 (Structure Theorem)** *Let $S$ be a (perturbed) point set in $\mathbb{R}^d$ contained in a bounding box $L^d$ and with minimum nonzero interdistance at least 8. Pick a random-shifted dissection of $L^d$. Then, with probability at least 0.9, there is an $m$-portal-respecting $r$-light Steiner $k$-VCSS for $S$ whose cost is at most $(1 + \mathcal{O}(\varepsilon))$-time the optimal $k$-VCSS for $S$, where $m = (\mathcal{O}(d \log L/\varepsilon))^{d-1}$ and $r = (\mathcal{O}(\sqrt{d}\, k/\varepsilon))^{d-1}$.*

The proof of the Structure Theorem follows from the aforementioned Patching Lemma by repeatedly patching the original graph in an appropriated order of facets, following the original approach of Arora. This part of the analysis is technical and subtle, and we only sketch it here; a reader interested in more detail is referred to Reference 24 or References 14, 36.

*Sketch of the proof.* The idea is to transform an optimal $k$-VCSS for $S$ into a $r$-light Steiner $k$-VCSS for $S$ of low cost by applying the Patching Lemma 2.3 to every facet that is crossed too often. Lemma 2.3 ensures that the resulting graph is a $r$-light Steiner $k$-VCSS for $S$. However, as its every application increases the cost of the resulting graph, it is crucial to show that the expected cost of the resulting graph is at most $(1 + \mathcal{O}(\varepsilon))$-time the optimal $k$-VCSS for $S$. If we prove this claim, then the lemma follows by applying Lemma 2.1 and by Markov inequality.

We bound the total cost of the new edges resulting from invoking the Patching Lemma by charging their cost to grid hyperplanes. For every facet in the dissection, we charge the cost of removing the excess of the edges crossing the facet to the grid hyperplane that contains the facet. We show that the expected cost charged to a grid hyperplane $\mathcal{H}$ is at most $\varepsilon\, t(\mathcal{H})/(2\sqrt{d})$, where $t(\mathcal{H})$ is the number of crossings of the hyperplane $\mathcal{H}$ by the optimal $k$-VCSS for $S$. Now, the result follows by the linearity of expectations and by the fact that $\sum_{\mathcal{H}} t(\mathcal{H})$ is at most $2\sqrt{d}$ times the cost of the optimal $k$-VCSS for $S$ (this result is obtained by well-known relation between the $\ell_1$ and $\ell_2$ norms).

Let us fix a grid hyperplane $\mathcal{H}$ perpendicular to some coordinate axis. Note that within the bounding box $L^d$, $\mathcal{H}$ forms an $L^{d-1}$ cube. We apply the Patching Lemma to all facets of the dissection that belong to $\mathcal{H}$. We first begin with the smallest facets and then consider the facets in the increasing order of their sizes. Let $c_j$ be the number of facets in $\mathcal{H}$ of side-length $L/2^j$ for which patching has been invoked. For $\ell \leq c_j$, let $t_{j,\ell}$ be the number of crossings of the $\ell$th facet of side-length $L/2^j$ for which patching has been applied. Observe that for the $\ell$th facet of side-length $L/2^j$ for which patching has been applied, the total cost of the new edges added by the Patching Lemma is upper bounded by $\mathcal{O}(k\,(L/2^j)\,(t_{j,\ell})^{1-1/(d-1)})$. Therefore, if the largest facet in the hyperplane $\mathcal{H}$ has side-length $L/2^i$, then the total cost of the new edges added by applying the Patching Lemma to all facets in $\mathcal{H}$ is upper bounded by as follows:

$$\mathcal{O}\left( \sum_{j=i}^{\log L} \sum_{\ell=1}^{c_j} k\,(L/2^j)\,(t_{j,\ell})^{1-1/(d-1)} \right). \tag{2.1}$$

Next, we study the expected cost as earlier, in which the expectation is taken over shifts chosen in the random-shifted dissection. Let us assume that the grid hyperplane $\mathcal{H}$ is perpendicular to the $s$th coordinate axis. Let us fix the random vector $\mathbf{a} = (a_1, \ldots, a_d)$ used to determine the random-shifted dissection in which all elements are fixed with the exception of $a_s$, which is kept random. We observe that the random shift in the dissection depends only on the value of $a_s$, and therefore, if $a_1, \ldots, a_{s-1}, a_{s+1}, \ldots, a_d$ are fixed, then the probability that the largest facet in the hyperplane $\mathcal{H}$ has side-length $L/2^i$ is $\mathcal{O}(2^i/L)$. Furthermore, one can show that the values of $c_j$ and $t_{j,\ell}$ are independent of $a_s$. Therefore, the expected cost of all edges added by applying patching to all facets in $\mathcal{H}$ is at most as follows:

$$\sum_{i=0}^{\log L} \mathcal{O}(2^i/L) \cdot \mathcal{O}\left( \sum_{j=i}^{\log L} \sum_{\ell=1}^{c_j} k\,(L/2^j)\,(t_{j,\ell})^{1-1/(d-1)} \right) \leq \mathcal{O}\left( \sum_{j=0}^{\log L} \sum_{\ell=1}^{c_j} k/2^j\,(t_{j,\ell})^{1-1/(d-1)} \sum_{i=0}^{j} 2^i \right)$$

$$= \mathcal{O}(k) \cdot \sum_{j=0}^{\log L} \sum_{\ell=1}^{c_j} (t_{j,\ell})^{1-1/(d-1)}.$$

As $t_{j,\ell} \geq r+1$, the bound above is maximized when each $t_{j,\ell} = r+1$, and therefore it is bounded by $\mathcal{O}(k) \cdot (r+1)^{1-1/(d-1)} \cdot \sum_{j=0}^{\log L} c_j$. Now, we need a good upper bound for $\sum_{j=0}^{\log L} c_j$. As each application of the Patching Lemma reduces the number of crossings of $\mathcal{H}$ by at least $r+1-k$, the definition of $t(\mathcal{H})$ yields

$$\sum_{j=0}^{\log L} c_j \leq \frac{t(\mathcal{H})}{r+1-k}.$$

Therefore, the expected cost of all edges added by applying patching to all facets in $\mathcal{H}$ is upper bounded by as follows:

$$\mathcal{O}(k) \cdot (r+1)^{1-1/(d-1)} \cdot \sum_{j=0}^{\log L} c_j \leq \mathcal{O}\left( \frac{k\,(r+1)^{1-1/(d-1)}\,t(\mathcal{H})}{r+1-k} \right).$$

We set $r = (\mathcal{O}(\sqrt{d}\,k/\varepsilon))^{d-1}$ to upperbound this by $\varepsilon\,t(\mathcal{H})/(2\sqrt{d})$. By our previous arguments, this implies that the expected cost of all edges added by applying the Patching Lemma (which results in a transformation of the graph into an $r$-light one) to an optimal $k$-VCSS for $S$ is at most $\varepsilon$ times the cost of the optimal $k$-VCSS for $S$.

Finally, we have to transform the graph into $m$-portal-respecting. We apply the construction presented in Lemma 2.1 with the value of $m = (\mathcal{O}(d\,\log L/\varepsilon))^{d-1}$. As, by Lemma 2.1, this construction increases in expectation, the cost of the graph by at most a factor of $\mathcal{O}(\varepsilon)$, the final result follows. ∎

## 2.3.5 Final Result: *Pseudoapproximation* Schemes for Multiconnectivity Problems

The results from the previous sections (Lemma 2.2 and Theorem 2.1) are summarized in the following theorem.

**Theorem 2.2** *Let $k$ and $d$ be any integers, $k, d \geq 2$, and let $\varepsilon$ be any positive real. Let $S$ be a set of $n$ points in $\mathbb{R}^d$. There is a randomized algorithm which finds a Steiner $k$-VCSS for $S$, whose cost is at most $(1 + \varepsilon)$-time the optimal $k$-VCSS for $S$, in time $n \cdot (\log n)^{(\mathcal{O}(\sqrt{d}\,k/\varepsilon))^{d-1}} \cdot 2^{(d\,k/\varepsilon)^{(\mathcal{O}(\sqrt{d}\,k/\varepsilon))^{d-1}}}$ with probability at least 0.9.*

*Furthermore, within the same running time, one can find a Steiner $k$-ECSS for $S$ whose cost is at most $(1 + \varepsilon)$-time the optimal $k$-ECSS for $S$. Also, all these algorithms can be derandomized in polynomial time.*

Observe that when all $d$, $k$, and $\varepsilon$ are constant, the running time of the randomized algorithm is $n \cdot (\log n)^{\mathcal{O}(1)}$. When $d$ is a constant and $k$ and $\varepsilon$ are arbitrary, then the running time is $n \cdot (\log n)^{(k/\varepsilon)^{\mathcal{O}(1)}} \cdot 2^{2^{(k/\varepsilon)^{\mathcal{O}(1)}}}$.

# 2.4 Polynomial-Time Approximation Scheme for Geometric Multiconnectivity Problems

The results from the previous section are certainly not fully satisfactory, and a natural question arises if we can obtain a similar result *without using Steiner points* in the solution. In this section, we discuss in details how one can modify the approach from Section 2.3 to obtain a PTAS for geometric multiconnectivity problems. Even if this method can be seen as a generalization of the approach developed initially by Arora [14], the details of the new construction are significantly different than those used for TSP and related problems. The material in this section is based on Reference 33, an updated and improved version of References 19, 25.

The main idea of the PTAS is similar to that from the previous section: We want to prove a result of the form similar to that from Structure Theorem 2.1. However this time, we want to make sure that no new Steiner points difficult to remove are created. We achieve this goal by aiming at a variant of the Structure Theorem that does not require the resulting graph to be $r$-light but only $r$-*locally-light*, see the definition in the following. The difference between these two requirements is insignificant for the dynamic programming phase, but it is critical in our analysis: As we show in our main theorem, there is always an almost optimal $k$-VCSS for a set of points in $\mathbb{R}^d$ that is $m$-portal-respecting and $r$-locally-light for small values of $m$ and $r$. Before we proceed on, we begin with introducing some new notation.

*Relevant crossings and vital edges:* A crossing of an edge with a region facet of side-length $W$ in a dissection is called *relevant* if it has exactly one endpoint in the region and its length is at most $2\sqrt{d}\,W$. For a given region $Q$ in a shifted dissection, any edge having exactly one endpoint in $Q$ is called *vital* (for $Q$).

*Special forms of geometric graphs: $r$-gray and $r$-locally-light graphs:* We say a geometric graph is $r$-*gray* (with respect to a shifted dissection) if for each region in the dissection there are at most $r$ relevant crossings. A graph is $r$-*locally-light* (with respect to a shifted dissection) if each region in the dissection has at most $r$ vital edges.

*Augmented TSTs:* A $k$th *power* of a graph $G$ is obtained by augmenting $G$ by the edges whose endpoints are connected by paths consisting of at most $k$ edges in $G$. For any set $S$ and $\ell$, an $\ell$-*augmented TST on* $S$ is either a clique on $S$ if $|S| \le 2\ell$, or the $\ell$th power of some TST on $S$ if $|S| \ge 2\ell + 1$.

## 2.4.1 Transformation Lemmata

In this section, we present a variant of the Structure Theorem designed to deal with the problem of finding a minimum-cost $k$-VCSS for a set of points in $\mathbb{R}^d$. Our goal is to obtain a similar claim as the Structure Theorem 2.1 but without the assumption that the promised graph has Steiner points. We prove this new Structure Theorem in three steps. We take an optimal solution for the minimum-cost $k$-VCSS problem, and we modify it to a suitable form to obtain a graph that is still $k$-VCSS and whose cost is just slightly larger than that of the minimum-cost. In the first two steps, we remove some number of edges (and thus, we do not increase the cost of the graph) to ensure that the resulting graph is first $r$-gray and then $r$-locally-light. In the third step, we add replacement of the removed edges to ensure that the obtained graph is $k$-VCSS. The first and the third steps are randomized, and they show that in expectation the cost of the resulting graph is at most $(1 + \mathcal{O}(\varepsilon))$ times the minimum-cost $k$-VCSS.

### 2.4.1.1 Local Decomposition Lemma

In this section, we discuss our first key result in the analysis, the so-called Local Decomposition Lemma. The Local Decomposition Lemma aims at reducing the number of *relevant* crossings of any given facet to at most $k$. This procedure is very similar in the spirit to the Patching Lemma 2.3. However, unlike the previously known approaches, the Local Decomposition Lemma *does not use any Steiner points*. The key feature of this construction is that it *only removes edges* and the decision which new edges should be inserted to ensure the connectivity requirements is *delayed*. Instead, a description of properties the new edges must satisfy is provided and these edges are inserted only at the very end of the algorithm (using the TST Covering Lemma 2.7).*

To streamline maintaining the connectivity properties of the missing edges, we always describe missing edges in a form of $k$-augmented TSTs. The idea is that to ensure that a set of points is $k$-connected, it is enough to maintain its TST and then observe that the $k$-augmented TST is $k$-connected. Furthermore, by controlling the cost of a minimum-cost TST for that set of points, we can also control an upper bound for a minimum-cost $k$-augmented TST for these points. This will be important in our analysis.

**Lemma 2.4 (Local Decomposition Lemma)** *Let $G$ be a Euclidean graph on a multiset $S$ of points in $\mathbb{R}^d$. Let $\mathcal{F}$ be a $(d-1)$-dimensional facet of side length $W$ in a dissection of the bounding box of $S$. If the edges of $G$ form $\ell$ relevant crossings of $\mathcal{F}$, then there exist a subgraph $G^*$ of $G$, and two disjoint subsets $S_1$ and $S_2$ of $S$, such that*

- *There are at most $2k^2$ relevant crossings of $\mathcal{F}$ in $G^*$*
- *There are a TST on $S_1$ and a TST on $S_2$ such that the cost of each is upper bounded by $\mathcal{O}(d\, W\, \ell^{1-\frac{1}{d}})$*
- *If $G$ is a $k$-VCSS on $S$, then the graph $H^*$ resulting from the graph $G^*$ by adding **any** $k$-augmented TST on $S_1$ and **any** $k$-augmented TST on $S_2$, is a $k$-VCSS on $S$*

**Remark 2.1** *There are three key differences between the Local Decomposition Lemma and the Patching Lemma 2.3: (i) The Local Decomposition Lemma does not introduce any new points to the obtained graph; (ii) it reduces only the number of relevant crossings, leaving the number of arbitrary crossings possibly arbitrarily large; and (iii) it does not produce a $k$-VCSS on $S$, but rather it says that one can build one by adding some additional edges.*

**Remark 2.2** *For a given TST $\mathcal{T}$ on $X$, it is easy to construct a $k$-augmented TST $\mathfrak{T}^{(k)}$ on $X$ such that the cost of $\mathfrak{T}^{(k)}$ is at most $\binom{k+1}{2} \leq 2k^2$ times larger than the cost of $\mathcal{T}$ and each hyperplane $\mathcal{H}$ (which does not contain any edge from $\mathcal{T}$) is crossed by the edges of $\mathfrak{T}^{(k)}$ at most $\binom{k+1}{2} \leq 2k^2$ times more than it is crossed by the edges of $\mathcal{H}$.*

*Proof.* We can assume $\ell > 2k^2$. We first construct the subgraph $G^*$ and the subsets $S_1$ and $S_2$ and then briefly argue about their properties.

Let $\mathcal{E}$ be the set of the $\ell$ edges of $G$ forming the $\ell$ *relevant* crossings with $\mathcal{F}$. We define $S_1 = \{x_1, \ldots, x_\ell\}$ as the set of endpoints of the edges in $\mathcal{E}$ in the first half-space induced by $\mathcal{F}$ and $S_2 = \{y_1, \ldots, y_\ell\}$ as the corresponding set of endpoints of these edges in the other half-space. Next, we define $G^*$. $G^*$ is obtained by removing from $G$ a subset of the edges in $\mathcal{E}$. Let $\mathbb{M}$ be a maximum cardinality subset of $\mathcal{E}$ such that no two edges in the subset are incident. Let $q = \min\{k, |\mathbb{M}|\}$. Then, we define the set $\mathcal{E}^*$ of edges in $\mathcal{E}$ that will remain in $G^*$ to consist of

---

* One can ask why do we delay inserting the new edges: For example, in a similar situation in Arora's PTAS for TSP [14], the new edges are inserted at once, as we also do in the analysis of the Structure Theorem 2.1. Note however that Arora [14] and others were always able to place the new edges on the facet for which the Patching Lemma is applied, which facilitates dealing with the new crossings. In the case discussed here, we do not want to create Steiner points and therefore we need to add new edges in arbitrary locations.

- The first $q$ edges of $\mathbb{M}$
- If $q < k$, then, in addition, for each endpoint $v$ of each edge from $\mathbb{M}$ we add to $\mathcal{E}^*$ $\min\{k-1,$ $deg_{\mathcal{E}}(v) - 1\}$ edges in $\mathcal{E}\backslash\mathbb{M}$ incident to $v$, where $deg_{\mathcal{E}}(v)$ is the number of edges in $\mathcal{E}$ incident to $v$

Now, the graph $G^*$ is obtained from $G$ by removing the edges in $\mathcal{E}\backslash\mathcal{E}^*$.

To complete the proof, we must show that $G^*$, $S_1$, and $S_2$ satisfy the properties promised in the lemma. Clearly, $\mathcal{E}^*$ is of size at most $2k^2$, and hence there are at most $2k^2$ relevant crossings of $\mathcal{F}$ in $G^*$. Furthermore, each of $S_1$ and $S_2$ consists of at most $\ell$ vertices that are contained in a bounding box of size $\mathcal{O}(\sqrt{d}\,W)$. (Indeed, as the vertices in $S_1$ and $S_2$ are endpoints of relevant crossings $\mathcal{F}$, their distance from $\mathcal{F}$ is bounded by $2\sqrt{d}\,W$.) Thus, there is a TST on each of $S_1$ and $S_2$ of total length smaller than $\mathcal{O}(d\,W\,\ell^{1-\frac{1}{d}})$ (Section 6 in [38]).* The remaining properties can be also easily shown, see [19,33] for details. ∎

### 2.4.1.2 Weak (too Weak) Version of Structure Theorem: Global Decomposition Lemma

With the above-mentioned Local Decomposition Lemma, we can provide a weak version of the Structure Theorem that uses similar arguments as those used in the proof of Theorem 2.1. As this formulation is too weak for our applications, our goal in the following sections will be to extend it to obtain a stronger result.

**Lemma 2.5 (Global Decomposition Lemma)** *Let $S$ be a (perturbed) point set in $\mathbb{R}^d$ contained in a bounding box $L^d$ and with minimum nonzero interdistance at least 8. Pick a random-shifted dissection of $L^d$. Then, there is an r-gray graph $G$ on $S$ and a collection $\mathbb{S}$ of (possible intersecting) subsets of $S$ such that*

- *The cost of $G$ is not larger than the minimum cost of k-VCSS for $S$*
- $r = (\mathcal{O}(k^2\,d^{3/2}/\varepsilon))^d$
- *There is a graph $H$ consisting of (possible nondisjoint) TSTs on every set $X \in \mathbb{S}$ whose expected (over the choice of the random-shifted dissection) total cost is $\mathcal{O}(\varepsilon/k^2)$ times the minimum cost of k-VCSS for $S$*
- *The graph resulting from $G$ by adding **any** k-augmented TSTs on each $X \in \mathbb{S}$ is a k-VCSS on $S$*

*Proof.* The proof of this result mimics the proof of the Structure Theorem 2.1 with the exception of a few modifications that are caused by a different form of the Local Decomposition Lemma 2.4. We take a minimum-cost $k$-VCSS $G_{\text{opt}}$ for $S$ and apply a sequence of the Local Decomposition Lemma to make this graph $r$-gray. As each application of the Local Decomposition Lemma only removes the edges from $G_{\text{opt}}$, the obtained graph $G$ is a subgraph of $G_{\text{opt}}$, and hence its cost is not larger than the minimum cost of $k$-VCSS for $S$. Furthermore, the resulting graph is $r$-gray by Lemma 2.4. This lemma ensures also that if we define $\mathbb{S}$ as the family of sets returned by all calls to the Local Decomposition Lemma, then by adding to $G$ any $k$-augmented TSTs on all $X \in \mathbb{S}$, we obtain a $k$-VCSS on $S$.

What remains to prove is that for the sets $X \in \mathbb{S}$, the total expected costs of minimum-cost TSTs on the sets $X \in \mathbb{S}$ is at most $\mathcal{O}(\varepsilon/k^2)$ times the cost of $G_{\text{opt}}$. The proof of this fact mimics the analysis of the Structure Theorem 2.1. We charge the cost of invoking the Local Decomposition Lemma to a facet contained in a grid hyperplane to that hyperplane. Then, a similar analysis implies that the expected cost of all minimum-cost TSTs on all sets $X \in \mathbb{S}$ resulting from applying the Local Decomposition Lemma to all facets contained in $\mathcal{H}$ is upper bounded by

---

* Note that we need here the assumption that each of $S_1$ and $S_2$ is included in a bounding box of size $\mathcal{O}(\sqrt{d}\,W)$. In contrast, in the Patching Lemma 2.3, the points could be arbitrarily far away from each other and thus, for example, there could be no TSP on $S_1$ of length $o(n)$.

$$\mathcal{O}\left(\frac{d \cdot (r+1)^{1-1/d} \cdot t(\mathcal{H})}{r+1-2\,k^2}\right).$$

Now, if we set $r = (\mathcal{O}(k^2\,d^{3/2}/\varepsilon))^d$, then the same arguments as those used in the proof of the Structure Theorem 2.1 imply that the expected (over the choice of the random-shifted dissection) total cost of minimum-cost TSTs on all sets $X \in \mathbb{S}$ is upper bounded by $\mathcal{O}(\varepsilon/k^2)$ times the minimum cost of $k$-VCSS for $S$. ∎

### 2.4.1.3 Filtering Lemma

The Global Decomposition Lemma 2.5 transforms an arbitrary Euclidean graph $G$ into an $r$-gray graph, so that certain properties of optimal $k$-vertex-connected graphs induced by these graphs are satisfied. There are however stronger requirements for the transformed graph to get a PTAS. Even if after applying the Global Decomposition Lemma each facet in an $r$-gray graph has only $\mathcal{O}(r)$ relevant crossings, many other (longer) crossings are possible. The following Filtering Lemma transforms any $r$-gray graph into an $r^*$-locally-light one by removing a set of edges of total small cost, with the parameter $r^*$ just slightly bigger than $r$.

**Lemma 2.6 (Filtering Lemma)** *Let $r \geq 1$ and let $S$ be a (perturbed) point set in $\mathbb{R}^d$ contained in a bounding box $L^d$ and with minimum nonzero interdistance at least $8$. For a given shifted dissection, let $G = (S, E)$ be any $r$-gray graph on $S$. Then, we can find a subgraph $G^*$ of $G$ that is $r^*$-locally-light for $r^* = \mathcal{O}(r\,d\,\log(d\,k/\varepsilon))$, and such that the total cost of the edges in $G\backslash G^*$ is at most $\mathcal{O}(\varepsilon/k^2) \cdot \mathrm{cost}(G)$.*

The proof of the Filtering Lemma explores the property that if a graph is $r$-gray, then for every region $Q$ of side length $L$ in the dissection there are at most $2\,d\,r$ vital edges for $Q$ the length of which is in the interval $(2^j\,\sqrt{d}\,L, 2^{j+1}\,\sqrt{d}\,L]$ for every value of $j$. This implies that if there are many vital edges crossing any single facet then most of them (all but a small number of the heaviest edges) have small cost. Therefore, one can transform any $r$-gray graph into an $r^*$-locally-light one by deleting some number of short edges whose total cost (by careful charging arguments) is low.

### 2.4.1.4 Traveling Salesman Tour Covering Lemma

In the previous subsections, we have transformed a Euclidean graph into the one that possesses fewer edges crossing each facet in the dissection. The key feature of the Global Decomposition Lemma and the Filtering Lemma is that after the graph transformations we are left with some (possible intersecting) sets of nodes that are to be connected in some way (either in pairs by edges or into $k$-augmented TSTs). The main reason of such construction was to postpone immediately connecting the nodes within each set because this could introduce many new crossings and might destroy the $r$-locally-lightness of the graph. The following TST Covering Lemma shows how to connect the nodes within each set without increasing the cost of the graph too significantly and without introducing too many crossings of any facet.

We need a definition of a *cover* of a superset that can be seen as a way of connecting multiple TSTs. Let $\mathbb{S}$ be a collection of (not necessarily disjoint) sets. A collection $\mathbb{S}^*$ is called a *cover* of $\mathbb{S}$ if (i) for every $X \in \mathbb{S}$ there is a $Y \in \mathbb{S}^*$ such that $X \subseteq Y$ and (ii) $\bigcup_{X \in \mathbb{S}} X = \bigcup_{Y \in \mathbb{S}^*} Y$. Now, we are ready to state the TST Covering Lemma.

**Lemma 2.7 (TST Covering Lemma)** *Let $S$ be a (perturbed) point set in $\mathbb{R}^d$ contained in a bounding box $L^d$ and with minimum nonzero interdistance at least $8$. Pick a random-shifted dissection of $L^d$. Let $\mathbb{S}$ be a collection of (possibly nondisjoint) subsets of $S$. Suppose there is a graph $G$ on $S$ that is a union of TSTs, one for each $X \in \mathbb{S}$, of total cost $\mathrm{cost}(G)$. Then, there is a graph $G^*$ such that*

- *$G^*$ is $r$-light with respect to the dissection, where $r = (\mathcal{O}(\sqrt{d}))^{d-1}$*
- *There is a cover $\mathbb{S}_{G^*}$ of $\mathbb{S}$ such that $G^*$ is the union of TSTs for each $Y \in \mathbb{S}_{G^*}$*
- *The expected (over the choice of the random-shifted dissection) cost of $G^*$ is at most $\mathcal{O}(\mathrm{cost}(G))$* ∎

The proof of this lemma is an extension of the PTAS for Euclidean TSP by Arora [14] and uses ideas similar to those underlined in the proof of the Structure Theorem 2.1. (Observe that as of now we need to find TSTs, the appearance of Steiner points in the approach of Arora [14] does not cause any problems.)

#### 2.4.1.5 Concluding: Structure Theorem for $k$-Vertex Connectivity

We conclude with a Structure Theorem for $k$-vertex connectivity that shows the existence of a low-cost locally-light graph. This theorem is obtained by combining Lemma 2.1, the Global Decomposition Lemma, the Filtering Lemma, and the TST Covering Lemma, when applied to a minimum-cost $k$-VCSS $G$ for the input point set.

**Theorem 2.3 (Structure Theorem II)** *Let $S$ be a (perturbed) point set in $\mathbb{R}^d$ contained in a bounding box $L^d$ and with minimum nonzero interdistance at least 8. Pick a random-shifted dissection of $L^d$. Then, with probability at least 0.9, there is an m-portal-respecting r-light $k$-VCSS for $S$ whose cost is at most $(1 + \mathcal{O}(\varepsilon))$-time the optimal $k$-VCSS for $S$, where $m = (\mathcal{O}(d \log L/\varepsilon))^{d-1}$ and $r = (\mathcal{O}(k^2 d^{3/2}/\varepsilon))^d \log(k/\varepsilon)$.*

### 2.4.2 Polynomial-Time Approximation Scheme for Euclidean $k$-Vertex and $k$-Edge Connectivity

Now, with the Structure Theorem II 2.3 at hand, we are ready to present the "real" PTAS for the minimum-cost $k$-VCSS problem in geometric graphs. In Section 2.3.2, we showed how to find a $(1+\mathcal{O}(\varepsilon))$-approximation of a minimum-cost $m$-portal-respecting $r$-light Steiner $k$-VCSS for a set of points, see Lemma 2.2. Similar result holds also for finding a minimum-cost $m$-portal-respecting $r$-locally-light $k$-VCSS for a set of points. The running time of the appropriated dynamic programming scheme is the same as that promised in Lemma 2.2, but this time we can even find an optimal solution [not an $(1+\mathcal{O}(\varepsilon))$-approximation, as in Lemma 2.2). Therefore, we can combine this result with the Structure Theorem II 2.3 to obtain the following result.

**Theorem 2.4** *Let $k$ and $d$ be any integers, $k, d \geq 2$, and let $\varepsilon$ be any positive real. Let $S$ be a set of $n$ points in $\mathbb{R}^d$. There is a randomized algorithm that in time $n \cdot (\log n)^{(k d/\varepsilon)^{\mathcal{O}(d)}} \cdot 2^{2^{(k d/\varepsilon)^{\mathcal{O}(d)}}}$ with probability at least 0.9 finds a $k$-VCSS for $S$ whose cost is at most $(1 + \varepsilon)$-time the optimal $k$-VCSS for $S$.*

*Furthermore, within the same running time, one can find a $k$-ECSS for $S$ whose cost is at most $(1+\varepsilon)$-time the optimal $k$-ECSS for $S$. Also, all these algorithms can be derandomized in polynomial time.*

When the parameters $\varepsilon$, $k$, and $d$ are constants, then the running time of the randomized algorithm is $n \cdot \log^{\mathcal{O}(1)} n$. When $d$ and $\varepsilon$ are constant and $k$ is arbitrary, the running time becomes $n \cdot (\log n)^{k^{\mathcal{O}(1)}} \cdot 2^{2^{k^{\mathcal{O}(1)}}}$; when $\varepsilon$ is arbitrary, it is $n \cdot (\log n)^{(1/\varepsilon)^{\mathcal{O}(1)}} \cdot 2^{2^{(1/\varepsilon)^{\mathcal{O}(1)}}}$. In particular, for a constant dimension $d$, our scheme leads to a PTAS for the minimum-cost $k$-VCSS and $k$-ECSS problems for all $k$ such that $k \leq (\log \log n)^c$ for certain positive constant $c < 1$.

## 2.5 Faster Polynomial-Time Approximation Scheme for Euclidean $k$-Edge Connected Spanning Sub-Multigraph and 2-Connected Graphs

Czumaj and Lingas [25] showed that the approximation schemes from Section 2.4 can be improved in the special case when $k = 2$ and for the minimum-cost $k$-ECSSM problem. The main source of the improvement is the observation that if we knew a graph/multigraph that contains an optimal or near optimal $k$-VCSS ($k$-ECSS, $k$-ECSSM), then we would be able to apply similar transformations as those described in Section 2.4 to transform this graph into an $r$-locally-light one. Comparing to the result from the Structure Theorem II 2.3, we would gain by not having to make the graph $m$-portal-respecting, because dynamic programming would not have to "guess" the locations of crossings of the facets. This would potentially eliminate term $m$ in the analysis (see Lemma 2.2) and thus greatly improve the running time.

A geometric graph $G$ on a set of points in $\mathbb{R}^d$ is called a *t-spanner* of $S$, $t \geq 1$, if for any pair of points $p, q \in S$, there is a path in $G$ from $p$ to $q$ of length at most $t$ times the distance between $p$ and $q$. Gudmundson et al. [39] showed that for any set $S$ of $n$ points in $\mathbb{R}^d$ and for any positive $\varepsilon$, in time $\mathcal{O}((d/\varepsilon)^{\mathcal{O}(d)} \cdot n + d \cdot n \cdot \log n)$, one can find a $(1 + \varepsilon)$-spanner of $S$ with maximum degree $(d/\varepsilon)^{\mathcal{O}(d)}$ and with the total cost at most $(d/\varepsilon)^{\mathcal{O}(d)} \cdot \text{MST}(S)$.

For a given multigraph $H$, the *graph induced* by $H$ is the graph obtained by reducing the multiplicity of each edge of $H$ to one. The following lemma formally describes the intuition that a $t$-spanner contains (implicitly) a $t$-approximation of the minimum-cost $k$-ECSSM.

**Lemma 2.8** *Let $G$ be a t-spanner for a point set $S$ in $\mathbb{R}^d$ and let $k$ be an arbitrary positive integer. Then, there exists a k-edge-connected multigraph $H$ on $S$ such that (i) the graph induced by $H$ is a subgraph of $G$, (ii) the total cost of $H$ is at most $t$ times larger than the minimum-cost k-edge-connected multigraph on $S$, and (iii) there are no parallel edges in $H$ of multiplicity exceeding $k$.*

Now, with a good spanner at hand and with Lemma 2.8, we can proceed with the approach sketched before. This approach is partly inspired by the recent use of spanners to speedup PTAS for Euclidean versions of TSP due to Rao and Smith [37]. The analysis relies on a series of transformations of a low cost and sparse $(1 + \mathcal{O}(\varepsilon))$-spanner for the input point set into an $r$-locally-light $k$-edge-connected *multigraph* spanning the input set and having nearly optimal cost. With some modifications of the analysis from the Structure Theorem II, one can get the following theorem.

**Theorem 2.5 (Structure Theorem III)** *Let $S$ be a (perturbed) set of $n$ points in $\mathbb{R}^d$ contained in a bounding box $L^d$ and with minimum nonzero interdistance at least 8. Let $G$ be a $(1 + \varepsilon)$-spanner for $S$ that has $n(d/\varepsilon)^{\mathcal{O}(d)}$ edges and has total cost $(d/\varepsilon)^{\mathcal{O}(d)} \cdot \text{MST}(S)$. Choose a shifted dissection uniformly at random. Then, one can transform $G$ into a graph $G^*$ on $S$ such that with probability (over the random choice of the shifted dissection) at least 0.9,*

- *$G^*$ is $r$-locally-light with respect to the shifted dissection, $r = k \, d^{\mathcal{O}(d)} + \mathcal{O}(k \, d^2 \log(d/\varepsilon)) + (d/\varepsilon)^{\mathcal{O}(d^2)}$*
- *There exists a k-edge-connected multigraph $\mathbb{H}$ which is a spanning subgraph of $\mathbb{G}$ with possible parallel edges (of multiplicity at most $k$) whose cost is upper bounded by $(1 + \mathcal{O}(\varepsilon))$ times the minimum-cost k-ECSSM for $S$*

*Moreover, the transformation can be done in time $n \cdot 2^{(\mathcal{O}(\sqrt{d}))^{d-1}} + n \cdot (d/\varepsilon)^{\mathcal{O}(d)} \log n$.*

Once we have the transformation defined in the Structure Theorem III 2.5, we can use dynamic programming, similar to that described in Lemma 2.2 and in Section 2.4.2, to obtain the following lemma.

**Lemma 2.9** *Let $S$ be a set of $n$ points in $\mathbb{R}^d$ contained in a bounding box $L^d$ and with minimum nonzero interdistance at least 8. Consider an arbitrary shifted dissection and assume that the $2^d$-ary dissection tree of $S$ is given. Let $G$ be an $r$-locally-light graph on $S$, where $r \geq 1$ is arbitrary. Then, a minimum-cost k-ECSSM $G^*$ on $S$ for which the induced graph is a subgraph of $G$ can be found in time $n \cdot 2^{d+(k\,r)^{\mathcal{O}(k\,r)}}$.*

If we combine the Structure Theorem III 2.5 with Lemma 2.9, we directly obtain the following theorem.

**Theorem 2.6** *Let $k$ and $d$ be any integers, $k, d \geq 2$, and let $\varepsilon$ be any positive real. Let $S$ be a set of $n$ points in $\mathbb{R}^d$. There is a randomized algorithm that in time $n \cdot \log n \cdot (d/\varepsilon)^{\mathcal{O}(d)} + n \cdot 2^{2^{(k^{\mathcal{O}(1)} \cdot (d/\varepsilon)^{\mathcal{O}(d^2)})}}$, with probability at least 0.9 finds a k-ECSSM for $S$ whose cost is at most $(1 + \varepsilon)$-time the optimal k-ECSSM for $S$. The algorithm can be derandomized in polynomial time.*

Observe that when all $d$, $k$, and $\varepsilon$ are constant, the running time of the randomized algorithm is $\mathcal{O}(n \log n)$. When $d$ and $k$ are constant and $\varepsilon$ is arbitrary, the running time becomes $n \log n \, (1/\varepsilon)^{\mathcal{O}(1)} + n \, 2^{2^{(1/\varepsilon)^{\mathcal{O}(1)}}}$. When $d$ and $\varepsilon$ are set to constants, then the running time is $\mathcal{O}(n \log n) + n \, 2^{2^{k^{\mathcal{O}(1)}}}$.

### 2.5.1 2-Connected Graphs Are Not Worse than 2-Connected Multigraphs

The algorithm presented in the previous section does not work for minimum-cost $k$-VCSS or $k$-ECSS problems. The reason is that no result similar to that from Lemma 2.8 holds. However, in the special case when $k = 2$, we still can use multigraph approach to obtain a fast PTAS for the minimum-cost 2-VCSS or 2-ECSS problems. Indeed, it is known than any 2-VCSS is also a 2-ECSSM. Therefore, the minimum-cost 2-ECSSM for a set of points is not bigger than the minimum-cost 2-VCSS for the same point set. The following theorem shows that actually, we can always quickly find a 2-VCSS (and hence also 2-ECSS) that has cost not larger than that of a 2-ECSSM.

**Lemma 2.10** [20,25] *A 2-edge-connected multigraph on a set of points in $\mathbb{R}^d$ can be transformed in linear time into a biconnected graph on the same set of points without increasing the total cost.*

In view of this result, we could find a $(1 + \varepsilon)$-approximation for the minimum-cost 2-VCSS problem by first running an algorithm for finding a $(1 + \varepsilon)$-approximation of the minimum-cost 2-ECSSM and then applying Lemma 2.10. By Theorem 2.6, such randomized algorithm for the minimum-cost 2-VCSS problem runs in time $n \cdot \log n \cdot (d/\varepsilon)^{\mathcal{O}(d)} + n \cdot 2^{2^{(\mathcal{O}(d/\varepsilon))^{\mathcal{O}(d^2)}}}$. However, as Czumaj and Lingas [25,34] proved, one can obtain further speedup by improving the dynamic programming scheme from Lemma 2.9 in the special case $k = 2$. For any set $S$ of $n$ points in $\mathbb{R}^d$ and for any Euclidean graph $G$ on $S$ that is $r$-locally-light with respect to some given shifted dissection, one can use dynamic programming to find in time $n \cdot 2^d \cdot r^{\mathcal{O}(r\,2^d)}$ a minimum-cost 2-edge-connected multigraph on $S$ for which the induced graph is a subgraph of $G$. This yields the following theorem.

**Theorem 2.7** *Let $d$ be any integer $d \geq 2$, and let $\varepsilon$ be any positive real. Let $S$ be a set of $n$ points in $\mathbb{R}^d$. There is a randomized algorithm which in time $n \cdot \log n \cdot (d/\varepsilon)^{\mathcal{O}(d)} + n \cdot 2^{(d/\varepsilon)^{\mathcal{O}(d^2)}}$, with probability at least 0.9 finds a 2-VCSS for $S$ whose cost is at most $(1 + \varepsilon)$-time the optimal 2-VCSS for $S$.*

*The same holds for the minimum-cost 2-ECSS problem; these algorithms can be derandomized in polynomial time.*

For constant $d$ and arbitrary $\varepsilon$, the running time of the randomized algorithm is $n \log n \, (1/\varepsilon)^{\mathcal{O}(1)} + 2^{(1/\varepsilon)^{\mathcal{O}(1)}}$.

## 2.6 Lower Bounds

The results discussed in previous sections show that various multiconnectivity problems have a PTAS. However, the obtained algorithms work in polynomial-time only for small values of $d$ and $k$. Are these results just a sign that our methods still need to be improved or they are inherent for the multiconnectivity problems?

As for now, we still do not know if there is a PTAS for large values of $k$ and, say, if we pick $k = \log n$ we do not know if the $k$-VCSS problem for geometric graphs on the plane (i.e., for $d = 2$) has a PTAS or does not. However, we know that we cannot obtain a PTAS for large values of $d$. Our basic tool is a powerful result of Trevisan [40] that connects the inapproximability of TSP in geometric graphs with the inapproximability of TSP in the so-called 1–2 graphs. A weighted undirected complete graph $G$ is a *1–2 graph* if each of its edges has weight either 1 or 2. It is called a *1–2–$\Delta$ graph* if it is a 1–2 graph and each of its vertices is incident to at most $\Delta$ edges of weight 1. It is easy to see that in every graph, TST has cost that is not smaller than the cost of a minimum-cost 2-VCSS. The following result showing that in 1–2 graphs TST and minimum-cost 2-VCSS coincide is central for our analysis.

**Lemma 2.11** *In every 1–2 graph, TST is a minimum-cost 2-VCSS.*

With this result, general inapproximability results for TST in 1–2 graphs proven by Trevisan [40] directly imply similar results for the 2-VCSS problem.

**Theorem 2.8 [19]** *There exist constants $\Delta_0 > 0$ and $\varepsilon > 0$ such that, given a 1–2–$\Delta_0$ graph G on n vertices, and given the promise that either its minimum-cost 2-VCSS H has cost n, or its cost is greater than or equal to $(1 + \varepsilon) n$, it is $\mathcal{NP}$-hard to distinguish which of the two cases holds. In particular, it is $\mathcal{NP}$-hard to approximate within $(1 + \varepsilon)$ the cost of a minimum-cost 2-VCSS of a 1–2–$\Delta_0$ graph.*

The next result is a direct application of Theorem 2.8 combined with classical results on metric embeddings.

**Theorem 2.9 [19]** *For any fixed $p \geq 1$, there exists a constant $\xi > 0$ such that it is $\mathcal{NP}$-hard to approximate within $1 + \xi$ the minimum-cost 2-connected graph spanning a set of n points in the $\ell_p$ metric in $\mathbb{R}^{\log n}$.*

**Corollary 2.1** *The minimum k-VCSS problem in graphs of maximum degree bounded by some constant is APX-hard and hence does not have a PTAS unless $\mathcal{P} = \mathcal{NP}$.*

One can easily modify the proofs of the theorems presented in this section to obtain similar inapproximability results for the problem of finding a minimum-cost $k$-edge-connected subgraph of a $k$-edge-connected graph.

# 2.7 Extensions to Other Related Problems

The results and techniques we discussed in the previous sections can be applied to various related problems.

## 2.7.1 Pseudoapproximations and Steiner *k*-Vertex Connected Spanning Subgraph/Edge Connected Spanning Subgraph

It is not hard to improve the pseudoapproximation result obtained in Theorem 2.2 by modifying the result from Theorem 2.6. We begin with finding a $k$-ECSSM whose cost is within $1 + \varepsilon$ of the minimum using the result from Theorem 2.6. Then, we can trivially transform this multigraph into a Steiner $k$-VCSS by placing $k - 1$ Steiner points on each input point (i.e., at the length zero from it) and forming a $k$-clique of zero cost out of the point and its associated $k - 1$ Steiner points. The cost of the resulting graph is within $(1 + \varepsilon)$ of the minimum-cost $k$-ECSSM for the input set, which, in turn, does not exceed $(1 + \varepsilon)$ times the minimum-cost $k$-VCSS on the input set. Such a Steiner $k$-VCSS can be found in (asymptotically) the same time as required by Theorem 2.6 to find the $k$-ECSSM, which is significantly better than the result in Theorem 2.2. The same approach works also for Steiner $k$-ECSS.

## 2.7.2 Steiner *k*-Connectivity–Real Approximation Schemes

The techniques described in the survey can be also used to derive efficient approximation schemes for Euclidean minimum-cost Steiner $k$-connectivity. In contrast to the result in Section 2.7.1, our goal is to find a Steiner $k$-VCSS (or $k$-ECSS) for a set of points $S$ in $\mathbb{R}^d$ whose cost is at most $(1 + \varepsilon)$ times the minimum-cost Steiner $k$-VCSS ($k$-ECSS, respectively) for $S$; so both, the solution found and the optimal solution, are allowed to use Steiner points.

The main difficulty with extending the result from Section 2.7.1 to a real PTAS for Steiner $k$-VCSS/ECSS is that the spanners used in the Structure Theorem III 2.5 and in the PTAS from Theorem 2.6 do not include Steiner points. Nevertheless, one can decompose optimal Steiner solutions for $k$-connectivity and combine this decomposition with the construction of banyans due to Rao and Smith [37].

The case of $k = 2$ is most interesting. Extending the work of Hsu and Hu [35], Czumaj and Lingas [25] showed a new structural characterization of minimum-cost Steiner biconnected graphs that lead to a decomposition of an optimal Steiner solution into minimum Steiner trees. This opened the possibility of using the so-called $(1 + \varepsilon)$-banyans [37], for the purpose of approximating the Euclidean minimum Steiner tree problem. As the result, Czumaj and Lingas [25] obtain a PTAS for Euclidean minimum-cost Steiner biconnectivity and Euclidean minimum-cost two-edge connectivity; the algorithms run in time $\mathcal{O}(n \log n)$ for any constant dimension and $\varepsilon$. For general $d$ and $\varepsilon$, the running time is

$$n \log n \, (d/\varepsilon)^{\mathcal{O}(d)} + n \, 2^{(d/\varepsilon)^{\mathcal{O}(d^2)}} + n \, 2^{2^{d^{\mathcal{O}(1)}}}.$$

### 2.7.3 Survivable Networks

Czumaj et al. [32] extended the analysis from previous sections (in particular, Theorem 2.7) to a more general problem of survivable networks. They considered the variant of the *survivable network design problem* in which for a given set $S$ of $n$ points in Euclidean space $\mathbb{R}^d$ and a connectivity requirement function $r : S \to \mathbb{N}$, the goal is to find a minimum-cost graph $G$ on $S$ such that for any pair of points $x, y \in S$, $G$ has $\min\{r(x), r(y)\}$ internally vertex-disjoint paths between $x$ and $y$. The two most basic (and of largest practical relevance) variants of this problem are those in which $r(x) \in \{0, 1\}$ and when $r(x) \in \{0, 1, 2\}$, for any point $x \in S$.

First, for the simplest case in which $r(x) \in \{0, 1\}$ for any point $x \in S$, that is, for the *Steiner tree problem,*[*] Czumaj et al. [32] designed a randomized algorithm that, for any constant $d$ and any constant $\varepsilon$, in time $\mathcal{O}(n \log n)$ finds a Steiner tree whose cost is at most $(1 + \varepsilon)$ times larger than the minimum. For general $d$ and $\varepsilon$, its running time is $n \log n \, (d/\varepsilon)^{\mathcal{O}(d)} + n \, 2^{(d/\varepsilon)^{\mathcal{O}(d^2)}} + n \, 2^{2^{d^{\mathcal{O}(1)}}}$.

Next, for the case when $r(x) \in \{0, 1, 2\}$ for any point $x \in S$ (this is the classical problem investigated thoroughly by Grötschel and Monma et al. [3,5,7,12,13]), Czumaj et al. [32] extended algorithm for the Steiner tree problem to design an algorithm that, for any constant $d$ and any constant $\varepsilon$, in time $\mathcal{O}(n \log n)$ finds a graph satisfying all the vertex connectivity requirements and having the cost at most $(1 + \varepsilon)$ times the minimum. When $d$ and $\varepsilon$ are allowed to be arbitrarily, its running time is $n \log n \, (d/\varepsilon)^{\mathcal{O}(d)} + n \, 2^{(d/\varepsilon)^{\mathcal{O}(d^2)}} + n \, 2^{2^{d^{\mathcal{O}(1)}}}$.

Finally, essentially the same techniques can be used to obtain a PTAS for the multigraph variant, in which the edge-connectivity requirements satisfy $r(x) \in \{0, 1, \dots, k\}$ and $k = \mathcal{O}(1)$.

All these approximation schemes are randomized, but they can be *derandomized* in a polynomial time.

### 2.7.4 Finding Low-Cost $k$-Vertex Connected Spanning Subgraph and $k$-Edge Connected Spanning Subgraph in Planar Graphs

Although the main focus of this survey is to present the results for Euclidean graphs, this line of research has been extended in the last decade to lead to a progress in designing approximation schemes for the 2-VCSS and 2-ECSS problem in *planar graphs* [26,27]. Similarly as for the TSP problem in planar graphs [23,41], the first step toward an efficient approximation scheme has been achieved for unweighted graphs. Czumaj et al. [27] showed that for every positive $\varepsilon$, for a given undirected graph planar $G$ with $n$ vertices, one can find in time $n^{\mathcal{O}(1/\varepsilon)}$ a 2-VCSS (or 2-ECSS) of $G$ whose total number of edges is at most $(1 + \varepsilon)$ times the minimum number of edges in any 2-VCSS (or 2-ECSS, respectively) of $G$; this gives a PTAS for the unweighted version of the 2-VCSS and 2-ECSS problem in planar graphs. In fact, the approximation

---

[*] Note that this variant of the Steiner tree problem is different from the Steiner tree problem considered by Arora [14], for which a PTAS is also known [14,22,36,37]. The variant considered in this survey requires that all locations of Steiner points are given in advance (they are the points $x \in S$ with $r(x) = 0$), whereas in the other variant, all points in $\mathbb{R}^d$ could be used as Steiner points.

scheme provided in Reference 27 works also for the weighted case, but then the running time becomes $n^{\mathcal{O}(\gamma/\varepsilon)}$, where $\gamma$ is the ratio of the total edge cost to the optimum solution cost.

Soon later, Berger et al. [26] modified the scheme from Reference 27 and obtained a *quasi-PTAS* for the 2-VCSS and 2-ECSS problem in planar graphs. Their algorithm runs in time $n^{\mathcal{O}(\log n \, \log(1/\varepsilon)/\varepsilon)}$ and finds a 2-VCSS (or 2-ECSS) of $G$ whose total cost is at most $(1 + \varepsilon)$ times the minimum-cost 2-VCSS (or 2-ECSS, respectively) of $G$. Furthermore, their algorithm can be extended to solve within the same runtime bounds the survivable network design problem in planar graphs in which $r(x) \in \{1, 2\}$ for any vertex.

The underlying techniques developed for the approximation schemes for the 2-VCSS and 2-ECSS problem in planar graphs were surprisingly similar to those used for geometric graphs: a combination of (new) separator theorems with dynamic programming, and then new constructions of *light spanners* for planar graphs. For more details, we refer interested readers to the original papers [26,27].

Later, Berger and Grigni have proposed a PTAS for $\{1, 2\}$-edge connectivity (in the spanning case) in planar multigraphs [28]. Shortly afterward, Borradaile et al. have designed the first PTASs for 1-VCSS (SMT) [30] and the two-edge connectivity survivable network [31] for planar multigraphs (for the survivable network design problem in planar graphs with edge-connectivity constraints in which $r(x) \in \{0, 1, 2\}$ for any vertex). Both PTASs run in $O(n \log n)$ time. The key idea in References 30, 31 is that the interaction between different parts of an optimal solution can be simplified while causing only a small increase in the cost of the solution. These results have been generalized to include bounded-genus graphs in Reference 29.

# 2.8 Final Comments

In the current paper, we surveyed recent approximation schemes for various variants of network design problems for geometric graphs. Our main goal was not only to show the results but also to demonstrate a variety of new techniques developed to coupe with these problems.

## 2.8.1 Interesting Open Questions

Our first open problem is whether there exists a PTAS for the Euclidean minimum-cost $k$-VCSS and $k$-ECSS problems for very large values of $k$. The techniques presented in this survey seem to work only for the values of $k$ up to $(\log \log n)^c$ for certain positive constant $c < 1$. What about large values of $k$?

Recently, Bartal and Gottlieb [42] designed an improved PTAS for the Euclidean TSP problem (in $d$ dimensions) that computes a $(1 + \varepsilon)$-approximation for the TSP in time $2^{(d/\varepsilon)^{O(d)}} n$ in the integer RAM model. For constant values of $d$ and $\varepsilon$, the running time of this algorithm is linear in $n$. It is an interesting open question if there is a linear-time PTAS for the Euclidean minimum-cost $k$-VCSS and $k$-ECSS problems; the problem is open even in the case $k = 2$.

The class of Euclidean graphs can be seen as a special class of *metrics with bounded doubling dimension* [43]. The doubling dimension of a metric is the smallest $D$ such that any ball of radius $2r$ can be covered using $2D$ balls of radius $r$; for example, a Euclidean graph in $\mathbb{R}^d$ can be represented as a graph in a metric with doubling dimension $O(d)$. In a series of papers [43–45], it has been shown that there is a PTAS for the TSP problem for metrics with bounded doubling dimension $D$. In particular, Gottlieb (see the full version of [45] at arxiv) shows how to find a $(1 + \varepsilon)$-approximation for the TSP in a metric of doubling dimension $D$ in time $\mathcal{O}(2^{(D/\varepsilon)^{D^2}} n) + (D/\varepsilon)^{O(D)} n \log^2 n$. With these results at hand, it is natural to expect that there are PTASs for the minimum-cost $k$-VCSS and $k$-ECSS problems in metrics with bounded doubling dimension. We conjecture that such results are true for constant values of $k$.

Another intriguing open problem is whether the minimum-cost $k$-VCSS and $k$-ECSS problems for planar graphs have a PTAS. Although this is known for the 2-ECSS problem in multigraphs [31], we conjecture that this result can be extended to PTASs for planar graphs for all values of $k \leq 5$ for both $k$-VCSS and $k$-ECSS problems (note that for $k > 5$, no planar graph can be $k$-vertex-connected).

Finally, and perhaps most importantly, how practical are the methods discussed in this survey? Even though, most probably, any direct implementation of the PTAS for $k$-connectivity problems would be inferior to the existing heuristic implementations discussed, [3,5,7,13], we believe that the techniques presented in this survey when combined with heuristics could lead to significant improvements in practical implementations.

# Acknowledgments

We thank Hairong Zhao, Christos Levcopoulos, Sanjeev Arora, and Luca Trevisan for their valuable comments and suggestions. We thank an anonymous referee for pointing out the relation between TST and 2-VCSS in 1-2-graphs.

# References

1. Ahuja, R.K., Magnanti, T.L., Orlin, J.B., and Reddy, M.R., Applications of network optimization, in *Handbooks in Operations Research and Management Science*, vol. 7: Network Models, Ball, M.O., Magnanti, T.L., Monma, C.L., and Nemhauser, G.L., (Eds.), Elsevier, North-Holland, Amsterdam, 1995, chap 1.
2. Goemans, M.X. and Williamson, D.P., The primal-dual method for approximation algorithms and its application to network design problems, in *Approximation Algorithms for $\mathcal{NP}$-Hard Problems*, Hochbaum, D.S., (Ed.), PWS Publishing Company, Boston, MA, 1996, chap 4.
3. Grötschel, M., Monma, C.L., and Stoer, M., Design of survivable networks, in *Handbooks in Operations Research and Management Science, vol. 7, Network Models*, Ball, M.O., Magnanti, T.L., Monma, C.L., and Nemhauser, G.L., (Eds.), Boston, MA, 1995, chap 10.
4. Schrijver, A., *Combinatorial Optimization Polyhedra and Efficiency*, Springer-Verlag, New York, 2003.
5. Stoer, M., *Design of Survivable Networks, volume 1531 of Lecture Notes in Mathematics*, Springer-Verlag, Berlin, Germany, 1992.
6. Mihail, M., Shallcross, D., Dean, N., and Mostrel, M., A commercial application of survivable network design: ITP/INPLANS CCS network topology analyzer, in *Proceedings of SODA*, Atlanta, GA, 1996, p. 279. ACM-SIAM, New York.
7. Monma, C.L. and Shallcross, D.F., Methods for designing communications networks with certain two-connected survivability constraints, *Oper. Res.*, 37(4), 531, 1989.
8. Du, D.Z. and Hwang, F.K., A proof of the Gilbert-Pollak conjecture on the Steiner ratio, *Algorithmica*, 7, 121, 1992.
9. Hwang, F.K., Richards, D.S., and Winter, P., *The Steiner Tree Problem, Vol 53 of Annals of Discrete Mathematics*, North-Holland, Amsterdam, 1992.
10. Borradaile, G., Klein, P. N., and Mathieu, C., A polynomial-time approximation scheme for Euclidean Steiner forest, *ACM Trans. Algorithms*, 11(3), 19, 1, 2015.
11. Gabow, H.N., Goemans, M.X., and Williamson, D.P., An efficient approximation algorithm for the survivable network design problem, *Math. Prog. Ser. B*, 82(1–2), 13, 1998.
12. Grötschel M. and Monma, C.L., Integer polyhedra arising from certain network design problems with connectivity constraints, *SIAM J. Disc. Math.*, 3(4), 502, 1990.
13. Grötschel, M., Monma, C.L., and Stoer, M., Computational results with a cutting plane algorithm for designing communication networks with low-connectivity constraints, *Oper. Res.*, 40(2), 309, 1992.
14. Arora, S., Polynomial time approximation schemes for Euclidean traveling salesman and other geometric problems, *J. ACM*, 45(5), 753, 1998.

15. Garey, M.R. and Johnson, D.S., *Computers and Intractability: A Guide to the Theory of NP-Completeness*, Freeman, New York, 1979.

16. Hochbaum, D.S., (Ed.), *Approximation Algorithms for $\mathcal{N}$P-Hard Problems*, PWS Publishing Company, Boston, MA, 1996.

17. Kortsarz, G. and Nutov, Z., Approximating $k$-node connected subgraphs via critical graphs, *SIAM J. Comput.*, 35(1), 247, 2005,

18. Khuller, S., Approximation algorithms for finding highly connected subgraphs, in *Approximation Algorithms for $\mathcal{N}$P-Hard Problems*, Hochbaum, D.S., (Ed.), PWS Publishing Company, Boston, MA, 1996, chap 6.

19. Czumaj, A. and Lingas, A., On approximability of the minimum-cost $k$-connected spanning subgraph problem, in *Proceedings of SODA*, Baltimore, MD, 1999, p. 281. ACM-SIAM, New York.

20. Frederickson, G.N. and JáJá, J., On the relationship between the biconnectivity augmentation and traveling salesman problem, *Theor. Comput. Sci.*, 19(2), 189, 1982.

21. Cheriyan, J. and Vetta, A., Approximation algorithms for network design with metric costs, in *Proceedings of STOC*, Baltimore, MD, 2005, p. 167. ACM, New York.

22. Mitchell, J.S.B., Guillotine subdivisions approximate polygonal subdivisions: A simple polynomial-time approximation scheme for geometric TSP, $k$-MST, and related problems, *SIAM J. Comput.*, 28(4), 1298, 1999.

23. Arora, S., Grigni, M., Karger, D., Klein, P., and Woloszyn, A., A polynomial-time approximation scheme for weighted planar graph TSP, in *Proceedings of SODA*, San Francisco, CA, 1998, p. 33. ACM-SIAM, New York.

24. Czumaj, A. and Lingas, A., A polynomial time approximation scheme for Euclidean minimum cost k-connectivity, in *Proceedings of ICALP*, Aalborg, Denmark, 1998, p. 682. Springer, Berlin, Germany.

25. Czumaj, A. and A. Lingas, A., Fast approximation schemes for Euclidean multi-connectivity problems, in *Proceedings of ICALP*, Geneva, Switzerland, 2000, p. 856. Springer, Berlin, Germany.

26. Berger, A., Czumaj, A., Grigni, M., and Zhao, H., Approximation schemes for minimum 2-connected spanning subgraphs in weighted planar graphs, in *Proceedings of ESA*, Palma de Mallorca, Spain, 2005, p. 472. Springer, Berlin, Germany.

27. Czumaj, A., Grigni, M., Sissokho, P., and Zhao, H., Approximation schemes for minimum 2-edge-connected and biconnected subgraphs in planar graphs, in *Proceedings of SODA*, 2004, p. 489. ACM-SIAM, New York.

28. Berger, A. and Grigni, M., Minimum weight 2-edge-connected spanning subgraphs in planar graphs, in *Proceedings of ICALP*, Eilat, Israel, 2007, p. 90. Springer, Berlin, Germany.

29. Borradaile, G., Demaine, E. D., and Tazari, S., Polynomial-time approximation schemes for subset-connectivity problems in bounded-genus graphs, *Algorithmica*, 68(2), 287, 2014.

30. Borradaile, G., Klein, P N., and Mathieu, C., An $O(n \log n)$ approximation scheme for Steiner tree in planar graphs, *ACM Trans. Algorithms*, 5(3), 1, 2009.

31. Borradaile, G. and Klein, P. N., The two-edge connectivity survivable network problem in planar graphs, in *Proceedings of ICALP (1)*, Reykjavik, Iceland, 2008, p. 485. Springer, Berlin, Germany.

32. Czumaj, A., Lingas, A., and Zhao, H., Polynomial-time approximation schemes for the Euclidean survivable network design problem, in *Proceedings of ICALP*, Malaga, Spain, 2002, p. 973. Springer, Berlin, Germany.

33. Czumaj, A. and Lingas, A., Polynomial time approximation schemes for Euclidean multi-connectivity problems. Manuscript, 2005, 45(5), 753–782, 2005.

34. Czumaj, A. and Lingas, A., Fast approximation schemes for Euclidean biconnectivity problems and for multi-connectivity problems. 2005.

35. Hsu, D.F. and Hu, X.-D., On short two-connected Steiner networks with Euclidean distance, *Networks*, 32(2), 133, 1998.

36. Arora, S., Approximation schemes for $\mathcal{NP}$-hard geometric optimization problems: A survey, *Math. Prog. Ser. B*, 97(1–2), 43, 2003.

37. Rao, S.B. and Smith, W.D., Approximating geometrical graphs via "spanners" and "banyans," in *Proceedings of STOC*, Dallas, TX, 1998, p. 540. ACM, New York.

38. Lawler, E.L., Lenstra, J.K., Rinnooy Kan, and A.H.G., Shmoys, D.B., (Eds.), *The Traveling Salesman Problem*, John Wiley & Sons, New York, 1985.

39. Gudmundson, J., Levcopoulos, C., and Narasimhan, G., Fast greedy algorithms for constructing sparse geometric spanners, *SIAM J. Comput.*, 31(5), 1479, 2002.

40. Trevisan, L., When Hamming meets Euclid: The approximability of geometric TSP and Steiner Tree, *SIAM J. Comput.*, 30(2), 475, 2000.

41. Grigni, M., Koutsouias, E., and Papadimitriou, C., An approximation scheme for planar graph TSP, in *Proceedings of FOCS*, Washington, DC, 1995, p. 640. IEEE Computer Society, Los Alamitos, CA.

42. Bartal, Y. and Gottlieb, L.-A., A linear time approximation scheme for Euclidean TSP, in *Proceedings of FOCS*, Berkeley, CA, 2013, p. 698. IEEE Computer Society, Los Alamitos, CA.

43. Talwar, K., Bypassing the embedding: Algorithms for low dimensional metrics, in *Proceedings of STOC*, 2004, p. 281. ACM, New York.

44. Bartal, Y., Gottlieb, L.-A., and Krauthgamer, R., The traveling salesman problem: Low-dimensionality implies a polynomial time approximation scheme, in *Proceedings of STOC*, 2012, p. 663. Also at http://arxiv.org/abs/1112.0699.

45. Gottlieb, L.-A., A light metric spanner, in *Proceedings of FOCS*, 2015, p. 759. Also at http://arxiv.org/abs/1505.03681.

# 3

# Dilation and Detours in Geometric Networks

Joachim Gudmundsson

Christian Knauer

## 3.1 Introduction

In the current chapter, we consider geometric networks and two important quality measures of such networks; namely, *dilation* and *detour*. A *geometric network* is an undirected graph the vertices of which are points in $d$-dimensional space, and the edges are straight-line segments connecting the vertices. The sites are usually located in the Euclidean plane, but other metrics and higher dimensions are also common. Geometric networks arise in many applications. Road networks, railway networks, telecommunication, pattern matching, bioinformatics—any collection of objects in space that have some connections between them can be modeled as a geometric network.

The weight of an edge $e = (u, v)$ in a geometric network $G = (S, E)$ on a set $S$ of $n$ points is the (usually Euclidean) distance between $u$ and $v$, which we denote by $d(u, v)$. The *graph distance* $d_G(u, v)$ between two vertices $u, v \in S$ is the length of the shortest path in $G$ connecting $u$ to $v$.

The current chapter will consider the problem of designing a "good" network and the dual problem, that is, evaluating how "good" a given network is. When designing a network for a given set $S$ of points, several criteria have to be taken into account. In particular, in many applications it is important to ensure a fast connection between every pair of points in $S$. For this, it would be ideal to have a direct connection between every pair of points; the network would then be a complete graph. In most applications, however, this is unacceptable due to the high costs. This leads to the concepts of dilation and detour of a graph, which we define next.

The *dilation* or *stretch factor* of $G$, denoted as $\Delta(G)$, is the maximum factor by which graph distance $d_G$ differs from the geometric distance $d$ between every pair of vertices.

**Definition 3.1** *Let $S$ be a set of points in $\mathbb{R}^d$, let $M = (\mathbb{R}^d, d_M)$ be a metric space on $\mathbb{R}^d$, and let $G$ be a graph in $M$ with vertex set $S$. For any two vertices $x, y \in S$, let $d_G(x, y)$ be the infimum length of all paths connecting $x$ to $y$ in $G$. We call*

$$\Delta_M^G(x, y) = \frac{d_G(x, y)}{d_M(x, y)}.$$

the M-dilation between x and y in G, and $\Delta_M(G) = \sup_{x,y \in S} \Delta_M^G(x, y)$ the M-dilation of G.

A graph $G = (S, E)$ with $\Delta_M^G(G) \le t$ is said to be a *t-spanner* of S.

The notion of dilation can be generalized to arbitrary connected sets $P \subset \mathbb{R}^d$. This measure is called *detour* and compares the length of a shortest path inside P between any two points $x, y \in P$ with their distance measured, for example, in the Euclidean metric on $\mathbb{R}^d$.

**Definition 3.2** *Let $P \subset \mathbb{R}^d$ be a connected set, and $M = (\mathbb{R}^d, d_M)$ be a metric space on $\mathbb{R}^d$. For any two points, $x, y \in P$, let $d_P(x, y)$ be the infimum length of all curves connecting x to y that are contained in P. We call*

$$\delta_M^P(x, y) = \frac{d_P(x, y)}{d_M(x, y)}$$

*the M-detour between x and y in P, and $\delta_M(P) = \sup_{x,y \in P} \delta_M^P(x, y)$ the M-detour of P.*

In some sense, the detour measures how much two metric spaces on $\mathbb{R}^d$—namely, M and $\mathbb{R}^d$ with the shortest path metric induced by P—resemble each other.

In the current chapter, we will mainly consider the case $M = \mathbb{E}^d$, the d-dimensional Euclidean space. We then write $\Delta(G)$ and $\delta(P)$ instead of $\Delta_{\mathbb{E}^d}(G)$ and $\delta_{\mathbb{E}^d}(P)$, respectively. Whenever we speak about the dilation or detour without specifying M, we refer to the case $M = \mathbb{E}^d$.

The chapter is organized as follows. In Section 3.2, we give an overview of the construction of *t*-spanners. Section 3.3 briefly considers the dual problem, namely, computing the dilation of a given graph. Then in Section 3.4, we turn our attention to the problem of computing the detour. Finally, in Section 3.5 we end this chapter by looking at structures with small dilation.

## 3.2 Constructing *t*-Spanners

The problem considered in this section is the construction of *t*-spanners given a set S of n points in $\mathbb{R}^d$ and a positive real value $t > 1$. The aim is to compute a *t*-spanner for S with certain desirable properties, such as

Size: Defined to be the number of edges in the graph.

Degree: The maximum number of edges incident to a vertex.

Weight: The weight of a Euclidean network G is the sum of the edge weights.

Spanner diameter (or simply Diameter): Defined as the smallest integer d such that for any pair of vertices u and v in S, there is a *t*-path in the graph (a path of length at most $t \cdot |uv|$) between u and v containing at most d edges.

There are trade-offs between different properties, for example, between the degree and the diameter [1]; a graph with constant degree will have diameter $\Omega(\log n)$. A further example is the trade-off between the diameter and the weight [2], that is, if the diameter of a Euclidean graph G is bounded by $O(\log n)$, then the weight of G is $\Omega(wt(MST(S)) \cdot \frac{\log n}{\log \log n})$, where $wt(MST(S))$ denotes the weight of the minimum spanning tree of S. Finally, there is also an $\Omega(n \log n)$ time lower bound in the algebraic computation tree model for computing any *t*-spanner for a given set of points S in $\mathbb{R}^d$ is $\Omega(n \log n)$ [3].

The most well-known *t*-spanners can be divided into three groups: $\Theta$-graphs, WSPD-graphs, and greedy-graphs. In Sections 3.2.1–3, we give the main idea of each of these, together with the known bounds. Throughout this section, it will be assumed that the set of input points is given in d-dimensional Euclidean space. For a more detailed description of the construction of *t*-spanners, see the extensive and thorough work by Narasimhan and Smid [4].

### 3.2.1 The $\Theta$-Graph

The $\Theta$-graph was discovered independently by Clarkson [5] and Keil [6]. Keil considered the graph in two dimensions, whereas Clarkson extended his construction to also include three dimensions. Althöfer et al. [7] defined the $\Theta$-graph for higher dimensions, and Ruppert and Seidel [8] improved the construction time to $O(n \log^{d-1} n)$. The general approach is stated in the following. Note that it is possible to cover $\mathbb{E}^d$ by $k$ simplicial cones of angular diameter $\theta$, where $k = O(1/\theta^{d-1})$ as defined in the algorithm.

---

**ALGORITHM   $\Theta$-Graph($S, t$)**

1.   Set $k := 2d! \left\lceil \sqrt{\frac{2(d-1)}{1-\cos\theta}} \right\rceil^{d-1}$ such that $t = \frac{1}{\cos\theta - \sin\theta}$ for $\theta = 2\pi/k$.
2.   Set $E := \emptyset$.
3.   **for** each point $u \in S$
4.        Consider $k$ cones $C_1, \ldots, C_k$ with angular diameter $\theta$ and apex at $u$ that cover $\mathbb{E}^d$.
5.        **for** each cone $C_i$
6.             Find the point $v$ within $C_i$ whose orthogonal projection onto the bisector of $C_i$ is closest to $u$.
7.             Add $(u, v)$ to $E$.
8.   **return** $G = (S, E)$.

---

A similar construction was already defined by Yao [9] in 1982, with the difference that for every point $u$ and every cone $C_i$, $u$ is connected to the closest point in $C_i$. Defining the edges as in the $\Theta$-graph algorithm has the advantage of faster computation.

**Theorem 3.1** *The $\Theta$-graph is a $t$-spanner of $S$ for $t = \frac{1}{\cos\theta - \sin\theta}$ with $O(\frac{n}{\theta^{d-1}})$ edges and can be computed in $O(\frac{n}{\theta^{d-1}} \log^{d-1} n)$ time using $O(\frac{n}{\theta^{d-1}} + n \log^{d-2} n)$ space.*

Even though the "out-degree" of each vertex is bounded by $k$, the "in-degree" could be linear. Moreover, in worst case, the weight and the diameter of the $\Theta$-graph can be $\Omega(n \cdot wt(MST(S)))$ and $n-1$, respectively. However, there are several variants of the $\Theta$-graph that improve these bounds.

*Sink-spanners*:   To obtain a spanner with constant degree, one can use the construction of *sink-spanners* by Arya et al. [1]. The basic idea is as follows. Start with a $\Theta$-graph that is a $\sqrt{t}$-spanner. Direct all the edges such that the out-degree is bounded by a constant for every vertex. To handle the vertices with high in-degree, replace each high degree node $q$ and its adjacent neighbors, that is, the star centered at $q$, with a bounded degree $\sqrt{t}$-*sink-spanner*. A $\sqrt{t}$-sink-spanner is a directed graph in which each point has a directed $\sqrt{t}$-spanner path to the center $q$. This is done in a way that may increase the dilation by a factor of $\sqrt{t}$, thus resulting in a $t$-spanner with degree $O(\frac{1}{(t-1)^{2d-2}})$.

**Theorem 3.2** *The sink spanner is a $t$-spanner of $S$ for $t = \frac{1}{(\cos\theta - \sin\theta)^2}$ with $O(\frac{n}{\theta^{d-1}})$ edges and can be computed in $O(\frac{n}{\theta^{d-1}} \log^{d-1} \frac{n}{\theta^{d-1}})$ time using $O(\frac{n}{\theta^{d-1}} + n \log^{d-2} n)$ space.*

The transformation from a directed $\sqrt{t}$-spanner with bounded out-degree to a $t$-spanner with bounded degree is called a *sink-spanner transformation*.

*Skip-list spanners*:   The idea is to generalize skip-lists [10] and apply them to the construction of spanners. Construct a sequence of subsets, as follows: Let $S_1 = S$. Let $i > 1$ and assume that we already have constructed the subset $S_i$. For each point in $S_i$, flip a fair coin. The set $S_{i+1}$ is defined as the set of all points of $S_i$ the coin flip of which produced heads. The construction stops if $S_{i+1} = \emptyset$. We have $\emptyset = S_{h+1} \subset S_h \subseteq S_{h-1} \subseteq \cdots \subseteq S_1 = S$. It holds that $h = O(\log n)$ with high probability and that $\sum_{i=1}^{h} |S_i| = O(n)$ with high probability. For each $1 \le i \le h$, construct a $\Theta$-graph $G(S_i)$. The union of the graphs $G(S_1), \ldots, G(S_h)$ is the skip-list spanner $G$. The skip-list spanner is a $t$-spanner having $O(n)$ edges and $O(\log n)$ spanner diameter with high probability.

*Ordered Θ-graphs*: A recent modification of the Θ-graph by Bose et al. [11] is the so-called Ordered Θ-graph that considers the order in which the points of $S$ are processed, that is, the graph is built incrementally by inserting and processing each point in some predefined order. When a new point is processed, it only considers the points in the graph that have already been processed. They show an ordering that guarantees that the degree is bounded by $O(k \log n)$ and that a random order gives a $t$-spanner for which the diameter is bounded by $O(\log n)$ with high probability.

*Gap-greedy*: The final variant is a combination of the Θ-graph and a greedy approach. A set of directed edges is said to satisfy the *gap* property if the sources of any two edges in the set are separated by a distance that is at least proportional to the length of the shorter of the two edges. Chandra et al. [12] showed that any directed graph $G$ that fulfills the *gap* property has weight $O(\log n \cdot wt(MST(S)))$. However, the gap property is limited in power. Lenhof et al. [13] showed that there exists a graph that satisfies the gap property and has weight $\Omega(\frac{\log n}{\log \log n} \cdot wt(MST(S)))$.

By using the previous idea, Arya and Smid [14] proposed an algorithm that uses the gap property to decide if an edge should be added to the $t$-spanner graph or not. They consider pairs of points in order of increasing distance, adding an edge $(p, q)$ if and only if it does not violate the gap property.

**Theorem 3.3** *Let $t = 1/(\cos \theta - \sin \theta - 2w)$ for some real numbers $0 < \theta < \pi/4$ and $0 \le w < (\cos \theta - \sin \theta)/2$. The gap-greedy algorithm produces a $t$-spanner $G$ of $S$ in time $O(n/\theta^{d-1} \log^d n)$ such that each vertex has degree $O(1/\theta^{d-1})$ and weight $O(\frac{1}{\theta^{d-1}} \cdot (1 + \frac{1}{w}) \log n \cdot wt(MST(S)))$.*

### 3.2.2 The Well-Separated Pair Decomposition Graph

The well-separated pair decomposition (WSPD) was developed by Callahan and Kosaraju [15]. A detailed description of the WSPD can be found in Volume 1, Chapter 4 by Smid in this handbook. The WSPD-graph was first described by Callahan and Kosaraju [16], but similar ideas were used earlier by Salowe [17,18] and Vaidya [19–21].

## ALGORITHM   WSPD-Graph(*S, t*)

1.  $E' := \emptyset$
2.  $G' := (S, E')$.
3.  $\{A_1, B_1\}, \ldots, \{A_m, B_m\} \leftarrow$ the well-separated pair decomposition of $S$ w.r.t. $s = \frac{4(t+1)}{(t-1)}$.
4.  **for** each well-separated pair $\{A_i, B_i\}$
5.      Let $a_i$ and $b_i$ be arbitrary points in $A_i$ and $B_i$, respectively.
6.      Add $(a_i, b_i)$ to $E'$.
7.  **return** $G' = (S, E')$.

The following theorem summarizes the properties.

**Theorem 3.4** *The WSPD-graph is a $t$-spanner for $S$ with $O(s^d \cdot n)$ edges and can be constructed in time $O(s^d n + n \log n)$, where $s = 4(t+1)/(t-1)$.*

There are modifications that can be made to obtain bounded diameter or bounded degree.

*Bounded diameter*: Arya et al. [22] showed how the construction algorithm can be modified such that the diameter of the graph is bounded by $2 \log n$. In the basic construction of a WSPD-graph, a graph is constructed by adding an edge for every well-separated pair in the WSPD. Instead of selecting an arbitrary point in each well-separated set, they choose a representative point by a search in the fair-split tree (see Vol. 1, Chap. 4), that is, for a node $u$ in the split tree, follow the path down the tree by always choosing the larger subtree. The point stored at the leaf in which the path ends is the representative point for $u$. This approach guarantees that the diameter of the constructed $t$-spanner is bounded by $2 \log n$.

*Bounded degree*: The main problem in the construction of the WSPD-graph is that a single point $v$ can be part of many well-separated pairs, and each of the pairs generates an edge with an endpoint at $v$. Arya et al. [1] suggest to keep only the shortest edge for each cone direction, thus combining the $\Theta$-graph approach with the WSPD-graph. The resulting spanner has bounded "out-degree" and by applying the sink-spanner transformation, a $t$-spanner of degree $O(\frac{1}{(t-1)^{2d-1}})$ is obtained.

### 3.2.3 The Greedy-Graph

The greedy algorithm was first presented in 1989 by Bern. Althöfer et al. [7] gave the first theoretical bounds, and since then, the greedy algorithm has been subject to considerable research [12,23–28]. The graph constructed using the greedy algorithm is called a greedy graph, and the general approach is given in the following.

---

### ALGORITHM  GREEDY-Graph(*S, t*)

1.  Construct the complete graph of $S$, denoted by $G = (S, E)$.
2.  $E' := \emptyset$
3.  $G' := (S, E')$.
4.  **for** each edge $(u, v) \in E$ in order of increasing weight
5.      **if** SHORTESTPATH$(G', u, v) > t \cdot d_G(u, v)$
6.          Add $(u, v)$ to $E'$.
7.  **return** $G' = (S, E')$.

---

Chandra et al. [12] proved that the maximum degree of the graph is bounded by a constant. The running time of the naïve implementation of GREEDY-GRAPH considered in their paper is $O(n^3 \log n)$. Das and Narasimhan [25] made a breakthrough in 1994 when they showed that a modified greedy graph could be constructed in $O(n \log^2 n)$ time. They detailed how to use clustering to speed up shortest path queries, by showing that approximate shortest path queries suffice to produce sparse spanners. However, their algorithm was not efficient as the clusters were not maintained efficiently and had to be frequently rebuilt. This problem was solved by Gudmundsson et al. [27] who developed techniques to efficiently perform clustering. The following theorem summarizes the known bounds.

**Theorem 3.5** *The greedy graph is a $t$-spanner of $S$ with $O(\frac{n}{(t-1)^d} \log(\frac{1}{t-1}))$ edges and $O(\frac{1}{(t-1)^d} \log(\frac{1}{t-1}))$ maximum degree and can be computed in time $O(\frac{n}{(t-1)^{2d}} \log n)$.*

*A note on the weight of a t-spanner*: Arya et al. [1] presented an algorithm that claimed to produce $t$-spanners of weight $O(wt(MST(S)))$. However, (Personal communication, 1998) showed that the claimed result was incorrect.

Das and Narasimhan [25] proved that the greedy-graph satisfies the so-called leapfrog property and claimed that any graph satisfying this property has weight $O(wt(MST(S)))$. At this moment, however, no complete proof of this claim has been published.

### 3.2.4 Experimental Studies

The first experimental study of the construction of $t$-spanners was performed by Navarro and Paredes [29] who presented four heuristics for point sets in high-dimensional metric space ($d = 20$) and showed by empirical methods that the running time was $O(n^{2.24})$, and the number of edges in the produced graphs was $O(n^{1.13})$. Sigurd and Zachariasen [30] considered the problem of constructing a minimum weight $t$-spanner of a given graph, but they only considered sparse graphs of small size, that is, graphs with at most 64 vertices and with average vertex degree 4 or 8. In the case in which the input points are given in the Euclidean plane, an extensive study of the main algorithms presented in Sections 3.2.1–3.2.3 was performed by Farshi and Gudmundsson [31].

## 3.3 Computing the Dilation of a Graph

The previous section considered the problem of constructing a graph for a given point set. In Section 3.2 we considered the problem, that is, given a graph $G$ compute $\Delta(G)$.

The problem of calculating the dilation of a given geometric graph can be solved by computing the All-Pairs-Shortest-Path of $G$. Running Dijkstra's algorithm—implemented using Fibonacci heaps—gives the dilation of $G$ in time $O(mn + n^2 \log n)$ using linear space, where $m$ is the number of edges in $G$. For a long time, there were no considerable improvements but in 2002, Langerman et al. [32] and Agarwal et al. [33] showed the first subquadratic bounds for any type of graph. They proved that the dilation of a planar polygonal path can be computed in $O(n \log n)$ expected time. The algorithm can be generalized to planar trees and cycles, with a randomized expected running time of $O(n \log^2 n)$, or $O(n \log^c n)$ worst case running time. This results holds in the plane and was later extended to three dimensions [34]. Agarwal et al. [34] also showed that in three dimensions, one can compute the dilation of a path, cycle, or tree in $O(n^{4/3+\epsilon})$ in randomized expected time. More details about their construction can be found in Section 3.4.2.

Eppstein and Wortman [35] presented an $O(n \log n)$-time algorithm for evaluating the dilation when the input graph $G$ is a star. Computing the shortest path between two points in a star obviously takes constant time, their idea is to identify $O(n)$ candidate pairs and prove that the pair deciding the dilation of $G$ is among these pairs. The point pairs are identified using two techniques, each generating $O(n)$ pairs.

Assume that $(x, y)$ is a pair of points in $G$ with dilation $\Delta(G)$.

In the case when the dilation of $G$ is high, that is, greater than 3, then it holds that $y$ is one of $x$'s $k$ nearest neighbors, for a constant $k$. The $k$ nearest neighbors of every point in $V$ may be reported in time $O(kn \log n)$ using the algorithm by Vaidya [20]. So the process of identifying the $O(n)$ candidate point pairs takes $O(n \log n)$ time.

In the case when the dilation of $G$ is low, that is, smaller than or equal to 3, then it holds that $x$ and $y$ must have almost the same distance to the center of $G$. Assume that the vertices of $V$ are sorted with respect to their distance from the center of $G$, $\langle v_1, \ldots, v_n \rangle$, and that $x = v_i$ and $y = v_j$. It holds that $|i - j| \le \ell$, where $\ell$ is a constant, and thus identifying $O(n)$ candidate point pairs requires $O(n \log n)$ time in this case.

### 3.3.1 Approximating the Dilation

For general geometric graphs, it seems unavoidable to test all the $\binom{n}{2}$ pairs of vertices that may decide the dilation of the graph. However, in the case when it suffices to approximate the dilation, this bound is no longer correct. Narasimhan and Smid [36] showed that $O(n/\epsilon^d)$ pairs of vertices are sufficient to test to give a good approximation. Their algorithm is very simple, and it is stated in the following. It is assumed that an algorithm $\text{ASP}_c(p, q, G)$ is given that takes a graph $G$ and two vertices $p$ and $q$ as input and computes a $c$-approximation of $\Delta_G(p, q)$, where $\Delta_G(p, q) = \frac{d_G(p,q)}{d(p,q)}$. Denote by $T(n, m, k)$ the running time of $\text{ASP}_c$, when given (i) a graph having $n$ vertices, $m$ edges and (ii) a sequence of $k$ $c$-approximate shortest path queries.

The algorithm takes a Euclidean graph $G = (V, E)$ and a real constant $\epsilon > 0$ as input. A WSPD of $V$ is computed with separation constant $4(1 + \epsilon)/\epsilon$. For each well-separated pair $\{A_i, B_i\}$, two arbitrary points $a_i \in A_i$ and $b_i \in B_i$ are chosen, and the dilation of $a_i$ and $b_i$ is estimated by taking the ratio between $\text{ASP}_c(a_i, b_i, G)$ and $d(a_i, b_i)$. The maximum over all the values is returned.

The main theorem can now be stated.

**Theorem 3.6** *Let G be a Euclidean graph in $\mathbb{E}^d$ and let $\epsilon$ be a real constant such that $0 < \epsilon \le 3$, one can compute a $((1 + \epsilon)c)$-approximate dilation of G in time*

$$O(n \log n) + T(n, m, n/\epsilon^d).$$

By using known data structures to answer (approximate) distance queries together with Theorem 3.6 gives an $O(n \log n)$ time $(1 + \epsilon)$-approximation algorithm for paths, cycles, and trees, an $O(n\sqrt{n})$ time $(1 + \epsilon)$-approximation algorithm for plane graphs [37], an $O(m + n(t^5/\epsilon^2)^d (\log n + (t/\epsilon)^d))$ time $(1 + \epsilon)$-approximation algorithm for $t$-spanners [38,39], and an $O(mn^{1/\beta} \log^2 n)$ expected time $O(2\beta(1 + \epsilon)^2)$-approximation algorithm for any Euclidean graph [40].

Note that any algebraic computation tree algorithm that computes a $c$-approximate dilation of a path or a cycle has running time $\Omega(n \log n)$ [36].

# 3.4 Detour

When a geometric network $G$ models an urban street system, the dilation is not necessarily an appropriate measure to assess the quality of $G$. As houses are spread everywhere along the streets, one has to take into account not only the vertices of $G$ but also all the points on its edges, that is, consider the detour of $G$. The detour is also of particular interest in various other applications:

- Analyzing on-line navigation strategies often involves estimating the detour of curves: The length of a path created by some robot must be compared with the shortest path connecting two points [41].
- When comparing the Fréchet distance $F(P, Q)$ between two curves $P, Q$ of detour at most $\kappa$ with their Hausdorff distance $H(P, Q)$ (see Reference 42 for the definition of $F$ and $H$), it turns out that (under some additional technical condition) $F(P, Q) \le (1 + \kappa)H(P, Q)$, whereas no such bound is known for general curves [43,44].

In Section 3.4.1, we describe properties of pairs of points with maximum detour for various scenarios, and in Section 3.4.2, we give algorithms for computing the detour in these scenarios. All the algorithms exploit the structural properties described earlier. The problem of constructing graphs of small detour that contain a prescribed finite point set will be considered in Section 3.4.3.

## 3.4.1 Structural Properties

We describe properties of pairs of points with maximum detour for the following scenarios:

- $P$ is a simple polygonal curve (possibly closed), or a simple tree in $\mathbb{R}^2$.
- $P$ is a simple polygon in $\mathbb{R}^2$.
- $P$ is a geometric graph in $\mathbb{R}^2$.

As already mentioned, the case of the detour of a planar polygonal chain $P$ is of particular interest in various applications. The problem of (approximately) computing $\delta(P)$ in that setting was first addressed in Ebbers-Baumann et al. [45]. The proposed algorithm exploits several structural properties of the problem.

Consider for instance an edge $e$ of $P$ with endpoints $r, s$, and let $q$ be a fixed point on $P$, such that $d_P(q, s) > d_P(q, r)$, cf, Figure 3.1 on page 60. The function $\delta^P(., q)$ takes on a unique maximum on $e$, and if $\beta := \cos \angle(q, p, s) = -\frac{||p-q||}{d_P(q,r)}$ holds, then this maximum is attained at $p$.

To see this, let us assume that $0 < \beta < \pi$. For $-||p - r|| \le t \le 0$, let $p(t)$ be the point on $\overline{rp}$ that has distance $|t|$ to $p$, and for $0 \le t \le ||p - s||$ let $p(t)$ be the point on $\overline{ps}$ that has distance $|t|$ to $p$. We have

$$f(t) := \delta^P(p(t), q) = \frac{d_P(p(t), q)}{||p(t) - q||} = \frac{t + d_P(p, q)}{\sqrt{t^2 + ||p - q||^2 - 2t||p - q|| \cos \beta}}$$

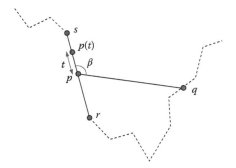

**FIGURE 3.1**   The detour $\delta^P(p(t), q)$ is larger than $\delta^P(p, q)$.

As $||p - q|| \cos \beta + d_P(p, q) > 0$, the derivative of $f(t)$ has a positive denominator and its numerator has the same sign as

$$n(t) := ||p - q|| \frac{||p - q|| + d_P(p, q) \cos \beta}{||p - q|| \cos \beta + d_P(p, q)} - t.$$

Thus, if $||p - q|| + d_P(p, q) \cos \beta$ is positive (resp. negative), the detour can be increased by moving $p$ toward $s$ (resp. $r$).

Another crucial property is that there is always a pair of points $(p', q')$ attaining the detour of a simple polygonal chain $P$, such that $p'$ and $q'$ are *covisible*, that is, $\overline{p'q'} \cap P = \{p', q'\}$. This can be seen as follows: Let $p', q' \in P$ attain the detour of $P$, that is, $\delta^P(p, q) = \delta(P)$, and $p = p_0, \ldots, p_k = q$ be the points of $P$ intersected by the segment $s = \overline{pq}$, ordered by their appearance on $s$, cf Figure 3.2.
Then

$$\delta^P(p, q) = \frac{d_P(p, q)}{||p - q||} \leq \frac{\sum_{i=0}^{k-1} d_P(p_i, p_{i+1})}{\sum_{i=0}^{k-1} ||p_i - p_{i+1}||} \leq \max_{0 \leq i < k} \frac{d_P(p_i, p_{i+1})}{||p_i - p_{i+1}||} = \max_{0 \leq i < k} \delta^P(p_i, p_{i+1})$$

that is, some covisible pair $p_i$ and $p_{i+1}$ attains the detour of $P$.

We can summarize the previous discussion in the following.

**Lemma 3.1** [45] *The detour of a simple polygonal chain $P$ in the plane is attained by a pair of points $(p, q)$ on $P$, where $p$ is a vertex of $P$, and $p$ and $q$ are covisible.*

Similar results were obtained for various other cases. Lemma 3.1 was generalized by Agarwal et al. [34] to the case of simple trees in the plane, and by Ebbers-Baumann et al. [46] it is shown that Lemma 3.1 also holds for simple polygons in the plane.

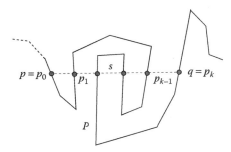

**FIGURE 3.2**   There is always a covisible pair of points attaining the detour.

**Lemma 3.2 [47]** *For any metric space $M$ on $\mathbb{R}^2$, the $M$-detour of a simple polygon $P$ in the plane with $\delta_M(P) > 1$ is attained by a pair of points $(p, q)$ on the boundary of $P$, with the property that $\overline{pq} \cap P = \{p, q\}$, and at least one of $p, q$ is a vertex of $P$.*

For the Euclidean metric, one can even show that every pair of points $(p, q)$ on the boundary that attains the detour of any nonconvex simple polygon $P$ must have the property that $\overline{pq} \cap P = \{p, q\}$.

It is easy to see that the detour of a *closed* simple polygonal curve is *not necessarily* attained at a vertex. Still, a somewhat weaker property was shown by Agarwal et al. [34] for this case:

**Lemma 3.3 [34]** *The detour of a closed simple polygonal curve $P$ of length $\ell$ in the plane is attained by a pair $(p, q)$ of points of $P$, such that, either one of them is a vertex of $P$, or $d_P(p, q) = \ell/2$.*

Moreover, Ebbers-Baumann et al. [46] observed that it is still the case, that a covisible pair of points attains the detour of $P$. This is in fact true for arbitrary connected simple straight-line graphs in the plane.

**Lemma 3.4 [46]** *The detour of a connected simple straight-line graph $P$ in the plane is attained by a pair of covisible points of $P$.*

Note that the *dilation* of a geometric graph is not necessarily attained at a covisible pair of vertices. Moreover, most of these properties fail to hold in higher dimensions. The detour of a polygonal curve in $\mathbb{R}^3$, for instance, is not necessarily attained at a vertex of the chain [34].

## 3.4.2 Algorithmic Questions

Based on the earlier work of Narasimhan and Smid [36], Grüne [48] has shown that there is an $\Omega(n \log n)$ lower bound in the algebraic decision tree model for computing the *dilation* of a monotone and hence simple planar polygonal curve. On the contrary, for the problem of computing the detour of such curves, no nontrivial lower bound is known.

However, as was shown by Agarwal et al. [34], computing the detour of a 3-dimensional polygonal path is as hard as Hopcroft's problem: Given a set $L$ of $n$ lines in $\mathbb{R}^2$ and a set $Q$ of $n$ points in $\mathbb{R}^2$, decide whether any line of $L$ contains any point of $Q$.

The idea is to reduce an instance of Hopcroft's problem to the problem of computing the detour of a 3-dimensional path. To this end, a 3-dimensional path $P_{L,Q}$ is built in such a way that $P_{L,Q}$ has infinite detour (i.e., it self-intersects), iff any line of $L$ contains any point of $Q$. By using techniques developed by Erickson [49], the construction is then modified to cover the case in which it is known in advance that the input chain is not self-intersecting.

The construction shows that, if there is an algorithm to compute $\delta(P)$ for a simple polygonal chain $P$ on $n$ vertices in $\mathbb{R}^3$ in $T(n)$ time, Hopcroft's problem can be solved in $O(n \log n + T(n))$ time. There is an abundance of evidence that suggests that Hopcroft's problem has an $\Omega(n^{4/3})$ lower bound [49] in any reasonable model of computation.

On the positive side, for arbitrary connected plane graphs $P$ on $n$ vertices, the detour can be computed in $O(n^2)$ time in the following way: Compute the shortest path distance for all pairs of vertices of $P$. As $P$ is planar this can be done in $O(n^2)$ time [50]. For every pair of edges $e, f$ of $P$, compute the detour $\delta^P(e, f) = \max\{\delta^P(x, y) \mid x \in e, y \in f\}$. This can be done in constant time per pair, as there are only four combinatorial different types of shortest paths going from points on $e$ to points on $f$.

The structural properties shown in the previous section can be exploited to obtain faster algorithms in some cases. The case in which $P$ is a planar polygonal chain without self-intersections was first studied by Ebbers-Baumann et al. [45], where the following result is shown.

**Theorem 3.7 [45]** *Let $P$ be a simple polygonal chain on $n$ vertices in the plane, and let $\epsilon$ be a positive constant. In $O(n \log n)$ time a pair of points $(p, q)$ on $P$ can be computed, such that $\delta(P) \le (1 + \epsilon)\delta^P(p, q)$.*

The problem of exactly computing $\delta(P)$ for a simple polygonal chain $P$ was independently considered by Langermann et al. and Agarwal et al. [32,33] (see also Reference 34 for a combined version).

The approach in Reference 34 is as follows: First, a deterministic $O(n \log n)$ time algorithm for the decision problem is developed, that is, an algorithm that decides on input $(P, \kappa)$ weather $\delta(P) \leq \kappa$. Note that according to Lemma 3.1, $\delta(P) \leq \kappa$ iff $\delta^P(q, p) \leq \kappa$ for all *vertices* $p \in P$ and all points $q \in P$.

This problem can be restated in a form that makes it amenable to range-searching techniques: Let $p_0$ be the first point of $P$. For a point $p \in P$, define the *weight* of $p$ to be $\omega(p) = d_P(p, p_0)/\kappa$. Let $C$ denote the cone $z = \sqrt{x^2 + y^2}$ in $\mathbb{R}^3$ and map each vertex $p = (p_x, p_y) \in P$ to the cone $C_p = C + (p_x, p_y, \omega(p))$, that is, translate the apex of $C$ (that is, the origin) to the point $(p_x, p_y, \omega(p))$. If $C_p$ is regarded as the graph of a bivariate function, which will also be denoted by $C_p$, then for any point $x \in \mathbb{R}^2$, $C_p(x) = |xp| + \omega(p)$. Map a point $q = (q_x, q_y) \in P$ to the point $\hat{q} = (q_x, q_y, \omega(q))$ in $\mathbb{R}^3$, cf Figure 3.3 on page 62. Now, for any point $q \in P$ and a vertex $p \in P$ that lies *before* $q$ on $P$ (in the sense that $d_P(p, q) = d_P(p_0, q) - d_P(p_0, p)$), $\delta^P(q, p) \leq \kappa$ if and only if $\hat{q}$ lies below the cone $C_p$:

$$\delta^P(q, p) \leq \kappa \Leftrightarrow \frac{d_P(p_0, q) - d_P(p_0, p)}{|qp|} \leq \kappa \Leftrightarrow \frac{d_P(p_0, q)}{\kappa} \leq |qp| + \frac{d_P(p_0, p)}{\kappa} \Leftrightarrow \omega(q) \leq C_p(q).$$

That is, $\delta^P(q, p) \leq \kappa$ iff $\hat{q}$ lies below the cone $C_p$, and consequently $\delta(P) \leq \kappa$ iff the polygonal chain $\hat{P} = \{\hat{p} \mid p \in P\}$ lies below the lower envelope of the cones $\{C_p \mid p$ is a vertex of $P\}$. By exploiting the covisibility property from Lemma 3.1, this condition can be verified in $O(n \log n)$ deterministic time.

By using a randomized technique of Chan [51] or parametric search [52], the decision procedure is then turned into an algorithm for actually computing $\delta(P)$ (parametric search incurs a polylogarithmic overhead).

**Theorem 3.8** [34] *Let $P$ be a simple polygonal chain on $n$ vertices in the plane. There is*

- *A deterministic algorithm to decide in $O(n \log n)$ time whether $\delta(P) \leq \kappa$, for any $\kappa > 0$*
- *A randomized algorithm to compute $\delta(P)$ in $O(n \log n)$ expected time*
- *A deterministic algorithm to compute $\delta(P)$ in $O(n \log^{O(1)} n)$ time.*

By using appropriate recursive partitioning schemes—based on Lemma 3.3 or the variant of Lemma 3.1 for trees—similar techniques can be applied to the case in which $P$ is a tree or a cycle.

**Theorem 3.9** [34] *Let $P$ be a simple closed polygonal chain or a plane tree on $n$ vertices in the plane. There is a randomized algorithm to compute $\delta(P)$ in $O(n \log^2 n)$ expected time.*

As shown by Grüne et al. [47,48], the approach of Reference 45 can be modified to handle the case in which $P$ is a simple polygon in the plane. This yields an efficient approximation algorithm.

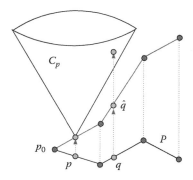

**FIGURE 3.3**   The chain $P$ lifted to $\mathbb{R}^3$.

**Theorem 3.10** [47,48] *Let P be a simple polygon on n vertices in the plane, and let $\epsilon$ be a positive constant. In $O(n \log n)$ time, a pair of points $(p, q)$ on the boundary of P can be computed, such that $\delta(P) \leq (1 + \epsilon)\delta^P(p, q)$.*

The fastest known algorithm for computing $\delta(P)$ for a simple polygon P *exactly* is similar to the brute-force approach described at the beginning of this section. Of course Lemma 3.2 plays a crucial role here. Moreover, the shortest path computation is more involved as shortest geodesic paths inside P have to be computed.

**Theorem 3.11** [47] *Let P be a simple polygon on n vertices in the plane. There is a deterministic algorithm to compute $\delta(P)$ in $O(n^2)$ time.*

As already mentioned, it is no longer true that the detour is attained at a vertex of P, when P is a simple polygonal chain in $\mathbb{R}^3$. This makes the 3-dimensional algorithm considerably more complicated, and less efficient, than its 2-dimensional counterpart.

**Theorem 3.12** [34] *Let P be a simple polygonal chain on n vertices in $\mathbb{R}^3$. There is a randomized algorithm to compute $\delta(P)$ in $O(n^{16/9+\epsilon} \log n)$ expected time for any $\epsilon > 0$.*

### 3.4.3 Low Detour Embeddings of Point Sets

Besides computing the detour of given graphs, the problem of constructing plane graphs of small detour that contain a given finite point set was also investigated.

**Definition 3.3 (Detour of a Point Set)** *The detour $\mathring{\delta}(P)$ of a finite point set P in the plane is the smallest possible detour of any finite plane graph that contains all points of P, that is,*

$$\mathring{\delta}(P) := \inf_{\substack{P \subset G \\ G \text{ finite, plane}}} \delta(G)$$

Even for a point set P of size 3, computing $\mathring{\delta}(P)$ is a nontrivial task. For the *dilation*, the optimum solution must be a triangulation, as an optimal solution only contains straight edges, and adding edges never increases the dilation. Still, it is not known how to efficiently compute the triangulation that minimizes the dilation of a given point set.

As a consequence of Lemmas 3.3 and 3.4, the detour of any *rational* point set P is bounded by two, as it can be embedded into a square grid, that is, $\mathring{\delta}(P) \leq 2$ for all $P \subseteq \mathbb{Q}^2$. A construction of Reference 46 shows that this can be improved:

**Theorem 3.13** [46] *There is a periodic, plane covering graph $G_\infty$ of detour 1.67784..., such that each finite set of rational points is contained in a finite part of a scaled copy of $G_\infty$.*

On the other side of the spectrum, Reference 46 gives an example of a point set P with detour $\mathring{\delta}(P) \geq \pi/2 = 1.57079...$. A subsequent work in References 53, 54 improved this lower bound by exhibiting a point set P with $\mathring{\delta}(P) > \pi/2$.

**Theorem 3.14** [46,54]

- *Let P be the vertex set of the regular n-gon on the unit circle. Then $\mathring{\delta}(P) \geq \pi/2$.*
- *Let $P = \{(x, y) \mid x, y \in \{-9, \ldots, 9\}\}$. Then $\mathring{\delta}(P) \geq (1 + 10^{-11})\pi/2$.*

# 3.5 Low-Dilation Networks

Besides computing the dilation of given graphs, the problem of constructing certain finite plane graphs $G = (V, E)$ of small dilation that contain a given finite point set $P$ is also interesting. There are several different variants of this problem, depending on weather $G$ may or may not contain Steiner-points, or if $G$ is restricted to belong to a certain class of graphs $\mathcal{G}$, such as triangulations and trees.

**Definition 3.4 (Dilation of a Point Set)** *Let $\mathcal{G}$ be a class of graphs and $P$ be a finite point set in the plane. The dilation $\dot{\Delta}^{\mathcal{G}}(P)$ of $P$ w.r.t. $\mathcal{G}$ is the smallest possible dilation of any finite plane graph $G = (P, E)$ in $\mathcal{G}$, that is,*

$$\dot{\Delta}^{\mathcal{G}}(P) := \min_{\substack{G=(P,E)\in\mathcal{G} \\ G \text{ finite, plane}}} \Delta(G).$$

If $\mathcal{G}$ is the class of all graphs, we can assume $G$ to be a triangulation, as an optimal solution only contains straight edges, and adding edges never increases the dilation. We omit the superscript $\mathcal{G}$ in this case.

## 3.5.1 Triangulations

A triangulation defining $\dot{\Delta}(P)$ is called a *minimum dilation triangulation* of $P$. So far, only little research has been conducted on minimum dilation triangulations. The complexity status of the problem is open. Most work upperbounds the dilation of certain types of triangulations. Chew [55] has shown that the rectilinear Delaunay triangulation has dilation at most $\sqrt{10}$. Dobkin et al. [56] gave a similar result for the Euclidean Delaunay triangulation. They show that its dilation can be bounded from above by $\left((1 + \sqrt{5})/2\right)\pi \approx 5.08$. This bound was further improved to $2\pi/(3\cos(\pi/6)) \approx 2.42$ by Keil and Gutwin [57,58]. Das and Joseph [59] generalized all these results by identifying two properties of planar graphs such that if $A$ is an algorithm that computes a planar graph from a given set of points and if all the graphs constructed by $A$ meet these properties, then the dilation of all the graphs constructed by $A$ is bounded by a constant.

*Exclusion and inclusion regions:* When investigating optimal triangulations, it is usually instructive to consider local properties of the edges in these triangulations. One important class of local properties that has been studied extensively, for example, for minimum weight triangulations are *exclusion regions*. They provide a necessary condition for the inclusion of an edge into an optimal triangulation: If $p$ and $q$ are two points in $P$, then the edge $e := \overline{pq}$ can only be contained in an optimal triangulation of $P$ if no other points of $P$ lie in (certain parts of) the exclusion region of $e$.

To obtain an exclusion region for the minimum dilation triangulation, one can observe the following: We know from Reference 57 that the dilation of the Delaunay triangulation of $P$ is bounded by $\gamma = 2\pi/(3\cos(\pi/6))$. Moreover, if we have an edge $e$ and two points $x, y$ on opposite sides of $e$ that are close to the center of $e$, then the dilation between $x$ and $y$ is large, because $e$ constitutes an obstacle that the shortest path between $x$ and $y$ has to surpass. In fact, if we can quantify this and show that the dilation between any pair of points in a certain region $D_{e,\gamma}$ that lie on opposing sides of $e$ is larger than $\gamma$, we can conclude that, if $D_{e,\gamma}$ contains such a pair of points, then $e$ cannot be contained in the minimum dilation triangulation of $P$, as the Delaunay triangulation gives a better dilation than any triangulation containing $e$.

The upper bound of Reference 57 can also be used to obtain a *sufficient* condition for the inclusion of an edge. More specifically, consider for two points $p, q \in P$ the ellipsoid $E_{p,q,\gamma}$ with foci $p$ and $q$ that is given by $E_{p,q,\gamma} = \{x \in \mathbb{R}^2 \mid |px| + |qx| \leq \gamma \cdot |pq|\}$. If $E_{p,q,\gamma}$ is empty, then the line segment $\overline{pq}$ has to be included in the minimum dilation triangulation of $P$, as otherwise the dilation between $p$ and $q$ would be larger than $\gamma$.

**Theorem 3.15 [60,61]** *Let* $\gamma = 2\pi/(3\cos(\pi/6))$ *and* $p, q \in P$.

1. *For any* $0 < \alpha < 1/(2\gamma)$, *the disk* $D_{\overline{pq},\alpha}$ *of radius* $\alpha|pq|$ *centered at the midpoint of* $\overline{pq}$ *is an exclusion region for the minimum dilation triangulation.*
2. *The ellipsoid* $E_{p,q,\gamma} = \{x \in \mathbb{R}^2 \,|\, |px| + |qx| \leq \gamma \cdot |pq|\}$ *is an inclusion region for the minimum dilation triangulation.*

*Regular n-gons:* Even for the vertex set $S_n = \{s_1, \ldots, s_n\}$ of a regular $n$-gon, it is not known how to efficiently compute a minimum dilation triangulation. There is however some additional understanding of the structure of optimal triangulations in that case. In particular, there is a simple lower bound for the dilation of $S_n$.

**Theorem 3.16 [61]** *Let* $n \geq 74$ *and assume that* $\max_{1 \leq i < j \leq n} ||s_i - s_j|| = 2$. *For any triangulation* $T$ *of* $S_n$ *and any maximum dilation pair* $s_x, s_y \in S_n$ *of* $T$, *we have that*

1. $\Delta(T) \geq \sqrt{2 - \sqrt{3}} + \sqrt{3}/2 \approx 1.3836$,
2. $|a - b| > 5n/12$, *and*
3. $||s_a - s_b|| > \left(\sqrt{6 + 3\sqrt{3}} + \sqrt{2 - \sqrt{3}}\right)/2 \approx 1.93185$.

This can be used to derive an efficient approximation algorithm that computes a triangulation the dilation of which is within a factor of $1 + O\left(1/\sqrt{\log n}\right)$ of the optimum.

**Theorem 3.17 [61]** *In* $O(n\sqrt{\log n})$, *a triangulation* $T^*$ *of* $S_n$ *can be computed, such that*

$$\Delta(T^*) \leq \left(1 + O\left(1/\sqrt{\log n}\right)\right) \dot{\Delta}(S_n).$$

### 3.5.2 Stars

The problem of computing a minimum dilation star of $P$, that is, a graph $G \in \mathcal{G}$ defining $\dot{\Delta}^P$ where $\mathcal{G}$ is a star, was considered for the first time by Eppstein and Wortman [35]. They proved the following:

**Theorem 3.18** *Let* $P$ *be a set of* $n$ *points in* $\mathbb{E}^2$, *one can compute a minimum dilation star of* $P$ *in* $O(n2^{\alpha(n)} \log n)$ *expected time, where* $\alpha$ *is the functional inverse of Ackermann's function [62].*

The algorithm works by iteratively selecting a random vertex $c$ in a region $R$, evaluating the dilation that would result from using $c$ as a center, computing the region $R$ that could contain a center yielding a lower dilation, and discarding the vertices outside $R$. By evaluating the dilation $\Delta_c$ of a given star with center at $c$ in $O(n \log n)$ time was discussed in Section 3.3.

The region $R$ is the intersection of $O(n)$ ellipses defined by the $O(n)$ pairs of points identified in Section 3.3, that is, for each of the pairs $v_i$ and $v_j$ the level set $f_{i,j}^{\leq \lambda} = \{x \in \mathbb{E}^2 | f_{i,j}(x) = \frac{|v_i x| + |x v_j|}{|v_i v_j|} \leq \lambda\}$ defines an ellipsoid with foci $v_i$ and $v_j$. The intersection of those $O(n)$ ellipses can be described by $O(n2^{\alpha(n)})$ arcs.

In each iteration, any vertex in $R$ will be removed with probability $1/2$; so the expected number of iterations is $O(\log n)$, resulting in an $O(n2^{\alpha(n)} \log n)$ expected time algorithm.

### 3.5.3 Small Spanners with Small Dilation

Aronov et al. [63] considered the problem of constructing a minimum dilation graph given the number of edges as a parameter. Any spanner of a set of $n$ points $S$ must have at least $n - 1$ edges, otherwise the graph would not be connected and the dilation would be infinite. The quantity $\Delta(S, k)$ is defined as:

$$\Delta(S, k) = \min_{\substack{V(G)=S \\ |E(G)|=n-1+k}} \Delta(G).$$

Thus $\Delta(S, k)$ is the minimum dilation one can achieve with a network on $S$ that has $n - 1 + k$ edges.

**Theorem 3.19** *For any $n$ and any $k$ with $1 \leq k \leq 2n$, there is a set $S$ of $n$ points such that any graph on $S$ with $n - 1 + k$ edges has dilation at least $\frac{2}{\pi} \cdot \lfloor \frac{n}{k+1} \rfloor - 1$.*

Consider a set $S$ of $n$ points $p_1, \ldots, p_n$ spaced equally on the unit circle, and let $o$ be the center of the circle. The first step is to prove a lower bound on any tree $T$ for $S$.

Let $x$ and $y$ be two points in $S$ and let $\gamma$ and $\gamma'$ be two paths from $x$ to $y$ avoiding $o$. The paths $\gamma$ and $\gamma'$ are *(homotopy) equivalent* if $\gamma$ and $\gamma'$ belong to the same homotopy class in the punctured plane $\mathbb{R}^2 \setminus \{o\}$. Let $\gamma_i$ be the unique path in $T$ from $p_i$ to $p_{i+1}$ (where $p_{n+1} := p_1$). Aronov et al. [63] prove that there must be at least one index $i$ for which $\gamma_i$ is not equivalent to the straight segment $p_i p_{i+1}$. As the path $\gamma_i$ must not only "go around" $o$ but must do so using points $p_j$ on the circle only it follows that $\Delta(S, 0) \geq \frac{2}{\pi} n - 1$.

Theorem 3.19 follows from the above-mentioned argument by letting $S$ consist of $k + 1$ copies of the earlier construction, that is, sets $S_i$, for $1 \leq i \leq k + 1$, each consisting of at least $\lfloor n/(k+1) \rfloor$ points. The points in $S_i$ are placed equally spaced on a unit-radius circle with center at $(2 \cdot i \cdot n, 0)$. The set $S$ is the union of $S_1, \ldots, S_{k+1}$.

In the same paper, Aronov et al. [63] give a matching upper bound. The algorithm constructs a graph $G = (S, E)$ with $n - 1 + k$ edges, and dilation $O(n/(k + 1))$ in $O(n \log n)$ time.

Let $m \leftarrow \lfloor (k + 5)/2 \rfloor$. Partition a minimum spanning tree $T$ of $S$ into $m$ disjoint connected subtrees, $T_1, \ldots, T_m$, each containing $O(n/m)$ points. The edges of each subtree is added to $E$. Next, consider a Delaunay triangulation of $S$. For each pair of subtrees $T_i$ and $T_j$, the shortest Delaunay edge (if any) is added to $E$. This completes the construction of $G = (S, E)$.

The number of edges in $G$ is at most $n - 1 + k$ and $\Delta(G)$ can be bounded by $O(n/(k + 1))$ [63].

## Acknowledgments

We would like to thank Michiel Smid for valuable comments on an earlier draft of this chapter.

## References

1. Arya, S., Das, G., Mount, D.M., Salowe, J.S., and Smid, M., Euclidean spanners: Short, thin, and lanky, in *Proceedings of STOC*, ACM-SIAM, 1995, p. 489.

2. Agarwal, P.K., Wang, Y., and Yin, P., Lower bound for sparse Euclidean spanners, in *Proceedings of SODA*, ACM-SIAM, 2005, p. 670. Society for Industrial and Applied Mathematics, Philadelphia, PA.

3. Chen, D.Z., Das, G., and Smid, M., Lower bounds for computing geometric spanners and approximate shortest paths, *Disc. Appl. Math.*, 110, 151, 2001.

4. Narasimhan, G. and Smid, M., *Geometric Spanner Networks*, Kluwer, Dordrecht, the Netherlands, 2006.

5. Clarkson, K.L., Approximation algorithms for shortest path motion planning, in *Proceedings of ACM Symposium on Computational Geometry*, ACM, 1987, p. 56.

6. Keil, J.M., Approximating the complete Euclidean graph, in *Proceedings of the Scandinavian Workshop on Algorithmic Theory*, LNCS, Springer, London, UK, 382, 1988, p. 208.

7. Althöfer, I., Das, G., Dobkin, D.P., Joseph, D., and Soares, J., On sparse spanners of weighted graphs, *Disc. Comput. Geom.*, 9(1), 81, 1993.

8. Ruppert, J. and Seidel, R., Approximating the $d$-dimensional complete Euclidean graph, in *Proceedings of the Canadian Conference on Computational Geometry*, Burnaby, Canada, 1991, p. 207.

9. Yao, A.C., On constructing minimum spanning trees in $k$-dimensional spaces and related problems, *SIAM J. Comput.*, 11, 721–736, 1982.

10. Pugh, W., Skip lists: A probabilistic alternative to balanced trees, *CACM*, 33(6), 668, 1990.

11. Bose, P., Gudmundsson, J., and Morin, P., Ordered theta graphs, *Comput. Geom. Theor. Appl.*, 28, 11, 2004.

12. Chandra, B., Das, G., Narasimhan, G., and Soares, J., New sparseness results on graph spanners, *Int. J. Comput. Geom. Appl.*, 5, 124, 1995.

13. Lenhof, H.-P., Salowe, J.S., and Wrege, D.E., New methods to mix shortest-path and minimum spanning trees, 1993. Manuscript.

14. Arya, S. and Smid, M., Efficient construction of a bounded-degree spanner with low weight, *Algorithmica*, 17, 33, 1997.

15. Callahan, P.B. and Kosaraju, S.R., A decomposition of multidimensional point sets with applications to $k$-nearest-neighbors and $n$-body potential fields, *JACM*, 42, 67, 1995.

16. Callahan, P.B. and Kosaraju, S.R., Faster algorithms for some geometric graph problems in higher dimensions, in *Proceedings of SODA*, ACM-SIAM, 1993, p. 291.

17. Salowe, J.S., Constructing multidimensional spanner graphs, *Int. J. Comput. Geom. Appl.*, 1(2), 99, 1991.

18. Salowe, J.S., Enumerating interdistances in space, *Int. J. Comput. Geom. Appl.*, 2(1), 49, 1992.

19. Vaidya, P.M., Minimum spanning trees in $k$-dimensional space, *SIAM J. Comput.*, 17, 572, 1988.

20. Vaidya, P.M., An $O(n \log n)$ algorithm for the all-nearest-neighbors problem, *Disc. Comput. Geom.*, 4, 101, 1989.

21. Vaidya, P.M., A sparse graph almost as good as the complete graph on points in $K$ dimensions, *Disc. Comput. Geom.*, 6, 369, 1991.

22. Arya, S., Mount, D.M., and Smid, M., Randomized and deterministic algorithms for geometric spanners of small diameter, in *Proceedings of FOCS*, IEEE, 1994, p. 703.

23. Chandra, B., Constructing sparse spanners for most graphs in higher dimensions, *Inf. Proceedings Lett.*, 51(6), 289, 1994.

24. Das, G., Heffernan, P., and Narasimhan, G., Optimally sparse spanners in 3-dimensional Euclidean space, in *Proceedings of the Symposium on Computational Geometry*, ACM, 1993, p. 53.

25. Das, G. and Narasimhan, G., A fast algorithm for constructing sparse Euclidean spanners, *Int. J. Comput. Geom. Appl.*, 7, 297, 1997.

26. Das, G., Narasimhan, G., and Salowe, J., A new way to weigh malnourished Euclidean graphs, in *Proceedings of SODA*, San Francisco, CA, 1995, p. 215.

27. Gudmundsson, J., Levcopoulos, C., and Narasimhan, G., Improved greedy algorithms for constructing sparse geometric spanners, *SIAM J. Comput.*, 31(5), 1479, 2002.

28. Soares, J., Approximating Euclidean distances by small degree graphs, *Disc. Comput. Geom.*, 11, 213, 1994.

29. Navarro, G. and Paredes, R., Practical construction of metric $t$-spanners, in *Proceedings of the Workshop on Algorithm Engineering and Experiments*, Springer, 2003, p. 69.

30. Sigurd, M. and Zachariasen, M., Construction of minimum-weight spanners, in *Proceedings of the European Symposium on Algorithms*, LNCS, 3221, Springer, 2004, p. 797.

31. Farshi, M. and Gudmundsson, J., Experimental study of geometric t-spanners, in *Proceedings of the 13th European Symposium on Algorithms*, LNCS, 2005.

32. Langerman, S., Morin, P., and Soss, M., Computing the maximum detour and spanning ratio of planar chains, trees and cycles, in *Proceedings of STACS*, LNCS, 2285, Springer, 2002, p. 250.

33. Agarwal, P.K., Klein, R., Knauer, C., and Sharir, M., Computing the detour of polygonal curves, TR B 02-03, Freie Universität Berlin, Fachbereich Mathematik und Informatik, 2002.

34. Agarwal, P.K., Klein, R., Knauer, C., Langerman, S., Morin, P., Sharir, M., and Soss, M., Computing the detour and spanning ratio of paths, trees and cycles in 2D and 3D, *Disc. Comput. Geom.*, 39(1–3), 17–37, 2008.

35. Eppstein, D. and Wortman, K.A., Minimum dilation stars, in *Proceedings of the Symposium on Computational Geometry*, ACM, 2005, p. 321.

36. Narasimhan, G. and Smid, M., Approximating the stretch factor of Euclidean graphs, *SIAM J. Comput.*, 30(3), 978, 2000.

37. Arikati, S.R., Chen, D.Z., Chew, L.P., Das, G., Smid, M., and Zaroliagis, C.D., Planar spanners and approximate shortest path queries among obstacles in the plane, in *Proceedings of the European Symposium on Algorithms*, LNCS, 1136, Springer, 1996, p. 514.

38. Gudmundsson, J., Levcopoulos, C., Narasimhan, G., and Smid, M., Approximate distance oracles for geometric graphs, in *Proceedings of SODA*, ACM-SIAM, 2002, p. 828.

39. Gudmundsson, J., Levcopoulos, C., Narasimhan, G., and Smid, M., Approximate distance oracles revisited, in *Proceedings of the International Symposium on Algorithms and Computation*, LNCS, 2518, Springer, 2002, p. 357.

40. Cohen, E., Fast algorithms for constructing $t$-spanners and paths with stretch $t$, *SIAM J. Comput.*, 28, 210, 1998.

41. Icking, C. and Klein, R., Searching for the kernel of a polygon: A competitive strategy, in *Proceedings of the ACM Symposium on Computational Geometry*, 1995, p. 258. ACM, Vancouver, Canada.

42. Alt, H. and Guibas, L.J., Discrete geometric shapes: Matching, interpolation, and approximation, in *Handbook of Computational Geometry*, Sack, J.R. and Urrutia, J. (Eds.), Elsevier Science Publishers, Elsevier, 2000, p. 121.

43. Alt, H., Knauer, C., and Wenk, C., Bounding the Fréchet distance by the Hausdorff distance, in *Abstracts European Workshop on Computational Geometry*, Freie Universität Berlin, Berlin, Germany, 2001, p. 166.

44. Alt, H., Knauer, C., and Wenk, C., Comparison of distance measures for planar curves, *Algorithmica*, Springer, 38(2), 45, 2004.

45. Ebbers-Baumann, A., Klein, R., Langetepe, E., and Lingas, A., A fast algorithm for approximating the detour of a polygonal chain, *Comput. Geom. Theor. Appl.*, Elsevier, 27, 123, 2004.

46. Ebbers-Baumann, A., Grüne, A., and Klein, R., On the geometric dilation of finite point sets, in *Proceedings of the International Symposium on Algorithms and Computation*, LNCS, 2906, Springer, 2003, p. 250.

47. Grüne, A., Klein, R., and Langetepe, E., Computing the detour of polygons, in *Abstracts of the European Workshop on Computational Geometry*, 2003, p. 61.

48. Grüne, A., Umwege in polygonen, Master Thesis, Universität Bonn, Bonn, Germany, 2002.

49. Erickson, J., New lower bounds for Hopcroft's problem, *Disc. Comput. Geom.*, 16, 389, 1996.

50. Henzinger, M., Klein, P.N., Rao, S., and Subramanian, S., Faster shortest-path algorithms for planar graphs, *JCSS*, 55(1), 3, 1997.

51. Chan, T.M., Geometric applications of a randomized optimization technique, *Disc. Comput. Geom.*, 22(4), 547, 1999.

52. Megiddo, N., Applying parallel computation algorithms in the design of serial algorithms, *JACM*, 30(4), 852, 1983.

53. Ebbers-Baumann, A., Grüne, A., and Klein, R., Geometric dilation of closed planar curves: A new lower bound, in *Abstracts of the European Workshop on Computational Geometry*, 2004, p. 123.

54. Dumitrescu, A., Grüne, A., and Rote, G., Improved lower bound on the geometric dilation of point sets, in *Abstracts of the European Workshop on Computational Geometry*, Berlin, Germany, 2005, p. 37.

55. Chew, L.P., There are planar graphs almost as good as the complete graph, *JCSS*, 39, 205, 1989.

56. Dobkin, D.P., Friedman, S.J., and Supowit, K.J., Delaunay graphs are almost as good as complete graphs, *Disc. Comput. Geom.*, 5, 399, 1990.

57. Keil, J.M. and Gutwin, C.A., The Delaunay triangulation closely approximates the complete Euclidean graph, in *Proceedings of the Workshop on Algorithms and Data Structures*, LNCS, 382, Springer, 1989, p. 47.

58. Keil, J.M. and Gutwin, C.A., Classes of graphs which approximate the complete Euclidean graph, *Disc. Comput. Geom.*, 7, 13, 1992.
59. Das, G. and Joseph, D., Which triangulations approximate the complete graph? in *Proceedings of the International Symposium on Optimal Algorithms*, LNCS, Springer, 401, Springer, 1989, p. 168.
60. Knauer, C. and Mulzer, W., An exclusion region for minimum dilation triangulations, in *Proceedings of the European Workshop on Computational Geometry*, 2005, p. 33.
61. Knauer, C. and Mulzer, W., Minimum dilation triangulations, TR B 05-06, Freie Universität Berlin, Fachbereich Mathematik und Informatik, 2005.
62. Ackermann, W., Zum Hilbertschen Aufbau der reellen Zahlen, *Math. Ann.*, 99, 118, 1928.
63. Aronov, B., de Berg, M., Cheong, O., Gudmundsson, J., Haverkort, H.J., Smid, M.H.M., and Vigneron, A., Sparse geometric graphs with small dilation, *Comput. Geom.*, 40(3), 207–219, 2008.

# 4

# The Well-Separated Pair Decomposition and Its Applications

Michiel Smid

## 4.1 Introduction

Computational geometry is concerned with the design and analysis of algorithms that solve problems on geometric data in $\mathbb{R}^d$, in which the dimension $d$ is a constant. A large part of the field has been devoted to problems that involve distances determined by pairs of points in a given point set. Given a set $S$ of $n$ points in $\mathbb{R}^d$, we may wish to compute a pair $p, q$ of distinct points in $S$ whose distance is minimum, the $k$ smallest distances among the $\binom{n}{2}$ pairwise distances, the nearest neighbor of each point of $S$, or the minimum spanning tree of $S$. Most problems of this type can be rephrased as a graph problem on the complete Euclidean graph on $S$, in which each edge $pq$ has a weight being the Euclidean distance $|pq|$ between $p$ and $q$. As the number of edges in this graph is $\Theta(n^2)$, many problems involving pairwise distances can trivially be solved in $O(n^2)$ time. Even though the complete Euclidean graph has size $\Theta(n^2)$, it can be represented in $\Theta(n)$ space: It is clearly sufficient to only store the points of $S$, because the weight of any edge can be computed in $O(1)$ time. This leads to the question whether distance problems can be solved in subquadratic time, possibly at the cost of obtaining an approximate solution. For many of these problems, subquadratic algorithms have indeed been designed; see for example, the books by Preparata and Shamos [1] and de Berg et al. [2], and the survey papers by Bern and Eppstein [3], Eppstein [4], and Smid [5]. Most of these algorithms, however, are tailored to the problem at hand.

Callahan and Kosaraju [6,7] devised the *well-separated pair decomposition (WSPD)* and showed that it can be used to solve a large variety of distance problems. Intuitively, a WSPD is a partition of the $\binom{n}{2}$ edges of the complete Euclidean graph into $O(n)$ subsets. Each subset in this partition is represented by two subsets $A$ and $B$ of the point set $S$, such that (i) all distances between points in $A$ and points in $B$ are approximately equal, (ii) all distances within the point set $A$ are much smaller than distances between $A$ and $B$, and (iii) all distances within the point set $B$ are much smaller than distances between $A$ and $B$. Thus, a WSPD can be regarded as a set of $O(n)$ edges that approximates the dense complete Euclidean graph.

Callahan and Kosaraju showed that the WSPD can be used to obtain optimal algorithms for solving the closest pair problem, the $k$ closest pairs problem, the all-nearest neighbors problem, and the approximate minimum spanning tree problem. After the publication of this influential paper, other researchers have shown that the WSPD can be used to solve many other problems. In this paper, we give an overview of several proximity problems that can be solved efficiently using the WSPD. We mention that the WSPD has also been a critical tool for the solution of several variants of the problem of constructing spanners; an overview of these results can be found in the chapter by Gudmundsson and Knauer [8] in this handbook. An extensive treatment of the WSPD and its applications are given in the book by Narasimhan and Smid [9].

The rest of this paper is organized as follows. In Section 4.2, we define the WSPD. In Section 4.3, we present an efficient algorithm for constructing a WSPD. In Section 4.4, we show that the WSPD can be used to obtain optimal algorithms for the closest pair problem, the $k$ closest pairs problem, and the all-nearest neighbors problem. In Section 4.5, we use the WSPD to obtain approximate solutions for the diameter problem, the spanner problem, the minimum spanning tree problem, and the problem of computing the $k$th closest pair. Finally, in Section 4.6, we mention some results on generalizing the WSPD to more general metric spaces.

## 4.2 Well-Separated Pairs

In the current section, we define the WSPD and prove one of its main properties in Lemma 4.1. As mentioned in Section 4.1, the WSPD was introduced by Callahan and Kosaraju [6,7]. Previously, however, similar ideas were used by Salowe [10,11] and Vaidya [12–14], who designed efficient algorithms for computing spanners, all-nearest neighbors, and $k$ closest pairs.

We start by defining the notion of two sets being well separated. For any set $X$ of points in $\mathbb{R}^d$, we denote its *bounding box* by $R(X)$. Thus, $R(X)$ is the smallest axes-parallel hyperrectangle that contains the set $X$.

**Definition 4.1** *Let $A$ and $B$ be two finite sets of points in $\mathbb{R}^d$, and let $s > 0$ be a real number. We say that $A$ and $B$ are* well separated *with respect to $s$, if there exist two disjoint balls $C_A$ and $C_B$, such that*

1. *$C_A$ and $C_B$ have the same radius.*
2. *$C_A$ contains $R(A)$.*
3. *$C_B$ contains $R(B)$.*
4. *The distance between $C_A$ and $C_B$ is at least $s$ times the radius of $C_A$.*

*The real number $s$ is called the* separation ratio.

If we are given the bounding boxes $R(A)$ and $R(B)$ of the sets $A$ and $B$, respectively, then we can test in $O(1)$ time whether these two sets are well separated.

In the next lemma, we prove the two properties of well-separated sets that were mentioned already in Section 4.1: If $A$ and $B$ are well separated with respect to a large separation ratio $s$, then (i) all distances between points in $A$ and $B$ are approximately equal and (ii) all distances between points in $A$ (resp. $B$) are much smaller than distances between $A$ and $B$. These two properties will be used repeatedly in the rest of this paper.

**Lemma 4.1** *Let $s > 0$ be a real number, let $A$ and $B$ be two sets in $\mathbb{R}^d$ that are well separated with respect to $s$, let $a$ and $a'$ be two points in $A$, and let $b$ and $b'$ be two points in $B$. Then, we have*

1. *$|aa'| \leq (2/s)|ab|$*
2. *$|a'b'| \leq (1 + 4/s)|ab|$*

*Proof.* Let $C_A$ and $C_B$ be disjoint balls of the same radius, say $\rho$, such that $R(A) \subseteq C_A$, $R(B) \subseteq C_B$, and the distance between $C_A$ and $C_B$ is at least $s\rho$. The first claim follows from the facts that $|aa'| \leq 2\rho$ and $|ab| \geq s\rho$. By combining the first claim with the triangle inequality, we obtain

$$|a'b'| \leq |a'a| + |ab| + |bb'| \leq (1 + 4/s)|ab|,$$

proving the second claim. ∎

**Definition 4.2** *Let S be a set of n points in $\mathbb{R}^d$, and let $s > 0$ be a real number. A WSPD for S, with respect to s, is a sequence*

$$\{A_1, B_1\}, \{A_2, B_2\}, \ldots, \{A_m, B_m\}$$

*of pairs of nonempty subsets of S, for some integer m, such that*

1. *For each i with $1 \leq i \leq m$, $A_i$ and $B_i$ are well separated with respect to s*
2. *For any two distinct points p and q of S, there is exactly one index i with $1 \leq i \leq m$, such that*
   a. *$p \in A_i$ and $q \in B_i$*
   b. *$p \in B_i$ and $q \in A_i$*

*The integer m is called the* size *of the WSPD.*

Observe that a WSPD always exists: If we let any two distinct points $p$ and $q$ of $S$ form a pair $\{\{p\}, \{q\}\}$, then the conditions in Definition 4.2 are satisfied. The size of this WSPD, however, is $\binom{n}{2}$. In the next section, we will give an algorithm that computes a WSPD whose size is only $O(n)$.

We remark that, for any set of $n$ points in $\mathbb{R}^d$, the size $m$ of any WSPD satisfies $m \geq n - 1$. An elegant proof of this fact, using linear algebra, was given by Graham and Pollak [15]; see also Volume 1, Chapter 9 in Aigner and Ziegler [16]. Moreover, for any set of $n$ points in $\mathbb{R}^d$ and for any WSPD, the total size $\sum_i (|A_i| + |B_i|)$ of all sets in the decomposition is $\Omega(n \log n)$; see Bollobás and Scott [17]. Callahan and Kosaraju have shown that, for some point sets, this summation is, in fact, $\Omega(n^2)$ for any WSPD.

## 4.3 Computing a Well-Separated Pair Decomposition

Let $S$ be a set of $n$ points in $\mathbb{R}^d$, and let $s > 0$ denote the separation ratio. The algorithm that constructs a WSPD for $S$ consists of two phases. In the first phase, a so-called split tree is constructed, which can be considered to be a hierarchical decomposition of the bounding box of $S$ into axes-parallel hyperrectangles. In the second phase, the split tree is used to actually compute the WSPD. The algorithm is due to Callahan and Kosaraju [7].

### 4.3.1 The Split Tree

The *split tree* $T(S)$ for the point set $S$ is a binary tree that is defined as follows:

1. If $n = 1$, then $T(S)$ consists of a single node storing the only element of $S$.
2. Assume that $n \geq 2$ and consider the bounding box $R(S) = \prod_{i=1}^{d} [\ell_i, r_i]$ of $S$. Let $i$ be the dimension such that $r_i - \ell_i$ is maximum and define $S_1 := \{p \in S : p_i \leq (\ell_i + r_i)/2\}$ and $S_2 := S \setminus S_1$. The split tree $T(S)$ for $S$ consists of a root whose two children are recursively defined split trees $T(S_1)$ and $T(S_2)$ for the sets $S_1$ and $S_2$, respectively. The root of $T(S)$ stores the bounding box $R(S)$.

Thus, the split tree $T(S)$ stores the points of $S$ at its leaves. Each internal node of $T(S)$ stores an axes-parallel hyperrectangle, which is the bounding box of the set of all points of $S$ that are stored in its subtree. Observe that the split tree is, in general, not balanced.

The aforementioned definition immediately leads to an $O(n^2)$-time algorithm for constructing the split tree. Callahan and Kosaraju show that, by using a divide-and-conquer approach, the split tree can in fact be constructed in $O(n \log n)$ time:

**Lemma 4.2** *The split tree for any set of n points in $\mathbb{R}^d$ can be constructed in $O(n \log n)$ time.*

### 4.3.2 Using the Split Tree to Compute Well-Separated Pairs

Consider the split tree $T = T(S)$ for the point set $S$. For any node $u$ of $T$, we denote by $S_u$ the subset of $S$ that is stored at the leaves of the subtree rooted at $u$. For any subset $X$ of $\mathbb{R}^d$, we denote by $L_{\max}(R(X))$ the length of a longest side of the bounding box $R(X)$ of $X$.

The following algorithm uses the split tree to compute a WSPD for $S$. Recall that $s$ denotes the separation ratio.

    *Step 1*: Initialize an empty queue $Q$. For each internal node $u$ of $T$, do the following: Let $v$ and $w$ be the two children of $u$. Insert the pair $(v, w)$ into $Q$.

    *Step 2*: Repeat the following until the queue $Q$ is empty: Take the first pair $(v, w)$ in $Q$ and delete it from $Q$. If the sets $S_v$ and $S_w$ are well separated with respect to $s$, then output the pair $\{S_v, S_w\}$. Otherwise, assume without loss of generality that $L_{\max}(R(S_v)) \leq L_{\max}(R(S_w))$. Let $w_1$ and $w_2$ be the two children of $w$. Insert the pairs $(v, w_1)$ and $(v, w_2)$ into the queue $Q$.

It is not difficult to see that the output of this algorithm is a WSPD of the point set $S$. Callahan and Kosaraju use a nontrivial packing argument to show that the number of pairs is $O(n)$.

**Lemma 4.3** *Given the split tree, the above-mentioned algorithm constructs, in $O(s^d n)$ time, a WSPD for S that consists of $O(s^d n)$ pairs.*

Observe that the WSPD is represented implicitly by the split tree: Each pair $\{A, B\}$ in the WSPD is represented by two nodes $v$ and $w$, such that $A = S_v$ and $B = S_w$. Thus, the entire WSPD can be represented using $O(s^d n)$ space.

By combining Lemmas 4.2 and 4.3, we obtain the following result:

**Theorem 4.1** [7] *Given a set S of n points in $\mathbb{R}^d$, and given a real number $s > 0$, a WSPD for S, with separation ratio s, consisting of $O(s^d n)$ pairs, can be computed in $O(n \log n + s^d n)$ time.*

In some applications, it is useful to have a WSPD in which each well-separated pair consists of two sets, at least one of which is a singleton set. For the WSPD $\{A_i, B_i\}$, $1 \leq i \leq m$, that is constructed by the algorithm given previously, Callahan [6] proved that

$$\sum_{i=1}^{m} \min(|A_i|, |B_i|) = O\left(s^d n \log n\right).$$

Thus, if we replace each pair $\{A_i, B_i\}$ (where we assume without loss of generality that $|A_i| \leq |B_i|$) by $|A_i|$ pairs $\{\{a\}, B_i\}$, $a \in A_i$, then we obtain the following result:

**Theorem 4.2** ([6]) *Let S be a set of n points in $\mathbb{R}^d$, and let $s > 0$ be a real number. In $O(s^d n \log n)$ time, a WSPD for S, with separation ratio s, can be constructed, such that each well-separated pair consists of two sets, at least one of which is a singleton set, and the total number of pairs is $O(s^d n \log n)$.*

## 4.4 Exact Algorithms for Proximity Problems

As we have mentioned earlier, the WSPD is an $O(n)$-size approximation of the set of $\Theta(n^2)$ distances determined by a set of $n$ points. In this section, we show that, despite the fact that the WSPD *approximates* all distances, it can be used to solve several proximity problems *exactly*. All results in this section are due to Callahan and Kosaraju [6,7].

Let $S$ be a set of $n$ points in $\mathbb{R}^d$, and let $s > 0$ be a real number. Consider the split tree $T$ and the corresponding WSPD for $S$, with separation ratio $s$, consisting of the pairs $\{A_i, B_i\}$, $1 \le i \le m$.

## 4.4.1 The Closest Pair Problem

In this problem, we want to compute a *closest pair* in $S$, that is, two distinct points $p$ and $q$ in $S$ such that $|pq|$ is minimum. Many algorithms are known that solve this problem optimally in $O(n \log n)$ time; see Smid [5]. We show that the WSPD "contains" a solution to the closest pair problem.

Let $(p, q)$ be a closest pair in $S$, and let $i$ be the index such that $p \in A_i$ and $q \in B_i$. If we assume that the separation ratio $s$ is a constant larger than two, then it follows from Lemma 4.1 that both $A_i$ and $B_i$ are singleton sets. Thus, by considering all pairs $\{A_j, B_j\}$, $1 \le j \le m$, such that both $A_j$ and $B_j$ are represented by leaves of the split tree, we obtain the closest pair in $O(m)$ time. Combining this with Theorem 4.1, we obtain the following result:

**Theorem 4.3** *Given a set $S$ of $n$ points in $\mathbb{R}^d$, a closest pair in $S$ can be computed in $O(n \log n)$ time.*

## 4.4.2 The $k$ Closest Pairs Problem

We next consider the problem of computing the $k$ *closest pairs* in $S$, for any given integer $k$ with $1 \le k \le \binom{n}{2}$. That is, we want to compute the $k$ smallest elements in the (multi)set of $\binom{n}{2}$ distances determined by pairs of points in $S$. Several algorithms have been designed that solve this problem optimally in $O(n \log n + k)$ time; see Smid [5]. As for the closest pair problem, we show that, once the WSPD is given, the $k$ closest pairs can be obtained in a simple way.

For each $i$ with $1 \le i \le m$, we denote by $|R(A_i)R(B_i)|$ the minimum distance between the bounding boxes $R(A_i)$ and $R(B_i)$ of $A_i$ and $B_i$, respectively. We assume, for ease of notation, that the pairs in the WSPD are numbered such that

$$|R(A_1)R(B_1)| \le |R(A_2)R(B_2)| \le \cdots \le |R(A_m)R(B_m)|.$$

(This ordering of the pairs is only used in the analysis, it is *not* computed by the algorithm.) The algorithm does the following:

*Step 1*: Compute the smallest integer $\ell \ge 1$, such that $\sum_{i=1}^{\ell} |A_i| \cdot |B_i| \ge k$.
*Step 2*: Compute the distance $r$ between the bounding boxes $R(A_\ell)$ and $R(B_\ell)$ of the sets $A_\ell$ and $B_\ell$, respectively.
*Step 3*: Compute the largest index $\ell'$ such that $|R(A_{\ell'})R(B_{\ell'})| \le (1 + 4/s)r$.
*Step 4*: Compute the set $L'$ consisting of all pairs $(p, q)$ for which there is an index $i$ with $1 \le i \le \ell'$, such that $p \in A_i$ and $q \in B_i$.
*Step 5*: Return the $k$ smallest distances determined by the pairs in the set $L'$.

**Lemma 4.4** *This algorithm computes the $k$ closest pairs in $S$.*

*Proof.* Let $(p, q)$ be one of the $k$ closest pairs, and let $j$ be the index such that $p \in A_j$ and $q \in B_j$. It suffices to prove that $(p, q)$ is an element of $L'$, that is, $j \le \ell'$. To prove this, assume that $j > \ell'$. Then

$$|pq| \ge |R(A_j)R(B_j)| > (1 + 4/s)r.$$

Let $L$ be the set consisting of all pairs $(x, y)$ for which there is an index $i$ with $1 \le i \le \ell$, such that $x \in A_i$ and $y \in B_i$. This set contains at least $k$ elements. By using Lemma 4.1, we have, for each pair $(x, y)$ in $L$,

$$|xy| \le (1 + 4/s)|R(A_i)R(B_i)| \le (1 + 4/s)|R(A_\ell)R(B_\ell)| = (1 + 4/s)r.$$

This contradicts our assumption that $(p, q)$ is one of the $k$ closest pairs. ∎

By using a linear-time (weighted) selection algorithm, the running time of the algorithm can be bounded by

$$O\left(m + \sum_{i=1}^{\ell'} |A_i| \cdot |B_i|\right) = O(m + |L'|).$$

Let $\delta$ be the $k$th smallest distance in $S$. Then it can be shown that $r \leq \delta$ and $|pq| \leq (1 + 4/s)^2\delta$, for any pair $(p, q)$ in $L'$. Hence, if we denote by $M$ the number of distances in the set $S$ that are at most $(1 + 4/s)^2\delta$, then the running time of the algorithm is $O(m + M)$. By using a counting technique, based on a grid with cells having sides of length $\delta/\sqrt{d}$, it can be shown that

$$M = O\left((1 + 4/s)^{2d}(n + k)\right).$$

We take the separation ratio $s$ to be equal to, say, one. Then, by combining our results with Theorem 4.1, we obtain the following theorem:

**Theorem 4.4** *Given a set $S$ of $n$ points in $\mathbb{R}^d$, and given an integer $k$ with $1 \leq k \leq \binom{n}{2}$, the $k$ closest pairs in $S$ can be computed in $O(n \log n + k)$ time.*

## 4.4.3 The All-Nearest Neighbors Problem

In this problem, we want to compute for each point $p$ of $S$ a *nearest neighbor* in $S$, that is, a point $q \in S \setminus \{p\}$ for which $|pq|$ is minimum. Vaidya [13] was the first to solve this problem optimally in $O(n \log n)$ time. In fact, his algorithm uses ideas that are very similar to the WSPD. In this section, we sketch the algorithm of Callahan and Kosaraju [7].

Let $p$ be a point of $S$ and let $q$ be its nearest neighbor. Let $i$ be the index such that $p \in A_i$ and $q \in B_i$. It follows from Lemma 4.1 that the set $A_i$ consists only of the point $p$. Hence, to solve the all-nearest neighbors problem, we only have to consider pairs of the WSPD, for which at least one of their sets is a singleton set. This observation does not lead to an efficient algorithm yet.

For any node $u$ of the split tree $T$, we define $F(u)$ to be the set of all points $p \in S$ such that $\{\{p\}, S_v\}$ is a pair in the WSPD, for some ancestor $v$ of $u$. (We consider $u$ to be an ancestor of itself.) Moreover, we define $N(u)$ to be the set of all points $p \in F(u)$, such that the distance from $p$ to the smallest ball containing $R(S_u)$ is at most equal to the smallest distance between $p$ and any other point of $F(u)$. Observe that $N(u) \subseteq F(u)$.

**Lemma 4.5** *The size of the set $N(u)$ is $O((s/(s-1))^d)$.*

*Proof.* Let $C$ be the smallest ball that contains $R(S_u)$. We claim that

1. For each $p \in N(u)$, the sets $\{p\}$ and $S_u$ are well separated.
2. $|pC| \leq |pq|$, for any two distinct points $p$ and $q$ of $N(u)$.

By combining these two claims with a generalization of the fact that any point can be the nearest neighbor of at most a constant number of other points, it can be shown that the size of $N(u)$ is $O((s/(s-1))^d)$.

To prove the first claim, let $p \in N(u)$. As $p \in F(u)$, there is an ancestor $v$ of $u$ such that the sets $\{p\}$ and $S_v$ are well separated. As $S_u$ is a subset of $S_v$, the sets $\{p\}$ and $S_u$ are well separated as well.

To prove the second claim, let $p$ and $q$ be two distinct points of $N(u)$. The definition of $N(u)$ implies that the distance between $p$ and $C$ is at most the smallest distance between $p$ and any other point of $F(u)$. In particular, a $q \in F(u)$, we have $|pC| \leq |pq|$. ∎

We assume from now on that the separation ratio $s$ is a constant larger than two. The sets $N(u)$, where $u$ ranges over all nodes of the split tree $T$, can be computed in a top–down fashion, in $O(n)$ total time. How do we use these sets? Let $p$ be any point of $S$, let $q$ be a nearest neighbor of $p$, and let $u$ be the leaf of the split tree that stores $q$. Let $i$ be the index such that $A_i = \{p\}$ and $q \in B_i$, and let $v$ be the ancestor of $u$ such that $B_i = S_v$. Then, $p \in F(u)$. Moreover, as $S_u = \{q\}$, the distance between $p$ and the smallest ball containing $R(S_u)$ is equal to $|pq|$, which is at most the distance between $p$ and any other point of $F(u)$. Therefore, we have $p \in N(u)$.

The earlier discussion, together with Theorem 4.1, leads to an algorithm that solves the all-nearest neighbors problem in $O(n \log n)$ time.

**Theorem 4.5** *Given a set $S$ of $n$ points in $\mathbb{R}^d$, the all-nearest neighbors problem can be solved in $O(n \log n)$ time.*

## 4.5 Approximation Algorithms for Proximity Problems

In the current section, we consider proximity problems for which no optimal exact algorithms are known. For each of these problems, we show that the WSPD leads to simple and fast approximation algorithms.

Let $S$ be a set of $n$ points in $\mathbb{R}^d$, and let $s > 0$ be a real number. We assume that we have already computed the split tree $T$ and the corresponding WSPD for $S$, with separation ratio $s$, consisting of the pairs $\{A_i, B_i\}$, $1 \le i \le m$. For each $i$ with $1 \le i \le m$, we choose an arbitrary element $a_i$ in $A_i$, and an arbitrary element $b_i$ in $B_i$.

### 4.5.1 The Diameter Problem

The *diameter* of $S$ is defined to be the largest distance between any two points of $S$. If the dimension $d$ is equal to two, the diameter can be computed in $O(n \log n)$ time; see Preparata and Shamos [1]. Ramos [18] obtained the same time bound for the three-dimensional case. It is not known if for dimensions larger than three, the diameter can be computed in $O(n \log^{O(1)} n)$ time. In this section, we show that the WSPD leads to a simple and efficient algorithm that approximates the diameter, for any constant dimension $d$.

Let $D$ be the diameter of $S$, and let $i$ be the index for which $|a_i b_i|$ is maximum. A straightforward application of Lemma 4.1 shows that $D/(1 + 4/s) \le |a_i b_i| \le D$. Hence, if we choose $s = 4(1 - \epsilon)/\epsilon$, then this result, together with Theorem 4.1, yields the following theorem.

**Theorem 4.6** *Given a set $S$ of $n$ points in $\mathbb{R}^d$, and given a real constant $0 < \epsilon < 1$, a $(1 - \epsilon)$-approximation to the diameter of $S$ can be computed in $O(n \log n)$ time.*

We remark that the same approximation factor can be achieved in only $O(n)$ time: Choose $O(1/\epsilon^{d-1}) = O(1)$ vectors such that, for any pair $p, q$ of distinct points in $\mathbb{R}^d$, one of these vectors makes an angle of at most $\epsilon$ with the vector $\overrightarrow{pq}$; see Volume 1, Chapter 5 in Narasimhan and Smid [9]. Then, for each of these vectors $\vec{v}$, compute an extreme point $a$ of $S$ in the positive direction $\vec{v}$, compute an extreme point $b$ of $S$ in the negative direction $-\vec{v}$, and compute the distance $|ab|$. The largest distance $|ab|$ obtained is a $(1 - \epsilon)$-approximation to the diameter of $S$. The running time of this algorithm is $O(n)$. For details, see Janardan [19].

### 4.5.2 The Spanner Problem

For a real number $t > 1$, a graph $G = (S, E)$ is called a *t-spanner* for the point set $S$, if for any two points $p$ and $q$ of $S$, we have $|pq|_G \le t|pq|$, where $|pq|_G$ denotes the length of a shortest path in $G$ between $p$ and $q$. Many algorithms are known that compute spanners; see the chapter by Gudmundsson and Knauer [8] in this handbook, and Narasimhan and Smid [9]. In this section, we show that the WSPD immediately gives a spanner for $S$ consisting of $O(n)$ edges. The construction is due to Callahan and Kosaraju [20].

**Lemma 4.6** *Assume that the separation ratio s is larger than four. Define $G = (S, E)$ to be the graph with edge set $E = \{a_i b_i : 1 \le i \le m\}$. Then, G is a t-spanner for S, where $t = (s + 4)/(s - 4)$.*

*Proof.* The proof is by induction. Consider two points $p$ and $q$ in $S$. If $p = q$, then obviously $|pq|_G \le t|pq|$. Assume that $p \ne q$. Moreover, assume that $|xy|_G \le t|xy|$, for all points $x$ and $y$ in $S$ for which $|xy| < |pq|$. Let $i$ be the index such that $p \in A_i$ and $q \in B_i$. By using Lemma 4.1, we obtain

$$|pq|_G \le |pa_i|_G + |a_i b_i|_G + |b_i q|_G$$
$$= |pa_i|_G + |a_i b_i| + |b_i q|_G$$
$$\le t|pa_i| + |a_i b_i| + t|b_i q|$$
$$\le (2t/s + (1 + 4/s) + 2t/s)|pq|$$
$$= t|pq|. \qquad\blacksquare$$

Observe that, if the separation ratio $s$ goes to infinity, the value $t = (s + 4)/(s - 4)$ converges to 1. Thus, Theorem 4.1 implies the following result.

**Theorem 4.7** *Given a set S of n points in $\mathbb{R}^d$ and given a real constant $t > 1$, a t-spanner for S, consisting of $O(n)$ edges, can be computed in $O(n \log n)$ time.*

### 4.5.3 The Greedy Spanner

Let $S$ be a set of $n$ points in $\mathbb{R}^d$ and let $t > 1$ be a real constant. We have seen in Theorem 4.7 that a $t$-spanner with $O(n)$ edges can be computed in $O(n \log n)$ time. In one of the earliest papers on geometric spanners, Althöfer et al. [21] introduced the following simple greedy algorithm for computing a $t$-spanner:

---

**ALGORITHM   GreedySpanner($S, t$)**

sort the $\binom{n}{2}$ pairs of distinct points in nondecreasing order of their
distances (breaking ties arbitrarily), and store them in list $L$;
$E := \emptyset$;
$G := (S, E)$;
**for each** pair $pq$ in $L$   (∗ consider pairs in sorted order ∗)
**do if** $|pq|_G > t|pq|$
   **then** $E := E \cup \{pq\}$;
        $G := (S, E)$
   **endif**
**endfor**;
return the graph $G$

---

It is obvious that the output of this algorithm is a $t$-spanner for the input set $S$. The following theorem uses the WSPD of $S$ to prove that the number of edges in the greedy spanner is $O(n)$.

**Theorem 4.8** *Let S be a set of n points in $\mathbb{R}^d$, let $t > 1$ be a real constant, and assume that the separation ratio s in the WSPD for S satisfies $t = (s + 4)/(s - 4)$. Then the number of edges in the greedy spanner for S is $O(n)$.*

*Proof.* The proof will follow from the claim that, for each pair $\{A_i, B_i\}$ in the WSPD for $S$, the greedy spanner contains at most one edge $pq$ with $p \in A_i$ and $q \in B_i$.

   We prove this claim by contradiction. Assume that the greedy spanner contains two distinct edges $pq$ and $p'q'$, with $p, p' \in A_i$ and $q, q' \in B_i$. We may assume without loss of generality that $pq$ is before $p'q'$

in the list $L$. Consider the moment when the algorithm examines the pair $p'q'$. At this moment, the edge $pq$ is already in the edge set $E$. Moreover, by Lemma 4.1, $|pp'| \leq (2/s)|p'q'| < |p'q'|$ and, thus, at this moment, we have $|pp'|_G \leq t|pp'|$. By the same argument, we have $|qq'|_G \leq t|qq'|$. It follows that, at this moment,

$$
\begin{aligned}
|p'q'|_G &\leq |p'p|_G + |pq| + |qq'|_G \\
&\leq t|p'p| + |pq| + t|qq'| \\
&\leq (2t/s + (1 + 4/s) + 2t/s)|p'q'| \\
&= t|p'q'|.
\end{aligned}
$$

As a result, the algorithm does not add $p'q'$ as an edge to the greedy spanner. This is a contradiction. ∎

By using geometric arguments, it can be shown that any two edges $pq$ and $pq'$ of the greedy spanner make a "large" angle. This implies that the maximum degree of the greedy spanner is $O(1)$. Moreover, using a lengthy analysis, it can be shown that the total edge length of the greedy spanner is within a constant factor of the total length of a minimum spanning tree. Proofs of these claims can be found in Narasimhan and Smid [9]. See also Smid [22].

## 4.5.4 The Minimum Spanning Tree Problem

In the two-dimensional case, the minimum spanning tree of $S$ can be computed in $O(n \log n)$ time, by using the fact that it is contained in the Delaunay triangulation of $S$; see Preparata and Shamos [1,11] and de Berg et al. [2]. For the three-dimensional case, the best known algorithm has a running time that is close to $O(n^{4/3})$; see Agarwal et al. [23]. Erickson [24] argues that it is unlikely that this running time can be improved considerably. In this section, we show that, in any constant dimension $d$, an approximation to the minimum spanning tree can be computed in $O(n \log n)$ time. The algorithm is due to Callahan and Kosaraju [20].

Let $t > 1$ be a real constant and consider an arbitrary $t$-spanner for $S$ having $O(n)$ edges. Observe that $G$ is a connected graph. Let $T$ be a minimum spanning tree of $G$.

**Lemma 4.7** *$T$ is a $t$-approximate minimum spanning tree of $S$.*

*Proof.* Let $T^*$ be a minimum spanning tree of $S$ and denote its total edge length by $|T^*|$. Number the edges of $T^*$ as $e_1, e_2, \ldots, e_{n-1}$. For each $i$ with $1 \leq i \leq n-1$, let $P_i$ be a $t$-spanner path (in $G$) between the endpoints of $e_i$ and denote the length of this path by $|P_i|$. Then,

$$
\sum_{i=1}^{n-1} |P_i| \leq \sum_{i=1}^{n-1} t|e_i| = t|T^*|.
$$

Let $G'$ be the subgraph of $G$ consisting of the union of the edges of all paths $P_i$, $1 \leq i \leq n-1$. Then $G'$ is a connected graph on the points of $S$, and its weight is at most $t|T^*|$. As the weight of $T$ is at most that of $G'$, the weight of $T$ is at most $t$ times the weight of $T^*$. ∎

As the graph $G$ contains $O(n)$ edges, its minimum spanning tree $T$ can be computed in $O(n \log n)$ time. By combining this with Theorem 4.7, we obtain the following result.

**Theorem 4.9** *Given a set $S$ of $n$ points in $\mathbb{R}^d$ and given a real constant $\epsilon > 0$, a $(1 + \epsilon)$-approximation to the minimum spanning tree of $S$ can be computed in $O(n \log n)$ time.*

### 4.5.5 The $k$th Closest Pair Problem

In this problem, we are given an integer $k$ with $1 \leq k \leq \binom{n}{2}$ and want to compute the $k$th smallest element in the (multi)set of $\binom{n}{2}$ distances determined by pairs of points in $S$. In the two-dimensional case, the best known algorithm for this problem has a running time that is close to $O(n^{4/3})$; see Katz and Sharir [25]. Again, Erickson [24] argues that it is unlikely that a significantly faster algorithm exists. We show that, for any constant dimension, there is a simple and efficient algorithm that approximates the $k$th closest pair. The results in this section are due to Bespamyatnikh and Segal [26].

Let $k$ be any integer with $1 \leq k \leq \binom{n}{2}$. As in Section 4.4.2, we assume that the pairs in the WSPD are numbered such that

$$|R(A_1)R(B_1)| \leq |R(A_2)R(B_2)| \leq \cdots \leq |R(A_m)R(B_m)|.$$

Let $\ell \geq 1$ be the smallest integer such that $\sum_{i=1}^{\ell} |A_i| \cdot |B_i| \geq k$, let $x$ be an arbitrary element of $A_\ell$, and let $y$ be an arbitrary element of $B_\ell$.

**Lemma 4.8** *If $\delta$ is the $k$th smallest distance in the set $S$, then $\delta/(1 + 4/s) \leq |xy| \leq (1 + 4/s)\delta$.*

*Proof.* As mentioned in Section 4.4.2, it can be shown that $|R(A_\ell)R(B_\ell)| \leq \delta$. If we combine this fact with Lemma 4.1, then we obtain

$$|xy| \leq (1 + 4/s)|R(A_\ell)R(B_\ell)| \leq (1 + 4/s)\delta.$$

Let $L$ be the set consisting of all pairs $(a, b)$ for which there is an index $i$ with $1 \leq i \leq \ell$, such that $a \in A_i$ and $b \in B_i$. Let $(a, b)$ be the pair in $L$ for which $|ab|$ is maximum, and let $i$ be the index such that $a \in A_i$ and $b \in B_i$. Observe that $i \leq \ell$. As $L$ has size at least $k$, we have $\delta \leq |ab|$. Therefore (again using Lemma 4.1),

$$\delta \leq |ab| \leq (1 + 4/s)|R(A_i)R(B_i)| \leq (1 + 4/s)|R(A_\ell)R(B_\ell)| \leq (1 + 4/s)|xy|. \qquad \blacksquare$$

Thus, using Theorem 4.1, we obtain the following result.

**Theorem 4.10** *Let $S$ be a set of $n$ points in $\mathbb{R}^d$, let $k$ be an integer with $1 \leq k \leq \binom{n}{2}$, let $\epsilon > 0$ be a constant, and let $\delta$ be the $k$th smallest distance in $S$. In $O(n \log n)$ time, a pair $(x, y)$ of points in $S$ can be computed for which $(1 - \epsilon)\delta \leq |xy| \leq (1 + \epsilon)\delta$.*

Significantly, computing a pair $(x, y)$ of points in $S$ such that $\delta \leq |xy| \leq (1 + \epsilon)\delta$ is more difficult. To compute such a pair, we use the WSPD of Theorem 4.2. Thus, the WSPD consists of pairs $\{A_i, B_i\}$, $1 \leq i \leq m$, where each set $A_i$ is a singleton set, say $A_i = \{a_i\}$, and $m = O(n \log n)$.

For each $i$ with $1 \leq i \leq m$, we define $d_i = \min\{|a_ib| : b \in B_i\}$ and $D_i = \max\{|a_ib| : b \in B_i\}$. Assume that the pairs in the WSPD are numbered such that $d_1 \leq d_2 \leq \cdots \leq d_m$. Let $\ell \geq 1$ be the smallest integer such that $\sum_{i=1}^{\ell} |B_i| \geq k$, and let $D = \max(D_1, D_2, \cdots, D_\ell)$.

**Lemma 4.9** *If $\delta$ is the $k$th smallest distance in the set $S$, then $\delta \leq D \leq (1 + 4/s)\delta$*

*Proof.* As $\sum_{i=1}^{\ell} |B_i| \geq k$, the pairs $\{A_i, B_i\}$ with $1 \leq i \leq \ell$, define at least $k$ distances. As $D$ is the largest among these distances, it follows that $\delta \leq D$.

Let $L = \{(a_i, b) : b \in B_i, i \geq \ell\}$. Then $d_\ell$ is the minimum distance of any element in $L$. As the size of $L$ is larger than $\binom{n}{2} - k$, it follows that $\delta \geq d_\ell$. Let $i$ be the index such that $D = D_i$. By Lemma 4.1, we have $D_i \leq (1 + 4/s)d_i$. Therefore, we have

$$D = D_i \leq (1 + 4/s)d_i \leq (1 + 4/s)d_\ell \leq (1 + 4/s)\delta. \qquad \blacksquare$$

We obtain the value of $D$ (which is an approximation to the $k$th smallest distance in $S$), by computing all values $d_i$ and $D_i$, $1 \leq i \leq m$. That is, for each point $a_i$, we compute its nearest and furthest neighbors in the set $B_i$. We will show how to use the split tree to solve this problem for the case when the points are in $\mathbb{R}^2$. The algorithm uses the following result.

**Lemma 4.10** *Let V be a set of N points in the plane. There exists a data structure that supports the following operations:*

1. *For any given query point $q \in \mathbb{R}^2$, report the nearest neighbor of $q$ in $V$. The query time is $O(\log^2 N)$.*
2. *For any given query point $q \in \mathbb{R}^2$, report the furthest neighbor of $q$ in $V$. The query time is $O(\log^2 N)$.*
3. *Insert an arbitrary point into the set $V$. The amortized insertion time is $O(\log^2 N)$.*

*Proof.* Consider the Voronoi diagram of the set $V$, which can be constructed in $O(N \log N)$ time. If we store this diagram, together with a point location data structure, then a nearest neighbor query can be answered in $O(\log N)$ time; see Preparata and Shamos [1,11] and de Berg et al. [2]. If we use the furthest point Voronoi diagram, then we obtain the same result for furthest neighbor queries. On the contrary, these Voronoi diagrams cannot be maintained efficiently under insertions of points. As nearest neighbor and furthest neighbor queries are *decomposable*, however, we can use the *logarithmic method* of Bentley and Saxe [27] to obtain the time bounds that are claimed in the lemma. ∎

We now show how Lemma 4.10 can be used to compute all values $d_i$ and $D_i$, $1 \leq i \leq m$. Consider the split tree $T$. Recall that for any node $u$, $S_u$ denotes the subset of $S$ that is stored at the leaves in the subtree of $u$. We store with each node $u$, a list $L_u$ consisting of all points $a_i$ such that $B_i = S_u$.

The algorithm traverses the split tree $T$ in postorder. During this traversal, the following invariant is maintained: If $u$ is a node that has been traversed, but none of its proper ancestors has been traversed yet, then $u$ stores the data structure $DS_u$ of Lemma 4.10 for the point set $S_u$.

The postorder traversal of $T$ does the following. Let $u$ be the current node in this traversal. If $u$ is a leaf storing the point, say, $p$, then we compute $d_i = D_i = |a_i p|$ for each point $a_i$ in $L_u$, and we build the data structure $DS_u$ of Lemma 4.10 for the singleton set $S_u = \{p\}$. Assume that the current node $u$ is not a leaf. Let $v$ and $w$ be the two children of $u$. By the invariant, $v$ and $w$ store the data structure $DS_v$ and $DS_w$ of Lemma 4.10 for the sets $S_v$ and $S_w$, respectively. Assume without loss of generality that $|S_v| \leq |S_w|$. We do the following: First, we discard the data structure $DS_v$. Then, we insert each element of $S_v$ into $DS_w$; this results in the data structure $DS_u$ storing all elements of $S_u$. Finally, for each element $a_i$ of $L_u$, we use $DS_u$ to find the nearest and furthest neighbors of $a_i$ in the set $S_u$ and compute the values of $d_i$ and $D_i$.

We analyze the total time of this algorithm. As the WSPD contains $O(n \log n)$ pairs, the total number of nearest neighbor and furthest neighbor queries is $O(n \log n)$. Consider any fixed point $p$ of $S$. If $p$ is inserted into a data structure, then it "moves" to a new set whose size is at least twice the size of the previous set containing $p$. As a result, $p$ is inserted $O(\log n)$ times. Thus, overall, the total number of insertions is $O(n \log n)$. Lemma 4.10 then implies that the running time of the algorithm is $O(n \log^3 n)$. If we combine this with Theorem 4.2, we obtain the following result.

**Theorem 4.11** *Let S be a set of n points in $\mathbb{R}^d$, let k be an integer with $1 \leq k \leq \binom{n}{2}$, let $\epsilon > 0$ be a constant, and let $\delta$ be the kth smallest distance in S. In $O(n \log^3 n)$ time, a pair $(x, y)$ of points in S can be computed for which $\delta \leq |xy| \leq (1 + \epsilon)\delta$.*

# 4.6 Generalization to Metric Spaces

All results in the previous sections are valid for Euclidean spaces $\mathbb{R}^d$, where $d$ is a constant. In recent years, the WSPD (and its applications) has been generalized to more general metric spaces.

Consider an arbitrary metric space $(S, \delta)$, where $S$ is a set of $n$ elements, and $\delta : S \times S \longrightarrow \mathbb{R}$ is the metric defined on $S$. For any two subsets $A$ and $B$ of $S$, we denote the minimum distance between any point in $A$ and any point in $B$ by $\delta(A, B)$, and we denote the diameter of $A$ by $D(A)$. If $s > 0$ is a real number, then we say that $A$ and $B$ are well separated with respect to $s$, if

$$\delta(A, B) \geq s \cdot \max(D(A), D(B)).$$

By using this generalized notion of being well separated, we define a WSPD for $S$ as in Definition 4.2.

Gao and Zhang [28] considered the problem of constructing a WSPD for the unit-disk graph metric: Let $S$ be a set of $n$ points in $\mathbb{R}^d$. The *unit-disk graph* is defined to be the graph with vertex set $S$, in which any two distinct points $p$ and $q$ are connected by an edge if and only $|pq| \leq 1$. If we define $\delta(p, q)$ to be the length of a shortest path between $p$ and $q$ in the unit-disk graph, then $(S, \delta)$ is a metric space. Observe that even though the unit-disk graph may have $\Theta(n^2)$ edges, it can be represented in $O(n)$ space: It suffices to store the points of $S$. Given any two points $p$ and $q$ of $S$, we can decide in $O(1)$ time if $p$ and $q$ are connected by an edge and, if so, compute its length $|pq|$. (If $p$ and $q$ are not connected by an edge, however, then a shortest path computation is needed to compute $\delta(p, q)$.) Gao and Zhang proved the following result:

**Theorem 4.12** *Let $S$ be a set of $n$ points in $\mathbb{R}^d$, and let $s > 1$ be a real number. Consider the unit-disk graph metric on $S$.*

1. *If $d = 2$, then a WSPD for $S$ with respect to $s$, consisting of $O(s^4 n \log n)$ pairs, can be computed in $O(s^4 n \log n)$ time.*
2. *If $d = 3$, then a WSPD for $S$, with respect to $s$, consisting of $O(n^{4/3})$ pairs, can be computed in $O(n^{4/3} \log^{O(1)} n)$ time.*
3. *If $d \geq 4$, then a WSPD for $S$, with respect to $s$, consisting of $O(n^{2-2/d})$ pairs, can be computed in $O(n^{2-2/d})$ time.*

Talwar [29] extended the WSPD to metric spaces whose *doubling dimension* is a constant. To define this notion, let $(S, \delta)$ be a metric space. A ball, with center $p \in S$ and radius $R$, is defined to be the set $\{q \in S : \delta(p, q) \leq R\}$. The *doubling parameter* of $S$ is defined to be the smallest integer $\lambda$ such that the following holds, for all real numbers $R > 0$: Every ball with radius $R$ can be covered by $\lambda$ balls of radius $R/2$. The doubling dimension of the metric space is defined to be $\log \lambda$. Observe that this generalizes Euclidean space $\mathbb{R}^d$, because the doubling dimension of $\mathbb{R}^d$ is proportional to $d$.

Many algorithms solving proximity problems in $\mathbb{R}^d$ are analyzed using a packing argument. If the doubling parameter $\lambda$ is small, then, in many cases, a similar analysis can be used to efficiently solve these problems.

Talwar showed how to compute a WSPD consisting of $O(s^{\log \lambda} n \log \Delta)$ pairs, where $\Delta$ is the aspect ratio, which is defined to be the ratio of the diameter and the closest pair distance. Har-Peled and Mendel [30] gave an improved construction and obtained the following result:

**Theorem 4.13** *Let $(S, \delta)$ be a metric space, let $n = |S|$, let $\lambda$ be the doubling parameter of $S$, and let $s > 1$ be a real number. There exists a randomized algorithm that constructs, in $O(\lambda n \log n + s^{\log \lambda} n)$ expected time, a WSPD for $S$, with separation ratio $s$, consisting of $O(s^{\log \lambda} n)$ pairs.*

Most results of the previous sections remain valid for metric spaces whose doubling parameter is bounded by a constant. Significantly, Har-Peled and Mendel have shown that this is not the case for the all-nearest neighbors problem: For every deterministic algorithm that solves this problem, there exists a metric space on $n$ points and doubling parameter $\lambda \leq 3$, such that this algorithm must examine all $\binom{n}{2}$ distances determined by these points. This implies that, in such metric spaces, it takes $\Omega(n^2)$ time to compute a minimum spanning tree. As the greedy spanner contains a minimum spanning tree, this also implies an $\Omega(n^2)$–time lower bound for computing this spanner. For more information, refer to Smid [22].

# References

1. F. P. Preparata and M. I. Shamos. *Computational Geometry: An Introduction.* Springer-Verlag, Berlin, Germany, 1988.
2. M. de Berg, O. Cheong, M. van Kreveld, and M. Overmars. *Computational Geometry: Algorithms and Applications.* Springer-Verlag, Berlin, Germany, 3rd edition, 2008.

3. M. Bern and D. Eppstein. Approximation algorithms for geometric problems. In D. S. Hochbaum, (Ed.), *Approximation Algorithms for NP-Hard Problems*, pp. 296–345. PWS Publishing Company, Boston, MA, 1997.

4. D. Eppstein. Spanning trees and spanners. In J.-R. Sack and J. Urrutia, (Eds.), *Handbook of Computational Geometry*, pp. 425–461. Elsevier Science, Amsterdam, the Netherlands, 2000.

5. M. Smid. Closest-point problems in computational geometry. In J.-R. Sack and J. Urrutia, (Eds.), *Handbook of Computational Geometry*, pp. 877–935. Elsevier Science, Amsterdam, the Netherlands, 2000.

6. P. B. Callahan. Dealing with higher dimensions: The well-separated pair decomposition and its applications. PhD thesis, Department of Computer Science, Johns Hopkins University, Baltimore, MD, 1995.

7. P. B. Callahan and S. R. Kosaraju. A decomposition of multidimensional point sets with applications to $k$-nearest-neighbors and $n$-body potential fields. *Journal of the ACM*, 42:67–90, 1995.

8. J. Gudmundsson and C. Knauer. Dilation and detours in geometric networks. In T. F. Gonzalez, (Ed.), *Handbook of Approximation Algorithms and Metaheuristics*. Taylor & Francis, London, UK, 2nd ed., 2017.

9. G. Narasimhan and M. Smid. *Geometric Spanner Networks*. Cambridge University Press, Cambridge, UK, 2007.

10. J. S. Salowe. Constructing multidimensional spanner graphs. *International Journal of Computational Geometry & Applications*, 1:99–107, 1991.

11. J. S. Salowe. Enumerating interdistances in space. *International Journal of Computational Geometry & Applications*, 2:49–59, 1992.

12. P. M. Vaidya. Minimum spanning trees in $k$-dimensional space. *SIAM Journal on Computing*, 17:572–582, 1988.

13. P. M. Vaidya. An $O(n \log n)$ algorithm for the all-nearest-neighbors problem. *Discrete & Computational Geometry*, 4:101–115, 1989.

14. P. M. Vaidya. A sparse graph almost as good as the complete graph on points in $K$ dimensions. *Discrete & Computational Geometry*, 6:369–381, 1991.

15. R. L. Graham and H. O. Pollak. On the addressing problem for loop switching. *Bell System Technical Journal*, 50:2495–2519, 1971.

16. M. Aigner and G. M. Ziegler. *Proofs from THE BOOK*. Springer-Verlag, Berlin, Germany, 3rd ed., 2004.

17. B. Bollobás and A. Scott. On separating systems. *European Journal of Combinatorics*, 28:1068–1071, 2007.

18. E. A. Ramos. An optimal deterministic algorithm for computing the diameter of a three-dimensional point set. *Discrete & Computational Geometry*, 26:233–244, 2001.

19. R. Janardan. On maintaining the width and diameter of a planar point-set online. *International Journal of Computational Geometry & Applications*, 3:331–344, 1993.

20. P. B. Callahan and S. R. Kosaraju. Faster algorithms for some geometric graph problems in higher dimensions. In *Proceedings of the 4th ACM-SIAM Symposium on Discrete Algorithms*, pp. 291–300. Society for Industrial and Applied Mathematics, Philadelphia, PA, 1993.

21. I. Althöfer, G. Das, D. P. Dobkin, D. Joseph, and J. Soares. On sparse spanners of weighted graphs. *Discrete & Computational Geometry*, 9:81–100, 1993.

22. M. Smid. The weak gap property in metric spaces of bounded doubling dimension. In *Efficient Algorithms, Essays Dedicated to Kurt Mehlhorn on the Occasion of His 60th Birthday*, volume 5760 of *Lecture Notes in Computer Science*, pp. 275–289, Springer-Verlag, Berlin, Germany, 2009.

23. P. K. Agarwal, H. Edelsbrunner, O. Schwarzkopf, and E. Welzl. Euclidean minimum spanning trees and bichromatic closest pairs. *Discrete & Computational Geometry*, 6:407–422, 1991.

24. J. Erickson. On the relative complexities of some geometric problems. In *Proceedings of the 7th Canadian Conference on Computational Geometry*, pp. 85–90. Université Laval, Quebec City, Canada, 1995.

25. M. J. Katz and M. Sharir. An expander-based approach to geometric optimization. *SIAM Journal on Computing*, 26:1384–1408, 1997.

26. S. Bespamyatnikh and M. Segal. Fast algorithms for approximating distances. *Algorithmica*, 33:263–269, 2002.

27. J. L. Bentley and J. B. Saxe. Decomposable searching problems I: Static-to-dynamic transformations. *Journal of Algorithms*, 1:301–358, 1980.

28. J. Gao and L. Zhang. Well-separated pair decomposition for the unit-disk graph metric and its applications. *SIAM Journal on Computing*, 35:151–169, 2005.

29. K. Talwar. Bypassing the embedding: Approximation schemes and compact representations of low dimensional metrics. In *Proceedings of the 36th ACM Symposium on the Theory of Computing*, pp. 281–290. Association for Computing Machinery, New York, 2004.

30. S. Har-Peled and M. Mendel. Fast construction of nets in low-dimensional metrics and their applications. *SIAM Journal on Computing*, 35:1148–1184, 2006.

# 5

# Covering with Unit Balls

Hossein
Ghasemalizadeh

Mohammadreza
Razzazi

## 5.1 Introduction

Geometric covering problems deals with covering some points in geometric space with geometric shapes. These problems have many applications such as finding the locations of emergency facilities to be fair for all the clients, placing the wireless sensors or antennas to cover some targets, and image processing [1]. Examples of geometric shapes that have been studied more are disks and squares that are unit balls in $(R^2, l_2)$ and $(R^2, l_\infty)$ metric spaces, respectively. The covering with unit balls problem is defined as follows: Given a set of points in $(R^d, l_p)$ metric space, find the minimum number of unit balls of that metric to cover all the points. The problem is NP-hard as it generalizes the covering by disks problem [2]. Different constraints have been added to the problem and new problems have been introduced. In the current chapter, we survey the approximation algorithms for the covering with unit balls problem and some of its variants.

### 5.1.1 Problems Definition

In the current section, we define the problems formally and in subsequent sections we discuss them.

*Covering with unit balls*: Given a set of points in $(R^d, l_p)$ metric space, find the minimum number of required unit balls of that metric to cover all the points.

*Partial covering*: Given a set of $n$ points, $P$, in $(R^d, l_p)$ metric space. For a given $0 < p \leq n$, find the minimum number of unit balls in the given metric space to cover at least $p$ points.

*Maximum covering*: Given a set $n$ of points, $P$, in $(R^d, l_p)$ metric space. For a given integer $m > 0$, find a set of $m$ unit balls in the given metric space to cover the maximum number of points from $P$.

*Cardinality constrained covering*: Given a set of $n$ points, $P$, in $(R^d, l_p)$ metric space. For a given integer $0 < L \leq n$, find the minimum number of unit balls of that metric to cover all the points provided that each unit balls contains at most $L$ points.

## 5.2 Covering with Unit Balls Problem

### 5.2.1 Background

Covering with unit balls problem is known as disk covering problem (in $(R^2, l_2)$ metric space) when the points are in the plane and the shapes are unit disks ((in $(R^2, l_\infty)$ metric space). The disk covering problem is NP-hard [2] and can be optimally solved using the idea of geometric separators in $O(2^{\sqrt{n}})$ [3]. In an arbitrary metric space, the problem is equal to the set cover problem which cannot be approximated under factor $O(lnn)$, unless $P = NP$ [4]. However, in $l_p$ metrics the problem admits Polynomial Time Approximation Schema (PTAS) that was first given by Hochbaum and Maass [1]. They used a shifting strategy by dividing the plane into height $h$ strips and solving the problem in each strip, separately. To solve the problem in each strip, they recursively used the same method. In the base case, in which the input points are restricted to a $h \times h$ square, they used a brute-force method to generate all possible ways of covering points with disks. Then these strips are shifted slightly and the process is repeated. The solution with minimum number of disks over all shifts is reported as the final solution. This results in a $(1 + 1/h)^2$-approximation algorithm which runs in $O(n^{4h^2+1})$ time. In Section 5.2.2, we describe this algorithm in more detail. Gonzalez [5] later proposed a PTAS with improved running time, by computing the optimal cover in each strip using dynamic programming, and then applying the shifting strategy to solve the general problem. Ghasemi et al. [6] proposed an algorithmic framework that gives a PTAS for a class of covering problem, provided that its prerequisites are satisfied. The class of problems involves covering point sets with bounded shapes in a space with fixed dimension. For a problem to be in this class, it must be possible to bound the maximum number of shapes to a polynomial number(two shapes are different if they cover a different subset of points) and it must be possible to cover all the points in a small box with a small number of shapes.

Some researchers have proposed approximation algorithms with better running times but higher approximation factors. Gonzalez [5] gave a $2^{d-1}$-approximation algorithm for covering with squares in $d$ dimensions which takes $O(dn + n \log OPT)$ time, where $OPT$ is the size of an optimal solution. In Reference 7, a $3 + \varepsilon$-approximation algorithm in $O(n)$ time was given for the covering with disks problem in the plane by restricting the centers of disks to be on a $1 \times 1$ grid. More recently, Fu et al. improved this bound to 2.83 with $O(n(\log n \log \log n)^2)$ time complexity [8]. The work of Bronniman and Goodrich also resulted in an $O(1)$-approximation $O(n^3 \log n)$-time algorithm [9].

### 5.2.2 PTAS for Covering with Unit Balls Problem

In this section, we explain the algorithm of Reference 1 to obtain a PTAS for covering with unit balls problem. For simplicity we describe the algorithm for covering the points in the plane with disks. In the disk covering problem, we want to cover a point set, $P$, with the minimum number of disks with radius $r$. Suppose there is an $\alpha$-approximation algorithm that solves this problem when the input points are located in a horizontal strip of height $2r \times l$, where $l$ is an input parameter which is a positive integer. We call this problem *strip problem* and the algorithm that solves it as *strip algorithm*. To solve the disk covering problem in the general case, we divide the plane into horizontal strips of height $2r$ and group every $l$ of them as a *master strip*. We call this a partitioned problem $P_1$ which consists of $m$ master strips $S_{11} \cdots S_{1m}$. We solve $P_1$ by solving each strip problem $S_{1i}$ $(i = 1, \ldots, m)$ using the strip algorithm and consider the union of the results as a solution for $P_1$. We, then, create another partitioned problem $P_2$, by shifting each master strip one strip down and solve it like $P_1$. Figure 5.1 shows two sample partitioned problems in a sample point set for $l = 3$, along with their master strips. We repeat this process $l$ times gathering solutions

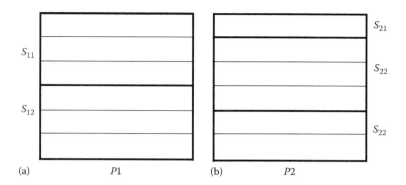

**FIGURE 5.1** Two partitioned problems in a sample point set for $l = 3$. In (a), the partitioned problem $P1$ has two master strips and in (b), the partitioned problem $P2$ has three master strips.

of partitioned problems $P_1$ to $P_l$ and report the solution with the minimum number of disks among $P_i$'s, $1 \leq i \leq l$, as the final solution of the disk covering problem. The shifting lemma [1] by using an averaging argument, guarantees that this process will produce an $\alpha(1 + \frac{1}{l})$ approximation of the optimal solution.

**Lemma 5.1** *Let $A_i$ be the solution of $P_i$, $1 \leq i \leq l$ and $A_{min}$ be the solution with the minimum number of disks among $A_i$'s. Let OPT be the solution that uses the minimum number of disks to cover P and let $|A_i|$ and $|OPT|$ denote the number of disks in $A_i$ and OPT, respectively. We have $|A_{min}| < \alpha \times (1 + \frac{1}{l})|OPT|$.*

*Proof.* By the definition of $|A_{min}|$ we have:

$$l \times |A_{min}| < \sum_{1 \leq i \leq l} |A_i| \tag{5.1}$$

Consider a master strip $S_{ik}$ of a partitioned problem $P_i$. Let $|A_{ik}|$ and $|OPT_{ik}|$ be the number of disk from $A_i$ and OPT in master strip $S_{ik}$. As $|A_i|$ is an $\alpha$-approximation algorithm for $P_i$ we have:

$$|A_{ik}| \leq \alpha \times |OPT_{ik}| \tag{5.2}$$

Let $D$ be a disk in OPT. $D$ will be located in at most two master strips of one of the partitioned problems and in other partitioned problems, it will be located inside one master strip. Consider $m$ as the number of master strips in partitioned problem $P_i$. Summing up over all master strips of all partitioned problems we have

$$\sum_{\substack{i=1,\ldots,l \\ k=0,\ldots,m}} |A_{ik}| \leq \alpha \sum_{\substack{i=0,\ldots,l \\ k=0,\ldots,m}} |OPT_{ik}| \leq \alpha(l|OPT| + |OPT|) \tag{5.3}$$

from 5.1 and 5.3 we have

$$l \times |A_{min}| \leq \alpha(l|OPT| + |OPT|)$$

$$|A_{min}| \leq \alpha \times \left(1 + \frac{1}{l}\right) |OPT| \qquad \blacksquare$$

To solve the strip problem, Hochbaum and Maass applied the above-mentioned procedure recursively, but by partitioning in the vertical direction. This results to squares of size $l2r \times l2r$. They cover the points in the squares with the minimum number of disks which can be done in $O(n^{4l^2+1})$ time. Putting all these together, Hochbaum and Maass [1] derived a $(1 + \frac{1}{l})^2$-approximation algorithm for the disk covering problem in time $O(l^4 n^{4l^2+1})$. Their algorithm can be extended to higher dimensions and arbitrary fixed shapes.

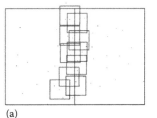 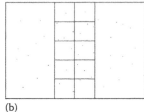

(a)                                              (b)

**FIGURE 5.2**  Every vertical line crosses at most $2l - 1$ squares in an equivalent covering. In (a), there is a covering in which its squares cross a vertical line more than $2l$ times, and in (b), there is another covering which covers all the points covered in (a) with $2l$ squares.

Gonzalez [5] solved the strip problem exactly using dynamic programming. By a packing argument, he proved that any vertical line in any optimal solution crosses at most $O(l)$ disks (Figure 5.2). By using this fact, one can generate all possible coverings by sweeping a line in the strip from left to right and at each event point, combining the coverings obtained before the sweep line with the set of disks intersecting the sweep line. The optimal solution is one of the valid coverings when the sweep line reaches the end point. The solution can be obtained in $O(n^{4l})$ time. By using this algorithm combined with the shifting lemma, Gonzalez derived a $(1 + \frac{1}{l})$-approximation algorithm for the disk covering problem in $O(ln^{4l})$ time.

# 5.3  Partial Covering Problem

## 5.3.1  Background

The partial covering problem was first considered by Gandhi et al. [10]. This problem in the plane is defined as follow: given $n$ points in the plane, cover at least $p$ of them with the minimum number of disks of fixed radius. To give a PTAS for this problem, they used the Hochbaum and Maass's approach [1] by applying the same partitioning and shifting strategy along with a dynamic programming algorithm to generate all possible ways of partitioning $p$ into smaller integers. They first used dynamic programming to obtain the optimal distribution of $p$ points in the master strips of a partitioned problem. To solve the problem in each master strip, they used the same dynamic programming to obtain the optimal distribution of $p$ point in the squares of a master strip. Although they did not give the running time of their algorithm, by analyzing their algorithm one can conclude that it runs in $O(\varepsilon^{-2}p^2n^{4\varepsilon^{-2}+2})$ time and has approximation ratio $O((1 + \varepsilon)^2)$. Ghasemalizadeh and Razzazi [15] proposed an improved algorithm for this problem. To solve the partial covering problem, they first exactly solved the problem when restricted to a horizontal strip of height $l$ using a dynamic programming algorithm, which is an extension of Gonzalez's algorithm [5]. Then, using the shifting lemma [1] and the idea of Gandhi et al. [10], they gave an improved solution to the problem. In Section 5.3.2, we describe the algorithm of Reference 15 for partial covering problem.

## 5.3.2  PTAS for Partial Covering Problem

For simplicity, we explain the algorithm in the plane and $l_\infty$ metric, which is covering the points with unit squares. The generalization to higher dimensions and other metrics is straightforward. The main algorithm in Reference 15 is a dynamic programming algorithm that is described somehow based on Reference 11. To solve the problem, first we solve it when the points are located in a height $l$ strip, and we call the algorithm *strip partial covering algorithm*. To explain the algorithm, we need the following definitions. Let $St$ be a height $l$ strip and $P$ be a set of $n$ point in $St$. Let $p \leq n$ be the number of points that should be covered in $St$. We sort the points in $P$ from left to right. Assume that no two points have the same $x$ coordinate and if so, we can use $y$ coordinate to order them. Let $p_i$ be the $i$'th point in the sorted points and $x_i$ be the $x$ coordinate of $p_i$. An *anchored* square is a square with one point on its left side and one (not necessarily distinct) point on its bottom side. Any square can be replaced with an anchored square that covers the same subset of points by shifting it to the right and up. So, from now on, we assume all squares are anchored squares. A $(k, i)$-*subcover* is a set (possibly empty) of squares that covers at least $k$ points from $p_1$ to $p_i$ and no square completely lies in the $x > x_i$ half plane. *The signature of a $(k, i)$-subcover* $(0 \leq k \leq p, 1 \leq i \leq n)$ is defined as the set of subcover squares (uniquely identified by their anchor points) that intersects with the line $x = x_i$. A $(k, i)$-subcover $c_1$ *dominates* a $(k, i)$-subcover $c_2$ if both have the same signature and $c_1$ contains equal or less number of squares than $c_2$.

By using the previous definitions, we may describe the dynamic programming algorithm for strip partial covering problem as follows: We denote each dynamic programming state as a tuple $((k, i), T)$, where $(k, i)$ stands for a certain $(k, i)$-subcover, and $T$ is the set of this subcover squares. The signature of the state $s$ is the signature of its corresponding subcover, and we denote it as $sig(s)$. Let $S_i$ denote the set of all states at iteration $i$. The result of the algorithm is a state $((p, n), T)$ in $S_n$.

The algorithm is a state-based iterative dynamic programming algorithm. In each iteration, first all possible states are generated from the previous iteration states and then the generated states are filtered to keep only nondominated subcovers.

Let $A_i$ be the set of all squares with $p_i$ as their left anchor. $A_i$ can contain at most $n$ squares, because we can have at most $n$ choices for the bottom side anchor. To compute $S_i$ from $S_{i-1}$, we iterate through all $((k, i - 1), T)$ states in $S_{i-1}$. For each $((k, i - 1), T)$ state, if it already covers $p_i$, we add $((k + 1, i), T)$ to $S_i$, otherwise we add both $((k, i), T)$ and $((k + 1, i), T \cup \{s\})$ to $S_i$ for each $s \in A_i$.

In the filtering step, if there were two states $((k_1, i), T_1)$ and $((k_2, i), T_2)$, which have the same signature and $k_2 = k_1$ and $|T_1| \leq |T_2|$, we remove the latter state from $S_i$. Gonzalez [5] has shown that any covering can be converted to a covering which has at most the same number of squares with the added property that any vertical line crosses at most $2l$ squares (Figure 5.2). Therefore, we only need to consider $(k, i)$-subcovers that their signatures have less than $2l$ squares. Hence, in the filtering phase, we remove all states that their signature has more than $2l - 1$ squares.

For a set $T$ of squares and a point $p$, we say $p$ is covered by $T$ and denote it by $p \in T$, if $p$ is covered by at least one of the squares in $T$. A $(k, i)$-subcover $c_1$ is *similar* to a $(k', i)$-subcover $c_2$, and we denote it by $c_1 \approx c_2$, if they both have the same number of squares, $k = k'$, and $sig(c_1) = sig(c_2)$.

Putting all these together, the strip partial covering algorithm is as follows:

## Strip Partial Covering Algorithm

*Inputs: A set P of n points in a horizontal strip with height l, a positive integer p.*
*Outputs: The minimum number of disks required to cover p points of P.*
*Definitions: A state $s = ((k, i), T)$ and the function sig(s) as defined in the text.*

```
1        S₀ = {((0, 0), ∅)}
2        //Generation Phase
3        For i = 1 to n do
4            Gᵢ = ∅
5            For each state ((k, i − 1), T) ∈ Sᵢ₋₁ do
6                If pᵢ ∈ T then
7                    Gᵢ = Gᵢ ∪ {((k + 1, i), T)}
8                else
9                    Gᵢ = Gᵢ ∪ {((k, i), T)}
10                   For each square s ∈ Aᵢ
11                       Gᵢ = Gᵢ ∪ {((k + 1, i), T ∪ {s})}
12       //Filtering Phase
13       //Filter states with more than 2l-1 squares
14       For each state ((k, i), T) ∈ Gᵢ do
```

15              If $|sig(s)| \leq 2l - 1$ then
16                  $S_i = S_i \cup \{((k, i), T)\}$
17          //Filter dominated states
18          For each state $s_1 = ((k_1, i), T_1) \in S_i$ do
19              For each state $s_2 = ((k_2, i), T_2) \in S_i$ do
20                  If $sig(s_1) == sig(s_2)$ then
21                      If $|T_1| \leq |T_2|$ and $k_2 == k_1$ then
22                          $S_i = S_i - \{s_2\}$
23          Return one of the $((p, n), T)$ states in $S_n$.

The correctness of the algorithm is proved in a series of lemmas in Reference 15 which leads to the following

**Lemma 5.2** *The strip partial covering algorithm correctly obtain the minimum number of squares required to cover at least p points of n point in height l strip for different values of $1 \leq p \leq n$ in $O(pn^{4l})$ time.*

To give an approximation algorithm for the partial covering problem in the general case, not restricted to a height $l$ strip, we use the shifting technique. We partition the plane into horizontal strips of height 1 and group every $l$ of them as a master strip. We call this a partitioned problem $P_1$ that consists of $k$ master strips $S_{11} \cdots S_{1k}$. To determine the number of points that should be covered in each master strip, we just enumerate all possible ways of distributing $p$ points among master strips using dynamic programming. Let $D(i, q)$ denote the minimum number of squares required to cover at least $q$ points in the first $i$ master strips in a given partitioned problem, and let $C(i, q)$ be the answer of the strip partial covering algorithm in master strip $i$ when $q$ is the number of points to be covered. $D(i, q)$ can be computed recursively:

$$D(i, q) \underset{1 \leq i \leq k, 0 \leq q \leq p}{=} \min_{0 \leq j \leq q} \{D(i - 1, q - j) + C(i, j)\}$$

We call this algorithm the *partial covering algorithm*. Note that in the strip partial covering algorithm, we solve the partial cover for all values of $q$, $1 \leq q \leq p$. So $C(i, q)$ can be computed by running the strip partial cover algorithm in master strip $i$ only once. To solve the partitioned problem $P_1$, we run the strip partial covering algorithm for each of its master strips, and then compute $D(i, q)$ values for all $i$ and $q$, $0 \leq i \leq k$ and $0 \leq q \leq p$. Overall, we need to run the strip partial cover algorithm $k$ times which, by considering only nonempty strips takes $O(n)$ time. Thus, by Lemma 5.2, the dynamic programming algorithm runs in time $O(pn^{4l+1})$ for the partitioned problem $P_1$. Applying the shifting process $l$ times, we have the following theorem:

**Theorem 5.1** *n points in the plane and a parameter $0 < \varepsilon < 1$ are given. There is a $(1 + \varepsilon)$-approximation algorithm for the partial covering problem in $O(\varepsilon^{-1} n^{4\varepsilon^{-1}+2})$ time, for every integer $0 \leq p \leq n$, which is the number of points to be covered.*

We can use disks instead of squares in the partial covering algorithm. Every two points determine two disks, so the total number of different disks (two disks are different if they cover different subsets of points) is $2n^2$. We should also compute the maximum number of disks that intersects a vertical line of height $l$ in the strip partial covering algorithm. This number can be obtained by considering the circumvented squares of the disks and it becomes at most $2\sqrt{2}l$. So the maximum number of different signatures to be considered is $(2n^2)^{2\sqrt{2}l}$ and the running time of the algorithm is $O(\varepsilon^{-1} pn^{4\sqrt{2}\varepsilon^{-1}+1})$. The generalization to $R^d$ and other $l_p$ metrics can be done as in Reference 11. In $R^d$, the algorithm runs in $O(\varepsilon^{-d+1} n^{O(d)\varepsilon^{-d+1}})$ time.

# 5.4 Maximum Covering Problem

## 5.4.1 Background

The dual of the partial covering problem is the problem of covering the maximum number of points when only $m$ disks are available. We will refer to this problem as the maximum covering problem. One solution for the maximum covering problem with a constant approximation factor is a greedy approach based on the method presented in Reference 12 for the set cover problem. Applying this method, we first find a disk center that covers the maximum number of points which can be done in $O(n^2)$ time [13]. Then, we remove the points located in the first disk and find the second disk which covers the maximum number of points using the same method. We continue this process until we find $m$ disks. The algorithm runs in $O(mn^2)$ time and its approximation factor is $(1 - \frac{1}{e})$ [12]. Berg et al. [14] gave the first $(1 - \varepsilon)$-approximation algorithm for the maximum covering problem. They call the problem $max(P, m)$, which is to cover the maximum number of points in the point set $P$ with $m$ unit disks. To solve the $max(P, m)$, they first define a grid with a specific spacing depending on $m$, and they demonstrate that for every arrangement of $m$ disks on the point set $P$, one can find a placement for the grid such that none of the grid lines intersects any of the disks. In other words, each of the disks lies completely in the grid cells. Suppose we know the placement of the grid and the number of the disks in every grid cell. In this case, the problem reduces to cover the maximum number of points with $j \leq m$ disks in a grid cell $C$; they call this problem $max(P, j, C)$. To find the optimum solution to $max(P, m)$, they examined different placements of the grid and in each placement and they checked different distribution of $m$ disks among the grid cells using dynamic programming. To solve $max(P, j, C)$, they first determined all possible locations for disks centers inside the cell $C$ which the number of them is $O(n^2)$. By considering all subsets of size $m$ of these centers, they found the subset which covers the most number of points. This exactly solves the problem and the algorithm running time for each cell is $n^{O(m)}$. To omit the dependency on $n$ in the running time of the algorithm, they use an $\varepsilon$-approximation in each cell to reduce the number of points in the cell to a constant number which depends on $\varepsilon$. By using this technique, they obtain a $(1 - \varepsilon)$-approximation algorithm which runs in time $(\frac{1}{\varepsilon})^{O(m)}$. The actual running time of their algorithm is $O(n\log n + n\varepsilon^{-6m+6}\log(\frac{1}{\varepsilon}))$ that is exponential with respect to $m$. However, considering $m$ as a constant, as the authors did, the running time of the algorithm can be evaluated as $O(n\log n)$. Ghasemalizadeh and Razzazi [15] proposed a PTAS for the maximum covering problem with polynomial running time with respect to both $m$ and $n$. The main idea is based on using the partial covering problem to find a near optimal solution and then by a careful modification of this solution generating a valid solution for the maximum covering problem. Their algorithm runs in $O((1 + \varepsilon)mn + \varepsilon^{-1}n^{4\sqrt{2}\varepsilon^{-1}+2})$ time for maximum covering in the plane with $m$ disks. In Section 5.4.2, we describe this algorithm.

## 5.4.2 PTAS for Maximum Covering Problem

In the current section, we describe the algorithm in Reference 15 for obtaining a PTAS for the *maximum covering* problem. We explain the algorithm in the plane and for unit disks ($l_2$ metric), the generalization to higher dimensions and other $l_p$ metrics is same as for the partial covering algorithm. We present an approximation algorithm for the problem whose running time is polynomial with respect to $m$ and $n$. We call the algorithm the *maximum covering* algorithm.

Let define $Max(P, m)$ as the value of an optimal solution, which is the maximum number of points where $m$ unit disks can cover from the set $P$. The main idea of the algorithm is as follows:

The dual of the maximum covering problem is the partial covering problem. Assume that there is an algorithm which solves the partial covering problem exactly, and we call it *ExactPartialCovering(P, p)*. It accepts a point set $P$ and a positive integer $p$ and it returns $c$ which is the minimum number of disks required to cover at least $P$ points of $P$. The maximum covering problem can be solved by several times calling the *ExactPartialCovering(P, p)* with different values of $p$. The maximum value of $p$ for which the

*ExactPartialCovering*$(P, p)$ returns $c = m$ is the optimal solution for the maximum covering problem. As we do not have the exact solution for the partial covering problem, we use the approximate solution for the partial covering problem to obtain an approximate solution for the maximum covering problem.

The maximum covering algorithm has two steps. In the first step, using an approximation algorithm for the partial covering problem, we generate a rough solution for the maximum covering problem which covers at least $Max(P, m)$ points with $c \geq m$ disks. In the second step of the algorithm, we remove $c - m$ disks with the smallest number of points from these $c$ disks to obtain a valid approximate solution for the maximum covering problem. In the following, we describe the algorithm in more details. We refer to the algorithm in Theorem 5.1 as *PartialCover*$(p, P, \varepsilon)$, which returns a $(1 + \varepsilon)$ approximation solution of the partial covering problem.

The maximum covering algorithm receives the point set $P$, the number of disks $m$, and the approximation factor $\varepsilon$ as input. In step 1, the algorithm does a binary search on the values of $p$, $1 \leq p \leq n$, and it calls the *PartialCover*$(p, P, \varepsilon)$ using these values of $p$. The *PartialCover*$(p, P, \varepsilon)$ in each call returns $c$ disks where $c$ is at most $(1 + \varepsilon)$ times more than the minimum number of disks that can cover $p$ points. In each call, the algorithm compares $c$ with $m$, and if $c$ is equal or less than $(1 + \varepsilon)m$, it increases $p$, otherwise it decreases it. It continues this process until it reaches a value of $p$ such that *PartialCover*$(p, P, \varepsilon)$ returns $c$ disks where $c - 1 \leq (1 + \varepsilon)m < c$. We call this value of $p$, $opt_m$. It can be proved that $opt_m$ is an upper bound for $Max(P, m)$. At the end of step 1, we have a solution with $c$ disks which covers at least $Max(P, m)$ points where $c - 1 \leq (1 + \varepsilon)m < c$.

In step 2, the algorithm acquires a valid solution for the maximum covering problem by removing some of these $c$ disks. First, it sorts the disks in increasing number of covered points, and then it removes $\lfloor \varepsilon m \rfloor + 1$ disks with smallest number of points. After removing the $\lfloor \varepsilon m \rfloor + 1$ disks, we obtain the desired solution with $m$ disks. The algorithm is illustrated as follows:

---

## Maximum Covering Algorithm

*Inputs: A set of n points P, a positive integer m, the approximation factor $\varepsilon$.*
*Outputs: A set of m disks.*
*Definitions: The method PartialCover(p,P,$\varepsilon$) as stated in the text.*

```
1     //Step 1.
2         b = 1 ; e = n ; p = 0;
3         while p ≤ n
4             p = ⌊ (b+e)/2 ⌋
5             disks = PartialCover(p,P,ε)
6             c = |disks|
7             if c − 1 ≤ (1 + ε)m < c then
8                 goto step 2
9             if c ≤ (1 + ε)m then
10                b = p
11            else
12                e = p
13        return disks //with m disks we can cover all the input points
14    //Step 2
15        Sort disks in increasing number of points they cover
16        Remove the first ⌊εm⌋ + 1 disks
17        return disks
```

---

The correctness of the algorithm will be proved in a series of lemmas that leads to the following theorem:

**Theorem 5.2** *There is a* $(1 - \frac{2\varepsilon}{1+\varepsilon})$*-approximation algorithm for the maximum covering problem for covering with disks in the plane that runs in* $O((1 + \varepsilon)mn + \varepsilon^{-1}n^{4\sqrt{2}\varepsilon^{-1}+2})$ *time.*

## 5.5 Cardinality Constrained Covering Problem

### 5.5.1 Background

In covering problems, geometric shapes represents facilities and the points represents clients to be served. As facilities are serving entities, it is natural to consider a capacity for the amount of their services. This calls for making covering problems capacitated. There are many problems in this category such as the capacitated facility location problem [16], the capacitated set covering problem [17], and the capacitated $k$-center problem [18]. In nongeometric settings, many approximation algorithms for capacitated facility location problems are developed. Wolsey [19] (see also Reference 17) used a greedy algorithm to derive a $\log(n)$-approximation algorithm for the capacitated set cover problem (and more generally, the submodular set cover problem). A different greedy algorithm that achieves the same bound was presented in Reference 18. Moreover in Reference 18, the classical approximation algorithm for the metric $k$-center problem was extended to develop a 10-approximation algorithm for the capacitated metric $k$-center problem. This bound was later improved to 6 by Khuller and Sussmann [20]. There are many attempts related to the soft-capacitated and hard-capacitated metric facility location problems. In the soft-capacitated case, one is allowed to create several copies of a facility, if required, but in the hard-capacitated case, only available facilities can be used. The best approximation ratios for the soft and hard capacitated metric facility location problems are 2 [21] and 5.83 [22], respectively. For the soft-capacitated vertex cover problem, the best bound is 2, which is achieved by both primal-dual approach [23] and a variant of randomized rounding called dependent rounding [24]. In the hard case and for arbitrary weights, Chuzhoy and Naor [17] showed that this problem is as hard as set cover to approximate. In the unweighted case, they presented a 3-approximation algorithm which later improved to a 2-approximation algorithm by Gandhi et al. [25] using randomized rounding.

In geometric settings, Arora et al. [26] gave a PTAS for the capacitated Euclidean $k$-median problem by changing the dynamic programming formulation of their PTAS for the Euclidean $k$-median problem. Agarwal and Procopiuc [27] considered the $L$-capacitated $k$-center problem in $(R^d, l_p)$ metric spaces, which is similar to the $k$-center problem but now each center can cover at most $L$ points. When $L = O(1)$ or $L = \Omega(n/k^{1-1/d})$, they gave an exact algorithm to solve the problem in $n^{O(k^{1-1/d})}$ time by partitioning the plane into horizontal strips and using a dynamic programming algorithm based on Gonzalez's algorithm [5] to solve the strip problem. They left solving the problem for any value of $L$ as an open problem. Having an $\alpha$-approximation algorithm for the uncapacitated geometric covering problem, one can use the general approach of Berman et al. [28] to convert it to an $\alpha + 1.357$-approximation algorithm for the capacitated case. Ghasemi and Razzazi [29] presented a PTAS for cardiniality constrained covering with unit balls in $(R^d, l_p)$ metric space. In the next section, we describe the algorithm of Reference 29 in more details in Section 5.5.2.

### 5.5.2 PTAS for Cardinality Constrained Covering with Unit Balls

For a point set $P$ and an integer $L$, let $CC(P, L)$ refer to the covering with unit balls problem with cardinality constrain and $UC(P)$ refer to this problem without any constraint. Let $U$ be the optimal solution for $UC(P)$. To obtain a solution for $CC(P, L)$, we can copy every unit ball $u \in U$ which covers more than $L$ points, to obtain a set of unit balls such that each of them covers at most $L$ points. We call this algorithm COPY and we can prove that the COPY algorithm is a 2-approximation algorithm for $CC(P, L)$. The optimal solution for $CC(P, L)$ may contain some balls that cover points from more than one unit ball $u \in U$. We call this unit balls *active* unit balls. Let define $k$-constrained $L$-capacitated problem as a problem that have at most $k$ active balls on each unit ball of $U$. We denote an instance of this problem by $CC^k(P, L, U)$. To derive a PTAS for this problem, we use the structure theorem:

**Theorem 5.3** *Given any fixed $\varepsilon > 0$ and any feasible solution $U$ for $UC(P)$, there exists an integer $k = \theta(1/\varepsilon)$ such that $|OPT_k| \leq (1 + \varepsilon).|OPT|$, where $|OPT_k|$ is the size of an optimal solution of $CC^k(P, L, U)$ and $|OPT|$ is the size of an optimal solution of $CC(P, L)$.*

The structure theorem tells us that to have a $(1 + 1/k)$-approximation solution, it is sufficient to find an optimal solution to the $k$-constrained problem. Feasible solutions of this problem have at most $k$ active unit balls on each unit ball of the underlying uncapacitated solution. As an active unit ball can cover at most $(L - 1)$ points from an uncapacitated unit ball, there are at most $k(L - 1)$ points from every uncapacitated unit ball that need to be covered carefully. These points are likely to be covered by active unit balls in the final solution. Other points can simply be covered using COPY algorithm, which only generates inactive unit balls. On the contrary, we do not know the position of these special points. Thus, we need to develop an algorithm that simultaneously finds these points and covers them with a minimum number of $L$-capacitated unit balls. From now on, we restrict ourselves to the case in which points lie in a width-bounded hyperstrip. The capacitated covering problem when input lies in a hyperstrip is called the capacitated strip covering problem. We can use the shifting strategy to remove this restriction.

For a given $\varepsilon > 0$, let $k = \theta(1/\varepsilon)$ be a constant. Assume that the points are all lying in a hyperstrip of width $2h$. Let $U$ be a solution for $UC(P)$. For a $u \in U$, define its covering requirement as follows:

$$r(u) = \begin{cases} kL + (\ell(u) \bmod L) & \ell(u) > kL \\ \ell(u) & \ell(u) \leq kL \end{cases}$$

Consider the following problem: find a minimum number of $L$-capacitated unit balls whose union covers $r(u)$ points from the points in $P(u)$, for each $u \in U$. We refer to this problem as the *capacitated group covering problem* (CGC). Suppose we have an algorithm CGC that optimally solves this problem. Now we can solve the capacitated strip covering problem using the following algorithm:

---

## Capacitated-Strip-Covering Algorithm

*Inputs P: A set of n points lying in a hyperstrip of width 2h, $\varepsilon > 0$, an integer $L > 0$*
*Outputs: An L-capacitated covering of P.*

| 1 | *Compute a $U \in UC(P)$ with the packing property.* |
|---|---|
| 2 | *Compute the k and $r(u)$ $\forall u \in U$ according to the above-mentioned definitions.* |
| 3 | *Solve the CGC problem.* |
| 4 | *Cover the remaining points in the unit balls of U using COPY algorithm.* |
| 5 | *Report the union of the solution of line 3 and the solution of line 4 as the final solution.* |

---

The CGC problem in a hyperstrip of width $2h$ strips is solved by a dynamic programming algorithm similar to Reference 5. Applying the shifting strategy in the capacitated case, we derive a $(1 + \varepsilon)^d$-approximation $n^{O(1/\varepsilon^d)}$-time algorithm for the problem in $d$ dimensions, assuming $d$ is a fixed constant.

## 5.6 Conclusion

In the current chapter we reviewed the algorithms for covering with unit balls problem and some of its variants. This problem is in NP-hard and it admits PTAS. The main technique to obtain PTAS for this problem is to divide the plane and solving the problem in several shifted divisions which is known as shifting lemma that was first introduced in Reference 1. We reviewed partial covering, maximum covering, and cardinality constrained covering as some variants of covering with unit balls problem. The partial covering problem is solved by using shifting technique along with a dynamic programming to enumerate partial covers. Maximum covering problem is solved by using a primal dual technique from the solution of the partial covering problem. To solve cardinality constrained covering problem, a structural theorem is proved to solve the problem in bounded height strips and shifting technique is used to solve the problem in general.

# Acknowledgment

The authors acknowledge Springer for permission to reuse some material from our paper. "An Improved Approximation Algorithm for the Most Points Covering Problem" published in Theory Comput Syst, 50:545–558, Springer (2011). We would also like to acknowledge Elsevier for allowing us to reprint portions for our paper. "A PTAS for the cardinality constrained covering with unit balls," that appeared in Theoretical Computer Science, 527, 50–60.

# References

1. Hochbaum, D.S., Maass, W., Approximation schemes for covering and packing problems in image processing and VLSI, *Journal of ACM,* 32, 130, 1985.
2. Fowler, R.J., Paterson, M.S., Tanimoto, S.L., Optimal packing and covering in the plane are NP-complete, *Information Processing Letters,* 3, 133, 1981.
3. Fu, B., Theory and application of width bounded geometric separator, *Lecture Notes in Computer Science,* 3884, 277, 2006.
4. Feige, U., A threshold of ln n for approximating set cover, *Journal of the ACM,* 45, 634, 1998.
5. Gonzalez, T.F., Covering a set of points in multidimensional space, *Information Processing Letters,* 40, 181, 1991.
6. Ghasemi, T., Ghasemalizadeh, H., Razzazi, M., An Algorithmic framework for solving geometric covering problems—with applications. *International Journal of Foundations of Computer Science,* 25(5), 623, 2014.
7. Franceschetti, M., Cook, M., Bruck, J., A geometric theorem for approximate disk covering algorithms, Report ETR035, Caltech, Pasadena, CA, 2001.
8. Fu, B., Chen, Z., Abdelguerfi, M., An almost linear time 2.8334—approximation algorithm for the disc covering problem, *Lecture Notes on Computer Science,* 4508, 317, 2007.
9. Bronnimann, H., Goodrich, M. T., Almost optimal set covers in finite vc-dimension, *Discrete and Computational Geometry,* 14, 463, 1995.
10. Gandhi, R., Khuller, S., Srinivasan, A., Approximation algorithms for partial covering problems, *Journal of Algorithms,* 53, 55, 2004.
11. Agarwal, P.K., Procopiuc, C.M., Exact and approximation algorithms for clustering, *Algorithmica,* 33, 201, 2002.
12. Khuller, S., Moss, A., Naor, J.S., The budgeted maximum coverage problem, *Information Processing Letters,* 70, 39, 1999.
13. Chazelle, B.M., Lee, D.T., On a circle placement problem, *Computing,* 36, 1, 1986.
14. Berg, M.D., Cabello, S., Har-Peled, S., Covering many or few points with unit disks, *Theory of Computing Systems,* 45, 446, 2009.
15. Ghasemalizadeh, H., Razzazi, M., An improved approximation algorithm for the most points covering problem, *Theory of Computing Systems,* 50(3), 545, 2012.
16. Shmoys, D.B., Tardos, É., Aardal, K., Approximation algorithms for facility location problems, *Proceedings of 29th Annual Symposium on Theory of Computing*, ACM, New York, pp. 265–274, 1997.
17. Chuzhoy, J., Naor, J., Covering problems with hard capacities, *SIAM Journal on Computing,* 36(2), 498, 2006.

18. Bar-Ilan, J., Kortsarz, G., Peleg, D., How to allocate network centers, *Journal of Algorithms,* 15, 385, 1993.

19. Wolsey, L.A., An analysis of the greedy algorithm for the submodular set covering problem, *Combinatorica,* 2, 385, 1982.

20. Khuller, S., Sussmann, Y., The capacitated k-center problem, *SIAM Journal on Discrete Mathematics,* 13, 403, 2000.

21. Mahdian M., Ye Y., Zhang J., A 2-approximation algorithm for the soft-capacitated facility location problem, *Approximation, Randomization, and Combinatorial Optimization. Algorithms and Techniques,* Lecture Notes in Computer Science, vol. 2764, Springer, Berlin, Germany, pp. 129–140, 2003.

22. Zhang, J., Chen, B., Ye, Y., A multiexchange local search algorithm for the capacitated facility location problem, *Mathematical Operations Research,* 30, 389, 2005.

23. Guha, S., Hassin, R., Khuller, S., Or, E., Capacitated vertex covering, *Journal of Algorithms,* 48, 257, 2003.

24. Gandhi, R., Khuller, S., Parthasarathy, S., Srinivasan, A., Dependent rounding and its applications to approximation algorithms, *Journal of the ACM,* 53(3), 324, 2006.

25. Gandhi, R., Halperin, E., Khuller, S., Kortsarz, G., Srinivasan, A., An improved approximation algorithm for vertex cover with hard capacities, *30th Annual International Colloquium on Automata, Languages and Programming,* Springer, Berlin, Germany, pp. 164–175, 2003.

26. Arora, S., Raghavan, P., Rao, S., Approximation schemes for euclidean k-medians and related problems, *Annual ACM Symposium on Theory of Computing,* 1998.

27. Agarwal, P.K., Procopiuc, C.M., Exact and approximation algorithms for clustering, *Algorithmica,* 33, 201, 2002.

28. Berman, P., Karpinski M., Lingas, A., Exact and approximation algorithms for geometric and capacitated set cover problems with applications, *Proceedings of the 16th Annual International Conference on Computing and Combinatorics,* Springer-Verlag, Nha Trang, Vietnam, pp. 226–234, 2010.

29. Ghasemi, T., Razzazi, M., A PTAS for the cardinality constrained covering with unit balls, *Theoretical Computer Science,* 527, 50, 2014.

# 6

# Minimum Edge-Length Rectangular Partitions

Teofilo F. Gonzalez

Si Qing Zheng

## 6.1 Introduction

In the current chapter, we discuss approximation algorithms for partitioning a rectangle with interior points. This problem is denoted by *RG-P* in which *RG* stands for *rectangle* and *P* stands for *points*. An instance of the *RG-P* problem is a set *P* of *n* points inside a rectangle *R* in the plane. A feasible solution is a *rectangular partition*, which consists of a set of (orthogonal) line segments that partition *R* into rectangles such that each of the *n* points in *P* is on a partitioning line segment. The objective of the *RG-P* problem is to find a rectangular partition whose line segments have least total length. The *RG-P* problem is an NP-hard problem [1].

A more general version of the problem is when *R* has interior rectilinear holes instead of points. This problem arises in Very Large Scale Integration (VLSI) design where it models the problem of partitioning a routing region into channels [2]. Approximation algorithms for this more general problem exist [3–6]. Levcopoulos' algorithms [4,5] are the ones with the smallest approximation ratio. His fastest algorithm [5] invokes as a subprocedure the divide-and-conquer algorithm for the *RG-P* problem developed by Gonzalez and Zheng [7], which is a preliminary version of the one discussed in this chapter.

The structure of an optimal rectangular partition can be very complex (Figure 6.1a). A restricted version of the *RG-P* problem is to only allow the introduction of line segments that recursively partition each subinstances into exactly two subinstances of the *RG-P* problem. A recursive partition with this property is given in Figure 6.1b. For a given rectangle, such a line segment is called a *cut* of the rectangle in Reference 7 and a *guillotine cut* in Reference 8. A rectangular partition, in which at every level there is a guillotine cut, is called a *guillotine partition*. By definition every guillotine partition is a rectangular partition, but the converse is not true. For example, the rectangular partition given in Figure 6.1a is not a guillotine partition. For the same instance, the structure of an optimal guillotine partition can be much simpler than the structure of an optimal rectangular partition. Furthermore, the concept of guillotine cut is useful for developing approximation algorithms for the *RG-P* problem using divide-and-conquer and dynamic programming techniques. The algorithm given in Reference 7 finds an approximation to the *RG-P* problem by producing a suboptimal guillotine partition via a divide-and-conquer algorithm, whereas the algorithm given by Du et al. [8] finds an approximation to the *RG-P* problem by producing an optimal guillotine partition via dynamic programming. Gonzalez and Zheng [7] algorithm takes $O(n \log n)$ time. Du et al. [8]

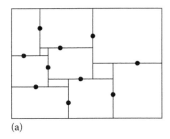

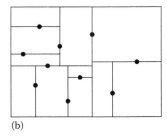

(a)                                                                          (b)

**FIGURE 6.1**    (a) Optimal rectangular partition. (b) Optimal guillotine partition.

algorithm takes $O(n^5)$ time to construct an optimal guillotine partition via dynamic programming. On the other hand, Du et al. [8] showed that the length of an optimal guillotine partition is at most twice the length of an optimal rectangular partition for the *RG-P* problem, but the approximation ratio for the divide-and-conquer algorithm given in Reference 7 is $3\sqrt{3}$. This ratio was reduced to 4 by using a slightly different divide-and-conquer procedure. A complex proof showing that the length of an optimal guillotine partition is within 1.75 times the length of an optimal rectangular partition is given by Gonzalez and Zheng [9]. Clearly, neither method dominates the other when one takes into account both the time complexity and the approximation ratio.

It is important to note that optimal and near-optimal guillotine partitions are not the only way one can generate solutions with a constant approximation ratio for the *RG-P* problem. Gonzalez and Zheng [10] developed an algorithm with approximation ratio of 3 for the *RG-P* problem that generates a rectangular partition that is not necessarily a guillotine partition. Both the time and approximation ratio for this algorithm are between the ones of the two algorithms mentioned earlier.

The *RG-P* problem was generalized to *d*-dimensional Euclidean space. In this generalized problem, *R* is a *d*-box and *P* is a set of points in *d*-dimensional space, and the objective is to find a set of orthogonal hyperplane segments of least total $(d-1)$-volume that include all points of *P*. This is the *d-dimensional RG-P problem*. Approximating an optimal *d*-box partition by suboptimal and optimal guillotine partitions based on divide-and-conquer and dynamic programming techniques has been established in References 11, 12, respectively.

In the current chapter, we present two approximation algorithms for the *RG-P* problems. For an *RG-P* instance $I$, we use $E(I)$ to represent the set of line segments in the solution generated by our algorithm, and $E_{opt}(I)$ the set of line segments in an optimal rectangular partition. We use $L(S)$ to represent the length of the line segments in set $S$. We first present an $O(n \log n)$-time divide-and-conquer approximation algorithm that generates suboptimal guillotine partitions with total edge length within four times the optimal rectangular partition value, that is for every $I$, $L(E(I)) \leq 4 \cdot L(E_{opt}(I))$. We then present a simple proof that shows that an optimal guillotine partition has total edge length that is within two times the optimal solution value, that is, for every $I$, $L(E(I)) \leq 2 \cdot L(E_{opt}(I))$. This proof is much simpler than the proof of Reference 8 for the same bound. We also outline the main ideas of the complex proof in Reference 9 that establishes that an optimal guillotine partition has edge length that is within 1.75 times an optimal rectangular partition value. One may improve Levcopoulos algorithm [5] by replacing the algorithm in Reference 7 by any of the aforementioned algorithms.

## 6.2  A Divide-and-Conquer Algorithm

An *RG-P* problem instance is given by $I = ((X, Y), P)$, in which $(X, Y)$ defines the rectangle $R$ (with height $Y$ and width $X$), and $P = \{p_1, p_2, \ldots, p_n\}$ is a nonempty set of points located inside $R$. Before we present our algorithm, we define some terms and define a way to establish lower bounds for the length of rectangular partitions.

A *partial rectangular partition* is a partition of $R$ into rectangles in which not all the points in $P$ are located along the partitioning line segments. Let $Q$ be any partial rectangular partition for problem instance $I$. A subset of the points in $P$ is said to be *assigned* to $Q$ if the following three conditions are satisfied:

1. Every rectangle $r$ in $Q$ with at least one point inside it has one such point assigned to it.
2. A rectangle $r$ in $Q$ without points inside it may be assigned at most one point on one of its sides.
3. Two rectangles with a common boundary cannot both have their assigned points on their common boundary.

An assignment of the points in $P$ to $Q$ is denoted by $A(Q)$.

Every rectangle $r$ with a point assigned to it is said to have *value, $v(r)$,* equal to the minimum of the height of $r$ and the width of $r$. The assignment $A(Q)$ of the partial rectangular partition $Q$ has value equal to the sum of the values of the rectangles that have an assigned point. This value is denoted by $v(A(Q))$.

We now establish that the edge length of a rectangular partition is at least equal to the value of any assignment for any partial rectangular partition of $R$.

**Lemma 6.1** *For every problem instance $I$, partial rectangular partition $Q$ and assignment $A(Q)$, a lower bound for $L(E(I))$ is given by $v(A(Q))$, that is, $v(A(Q)) \leq L(E(I))$. In particular, $v(A(Q)) \leq L(E_{OPT}(I))$.*

*Proof.* Consider any rectangle $r$ with one or more points in it. By definition, one of these points is assigned to the rectangle. Clearly, any partition into rectangles must include line segments inside $r$ with length greater or equal to the minimum of the height of $r$ or the width of $r$. This is equal to the value $v(r)$.

Consider now any rectangle $r$ without points inside, but with one assigned point. The point assigned to $r$ must be located on one of its sides. Lets say it is on side $s$. By definition, any rectangle with a common boundary to side $s$ of rectangle $r$ does not have its assigned point on this common boundary. Therefore, any rectangular partition that covers the assigned point of $r$ must include line segments in $r$ with length at least equal to the minimum of the height of $r$ or the width of $r$. This is equal to the value $v(r)$.

As $v(A(Q))$ is equal to the sum of the values of the rectangles that have an assigned point, the above-mentioned arguments establish that $v(A(Q)) \leq L(E(I))$, that is $v(A(Q))$ is a lower bound for $L(E(I))$. ∎

We now define our divide-and-conquer procedure to generate a rectangular partition. From this rectangular partition, we identify a partial rectangular partition and an assignment of a subset of the points $P$. By applying Lemma 6.1, we have a lower bound for an optimal rectangular partition. Then we show that the length of the edges in the solution generated is within four times the lower bound provided by the assignment.

Assume without loss of generality that we start with a nonempty problem instance such that $Y \leq X$. Procedure $DC$ introduces a *mid-cut* or an *end-cut*, depending on whether or not the rectangle has points to the left and also to the right of the center of $R$. The cut is along a vertical line that partitions $R$ into $R_l$ and $R_r$. The set of points in $P$ that are not part of the cut are partitioned into $P_l$ and $P_r$ depending on whether they are inside of $R_l$ or $R_r$. A *mid-cut* is a vertical line segment that partitions $R$ and includes the center of the rectangle. An *end-cut* is a vertical line segment that partitions $R$ and includes either the "rightmost" or the "leftmost" points in $P$, depending whether or not there are points to the left of the center of $R$. Procedure $DC$ is then applied recursively to the nonempty resulting subproblems, that is, $I_l = ((X_l, Y), R_l)$ and $I_r = ((X_r, Y), R_r)$, where $X_l$ and $X_r$ represent the length along the x-axis of the two resulting subinstances ($I_l$ and $I_r$), respectively. Note that when a mid-cut is introduced, it must be that both $P_l$ and $P_r$ end up being nonempty. For an end-cut, at least one of these sets will be empty. When both sets of points are empty, the cut is called *terminal end-cut*.

It is easy to verify that Figure 6.2 represents all the possible outcomes of one step in the recursive process of our procedure. A subinstance without interior points is represented by a rectangle filled with diagonal line segments.

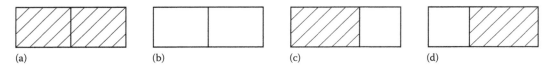

(a)                    (b)                    (c)                    (d)

**FIGURE 6.2**  Sub instances generated by Procedure *DC*: (a) A terminal end-cut is introduced, $P_l = \emptyset$ and $P_r = \emptyset$, (b) A mid-cut is introduced, $P_l \neq \emptyset$ and $P_r \neq \emptyset$, (c) A non-terminal end-cut is introduced, $P_l = \emptyset$ and $P_r \neq \emptyset$, and (d) A non-terminal end-cut is introduced, $P_l \neq \emptyset$ and $P_r = \emptyset$.

Our lower bound function, $LB(I)$, is defined from a partial rectangular partition of the rectangular partition generated by Procedure *DC*. The partial rectangular partition is the rectangular partition generated by Procedure *DC*, except that all the terminal end-cuts are not included. The association for the partial rectangular partition is defined as follows. Every rectangle with points inside it is assigned one of the points inside it. Rectangles without points inside them resulting from a nonterminal end-cut have one of the points along the end-cut assigned to them. By using Lemma 6.1, the value of this assignment is a lower bound for the length of an optimal rectangular partition. A recursive definition for the value of the earlier assignment is given by the following recurrence relation $LB(I)$:

$$LB(I) = \begin{cases} Y & \text{(a) A terminal end-cut is introduced, } P_l = \emptyset \text{ and } P_r = \emptyset. \\ LB(I_l) + LB(I_r) & \text{(b) A mid-cut is introduced, } P_l \neq \emptyset \text{ and } P_r \neq \emptyset. \\ min\{Y, X_l\} + LB(I_r) & \text{(c) A non-terminal end-cut is introduced, } P_l = \emptyset \text{ and } P_r \neq \emptyset. \\ LB(I_l) + min\{Y, X_r\} & \text{(d) A non-terminal end-cut is introduced, } P_l \neq \emptyset \text{ and } P_r = \emptyset. \end{cases}$$

Let $L(E_{DC}(I))$ denote the total length of the set $E_{DC}(I)$ of line segments introduced by Procedure *DC*. We define $USE(I)$ to be the length of the line segments introduced during the first call to Procedure $DC(I)$. That is, when $P = \emptyset$ then $USE(I) = 0$; otherwise, $USE(I) = L(E_{DC}(I)) - L(E_{DC}(I_l)) - L(E_{DC}(I_r))$.

Assume $X \geq Y$. A problem instance $I = ((X, Y), P)$ is said to be *regular* if $X \leq 2Y$, and *irregular* otherwise (i.e., $X > 2Y$). We define the *carry function* $C$ for a problem instance $I$ as

$$C(I) = \begin{cases} 3Y & \text{if } I \text{ is irregular,} \\ X + Y & \text{if } I \text{ is regular.} \end{cases}$$

One may visualize the analysis of our approximation algorithm as follows. Whenever a line segment (*mid-cut* or *end-cut*) is introduced by Procedure *DC*, it is colored red, and when a lower bound from $LB(I)$ is "identified" we draw an "invisible" blue line segment with such length. Our budget is four times the length of the blue line segments that we must use to pay for all the red line segments. In other words, the idea is to bound the sum of the length of all the red segments by four times the one of the blue segments. The length of the red line segments introduced at previous recursive invocations of Procedure *DC* that have not yet been accounted by previously identified blue segments is bounded above by the carry function $C$, defined previously. In other words, the carry value is the maximum edge length (length of red segments) we could possibly owe at this point. Before proving our result, we establish some preliminary bounds.

**Lemma 6.2**  *For any problem instance $I$ such that $X \geq Y$, $2Y \leq C(I) \leq 3Y$.*

*Proof.* The proof follow from the definition of the carry function.  ∎

**Lemma 6.3**  *For every problem instance $I$, $L(E_{DC}(I)) + C(I) \leq 4 \cdot LB(I)$.*

*Proof.* The proof is by contradiction. Let $I$ be a problem instance with the least number of points $P$ that does not satisfy the lemma, that is,

$$L(E_{DC}(I)) + C(I) > 4 \cdot LB(I). \tag{6.1}$$

Assume without loss of generality that $Y \leq X$. There are three cases depending on the cut introduced by Procedure *DC* when it is initially invoked with $I$.

*Case 1*: The procedure introduces a terminal end-cut.

As $Y \leq X$ and $P_l = P_r = \emptyset$, we know that $LB(I) = Y$. From the way the procedure operates, we know $L(E_{DC}(I)) = USE(I) = Y$. Substituting in Equation 6.1, we know that $C(I) > 3Y$. But this contradicts Lemma 6.2. So it cannot be that the algorithm introduces a terminal end-cut when presented with instance $I$.

*Case 2*: The procedure introduces a mid-cut.

As a *mid-cut* is introduced, both $P_l$ and $P_r$ must be nonempty and as both $I_l$ and $I_r$ have fewer points than $P$, we know they satisfy the conditions of the lemma. Combining the conditions of the lemma for $I_l$ and $I_r$, we know that

$$L(E_{DC}(I_l)) + L(E_{DC}(I_r)) + C(I_l) + C(I_r) \leq 4LB(I_l) + 4LB(I_r)$$

By definition, $L(E_{DC}(I)) = L(E_{DC}(I_l)) + L(E_{DC}(I_r)) + USE(I)$ and $LB(I) = LB(I_l) + LB(I_r)$. As $Y \leq X$, we know $USE(I) = Y$. Substituting these equations in Equation 6.1, we have

$$Y + C(I) > C(I_l) + C(I_r). \tag{6.2}$$

There are two cases depending on whether $I$ is regular or irregular.

*Subcase 2.1*: $I$ is regular.

By definition, $C(I) = X + Y$. Substituting in Equation 6.2,

$$Y + C(I) = X + 2Y > C(I_l) + C(I_r).$$

As $I$ is regular and the procedure introduces a mid-cut, both $I_1$ and $I_2$ must also be regular. Therefore, $X + 2Y = C(I_l) + C(I_r)$. A contradiction. So it cannot be that $I$ is regular.

*Subcase 2.2*: $I$ is irregular.

As $Y \leq X$ and $I$ is irregular, we know that $Y + C(I) = 4Y$. Substituting in Equation 6.2, we have $4Y > C(I_l) < 2Y + C(I_r)$. As $I$ is irregular and the algorithm introduces a mid-cut, it must be that $X_l = X_r \geq Y$. But by Lemma 6.2, $C(I_l) \geq 2Y$ or $C(I_r) \geq 2Y$. A contradiction. So it cannot be that the procedure introduces a mid-cut.

*Case 3*: The procedure introduces a nonterminal end-cut.

When a nonterminal end-cut is introduced, exactly one of the two resulting subproblems has no interior points (Figure 6.2c and d). Assume without loss of generality that $I_r$ has no interior points. From the lower bound function and the procedure, we know that

$$LB(I) = LB(I_l) + min\{Y, X_r\} \text{ and } L(E_{DC}(I)) = L(E_{DC}(I_l)) + USE(I).$$

As instance $I_l$ has fewer points than $I$, it then follows that $L(E_{DC}(I_l)) + C(I_l) \leq 4LB(I_l)$. Clearly, $USE(I) = Y$. Substituting these inequalities in Equation 6.1, we know that

$$Y + C(I) > C(I_l) + 4 \, min\{Y, X_r\}. \tag{6.3}$$

By Lemma 6.2, we know $C(I) \leq 3Y$ and $C(I_l) > 0$. So Equation 6.3 is $4Y > 4min\{Y, X_r\}$. Therefore, it cannot be that $Y \leq X_r$ as otherwise there is a contradiction. It must then be that $X_r < Y$. Substituting in Equation 6.3, we have

$$Y + C(I) > C(I_l) + 4X_r. \tag{6.4}$$

Instance $I$ is regular because $X_r < Y$ and $X_r \geq \frac{X}{2}$ implies $X \leq 2Y$. Substituting $C(I) = X + Y$ in Equation 6.4, we know

$$X + 2Y > C(I_l) + 4X_r. \tag{6.5}$$

If $I_l$ is regular, then substituting $C(I_l)$ in Equation 6.5, we know $X + 2Y > X_l + Y + 4X_r$. As $X_l + X_r = X$, Equation 6.5 becomes $Y > 3X_r$. But we know that $X_r \geq X/2$ and $X \geq Y$. So $X_r \geq X/2$. A contradiction. So it must be that $I_l$ is irregular.

On the other hand if $I_l$ is irregular, then as $X_l \leq Y$ substituting $C(I_l)$ in Equation 6.5, we know $X + 2Y > 3X_l + 4X_r$. As $X_l + X_r = X$, we know $2Y > 2X + X_r$. But we know $X \geq Y$ and $X_r > 0$. A contradiction.

This completes the proof of this case and the lemma. ∎

We establish the main result (Theorem 6.1), show that the approximation bound is tight (Theorem 6.2), and explain implementation details needed to establish the time complexity bound for procedure $DC$ (Theorem 6.3).

**Theorem 6.1** *For any instance of the RG-P problem, algorithm DC generates a solution such that* $L(E_{DC}(I)) \leq 4 \cdot L(E_{opt}(I))$.

*Proof.* The proof follows from Lemmas 6.2 and 6.3. ∎

We now show that the approximation bound is asymptotically tight, that is, $L(E_{DC}(I))$ is about $4L(E_{opt}(I))$. The problem instance we use to establish this result has the property that $LB(I) = L(E_{opt}(I))$. Our approach is a standard one that begins with small problem instances and then combines them to build larger ones. As the problem instances becomes larger, the approximation ratio for the solution generated by Procedure $DC$ increases. The analysis just needs to take into consideration a few steps performed by the procedure and the previous analysis for the smaller problems instances.

We define problem instances $P_i$, for $i \geq 0$ as follows: Problem instance $P_i$ consists of a rectangle of size $2^i$ by $2^i$. The instance $P_1$ contains two points. Figure 6.3a and b depicts $E_{DC}(P_1)$ and $E_{opt}(P_1)$, respectively. Clearly, $L(E_{DC}(P_1)) = 4$ and $L(E_{opt}(P_1)) = 2$. In this case, the approximation ratio is 2. Instance $P_2$ is a combination of four instances of $P_1$. Figure 6.3c and d depicts $E_{DC}(P_2)$ and $E_{opt}(P_2)$, respectively. Clearly, $L(E_{DC}(P_2)) = 2^3 + 4L(E_{DC}(P_1)) = 24$ and $L(E_{opt}(P_2)) = 4L(E_{opt}(P_1)) = 8$. The ratio is 3. Problem

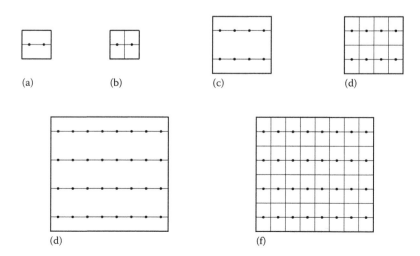

(a)                    (b)                    (c)                    (d)

(d)                                        (f)

**FIGURE 6.3**    (a) $E_{DC}(P_1)$, (b) $E_{opt}(P_1)$, (c) $E_{DC}(P_2)$, (d) $E_{opt}(P_2)$, (e) $E_{DC}(P_3)$, and (f) $E_{opt}(P_3)$.

instance $P_3$ combines four instances of $P_2$ as shown in Figure 6.3c. Figure 6.3e and f depicts $E_{DC}(P_3)$ and $E_{opt}(P_3)$, respectively. Clearly, $L(E_{DC}(P_3)) = 2^4 + 4L(E_{DC}(P_2)) = 112$ and $L(E_{opt}(P_3)) = 4L(E_{opt}(P_2)) = 32$. The ratio is $112/32 = 3.5$. To construct $P_i$, we apply the same combination using $P_{i-1}$. Note that when applying our procedure to $P_i$ it always introduce mid-cuts, except when presented $P_0$ in which case it introduced a terminal end-cut. It is simple to see that our approximation algorithm introduces cuts with length $L(E_{DC}(P_i)) = 2^{i+1} + 4L(E_{DC}(P_{i-1}))$, for $i > 1$ and $L(E_{DC}(P_1)) = 4$. An optimal solution, in this case, is identical to the lower bound function which is $L(E_{opt}(P_i)) = 4L(E_{opt}(P_{i-1}))$, for $i > 1$ and $L(E_{opt}(P_1)) = 2$.

Solving the above-mentioned recurrence relations, we know that

$$L(E_{DC}(P_i)) = \sum_{j=0}^{i-2} 2^{2i-j-1} + 2^{2i-2}L(E_{DC}(P_1)) = 2^{2i+1} - 2^{i+1}, \text{and}$$

$$L(E_{opt}(P_i)) = 2^{2(i-1)}L(E_{opt}(P_1)) = 2^{2i-1}$$

Therefore, the approximation ratio is $L(E_{DC}(P_i))/L(E_{opt}(Pi)) = 4 - 1/2^{i-2}$.

**Theorem 6.2** *There are problem instances for which procedure DC generates a solution such that $L(E_{DC}(I))$ tends to $4 \cdot L(E_{opt}(I))$.*

*Proof.* By the aforementioned discussion. ∎

Procedure $DC$ can be easily modified so that for problem instances with "many" points along the same line, it will introduce a line to cover all the points provided that the line segment has length close to $Y$. For the problem instance given earlier, the modified algorithm will generate a better solution decreasing substantially the approximation ratio. However, as pointed out in Reference 11, there are problem instances for which the modified procedure will generate solutions with a ratio of about $4LB(I)$. The idea is to perturb the points slightly this way, no two points will belong to the same vertical or horizontal line.

The time complexity $T(n)$ for Procedure $DC$ when operating on an instance with $n = |P|$ points is given by the recurrence relation $T(n) \leq T(n - i - 1) + T(i) + cn$, for $1 \leq i < n$ and some constant $c$. However, it is possible to implement the procedure so that it takes $O(n \log n)$ time [5]. In what follows, we briefly describe one of the two implementations given in Reference 5 with the $O(n \log n)$ bound. To simplify the presentation, assume that no two points can be covered by the same vertical or horizontal line. The idea is to change the procedure so that the time complexity term $cn$ becomes $cf(min\{i, n-i-1\})$, for some function $f()$ that we specify in the following. Note that this requires a preprocessing step that takes $O(n \log n)$ time. For certain functions $f()$, this reduces the overall time complexity bound $O(n \log n)$.

In the preprocessing step, we create a multilinked structure in which there is a record (data node) for each point. The record contains the $x$ and $y$ coordinate values of the point. We also include the rank of each point with respect to their $x$ and $y$ values. For example, if the rank of a point is $(i, j)$ then it is the $i$th smallest point with respect to its $x$ value and the $j$th smallest point with respect to its $y$ value. Note that this ranking is with respect to the initial set of points. We also have all of these records in two circular lists, (1) one sorted with respect to the $x$ values and (2) the other sorted with respect to the $y$ values. It is simple to see that this multilinked structure can be constructed in $O(n \log n)$ time.

In each recursive invocation of procedure $DC$, we need to construct $P_l$ and $P_r$ from $P$, and assume that $X \geq Y$. This construction can be implemented to take $O(min\{|P_l|, |P_r|\})$ by scanning the doubly linked lists for the $x$ values from both ends (alternating one step from each end) until all the points at one end are $P_l$ and the other ones are on $P_r$. Assume that $n \geq |P| = m > |P_l| = m - q \geq |P_r| = q$, and $1 < q \leq m/2$. Clearly, the above-mentioned procedure can be used to identify the points in $P_r$ and then remove the points in $P_r$ from $P$ in $O(q)$ time. The remaining points are $P_l$ and are represented according to our structure. But now the problem is that we need to construct a multilinked data structure for the points in $P_l$. These points are already sorted by their $x$ values, but not by their $y$ values. The algorithm given in Reference 13 can sort any $q$ integers in the range $[1, n]$ in $O(q \log \log_q n)$ time. Once we have them sorted we can construct the multilinked data structure for $P_l$. This takes $O(q \log \log_q n)$ time. Let $T_n(m)$ be the total time required by procedure $DC$ when the initial invocation involved a problem with $n$ points and we now have a problem with $m$ points. Clearly, $T_n(1) = O(1)$ and $T_n(m) = \max\{O(q \log \log_q n) + T_n(q) + T_n(m - q) | 1 \le q \le m/2\}$ for $m > 1$. By the analysis of [5], $T_n(n) = O(n \log n)$. This result is summarized in the following theorem:

**Theorem 6.3** *Procedure DC and its preprocessing step can be implemented to take $O(n \log n)$ time.*

*Proof.* By the aforementioned discussion. ∎

# 6.3 Dynamic Programming Approach

The divide-and-conquer algorithm $DC$ presented in the previous section introduces guillotine cuts by following a set of simple rules, which makes the algorithm run very efficiently. But such guillotine cuts do not form an optimal guillotine partition. It seems natural that using an optimal guillotine partition would generate a better solution. But how fast can one generate an optimal guillotine partition? By the recursive nature of guillotine partitions, it is possible to construct an optimal guillotine partition in polynomial time. In this section, we analyze an approximation algorithm based on this approach. As we shall see the algorithm has a smaller approximation ratio, but it takes longer to generate its solution.

## 6.3.1 Algorithm

Let $I = (R = (X, Y), P)$ be any problem instance and let the $x$- and $y$-interval define the rectangle $R_{x,y}$, which is part of rectangle $R$. We use $g(R_{x,y})$ to represent the length of an optimal guillotine partition for $R_{x,y}$.

First, it is important to esablish that there is always an optimal guillotine partition such that all its line segments include at least one point from $P$ inside them (excluding those at its endpoints). This is based on the observation that given any optimal partition that does not satisfied this property it can either be transformed to one that does satisfy the property or one can establish that it is not an optimal guillotine partition. The idea is to move each horizontal (vertical) guillotine cuts without points from $P$ inside them either to the left or right (up or down) without increasing the total edge length. Note that when the previous operation is perfomed, some vertical (horizontal) segments need to be extended and some need to be retracted to preserve a guillotine partition. If the total edge length decreases, then we know that it is not an optimal guillotine partition and when it remains unchanged after making all the transformations, it becomes an optimal guillotine partition in which all its guillote cuts include at least one point from $P$.

By applying the above-mentioned argument, we know that for any rectangle $R_{x,y}$ one can easily compute $g(R_{x,y})$ recursively by selecting the best solution obtained by trying all $2n$ guillotine cuts (that include a point from $P$) and then solving recursively the two resulting problem instances. It is simple to show that there are $O(n^4)$ different $g$ that are needed to be computed. By using dynamic programming and the aforementioned recurrence relation the length of an optimal guillotine partition can be constructed in $O(n^5)$ time. By recording at each step a guillotine cut forming an optimal solution and using the $g$ values, an optimal guillotine partition can be easily constructed within the same time complexity bound. The following theorem follows the earlier arguments.

**Theorem 6.4** *An optimal guillotine partition for the RG-P problem can be constructed in $O(n^5)$ time.*

## 6.3.2 Approximation Bound

The set of line segments in a feasible rectangular partition of $I$ is denoted by $E(I)$, and a set of line segments in a minimum length guillotine partition is denoted by $E_G(I)$. We use $L(E)$ to represent the length of the edges in a rectangular partition $E$. In what follows, we show that $L(E_G(I)) \leq 2L(E(I))$ by introducing a set of vertical and horizontal line segments $A(I)$ such that $E(I) \cup A(I)$ is a guillotine partition and $L(E(I) \cup A(I)) \leq 2L(E(I))$. The approximation ratio follows from the fact that $L(E_G(I)) \leq L(E(I) \cup A(I))$. The set $A(I)$ of additional segments are introduced by the transformation Procedure *TR1*. Before we present this procedure and its analysis, we define some terms.

A horizontal (vertical) line segment that partitions rectangle $R$ into two rectangles is called a *horizontal (vertical) cut*. We say that $E(I)$ has a *horizontal guillotine cut*, if there is a horizontal cut, $l$, such that $L(E(I) \cap l) = X$. A rectangular partition $E(I)$ has a *half horizontal overlapping cut* if there is a horizontal cut $l$ such that $L(E(I) \cap l) \geq 0.5X$. *Vertical guillotine cuts* are defined similarly.

Suppose $R$ is partitioned by a vertical or horizontal cut into two rectangles, $R_l$ and $R_r$. With respect to this partition we define $E(I_l)$ and $E(I_r)$ as the subset of line segments of $E(I)$ inside $R_l$ and $R_r$, respectively. We use $E_h(I)$ ($E_v(I)$) to denote all the horizontal (vertical) line segments in $E(I)$. With respect to $A(I)$, we define similarly $A_h(I)$ and $A_v(I)$.

Assume that $E(I)$ is nonempty. The idea behind Procedure *TR1* is to either introduce horizontal or vertical line segments at each step and then apply recursively the procedure to the nonempty problem instances. When a horizontal line segment is introduced, it is added along a half horizontal overlapping cut. The two resulting problems will not have any of these segments inside them. So at this step the added horizontal segments have length at most equal to the ones of the half horizontal overlapping cut.

Since there are problem instances without half horizontal overlapping cuts, Procedure *TR1* checks to see if there is a vertical guillotine cut. In this case, we just partition the rectangle along this cut. Clearly, there are no additional line segments for this case.

There are rectangular partitions without a half horizontal overlapping cut or a vertical guillotine cut. In this case, Procedure *TR1* introduces a *mid-cut*, which is just a vertical line that partitions the rectangle along its center. The problem now is that this mid-cut does not necessarily overlap with any of the segments in $E_v(I)$. Our approach is to remember this fact and later on identify a set of line segments in $E_v(I)$ that will account for the length of this mid-cut. To remember this fact we will color the right side of rectangle $R_l$. As we proceed in the recursive process different parts of this colored side will be inherited by smaller rectangles that will get their right side colored. At some point later on when we pay for the segment represented by a colored right side of a rectangle, it will no longer appear in recursive calls resulting from the one for this rectangle. The budget in this case is two times the total length of the vertical line segments in $E_v(I)$. This budget should be enough to pay for the total length of the vertical line segments introduced during the transformation.

In order to be able to show that $A_v(I) \leq E_v(I)$, it must be that the rectangles in terminal recursive call should not have their right side colored. Before we establish this result, we formally present Procedure *TR1* that defines the way in which colors are inherited in the recursive calls.

---

**Procedure** *TR1*$(I = (R = (X, Y), E(I)))$;
    **case**
      :$E(I)$ is empty: **return**;
      :There is a half horizontal overlapping cut in $E(I)$:
          Partition the rectangle $R$ along one such cut;
          If the right side of $R$ is colored, the right sides of rectangles $R_l$ and $R_r$ remain colored;
      :There is a vertical guillotine cut in $E(I)$:
          Partition $R$ along one such cut;
          Remove the color, if any, of the right side of rectangle $R_r$;
     : **else**:
          Partition $R$ by a vertical cut intersecting the center of $R$; // introduce a mid-cut
          If the right side of $R$ is colored, the right side of $R_r$ remains colored;
          Color the right side of $R_l$;
    **endcase**

Apply *TR1* recursively to $(I_l = (R_l, E(I_l)))$;
Apply *TR1* recursively to $(I_r = (R_r, E(I_r)))$;
**end of Procedure TR1**

It is important to remember that Procedure *TR1* is only used to establish our approximation bound. The right side of rectangle $R$ is not colored in the initial invocation to Procedure *TR1*. Before establishing the approximation ratio of two in Theorem 6.5, we prove the following lemmas:

**Lemma 6.4** *Every invocation of Procedure TR1 $(I, E(I))$ satisfies the following conditions:*

(a) *If $E(I)$ is empty, then the right side of $R$ is not colored.*
(b) *If the right side of $R$ is colored, then $E(I)$ does not have a horizontal guillotine cut.*

*Proof.* Initially, the right side of $R$ is not colored, so the the first invocation $(I, E(I))$ satisfies conditions (a) and (b). We now show that if upon entrance to the procedure $(I, E(I))$ the two conditions (a) and (b) were satisfied, then the invocations made directly from it will also satisfy (a) and (b). There are three cases depending on the type of cut introduced by procedure *TR1*, which are described in the following:

*Case 1*: Procedure *TR1* partitions $R$ along a half horizontal overlapping cut.
     First consider the subcase when $E(I)$ has a horizontal guillotine cut. From (b) we know that the right side of $R$ is not colored, and the algorithm does not color the right side of $R_l$ nor $R_r$. Therefore, the two invocations made directly from this call satisfy properties (a) and (b).
     On the other hand, when $E(I)$ does not have a horizontal guillotine cut, then the right side of $R$ may be colored. As $E(I)$ does not have a horizontal guillotine cut, we know that there is at least one vertical line segment on each side of the half horizontal overlapping cut, so $E(I_l)$ and $E(I_r)$ must be nonempty. As neither of these two partitions has a horizontal guillotine cut, each invocation made directly by our procedure satisfy properties (a) and (b). This completes the proof for this case.

*Case 2*: Procedure *TR1* partitions $R$ along a vertical guillotine cut.
     As the algorithm does not color the right side of $R_l$ and $R_r$, then the invocations made directly by the procedure satisfy (a) and (b). This completes the proof of this case.

*Case 3*: Procedure *TR1* introduces a mid-cut.
     As $E(I)$ does not have a half horizontal overlapping cut and there is no vertical guillotine cut, then each of the resulting rectangular partitions has at least one vertical line segment and there are no horizontal guillotine cuts in the two resulting problems. Therefore, both of the resulting problem instances are not empty and do not have a horizontal guillotine cut. This implies that both problem instances satisfy (a) and (b). This completes the proof for this case and the lemma. ∎

**Lemma 6.5** *For any nonempty rectangular partition $E(I)$ of any instance $I$ of the RG-P problem, procedure TR1 generates a set $A(I)$ of line segments such that $L(A_h(I)) \leq L(E_h(I))$ and $L(A_v(I)) \leq L(E_v(I))$.*

*Proof.* First we show that $L(A_h(I)) \leq L(E_h(I))$. This is simple to prove because horizontal cuts are only introduced over half horizontal overlapping cuts. Each time a horizontal cut is introduced, the segments added to $A_h(I)$ have length that is at most equal to the length of the segments in the half horizontal cut in $E_h(I)$. As these line segments are located on the boundary of the two resulting instances, these line segments will not account for other segments in $A_h(I)$ and thus $L(A_h(I)) \leq L(E_h(I))$.

Let us now establish that $L(A_v(I)) \leq L(E_v(I))$. It is simple to verify that Procedure *TR1* does not color a side more than once, and all empty rectangular partitions do not have their sides colored (Lemma 6.4). Every invocation of Procedure *TR1* with a nonempty rectangular partition $R$ generates two problem instances whose total length of their colored right sides is at most the length of the right side of $R$, if it is colored, plus the length of the vertical line segment introduced. The only exception is when the procedure introduces a cut along an existing vertical guillotine cut. In this case, if the right side of $R$ is colored, the right sides of two resulting partitions will not be colored. So the cost of a line segment previously

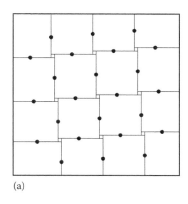
(a)

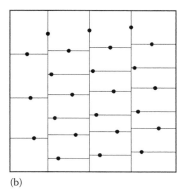
(b)

**FIGURE 6.4** (a) Optimal rectangular partition. (b) Optimal guillotine partition.

introduced in $A(I)$, which is recorded by the fact that the right side of $R$ colored, is charged to the existing guillotine cut and such cut will not be charged another segment again. Therefore, $L(A_v(I)) \leq L(E_v(I))$. This concludes the proof of the lemma. ∎

**Theorem 6.5** *The length of an optimal guillotine partition is at most twice the length of an optimal rectangular partition, that is, $L(E_G(I)) \leq 2L(E_{opt}(I))$.*

*Proof.* Apply procedure *TR1* to any rectangular partition $E(I)$. By Lemma 6.5 we know that $L(E(I) \cup A(I)) \leq 2L(E_h(I)) + 2L(E_v(I)) = 2L(E(I))$. As $E(I) \cup A(I)$ is a guillotine partition, $L(E_G(I)) \leq L(E(I) \cup A(I))$. Hence, $L(E_G(I)) \leq 2L(E(I))$. ∎

It is simple to find a problem instance $I$ such that $L(E_G(I))$ is about 1.5 $L(E_{opt}(I))$ [8]. One of such problem instances has the distribution of points shown in Figure 6.4. As the number of points increases, the ratio $\frac{L(E_G(I))}{L(E_{opt}(I))}$ approaches 1.5.

## 6.3.3 Improved Approximation Bound

In the current section, we describe the idea behind the complex proof given in Reference 9 that establishes the fact that $L(E_G(I)) \leq 1.75L(E_{opt}(I))$. The proof is based on a recursive transformation procedure *TR2* that, when performed on any rectangular partition $E(I)$, generates a set $A(I)$ of line segments such that $E(I) \cup A(I)$ forms a guillotine partition (of course $A(I) \cap E(I) = \emptyset$). Without loss of generality, assume that $L(E_v(I)) \leq L(E_h(I))$. The transformation is performed in such a way that

$$L(A_v(I)) \leq L(E_v(I)), \tag{6.6}$$

and

$$L(A_h(I)) \leq 0.5 \cdot L(E_h(I)). \tag{6.7}$$

Then, we have

$$
\begin{aligned}
L(E_G(I)) &\leq L(A(I) \cup E(I)) \\
&= L(A_h(I)) + L(A_v(I)) + L(E(I)) \\
&\leq 0.5 \cdot L(E_h(I)) + L(E_v(I)) + L(E(I)) \\
&= 0.5 \cdot L(E_v(I)) + 1.5L(E(I)) \\
&\leq 1.75 \cdot L(E(I)).
\end{aligned}
$$

To satisfy Equation 6.6, the notion of *vertical separability* is introduced. Denote the $x$-coordinate of a vertical segment $l$ by $x(l)$. A vertical cut $l$ is *left* (*right*) *covered* by $E_v(I)$ if for every point $p$ on $l$ there exists a line segment $l'$ in $E_v(I)$ such that $x(l') \leq x(l)$ ($x(l') \geq x(l)$), and there is a point $p'$ on $l'$ with $x(p') = x(p)$.

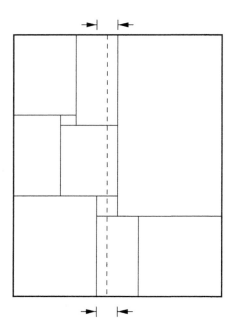

**FIGURE 6.5**  A vertically separable $E(I)$. $E(I)$ has no guillotine cut. The dashed segment is a vertical through cut of $E(I)$. In fact, any vertical cut in the region marked by the vertical segments outside of rectangular boundary is a vertical through cut for $E(I)$.

A vertical cut is called a *vertical through cut* if it is both left and right covered by $E_v(I)$. The set of segments $E(I)$ is said *vertically separable* if there exists at least one vertical through cut. Note that a vertical guillotine cut is also a vertical through cut, but the converse is not necessarily true. Figure 6.5 shows a vertically separable $E(I)$ and a vertical through cut. When a new vertical line segment is introduced by *TR2*, it is ensured to be a portion of a vertical through cut.

To satisfy Equation 6.7, horizontal segments are carefully introduced by *TR2* according to the structure of $E(I)$. When there is neither guillotine cut and nor vertical through cut in $E(I)$, a three-step subprocedure *HVH_CUT* is invoked to introduce new segments. Procedure *TR2* is given in the following:

---

**Procedure** *TR2(E(I))*;
    **case**
      :$E(I)$ is empty:
        **return**;
      :$E(I)$ has a guillotine cut $l$:
        Partition $E(I)$ along $l$ into $E(I_1)$ and $E(I_2)$;
        Recursively apply *TR2* to $E(I_1)$ and $E(I_2)$;
      :$E(I)$ is vertically separable:
        Let $l$ be any vertical through cut;
        Partition $E(I)$ along $l$ into $E(I_1)$ and $E(I_2)$;
        Recursively apply *TR2* to $E(I_1)$ and $E(I_2)$;
      :**else**:
        Use Procedure *HVH_CUT* to partition $E(I)$ into $E(I_1), E(I_2) \cdots E(I_{q+1})$;
        Recursively apply *TR2* to each $E(I_i)$;
    **endcase**
  **end of Procedure TR2**

---

Procedure *HVH_CUT* is outlined in the following:

---

**Procedure** *HVH_CUT(E(I))*;

1. Introduce a carefully selected set of $q$ horizontal cuts that partition $E(I)$ into $q + 1$ vertically separable rectangular subpartitions.
2. In each resulting partition of Step 1, introduce either the leftmost or the rightmost vertical through cut.
3. Divide each resulting partition of Step 2 into two rectangular subpartitions by a horizontal guillotine cut if such a cut exists.

**end of Procedure HVH_CUT**

---

Let $H(I)$ be the set of horizontal cuts introduced in Step 1, $V(I)$ the set of vertical through cuts chosen in Step 2, and $H_3(I)$ the horizontal guillotine cuts found in Step 3. Define $H_1(I) = H(I) \cap E_h(I)$ and $H'_1(I) = H(I) - E_h(I)$. Note that $H_3(I) \subset E_h(I)$. The sets $H(I)$ and $V(I)$ are selected in such a way

$$L(H'_1(I)) \leq 0.5 \cdot qX \tag{6.8}$$

and

$$L(H_1(I) \cup H_3(I)) \geq qX. \tag{6.9}$$

It is quite complex to establish such that $H(I)$ and $V(I)$ always exist [9]. We omit additional details of the procedure *HVH_CUT* and the related proofs. Vertical though cuts introduced by *TR2* are carefully selected to satisfy Equation 6.6 and ensure Equation 6.9. As all horizontal cuts introduced by *HVH_CUT* satisfy Equations 6.8 and 6.9, and all horizontal cuts introduced not by invocations of *HVH_CUT* are horizontal guillotine cuts in their respective subrectangular boundaries, Equation 6.7 is satisfied. Consequently, $L(E_G(I)) \leq 1.75 \cdot L(E(I))$. As $E(I)$ is an arbitrary rectangular partition, we have $L(E_G(I)) \leq 1.75 \cdot L(E_{opt}(I))$. The next theorem sums up the discussion.

**Theorem 6.6** *The length of an optimal guillotine partition is at most 1.75 times the length of an optimal rectangular partition, that is, $L(E_G(I)) \leq 1.75L(E_{opt}(I))$.*

*Proof.* The full details of the proof, whose outline is given earlier, appear in Reference 9. ∎

## 6.4 Concluding Remarks

In this chapter, we considered the *RG-P* problem. We presented a fast divide-and-conquer approximation algorithm with approximation ratio 4. This ratio is smaller than the $3 + \sqrt{3}$ ratio of a similar divide-and-conquer approximation algorithm given [7]. We also examined in detail the approach of approximating optimal rectangular partitions via optimal guillotine partitions. Optimal guillotine partitions can be constructed in polynomial time using dynamic programming [3]. For this approach, we presented a proof that the approximation ratio is at most 2. Our proof is simpler than the proof of for the same bound given in Reference 3. We presented the idea behind a complex proof given in Reference 9 that establishes that the approximation ratio is at most 1.75 when approximating optimal rectangular partitions via optimal guillotine partitions. Both proofs are based on the technique of recursively transforming a rectangular partition into a guillotine partition by introducing additional line segments. The difference is that the additional segments in the transformation for the approximation ratio of 1.75 are more carefully selected than those introduced by the transformation for bound of 2. Whether or not one can further reduce this bound remains a challenging open problem.

For the *RG-P* problem, the partitions obtained by the divide-and-conquer algorithms (the one of Reference 7 and the one presented in this chapter) and the dynamic programming algorithm are guillotine partitions formed by recursive guillotine cuts. The approach examined in this chapter can be referred to as approximating optimal rectangular partitions via optimal and suboptimal guillotine partitions. The first approximation algorithm for this problem based on suboptimal guillotine partitions appeared in Reference 7. Subsequently, optimal guillotine partitions were used in Reference 8. Both of these approaches were generalized to multidimensional space in References 11, 12.

The term "guillotine cut" was introduced in the 60s in the context of cutting stock problems. In the context of rectangular partitions, it was first used in Reference 8. As pointed out in Reference 14, the concept of guillotine partition has been generalized into a powerful general approximation paradigm for solving optimization problems in different settings. The guillotine partition algorithms for the *RG-P* problem were among the first that manifested the power of this paradigm.

# References

1. Lingas, A., Pinter, R.Y., Rivest, R.L., and Shamir, A., Minimum edge length partitioning of rectilinear polygons, in *Proceedings of the 20th Allerton Conference on Communication, Control, and Computing,* Monticello, IL, 1982.
2. Rivest, R.L., The "PI" (placement and interconnect) system, in *Proceedings of the Design Automation Conference,* 1982.
3. Du, D.Z. and Chen, Y.M., On fast heuristics for minimum edge length rectangular partition, Technical Report, Mathematical Sciences Research Institute, UC Berkeley, Berkeley, CA, 03618-86, 1986.
4. Levcopoulos, C., Minimum length and thickest–first rectangular partitions of polygons, in *Proceedings of the 23rd Allerton Conference on Communication, Control, and Computing,* Monticello, IL, 1985.
5. Levcopoulos, C., Fast heuristics for minimum length rectangular partitions of polygons, in *Proceedings of the 2nd ACM Symposium on Computational Geometry,* 1986.
6. Lingas, A., Heuristics for minimum edge length rectangular partitions of rectilinear figures, in *Proceedings of the 6th GI-Conference,* Dortmund, Germany, LNCS, 195, Springer-Verlag, 1983.
7. Gonzalez, T.F. and Zheng, S.Q., Bounds for partitioning rectilinear polygons, in *Proceedings of ACM Symposium on Computational Geometry,* 1985, p. 281.
8. Du, D.Z., Pan, L.Q., and Shing, M.T., Minimum edge length guillotine rectangular partition, Technical Report, MSRI 02418-86, 1986.
9. Gonzalez, T.F. and Zheng, S.Q., Improved bounds for rectangular and guillotine partitions, *J. Symb. Comput.,* 7, 591, 1989.
10. Gonzalez, T.F. and Zheng, S.Q., Approximation algorithms for partitioning rectilinear polygons with interior points, *Algorithmica,* 5, 11, 1990.
11. Gonzalez, T.F., Razzazi, M., and Zheng, S.Q., An efficient divide-and-conquer algorithm for partitioning into *d*-boxes, *Int. J. Comput. Geom. Appl.,* 3(4) 417, 1993. (condensed version appeared in *Proceedings of the 2nd Canadian Conference on Computational Geometry,* 1990, p. 214).
12. Gonzalez, T.F., Razzazi, M., Shing, M., and Zheng, S.Q., On optimal *d*-guillotine partitions approximating hyperrectangular partitions, *Comput. Geom. Theor. Appl.,* 4(1), 1, 1994.
13. Kirkpatrick, D.G., An upper bound for sorting integers in restricted ranges, in *Proceedings of the 18th Allerton Conference on Communication, Control, and Computing,* Monticello, IL, 1980.
14. Cardei, M., Cheng, X., Cheng, X., and Du, D.Z., A tale on guillotine cut, in *Proceedings of Novel Approaches to Hard Discrete Optimization,* ON, Canada, 2001.

# 7

# Automatic Placement of Labels in Maps and Drawings

Konstantinos
G. Kakoulis

Ioannis G. Tollis

## 7.1 Introduction

Traditionally, the placement of labels has been central in the framework of cartography. Today, due to the significant growth in the area of information visualization the placement of labels may be seen through a wider scope. Labels are textual descriptions, and thus another means to convey information or clarify the meaning of complex structures. The problem of automatic label placement is important [1], and has applications in many areas including information visualization [2], cartography [3], geographic information science (GIS) [4], and graph drawing [5,6].

For centuries, cartographers have manually perfected the art of placing names corresponding to graphical features of maps. The computer revolution has changed many of the long-established practices in map production. However, one aspect of cartography that has not been fully automated is the name placement or lettering process. Even today, this aspect of map production requires demanding manual work to produce high-quality results. There are two main reasons for that: (1) It is very difficult to quantify all the characteristics of a good label placement as they reflect human visual perception, intuition, and experience, which have been perfected through the centuries by cartographers who have elevated the placement of labels into an art and (2) most of the optimization problems associated with the placement of labels are computationally hard. A practical approach to produce high-quality map labels might be a semiautomated interactive name placement system. Yet, fully automated name placement may be applied in the context of online GIS- or internet-based map search and special-purpose technical maps. Significant progress has been made toward automating the label placement process in map production in the last three decades.

Today, dynamic and interactive maps are gaining enormous popularity and there are numerous application fields and devices on which they are used. Dynamic digital maps, in which users can navigate by continuously zooming, panning, or rotating, opened up a new era in cartography and GIS, from professional applications to personal mapping services on mobile devices. Even though much effort has been placed in map labeling, the dynamic map labeling has received surprisingly little attention. Furthermore, commercial online maps and navigation devices use inferior algorithms to place labels [7].

A key task in information visualization is the annotation of the information through label placement. Graph drawings visualize information by constructing geometric representations of conceptual structures that are modeled by graphs, networks, or hypergraphs. Graphs may be used to represent any information that can be modeled as objects and connections between those objects. The need for displaying labels, related to graphical features of a graph drawing, is essential in many applications, for example, in software engineering (call graphs, class hierarchies), database systems (entity-relationship diagrams), VLSI (symbolic layout), electronic systems (block diagrams, circuit schematics), and project management (PERT diagrams, organization charts) [5,6]. The label placement problem is well studied in the context of the graph drawing [2].

In the following sections, we present the labeling problem not only in its traditional form (i.e., cartography), but also in the context of information visualization (e.g., graph drawings).

# 7.2 The Label Placement Problem

In the label placement problem, we assign overlap-free labels to each graphical feature of a given map or drawing. The goal is to communicate the information for each graphical feature via text labels in the best possible way, by positioning the labels in the most appropriate place. Each label position that is part of a final label assignment is associated with a cost which reflects the severity of the violation of the basic label placement rules established mainly by cartographers. Thus, the label placement problem can be viewed as an optimization problem where the objective is to find a label assignment of minimum total cost.

Good label placement aids in conveying the information that labels represent and enhances the esthetics of the input map or drawing. A critical task in the labeling process is to decide the best position for a label with respect to its corresponding graphical feature. What constitutes a good label assignment has its roots in the art of cartography. For many centuries, cartographers have perfected the art of placing labels on maps.

It is difficult to quantify all the characteristics of a good label placement, because they reflect human visual perception and intuition. It is trivial to place a label when its associated object is isolated. The real difficulty arises when the freedom to place a label is restricted by the presence (in close proximity) of other objects of the drawing. Then, we must consider not only the position of a label with respect to its associated object, but also how it relates to other labels and objects in the surrounding area.

In a successful label assignment, labels must be positioned such that they are legible and follow basic esthetic quality criteria [8,9]:

*Clarity:* A label must not overlap other labels or graphical features.

*Readability:* A label must have legible size.

*Associativity:* A label must be easily identified with exactly one graphical feature.

*Quality:* A label must be placed in the most preferred position, among all legible positions.

In the production of geographical maps, we rank label positions according to rules developed through years of experience of manual placement. These rules typically capture the esthetic quality of label positions. When the graphical objects to be labeled belong to a technical map or drawing, then, usually, a different set of rules govern the preferred label positions. These rules depend on the particular application. It is important to emphasize here that for drawings other than geographical maps, the user must be able to customize the rules of label quality to meet specific needs and/or expectations. Therefore, any successful labeling algorithm must have the capability to take into account the user's preferences.

## 7.3 Label Placement Algorithms for Maps and Drawings

Most of the research addressing the labeling problem has been focused on labeling graphical features of geographical and technical maps. The label placement problem is typically partitioned into three tasks: (a) labeling points (e.g., cities), (b) labeling lines (e.g., roads or rivers), and (c) labeling areas (e.g., lakes or oceans).

The general labeling problem, the *graphical feature label placement* (GFLP) problem (where a graphical feature can be a node, edge, or area), has been addressed primarily in the context of cartography; however, it has direct applications in the area of graph drawing [10–17].

The problem of assigning labels to a set of points or nodes, the *node label placement* (NLP) problem, has been studied in depth [3,18–24].

The problem of assigning labels to a set of lines or edges, the *edge label placement* (ELP) problem, has been addressed in the context of geographical and technical maps [13,24–26] and graph drawing [27,28].

The problem of assigning labels to areas, also known as the *area label placement* (ALP) problem, has been addressed in References 8–10, 14, 29–33.

In many practical applications, each graphical feature may have more than one label. The need for assigning multiple labels is necessary not only when objects are large or long, but also when it is necessary to display different attributes of an object. This problem is known as the *multiple label placement* (MLP) problem and has been addressed in References 14, 31, 34–37.

The labeling process is not allowed to modify the underlying geometry of geographical and technical maps that are fixed. However, one can modify a graph drawing to accommodate the placement of labels. The problem of modifying a graph drawing to open up space that can be used to assign labels without compromising the quality of the drawing, the so-called opening space label placement (OSLP) problem, has been addressed in References 38, 39.

In special-purpose technical drawings and medical atlases, it is common to place the labels on the boundary of the drawing and to connect them to the features they describe by curves. The problem of placing labels on the boundary of a drawing that contains the graphical features, known as the *boundary label placement* (BLP) problem, is studied in References 40–51.

Dynamic maps, in which users can navigate by continuously zooming, panning, or rotating, are essential in online map services and portable navigation devices. The problem of placing labels, at any scale, related to graphical features of dynamic maps, the *dynamic label placement* (DLP) problem, is addressed in References 7, 52–67.

The problem of labeling moving points on a static map, which has applications in annotating radar screens, is addressed in References 68–70.

The problem of annotating 3D scenes, which has many applications in virtual and augmented reality systems and relates to dynamic labeling, is addressed in References 71–77.

In References 78–80, algorithms that combine the layout and labeling process of orthogonal drawings of graphs are presented.

Finally, an alternative approach for displaying edge labels is presented in Reference 81. Each edge is replaced by its corresponding edge label in the drawing. The label font starts out larger from the source node and shrinks gradually until it reaches the destination node. The tapered label also indicates the direction of the edge.

It is worth noting that the NLP [19,82,83], ELP [84], BLP [42], OSLP [85], and DLP [7,52,54] problems are NP-hard. As automatic labeling is a very difficult problem, we rely on heuristics to provide practical solutions for real-world problems.

A number of types of algorithms have been used to solve the labeling problem such as greedy algorithms [3,20], exhaustive search algorithms [11,12,14], algorithms that simulate physical models (i.e., simulated annealing [13]), algorithms that reduce the labeling problem to a variant of 0-1 integer programming [21,22,24], algorithms that restrict the labeling problem to a variant of the 2-SAT problem [19,23], approximation algorithms [86], tabu search [87], genetic algorithms [88,89], and algorithms that transform the labeling problem into a matching problem [16,28,35].

## 7.3.1 Graphical Feature Label Placement

The general labeling problem (i.e., the labeling of points, lines, and areas), also known as the GFLP problem, has been addressed in the context of cartography and graph drawing [10–17].

Most algorithms for the GFLP problem are based on local and exhaustive search [10,11,13–15]. These algorithms first create an initial label assignment in which conflicts between labels are allowed. Then conflicts are resolved by repositioning assigned labels until all conflicts are resolved, or no further improvement can be achieved. One disadvantage of these algorithms is that they have the tendency to get trapped in local optima; in addition, they use inferior optimization techniques [3,24]. Actually, these methods use a rule-based approach to evaluate good label placement and variants of depth-first search to explore different labeling configurations. Edmondson et al. [13] improved considerably the optimization process by applying simulated annealing, which allows occasionally single label moves that worsen the quality of the labeling in the hope of avoiding bad labelings that happen to be locally optimal. Furthermore, this technique separates the cartographic knowledge needed to recognize the best label positions from the optimization procedure needed to find them.

All of the above-mentioned techniques for the general labeling problem start with a rather small initial set of potential label positions from which they derive a final label assignment. The performance of these techniques decreases considerably when the number of potential label positions increases.

Kakoulis and Tollis [16] transformed the labeling problem into a matching problem. The general framework of this technique is flexible and can be adjusted for particular labeling requirements. In Figure 7.1, the labeling assignment is produced by this technique. The basic steps for this labeling technique are given in the following:

1. A set of discrete potential label solutions for each graphical feature is carefully selected.
2. This set of labels is reduced by removing heavily overlapping labels. The remaining labels are assigned to groups, such that, if two labels overlap then they belong to the same group.
3. Finally, labels are assigned by solving a variant of the matching problem, where at most one label position from each group is part of the solution.

Next, the three main steps of the basic labeling algorithm are presented in detail.

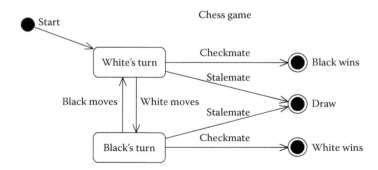

**FIGURE 7.1** An UML state diagram that makes use of labels for states and transitions as well as the title of the diagram.

### 7.3.1.1 Creating Discrete Label Positions

To find a set of discrete label positions for each graphical feature, a number of heuristics can be used.

For points, a number of label positions that touch their corresponding point are defined. In most algorithms, a set of four or eight potential label positions is associated with each point [3] (Figure 7.2).

In the framework of cartography, area labels must follow the general shape of their corresponding area and must be inside the boundaries of the area (Figure 7.3). For each area, a number of potential label positions is defined according to the techniques described in References 14, 31–33.

In graph drawings, edge labels touch the edges that they belongs to (Figure 7.4a–c). Label positions that intersect their corresponding edge are acceptable but not preferable (Figure 7.4). Simple heuristics

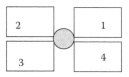
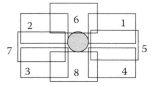

**FIGURE 7.2**  Typical potential label positions for a point or node. Lower values indicate more desirable positions.

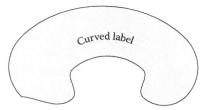

**FIGURE 7.3**  Typical area label.

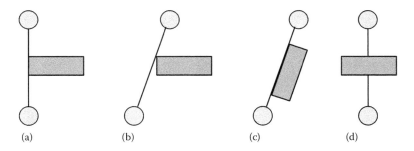

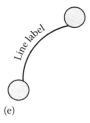

**FIGURE 7.4**  Label positions for lines and edges.

for finding an initial set of edge label positions can be found in References 16, 28. In the framework of cartography, linear labels must follow the shape of the line (Figure 7.4). For each linear feature, a number of potential label positions is defined according to the techniques described in References 11, 13, 24–26, 31, 33, 90, 91.

### 7.3.1.2 Refining the Set of Potential Labels

The size of the initial set of label positions must be kept reasonably small as it affects the performance of any labeling algorithm.

In order to reduce the set of label positions, an intersection graph $G_i$ is first created, where each label position is a node in $G_i$, and if two label positions intersect then there is an edge in $G_i$ connecting their corresponding nodes.

Then, in a preprocessing step, the set of potential labels is refined by removing label positions that add to the complexity of the problem without potentially improving the solution or by assigning labels in obvious cases. Some simple but efficient heuristics for refining the set potential labels are given in the following (for more details see References 16, 28):

- If a graphical feature $F$ has only one potential label, then assign that label to $F$.
- If a label position $l$ of $F$ is free of overlaps, then any less desirable label position of $F$ is removed.
- If two pairs of label positions of two different graphical features overlap, then both graphical features may have a label free of overlaps assigned to them. Hence, any other label position of these two graphical features can be safely removed from the set of potential labels.

The remaining labels (nodes in $G_i$) are assigned to groups, such that, if two labels overlap then they belong to the same group, whereas heavily overlapping labels are removed. The goal of the third step of the algorithm is to select at most one label from each group as part of the solution. This way, the algorithm will produce a label assignment free of overlaps.

In principle, the set of potential label positions is further refined by removing heavily overlapping labels while maintaining a large number of potential labels for each object $f$ by keeping track of the number of labels associated with $f$. In the end, the goal is to reduce the intersection graph into a set of disconnected subgraphs.

A simple and very successful (according to experiments) technique for removing overlapping labels is as follows: If a subgraph $c$ of $G_i$ must be split, then the node with the highest degree is removed from $c$, unless that node corresponds to a label position of some object with very few label positions. In that case, the next highest degree node from $c$ is removed. This process is repeated until either $c$ is split into at least two disjoint subgraphs, or $c$ is complete.

### 7.3.1.3 Matching Labels to Graphical Features

First, a bipartite *matching* graph $G_m(V_f, V_c, E_m)$ is created as follows: (i) Each node in $V_f$ corresponds to a graphical feature; (ii) each node in $V_c$ corresponds to a group of overlapping labels; and (iii) each edge $(i, j)$ in $E_m$ connects a node $i$ in $V_f$, to a node $j$ in $V_c$, if and only if the graphical feature that corresponds to $i$ has a label position that is a member of the group that corresponds to $j$.

The cost of assigning label $l$ to graphical feature $f$ can be the weight of edge $(f, l)$ in $G_m$. Therefore, a maximum cardinality minimum weight matching for graph $G_m$ will give us an optimal (maximum number of labels with minimum cost) label assignment with no overlaps with respect to the *reduced* set of label positions. Figure 7.5 shows a label assignment produced by the matching technique for the GFLP problem.

## 7.3.2 Node Label Placement

The problem of assigning labels to a set of points or nodes, also known as the NLP problem, has been extensively studied in the context of automated cartography, and many successful algorithmic approaches

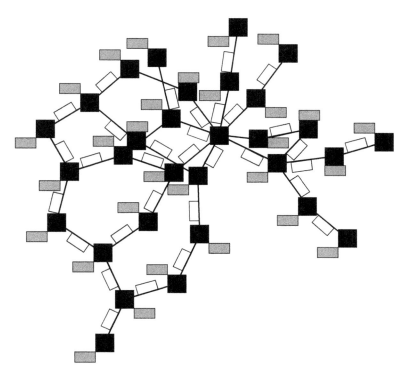

**FIGURE 7.5** A force-directed drawing where labels are positioned by the matching technique for the GFLP problem. The gray boxes are node labels and the white boxes are edge labels.

have been introduced [3,18–24]. Furthermore, any of the techniques for solving the general labeling problem (Section 7.3.1) can be applied successfully in solving the NLP problem.

Algorithms based on local and exhaustive search [11,12,14] and simulated annealing [13] are well suited for solving the NLP problem. Experimental results [3] have shown that simulated annealing outperforms all algorithms based on local and exhaustive search. In addition, simulated annealing is one of the easiest algorithms to implement.

These algorithms start with a rather small initial set of potential label positions from which they derive a final label assignment. This is because the size of the initial set of label positions plays a critical role in the performance of these algorithms. This precondition works well when solving the NLP problem. For example, each point is given at most four or eight potential label positions (Figure 7.2).

Approximation algorithms for restricted versions of the NLP problem are presented in References 18, 19. Specifically, the approach of 19 assigns labels of *equal* size to all points while attempting to maximize the size of the assigned labels. The work in Reference 23 improves the results in Reference 18 by using heuristics. A similar approach has been taken in Reference 18. In effect, finding the maximum label size is equivalent to finding the smallest factor by which the map has to be zoomed out such that each point has a label assigned to it. These are very interesting results. However, it is not clear how these techniques can be modified to solve real-world problems, including the labeling of graphical features of graph drawings, where the label size is usually predefined and the labels are not necessarily of equal size.

Another approach to solve the NLP problem is based on the sliding model, where sliding labels can be attached to the point in which they label anywhere on their boundary. This model was first introduced in Reference 20 who gave an iterative algorithm that uses repelling forces between labels to eventually find a placement of labels. A polynomial time approximation scheme and a fast factor-2 approximation algorithm for maximizing the number of points that are labeled by axis-parallel sliding rectangular labels of common height, based on the sliding model, are presented in Reference 92.

### 7.3.3 Edge Label Placement

The problem of assigning labels to a set of lines or edges, also known as the ELP problem, has been addressed in the context of geographical and technical maps [11,13,24–26,31,33,90,91] and graph drawing [27,28]. Furthermore, any of the techniques for solving the general labeling problem (Section 7.3.1) can be applied in solving the ELP problem.

In the context of geographical and technical maps, edges are linear features. Labels should be placed along (or inside) rivers, boundaries, roads, or linear features. If the linear feature is curved, the shape of the label must follow the curvature of the linear feature (Figure 7.4). The positioning of linear labels has the greatest degree of freedom as labels can be placed almost anywhere along the linear feature; thus, cartographers have focused their effort in finding the right shape of a linear label.

However, in the context of graph drawing, placing labels to edges is a more complicated process. Edges are not necessarily long; they are usually straight lines or polygonal chains, and they have to follow user preferences and specifications. For example, an edge label might be related to the source node of the edge; thus, it must be placed closer to the source node rather than the target node to avoid a misleading label assignment.

In Reference 27, a labeling system is presented that includes a very functional interface and labeling engine that addresses the ELP problem in the context of a graph drawing editor. It is noteworthy that the interface of that system allows the user to set the labeling preferences interactively.

Kakoulis and Tollis [28] proposed a fast and simple technique for solving the problem of positioning text or symbol labels corresponding to edges of a graph drawing (Figure 7.6 for an example). This technique is based on the matching technique for solving the general labeling problem presented in Section 7.3.1 and works best for labels that are parallel to the horizontal axis and have approximately equal height and arbitrary width. The main idea of this technique is explained the following:

First, a set of label positions is produced. Next, label positions are grouped such that each label position that is part of a group overlaps every other label position that belongs to the same group. Then, edges are matched to label positions by allowing at most one label position from each group to be part of a label assignment by using a fast matching heuristic.

The initial set of label positions is carefully constructed by dividing the input drawing into consecutive horizontal strips of equal height. The height of each strip is equal to the height of the labels. Next, a set of label positions for each edge is found. Each label position must be inside a horizontal strip. Labels are slid inside each horizontal strip until a label touches its edge. A label position is considered only if it

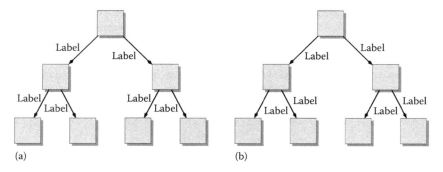

(a)                                                          (b)

**FIGURE 7.6**  A hierarchical drawing with edge labels produced by the fast ELP algorithm. (a) The preferred label positions are to the left side of each edge. (b) The preferred label positions are to the right side of each edge.

does not overlap any other graphical feature except its corresponding edge and other labels that belong to the same horizontal strip. Next, an intersection graph $G_i$ and a bipartite *matching* graph $G_m$ are created (Section 7.3.1 for details). Finally, a maximum cardinality matching of graph $G_m$ assigns labels to the maximum number of edges with respect to the initial set of label positions.

A fast heuristic that solves the maximum cardinality matching problem for the matching graph $G_m$ in linear time is presented in Reference 28. This heuristic takes advantage of the simple structure of the matching graph $G_m$. Each node of $G_m$, which corresponds to a group of overlapping labels, has degree at most two. As, by construction, each label position overlaps at most one other label position, it is worth noting that the technique described earlier is especially suited for straight-line and hierarchical drawings.

### 7.3.4 Area-Label Placement

The problem of assigning labels to areas, also known as the ALP problem, has been addressed in the context of geographical and technical maps [8–10,14,29–33]. It is worth noting that areal labels are the first to be placed in a map, since they are less flexible than node and edge (line) labels.

The accepted practice for placing a label associated to an area, especially in the framework of cartography, is to have the label span the entire area and conform to its shape, as shown in Figure 7.3. Most of the research addressing the ALP problem has been focused in finding the right shape and position of areal labels. The most common practice is to place an areal label on the skeleton (medial axis) of the area, which represents the geometric shape of the area as a polyline. A technique for assigning rectangular labels within a polygon is described in Reference 33.

### 7.3.5 Multiple Labels Placement

Many algorithms exist for the labeling problem; however, very little work has been directed toward positioning many labels per graphical feature in a map or drawing [14,31,34–37].

In map production, the need for multiple labels usually arises when a graphical feature covers a large area of a map. It is mostly related to labeling linear graphical features (e.g., streets, rivers, or boundary lines of areas). For example, if the graphical feature to be labeled is a road in a map, a text label with the name of this road must be placed in regular intervals. This problem is referred to as the street signing problem (i.e., one street sign close to each intersection). The need for assigning multiple labels is necessary not only when graphical features are large or long, but also when it is necessary to display different attributes of an object.

In existing automated name placement systems for geographic maps, simple techniques have been utilized to address the multiple labels per graphical feature problem. Specifically, each feature to be labeled is partitioned into as many pieces as the number of labels for that feature. Then, labeling algorithms for single label per graphical feature may be applied to the new set of partitioned graphical features. Most labeling algorithms are based on local or exhaustive search. Thus, their performance (running time and quality of solutions) is sensitive to the size of the graphical features to be labeled and to the density of the drawing. Clearly, if each graphical feature in a drawing is associated with $i$ labels then the size of the problem becomes $i$ times larger; therefore, techniques based on local or exhaustive search might be slow even for small instances.

In References 36, 37, algorithms for labeling points with two labels for each point are presented. In References 34, 35, the MLP problem is treated in the context of graph drawing. A framework for evaluating the quality of label positions is presented. In addition, two algorithmic schemes are presented: (1) A simple and practical iterative technique and (2) a flow-based technique, which is an extension of the matching technique presented in Section 7.3.1. In the following section, these techniques will be presented in detail.

### 7.3.5.1 An Algorithm for the MLP Problem

In the production of geographical maps, label positions are ranked according to rules developed through years of experience with manual placement. These rules typically capture the esthetic quality of label positions. For technical maps or drawings, one must be able to customize the rules of label quality to meet a user's specific needs and/or expectations. Furthermore, when graphical features are associated with more than one label, then one must take into account how labels for the same graphical feature influence each other.

Next, some constraints are presented that may be used to ensure that each label is unambiguous and easily read and recognized, when more than one label is associated with a graphical feature. These constraints can be divided into three general categories: (1) *proximity*, (2) *partial order*, and (3) *priority*. To illustrate the three sets of constraints, we will use as an example the labeling of a single edge with two labels. Label $l_s$ is associated with the source node and label $l_t$ is associated with the target node.

*Proximity:* Label $l_s$ ($l_t$) must be in close proximity with the source (target) node to avoid ambiguity. Therefore, it is necessary to define a maximum distance from the source (target) node that label $l_s$ ($l_t$) must be positioned.

*Partial order:* A label associated with the source (target) node must be closer to the source (target) node than the other label to avoid ambiguity. Thus, in many cases, it is appropriate to define a partial order among the labels according to some invariants (e.g., $x$ or $y$ axis, distance from a fixed point).

*Priority:* In many cases, it is impossible to assign all labels associated with a graphical feature due to the density of the drawing. Then, the user might prefer to have the important labels assigned first and then might assign the rest of the labels if there is available space.

A simple iterative algorithm for solving the MLP problem consists of a main loop, which is executed as many times as the number of labels per graphical feature. For example, at the $i$th execution of the loop, the $i$th label is assigned to each graphical feature. Labeling algorithms for single label per graphical feature may be used at each execution of the loop. One can refine this technique by first finding a set of label positions before entering the loop and then executing only the step of positioning labels inside the loop.

This technique can take into account all three sets of constraints (proximity, partial order, priority). Thus, it leads to very practical solutions for the MLP problem. This iterative approach is especially suited for the labeling algorithms presented in References 16, 28, because they first find a set of label positions, and then they produce a label assignment in a single step without any repositioning of labels. Figure 7.7 shows label assignments produced by the simple iterative algorithm.

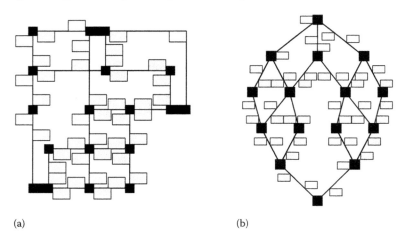

(a)                                               (b)

**FIGURE 7.7**   An orthogonal (a) and a hierarchical (b) drawing with two labels per edge, positioned by the iterative algorithm.

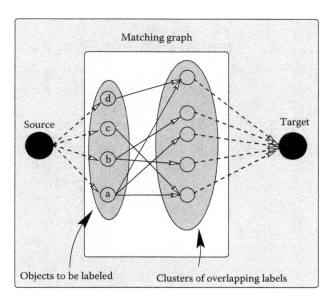

**FIGURE 7.8** The flow graph.

The matching technique presented in Section 7.3.1 can be further extended to support the placement of more than one label per graphical feature of a graph drawing in a noniterative fashion. It can solve the MLP problem by using flow-based techniques.

First, the matching graph $G_m$ is created (Section 7.3.1 for more details). Next, the matching graph $G_m$ is transformed into a flow graph $G_{flow}$ by inserting a source node $s$ and a target node $t$ (Figure 7.8). Then, capacities to each edge of the flow graph $G_{flow}$ are assigned in the following way: (1) Each edge of the original matching graph has capacity one, (2) each edge incident to the target node has capacity one, and (3) each edge $(s, v)$ incident to the source node has capacity equal to the number of labels associated with the graphical feature of the input graph that is represented by node $v$ in $G_m$.

Finally, a maximum flow of graph $G_{flow}$ will produce a maximum cardinality label assignment with respect to the set of labels encoded in the matching graph in $O(n\, m\, \log\, n)$ time. One point that needs to be emphasized is that the framework just described can take into account the cost of a label assignment with respect to priority, proximity, and esthetic criteria.

Experimental results have shown that both algorithms perform essentially the same with respect to the success rate of assigning labels, even though the matching technique produces a slightly better quality of label assignments.

## 7.3.6 Boundary Label Placement

Boundary labeling addresses the problem of placing large labels on the boundary of a rectangle that contains the graphical features, also known as the BLP problem. Each feature is connected to its associated label through a curve, called *leader*. A leader connects its label to an *anchor* point which marks a point of the graphical feature to be labeled. Clearly, the leaders should not intersect with each other to avoid confusion. Labels can be attached to one, two, and four sides of the rectangle by using different types of leaders. Efficient boundary labeling algorithms aim at producing overlap-free labels, leader-bend minimization, and leader-length minimization. Generally speaking, the boundary labeling problem is simpler than the general labeling problem. However, when labels are placed to more than one side of the bounding rectangle, then for many of the variations of the boundary labeling problem the optimization process is computationally hard [42].

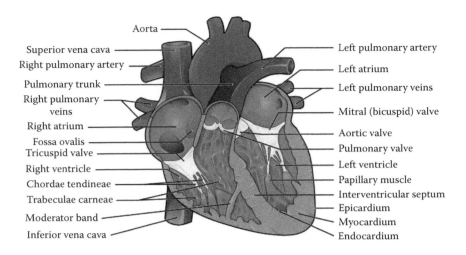

Aorta

Superior vena cava

Right pulmonary artery

Pulmonary trunk

Right pulmonary
veins

Right atrium

Fossa ovalis
Tricuspid valve

Right ventricle

Chordae tendineae

Trabeculae carneae

Moderator band

Inferior vena cava

Left pulmonary artery

Left atrium

Left pulmonary veins

Mitral (bicuspid) valve

Aortic valve
Pulmonary valve

Left ventricle

Papillary muscle

Interventricular septum

Epicardium

Myocardium

Endocardium

**FIGURE 7.9**    Anterior view of the heart.

In practice, large labels are common in technical drawings and illustrations in medical atlases. For example, a medical map depicting the anatomy of the human heart is shown in Figure 7.9[*]. If the labels were placed next to the features they describe, then certain parts of the map would have been obscured. Therefore, it is desirable to keep the map as clear as possible by placing the labels on the boundary of the map.

Bekos [93] was the first to study in depth the boundary labeling problem. Bounady labeling models can be classified by the shape of the leaders and the sides of the rectangle at which labels can be placed. The basic types of leaders are described in the following (Figure 7.10):

*s-leader:* The leader is a straight line.

*po-leader:* The first-line segment of a leader is parallel (*p*) to the side of the rectangle and the second-line segment is orthogonal (*o*) to that side.

*opo-leader:* The first-line segment of a leader is orthogonal (*o*) to the side of the rectangle, the second-line segment is parallel (*p*) to that side, and the third-line segment is orthogonal (*o*). The two bends are inside a strip *S* next to the label. The strip *S* is called the track-routing area of the rctangle.

*do-leader:* The first-line segment of a leader is diagonal (*d*) to the side of the rectangle and the second-line segment is orthogonal (*o*) to that side.

Leaders in automated map labeling were probably used for the first time in References 94, 95. Fekete et al. [46] presented the first algorithm for labeling dense maps by extending the infotip mechanism, which supplies the user with additional information about screen objects whenever the mouse pointer rests a certain amount of time in their vicinity (Figure 7.11). In that algorithm, a circle of fixed radius is drawn around the current cursor position, the so-called focus circle, and the points that fall into the circle are labeled by axis-parallel rectangles that contain the names associated with the points. Labels are placed in two stacks to the left and the right of the circle.

For drawing the leaders, they present two main approaches. In the first approach, they order the labels within each stack according to the vertical order of the corresponding sites. In this approach, leaders may cross. In the second approach, to connect a point with its label, they use a nonorthogonal leader that

---

[*] This figure is licensed under a Creative Commons Attribution 4.0 International License. You can also download for free at http://cnx.org/contents/14fb4ad7-39a1-4eee-ab6e-3ef2482e3e22@8.24.

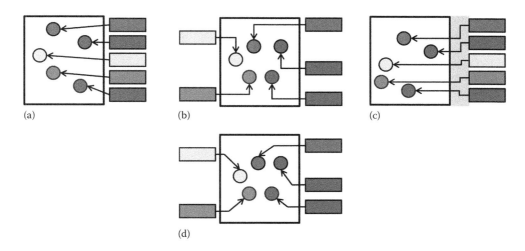

**FIGURE 7.10** Illustration of leader types: (a) *s*-leaders, (b) *po*-leaders, (c) *opo*-leaders, and (d) *do*-leaders.

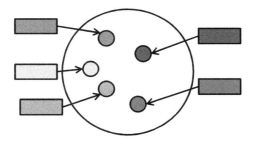

**FIGURE 7.11** A focus style boundary labeling.

consists of two line segments: one radially from the point to its projection on the focus circle and one from there to the midpoint of the left edge of the corresponding label. This approach guarantees that no two leaders intersect. In the worst case, they may overlap within the focus circle.

Ali et al. [40] introduced heuristics to label 3D illustrations using *s* and *po* style leaders on the boundary of a rectangle or a circle (focus labeling). Layout-specific algorithms determine initial positions for labels. Then, label overlaps are eliminated by using spring-embedding techniques [6]. Possible leader intersections are resolved by interchanging label positions.

Bekos et al. [44] presented efficient labeling algorithms for *s*-, *po*-, and *opo*-type leaders. As objective functions, they considered minimizing the number of bends and the total leader length. The basic framework for producing *opo* style boundary labels, in $O(n\log n)$ time, is as follows:

1. Labels are stacked, in increasing order with respect to the $y$-coordinate of their corresponding anchor points, next to one side of the bounding rectangle. Labels are placed on top of each other.
2. Then, each anchor point is connected with a horizontal segment to the side of the bounding rectangle.
3. Finally, the track-routing area is used to lay out the remaining parts of the leaders from the side of the bounding rectangle to the label sides (Figure 7.10). A simple way to determine the $x$-coordinate of the vertical leader segments, which are placed in the track-routing area, is to offset from the boundary the index of the anchor points in a bottom to top ordering.

An $O(n^2)$-time dynamic programming algorithm minimizes the total number of leader bends. To attach labels to all sides of the bounding rectangle $R$, they partition $R$ into four disjoint regions such that the algorithms for *opo* type leaders can be applied to each region separately. In addition, the number of labels for each side of $R$ is predefined.

Many variations of the boundary labeling problem have been studied in depth in the recent years: *po*-labeling [45,50], *do*-labeling [45], dynamic *po*-labeling, in which the user can zoom and pan the map view [51], flexible label positions [49], labeling focus regions using *s*-leaders and leaders based on Bezier curves [47], vertical *s*-labeling for panorama images [48], and multistack labeling [43]. Finally, Barth et al. [41] presented a formal user study on the readability of boundary labeling.

It is worth noting that when labels can be placed to more than one sides of a bounding rectangle $R$, most algorithms use a rather naive way to decide on which side of $R$ each label will be placed. Actually, this problem corresponds to the well-known partition problem, which is NP-complete [96]. Usually, in illustrations of maps, the number of boundary labels to be placed is small and optimal partition of these labels is not that important. However, when many objects have to be labeled (e.g., technical maps [95]), finding an optimal partition of the label positions is essential.

# 7.4 Label Placement Algorithms for Drawings

Automatic labeling is a very difficult problem, and as we rely on heuristics to solve it, there are cases where the best methods available do not always produce an acceptable or legible solution, even if one exists. Furthermore, there are cases where no feasible solution exists. Given a specific drawing and labels of a fixed size, it might be impossible to assign labels without violating any of the basic rules of a good label assignment (e.g., label-to-label overlap, legibility, and unambiguous assignment). These cases often appear in practical applications when drawings are dense, labels are oversized, or the label assignment must meet the minimum requirements set by the user (e.g., font size or preference of label placement).

To solve the labeling problem where the best solution that we can obtain is either incomplete or not acceptable, one must modify the drawing. This approach cannot be applied in drawings that represent geographical or technical maps where the underlying geometry is fixed by definition. However, the coordinates of nodes and edges of a given graph drawing or diagram can be changed, as it is the result of an algorithm that draws the graph.

Generally speaking, there can be two algorithmic approaches in modifying the layout of a graph drawing as follows:

- Modify the existing layout of a graph drawing to make room for the placement of labels.
- Produce a new layout of a graph drawing that integrates the layout and labeling processes.

In References 78–80, algorithms that combine the layout and labeling process of orthogonal drawings of graphs are presented. In Reference 80, the authors studied the problem of computing a grid drawing of an orthogonal representation of a graph with labeled nodes and minimum total edge length. They showed an integer linear programming formulation of the problem and present a branch-and-cut-based algorithm that combines compaction and labeling techniques. The work in Reference 78 made a further step in the direction defined in Reference 80 by integrating the topology-shape-metrics approach with algorithms for edge labeling. In Reference 79, an approach to combining the layout and labeling process of orthogonal drawings is presented. Labels are modeled as dummy nodes and the topology-shape-metrics approach is applied to compute an orthogonal drawing where the dummy nodes are constrained to have fixed size.

The problem of modifying a graph drawing to open up space that can be used to assign labels without compromising the quality of the drawing, the so-called OSLP problem, is an optimization problem. The goal is to minimize the extra area needed to resolve overlaps while preserving the constraints and esthetics of the drawing. In any approach to solve the OSLP problem, the following two things are of particular importance:

- The embedding of the drawing must not change as the drawings before and after opening extra space should look similar. The reason is
  - We want to maintain the mental map of the drawing.
  - We do not want to reduce the quality of the drawing by violating any of the layout constraints.
- Opening up the minimum amount of space.

The notion of the mental map of a drawing has been introduced in Reference 97. By preserving the user's mental map of a drawing, we generally mean that a user who has studied a drawing will not have to reinvest significant time and effort to visualize or understand the modified drawing.

In References 38, 85, algorithms that modify an existing layout of a graph drawing to make room for the placement of labels are presented.

Hu [38] presented an algorithm that modifies an existing force-directed graph drawing to resolve edge label overlaps. Next, to better understand this algorithm, we describe the most influential framework for creating force-directed graph drawings introduced by Eades [98]. A graph is modeled as a physical system where the nodes are rings and the edges are springs that connect pairs of rings. The rings are placed in some initial layout and the spring forces move the system to a local minimum of energy. Attractive forces are applied only between pairs of adjacent nodes. Repulsive forces act between every pair of nodes to avoid node overlaps. A node will move to its new location based on the sum of those attractive and repulsive forces applied to the node.

It is worth noting that force-directed graph drawings are esthetically pleasing, exhibit symmetries, and produce well-balanced distribution of nodes and few edge crossings. They have applications in many areas of visualization of complex structures, such as, bioinformatics, enterprise networking, knowledge representation, system management, and mesh visualization.

In Reference 38, label overlaps are resolved by applying an algorithm based on the techniques used to produce force-directed drawings. Generally speaking, attractive and repulsive forces are applied to push overlapping objects apart. This algorithm iteratively moves the nodes and labels to remove overlaps, while keeping the relative positions between them as close to those in the original layout as possible, and edges as straight as possible. Because force-directed drawings are often unpredictable, the final drawing might not preserve all the mental map criteria with respect to the initial drawing.

Kakoulis and Tollis [85] presented algorithms that modify an existing orthogonal graph drawing by inserting extra space to accommodate the placement of overlap-free node and edge labels. First, a label assignment is computed, where overlaps are allowed, using existing techniques. Then, the drawing is modified by applying polynomial time algorithms based on plane sweep and minimum flow techniques to find the extra space needed to eliminate label overlaps, while preserving the mental map of the drawing. Furthermore, they showed that the OSLP problem is NP-Complete even for simple cases. In the next section, these algorithms are presented in detail.

## 7.4.1 Opening Space Label Placement

First, we give some necesary definitions. In a graph drawing, nodes are represented by symbols, such as circles or boxes, and each edge is represented by a simple open curve. A *polyline* drawing maps each edge into a polygonal chain. An *orthogonal* drawing is a polyline drawing in which each edge consists only of horizontal and vertical segments. An *orthogonal representation* captures the notion of the orthogonal shape of planar orthogonal drawings by taking into account bends along each edge and angles around each node, but disregarding edge lengths. An orthogonal representation describes an equivalence class of orthogonal drawings with a similar shape; more details can be found in References 5, 99. Two planar orthogonal drawings have the same orthogonal representation if each of the following rules is true:

- Both drawings have the same sequence of left and right turns (bends) along the edges.
- Two edges incident at a common node determine the same angle.

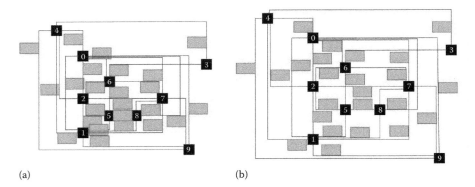

(a)                                                    (b)

**FIGURE 7.12**    (a) An orthogonal drawing with overlapping edge labels and (b) edge label overlaps have been resolved by inserting extra space using our flow technique.

Orthogonal drawings have been studied extensively [5] because they produce drawings with high clarity and readability, as they contain only horizontal and vertical lines. They have many applications in key areas, including software engineering, database design, circuit schematics, and logic diagrams.

For orthogonal drawings, we can preserve the mental map of a drawing by requiring that the orthogonal representation of a drawing remains the same after opening space. The drawings (a,b) in Figure 7.12 have the same orthogonal representation. It is obvious that both drawings have similar shapes, and the mental map between those two drawings is similar.

Now, a practical solution for the OSLP problem is presented. Given an orthogonal drawing $\Gamma$, first an edge label assignment is found where overlaps are allowed by using existing techniques [13,16]. Then, $\Gamma$ is modified by locally blowing up part of the drawing (i.e., inserting rows and/or columns), so that the existing label assignment becomes overlap-free while preserving the orthogonal representation of the drawing. Next, we expand on the latter step of the technique that resolves label overlaps.

### 7.4.1.1 Resolving Overlaps

Generally speaking, a two-phase technique is used to resolve overlaps. During Phase 1, overlaps are resolved in the $x$-direction. Analogously, in Phase 2, overlaps are resolved in the $y$-direction. First, a partial order of all graphical features in the drawing (nodes, edges, and labels) is defined. Next, plane sweep and flow techniques are used to find the minimum width and/or height needed to push apart objects in the drawing, so that all label overlaps are eliminated.

The OSLP problem has many similarities with the compaction problem in VLSI layout [100,101]. Lengauer [101] has pointed out that what makes the compaction problem computationally hard is that it is difficult to decide the partial order of two objects that overlap. Once a partial order for all objects is given, the compaction problem is solved in polynomial time if, in addition, objects are allowed to move only in one direction [100].

Next, it is shown that if the partial order of overlapping objects is given and labels in the drawing are allowed to move only in one direction (horizontal or vertical), then the minimum area needed to resolve overlaps can be found in polynomial time while preserving the orthogonal representation of the input drawing. We will refer to it as the *one-dimensional separation* problem.

We note the partial order of two objects $i$ and $j$ as $i \prec j$ when object $i$ is to the left (respectively bottom) of $j$. In the one-dimensional separation problem, parts of the drawing are not allowed to be swapped to preserve the partial order, and insertion of new bends or crossings is not allowed to preserve the orthogonal representation. Generally speaking, the only action that is allowed is stretching the edges in the drawing to allocate space that allows overlapping labels to separate. Furthermore, the following assumptions are made to simplify the description: Labels are isothetic rectangles; the drawing is planar, if the drawing has crossings, then a dummy node is inserted at each intersection to augment the input drawing into a planar one; we consider the separation of label overlaps in the $x$-direction, that is, we increase only the width of the drawing to resolve overlaps.

Next, plane sweep and flow techniques are presented for solving the one-dimensional separation problem.

### 7.4.1.2 The Plane Sweep Method

To solve the one-dimensional separation problem, a partial order of all objects in the drawing (nodes, edges, and labels) is defined. Then, enough extra columns (horizontal space) are inserted between consecutive objects in the partial order, such that these objects are separated. Finally, a compaction step is performed in the drawing.

Before the assignment of the partial order of labels, one needs to decide if overlaps will be resolved by increasing the width or height of the drawing. Actually, in some cases, overlaps can be resolved only by increasing the height (width) of the drawing. Usually, the height (width) is increased when labels have a greater overlap in the horizontal (vertical) direction.

The task of defining the partial order of the objects in the drawing is not trivial, as an optimal partial order will produce an optimal solution for the OSLP problem. To define the partial order of the labels, each label is associated with a face and an edge segment of the drawing by following some simple rules: If a label overlaps more than one face of the drawing then the label is placed in the face where it overlaps the minimum area of other objects; labels that overlap the boundaries of the face that they belong to are retracted inside that face; a label is associated with one edge segment.

Next, each label is associated with a vertical line segment (e.g., the left or right vertical side of the rectangle corresponding to the label). Then, nodes and vertical edge segments that are connected and shared the same column in the drawing are grouped into single vertical objects, to preserve the orthogonal representation of the drawing. Next, the plane is swept from left to right, and each time a vertical object is encountered for the first time, it is placed at the end of a queue. The order in which each object is placed in the queue reveals its position in the partial order.

Then, a plane sweep of the drawing is performed from left to right with a vertical line. Each time the sweep line touches a vertical segment corresponding to a label $l$, columns are inserted to separate $l$ from all objects to the right of the sweep line (i.e., all objects that follow $l$ in the partial order are pushed to the right). This ensures that all overlaps will be eliminated.

Finally, a compaction step is performed on the resulting drawing to minimize the width of the drawing. There are efficient techniques that optimally perform one-dimensional compaction (i.e., compaction in one direction for objects with a predefined partial order) in polynomial time [100–102]. These techniques are based on the longest path method. It is trivial to conclude that the plane sweep algorithm runs in $O(n + k + l)$ time, where $n$ is the number of vertices, $k$ is the number of vertical edge segments of the orthogonal representation, and $l$ is the number of edge labels.

One weak point of the longest path method is that it has the tendency to push all objects toward one side. Thus, it produces not only drawings that have long edges, but also drawings that are very dense on one side, while space on the other side remains unused. The previous technique, even though it is very attractive as it is fast and simple, produces esthetically inferior drawings. However, the esthetic quality of a drawing is critical; thus, one must preserve the mental map of the drawing after inserting extra space to resolve label overlaps. In the next section, a method is presented for solving the one-dimensional separation problem that produces not only minimum width, but also minimum edge length and evenly, spaced drawings based on flow techniques.

### 7.4.1.3 The Flow Method

In this section, minimum flow techniques are used to find the minimum width needed to eliminate label overlaps in an orthogonal drawing $\Gamma$, given a partial order of overlapping objects, while preserving the orthogonal representation of $\Gamma$. The case of finding minimum height is analogous.

Generally speaking, a directed acyclic graph $G_{flow}$ transfers flow from the top to the bottom of the drawing to insert extra vertical space to resolve horizontal overlaps. Intuitively, if two objects overlap, then we must push between them at least as much flow as the amount of their overlap in the $x$-direction. $G_{flow}$ captures all possible routes that flow can be pushed such that all horizontal overlaps are resolved. As the orthogonal representation of the drawing must be preserved, space is inserted in such a way that the path of inserted flow does not intersect any vertical edge segment. It is important to note that the space opened at the top of the drawing can be reused by objects below, and the goal is to maximize the reuse of the space.

First, the input drawing is decomposed into vertical edge segments, nodes, and labels. Next, all horizontal edge segments are removed. Then, nodes and edge segments that are connected and share the same column in the drawing are grouped into single vertical objects, to preserve the orthogonal representation of the drawing.

Next, the partial order of the objects (vertical objects, nodes, and labels) is obtained from the decomposed drawing by performing a plane sweep similar to the one described in Section 7.4.1.2.

Then, a *separation visibility graph* $G_{sv}$ is created as follows: For each vertical segment and node of the decomposed input drawing, a node is inserted in $G_{sv}$. For each edge label $l$, which is a rectangle, two nodes and two horizontal edges are inserted in $G_{sv}$. The nodes correspond to the vertical sides of $l$ and the edges to the horizontal sides of $l$. Next, edges are inserted in $G_{sv}$ by expanding (to the left and to the right) the horizontal sides of each node in $G_{sv}$ until they touch another node of $G_{sv}$. Each edge $e$ of $G_{sv}$ is assigned a weight, which represents the minimum distance the two objects connected with edge $e$ must be kept apart to avoid overlaps.

Next, a *flow* graph $G_{flow}$ is created from graph $G_{sv}$ by adding into $G_{flow}$: (i) A source node $s$ and a target node $t$, (ii) a node for each face of $G_{sv}$, and (iii) an edge for each pair of neighboring faces of $G_{sv}$ that share a horizontal edge segment. For each edge in $G_{flow}$, lower and upper capacities are assigned, which is equal to a pair of weights for the only edge in $G_{sv}$ it intersects.

The minimum flow of $G_{flow}$ gives the minimum width of the drawing, in $O(m \log n \, (m + n \log n))$ time, such that all label overlaps are resolved in one direction and the orthogonal representation of the drawing is preserved.

Figure 7.13 illustrates the ordered set of objects corresponding to the input drawing with overlapping labels shown in Figure 7.13. The corresponding separation visibility graph is shown in Figure 7.14.

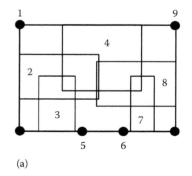

(a)

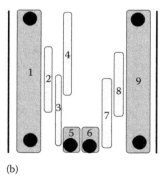

(b)

**FIGURE 7.13** (a) An input drawing with label overlaps and (b) its ordered vertical objects.

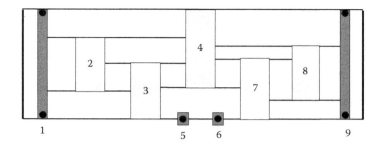

**FIGURE 7.14** The separation visibility graph of the ordered vertical objects as shown in Figure 7.13.

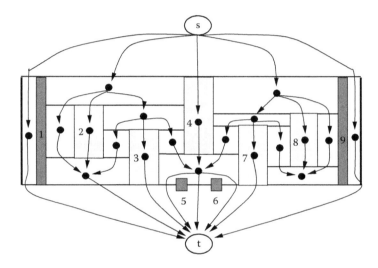

**FIGURE 7.15** The flow graph of the separation visibility graph as shown in Figure 7.14.

Notice that two vertical line segments (nodes in $G_{sv}$) are inserted, one to the left and one to the right of the set of ordered objects. By adding these two objects, each face of the separation visibility graph becomes a rectangle. Finally, the corresponding flow graph $G_{flow}$ is shown in Figure 7.15.

There are cases where overlaps of labels can be resolved by inserting extra space in both or only in one direction. For the orthogonal drawing shown in Figure 7.12, overlaps are resolved by applying the flow technique and inserting horizontal and vertical extra space (Figure 7.12). Furthermore, for the orthogonal drawing shown in Figure 7.13, by running the algorithm once for each direction, we get better results. In Figure 7.16, we resolve all overlaps in the x-direction; in Figure 7.16, we resolve all overlaps in the y-direction; and in Figure 7.16, we resolve overlaps of two objects in the direction that they intersect less. For example, if two objects overlap more horizontally than vertically, then we resolve overlaps by increasing the height of the drawing.

It is worth noting that the flow technique, in addition to resolving overlaps, can produce high-quality label placement with respect to user's preferences and esthetic criteria. For example, to keep or place an edge label close to its associated node, we could assign a constant value to the upper and lower capacity of the edge in the flow graph that separates the node from each associated edge label.

These methods can be trivially extended to assign labels to nodes and edges of an orthogonal drawing.

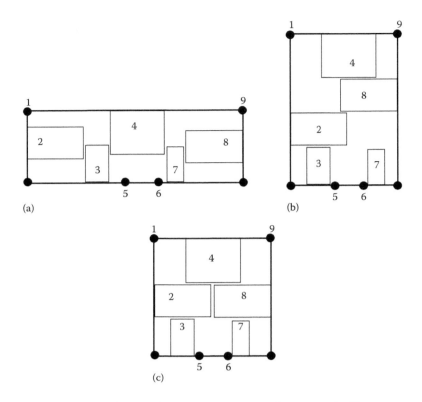

**FIGURE 7.16**  For the orthogonal drawing shown in Figure 7.13, overlaps are resolved by increasing: (a) only the width, (b) only the height, and (c) the width and height.

## 7.5 Label Placement Algorithms for Dynamic Maps and Drawings

Dynamic maps, in which users can navigate by continuously zooming, panning, or rotating, are essential in online map services and portable navigation devices. Dynamic labeling is the problem of placing labels, at any scale, related to graphical features of dynamic maps, also known as the DLP problem. Even though much effort has focused on automating and optimizing the placement of labels in maps and drawings, the dynamic map labeling has received surprisingly little attention. Modern dynamic displays demand interactive and fast labeling.

The problem of labeling dynamic maps is computationally hard [7,52,54]. It is worth noting that the dynamic map-labeling problem can be viewed as the traditional map labeling with the addition of scale as a third dimension. Generally speaking, dynamic labeling is probably a harder problem to be solved than traditional map labeling for two main reasons: (1) dynamic labeling has an extra dimension that adds complexity to the problem and makes harder to extend traditional labeling techniques to dynamic labeling and (2) instantaneous placement of labels is essential to support interaction; thus, real-time algorithms for the dynamic labeling problem need to be devised.

Petzold et al. [60,61] presented an efficient dynamic labeling technique which supports zooming and panning in real time. They use a computationally intensive preprocessing step to generate a data structure, the so-called reactive conflict graph, that represents possible label overlaps for all scales. For any fixed scale and map region, an overlap-free labeling can be computed quickly using heuristics that exploit the reactive conflict graph. Continuous movement and zooming, however, are not explicitly supported by their methods and may lead to sudden discrete changes of label positions. Poon and Shin [62] presented an algorithm for labeling points that precomputes solutions for a number of scales and interpolation

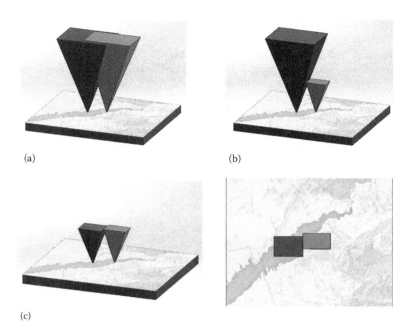

**FIGURE 7.17** The dymanic labeling framework: (a) the selectable ranges of the labels, (b) the active ranges of the labels, and (c) the largest zoom scale where both labes will be placed.

between these produces solutions for any scale. However, popping and jumping effects can occur during zooming.

Been et al. [52] introduced a formal model for the dynamic labeling problem. First, a set of desired characteristics for providing consistency in dynamic map labeling are presented:

1. Labels should not vanish when zooming in or appear when zooming out.
2. The position and size of a label should change continuously under the pan and zoom operations.
3. Labels should not vanish or appear during panning (except for sliding in or out of the view area).
4. The labeling should not depend on the user's navigation history.

Then, a labeling framework is defined to address the additional dimension of scale. Each label is represented by a 3D-solid, with scale as the third dimension (Figure 7.17). For each solid (label), a single-scale interval is defined, called its *active range*, which corresponds to the scales at which the label will be selected. The active range of a label begins at zero and extends just to the point where its solid would overlap the first solid of a higher priority label. At that top end of the active range the solid is truncated, thus eliminating that label from view at all higher scales (Figure 7.17). By representing the labels as 3D-solids, the DLP problem is transformed to the *active range optimization* (ARO) problem where the goal is to select active ranges so that no two truncated solids intersect and the sum of the heights of the active ranges is maximized. It is worth noting that the label size is proportional to scale, which means the screen size of the label is invariant under zooming.

Finally, a two-phase approximation algorithm for solving the dynamic labeling problem, which is an extension of the work in References 60, 61, is presented. In the preprocessing phase, the active ranges of all potential labels are computed. In the interaction phase, a filtering is applied to retrieve the precomputed valid labels (i.e., labels that the current scale is within their active range). It is worth noting that no label conflict computations are performed during the interaction phase, which makes the algorithm fast and efficient. The work in Reference 7 improves aspects of this dynamic labeling framework.

Mote [58] provided a fast and efficient method for real-time point feature labeling on dynamic maps without preprocessing. He subdivides the map space into a trellis structure of rows and columns; each trellis cell is associated with the features within its boundaries, significantly limiting the search for label conflicts. In addition, a weighted scheme based on label priority and esthetic preference is used to resolve label conflicts.

Dynamic labeling has received considerable attention in recent years. Most work covers maps that allow panning and zooming [56,59,63,66]. The problem of labeling rotated maps is addressed in References 53–55, 67. In Reference 54, the dynamic labeling framework, presented earlier, was adapted to rotating maps and efficient approximation algorithms were provided. The problem of labeling moving points on a static map, which has applications in annotating radar screens, is studied in References 68–70.

Dynamic labeling of linear features through embedded street labels and billboards is addressed in References 57, 64–66. Finally, annotation of 3D scenes has many applications in virtual and augmented reality systems and relates to dynamic labeling [71–74,76,77].

## 7.6 Reaserch Issues and Summary

The computer revolution has changed many of the long-established practices in map production. Since Yoeli [9] first proposed that map label placement could be automated through an interactive process based on cartographic best practices, researchers continue to apply new techniques while building on existing knowledge. One aspect of cartography that still demands intensive manual work is the label placement process to produce high-quality maps, as the automatic label placement process has been proven to be a very difficult problem. As Freeman points out [103], the ultimate test for an automatic map-labeling system is how well its results compare with those produced by an expert cartographer. A practical approach to producing a high-quality map labels might be a semi-automated interactive name placement system. That system might produce an initial label placement which could be improved manually by cartographers. An interactive map-labeling framework for such systems is presented in Reference 104. Certainly, more research in this area is promising.

Labeling of dynamic maps, in which users can navigate by continuously zooming, panning, or rotating, has received little attention until recently. This area of research offers many interesting and challenging problems: (1) As it is noted in Reference 7, commercial on-line maps and navigation devices use inferior algorithms to place labels. Thus, further improvement of the existing state of the art labeling techniques and incorporation into real-world applications is essential; (2) an important challenge is to devise an algorithmic framework that combines map rotation with zooming and panning; (3) labeling dynamic 3D scenes (e.g., video games) presents another challenge for researchers; and (4) ARO in the dynamic labeling framework introduced in Reference 52 presents many new challenges, both of theoretical and practical importance.

In the context of graph drawing, an interesting and untackled problem is that of placing labels to graphical features of 3D drawings. This is quite a challenging problem due to the occlusion problems involved.

## References

1. Bernard Chazelle et al. Application challenges to computational geometry: CG impact task force report. In B. Chazelle, J. E. Goodman, and R. Pollack, Eds., *Advances in Discrete and Computational Geometry*, volume 223 of *Contemporary Mathematics*, pp. 407–463. American Mathematical Society, Providence, RI, 1999.
2. K. G. Kakoulis and I. G. Tollis. Labeling algorithms. In R. Tamassia, Ed., *Handbook of Graph Drawing and Visualization*, pp. 489–515. Chapman and Hall/CRC Press, Boca Raton, FL, 2013.
3. J. Christensen, J. Marks, and S. Shieber. An empirical study of algorithms for point feature label placement. *ACM Transactions on Graphics*, 14(3):203–232, 1995.

4. H. Freeman. Computer name placement. In D. J. Maguire, M. F. Goodchild, and D. W. Rhind, Eds., *Geographical Information Systems: Principles and Applications*, pp. 445–456. Longman, London, UK, 1991.

5. G. Di Battista, P. Eades, R. Tamassia, and I. G. Tollis. *Graph Drawing: Algorithms for the Visualization of Graphs*. Prentice Hall, Upper Saddle River, NJ, 1999.

6. I. G. Tollis and K. G. Kakoulis. Graph drawing. In A. B. Tucker and T. Gonzalez, Eds., *Computing Handbook: Computer Science and Software Engineering, Third Edition*, pp. 14:1–21. Chapman and Hall/CRC Press, Boca Raton, FL, 2014.

7. K. Been, M. Nöllenburg, S. Poon, and A. Wolff. Optimizing active ranges for consistent dynamic map labeling. *Computational Geometry*, 43(3):312–328, 2010.

8. E. Imhof. Positioning names on maps. *The American Cartographer*, 2(2):128–144, 1975.

9. P. Yoeli. The logic of automated map lettering. *The Cartographic Journal*, 9(2):99–108, 1972.

10. J. Ahn and H. Freeman. A program for automatic name placement. *Cartographica*, 21(2 & 3): 101–109, 1984.

11. J. S. Doerschler and H. Freeman. A rule based system for dense map name placement. *Communications of ACM*, 35(1):68–79, 1992.

12. L. R. Ebinger and A. M. Goulete. Noninteractive automated names placement for the 1990 decennial census. *Cartography and Geographic Information Systems*, 17(1):69–78, 1990.

13. S. Edmondson, J. Christensen, J. Marks, and S. Shieber. A general cartographic labeling algorithm. *Cartographica*, 33(4):321–342, 1997.

14. H. Freeman and J. Ahn. On the problem of placing names in a geographical map. *International Journal of Pattern Recognition and Artificial Intelligence*, 1(1):121–140, 1987.

15. C. B. Jones. Cartographic name placement with prolog. *IEEE Computer Graphics and Applications*, 9(5):36–47, 1989.

16. K. G. Kakoulis and I. G. Tollis. A unified approach to automatic label placement. *International Journal of Computational Geometry and Applications*, 13(1):23–60, 2003.

17. F. Wagner and A. Wolff. A combinatorial framework for map labeling. In S. Whitesides, Ed., *Graph Drawing (Proceedings of the GD 1998)*, volume 1547 of *Lecture Notes in Computer Science*, pp. 316–331. Springer, Berlin, Germany, 1998.

18. S. Doddi, M. V. Marathe, A. Mirzaian, B. M. Moret, and B. Zhu. Map labeling and its generalizations. In *Proceedings of the 8th ACM-SIAM Symposium on Discrete Algorithms*, pp. 148–157, 1997.

19. M. Formann and F. Wagner. A packing problem with applications to lettering of maps. In *Proceedings of the 7th Annual ACM Symposium on Computational Geometry*, pp. 281–288, 1991.

20. S. A. Hirsch. An algorithm for automatic name placement around point data. *The American Cartographer*, 9(1):5–17, 1982.

21. T. Strijk, B. Verweij, and K. Aardal. Algorithms for maximum independent set applied to map labelling. Technical Report UU-CS-2000-22, Department of Computer Science, Utrecht University, Utrecht, the Netherlands, 2000.

22. B. Verweij and K. Aardal. An optimisation algorithm for maximum independent set with applications in map labelling. In J. Nešetřil, Ed., *7th Annual European Symposium on Algorithms (Proceedings of the ESA 1999)*, pp. 426–437. Springer, Berlin, Germany, 1999.

23. F. Wagner and A. Wolff. Map labeling heuristics: Provably good and practically useful. In *Proceedings of the 11th Annual ACM Symposium on Computational Geometry*, pp. 109–118, 1995.

24. S. Zoraster. The solution of large 0–1 integer programming problems encountered in automated cartography. *Operation Research*, 38(5):752–759, 1990.

25. D. H. Alexander and C. S. Hantman. Automating linear text placement within dense feature networks. In *Proceedings of Auto-Carto 12*, pp. 311–320. ACSM/ASPRS, Bethesda, MD, 1995.

26. A. Wolff, L. Knipping, M. van Kreveld, T. Strijk, and P. K. Agarwal. A simple and efficient algorithm for high-quality line labeling. In D. Martin and F. Wu, Eds., *7th Annual GIS Research UK Conference (Proceedings of the GISRUK 1999)*, pp. 146–150, 1999.

27. U. Doğrusöz, K. G. Kakoulis, B. Madden, and I. G. Tollis. On labeling in graph visualization. *Special Issue on Graph Theory and Applications, Information Sciences Journal*, 177(12):2459–2472, 2007.

28. K. G. Kakoulis and I. G. Tollis. An algorithm for labeling edges of hierarchical drawings. In G. Di Battista, Ed., *Graph Drawing (Proceedings of the GD 1997)*, volume 1353 of Lecture Notes in Computer Science, pp. 169–180. Springer, Berlin, Germany, 1997.

29. M. Barrault. A methodology for placement and evaluation of area map labels. *Computers, Environment and Urban Systems*, 25(1):33–52, 2001.

30. D. Dorschlag, I. Petzold, and L. Plumer. Placing objects automatically in areas of maps. In *Proceedings of the 21st International Cartographic Conference*, 2003.

31. H. Freeman. An expert system for the automatic placement of names on a geographic map. *Information Sciences*, 45:367–378, 1988.

32. I. Pinto and H. Freeman. The feedback approach to cartographic area text placement. In P. Perner, P. Wang, and A. Rosenfeld, Eds., *Advances in Structural and Syntactical Pattern Recognition*, volume 1121 of *Lecture Notes in Computer Science*, pp. 341–350. Springer, Berlin, Germany, 1996.

33. J. W. van Roessel. An algorithm for locating candidate labeling boxes within a polygon. *The American Cartographer*, 16(3):201–209, 1989.

34. K. G. Kakoulis and I. G. Tollis. On the multiple label placement problem. In *Proceedings of the 10th Canadian Conference on Computational Geometry*, pp. 66–67, 1998.

35. K. G. Kakoulis and I. G. Tollis. Algorithms for the multiple label placement problem. *Computational Geometry*, 35(3):143–161, 2006.

36. Z. Qin, A. Wolff, Y. Xu, and B. Zhu. New algorithms for two-label point labeling. In M. S. Paterson, Ed., *8th Annual European Symposium on Algorithms (Proceedings of the ESA 2000)*, pp. 368–380. Springer, Berlin, Germany, 2000.

37. B. Zhu and C. Poon. Efficient approximation algorithms for multi-label map labeling. In A. Aggarwal and C. Pandu Rangan, Eds., *10th Annual International Symposium on Algorithms and Computation (Proceedings of the ISAAC 1999)*, volume 1741 of Lecture Notes in Computer Science, pp. 143–152. Springer, Berlin, Germany, 1999.

38. Y. Hu. Visualizing graphs with node and edge labels. *arXiv preprint arXiv:0911.0626*, 2009.

39. K. G. Kakoulis and I. G. Tollis. Placing edge labels by modifying an orthogonal graph drawing. In U. Brandes and S. Cornelsen, Eds., *Graph Drawing (Proceedings of the GD 2010)*, volume 6502 of Lecture Notes in Computer Science, pp. 395–396. Springer, Berlin, Germany, 2010.

40. K. Ali, K. Hartmann, and T. Strothotte. Label layout for interactive 3D illustrations. *Journal of the WSCG*, 13:1–8, 2005.

41. L. Barth, A. Gemsa, B. Niedermann, and M. Nöllenburg. On the readability of boundary labeling. In E. Di Giacomo and A. Lubiw, Eds., *Graph Drawing and Network Visualization (Proceedings of GD 2015)*, pp. 515–527. Springer, Berlin, Germany, 2015.

42. M. A. Bekos, M. Kaufmann, D. Papadopoulos, and A. Symvonis. Combining traditional map labeling with boundary labeling. In I. Černá, T. Gyimóthy, J. Hromkovič, K. Jefferey, R. Královič, M. Vukolić, and S. Wolf, Eds., *37th International Conference on Current Trends in Theory and Practice of Computer Science (Proceedings of the SOFSEM 2011)*, pp. 111–122. Springer, Berlin, Germany, 2011.

43. M. A. Bekos, M. Kaufmann, K. Potika, and A. Symvonis. Multi-stack boundary labeling problems. In S. Arun-Kumar and N. Garg, Eds., *26th International Conference on Foundations of Software Technology and Theoretical Computer Science (Proceedings of the FSTTCS 2006)*, pp. 81–92. Springer, Berlin, Germany, 2006.

44. M. A. Bekos, M. Kaufmann, A. Symvonis, and A. Wolff. Boundary labeling: Models and efficient algorithms for rectangular maps. *Computational Geometry*, 36(3):215–236, 2007.

45. M. Benkert, H. Haverkort, M. Kroll, and M. Nöllenburg. Algorithms for multi-criteria boundary labeling. *Journal of Graph Algorithms and Applications*, 13(3):289–317, 2009.

46. J. Fekete and C. Plaisant. Excentric labeling: Dynamic neighborhood labeling for data visualization. In *Proceedings of the SIGCHI Conference on Human Factors in Computing Systems*, pp. 512–519, 1999.

47. M. Fink, J. H. Haunert, A. Schulz, J. Spoerhase, and A. Wolff. Algorithms for labeling focus regions. *IEEE Transactions on Visualization and Computer Graphics*, 18(12):2583–2592, 2012.

48. A. Gemsa, J. Haunert, and M. Nöllenburg. Multirow boundary-labeling algorithms for panorama images. *ACM Transactions on Spatial Algorithms and Systems*, 1(1):1:1–1:30, 2015.

49. Z. Huang, S. Poon, and C. Lin. Boundary labeling with flexible label positions. In S. P. Pal and K. Sadakane, Eds., *8th International Workshop on Algorithms and Computation* (*Proceedings of the WALCOM 2014*), pp. 44–55. Springer, Berlin, Germany, 2014.

50. P. Kindermann, B. Niedermann, I. Rutter, M. Schaefer, A. Schulz, and A. Wolff. Two-sided boundary labeling with adjacent sides. In F. Dehne, R. Solis-Oba, and J. Sack, Eds., *13th International Symposium on Algorithms and Data Structures* (*Proceedings of the WADS 2013*), pp. 463–474. Springer, Berlin, Germany, 2013.

51. M. Nöllenburg, V. Polishchuk, and M. Sysikaski. Dynamic one-sided boundary labeling. In *Proceedings of the 18th SIGSPATIAL International Conference on Advances in Geographic Information Systems*, pp. 310–319, 2010.

52. K. Been, E. Daiches, and C. Yap. Dynamic map labeling. *IEEE Transactions on Visualization and Computer Graphics*, 12(5):773–780, 2006.

53. A. Gemsa, B. Niedermann, and M. Nöllenburg. Trajectory-based dynamic map labeling. In L. Cai, S. Cheng, and T. Lam, Eds., *24th International Symposium on Algorithms and Computation* (*Proceedings of the ISAAC 2013*), pp. 413–423. Springer, Berlin, Germany, 2013.

54. A. Gemsa, M. Nöllenburg, and I. Rutter. Consistent labeling of rotating maps. In F. Dehne, J. Iacono, and J. Sack, Eds., *12th International Symposium on Algorithms and Data Structures* (*Proceedings of the WADS 2011*), pp. 451–462. Springer, Berlin, Germany, 2011.

55. A. Gemsa, M. Nöllenburg, and I. Rutter. Evaluation of labeling strategies for rotating maps. *Journal of Experimental Algorithmics*, 21(1):1.4:1–1.4:21, 2016.

56. C. Liao, C. Liang, and S. Poon. Approximation algorithms on consistent dynamic map labeling. In J. Chen, J. E. Hopcroft, and J. Wang, Eds., *8th International Workshop on Frontiers in Algorithmics* (*Proceedings of the FAW 2014*), pp. 170–181. Springer, Berlin, Germany, 2014.

57. S. Maass and J. Düllner. Embedded labels for line features in interactive 3D virtual environments. In *Proceedings of the 5th International Conference on Computer Graphics, Virtual Reality, Visualisation and Interaction in Africa*, pp. 53–59, 2007.

58. K. Mote. Fast point-feature label placement for dynamic visualizations. *Information Visualization*, 6(4):249–260, 2007.

59. K. Ooms, W. Kellens, and V. Fack. Dynamic map labeling for users. In *24th International Cartographic Conference* (*Proceedings of the ICC 2009*), Santiago, Chile, 2009.

60. I. Petzold, G. Gröger, and L. Plümer. Fast screen map labeling—Data structures and algorithms. In *21st International Cartographic Conference* (*Proceedings of the ICC 2003*), Durban, South Africa, 2003.

61. I. Petzold, H. Plümer, and M. Heber. Label placement for dynamically generated screen maps. In *19th International Cartographic Conference* (*Proceedings of the ICC 1999*), Ottawa, Canada, 1999.

62. S. Poon and C. Shin. Adaptive zooming in point set labeling. In M. Liśkiewicz and R. Reischuk, Eds., *15th International Symposium on Fundamentals of Computation Theory* (*Proceedings of the FCT 2005*), pp. 233–244. Springer, Berlin, Germany, 2005.

63. N. Schwartges, J. Haunert, A. Wolff, and D. Zwiebler. Point labeling with sliding labels in interactive maps. In J. Huerta, S. Schade, and C. Granell, Eds., *Connecting a Digital Europe Through Location and Place*, pp. 295–310. Springer, Berlin, Germany, 2014.

64. N. Schwartges, B. Morgan, J. Haunert, and A. Wolff. Labeling streets along a route in interactive 3D maps using billboards. In F. Bacao, Y. Santos, and M. Painho, Eds., *AGILE 2015: Geographic Information Science as an Enabler of Smarter Cities and Communities*, pp. 269–287. Springer, Berlin, Germany, 2015.

65. N. Schwartges, A. Wolff, and J. Haunert. Labeling streets in interactive maps using embedded labels. In *Proceedings of the 22nd ACM SIGSPATIAL International Conference on Advances in Geographic Information Systems*, pp. 517–520, 2014.

66. M. Vaaraniemi, M. Treib, and R. Westermann. Temporally coherent real-time labeling of dynamic scenes. In *Proceedings of the 3rd International Conference on Computing for Geospatial Research and Applications*, pp. 17:1–17:10, 2012.

67. Y. Yokosuka and K. Imai. Polynomial time algorithms for label size maximization on rotating maps. In *Proceedings of the 25th Canadian Conference on Computational Geometry*, pp. 187–192, 2013.

68. K. Buchin and D. H. P. Gerrits. Dynamic point labeling is strongly pspace-complete. *International Journal of Computational Geometry and Applications*, 24(4):373–395, 2014.

69. M. de Berg and D. H. P. Gerrits. Approximation algorithms for free-label maximization. *Computational Geometry*, 45(4):153–168, 2012.

70. M. de Berg and D. H. P. Gerrits. Labeling moving points with a trade-off between label speed and label overlap. In H. L. Bodlaender and G. F. Italiano, Eds., *21st Annual European Symposium on Algorithms* (*Proceedings of the ESA 2013*), pp. 373–384. Springer, Berlin, Germany, 2013.

71. R. Azuma and C. Furmanski. Evaluating label placement for augmented reality view management. In *Proceedings of the 2nd IEEE/ACM International Symposium on Mixed and Augmented Reality*, pp. 66–75, 2003.

72. B. Bell, S. Feiner, and T. Höllerer. View management for virtual and augmented reality. In *Proceedings of the 14th Annual ACM Symposium on User Interface Software and Technology*, pp. 101–110, 2001.

73. L. Čmolík and J. Bittner. Layout-aware optimization for interactive labeling of 3D models. *Computers and Graphics*, 34(4):378–387, 2010.

74. R. Grasset, T. Langlotz, D. Kalkofen, M. Tatzgern, and D. Schmalstieg. Image-driven view management for augmented reality browsers. In *IEEE International Symposium on Mixed and Augmented Reality* (*Proceedings of ISMAR 2012*), pp. 177–186, 2012.

75. S. Pick, B. Hentschel, I. Tedjo-Palczynski, M. Wolter, and T. Kuhlen. Automated positioning of annotations in immersive virtual environments. In *Proceedings of the Joint Virtual Reality Conference of EGVE—EuroVR—VEC*, pp. 1–8, Eurographics Association, 2010.

76. T. Stein and X. Décoret. Dynamic label placement for improved interactive exploration. In *Proceedings of the 6th International Symposium on Non-photorealistic Animation and Rendering*, pp. 15–21, 2008.

77. M. Vaaraniemi, M. Freidank, and R. Westermann. Enhancing the visibility of labels in 3D navigation maps. In J. Pouliot, S. Daniel, F. Hubert, and A. Zamyadi, Eds., *Progress and New Trends in 3D Geoinformation Sciences*, pp. 23–40. Springer, Berlin, Germany, 2013.

78. C. Binucci, W. Didimo, G. Liotta, and M. Nonato. Orthogonal drawings of graphs with vertex and edge labels. *Computational Geometry*, 32(2):71–114, 2005.

79. R. Di Battista, W. Didimo, M. Patrignani, and M. Pizzonia. Orthogonal and quasi-upward drawings with vertices of prescribed size. In J. Kratochvil, Ed., *Graph Drawing* (*Proceedings of the GD 1999*), volume 1731 of *Lecture Notes in Computer Science*, pp. 297–310. Springer, Berlin, Germany, 1999.

80. G. W. Klau and P. Mutzel. Combining graph labeling and compaction. In J. Kratochvil, Ed., *Graph Drawing* (*Proceedings of the GD 1999*), volume 1731 of Lecture Notes in Computer Science, pp. 27–37. Springer, Berlin, Germany, 1999.

81. P. C. Wong, P. Mackey, K. Perrine, J. Eagan, H. Foote, and J. Thomas. Dynamic visualization of graphs with extended labels. In *IEEE Symposium on Information Visualization* (*Proceedings of the INFOVIS 2005*), pp. 73–80, 2005.

82. T. Kato and H. Imai. The NP-Completeness of the character placement problem of 2 or 3 degrees of freedom. In *Record of Joint Conference of Electrical and Electronic Engineers in Kyushu*, pp. 11–18, in Japanese, 1988.

83. J. Marks and S. Shieber. The computational complexity of cartographic label placement. Technical Report 05-91, Harvard University, Cambridge, MA, 1991.

84. K. G. Kakoulis and I. G. Tollis. On the complexity of the edge label placement problem. *Computational Geometry*, 18(1):1–17, 2001.

85. K. G. Kakoulis and I. G. Tollis. Modifying orthogonal drawings for label placement. *Algorithms*, 9(2):22, 2016.

86. S. Poon, C. Shin, T. Strijk, T. Uno, and A. Wolff. Labeling points with weights. *Algorithmica*, 38(2):341–362, 2004.

87. M. Yamamoto, G. Camara, and L. Lorena. Tabu search heuristic for point-feature cartographic label placement. *GeoInformatica*, 6(1):77–90, 2002.

88. S. van Dijk, D. Thierens, and M. de Berg. On the design of genetic algorithms for geographical applications. In *Proceedings of the 1st Annual Conference on Genetic and Evolutionary Computation*, pp. 188–195. Morgan Kaufmann Publishers, 1999.

89. O. V. Verner, R. L. Wainwright, and D. A. Schoenefeld. Placing text labels on maps and diagrams using genetic algorithms with masking. *INFORMS Journal on Computing*, 9(3):266–275, 1997.

90. M. Barrault and F. Lecordix. An automated system for linear feature name placement which complies with cartographic quality criteria. In *Proceedings of Auto-Carto 12*, pp. 321–330. ACSM/ASPRS, Bethesda, MD, 1995.

91. F. Chirié. Automated name placement with high cartographic quality: City street maps. *Cartography and Geographic Information Science*, 27(2):101–110, 2000.

92. M. van Kreveld, T. Strijk, and A. Wolff. Point labeling with sliding labels. *Computational Geometry*, 13:21–47, 1999.

93. M. A. Bekos. *Map Labeling Algorithms with Application in Graph Drawing and Cartography*. PhD thesis, National Technical University of Athens, Athens, Greece, 2008.

94. H. Freeman, S. Marrinan, and H. Chitalia. Automated labeling of soil survey maps. In *Proceedings of the ASPRS-ACSM Annual Convention, Baltimore*, Vol. 1, pp. 51–59, 1996.

95. S. Zoraster. Practical results using simulated annealing for point feature label placement. *Cartography and Geographic Information Systems*, 24(4):228–238, 1997.

96. M. R. Garey and D. S. Johnson. *Computers and Intractability: A Guide to the Theory of NP-Completeness*. W. H. Freeman, New York, 1979.

97. K. Misue, P. Eades, W. Lai, and K. Sugiyama. Layout adjustement and the mental map. *Journal of Visual Languages and Computing*, 6:183–210, 1995.

98. P. Eades. A heuristic for graph drawing. *Congressus Numerantium*, 42:149–160, 1984.

99. R. Tamassia. On embedding a graph in the grid with the minimum number of bends. *SIAM Journal on Computing*, 16(3):421–444, 1987.

100. J. Doenhardt and T. Lengauer. Algorithmic aspects of one dimentional layout compaction. *IEEE Transactions on Compute-Aided Design of Integrated Circuits and Systems*, 6(5):863–879, 1987.

101. T. Lengauer. *Combinatorial Algorithms for Integrated Circuuit Layout*. John Wiley & Sons, New York, 1990.

102. M. Y. Hsueh. *Symbolic Layout and Compaction of Integrated Circuits.* PhD thesis, University of California at Berkeley, Berkeley, CA, 1979.

103. H. Freeman. Automated cartographic text placement. *Pattern Recognition Letters*, 26(3):287–297, 2005.

104. H. A. D. do Nascimento and P. Eades. User hints for map labeling. *Journal of Visual Languages and Computing*, 19(1):39–74, 2008.

# 8

# Complexity, Approximation Algorithms, and Heuristics for the Corridor Problems

Teofilo F. Gonzalez

Arturo Gonzalez-Gutierrez

## 8.1 Introduction

In the current chapter, we examine complexity issues and approximation algorithms for the Minimum-Length Corridor (MLC) problem and its variants. These problems have applications in several fields of study including VLSI (Very Large Scale Integration) routing, transportation systems, and so on. Specifically, we discuss the NP-hardness (Nondeterministic Polynomial time-hardness) of the MLC problem and present a hierarchy of related problems, most of which remain NP-hard. We present a general framework for a class of polynomial-time constant ratio-approximation algorithms for the MLC-R, which is the MLC problem in which all the objects are rectangles. The approximation algorithms are based on the restriction and relaxation methodologies. Finally, we present new experimental analysis of our approximation algorithms for families of problem instances.

The MLC problem is defined as follows. Given a rectangular boundary $F^*$ partitioned into rectilinear polygons ($R = \{R_1, R_2, \ldots, R_r\}$), a corridor is a set of connected line segments $T$ each of which must lie along the line segments that form the rectangular boundary $F$ and/or the boundary of the rectilinear

---

* Throughout this chapter, we assume that all the rectangles and rectilinear polygons have their boundaries orthogonal to the x or y axis, that is, they are all orthogonal rectangles or orthogonal rectilinear polygons.

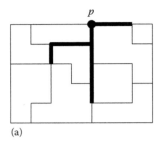
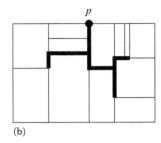

**FIGURE 8.1**    Optimal corridors: (a) MLC and $p$-MLC instance and (b) MLC-R and $p$-MLC-R instance.

polygons $R$. The line segments in $T$ form a tree and include at least one point from $F$ and at least one point from the boundary of each of the rectilinear polygons in $R$. The sum of the length of the line segments in $T$ is called the *edge-length* or simply the *length* of $T$ and is denoted by $L(T)$.

One may view the pair $(F, R)$ as a floorplan with $r$ rooms, and the set $T$ of line segments as a *corridor* connecting the rooms. Figure 8.1 shows two problem instances and their corresponding optimal corridors, represented by thick lines. Any corridor may be used to connect all the rooms to the outside of $F$ through any of the points along the rectangular boundary $F$.

The MLC problem and its variants have applications when laying optical fiber for data communication or wires for electrical connection in floorplans. In these applications, the rectangular boundary $F$ corresponds to the floorplan, the partition $R$ of rectilinear polygons corresponds to the configuration of the floorplan into individual rooms, and a corridor corresponds to the placement of the optical fiber or wires that provide data communication or power to all the rooms. The minimum edge-length corridor corresponds to the minimum length optical fiber or wire needed to provide connectivity. Other applications include the laying of water, sewer, and electrical lines on parcels in housing developments or farming regions, and laying of wires for power or signal communication in circuit layout design.

A restricted version of the MLC problem is when all the rooms are rectangles (MLC-R problem). Figure 8.1b represents an instance of the MLC-R problem, but not the instance of Figure 8.1a.

It is convenient to view the geometric representation of our problems as a graph. There is a vertex for every distinct point located at the intersection of two orthogonal line segments (sides of rooms or the boundary $F$). A vertical (resp. horizontal) line segment in the instance is called an *edge* of the graph if its endpoints are vertices as identified above. Thus, every instance $I$ of the MLC-R problem may be viewed as the graph $G(I) = (V, E, w)$, where the set $V$ of vertices and the set $E$ of edges are as defined earlier, and the weight $w(e)$ of an edge $e \in E$ corresponds to the length of the line segment represented by the edge. We use $V(R_i)$ to represent all the vertices along the sides of rectangle $R_i$.

An *access point* is a vertex located on the boundary $F$. The $p$-MLC ($p$-MLC-R) problem is a restricted version of the MLC (resp. MLC-R) problem in which the input has additionally an access point $p$ and the solution must include point $p$. Figure 8.1 shows an instance for the $p$-MLC ($p$-MLC-R) problem as well as its optimal corridor represented by thick lines. The solution to any instance of the MLC (respectively, MLC-R) problem can be obtained by finding a corridor for the $p$-MLC (respectively, $p$-MLC-R) problem at each access point $p$, and then selecting the best of these corridors.

The MLC problem was initially posted as an open problem in the Proceedings of the 13th Canadian Conference on Computational Geometry (CCCG 2001) [1] and subsequently Eppstein [2] introduced the MLC-R problem. The question as to whether or not the decision version of each of these problems is NP-complete is raised in the previous two references. Gonzalez-Gutierrez and Gonzalez [3] and Bodlaender et al. [4] proved independently and simultaneously the NP-completeness of the MLC problem. The decision versions of the MLC-R and $p$-MLC-R problems are shown to be NP-complete in Gonzalez-Gutierrez and Gonzalez [3].

Gonzalez-Gutierrez and Gonzalez [5] established the first constant ratio-approximation algorithm for the MLC-R problem. It is still an open question whether or not there are constant ratio polynomial time-approximation algorithms for the MLC problem. In this chapter, we discuss complexity issues and approximation algorithms for the MLC-R problem and its variations. We also examine empirical evaluations of these algorithms.

## 8.2 Related Problems

A corridor is called a *partial corridor* if there is at least one room that is not reached by the corridor, that is, at least one room is not *exposed* to the corridor. A more general version of our corridor problems allows a set or forest of partial corridors rather than just a corridor, with the partial corridors connecting all the rooms to access points. We refer to these problem as the $MLC_f$ and $MLC_f$-R problems.

The Multiple Access (MA) points versions of MA-$MLC_f$ and MA-$MLC_f$-R restrict the partial corridors to be rooted at a given set of access points. When the top-right corner of $F$ is the only access point, the MA-$MLC_f$ and MA-$MLC_f$-R problems are referred to as the *top-right access* (TRA) point versions or simply the *TRA-MLC* and *TRA-MLC-R* problems.

Gonzalez-Gutierrez and Gonzalez [3] proved that the most fundamental of the problems so far discussed, TRA-MLC-R, is NP-complete. From this result and simple reductions, one can show that all the remaining problems discussed so far are NP-complete.

The generalization of the MLC-R problem to graphs is referred to as the Network-MLC (or simply N-MLC) problem. The input to this problem is a connected undirected edge-weighted graph, and the objective is to find a tree with least total edge-weight such that every cycle in the graph has at least one of its vertices in the tree. In graph theoretic terms, the set of vertices that break all the cycles in a graph is called a *feedback node set* (FNS). One may redefine the N-MLC problem by requiring that the vertices in the corridor form a FNS for the graph. Thus, this problem can also be referred to as the *tree FNS* problem.

The *Group Steiner tree* (GST) problem was introduced by Reich and Widmayer [6] to model VLSI applications. This problem may be viewed as a generalization of the MLC problem. The input to the GST problem is an edge-weighted connected graph $G = (V, E, w)$, where $w : E \rightarrow \mathbb{R}^+$ is an edge-weight function; a non empty subset $S$, $S \subseteq V$, of *terminals*; and a partition $\{S_1, S_2, \ldots, S_k\}$ of $S$. The objective is to find a tree $T(S) = (V', E')$, where $E' \subseteq E$ and $V' \subseteq V$, such that at least one terminal from each set $S_i$ is in the tree $T(S)$ and the total edge-length $\sum_{e \in E'} w(e)$ is minimized.

The graph Steiner tree (ST) problem is a special case of the GST problem where for each set $S_i, |S_i| = 1$. Karp [7] proved that the decision version of the ST problem is NP-complete. As the GST problem includes the ST problem, the decision version of the GST problem is also NP-complete. Approximation algorithms for the GST problem are given in References 8–10. There is a simple and straightforward reduction from the MLC problem to the GST problem, which can be used to show that any constant ratio-approximation algorithm for the GST problem is a constant ratio-approximation algorithm for the MLC problem. However, there is no known constant ratio-approximation algorithm for the GST problem. Safra and Schwartz [11] established inapproximability results for the 2D version of the GST problem when each set is connected and the sets are allowed to intersect; but again, these results do not seem to carry over to the MLC problem. Other authors [11–13] defined a more general version of the GST problem where $\{S_1, S_2, \ldots, S_k\}$ of $S$ is not a partition, but each $S_i \subseteq S$, that is, a vertex may be in more than one set $S_i$. This version of the GST problem is referred to as the *tree errand cover* (TEC) problem, and it was studied by Slavik [12,13]. Slavik developed an approximation algorithm, based on the relaxation methodology, to generate solutions to the TEC problem with approximation ratio $2d$, where $d$ is the maximum number of vertices in any of the $S_i$ sets.

As we have seen, our problems are restricted versions of more general ones reported in the literature. But previous results for those problems do not establish NP-completeness results, inapproximability results, nor constant ratio-approximation algorithms for our MLC problems.

# 8.3 Parameterized Algorithm and Design Principles

In the current section, we outline approximation algorithms for the $p$-MLC-R problem. These algorithms can be easily used to generate provably constant ratio-approximation algorithms for the MLC-R problem. The approximation algorithms are based on the restriction and relaxation-approximation methodologies.

## 8.3.1 General Framework

An approach to generate a suboptimal solution to the $p$-MLC-R problem is to simply reduce it to the TEC problem by defining an errand that includes only point $p$ and an errand for each rectangle $R_j$ that includes the vertices in $V(R_j)$. Then one uses Slavik's [12,13] approximation algorithm that is bounded by $2d$, where $d$ is the maximum number of vertices on any of the rectangles. The only reason this approach does not provide a constant ratio approximation is that $d$ is not bounded above by a constant.

Our parameterized approximation algorithm for the $p$-MLC-R problem uses the same approach but circumvents the drawback by limiting the number of points identified at each rectangle. This is accomplished by defining a selector function $S$ that identifies a subset of vertices from each rectangle $R_j$ that are referred to as *critical points*. The $p$-MLC-$R_S$ is exactly similar to the $p$-MLC-R problem, except that every feasible corridor must include a critical point from each rectangle. Then Slavik's approximation algorithm for the TEC problem [12,13] is used to generate a corridor for the $p$-MLC-$R_S$ problem instance.

When the maximum number of critical points identified for each rectangle is $k_S$, our corridor has length at most $2k_S$ times the length of an optimal corridor for the $p$-MLC-$R_S$ problem instance. The approximation ratio of our algorithm for the $p$-MLC-R problem also depends on the ratio ($r_S$) between optimal solutions for $p$-MLC-$R_S$ and $p$-MLC-R problem instances. The approximation ratio for the $p$-MLC-$R_S$ problem of our parameterized algorithm is $2k_Sr_S$.

In order for our algorithm to generate a constant ratio approximation, the selector function $S$ must identify from each rectangle a set of critical points that is bounded above by a constant. For example, the selector function that identifies from each rectangle its corners as critical points has $k_S = 4$. However, as we shall see later on, the ratio $r_S$ is only bounded above by a constant for some selector functions $S$.

We begin by discussing several simple selector functions for which one cannot bound $r_S$ by a constant. These negative results shed some light on design principles that may result in constant ratio-approximation algorithms. Then we discuss two constant ratio-approximation algorithms arising from the above-mentioned design principles.

Formally, given an instance $I$ of the $p$-MLC-R problem and a selector function $S$, we use $I_S$ to denote the instance of the corresponding $p$-MLC-$R_S$ problem. The instance of the TEC problem, denoted by $J_S$, is constructed from the instance $I_S$ of the $p$-MLC-$R_S$ problem using the approach discussed earlier, but limiting the errand at each rectangle to the critical points of the rectangle. Clearly, every feasible solution to the $p$-MLC-$R_S$ problem instance $I_S$ is also a feasible solution to the instance $J_S$ of the TEC problem, and vice versa. Furthermore, the objective function value of every feasible solution to both problems is identical. Slavik's algorithm applied to the instance $J_S$ of the TEC problem generates a solution $T(J_S)$ from which we construct a corridor $T(I)$ with edge-length $t(I)$ for the $p$-MLC-R problem. We call our approach the parameterized algorithm $Alg(S)$, where $S$ is the parameter.

As Slavik's approximation algorithm is based on relaxation techniques and we apply it to a restricted version of the $p$-MLC-R problem, we say that our approximation algorithm is based on restriction and relaxation approximation techniques.

Let $OPT(I_S)$ be an optimal corridor for $I_S$ and let $opt(I_S)$ be its edge-length. Theorem 8.1 establishes a general approximation ratio for our parameterized algorithm $Alg(S)$. It is simple to see that the total edge-length of an optimal solution of the instance $I_S$, $opt(I_S)$, corresponding to the $p$-MLC-$R_S$ problem, is at least as large as the total edge-length of an optimal solution of the instance $I$, $opt(I)$, of the $p$-MLC-R problem. We define the ratio between $opt(I_S)$ and $opt(I)$ as $r_S$ (with $r_S \geq 1$). In other words, one needs to prove that $opt(I_S) \leq r_S \times opt(I)$ for every instance $I$ of the $p$-MLC-R problem to use the following theorem.

**Theorem 8.1** *Parameterized algorithm Alg(S) generates for every instance $I$ of the $p$-MLC-R problem a corridor $T(I)$ of length $t(I)$ at most $2k_S \times r_S$ times $opt(I)$, provided that $opt(I_S) \leq r_S \times opt(I)$.*

For the previous approach to yield a constant ratio-approximation algorithm, we need both $k_S$ and $r_S$ to be bounded above by constants. For example, $S(4C)$ selects from each rectangle $R_j$ its four corner points, so $k_S$ is four. However, in order for our parameterized algorithm $Alg(S(4C))$ to be a constant ratio-approximation algorithm for the $p$-MLC-R problem, we need to show that $opt(I_{S(4C)}) \leq r_{S(4C)} \times opt(I)$, for some $r_{S(4C)}$ bounded above by a constant.

For most selector functions $S$, proving that $opt(I_S) \leq r_S \times opt(I)$, for every instance $I$ of the $p$-MLC-R problem, seems impossible because we do not have the optimal solutions at hand. Instead we establish a bound for all corridors. That is, we prove that for every corridor $T(I)$ with edge-length $t(I)$, there is a corridor for $I_S$ denoted by $T(I_S)$ with edge-length $t(I_S) \leq r_S \times t(I)$. Applying this to $T(I) = OPT(I)$, we know that there is a corridor $T(I_S)$ such that $t(I_S) \leq r_S \times opt(I)$. As $opt(I_S) \leq t(I_S)$, we know that $opt(I_S) \leq r_S \times opt(I)$. We discuss several selector functions in the next two subsections.

## 8.3.2 No Constant Ratio Approximations

In this subsection we show that for a set of selector functions, $r_S$ cannot be bounded above by a constant. When one incorporates them into our parameterized algorithm, $Alg(S)$ does not result in constant ratio approximations.

### 8.3.2.1 Selecting a Subset with the Four Corners

Consider the selector function $S(4C)$ that identifies from each rectangle $R_i$ its four corners $C(R_i)$, that is, the four critical points for $R_i$ are its corner points. Clearly, $k_{S(4C)} = 4$. We now show that $r_{S(4C)}$ cannot be bounded above by a constant, and thus the resulting parameterized algorithm $Alg(S(4C))$ is not a constant ratio-approximation algorithm.

To establish our claim, we give a family of problem instances with parameter $j$ such that $r_{S(4C)}$ is proportional to $j$, and $j$ can be made arbitrarily large. Consider the $j$-layer family of instances $I(j)$ of the $p$-MLC-R problem represented in Figure 8.2. The rectangle $F$ has width 4 and height $\epsilon \ll 4$.

The point $p$ is the top-right corner of $F$. The rectangle $F$ is partitioned into $j$ layers of rectangles. Layer $i = 1$ is formed by two rectangles of size $(\frac{\epsilon}{2} - \delta) \times 2$, above three rectangles with (very tiny) heigth $\delta$ and width 1, 2, and 1, respectively (see the bottom part of Figure 8.2). The rectangle in layer 1 with height $\delta$ and width 2 is the shaded one. Layer $i + 1$ consists of two copies of layer $i$ scaled by 50% laid side by side and placed on top of layer $i$. In other words, layer 2 has four rectangles of size $(\frac{\epsilon}{4} - \frac{\delta}{2}) \times 1$, above two sets

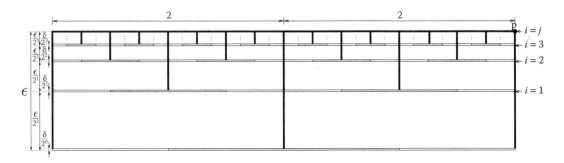

**FIGURE 8.2**   Optimal solution $OPT(I(j))$ for the family of instances $I(j)$ of the $p$-MLC-R problem.

of three rectangles each with (very tiny) height $\frac{\delta}{2}$ and width $\frac{1}{2}$, 1, and $\frac{1}{2}$, respectively. The two rectangles in layer 2 with height $\frac{\delta}{2}$ and width 1 are the shaded ones.

An optimal solution $OPT(I(j))$ of the $j$-layer family of problem instances $I(j)$ is given by the thick black lines in Figure 8.2. The total edge-length of the optimal corridor $OPT(I(j))$ is $opt(I(j)) = 4 + (j+2)(\epsilon - \delta)$. It is simple to show that in an optimal solution for instance $I_{S(4C)}(j)$ of the $p$-MLC-$R_{S(4C)}$ problem, denoted by $OPT(I_{S(4C)}(j))$, there must be a segment of length 1 to connect a corner point of the shaded rectangle in layer 1, there must be two segments of length $\frac{1}{2}$ to connect a corner point of the two shaded rectangles in layer 2, and so on. Furthermore, all of these segments must be distinct. Therefore, the total length of $OPT(I_{S(4C)}(j))$ is $opt(I_{S(4C)}(j)) > j$. Making $\delta$ and $\epsilon$ approach zero, the ratio $r\{S(4C)\}$ is about $\frac{j}{4}$, and $j$ can be made arbitrarily large.

Let $S(F4C)$ be a function that identifies fewer corners than $S(4C)$ for each rectangle. It is simple to see that by using the same example, our parameterized algorithm $Alg(S(F4C))$ has an approximation ratio that cannot be bounded above by any constant, as $r_{S(F4C)}$ is again proportional to $j$.

### 8.3.2.2  Selecting $k$ Points at Random

Randomization is a powerful technique to generate near-optimal solutions to some problems. Lets apply it to restrict the set of vertices that belong to rectangle $R_i$. Consider the selector function $S(R_k)$ that identifies randomly at most $k \geq 1$ critical points among the vertices of each rectangle. If $k = 7$, the example given in Figure 8.2 will have $opt(I_{S(R_k)}) = opt(I(j))$ because every rectangle has at most seven vertices. Actually, one can show that $opt(I_{S(R_k)}) = opt(I(j))$ holds even when $k = 5$. However, there is a large set of problem instances for which the parameterized algorithm $Alg(S(R_k))$ does not generate solutions with an expected approximation ratio bounded above by a constant. These instances include the ones given in Figure 8.2 after partitioning the rectangles adjacent to all of the shaded rectangles (left and right sides rectangles) into $k$ (stacked) rectangles.

### 8.3.2.3  Selecting One Special Point

Considering the family of instances given in Figure 8.2, the only "good" selector functions $S$ for our parameterized algorithm are those that include for every shaded rectangle its middle point as a critical point. We call these points *special points*, and we formally define them shortly. For the definition of special point, assume that "rectangle" $R_0$, which is just point $p$, is included in $R$, and $p$ is said to be a corner of $R_0$. Depending on the selector function $S$, a subset of the corner points $C(R_i)$ is called the fixed points $F(R_i)$ of rectangle $R_i$. Now, the set of critical points consists of the union of the disjoint sets of fixed points and special points. The middle point of each shaded rectangle in Figure 8.2 has the *minimum connectivity distance property*. By this we mean, in very general terms, that if given all partial corridors that do not include a vertex from rectangle $R_i$, but include vertices from all other rectangles, then a special point of $R_i$ is a vertex in rectangle $R_i$ that is not in $F(R_i)$, and the maximum edge-length needed to connect it to each one of the partial corridors is least possible. Finding special points in this way is in general time consuming. Moreover, this definition is not valid for all problem instances as the set of partial corridors, in which every corridor includes vertices from all the rectangles except from $R_i$, may be empty.

In what follows, we define special points precisely for all problem instances in a way that is computationally easy to identify a set of special points for each rectangle. Special points are identified using an upper bound on the connectivity distance. Given that we have selected a set $F(R_i)$ of fixed points for each rectangle $R_i$, we define a special point as follows. Let $u \in V(R_i)$ and let $T_u$ be a tree of shortest paths rooted at $u$ to all other vertices $(\bigcup_{j \neq i} V(R_j))$ along the sides of rectangle $F$ and the sides of the rooms. Let $SP(u, v)$ be the length of the (shortest) path from vertex $u$ to vertex $v$ along $T_u$. Let $FP(u, R_j)$ be the length of the (shortest) path from point $u \in V(R_i)$ to the "farthest" vertex of rectangle $R_j$ along $T_u$, for $i \neq j$, that is, $FP(u, R_j) = max_{v \in V(R_j)}\{SP(u, v)|u \in V(R_i), j \neq i\}$. In other words, the edge-length needed to connect vertex $u$ of room $R_i$ to any corridor through the connection of room $R_j$ is at most $FP(u, R_j)$. We define the connectivity distance $CD(u, R)$ of vertex $u$ in room $R_i$ as $min_{j \neq i}\{FP(u, R_j)|R_j \in R\}$. If $F(R_i) \subset V(R_i)$, we define the connectivity distance $CD(R_i, R)$ of room $R_i$ as $min_{u \in V(R_i) \backslash F(R_i)}\{CD(u, R)\}$. In other words,

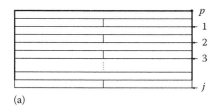 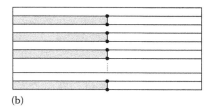

(a)                                                                 (b)

**FIGURE 8.3**  Family of instances $I(j)$ of the $p$-MLC-R problem: (a) Optimal corridor of length $opt(I(j))$ and (b) 2 $SpP$ candidates for shaded rectangles.

$CD(R_i, R)$ is the edge-length needed to connect some specific vertex in $V(R_i) \backslash F(R_i)$ to any corridor through the connection of another room. The special point of $R_i$ is a vertex $u \in V(R_i) \backslash F(R_i)$ such that $CD(u, R) = CD(R_i, R)$. Notice that there may be more than one point satisfying this condition, in which case we select any of these points as the special point. When $F(R_i) = V(R_i)$, then there is no special point. It is important to remember that for the definition of special point, $R_0$ which is simply $p$ is included in $R$.

Consider now the selector function $S(+)$ that identifies one special point from each room $R_i$. The special point for $R_i$ is referred to as $SpP_i$. In this case, $F(R_i)$ is the empty set. For the problem instance given in Figure 8.2, $opt(I(j)) = opt(I_{S(+)}(j))$. However, as we shall see shortly, this property does not hold in general. Consider the $j$-layer family of instances $I(j)$ of the $p$-MLC-R problem given in Figure 8.3, where the height $h$ of $F$ is very small compared with its width $w$.

The rectangle $F$ is partitioned into $j$ layers of rectangles. Each layer is formed by one rectangle of size $w \times \delta$ on top of two rectangles, each of size $\frac{w}{2} \times \delta$, where $2j\delta = h$. An optimal solution is represented by the thick black lines in Figure 8.3a, and its total edge-length $opt(I(j)) = w + 2(h - \delta)$. The special point of each shaded rectangle in Figure 8.3b is either its top-right or bottom-right corner. An optimal solution for the instance $I_{S(+)}$ must include a segment from a rightmost corner of each shaded rectangle to the left- or right-hand side of rectangle $F$. The total edge-length of the optimal solution for the instance $I_{S(+)}$ is $opt(I_{S(+)}) = j\frac{w}{2} + h - \delta$. Making $\delta$ and $h$ approach zero, the ratio $\frac{opt(I_{S(+)})}{opt(I(j))}$ is about $\frac{j}{2}$, and $j$ can be made arbitrarily large. Therefore, $r_{S(+)}$ is not bounded above by any constant. Thus, the restriction $S(+)$ does not result in constant ratio approximation.

#### 8.3.2.4 Selecting Several Special Points

The selector function $S(K+)$ consisting of $k > 1$ special points from each rectangle does not result in constant ratio approximations. There are many instances that show that the ratio $r_{S(K+)}$ is not bounded by any constant. These instances include the problem instances given in Figure 8.3 after stacking $k$ rectangles at every rectangle on the right side of each of the shaded rectangles.

#### 8.3.2.5 Selecting Two Adjacent Corners and One Special Point

The case for $S(2AC+)$ having two adjacent corners and a special point as critical points does not result in a constant ratio approximation, that is, $k_{S(2AC+)}$ is 3, but $r_{S(2AC+)}$ is not bounded above by a constant. The same family of instances of the $p$-MLC-R problem given in the previous case can be used to establish this claim. Obviously, the case for $S$ consisting of one corner and one special point does not result in constant ratio approximations.

### 8.3.3 Constant Ratio Approximations

First, let us identify some important characteristics of selector functions that will enable us to come up with ones for which $r_S$ is bounded above by a constant. Our approach is motivated by two facts. First, the class of instances given in Figure 8.2 requires the selection of one special point from each rectangle $R_i$, as otherwise the resulting ratio will not be bounded by a constant. Second, the two last family of instances of the $p$-MLC-R problem given in Figure 8.3 require the selection of two opposite corners from each rectangle, as otherwise the resulting ratio will not be bounded by a constant. In the following Sections 8.3.3.1 and 8.3.3.2 we discuss the two most promising approaches based on the previous two design principles.

### 8.3.3.1 Selecting Two Opposite Corners and a Special Point

The selector function is called $S(2OC+)$. Without loss of generality, we select the top-right and bottom-left corners as the two opposite corners. Clearly $k_{S(2OC+)} = 3$, and we prove in Reference 5 that $r_{S(2OC+)} = 5$. The proof of this property is quite complex. We give an example to illustrate the main idea behind the proof that we follow to prove that $r_{S(2OC+)}$ is bounded above by a constant. Consider the instance problem $I$ of the $p$-MLC-R problem and its solution $T(I)$, given in Figure 8.4a. The selector function $S(2OC+)$ selects uniformly the top-right and bottom-left corners as well as a special point as critical points for each rectangle. Each rectangle is connected through at least one of its critical points except for the two (shaded) rectangles given in Figure 8.4b. It is easy to see that the solution $T(I)$ is a partial corridor for the instance $I_{S(2OC+)}$ of the $p$-MLC-R$_{S(2OC+)}$ problem, as the two (shaded) rectangles are not connected through at least one of their critical points. Notice that the (shaded) rectangle located on the top-left corner of $F$ has two points located along its right edge with the same minimum connectivity distance value. These two points are the candidates to be special points, and the algorithm may select any of them. Let us say that in this case, we select as the special point the top-most point. In order for the two (shaded) rectangles to be connected, we need to extend the solution $T(I)$. The problem instance $I_{S(2OC+)}$ of the $p$-MLC-R$_{S(2OC+)}$ problem and its solution $T(I_{S(2OC+)})$ are given in Figure 8.4c. Clearly for this example, the total length of the corridor $T(I_{S(2OC+)})$ is less than $r_{S(2OC+)} = 5$ times the total length of the solution T(I). But in general, it does not seem possible to improve on this result.

By Theorem 8.1, this results in an approximation ratio 30 for the parameterized algorithm.

**Theorem 8.2** *For every instance $I$ of the $p$-MLC-R problem, the parameterized algorithm $Alg(S(2OC+))$ generates, in polynomial time with respect to the total number $m$ of rectangles in the partition of $F$, a corridor of length at most $30 \times opt(I)$.*

### 8.3.3.2 Selecting the Four Corners and a Special Point

The critical points in $S(4C+)$ for each rectangle $R_i \in R$ are its four corners and a special point. For this case $k_{S(4C+)} = 5$, and we show in Reference 5 that $r_{S(4C+)} = 3$. Therefore, the approximation ratio of the parametrized algorithm is 30 as in the case of $S(2OC+)$. We claim without stating any further details that the approximation ratio of the parameterized algorithm $Alg(S(4C+))$ is 30. The proof is more complex than the one for $S(2OC+)$, but its main idea is similar to the one for the previous selector function.

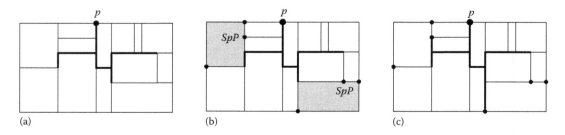

(a)                          (b)                          (c)

**FIGURE 8.4** Constructing a corridor $T(I_{S(2OC+)})$ from $T(I)$: (a) Instance $I \in p$-MLC-R, and its solution $T(I)$, (b) partial corridor T(I) for instance $I_{S(2OC+)} \in p$-MLC-R$_{S(2OC+)}$, and (c) instance $I_{S(2OC+)}$, and its solution $T(I_{S(2OC+)})$.

# 8.4 Heuristics for Solving the MLC-R Problem

In the current section, we implement our approximation algorithms for the $p$-MLC-R$_S$ problem and examine empirically their performance. We begin by formulating our problem as minimum cost integer multicommodity flow problem. Then by relaxing the integer constraints, we formulate it as a linear programming (LP) problem that we solve by using the Gurobi LP package [14]. In Section 8.4.1, we discuss a sophisticated rounding technique to obtain the corridor for the $p$-MLC-R$_S$ problem.

Let $G(I) = (V, E, w)$ be the graph for the instance $I$ of the $p$-MLC-R$_S$ problem as defined in Section 8.1 and $S$ be a selector function. Now let us define the directed graph $H = (V, A)$, in which $A$ is the set of directed edges $(i, j)$ and $(j, i)$ for every edge $\{i, j\}$ in $E$. The two arcs associated with edge $e \in E$ are referred to by $a_1(e)$ and $a_2(e)$. A Corridor is defined by a subgraph $G' = (V', E')$, where $V' \subset V$, $E' \subset E$, and for every rectangle $R_k$ there is a path from vertex $p$ to a critical point defined by $S$ on $R_k$ on $G'$. The Corridor $G'$ may also be specified by the weight function $C(e) = 1$ for every edge in $E'$, and $C(e) = 0$ for every edge in $E \setminus E'$. The variable $y_i$ for vertex $i \in V$ is one if there is a path from vertex $p$ to vertex $i$ in $G'$, and zero otherwise.

The minimum cost integer multicommodity flow objective function is to find the corridor (specified by the weight function $C$) such that $\sum_{e \in E} w(e) \times C(e)$ is least possible. The value $C(e)$ will be interpreted as the capacity of the arcs $a_1(e)$ and $a_2(e)$ in $H$. Now we have to define the constraints for our problem. Commodity $k$ is associated with rectangle $R_k$. The first set of constraints is to find a flow function $f_k$ in $H$ (subject to the typical flow conservation rules) that sends 1 unit of flow from $p$ to one of the critical points of $R_k$. The flow along each directed arc $a_1(e)$ and $a_2(e)$ denoted by $f_k(a_1(e))$ and $f_k(a_2(e))$ of the two-directional edges associated with edge $e$ must be such that $f_k(a_1(e)) \leq C(e)$ and $f_k(a_2(e)) \leq C(e)$. Note that our constrains differ from the classical multicommodity flow problem. Instead of requiring that the sum of the flows from all commodities along a given edge to be at most the capacity of the arc, we just require that the flow of each commodity along each arc be at most the capacity of the arc. The variable $y_i$ defines the maximum flow of a commodity from vertex $p$ to vertex $i$.

By using standard techniques, one can formulate our minimum cost integer multicommodity flow problem as an integer LP problem. But relaxing the integer constraints (on the flows) we obtain a LP problem that we solve via the Gurobi optimization package [14]. The resulting function $C(e)$ is a fractional corridor that we need to transform into a valid corridor. The fractional corridor defines fractional values for the $y_i$ values. In Section 8.4.1, we outline our sophisticated rounding strategy.

## 8.4.1 Rounding Scheme

The solution generated by the LP model consists of a set of real values for the capacities $C(e)$ and the $y_i$ values for every vertex (the flow value from the access point $p$ to vertex $i$). Our rounding scheme will refine the set of critical points $CP(R_k)$ for some rectangles and then will resolve the LP program. After repeating this process several times, we use the solution to the last LP problem to generate the final corridor.

Before we proceed with the rounding scheme, we introduce additional notation. The current set of critical points for $R_k$ is denoted by $CSCP(R_k)$. Let $\overline{y}_k = \frac{\sum_{i \in CSCP(R_k)} y_i}{|CSCP(R_k)|}$ be the average flow going in through the vertices in $CSCP(R_k)$ of rectangle $R_k$. We use $R(i)$ to denote the set of rectangles where vertex $i$ is a current critical point.

Algorithm 1 specifies in detail one iteration of our refinement process.

Once we have refined our critical points and solved the corresponding LP problems several times, we use the final solution to the LP problem to generate our corridor. The idea is to select vertex $p$ and all the vertices with above average flow from the access point $p$. We use all of these points as the terminal points for a ST problem. Then we just construct a ST as follows.

We construct a complete graph $Q$ where each vertex $i$ is a vertex in $G$ that satisfies the *average flow condition* $y_i \geq (\overline{y}_k - \epsilon)$ for at least one of the rectangles $R_k \in R(i)$. The edge length of $e = \{i, j\}$ is equal to the length of the shortest path joining the vertices $i$ and $j$ through the original graph $G = (V, E, w)$. Thus,

## ALGORITHM 8.1    Refinement Process

1: **for** every vertex $i$ **do**
2:     Let $R(i) = \{k | i \in CP(R_k)\}$
3: **end for**
4: **for** each rectangle $R_k$ **do**
5:     $CSCP(R_k) \leftarrow CP(R_k)$
6:     $\bar{y}_k = \frac{\sum_{i \in CSCP(R_k)} y_i}{|CSCP(R_k)|}$
7:     **for** every vertex $i \in CSCP(R_k)$ **do**
8:         **if** $y_i \leq \bar{y}_k - \epsilon$ **then**
9:             Assign label $Y$ to $l_{i,k}$
10:        **else**
11:            Assign label $N$ to $l_{i,k}$
12:        **end if**
13:    **end for**
14:    **if** more than one of the critical points in $R_k$ are labeled $N$ **then**
15:        Change all labels $N$ to $M$ for $R_k$
16:    **end if**
17: **end for**
18: **while** there is a vertex $i$ with all labels $Y$ or $M$ in $R(i)$ **do**
19:    Delete vertex $i$ from every set $CSCP(R_k)$
20:    **while** there is only one vertex in some $CSCP(R_k)$ with label $M$ **do**
21:        Change to $N$ the label of such vertex in $CSCP(R_k)$
22:    **end while**
23: **end while**
24: **for** each rectangle $R_k$ **do**
25:    $CP(R_k) \leftarrow CSCP(R_k)$
26: **end for**

$Q$ is a complete graph that interconnects all its vertices. Let $q$ be the number of vertices in $Q$. For $Q$, we compute its corresponding minimum-cost spanning tree.

In the final step, we just delete leaves from the minimum-cost spanning tree that do not affect connectivity in our original problem. The resulting tree is the approximation solution to our $p$-MLC-R problem instance.

## 8.5  Experimental Evaluation

We perform an experimental evaluation of our LP based algorithm and the rounding scheme discussed in the previous section. The LP problems were solved by the Gurobi LP Solver [14]. We run our algorithm with the following two selector functions: (a) selecting for each rectangle the top-left and bottom-right corners as well as a special point as critical points ($S(2OC+)$) and (b) selecting all the vertices of each rectangle as critical points ($S(AllP)$).

All the instances used in our evaluation have a room configuration that may be viewed as a guillotine partition, that is, the boundary is partitioned by a horizontal or vertical cut into two rectangles and each of the resulting rectangles may be a room or a guillotine partition.

All the instances used in our evaluation fall into two categories: nonuniform and uniform. The difference between them is that in the uniform case, there are bounds on the aspect ratio of the rooms, but the aspect ratio is not limited for the nonuniform case. The number of rectangles were between 5 and 400.

All the evaluations were performed on a Dell Computer with an Intel Core I5-2400 CPU machine operating at 3.10 GHz X 4 (Cores) with 8 GB of RAM memory under Fedora21 (x86 - 64).

### 8.5.1 Nonuniform Instances

For this case, we run our code with 5 to 230 rooms. To compute the approximation ratio of the solutions generated, we used the value generated for the LP formulation as the optimal solution value for the corridor problem. Note that in general, the value of the optimal solution to our corridor problem is larger. For all the instances, the actual approximation ratio was not more than two. Moreover, the approximation ratio using the two selector functions is about the same (Figure 8.5a).

Significantly enough, the actual running-time used for both selector functions ($S(2OC+)$ and $S(AllP)$) is almost the same (Figure 8.5b). The actual running-time for the largest instances with $r = 220$ and $230$ is around 15 and 20 hours, respectively, for both selector functions.

In Figure 8.5, we present the total-edge length for both selector functions. For each selector function, we also present the lower bound for the total-edge length produced by the LP solver.

For this family of problem instances, in which the size of some rectangles can end up being almost as large as the rectangular boundary, combined with many small size rectangles, the total edge-length may decrease when the number of rectangles grows. For example, you can see this anomaly for 90 rectangles in Figure 8.5c. Figure 8.6 corresponds to the problem instances with 80 and 90 rectangles, respectively. Note that both boundaries have the same height and width, but the total edge-length decreases from $21,415$ to $6473$ length units for the selector function $S(AllP)$.

It is interesting to note that the selector function is not a decisive factor that guarantees a better solution. The results in Figure 8.5c show that the total edge-lengths for both the $S(2OC+)$ and $S(AllP)$ are similar. On the other hand, the lower bounds provided by the LP Solver for both selector functions are about the same.

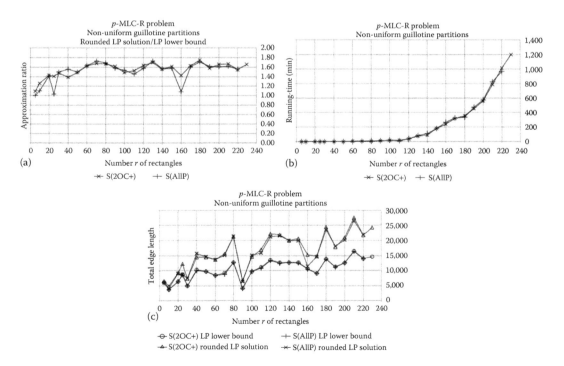

**FIGURE 8.5**  Experimental results for nonuniform guillotine partitions: (a) approximation ratio, (b) running-time, and (c) total edge length.

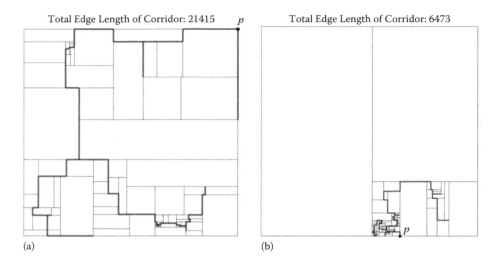

**FIGURE 8.6**    A $p$-MLC-R$_{S(AllP)}$ problem instance: (a) $r = 80$ and (b) $r = 90$.

## 8.5.2   Uniform Guillotine Partitions

In this case, we generated slightly larger instances in which the number $r$ of rectangles was in the range 5 and 400. The approximation ratio is computed as in the previous subsection. Figure 8.7a presents the resulting approximation ratio, which is again not more than 2 for both selector functions used. The largest instance for 400 rectangles takes about 8 days for both selector functions ($S(2OC+)$ and $S(AllP)$). This is why we could not run larger instances (Figure 8.7b). The resulting total edge-length of the corridors generated is also invariant with respect to the selector function used (Figure 8.7c). As expected, the optimal

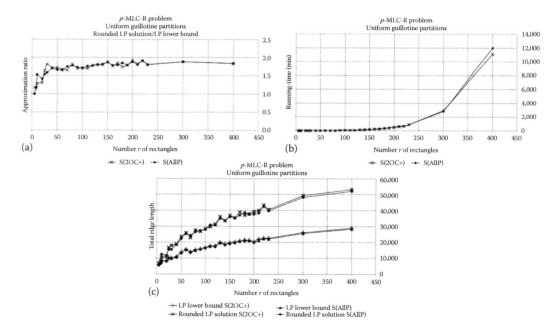

**FIGURE 8.7**   Experimental results for uniform guillotine partitions: (a) approximation ratio, (b) running-time, and (c) total edge-length.

solution to the LP problems is the same for the selector functions $S(2OC+)$ and $S(AllP)$. Because of the relative uniformity of the instances generated, in most of the cases, both the total edge-length and the lower bound grow as the number of rectangles grows.

## 8.6 Conclusion

We have developed a constant ratio-approximation algorithm for the TRA-MLC-R problem [5]. However, the approximation ratio is large (30). The reason for this is mainly due to the approximation ratio of Slavik's algorithm for the TEC problem. But note that it is not for graphs limited to 2D. Our experimental evaluation indicates that the ratio is at just 2. It is likely then that one can develop an algorithm with an approximation ratio of 2 for the TRA-MLC-R problem.

Another interesting open problem is to either develop a constant ratio-approximation algorithm or establish constant-ratio inapproximability results for the TRA-MLC problem. We believe an inapproximability result is more likely for this case.

## References

1. E. D. Demaine and J. O'Rourke. Open problems from CCCG 2000. In *Proceedings of the 13th Canadian Conference on Computational Geometry (CCCG 2001)*, pp. 185–187, 2001.
2. D. Eppstein. Some open problems in graph theory and computational geometry. PDF file (3.89 MB), World Wide Web, http://www.ics.uci.edu/~eppstein/200-f01.pdf, November 2001.
3. A. Gonzalez-Gutierrez and T. F. Gonzalez. Complexity of the minimum-length corridor problem. *Computational Geometry*, 37(2):72–103, 2007.
4. H. L. Bodlaender, C. Feremans, A. Grigoriev, E. Penninkx, R. Sitters, and T. Wolle. On the minimum corridor connection problem and other generalized geometric problems. *Computational Geometry*, 42(9):939–951, 2009.
5. A. Gonzalez-Gutierrez and T. F. Gonzalez. Approximating corridors and tours via restriction and relaxation techniques. *ACM Transactions on Algorithms*, 6(3):56:1–56:36, 2010. doi:10.1145/1798596.1798609.
6. G. Reich and P. Widmayer. Beyond Steiner's problem: A VLSI oriented generalization. In *WG '89: Proceedings of the Fifteenth International Workshop on Graph-theoretic Concepts in Computer Science*, pp. 196–210. Springer-Verlag, New York, 1990.
7. R. M. Karp. Reducibility among combinatorial problems. *Complexity of Computer Communications, Proceedings of a Symposium on the Complexity of Computers Communications*, March 20–22, IBM Thomas J. Watson Center, Yorktown Heights, NY, R. E. Miller and J. W. Thatcher (Eds.), pp. 85–103. Plenum Press, New York, 1972.
8. C. D. Bateman, C. S. Helvig, G. Robins, and A. Zelikovsky. Provably good routing tree construction with multi-port terminals. In *ISPD '97: Proceedings of the 1997 International Symposium on Physical Design*, pp. 96–102. ACM Press, New York, 1997.
9. C. S. Helvig, G. Robins, and A. Zelikovsky. An improved approximation scheme for the group Steiner problem. *Networks*, 37(1):8–20, 2001.
10. E. Ihler. Bounds on the quality of approximate solutions to the group Steiner problem. In *WG '90: Proceedings of the 16th International Workshop on Graph-theoretic Concepts in Computer Science*, pp. 109–118. Springer-Verlag, New York, 1991.
11. S. Safra and O. Schwartz. On the complexity of approximating TSP with neighborhoods and related problems. *Computational Complexity*, 14(4):281–307, 2006. doi:10.1007/s00037-005-0200-3.

12. P. Slavik. The errand scheduling problem. Technical Report 97-02, Department of Computer Science and Engineering, University of New York at Buffalo, 1997. http://www.cse.buffalo.edu/tech-reports/.

13. P. Slavik. Approximation algorithms for set cover and related problems. PhD Thesis 98-06, Department of Computer Science and Engineering, University of New York at Buffalo, 1998. http://www.cse.buffalo.edu/tech-reports/.

14. Gurobi Optimization Inc. Gurobi optimizer reference manual, 2016. http://www.gurobi.com.

# Approximate Clustering

Ragesh Jaiswal

Sandeep Sen

## 9.1 Introduction

Clustering is the task of partitioning a given set of objects so that similar objects belong to the same group. This general idea is extremely useful in unsupervised learning in which little to no prior knowledge about the data is available. Clustering is usually the first data-analysis technique employed when analyzing big data. The importance of clustering and its applications may be found in a number of books [1] on the subject, and we avoid the detailed discussion here. Instead, we focus on formulating clustering and analyzing some mathematically well-defined problems. The first issue that we need to address when formulating the clustering problem is how is the data represented? In case it is possible to study and measure the defining properties and attributes of the objects that need to be clustered, then individual objects may be represented as points in some Euclidean space $\mathbb{R}^d$, where the dimension $d$ denotes the number of attributes that define objects. In this scenario, the similarity or dissimilarity between objects may be defined as a function of the Euclidean distance between the points corresponding to the objects. In many situations, representing data as points in Euclidean space may not be possible, and all that is known is how similar or dissimilar a pair of objects is. As the goal of clustering is to group similar objects in the same group, this pairwise similarity/dissimilarity information suffices. In such cases, what is given is a distance function $D : \mathcal{X} \times \mathcal{X} \to \mathbb{R}^+$, where $\mathcal{X}$ denotes the set of objects. In most practical scenarios, $(\mathcal{X}, D)$ is a *metric*. Any metric $(\mathcal{X}, D)$ satisfies the following three properties: (i) $\forall x \in \mathcal{X}, D(x, x) = 0$, (ii) $\forall x, y \in \mathcal{X}, D(x, y) = D(y, x)$ (symmetry), and (iii) $\forall x, y, z \in \mathcal{X}, D(x, z) \leq D(x, y) + D(y, z)$ (triangle inequality). The Euclidean setting may be regarded as a special case of the metric setting where $\mathcal{X} = \mathbb{R}^d$ and where the distance $D(x, y)$ between two points $x, y \in \mathcal{X}$ is the Euclidean distance (or squared Euclidean distance).

Given any set of objects $X \subseteq \mathcal{X}$, the goal of clustering is to partition $X$ into sets $X_1, \ldots, X_k$ (these sets are called *clusters*) for some $k$ such that for any pair $(x, y)$ of data points in the same cluster, $D(x, y)$ is small and for any pair of points in different clusters, $D(x, y)$ is large. In many scenarios, the number of clusters $k$ is known. In such scenarios, it makes sense to define the clustering problem in one of the following ways:

- *Clustering problem 1*: Partition the given dataset $X$ into $k$ clusters $X_1, \ldots, X_k$ such that the following cost function is minimized: $\Psi_{sum}(X_1, \ldots, X_k) \overset{def.}{=} \sum_{i=1}^{k} \sum_{x,y \in X_i} D(x, y)$.
- *Clustering problem 2*: Partition the given dataset $X$ into $k$ clusters $X_1, \ldots, X_k$ such that the following cost function is minimized: $\Psi_{max}(X_1, \ldots, X_k) \overset{def.}{=} \sum_{i=1}^{k} \max_{x,y \in X_i} D(x, y)$.

Note that the goal of minimizing the above-mentioned objective functions is aligned with the goal of putting similar objects in the same cluster and different objects in different clusters[*]. So, these could be the problems most relevant for clustering. However, the previous problem statements are in terms of partitions of the given dataset $X$. Partitions are complex objects (in terms of description), and it would nicer if there was an alternate formulation in which the solution is much simpler to describe compared with a partition. Consider the following idea: Suppose we are given $k$ points $\{c_1, \ldots, c_k\} \subseteq \mathcal{X}$ (let us call them *centers*) such that each of these centers *represent* a cluster. That is, all points that have the closest centers as $c_i$ are in the cluster $X_i$. This is also known as *Voronoi partitioning*. Note that as $D$ is a metric, all points in cluster $X_i$ defined as earlier will be close to each other. So, these centers may be used as compact representation of the clusters that they define, and then the clustering goal would be to find good centers. This motivates the following clustering problems that we will study in the remainder of this chapter.

- *Clustering problem 3 (Metric-k-median)*: Given dataset $X \subseteq \mathcal{X}$, find $k$ centers $C = \{c_1, \ldots, c_k\} \subseteq X$ such that the following cost function is minimized: $\Psi(C, X) \overset{def.}{=} \sum_{x \in X} \min_{c \in C} D(x, c)$.
- *Clustering problem 4 (Metric-k-center)*: Given dataset $X \subseteq \mathcal{X}$, find $k$ centers $C = \{c_1, \ldots, c_k\} \subseteq X$ such that the following cost function is minimized: $\Psi_{max}(C, X) \overset{def.}{=} \max_{x \in X}\{\min_{c \in C} D(x, c)\}$.

In the Euclidean setting where $\mathcal{X} = \mathbb{R}^d, D(x, y) = ||x - y||$ ($||.||$ denotes Euclidean distance), and $C$ is allowed to be any set of $k$-points from $\mathcal{X}$, these problems are called the Euclidean-$k$-median and Euclidean-$k$-center problem. Furthermore, $D(x, y) = ||x - y||^2$, the problem is called the Euclidean-$k$-means problem. The Euclidean-$k$-means problem is one of the most studied clustering problems and is popularly known as the $k$-means problem. For the sake of clarity, we give the description of these problems in the following:

- *Clustering problem 5 (Euclidean-k-median)*: Given dataset $X \subseteq \mathbb{R}^d$, find $k$ centers $C = \{c_1, \ldots, c_k\} \subseteq \mathbb{R}^d$ such that the following cost function is minimized: $\phi(C, X) \overset{def.}{=} \sum_{x \in X} \min_{c \in C} ||c - x||$.
- *Clustering problem 6 (Euclidean-k-center)*: Given dataset $X \subseteq \mathbb{R}^d$, find $k$ centers $C = \{c_1, \ldots, c_k\} \subseteq \mathbb{R}^d$ such that the following cost function is minimized: $\phi_{max}(C, X) \overset{def.}{=} \max_{x \in X}\{\min_{c \in C} ||c - x||\}$.
- *Clustering problem 7 (k-means)*: Given dataset $X \subseteq \mathbb{R}^d$, find $k$ centers $C = \{c_1, \ldots, c_k\} \subseteq \mathbb{R}^d$ such that the following cost function is minimized: $\Phi(C, X) \overset{def.}{=} \sum_{x \in X} \min_{c \in C} ||c - x||^2$.

Clustering problems 3, 4, 5, 6, and 7 together are known as *center-based clustering* problems. As they are related to Voronoi partitioning, a brute-force algorithm may end up generating all possible Voronoi partitioning of $n$ points in $d$ dimensions. It is known [2] that there are $O(n^{kd})$ possible partitioning defined by $k$ centers. Clearly, this would be completely impractical. We will be discussing only these center-based clustering problems in the remainder of this chapter. The corresponding weighted version of the above-mentioned problems are also important for many applications, in particular, for many facility location problems. Each point has an associated positive real number, and the objective function is weighted by the corresponding weight for each of the terms. The weighted versions can sometimes be solved by the same algorithms with some additional adjustments in the analysis. In this chapter, we will address the unweighted versions for ease of exposition.

## 9.2 The $k$-Center Problem

The two versions of the problem that we consider in this section is the Metric-$k$-center problem (problem 4) and the Euclidean-$k$-center problem (problem 6). For both these versions of the $k$-center problem, even approximating the solution to within a fixed constant factor is NP-hard [3–5]. For the Euclidean-$k$-center problem, this approximation factor has been shown to be 1.822, and for the Metric

---

[*] However, note that the distances *across* clusters has no contribution in the cost function.

version, the approximation factor is known to be 2. Bern and Eppstein [6] gave a nice summary of these results in their book chapter about approximation algorithms for geometric problems.

*Sequential-selection*:  Here, we will focus on the simple 2-factor approximation algorithm for the Metric-$k$-center problem (and hence also the Euclidean-$k$-center) that is based on iteratively picking $k$ centers such that the choice of the $i$th center depends on all the previously chosen centers. Let us call such algorithms *sequential-selection* algorithms. We will see more such sequential-selection algorithms in this chapter. The nice property of these algorithms is that they are extremely simple (and hence easy to implement and test) and in many cases run very fast in practice. When solving the $k$-center problem, the high-level objective is to find $k$ centers that are well distributed within the given data points. Solutions in which two centers are very close to each other may not optimize the objective function well. Consider a very simple algorithm that iteratively picks $k$ centers that are far apart from each other. This algorithm is called the *Farthest First Traversal* algorithm [5]. This algorithm is used as an effective heuristic for many different problems such as the Traveling Salesman Problem. For the Metric-$k$-center problem, we will show that this algorithm gives a 2-factor approximation. The hardness of approximation guarantee discussed in the previous paragraphs tells us that this is the best approximation guarantee that can be achieved unless $P = NP$.

> (*Farthest first-traversal*): Let $X$ denote the given points and $D(.,.)$ be the distance function. Pick the first center arbitrarily from the given points. After having picked $(i - 1)$ centers denoted by $C_{i-1}$, pick a point $p \in X$ to be the $i$th center $c_i$ such that $p$ is the farthest point from all centers in $C_{i-1}$. That is, $c_i = \arg\max_{x \in X} \left[ \min_{c \in C_{i-1}} D(x, c) \right]$.

The next theorem shows that the above-mentioned algorithm gives a 2-factor approximation.

**Theorem 9.1** *For any metric $(\mathcal{X}, D)$ and any dataset $X \subseteq \mathcal{X}$, let $C$ be the $k$ centers produced by the farthest first-traversal algorithm and let $C^*$ denote any optimal solution. Then $\Psi_{max}(C, X) \leq 2 \cdot \Psi_{max}(C^*, X)$.*

*Proof.* Let $y = \arg\max_{x \in X}\{\min_{c \in C} D(x, c)\}$. So, $y$ is the point that is farthest from the set of centers $C$ and let $R = \min_{c \in C} D(y, c)$ be this farthest distance. Consider the set $C' = C \cup \{y\}$. First, observe that for every point $x \in \mathcal{X}$, $D(x, C) \leq R$. Moreover, note that due to the manner in which the centers in the set $C$ are picked, each pair of points in the set $C'$ have distance at least $R$. Now, at least two points in the set $C'$ share the same nearest center in the set $C^*$. Let these points be $p$ and $q$ and the center be $c^*$. Then by triangle inequality, we have $D(p, c^*) + D(q, c^*) \geq D(p, q) \geq R$. So, we have $\max(D(p, c^*), D(q, c^*)) \geq R/2$ that further gives $\Psi_{max}(C^*, X) \geq \Psi_{max}(C, X)/2$. ∎

Note that the approximation factor of 2 given by the aforementioned algorithm holds for both Metric and Euclidean versions of the $k$-center problem. As mentioned earlier, it has been shown that one cannot get a better approximation than 2 for the Metric version of the problem unless $P = NP$. However, for the Euclidean version, the known lower bound on the approximation factor is 1.822. So, the following problem remains open to the best of our knowledge: Is it possible to design an efficient algorithm for the Euclidean $k$-center problem that gives approximation guarantee better than 2? Is it possible to argue that one cannot find an efficient algorithm with approximation guarantee $c > 1.822$ unless $P = NP$?

## 9.3 The $k$-Means/Median Problem

We discuss the $k$-Means (problem 7) and $k$-Median (problems 3 and 5) problems together in this section. The reason for similar treatment of $k$-means and $k$-median is because the statements of the these problems are similar. Recall that in the $k$-means problem, the goal is to minimize the sum of squared distances, whereas the goal in the $k$-median problem is to minimize the sum of distances. However, there are some crucial differences that one should note between these problems. One key difference due to squared versus nonsquared distance is that in the Euclidean setting, the 1-median problem, also known as the

*Geometric Median* problem or a special case of the *Fermat–Weber* problem, is a problem of unknown difficulty* whereas the 1-means problem is easy. In fact, there is a closed form expression for the solution of any given 1-means problem. The geometric mean of all the given points is the optimal solution for the problem. This is evident from the following well-known fact.

**Fact 9.1** *For any set $X \subseteq \mathbb{R}^d$ and any point $p \in \mathbb{R}^d$, we have*

$$\sum_{x \in X} ||x - p||^2 = \sum_{x \in X} ||x - \mu(X)||^2 + |X| \cdot ||p - \mu(X)||^2,$$

*where $\mu(X) \stackrel{def.}{=} \frac{\sum_{x \in X} x}{|X|}$ is the geometric mean (or centroid) of the points in the set X.*

Another, important difference is due to the triangle inequality that is satisfied when distance is being considered as in the $k$-median problem as opposed to squared distance that does not satisfy the triangle inequality[†]. This happens to be one of the main reasons why the techniques of Arora et al. [7] that work for the Euclidean $k$-median problem cannot be made to work for the $k$-means problem.

*Hardness:* The $k$-means and the Euclidean $k$-median problem are simple in one dimension (i.e., $d = 1$). There is a simple dynamic programming approach that efficiently solves the 1-dimensional version of these problems. However, the planar version ($d = 2$) and higher dimensional version ($d > 2$) of both the $k$-means [8–10] and $k$-median [4] have been shown to be NP-hard. So, the next natural question is how hard are these problems to approximate? For the $k$-means problem, Awasthi et al. [11] showed that there exists a constant $0 < c < 1$ such that it is NP-hard to approximate the $k$-means problem to a factor better than $(1 + c)$. For the Euclidean $k$-median problem, Guruswami and Indyk [12] showed that under standard complexity theoretic assumptions, one cannot obtain an efficient algorithm that gives $(1 + \varepsilon)$ approximation for an arbitrary small $\varepsilon$. Similarly for the metric $k$-median problem, it was shown by Jain et al. [13] that there is no efficient algorithm that achieves an approximation factor better than $(1 + 2/e)$ conditioned on some standard complexity theoretic assumption.

*Approximation algorithms:* On the positive side, there are efficient constant factor approximation algorithms for the $k$-means/median problems. More specifically, for the $k$-median problem (Metric and Euclidean versions), the best known efficient constant factor approximation algorithm is by Arya et al. [14] who gave a $(3 + \varepsilon)$-factor approximation algorithm. Similarly, for the $k$-means algorithm, the best known efficient constant factor approximation algorithm is by Kanungo et al. [15] who gave a $(9+\varepsilon)$-factor approximation algorithm[‡]. Both the above-mentioned algorithms are based on a technique known as *local search*. We discuss these techniques further in the next few subsections.

As far as approximation schemes are concerned, there are PTAS (Polynomial Time Approximation Scheme)[§] for these problems for special cases when either $k$ or the dimension $d$ is a fixed constant. Let us first discuss the special case in which the dimension $d$ is a fixed constant. In a significant technical breakthrough, Arora et al. [7] gave a PTAS for the Euclidean $k$-median problem when $d$ is a constant. The running time of this algorithm was later improved by Kolliopoulos and Rao [16]. Note that the dynamic programming-based approach of References 7, 16 did not extend to the $k$-means problem for reasons mentioned earlier and until recently the problem of obtaining a PTAS for $k$-means in constant dimension was open. In recent developments, Addad et al. [17] and Friggstad et al. [18] have obtained PTAS for

---

* Even though simple approximation algorithms exist.
[†] The triangle inequality says that $\forall p, q, r \in \mathbb{R}^d, ||p - q|| + ||q - r|| \geq ||p - q||$. Note that even though the sqaured Euclidean distance does not satisfy the triangle inequality, it does satisfy an *approximate* version $\forall p, q, r \in \mathbb{R}^d, 2 \cdot ||p - q||^2 + 2 \cdot ||q - r||^2 \geq ||p - q||^2$.
[‡] The running time of the algorithms in References 14 and 15 is polynomial in the input parameters and in $1/\varepsilon$.
[§] PTASs are $(1 + \varepsilon)$-factor approximation algorithms for any given $\varepsilon$. The running time of such algorithms is polynomial in the input parameters but can be exponential in $1/\varepsilon$.

the $k$-means problem in constant dimension. For the case when $k$ is a fixed constant, PTAS for the $k$-means problem has been known. Kumar et al. [19] gave the first PTAS for the $k$-means problems under the assumption that $k$ is a fixed constant. Using different techniques, Feldman et al. [20] and Jaiswal et al. [21,22] gave algorithms with improved running time.

*Techniques*: In the remainder of this section, we discuss the $k$-median and $k$-means problem with respect to the three main techniques or approaches that are used to obtain approximate solutions for these problems: (i) *linear program (LP)*, (ii) *local search*, and (iii) *sampling*. We discuss these three approaches in the next three subsections. We will discuss the sampling-based approach more elaborately because of the simplicity of the ideas involved that makes the resulting algorithms more practical.

## 9.3.1 Linear Program and Rounding

Consider the Metric-$k$-median problem. Note that the centers $C$ are supposed to be a subset of the given point set (i.e., $C \subseteq X \subseteq \mathcal{X}$). This allows us to write a simple LP for this problem. We will denote the given $n$ points with integers $1, \ldots, n$ and the distance between points $i$ and $j$ with $D(i,j)$. Let $y_i$ denote whether the point $i$ is chosen as a center (i.e., $y_i = 1$ when point $i$ is chosen as a center, otherwise 0). Similarly, let $x_{ij}$ denote whether point $j$ is assigned to center $i$. We can write the following program with respect to these variables:

$$\text{Minimize:} \quad \sum_{i,j} D(i,j) \cdot x_{ij}$$

$$\text{Subject to:} \quad \forall j, \sum_i x_{ij} = 1 \quad \text{and} \quad \forall i, j, x_{ij} \leq y_i \quad \text{and} \quad \sum_i y_i \leq k \quad \text{and} \quad \forall i, j, y_i, x_{ij} \in \{0, 1\}$$

The main idea in obtaining approximation algorithm using the previous integer LP is to solve a *relaxed* version of the program (i.e., relax the last two integer constraints to $0 \leq x_{ij} \leq 1, 0 \leq y_i \leq 1$) and then *round* the noninteger solution to obtain an approximate solution*. Lin and Vitter [23] developed a bicriteria rounding technique that gave a *pseudoapproximation* algorithm in which the number of centers produced is $2k$ and the cost is at most four times the optimal cost (with respect to $k$ centers). Such algorithms are also called *bicriteria* approximation algorithms. Charikar et al. [24] gave a rounding algorithm that gives an approximation guarantee of $6\frac{2}{3}$. This was later improved to 3.25 by Charikar and Li [25].

## 9.3.2 Local Search

*Local search* is another very highly explored technique in the context of the center-based clustering problems. This technique has given some very interesting results. The main idea of local search can be summarized as follows:

> (*Local search*): Start with a set of $k$ centers, and in each step update the set of $k$ centers by making a *local* change. Perform these local changes interatively until the set of centers satisfy some *stability* condition. The local change step is of the following nature: Suppose at the start of the step the set of centers is $C$, then in the current step update the centers from $C$ to $C'$ by performing one (single-swap) or few (multiswap) *swaps* such that $C'$ is better than $C$.

Note that due to the nature of the local search algorithm, the centers that are produced by the algorithm are either from among the given points or from among a set of *possible* centers (that may be chosen in a preprocessing step based on the input points). Arya et al. [14] showed that multiswap local search gives a $(3 + \varepsilon)$-factor approximation algorithm for the metric $k$-median problem with polynomial running time.

---

* Note that it is known that the *integrality gap* of the above-mentioned LP relaxation is at least 2. It is also known from the work of Archer et al. [26] that the integrality gap is upper bounded by 3.

Using similar ideas, Kanungo et al. [15] showed that the multiswap local search gives a $(9 + \varepsilon)$-factor approximation for the $k$-means problem. In more recent developments, Addad et al. [17] and Friggstad et al. [18] showed that the local search technique gives a PTAS[*] for the $k$-means problem for the case when $d$ is a fixed constant.

### 9.3.3 Sampling

Despite all the development in approximation algorithms for the clustering problems that we have discussed here, the algorithms that are used to solve some of these problems in practice are heuristics without any theoretical guarantees. The reason for using such heuristics over algorithms with guarantees is that these heuristics are extremely simple (so implementation and debugging is simple), and they produce very good results on real world input data. For instance, the heuristic that is used to solve the $k$-means problem in practice is called the *Lloyd's* algorithm. The algorithm works by starting with $k$ centers picked arbitrarily, and then in a sequence of steps *improving* the $k$ centers. A single improvement step consists of two parts: (1) perform a Voronoi partition with respect to the $k$ centers and (2) update the $k$ centers to the centroid of the $k$ Voronoi partitions. This procedure does not give any approximation guarantees. Figure 9.1 gives a simple example to show this. However, if started with $k$ centers with cost at most $c$ times the optimal, then the Lloyd's algorithm trivially gives a $c$ factor approximation as the procedure can only improve the solution. This encourages development of *seeding* algorithms that are fast and simple and that give some approximation guarantee. Such an algorithm, followed by Lloyd's procedure, would provide a "best of theoretical and practical worlds" scenario.

Sampling-based techniques give rise to such extremely simple and practical seeding algorithms. Note that even though we are calling them seeding algorithms, these algorithms are approximate clustering algorithms by themselves as they produce $k$ centers with provably bounded cost. One of the most popular seeding algorithm is called the $k$-means++ seeding algorithm that in essence is a randomized *sequential-selection* algorithm. This simple sampling procedure also called the $D^2$-sampling procedure. We will use $k$-means++ and $D^2$-sampling interchangeably in this chapter. The description of this algorithm is given as follows.

> (k-*means++ seeding or* $D^2$-*sampling*): Let $X$ denote the given points and $d(.,.)$ be the distance function. Pick the first center randomly from the given points. After having picked $(i - 1)$ centers denoted by $C_{i-1}$, pick a point $p \in X$ to be the $i$th center with probability proportional to $\min_{c \in C_{i-1}} d(x, c)$.

In the context of the $k$-means problem, the distance function $d$ is the squared Euclidean distance (hence the name $D^2$-sampling). This sampling procedure runs fast in practice and has a worst-case running time of $O(nkd)$. Figure 9.2 demonstrates the sampling procedure on a simple two-dimensional problem

**FIGURE 9.1**  A simple one dimensional example for showing that Lloyd's algorithm may not give any approximation guarantees. Consider four points as shown earlier and let their labels also be their x coordinates. For $k = 3$, the optimal centers are located at $a$, $b$, and $(c + d)/2$ with cost $(d - c)^2/2$. However, if the Lloyd's algorithm picks $a, c, d$ as the initial three centers, then the algorithm outputs centers $(a + b)/2, c, d$ with cost $(b - a)^2/2$ which may be arbitrarily bad compared with the optimal.

---

[*]  As explained earlier, a PTAS is a $(1 + \varepsilon)$-approximation in which the running time is polynomial in the input parameters but is allowed to be exponential in $1/\varepsilon$.

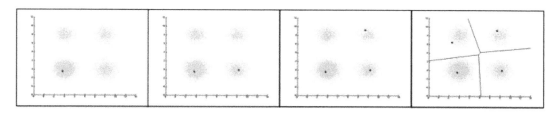

**FIGURE 9.2** The figure shows a two-dimensional data being clustered into four clusters using the sampling algorithm. The algorithm samples the centers in four simple steps. The first center is chosen randomly. For the second center, the points that are further away from the first center is given more priority and so on.

instance with four clusters. Arthur and Vassilvitskii [27] showed that the above-mentioned sampling procedure gives an $O(\log k)$-factor approximation guarantee in expectation. The formal statement of their result is given in the following

**Theorem 9.2 (Theorem 3.1 in Reference 27)** *Let $X \subseteq \mathbb{R}^d$ denote the data points, and let $C$ denote the set of $k$ centers produced by the $k$-means++ seeding algorithm. Then we have $\mathbf{E}[\Phi(C, X)] \leq 8(\ln k + 2) \cdot \Phi_{OPT}(X)$, where $\Phi_{OPT}(X)$ denotes the optimal $k$-means cost for the input $X$.*

The previous result generalizes for cost functions of the following form:

$$\Phi^\ell(C, X) = \sum_{x \in X} \min_{c \in C} ||x - c||^\ell \tag{9.1}$$

Note that the $k$-median cost function corresponds to $\ell = 1$. For such cost functions, we will use $D^\ell$-sampling instead of $D^2$-sampling. The more general result shows approximation guarantee for the $D^\ell$-sampling procedure.

**Theorem 9.3 (Theorem 5.1 in Reference 27)** *Let $X \subseteq \mathbb{R}^d$ denote the data points, and let $C$ denote the set of $k$ centers produced by the $D^\ell$-sampling algorithm. Then we have $\mathbf{E}[\Phi(C, X)] \leq 2^{2\ell}(\ln k + 2) \cdot \Phi^\ell_{OPT}(X)$, where $\Phi^\ell_{OPT}(X)$ denotes the optimal cost for the input $X$ (defined in Equation 9.1).*

The aforementioned results are obtained by making clever use of the next two lemmas in an induction-based argument. These lemmas form the foundation for analyzing such sampling algorithms. We state these lemmas in the context of the $k$-means objective function and $D^2$-sampling (i.e., $\ell = 2$).

**Lemma 9.1 (Lemma 3.1 in Reference 27)** *Let $A$ denote points in any optimal $k$-means cluster of data $X$. Let $C$ denote a randomly chosen center from $A$. Then $\mathbf{E}[\Phi(C, A)] \leq 2 \cdot \Delta(A)$. Here, $\Delta(A)$ denotes the optimal 1-means cost of $A$.*

Let $X_1, \ldots, X_k$ denote the optimal clusters for any dataset $X$. From Fact 9.1, we get that $OPT(X) = \sum_{i=1}^k \Delta(X_i)$. The above-mentioned lemma says that if a center is chosen uniformly at random from a set $X_i$ of points, then this center gives a 2-factor approximation to the 1-means problem for the point set $X_i$. This means that if a set $C$ of $k$ points sampled uniformly at random from the dataset $X$ happen to belong to the $k$ optimal clusters $X_1, \ldots, X_k$, then $C$ is a 2-approximate solution in expectation. However, some thinking reveals that the event that $k$ uniformly sampled points belong to the $k$ optimal clusters might be very unlikely for certain datasets. For example, consider a scenario where $k = 2$ and the dataset $X$ has one large cluster and another extremely small cluster that is very far from the larger cluster. There is a very small chance that uniformly sampled points will belong to the small cluster. On the other hand, if points are sampled using $D^2$-sampling, then the event that points are picked from both large and small clusters is more likely. The first uniformly sampled point will most likely belong to the large cluster. Now, as we are picking the second point with $D^2$-sampling, a point that is further away from the first point is more likely to be picked as compared with a point that is closer. Recall that given data $X$ and center set $C$, sampling a center with $D^2$-sampling means to pick a point from $x \in X$ with probability proportional to the squared Euclidean distance of $x$ from its nearest center in $C$. So, the second point is more likely to be picked from the smaller cluster. However, in this case, we cannot use Lemma 9.1 as the second sampled point is not a *uniform* sample from the smaller cluster. The next lemma helps out by making a claim similar to Lemma 9.1 but with respect to a center chosen with $D^2$-sampling.

**Lemma 9.2 (Lemma 3.2 in Reference 27)** *Let $A$ denote any optimal $k$-means cluster of $X$, and let $C$ denote an arbitrary set of centers. Let $c$ denote a center chosen using $D^2$-sampling with respect to center set $C$. Then $\mathbf{E}[\Phi(C \cup \{c\}, A)|c \in A] \leq 8 \cdot \Delta(A)$. Here $\Delta(A)$ denotes the optimal $1$-means cost of $A$.*

The above-mentioned lemma implies that the conditional expectation of the cost of $A$ with respect to the points $C \cup \{c\}$ is at most eight times optimal given that the sampled point is from the set $A$. So, as long as the sampled point happens to be from the set $A$, we get that the set of points $C \cup \{c\}$ gives a constant factor approximation with respect to the points in the set $A$. Arthur and Vassilvitskii [27] gave a clever induction-based argument using Lemmas 9.1 and 9.2 and showed a $O(\log k)$ factor approximation in expectation.

Let us try to use Lemma 9.2 in a weaker but simpler manner to get more intuition into the approximation analysis of the sampling procedure. At a high level, what we can say is that during $D^2$-sampling, if a chosen center belongs to some optimal cluster $A$, then we get a constant factor approximation with respect to that cluster. In some sense we can say that $A$ gets "covered" when this happens. So, in the event that all optimal clusters get covered, we get a constant factor approximation guarantee. However, the probability of the event that all optimal clusters get covered by picking $k$ centers may be very small. This can be seen as follows: Consider sampling the $i$th center after a set of $(i - 1)$ centers $C_{i-1}$ have already been chosen with $D^2$-sampling. Suppose the probability that the point sampled next is from an "uncovered" cluster is small, say at most $1/2$ and this is true for all $i$. Then the probability of covering all clusters by picking $k$ centers will be at most $1/2^k$. The next lemma shows that in any step of $D^2$-sampling, the probability of sampling the next point from a currently uncovered cluster cannot be too small unless the current set of centers already give a constant factor approximation (so picking more points will make the solution only better).

**Lemma 9.3 (Lemma 2.7 in Reference 28)** *Let $X$ denote the dataset and $X_1, \ldots, X_k$ denote the $k$ optimal $k$-means clusters. Let $C$ denote any set of centers. Let $I$ denote the subset of indices of clusters that are covered w.r.t. center set $C$. That is, for every $i \in I$, we have $\Phi(C, X_i) \leq c \cdot \Delta(X_i)$ for some fixed constant $c$. Let $x \in X_j$ denote the next point that is sampled with $D^2$-sampling. Then at least one of following statements is true: (i) $\mathbf{Pr}[j \notin I] \geq 1/2$, (ii) $\Phi(C, X) \leq (2c) \cdot \Phi_{OPT}(X)$.*

*Proof.* Suppose the first statement is false, then we will show that the second statement will be true. Let $X_c = \cup_{i \in I} X_i$ and $X_u = X \setminus X_c$. Then we have $1/2 > \mathbf{Pr}[j \notin I] = \frac{\Phi(C, X_u)}{\Phi(C, X_u) + \Phi(C, X_c)} \Rightarrow \Phi(C, X_c) > \Phi(C, X_u)$. Using this inequality, we get $\Phi(C, X) = \Phi(C, X_u) + \Phi(C, X_c) < 2 \cdot \Phi(C, X_c) \leq (2c) \cdot \sum_{i \in I} \Delta(X_i) \leq (2c) \cdot \sum_{i=1}^{k} \Delta(X_i) \leq (2c) \cdot \Phi_{OPT}(X)$. ∎

The above-mentioned lemma implies that the $k$-means++ algorithm gives constant approximation with probability at least $\frac{1}{2^{O(k)}}$. The lemma also implies that if we sample $O(k \log k)$ centers with $D^2$-sampling (instead of $k$ centers) and compare the cost with the optimal cost with respect to $k$ centers, then we get a constant factor *pseudoapproximation*. This observation was made by Ailon et al. [28]. With a little more work using martingale analysis, Aggarwal et al. [29] showed that $D^2$-sampling gives a constant pseudoapproximation even when only $O(k)$ centers are sampled.

Arthur and Vassilvitskii [27] had showed that $D^2$-sampling procedure gives an $O(\log k)$ approximation in expectation. They also support their upper bound on the approximation guarantee with a matching lower bound of $\Omega(\log k)$. That is, they give a dataset $X$ such that the $k$-means++ seeding algorithm when executed on $X$ gives a solution with expected cost $\Omega(\log k)$ times the optimal.

After the initial results on the analysis of the sampling algorithm by Arthur and Vassilvitskii [27], there has been a huge volume of work on understanding the $D^2$-sampling technique. We give a brief review of all these developments in the next few paragraphs.

*Pseudoapproximation*: Unlike the Lloyd's algorithm in which the number of clusters is used crucially in the algorithm, the sampling algorithm uses it just as a termination condition. That is, it terminates when it has sampled $k$ centers. An interesting property of the sampling algorithm is that if one considers the first $i$ centers $C_i$ that are sampled, then $C_i$ gives a $O(\log i)$ factor approximation for the $i$-means problem for the dataset. In this scenario, it makes sense to consider sampling more than $k$ centers and then compare the solution with the optimal solution with respect to $k$ centers. Such algorithms are known as *pseudoapproximation* algorithms. We have already mentioned the results of Ailon et al. [28] and Aggarwal et al. [29] who showed constant pseudoapproximation with $O(k \log k)$ and $O(k)$ centers, respectively. In a more recent work, Wei [30] showed that if $\beta k$ centers are sampled for any constant $\beta > 1$, then we get a constant factor pseudoapproximation in expectation.

*Lower bound on approximation*: The results of Arthur and Vassilvitskii [27] left a natural open question of whether the $k$-means++ seeding gives better than $O(\log k)$ approximation (say constant factor) with probability that is not too small (say $poly(1/k)$). Brunsch and Röglin [31] showed that this is not possible. More specifically, they created an instance in which the sampling procedure gives approximation ratio of $(2/3 - \varepsilon) \log k$ with probability that is exponentially small in $k$. However, the instance that they gave was a high-dimensional instance and they left an important open problem of analyzing the sampling procedure for small-dimensional instances. In fact, Brunsch and Röglin [31] conjectured that for constant dimensional instances, the sampling procedure gives a constant factor approximation with not too small probability. This conjecture was refuted by Bhattacharya et al. [32] who gave a two-dimensional instance such that the sampling procedure gives an $O(\log k)$ approximation on that instance with probability that is exponentially small in $k$.

*$D^2$-sampling under separation*: Qualitatively, the $D^2$-sampling procedure samples points that are far from each other. Consider the case when the optimal clusters are "separated" in some sense. If a center $c$ from one of the optimal clusters $X_i$ have been picked, then the probability of choosing another center from the same cluster would be low as the other points in $X_i$ will typically be closer to $c$ as compared with points from other clusters. So, in such cases, it is likely that the sampling procedure will pick one center from each of the optimal cluster, and given this, we know from Lemma 9.2 that the solution will be good. To formulate such results, we first need to define an appropriate notion of separation of clusters. One popular notion of separation was defined by Ostrovsky et al. [33]. This was in terms of the gap between the cost of the optimal solution with respect to $k$ centers and that with respect to $k - 1$ centers. If the ratio of the latter to the former is large, then this implies some kind of separation between clusters as clustering into $k$ cluster is much better than clustering into $k - 1$ clusters. Jaiswal and Garg [34] showed that under this kind of separation guarantee, the $k$-means++ seeding procedure indeed gives a constant approximation with probability $\Omega(1/k)$. Another notion of separation was given by Balcan et al. [35] and is known as *approximation stability*. A dataset is said to satisfy approximation stability iff for every clustering with cost near the optimal cost is "close" to the target optimal clustering. The *closeness* is in terms of the number of points that need to be reassigned to resemble the optimal clustering. Agrawal et al. [36] showed that if the dataset $X$ satisfies approximation stability and if all optimal clusters have some minimum size, then the $D^2$-sampling algorithm gives a constant factor approximation with probability $\Omega(1/k)$.

*PTAS using $D^2$-sampling*: PTAS is an algorithm that takes as input an error parameter $\varepsilon$ in addition to the problem and outputs a solution with an approximation guarantee of $(1 + \varepsilon)$. The running time of such an algorithm should be polynomial in the input parameters (but is allowed to be exponential in the error parameter $\varepsilon$). Such algorithms allow us to obtain solutions with cost that is arbitrarily close to the optimal cost. As we noted earlier, while discussing the hardness for the $k$-means problem (and also the $k$-median problem), such algorithms are not possible. However, when $k$ or $d$ is not part of the input and is a fixed constant, then there do exist PTAS for these problems. We noted earlier that a local search-based algorithm is a PTAS under the assumption that $d$ is not part of the input. Here, we will see a simple $D^2$-sampling-based PTAS under the assumption that $k$ is not part of the input. These are results from Jaiswal et al. [21,22]. We will discuss the $k$-means problem and later note that the techniques generalize to $k$-median and other settings. The main lemma that we will use in the discussion is the following lemma by Inaba et al. [2].

**Lemma 9.4** [2] *Let $S$ be a set of points obtained by independently sampling $M$ points with replacement uniformly at random from a point set $X \subset \mathbb{R}^d$. Then for any $\delta > 0$,*

$$\mathbf{Pr}\left[\Phi(\{\Gamma(S)\}, X) \leq \left(1 + \frac{1}{\delta M}\right) \cdot \Delta(X)\right] \geq (1 - \delta).$$

*Here $\Gamma(S)$ denotes the geometric centroid of the set $S$. That is $\Gamma(S) = \frac{\sum_{s \in S} s}{|S|}$*

Fact 9.1 tells us that the centroid of any point set $X \subset \mathbb{R}^d$ is the optimal 1-means solution for $X$. What the above-mentioned lemma tells us is that the centroid of a uniform sample from any point set *approximates* the 1-means solution to within a factor of $(1 + \varepsilon)$ if the size of the sample is $\Omega(1/\varepsilon)$. This implies that there exits at least one such subset among the possible $n^{O(1/\varepsilon)}$ subsets for which this holds. Worah and Sen [37] showed how to find such a subset using efficient derandomization. The situation here is different as we do not even know the clusters.

Let $X$ denote any dataset and let $X_1, \ldots, X_k$ denote an optimal $k$-means clusters of $X$. Suppose we manage to isolate uniform samples $S_1, \ldots, S_k$ from $X_1, \ldots, X_k$, respectively, such that $\forall i, |S_i| = \Omega(1/\varepsilon)$. Then, from the previous lemma, we get that the centers $(c_1, \ldots, c_k)$ where $c_i = \Gamma(S_i)$ is a $(1 + \varepsilon)$-approximate solution. Let us now try to think about how we can possibly perform this task. Without loss of generality, let us assume that $X_1$ is the largest cluster (in terms of the number of points). Isolating a uniform sample from $X_1$ is simple may be done in the following manner:

> Sample a set $U$ uniformly at random (with replacement) from $X$ such that $|U| = poly(k/\varepsilon)$. Let $P_1, \ldots, P_m$ denote all the $\binom{|U|}{\lceil 1/\varepsilon \rceil}$ subsets[*] of size $\lceil \frac{1}{\varepsilon} \rceil$ of the set $U$. As $X_1$ is well represented in the sample $U$ (note that at least $\frac{1}{k} \cdot |U| = \Omega(\frac{1}{\varepsilon})$ elements in $U$ will belong to $X_1$ in expectation), it will not be too difficult to argue that at least one of the subsets $P_1, \ldots, P_m$ will represent a uniform sample from $X_1$. $S_1$ denotes that subset.

Note that even though we are not able to isolate a single uniform sample $S_1$ from $X_1$, we are able to produce a *list* of samples such that one of them is such a sample. This is sufficient for developing an algorithm. What we can do is calculate the centroid of each of the subsets and try each of these centers as a possible first center. The next question is, how do we compute the uniform samples $S_2, \ldots, S_k$ from $X_2, \ldots, X_k$? Note that uniformly sampling from $X$ and trying all possible subsets (as we did for $S_1$) may not work as $X_2, \ldots, X_k$ may be tiny clusters and a uniformly sampled set from $X$ may not have adequate representatives from $X_2, \ldots, X_k$. This is where $D^2$-sampling becomes helpful. Suppose at this stage, we sample a set $U$ with $D^2$-sampling. We can now argue that there is a good chance that $U$ has adequate representation from $X_2, \ldots, X_k$ and so at least one of its subsets of size $\lceil \frac{1}{\varepsilon} \rceil$ has points entirely from one of $X_2, \ldots, X_k$. Moreover, we can argue that if the size of $U$ is chosen carefully, at least one of the subsets of size $\lceil \frac{1}{\varepsilon} \rceil$ will resemble a uniform sample from one of $X_2, \ldots, X_k$. Hence, the centroid of that set will be a good addition to the set of centers from Lemma 9.4. We continue this argument to show that one set of $k$

---

[*] This can be bounded by $2^{O\left(\frac{1}{\varepsilon} \log \frac{|U|}{\varepsilon}\right)}$ using the inequality $\binom{x}{y} \leq \left(\frac{ex}{y}\right)^y$.

---

Find-$k$-means$(X, k, \varepsilon)$

   - Let $N = \Theta(\frac{k}{\epsilon^2})$, $M = \Theta(\frac{1}{\varepsilon})$ and initialize $\mathcal{L}$ to $\emptyset$.

   - Repeat $2^{2k}$ times: Make a call to Sample-centers$(X, k, \epsilon, 0, \{\})$.

   - Return the set of $k$ centers from $\mathcal{L}$ that has least cost.

Sample-centers$(X, k, \epsilon, i, C)$

     (1) If $(i = k)$ then add $C$ to the set $\mathcal{L}$.

     (2) else

        (a) Sample a multi-set $S$ of $N$ points with $D^2$-sampling (w.r.t. centers $C$)

        (b) For all subsets $T \subset S$ of size $M$:

            (i) $C \leftarrow C \cup \{\Gamma(T)\}$.

            (ii) Sample-centers$(X, k, \epsilon, i + 1, C)$

---

**FIGURE 9.3**    $(1 + \varepsilon)$-approximation algorithm for $k$-means. Note that $\Gamma(T)$ denotes the centroid of $T$.

centers will have a good center from each of the $k$-optimal clusters. As this is a branching algorithm with a branch factor of $2^{\tilde{O}(1/\varepsilon)}$ and a depth of $k$, the running time of the algorithm is $O(nd \cdot 2^{\tilde{O}(k/\varepsilon)})^*$. The detailed algorithm is given in Figure 9.3. We give an analysis of this algorithm showing $(1 + \varepsilon)$-approximation in sufficient detail in the following. Following is the main result that we show:

**Theorem 9.4** *Let $0 < \varepsilon \leq 1/2$, $k$ be a positive integer, and $X \subseteq \mathbb{R}^d$. Then* Find-$k$-Means$(X, k, \varepsilon)$ *runs in time $O(|X|d \cdot 2^{\tilde{O}(k/\varepsilon)})$ and gives a $(1 + \varepsilon)$-approximation to the $k$-means problem.*

The algorithm Find-$k$-Means maintains a set $C$ of centers, that is initially empty. Every recursive call to the function Sample-centers increments the size of $C$ by one. In step (2) of the subroutine Sample-centers, it *tries out* various candidates that can be added to $C$. First, it samples a multiset $S$ of size $N = O(k/\varepsilon^3)$ by $D^2$-sampling from $X$ with respect to center set $C$. Then it considers all subsets of $S$ of size $M = O(1/\varepsilon)$ - for each such subset it adds the centroid of the subset to $C$ and makes a recursive call to itself. Thus, each invocation of Sample-centers makes precisely $\binom{N}{M}$ recursive calls to itself. It will be useful to think of the execution of this algorithm as a tree $\mathcal{T}$ of depth $k$. Each node in $\mathcal{T}$ can be labeled with a set $C$ – it corresponds to the invocation of Sample-centers with this set as $C$ (and $i$ being the depth of this node). The children of a node denote the recursive calls made by the corresponding invocation of Sample-centers. Finally, the leaves denote the set of candidate centers produced by the algorithm.

Let $X_1, \ldots, X_k$ denote the optimal $k$-means clustering of the given dataset $X$ and let $OPT(X)$ denote the optimal $k$-means cost for dataset $X$. As defined before, let $\Delta(R)$ denote the optimal 1-means cost for any dataset $R$. Note that $OPT(X) = \sum_{r=1}^{k} \Delta(X_r)$. A node $v$ at depth $i$ in the execution tree $\mathcal{T}$ corresponds to a set $C$ of size $i$. Let us call this set $C_v$. Our proof will argue inductively that for each $i$, there will be a node $v$ at depth $i$ such that the centers chosen so far in $C_v$ are good with respect to a subset of $i$ distinct clusters among $X_1, \ldots, X_k$ or the entire set $X$. More specifically, we will argue that the following invariant $P(i)$ is maintained during the recursive calls to **Sample-centers**:

---

  * Here $\tilde{O}$ hides logarithmic factors in $k$ and $\varepsilon$. Moreover, the branching structure will be more clear when we discuss the algorithm in detail in the following.

$P(i)$: With probability at least $\frac{1}{2^{i-1}}$, there is a node $v_i$ at depth $(i-1)$ in the tree $\mathcal{T}$ such that either **(a)** $\Phi(C_{v_i}, X) \leq (1+\varepsilon) \cdot OPT(X)$, or **(b)** there is a set of $(i-1)$ distinct clusters $X_{j_1}, X_{j_2}, \ldots, X_{j_{i-1}}$ such that $\Phi(C_{v_i}, X_{j_1} \cup \cdots \cup X_{j_{i-1}}) \leq (1+\frac{\varepsilon}{2}) \cdot \sum_{r=1}^{i-1} \Delta(X_{j_r})$.

Theorem 9.4 follows from the earlier as $P(k+1)$ holds and $2^{2k}$ repeated calls to the `Sample-centers` procedure are made and the best set of $k$ centers picked. We prove the invariant using induction. Note that $P(1)$ is trivially true as for the root $v$ of the tree, $C_v$ is the empty set. Now assume that $P(i)$ holds for some $i \geq 1$. We will show that $P(i+1)$ also holds. For this, we condition on the event $P(i)$ (that happens with probability at least $2^{i-1}$). Note that if $\Phi(C_i, X) \leq (1+\varepsilon) \cdot OPT(X)$, then $P(i+1)$ is trivially true. So, for the rest of the discussion, we will assume that this does not hold. Let $v_i$ and $X_{j_1}, \ldots, X_{j_{i-1}}$ be as guaranteed by the invariant $P(i)$. For ease of notation and without loss of generality, let us assume that the index $j_r$ is $r$, and let us call $C_{v_i}$ as just $C_i$. So, we have good approximation with respect to points in $X_1 \cup \cdots X_{i-1}$, and these cluster, may be thought of as "covered" clusters. Note that the probability of $D^2$-sampling a point from one of the uncovered clusters $X_i, \ldots, X_k$, say $X_r$, is equal to $\frac{\Phi(C_i, X_r)}{\Phi(C_i, X)}$. Let $\bar{i} \in \{i, i+1, \ldots, k\}$ be the index for which $\Phi(C_i, X_{\bar{i}})$ is the largest. We can break the analysis into the following two parts—(i) $\frac{\Phi(C_i, X_{\bar{i}})}{\Phi(C_i, X)} \leq \frac{\varepsilon}{2k}$, and (ii) $\frac{\Phi(C_i, X_{\bar{i}})}{\Phi(C_i, X)} > \frac{\varepsilon}{2k}$. In the former case, we will argue that $\Phi(C_i, X) \leq (1+\varepsilon) \cdot OPT(X)$, that is, the current set of centers already gives $(1+\varepsilon)$-approximation for the entire dataset, and hence $P(i+1)$ holds. In the latter case, we will argue that sufficient number of points will be sampled from $X_{\bar{i}}$ when sampling is done using $D^2$-sampling, and the added center corresponding to at least one of the branches incident on $v_i$ in the tree $\mathcal{T}$ will be a good center for $X_{\bar{i}}$ with probability at least $1/2$. Hence, $P(i+1)$ holds again. These two cases are discussed in the next two lemmas.

**Lemma 9.5** *Let $0 < \varepsilon \leq 1/2$. If $\frac{\Phi(C_i, X_{\bar{i}})}{\Phi(C_i, X)} \leq \frac{\varepsilon}{2k}$, then $\Phi(C_i, X) \leq (1+\varepsilon) \cdot \sum_{r=1}^{k} \Delta(X_r) = (1+\varepsilon) \cdot OPT(X)$.*

*Proof.* As $\frac{\Phi(C_i, X_{\bar{i}})}{\Phi(C_i, X)} \leq \frac{\varepsilon}{2k}$, we have $\frac{\sum_{r=i}^{k} \Phi(C_i, X_r)}{\sum_{r=1}^{i-1} \Phi(C_i, X_r) + \sum_{r=i}^{k} \Phi(C_i, X_r)} \leq \frac{\varepsilon}{2}$ which implies $\sum_{r=i}^{k} \Phi(C_i, X_r) \leq \frac{\varepsilon/2}{1-\varepsilon/2} \cdot \sum_{r=1}^{i-1} \Phi(C_i, X_r)$. We use this inequality to prove the lemma in the following manner:

$$\Phi(C_i, X) = \sum_{r=1}^{i-1} \Phi(C_i, X_r) + \sum_{r=i}^{k} \Phi(C_i, X_r) \leq \frac{1}{1-\frac{\varepsilon}{2}} \cdot \sum_{r=1}^{i-1} \Phi(C_i, X_r)$$

$$\leq \frac{1+\frac{\varepsilon}{2}}{1-\frac{\varepsilon}{2}} \cdot \sum_{r=1}^{i-1} \Delta(X_r) \leq (1+\varepsilon) \cdot OPT(X).$$

Note that the second to last inequality follows from the invariant. ■

The previous lemma handles case (i). Let us now discuss case (ii). So, we have $\frac{\Phi(C_i, X_{\bar{i}})}{\Phi(C_i, X)} > \frac{\varepsilon}{2k}$. What this essentially says is that the set $S$ in the `Find-k-Means` algorithm that is obtained by $D^2$-sampling with respect to center set $C_i$ will have good representation from $X_{\bar{i}}$ as long as $N = \Theta(k/\varepsilon^2)$ (as the expected number of points in the set $S$ from $X_{\bar{i}}$ will be $\Omega(1/\varepsilon)$). So, the important question now is that if we consider all possible subsets of $S$ of size $M = \Theta(1/\varepsilon)$, are we guaranteed that at least one of these subsets will be a uniform sample from $X_{\bar{i}}$? If this were true, then we can straightaway apply Inaba et al.'s lemma (Lemma 9.4) and argue that the centroid of such a sample will "cover" $X_{\bar{i}}$ and so $P(i+1)$ holds. However, this may not be true as there may be points in $X_{\bar{i}}$ that are very close to centers in $C_i$ and hence have very small probability of being sampled. To handle this issue, we partition the set $X_{\bar{i}}$ into two sets, $X_{\bar{i}}^{near}$ and $X_{\bar{i}}^{far}$ denoting the points that are near to the centers in $C_i$ and those that are far, respectively. More specifically, a point $x \in X_{\bar{i}}$ belongs to $X_{\bar{i}}^{far}$ iff $\frac{\Phi(C_i, \{x\})}{\Phi(C_i, X_{\bar{i}})} > \frac{\varepsilon}{16} \cdot \frac{1}{|X_{\bar{i}}|}$. This means that given that the sampled point is from $X_{\bar{i}}$, the conditional probability of sampling any given point $x \in X_{\bar{i}}^{far}$ is at least $\frac{\varepsilon}{16}$ fraction of the probability if it were sampled uniformly from $X_{\bar{i}}$. So, what we can argue for points in $X_{\bar{i}}^{far}$ is that if $N = |S|$ is chosen carefully, there is a good chance that one of the subsets of $S$ will be a uniform sample from $X_i^{far}$ and given this, the center corresponding to this subset will provide a $(1 + \frac{\varepsilon}{4})$ approximation for the set $X_i^{far}$. We omit the details of this analysis that may be found in Reference 21. As for points in $X_i^{near}$, we will argue that the cost of these points with respect to centers in $C_i$ is small (at most $\frac{\varepsilon}{4} \cdot \Delta(X_{\bar{i}})$), and so they do not contribute to the overall cost (even if no further centers are chosen). So, given that a good center for $X_i^{far}$ is chosen (this happens with high probability), the overall cost of $\bar{X}_i$ will be at most $(1 + \frac{\varepsilon}{4}) \cdot \Delta(X_i^{far}) + \frac{\varepsilon}{4} \cdot \Delta(X_{\bar{i}}) \leq (1 + \frac{\varepsilon}{2}) \cdot \Delta(X_{\bar{i}})$. The details of this part of the analysis are given in the next lemma that is a combination of Lemmas 2 and 3 in Reference 22).

**Lemma 9.6** $\Phi(C_i, X_i^{near}) \leq \frac{\varepsilon}{4} \cdot \Delta(X_{\bar{i}})$.

*Proof.* Let $D(p, q)$ denote the squared Euclidian distance between $p$ and $q$. Let $c_{\bar{i}}$ denote the centroid of $X_{\bar{i}}$. For any point $x \in X_i^{near}$ let $c_x = \arg\min_{c \in C_i}\{D(c, x)\}$. We have

$$\frac{\varepsilon}{16} \cdot \frac{1}{|X_{\bar{i}}|} \geq \frac{\Phi(C_i, \{x\})}{\Phi(C_i, X_{\bar{i}})} \geq \frac{D(x, c_x)}{\Phi(c_x, X_{\bar{i}})} \stackrel{Fact\ 9.1}{=} \frac{D(x, c_x)}{\Delta(X_{\bar{i}}) + |X_{\bar{i}}| \cdot D(c_x, c_{\bar{i}})} \geq \frac{D(x, c_x)}{\Delta(X_{\bar{i}}) + 2|X_{\bar{i}}| \cdot (D(x, c_x) + D(x, c_{\bar{i}}))}.$$

The last inequality uses the fact that for any three points $p, q, r, D(p, q) \leq 2(D(p, r) + D(r, q))$. Rearranging the terms of the previous inequality, we get that: $D(x, c_x) \leq \frac{\varepsilon/16}{1 - \varepsilon/16} \cdot \left(\frac{\Delta(X_{\bar{i}})}{|X_{\bar{i}}|} + D(x, c_{\bar{i}})\right) \leq \frac{\varepsilon}{8} \cdot \left(\frac{\Delta(X_{\bar{i}})}{|X_{\bar{i}}|} + D(x, c_{\bar{i}})\right)$. Using this, we can get a bound on $\Phi(C_i, X_i^{near})$ as follows:

$$\Phi(C_i, X_i^{near}) = \sum_{x \in X_i^{near}} D(x, C_i) \leq \sum_{x \in X_i^{near}} \frac{\varepsilon}{8} \cdot \left(\frac{\Delta(X_{\bar{i}})}{|X_{\bar{i}}|} + D(x, c_{\bar{i}})\right) \leq \frac{\varepsilon}{4} \cdot \Delta(X_{\bar{i}}).$$

This completes the proof of the lemma. ∎

This $D^2$-sampling technique generalizes for the Euclidean $k$-median problem, and we have a similar result as mentioned earlier. Significantly, these techniques generalize for distance measures other than Euclidean and squared Euclidean distance such as the Mahalanobis distance and $\mu$-similar Bregman divergences (see Reference 38 for discussion on these distance functions). The analysis uses only simple properties of the distance measure such as (approximate) symmetry and (approximate) triangle inequality that causes the analysis to carry over to these settings.

The $D^2$-sampling technique also extends to the setting of *constrained clustering* that is relevant in many machine-learning applications. In many clustering scenarios, optimizing the $k$-means cost function is not the only objective. Depending on the clustering context, there may be additional constraints. For example, consider the *r-gather problem* in which the additional constraint is that each cluster should have at least $r$ points. Ding and Xu [39] gave a general framework in which such algorithms can be studied and also gave a $(1 + \varepsilon)$-approximation algorithm (not based on $D^2$-sampling). Note that in such scenarios, the separation between optimal clusters is not as well defined as in the classical $k$-means problem (in which the clusters are optimal Voronoi partitions). However, even under such scenarios the algorithm based on $D^2$-sampling can be made to work with minimal changes in the overall outline. This is one of the main results by Bhattacharya et al. [40]. Let us go back to the outline of the algorithm and see the main issue. After finding good centers for clusters $X_1, \ldots, X_i$, we sample point set $U$ using $D^2$-sampling. This is done to isolate a uniform sample from one of the remaining clusters $X_{i+1}, \ldots, X_k$. However, some of the points of these clusters may be very close to the $i$ good centers that we have already picked and hence will have low chance of being sampled. This is handled by taking appropriate copies of the already chosen centers in the set $U$ (which act as proxies for points in remaining clusters that are close to the already chosen centers). The analysis becomes slightly tricky than before but it goes through without changing the basic outline of the algorithm for the classical $k$-means problem as in Reference 21.

*Variations of* k-*means++ seeding*: Due to its simplicity, many practitioners have been interested in the $D^2$-sampling technique and have used it in various data-mining tasks. A number of optimizations have also been suggested that make the sampling procedure more efficient in various contexts. One such context is parallel computing. With advancement in parallel architectures, there is a move toward developing parallel algorithms for popular data-mining tasks such as clustering. It would be nice if we can come up with a parallel version of the $k$-means++ seeding algorithm. However, we note that the sampling algorithm is inherently sequential. This is because the sampling of the $i$th center depends on the previous $i-1$ centers. So, as a minimum of $k$ sequential steps are required, parallelism cannot give arbitrary running time improvements. Bahmani et al. [41] gave a slightly modified version of the algorithm that they called the $k$-means|| ($k$-means *parallel*). The main idea is that in the $i$th iteration, instead of sampling a single point, sample $\ell$ points (for some $\ell > 1$) in parallel and run this algorithm for $t$ iterations (for some $t < k$). Finally, the sampled points are *pruned* to obtain $k$ centers. On the experimental side, it was shown that the behavior of this algorithm is similar to that of the $k$-means++ even when $t$ is very small, and on the theoretical side, they showed a constant approximation guarantee under some reasonable assumption on the lower bound on $t$.

The $D^2$-sampling algorithm runs in $k$ iterations, one center being picked in each iteration. Note that the running time of each of these iterations is $O(nd)$. This is because, at the start of every iteration, we need to compute the distribution from which the next center will be sampled, and for this we need to compute the distance of every point to its nearest center (from among the centers chosen until that iteration). Bachem et al. [42] suggested a simple Monte Carlo Markov Chain-based technique to improve the running time in every iteration. Their algorithm is called the $k$-MC$^2$ algorithm. The main idea is that to sample $x \in X$ with probability $p(x)$, we perform a bounded-length random walk on a Markov chain where the states denote points in $X$ and we make a transition from $x$ to $y$ with probability $\min(1, p(y)/p(x))$. Note that $p(x), p(y)$ is proportional to the squared distance of $x$ and $y$ from their nearest centers, respectively. As the transition only requires the ratio of these quantities, we do not need to compute the distance of every point from its nearest center. So, the running time in every iteration of the sampling algorithm is proportional to the length of the random walk performed. This may be much smaller than $n$ in many different contexts. In a follow-up paper, Bachem et al. [43] gave an improvement in which we can put an upper bound on the length of the random walk at a small cost to the approximation guarantee.

*Streaming algorithm*: In typical data-mining tasks where these clustering algorithms are typically used, the data are so large that the standard random access model of computation (in which data are assumed to be in the memory) is not appropriate. In such large-data scenarios, the algorithm typically gets to see the data by making one pass (or a few passes) over the data. This model of computation is called the *streaming model* as the data are accessible as a *stream*. The question in this context is whether we can solve the center-based clustering problems in the streaming setting in which the memory available is small compared with the number of data points. The trade-off of interest in this setting is that between the approximation guarantee and the amount of space required by the streaming algorithm. There is a standard method of using classical approximation algorithms to design approximation algorithms in the streaming setting where the memory is limited. Let $A, B$ be an approximation algorithms for the *weighted k-means problem* that gives an approximation factor of $\alpha, \beta$, respectively, and use linear amount of memory. The weighted $k$-means problem is a simple generalization of the $k$-means problem where each point has an associated weight and the goal is to minimize the sum of squared distances multiplied by the weights (instead of just the squared distances as in $k$-means). By using $A$ and $B$, we will design a streaming algorithm that uses $O(k\sqrt{n})$ memory. Imagine having a bucket that can store at most $\sqrt{n}$ data points. As we make a pass over the data, we keep filling this bucket. Once the bucket is full, we execute $A$ on the data points in the bucket and find $k$ centers. We empty the bucket and keep these weighted $k$ centers separately, the weight of a center being the number of points in the bucket for which this center is the closest. We continue with the data stream, repeating the same procedure as the previous one. Once all the $n$ points have been seen, we go back to the weighted $O(k\sqrt{n})$ centers that have been saved and use algorithm $B$ to cluster these weighted centers and obtain the final $k$ centers. Note that the total memory used is $O(k\sqrt{n})$ and the algorithm just makes one pass over the data. Guha et al. [44] showed that the approximation guarantee obtained by using algorithms $A$ and $B$ in such a manner (that is, using $B$ on the output of $A$) is $2\alpha + 4\beta(2\alpha + 1)$. Also, algorithm $A$ may be a pseudoapproximation algorithm. Ailon et al. [28] used the sampling-based pseudoapproximation algorithm in the above-mentioned general technique to obtain a streaming algorithm that for any constant $\varepsilon > 0$ uses $O(n^\varepsilon)$ space and gives $O(c^{1/\varepsilon} \log k)$ approximation guarantee for some constant $c$. Note that this is not the best trade-off that is known between space and approximation guarantee and there are better tradeoffs known [45] using other techniques.

## 9.4 Other Topics on Clustering

Even tough the title of this chapter is Approximate Clustering, there are many important topics on clustering that we did not discuss. The simple reason for this is that the literature on clustering has now become so vast that a comprehensive coverage in a book chapter is not possible. There are a few books and surveys [1,46–49] on various clustering topics that the reader may find useful. In this chapter, we discussed center-based clustering problems such as $k$-center/means/median and even within this topic, our main focus was on sampling-based sequential selection algorithms. Given that we have left out discussions on many clustering topics, it will be useful to mention at least a few of them here.

It is important to note that the center-based clustering problems provide just one way to partition the data into meaningful clusters. These problems make sense in contexts in which the target clusters satisfy *locality property*, meaning that the points within the same cluster are all close to each other (or *tightly packed* in some sense). However, there are other contexts in which the meaningful clusters do not satisfy such properties, and in these scenarios, modeling the clustering problem in terms of center-based problems such as $k$-center/means/median may not give the best results. Figure 9.4a shows one such scenario. In such contexts, a *density-based model* makes much more sense. The main idea is to keep neighbouring points in the same cluster even if all points within a cluster are not close to each other as in the aforementioned scenario. DBSCAN [50] is a density-based algorithm that is widely used such scenarios.

Data clustering can also be modeled as a graph theoretic problem in the following manner: For a given dataset along with similarity measure, we can define a graph in which the vertices correspond to the data points, and there is an edge between vertices $x$ and $y$ if similarity exceeds some threshold and the weight of the edge equals the similarity. Modeling in terms of a graph also makes sense in contexts in which the similarity/dissimilarity between data points is known at a coarse level such as for every pair $x, y$ all that is known is whether $x$ and $y$ are similar or not. The high-level goal is to find a clustering of the vertices such that the sum of weight of the edges that go across clusters is minimized (or in other words find *cuts* with

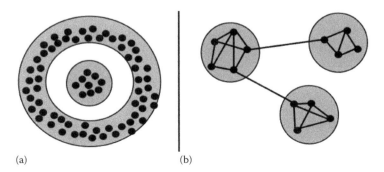

(a) (b)

**FIGURE 9.4** (a) shows a simple two-dimensional dataset in which the two meaningful clusters is unlikely to be found using center-based clustering algorithms and (b) shows a simple graph clustering problem.

small capacity). Figure 9.4b shows a simple graph clustering problem in which all edge weights are 1. Note that in this formulation, the number of clusters $k$ may not be given as input. *Spectral techniques* [51] are popularly used for such cut problems in graphs. Correlation clustering [52] may be described as a more general version of the graph clustering problem discussed in the previous paragraph. Here, the edges in the graph may have positive weight (indicating similarity) or negative weight (indicating dissimilarity). The goal is to find a partition of the vertices into clusters such that sum of weights of positive edges within clusters minus the sum of weights of negative edges across clusters is maximized.

# References

1. A. K. Jain and R. C. Dubes. *Algorithms for Clustering Data*. Prentice-Hall, Upper Saddle River, NJ, 1988.
2. M. Inaba, N. Katoh, and H. Imai. Applications of weighted Voronoi diagrams and randomization to variance-based $k$-clustering: (extended abstract). In *Proceedings of the Tenth Annual Symposium on Computational geometry (SoCG'94)*, ACM, New York, pp. 332–339, 1994.
3. T. Feder and D. Greene. Optimal algorithms for approximate clustering. In *Proceedings of the 20th ACM Symposium on Theory of Computing*, ACM, New York, pp. 434–444, 1988.
4. N. Megiddo and K. J. Supowit. On the complexity of some common geometric location problems. *SIAM Journal of Computing*, 13:182–196, 1984.
5. T. F. Gonzalez. Clustering to minimize the maximum intercluster distance. *Theoretical Computer Science*, 38:293–306, 1985.
6. M. Bern and D. Eppstein. Approximation algorithms for geometric problems. *Approximation Algorithms for NP-hard Problems*, D. Hochbaum, Ed., PWS Publishing, Boston, MA, pp. 296–345, 1996.
7. S. Arora, P. Raghavan, and S. Rao. Approximation schemes for Euclidean $k$-medians and related problems. In *Proceedings of the Thirtieth Annual ACM Symposium on Theory of Computing (STOC '98)*. ACM, New York, pp. 106–113, 1998.
8. S. Dasgupta. The hardness of $k$-means clustering. Technical Report CS2007-0890, Department of Computer Science and Engineering, University of California, San Diego, CA, 2007.
9. A. Vattani. The hardness of $k$-means clustering in the plane. Technical report, Department of Computer Science and Engineering, University of California, San Diego, CA, 2009.
10. M. Mahajan, P. Nimbhorkar, and K. Varadarajan. The planar $k$-means problem is NP-hard. *Theoretical Computer Science*, 442:13–21, 2012.
11. P. Awasthi, M. Charikar, R. Krishnaswamy, and A. K. Sinop. The hardness of approximation of Euclidean $k$-means. In *the 31st International Symposium on Computational Geometry (SoCG'15)*, Schloss Dagstuhl–Leibniz-Zentrum fuer Informatik, Dagstuhl, Germany, pp. 754–767, 2015.
12. V. Guruswami and P. Indyk. Embeddings and non-approximability of geometric problems. In *Proceedings of the fourteenth annual ACM-SIAM symposium on Discrete algorithms (SODA'03)*, Society for Industrial and Applied Mathematics, Philadelphia, PA, pp. 537–538, 2003.
13. K. Jain, M. Mahdian, and A. Saberi. A new greedy approach for facility location problems. In *Proceedings of the Thiry-fourth Annual ACM Symposium on Theory of Computing (STOC'02)*, ACM, New York, pp. 731–740, 2002.
14. V. Arya, N. Garg, R. Khandekar, A. Meyerson, K. Munagala, and V. Pandit. Local search heuristics for $k$-median and facility location problems. *SIAM Journal on Computing*, 33(3):544–562, 2004.
15. T. Kanungo, D. M. Mount, N. S. Netanyahu, C. D. Piatko, R. Silverman, and A. Y. Wu. A local search approximation algorithm for $k$-means clustering. *Computational Geometry*, 28(2):89–112, 2004.
16. S. G. Kolliopoulos and S. Rao. A nearly linear-time approximation scheme for the Euclidean $k$-median problem. *SIAM Journal on Computing*, 37:757–782, 2007.

17. V. Cohen-Addad, P. N. Klein, and C. Mathieu. Local search yields approximation schemes for *k*-means and *k*-median in Euclidean and minor-free metrics. In *Proceedings of the 57th Annual IEEE Symposium on Foundations of Computer Science (FOCS'16)*, 2016.

18. Z. Friggstad, M. Rezapour, and M. R. Salavatipour. Local search yields a PTAS for *k*-means in doubling metrics. In *Proceedings of the 57th Annual IEEE Symposium on Foundations of Computer Science (FOCS'16)*, 2016.

19. A. Kumar, Y. Sabharwal, and S. Sen. Linear-time approximation schemes for clustering problems in any dimensions. *Journal of the ACM*, 57(2):5:1–5:32, 2010.

20. D. Feldman, M. Monemizadeh, and C. Sohler. A PTAS for *k*-means clustering based on weak coresets. In *Symposium on Computational Geometry*, ACM, New York, pp. 11–18, 2007.

21. R. Jaiswal, A. Kumar, and S. Sen. A simple $D^2$-sampling based PTAS for *k*-means and other clustering problems. *Algorithmica*, 70(1):22–46, 2014.

22. R. Jaiswal, M. Kumar, and P. Yadav. Improved analysis of $D^2$-sampling based PTAS for *k*-means and other clustering problems. *Information Processing Letters*, 115(2):100–103, 2015.

23. J.-H. Lin and J. S. Vitter. Approximation algorithms for geometric median problems. *Information Processing Letters*, 44(5):245–249, 1992.

24. M. Charikar, S. Guha, E. Tardos, and D. B. Shmoys. A constant-gactor approximation algorithm for the *k*-median problem. *Journal of Computer and System Sciences*, 65(1):129–149, 2002.

25. M. Charikar and S. Li. A dependent LP-rounding approach for the *k*-median problem. In *Proceedings of the 39th International Colloquium Conference on Automata, Languages, and Programming—Volume Part I (ICALP'12)*, Lecture Notes in Computer Science, vol. 7391, Springer, Berlin, Germany, pp. 194–205, 2012.

26. A. Archer, R. Rajagopalan, and D. B. Shmoys. Lagrangian relaxation for the *k*-median problem: New insights and continuity properties. In *Proceedings of the 11th Annual European Symposium on Algorithms (ESA'03)*, Lecture Notes in Computer Science, vol. 2832, Springer, Berlin, Germany, pp. 31–42, 2003.

27. D. Arthur and S. Vassilvitskii. *k*-means++: The advantages of careful seeding. In *Proceedings of the Eighteenth Annual ACM-SIAM Symposium on Discrete Algorithms*, SODA '07, Society for Industrial and Applied Mathematics, Philadelphia, PA, pp. 1027–1035, 2007.

28. N. Ailon, R. Jaiswal, and C. Monteleoni. Streaming *k*-means approximation. In *NIPS*, Curran Associates, pp. 10–18, 2009.

29. A. Aggarwal, A. Deshpande, and R. Kannan. Adaptive sampling for *k*-means clustering. In *Approximation, Randomization, and Combinatorial Optimization. Algorithms and Techniques*, volume 5687 of Lecture Notes in Computer Science, Berkeley, CA, pp. 15–28. Berkeley, Springer, Berlin, Germany, 2009.

30. D. Wei. A constant-factor bi-criteria approximation guarantee for *k*-means++. In *NIPS*, 2016.

31. T. Brunsch and H. Röglin. A bad instance for *k*-means++. *Theoretical Computer Science*, 505:19–26, 2013.

32. A. Bhattacharya, R. Jaiswal, and N. Ailon. Tight lower bound instances for *k*-means++ in two dimensions. *Theoretical Computer Science*, 634:55–66, 2016.

33. R. Ostrovsky, Y. Rabani, L. J. Schulman, and C. Swamy. The effectiveness of Lloyd-type methods for the *k*-means problem. *Journal of the ACM*, 59(6):28:1–28:22, 2013.

34. R. Jaiswal and N. Garg. Analysis of *k*-means++ for separable data. In *Approximation, Randomization, and Combinatorial Optimization. Algorithms and Techniques, in:* Lecture Notes in Computer Science, vol. 7408, Springer, Berlin, Germany, pp. 591–602, 2012.

35. M.-F. Balcan, A. Blum, and A. Gupta. Clustering under approximation stability. *Journal of the ACM*, 60(2):8:1–8:34, 2013.

36. M. Agarwal, R. Jaiswal, and A. Pal. *k*-means under approximation stability. *Theoretical Computer Science*, 588:37–51, 2015.

37. P. Worah and S. Sen. A linear time deterministic algorithm to find a small subset that approximates the centroid. *Information Processing Letters*, 105(1):17–19, 2007.

38. M. R. Ackermann and J. Blömer. Coresets and approximate clustering for Bregman divergences. In *Proceedings of the Twentieth Annual ACM-SIAM Symposium on Discrete Algorithms (SODA'09)*, 2009.

39. H. Ding and J. Xu. A unified framework for clustering constrained data without locality property. In *Proceedings of the Twenty-Sixth Annual ACM-SIAM Symposium on Discrete Algorithms (SODA'15)*, Society for Industrial and Applied Mathematics, Philadelphia, PA, pp. 1471–1490, 2015.

40. A. Bhattacharya, R. Jaiswal, and A. Kumar. Faster algorithms for the constrained $k$-means problem. In *33rd Symposium on Theoretical Aspects of Computer Science (STACS 2016)*, Schloss Dagstuhl–Leibniz-Zentrum fuer Informatik, Dagstuhl, Germany, pp. 16:1–16:13, 2016.

41. B. Bahmani, B. Moseley, A. Vattani, R. Kumar, and S. Vassilvitskii. Scalable k-means++. *Proceedings of the VLDB Endowment*, 5(7):622–633, 2012.

42. O. Bachem, M. Lucic, S. H. Hassani, and A. Krause. Approximate $k$-means++ in sublinear time. In *The 30th AAAI Conference on Artificial Intelligence*, 2016.

43. O. Bachem, M. Lucic, S. H. Hassani, and A. Krause. Fast and provably good seedings for $k$-means. In *NIPS*, 2016.

44. S. Guha, A Meyerson, N. Mishra, R. Motwani, and L. O'Callaghan. Clustering data streams: Theory and practice. *IEEE Transactions on Knowledge and Data Engineering*, 15(3):515–528, 2003.

45. M. Charikar, L. O'Callaghan, and R. Panigrahy. Better streaming algorithms for clustering problems. In *Proceedings of the Thirty-fifth Annual ACM Symposium on Theory of Computing (STOC '03)*. ACM, New York, pp. 30–39, 2003.

46. J. A. Hartigan. *Clustering Algorithms*. Wiley, New York, 1975.

47. A. K. Jain, M. N. Murty, and P. J. Flynn. Data clustering: A review. *ACM Computing Surveys*, 31(3):264–323, 1999.

48. P. Berkhin. A survey of clustering data mining techniques. In: *Grouping Multidimensional Data*, Springer, Berlin, Germany, pp. 25–71, 2006.

49. A. K. Jain. Data clustering: 50 years beyond $k$-means. *Pattern Recognition Letters*, 31(8):651–666, 2010.

50. M. Ester, H. Kriegel, J. Sander, and X. Xu. A density-based algorithm for discovering clusters in large spatial databases. In *Proceedings of KDD*, AAAI Press, pp. 226–231, 1996.

51. U. Luxburg. A tutorial on spectral clustering. *Statistics and Computing*, 17(4):395–416, 2007.

52. N. Bansal, A. Blum, and S. Chawla. Correlation clustering. *Machine Learning*, 56:89, 2004.

# Maximum Planar Subgraph

Gruia Călinescu

Cristina G. Fernandes*

## 10.1 Introduction

In graph theory, questions related to planarity always played an important role. The main subject of this chapter is the following problem, which is denoted here as MAXIMUM WEIGHT PLANAR SUBGRAPH: given a graph $G$ with a nonnegative weight defined for each edge, find a planar subgraph of $G$ of maximum weight, in which the weight of a subgraph is simply the sum of the weights of the edges in the subgraph. Its unweighted version, denoted as MAXIMUM PLANAR SUBGRAPH, consists of: given a simple graph $G$, find a planar subgraph of $G$ with the maximum number of edges.

These problems have applications in circuit layout [1–6], facility layout [6,7], and graph drawing [8–10] (the process of drawing a graph, usually on a two-dimensional medium, as "nicely" as possible, in which "nicely" is defined by the application). For example, graph drawing of nonplanar graphs can start by drawing a planar subgraph [11,12], and then drawing the remaining edges. This method employs planarization, the process of obtaining a large planar subgraph from a nonplanar one [13–15].

On the contrary, MAXIMUM PLANAR SUBGRAPH is NP-hard [16,17]. In fact, MAXIMUM PLANAR SUBGRAPH is known to be Max SNP-hard [18], which means there is an $\epsilon > 0$ such that no approximation algorithm achieves a ratio of $1 - \epsilon$ for it, unless P = NP. This is true even if the input is a cubic graph [19]. The largest $\epsilon$ for which this result is known however is tiny, making $1 - \epsilon$ far from the best approximation ratios known for the problem.

On the positive side, for years, the best known approximation algorithm for MAXIMUM PLANAR SUBGRAPH was the trivial one that produces a spanning tree of the (connected) input graph. By using Euler's formula, it is easy to see that this algorithm achieves a ratio of $1/3$. There were several heuristics proposed for MAXIMUM PLANAR SUBGRAPH and MAXIMUM WEIGHT PLANAR SUBGRAPH, but most of them either have a ratio of $1/3$ or were not approximation algorithms at all [20]. In particular, the natural MST algorithm for MAXIMUM WEIGHT PLANAR SUBGRAPH, that outputs a maximum weight forest of the given graph, has also a ratio of $1/3$. Only in the 90s, the $1/3$ threshold was broken, by algorithms that use

*Work partially supported by the Brazilian agencies CNPq (Proc. 308523/2012-1 and 456792/2014-7) and FAPESP (Proc. 2013/03447-6).

*triangular structure* (a graph whose all blocks are edges or triangles) [18,21]. Since then, to our knowledge, no better approximation algorithms appeared in the literature for these problems.

Specifically, the best known approximation algorithm for MAXIMUM PLANAR SUBGRAPH is the one that outputs a triangular structure of the input graph with the maximum number of edges. Such a triangular structure can be computed in polynomial time by a sophisticated algorithm for polymatroid matching. The approximation ratio of the resulting algorithm is 4/9 [18]. Its analysis and implementation however are intricate. A simpler and faster greedy algorithm that also produces a triangular structure achieves a ratio of 7/18 [18].

As for MAXIMUM WEIGHT PLANAR SUBGRAPH, the best approximation ratio known is $1/3+1/72$ [21]. Triangular structures are related to 3-restricted Steiner trees, which are used to achieve good approximation ratios for the famous MINIMUM STEINER TREE problem (Vol. 1, Chap. 36). The ratio of $1/3+1/72$ for MAXIMUM WEIGHT PLANAR SUBGRAPH is achieved by an algorithm that follows closely an algorithm by Berman and Ramaiyer [22] for MINIMUM STEINER TREE. The output is also a triangular structure. A randomized pseudo-polynomial algorithm that computes a maximum weight triangular structure in a given weighted graph can be derived from a result of Camerini et al. [23]. This algorithm can be converted, using the standard method of rounding and scaling (as in Vol. 1, Chaps. 8 and 9, or in Reference 24, pp. 135–137), to a $(1 - \epsilon)$-approximation algorithm for computing a maximum weight triangular structure. A $(1 - \epsilon)$-approximation parallel algorithm with polylogarithmic running time also follows from a result by Prömel and Steger [25]. A maximum weight triangular structure also achieves a $1/3 + 1/72$ ratio, however we know of no direct proof for this. The only proof we know is based on the algorithm that mimics Berman and Ramaiyer's.

In the current chapter, we describe the main approximation algorithms known for both MAXIMUM PLANAR SUBGRAPH and MAXIMUM WEIGHT PLANAR SUBGRAPH. We also discuss the relation between these problems and MINIMUM STEINER TREE.

The chapter is organized as follows. Section 10.2 analyzes the MST algorithm. Section 10.3 is about triangular structures, their relation to 3-restricted Steiner trees, and the above-mentioned algorithms that produce triangular structures. Section 10.4 presents the analysis of some of these triangular structure algorithms. Section 10.5 shows the results of applying the same methods to the MAXIMUM OUTERPLANAR SUBGRAPH problem, in which the output graph must be outerplanar (i.e., can be drawn in the plane with all vertices on the boundary of the outer face). Due to the importance of these problems, exact algorithms with exponential worst-case running time have been proposed, and heuristics have been analyzed experimentally. We discuss these practical approaches in Section 10.6 together with related problems and results and ideas for improved approximation ratios.

## 10.2 Analysis of the MST Algorithm

The MST algorithm, given a graph $G$ and a nonnegative weight for each of its edges, outputs a maximum weight forest of $G$. We know two ways to demonstrate that this algorithm has a ratio of 1/3 for MAXIMUM WEIGHT PLANAR SUBGRAPH. One of them uses an idea of Korte and Hausmann [26]. The other one uses a well-known theorem of Nash-Williams involving the following concept.

The *arboricity* of a graph is the minimum number of spanning forests into which its edge set can be partitioned. Nash-Williams [27,28] proved the following classic theorem about it.

**Theorem 10.1** *The arboricity of a graph $G$ is the maximum, over all subgraphs $H$ of $G$ with at least two vertices, of*

$$\left\lceil \frac{|E(H)|}{|V(H)| - 1} \right\rceil.$$

For a function $w$ defined on the edge set of a graph $G$ and a set $S$ of edges of $G$, we use $w(S)$ to denote the sum of the weights of the edges in $S$ and, for a subgraph $H$ of $G$, we use $w(H)$ instead of $w(E(H))$. The next lemma is the key in the analysis of the approximation ratio of the MST algorithm.

**Lemma 10.1** *Let $\mathcal{F}$ be a family of graphs closed under taking subgraphs, such that, for some positive integer $c$, $|E(G)| \leq c(|V(G)| - 1)$ for all $G$ in $\mathcal{F}$. Let $w$ be a nonnegative weight function on the edges of some $G$ in $\mathcal{F}$, and $F$ be a maximum weight forest in $G$. Then, $w(F) \geq \frac{1}{c}w(G)$.*

*First proof.* Let $G$ be a graph in $\mathcal{F}$ and $w$ be a nonnegative weight function on its edges. We have $|E(H)| \leq c(|V(H)| - 1)$ for any subgraph $H$ of $G$, because $\mathcal{F}$ is closed under taking subgraphs. Thus, by Theorem 10.1, the arboricity of $G$ is at most $c$. This means that the edge set of $G$ can be partitioned into $c$ forests $F_1, \ldots, F_c$. Clearly, we have $w(F_1) + \cdots + w(F_c) = w(G)$. But $w(F) \geq w(F_i)$ for all $i$, and this implies that $w(G) = w(F_1) + \cdots + w(F_c) \leq c \cdot w(F)$. ∎

*Second proof.* Let $G$ be a graph in $\mathcal{F}$ with a nonnegative weight for each of its edges. Use Kruskal's algorithm to construct a maximum weight forest $F$. That is, let $e_1, \ldots, e_m$ be the edges of $G$ in nonincreasing order of weight. Start with $F_0 := \emptyset$. For $i = 1, \ldots, m$, if the addition of $e_i$ to $F_{i-1}$ creates a cycle, let $F_i := F_{i-1}$, otherwise let $F_i := F_{i-1} \cup \{e_i\}$. At the end, let $F := F_m$.

For each $i$, let $E_i := \{e_1, \ldots, e_i\}$ and let $w_i$ be the weight of $e_i$. Put $w_{m+1} := 0$. By rearranging the terms,

$$w(F) = \sum_{i=1}^{m} |F_i|(w_i - w_{i+1}), \text{ and } w(G) = \sum_{i=1}^{m} |E_i|(w_i - w_{i+1}).$$

It is therefore enough to show that $|E_i| \leq c|F_i|$ for $i = 1, \ldots, m$.

For each $i$, let $p_1, \ldots, p_k$ be the number of vertices in the components of $F_i$. Of course, $|F_i| = \sum_{j=1}^{k}(p_j - 1)$. Any edge of $E_i$ must have its two endpoints in the same component of $F_i$. (Otherwise, the edge could have been selected by the algorithm, merging two components of $F_i$.) We associate each edge of $E_i$ with the component of $F_i$ that contains both of its endpoints.

The edges of $E_i$ associated with a component of $F_i$ are a subset of the edges of the subgraph of $G$ induced by the vertices of this component. Thus, the number of edges associated with the $j$th component of $F_i$ is at most $c(p_j - 1)$, because the subgraph of $G$ induced by $F_i$ is in $\mathcal{F}$, as $\mathcal{F}$ is closed under taking subgraphs. But then $|E_i| \leq \sum_{j=1}^{k} c(p_j - 1) = c|F_i|$. ∎

The approximation ratio of 1/3 for the MST algorithm for MAXIMUM PLANAR SUBGRAPH follows from the next corollary.

**Corollary 10.1 (See Reference 1, Theorem 3.3)** *Let $P$ be a simple planar graph with a nonnegative weight function $w$ defined on its edges. If $F$ is a maximum weight forest in $P$, then $w(F) \geq \frac{1}{3}w(P)$.*

*Proof.* By Euler's formula, $|E(P)| \leq 3(|V(P)| - 1)$. Then the corollary follows from applying Lemma 10.1 to the family of simple planar graphs, with $c = 3$. ∎

Indeed, if $F$ is a maximum weight forest in a graph $G$ with a weight function $w$ on its edges and $P$ is any planar subgraph of $G$, then $w(F) \geq \frac{1}{3}w(P)$ by Corollary 10.1. This implies that the MST algorithm has a ratio of at least 1/3 for MAXIMUM WEIGHT PLANAR SUBGRAPH. The 1/3 bound is tight, as is the case when the input is an unweighted triangulated planar graph.

If one uses a Kruskal-like heuristic testing for planarity before adding an edge, the output contains a maximum weight forest and therefore this algorithm also has an approximation ratio of at least 1/3. Tight examples are known [1,20].

A graph is said to be *outerplanar* if it can be drawn in the plane with all vertices on the boundary of the outer face. Consider the MAXIMUM WEIGHT OUTERPLANAR SUBGRAPH problem: Given an edge-weighted graph $G$, find an outerplanar subgraph of $G$ of maximum weight. This problem is known to be NP-hard [24, p. 197]. The MST algorithm has a ratio of 1/2 for MAXIMUM WEIGHT OUTERPLANAR SUBGRAPH, as we show in the following using the same methods.

**Corollary 10.2** *Let $P$ be an outerplanar graph with a nonnegative weight function $w$ defined on its edges. Then, a maximum weight forest $F$ of $P$ satisfies $w(F) \geq \frac{1}{2}w(P)$.*

*Proof.* Any subgraph of an outerplanar graph is outerplanar, and, by Euler's formula, $|E(P)| \leq 2(|V(P)| - 1)$ for any outerplanar graph $P$ [29, Cor. 11.9]. Thus, it is enough to apply Lemma 10.1 with $c = 2$ to the family of all outerplanar graphs. ∎

Let $P$ be any outerplanar subgraph of the graph $G$ with a weight function $w$ on its edges and $F$ be a maximum weight forest. By Corollary 10.2, we deduce that $w(F) \geq \frac{1}{2}w(P)$, and the approximation ratio of $1/2$ for MAXIMUM WEIGHT OUTERPLANAR SUBGRAPH follows.

## 10.3 Triangular Structures and Better Approximations

A *triangular cactus* is a graph whose cycles (if any) are triangles and such that all edges appear in some cycle (Figure 10.1). A *triangular cactus in a graph G* is a subgraph of $G$ that is a triangular cactus.

A *triangular structure* is a graph whose cycles (if any) are triangles (Figure 10.2). A *triangular structure in a graph G* is a subgraph of $G$ that is a triangular structure. Note that every triangular cactus is a triangular structure, but not vice versa. Observe also that every triangular structure is planar, as it does not contain cycles of length greater than three.

We start by presenting two algorithms for MAXIMUM PLANAR SUBGRAPH: algorithm A (the greedy one) and algorithm B (the currently best one).

Given a graph $G = (V, E)$ and $E' \subseteq E$, we denote by $G[E']$ the *spanning* subgraph of $G$ induced by $E'$, that is, the graph $(V, E')$. (Note that this is not the usual definition of subgraph induced by an edge set, as we require the subgraph to be spanning.)

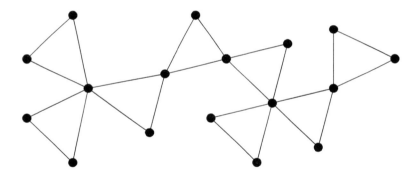

**FIGURE 10.1**　A (connected) triangular cactus.

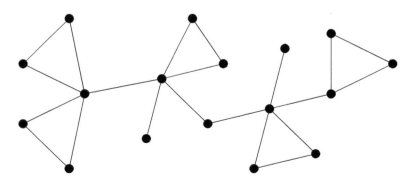

**FIGURE 10.2**　A (connected) triangular structure.

The approximation ratio of algorithm A for MAXIMUM PLANAR SUBGRAPH, as we will see, is $7/18 = 0.3888\ldots$. Algorithm A produces a triangular structure in the given graph $G$. As seen in the following, it consists of two phases. First, it greedily constructs a maximal triangular cactus $S_1$ in $G$. Second, it extends $S_1$ to a triangular structure $S_2$ in $G$ by adding as many edges as possible to $S_1$ without forming any new cycles.

---

## ALGORITHM 10.1    A (G)

1      $E_1 \leftarrow \emptyset$
2      **while** there is a triangle $\Delta$ in $G$ whose vertices are in different components of $G[E_1]$ **do**
3           $E_1 \leftarrow E_1 \cup E(\Delta)$
4      $E_2 \leftarrow E_1$
5      **while** there is an edge $e$ in $G$ whose endpoints are in different components of $G[E_2]$ **do**
6           $E_2 \leftarrow E_2 \cup \{e\}$
7      $S_2 \leftarrow G[E_2]$
8      **return** $S_2$

---

Note that $S_2$ is indeed a triangular structure in $G$ and therefore is a planar subgraph of $G$. It is easy to see that algorithm A is polynomial. Moreover, it can be implemented to run in linear time for graphs with bounded degree [18]. Its approximation ratio will be analyzed in the next section. Now we present the second algorithm for MAXIMUM PLANAR SUBGRAPH.

Algorithm B is very similar to algorithm A. The only difference is that, in the first phase, it finds a triangular cactus with the maximum number of triangles. This can be done in polynomial time using an algorithm for the so-called GRAPHIC MATROID PARITY problem [30,31] as a subroutine. In the following pseudocode, MAXTRICACTUS denotes a routine that uses such an algorithm to obtain a triangular cactus in $G$ with maximum number of triangles. See Reference 18 for the details on how this routine does that.

---

## ALGORITHM 10.2    B (G)

1      $E_1 \leftarrow \text{MAXTRICACTUS}(G)$
2      $E_2 \leftarrow E_1$
3      **while** there is an edge $e$ in $G$ whose endpoints are in different components of $G[E_2]$ **do**
4           $E_2 \leftarrow E_2 \cup \{e\}$
5      $S_2 \leftarrow G[E_2]$
6      **return** $S_2$

---

Again it is clear that $S_2$ is a triangular structure in $G$ and therefore is a planar subgraph of $G$. Gabow and Stallmann [30] described an algorithm for GRAPHIC MATROID PARITY that runs in time $O(m'n'\log^6 n')$, where $m'$ and $n'$ are the number of edges and vertices, respectively, in the input graph for this problem. By using this algorithm, one can get an implementation for MAXTRICACTUS that runs in time $O(m^{3/2}n\log^6 n)$, where $m$ is the number of edges in the input graph and $n$ is the number of vertices. So line 1 can be implemented to run in polynomial time, and again the whole algorithm has a polynomial-time implementation.

As for the weighted case, there is no polynomial-time algorithm known for GRAPHIC MATROID PARITY. However, Camerini et al. [23] proposed a randomized pseudo-polynomial algorithm to solve the MATROID PARITY problem. By using the standard method of rounding and scaling (as in Vol. 1, Chaps. 8 and 9, or in Reference 24, pp. 135–137), one can use their algorithm to obtain, in a weighted graph, for every $\epsilon > 0$, a triangular structure whose weight is at least $1 - \epsilon$ times the weight of a maximum weight triangular structure. The running time of this procedure is polynomial in the size of the input and in $1/\epsilon$.

Curiously, we do not know a direct way to prove that a maximum weight triangular structure achieves a ratio greater than 1/3 for MAXIMUM WEIGHT PLANAR SUBGRAPH. The only way we know how to prove that is by analyzing another algorithm—a greedy one, inspired in an algorithm by Berman and Ramaiyer [22] for MINIMUM STEINER TREE. This algorithm takes advantage of an interesting relation between triangular structures and a particular type of Steiner trees. To explain a bit of this relation, we need to introduce some notation on Steiner trees.

Let $G$ be a connected graph and $R$ be a set of vertices of $G$, usually called *terminals*. Each vertex not in $R$ is called a *Steiner vertex*. A *Steiner tree* is a tree in $G$ containing all terminals. We may assume all leaves in a Steiner tree are terminals. In the MINIMUM STEINER TREE problem, we are given a graph $G$, a nonnegative weight for each edge of $G$, and a set $R$ of vertices of $G$, and we want to find a minimum weight Steiner tree in $G$.

A *full component* of a Steiner tree $T$ is a maximal subtree of $T$ whose internal vertices are all Steiner vertices. For an integer $k$, a Steiner tree is $k$-restricted if all of its full components have at most $k$ leaves.

Berman and Ramaiyer [22] proposed an approximation algorithm for MINIMUM STEINER TREE that produces a 3-restricted Steiner tree of weight at most 11/6 times the minimum weight of a Steiner tree. Their algorithm works for larger values of $k$, producing $k$-restricted Steiner trees having better approximation ratio. The 3-restricted version has been applied to MAXIMUM WEIGHT PLANAR SUBGRAPH, and we give the intuition of this approach in the following.

There is a close relation between triangular structures in a graph $G$ and 3-restricted Steiner trees in an auxiliary graph $H$ defined as follows. The set of vertices of $H$ is $V(G)$ plus a new vertex for each triangle in $G$. The edge set of $H$ is $E(G)$ plus three new edges incident to each triangle vertex. The edges incident to a triangle vertex have as the other endpoints the three vertices of the triangle in $G$. This completes the description of graph $H$. Let the set of terminals $R$ be $V(G)$.

Now, suppose we are given a nonnegative weight for each edge in $G$ and let $M$ be 10 times the largest such weight. In $H$, an edge of $e \in E(G)$ has weight $w'(e) = M - w(e)$. For a triangle $xyz$ of $G$, if $u$ is the new vertex of $H$ corresponding to $xyz$, then we set $w'(xu) = w'(yu) = w'(zu) = 2M/3 - (w(xy) + w(xz) + w(yz))/3$. It is easy to check that a triangular structure of weight $W$ in $G$ corresponds to a 3-restricted Steiner tree in $H$ of weight $(n - 1)M - W$, where $n = |V(G)|$. Thus, a maximum weight triangular structure in $G$ corresponds to a minimum weight 3-restricted Steiner tree in $H$.

The above-mentioned reduction does not preserve approximation ratios. Nevertheless, some algorithms designed for MINIMUM STEINER TREE can be adapted to MAXIMUM WEIGHT PLANAR SUBGRAPH. In particular, the Berman and Ramaiyer algorithm is used by Reference 21 to break the 1/3 threshold for MAXIMUM WEIGHT PLANAR SUBGRAPH.

# 10.4 Analysis of the Triangular Structure Algorithms

We start by settling the approximation ratio of algorithm A.

**Theorem 10.2** *Algorithm A has approximation ratio of 7/18 for* MAXIMUM PLANAR SUBGRAPH.

*Proof.* First, let us show that the approximation ratio is at least 7/18. Without loss of generality, we may assume $G$ is connected and has at least three vertices. Observe that the number of edges in $S_2$ is the number of edges in a spanning tree of $G$ plus the number of triangles in $S_1$. So it suffices to count the number of triangles in $S_1$.

Let $H$ be a plane embedding of a maximum planar subgraph of $G$. Let $n \geq 3$ be the number of vertices in $G$, and $t \geq 0$ be such that $3n - 6 - t$ is the number of edges in $H$. We can think of $t$ as the number of edges missing for $H$ to be a triangulated plane graph. The number of triangular faces in $H$ is at least $2n - 4 - 2t$. (This is a lower bound on the number of triangular faces of $H$ as if $H$ were triangulated, it would have $2n - 4$ triangular faces, and each missing edge can destroy at most two of these triangular faces.)

Let $k$ be the number of components of $S_1$ each with at least one triangle, and let $p_1, p_2, \ldots, p_k$ be the number of triangles in each of these components. Let $p = \sum_{i=1}^{k} p_i$. We will prove that $p$, the number of triangles in $S_1$, is at least a constant fraction of $n - 2 - t$. Note that if a triangle cannot be added to $S_1$, it is because two of its vertices are in the same component of $S_1$. Hence, one of its edges has its two endpoints in the same component of $S_1$. This means that, at the end of the first phase, every triangle in $G$ must have some two vertices in the same component of $S_1$. In particular, every triangular face in $H$ must have some two vertices in the same component of $S_1$, and therefore one of its three edges must be in the subgraph of $H$ induced by the vertices in a component of $S_1$. Thus, we can associate with each triangular face $F$ in $H$ an edge $e$ in $F$ whose endpoints are in the same component of $S_1$. But any edge $e$ in $H$ lies in at most two triangular faces of $H$, so $e$ could have been chosen by at most two triangular faces of $H$. It follows that the number of triangular faces in $H$ is at most twice the number of edges in $H$ whose endpoints are in the same component of $S_1$.

Let $H'$ be the subgraph of $H$ induced by the edges of $H$ whose endpoints are in the same component of $S_1$. Note that $p_i \geq 1$, for all $i$, and that the number of vertices in the $i$th component of $S_1$ is $2p_i + 1 \geq 3$. As $H'$ is planar, $H'$ has at most $\sum_{i=1}^{k} (3(2p_i + 1) - 6) = 6p - 3k$ edges. By the observation at the end of the previous paragraph, $2(6p - 3k) \geq 2|E(H')| \geq$ (number of triangular faces in $H$) $\geq 2n - 4 - 2t$. From this, we have

$$p \geq \frac{n - 2 - t + 3k}{6} \geq \frac{n - 2 - t}{6}.$$

Therefore, the number of triangles in $S_1$ is at least $\frac{n-2-t}{6}$, and the ratio between the number of edges in $S_2$ and the number of edges in $H$ is at least

$$\frac{n - 1 + \frac{n-2-t}{6}}{3n - 6 - t} = \frac{7n - 8 - t}{18n - 36 - 6t} \geq \frac{7}{18},$$

as $t \geq 0$. This completes the proof that the approximation ratio of algorithm A is at least 7/18.

Now, we will prove that the approximation ratio is at most 7/18. Let $S$ be any connected triangular cactus with $p > 0$ triangles. Note that $S$ has $2p + 1 \geq 3$ vertices. Let $S'$ be any triangulated plane supergraph of $S$ on the same set of vertices ($S'$ can be obtained from $S$ by adding edges to $S$ until it becomes triangulated). As $S'$ is triangulated, $S'$ has $2(2p + 1) - 4 = 4p - 2$ (triangular) faces. For each face of $S'$, add a new vertex in the face and adjacent to all vertices on the boundary of that face. Let $G$ be the new graph. Observe that $G$ is a triangulated plane graph and has $(2p + 1) + (4p - 2) = 6p - 1$ vertices. This means that $G$ has $3(6p - 1) - 6 = 18p - 9$ edges. With $G$ as input for algorithm A, in the first phase it can produce $S_1 = S$, and $S_2$ can be $S$ plus one edge for each of the new vertices (the vertices in $G$ not in $S$). The number of edges in $S$ is $3p$. Hence, $S_2$ can have $3p + (4p - 2) = 7p - 2$ edges, whereas $G$ has $18p - 9$ edges. Thus, the ratio between the number of edges in $S_2$ and the number of edges in $G$ is

$$\frac{7p - 2}{18p - 9}.$$

By choosing $p$ as large as we wish, we get a ratio as close to 7/18 as we want. ∎

Algorithm B has an approximation ratio of $4/9 = 0.444\ldots$ The proof that its ratio is at least 4/9 is a consequence of a result on triangular structures in planar graphs.

For a graph $H$, denote by $\mathrm{mts}(H)$ the number of edges in a triangular structure in $H$ with the maximum number of edges. Then, we have the following:

**Theorem 10.3** *If $H$ is a planar graph, then $\mathrm{mts}(H)/|E(H)| \geq 4/9$.*

This theorem however has a long and complicated proof [18] (Raz also announced a proof [Personal communication, 1996]), so we chose to present here a weaker version of the result that has a nice proof.

**Theorem 10.4** *If H is a planar graph, then* $\mathrm{mts}(H)/|E(H)| \geq 0.4$.

*Proof.* The theorem is easily verified if $H$ has fewer than three vertices, so let us assume that $H$ has $n \geq 3$ vertices. We may furthermore assume that $H$ is connected. Embed $H$ in the plane without crossings. Let $t$ be such that $|E(H)| = 3n - 6 - t$. Clearly, $t \geq 0$.

Now let $J$ be any triangular cactus obtained by choosing triangular faces of $H$ until no more can be added; say the final $J$ has $k$ components. Let $p$ be the number of triangles in $J$. As in the proof of Theorem 10.2, if we count twice every edge in $H$ whose endpoints are in the same component of $J$, we will "cover" every triangular face of $H$. In fact, each triangular face of $J$ will be covered three times, by the three edges bounding the face. Let $s$ be the number of edges in $H$ whose endpoints are in the same component of $J$. Let $l$ be the number of triangular faces in $H$. As the $p$ triangles in $J$ are covered three times, we have $(l - p) + 3p = l + 2p \leq 2s$. As in Theorem 10.2, we have $s \leq 6p - 3k$ and $l \geq 2n - 4 - 2t$.

It follows that $2n - 4 - 2t + 2p \leq l + 2p \leq 2s \leq 2(6p - 3k)$, so that

$$p \geq \frac{2n - 4 - 2t + 6k}{10} = \frac{n - 2 - t + 3k}{5} \geq \frac{n - 2 - t}{5}.$$

As $\mathrm{mts}(H) \geq (n - 1) + p$, we have

$$\frac{\mathrm{mts}(H)}{|E(H)|} \geq \frac{n - 1 + \frac{n-2-t}{5}}{3n - 6 - t} = \frac{6n - 7 - t}{15n - 30 - 5t} \geq \frac{2}{5} = 0.4. \qquad \blacksquare$$

Now we proceed with the analysis of algorithm B.

**Theorem 10.5** *Algorithm B has approximation ratio of 4/9 for* MAXIMUM PLANAR SUBGRAPH.

*Proof.* To show that the approximation ratio of algorithm B is at least 4/9, it suffices to apply Theorem 10.3 to a maximum planar subgraph of the input graph of algorithm B. (Theorem 10.4 implies a lower bound of 0.4 on the approximation ratio.)

Now, let us prove that the approximation ratio is at most 4/9. Let $G'$ be any triangulated plane graph on $n'$ vertices. Call $V'$ the vertex set of $G'$. As $G'$ is triangulated, $G'$ has $2n' - 4$ (triangular) faces. For each face of $G'$, add a new vertex in the face, adjacent to all three vertices on the boundary of that face. Let $G$ be the new graph and let $V$ be the vertex set of $G$.

Observe that $G$ is a triangulated plane graph and has $n' + (2n' - 4) = 3n' - 4$ vertices. Therefore, $G$ has $3(3n' - 4) - 6 = 9n' - 18$ edges. Let $S$ be a maximum triangular structure in $G$. Any edge in $G$ has at least one endpoint in $V'$. Moreover, $|V'| = n'$. Therefore, a maximum matching in $G$ has at most $n'$ edges (each with at least one distinct endpoint in $V'$). The following lemma is observed in Reference 32, p. 440.

**Lemma 10.2** *If S is a triangular structure with t triangles in a given graph G, then there is a matching in G of size t.*

By using Lemma 10.2, we conclude that $S$ has at most $n'$ triangles. Recall that $S$, being a triangular structure, is a spanning tree of $G$ plus one edge per triangle in $S$, which implies that $S$ has at most $(3n' - 5) + n' = 4n' - 5$ edges. Furthermore, $G$ has $9n' - 18$ edges. Therefore, the ratio between the number of edges in $S$ and the number of edges in $G$ is

$$\frac{4n' - 5}{9n' - 18}.$$

By choosing $n'$ as large as we wish, we get a ratio as close to 4/9 as we want. $\qquad \blacksquare$

Significantly, somehow Theorem 10.3 does not extend to the weighted case. Let $\mathrm{mwts}(G)$ denote the weight of a maximum weight triangular structure in a weighted graph $G$. Then, we have the following.

**Theorem 10.6** *If H is a planar graph, then* $\mathrm{mwts}(H)/w(H) \geq 1/3 + 1/72$.

This theorem however has a long and complicated proof [21] based on the algorithm adapted from the Berman and Ramaiyer MINIMUM STEINER TREE algorithm. The bound is not tight, and in fact we conjecture the following.

**Conjecture 10.1** *If H is a planar graph, then* mwts($H$)/$w(H) \geq 5/12$.

The next theorem shows that Conjecture 10.1 is as strong as possible in the sense that it does not hold with a constant smaller than 5/12. Conjecture 10.1, if true, would imply that the algorithm that produces a maximum weight triangular structure in the input weighted graph has an approximation ratio of 5/12.

**Theorem 10.7** *For any $\epsilon > 0$, there is a planar graph H and a weight function w defined on its edges for which* mwts($H$)/$w(H) \leq 5/12 + \epsilon$.

*Proof.* Let $P$ be any triangulated plane graph with $n \geq 3$ vertices. By Euler's formula, $P$ has $3n - 6$ edges. Let the weight of each edge of $P$ be 2. In each of the $2n - 4$ faces of $P$, add a new vertex and three edges adjacent to this new vertex and the three vertices of $P$ that define the face. These new edges have weight 1. Denote by $H$ the plane graph obtained this way and by $w$ the weight function defined on its edges.

Observe that $H$ has $3n - 4$ vertices, is triangulated, and has total weight $2 \cdot [3n - 6] + 1 \cdot [3(2n - 4)] = 12n - 24$ (Figure 10.3). Let us prove that a maximum weight triangular structure in $H$ has weight at most $5/12 + o(1)$ of the weight of $H$.

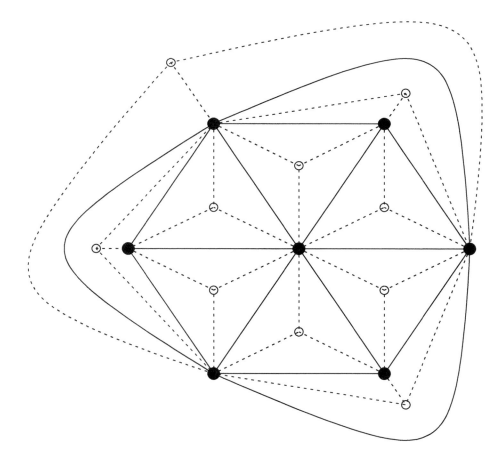

**FIGURE 10.3** An example of graph $H$. The solid edges show graph $P$. Each of them has weight 2. The dotted part shows the extra vertices and edges in $H$. The dotted edges have weight 1.

Consider an arbitrary triangular structure $S$ in $H$. Let $S_1$ be the set of edges of weight 1 in $S$ and $S_2$ be the remaining edges of $S$ (all of weight 2).

First, we may assume $S$ connects $V(H)$, as this can be obtained without decreasing the weight of $S$. Second, we may assume that $S_2$ connects $V(P)$. Indeed, assume this is not the case, and let $V_1, \ldots, V_k$ be the connected components of $P[S_2]$. Then there are two such components, say $V_1$ and $V_2$, such that two edges $e_1$ and $e_2$ in $S_1$ form a path of length two from a vertex $v_1$ in $V_1$ to a vertex $v_2$ in $V_2$. Indeed this is the only way to connect vertices of $P$ with edges from $S_1$. The middle point of this path sits in a face of $P$, so let $v_3$ be the third vertex of $P$, besides $v_1$ and $v_2$, bordering this face. By interchanging, if necessary, indexes 1 and 2, we may assume that $v_3 \notin V_2$. Then the weight-2 edges $v_1 v_2$ and $v_2 v_3$ do not belong to $S_2$. As $S$ only contains cycles of length three, we deduce that $e_2$ (which is incident to $v_2$) does not belong to any cycle of $S$. Thus, replacing $e_2$ by $v_1 v_2$ results in a triangular structure of larger weight.

Now we divide the edges of $S_2$ into two sets: set $X$, containing edges that belong to triangles of $S_2$, and set $Y$, containing the remaining edges. Removing exactly one edge from each triangle of $S_2$ leaves us with a spanning tree of $P$, and therefore $2|X|/3 + |Y| = n - 1$. Each triangle of $S$ that is not in $X$ must contain one edge from $Y$, or we have cycles of length 4. Now recall that $S_2$ connects $P$. Thus, for every vertex of $H$ not in $P$, we have in $S$ exactly one or two edges, and if we have two edges, we have a triangle of $S$ with one edge in $Y$. One edge of $Y$ can appear in at most one triangle, as otherwise we get a cycle of length four. In conclusion, we have

$$w(S) \leq (|X| + |Y|) \cdot 2 + (2n - 4) \cdot 1 + |Y| \cdot 1 \leq 3(\frac{2}{3}|X| + |Y|) + 2n - 4 = 3(n - 1) + 2n - 4 = 5n - 7.$$

Therefore, as $n$ goes to $\infty$, the ratio $\mathrm{mwts}(H)/w(H)$ approaches 5/12 and the theorem follows. ∎

A consequence of Theorem 10.7 is that the approximation ratio of any algorithm for MAXIMUM WEIGHT PLANAR SUBGRAPH that produces a triangular structure in the input graph is at most 5/12. In particular, the approximation ratio of the algorithm that mimics Berman and Ramaiyer's and the one that produces a maximum weight triangular structure in the input weighted graph is at most 5/12.

## 10.5 Outerplanar Subgraphs

A triangular structure is an outerplanar graph. We will show that a maximum weight triangular structure in a weighted graph $G$ has weight at least two-thirds of the weight of a maximum weight outerplanar subgraph of $G$. This implies a $(2/3 - \epsilon)$-approximation algorithm for MAXIMUM WEIGHT OUTERPLANAR SUBGRAPH and, for the unweighted version of the problem, a 2/3-approximation algorithm (via algorithm B, described in Section 10.3).

The proof of the next key lemma takes a few pages.

**Lemma 10.3** *In any maximal outerplanar graph $P$, there are at most three (not necessarily distinct) triangular structures in $P$ such that each edge of $P$ appears in exactly two of them.*

*Proof.* If $P$ has fewer than three vertices, then $P$ and the empty graph are triangular structures. Let us assume $P$ has at least three vertices. Embed $P$ in the plane as a triangulation of a polygon. Every maximal outerplanar graph with at least three vertices is a triangulation of a polygon. That is, the boundary of the exterior face of $P$ is a Hamiltonian cycle $H$ and each interior face is triangular [29, p. 106]. Let $E$ be the edge set of $P$, $b$ be the exterior face of $P$, and $F$ be the set of faces of $P$ other than $b$.

Let $D$ be the dual multigraph of $P$. Let us call the vertices of $D$ (which are faces of $P$) *nodes*, and the edges of $D$, *arcs*. All nodes of $D$ but $b$ have degree three. Also, the edges in the Hamiltonian cycle $H$ correspond to the arcs incident to $b$ in $D$.

Let $D'$ be the graph obtained from $D$ by subdividing each arc incident to $b$, and then removing $b$. See Figure 10.4 for an example.

**Lemma 10.4** *$D'$ is a tree all of whose internal nodes have degree three.*

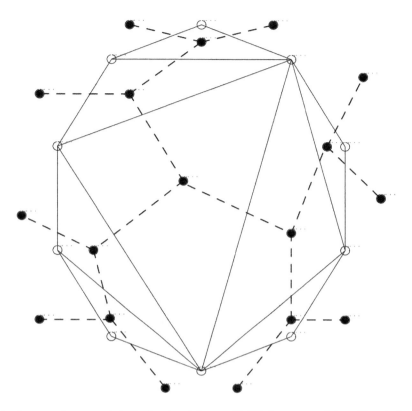

**FIGURE 10.4** An outerplanar graph (solid lines) and the tree $D'$ (filled vertices and dotted lines) obtained from its dual.

*Proof.* First, let us prove that $D'$ has no cycle. It is enough to show that any cycle in $D$ contains $b$. A cycle in $D$ corresponds to a cut in $P$ [33, p. 143, ex. 9.2.3]. As $H$ is a Hamiltonian cycle, any cut in $P$ contains at least two edges of $H$, which correspond to arcs incident to $b$. Therefore, any cycle in $D$ contains at least two arcs incident to $b$, so it contains $b$.

Second, let us prove that $D'$ is connected. If $D'$ were not connected, there would be two nodes $u$ and $v$ in different components of $D'$. Let us argue that we can choose $u$ and $v$ to be nodes in $V(D)$. If $u$ were not a node in $V(D)$, then it would be a node that originated from the subdivision of an arc incident to $b$, and thus it would have degree one in $D'$. Change $u$ to its unique neighbor in $D'$. Do the same for $v$. Note that the new $u$ and $v$ must still be in different components of $D'$, as they are in the same component as the initial $u$ and $v$, respectively. So we can assume $u$ and $v$ are nodes of $D$. And because $D$ is connected, there is a path in $D$ between $u$ and $v$. For this path not to exist in $D'$, it has to go through node $b$. But this implies that $b$ would be a cut vertex in $D$. If $b$ were a cut vertex in $D$, then there would be a *minimal* cut in $D$ containing exactly a proper subset of the arcs incident to $b$ (consider the set of edges going from $b$ to one of the components of $D$ after the removal of $b$). A minimal cut in $D$ corresponds to a cycle in $P$ [33, p. 143, ex. 9.2.3]. This implies that there would be a cycle in $P$ whose edges are a proper subset of the edges of $H$, a contradiction (a proper subset of the edges of any cycle induces an acyclic graph, as it is enough to remove one edge of a cycle to be left with a path that is acyclic).

Therefore, $D'$ is, in fact, a tree. Recall that all nodes of $D$ but $b$ have degree three. Before removing $b$, we subdivided all of the arcs incident to $b$. The new nodes have degree one in $D'$, and all others have the same degree as in $D$, that is, three. ∎

A *special 3-coloring of* $D'$ is a partition of the set of nodes of $D'$ into three sets $\{X_1, X_2, X_3\}$ (each set referred to as a *color class*) such that

1. Adjacent nodes have different colors.
2. For $i = 1, 2, 3$, if we remove all nodes of color $i$, in the resulting graph there is a path from any node to a leaf in $D'$.

**Lemma 10.5** *There is a special 3-coloring of* $D'$.

*Proof.* Root $D'$ at one of its leaves. In the rooted $D'$, all internal nodes except the root have two children. Color the root and its unique child with distinct colors. Now, start at level $i = 1$. If there are nodes in level $i + 1$, for each node in level $i$ with children, give its children distinct colors that differ from its own color. Proceed to level $i + 1$.

Clearly, adjacent nodes get distinct colors.

Suppose we remove all nodes of color $i$, for some $i$. Let $j$ and $k$ be the two remaining colors. Consider a remaining node. Either it is a leaf in $D'$, and there is nothing to prove, or it is an internal node of the rooted tree $D'$ different from the root (as the root is a leaf of $D'$). Any internal node colored $j$ that is not the root has a child of color $k$, and vice versa. This means that there is a path from any node to a leaf in $D'$.                                                                                                                     ∎

Given a special 3-coloring $\{X_1, X_2, X_3\}$ of $D'$, we now describe three triangular structures $S_1, S_2$, and $S_3$ as required by Lemma 10.3. Establish the natural one-to-one correspondence between edges of $P$ and arcs of $D'$: each edge of $P$ corresponds to the unique arc of $D'$ that "crosses" it.

Let $S_i$ be the set of edges in $P$ whose corresponding arc in $D'$ has an endpoint of color $i$ (i.e., in $X_i$). (An arc has either one or zero endpoints of a given color.)

**Lemma 10.6** $S_i$ *is a triangular structure in* $P$.

*Proof.* Suppose that there is a cycle $C$ in $S_i$ that is not a triangle. Cycle $C$ partitions the set of faces of $P$ into two sets $F_0 \ni b$ (outside $C$) and $F_1 \subseteq F$ (inside $C$), $F$ being the set of faces of $D$ other than $b$.

As $C$ is not a triangle, $|F_1| \geq 2$. (If $|F_1| = 1$, then $C$ would be the boundary of the unique face in $F_1 \subseteq F$, which is a triangle.)

A cycle in $P$ corresponds to a minimal cut in $D$ [33, p. 143, ex. 9.2.3]. So $F_1$ must induce a connected subgraph of $D$. As $D'$ differs from $D$ only in arcs incident to $b$, $F_1$ induces the same connected subgraph in $D'$. Also, $F_1$ consists only of internal nodes of $D'$.

Thus, $F_1$ induces a connected subgraph of $D'$ with at least two nodes. Hence, in the given special 3-coloring of $D'$, not all nodes of $F_1$ can be of color $i$; there is a node $d$ in $F_1$ of color $j \neq i$.

All leaves of $D'$ are outside $C$. As $C$ is in $S_i$, each edge of $C$ corresponds to an arc in $D'$ with one endpoint of color $i$. Removing the nodes of $D'$ of color $i$ hence eliminates all arcs with one endpoint inside $C$ and one outside. Thus, after removing from $D'$ the nodes (faces of $P$) of color $i$, there can be no path from node $d$ to a leaf of $D'$. This is a contradiction to the fact that we have a special 3-coloring.                       ∎

Clearly, $S_1, S_2$, and $S_3$ satisfy the statement of Lemma 10.3: each edge of $P$ appears in exactly two of them.                                                                                                                                                     ∎

**Theorem 10.8** *Let* $G$ *be a graph with a nonnegative weight function* $w$ *on its edges and let* $S$ *be a maximum weight triangular structure in* $G$. *Then* $w(S) \geq \frac{2}{3} w(P)$ *for any outerplanar subgraph* $P$ *of* $G$.

*Proof.* By adding edges possibly not in $G$, extend $P$ to a maximal outerplanar graph $\overline{P}$. For any edge $e$ in $\overline{P}$ but not in $G$, let $w(e) = 0$. Clearly, $w(\overline{P}) \geq w(P)$.

Let $S_1, S_2$, and $S_3$ be three triangular structures in $\overline{P}$ as given by Lemma 10.3. Each edge of $\overline{P}$ appears in exactly two of these triangular structures. Then $w(S_1) + w(S_2) + w(S_3) = 2w(\overline{P}) \geq 2w(P)$. Moreover, $w(S) \geq w(S_i)$, $i = 1, 2, 3$. Therefore, $2w(P) \leq 3w(S)$, implying that $w(S) \geq \frac{2}{3} w(P)$.                                     ∎

As there are outerplanar graphs $H_i$ with $2i$ vertices and $3i - 2$ edges that do not have any triangle (all faces except the outer one having size four), the previous theorem is tight. Thus, we conclude:

**Corollary 10.3** *Algorithm B, described in Section 10.3, has approximation ratio of 2/3 for the unweighted version of* MAXIMUM WEIGHT OUTERPLANAR SUBGRAPH. *There is a polynomial-time $(2/3 - \epsilon)$-approximation algorithm for* MAXIMUM WEIGHT OUTERPLANAR SUBGRAPH.

# 10.6 Practical Results and Discussion

Exact algorithms for MAXIMUM PLANAR SUBGRAPH have been proposed. Foulds and Robinson [4] presented a branch and bound algorithm successful only on small dense graphs. Cimikowski [34] also proposed a branch and bound algorithm. Jünger and Mutzel [9,35] presented a branch-and-cut methodology with promising results on unweighted instances. New facets for this method have been proposed by Hicks [36].

Heuristics without proven approximation ratio have been proposed and analyzed experimentally by Cacceta and Kusumah [37] and Poranen [38,39]. In particular, Poranen [39] proposed modified greedy algorithms that improve on experiments over algorithm A from Section 10.3. Precisely, triangles are added also when two of the three vertices are in the same component of $G[E_1]$ (the third being in another component) if the two vertices are adjacent in $E_1$. Poranen [39] proved that this modified algorithm also has approximation ratio at least 7/18 and conjectures it has approximation ratio 4/9. We have examples showing that in fact the 7/18 ratio is tight for his algorithms. Indeed, take an arbitrary triangulated plane graph $H$ with $q$ vertices (choose $q$ odd). Place one new vertex in each of the $2q - 4$ faces of $H$, making it adjacent to the three vertices in the face. The resulting (planar) graph has $3q - 4$ vertices and $9q - 18$ edges, as it is triangulated. Add another triangular cactus spanning the original $q$ vertices so that its edges have endpoints apart, forming no triangle with the plane graph $H$. Poranen's first algorithm can pick this last triangular cactus and no other triangle, resulting in a subgraph with $3(q - 1)/2$ edges for the cactus plus $2q - 4$ edges, one for each vertex placed in a face of $H$. This leads to a ratio that converges (when $q$ tends to infinity) to 7/18. In a similar way, one can describe a worst case example for another of his algorithms. Poranen reports that his modified greedy algorithms are a slightly better start for simulated annealing heuristics. Recently, Chimani et al. [40] also experimentally compared heuristics for MAXIMUM PLANAR SUBGRAPH.

For certain classes of inputs, better algorithms are known. Kühn et al. [41] showed that graphs having large minimum degree contain large planar subgraphs. For example, if the minimum degree is at least $1500\sqrt{n}/\epsilon^2$, then every graph $G$ with $n = |V(G)|$ sufficiently large has a planar subgraph with $(2 - \epsilon)n$ edges. Their proofs can be converted in polynomial-time algorithms. For a more up-to-date view of this line of algorithms [42].

If the input graph is a weighted clique whose edge weights satisfy the triangle inequality, the output of the algorithm that mimics Berman and Ramaiyer's [22] has approximation ratio at least 3/8 [21].

The related problem of GRAPH THICKNESS (given $G$, partition the edges of $G$ into as few planar subgraphs as possible) has been proven to be NP-hard [43]. A 3-approximation for GRAPH THICKNESS is trivial, via arboricity as in the proof of Lemma 10.1. Finding better approximation algorithms seems very hard.

We conclude with a discussion on improving the approximation ratio for MAXIMUM PLANAR SUBGRAPH and MAXIMUM WEIGHT PLANAR SUBGRAPH. First we remark that with the current methods adapted from MINIMUM STEINER TREE, MAXIMUM WEIGHT PLANAR SUBGRAPH is harder than MAXIMUM PLANAR SUBGRAPH, as opposed to many problems mentioned in Reference 44.

For MAXIMUM WEIGHT PLANAR SUBGRAPH, there is still room for improvement with triangular structures. Conjecture 10.1, for instance, (or some weaker version of it with a bound greater than $1/3 + 1/72$) would already imply a better ratio for MAXIMUM WEIGHT PLANAR SUBGRAPH.

A natural approach is to consider structures with larger blocks. For example, what is the approximation ratio of an algorithm that produces a "large" structure whose blocks have up to four vertices? (Note that such structures are planar.) On the contrary, finding such a structure with the maximum number of edges is NP-hard. (One can reduce packing of triangles in graphs of maximum degree bounded by four to this problem.) The approximation algorithm of Berman and Ramaiyer's [22] and the relative greedy algorithm of Zelikovsky [45] can be adapted to use planar structures with blocks of size at most $k$, for an arbitrary fixed $k$, and run in time polynomial in $n^k$, where $n$ denotes, as usual, the number of vertices of the input graph.

For a graph $H$, denote by $\text{mts}_k(H)$ the maximum number of edges in a subgraph of $H$ whose blocks have size at most $k$. Theorem 10.3 states that, if $H$ is planar, $\text{mts}_3(H)/|E(H)| \geq 4/9$. Călinescu and Fernandes [46] prove that, as $k$ tends to infinity, $\text{mts}_k(H)/|E(H)|$ tends to $1/2$. This convergence also holds for the weighted version, although the convergence rate is slower. The corresponding ratios, for outerplanar and weighted outerplanar graphs, tend to 1.

These results are analogous to computing the Steiner ratios for $k$-restricted Steiner trees, which tend to 1 as $k$ goes to $\infty$ [47,48] (Vol. 1, Chap. 36). As $\text{mts}_k(H)/|E(H)|$ tends to $1/2$ instead of 1, we believe that algorithm B is better than possible adaptations of Berman and Ramaiyer's [22] or Zelikovsky's relative greedy algorithm [45]. This type of situation happens in fact for MINIMUM STEINER TREE, in which computing an almost optimal 3-restricted Steiner tree (as done by Prömel and Steger [25]) is better than both Berman and Ramaiyer's [22] and the relative greedy approximation algorithm of Zelikovsky [45].

All known MINIMUM STEINER TREE algorithms with better ratio than Prömel and Steger [25] (such as the one by Robins and Zelikovsky [49], and the best known obtained by Byrka et al. [50]) use not only the notion of "gain" (as defined in Reference 48), but also the notion of "loss." However, the notion of "loss" does not translate easily to MAXIMUM PLANAR SUBGRAPH.

# References

1. Dyer, M.E., Foulds, L.R., and Frieze, A.M., Analysis of heuristics for finding a maximum weight planar subgraph, *European Journal of Operations Research* 20, 102, 1985.
2. Eades, P., Foulds, L., and Giffin, J., An efficient heuristic for identifying a maximal weight planar subgraph, *Combinatorial Mathematics IX, Lectures Notes in Mathematics* 952, p. 239. Springer-Verlag, Berlin, Germany, 1982.
3. Foulds, L.R., *Graph Theory Applications*, Springer-Verlag, New York, 1992.
4. Foulds, L.R. and Robinson, D.F., Graph-theoretic heuristics for the plant layout problem, *International Journal of Production Research* 12, 27, 1978.
5. Krejcirik, K., Computer aided plant layout, *Computer Aided Design* 2, 7, 1969.
6. Mutzel, P., *The Maximum Planar Subgraph Problem*, PhD Thesis, Universität zu Köln, 1994.
7. Marek-Sadowska, M., Planarization algorithms for integrated circuits engineering, in *Proceedings of the IEEE International Symposium on Circuits and Systems*, p. 919. IEEE Press, Hoboken, NJ, 1978.
8. Di Battista, G., Eades, P., Tamassia, R., and Tollis, I.G., *Graph Drawing: Algorithms for the Visualization of Graphs*, Prentice-Hall, Englewood, NJ, 1998.
9. Jünger, M. and Mutzel, P., Maximum planar subgraphs and nice embeddings: practical layout tools, *Algorithmica* 16 (1), 33, 1996. (Special issue on graph drawing.)
10. Liebers, A., Planarizing graphs—A survey and annotated bibliography, *Journal of Graph Algorithms and Applications* 5 (1), 1, 2001.
11. Batini, C., Nardelli, E., Talamo, M., and Tamassia, R., A graph-theoretic approach to aesthetic layout of information system diagrams, in *Proceedings of the 10th International Workshop on Graph Theoretic Concepts in Computer Science*, p. 9. Springer-Verlag, Berlin, Germany, 1984.
12. Tamassia, R., Di Battista, G., and Batini, C., Automatic graph drawing and readability of diagrams, *IEEE Transactions on Systems, Man and Cybernetics* 18, 61, 1988.
13. Chiba, N., Nishizeki, T., and Shirakawa, I., An algorithm for maximal planarization of graphs, in *Proceedings of the IEEE International Symposium on Circuits and Systems*, p. 649. IEEE Press, Hoboken, NJ, 1979.

14. Jayakumar, R., Thulasiraman, K., and Swamy, M.N.S., On maximal planarization of nonplanar graphs, *IEEE Transactions on Circuits and Systems*, CAS-33, 843, 1986.

15. Ozawa, T. and Takahashi, H., A graph-planarization algorithm and its applications to random graphs, in *Graph Theory and Algorithms*, *Lectures Notes in Computer Science* 108, p. 95. Springer-Verlag, Berlin, Germany, 1981.

16. Liu, P.C. and Geldmacher, R.C., On the deletion of nonplanar edges of a graph, in *Proceedings of the 10th Southeastern Conference on Combinatorics, Graph Theory, and Computing*, p. 727. Utilitas Mathematica Pub., Winnipeg, Canada, 1977.

17. Yannakakis, M., Node- and edge-deletion NP-complete problems, in *Proceedings of ACM Symposium on Computational Geometry*, p. 253. ACM Media, New York, 1978.

18. Călinescu, G., Fernandes, C.G., Finkler, U., and Karloff, H., A better approximation algorithm for finding planar subgraphs, *Journal of Algorithms* 27, 269, 1998.

19. Faria, L., de Figueiredo, C.M.H., and Mendonça, C.F.X., On the complexity of the approximation of nonplanarity parameters for cubic graphs, *Discrete Applied Mathematics* 141, 119, 2004.

20. Cimikowski, R., An analysis of heuristics for graph planarization, *Journal of Information & Optimization Sciences* 18 (1), 49, 1997.

21. Călinescu, G., Fernandes, Karloff, H., and Zelikovsky, A., A new approximation algorithm for finding heavy planar subgraphs, *Algorithmica* 36 (2), 179, 2003.

22. Berman, P. and Ramaiyer, V., Improved approximations for the Steiner tree problem, *Journal of Algorithms* 17, 381, 1994.

23. Camerini, P.M., Galbiati, G., and Maffioli, F., Random pseudo-polynomial algorithms for exact matroid problems, *Journal of Algorithms* 13, 258, 1992.

24. Garey, M.R. and Johnson, D.S., *Computers and Intractability: A Guide to the Theory of NP-Completeness*, W.H. Freeman and Company, New York, 1979.

25. Prömel, M.J. and Steger, A., A new approximation algorithm for the Steiner tree problem with performance ratio 5/3, *Journal of Algorithms* 36, 89, 2000.

26. Korte, B. and Hausmann, D., An analysis of the greedy heuristic for independence systems, *Annals of Discrete Mathematics* 2, 65, 1978.

27. Chen, B., Matsumoto, M., Wang, J., Zhang, Z., and Zhang, J., A short proof of Nash-Williams' theorem for the arboricity of a graph, *Graphs and Combinatorics* 10, 27, 1994.

28. Nash-Williams, C.S.J.A., Decomposition of finite graphs into forests, *Journal of London Mathematical Society* 39, 12, 1964.

29. Harary, F., *Graph Theory*, Addison-Wesley Publishing Company, Boston, MA, 1972.

30. Gabow, H.N. and Stallmann, M., Efficient algorithms for graphic matroid intersection and parity, in *12th Colloquium on Automata, Language and Programming*, *Lectures Notes in Computer Science* 194, p. 210. Springer-Verlag, Berlin, Germany, 1985.

31. Szigeti, Z., On the graphic matroid parity problem, *Journal of Combinatorial Theory, Series B* 88 (2), 247, 2003.

32. Lovász, L. and Plummer, M.D., *Matching Theory*, Elsevier Science, Amsterdam, the Netherlands, 1986.

33. Bondy, J.A. and Murty, U.S.R., *Graph Theory with Applications*, MacMillan Press, Basingstoke, UK, 1976.

34. Cimikowski, R., Branch-and-bound techniques for the maximum planar subgraph problem, *International Journal of Computer Mathematics* 53, 135, 1994.

35. Jünger, M. and Mutzel, P., Solving the maximum weight planar subgraph problem, in *Proceedings of the 3rd Integer Programming and Combinatorial Optimization Conference*, p. 479. Springer-Verlag, Berlin, Germany, 1993.

36. Hicks, I.V., New facets for the planar subgraph polytope, *Networks*, 51 (2), 120, 2008.

37. Cacceta, L. and Kusumah, Y.S., A new heuristic for facility layout design, in *Proceedings of Optimization, Techniques and Applications*, p. 287. Curtin University of Technology, Bentley, Australia, 1998.
38. Poranen, T., A simulated annealing algorithm for the maximum planar subgraph problem, *International Journal of Computer Mathematics* 81 (5), 555, 2004.
39. Poranen, T., Two new approximation algorithms for the maximum planar subgraph problem, *Acta Cybernetica* 18 (3), 503, 2008.
40. Chimani, B., Klein, K., and Wiedera, T., A note on the practicality of maximal planar subgraph algorithms, in *Proceedings of the 24th International Symposium on Graph Drawing & Network Visualization*, 2016. Retrieved from http://arxiv.org/pdf/1608.07505v1.pdf.
41. Kühn, D., Osthus, D., and Taraz, A., Large planar subgraphs in dense graphs, *Journal of Combinatorial Theory, Series B* 95 (2), 263, 2005.
42. Allen, P., Skokan, J., and Würfl, A., Maximum planar subgraphs in dense graphs, *Electronic Journal of Combinatorics* 20 (3), P1, 2013.
43. Mansfield, A., Determining the thickness of graphs is NP-Hard, *Mathematical Proceedings of the Cambridge Philosophical Society*, 93, 9, 1983.
44. Crescenzi, P., Silvestri, R., and Trevisan, L., To weight or not to weight: where is the question? in *Proceedings of the 4th Israel Symposium on Theory of Computing and Systems*, p. 68. IEEE Computer Society Press, Los Alamitos, CA, 1996.
45. Zelikovsky, A., Better approximation bounds for the network and Euclidean Steiner tree problems, Department of Computer Science, University of Virginia, CS-96-06, 1996.
46. Călinescu, G. and Fernandes, C.G., On the *k*-structure ratio in planar and outerplanar graphs, *Discrete Mathematics & Theoretical Computer Science* 10 (3), 135, 2008.
47. Du, D.-Z., Zhang, Y.-J., and Feng, Q., On better heuristic for Euclidean Steiner minimum trees, in *Proceedings of the 32nd Annual IEEE Symposium on Foundations of Computer Science*, p. 431. IEEE Computer Society Press, Los Alamitos, CA, 1991.
48. Zelikovsky, A., An 11/6-approximation algorithm for the network Steiner problem, *Algorithmica* 9, 463, 1993.
49. Robins, G. and Zelikovsky, A., Tighter bounds for graph Steiner tree approximation, *SIAM Journal on Discrete Mathematics* 19 (1), 122, 2005.
50. Byrka, J., Grandoni, F., Rothvoß, T., and Sanitá, L., Steiner tree approximation via iterative randomized rounding, *Journal of the ACM* 60 (1), 1, 2013.

# Disjoint Paths and Unsplittable Flow

Stavros G. Kolliopoulos

## 11.1 Introduction

Finding disjoint paths in graphs is a problem that has attracted considerable attention from at least three perspectives: graph theory, VLSI design, and network routing/flow. The corresponding literature is extensive. In the current chapter, we focus mostly on results on offline approximation algorithms for problems on general graphs as influenced from the network flow perspective. Surveys examining the underlying graph theory, combinatorial problems in VLSI, and disjoint paths on special graph classes can be found in References 1–8. We sporadically mention some results on fixed-parameter tractability, but this is not an aspect this survey covers at any depth.

An instance of *edge-disjoint (vertex-disjoint) paths* consists of a graph $G = (V, E)$ and a multiset $\mathcal{T} = \{(s_i, t_i) : s_i \in V, t_i \in V, \ i = 1, \ldots, k\}$ of $k$ source-sink pairs. Any source or sink is called a *terminal*. An element of $\mathcal{T}$ is also called a *commodity*. In the decision problem, one asks whether there is a a set of edge-disjoint (vertex-disjoint) paths $P_1, P_2, \ldots, P_k$, where $P_i$ is an $s_i$-$t_i$ path, $i = 1, \ldots, k$. The graph $G$ can be either directed or undirected. Typically, a terminal may appear in more than one pair in $\mathcal{T}$. For vertex-disjoint paths, one requires that the terminal pairs are mutually disjoint. We abbreviate the edge-disjoint paths problem by EDP and vertex-disjoint paths by VDP. The notation introduced will be used throughout the chapter to refer to an input instance. We will also denote $|V|$ by $n$ and $|E|$ by $m$ for the corresponding graph.

Based on whether $G$ is directed or undirected and the edge- or vertex-disjointness condition, one obtains four basic problem versions. The following polynomial-time reductions exist among them. Any undirected problem can be reduced to its directed counterpart by replacing an undirected edge with an appropriate gadget; both reductions maintain planarity. See Reference 9 and Chapter 28 of [4] for details. An edge-disjoint problem can be reduced to its vertex-disjoint counterpart by replacing $G$ with its line graph (or digraph as the case may be). Directed vertex-disjoint paths reduce to directed edge-disjoint paths by replacing every vertex with a pair of new vertices connected by an edge. There is no known reduction from a directed to an undirected problem. These transformations can serve for translating approximation guarantees or hardness results from the edge-disjoint to the vertex-disjoint setting and vice versa.

The *unsplittable flow* problem (UFP) is the generalization of EDP in which every edge $e \in E$ has a positive capacity $u_e$, and every commodity $i$ has a demand $d_i > 0$. The demand from $s_i$ to $t_i$ has to be routed in an unsplittable manner, that is, along a single path from $s_i$ to $t_i$. For every edge $e$, the total demand routed through that edge should be at most $u_e$. We will sometimes refer to a capacitated graph as a *network*. In a similar manner, a vertex-capacitated generalization of vertex-disjoint paths can be defined. UFP was introduced in the PhD thesis of Kleinberg [8]. Versions of the problem had been studied before though not under the UFP moniker [10,11].

If one relaxes the requirement that every commodity should use exactly one path, one obtains the *multicommodity flow* problem which is solvable in polynomial time, for example, through linear programming. When all the sources of a multicommodity flow instance coincide at a vertex $s$ and all the sinks at a vertex $t$, we obtain the classical *s-t flow* problem, whose maximization version is the well-studied *maximum flow* problem. The relation between UFP and multicommodity flow is an important one to which we shall return often in this survey. We will denote a solution to either problem as a flow vector $f$ defined on the edges or the paths of $G$ as appropriate.

*Complexity of disjoint-path problems:* For general $k$, all four basic problems are NP-complete. The undirected VDP was shown to be NP-complete by Knuth in 1974 [12], via a reduction from SAT, and by Lynch [13]. This implies the NP-completeness of directed VDP and directed EDP. Even, Itai and Shamir [14] showed that both problems remain NP-complete on directed acyclic graphs (DAGs). In the same paper the undirected EDP was shown NP-complete even when the multiset $\mathcal{T}$ contains only two distinct pairs of terminals. In the case when $s_1 = s_2 = \cdots = s_k$ all four versions are in $P$ as special cases of maximum flow. For planar graphs, Lynch's reduction [13] shows NP-completeness for undirected VDP; Kramer and van Leeuwen [15] showed that undirected EDP is NP-complete. The NP-completeness of the directed planar versions follows.

For fixed $k$, the directed versions are NP-complete even for the case of two pairs with opposing source-sinks, that is, $(s, t)$ and $(t, s)$ [16].[*] Undirected VDP, and by implication EDP as well, can be solved in polynomial time [17]. This is an outcome of the celebrated project of Robertson and Seymour on graph minors. See Reference 18 for an informal description of the highly impractical Robertson-Seymour algorithm, which runs in $f(k) \cdot n^3$ for an immense, but computable, function $f(k)$. By an easy reduction, the Robertson-Seymour algorithm also works when terminal pairs are not mutually disjoint and one wants the output to consist of internally-disjoint paths. A decision problem is *fixed-parameter tractable* (FPT) parameterized by $t$ if it can be solved in time $f(t)n^{O(1)}$ for a computable function $f$ that depends only on $t$. Therefore, the Robertson-Seymour algorithm implies that the decision version of disjoint paths in undirected graphs is FPT when parameterized by $k$. It is notable that for fixed $k$, VDP, and by consequence EDP, can be solved on DAGs by a fairly simple polynomial-time algorithm [16]. Earlier polynomial-time algorithms for $k = 2$ include the one by Perl and Shiloach on DAGs [9] and the ones derived independently by Seymour [19], Shiloach [20], and Thomassen [21] for VDP on general undirected graphs.

For planar graphs and fixed $k$, the directed VDP is in P [22], whereas the complexity of the edge-disjoint case is open, even for $k = 2$. Schrijver's algorithm runs in $n^{O(k)}$ but Cygan et al. [23] showed that the problem is in fact FPT with an algorithm that runs in $2^{2^{O(k)}} \cdot n^{O(1)}$. These algorithms solve the decision problems. A few polynomial-time algorithms are known for optimization versions. When the input graph is a tree, Garg et al. gave a polynomial-time algorithm to maximize the number of pairs that can be connected by edge-disjoint paths [24]. The algorithm extends for vertex-disjoint paths (N. Garg, personal communication, July 2005). By total unimodularity, maximizing the number of pairs that can be connected by edge-disjoint paths is polynomial-time solvable on *di-trees* as well, that is, directed graphs

---

[*] The NP-completeness proof holds for a sparse graph with $m = \Theta(n)$; this observation has consequences for hardness of approximation proofs in References 25–27.

in which there is a unique directed path from $s_i$ to $t_i$, for all $i$; a reduction to a minimum-cost circulation problem is also possible in this case [28]. Reducing directed vertex-disjoint paths to EDP maintains the di-tree property, hence the maximization version of the former problem is polynomial-time solvable as well. Observe that directed out- and in-trees are special cases of di-trees. Additional complexity results for special graph classes can be found in References 29–32. For a comprehensive complexity classification up to 2009 [33].

*Optimization versions*: Two basic NP-hard optimization problems are associated with unsplittable flow and hence with EDP. Given a UFP instance, an *unsplittable flow solution* or simply a *routing* is a selection of $k' \leq k$ paths, one each for a subset $\mathcal{T}' \subseteq \mathcal{T}$ of $k'$ commodities. For every commodity, $(s_i, t_i) \in \mathcal{T}'$ demand $d_i$ is routed along the corresponding path. Any routing can be expressed as a flow vector $f$; the flow $f_e$ through edge $e$ equals the sum of the demands using $e$. A *feasible* routing is one that respects the capacity constraints. In the *maximum-demand* optimization problem, one seeks a feasible routing of a subset $\mathcal{T}'$ of commodities such that $\sum_{i \in \mathcal{T}'} d_i$ is maximized. The *congestion* of a routing $f$ is defined as $\max_{e \in E}\{\max\{f_e/u_e, 1\}\}$. Note that the events $f_e < u_e$ and $f_e = u_e$ are equivalent for this definition. In the *minimum-congestion* optimization problem, one seeks a routing of all $k$ commodities that minimize the congestion, that is, one seeks the minimum $\lambda \geq 1$ such that all $k$ commodities can be feasibly routed if all the capacities are multiplied by $\lambda$. From now on, we use MEDP to denote the maximum-demand optimization version of EDP. Similarly, MVDP for the vertex-disjoint maximization problem. Some other objective functions of interest will be defined in Section 11.3.

*Fractional solutions and flow-cut gaps*: We present now some background on multicommodity flow.

*LP-rounding (linear program) algorithms*: As mentioned, multicommodity flow is an efficiently-solvable relaxation of EDP. Hence, it is no accident that multicommodity flow theory has played such an important part in developing algorithms for disjoint-path problems. This brings us to the standard linear programming formulation for the optimization version of multicommodity flow. Let $\mathcal{P}_i$ denote the set of paths from $s_i$ to $t_i$. Set $\mathcal{P} := \bigcup_{i=1}^{k} \mathcal{P}_i$. Consider the following LP formulation for *maximum multicommodity flow*:

$$\text{maximize} \sum_{P \in \mathcal{P}} f_P \qquad\qquad \text{(LP-MCF)}$$

$$\sum_{P \in \mathcal{P}_i} f_P \leq d_i \qquad\qquad \text{for } i = 1, \ldots, k$$

$$\sum_{P \in \mathcal{P} \,:\, P \ni e} f_P \leq u_e \qquad\qquad \text{for } e \in E$$

$$f_P \geq 0 \qquad\qquad \text{for } P \in \mathcal{P}$$

The number of variables in the LP is exponential in the size of the graph. By using flow variables defined on the edges, one can write an equivalent LP of polynomial size. We choose to deal with the more elegant flow-path formulation. We observe that adding the constraint $f_P \in \{0, d_i\}, \forall P \in \mathcal{P}_i$, to (LP-MCF) turns it into an exact formulation for maximum-demand UFP. We call an LP solution for the optimization problem of interest *fractional*. A similar LP, corresponding to the *concurrent flow* problem, can be written for minimizing congestion. See Reference 34 for details. Our presentation in this section focuses on edge-capacitated graphs in which the dual problems concern the computation of minimum (fractional) edge cuts. Similarly, one can write the multicommodity flow LP for vertex-capacitated graphs in which the dual objects of interest are vertex cuts.

Several early approximation algorithms for UFP, and more generally, integer multicommodity flow, work in two stages. First, a fractional solution $f$ is computed. Then $f$ is rounded to an unsplittable solution $\hat{f}$ through procedures of varying intricacy, most commonly by randomized rounding as shown by

Raghavan and Thompson [35]. The randomized rounding stage can usually be derandomized using the method of conditional probabilities [36–38]. The derandomization component has gradually become very important in the literature for two reasons. First, through the key work of Srinivasan [39,40] on pessimistic estimators, good deterministic approximation algorithms were designed even in cases in which the success probability of the randomized experiment was small. See References 41, 42 for applications to disjoint paths. Second, in some cases, the above-mentioned two-stage scheme can be implemented rather surprisingly without solving first the LP. Instead, one designs directly a suitable Langrangean relaxation algorithm implementing the derandomization part. See the work of Young [43] and Volume 1, Chapter 4.

We note that some of the approximation ratios obtained through the LP-rounding method can nowadays be matched (or surpassed) by simple combinatorial algorithms. By combinatorial, one usually means algorithms restricted to ordered ring operations as opposed to ordered field ones. Two distinct greedy algorithms for MEDP were given by Kleinberg [8] (see also Reference 44), and Kolliopoulos and Stein [45] (see also Reference 46). A lot of the subsequent work on combinatorial algorithms for general graphs uses these two approaches as a basis. Still the influence of rounding methods on the development of algorithms for disjoint-path problems can hardly be overstated. See Volume 1, Chapter 7 in this volume for further background on LP-based approximation algorithms.

*Approximate max-flow/min-cut theorems*:  One of the first results on disjoint paths and in fact one of the cornerstones of graph theory is Menger's Theorem [47]: an undirected graph is $k$ vertex-connected if and only if there are $k$ vertex-disjoint paths between any two vertices. The edge analog holds as well and the min–max relation behind the theorem has resurfaced in a number of guises, most notably as the max-flow/min-cut theorem for $s$-$t$ flows. Let $G = (V, E)$ be undirected. For $S \subseteq V$, define $\delta_G(S) := \{\{u, v\} \in E : u \in S$ and $v \in V \setminus S\}$. Similarly, dem($S$) is the sum of all demands over commodities, which are separated by the cut $\delta_G(S)$. A necessary condition for the existence of a feasible fractional solution to (LP-MCF) that satisfies all demands is the *cut condition*:

$$\sum_{e \in \delta_G(S)} u_e \geq \mathrm{dem}(S), \text{ for each } S \subset V.$$

Define the quantity $\frac{\sum_{e \in \delta_G(S)} u_e}{\mathrm{dem}(S)}$ as the *sparsity* of the cut $\delta_G(S)$, and let $\lambda$ be the sparsity of the minimum sparsity cut. The cut condition is equivalent to $\lambda \geq 1$. For $s$-$t$ flows, the max-flow/min-cut theorem [48–50] yields that the cut condition is also sufficient. For undirected multicommodity flow, Hu showed that the cut condition is sufficient for $k = 2$ [51]. It fails in general for $k \geq 3$. For directed multicommodity flows, there are simple examples with $k = 2$, for which the directed analog of the cut condition holds but the demands cannot be satisfied fractionally [4]. For undirected EDP, already for $k = 2$, the cut condition is not sufficient for a solution to exist [1]. Let $\phi > 0$ be the largest quantity so that a (fractional) multicommodity flow exists that respects capacities and routes the scaled demand values $\phi d_i$ for all $i$. The *concurrent-flow/cut gap* of an instance is the quantity $\frac{\lambda}{\phi}$. Starting with the seminal work of Leighton and Rao [52], a lot of effort has been spent on determining the concurrent-flow/cut gap in a variety of settings. Among several other results, an optimal upper bound of $O(\log k)$ has been established for general undirected graphs both for edge as well as vertex cuts [53–55].

A *multicut* in an undirected capacitated graph $G = (V, E)$ is a subset of edges $F \subseteq E$, such that if all edges in $F$ are deleted, no two vertices from a pair $(s_i, t_i)$ $i = 1, \ldots, k$ are in the same connected component of the remaining graph. The *maximum-multiflow/cut gap* of an instance is the ratio between the values of the minimum-capacity multicut and the maximum (fractional) multicommodity flow. The latter quantity is simply the optimum of the (LP-MCF) formulation. Garg et al. [56] showed that the maximum-multiflow/cut gap in an undirected graph is $O(\log k)$, and this is existentially tight. See References 34, 57, and 58 for surveys of the results in this area and their applications to approximation algorithms. In contrast to the undirected case, for both types of flow-cut gaps there are polynomial lower bounds in directed graphs [59,60]. Polylogarithmic bounds exist for the problem in which one wants to separate sets as opposed to pairs of vertices [61]. This includes the case in which demands are symmetric, that is, there are commodities for both ordered pairs $(s_i, t_i)$ and $(t_i, s_i)$.

The outline of the current chapter is as follows. In Section 11.2, we present hardness of approximation results and (mostly greedy) algorithms for MEDP and MVDP. In Section 11.3, we examine algorithms for the various optimization versions of UFP, properties of the fractional relaxation, packing integer programs (PIPs), and UFP-specific hardness results. Section 11.4 outlines developments specific to undirected graphs that have for the most part occurred at the intersection of approximation algorithms and structural graph theory. In Section 11.5, we present results on some variants of the basic problems. Unless mentioned otherwise, all of the algorithms we describe for directed graphs work also on undirected graphs.

## 11.2 Disjoint Paths

In the current section, we examine the problem of finding a maximum-size set of edge-disjoint paths, mostly from the perspective of combinatorial algorithms. We defer the discussion of the LP-rounding algorithms and the integrality gaps of the linear relaxations until Section 11.3, in which we examine them in the more general context of UFP, similarly for some key results on expander graphs and hardness bounds particular to UFP. Table 11.1 summarizes the known positive and negative results for MEDP and MVDP on general graphs, including bounds on the integrality gap of the multicommodity flow relaxation.

*Hardness results*: Ma and Wang [67] showed via the PCP theorem that MVDP and MEDP on directed graphs cannot be approximated within $2^{O(\log^{1-\varepsilon} n)}$, unless NP $=$ DTIME($2^{\text{polylog}(n)}$). Guruswami et al. [25] showed that on directed graphs it is NP-hard to obtain an $O(n^{1/2-\varepsilon})$ approximation for any fixed $\varepsilon > 0$. They gave a gap-inducing reduction from the two-pair decision problem to EDP on a sparse graph with $\Theta(n)$ edges. As this EDP problem reduces to a vertex-disjoint path instance on a graph with $N = \Theta(n)$ vertices, we obtain that is NP-hard to approximate MVDP on graphs with $N$ vertices within $O(N^{1/2-\varepsilon})$, for any fixed $\varepsilon > 0$. See Chapter 17 in the 1st edition of this handbook for background on the PCP theorem and the theory of inapproximability. Chalermsook et al. [68] showed that it is NP-hard to obtain an $n^{1/2-\varepsilon}$ approximation for MEDP on DAGs. MEDP on undirected graphs is much less understood. It was first shown MAX SNP-hard in Reference 24. Improving upon an earlier result [69], Andrews et al. [65] showed that, for any constant $\varepsilon > 0$, there is no $\log^{1/2-\varepsilon} n$ approximation algorithm unless NP $\subseteq$ ZPTIME($n^{\text{polylog}(n)}$). ZPTIME($n^{\text{polylog}(n)}$) is the set of languages that have randomized algorithms that always give the correct answer in expected running time $n^{\text{polylog}(n)}$. The same hardness result holds for MVDP on undirected graphs. Even when congestion $1 \leq c \leq \alpha \log \log n / \log \log \log n$ is allowed for some constant $\alpha > 0$, Reference 65 shows that MEDP and MVDP are $\log^{\Omega(1/c)} n$-hard to approximate. For directed graphs, there is a constant $0 < \lambda < 1/4$ such that approximating MEDP while allowing congestion $1 \leq c \leq \log^{\lambda} n$ is hard to approximate within a factor $\Omega(n^{1/c})$, unless NP $\subseteq$ ZPTIME($n^{\text{polylog}(n)}$) [70]. We return to disjoint paths with congestion in Section 11.4.

*Greedy algorithms*: The first approximation algorithm analyzed in the literature for MEDP on general graphs seems to be the online *Bounded Greedy Algorithm (BGA)* in the PhD thesis of Kleinberg [8];

**TABLE 11.1**  Known Upper and Lower Bounds for Disjoint Paths on General Graphs in Terms of $m$ and $n$. The Best Upper Bounds on the Integrality Gap for MEDP are $O\left(\min\{\sqrt{m}, n^{4/5}\}\right)$ on Directed Graphs [41,66] and $O(\sqrt{n})$ on Undirected Graphs [64]. For MVDP the Corresponding Upper Bound is $O(\sqrt{n})$ [45] on Both Types of Graphs

| Problem | Approximation Ratio | Integrality Gap | Hardness |
|---|---|---|---|
| Directed MEDP | $O(\min\{\sqrt{m}, n^{2/3} \log^{1/3} n\})$ [41,45,62,63] | $\Omega(\sqrt{n})$ [24] | $\Omega(n^{1/2-\varepsilon})$ [25] |
| Undirected MEDP | $O(\sqrt{n})$ [64] | $\Omega(\sqrt{n})$ [24] | $\log^{1/2-\varepsilon} n$ [65] |
| Directed MVDP | $O(\sqrt{n})$ [42,45] | $\Omega(\sqrt{n})$ [24] | $\Omega(n^{1/2-\varepsilon})$ [25] |
| Undirected MVDP | $O(\sqrt{n})$ [42,45] | $\Omega(\sqrt{n})$ [24] | $\log^{1/2-\varepsilon} n$ [65] |

see also Reference 44. The algorithm is parameterized by a quantity $L$. The terminal pairs are examined in one pass. When $(s_i, t_i)$ is considered, check if $s_i$ can be connected to $t_i$ by a path of length at most $L$. If so, route $(s_i, t_i)$ on such a path $P_i$. Delete $P_i$ from $G$ and iterate. To simplify the analysis, we assume that the last terminal pair is always routed if all the previous pairs have been rejected.

The idea behind the analysis of BGA [8] is very simple but it has influenced later work such as References 45, 66, 71. Informally it states that *in any graph there cannot be too many long paths that are edge-disjoint*. In Reference 8, the algorithm was shown to achieve a $(2L + 1)$-approximation if $L = \max\{diam(G), \sqrt{m}\}$. Several people quickly realized that the analysis can be slightly altered to obtain an $O(\sqrt{m})$-approximation. We provide such an analysis with $L = \sqrt{m}$. The first published $O(\sqrt{m})$ approximation for MEDP was given by Srinivasan using LP-rounding methods [41].

Let $\mathcal{O}$ be a maximum-cardinality set of edge-disjoint paths connecting pairs of $\mathcal{T}$. Let $\mathcal{B}$ be the set of paths output by BGA and $\mathcal{O}_u \subset \mathcal{O}$ be the set of paths corresponding to terminal pairs unrouted by the BGA. We have that

$$|\mathcal{O}| - |\mathcal{O}_u| = |\mathcal{O} \setminus \mathcal{O}_u| \le |\mathcal{B}|. \tag{11.1}$$

One tries to relate $|\mathcal{O}_u|$ to $|\mathcal{B}|$. This is done by observing that a commodity $l$ routed in $\mathcal{O}_u$ was not routed in $\mathcal{B}$ because one of two things happened: (i) no path of length shorter than $L$ exists or (ii) the existing paths from $s_l$ to $t_l$ were blocked by (intersect on at least one edge with) paths selected earlier in $\mathcal{B}$. The paths in $\mathcal{O}_u$ can thus be partitioned into the two corresponding subsets $\mathcal{O}_1$ and $\mathcal{O}_2$. $\mathcal{O}_1$ contains paths blocked by a path in $\mathcal{B}$ and has size at most $L|\mathcal{B}|$, as the elements of $\mathcal{B}$ are edge-disjoint paths of length at most $L$. The second set $\mathcal{O}_2 := \mathcal{O}_u \setminus \mathcal{O}_1$, consists of disjoint paths longer than $L$, hence $|\mathcal{O}_2| < m/L$. Therefore

$$|\mathcal{O}_u| < \frac{m}{L} + L|\mathcal{B}| = \sqrt{m} + \sqrt{m}|\mathcal{B}| \le 2\sqrt{m}|\mathcal{B}|. \tag{11.2}$$

Adding inequalities (11.1) and (11.2) yields that the BGA is an $O(\sqrt{m})$-approximation algorithm. In Section 11.3, we return to the performance of the BGA on expander graphs.

The astute reader has noticed that the idea used in the above-mentioned analysis is an old one. It goes back to the blocking flow method of Dinitz [72] for the maximum $s$-$t$ flow problem as applied to unit-capacity networks by Even and Tarjan [73]. A *blocking flow* is a flow that cannot be augmented without rerouting. The blocking flow method iterates over the residual graph. In every iteration, a blocking flow is found over the subgraph of the residual graph that contains the edges on a shortest path from $s$ to $t$. At the end of an iteration, the distance from $s$ to $t$ in the new residual graph can be shown to have increased by at least one. When the distance becomes larger than $L$, the number of edge-disjoint paths from $s$ to $t$ is $O(\min\{m/L, n^2/L^2\})$ and this bounds also the remaining number of augmentations required by the algorithm [73].

Kolliopoulos and Stein [45] made the connection with the blocking flow idea explicit and proposed the offline Greedy_Path algorithm, from now on called simply *the Greedy algorithm*. The motivation behind the Greedy algorithm was the following: What amount of residual flow has survived if one is never allowed to reroute the flow sent along shortest paths at a given iteration? In every iteration, Greedy picks the unrouted $(s_i, t_i)$ pair such that the length of the shortest path $P_i$ from $s_i$ to $t_i$ is minimized. The pair is routed using $P_i$; the edges of $P_i$ are deleted from the graph. In the original paper on the Greedy algorithm, it was shown to output a solution of size $\Omega(\max\{OPT/\sqrt{m_0}, OPT^2/m_0, OPT/d_0\})$, where $OPT$ is the optimum, $m_0$ is the minimum number of edges used in an optimal solution, and $d_0$ is the minimum average length of the paths in an optimal solution [45]. In particular, when the terminals are disjoint and there exists an acyclic optimal solution, one can show using a result in Reference 74 that $m_0 = O(n^{3/2})$. By using the above-mentioned BGA notation and analysis from we obtain the following [66].

**Lemma 11.1** *Consider the restriction of the Greedy algorithm that as soon as the minimum shortest path length among the unrouted pairs exceeds L selects one more path and terminates. The approximation guarantee is at most* $\max\{L, |\mathcal{O}_2|\}$.

The analysis in Reference 45 used the fact that $|\mathcal{O}_2| \leq m/L$. This was extended by Chekuri and Khanna [66]:

**Theorem 11.1 [66]:** *Using the notation defined earlier* $|\mathcal{O}_2| = O(n^2/L^2)$ *for undirected simple graphs and* $|\mathcal{O}_2| = O(n^4/L^4)$ *for the directed case.*

The theorem together with Lemma 11.1 and Reference 45 yields immediately that the Greedy algorithm achieves an $O(\min\{\sqrt{m}, n^{2/3}\})$-approximation for undirected MEDP and an $O(\min\{\sqrt{m}, n^{4/5}\})$ for directed MEDP. Varadarajan and Venkataraman [63] improved the bound for directed graphs to $O(\min\{\sqrt{m}, n^{2/3} \log^{1/3} n\})$, again for the Greedy algorithm. Their argument shows the existence of a cut of size $O((n^2/L^2) \log(n/L))$ that separates all terminal pairs $(s_i, t_i)$ lying at distance $L \geq \log n$ or more. This brings us almost full circle back to the Even-Tarjan bound [73] for $s$-$t$ flows. The latter argument demonstrates the existence of a cut of size $O(n^2/L^2)$ when the source is at distance $L$ or more from the sink. Reference 66 demonstrates an infinite family of undirected and directed acyclic instances on which the approximation ratio achieved by the Greedy algorithm is $\Omega(n^{2/3})$. New ideas are thus required to bring the approximation for directed graphs down to $O(\sqrt{n})$ which in Reference 66 is conjectured to be possible. This conjecture is still open.

Because of the $d_0$-approximation outlined earlier, one can assume without loss of generality that all shortest $s_i - t_i$ paths have length $\Omega(\sqrt{n})$. Then a counting argument shows that there is a vertex $u$ such that at least $\Omega(OPT/\sqrt{n})$ paths in the optimal solution go through this "congested" vertex $u$. We guess $u$ and concentrate on finding an $O(1)$-approximation to the maximum-size *u-solution,* to our original EDP instance: this consists only of paths going through $u$. By using this approach Chekuri et al. [64] and independently Nguyen [75] obtained $O(\sqrt{n})$-approximation algorithms for EDP on undirected graphs and DAGs. The results of Reference 64 establish matching tight bounds on the integrality gap.

We now sketch the proof of Theorem 11.1 for the undirected case as given by Chekuri and Khanna [66]. The theorem holds for the fractional solution as well, that is, the value $\nu$ of the maximum fractional multicommodity flow connecting terminals at distance more than $L$. Call a vertex of $G$ *high-degree* if its degree is more than $6n/L$ and *low-degree* otherwise. The total capacity incident to low-degree vertices is $O(n^2/L)$. We claim that every $s_i - t_i$ path, $(s_i, t_i) \in \mathcal{T}$, must contain at least $L/6$ of the low-degree vertices. Therefore $\nu$, the sum of flow values over the paths used in the fractional solution, is $O(n^2/L^2)$. To prove the claim consider a breadth-first search tree rooted at $s_i$ and let *layer* $L_j$ be the set of vertices at distance $j$ from $s_i$. We will show something stronger: There are at least $L/6$ layers among the first $L$ consisting only of low-degree vertices. Partition the layers into blocks of three contiguous layers and let $B_j$ denote the block made up of layers $L_{3j+1}, L_{3j+2}, L_{3j+3}$. Discard the blocks that contain at least one layer consisting entirely of low-degree vertices. If $L/6$ or more blocks are discarded, we are done. So assume that we are left with at least $L/6$ blocks. The blocks are disjoint so at least one of the remaining blocks, call it $B_*$, must contain at most $6n/L$ vertices. Consider a high-degree vertex in the middle layer of $B_*$. By the breadth-first search property all its neighbors must be within $B_*$ itself, a contradiction. This completes the proof of Theorem 11.1.

*Vertex-disjoint paths:* The Greedy algorithm, with the obvious modification, connects a set of terminal pairs of size $\Omega(\max\{OPT/\sqrt{n_0}, OPT^2/n_0, OPT/d_0\})$ [45]. Here $n_0$ denotes the minimum size of a set of vertices used in the optimal solution and $d_0$ the minimum average path length in an optimal solution. By the hardness result of Reference 25, this result is asymptotically tight on directed graphs, unless $P = NP$.

# 11.3 Unsplittable Flow

We start with some additional definitions. We assume that a UFP instance satisfies the *no-bottleneck assumption* (NBA)*: $d_{max} := \max_{i=1,\dots,k} d_i \leq u_{min} := \min_{e \in E} u_e$, that is, any commodity can be routed through any of the edges. This assumption is common in the literature and we will mention it explicitly when NBA is not met. In the *weighted* UFP, commodity $i$ has an associated weight (profit) $w_i > 0$; one wants to route feasibly a subset of commodities with maximum total weight. Note that maximizing demand reduces to maximizing the weight: simply set $w_i := d_i$, $i = 1, \dots, k$. Another objective function of interest in addition to maximizing demand and minimizing congestion is routing in the *minimum number of rounds*. A round corresponds to a set of commodities that can be routed feasibly, hence one seeks a minimum-size partition of the set of commodities into feasible unsplittable flow solutions. A *uniform capacity unsplittable flow problem* (UCUFP) is a UFP in which all edges of the input graph have the same capacity value.

*Randomized rounding and UFP*: Some of the approximation ratios achieved by LP-rounding that we are about to present are currently also obtainable with simple greedy algorithms. See the paragraph on combinatorial algorithms in the following. Nevertheless LP-rounding algorithms have the advantage that they are analyzed with respect to the existentially weak optima of the linear relaxations. In addition, their analysis yields upper bounds on the respective integrality gaps. An implementation study comparing the actual performance of the LP-based versus the more combinatorially-flavored algorithms would be of interest. For an in-depth survey of randomization for routing problems, see Reference 76.

Minimizing congestion. The best known algorithm for congestion is also perhaps the best known example of the randomized rounding method of Raghavan and Thompson [35]. A fractional solution $f$ to the concurrent flow problem is computed and then one path is selected independently for every commodity from the following distribution: Commodity $i$ is assigned to path $P \in \mathcal{P}_i$ with probability $f_P/d_i$. An application of the Chernoff bound [77] shows that with high probability the resulting congestion is $O(\log n / \log \log n)$ times the fractional optimum. The process can be derandomized using the method of conditional probabilities [38]. Young [43] shows how to construct the derandomized algorithm without having first obtained the fractional solution.

The analysis of the performance guarantee cannot be improved. Leighton et al. [78] provide an instance on a directed graph on which a fractional solution routes at most $1/\log^c n$ flow per edge, for any constant $c > 0$, while any unsplittable solution incurs congestion $\Omega(\log n / \log \log n)$. If the unsplittable solution uses only paths with nonzero fractional flow, the lower bound holds for both undirected and directed instances with optimal UFP congestion 1 [78,79]. Trivially, it is NP-hard to approximate congestion within better than 2 in the case of EDP; this would solve the decision problem. For directed graphs, the hardness results were improved in References 80, 81 and finally, Chuzhoy et al. [70] showed a tight $\Omega(\log n / \log \log n)$ bound, assuming that NP $\not\subseteq$ BPTIME($n^{O(\log \log n)}$). For undirected graphs, Andrews and Zhang [82] showed that congestion cannot be approximated within $(\log \log n)^{1-\varepsilon}$, for any constant $\varepsilon > 0$, unless NP $\subseteq$ ZPTIME($n^{polylog(n)}$). Improving upon this bound, Andrews et al. [65] showed that minimizing congestion in undirected graphs is hard to approximate within $\Omega(\log \log n / \log \log \log n)$, assuming NP $\not\subseteq$ ZPTIME($n^{polylog(n)}$).

Maximizing congestion. Srinivasan published the first $O(\sqrt{m})$-approximation for MEDP and more generally, maximum-demand UCUFP in Reference 41. The first nontrivial $O(\sqrt{m} \log m)$-approximation for (weighted) UFP was published in the IPCO version of Reference 45. Simultaneously and independently, Baveja and Srinivasan refined the results in Reference 41 to obtain an $O(\sqrt{m})$-approximation for weighted UFP; this work was published in Reference 42. The Baveja-Srinivasan methods extend the earlier key work of Srinivasan on LP-rounding methods for approximating PIPs [39,40]. We outline now some

---

* In early literature it was called the *balance condition*.

of the ideas in References 39, 41, 42. The algorithm computes first a fractional solution $f$ to the weighted modification of the (LP-MCF) relaxation that has the same constraints as (LP-MCF) and objective function $\sum_{i=1}^{k} w_i \sum_{P \in \mathcal{P}_i} f_P$. We call this relaxation from now on (LP-WMCF). The rounding method has two phases. First, a randomized rounding experiment is analyzed to show that it produces with positive probability a near-optimal feasible unsplittable solution. Second, the experiment is derandomized yielding a deterministic polynomial-time algorithm for computing a feasible near-optimal solution. Let $y_*$ be the fractional optimum.

One starts by scaling down every variable $f_P$ by an appropriate parameter $\alpha > 1$. This is done to boost the probability that after randomized rounding all edge capacities are met. Let $B_i$ denote the event that in the unsplittable solution, the capacity of the edge $e_i \in E$ is violated. Let $B_{m+1}$ denote the event that the routed demand will be less than $y_*/(\beta\alpha)$, for some $\beta > 1$. The quantity $\beta\alpha$ is the targeted approximation ratio. The randomized rounding method of Raghavan and Thompson in the context of UFP works by bounding the probability of the "bad" event $\bigcup_{i=1}^{m+1} B_i$ by $\sum_{i=1}^{m+1} Pr(B_i)$. Srinivasan [41] and later Srinivasan and Baveja [42] exploited the fact that the events $\overline{B_i}$ are *positively correlated*: If it is given that a routing respects the capacities of the edges in some $S \subset E$, the conditional probability that for $e_i \in E \setminus S$, $\overline{B_i}$ occurs is at least $Pr(\overline{B_i})$. Mathematically this is expressed via the FKG inequality due to Fortuin, Ginibre, and Kasteleyn (see Reference 83, Vol. 1, Chap. 6). By using the positive correlation property, Baveja and Srinivasan obtained a better upper bound on $Pr(\bigcup_{e_i \in E} B_i)$ than the naive union bound and therefore can prove the existence of an unsplittable solution while using a better, that is, smaller, $\beta\alpha$ scaling factor than traditional randomized rounding. The second ingredient of Srinivasan's method [39,40] is to design an appropriate pessimistic estimator to constructively derandomize the method. Such an estimator is shown for UFP as well in Reference 42. The by-now standard derandomization approach of Raghavan [38] fails as it relies precisely on the probability $Pr(\bigcup_{i=1}^{m+1} B_i)$ being upper-bounded by $\sum_{i=1}^{m+1} Pr(B_i)$.

Let $d$ denote the *dilation* of the optimal fractional solution $f$, that is, the maximum number of edges on any flow-carrying path. The Baveja-Srinivasan algorithm computes a solution to weighted UFP of value

$$\Omega(\max\{(y_*)^2/m, y_*/\sqrt{m}, y_*/d\}), \tag{11.3}$$

The corresponding upper bounds on the integrality gap of (LP-WMCF) follow. The analysis of Reference 39 was simplified by Srinivasan [84] by using randomized rounding followed by alteration. Here the outcome of the random experiment is allowed to violate some constraints. It is then altered in a greedy manner to achieve feasibility. The problem-dependent alteration step should be analyzed to quantify the potential degradation of the performance guarantee. This method was applied to UFP in Reference 85.

Integrality gaps for weighted MEDP and MVDP. In *weighted* MEDP or MVDP, one wants to maximize the total weight of the paths that can be feasibly routed. For weighted MVDP the bounds of (11.3) hold with $n$ in place of $m$ [42,45]. In LP-based algorithms in which the selection of the fractional paths that respect capacities and the rounding are two distinct stages, it is possible in the vertex-disjoint case to accommodate the more general setting in which different commodities may share terminals, that is, when one requires that the internal vertices of a chosen path are not used in any other. We show how one can obtain an $O(\sqrt{n+\mu})$-approximation in this case, where $\mu$ denotes the sum of the multiplicities of the terminals that belong to more than one pairs. Clearly, $\mu \leq 2k$. We restrict the discussion to undirected graphs, but it is easy to extend it to the directed case. Let $G$ be the input graph and $\mathcal{F}$ be the set of paths that support a feasible fractional solution for weighted MVDP. We will set up the rounding stage on a modified graph $G'$ with a modified set of paths $\mathcal{F}'$. Consider a terminal $s$ that has multiplicity $l := \mu(s)$, that is, belongs to the pairs of $l > 1$ commodities. We create a new graph $G'$ where $s$ is replaced by $l$ vertices $s^1, \ldots, s^l$ where each is connected to the neighborhood of $s$ in $G$. Fractional paths in $\mathcal{F}$ that correspond to one of the $l$ commodities originating at $s$ are each mapped to a path in $\mathcal{F}'$ that originates at the corresponding $s^i$. Let $P$ be a path in $\mathcal{F}$ that uses $s$ as an internal vertex. In $\mathcal{F}'$ obtain $P'$ from $P$ by splicing out $s$ and replacing it by a subpath through $s^1, \ldots, s^l$. For each $i$, set up the vertex-capacity constraint for $s^i$ so that either one can route a commodity originating from $s^i$ or one can use it as an internal vertex in at most one path in $\mathcal{F}'$. Producing $G', \mathcal{F}'$ and enforcing the new vertex capacity constraints requires simply writing the corresponding PIP in which the columns correspond to the paths in $\mathcal{F}'$, see the rounding algorithms for VDP in Reference 45 and the upcoming discussion of PIPs in the following.

In addition to the upper bounds on the integrality gap of (LP-WMCF) given by (11.3), the integrality gap for weighted MEDP is $O(\sqrt{n})$ on undirected graphs [64] and $O(n^{4/5} \log n)$ on directed graphs [66]. The gap is known to be at least $k/2$ for the unit-weight case by an example in a grid-like planar graph with $k = \Theta(\sqrt{n})$, for both MEDP and MVDP [24]. The best known lower bounds for maximum-demand and weighted UFP are summarized in Table 11.2. We provide some additional negative results for UFP in the upcoming paragraph on combinatorial algorithms.

Minimizing the number of rounds. Aumann and Rabani [87] (see also Reference 8) show that a $\rho$-approximation for maximum demand translates to an $O(\rho \log n)$ guarantee for the number of rounds objective. Reference 42 provides improvements when all edge capacities are unit. Let $\chi(\mathcal{T})$ be the minimum number of rounds. In deterministic polynomial-time one can feasibly "route in rounds," the number of rounds being the minimum of (i) $O(\chi(\mathcal{T}) d^\delta \log n + d(y_* + \log n))$ for any fixed $\delta \in (0,1)$, (ii) $O(\eta^{-1} d(y_* + \log n))$, if for all $i$, $d_i \geq \eta$, and (iii) $O\left(\chi(\mathcal{T})\sqrt{m\left(1 + (\log n)/\chi(\mathcal{T})\right)}\right)$ [42]. Minimizing the number of rounds for UFP is related to wavelength assignment in optical networks. Connections routed in the same round can be viewed as being assigned the same wavelength. There is extensive literature dealing with *path coloring* as this problem is often called; usually the focus is on special graph classes. See Reference 88, Volume 1, Chapter 2 for an introduction to this area.

UFP with small demands. Versions of UFP have been studied that impose stronger restrictions on $d_{\max}$ than just NBA. In UFP *with small demands*, one assumes that $d_{\max} \leq u_{\min}/B$, for some $B > 1$. Various improved bounds that depend on $B$ exist, some obtainable via combinatorial algorithms. In the rather arbitrarily named *high-capacity* UFP, $B = \Omega(\log m)$. An optimal deterministic $O(\log n)$-competitive online algorithm was obtained by Awerbuch et al. [11]. It maintains length functions for the edges that are exponential in the current load. This idea was introduced for multicommodity flow in Reference 89 and heavily used thereafter [43,90–92]. Raghavan [38] showed that standard randomized rounding achieves with high probability an $O(1)$-approximation for maximum weight with respect to the fractional optimum. Similarly, one obtains that the high-capacity UFP admits an $O(1)$-approximation for congestion. For general $B > 1$, various bounds that depend on $B$ exist, some obtainable via combinatorial algorithms. Baveja and Srinivasan obtained an $O(t^{1/\lfloor B \rfloor})$-approximation where $t$ is the maximum total capacity used by a commodity along a path in a fractional solution. Under the standard assumption that demands have been scaled to lie in $[0,1]$, $t \leq d$. Azar and Regev gave a combinatorial algorithm that achieves an $O(BD^{1/B})$-approximation where $D$ is the maximum length of a path used in the optimal solution. Kolman and Scheideler [71] investigated the approximability in terms of a different network measure,

**TABLE 11.2** Known Upper and Lower Bounds Involving $m$ and $n$ for the Maximization Versions of UFP on General Graphs, with (w) or without (w/o) NBA. The Best Known Upper Bound on the Integrality Gap for Directed Max-Weight with NBA is $O(\min\{\sqrt{m}, n^{4/5} \log n\})$ [42,66]. For Undirected Max-Weight with NBA a Matching $O(\sqrt{n})$ Upper Bound on the Integrality Gap Was Given in Reference 64

| UFP Problem | NBA | Approximation Ratio | Integrality Gap | Hardness |
|---|---|---|---|---|
| Directed max-demand | w/o | $O(\min\{\sqrt{m}, n^{4/5}\})$ [86] | $\Omega(\sqrt{n})$ [24]* | $\Omega(n^{1/2-\varepsilon})$ [25]* |
| Undirected max-demand | w/o | $O(\min\{\sqrt{m}, n^{2/3}\})$ [86] | $\Omega(\sqrt{n})$ [24]* | $\Omega(n^{1/2-\varepsilon})$ [27] |
| Directed max-weight | w | $O(\min\{\sqrt{m}, n^{4/5} \log n\})$ [26,42,66] | $\Omega(\sqrt{n})$ [24] | $\Omega(n^{1/2-\varepsilon})$ [25] |
| Undirected max-weight | w | $O(\sqrt{n})$ [64] | $\Omega(\sqrt{n})$ [24] | $\log^{1/2-\varepsilon} n$ [65] |
| Directed max-weight | w/o | $O(\sqrt{m} \log(2 + \frac{d_{\max}}{u_{\min}}))$ [26] | $\Omega(n)$ [85] | $\Omega(n^{1-\varepsilon})$ [26] |
| Undirected max-weight | w/o | $O(\sqrt{m} \log(2 + \frac{d_{\max}}{u_{\min}}))$ [26] | $\Omega(n)$ [85] | $\Omega(n^{1-\varepsilon})$ [27] |

*Note:* Lower bounds with a (*) apply also with NBA, in fact they carry over from MEDP.

cf the upcoming discussion on the concept of the flow number. See References 26, 42, 45, 71, 85 for further bounds and details. Interesting results for the *half-disjoint* case, that is, when $B = 2$, include the following: a polynomial-time algorithm on undirected graphs for the decision version of UFP for fixed $k$ [93], a poly-logarithmic approximation for MEDP on undirected planar graphs [94], and an $O(\sqrt{n})$-approximation for MEDP on directed graphs [75]. See also the discussion on Disjoint Paths with congestion in Section 11.4.

*Packing integer programs and UFP:*  Given $A \in [0,1]^{M \times N}$, $b \in [1, \infty)^M$, and $c \in [0,1]^N$ with $\max_j c_j = 1$, a *PIP* $\mathcal{P} = (A, b, c)$ seeks to maximize $c^T \cdot x$ subject to $x \in \mathbb{Z}_+^N$ and $Ax \leq b$. Let $B$ and $\zeta$ denote respectively $\min_i b_i$, and the maximum number of nonzero entries in any column of $A$. The restrictions on the values in $A, b, c$ are without loss of generality; arbitrary nonnegative values can be scaled appropriately [39]. When $A \in \{0,1\}^{M \times N}$, we say that we have a $(0,1)$-*PIP*. The best guarantees known for PIPs are due to Srinivasan; those for $(0,1)$-PIPs are better than those known for general PIPs by as much as an $\Omega(\sqrt{M})$ factor [39,40].

As witnessed by the (LP-WMCF) relaxation, weighted UFP is a packing problem, albeit one with an exponential number of variables. Motivated by UFP, Kolliopoulos and Stein [45] defined the class of *column-restricted PIPs (CPIPs):* these are the PIPs in which all nonzero entries of column $j$ of $A$ have the same value $\rho_j$, for all $j$. Observe that a CPIP generalizes Knapsack. If one obtains the fractional solution $f$ to (LP-WMCF), one can formulate, at a loss of a $\log m$ factor, the *rounding problem* as a polynomial-size CPIP in which the columns of $A$ correspond to the paths used in the fractional solution and the rows correspond to edges in the graph, hence to capacity constraints. The column value $\rho_j$ equals the demand $d_j$ of the commodity corresponding to the path represented by the column. In combination with improved bounds for CPIPs, this approach yielded the $O(\sqrt{m} \log m)$-approximation for weighted UFP [45] mentioned previously.

A result of independent interest in Reference 45 shows that approximating a family of column-restricted PIPs can be reduced in an approximation-preserving fashion to approximating the corresponding family of $(0,1)$-PIPs. This result is obtained constructively via the *grouping-and-scaling* technique which first appeared in Reference 95 in the context of single-source UFP. Let $z_*$ be the fractional optimum. For a general CPIP, the reduction of Reference 45 using the bounds for $(0,1)$-PIPs in References 39, 40 translates to the existence of an integral solution of value $\Omega \left( \max \left\{ \frac{z_*}{M^{1/(\lfloor B \rfloor + 1)}}, \frac{z_*}{\zeta^{1/\lfloor B \rfloor}}, z_* \left( \frac{z_*}{M} \right)^{1/\lfloor B \rfloor} \right\} \right)$.

Baveja and Srinivasan [96] improved the dilation bound for CPIPs to $\Omega(\frac{z_*}{t^{1/\lfloor B \rfloor}})$ where $t \leq \zeta$ is the maximum column sum of $A$. Notably, Baveja and Srinivasan [96] treat CPIPs as a special case of an abstract generalization of UFP that they call *low-congestion routing problem (LCRP)*. LCRP is a PIP in which it is convenient to think of the columns of a given matrix $A$ as corresponding to "paths," even though there is no underlying graph. Columns are partitioned into $k$ groups: the set of variables is $\{z_{u,v} : u \in [k], v \in [l_u]\}$, where $l_u$ are given integers. In addition to the column-restricted packing constraints $Az \leq b$, at most one variable from each group can be set to 1, that is, $\forall u, \sum_{v \in [l_u]} z_{u,v} \leq 1$. The objective is to maximize $\sum_{u \in [k]} w_u \sum_{v \in [l_u]} z_{u,v}$, for a vector $w \in \mathbb{R}_+^k$. Let $m$ denote the number of rows of $A$, $z_*$ the fractional optimum of an LCRP, and $\zeta$ the maximum number of nonzero entries in any column of $A$, Baveja and Srinivasan [42] show how to obtain an integer solution to LCRP of value

$$\Omega(\max\{(z_*)^2/m, z_*/\sqrt{m}, z_*/\zeta\}) \tag{11.4}$$

that is, LCRP can be approximated as well as its special case of weighted UFP. In Reference 96, they showed the $O(t^{1/\lfloor B \rfloor})$-approximation mentioned earlier for LCRPs, and as a consequence for weighted UFP as well. Observe that the approach of Reference 45 to UFP was the opposite to the one of References 42, 96. In Reference 45, the authors reduced the rounding stage of a UFP algorithm to a CPIP, at a loss of a $\log m$ factor. Baveja and Srinivasan attacked directly LCRPs, saved the $\log m$ factor from the UFP approximation ratio, and then obtained results for CPIPS as corollaries.

Chekuri et al. [97] translated in a convenient way the grouping-and-scaling approach of Reference 45 so that it can be used as a black box for UFP problems: The integrality gap for instances with NBA is within a constant factor of the integrality gap for unit-demand instances. Chekuri et al. [98] investigated CPIPs without NBA and among other results established an $O(L)$-approximation for CPIPs with at most $L$ nonzero entries per column.

*Combinatorial algorithms*: For UFP with polynomially bounded demands and without NBA Guruswami et al. [25] gave a simple randomized algorithm that achieves an $O(\sqrt{m}\log^{3/2} m)$-approximation and generalized the Greedy algorithm for MEDP [45] (cf Section 11.2) to UFP, to obtain an $O(\sqrt{m}\log^2 m)$-approximation. Azar and Regev [26] provided the first strongly-polynomial algorithm for weighted UFP with NBA that achieves an $O(\sqrt{m})$-approximation. For weighted UFP without NBA, they obtained a strongly-polynomial $O(\sqrt{m}\log(2 + \frac{d_{\max}}{u_{\min}}))$-approximation algorithm. By a reduction from the two-pair decision problem, it is NP-hard to obtain an $O(n^{1-\varepsilon})$-approximation for weighted UFP on directed graphs without NBA, for any fixed $\varepsilon > 0$ [26]. The lower bound applies with all the commodities sharing the same source and the weights being such that the objective function is the cardinality of the set of commodities that can be feasibly routed. To quantify the effect of $d_{\max}/u_{\min}$, Azar and Regev showed via a different reduction that it is NP-hard to obtain an $O\left(n^{1/2-\varepsilon}\sqrt{\log(2 + \frac{d_{\max}}{u_{\min}})}\right)$ ratio [26]. For weighted UFP without NBA the integrality gap of the multicommodity flow relaxation is $\Omega(n)$ even when the input graph is a path [85].

Guruswami et al. [25] considered the *integral splittable flow (ISF)* problem in which one allows the flow for a commodity to be split along more than one path but each of these paths must carry an integral amount of flow. The objective function is to maximize the total weight of the commodities for which the entire demand has been routed. This problem is NP-hard on both directed and undirected graphs, even with just two sources and sinks [14]. Hardness results for MEDP trivially carry over to ISF. Guruswami et al. [25] observe that there is an approximation-preserving reduction from maximum independent set to ISF, therefore the latter problem cannot be approximated on undirected graphs within $m^{1/2-\varepsilon}$ unless NP = ZPP. Generalizing the Greedy algorithm of Reference 45, they showed that ISF with polynomially-bounded demands is approximable within a factor of $O(\sqrt{md_{max}}\log^2 m)$ [25]. Another of the few known polynomial lower bounds for undirected graphs was also given in Reference 25: it is NP-hard to approximate the maximum-cardinality objective of vertex-capacitated UFP within a factor of $n^{1/2-\varepsilon}$.

Further progress in terms of greedy algorithms was achieved by Kolman and Scheideler [71] and Kolman [86]. Recall the BGA algorithm from Section 11.2. Kolman and Scheideler [71] proposed the *careful BGA*, parameterized by $L$. The commodities are ordered according to their demands, starting with the largest. Commodity $i$ is accepted if there is a feasible path $P$ for it such that, after routing $i$, the total flow is larger than half their capacity on at most $L$ edges of $P$. Let $\mathcal{B}_1$ be the solution thus obtained and $\mathcal{B}_2$ the solution consisting simply of the largest demand routed on any path. The output is $\mathcal{B} := \max\{\mathcal{B}_1, \mathcal{B}_2\}$. In Reference 71, the careful BGA is shown to achieve an $O(\sqrt{m})$-approximation for maximum-demand UFP without NBA. Generalizing the above-mentioned Theorem 11.1 to maximum-demand UFP, Kolman showed that the careful BGA achieves an $O(\min\{\sqrt{m}, n^{2/3}\})$-approximation on undirected networks and $O(\min\{\sqrt{m}, n^{4/5}\})$-approximation on directed networks, even without NBA [71]. Currently these are the best published bounds for maximum-demand UFP; previously they had been shown for maximum-demand UCUFP in Reference 66. By using the grouping-and-scaling translation from unit to arbitrary demands, Chekuri et al. [64] obtained an LP-based $O(\sqrt{n})$-approximation for weighted UFP on undirected graphs and DAGs. Table 11.2 summarizes the known bounds that involve $m$ and $n$ for the maximization versions of UFP, including bounds on the integrality gap of the multicommodity-flow based relaxation.

*Exploiting the network structure*: Existing approximation guarantees for UFP are rather weak, and on directed graphs, one cannot hope for significant improvements, unless P = NP. A different line of work has aimed for approximation ratios depending on parameters other than $n$ and $m$. This type of work was originally motivated in part by popular hypercube-derived interconnection networks [99]. Theoretical advances on these networks are typically facilitated by their rich expansion properties. A graph $G = (V, E)$ is an $\alpha$-expander if for every set $X$ of at most half the vertices, the number of edges leaving $X$ is at least $\alpha|X|$. Concluding a long line of research, Frieze [100] showed that in any $r$-regular graph with sufficiently strong expansion properties and $r$ a sufficiently large constant, *any* $\Omega(n/\log n)$ vertex pairs can be connected via edge-disjoint paths. See Reference 100 for references on the long history of the topic and the precise underlying assumptions. In such an expander, the median distance between pairs of vertices is $O(\log n)$, hence the result of Frieze is within a constant factor of optimal. This basic property, that expanders are rich in short edge-disjoint paths, has been exploited in various guises in the literature. Results for fractional multicommodity flows along short paths were first given by Leighton and Rao [52].

Kleinberg and Rubinfeld analyzed the BGA on expanders [101]. In the light of the above-mentioned Frieze's result, the BGA achieves an $O(\log n)$-approximation. In Reference 101, it was also shown that for UCUFP, one can efficiently compute a fractional solution that routes at least half the maximum demand with dilation $d = O(\Delta^2 \alpha^{-2} \log^3 n)$. Here $\Delta$ denotes the maximum degree of the (arbitrary) input graph. The bound on $d$ was improved in Reference 71. Kolman and Scheideler introduced a new network measure, the *flow number* $F_{G,u}$, and showed that (in undirected graphs) there is always a $(1 + \varepsilon)$-approximate fractional flow of dilation $O(F_{G,u}/\varepsilon)$. The flow number is defined on the basis of the solution to a product multicommodity flow problem on $G$ and is computable in polynomial time. If $u_{\min} \geq 1$, $F_{G,u}$ is always $\Omega(\alpha^{-1})$ and $O(\Delta \alpha^{-1} \log n)$ [71]. The BGA examining the demands in nonincreasing order and with $L := 4F_{G,u}$ achieves an $O(F_{G,u})$ approximation for UFP on undirected graphs with NBA [71]. For UFP with small demands Kolman and Scheideler gave an $O(u_{\min}(F_{G,u}^{1/u_{\min}} - 1))$ guarantee for integral $u_{\min} \geq 1$, which is $O(\log F_{G,u})$ if $u_{\min} \geq \log F_{G,u}$. It is NP-hard to approximate maximum-demand UFP on directed graphs with $F_{G,u} = n^\gamma$, $0 < \gamma \leq 1/2$, within $F_{G,u}^{1-\varepsilon}$ [71]. Chakrabarti et al. [85] provide an $O(F_G \log n)$-approximation for weighted undirected UFP where $F_G$ is a definition of the flow number concept of Reference 71 made independent of capacities. $F_G$ and $F_{G,u}$ coincide on uniform-capacity networks but are otherwise incomparable. Notably Reference 85 presents an $O(\sqrt{\Delta \log n})$-approximation for UCUFP on $\Delta$-regular graphs with sufficiently strong, in the sense of Reference 100, expansion properties.

A special case that has received considerable attention is UFP *on a path* in which the input graph is a path. Commodities have weights and the objective is to maximize the total weight of the commodities that can be feasibly routed. This formulation can model bandwidth allocation on a single link in which every user requests an amount of bandwidth for a given time window and the capacity of the link changes over time, under the assumption that the time breakpoints are integer quantities. UFP on a path remains NP-hard as Knapsack reduces to UFP on a single edge. The currently best polynomial-time approximation is a $(2 + \varepsilon)$-approximation (without NBA), in time $n^{O(1/\varepsilon^4)}$ [102]. Bansal et al. [103] obtained a $(1 + \varepsilon)$-approximation in quasi polynomial time. An improved QPTAS and some better results for special cases are given in Reference 104. The first constant-factor approximations were given in Reference 85 with NBA and without NBA in Reference 105. UFP *on trees* has also been studied, but so far a constant-factor approximation remains elusive, see References 98, 106. The only hardness result known is that the problem is APX-hard on capacitated trees, even when all the demands are unit [24]. Recently the $O(\log^2 n)$-approximation of Reference 98 was extended to the case where the objective function is submodular as opposed to linear [107].

Another interesting case is UFP on a cycle which is commonly called *the ring loading problem*. In fact this is the first UFP problem studied in the literature [10], before the term "unsplittable" was coined in Reference 62. The input is an undirected cycle with vertices numbered clockwise along the ring and demands $d_{ij} \geq 0$ for each pair of vertices $i < j$. The task is to route all demands unsplittably, that is demand $d_{ij}$ needs to be routed either in clockwise or in counterclockwise direction. The objective is to minimize the maximum load on an edge of the ring. Let $L^*$ be the optimal *split* (fractional) load. In a landmark paper, Schrijver et al. [108] showed one can always achieve maximum load at most $L^* + \frac{3}{2}D$, where $D$ is the maximum demand and conjectured that $L^* + D$ is achievable. Skutella [109] proved that any split routing can be turned into an unsplittable one while increasing the load on any edge by at most $\frac{19}{14}D$ and also disproved the above-mentioned conjecture of Reference 108.

*Single-Source Unsplittable Flow*: Constant approximation guarantees exist for the case in which all commodities share the same source, the so-called *single-source* UFP (SUFP). In contrast to single-source EDP, SUFP is strongly NP-complete [62]. The version of SUFP with costs has also been studied. In the latter problem, every edge $e \in E$, has a nonnegative cost $c_e$. The cost of an unsplittable flow solution is $\sum_{e \in E} c_e f_e$.

The first constant-factor approximations for all the three main objectives (minimizing congestion, maximizing demand and minimizing the number of rounds) were given by Kleinberg [62]. The factors were improved by Kolliopoulos and Stein [95] in which the first approximation for minimizing congestion without NBA was also given. The grouping-and-scaling technique of Reference 95 partitions the original problem into a collection of independent subproblems, each of them with demands in a specified range. The fractional solution is then used to assign capacities to each subproblem. The technique is in general useful for translating within constant factors integrality gaps obtained for unit demand instances to arbitrary demand instances. It found further applications, for example, in approximating CPIPs [45,96], and weighted UFP on trees [97]. See also the problems treated in Reference 110. The currently best constant factors for SUFP were obtained by Dinitz, Garg and Goemans [111], though none of them is known to be best possible under some complexity-class separation assumption. Our understanding seems to be better for congestion. The 2-approximation in Reference 111 is best possible if the fractional congestion is used as a lower bound. No ratio better than $3/2$ is possible unless P $=$ NP. The lower bound comes from minimizing makespan on parallel machines with allocation restriction [112] which reduces in an approximation-preserving manner to minimum-congestion SUFP. Significantly, Svensson [113] has given for the latter problem a polynomial-time algorithm that estimates the optimal makespan within a factor $33/17 + \epsilon$. This scheduling problem is also a special case of the generalized assignment problem for which a simultaneous $(2, 1)$-approximation for makespan and assignment cost exists [114]. Naturally one wonders whether a simultaneous $(2, 1)$-approximation for congestion and cost is possible for SUFP. This is an outstanding open problem. More precisely, given a fractional solution $f^*$ the conjecture is that there is there is an unsplittable solution with the same cost such that for *every* edge $e$ the flow through it is at most $f_e^* + d_{max}$. The currently best trade-off is a $(3, 1)$-approximation algorithm due to Skutella [115] which cleverly builds on the earlier $(3, 2)$-approximation in Reference 95. Erlebach and Hall [116] showed that it is NP-hard to obtain a $(2 - \varepsilon, 1)$-approximation, for any fixed $\varepsilon > 0$. Experimental evaluations of algorithms for congestion can be found in References 117, 118. For SUFP without NBA, the $O(n^{1-\varepsilon})$-hardness result of Azar and Regev [26] mentioned earlier for the cardinality objective holds on directed graphs. It was extended to undirected planar graphs by Shepherd and Vetta [27], always under the assumption that P $\neq$ NP. For a small demand regime, the lower bound for the cardinality objective becomes $\Omega(n^{1/2-\varepsilon}\sqrt{\log(d_{max}/u_{min})})$, with $d_{max}/u_{min} > 1$, for both directed and undirected graphs [27]. A lower bound of $\Omega(n^{1/2-\varepsilon})$ holds also for the maximum demand objective in both directed and undirected graphs [27]. Finally, we refer the interested reader to the excellent survey by Shepherd [119] on single-sink problems, including but not limited to unsplittable flow, which have been motivated by telecommunications networks.

# 11.4 Results Specific to Undirected Graphs

The most interesting developments over the last decade have taken place mostly for undirected graphs. Chekuri et al. [94,110,120] introduced the framework of well-linked decompositions which brought into approximation algorithms of some of the breakthrough insights from Topological Graph Theory and in particular the Graph Minors Project of Robertson and Seymour. This fertile exchange has produced a considerable body of work for problems such as All-or-Nothing Multicommodity Flow, Disjoint Paths with Congestion, and Disjoint Paths on Planar Graphs, which we survey in this section. The story has recently come full-circle with some of the new algorithmic ideas contributing to a significant improvement

[121,122] of the bounds in the Grid Minor Theorem, whose first version was given by Robertson and Seymour [123]. All graphs in this section are undirected unless mentioned otherwise. We denote by tw($G$) the treewidth of $G$, a key parameter in structural graph theory. For informative yet accessible discussions of treewidth, the reader is referred to the surveys in References 124, 125. Informally, a graph $G$ with tw($G$) $\leq k$ can be recursively partitioned via "balanced" vertex separators of size at most $k + 1$. Moreover, if $k$ is a small constant then $G$ is "tree-like."

*Well-linked sets*: The notion of a well-linked set was introduced by Reed [126] (see also Reference 127 for a similar definition) in an attempt to capture the concept of a highly connected graph in a global manner. The standard definition of vertex $k$-connectivity in a sense focuses on local properties: a $(k + 1)$-cutset may disconnect the graph but not "shatter" it globally. Given a graph $G$, a set $X \subseteq V(G)$ is *well-linked* if for every pair $A, B \subseteq X$ such that $|A| = |B|$, there exists a set of $|A|$ vertex-disjoint paths from $A$ to $B$ in $G$. The *well-linked number* of $G$ denoted wl($G$) is the size of the largest well-linked set in $G$. Reed [126] showed that tw($G$) $+ 1 \leq$ wl($G$) $\leq 4($tw($G$) $+ 1)$.

Chekuri et al. [110] generalized the definition of a well-linked set and connected it to the concurrent-flow/cut gap. Given $G = (V, E)$, let $X \subseteq V$ and $\pi : X \to [0, 1]$ a weight function on $X$. $X$ is $\pi$-*edge-well-linked* in $G$ if $|\delta_G(S)| \geq \pi(X \cap S)$ for all $S \subset V$ such that $\pi(X \cap S) \leq \pi(X \cap (V \setminus S))$. If $\pi(u) = \alpha$ for all $u \in V$, we say that $X$ is $\alpha$-edge-well-linked; $X$ is *edge-well-linked* if it is 1-edge-well-linked. For vertex-disjoint paths or more generally, for vertex-capacitated problems, $\pi$-vertex-well-linked sets are the natural choice; they are defined similarly with vertex cuts in the place of edge-cuts. Reed's previous definition corresponds to 1-vertex-well-linked sets. We focus on edge-well-linkedness. The decomposition framework of Chekuri et al. [110] established that given an instance $(G, \mathcal{T})$ of MEDP, one can decompose it in polynomial-time into vertex-disjoint subinstances $(G_1, \mathcal{T}_1), \ldots, (G_l, \mathcal{T}_l)$ where for every pair $s_j, t_j \in \mathcal{T}_i$, $s_j, t_j \in V(G_i)$ and if $X_i$ is the set of terminals in $\mathcal{T}_i$, then $X_i$ is $\pi_i$-well-linked in $G_i$ for some appropriate weight function $\pi_i$. Moreover, this happens only at a polylogarithmic loss, that is, $\sum_{i=1}^{l} \pi_i(X_i) = \Omega(OPT/\log^2 k)$. One can boost well-linkedness, that is, Chekuri et al. [120] showed that given a $\pi$-well-linked set $X$, there is a subset $X' \subset X$ such that $X'$ is $\alpha$-well-linked for some constant $\alpha \leq 1$, and $|X'| = \Omega(\pi(X))$. The importance of well-linkedness stems from the fact that as long as one is interested in fractional routings, in an $\alpha$-well-linked set $X$, any matching can be routed with congestion $O(\beta(G)/\alpha)$ where $\beta(G)$ is the concurrent-flow/cut gap for product multicommodity flow in $G$. The $\beta(G)$ factor in the congestion can be saved with the stronger notion of flow-well-linked sets, see Reference 110 for details. This decomposition framework has been used as a black-box in many later papers, including several of those that we survey in the following. At a loss of a polylogarithmic factor in the total flow one can assume that the input is well-linked. Obtaining integral feasible routings or bringing congestion down to a small constant require numerous further powerful ideas that exceed the scope of our survey. This a complex body of work which reaps dividends from sustained interaction with Topological Graph Theory.

*All-or-Nothing Multicommodity Flow*: *All-or-Nothing Multicommodity Flow* (AN-MCF) is a relaxation of MEDP. A subset $M'$ of the pairs $\{(s_1, t_1), \ldots, (s_k, t_k)\}$ is *routable* if there is a feasible (fractional) multicommodity flow that routes one unit of flow from $s_i$ to $t_i$ for every pair that belongs to $M'$. The objective is to maximize the cardinality of $M$. Although the decision version of AN-MCF is in P via linear programming, AN-MCF is NP-hard and APX-hard to approximate, even in capacitated trees [24]. The best known approximations that are known are $O(\log^2 k)$ in edge-capacitated graphs [110] and $O(\log^4 k \log n)$ with congestion $(1 + \varepsilon)$ in vertex-capacitated graphs [110]. These are improved to $O(\log k)$ and $O(\log^2 k \log n)$ for planar graphs [110]. The grouping-and-scaling technique of Reference 45 (see also Reference 97) can be used to extend the results to arbitrary demands with NBA. We note that the original paper of Chekuri et al. [120] on AN-MCF used the oblivious routing techniques of Räcke [128] to implement the (implicit) decomposition into well-linked sets. The general framework and its computation via near-optimal separators were made explicit in Reference 110. All bounds above hold also against the optimal LP solution without congestion. This is in contrast to the $\Omega(\sqrt{n})$ lower bound on the integrality gap for MEDP. The approximation ratios in Reference 120 hold also for the weighted version of the problem.

In Reference 97, a 4-approximation was given for the weighted version on trees, whereas the cardinality version is 2-approximable [24]. The known hardness results for AN-MCF are the same as for MEDP, that is, AN-MCF is hard to approximate within $\log^{1/2-\varepsilon}$. Even with congestion $c$, AN-MCF is $\log^{\Omega(1/c)} n$-hard to approximate for undirected graphs, and $\Omega(n^{1/c})$, for directed graphs. See Section 11.2 for details. For AN-MCF with congestion $2 \le c \le O(\log \log n / \log \log \log n)$ the integrality gap of the multicommodity flow relaxation is $\Omega(\frac{1}{c^2}(\frac{\log n}{(\log \log n)^2})^{1/(c+1)})$ [65].

Recently, there has been some progress for directed graphs when demand pairs are *symmetric*: for each routable pair $(s_i, t_i)$, at least one unit of flow should be routed both from $s_i$ to $t_i$ and from $t_i$ to $s_i$. For AN-MCF with symmetric demand pairs on vertex-capacitated directed graphs, an $O(\log^2 k)$-approximation with constant vertex congestion was given in Reference 129.

*Disjoint Paths with Congestion:* In the *Edge-Disjoint Paths with Congestion* problem (EDPWC), we are given as an additional input an integer $c$. The objective is to route the maximum number of demand pairs so that the maximum edge congestion is $c$. For $c = 1$ we obtain MEDP. For $c > 1$, EDPWC is a special case of UFP with small demands, which we examined in Section 11.3. We review some of the basic results for the latter problem. Allowing $c = \Omega(\log n / \log \log n)$ the classical randomized rounding of Raghavan and Thompson [35] gives a constant-factor approximation. For congestion $c \ge 1$, there is an $O(d^{1/\lfloor c \rfloor})$-approximation and a matching upper bound on the integrality gap, where $d$ is the maximum length of a flow path in the optimal fractional solution [96]. For combinatorial algorithms, see also Reference 26 for a matching bound and Reference 75 for an $O(\sqrt{n})$-approximation for the case $c = 2$ and all demands being unit. All these results hold also for directed graphs and in fact they are tight under complexity-theoretic assumptions, see Sections 11.2 and 11.3.

For the undirected case, the interest in routing with low congestion $c \ge 2$ is motivated by the large gap between the existing upper and lower bounds for the approximability of MEDP. In addition, for any congestion $2 \le c \le O(\log \log n / \log \log \log n)$ the integrality gap of the multicommodity flow relaxation of EDPWC is $\Omega(\frac{1}{c}(\frac{\log n}{(\log \log n)^2})^{1/(c+1)})$ [65]. Andrews [130] gave a $O(\log^{61} n)$-approximation with congestion $O((\log \log n)^6)$. Chuzhoy [131] improved this to an $O(\log^{22.5} k \log \log k)$ approximation with congestion 14. Chuzhoy and Li [132] gave a randomized algorithm that achieves $O(\text{polylog } k)$ approximation with congestion 2. All these algorithms round the solution to the multicommodity flow relaxation, therefore when congestion 2 is allowed the integrality gap improves from $O(\sqrt{n})$ to polylogarithmic. Before that, Kawarabayashi and Kobayashi [133] had given an $O(n^{3/7} \text{ polylog } n)$-approximation for congestion $c = 2$. For planar graphs, improved bounds had been obtained earlier. Chekuri et al. had shown an $O(\log k)$-approximation with congestion 2 [110] and then an $O(1)$-approximation with congestion 4 [134]. Seguin-Charbonneau and Shepherd [135] showed a constant-factor approximation with congestion 2. For *Vertex-Disjoint Paths with Congestion,* Chekuri and Ene [136] gave an $O(\text{polylog } k)$-approximation with congestion 51. Chekuri and Chuzhoy have announced an $O(\text{polylog } k)$ approximation with congestion 2 [137]. Simple modifications to the reductions in Reference 65 show that the $\log^{\Omega(1/c)} n$-hardness result for EDPWC holds also for Vertex-Disjoint Paths, assuming NP $\not\subseteq$ ZPTIME($n^{\text{polylog}(n)}$). In contrast to these intractability results, when $k \le \delta(\log \log n)^{2/15}$, where $\delta$ is an appropriate constant, Kleinberg showed that for $c = 2$ the decision version is in P both for edge- and vertex-capacitated undirected graphs [93]. Finally, Chekuri et al. [138] gave a poly logarithmic approximation with congestion 5 for MVDP with symmetric demands on planar digraphs.

*Disjoint Paths without Congestion:* An interesting result by Rao and Zhu [139] gives an $O(\text{polylog } n)$ approximation for MEDP (with congestion 1), if the global min cut has size $\Omega(\log^5 n)$. This result is not comparable to the above but the ideas have proved valuable for References 130–133.

For MVDP (with congestion 1) Chuzhoy et al. [140] gave recently an $O(n^{9/19} \text{ polylog } n)$-approximation on planar graphs. Significantly, this is the first result on planar graphs that improves upon the $O(\sqrt{n})$-approximation which the Greedy algorithm achieves on general graphs [45]. Recall that on the negative side, there is no $\log^{1/2-\varepsilon} n$ approximation algorithm on general undirected graphs unless

NP $\subseteq$ ZPTIME($n^{\text{polylog}(n)}$) [65]. The only hardness result specific to planar graphs is APX-hardness on grid graphs [141].

Several other improved approximation ratios exist for special graph classes. See the references contained in References 140, 142. Notably, for graphs of treewidth $r$, improving upon Reference 143, Ene et al. [144] obtained an $O(r^3)$-approximation for MEDP and a similar approximation for MVDP as a function of pathwidth. These guarantees hold with respect to the fractional solution. The standard gap example of Reference 24 yields an $\Omega(r)$ lower bound on the integrality gap. Chekuri et al. [145] asked whether there is a matching upper bound. It should be noted that EDP remains NP-complete even in graphs of treewidth $r = 2$ [31]. Assuming P $\neq$ NP, this rules out the existence of a fixed-parameter algorithm for EDP parameterized by treewidth. For a characterization of the fixed-parameter tractability of VDP, see Reference 146.

## 11.5 Further Variants of the Basic Problems

Length-bounded flows are single- or multi commodity flow problems in which every path used must obey a length constraint. See References 147, 148 for background on this topic, including results on the associated flow/cut gaps. In the *bounded-length* EDP (BLEDP), an additional input parameter $M$ is specified. One seeks a set of disjoint $s_i$-$t_i$ paths under the constraint that the length of each path is at most $M$. In $(s, t)$-BLEDP all the pairs share the same source $s$ and sink $t$. Cases that are used to be tractable become NP-complete with the length constraint. Both in the vertex and the edge-disjoint case, $(s, t)$-BLEDP is NP-complete on undirected graphs even for fixed $M \geq 4$ [149]. For variable $M$ and fixed $k$, the problems remain NP-complete [150]. In the optimization version BLMEDP, one seeks a maximum-cardinality set of edge-disjoint paths that satisfy the length constraint. It is NP-hard to approximate $(s, t)$-BLMEDP within $O(n^{1/2-\varepsilon})$ on directed graphs and, unless NP = ZPP, BLMEDP cannot be approximated in polynomial time within $O(n^{1/2-\varepsilon})$ on undirected graphs, for any fixed $\varepsilon > 0$ [25]. On the positive side it is easy to obtain an $O(\sqrt{m})$-approximation for BLMEDP. For the paths in the optimal solution with length at most $M' := \min\{\sqrt{m}, M\}$, the BGA with parameter $L = M'$ achieves an $O(M')$-approximation. This is because, in the notation of the BGA analysis in Section 11.2, the set $\mathcal{O}_2$ is empty. On the other hand, there are at most $\sqrt{m}$ edge-disjoint paths of length more than $\sqrt{m}$. See Reference 25 for other algorithmic results. For the *bounded-length weighted* UFP with NBA, the results in References 41, 45, 85 yield $O(M)$-approximations. By using the ellipsoid algorithm, one can find an optimal fractional solution whose support contains only paths that satisfy the length constraint. By using their result on CPIPs without NBA, Chekuri et al. [98] showed that the same $O(M)$ ratio holds for bounded-length weighted UFP without NBA. Significantly, the integrality gap of the natural relaxation for the maximum integral length-bounded $s$-$t$ flow is $\Omega(\sqrt{n})$ even for directed or undirected planar graphs [148].

In transportation logistics a commodity may be splittable in different containers, each of them to be routed along a single path. One wishes to bound the number of containers used. This motivates the *K-splittable flow problem*, a relaxed version of UFP in which a commodity can be split along *at most $K \geq 1$* paths, $K$ an input parameter. This problem was introduced and first studied by Baier et al. [151]. Clearly for $K = m$, it reduces to solving the fractional relaxation; it is NP-complete for $K = 2$. See References 79, 152, 153 for a continuation of the work in Reference 151. The author observed in Reference 154 that the single-source 2-splittable flow problem admits a simultaneous $(2, 1)$-approximation for congestion and cost. This was improved and generalized by Salazar and Skutella to a $(1 + 1/K + 1/(2K - 1), 1)$-approximation for $K$-splittable flows [155].

Finally, a problem in a sense complementary to $K$-splittable flow is the *multiroute flow* in which for reliability purposes the flow of a commodity *has to* be split along a given number of edge-disjoint paths. Observe that in the $K$-splittable flow, there is no edge-disjointness requirement. Given a source $s$ and a sink $t$, and an integer $K \geq 1$, an *elementary $K$-route flow* is a set of $K$ edge-disjoint paths between $s$ and $t$. A *K-route flow* is a nonnegative linear combination of elementary $K$-route flows. The multiroute flow problem consists of finding a maximum $K$-route flow that respects the edge capacities. It was introduced by Kishimoto and Tagauchi [156] (see Reference 157) and is solvable in polynomial time. See Reference 158 for simplifications and extensions of the basic $s$-$t$ multiroute flow theory. A dual object of interest to a $K$-route flow is a *K-route cut*: a subset of the edges whose removal leaves at most $K - 1$ edge-disjoint paths between $s$ and $t$. We note that in the early literature [157,158] a different definition of cut was considered for the purpose of a flow/cut duality result. It is a natural question to study multiroute flows in the multicommodity setting in conjunction with the $K$-route flow/$K$-route cut gap. A sample of this work can be found in References 159–163. Martens [164] gives a greedy algorithm for a related multicommodity problem which he calls the *k-disjoint flow* problem. For every commodity $i$, one seeks $K_i$ edge-disjoint paths and the flow has to be perfectly balanced: On each path, it must be equal to $d_i/K_i$. The objective function is to maximize the sum of routed demands subject to the capacity constraints and the approximation ratio achieved is $O(K_{max}\sqrt{m}/K_{min})$ [164]. A single-commodity $(s, t)$-flow of value $F$ is $K$-*balanced* if every edge carries at most $F/K$ units of flow. Significantly, an acyclic $(s, t)$-flow is $K$-balanced if and only if it is a $K$-route flow [157,158,165].

## Acknowledgments for the First Edition

Thanks to Chandra Chekuri, Sanjeev Khanna, Maren Martens, Martin Skutella, and Cliff Stein for valuable comments and suggestions. Thanks to Naveen Garg for a clarification on Reference 24, to Aris Pagourtzis for pointing out Reference 88, and to Lex Schrijver for information on EDP on planar graphs.

## Acknowledgments for the Second Edition

Thanks to Chandra Chekuri, Petr Kolman, Bruce Shepherd, and Dimitrios Thilikos for valuable comments and suggestions.

## References

1. A. Frank. Packing paths, cuts and circuits—A survey. In B. Korte, L. Lovász, H. J. Prömel, and A. Schrijver, Eds., *Paths, Flows and VLSI-Layout*, pp. 49–100. Springer-Verlag, Berlin, Germany, 1990.
2. A. Frank. Connectivity and network flows. In R. Graham, M. Grötschel, and L. Lovász, Eds., *Handbook of Combinatorics*, pp. 111–177. North-Holland, Amsterdam, the Netherlands, 1995.
3. A. Schrijver. Homotopic routing methods. In B. Korte, L. Lovász, H. J. Prömel, and A. Schrijver, Eds., *Paths, Flows and VLSI-Layout*. Springer-Verlag, Berlin, Germany, 1990.
4. A. Schrijver. *Combinatorial Optimization: Polyhedra and Efficiency*. Springer-Verlag, Berlin, Germany, 2003.
5. R. H. Möhring, D. Wagner, and F. Wagner. VLSI network design: A survey. In M. O. Ball, T. L. Magnanti, C. L. Monma, and G. L. Nemhauser, Eds., *Handbooks in Operations Research/ Management Science, Volume on Networks*, pp. 625–712. North-Holland, Amsterdam, the Netherlands, 1995.
6. H. Ripphausen-Lipa, D. Wagner, and K. Weihe. Survey on efficient algorithms for disjoint paths problems in planar graphs. In W. Cook, L. Lovász, and P. D. Seymour, Eds., *DIMACS-Series in Discrete Mathematics and Theoretical Computer Science, Volume 20 on the "Year of Combinatorial Optimization,"* pp. 295–354. AMS, Providence, RI, 1995.
7. R. H. Möhring and D. Wagner. Combinatorial topics in VLSI design, annotated bibliography. In M. Dell'Amico, F. Maffioli, and S. Martello, Eds., *Annotated Bibliographies in Combinatorial Optimization*, pp. 429–444. Wiley, New York, 1997.

8. J. M. Kleinberg. Approximation algorithms for disjoint paths problems. PhD thesis, MIT, Cambridge, MA, May 1996.

9. Y. Perl and Y. Shiloach. Finding two disjoint paths between two pairs of vertices in a graph. *Journal of the ACM*, 25:1–9, 1978.

10. S. Cosares and I. Saniee. An optimization problem related to balancing loads on SONET rings. *Telecommunications Systems*, 3:165–181, 1994. Prelim. version as Technical Memorandum. Bellcore, Morristown, NJ, 1992.

11. B. Awerbuch, Y. Azar, and S. Plotkin. Throughput-competitive online routing. In *Proceedings of the 34th Annual IEEE Symposium on Foundations of Computer Science*, pp. 32–40. IEEE Computer Society, 1993.

12. R. M. Karp. On the computational complexity of combinatorial problems. *Networks*, 5:45–68, 1975.

13. J. F. Lynch. The equivalence of theorem proving and the interconnection problem. *ACM SIGDA Newsletter*, 5:31–36, 1975.

14. S. Even, A. Itai, and A. Shamir. On the complexity of timetable and multicommodity flow problems. *SIAM Journal on Computing*, 5:691–703, 1976.

15. M. R. Kramer and J. van Leeuwen. The complexity of wire-routing and finding minimum-area layouts for arbitrary VLSI circuits. In F. P. Preparata, Ed., *VLSI Theory*, volume 2 of *Advances in Computing Research*, pp. 129–146. JAI Press, Greenwich, CT, 1984.

16. S. Fortune, J. Hopcroft, and J. Wyllie. The directed subgraph homeomorphism problem. *Theoretical Computer Science*, 10:111–121, 1980.

17. N. Robertson and P. D. Seymour. Graph Minors XIII. The disjoint paths problem. *Journal of Combinatorial Theory B*, 63:65–110, 1995.

18. D. Bienstock and M. A. Langston. Algorithmic implications of the Graph Minor Theorem. In M. O. Ball, T. L. Magnanti, C. L. Monma, and G. L. Nemhauser, Eds., *Handbook in Operations Research and Management Science 7: Network models*. North-Holland, Amsterdam, the Netherlands, 1995.

19. P. D. Seymour. Disjoint paths in graphs. *Discrete Mathematics*, 29:293–309, 1980.

20. Y. Shiloach. A polynomial solution to the undirected two paths problem. *Journal of the ACM*, 27:445–456, 1980.

21. C. Thomassen. 2-Linked graphs. *European Journal of Combinatorics*, 1:371–378, 1980.

22. A. Schrijver. Finding $k$ disjoint paths in a directed planar graph. *SIAM Journal on Computing*, 23:780–788, 1994.

23. M. Cygan, D. Marx, M. Pilipczuk, and M. Pilipczuk. The planar directed $k$-vertex-disjoint paths problem is fixed-parameter tractable. In *54th Annual IEEE Symposium on Foundations of Computer Science, FOCS 2013*, Berkeley, CA, October 26–29, pp. 197–206, 2013.

24. N. Garg, V. Vazirani, and M. Yannakakis. Primal-dual approximation algorithms for integral flow and multicut in trees. *Algorithmica*, 18:3–20, 1997. Prelim. version in ICALP 93.

25. V. Guruswami, S. Khanna, R. Rajaraman, B. Shepherd, and M. Yannakakis. Near-optimal hardness results and approximation algorithms for edge-disjoint paths and related problems. *Journal of Computer and System Sciences*, 67:473–496, 2003. Prelim. version in STOC 99.

26. Y. Azar and O. Regev. Combinatorial algorithms for the unsplittable flow problem. *Algorithmica*, 44(1):49–66, 2006.

27. F. B. Shepherd and A. Vetta. The inapproximability of maximum single-sink unsplittable, priority and confluent flow problems. *CoRR*, abs/1504.00627, 2015.

28. M. C. Costa, L. Létocart, and F. Roupin. A greedy algorithm for multicut and integral multiflow in rooted trees. *Operations Research Letters*, 31:21–27, 2003.

29. M. Middendorf and F. Pfeiffer. On the complexity of the disjoint paths problems. *Combinatorica*, 13(1):97–107, 1993.

30. J. Vygen. NP-completeness of some edge-disjoint paths problems. *Discrete Applied Mathematics*, 61(1):83–90, 1995.

31. T. Nishizeki, J. Vygen, and X. Zhou. The edge-disjoint paths problem is NP-complete for series-parallel graphs. *Discrete Applied Mathematics*, 115(1–3):177–186, 2001.

32. D. Marx. Eulerian disjoint paths problem in grid graphs is NP-complete. *Discrete Applied Mathematics*, 143(1–3):336–341, 2004.

33. G. Naves and A. Sebő. Multiflow feasibility: An annotated tableau. In W. Cook, L. Lovász, and J. Vygen, Eds., *Research Trends in Combinatorial Optimization: Bonn 2008*, pp. 261–283. Springer, Berlin, Germany, 2009.

34. V. V. Vazirani. *Approximation Algorithms*. Springer-Verlag, Berlin, Germany, 2001.

35. P. Raghavan and C. D. Thompson. Randomized rounding: A technique for provably good algorithms and algorithmic proofs. *Combinatorica*, 7:365–374, 1987.

36. P. Erdős and J. L. Selfridge. On a combinatorial game. *Journal of Combinatorial Theory A*, 14:298–301, 1973.

37. J. Spencer. *Ten Lectures on the Probabilistic Method*. SIAM, Philadelphia, PA, 1987.

38. P. Raghavan. Probabilistic construction of deterministic algorithms: Approximating packing integer programs. *Journal of Computer and System Sciences*, 37:130–143, 1988.

39. A. Srinivasan. Improved approximations guarantees for packing and covering integer programs. *SIAM Journal on Computing*, 29:648–670, 1999. Prelim. version in STOC 95.

40. A. Srinivasan. An extension of the Lovász Local Lemma and its applications to integer programming. In *Proceedings of the 7th ACM-SIAM Symposium on Discrete Algorithms*, pp. 6–15. ACM/SIAM, 1996.

41. A. Srinivasan. Improved approximations for edge-disjoint paths, unsplittable flow and related routing problems. In *Proceedings of the 38th Annual IEEE Symposium on Foundations of Computer Science*, pp. 416–425. IEEE Computer Society, 1997.

42. A. Baveja and A. Srinivasan. Approximation algorithms for disjoint paths and related routing and packing problems. *Mathematics of Operations Research*, 25:255–280, 2000.

43. N. E. Young. Randomized rounding without solving the linear program. In *Proceedings of the 6th ACM-SIAM Symposium on Discrete Algorithms*, pp. 170–178. ACM/SIAM, 1995.

44. J. M. Kleinberg and É. Tardos. Disjoint paths in densely-embedded graphs. In *Proceedings of the 36th Annual IEEE Symposium on Foundations of Computer Science*, pp. 52–61. IEEE Computer Society, 1995.

45. S. G. Kolliopoulos and C. Stein. Approximating disjoint-path problems using packing integer programs. *Mathematical Programming A*, 99:63–87, 2004. Prelim. version in IPCO 98.

46. S. G. Kolliopoulos. Exact and approximation algorithms for network flow and disjoint-path problems. PhD thesis, Dartmouth College, Hanover, NH, August 1998.

47. K. Menger. Zur allgemeinen kurventheorie. *Fundamenta Mathematicae*, 10:96–115, 1927.

48. L. R. Ford and D. R. Fulkerson. Maximal flow through a network. *Canadian Journal of Mathematics*, 8:399–404, 1956.

49. D. R. Fulkerson and G. B. Dantzig. Computation of maximum flow in networks. *Naval Research Logistics Quarterly*, 2:277–283, 1955.

50. P. Elias, A. Feinstein, and C. E. Shannon. A note on the maximum flow through a network. *IRE Transactions on Information Theory*, 2:117–199, 1956.

51. T. C. Hu. Multi-commodity network flows. *Operations Research*, 11:344–360, 1963.

52. T. Leighton and S. Rao. Multicommodity max-flow min-cut theorems and their use in designing approximation algorithms. *Journal of the ACM*, 46:787–832, 1999. Prelim. version in FOCS 88.

53. Y. Aumann and Y. Rabani. An $O(\log k)$ approximate Min-Cut Max-Flow theorem and approximation algorithm. *SIAM Journal on Computing*, 27:291–301, 1998.

54. N. Linial, E. London, and Y. Rabinovich. The geometry of graphs and some of its algorithmic applications. *Combinatorica*, 15:215–246, 1995.

55. U. Feige, M. T. Hajiaghayi, and J. R. Lee. Improved approximation algorithms for minimum weight vertex separators. *SIAM Journal on Computing*, 38(2):629–657, 2008.

56. N. Garg, V. Vazirani, and M. Yannakakis. Approximate max-flow min-(multi)cut theorems and their applications. *SIAM Journal on Computing*, 25:235–251, 1996. Prelim. version in STOC 93.

57. D. B. Shmoys. Cut problems and their applications to Divide and Conquer. In D. S. Hochbaum, Ed., *Approximation Algorithms for NP-Hard Problems*, pp. 192–231. PWS, Boston, MA, 1997.

58. M. C. Costa, L. Létocart, and F. Roupin. Minimal multicut and maximum integer multiflow: A survey. *European Journal of Operational Research*, 162:55–69, 2005.

59. M. E. Saks, A. Samorodnitsky, and L. Zosin. A lower bound on the integrality gap for minimum multicut in directed networks. *Combinatorica*, 24(3):525–530, 2004.

60. J. Chuzhoy and S. Khanna. Polynomial flow-cut gaps and hardness of directed cut problems. *Journal of the ACM*, 56(2):6:1–6:28, 2009.

61. P. N. Klein, S. A. Plotkin, S. Rao, and É. Tardos. Approximation algorithms for Steiner and directed multicuts. *Journal of Algorithms*, 22(2):241–269, 1997.

62. J. M. Kleinberg. Single-source unsplittable flow. In *Proceedings of the 37th Annual IEEE Symposium on Foundations of Computer Science*, pp. 68–77. IEEE Computer Society, October 1996.

63. K. Varadarajan and G. Venkataraman. Graph decomposition and a greedy algorithm for edge-disjoint paths. In *Proceedings of the 15th ACM-SIAM Symposium on Discrete Algorithms*, pp. 379–380. SIAM, 2004.

64. C. Chekuri, S. Khanna, and F. B. Shepherd. An $O(\sqrt{n})$ approximation and integrality gap for disjoint paths and unsplittable flow. *Theory of Computing*, 2(7):137–146, 2006.

65. M. Andrews, J. Chuzhoy, V. Guruswami, S. Khanna, K. Talwar, and L. Zhang. Inapproximability of edge-disjoint paths and low congestion routing on undirected graphs. *Combinatorica*, 30(5): 485–520, 2010.

66. C. Chekuri and S. Khanna. Edge-disjoint paths revisited. *ACM Transactions on Algorithms*, 3(4), 2007. Prelim. version in SODA 2003.

67. B. Ma and L. Wang. On the inapproximability of disjoint paths and minimum Steiner forest with bandwidth constraints. *Journal of Computer and System Sciences*, 60:1–12, 2000.

68. P. Chalermsook, B. Laekhanukit, and D. Nanongkai. Pre-reduction graph products: Hardnesses of properly learning DFAs and approximating EDP on DAGs. In *55th IEEE Annual Symposium on Foundations of Computer Science, FOCS 2014*, Philadelphia, PA, October 18–21, pp. 444–453, 2014.

69. M. Andrews and L. Zhang. Hardness of the undirected edge-disjoint paths problem. In *Proceedings of the 37th annual ACM Symposium on Theory of Computing*, pp. 276–283. ACM, 2005.

70. J. Chuzhoy, V. Guruswami, S. Khanna, and K. Talwar. Hardness of routing with congestion in directed graphs. In *Proceedings of the 39th Annual ACM Symposium on Theory of Computing*, San Diego, CA, June 11–13, pp. 165–178, 2007.

71. P. Kolman and C. Scheideler. Improved bounds for the unsplittable flow problem. *Journal of Algorithms*, 61(1):20–44, 2006. Prelim. version in SODA 02.

72. E. A. Dinitz. Algorithm for solution of a problem of maximum flow in networks with power estimation. *Soviet Mathematics Doklady*, 11:1277–1280, 1970.

73. S. Even and R. E. Tarjan. Network flow and testing graph connectivity. *SIAM Journal on Computing*, 4:507–518, 1975.

74. D. R. Karger and M. S. Levine. Finding maximum flows in simple undirected graphs seems easier than bipartite matching. In *Proceedings of the 30th Annual ACM Symposium on Theory of Computing*, 1998.

75. T. Nguyen. On the disjoint paths problem. *Operations Research Letters*, 35(1):10–16, 2007.

76. A. Srinivasan. A survey of the role of multicommodity flow and randomization in network design and routing. In P. M. Pardalos, S. Rajasekaran, and J. Rolim, Eds., *Randomization Methods in Algorithm Design, Proceedings of a DIMACS Workshop*, Princeton, NJ, December 12–14, volume 43 of *DIMACS Series in Discrete Mathematics and Theoretical Computer Science*, pp. 271–302. DIMACS/AMS, 1997.

77. H. Chernoff. A measure of the asymptotic efficiency for tests of a hypothesis based on sum of observations. *The Annals of Mathematical Statistics*, 23:493–509, 1952.

78. F. T. Leighton, S. Rao, and A. Srinivasan. Multicommodity flow and circuit switching. In *Hawaii International Conference on System Sciences*, pp. 459–465. IEEE Computer Society, 1998.

79. M. Martens and M. Skutella. Flows on few paths: Algorithms and lower bounds. *Networks*, 48(2): 68–76, 2006.

80. J. Chuzhoy and S. Naor. New hardness results for congestion minimization and machine scheduling. In *Proceedings of the 36th Annual ACM Symposium on Theory of Computing*, pp. 28–34. ACM, 1994.

81. M. Andrews and L. Zhang. Almost-tight hardness of directed congestion minimization. *Journal of the ACM*, 55(6), 27, 2008.

82. M. Andrews and L. Zhang. Hardness of the undirected congestion minimization problem. In *Proceedings of the 37th annual ACM Symposium on Theory of Computing*, pp. 284–293. ACM, 2005.

83. N. Alon and J. Spencer. *The Probabilistic Method*, 2nd ed. John Wiley and Sons, New York, 2000.

84. A. Srinivasan. New approaches to covering and packing problems. In *Proceedings of the 12th ACM-SIAM Symposium on Discrete Algorithms*, pp. 567–576. ACM/SIAM, 2001.

85. A. Chakrabarti, C. Chekuri, A. Gupta, and A. Kumar. Approximation algorithms for the unsplittable flow problem. *Algorithmica*, 47(1):53–78, 2007. Prelim. version in APPROX 02.

86. P. Kolman. A note on the greedy algorithm for the unsplittable flow problem. *Information Processing Letters*, 88:101–105, 2003.

87. Y. Aumann and Y. Rabani. Improved bounds for all-optical routing. In *Proceedings of the 6th ACM-SIAM Symposium on Discrete Algorithms*, pp. 567–576. ACM/SIAM, 1995.

88. S. Stefanakos. On the design and operation of high-performance optical networks. PhD thesis, ETH Zurich, No. 15691, 2004.

89. F. Shahrokhi and D. W. Matula. The maximum concurrent flow problem. *Journal of the ACM*, 37:318–334, 1990.

90. T. Leighton, F. Makedon, S. Plotkin, C. Stein, É. Tardos, and S. Tragoudas. Fast approximation algorithms for multicommodity flow problems. *Journal of Computer and System Sciences*, 50: 228–243, 1995. Prelim. version in STOC 91.

91. S. Plotkin, D. B. Shmoys, and É. Tardos. Fast approximation algorithms for fractional packing and covering problems. *Mathematics of Operations Research*, 20:257–301, 1995.

92. N. Garg and J. Könemann. Faster and simpler algorithms for multicommodity flow and other fractional packing problems. In *Proceedings of the 39th Annual IEEE Symposium on Foundations of Computer Science*, pp. 300–309. IEEE Computer Society, 1998.

93. J. M. Kleinberg. Decision algorithms for unsplittable flow and the half-disjoint paths problem. In *Proceedings of the 30th Annual ACM Symposium on Theory of Computing*, pp. 530–539. ACM, 1998.

94. C. Chekuri, S. Khanna, and F. B. Shepherd. Edge-disjoint paths in planar graphs. In *Proceedings of the 45th Annual IEEE Symposium on Foundations of Computer Science*, pp. 71–80. IEEE Computer Society, 2004.

95. S. G. Kolliopoulos and C. Stein. Approximation algorithms for single-source unsplittable flow. *SIAM Journal on Computing*, 31:919–946, 2002. Prelim. version in FOCS 97.

96. A. Baveja and A. Srinivasan. Approximating low-congestion routing and column-restricted packing problems. *Information Processing Letters*, 74:19–25, 2000.

97. C. Chekuri, M. Mydlarz, and F. B. Shepherd. Multicommodity demand flow in a tree and packing integer programs. *ACM Transactions on Algorithms*, 3(3):27, 2007.

98. C. Chekuri, A. Ene, and N. Korula. Unsplittable flow in paths and trees and column-restricted packing integer programs. In *Approximation, Randomization, and Combinatorial Optimization. Algorithms and Techniques, 12th International Workshop, APPROX 2009, and 13th International Workshop, RANDOM 2009, Proceedings*, Berkeley, CA, August 21–23, pp. 42–55, 2009.

99. C. Scheideler. *Universal Routing Strategies for Interconnection Networks*, volume 1390 of *LNCS*. Springer-Verlag, Berlin, Germany, 1998.

100. A. M. Frieze. Edge-disjoint paths in expander graphs. *SIAM Journal on Computing*, 30:1790–1801, 2001. Prelim. version in SODA 00.

101. J. M. Kleinberg and R. Rubinfeld. Short paths in expander graphs. In *Proceedings of the 37th Annual IEEE Symposium on Foundations of Computer Science*, pp. 86–95. IEEE Computer Society, 1996.

102. A. Anagnostopoulos, F. Grandoni, S. Leonardi, and A. Wiese. A mazing $2 + \epsilon$ approximation for unsplittable flow on a path. In *Proceedings of the Twenty-Fifth Annual ACM-SIAM Symposium on Discrete Algorithms, SODA 2014*, Portland, OR, January 5–7, pp. 26–41, 2014.

103. N. Bansal, A. Chakrabarti, A. Epstein, and B. Schieber. A quasi-PTAS for unsplittable flow on line graphs. In *Proceedings of the 38th Annual ACM Symposium on Theory of Computing*, Seattle, WA, May 21–23, pp. 721–729, 2006.

104. J. Batra, N. Garg, A. Kumar, T. Mömke, and A. Wiese. New approximation schemes for unsplittable flow on a path. In *Proceedings of the Twenty-Sixth Annual ACM-SIAM Symposium on Discrete Algorithms, SODA 2015*, San Diego, CA, January 4–6, pp. 47–58, 2015.

105. P. S. Bonsma, J. Schulz, and A. Wiese. A constant-factor approximation algorithm for unsplittable flow on paths. *SIAM Journal on Computing*, 43(2):767–799, 2014.

106. Z. Friggstad and Z. Gao. On linear programming relaxations for unsplittable flow in trees. In *Approximation, Randomization, and Combinatorial Optimization. Algorithms and Techniques, APPROX/RANDOM 2015*, Princeton, NJ, August 24–26, pp. 265–283, 2015.

107. A. Adamaszek, P. Chalermsook, A. Ene, and A. Wiese. Submodular unsplittable flow on trees. In *Integer Programming and Combinatorial Optimization—18th International Conference, IPCO 2016, Proceedings*, Liège, Belgium, June 1–3, pp. 337–349, 2016.

108. A. Schrijver, P. D. Seymour, and P. Winkler. The ring loading problem. *SIAM Journal on Discrete Mathematics*, 11(1):1–14, 1998.

109. M. Skutella. A note on the ring loading problem. *SIAM Journal on Discrete Mathematics*, 30(1): 327–342, 2016.

110. C. Chekuri, S. Khanna, and F. B. Shepherd. Multicommodity flow, well-linked terminals, and routing problems. In *Proceedings of the 37th annual ACM Symposium on Theory of Computing*, pp. 183–192. ACM, 2005.

111. Y. Dinitz, N. Garg, and M. X. Goemans. On the single-source unsplittable flow problem. *Combinatorica*, 19:1–25, 1999. Prelim. version in FOCS 98.

112. J. K. Lenstra, D. B. Shmoys, and É. Tardos. Approximation algorithms for scheduling unrelated parallel machines. *Mathematical Programming A*, 46:259–271, 1990.

113. O. Svensson. Santa claus schedules jobs on unrelated machines. *SIAM Journal on Computing*, 41(5):1318–1341, 2012.

114. D. B. Shmoys and É. Tardos. An approximation algorithm for the generalized assignment problem. *Mathematical Programming A*, 62:461–474, 1993.

115. M. Skutella. Approximating the single-source unsplittable min-cost flow problem. *Mathematical Programming B*, 91:493–514, 2002. Prelim. version in FOCS 2000.

116. T. Erlebach and A. Hall. NP-hardness of broadcast sheduling and inapproximability of single-source unsplittable min-cost flow. *Journal of Scheduling*, 7:223–241, 2004. Prelim. version in SODA 02.

117. S. G. Kolliopoulos and C. Stein. Experimental evaluation of approximation algorithms for single-source unsplittable flow. In G. Cornuéjols, R. E. Burkard, and G. J. Woeginger, Eds., In *Proceedings of the 7th Conference on Integer Programming and Combinatorial Optimization*, volume 1610 of *Lecture Notes in Computer Science*, pp. 328–344. Springer-Verlag, Berlin, Germany, June 1999.

118. J. Du and S. G. Kolliopoulos. Implementing approximation algorithms for the single-source unsplittable flow problem. *ACM Journal of Experimental Algorithmics*, 10:2.3:1–2.3:21, 2005.

119. F. B. Shepherd. Single-sink multicommodity flow with side constraints. In W. Cook, L. Lovász, and J. Vygen, Eds., *Research Trends in Combinatorial Optimization: Bonn 2008*, pp. 261–283. Springer, Berlin, Germany, 2009.

120. C. Chekuri, S. Khanna, and F. B. Shepherd. The all-or-nothing multicommodity flow problem. *SIAM Journal on Computing*, 42(4):1467–1493, 2013. Prelim. version in STOC 04.

121. C. Chekuri and J. Chuzhoy. Polynomial bounds for the grid-minor theorem (extended abstract). In *Proceedings of the Symposium on Theory of Computing (STOC)*, New York, pp. 60–69, 2014.

122. J. Chuzhoy. Excluded grid theorem: Improved and simplified. In *Proceedings of the Forty-Seventh Annual ACM on Symposium on Theory of Computing, STOC 2015*, Portland, OR, June 14–17, pp. 645–654, 2015.

123. N. Robertson and P. D. Seymour. Graph minors. V. Excluding a planar graph. *Journal of Combinatorial Theory, Series B*, 41(1):92–114, 1986.

124. B. A. Reed. Algorithmic aspects of tree width. In B. A. Reed and C. L. Sales, Eds., *Recent Advances in Algorithms and Combinatorics*, pp. 85–107. Springer, New York, 2003.

125. D. J. Harvey and D. R. Wood. Parameters tied to treewidth. *Journal of Graph Theory*, 84:364–385.

126. B. A. Reed. Tree width and tangles: A new connectivity measure and some applications. In R. A. Bailey, Ed., *Surveys in Combinatorics*, volume 241 of *London Mathematical Society Lecture Note Series*, pp. 87–162. Cambridge University Press, Cambridge, UK, 1997.

127. N. Robertson, P. D. Seymour, and R. Thomas. Quickly excluding a planar graph. *Journal of Combinatorial Theory, Series B*, 62(2):323–348, 1994.

128. H. Räcke. Minimizing congestion in general networks. In *43rd Symposium on Foundations of Computer Science (FOCS 2002)*, Proceedings, Vancouver, Canada, November 16–19, pp. 43–52, 2002.

129. C. Chekuri and A. Ene. The all-or-nothing flow problem in directed graphs with symmetric demand pairs. *Mathematical Programming*, 154(1–2):249–272, 2015.

130. M. Andrews. Approximation algorithms for the edge-disjoint paths problem via Räcke decompositions. In *51st Annual IEEE Symposium on Foundations of Computer Science, FOCS 2010*, Las Vegas, NV, October 23–26, pp. 277–286, 2010.

131. J. Chuzhoy. Routing in undirected graphs with constant congestion. In *Proceedings of the 44th Symposium on Theory of Computing Conference, STOC 2012*, New York, May 19–22, pp. 855–874, 2012.

132. J. Chuzhoy and S. Li. A polylogarithmic approximation algorithm for edge-disjoint paths with congestion 2. In *53rd Annual IEEE Symposium on Foundations of Computer Science, FOCS 2012*, New Brunswick, NJ, October 20–23, pp. 233–242, 2012.

133. K.-I. Kawarabayashi and Y. Kobayashi. Breaking $O(n^{1/2})$-approximation algorithms for the edge-disjoint paths problem with congestion two. In *Proceedings of the 43rd ACM Symposium on Theory of Computing, STOC 2011*, San Jose, CA, June 6–8, pp. 81–88, 2011.

134. C. Chekuri, S. Khanna, and F. B. Shepherd. Edge-disjoint paths in planar graphs with constant congestion. *SIAM Journal on Computing*, 39(1):281–301, 2009.

135. L. Seguin-Charbonneau and F. B. Shepherd. Maximum edge-disjoint paths in planar graphs with congestion 2. In *IEEE 52nd Annual Symposium on Foundations of Computer Science, FOCS 2011*, Palm Springs, CA, October 22–25, pp. 200–209, 2011.

136. C. Chekuri and A. Ene. Poly-logarithmic approximation for maximum node disjoint paths with constant congestion. In *Proceedings of the Twenty-Fourth Annual ACM-SIAM Symposium on Discrete Algorithms, SODA 2013*, New Orleans, LA, January 6–8, pp. 326–341, 2013.

137. C. Chekuri and J. Chuzhoy. Half-integral all-or-nothing flow. Unpublished manuscript.

138. C. Chekuri, A. Ene, and M. Pilipczuk. Constant congestion routing of symmetric demands in planar directed graphs. In *43rd International Colloquium on Automata, Languages, and Programming, ICALP 2016*, Rome, Italy, July 11–15, pp. 7:1–7:14, 2016.

139. S. Rao and S. Zhou. Edge disjoint paths in moderately connected graphs. *SIAM Journal on Computing*, 39(5):1856–1887, 2010.

140. J. Chuzhoy, D. H. K. Kim, and S. Li. Improved approximation for node-disjoint paths in planar graphs. In *Proceedings of the 48th Annual ACM SIGACT Symposium on Theory of Computing, STOC 2016*, Cambridge, MA, June 18–21, pp. 556–569, 2016.

141. J. Chuzhoy and D. H. K. Kim. On approximating node-disjoint paths in grids. In *Approximation, Randomization, and Combinatorial Optimization. Algorithms and Techniques, APPROX/RANDOM 2015*, Princeton, NJ, August 24–26, pp. 187–211, 2015.

142. T. Erlebach. Approximation algorithms for edge-disjoint paths and unsplittable flow. In E. Bampis, K. Jansen, and C. Kenyon, Eds., *Efficient Approximation and Online Algorithms: Recent Progress on Classical Combinatorial Optimization Problems and New Applications*, pp. 97–134. Springer, Berlin, Germany, 2006.

143. C. Chekuri, G. Naves, and F. B. Shepherd. Maximum edge-disjoint paths in *k*-sums of graphs. In *Automata, Languages, and Programming—40th International Colloquium, ICALP 2013, Proceedings, Part I*, Riga, Latvia, July 8–12, pp. 328–339, 2013.

144. A. Ene, M. Mnich, M. Pilipczuk, and A. Risteski. On routing disjoint paths in bounded treewidth graphs. In *15th Scandinavian Symposium and Workshops on Algorithm Theory, SWAT 2016*, Reykjavik, Iceland, June 22–24, pp. 15:1–15:15, 2016.

145. C. Chekuri, S. Khanna, and F. B. Shepherd. A note on multiflows and treewidth. *Algorithmica*, 54(3):400–412, 2009.

146. D. Marx and P. Wollan. An exact characterization of tractable demand patterns for maximum disjoint path problems. In *Proceedings of the Twenty-Sixth Annual ACM-SIAM Symposium on Discrete Algorithms, SODA 2015*, San Diego, CA, January 4–6, pp. 642–661, 2015.

147. G. Baier. Flows with path restrictions. PhD thesis, TU Berlin, 2003.

148. G. Baier, T. Erlebach, A. Hall, E. Köhler, P. Kolman, O. Pangrác, H. Schilling, and M. Skutella. Length-bounded cuts and flows. *ACM Transactions on Algorithms*, 7(1):4, 2010.

149. A. Itai, Y. Perl, and Y. Shiloach. The complexity of finding maximum disjoint paths with length constraints. *Networks*, 12:277–286, 1982.

150. C. Li, T. McCormick, and D. Simchi-Levi. The complexity of finding two disjoint paths with min-max objective function. *Discrete Applied Mathematics*, 26:105–115, 1990.

151. G. Baier, E. Köhler, and M. Skutella. The *k*-splittable flow problem. *Algorithmica*, 42(3–4):231–248, 2005.

152. R. Koch and I. Spenke. Complexity and approximability of *k*-splittable flows. *Theoretical Computer Science*, 369(1–3):338–347, 2006.

153. R. Koch, M. Skutella, and I. Spenke. Maximum *k*-splittable *s, t*-flows. *Theory of Computing Systems*, 43(1):56–66, 2008.

154. S. G. Kolliopoulos. Minimum-cost single-source 2-splittable flow. *Information Processing Letters*, 94:15–18, 2005.

155. F. Salazar and M. Skutella. Single-source k-splittable min-cost flows. *Operations Research Letters*, 37(2):71–74, 2009.

156. W. Kishimoto and M. Takeuchi. On *m*-route flows in a network. *IEICE Transactions*, J-76-A(8):1185–1200, 1993. In Japanese.

157. W. Kishimoto. A method for obtaining the maximum multiroute flows in a network. *Networks*, 27(4):279–291, 1996.

158. C. C. Aggarwal and J. B. Orlin. On multiroute maximum flows in networks. *Networks*, 39(1):43–52, 2002.

159. A. Bagchi, A. Chaudhary, and P. Kolman. Short length Menger's Theorem and reliable optical routing. *Theoretical Computer Science*, 339:315–332, 2005. Prelim. version in SPAA 03 (revue paper).

160. H. Bruhn, J. Cerný, A. Hall, P. Kolman, and J. Sgall. Single source multiroute flows and cuts on uniform capacity networks. *Theory of Computing*, 4(1):1–20, 2008.

161. C. Chekuri and S. Khanna. Algorithms for 2-route cut problems. In *Automata, Languages and Programming, 35th International Colloquium, ICALP 2008, Proceedings, Part I: Tack A: Algorithms, Automata, Complexity, and Games*, Reykjavik, Iceland, July 7–11, pp. 472–484, 2008.

162. P. Kolman and C. Scheideler. Approximate duality of multicommodity multiroute flows and cuts: Single source case. In *Proceedings of the Twenty-Third Annual ACM-SIAM Symposium on Discrete Algorithms, SODA 2012*, Kyoto, Japan, January 17–19, pp. 800–810, 2012.

163. P. Kolman and C. Scheideler. Towards duality of multicommodity multiroute cuts and flows: Multilevel ball-growing. *Theory of Computing Systems*, 53(2):341–363, 2013.

164. M. Martens. A simple greedy algorithm for the k-disjoint flow problem. In *Theory and Applications of Models of Computation, 6th Annual Conference, TAMC 2009, Proceedings*, Changsha, China, May 18–22, pp. 291–300, 2009.

165. A. Bagchi, A. Chaudhary, P. Kolman, and J. Sgall. A simple combinatorial proof of duality of multiroute flows and cuts. Technical Report 2004-662, Charles University, Prague, Czech Republic, 2004.

# 12

# The *k*-Connected
# Subgraph Problem

Zeev Nutov

## 12.1  Introduction

Two paths in a graph are **internally disjoint** if no internal node of one of the paths belongs to the other. A graph is *k*-**connected** if it contains *k* pairwise internally disjoint paths from every node to any other node. We survey approximation algorithms for the *k*-Connected Subgraph problem, formally defined as follows.

---

*k*-Connected Subgraph

*Input:* A directed/undirected graph $\hat{G} = (V, \hat{E})$ with edge costs $\{c_e : e \in \hat{E}\}$ and a positive integer *k*.
*Output:* A minimum-cost *k*-connected spanning subgraph of $\hat{G}$.

---

In the related *k*-Edge-Connected Subgraph problem, the paths are required to be only edge disjoint. For undirected graphs, both problems are NP-hard for $k = 2$ (the case $k = 1$ is the Minimum Spanning Tree problem), even when all edges in $\hat{G}$ have unit costs. This is as any feasible solution with $|V|$ edges is a Hamiltonian cycle. For directed graphs, the problem is NP-hard already for $k = 1$, by a similar reduction.

We may assume that the input graph $\hat{G}$ is complete, by assigning infinite costs to "forbidden" edges. Under this assumption, we consider four types of edge costs:

- {0, 1}-*costs*: Here we are given a graph $G$, and the goal is to find a minimum size augmenting edge set $J$ of new edges (any edge is allowed) such that $G \cup J$ is *k*-connected.
- {1, ∞}-*costs*: Here we seek a *k*-connected spanning subgraph of $\hat{G}$ with minimum number of edges.
- *Metric costs*: The costs satisfy the triangle inequality $c_{uv} \leq c_{uw} + c_{wv}$ for all $u, w, v \in V$.
- *General costs*: The costs are arbitrary nonnegative integers or ∞.

**TABLE 12.1**　Known Approximability Status of $k$-Connected Subgraph Problems

| Costs | Node-Connectivity | | Edge-Connectivity | |
| --- | --- | --- | --- | --- |
| | Undirected | Directed | Undirected | Directed |
| $\{0,1\}$ | $\min\{2, 1 + \frac{k^2}{2\mathrm{opt}}\}$ [5,6] | in P [5] | in P [7] | in P [8] |
| $\{1,\infty\}$ | $1 - \frac{1}{k} + \frac{n}{\mathrm{opt}} \leq 1 + \frac{1}{k}$ [9] ([10]) | $1 - \frac{1}{k} + \frac{2n}{\mathrm{opt}} \leq 1 + \frac{1}{k}$ [9] ([10]) | $1 + \frac{1}{2k} + O\left(\frac{1}{k^2}\right)$ [11] | $1 + \frac{1}{k}$ [12] |
| | | | $1 + \Omega\left(\frac{1}{k}\right)$ [13] | $1 + \Omega\left(\frac{1}{k}\right)$ [13] |
| Metric | $2 + (k-1)/n$ [14] | $2 + k/n$ [14] | 2 [15] | 2 [15] |
| General | $O\left(\ln\frac{n}{n-k} \cdot \ln k\right)$ [16] | $O\left(\ln\frac{n}{n-k} \cdot \ln k\right)$ [16] | 2 [15] | 2 [15] |
| | 6 if $n \geq k^3$ [17] ([18]) | | | |

Here and everywhere, references in parenthesis give a simplified proof and/or a slight improvement needed to achieve the approximation ratio stated.

In this survey, we consider only the *node-connectivity* version, when the paths are internally disjoint. We survey only *approximation algorithms* (see References 1 and 2 for polynomially solvable cases and References 3 and 4 for previous surveys), with the currently best known approximation ratios, as summarized in Table 12.1, in which for comparison we also included results for edge-connectivity.

We mention some additional results not appearing in Table 12.1. The best known ratios for $k$-Connected Subgraph with small values of $k$ are as follows: $\lceil(k+1)/2\rceil$ if $k \leq 8$ for undirected graphs and $k + 1$ if $k \leq 6$ for directed graphs [14,19,20]. For $\{0,1\}$-costs, the complexity status of the problem is not known for undirected graphs, but for any constant $k$, an optimal solution can be computed in polynomial time [21]. When $\hat{G}$ contains a spanning $(k-1)$-connected subgraph of cost 0, the $\{0,1\}$-costs case can be solved in polynomial time for any $k$ [22]. In the case of $\{1,\infty\}$-costs, directed 1-Connected Subgraph admits ratio 3/2 [23], and undirected $k$-Edge-Connected Subgraph admits ratio 4/3 for $k = 2$ [24] and ratio $1 + \frac{2}{k+1}$ for $3 \leq k \leq 6$ [9]. In the case of metric costs, both 2-Connected Subgraph and 2-Edge-Connected Subgraph admit ratio 3/2 [25]. In the case of $\{0,1,\infty\}$-costs when $\hat{G}$ contains a spanning tree of cost 0, 2-Edge-Connected Subgraph admits ratio 3/2 [26].

We will assume that $n \geq k + 2$ and that the input graph $\hat{G}$ is simple. This is justified as follows. If $n \leq k+1$, then an easy argument shows that an optimal solution is obtained by taking the cheapest $k - n$ parallel edges from any node to the other; thus, this case can be solved in polynomial time. If $n \geq k + 1$, then it is known that any $k$-connected graph has a simple $k$-connected spanning subgraph. Thus, in this case for any set of parallel edges of the input graph $\hat{G}$ we can keep only the cheapest one. The case $k = n-2$ can also be solved in polynomial time. It is not hard to see that a graph $G$ is $(n-2)$-connected if and only if each node has degree (indegree and outdegree, in the case of directed graphs) at least $n - 2$. The problem of finding the cheapest subgraph that satisfies some prescribed degree bounds can be solved in polynomial time [27]. This gives a polynomial time algorithm for $k = n - 2$.

Here is some notation used. An edge from $u$ to $v$ is denoted by $uv$. A $uv$-**path** is a path from $u$ to $v$. For arbitrary sets $A, B$ of nodes and edges (or graphs), $A \backslash B$ is the set (or graph) obtained by deleting $B$ from $A$, in which deletion of a node implies also deletion of all the edges incident to it; similarly, $A \cup B$ is the set (graph) obtained by adding $B$ to $A$. For real values $\{x_u : u \in U\}$, let $x(U) = \sum_{u \in U} x_u$ and $\max(U) = \max_{u \in U} x_u$.

We denote $n = |V|$ and assume that $n \geq k + 2$ (otherwise the problem is trivial).

*Organization.*

In the next Section 12.2, we give some reductions between directed and undirected versions of our problem and related problems. In Section 12.3, we describe the algorithm of Reference 9 for the $\{1,\infty\}$-costs case. In Section 12.4, we give a Biset-LP formulation of the problem and some properties of relevant biset functions. In Section 12.5, we consider metric costs. In Section 12.6, we describe a 6-approximation algorithm of Reference 17 for undirected graphs with $n = \Omega(k^3)$. In Section 12.7, we discuss the relation between $k$-Connected Subgraph and the augmentation version of the problem when the input graph $\hat{G}$ has a $(k-1)$-connected spanning subgraph of cost zero, and in Section 12.8, we describe an $O\left(\ln \frac{n}{n-k}\right)$-approximation algorithm for the augmentation problem. We conclude in Section 12.9 with some open problems in the field.

## 12.2 Relation between Directed and Undirected Problems

We mention several relevant related problems. Let $\kappa_G(s,t)$ denote the maximum number of internally disjoint $st$-paths in a graph $G$. We say that $G$ is $k$-**inconnected to** $s$ if $\kappa_G(v,s) \geq k$ for all $v \in V \setminus \{s\}$. The corresponding min-cost connectivity problem is $k$-Inconnected Subgraph. In the related edge-connectivity problem $k$-Edge-Inconnected Subgraph, the paths are required only to be edge disjoint. For directed graphs, these problems can be solved in polynomial time, see References 28 and 29 for the edge-connectivity case, Reference 30 for the node connectivity case, and References 1 and 31 for various generalizations. The following reduction shows that undirected problems are not much harder to approximate than the directed ones.

**Lemma 12.1** *For the four problems* $k$-Connected Subgraph, $k$-Inconnected Subgraph, $k$-Edge-Connected Subgraph, *and* $k$-Edge-Inconnected Subgraph, *ratio* $\rho$ *for directed graphs implies ratio* $2\rho$ *for undirected graphs.*

*Proof.* Given an undirected instance $\mathcal{I} = (\hat{G}, c, k)$ of one of the four problems in the lemma, obtain a "bidirected" instance $\mathcal{I}' = (\hat{G}', c', k)$ by replacing every edge $e = uv$ of $\hat{G}$ by the two opposite directed edges $uv, vu$ each of the same cost as $e = uv$. Then compute a $\rho$ approximate solution $J'$ to $\mathcal{I}'$ and output its underlying graph $J$. It is easy to see that $J$ is a feasible solution for $\mathcal{I}$. Furthermore, if $G$ is an arbitrary subgraph of $\hat{G}$, and $G'$ is the corresponding bidirected subgraph of $\hat{G}'$, then $c'(G') = 2c(G)$, and $G$ is a feasible solution for $\mathcal{I}$ if $G'$ is a feasible solution for $\mathcal{I}'$. Thus, $\text{opt}' \leq 2\text{opt}$, in which opt and opt' denote the optimal solution value of $\mathcal{I}$ and $\mathcal{I}'$, respectively. Consequently, $c(J) \leq c'(J') \leq \rho \cdot \text{opt}' \leq 2\rho \cdot \text{opt}$. ∎

Here is a typical application of Lemma 12.1. As the directed $k$-Inconnected Subgraph problem can be solved in polynomial time [30], we get from Lemma 12.1 ratio 2 for undirected $k$-Inconnected Subgraph. Another immediate application is as follows. As a directed graph $G = (V, E)$ is $k$-edge-connected iff both $G$ and the reverse graph of $G$ are $k$-edge-inconnected to $s$ for some $s \in V$, we get ratio 2 for the directed $k$-Edge-Connected Subgraph problem. From Lemma 12.1, we also get ratio 2 for the undirected $k$-Edge-Connected Subgraph problem, as an undirected graph $G$ is $k$-edge-connected if and only if $G$ is $k$-edge-inconnected to some node $s$. This method does not work directly for the $k$-Connected Subgraph problem, as a graph which is $k$-inconnected to $s$ may not be $k$-connected. However, many algorithms for $k$-Connected Subgraph use an extension of this method.

The following reduction shows that for $k$-Connected Subgraph instances with "high" values of $k$, the undirected variant cannot be much easier than the directed one.

**Theorem 12.1 (Lando & Nutov [32])** *Let* $\rho_{\text{dir}}(k, n)$ *and* $\rho_{\text{und}}(k, n)$ *denote the best possible approximation ratio for the directed and undirected $k$-Connected Subgraph problem on graphs on $n$ nodes, respectively. Then* $\rho_{\text{dir}}(k, n) \leq \rho_{\text{und}}(k + n, 2n)$.

For even $n$ and $k \geq n/2 + 1$, the inequality in Theorem 12.1 can be written as $\rho_{\text{dir}}(k - n/2, n/2) \leq \rho_{\text{und}}(k, n)$. Combining with Lemma 12.1 gives $\rho_{\text{dir}}(k - n/2, n/2) \leq \rho_{\text{und}}(k, n) \leq 2\rho_{\text{dir}}(k, n)$. Loosely speaking, this means that for "high" values of $k$, say $k = n - o(n)$, the approximability of directed and undirected variants of the $k$-Connected Subgraph problem is the same, up to a constant factor.

In the rest of this section, we prove Theorem 12.1. An ordered pair $(S, T)$ of disjoint subsets of $V$ is called a **setpair**. For $s, t \in V$ we say that $(S, T)$ an *st*-**setpair** if $s \in S$ and $t \in T$. Let $d_G(S, T)$ denote the number of edges in $G$ that go from $S$ to $T$. Then the node connectivity version of Menger's Theorem can be formulated as follows.

**Lemma 12.2** *Let $G = (V, E)$ be a (directed or undirected) graph. Then for any $s, t \in V$*

$$\kappa_G(s, t) = \min\{d_G(S, T) + |V| - (|S| + |T|) : (S, T) \text{ is an st-setpair in } G\}$$

*Furthermore, if $st \notin E$, then the minimum is attained for an st-setpair $(S, T)$ with $d_G(S, T) = 0$.*

By using Lemma 12.2, it is not hard to prove the following known statement.

**Lemma 12.3** *A graph $G = (V, E)$ with $|V| \geq k + 1$ is $k$-connected if and only if $\kappa_G(s, t) \geq k$ for all $s, t \in V$ with $st \notin E$.*

Let $G = (V, E)$ be a directed graph. The **bipartite graph of** $G$ has node set $V' \cup V''$ in which $V', V''$ are copies of $V$, and edge set $\{u'v'' : uv \in E\}$, in which for $v \in V$ we denote by $v'$ and $v''$ the copies of $v$ in $V'$ and $V''$, respectively. The **padded graph of** $G$ is obtained by adding to the bipartite graph of $G$ a **padding edge set** of cliques on each of $V'$ and $V'$ and the matching $M = \{v'v'' : v \in V\}$. Given an instance $\hat{G} = (V, \hat{E}), c, k$ of directed $k$-Connected Subgraph, obtain an instance $\hat{H}, c', k + n$, of undirected $k$-Connected Subgraph in which $\hat{H}$ is the padded graph of $\hat{G}$, and the costs are: $c'(u'v'') = c(uv)$ if $uv \in \hat{E}$ and the padding edges have cost 0. Note that if $G$ is a spanning subgraph of $\hat{G}$ then the padded graph $H$ of $G$ is a spanning subgraph of $\hat{H}$, and $G$ and $H$ have the same cost. Moreover, note that the set of nonadjacent node pairs in $H$ is $\{\{s', t''\} : s, t \in V, st \notin E\}$. Thus, the following lemma combined with Lemma 12.3 finishes the proof of Theorem 12.1 (for the proof see Reference 33).

**Lemma 12.4 (Lando & Nutov [32] ([33]))** *If $H$ is the padded graph of a directed graph $G = (V, E)$ on $n$ nodes, then $\kappa_H(s', t'') = \kappa_G(s, t) + n$ for all $s, t \in V$. Thus, $G$ is $k$-connected if and only if $H$ is $(k + n)$-connected.*

## 12.3 Min-Size $k$-Connected Subgraph ($\{1, \infty\}$-Costs)

Here we consider the $\{1, \infty\}$-costs $k$-Connected Subgraph problem, often called the Min-Size $k$-Connected Subgraph problem. We present a slight improvement due to Reference 10 of the ratio $1 + \frac{1}{k}$ of Cheriyan and Thurimella [9].

**Theorem 12.2 (Cheriyan & Thurimella [9] ([10]))** Min-Size $k$-Connected Subgraph *admits the following approximation ratios:* $1 - \frac{1}{k} + \frac{n}{\text{opt}} \leq 1 + \frac{1}{k}$ *for undirected graphs and* $1 - \frac{1}{k} + \frac{2n}{\text{opt}} \leq 1 + \frac{1}{k}$ *for directed graphs.*

An edge $e$ of a $k$-connected graph $G$ is said to be **critical** if $G \setminus \{e\}$ is not $k$-connected. One of the most important theorems in theory of $k$-connected graphs is the following.

**Theorem 12.3 (Mader's Undirected Critical Cycle Theorem [34])** *In a $k$-connected undirected graph $G$, any cycle of critical edges has a node $v$ whose degree in $G$ is $k$.*

Mader [35] also formulated and proved a similar theorem for directed graphs. An even length sequence of directed edges $C = (v_1v_2, v_3v_2, v_3v_4, \ldots, v_{2q-1}v_{2q}, v_1v_{2q})$ of a directed graph $G$ is called an **alternating cycle**; the nodes $v_1, v_3, \ldots, v_{2q-1}$ are $C$-**out nodes**, and $v_2, v_4, \ldots, v_{2q}$ are $C$-**in nodes**.

**Theorem 12.4 (Mader's Directed Critical Cycle Theorem [35])** *In a $k$-connected directed graph $G$, any alternating cycle $C$ of critical edges contains a $C$-in node whose indegree in $G$ is $k$, or a $C$-out node whose outdegree in $G$ is $k$.*

**Definition 12.1** *A graph $G = (V, E)$ is an $\ell$-**edge-cover** if it has minimum degree $\geq \ell$ if $G$ is undirected, and minimum outdegree $\geq \ell$ and minimum indegree $\geq \ell$ if $G$ is directed.*

It is not hard to see that a directed graph $G$ has no alternating cycle iff its bipartite graph $G'$ is a forest. From Theorems 12.3 and 12.4, it is easy to deduce the following corollary, also due to Mader [34,35].

**Corollary 12.1** *Let $G = (V, I \cup F)$ be a k-connected graph such that I is a $(k-1)$-edge-cover and all the edges in F are critical. If G is an undirected graph then $(V, F)$ is a forest, and if G is a directed graph then the bipartite graph of $(V, F)$ is a forest.*

The following algorithm from Reference 9 achieves the desired ratio for both directed and undirected graphs.

---

## ALGORITHM 12.1   Min-Size $k$-Connected Subgraph($\hat{G}, k$)

1 find a minimum size $(k-1)$-edge-cover $I \subseteq \hat{E}$
2 find an inclusionwise minimal edge set $F \subseteq \hat{E} \setminus I$ such that $(V, I \cup F)$ is $k$-connected
3 return $I \cup F$

---

The problem of finding the cheapest $\ell$-edge-cover can be solved in polynomial time (cf. [27]), and it is also not hard to see that Step 2 of the algorithm can be implemented in polynomial time [9]. To show the approximation ratio we will use the following theorem, to be proved later.

**Theorem 12.5 (Nutov [10])** *Let $G = (V, E)$ be an undirected graph with edge costs $\{c_e : e \in E\}$ and minimum degree $\geq k$ and let $I \subseteq E$ be a minimum cost $\ell$-edge cover in G, $1 \leq \ell \leq k-1$. If G is bipartite, then $c(F) \leq \frac{\ell}{k} \cdot c(E)$. If G is k-edge-connected, then $c(I) \leq \frac{\ell + 1/n}{k} \cdot c(E)$ and $c(I) \leq \frac{\ell}{k} \cdot c(E)$ if $\ell|V|$ is even.*

Let us consider directed graphs. Let $I'$ and $F'$ be the "bipartite" edge sets in the bipartite graph $G'$ of $G$ that corresponds to $I$ and $F$, respectively. Then $I$ is an $\ell$-edge-cover if and only if $I'$ is an $\ell$-edge-cover. Thus, $|I| = |I'| \leq \frac{k-1}{k}\mathrm{opt}$, by Theorem 12.5. On the other hand, by Corollary 12.1, $F'$ is a forest, hence $|F| = |F'| \leq 2n - 1$. Consequently, $\frac{|I|+|F|}{\mathrm{opt}} \leq 1 - \frac{1}{k} + \frac{2n-1}{\mathrm{opt}}$.

Let us consider undirected graphs. If $(k-1)n$ is even or if $\mathrm{opt} \geq \frac{kn}{2} + \frac{k}{2(k-1)}$, then $|I| \leq \frac{k-1}{k}\mathrm{opt}$, by Theorem 12.5. By Corollary 12.1, $F$ is a forest, hence $|F| \leq n - 1$. Consequently, $\frac{|I|+|F|}{\mathrm{opt}} \leq 1 - \frac{1}{k} + \frac{n-1}{\mathrm{opt}}$. If $(k-1)n$ is odd and $\mathrm{opt} < \frac{kn}{2} + 1$, then an optimal solution is $k$-regular and hence $|I| \leq \frac{(k-1)n+1}{2} \leq \left(1 - \frac{1}{k}\right)(\mathrm{opt} + 1)$. Combining we get $\frac{|I|+|F|}{\mathrm{opt}} \leq 1 - \frac{1}{k} + \frac{1-1/k}{\mathrm{opt}} + \frac{n-1}{\mathrm{opt}} < 1 - \frac{1}{k} + \frac{n}{\mathrm{opt}}$.

Before proving Theorem 12.5, let us give two additional applications of Theorem 12.5.

*Max-Connectivity m-Subgraph.*

Here we seek a maximum connectivity $k^*$ spanning subgraph $G$ of $\hat{G}$ with at most $m$ edges. Note that if we apply Algorithm 12.1 with $k$ replaced by $k - 1$, then from Theorem 12.5 we get: $|I| \leq \frac{k-2}{k}\mathrm{opt}$ and $|F| \leq \frac{2}{k}\mathrm{opt}$ for directed graphs, and $|I| \leq \frac{k-2+1/n}{k}\mathrm{opt}$ and $|F| \leq \frac{2-2/n}{k}\mathrm{opt}$ for undirected graphs. Thus, the algorithm returns a $(k-1)$-connected spanning subgraph $G$ with at most opt edges. We can apply this algorithm to find the maximum integer $k$ for which the algorithm returns a subgraph with at most $m$ edges. Then $k \geq k^* - 1$, hence we obtain a $(k^* - 1)$-connected spanning subgraph with at most $m$ edges. Note that this is tight, as the problem is NP-hard.

*$\beta$-metric costs.*

Here for some $\frac{1}{2} \leq \beta < 1$ the costs satisfy the $\beta$-triangle inequality $c_{uv} \leq \beta(c_{uw} + c_{wv})$ for all $u, w, v \in V$. When $\beta = 1/2$, we have the min-size version of the problem. If we allow $\beta = 1$, then we get metric costs. In Reference 36, it is proved that $c(F) \leq \frac{2\beta}{k(1-\beta)}\mathrm{opt}$. If $(k-1)n$ is even, or if there exists an optimal solution with at least $\frac{kn}{2} + \frac{k}{2(k-1)} \leq \frac{kn}{2} + 1$ edges, then Theorem 12.5 gives the bound $c(I) \leq \left(1 - \frac{1}{k}\right)$ opt; else, $c(I) \leq \left(1 - \frac{1}{k} + \frac{1}{kn}\right)$ opt. Consequently, we get ratio $1 - \frac{1}{k} + \frac{1}{kn} + \frac{2\beta}{k(1-\beta)}$ for this version of the problem.

In the rest of this section, we prove Theorem 12.5. Let for $A \subseteq V$ let $\delta(A)$ denote the set of edges in $E$ with exactly one endnode in $A$. Moreover, let $\zeta(A)$ denote the set of edges in $E$ with at least one endnode in $A$. For $x \in \mathbb{R}^E$, the $\ell$-**edge-cover polytope** $P_{\text{cov}}(G, \ell)$ is defined by the constraints

$$
\begin{aligned}
0 \leq x_e \leq 1 & \qquad e \in E \\
x(\delta(v)) \geq \ell & \qquad v \in V \\
x(\zeta(A)) - x(F) \geq (\ell|A| - |F| + 1)/2 & \qquad A \subseteq V, F \subseteq \delta(A), \ell|A| - |F| \text{ odd}
\end{aligned}
$$

The **fractional $\ell$-edge-cover polytope** $P^f_{\text{cov}}(G, \ell)$ is defined by the first two sets of constraints. It is known that if $G$ is bipartite then $P^f_{\text{cov}}(G, \ell) = P_{\text{cov}}(G, \ell)$ (cf. [27], (31.7) on page 340). The **fractional $k$-edge-connectivity polytope** $P^f_{\text{con}}(G, k)$ is defined by the constraints

$$
\begin{aligned}
0 \leq x_e \leq 1 & \qquad e \in E \\
x(\delta(A)) \geq k & \qquad \emptyset \neq A \subset V
\end{aligned}
$$

We will prove the following statement that implies Theorem 12.5.

**Lemma 12.5** *Let $G = (V, E)$ be an undirected graph with costs $\{c_e : e \in E\}$ and minimum degree $\geq k \geq 2$ and let $1 \leq \ell \leq k - 1$. If $G$ is bipartite and $x \in P^f_{\text{cov}}(G, k)$, then $\frac{\ell}{k}x \in P_{\text{cov}}(G, \ell)$. If $G$ is $k$-edge-connected and $x \in P^f_{\text{con}}(G, k)$, then $\frac{\ell+1/n}{k}x \in P_{\text{cov}}(G, \ell)$ and $\frac{\ell}{k}x \in P_{\text{cov}}(G, \ell)$ if $\ell n$ is even.*

*Proof.* Clearly, if $x \in P^f_{\text{cov}}(G, k)$, then $\frac{\ell}{k} \in P^f_{\text{cov}}(G, \ell)$. If $G$ is bipartite, then $P^f_{\text{cov}}(G, \ell) = P_{\text{cov}}(G, \ell)$ and thus $\frac{\ell}{k}x \in P^f_{\text{cov}}(G, \ell)$. Assume that $G$ is $k$-edge-connected and let $x \in P^f_{\text{con}}(G, k)$. We will show that $\mu \cdot x \in P_{\text{cov}}(G, \ell)$, in which $\mu = \frac{\ell}{k}$ if $\ell n$ is even and $\mu = \frac{\ell+1/n}{k}$ otherwise. Note that $\frac{\ell}{k}x$ satisfies the first two sets of constraints in the definition of $P_{\text{cov}}(G, \ell)$. Thus, we only need to prove that

$$
\mu(x(\zeta(A)) - x(F)) \geq (\ell|A| - |F| + 1)/2 \qquad A \subseteq V, F \subseteq \delta(A), \ell|A| - |F| \text{ odd}
$$

Note that $x(\zeta(A)) = \frac{1}{2}(\sum_{v \in A} x(\delta(v)) + x(\delta(A))) \geq \frac{1}{2}(k|A| + x(\delta(A))$. Substituting, multiplying by 2, and rearranging terms, we obtain that it is sufficient to prove that

$$
|A|(\mu k - \ell) + (|F| - \mu x(F)) + \mu(x(\delta(A)) - x(F)) \geq 1 \qquad A \subseteq V, F \subseteq \delta(A), \ell|A| - |F| \text{ odd}
$$

If $A = V$, then $F = \delta(A) = \emptyset$. Then the above-mentioned condition is void if $\ell|V|$ is even, and it reduces to the condition $n(\mu k - \ell) \geq 1$ otherwise, which holds as equality for $\mu = \frac{\ell+1/n}{k}$.

Suppose that $A$ is a proper subset of $V$. Then substituting $\mu = \frac{\ell}{k}$, multiplying both sides by $k$, and observing that $x(F) \leq |F|$, we obtain that it is sufficient to prove that

$$
|F|(k - \ell) - \ell x(F) + \ell x(\delta(A)) \geq k
$$

If $|F| \geq \frac{k}{k-\ell}$, then this is so as $x(\delta(A)) \geq x(F)$. If $k \geq 2\ell$, then this is so as $|F| \geq x(F)$ and $x(\delta(A)) \geq k$. The remaining case is $|F| < \frac{k}{k-\ell}$ and $k < 2\ell$. Then as $|F| \geq x(F)$ and $x(\delta(A)) \geq k$

$$
|F|(k - \ell) - \ell x(F) + \ell x(\delta(A)) \geq |F|(k - 2\ell) + k\ell \geq \frac{k}{k - \ell}(k - 2\ell) + k\ell = k - \frac{k\ell}{k - \ell} + k\ell \geq k
$$

This concludes the proof of the lemma.                                                                     ∎

## 12.4 Biset Functions

In Lemma 12.2, we formulated Menger's Theorem in terms of setpairs. It would be more convenient to consider instead of a setpair $(A, B)$ the pair of sets $(A, V\backslash B)$ called a "biset."

**Definition 12.2** *An ordered pair* $\mathbb{A} = (A, A^+)$ *of subsets of V with* $A \subseteq A^+$ *is called a* **biset**; *A is the* **inner part** *and* $A^+$ *is the* **outer part** *of* $\mathbb{A}$, *and* $\partial\mathbb{A} = A^+\backslash A$ *is the* **boundary** *of* $\mathbb{A}$. *The* **co-set** *of* $\mathbb{A}$ *is* $A^* = V\backslash A^+$; *the* **co-biset** *of* $\mathbb{A}$ *is* $\mathbb{A}^* = (A^*, V\backslash A)$. *We say that* $\mathbb{A}$ *is* **void** *if* $A = \emptyset$, **co-void** *if* $A^+ = V$, *and* $\mathbb{A}$ *is* **proper** *otherwise. Let* $\mathcal{V}$ *denote the family of bisets over V.*

A **biset function** assigns to every $\mathbb{A} \in \mathcal{V}$ a real number; in our context, it will always be an integer.

**Definition 12.3** *An* **edge covers a biset** $\mathbb{A}$ *if it goes from A to* $A^*$. *For a biset* $\mathbb{A}$ *and an edge-set/graph J, let* $\delta_J(\mathbb{A})$ *denote the set of edges in J covering* $\mathbb{A}$ *and let* $d_J(\mathbb{A}) = |\delta_J(\mathbb{A})|$. *The* **residual function** *of a biset function f w.r.t. a partial f-cover J is defined by* $f^J(\mathbb{A}) = f(\mathbb{A}) - d_J(\mathbb{A})$ *for all* $\mathbb{A} \in \mathcal{V}$. *We say that an edge set/graph J covers a biset function f, or that J is an f-***cover***, if* $d_J(\mathbb{A}) \geq f(\mathbb{A})$ *for all* $\mathbb{A} \in \mathcal{V}$.

We say that $\mathbb{A}$ is an *st*-**biset** if $s \in A$ and $t \in A^*$. In biset terms, Menger's Theorem (Lemma 12.2) is

$$\kappa_G(s, t) = \min\{|\partial\mathbb{A}| + d_G(\mathbb{A}) : \mathbb{A} \text{ is an } st\text{-biset}\}$$

Thus, $\kappa_G(s, t) \geq k$ iff $d_G(\mathbb{A}) \geq k - |\partial\mathbb{A}|$ for every *st*-biset $\mathbb{A}$. Consequently, *G* is *k*-connected iff *G* covers the *k*-**connectivity biset function** $f_{k\text{-CS}}$ defined by

$$f_{k\text{-CS}}(\mathbb{A}) = \begin{cases} k - |\partial\mathbb{A}| & \text{if } \mathbb{A} \text{ is proper} \\ 0 & \text{otherwise} \end{cases}$$

We thus will often consider the following generic problem:

---

**Biset-Function Edge-Cover**

*Input:* A graph $\hat{G} = (V, \hat{E})$ with edge costs $\{c_e : e \in \hat{E}\}$ and a biset function $f$ on $V$.
*Output:* A minimum cost edge-set $E \subseteq \hat{E}$ that covers $f$.

---

Here $f$ may not be given explicitly, and an efficient implementation of algorithms requires that certain queries related to $f$ can be answered in time polynomial in $n$. We will not consider implementation details. In the applications discussed here, relevant polynomial time oracles are available via min-cut computations. In particular, we have a polynomial time separation oracle for the LP-relaxation due to Frank and Jordán [5]:

$$\tau(f) = \min \quad c \cdot x$$
**(Biset-LP)** $$\text{s.t.} \quad x(\delta_{\hat{E}}(\mathbb{A})) \geq f(\mathbb{A}) \quad \forall \mathbb{A} \in \mathcal{V}$$
$$0 \leq x_e \leq 1 \quad \forall e \in E$$

This LP is particularly useful if the biset function $f$ has good uncrossing/supermodularity properties. To state these properties, we need to define the intersection and the union of bisets.

**Definition 12.4** *The* **intersection** *and the* **union** *of two bisets* $\mathbb{A}, \mathbb{B}$ *are defined by* $\mathbb{A}\cap\mathbb{B} = (A\cap B, A^+\cap B^+)$ *and* $\mathbb{A}\cup\mathbb{B} = (A\cup B, A^+\cup B^+)$. *The biset* $\mathbb{A}\backslash\mathbb{B}$ *is defined by* $\mathbb{A}\backslash\mathbb{B} = (A\backslash B^+, A^+\backslash B)$. *We say that* $\mathbb{B}$ **contains** $\mathbb{A}$ *and write* $\mathbb{A} \subseteq \mathbb{B}$ *if* $A \subseteq B$ *and* $A^+ \subseteq B^+$. *We say that* $\mathbb{A}, \mathbb{B}$ **intersect** *if* $A\cap B \neq \emptyset$, *and* **cross** *if* $A\cap B \neq \emptyset$ *and* $A^+ \cup B^+ \neq V$.

The following properties of bisets are easy to verify.

**Fact 12.1** *For any bisets* $\mathbb{A}, \mathbb{B}$ *the following holds. If a directed/undirected edge $e$ covers one of* $\mathbb{A} \cap \mathbb{B}, \mathbb{A} \cup \mathbb{B}$, *then $e$ covers one of* $\mathbb{A}, \mathbb{B}$; *if $e$ is an undirected edge, then if $e$ covers one of* $\mathbb{A} \setminus \mathbb{B}, \mathbb{B} \setminus \mathbb{A}$, *then $e$ covers one of* $\mathbb{A}, \mathbb{B}$. *Furthermore,* $|\partial \mathbb{A}| + |\partial \mathbb{B}| = |\partial(\mathbb{A} \cap \mathbb{B})| + |\partial(\mathbb{A} \cup \mathbb{B})| = |\partial(\mathbb{A} \setminus \mathbb{B})| + |\partial(\mathbb{B} \setminus \mathbb{A})|$.

For a biset function $f$ and bisets $\mathbb{A}, \mathbb{B}$, the **supermodular inequality** is

$$f(\mathbb{A} \cap \mathbb{B}) + f(\mathbb{A} \cup \mathbb{B}) \geq f(\mathbb{A}) + f(\mathbb{B})$$

We say that a biset function $f$ is **supermodular** if the supermodular inequality holds for all $\mathbb{A}, \mathbb{B} \in \mathcal{V}$ and **modular** if the supermodular inequality holds as equality for all $\mathbb{A}, \mathbb{B} \in \mathcal{V}$. $f$ is **symmetric** if $f(\mathbb{A}) = f(\mathbb{A}^*)$ for all $\mathbb{A} \in \mathcal{V}$. From Fact 12.1, one can deduce the following.

- For any (directed or undirected) graph $G$ the function $-d_G(\cdot)$ is supermodular.
- The function $|\partial(\cdot)|$ is modular, and for any $R \subseteq V$, the function $|A \cap R|$ is modular.

Two additional important types of biset functions are given in the following definition.

**Definition 12.5** *A biset function $f$ is* **intersecting/crossing supermodular** *if the supermodular inequality holds whenever* $\mathbb{A}, \mathbb{B}$ *intersect/cross.*

Biset-Function Edge-Cover with intersecting supermodular $f$ admits a polynomial time algorithm that for directed graphs computes an $f$-cover of cost $\tau(f)$ [31]; for undirected graphs the cost is at most $2\tau(f)$, by the "bidirection" reduction from Lemma 12.1. Intersecting supermodular functions usually arise by zeroing a supermodular function on void bisets, whereas crossing supermodular functions arise by zeroing a supermodular function on non-proper bisets. For example, the $k$-connectivity function $f_{k\text{-CS}}$ is crossing supermodular, as it is obtained by zeroing the modular function $k - |\partial \mathbb{A}|$ on non-proper bisets.

## 12.5 Metric Costs

For metric costs, Khuller and Raghavachari [37] obtained ratio $2 + \frac{2(k-1)}{n}$ for undirected graphs. We survey a result of Reference 14, which gives an improved ratio for undirected graphs, and a similar ratio for directed graphs.

**Theorem 12.6 (Kortsarz & Nutov [14])** $k$-Connected Subgraph *with metric costs admits the following approximation ratios:* $(2 + \frac{k-1}{n})$ *for undirected graphs and* $(2 + \frac{k}{n})$ *for directed graphs.*

Let $R \subseteq V$ with $|R| \geq k$. A $k$-**fan from** $v$ **to** $R$ is a set of $k$ distinct paths (possibly of length 0) from $v$ to $R$ such that any two of them have only the node $v$ in common. Let us say that a graph $G$ is $k$-**fan-connected to** $R$ if $G$ has a $k$-fan from any $v \in V$ to $R$. In a similar way, we define a $k$-**fan from** $R$ **to** $v$ and say that $G$ is $k$-**fan-connected from** $R$ if the paths go from $R$ to $v$. Note that the problem of finding a minimum-cost spanning subgraph that is $k$-fan-connected to $R$ is equivalent to the $k$-Inconnected Subgraph problem in the graph obtained from $\hat{G}$ by adding a new node $s$ and edges of cost 0 from every $v \in R$ to $s$. Let the $k$-**fan-connectivity function** $f_{k,R}$ be obtained by zeroing the modular function $k - |\partial \mathbb{A}| - |A \cap R|$ on void bisets. Then $f_{k,R}$ is intersecting supermodular. By Menger's Theorem, $G$ is $k$-fan-connected to $R$ iff $d_G(\mathbb{A}) \geq f_{k,R}(\mathbb{A})$ for all $\mathbb{A} \in \mathcal{V}$. Note that if $\mathbb{A}$ is co-void then $|\partial \mathbb{A}| + |A \cap R| \geq |R| \geq k$, hence $f_{k,R}(\mathbb{A}) \leq 0$. Summarizing, we have:

**Lemma 12.6** *If* $|R| \geq k$, *then* $f_{k,R}$ *is intersecting supermodular and* $f_{k\text{-CS}}(\mathbb{A}) - |A \cap R| \leq f_{k,R}(\mathbb{A}) \leq f_{k\text{-CS}}(\mathbb{A})$ *for all* $\mathbb{A} \in \mathcal{V}$. *Thus, if an edge set $J$ covers* $f_{k,R}$ *then* $A \cap R \neq \emptyset$ *whenever* $f^J_{k\text{-CS}}(\mathbb{A}) > 0$.

An undirected edge set (or a graph) $S$ is a **star** if all its edges are incident to the same node, called the **center** of the star. In the case of directed graphs, $S$ should be either an **outstar**—all edges in $S$ leave the center, or an **instar**—all edges in $S$ enter the center. An $\ell$-**star** is a star with $\ell$ leaves. For a star $S$, let max$(S)$ denote the maximum cost of an edge in $S$. The algorithm for undirected graphs is given in Algorithm 12.2.

---

## ALGORITHM 12.2 Undirected Metric $k$-Connected Subgraph$(\hat{G}, c, k)$

1 find a $(k-1)$-star $(R, S)$ in $\hat{G}$ for which $c(S) + (k-2)$ max$(S)$ is minimal
2 find a 2-approximate $k$-fan-connected to $R$ spanning subgraph $(V, I)$ of $\hat{G}$
3 find an inclusion minimal edge set $F$ on $R$ such that $(V, I \cup F)$ is $k$-connected
4 return $I \cup F$

---

Note that by Lemma 12.6, an edge set $F$ as in step 3 exists. Let $e_S$ denote the maximum-cost edge of $S$, so $c(e_S) = $ max$(S)$. The approximation ratio follows from the following lemma, which shows that $c(F) \leq \frac{k-1}{n} \cdot $ opt.

**Lemma 12.7** (i) $c(F) \leq c(S) + (k-2)$ max$(S)$.
    (ii) *There exists a* $(k-1)$-*star $S$ such that* $c(S) + (k-2)$ max$(S) \leq \frac{k-1}{n} \cdot $ opt.

*Proof.* We prove (i). By Corollary 12.1, $F$ is a forest. Let $z$ be the center of $S$ and for $v \in R$ denote $w_v = c_{zv}$ (where $w_z = 0$). Note that $w(R) + (k-2) \max_{v \in R} w_v = c(S) + (k-2)$ max$(S)$. As the costs are metric,
$$c(F) \leq \sum_{uv \in F}(w_u + w_v) = \sum_{v \in R} d_F(v)w_v.$$ However, it is easy to see that for any non-negative node weights $\{w_v : v \in R\}$, the maximum of $\sum_{v \in R} d_F(v)w_v$ is attained when $F$ is a star on $R$ centered at the maximum weight node. Thus, $\sum_{v \in R} d_F(v)w_v \leq w(R) + (k-2) \max_{v \in R} w_v \leq c(S) + (k-2)$ max$(S)$.

We prove (ii). Let $p = $ opt$/n$. By an averaging argument, there exists a $k$-star $S'$ of cost $c(S') \leq 2p$. Obtain a star $S$ from $S'$ by deleting the maximum-cost edge of $S'$. Then $c(S) + c(e_S) \leq c(S') \leq 2p$ and $c(e_S) \leq c(S')/2 \leq p$. Consequently, $c(S) + (k-2)c(e_S) = c(S) + c(e_S) + (k-3)c(e_S) \leq (k-1)p$. ∎

The algorithm for directed graphs is as follows.

---

## ALGORITHM 12.3 Directed Metric $k$-Connected Subgraph$(\hat{G}, c, k)$

1 find a $(k-1)$-outstar $(R^{out}, S^{out})$ and a $(k-1)$-instar $(R^{in}, S^{in})$ with common center such that
   $c(S^{out}) + c(S^{in}) + (k-1) ($max$(S^{out}) + $max$(S^{in}))$ is minimal
2 find a 2-approximate subgraph $(V, I)$ that is $k$-fan-connected to $R^{out}$ and $k$-fan-connected from $R^{in}$
3 find an inclusion minimal edge set $F$ from $R^{out}$ to $R^{in}$ such that $(V, I \cup F)$ is $k$-connected
4 return $I \cup F$

---

The following directed counterpart of Lemma 12.7 (the proof is omitted) implies that $c(F) \leq \frac{k}{n}$opt.

**Lemma 12.8** (i) $c(F) \leq c(S^{out}) + c(S^{in}) + (k-1) ($max$(S^{out}) + $max$(S^{in}))$.
    (ii) *There exists a* $(k-1)$-*outstar* $(R^{out}, S^{out})$ *and a* $(k-1)$-*instar* $(R^{in}, S^{in})$ *with common center such that* $c(S^{out}) + c(S^{in}) + (k-1) ($max$(S^{out}) + $max$(S^{in})) \leq \frac{k}{n} \cdot $ opt.

## 12.6 Ratio 6 for Undirected Graphs with $n = \Omega\left(k^3\right)$

This section is devoted for the proof of the following result.

**Theorem 12.7 (Cheriyan & Végh [17] ([18]))** *Undirected* biset-function edge-cover *with symmetric crossing supermodular $f$ admits ratio 6 provided that* $n \geq \gamma(2\gamma + 1)^2 + 2\gamma + 1$, *where* $\gamma = \gamma_f = \max_{f(\mathbb{A}) > 0} |\partial \mathbb{A}|$.

In the $k$-Connected Subgraph problem, the function $f_{k\text{-CS}}$ has $\gamma = k - 1$. This implies that undirected $k$-Connected Subgraph admits ratio 6 provided that $n \geq (k - 1)(2k - 1)^2 + 2k - 1$. Better bounds can be obtained under stronger assumptions on $f$, for example, for $f = f_{k\text{-CS}}$ [18] gives the bound $n \geq k(k - 1)(k - 1.5) + k$.

In the rest of this section, we prove Theorem 12.7. We need some definitions. Let us say that bisets $\mathbb{A}, \mathbb{B}$ **co-cross** if $A \backslash \mathbb{B}$ and $\mathbb{B} \backslash A$ are both non-void, and that $\mathbb{A}, \mathbb{B}$ **mesh** if they do not cross nor co-cross. One can verify that $\mathbb{A}, \mathbb{B}$ mesh iff one of the following holds: $A \subseteq \partial\mathbb{B}$, or $A^* \subseteq \partial\mathbb{B}$, or $B \subseteq \partial\mathbb{A}$, or $B^* \subseteq \partial\mathbb{A}$.

A biset function $f$ is **positively intersecting supermodular** if the supermodular inequality holds whenever $\mathbb{A}, \mathbb{B}$ intersect and $f(\mathbb{A}) > 0$ and $f(\mathbb{B}) > 0$. $f$ is **positively skew-supermodular** if the supermodular inequality or the **co-supermodular inequality** $f(\mathbb{A} \backslash \mathbb{B}) + f(\mathbb{B} \backslash \mathbb{A}) \geq f(\mathbb{A}) + f(\mathbb{B})$ holds whenever $f(\mathbb{A}) > 0$ and $f(\mathbb{B}) > 0$. Each of the corresponding Biset-Function Edge-Cover problems, in which $f$ is positively intersecting supermodular, or when $f$ is positively skew-supermodular, admits ratio 2 [30,38].

The idea of the proof is to cover sequentially three functions dominated by $f$; one intersecting supermodular, the second positively intersecting supermodular, and the last positively skew-supermodular.

**Lemma 12.9** *Let $f$ be a symmetric crossing supermodular biset function. If $\mathbb{A}, \mathbb{B}$ are non-meshing bisets, then the supermodular or the co-supermodular inequality holds for $\mathbb{A}, \mathbb{B}$, and $f$.*

*Proof.* If $\mathbb{A}, \mathbb{B}$ cross then the supermodular inequality holds for $\mathbb{A}, \mathbb{B}$. Assume that $\mathbb{A}, \mathbb{B}$ co-cross. Then $\mathbb{A}$ and $\mathbb{B}^*$ cross, and thus the supermodular inequality holds for $\mathbb{A}, \mathbb{B}^*$, and $f$. Note that (i) $\mathbb{A} \backslash \mathbb{B} = \mathbb{A} \cap \mathbb{B}^*$; (ii) $\mathbb{A} \cup \mathbb{B}^*$ is the co-biset of $\mathbb{B} \backslash \mathbb{A}$, hence $f(\mathbb{A} \cup \mathbb{B}^*) = f(\mathbb{B} \backslash \mathbb{A})$, by the symmetry of $f$. Thus, we get $f(\mathbb{A} \backslash \mathbb{B}) + f(\mathbb{B} \backslash \mathbb{A}) = f(\mathbb{A} \cap \mathbb{B}^*) + f(\mathbb{A} \cup \mathbb{B}^*) \geq f(\mathbb{A}) + f(\mathbb{B}^*) = f(\mathbb{A}) + f(\mathbb{B})$. ∎

From Lemma 12.9, we obtain a sufficient condition for a symmetric crossing supermodular biset function to be positively skew-supermodular.

**Corollary 12.2** *Let $f$ be a symmetric crossing supermodular biset function. If $f(\mathbb{C}) \leq 0$ holds for every biset $\mathbb{C}$ with $|C| \leq \gamma$, then $f$ is positively skew-supermodular.*

*Proof.* Let $\mathbb{A}, \mathbb{B}$ be bisets with $f(\mathbb{A}) > 0$ and $f(\mathbb{B}) > 0$. We claim that $\mathbb{A}, \mathbb{B}$ do not mesh, and thus by Lemma 12.9, the supermodular or the co-supermodular inequality holds for $\mathbb{A}, \mathbb{B}$ and $f$, as required.

Suppose to the contrary that $\mathbb{A}, \mathbb{B}$ mesh. Then $A \subseteq \partial\mathbb{B}$, or $A^* \subseteq \partial\mathbb{B}$, or $B \subseteq \partial\mathbb{A}$, or $B^* \subseteq \partial\mathbb{A}$. If $A \subseteq \partial\mathbb{B}$ holds, then $|A| \leq |\partial\mathbb{B}| \leq \gamma$, and thus $f(\mathbb{A}) \leq 0$. If $A^* \subseteq \partial\mathbb{B}$ holds, then $|A^*| \leq |\partial\mathbb{B}| \leq \gamma$, and thus by the symmetry of $f$ we get $f(\mathbb{A}) = f(\mathbb{A}^*) \leq 0$. In both cases, we obtain a contradiction to the assumption $f(\mathbb{A}) > 0$. The contradiction for the cases $B \subseteq \partial\mathbb{A}$ or $B^* \subseteq \partial\mathbb{A}$ is obtained in a similar way. ∎

In what follows, for a biset function $f$ on $V$ and $R \subseteq V$ let $f_R$ be defined by

$$f_R(\mathbb{A}) = \begin{cases} f(\mathbb{A}) & \text{if } A \cap R = \emptyset \\ 0 & \text{otherwise} \end{cases}$$

**Lemma 12.10** *If $f$ is crossing supermodular and $|R| \geq 2\gamma + 1$, then $f_R$ is positively intersecting supermodular.*

*Proof.* Let $\mathbb{A}, \mathbb{B}$ be intersecting bisets with $f_R(\mathbb{A}), f_R(\mathbb{B}) > 0$. Then $A \cap R = B \cap R = \emptyset$, and thus $(A \cap B) \cap R = (A \cup B) \cap R = \emptyset$. Consequently, $f_R(\mathbb{A}) = f(\mathbb{A})$, $f_R(\mathbb{B}) = f(\mathbb{B})$, $f_R(\mathbb{A} \cap \mathbb{B}) = f(\mathbb{A} \cap \mathbb{B})$, and $f_R(\mathbb{A} \cup \mathbb{B}) = f(\mathbb{A} \cup \mathbb{B})$. Furthermore, $\mathbb{A}, \mathbb{B}$ cross, as $|R| \geq 2\gamma + 1$. Thus, as $f$ is crossing supermodular

$$f_R(\mathbb{A}) + f_R(\mathbb{B}) = f(\mathbb{A}) + f(\mathbb{B}) \leq f(\mathbb{A} \cap \mathbb{B}) + f(\mathbb{A} \cup \mathbb{B}) = f_R(\mathbb{A} \cap \mathbb{B}) + f_R(\mathbb{A} \cup \mathbb{B})$$

Consequently, the supermodular inequality holds for $\mathbb{A}, \mathbb{B}$ and $f_R$ whenever $f_R(\mathbb{A}), f_R(\mathbb{B}) > 0$. ∎

For a biset function $f$ and an integer $p$, let us denote $U(f, p) = \bigcup\{A : f(\mathbb{A}) > 0, |A| \leq p\}$. Note that

- $f(\mathbb{A}) \leq 0$ whenever $|A| \leq p$ and $A \setminus U(f, p) \neq \emptyset$.
- If $I$ covers $f_R$ then $f^I(\mathbb{A}) \leq 0$ whenever $A \cap R = \emptyset$, namely, $R \cap A \neq \emptyset$ whenever $f^I(\mathbb{A}) > 0$.

This implies that if $R \subseteq V \setminus U(f, p)$ and if $I$ covers $f_R$, then $f^I(\mathbb{A}) \leq 0$ whenever $|A| \leq p$. Combining with Lemma 12.10(ii) we get

**Corollary 12.3** *Let $f$ be symmetric crossing supermodular, let $R \subseteq V \setminus U(f, \gamma)$, and let $I$ be an $f_R$-cover. Then $f^I$ is positively skew-supermodular.*

Corollary 12.3 gives a "cheap" method to "convert" a symmetric crossing supermodular function $f$ into a positively skew-supermodular function: just find $R \subseteq V \setminus U(f, \gamma)$ with $|R| = 2\gamma + 1$ and compute a 2-approximate cover $I$ of $f_R$ – the residual function $f^I$ of $f$ will be positively crossing supermodular. The difficulty is that such $R$ may not exist, for example, if $f = f_{k\text{-CS}}$ then $U(f, \gamma) = V$. We thus find some "cheap" edge set $I'$ such that $|U(g, \gamma)| \leq n - (2\gamma + 1)$ will hold for the residual function $g = f^{I'}$. The next lemma shows that such $I'$ can be a cover of the function $f_S$ for arbitrary $S \subseteq V$, provided that $n$ is not too small.

**Lemma 12.11** *Let $g$ be a symmetric crossing supermodular biset function and let $\mathcal{F} = \{\mathbb{A} : g(\mathbb{A}) > 0, |A| \leq p\}$. Let $S \subseteq V$ such that $S \cap A \neq \emptyset$ for all $\mathbb{A} \in \mathcal{F}$. Then $|U(g, p)| \leq (2\gamma + 1)|S|p$.*

*Proof.* Let $\mathcal{F}' = \{\mathbb{A}_1, \ldots, \mathbb{A}_\ell\}$ be a minimum size subfamily of $\mathcal{F}$ with $\bigcup_{A \in \mathcal{F}} A = \bigcup_{A \in \mathcal{F}'} A$. It is sufficient to show that $|\mathcal{F}'| \leq (2\gamma + 1)|S|$. By the minimality of $|\mathcal{F}'|$, for every $\mathbb{A}_i \in \mathcal{F}'$ there is $v_i \in A_i$ such that $v_i \notin A_j$ for every $j \neq i$. For every $i$, let $\mathbb{C}_i$ be an inclusion minimal member of the family $\{\mathbb{A} \in \mathcal{F} : \mathbb{A} \subseteq \mathbb{A}_i, v_i \in A\}$. By Corollary 12.2, $\mathcal{F}$ has the following property: $\mathbb{A} \setminus \mathbb{B} \in \mathcal{F}$ or $\mathbb{B} \setminus \mathbb{A} \in \mathcal{F}$ whenever none of $\mathbb{A} \setminus \mathbb{B}, \mathbb{B} \setminus \mathbb{A}$ is void. Thus, the minimality of $\mathbb{C}_i$ implies that one of the following holds for any $i \neq j$:

- $v_i \in \partial \mathbb{C}_j$ or $v_j \in \partial \mathbb{C}_i$
- $\mathbb{C}_i = \mathbb{C}_i \setminus \mathbb{C}_j$ or $\mathbb{C}_j = \mathbb{C}_j \setminus \mathbb{C}_i$

Construct an auxiliary directed graph $\mathcal{J}$ on node set $\mathcal{C} = \{\mathbb{C}_1, \mathbb{C}_2, \ldots, \mathbb{C}_\ell\}$. Add an arc $\mathbb{C}_i\mathbb{C}_j$ if $v_i \in \partial \mathbb{C}_j$. The indegree in $\mathcal{J}$ of a node $\mathbb{C}_i$ is at most $|\partial \mathbb{C}_i| \leq \gamma$. Thus, every subgraph of the underlying graph of $\mathcal{J}$ has a node of degree $\leq 2\gamma$. A graph is $d$-degenerate if every subgraph of it has a node of degree $\leq d$. It is known that any $d$-degenerate graph is $(d + 1)$-colorable. Hence, $\mathcal{J}$ is $(2\gamma + 1)$-colorable, so its node set can be partitioned into $2\gamma + 1$ independent sets. The bisets in each independent set are inner part disjoint, hence their number is at most $|S|$. Consequently, $\ell \leq (2\gamma + 1)|S|$, as claimed. ∎

Let $I'$ be a cover of $f_S$ in which $|S| = 2\gamma + 1$ and let $g = f^{I'}$ be the residual function of $f$ w.r.t. $I'$. Note that $A \cap S \neq \emptyset$ whenever $g(\mathbb{A}) > 0$, thus Lemma 12.11 gives the bound $|U(g, \gamma)| \leq \gamma(2\gamma + 1)^2$. Thus, if $n \geq \gamma(2\gamma + 1)^2 + 2\gamma + 1$, then there *exists* $R \subseteq V \setminus U(g, \gamma)$ with $|R| = 2\gamma + 1$.

However, no polynomial time algorithm for *finding* $R$ as earlier is known. This can be resolved as follows. The iterative rounding algorithm of Reference 38, when applied on an *arbitrary* biset function $h$, either returns a 2-approximate cover $J$ of $h$, or a **failure certificate**: a pair $\mathbb{A}, \mathbb{B}$ of bisets with $h(\mathbb{A}) > 0$ and $h(\mathbb{B}) > 0$ for which both the supermodular and the co-supermodular inequality does not hold. In the case when $h = g^I$, in which $g$ is symmetric supermodular and $I$ is a cover of $g_R$ with $|R| \geq 2\gamma + 1$, this can happen only if $\mathbb{A}, \mathbb{B}$ mesh (by Corollary 12.2) and $A \cap R, B \cap R$ are both non-empty (as $A \cap R \neq \emptyset$ whenever $g^I(\mathbb{A}) > 0$). As $g^I$ is symmetric, then by interchanging the roles of $\mathbb{A}, \mathbb{A}^*, \mathbb{B}, \mathbb{B}^*$, we can assume that our failure certificate $\mathbb{A}, \mathbb{B}$ satisfies $A \subseteq \partial \mathbb{B}$. Then $A \subseteq U(g, \gamma)$ and $A \cap R \neq \emptyset$, and we can apply the same procedure by choosing a different $R$ from a smaller range set that excludes $A$. Formally, the algorithm is as follows:

---

## ALGORITHM 12.4   Crossing Supermodular Edge-Cover$(\hat{G}, c, f)$
## (Assume $n \geq \gamma(2\gamma + 1)^2 + 2\gamma + 1$)

1   choose $S \subseteq V$ with $|S| = 2\gamma + 1$ and find a 2-approximate cover $I' \subseteq \hat{E}$ of $f_S$
2   $g \leftarrow f^{I'}, \hat{E} \leftarrow \hat{E} \setminus I', U \leftarrow \emptyset, J \leftarrow \text{NIL}$
3   **while** $J = \text{NIL}$ **do**
4   $\quad$ choose $R \subseteq V \setminus U$ with $|R| = 2\gamma + 1$ and find a 2-approximate cover $I \subseteq \hat{E}$ of $g_R$
5   $\quad$ apply the algorithm of Reference 38 on $g^I$ and $\hat{E} \setminus I$
    $\quad\quad$ - if the algorithm returns a failure certificate $\mathbb{A}, \mathbb{B}$ with $A \subseteq \partial\mathbb{B}$ then do $U \leftarrow U \cup A$
    $\quad\quad$ - else, $J \leftarrow$ a 2-approximate cover of $g^I$ computed by the algorithm of Reference 38
6   **return** $I' \cup I \cup J$

---

Clearly, the algorithm computes a feasible solution. Ratio 6 follows from the fact that the solution is a union of 3 edge sets such that each of them is a 2-approximate cover of a function dominated by $f$.

## 12.7 Biset Families and $k$-Connectivity Augmentation Problems

Any $\{0, 1\}$-valued biset function $f$ bijectively corresponds to the **biset family** $\mathcal{F} = \{\mathbb{A} \in \mathcal{V} : f(\mathbb{A}) = 1\}$. We thus use for biset families the same terminology and notation as for biset functions, for example, we say that $J$ covers $\mathcal{F}$ if $d_J(\mathbb{A}) \geq 1$ for all $\mathbb{A} \in \mathcal{F}$, in the Biset-Family Edge-Cover problem we seek a minimum-cost edge-set $E \subseteq \hat{E}$ that covers $\mathcal{F}$, and $\tau(\mathcal{F})$ is the optimal value of the Biset-LP for covering $\mathcal{F}$.

Let $k$-Connectivity Augmentation be the restriction of $(k + 1)$-Connected Subgraph to instances when the input graph contains a $k$-connected spanning subgraph $G$ of cost 0.

---

$k$-Connectivity Augmentation

*Input:* An integer $k$, a $k$-connected graph $G = (V, E)$, and an edge set $\hat{E}$ with costs $\{c_e : e \in \hat{E}\}$.
*Output:* A min-cost augmenting edge set $J \subseteq \hat{E}$ such that $G \cup J$ is $(k + 1)$-connected.

---

Given an instance of $k$-Connectivity Augmentation let us say that $\mathbb{A} \in \mathcal{V}$ is a **tight biset** if $d_G(\mathbb{A}) = 0$, $|\partial\mathbb{A}| = k$, and $\mathbb{A}$ is proper. By Lemma 12.2, $k$-Connectivity Augmentation is equivalent to the Biset-Family Edge-Cover problem with $\mathcal{F} = \{\mathbb{A} \in \mathcal{V} : d_G(\mathbb{A}) = 0, |\partial\mathbb{A}| = k, \mathbb{A} \text{ is proper}\}$ being the family of tight bisets.

Suppose that $k$-Connectivity Augmentation admits ratio $\rho(k)$, and consider the following algorithm:

---

## ALGORITHM 12.5   Sequential Augmentation$(\hat{G}, c, k)$

1   $E \leftarrow \emptyset$
2   **for** $\ell = 0$ **to** $\ell = k - 1$ **do**
3   $\quad$ find a $\rho(\ell)$-approximate solution $J$ to $\ell$-Connectivity Augmentation instance $G = (V, E), \hat{E}$
4   $\quad$ $E \leftarrow E \cup J, \hat{E} \leftarrow \hat{E} \setminus J$
5   **return** $E$

---

Clearly, the algorithm computes a feasible solution for $k$-Connected Subgraph and has ratio $\sum_{\ell=0}^{k-1} \rho(\ell)$. One can show a better ratio if the ratio $\rho(\ell)$ is w.r.t. the Biset-LP. Let us say that Biset-Function Edge-Cover **admits LP-ratio** $\rho$ if there exists a polynomial time algorithm that computes an $f$-cover of cost at most $\rho \cdot \tau(f)$. A similar terminology is used for $k$-Connected Subgraph and $k$-Connectivity Augmentation instances, meaning that the ratio $\rho$ is w.r.t. the Biset-LP for these problems.

**Lemma 12.12** *If $k$-Connectivity Augmentation admits LP-ratio $\rho(k)$, then $k$-Connected Subgraph admits*

$$LP\text{-ratio } \sum_{\ell=0}^{k-1} \frac{\rho(\ell)}{k-\ell}; \text{ thus if } \rho(\ell) \text{ is a nondecreasing function of } \ell, \text{ then the ratio is bounded by } \rho(k-1)H(k).$$

*Proof.* At iteration $\ell$ of Algorithm 12.5, we cover the family $\mathcal{F}_\ell = \{\mathbb{A} \in \mathcal{V} : d_G(\mathbb{A}) = 0, |\partial \mathbb{A}| = \ell, \mathbb{A} \text{ is proper}\}$ by an edge set $J_\ell$ of cost $c(J_\ell) \leq \rho(\ell)\tau(\mathcal{F}_\ell)$. Let $x$ be a feasible solution to the Biset-LP for $f_{k\text{-CS}}$. Then $x(\delta(\mathbb{A})) \geq k - |\partial\mathbb{A}| = k - \ell$ for every $\mathbb{A} \in \mathcal{F}$, hence $\frac{x}{k-\ell}$ is a feasible solution to the Biset-LP for covering $\mathcal{F}_\ell$. Thus, $\tau(\mathcal{F}_\ell) \leq \frac{1}{k-\ell}\tau(f_{k\text{-CS}})$. Consequently, $c(J_\ell) \leq \rho(\ell)\tau(\mathcal{F}_\ell) \leq \rho(\ell) \cdot \frac{1}{k-\ell}\tau(f_{k\text{-CS}})$, and thus $c(E) \leq \sum_{\ell=0}^{\ell=k-1} c(J_\ell) \leq \tau(f_{k\text{-CS}}) \sum_{\ell=0}^{\ell=k-1} \frac{\rho(\ell)}{k-\ell}$, as claimed. ∎

We now establish some properties of tight bisets that will be used later.

**Definition 12.6** *A biset family $\mathcal{F}$ is **intersecting/crossing** if $\mathbb{A} \cap \mathbb{B}, \mathbb{A} \cap \mathbb{B} \in \mathcal{F}$ whenever $\mathbb{A}, \mathbb{B}$ intersect/cross. A crossing family is $p$-**crossing** if whenever $\mathbb{A}, \mathbb{B} \in \mathcal{F}$ intersect and $|A \cup B| \leq n - p - 1$ holds, $\mathbb{A}$ and $\mathbb{B}$ cross.*

For a biset family $\mathcal{F}$, the **co-family** of $\mathcal{F}$ is defined to be $\mathcal{F}^* = \{\mathbb{A}^* : \mathbb{A} \in \mathcal{F}\}$. We will use the following property of the family of tight bisets.

**Lemma 12.13** *Let $\mathcal{F} = \{\mathbb{A} \in \mathcal{V} : d_G(\mathbb{A}) = 0, |\partial\mathbb{A}| = k, \mathbb{A} \text{ is proper}\}$ be the family of tight bisets of a $k$-connected (directed or undirected) graph $G$. Then $\mathcal{F}$ and $\mathcal{F}^*$ are both $k$-crossing.*

*Proof.* We will show that $\mathcal{F}$ is $k$-crossing; the proof for $\mathcal{F}^*$ is similar. Let $\mathbb{A}, \mathbb{B} \in \mathcal{F}$. Note that by Fact 12.1 $d_G(\mathbb{A} \cap \mathbb{B}) = d_G(\mathbb{A} \cup \mathbb{B}) = 0$.

If $\mathbb{A}, \mathbb{B}$ cross, then none of $\mathbb{A} \cap \mathbb{B}, \mathbb{A} \cup \mathbb{B}$ is void or co-void, thus $|\partial(\mathbb{A} \cap \mathbb{B})| \geq k$ and $|\partial(\mathbb{A} \cup \mathbb{B})| \geq k$, as $G$ is $k$-connected. Thus by Fact 12.1 $k + k = |\partial\mathbb{A}| + |\partial\mathbb{B}| = |\partial(\mathbb{A} \cap \mathbb{B})| + |\partial(\mathbb{A} \cup \mathbb{B})| \geq k + k$. Hence $|\partial(\mathbb{A} \cap \mathbb{B})| = k$ and $|\partial(\mathbb{A} \cup \mathbb{B})| = k$, which implies $\mathbb{A} \cap \mathbb{B}, \mathbb{A} \cup \mathbb{B} \in \mathcal{F}$. Thus $\mathcal{F}$ is a crossing family.

Now let $\mathbb{A}, \mathbb{B}$ intersect and satisfy $|A \cup B| \leq n - k - 1$. Note that $\mathbb{A} \cap \mathbb{B}$ is not co-void, and thus $|\partial(\mathbb{A} \cap \mathbb{B})| \geq k$, as $G$ is $k$-connected. If $\mathbb{A}, \mathbb{B}$ do not cross, then $|\partial(\mathbb{A} \cup \mathbb{B})| = n - |A \cup B| \geq k + 1$, and thus $|\partial(\mathbb{A} \cap \mathbb{B})| = |\partial\mathbb{A}| + |\partial\mathbb{B}| - |\partial(\mathbb{A} \cup \mathbb{B})| \leq k - 1$, contradicting that $G$ is $k$-connected. ∎

## 12.8 LP-Ratio $O\left(\ln \frac{n}{n-k}\right)$ for $k$-Connectivity Augmentation

Let $q$ be a parameter eventually set to $q = \lceil \frac{n-k}{2} \rceil$ and let $\mu = \left\lfloor \frac{n}{q+1} \right\rfloor \leq \frac{2n}{n-k}$. We survey the following result:

**Theorem 12.8 (Nutov [16])** *Directed $k$-Connectivity Augmentation admits LP-ratio $H(\mu) + 2$.*

Note that combining Theorem 12.8 with Lemma 12.12 gives ratio $H(k)(H(\mu) + 2) = O\left(\ln k \cdot \ln \frac{n}{n-k}\right)$ for directed $k$-Connected Subgraph, and via Lemma 12.1 also for undirected $k$-Connected Subgraph.

Recall that the $k$-Connectivity Augmentation problem is equivalent to covering the family $\mathcal{F} = \{\mathbb{A} : d_G(\mathbb{A}) = 0, |\partial\mathbb{A}| = k, \mathbb{A} \text{ is proper}\}$ of tight bisets, and that by Lemma 12.13, both $\mathcal{F}$ and $\mathcal{F}^*$ are $k$-crossing. In the rest of this section, we will prove the following general theorem that implies Theorem 12.8.

**Theorem 12.9 (Nutov [16])** Biset-Family Edge-Cover *such that $\mathcal{F}$ and $\mathcal{F}^*$ are both $k$-crossing admits LP-ratio $O(\ln \mu)$. Moreover, if $|\partial\mathbb{A}| \geq k$ for all $\mathbb{A} \in \mathcal{F}$ then the problem admits LP-ratio $H(\mu) + 2$.*

As any crossing family $\mathcal{F}$ is $2\gamma$-crossing, in which $\gamma = \max_{\mathbb{A}\in\mathcal{F}} |\partial\mathbb{A}|$, we get that Biset-Family Edge-Cover with arbitrary crossing $\mathcal{F}$ admits LP-ratio $O\left(\ln \frac{n}{n-2\gamma}\right)$.

In the rest of this section, we prove Theorem 12.9. We need the following definition from Reference 39:

**Definition 12.7** *The inclusionwise minimal members of a biset family $\mathcal{F}$ are called $\mathcal{F}$-**cores**, or simply* **cores**, *if $\mathcal{F}$ is clear from the context. Let $\mathcal{C}(\mathcal{F})$ denote the family of $\mathcal{F}$-cores. For an $\mathcal{F}$-core $\mathbb{C} \in \mathcal{C}(\mathcal{F})$, the* **halo-family** *$\mathcal{F}(\mathbb{C})$ of $\mathbb{C}$ is the family of those members of $\mathcal{F}$ that contain $\mathbb{C}$ and contain no $\mathcal{F}$-core distinct from $\mathbb{C}$.*

The following lemma summarizes the properties of halo-families that we need.

**Lemma 12.14** *For any crossing biset family $\mathcal{F}$ the following holds.*

(i) *For any $\mathcal{F}$-core $\mathbb{C}$, $\mathcal{F}(\mathbb{C})$ is a crossing family and $\mathcal{F}(\mathbb{C})^* = \{\mathbb{A}^* : \mathbb{A} \in \mathcal{F}(\mathbb{C})\}$ is an intersecting family.*

(ii) *Bisets $\mathbb{A}_1$ and $\mathbb{A}_2$ that belong to distinct halo-families do not cross; thus no edge can cover both $\mathbb{A}_1$ and $\mathbb{A}_2$.*

(iii) *For any $\mathcal{F}$-core $\mathbb{C}$, if $J$ is an inclusion minimal edge set that covers $\mathcal{F}(\mathbb{C})$ then $\mathcal{C}(\mathcal{F}^J) = \mathcal{C}(\mathcal{F}) \setminus \{\mathbb{C}\}$.*

*Proof.* We prove (i). Let $\mathbb{A}, \mathbb{B} \in \mathcal{F}(\mathbb{C})$ cross. As $\mathbb{A}\cap\mathbb{B}, \mathbb{A}\cup\mathbb{B} \in \mathcal{F}$. As $\mathbb{A}\cap\mathbb{B} \subseteq \mathbb{A} \subseteq \mathbb{A}\cup\mathbb{B}$ and $\mathbb{A} \in \mathcal{F}(\mathbb{C})$, $\mathbb{A}\cap\mathbb{B} \in \mathcal{F}(\mathbb{C})$ and $\mathbb{C} \subseteq \mathbb{A}\cup\mathbb{B}$. We claim that $\mathbb{A}\cup\mathbb{B}$ contains no core $\mathbb{C}'$ distinct from $\mathbb{C}$. Otherwise, as none of $\mathbb{A}, \mathbb{B}$ can contain $\mathbb{C}'$, we must have that $\mathbb{C}', \mathbb{A}$ cross or $\mathbb{C}', \mathbb{B}$ cross, so $\mathbb{C}'\cap\mathbb{A} \in \mathcal{F}$ or $\mathbb{C}'\cap\mathbb{B} \in \mathcal{F}$; this contradicts that $\mathbb{C}'$ is a core. Thus, $\mathcal{F}(\mathbb{C})$ is a crossing family. We prove that $\mathcal{F}(\mathbb{C})^*$ is an intersecting family. Let $\mathbb{A}, \mathbb{B} \in \mathcal{F}(\mathbb{C})^*$ intersect. Then $\mathbb{C} \subseteq \mathbb{A}^* \cap \mathbb{B}^*$ so $\mathbb{A}, \mathbb{B}$ cross. Thus, as $\mathcal{F}(\mathbb{C})$ is a crossing family, we get that $\mathbb{A}\cap\mathbb{B}, \mathbb{A}\cup\mathbb{B} \in \mathcal{F}(\mathbb{C})^*$.

We prove (ii). Let $\mathbb{A}_1 \in \mathcal{F}(\mathbb{C}_1)$ and $\mathbb{A}_2 \in \mathcal{F}(\mathbb{C}_2)$ cross, for $\mathbb{C}_1, \mathbb{C}_2 \in \mathcal{C}(\mathcal{F})$. Then $\mathbb{A}_1 \cap \mathbb{A}_2 \in \mathcal{F}$, so $\mathbb{A}_1 \cap \mathbb{A}_2$ contains some $\mathcal{F}$-core $\mathbb{C}$. We have $\mathbb{C} = \mathbb{C}_1$ as $\mathbb{C} \subseteq \mathbb{A}_1$ and $\mathbb{C} = \mathbb{C}_2$ as $\mathbb{C} \subseteq \mathbb{A}_2$, hence $\mathbb{C}_1 = \mathbb{C}_2$.

Part (iii) follows from part (ii), as every $e \in J$ covers some biset in $\mathcal{F}(\mathbb{C})$ (by the minimality of $J$) and thus by (ii) cannot cover a core distinct from $\mathbb{C}$. ∎

We need the following theorem for the proof of Theorem 12.9.

**Theorem 12.10 (Fakcharoenphol & Laekhanukit [40] ([16,41]))** *Directed* Biset-Family Edge-Cover *with crossing $\mathcal{F}$ admits a polynomial time algorithm that given $\mathcal{C} \subseteq \mathcal{C}(\mathcal{F})$ computes $J \subseteq \hat{E}$ such that $\mathcal{C}(\mathcal{F}^J) = \mathcal{C}(\mathcal{F}) \setminus \mathcal{C}$ and $c(J) \leq H(|\mathcal{C}|) \cdot \tau(\mathcal{F})$. In particular,* Biset-Family Edge-Cover *with crossing $\mathcal{F}$ admits ratio $H(|\mathcal{C}(\mathcal{F})|)$.*

*Proof.* Consider the following algorithm. Start with a partial solution $J = \emptyset$. While $|\mathcal{C} \cap \mathcal{C}(\mathcal{F}^J)| \geq 1$ continue with iterations. At iteration $i$, compute for each $\mathbb{C} \in \mathcal{C} \cap \mathcal{C}(\mathcal{F}^J)$ an optimal inclusion minimal edge cover $J_\mathbb{C}$ of the family $\mathcal{F}^J(\mathbb{C})$ (the halo-family of $\mathbb{C}$ in $\mathcal{F}^J$); then add to $J$ a minimum-cost edge set $J_i$ among the edge sets $\{J_\mathbb{C} : \mathbb{C} \in \mathcal{C} \cap \mathcal{C}(\mathcal{F}^J)\}$. By part (i) of Lemma 12.14, each $J_\mathbb{C}$ can be computed in polynomial time and $c(J_\mathbb{C}) = \tau(\mathcal{F}^J(\mathbb{C}))$. By part (ii), $\sum_{\mathbb{C}\in\mathcal{C}(\mathcal{F})} c(J_\mathbb{C}) \leq \tau(\mathcal{F})$. Thus, there is $\mathbb{C} \in \mathcal{C}$ such that $c(J_\mathbb{C}) \leq \tau(\mathcal{F})/|\mathcal{C}|$. By part (iii), at iteration $i$ we have $|\mathcal{C}(\mathcal{F}^J) \cap \mathcal{C}| \leq |\mathcal{C}| - i + 1$. Thus, $c(J_i) \leq \tau(\mathcal{F})/(|\mathcal{C}| - i + 1)$ at iteration $i$, and the statement follows. ∎

**Definition 12.8** *A biset family $\mathcal{F}$ is* **intersection closed** *if $\mathbb{A} \cap \mathbb{B} \in \mathcal{F}$ for any intersecting $\mathbb{A}, \mathbb{B} \in \mathcal{F}$. An intersection closed $\mathcal{F}$ is $q$-**semi-intersecting** if $|A| \leq q$ for every $\mathbb{A} \in \mathcal{F}$, and if $\mathbb{A} \cup \mathbb{B} \in \mathcal{F}$ for any intersecting $\mathbb{A}, \mathbb{B} \in \mathcal{F}$ with $|A \cup B| \leq q$. The $q$-**truncated family** of a biset family $\mathcal{F}$ is $\mathcal{F}_{\leq q} := \{\mathbb{A} \in \mathcal{F} : |A| \leq q\}$.*

We obtain a $q$-semi-intersecting family from a $k$-crossing family as follows.

**Lemma 12.15** *If $\mathcal{F}$ is k-crossing and $q \leq \frac{n-k}{2}$ then the q-truncated family $\mathcal{F}_{\leq q}$ of $\mathcal{F}$ is q-semi-intersecting.*

*Proof.* Let $\mathbb{A}, \mathbb{B} \in \mathcal{F}_{\leq q}$ intersect. Then $|A \cup B| \leq |A| + |B| - 1 \leq 2q - 1 \leq n - k - 1$. Thus, $\mathbb{A} \cap \mathbb{B}, \mathbb{A} \cup \mathbb{B} \in \mathcal{F}$, as $\mathcal{F}$ is k-crossing. Hence, $\mathbb{A} \cap \mathbb{B} \in \mathcal{F}_{\leq q}$ (as $|A \cap B| \leq |A| \leq q$), and $\mathbb{A} \cup \mathbb{B} \in \mathcal{F}_{\leq q}$ if $|A \cup B| \leq q$. ∎

Let $\nu(\mathcal{F})$ denote the maximum number of pairwise inner part disjoint bisets in $\mathcal{F}$.

**Theorem 12.11 (Nutov [16])** *Directed Biset-Family Edge-Cover with q-semi-intersecting $\mathcal{F}$ admits a polynomial time algorithm that computes an edge-set $J \subseteq \hat{E}$ such that $\nu(\mathcal{F}^J) \leq \mu$ and $c(J) \leq \tau(\mathcal{F})$.*

Theorem 12.11 will be proved later. From Theorem 12.11 and Lemma 12.15 we have the following.

**Corollary 12.4** *Directed Biset-Family Edge-Cover with k-crossing $\mathcal{F}$ admits a polynomial time algorithm that for $q = \lceil \frac{n-k}{2} \rceil$ computes $J \subseteq E$ such that $\nu(\mathcal{F}_{\leq q}^J) \leq \mu$ and $c(J) \leq \tau(\mathcal{F}_{\leq q})$.*

A slightly weaker version of the following statement was proved in Reference 42.

**Lemma 12.16 ([42])** *If both $\mathcal{F}$ and $\mathcal{F}^*$ are k-crossing, then $|\mathcal{C}(\mathcal{F})| \leq \nu(\mathcal{F}_{\leq q}) + \nu(\mathcal{F}_{\leq q}^*) + \mu^2 H(\mu)$.*

*Proof.* We show that if $\mathcal{C} \subseteq \mathcal{F}$ has no two bisets that cross then $|\mathcal{C}| \leq \nu(\mathcal{F}_{\leq q}) + \nu(\mathcal{F}_{\leq q}^*) + \mu^2 H(\mu)$. Let $\mathcal{B} = \{\mathbb{A} \in \mathcal{C} : |A|, |A^*| \geq q + 1\}$. Clearly, $|\mathcal{C}| \leq |\mathcal{C}_{\leq q}| + |\mathcal{C}_{\leq q}^*| + |\mathcal{B}|$. Note that $|\mathcal{C}_{\leq q}| \leq \nu(\mathcal{F}_{\leq q})$, as $\mathcal{F}_{\leq q}$ is intersection closed, by Lemma 12.15. Similarly, $|\mathcal{C}_{\leq q}^*| \leq \nu(\mathcal{F}_{\leq q}^*)$. To see that $|\mathcal{B}| \leq \mu^2 H(\mu)$, note that

(i) $\Delta(\mathcal{B}) \leq \mu$, where $\Delta(\mathcal{B})$ is the maximum degree in the hypergraph $\mathcal{B}^{in}$ formed by the inner parts of the bisets in $\mathcal{B}$. This is so as no two bisets in $\mathcal{B}$ cross, and thus for any $v \in V$ the sets in the family $\{A^* : \mathbb{A} \in \mathcal{B}, v \in A\}$ of $\mathcal{B}^*$ are pairwise disjoint; hence their number is at most $\nu(\mathcal{B}^*) \leq \left\lfloor \frac{n}{q+1} \right\rfloor = \mu$.

(ii) The hypergraph $\mathcal{B}^{in}$ has a hitting-set $U$ of size $|U| \leq \mu H(\Delta(\mathcal{B})) \leq \mu H(\mu)$. This is so as this hypergraph has a fractional hitting-set $h$ of value $\mu$ defined by $h(v) = \frac{1}{q+1}$ for all $v \in V$.

As $|\mathcal{B}| \leq |U| \cdot \Delta(\mathcal{B})$ for any hitting-set $U$ of $\mathcal{B}^{in}$, the bound $|\mathcal{B}| \leq \mu^2 H(\mu)$ follows. ∎

The algorithm is as follows.

---

## ALGORITHM 12.6   Directed-Cover1 $(\mathcal{F}, \hat{G}, c)$ ($\mathcal{F}, \mathcal{F}^*$ are both k-crossing)

1 **compute** $J_1 \subseteq E$ with $\nu(\mathcal{F}_{\leq q}^{J_1}) \leq \mu$ and $c(J_1) \leq \tau(\mathcal{F}_{\leq q})$ using the algorithm from **Corollary 12.4**; **compute** a similar edge set $J_1^* \subseteq E$ for the family $\mathcal{F}_{\leq q}^*$.
2 **compute** $J_2 \subseteq E$ covering $\mathcal{F}^{J_1 \cup J_1^*}$ using the algorithm from **Theorem 12.10**.
3 **return** $J = J_1 \cup J_1^* \cup J_2$.

---

By Lemma 12.16, $\left| \mathcal{C}\left(\mathcal{F}^{J_1 \cup J_1^*}\right) \right| = O(\mu^2 \ln \mu)$ and thus $c(J_2) = \tau(\mathcal{F}) O(\ln \mu)$. Consequently, the cost of the solution computed is bounded by $\tau(\mathcal{F})(c(J_1) + c(J_1^*) + c(J_2)) \leq \tau(\mathcal{F})(1 + 1 + O(\ln \mu)) = O(\ln \mu)$.

In the case when $|\partial \mathbb{A}| \geq k$ for all $\mathbb{A} \in \mathcal{F}$, we get a slightly better ratio $H(\mu) + 2$ by the following observation.

**Lemma 12.17** *Let $\mathcal{F}$ be a biset family such that $\mathcal{F}^*$ is k-crossing and $|\partial \mathbb{A}| \geq k$ for all $\mathbb{A} \in \mathcal{F}$. Let $q = \lceil \frac{n-k}{2} \rceil$. If $\mathcal{F}_{\leq q} = \emptyset$ then $\mathcal{F}^*$ is an intersecting family;*

*Proof.* If $|A \cup B| \leq n - k - 1$, then $\mathbb{A} \cap \mathbb{B}, \mathbb{A} \cup \mathbb{B} \in \mathcal{F}^*$, as $\mathcal{F}^*$ is k-regular. Assume that $|A \cup B| \geq n - k$. Then $\max\{|A|, |B|\} \geq \frac{n-k}{2}$, say $|A| \geq \frac{n-k}{2}$. This implies $|A^*| \leq n - k - |A| \leq \frac{n-k}{2}$, so $\mathbb{A}^* \in \mathcal{F}_{\leq q}$, contradicting that $\mathcal{F}_{\leq q} = \emptyset$. ∎

---

## ALGORITHM 12.7   Directed-Cover2($\mathcal{F},\hat{G},c$) ($\mathcal{F}$ and $\mathcal{F}^*$ are both $k$-crossing, $|\partial\mathbb{A}| \geq k$ for all $\mathbb{A} \in \mathcal{F}$)

1  compute $J_1 \subseteq \hat{E}$ with $v(\mathcal{F}^{J_1}_{\leq q}) \leq \mu$ and $c(J_1) \leq \tau(\mathcal{F}_{\leq q})$ using the algorithm from **Corollary 12.4**.
2  compute $J_2 \subseteq \hat{E}$ covering $\mathcal{F}^{J_1}_{\leq q}$ with $c(J_2) \leq H(\mu)\tau\left(\mathcal{F}^{J_1}\right)$ using the algorithm from **Theorem 12.10**.
3  compute $J_3 \subseteq \hat{E}$ covering $\mathcal{F}^{J_1 \cup J_2}$ with $c(J_3) \leq \tau\left(\mathcal{F}^{J_1 \cup J_2}\right)$ using the algorithm from **Lemma 12.17**.
4  return $J_1 \cup J_2 \cup J_3$.

---

## ALGORITHM 12.8   $q$-Semi-Intersecting Family Edge-Cover($\mathcal{F},G,c$)

1  $J \leftarrow \emptyset, y \leftarrow 0, \mathcal{L} \leftarrow \emptyset$.
2  **while** $v(\mathcal{F}^J) \geq 1$ **do**
3  $\quad$ add some $\mathbb{C} \in \mathcal{C}(\mathcal{F}^J)$ to $\mathcal{L}$
4  $\quad$ raise $y_{\mathbb{C}}$ until the dual constraint of some $e \in \delta_{\hat{E}\setminus J}(\mathbb{C})$ becomes tight and add $e$ to $J$
5  Let $e_1, \ldots, e_j$ be the order in which the edges were added to $J$
6  **for** $i = j$ **downto** 1 **do**
7  $\quad$ if $J \setminus \{e_i\}$ covers the family $\mathcal{F}' = \{\mathbb{A} \in \mathcal{F} : \mathbb{A} \subseteq \mathbb{B} \text{ for some } \mathbb{B} \in \mathcal{L}\}$ then do $J \leftarrow J \setminus \{e_i\}$
8  return $J$

---

Relying on Corollary 12.4, Theorem 12.10, and Lemma 12.17, it is easy to see that Algorithm 12.7 computes a feasible solution of cost at most $\tau(\mathcal{F})(H(\mu) + 2)$.

In the rest of this section, we prove Theorem 12.11. Consider the dual program of the Biset-LP for covering $\mathcal{F}$ and the primal-dual Algorithm 12.8 for covering $\mathcal{F}$.

$$\max\left\{\sum_{\mathbb{A}\in\mathcal{V}} y_{\mathbb{A}} : \sum_{\delta(\mathbb{A})\ni e} y_{\mathbb{A}} \leq c_e \; \forall e \in E, \; y_{\mathbb{A}} \geq 0 \; \forall \mathbb{A} \in \mathcal{V}\right\}$$

Let $I$ denote the set of edges in $J$ right before the reverse-delete phase (steps 5,6, and 7). Note that $I$ covers $\mathcal{F}$, but in the reverse-delete phase we care to cover just the subfamily $\mathcal{F}'$ of $\mathcal{F}$. In fact, the algorithm coincides with a standard primal-dual algorithm for covering the biset family $\mathcal{F}'$. We will show that $\mathcal{F}'$ is an intersecting biset family and conclude that $c(J) = \tau(\mathcal{F}') \leq \tau(\mathcal{F})$. In what follows, let $\mathcal{M}$ denote the family of inclusionwise maximal members of $\mathcal{L}$, and for an $\mathcal{F}^J$-core $\mathbb{C}_i$ let $\mathcal{M}_i$ denote the family of bisets in $\mathcal{M}$ that intersect with $\mathbb{C}_i$, and $\mathbb{B}_i$ the union of $\mathbb{C}_i$ and the bisets in $\mathcal{M}_i$.

Note that each family $\mathcal{M}_i$ is non-empty, as $\mathbb{C}_i$ is covered by some edge $e \in I \setminus J$, and as any edge $e \in I$ covers some $\mathbb{A} \in \mathcal{L}$. Let us say that a biset family $\mathcal{L}$ is **laminar** if for any $\mathbb{A}, \mathbb{B} \in \mathcal{L}$ that intersect $\mathbb{A} \subseteq \mathbb{B}$ or $\mathbb{B} \subseteq \mathbb{A}$ holds. In the following lemma, we establish some properties of the families $\mathcal{L}$ and $\mathcal{F}'$.

**Lemma 12.18** *At the end of the algorithm the following holds*

(i) *$\mathcal{L}$ is a laminar biset family and $\mathcal{F}'$ is an intersecting biset family.*
(ii) *For any $\mathbb{A} \in \mathcal{M}$ there is a unique edge $e_{\mathbb{A}}$ in $I$ that covers $\mathbb{A}$, and $e_{\mathbb{A}} \in J$. Furthermore, if $\mathbb{A}$ and an $\mathcal{F}^J$-core $\mathbb{C}$ intersect, then $\delta_J(\mathbb{A} \cap \mathbb{C}) = \{e_{\mathbb{A}}\}$.*

*Proof.* We prove (i). Let $\mathbb{A}_1, \mathbb{A}_2 \in \mathcal{L}$ intersect in which $\mathbb{A}_1$ was added to $\mathcal{L}$ before $\mathbb{A}_2$. When $\mathbb{A}_1$ was added to $\mathcal{L}$, we had $\mathbb{A}_1 \in \mathcal{C}(\mathcal{F}^J)$ and $\mathbb{A}_2 \in \mathcal{F}^J$. Thus, $\mathbb{A}_1 \cap \mathbb{A}_2 = \mathbb{A}_1$ (namely, $\mathbb{A}_1 \subseteq \mathbb{A}_2$) by the minimality of $\mathbb{A}_1$ and as $\mathcal{F}$ (and thus, also $\mathcal{F}^J$) is intersection closed. This implies that $\mathcal{L}$ is laminar. We show that $\mathcal{F}'$ is an intersecting biset family. Let $\mathbb{A}_1, \mathbb{A}_2 \in \mathcal{F}'$ intersect. Then, as $\mathcal{L}$ is laminar, $\mathbb{A}_1 \cup \mathbb{A}_2 \subseteq \mathbb{B}$ for some $\mathbb{B} \in \mathcal{L}$.

Thus, $\mathbb{A}_1 \cup \mathbb{A}_2 \in \mathcal{F}$, as $|A_1 \cup A_2| \le |B| \le q$ and as $\mathcal{F}$ is $q$-semi-intersecting. This implies $\mathbb{A}_1 \cup \mathbb{A}_2 \in \mathcal{F}'$, and clearly $\mathbb{A}_1 \cap \mathbb{A}_2 \in \mathcal{F}'$ as $\mathbb{A}_1 \cap \mathbb{A}_2 \subseteq \mathbb{B}$ and as $\mathcal{F}$ is intersection closed.

We prove (ii). Let $e_{\mathbb{A}}$ be the edge that was added to $J$ at step 4 of the algorithm after $\mathbb{A}$ was added to $\mathcal{L}$ at step 3 (the first edge that covered $\mathbb{A}$). After $\mathbb{A}$ was added to $\mathcal{L}$, no biset that intersects with $\mathbb{A}$ was added to $\mathcal{L}$, as $\mathbb{A} \in \mathcal{M}$ and as $\mathcal{L}$ is laminar. Thus, edges added to $J$ after $e_{\mathbb{A}}$ do not cover $\mathbb{A}$, as their tails are in $V \setminus A$. Consequently, $e_{\mathbb{A}}$ is the unique edge in $I$ that covers $\mathbb{A}$, and thus, $e_{\mathbb{A}} \in J$. Now suppose that $\mathbb{A}$ and an $\mathcal{F}^J$-core $\mathbb{C}$ intersect. Then $\mathbb{A} \cap \mathbb{C} \in \mathcal{F}'$, as $\mathcal{F}$ is intersection closed and as $\mathbb{A} \cap \mathbb{C} \subseteq \mathbb{A}$. Thus, $\delta_J(\mathbb{A} \cap \mathbb{C}) \ne \emptyset$. Let $e \in \delta_J(\mathbb{A} \cap \mathbb{C})$. Then $e$ covers $\mathbb{A}$, as $e$ covers $\mathbb{A}$ or $\mathbb{C}$ by Fact 12.1, but $e$ cannot cover $C$ as $e \in J$ and $J$ does not cover $\mathbb{C}$. Thus, $e = e_{\mathbb{A}}$ for any $e \in \delta_J(\mathbb{A} \cap \mathbb{C})$, namely, $\delta_J(\mathbb{A} \cap \mathbb{C}) = \{e_{\mathbb{A}}\}$. ∎

**Lemma 12.19** *If $\mathcal{F}$ is $q$-semi-intersecting then at the end of the algorithm the following holds*

(i) $|\delta_J(\mathbb{A})| = 1$ *for any $\mathbb{A} \in \mathcal{L}$.*
(ii) *The sets $B_i$ are pairwise disjoint and each of them has size $\ge q + 1$.*

*Proof.* For part (i), let $\mathbb{A} \in \mathcal{L}$ and suppose to the contrary that there are $e_1, e_2 \in \delta_J(\mathbb{A})$ with $e_1 \ne e_2$. For $i = 1, 2$ let $\mathbb{A}_i$ be some biset in $\mathcal{F}'$ that became uncovered when $e_i$ was considered for deletion at step 7. Note that $\delta_J(\mathbb{A}_i) = \{e_i\}$ and that $\mathbb{A} \subseteq \mathbb{A}_i$, as the edges in $J$ were considered for deletion in the reverse order. Thus, $\mathbb{A} \subseteq \mathbb{A}_1 \cap \mathbb{A}_2$, and by Lemma 12.18(i) $\mathbb{A}_1 \cup \mathbb{A}_2 \in \mathcal{F}'$. Consequently, there is $e \in \delta_J(\mathbb{A}_1 \cup \mathbb{A}_2)$, hence $e \in \delta_J(\mathbb{A}_1)$ or $e \in \delta_J(\mathbb{A}_2)$, by Fact 12.1. Thus, $e = e_1$ or $e = e_2$. As the tail of each of $e_1, e_2$ is in $A \subseteq A_1 \cap A_2$, so is the tail of $e$. The head of $e$ is in $A_1^* \cap A_2^*$. This gives the contradiction $e \in \delta_J(\mathbb{A}_1) \cap \delta_J(\mathbb{A}_2)$.

We prove part (ii). Let $\mathbb{C}_i, \mathbb{C}_j$ be distinct $\mathcal{F}^J$-cores. Note that no two bisets in $\mathcal{M}$ intersect (as $\mathcal{L}$ is laminar) and that $C_i \cap C_j = \emptyset$ (as $\mathcal{F}$ is intersection closed). Thus, to prove that $B_i \cap B_j = \emptyset$, it is sufficient to prove that $\mathcal{M}_i \cap \mathcal{M}_j = \emptyset$. Suppose to the contrary that there is $\mathbb{A} \in \mathcal{M}_i \cap \mathcal{M}_j$. By Lemma 12.18(ii), the tail of $e_{\mathbb{A}}$ is both in $A \cap C_i$ and $A \cap C_j$. This contradicts $C_i \cap C_j = \emptyset$. We prove that $|B_i| \ge q + 1$. Note that $|B_i| \le q$ implies $\mathbb{B}_i \in \mathcal{F}$, as $\mathcal{F}$ is $q$-semi-intersecting. Thus, to prove that $|B_i| \ge q + 1$, it is sufficient to prove that $\delta_I(\mathbb{B}_i) = \emptyset$, as this implies $\mathbb{B}_i \notin \mathcal{F}$ (as $I$ covers $\mathcal{F}$). Suppose to the contrary that there is $e \in \delta_I(\mathbb{B}_i)$. Then there is a biset $\mathbb{A} \in \mathcal{M}$ whose inner part contains the tail of $e$, and we must have $\mathbb{A} \in \mathcal{M}_i$, by the definition of $\mathbb{B}_i$ and as no two bisets in $\mathcal{M}$ intersect. As $e$ covers the biset $\mathbb{B}_i$ that contains $\mathbb{A}$, $e$ covers $\mathbb{A}$, and thus, $e = e_{\mathbb{A}}$ and $\delta_J(\mathbb{A} \cap \mathbb{C}_i) = \{e_{\mathbb{A}}\}$, by Lemma 12.18(ii). The edge $e_{\mathbb{A}}$ has its tail in $C_i$ and covers the biset $\mathbb{B}_i$ that contains $\mathbb{C}_i$. Consequently, $e_{\mathbb{A}}$ covers $\mathbb{C}_i$, contradicting that $\mathbb{C}_i \in \mathcal{F}^J$. ∎

Lemma 12.19(ii) implies $\nu(\mathcal{F}^J) \le \lfloor n/(q+1) \rfloor$. To see that $c(J) = \tau(\mathcal{F}')$ let $x \in \{0, 1\}^F$ be the characteristic vector of $J$ and $y$ the dual solution produced by the algorithm. It is easy to see that $x$ and $y$ are feasible solutions for the primal and dual LPs, respectively, and that the Primal Complementary Slackness Conditions hold for $x$ and $y$. The Dual Complementary Slackness Conditions are as follows: $|\delta_J(\mathbb{A})| = 1$ whenever $y_{\mathbb{A}} > 0$, and they hold by Lemma 12.19(i), as $\{\mathbb{A} : y_{\mathbb{A}} > 0\} \subseteq \mathcal{L}$.

This concludes the proof of Theorem 12.11, and thus, the proof of Theorem 12.9 is also complete.

## 12.9 Open Problems

In the current section, we list some open problems in the field.

*k-Connected Subgraph with $\{0, 1\}$ costs.*

Can this problem be solved in polynomial time? A polynomial time algorithm is known for the augmentation version of increasing the connectivity by 1.

*Min-Size k-Connected Subgraph ($\{1, \infty\}$-costs).*

The best known ratios for this problem are $1 - \frac{1}{k} + \frac{n}{\mathrm{opt}} \le 1 + \frac{1}{k}$ for undirected graphs and $1 - \frac{1}{k} + \frac{2n}{\mathrm{opt}} \le 1 + \frac{1}{k}$ for directed graphs. In both cases, the ratio is better than $1 + \frac{1}{k}$, unless an optimal solution is a $k$-regular graph. Can the ratio $1 + \frac{1}{k}$ be improved? Can it be improved for $k = 2$? Such improvements are known in the undirected edge-connectivity case.

*k-Connected Subgraph with general costs.*

The best known ratios are as follows: $O\left(\ln k \ln \frac{n}{n-k}\right)$, for undirected graphs with $n = \Omega(k^3)$ ratio 6 is achievable, and for the augmentation version of increasing the connectivity by 1 ratio $O\left(\ln \frac{n}{n-k}\right)$ is known. One main open question here is whether the later problem admits a constant ratio. Another open question is whether the ratio $O\left(\ln \frac{n}{n-k}\right)$ extends to the general version of the problem.

# References

1. A. Frank. *Connections in Combinatorial Optimization.* Oxford University Press, Oxford, UK, 2011.
2. A. Frank and T. Kiraly. A survey on covering supermodular functions. In W. Cook, L. Lovász, and J. Vygen, Eds., *Research Trends in Combinatorial Optimization*, pp. 87–126. Springer, Berlin, Germany, 2009.
3. S. Khuller. Approximation algorithms for finding highly connected subgraphs. In D. Hochbaum, Ed., *Approximation Algorithms for NP-hard Problems*, chapter 6, pp. 236–265. PWS, Boston, MA, 1995.
4. G. Kortsarz and Z. Nutov. Approximating minimum cost connectivity problems. In T. Gonzalez, Ed., *Approximation Algorithms and Metaheuristics*, chapter 58. Chapman & Hall, Boca Raton, FL, 2007.
5. A. Frank and T. Jordán. Minimal edge-coverings of pairs of sets. *J. Comb. Theor. B*, 65:73–110, 1995.
6. B. Jackson and T. Jordán. A near optimal algorithm for vertex connectivity augmentation. In *ISAAC*, pp. 313–325, 2000.
7. T. Watanabe and A. Nakamura. Edge-connectivity augmentation problems. *Comput. Syst. Sci.*, 35(1):96–144, 1987.
8. A. Frank. Augmenting graphs to meet edge-connectivity requirements. *SIAM J. Disc. Math.*, 5(1):25–53, 1992.
9. J. Cheriyan and R. Thurimella. Approximating minimum-size $k$-connected spanning subgraphs via matching. *SIAM J. Comput.*, 30(2):528–560, 2000.
10. Z. Nutov. Small $\ell$-edge-covers in $k$-connected graphs. *Disc. Appl. Math.*, 161(13–14):2101–2106, 2013.
11. H. Gabow and S. Gallagher. Iterated rounding algorithms for the smallest $k$-edge connected spanning subgraph. *SIAM J. Comput.*, 41(1):61–103, 2012.
12. B. Laekhanukit, S. Gharan, and M. Singh. A rounding by sampling approach to the minimum size $k$-arc connected subgraph problem. In *ICALP (1)*, pp. 606–616, 2012.
13. H. Gabow, M. Goemans, E. Tardos, and D. Williamson. Approximating the smallest $k$-edge connected spanning subgraph by LP-rounding. *Networks*, 54(4):345–357, 2009.
14. G. Kortsarz and Z. Nutov. Approximating node-connectivity problems via set covers. *Algorithmica*, 37:75–92, 2003.
15. S. Khuller and U. Vishkin. Biconnectivity approximations and graph carvings. *J. Assoc. Comput. Mach.*, 41(2):214–235, 1994.
16. Z. Nutov. Approximating minimum-cost edge-covers of crossing biset-families. *Combinatorica*, 34(1):95–114, 2014.
17. J. Cheriyan and L. Végh. Approximating minimum-cost $k$-node connected subgraphs via independence-free graphs. *SIAM J. Comput.*, 43(4):1342–1362, 2014.
18. T. Fukunaga, Z. Nutov, and R. Ravi. Iterative rounding approximation algorithms for degree-bounded node-connectivity network design. *SIAM J. Comput.*, 44(5):1202–1229, 2015.

19. V. Auletta, Y. Dinitz, Z. Nutov, and D. Parente. A 2-approximation algorithm for finding an optimum 3-vertex-connected spanning subgraph. *J. Algorithms*, 32(1):21–30, 1999.

20. Y. Dinitz and Z. Nutov. A 3-approximation algorithm for finding optimum 4, 5-vertex-connected spanning subgraphs. *J. Algorithms*, 32(1):31–40, 1999.

21. B. Jackson and T. Jordán. Independence free graphs and vertex connectivity augmentation. *J. Comb. Theor. B*, 94:31–77, 2005.

22. L. Végh. Augmenting undirected node-connectivity by one. *SIAM J. Disc. Math.*, 25(2):695–718, 2011.

23. A. Vetta. Approximating the minimum strongly connected subgraph via a matching lower bound. In *SODA*, pp. 417–426, 2001.

24. A. Sebo and J. Vygen. Shorter tours by nicer ears: 7/5-approximation for the graph-TSP, 3/2 for the path version, and 4/3 for two-edge-connected subgraphs. *Combinatorica*, 2014.

25. G. Fredrickson and J. Jájá. On the relationship between the biconnectivity augmentation and traveling salesman problem. *Theor. Comput. Sci.*, 19(2):189–201, 1982.

26. G. Kortsarz and Z. Nutov. A simplified 1.5-approximation algorithm for augmenting edge-connectivity of a graph from 1 to 2. *ACM Trans. Algorithms*, 12(2):23, 2016.

27. A. Schrijver. *Combinatorial Optimization, Polyhedra and Efficiency*. Springer-Verlag, Berlin, Germany, 2003.

28. J. Edmonds. Matroid intersection. *Ann. Disc. Math.*, 4:185–204, 1979.

29. J. Edmonds and R. Giles. A min–max relation for submodular functions on graphs. *Ann. Disc. Math.*, 1:185–204, 1977.

30. A. Frank and É. Tardos. An application of submodular flows. *Linear Algebra Appl.*, 114/115: 329–348, 1989.

31. A. Frank. Rooted *k*-connections in digraphs. *Disc. Appl. Math.*, 157(6):1242–1254, 2009.

32. Y. Lando and Z. Nutov. Inapproximability of survivable networks. *Theor. Comput. Sci.*, 410(21–23):2122–2125, 2009.

33. Z. Nutov. Node-connectivity survivable network problems. In T. Gonzalez, Ed., *Approximation Algorithms and Metaheuristics*, chapter 13, Chapman & Hall, 2018.

34. W. Mader. Ecken vom grad *n* in minimalen n-fach zusammenhängenden graphen. *Arch. Math.*, 23:219–224, 1972.

35. W. Mader. Minimal *n*-fach in minimalen n-fach zusammenhängenden digraphen. *J. Comb. Theor. B*, 38:102–117, 1985.

36. J. David and Z. Nutov. Approximating survivable networks with $\beta$-metric costs. *J. Disc. Algorithms*, 9(2):170–175, 2011.

37. S. Khuller and B. Raghavachari. Improved approximation algorithms for uniform connectivity problems. *J. Algorithms*, 21:434–450, 1996.

38. L. Fleischer, K. Jain, and D. Williamson. Iterative rounding 2-approximation algorithms for minimum-cost vertex connectivity problems. *J. Comput. Syst. Sci.*, 72(5):838–867, 2006.

39. G. Kortsarz and Z. Nutov. Approximating *k*-node connected subgraphs via critical graphs. *SIAM J. Comput.*, 35(1):247–257, 2005.

40. J. Fakcharoenphol and B. Laekhanukit. An $O(\log^2 k)$-approximation algorithm for the *k*-vertex connected spanning subgraph problem. *SIAM J. Comput.*, 41(5):1095–1109, 2012.

41. J. Cheriyan and B. Laekhanukit. Approximation algorithms for minimum-cost *k*-(*s, t*) connected digraphs. *SIAM J. Disc. Math.*, 27(3):1450–1481, 2013.

42. Z. Nutov. Approximating subset *k*-connectivity problems. *J. Disc. Algorithms*, 17:51–59, 2012.

# 13

# Node-Connectivity Survivable Network Problems

Zeev Nutov

## 13.1 Introduction

We survey approximation algorithms and hardness of approximation results for the Survivable Network problem in which we seek a low edge-cost directed/undirected subgraph that satisfies prescribed connectivity demands w.r.t. a given "connectivity measure." These problems include the following well-known problems: Minimum Spanning Tree, Traveling Salesman Problem, Steiner Tree, Steiner Forest, and their directed variants. These are examples of *low connectivity* Survivable Network problems. In this survey, we will consider *high-connectivity* Survivable Network problems; some examples are Min-Cost $k$-Flow, $k$-Inconnected Subgraph, $k$-Connected Subgraph, and Rooted Survivable Network. See previous surveys on such problem in References 1 and 2.

Many common connectivity measures can be defined by the following unified framework. Let $G = (V, E)$ be a (possibly directed) graph. For $Q \subseteq V$, the $Q$-**connectivity** $\lambda_G^Q(s, t)$ of a node pair $(s, t)$ is the maximum number of $st$-paths such that no two of them have an edge or a node in $Q \setminus \{s, t\}$ in common. The case $Q = \emptyset$ is the case of **edge-connectivity**, and we use the notation $\lambda_G(s, t) := \lambda_G^\emptyset(s, t)$; the case $Q = V$ is the case of **node-connectivity**, and we use the notation $\kappa_G(s, t) := \lambda_G^V(s, t)$. Even more generally, given node capacities $\{q_v : v \in V\}$, the $q$-**connectivity** $\lambda_G^q(s, t)$ is the maximum number of pairwise edge disjoint $st$-paths such that for every $v \in V \setminus \{s, t\}$ at most $q_v$ of the paths contain $v$; $Q$-connectivity is the particular case when $q_v \in \{1, \infty\}$ for all $v \in V$ and $Q = \{v \in V : q_v = 1\}$. We will consider mainly the node-connectivity case $Q = V$, when the paths are required to be pairwise internally node disjoint. However, most algorithms presented can be adjusted to the $q$-connectivity case with the same approximation ratio.

Given positive integral connectivity **demands** $\{r_{st} \geq 1 : st \in D\}$ over a set $D$ of demand pairs on $V$ and $Q \subseteq V$, we say that $G$ **satisfies** $r$ if $\lambda_G^Q(s,t) \geq r_{st}$ for all $st \in D$. Our problem is:

---

Survivable Network

*Input:* A (multi)graph $\hat{G} = (V, \hat{E})$ with edge costs $\{c_e : e \in \hat{E}\}$, $Q \subseteq V$, and connectivity demands $\{r_{st} > 0 : st \in D\}$ on a set $D \subseteq V \times V$ of demand pairs.
*Output:* A minimum cost subgraph of $\hat{G}$ that satisfies $r$.

---

Assume that the input numbers are integers bounded by a polynomial in $n = |V|$. A node is a **terminal** if it belongs to some demand pair. Let $T$ denote the set of terminals. An important $Q$-connectivity measure is **element-connectivity**, when $Q \subseteq V \setminus T$. Let $k$-Survivable Network be the restriction of Survivable Network to instances with demands at most $k$, and unless stated otherwise $k = \max\limits_{st \in D} r_{st}$ is the maximum demand.

We may assume that the input graph $\hat{G}$ is complete, by assigning infinite costs to "forbidden" edges. Under this assumption, we consider four types of edge costs:

- $\{0,1\}$-*costs*: Here we are given a graph $G$, and the goal is to find a minimum size augmenting edge set $J$ of new edges (any edge is allowed, and parallel edges are allowed) such that $G \cup J$ satisfies $r$.
- $\{1,\infty\}$-*costs*: Here we seek a subgraph of $\hat{G}$ with minimum number of edges that satisfies $r$.
- *Metric costs*: The costs satisfy the triangle inequality $c_{uv} \leq c_{uw} + c_{wv}$ for all $u, v, w \in V$.
- *General costs*: The costs are arbitrary nonnegative integers or $\infty$.

For each type of costs, we have the following types of demands:

- *Uniform demands*: $r_{st} = k$ for all $s, t \in V$; this is the $k$-Connected Subgraph problem.
- *Rooted demands*: All pairs in $D$ share the same node $s$; this is the Rooted Survivable Network problem. If the demands are $r_{ts} = k$ for all $t \in T \setminus \{s\}$, we get **rooted subset uniform demands** and the Subset $k$-Inconnected Subgraph problem.
- *Subset uniform demands*: $r_{st} = k$ for all $s, t \in T$; this is the Subset $k$-Connected Subgraph problem.
- *Arbitrary (nonnegative) demands*.

We mention two important cases of Rooted Survivable Network. In the Min-Cost $k$-Flow problem $|D| = 1$. This problem admits a polynomial time algorithm for both directed and undirected graphs cf. [3,4]. Another case is the $k$-Inconnected Subgraph problem where the demands are $r_{ts} = k$ for every $t \in V \setminus \{s\}$; this is a particular case of the Subset $k$-Inconnected Subgraph problem, when $T = V$. For directed graphs, $k$-Inconnected Subgraph can be solved in polynomial time [5] (see also [6]).

Survivable Network admits a trivial approximation ratio $\min\{|D|, O(n)\}$. Ratio $|D|$ is obtained by applying the above-mentioned Min-Cost $k$-Flow algorithm for each pair in $D$ and taking the union of the $|D|$ computed edge sets. Ratio $O(n)$ is obtained by randomized rounding of an LP-relaxation, see Section 13.2.

In Table 13.1, we list famous particular cases of 1-Survivable Network. For a survey on the $k$-Connected Subgraph problem (the case of uniform demands) see [21]. For the metric costs case, see the original paper of Cheriyan and Vetta [22]. Here we consider all the other variants of the problem. We focus on *node-connectivity* and $k$ arbitrary, although there are many interesting results for edge-connectivity and element-connectivity, as well as for small requirements. We survey only *approximation algorithms* (for polynomially solvable cases see [3,23]), with the currently best known approximation ratios, as summarized in Table 13.2, in which, for comparison, we also included results for edge-connectivity.

**TABLE 13.1** Known Approximability Status of 1-Survivable Network Problems

| Problem | Demands | Approximability |
|---|---|---|
| Directed/undirected Shortest Path (Min-Cost 1-Flow) | Rooted, $|D| = 1$ | in P |
| Minimum Spanning Tree (undirected 1-Connected Subgraph) | Uniform/rooted, $T = V$ | in P |
| Steiner Tree (undirected Subset 1-Connected Subgraph) | Rooted/subset uniform | $\ln 4 + \varepsilon$ [7] ([8]) |
| Steiner Forest (undirected 1-Survivable Network) | General | 2 [9] ([10]) |
| Minimum Arborescence (directed 1-Inconnected Subgraph) | Rooted, $T = V$ | in P |
| Strongly Connected Subgraph (directed 1-Connected Subgraph) | Uniform | 2 for general costs <br> 3/2 for $\{1, \infty\}$-costs [11] |
| Directed Steiner Tree (directed Rooted 1-Survivable Network) | Rooted uniform | $O\left(\ell^3 \lvert T \rvert^{2/\ell}\right)$ in $O\left(\lvert T \rvert^{2\ell} n^\ell\right)$ time [12–15] <br> $\Omega(\ln^2 n)$ [16] |
| Directed Steiner Forest (directed 1-Survivable Network) | General | $n^\varepsilon \cdot \min\left\{\lvert D \rvert^{1/2}, n^{2/3}\right\}$ [17,18] ([19]) <br> $\Omega\left(2^{\ln^{1-\epsilon} n}\right)$ [20] |

References in parenthesis give a simplified proof, or a slight improvement needed to achieve the approximation ratio/threshold stated.

**TABLE 13.2** Known Approximability Status of Survivable Network Problems

| $c,r$ | Node-Connectivity | | Edge-Connectivity | |
|---|---|---|---|---|
| | Undirected | Directed | Undirected | Directed |
| $\{0,1\}$,R,RU | $O(\min\{\ln^2 k, \ln n\})$ [24] <br> $\Omega(\ln n)$ [25] | $O(\ln n)$ [24,25] <br> $\Omega(\ln n)$ [26] | in P [26] | $O(\ln n)$ [24] <br> $\Omega(\ln n)$ [26] |
| $\{0,1\}$,SU | $\frac{\lvert T \rvert \ln k}{\lvert T \rvert - k} O(\min\{\ln^2 k, \ln n\})$ [27] <br> $\Phi$ [28,29] | $\frac{\lvert T \rvert \ln k}{\lvert T \rvert - k} \cdot O(\ln n)$ [27] <br> $\Phi$ [28,29] | in P [26] | $O(\ln n)$ [24] <br> $\Omega(\ln n)$ [26] |
| $\{0,1\}$,G | $k \cdot O(\min\{\ln^2 k, \ln n\})$ [30,24] <br> $\Phi$ [28,29] | $O(k \ln n)$ [24] <br> $\Phi$ [28,29] | in P [26] | $O(\ln n)$ [24] <br> $\Omega(\ln n)$ [26] |
| $\{1,\infty\}$,R | $O(k^2)$ [31], $\Phi$ [32] | $\Phi$ [32] | 2 [33] | $\Phi$ [32] |
| $\{1,\infty\}$,RU | $O(k \ln k)$ [31], $\Phi$ [32] | $\Phi$ [32] | 2 [33] | $\Phi$ [32] |
| $\{1,\infty\}$,SU | $\frac{\lvert T \rvert}{\lvert T \rvert - k} O(k \ln k)$ [27], $\Phi$ [28] | $\Phi$ [20] | 2 [33] | $\Phi$ [20] |
| $\{1,\infty\}$,G | $O(k^3 \ln \lvert T \rvert)$ [34], $\Phi$ [28] | $\Phi$ [20] | 2 [33] | $\Phi$ [20] |
| MC,R | $O(\ln k)$ [22] | $\Phi$ [32] | 2 [33] | $\Phi$ [32] |
| MC,RU,SU | 24 [22] | $\Phi$ [32] | 2 [33] | $\Phi$ [32] |
| MC,G | $O(\ln k)$ [22] | $\Phi$ [20] | 2 [33] | $\Phi$ [20] |
| GC,R | $O(k^2)$ [31], $\Phi$ [32] | $\Phi$ [32] | 2 [33] | $\Phi$ [32] |
| GC,RU | $O(k \ln k)$ [31], $\Phi$ [32] | $\Phi$ [32] | 2 [33] | $\Phi$ [32] |
| GC,SU | $\frac{\lvert T \rvert}{\lvert T \rvert - k} O(k \ln k)$ [27], $\Phi$ [28] | $\Phi$ [20] | 2 [33] | $\Phi$ [20] |
| GC,G | $O(k^3 \ln \lvert T \rvert)$ [34], $\Phi$ [28] | $\Phi$ [20] | 2 [33] | $\Phi$ [20] |

MC and GC stand for metric and general costs, R, RU, SU, and G stand for rooted, rooted uniform, subset uniform, and general demands, respectively. All problems admit ratio $\min\{\lvert D \rvert, O(n)\}$. Here $\Phi = 2^{\ln^{1-\epsilon} n}$, meaning that the problem cannot be approximated within $2^{\ln^{1-\epsilon} n}$ for any fixed $\epsilon > 0$ (unless NP has quasipolynomial time algorithms).

The approximation lower bounds results in the tables are given in terms of $n$. In terms of the parameters $k$ and $|D|$, the following lower bounds are established in Reference 35 for rooted demands: $k^{1/2-\epsilon}$ for directed graphs, $k^{1/10-\epsilon}$ for undirected graphs, and $|D|^{1/4-\epsilon}$ for both directed and undirected graphs. For undirected graphs and arbitrary demands, Reference 35 shows a better bound $k^{1/6-\epsilon}$.

The reader may observe that most of the results for the edge-connectivity case follow from four basic algorithms. In the undirected case, we have ratio 2 for almost all variants by the seminal work of Jain [33], except that in the case of $\{0,1\}$-costs, we have a polynomial time algorithm by the work of Frank [26]. In the directed case, we have the trivial ratio $\min\{|D|, O(n)\}$ for almost all variants, except that in the case of $\{0,1\}$-costs ratio, $O(\ln n)$ is achievable using Set-Cover techniques [24]. On the contrary, node-connectivity Survivable Network problems are more complicated, and often one has to decompose the problem into a "small" number of subproblems on which traditional methods can be applied.

We mention some additional results not appearing in the tables. Survivable Network admits ratio 2 for $k = 2$ [36], but no constant ratio is known for $k = 3$. Element-connectivity Survivable Network admits ratio 2 [36]; in the case of $\{0,1\}$-costs, the problem is NP-hard for $k = 2$ [37] and admits ratios $7/4$ for arbitrary demands and $3/2$ for $\{0,1,2\}$ or $\{0,k\}$ demands [29]. When the sum of the demands is a constant, directed Rooted Survivable Network admits a polynomial time algorithm [32]. In the (undirected) Multiroot Connectivity problem, the demands are $\max\{r_s, r_t\}$ for given nonnegative integers $\{r_v : v \in V\}$. The best known ratios for this problem are $2(k-1)$ for general costs, 3 for metric costs, and 2 for $\{1, \infty\}$-costs [38].

Here is some notation used. An edge from $u$ to $v$ is denoted by $uv$. A $uv$-**path** is a path from $u$ to $v$. For arbitrary sets $A, B$ of nodes and edges (or graphs) $A \setminus B$ is the set (or graph) obtained by deleting $B$ from $A$, where deletion of a node implies also deletion of all the edges incident to it; similarly, $A \cup B$ is the set (graph) obtained by adding $B$ to $A$. For real values, $\{x_u : u \in U\}$ let $x(U) = \sum_{u \in U} x_u$ and $\max(U) = \max_{u \in U} x_u$.

*Organization.*

In Section 13.2, we give some reductions between various versions of our problem. In Section 13.3, we describe an $O(k^3 \ln n)$-approximation algorithm for undirected Survivable Network. In Section 13.4, we give a Biset-LP formulation of Survivable Network problems and some properties of relevant biset families. In Section 13.5, we consider rooted and subset uniform demands, whereas in Section 13.6, we consider the $\{0,1\}$-costs case. We conclude in Section 13.7 with some open problems.

## 13.2 Some Simple Reductions

Note that for directed graphs, the node-connectivity and the edge-connectivity cases have often the same approximability in Table 13.2. The following statement shows that this is not a coincidence.

**Lemma 13.1** *For directed graphs, if edge-connectivity Survivable Network admits ratio $\rho(n)$, then $q$-connectivity Survivable Network admits ratio $\rho(2n)$.*

*Proof.* Given a $q$-connectivity instance, obtain an edge-connectivity instance by a standard conversion of node capacities to edge-capacities: Replace every $v \in V$ by two nodes $v^{out}, v^{in}$ connected by $q_v$ edges from $v^{in}$ to $v^{out}$ of cost 0 each, and redirect the heads of the edges in $\hat{E}$ entering $v$ to $v^{in}$ and the tails of the edges in $\hat{E}$ leaving $v$ to $v^{out}$, keeping costs. The demands are $r'(s^{out}, t^{in}) = r(s, t)$. Then $J$ is a feasible solution to the $q$-connectivity instance iff $J' = \{u^{out}v^{in} : uv \in E\} \cup \{v^{in}v^{out} : v \in V\}$ is a feasible solution to the edge-connectivity instance, and clearly both solutions have the same cost.    ∎

As an application of Lemma 13.1, we show ratio $O(n)$ for edge-connectivity Survivable Network, and via Lemma 13.1 obtain the same ratio for $q$-connectivity. For $A \subseteq V$ let $f_r(A) = \max\{r_{st} : s \in A, t \in V \setminus A\}$, where the maximum or a sum taken over the empty set is defined to be 0. Let $\delta(A)$ denote the set of edges in $\hat{E}$ going from $A$ to $V \setminus A$. Consider the **Cut-LP** for directed edge-connectivity Survivable Network

$$\min\left\{\sum_{e\in\hat{E}}c_e x_e : \sum_{e\in\delta(A)}x_e \geq f_r(A)\ \forall A\subseteq V,\ 0\leq x_e\leq 1\ \forall e\in E\right\}$$

This LP can be solved in polynomial time by the ellipsoid method. As the number of constraints in the LP is $2^n$, we can get (expected) ratio $O(\ln 2^n) = O(n)$ via randomized rounding; the computed solution will be feasible with high probability.

Note that in Table 13.2, the $\{1,\infty\}$-costs case has the same approximability as the general costs case. This is also not a coincidence, as for problems we consider, the general costs case can be reduced to the $\{1,\infty\}$-costs case with a small loss in the approximation ratio, as follows. Assume that the set of 0 cost edges is not a feasible solution. Multiply the costs by $M = n^2/\varepsilon$, then assign cost 1 to 0 cost edges, and replace every edge $e = uv$ by a $uv$-path of length $c_e$ of 1 cost edges. Any feasible solution of cost $C$ is transformed into a solution of cost between $MC$ and $MC + n^2 = M(C+\varepsilon)$, which causes arbitrarily small loss in the ratio.

We now show that for high values of $k$, the directed and undirected variants of Survivable Network have similar approximability. Let $\rho_{\text{dir}}(k,n)$ and $\rho_{\text{und}}(k,n)$ denote the best possible approximation ratio for the directed and undirected $k$-Survivable Network problem on graphs on $n$ nodes, respectively.

**Lemma 13.2** $\rho_{\text{und}}(k,n) \leq 2\rho_{\text{dir}}(k,n)$.

*Proof.* Given an undirected Survivable Network instance $\mathcal{I} = (\hat{G}, Q, c, r)$, obtain a "bidirected" instance $\mathcal{I}' = (\hat{G}', Q, c', r)$ of directed Survivable Network by replacing every edge $uv$ of $\hat{G}$ by the two opposite directed edges $uv, vu$ each of the same cost as $uv$. Then compute a $\rho$ approximate solution $J'$ to $\mathcal{I}'$ and output its underlying graph $J$. It is easy to see that $J$ is a feasible solution for $\mathcal{I}$. Furthermore, if $G$ is an arbitrary subgraph of $\hat{G}$, and $G'$ is the corresponding bidected subgraph of $\hat{G}'$, then $c'(G') = 2c(G)$, and $G$ is a feasible solution for $\mathcal{I}$ if $G'$ is a feasible solution for $\mathcal{I}'$. Thus, $\text{opt}' \leq 2\text{opt}$, where opt and opt$'$ denote the optimal solution value of $\mathcal{I}$ and $\mathcal{I}'$, respectively. Consequently, $c(J) \leq c'(J') \leq \rho\cdot\text{opt}' \leq 2\rho\cdot\text{opt}$. $\blacksquare$

In the reduction in the proof of Lemma 13.2, it is also possible to "bidirect" the demands.

A pair $(A, B)$ of disjoint subsets of $V$ is called a **setpair**; $(A, B)$ is an $st$-**setpair** if $s \in A$ and $t \in B$. For a graph $G = (V, E)$, let $d_G(A, B) = d_E(A, B)$ denote the number of edges in $G$ going from $A$ to $B$. Recall that $\kappa_G(s, t)$ denotes the maximum number of pairwise internally disjoint $st$-paths in $G$. If $(A, B)$ is an $st$-setpair then $\kappa_G(s, t) \leq d_G(A, B) + |V \setminus (A \cup B)|$, and we say that $(A, B)$ is $st$-**tight** in $G$ if $\kappa_G(s, t) = d_G(A, B) + |V \setminus (A \cup B)|$. The node-connectivity version of Menger's Theorem states that for any $s, t \in V$ an $st$-tight setpair exists, namely

$$\kappa_G(s, t) = \min\{d_G(A, B) + n - |A \cup B| : (A, B) \text{ is an } st\text{-setpair in } G\}$$

**Theorem 13.1 (Lando & Nutov [39])** $\rho_{\text{dir}}(k, n) \leq \rho_{\text{und}}(k + n, 2n)$, *and this is so also if both directed and undirected problems are restricted to rooted demands, or to rooted uniform demands, or to $\{0, 1\}$-costs.*

In the rest of this section, we prove Theorem 13.1. Let $G = (V, E)$ be a directed graph. The **bipartite graph of** $G$ has node set $V' \cup V''$ where $V', V''$ are copies of $V$, and edge set $\{u'v'' : uv \in E\}$, where for $v \in V$ we denote by $v'$ and $v''$ the copies of $v$ in $V'$ and $V''$, respectively. The **padded graph of** $G$ is obtained by adding to the bipartite graph of $G$ a **padding edge set** of cliques on each of $V'$ and $V''$, and the matching $M = \{v'v'' : v \in V\}$ (Figure 13.1). Given an instance $\hat{G}, c, r$ of directed $k$-Survivable Network, obtain an instance of undirected $(k + n)$-Survivable Network $\hat{H}, c', r'$, where $\hat{H}$ is the padded graph of $\hat{G}$. The demands and the costs are derived via the natural bijection between ordered node pairs $(u, v)$ of $\hat{G}$ and $(u', v'')$ of $\hat{H}$, while keeping costs but shifting the demands by $n$; namely, the demand of the pair $(s', t'')$ is $r_{st} + n$ and the costs are $c'(u'v'') = c(uv)$ whenever $uv \in \hat{E}$ while the padding edges have cost 0. Note that if $G$ is a spanning subgraph of $\hat{G}$, then the padded graph $H$ of $G$ is a spanning subgraph of $\hat{H}$, and $G$ and $H$ have the same cost. Thus, to finish the proof of Theorem 13.1, it is sufficient to prove the following.

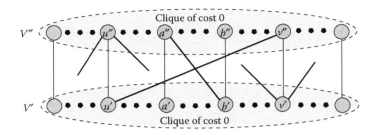

**FIGURE 13.1** The padded bipartite graph; edges in $M$ are shown by thin lines.

**Lemma 13.3 ([39])** *If $H$ is the padded graph of a directed graph $G = (V, E)$ on $n$ nodes, then $\kappa_H(s', t'') = \kappa_G(s, t) + n$ for all $s, t \in V$.*

*Proof.* Let $(A, B)$ be an $st$-tight setpair in $G$. Then, $(A', A'')$ is an $s't''$-setpair in $H$, where $A'$ is the copy of $A$ in $V'$ and $B''$ is the copy of $B$ in $V''$. Note that $d_G(A, B) = d_H(A', B'')$. This implies $\kappa_H(s', t'') \le d_H(A', B'') + 2n - |A' \cup B''| = \kappa_G(s, t) + n$. We prove the inverse inequality. Let $(A' \cup A'', B' \cup B'')$ be an $s't''$-tight setpair in $H$ with $|A'| + |A''| + |B'| + |B''|$ minimal, where $A', B' \subseteq V'$ and $A'', B'' \subseteq V''$. We claim that $A'' = \emptyset$ and $B' = \emptyset$; otherwise, if say there is $a'' \in A''$, then excluding $a''$ from $A''$ decreases $A''$ by exactly 1 and decreases $d_H(A' \cup A'', B' \cup B'')$ by at least $|B''| \ge 1$. Furthermore, no node has a copy both in $A'$ and $B''$, by a similar argument. Hence, $A'$ is a copy in $V'$ of some $A \subseteq V$, $B''$ is a copy in $V''$ of some $B \subseteq V$, and $A$ and $B$ are disjoint; thus, $(A, B)$ is an $st$-setpair. Consequently, $\kappa_H(s', t'') = d_H(A', B'') + 2n - |A' \cup B''| = d_G(A, B) + n + n - |A \cup B| \ge n + \kappa_G(s, t)$. ∎

# 13.3 General Demands and Costs

As for directed graphs, Survivable Network has strong inapproximability results already for small values of $k$, we will focus on undirected graphs and on variants that have good ratios for small values of $k$.

**Theorem 13.2 (Chuzhoy & Khanna [34])** *Undirected Survivable Network admits ratio $O(k^3 \ln |T|)$.*

The idea is to decompose the problem into $O(k^3 \ln |T|)$ element connectivity Survivable Network problems.

**Definition 13.1** *A set family $\mathcal{F}$ on a groundset $T$ is $k$-**resilient** if for any pair $(K, L)$ of disjoint subsets of $T$ with $|L| = 2$ and $|K| \le k$, there exists $S \in \mathcal{F}$ such that $L \subseteq S$ and $K \subseteq T \setminus S$. Let $\rho_n(k)$ denote the minimum size of a $k$-resilient set-family on a groundset of $n$ elements.*

**Lemma 13.4** *Suppose that for a Survivable Network instance we are given on the set $T$ of terminals a $(k-1)$-resilient family $\mathcal{F}$ of size $\rho$. Then such instance admits ratio $2\rho$.*

*Proof.* For each $T_i \in \mathcal{F}$, compute a 2-approximate solution $E_i$ using the algorithm of Reference 36 to the instance of element connectivity Survivable Network with $Q_i = V \setminus T_i$ and the demands restricted to the pairs in $T_i$. Let $E = \bigcup_{i=1}^{\rho} E_i$ be the union of the solutions computed. Any feasible solution to the Survivable Network instance is also feasible for each element connectivity instance; hence, $c(E)$ is at most $2\rho$ times the optimal solution value of the Survivable Network instance. To prove that $G = (V, E)$ is a feasible solution, we show that for any $st \in D$ and any $st$-setpair $(A, B)$, $d_E(A, B) + n - |A \cup B| \ge r_{st}$. If $n - |A \cup B| \ge k$ we are done, so assume that $n - |A \cup B| \le k - 1$. Let $T_i \in \mathcal{F}$ satisfy $\{s, t\} \subseteq T_i$ and $V \setminus (A \cup B) \subseteq T \setminus T_i$, so $V \setminus (A \cup B) \subseteq V \setminus T_i = Q_i$. Then $d_{E_i}(A, B) + n - |A \cup B| \ge r_{st}$, as $E_i$ is a feasible solution to the corresponding element-connectivity Survivable Network instance. However, $E_i \subseteq E$, and thus $d_E(A, B) \ge d_{E_i}(A, B)$. Consequently, $d_E(A, B) + n - |A \cup B| \ge r_{st}$, as required. ∎

We now show that there exists a $k$-resilient family of size $O(k^3 \ln |T|)$ and how to find such a family.

**Theorem 13.3 (Chuzhoy & Khanna [34])** $\rho_n(k) = O(k^3 \ln n)$, and there exists a randomized polynomial time algorithm that computes a $k$-resilient family within this bound, with high probability.

*Proof.* If $n \le 3k+1$, then the subsets of $T$ of size 2 form a $k$-resilient family of size $O(k^2)$, so assume that $n \ge 3k+2$. Moreover, by augmenting the groundset $T$ by a set of $k$ "dummy" elements and revising the notation to $n \leftarrow n+k$, we may consider sets $K$ of size exactly $k$.

We will seek a "small" resilient family in which all sets have the same size $p$. We cast this problem as a Hitting-Set problem in a hypergraph $(\mathcal{V}, \mathcal{E})$ where $\mathcal{V}$ is the family of the subsets of $T$ of size $p$ and $\mathcal{E}$ is the family of pairs $(K, L)$ of disjoint subsets of $T$ with $|L| = 2$ and $|K| = k$, namely,

$$\mathcal{V} = \{S : S \subseteq T, |S| = p\} \quad \mathcal{E} = \{(K, L) : K, L \subseteq V, K \cap L = \emptyset, |L| = 2, |K| = k\}$$

We say that $S \in \mathcal{V}$ **hits** $(K, L) \in \mathcal{E}$ if $L \subseteq S$ and $K \subseteq T \setminus S$; $\mathcal{F} \subseteq \mathcal{V}$ is a **hitting set** of $\mathcal{E}$ if for every $(K, L) \in \mathcal{E}$ there is $S \in \mathcal{V}$ that hits $(K, L)$. By Definition 13.1, $\mathcal{F} \subseteq \mathcal{V}$ is $k$-resilient if and only if $\mathcal{F}$ is a hitting set of $\mathcal{E}$. Thus, $\rho_n(k)$ is bounded by the minimum size of a hitting set of $\mathcal{E}$. A **fractional hitting set** of $\mathcal{E}$ is a function $h : \mathcal{V} \longrightarrow [0, 1]$ such that $\sum \{h(S) : S \text{ hits } (K, L)\} \ge 1$ for every $(K, L) \in \mathcal{E}$; the value of $h$ is $\sum_{S \in \mathcal{V}} h(S)$. It is known that if $\mathcal{E}$ has a fractional hitting set of value $\tau$, then $\mathcal{E}$ has a hitting set of size $O(\tau \ln |\mathcal{E}|)$. Our next goal is to show that $\mathcal{E}$ has a low value fractional hitting set, and to bound $\ln |\mathcal{E}|$.

**Claim 13.1** $|\mathcal{V}| = \binom{n}{p}$, $|\mathcal{E}| = \binom{n}{k}\binom{n-k}{2}$, and the size of each hyperedge $(K, L) \in \mathcal{E}$ is $s = \binom{n-k-2}{p-2}$.

*Proof.* $|\mathcal{V}|$ equals the number $\binom{n}{p}$ of choices of $p$ elements from $n$ elements. $|\mathcal{E}|$ equals the number $\binom{n}{k}$ of choices of a set $K$ of size $k$ multiplied by the number $\binom{n-k}{2}$ of choices of a pair $L$ from the set $V \setminus K$ of size $n - k$. For every $(K, L) \in \mathcal{E}$, the size of the set $\{S \subseteq V : |S| = p, L \subseteq S, K \subseteq V \setminus S\}$ equals the number $\binom{n-k-2}{p-2}$ of choices of the set $S \setminus L$ of size $p - 2$ from the set $T \setminus (K \cup L)$ of size $n - k - 2$. ∎

We have $\left(\frac{n}{k}\right)^k \le \binom{n}{k} \le \left(\frac{n}{k}\right)^k \cdot e^k$. In particular, $\ln |\mathcal{E}| = O(k \ln n)$.

Assigning value $1/s$ to each $S \in \mathcal{V}$ gives a fractional hitting set of value $|\mathcal{V}|/s$. Denote $m = n - k$. Then

$$\frac{|\mathcal{V}|}{s} = \frac{\binom{n}{p}}{\binom{m-2}{p-2}} = \frac{n!}{p!(n-p)!} \cdot \frac{(p-2)!(m-p)!}{(m-2)!} = \frac{m(m-1)}{p(p-1)} \cdot \frac{n!}{(n-p)!} \cdot \frac{(m-p)!}{m!} \le \frac{m^2}{(p-1)^2} \prod_{i=1}^{p} \frac{n-i+1}{m-i+1}$$

Note that for $1 \le i \le p$ we have $\frac{n-i+1}{m-i+1} = 1 + \frac{n-m}{m-i+1} \le 1 + \frac{k}{n-k-p}$. Let us choose $p$ such that $\frac{k}{n-k-p} = \frac{1}{p}$, so $p = \frac{n-k}{k+1}$; assume that $p$ is an integer, as adjustment to floors and ceilings only affects by a small amount the constant hidden in the $O(\cdot)$ term. As $(1 + 1/p)^p \le e$, we obtain

$$\prod_{i=1}^{p} \frac{n-i+1}{m-i+1} \le \left(1 + \frac{1}{p}\right)^p \le e$$

As we assume that $n \ge 3k+2$, we have $\frac{n-k}{k+1} \ge 2$, and thus $\frac{m}{p-1} \le 2(k+1)$. Consequently, we get that $\frac{|\mathcal{V}|}{s} \cdot \ln |\mathcal{E}| = O(k^3 \ln n)$. This implies that a standard greedy algorithm for Hitting Set produces a $k$-resilient family of size $O\left(k^3 \ln n\right)$. There is some difficulty to implement this algorithm in time polynomial in $n$; thus we use a randomized algorithm for Hitting-Set, by rounding each entry to 1 with probability determined by our fractional hitting set. It is known that repeating this rounding $2\lceil \ln |\mathcal{E}| \rceil$ times gives a hitting set w.h.p., and clearly its expected size is $2\lceil \ln |\mathcal{E}| \rceil$ times the value of the fractional hitting set. In our case, the value of every $S \in \mathcal{V}$ is $1/s = 1/\binom{n-k-2}{p-2}$. Thus, we just need to sample a set $S$ of size $p = \frac{n-k}{k+1}$ with probability $1/s$, independently, $2\lceil \ln |\mathcal{E}| \rceil = O(k \ln n)$ times. ∎

## 13.4 Biset Functions and Survivable Network Augmentation Problems

### 13.4.1 Biset Formulation of Survivable Network Problems

In Section 13.2, we formulated Menger's Theorem in terms of setpairs. It would be more convenient to consider instead of a setpair $(A, B)$ the pair of sets $(A, V \backslash B)$ called a "biset," defined as follows.

**Definition 13.2** *An ordered pair* $\mathbb{A} = (A, A^+)$ *of subsets of V with* $A \subseteq A^+$ *is called a* **biset**; $A$ *is the* **inner part** *and* $A^+$ *is the* **outer part** *of* $\mathbb{A}$, *and* $\partial \mathbb{A} = A^+ \backslash A$ *is the* **boundary** *of* $\mathbb{A}$. *The* **co-set** *of* $\mathbb{A}$ *is* $A^* = V \backslash A^+$; *the* **co-biset** *of* $\mathbb{A}$ *is* $\mathbb{A}^* = (A^*, V \backslash A)$. *Let* $\mathcal{V}$ *denote the family of bisets over V.*

A **biset function** assigns to every $\mathbb{A} \in \mathcal{V}$ a real number; in our context, it will always be an integer.

**Definition 13.3** *An* **edge covers a biset** $\mathbb{A}$ *if it goes from A to* $A^*$. *For a biset* $\mathbb{A}$ *and an edge-set/graph J let* $\delta_J(\mathbb{A})$, *denote the set of edges in J covering* $\mathbb{A}$ *and let* $d_J(\mathbb{A}) = |\delta_J(\mathbb{A})|$. *The* **residual function** *of a biset function f w.r.t. an edge set J is defined by* $f^J(\mathbb{A}) = f(\mathbb{A}) - d_J(\mathbb{A})$ *for all* $\mathbb{A} \in \mathcal{V}$. *We say that an edge set/graph J* **covers** *a biset function f, or that J is an f-***cover**, *if* $d_J(\mathbb{A}) \geq f(\mathbb{A})$ *for all* $\mathbb{A} \in \mathcal{V}$.

We say that $\mathbb{A}$ is an $st$-**biset** if $s \in A$ and $t \in A^*$. By the node-capacitated version of Menger's Theorem, we have that if a (directed or undirected) graph $G = (V, E)$ has node capacities $\{q(v) : v \in V\}$, then

$$\lambda_G^q(s, t) = \min\{q(\partial \mathbb{A}) + d_G(\mathbb{A}) : \mathbb{A} \text{ is an } st\text{-biset}\}$$

Thus, $\lambda_G^q(s, t) \geq r_{st}$ if and only if $d_G(\mathbb{A}) \geq r_{st} - q(\partial \mathbb{A})$ for every $st$-biset $\mathbb{A}$. Consequently, we get that $G$ satisfies $r$ if and only if $G$ covers the biset function defined by $f_r^q(\mathbb{A}) = \max_{st \in \delta_D(\mathbb{A})} r_{st} - q(\partial \mathbb{A})$ for all $\mathbb{A} \in \mathcal{V}$, where we treat demand pairs as edges and denote by $\delta_D(\mathbb{A})$ the set of demand pairs that cover $\mathbb{A}$. In the $Q$-connectivity case, we have $q(v) = 1$ if $v \in Q$ and $q(v) = \infty$ otherwise, so $G$ satisfies $r$ iff $G$ covers the biset function $f_r = f_r^Q$ defined by

$$f_r(\mathbb{A}) = \begin{cases} \max_{st \in \delta_D(\mathbb{A})} r_{st} - |\partial \mathbb{A}| & \text{if } \partial \mathbb{A} \subseteq Q \\ 0 & \text{otherwise} \end{cases}$$

We thus often will consider the following generic problem:

---

**Biset-Function Edge-Cover**

*Input:* A graph $\hat{G} = (V, \hat{E})$ with edge costs $\{c_e : e \in \hat{E}\}$ and a biset function $f$ on $V$.
*Output:* A minimum cost edge-set $E \subseteq \hat{E}$ that covers $f$.

---

Here $f$ may not be given explicitly, and an efficient implementation of algorithms requires that certain queries related to $f$ can be answered in time polynomial in $n$. We will not consider implementation details. In most applications, relevant polynomial time oracles are available via min-cut computations. In particular, we have a polynomial time separation oracle for the LP-relaxation due to Frank and Jordan [40]:

$$\tau(f) = \min \quad c \cdot x$$

$$\textbf{(Biset-LP)} \qquad \text{s.t.} \quad x(\delta_E(\mathbb{A})) \geq f(\mathbb{A}) \quad \forall \mathbb{A} \in \mathcal{V}$$

$$0 \leq x_e \leq 1 \qquad \forall e \in E$$

Note that any 0, 1-valued biset function $f$ corresponds to the **biset family** $\mathcal{F} = \{\mathbb{A} \in \mathcal{V} : f(\mathbb{A}) = 1\}$. We thus use for biset families the same terminology and notation as for biset functions, for example, $\tau(\mathcal{F})$ is the optimal value of the Biset-LP for covering $\mathcal{F}$, and $\mathcal{F}^J = \{\mathbb{A} \in \mathcal{F} : \delta_J(\mathbb{A}) = \emptyset\}$ denotes the residual family of $\mathcal{F}$. In the Biset-Family Edge-Cover problem, we seek a minimum cost edge-set $E \subseteq \hat{E}$ that covers $\mathcal{F}$.

## 13.4.2 The Backwards Augmentation Method

We now describe the **backwards augmentation method** due to Reference 41 for reducing a Biset-Function Edge-Cover problem to $\max(f) = \max_{\mathbb{A} \in \mathcal{V}} f(\mathbb{A})$ instances of the Biset-Family Edge-Cover problem, but with a loss of only $O(\ln \max(f))$ in the approximation ratio. Let us say that Biset-Function Edge-Cover **admits LP-ratio** $\rho$ if there exists a polynomial time algorithm that computes an $f$-cover of cost at most $\rho \cdot \tau(f)$. Let $H(k) = \sum i = 1^k 1/i$ denote the $k$th Harmonic number.

**Lemma 13.5** *Suppose that for a* Biset-Function Edge-Cover *instance* $(f, \hat{G}, c)$, *the following holds: For any* $E \subseteq \hat{E}$, *the* Biset-Family Edge-Cover *instance* $(\mathcal{F}, \hat{G} \setminus E, c)$ *with* $\mathcal{F} = \{\mathbb{A} \in \mathcal{V} : f^E(\mathbb{A}) = \max(f^E)\}$ *admits LP-ratio* $\rho$. *Then such* Biset-Function Edge-Cover *instance admits LP-ratio* $\rho \cdot H(k)$, *where* $k = \max(f)$.

*Proof.* Let $x$ be an optimal solution for the Biset-LP for $f$. Consider the following algorithm.

---

## ALGORITHM 13.1   Backwards Augmentation($f, \hat{G}, c$)

1  $E \leftarrow \emptyset$
2  **while** $\max(f^E) > 0$ **do**
3  |  find a cover $J \subseteq \hat{E}$ of the family $\mathcal{F} = \{\mathbb{A} \in \mathcal{V} : f^E(\mathbb{A}) = \max(f^E)\}$ of cost $c(J) \leq \rho\tau(\mathcal{F})$
4  |  $E \leftarrow E \cup J, \hat{E} \leftarrow \hat{E} \setminus J$
5  return $E$

---

After each iteration $\max(f^E)$ decreases by at least 1, hence the algorithm terminates and computes a feasible solution. Consider some iteration with $\max(f^E) = \ell$. Note that $\frac{x}{\ell}$ is a feasible solution to the Biset-LP for the family $\mathcal{F}$ covered at this iteration. This implies $\tau(\mathcal{F}) \leq \frac{\tau(f)}{\ell}$, and hence $c(J) \leq \rho\tau(\mathcal{F}) \leq \rho\frac{\tau(f)}{\ell}$. Consequently, at the end of the algorithm $c(E) \leq \rho\tau(f) \sum_{\ell=k}^{1} \frac{1}{\ell} = \rho\tau(f)H(k)$.  ∎

Applying the backward augmentation method on a Survivable Network instance means that at each iteration, we cover all bisets with maximum residual demand. This is equivalent to increasing the connectivity by 1 between demand pairs for which $r_{st} - \lambda_G^Q(s, t)$ is maximum, where $G = (V, E)$ is the partial solution computed so far. Then we get the following problem that is a restriction of Survivable Network to instances when $\hat{G}$ contains a subgraph $G = (V, E)$ of cost 0, and the demands are $r_{st} = \lambda_G^Q(s, t) + 1$ for all $st \in D$.

---

Survivable Network Augmentation

*Input:* A graph $G$, and edge set $\hat{E}$ with costs $\{c_e : e \in \hat{E}\}$ and a set $D \subseteq V \times V$ of demand pairs.
*Output:* A minimum cost edge-set $J \subseteq \hat{E}$ such that $\lambda_{G \cup J}^Q(s, t) \geq \lambda_G^Q(s, t) + 1$ for all $st \in D$.

---

**Definition 13.4** *A ts-biset* $\mathbb{A}$ *is ts-**tight** in a graph* $G$ *if* $|\partial\mathbb{A}| + d_G(\mathbb{A}) = \kappa_G(t, s)$; *in the case of Q-connectivity, we also require* $\partial\mathbb{A} \subseteq Q$. *Given an instance of* Survivable Network Augmentation, *we say that* $\mathbb{A}$ *is **tight** if it is ts-tight for some* $ts \in D$.

By Menger's Theorem we have

**Fact 13.1** *An edge set $J \subseteq \hat{E}$ is a feasible solution to the* Survivable Network Augmentation *problem if and only if $J$ covers the family of tight bisets.*

### 13.4.3 Properties of Tight Bisets

By Fact 13.1, the Survivable Network Augmentation problem is equivalent to the Biset-Family Edge-Cover problem with $\mathcal{F}$ being the family of tight bisets; in this section, we will establish some properties of this family.

**Definition 13.5** *The **intersection** and the **union** of two bisets $\mathbb{A}, \mathbb{B}$ are defined by $\mathbb{A} \cap \mathbb{B} = (A \cap B, A^+ \cap B^+)$ and $\mathbb{A} \cup \mathbb{B} = (A \cup B, A^+ \cup B^+)$. The biset $\mathbb{A} \setminus \mathbb{B}$ is defined by $\mathbb{A} \setminus \mathbb{B} = (A \setminus B^+, A^+ \setminus B)$. We say that $\mathbb{B}$* **contains** $\mathbb{A}$ *and write $\mathbb{A} \subseteq \mathbb{B}$ if $A \subseteq B$ and $A^+ \subseteq B^+$.*

The following properties of bisets are easy to verify (Figure 13.2).

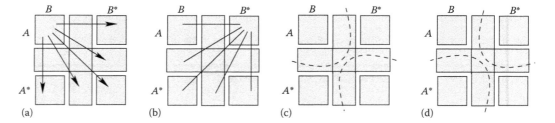

| $B$ $B^*$ | $B$ $B^*$ | $B$ $B^*$ | $B$ $B^*$ |

(a)    (b)    (c)    (d)

**FIGURE 13.2**    Illustration to Fact 13.2; the sets $A, \partial A, A^*$ are the "rows" and $B, \partial B, B^*$ are the "columns" of a $3 \times 3$ "matrix." (a) Types of directed edges that cover $\mathbb{A} \cap \mathbb{B}$. (b) Types of undirected edges that cover $\mathbb{A} \backslash \mathbb{B}$. (c) Nodes appearing in $\partial(\mathbb{A} \cap \mathbb{B})$ and in $\partial(\mathbb{A} \cup \mathbb{B})$. (d) Nodes appearing in $\partial(\mathbb{A} \backslash \mathbb{B})$ and in $\partial(\mathbb{B} \backslash \mathbb{A})$.

**Fact 13.2** *For any bisets $\mathbb{A}, \mathbb{B}$ the following holds. If a directed/undirected edge $e$ covers one of $\mathbb{A} \cap \mathbb{B}, \mathbb{A} \cup \mathbb{B}$, then $e$ covers one of $\mathbb{A}, \mathbb{B}$ (Figure 13.2); if $e$ is an undirected edge, then if $e$ covers one of $\mathbb{A} \setminus \mathbb{B}, \mathbb{B} \setminus \mathbb{A}$, then $e$ covers one of $\mathbb{A}, \mathbb{B}$. Furthermore, $|\partial \mathbb{A}| + |\partial \mathbb{B}| = |\partial(\mathbb{A} \cap \mathbb{B})| + |\partial(\mathbb{A} \cup \mathbb{B})| = |\partial(\mathbb{A} \setminus \mathbb{B})| + |\partial(\mathbb{B} \setminus \mathbb{A})|$.*

For a biset function $f$ and bisets $\mathbb{A}, \mathbb{B}$ the **supermodular inequality** is:

$$f(\mathbb{A} \cap \mathbb{B}) + f(\mathbb{A} \cup \mathbb{B}) \geq f(\mathbb{A}) + f(\mathbb{B})$$

A biset function $f$ is: **supermodular** if the supermodular inequality holds for all $\mathbb{A}, \mathbb{B} \in \mathcal{V}$, **submodular** if $-f$ is supermodular, and **modular** if $f$ is both submodular and supermodular. $f$ is **symmetric** if $f(\mathbb{A}) = f(\mathbb{A}^*)$ for all $\mathbb{A} \in \mathcal{V}$. It is easy to see that for symmetric $f$, if the supermodular inequality holds for $\mathbb{A}, \mathbb{B}^*$ then $f(\mathbb{A} \setminus \mathbb{B}) + f(\mathbb{B} \setminus \mathbb{A}) \geq f(\mathbb{A}) + f(\mathbb{B})$. From Fact 13.2, one can deduce the following.

- For any (directed or undirected) graph $G$, the function $d_G(\cdot)$ is submodular, and if $G$ is an undirected graph, $d_G(\cdot)$ is symmetric.
- The function $|\partial(\cdot)|$ is modular.

The following lemma from Reference 30 gives a general "uncrossing property" of tight bisets, namely, a useful characterization of those pairs $\mathbb{A}, \mathbb{B} \in \mathcal{F}$ for which $\mathbb{A} \cap \mathbb{B}, \mathbb{A} \cup \mathbb{B} \in \mathcal{F}$ or $\mathbb{A} \setminus \mathbb{B}, \mathbb{B} \setminus \mathbb{A} \in \mathcal{F}$ holds.

**Lemma 13.6** *Let $\mathbb{A}, \mathbb{B}$ be bisets in a graph $G$ such that $\mathbb{A}$ is $aa'$-tight, $\mathbb{B}$ is $bb'$-tight, and $\kappa_G(a, a') \geq \kappa_G(b, b')$.*

(i)    *If $\mathbb{A} \cap \mathbb{B}, \mathbb{A} \cup \mathbb{B}$ are both $aa'$-bisets then they are both $aa'$-tight. Otherwise, the following holds.*
   (a)    *If $\mathbb{A} \cap \mathbb{B}$ is an $aa'$-biset and $\mathbb{A} \cup \mathbb{B}$ is a $bb'$-biset then $\mathbb{A} \cap \mathbb{B}$ is $aa'$-tight and $\mathbb{A} \cup \mathbb{B}$ is $bb'$-tight.*
   (b)    *If $\mathbb{A} \cap \mathbb{B}$ is a $bb'$-biset and $\mathbb{A} \cup \mathbb{B}$ is an $aa'$-biset then $\mathbb{A} \cap \mathbb{B}$ is $bb'$-tight and $\mathbb{A} \cup \mathbb{B}$ is $aa'$-tight.*

(ii)   *Suppose that G is an undirected graph. If $\mathbb{A} \setminus \mathbb{B}$ is an aa′ biset and $\mathbb{B} \setminus \mathbb{A}$ is an a′a-biset then $\mathbb{A} \setminus \mathbb{B}$ is aa′-tight and $\mathbb{B} \setminus \mathbb{A}$ is a′a-tight. Otherwise, the following holds.*

   (a)   *If $\mathbb{A} \setminus \mathbb{B}$ is an aa′-biset and $\mathbb{B} \setminus \mathbb{A}$ is a bb′-biset then $\mathbb{A} \setminus B$ is aa′-tight and $\mathbb{B} \setminus \mathbb{A}$ is bb′ tight.*

   (b)   *If $\mathbb{A} \setminus \mathbb{B}$ is a b′b-biset and $\mathbb{B} \setminus \mathbb{A}$ is an a′a-biset then $\mathbb{A} \setminus B$ is b′b-tight and $\mathbb{B} \setminus \mathbb{A}$ is a′a tight.*

*Proof.* Let us use the notation $\kappa(s,t) = \kappa_G(s,t)$ and $\psi(\mathbb{A}) = |\partial \mathbb{A}| + d_G(\mathbb{A})$. Note that $\psi$ is submodular, namely, $\psi(\mathbb{A}) + \psi(\mathbb{B}) \geq \psi(\mathbb{A} \cap \mathbb{B}) + \psi(\mathbb{A} \cup \mathbb{B})$ for all $\mathbb{A}, \mathbb{B} \in \mathcal{V}$. Also note that for any $st$-biset $\mathbb{C}$ we have $\psi(\mathbb{C}) \geq \kappa(s,t)$ and $\kappa(s,t) = \psi(\mathbb{C})$ iff $\mathbb{C}$ is $st$-tight. If $\mathbb{A} \cap \mathbb{B}, \mathbb{A} \cup \mathbb{B}$ are both $aa′$-bisets then:

$$\kappa(a,a′) + \kappa(a,a′) \geq \kappa(a,a′) + \kappa(b,b′) = \psi(\mathbb{A}) + \psi(\mathbb{B}) \geq \psi(\mathbb{A} \cap \mathbb{B}) + \psi(\mathbb{A} \cup \mathbb{B}) \geq \kappa(a,a′) + \kappa(a,a′).$$

Hence, equality holds everywhere, so $\mathbb{A} \cap \mathbb{B}, \mathbb{A} \cup \mathbb{B}$ are both $aa′$-tight (and $\kappa(a,a′) = \kappa(b,b′)$).

In case (ia) we have $\kappa(a,a′) + \kappa(b,b′) = \psi(\mathbb{A}) + \psi(\mathbb{B}) \geq \psi(\mathbb{A} \cap \mathbb{B}) + \psi(\mathbb{A} \cup \mathbb{B}) \geq \kappa(a,a′) + \kappa(b,b′)$. Hence, equality holds everywhere, so $\mathbb{A} \cap \mathbb{B}$ is $aa′$-tight and $\mathbb{A} \cup \mathbb{B}$ is $bb′$-tight.

In case (ib) we have $\kappa(a,a′) + \kappa(b,b′) = \psi(\mathbb{A}) + \psi(\mathbb{B}) \geq \psi(\mathbb{A} \cap \mathbb{B}) + \psi(\mathbb{A} \cup \mathbb{B}) \geq \kappa(b,b′) + \kappa(a,a′)$. Hence, equality holds everywhere, so $\mathbb{A} \cap \mathbb{B}$ is $aa′$-tight and $\mathbb{A} \cup \mathbb{B}$ is $bb′$-tight.

We prove (ii). If $G$ is undirected then $\psi$ is symmetric, and thus $\psi(\mathbb{A}) + \psi(\mathbb{B}) \geq \psi(\mathbb{A} \setminus \mathbb{B}) + \psi(\mathbb{B} \setminus \mathbb{A})$. Also note that $\kappa(s,t) = \kappa(t,s)$ for all $s, t \in V$. If $\mathbb{A} \setminus \mathbb{B}$ is an $aa′$ biset and $\mathbb{B} \setminus \mathbb{A}$ is an $a′a$-biset then

$$\kappa(a,a′) + \kappa(a′,a) \geq \kappa(a,a′) + \kappa(b,b′) = \psi(\mathbb{A}) + \psi(\mathbb{B}) \geq \psi(\mathbb{A} \setminus \mathbb{B}) + \psi(\mathbb{B} \setminus \mathbb{A}) \geq \kappa(a,a′) + \kappa(a′,a).$$

Hence, equality holds everywhere, so $\mathbb{A} \setminus \mathbb{B}$ is $aa′$-tight and $\mathbb{B} \setminus \mathbb{A}$ is $a′a$-tight (and $\kappa(a,a′) = \kappa(b,b′)$).

In case (iia) we have $\kappa(a′,a) + \kappa(b′,b) = \psi(\mathbb{A}) + \psi(\mathbb{B}) \geq \psi(\mathbb{A} \setminus \mathbb{B}) + \psi(\mathbb{B} \setminus \mathbb{A}) \geq \kappa(a′,a) + \kappa(b′,b)$. Hence, equality holds everywhere, so $\mathbb{A} \setminus \mathbb{B}$ is $b′b$-tight and $\mathbb{B} \setminus \mathbb{A}$ is $a′a$-tight.

In case (iib) we have $\kappa(a,a′) + \kappa(b,b′) = \psi(\mathbb{A}) + \psi(\mathbb{B}) \geq \psi(\mathbb{A} \setminus \mathbb{B}) + \psi(\mathbb{B} \setminus \mathbb{A}) \geq \kappa(a′,a) + \kappa(b′,b)$. Hence, equality holds everywhere, so $\mathbb{A} \setminus \mathbb{B}$ is $aa′$-tight and $\mathbb{B} \setminus \mathbb{A}$ is $bb′$-tight. ∎

Inclusion minimal members of a biset family $\mathcal{F}$ are called $\mathcal{F}$-**cores**, or simply **cores**, if $\mathcal{F}$ is clear from the context. Let $\mathcal{C}(\mathcal{F})$ denote the family of $\mathcal{F}$-cores. From Lemma 13.6, we have the following:

**Corollary 13.1** *Let $\mathbb{A}, \mathbb{B}$ be distinct cores of the family of tight bisets, where $\mathbb{A}$ is aa′-tight and $\mathbb{B}$ is bb′-tight, and $aa′, bb′ \in D$. If $A \cap B \neq \emptyset$, $\{a,a′\} \cap \partial \mathbb{B} \neq \emptyset$ or $\{b,b′\} \cap \partial \mathbb{A} \neq \emptyset$, and if $a′ = b′$, $a \in \partial \mathbb{B}$ or $b \in \partial \mathbb{A}$.*

## 13.5 Rooted and Subset Uniform Demands

Let us say that a graph $G$ is $k$-$(T,s)$-**connected** if $\kappa_G(t,s) \geq k$ for all $t \in T$. In the $k$-$(T,s)$-Connectivity Augmentation problem, the goal is to augment a $k$-$(T,s)$-connected graph $G$ by a minimum cost edge set $J$ such that $G \cup J$ is $(k+1)$-$(T,s)$-connected. We survey the following result.

**Theorem 13.4 (Nutov [31,42])**   *Undirected $k$-$(T,s)$-Connectivity Augmentation admits LP-ratio $O(k)$.*

Theorem 13.4 can be applied to compute a solution to Rooted Survivable Network in $k$ iterations, where at iteration $\ell = 1, \dots, k$, we increase the connectivity between nodes in $T$ and $s$ from $\ell - 1$ to $\ell$. For general rooted demands, we get ratio $\sum_{\ell=1}^{k} O(\ell) = O(k^2)$. For demands $r_{st} = k$ for all $t \in T$, this is equivalent to the backward augmentation method, so we get ratio $O(k \ln k)$ in this case. Summarizing, we have

**Corollary 13.2**   *Undirected Rooted Survivable Network with demands $r_{st} = k$ for all $t \in T$ admits LP-ratio $O(k \ln k)$. For rooted general demands, the problem admits LP-ratio $O(k^2)$.*

We now prove Theorem 13.4. The idea of the proof is similar to the one of Theorem 13.2 to decompose the problem into $O(k)$ element-connectivity Survivable Network problems. We will prove a more general result stated in biset families terms. The family of tight bisets in our problem is

$$\mathcal{F} = \{\mathbb{A} \in \mathcal{V} : \mathbb{A} \text{ is a } ts\text{-biset for some } t \in T, |\partial\mathbb{A}| + d_G(\mathbb{A}) = k\}$$

**Definition 13.6** *A biset family $\mathcal{F}$ is **uncrossable** if $\mathbb{A} \cap \mathbb{B}, \mathbb{A} \cap \mathbb{B} \in \mathcal{F}$, or $\mathbb{A} \setminus \mathbb{B}, \mathbb{B} \setminus \mathbb{A} \in \mathcal{F}$ for any $\mathbb{A}, \mathbb{B} \in \mathcal{V}$. Let us say that bisets $\mathbb{A}, \mathbb{B}$: $T$-**intersect** if $A \cap B \cap T \neq \emptyset$, and $T$-**co-cross** if $A \cap B^* \cap T \neq \emptyset$ and $B \cap A^* \cap T \neq \emptyset$. A biset family $\mathcal{F}$ is $T$-**uncrossable** if $A \cap T \neq \emptyset$ for all $\mathbb{A} \in \mathcal{F}$ and if for any $\mathbb{A}, \mathbb{B} \in \mathcal{F}$ the following holds: $\mathbb{A} \cap \mathbb{B}, \mathbb{A} \cup \mathbb{B} \in \mathcal{F}$ if $\mathbb{A}, \mathbb{B}$ $T$-intersect, and $\mathbb{A} \setminus \mathbb{B}, \mathbb{B} \setminus \mathbb{A} \in \mathcal{F}$ if $\mathbb{A}, \mathbb{B}$ $T$-co-cross.*

**Lemma 13.7** *If $G$ is a $k$-$(T, s)$-connected undirected graph, then the family $\mathcal{F}$ of tight bisets is $T$-uncrossable.*

*Proof.* Let $\mathbb{A}, \mathbb{B} \in \mathcal{F}$. If $\mathbb{A}, \mathbb{B}$ $T$-intersect then $\mathbb{A} \cap \mathbb{B}, \mathbb{A} \cup \mathbb{B} \in \mathcal{F}$ by Lemma 13.6(i). If $\mathbb{A}, \mathbb{B}$ $T$-co-cross then $\mathbb{A} \setminus \mathbb{B}, \mathbb{B} \setminus \mathbb{A} \in \mathcal{F}$ by Lemma 13.6(iia). ∎

Recall that by Fact 13.1, any Survivable Network Augmentation problem is equivalent to the corresponding Biset-Family Edge-Cover problem with $\mathcal{F}$ being the family of tight bisets. By Lemma 13.7, in the $k$-$(T, s)$-Connectivity Augmentation problem, the family of tight bisets of the input graph $G$ is $T$-uncrossable. Also note that $|\partial\mathbb{A}| \leq k$ for every tigh biset $\mathbb{A}$ (see Definition 13.4). Thus, the following theorem implies Theorem 13.4.

**Theorem 13.5 (Nutov [31,42])** *Undirected Biset-Family Edge-Cover with $T$-uncrossable biset family $\mathcal{F}$ admits LP-ratio $\frac{4}{3}(4\gamma + \lceil\log_2\lfloor\gamma/2\rfloor\rceil) + 2 = O(\gamma)$, where $\gamma = \max_{\mathbb{A} \in \mathcal{F}} |\partial\mathbb{A} \cap T|$.*

In the rest of this section, we prove Theorem 13.5. For an $\mathcal{F}$-core $\mathbb{C} \in \mathcal{C}(\mathcal{F})$, the **halo-family** $\mathcal{F}(\mathbb{C})$ of $\mathbb{C}$ is the family of those members of $\mathcal{F}$ that contain $\mathbb{C}$ and contain no $\mathcal{F}$-core distinct from $\mathbb{C}$.

**Lemma 13.8** *Let $\mathcal{F}$ be a $T$-uncrossable biset family, and let $p = \min_{\mathbb{A} \in \mathcal{F}} |A \cap T|$. Then there exists a polynomial time algorithm that computes a partition $\Pi$ of $\mathcal{C}(\mathcal{F})$ with at most $2\lfloor\gamma/p\rfloor + 1$ parts such that for each $\mathcal{C} \in \Pi$, the family $\bigcup_{\mathbb{C} \in \mathcal{C}} \mathcal{F}(\mathbb{C})$ is uncrossable. Furthermore, if $p \geq \gamma + 1$, then $\mathcal{F}$ is uncrossable.*

*Proof.* It is easy to see that if $p \geq \gamma + 1$ then any $\mathbb{A}, \mathbb{B} \in \mathcal{F}$ must $T$-intersect or $T$-co-cross; thus, $\mathcal{F}$ is uncrossable in this case. We prove the first statement. For $\mathbb{C}_i \in \mathcal{C}(\mathcal{F})$, let $\mathbb{B}_i = \bigcup_{\mathbb{A} \in \mathcal{F}(\mathbb{C}_i)} \mathbb{A}$ be the union of the bisets in the halo family of $\mathbb{C}_i$, namely, $\mathbb{B}_i$ is the inclusionwise maximal biset in $\mathcal{F}(\mathbb{C}_i)$. Note that as $\mathcal{F}$ is $T$-uncrossable, then for any $\mathbb{A}_i \in \mathcal{F}(\mathbb{C}_i)$ and $\mathbb{A}_j \in \mathcal{F}(\mathbb{C}_j)$ we have

(i) $\mathbb{A}_i, \mathbb{A}_j$ $T$-intersect if and only if $i = j$.
(ii) If $C_i \cap B_j^* \cap T$ and $C_j \cap B_i^* \cap T$ are both nonempty, then $\mathbb{A}_i, \mathbb{A}_j$ $T$-co-cross.

Construct an auxiliary directed graph $\mathcal{J}$ as follows. The node set of $\mathcal{J}$ is $\mathcal{C}(\mathcal{F})$. Add an arc $\mathbb{C}_i\mathbb{C}_j$ if $C_i \cap T \subseteq \partial\mathbb{B}_j$. The indegree of every node in $\mathcal{J}$ is at most $\lfloor\gamma/p\rfloor$, by (i). This implies that every subgraph of the underlying graph of $\mathcal{J}$ has a node of degree $2\lfloor\gamma/p\rfloor$. A graph is $d$-degenerate if every subgraph of it has a node of degree $\leq d$. It is known that any $d$-degenerate graph can be colored with $d + 1$ colors, in polynomial time. Hence, $\mathcal{J}$ is $(2\lfloor\gamma/p\rfloor + 1)$-colorable, and such a coloring can be computed in polynomial time. Consequently, we can compute in polynomial time a partition $\Pi$ of $\mathcal{C}(\mathcal{F})$ into at most $2\lfloor\gamma/p\rfloor + 1$ independent sets. For each independent set $\mathcal{C} \in \Pi$, the family $\bigcup_{\mathbb{C} \in \mathcal{C}} \mathcal{F}(\mathbb{C})$ is uncrossable, by (ii). Hence, $\Pi$ is a partition as required, and the proof of the lemma is complete. ∎

The best known ratio for covering an uncrossable biset family is 2. However, ratio 4/3 is known for the case when the uncrossable family is a union of its halo families [43].

Consider the following algorithm.

---

## ALGORITHM 13.2  *T*-Uncrossable Edge-Cover($\hat{G}, c, \mathcal{F}$)

1  $J \leftarrow \emptyset$
2  **while** $p := \min\limits_{\mathbb{A} \in \mathcal{C}(\mathcal{F}^J)} |A \cap T| \leq \gamma/2$ **do**
3      find a partition $\Pi$ of $\mathcal{C}(\mathcal{F}^J)$ as in Lemma 13.8 with at most $2\lfloor \gamma/p \rfloor + 1$ parts
4      for every $\mathcal{C} \in \Pi$ find a 4/3-approximate edge-cover $J_\mathcal{C}$ of the family $\bigcup\limits_{\mathbb{C} \in \mathcal{C}} \mathcal{F}^J(\mathbb{C})$
5      for every $\mathcal{C} \in \Pi$ do: $J \leftarrow J \cup J_\mathcal{C}$
6  find a 2-approximate edge-cover of $J'$ of $\mathcal{F}^J$ and add $J'$ to $J$
7  **return** $J$

---

Let $p_i$ denote the value of $p$ at the beginning of iteration $i$ in the while loop. Initially, $p_1 \geq 1$. Note that for any $T$-uncrossable family $\mathcal{F}$, if an $\mathcal{F}$-core $\mathbb{C}$ and $\mathbb{A} \in \mathcal{F}$ $T$-intersect, then $\mathbb{C} \subseteq \mathbb{A}$; this implies that if $I$ is an edge set that covers all halo families of $\mathcal{F}$, every $\mathcal{F}^I$-core $\mathbb{A}$ contains at least two $\mathcal{F}$-cores. From this, it follows that $p_i \geq 2p_{i-1}$ for all $i$. Thus, the number of iterations in the while loop is at most $\lceil \log_2(\lfloor \gamma/2 \rfloor + 1) \rceil$. Consequently, the number of uncrossable biset families covered in the while loop is bounded by

$$\sum_{i=0}^{\lceil \log_2(\lfloor \gamma/2 \rfloor + 1) \rceil} (2\lfloor \gamma/2^i \rfloor + 1) = 1 + \lceil \log_2(\lfloor \gamma/2 \rfloor + 1) \rceil + 2\gamma \sum_{i=0}^{\lceil \log_2(\lfloor \gamma/2 \rfloor + 1) \rceil} (1/2)^i \leq 4\gamma + \lceil \log_2 \lfloor \gamma/2 \rfloor \rceil$$

Thus, the approximation ratio is bounded by $\frac{4}{3}(4\gamma + \lceil \log_2 \lfloor \gamma/2 \rfloor \rceil) + 2$, as claimed in Theorem 13.5.

Let us now consider the Subset $k$-Connected Subgraph problem. Let us say that $G$ is $k$-$T$-**connected** if $\kappa_G(s, t) \geq k$ for all $s, t \in T$. In the $k$-$T$-Connectivity Augmentation problem, the goal is to augment a $k$-$T$-connected graph $G$ by a minimum cost edge set $J$ such that $G \cup J$ is $(k+1)$-$T$-connected. The ratios for this problem can derived from those of the $k$-$(T, s)$-Connectivity Augmentation problem via the following relation (the proof is omitted).

**Theorem 13.6 (Nutov [27])**  *For* $|T| > k$, *if* $k$-$(T, s)$-Connectivity Augmentation *admits ratio (LP-ratio)* $\rho$ *then* $k$-$T$-Connectivity Augmentation *admits the following ratios (LP-ratios):*

(i)  $b(\rho + k) + O(\mu^2 \ln \mu)$, *where* $b = 1$ *for undirected graphs and* $b = 2$ *for directed graphs, and* $\mu = \frac{T}{|T| - k}$.

(ii)  $\rho \cdot O(\mu \ln k)$ *for both directed and undirected graphs; furthermore, this is so also for* $\{0, 1\}$-*costs.*

For $|T| > k$, the best known values of $\rho$ on undirected graphs are $O(k)$ for arbitrary costs and $O(\ln k)$ for $\{0, 1\}$-costs. For directed graphs $\rho = |T|$ for arbitrary costs and $\rho = O(\log |T|)$ for $\{0, 1\}$-costs. These are LP-ratios, so the backward augmentation method can be applied. Thus, Theorem 13.6 implies the following.

**Corollary 13.3**  *For* $|T| > k$, $k$-$T$-Connectivity Augmentation *admits the following LP-ratios.*

- *For undirected graphs:* $O(k + \mu^2 \ln \mu)$ *for arbitrary costs, and* $O(\mu \ln^2 k)$ *for* $\{0, 1\}$-*costs,* $\mu = \frac{T}{|T| - k}$.
- *For directed graphs:* $2(|T| + k) + O(\mu^2 \ln \mu)$ *for arbitrary costs, and* $O(\mu \ln k \ln |T|)$ *for* $\{0, 1\}$-*costs.*

*For* Subset $k$-Connected Subgraph, *the ratios are larger by a factor of* $O(\log k)$.

# 13.6 {0,1}-Costs Survivable Network Problems

Here we consider the {0, 1}-Costs Survivable Network problem that can be formulated as follows.

---

{0, 1}-Costs Survivable Network

*Input:* A directed/undirected graph $G = (V, E)$, $Q \subseteq V$, and connectivity demands $\{r_{st} : st \in D\}$.
*Output:* A minimum size set $J$ of new edges (any edge is allowed) such that $G \cup J$ satisfies $r$.

---

**Theorem 13.7 (Kortsarz & Nutov [24], Nutov [30])** *Suppose that for an instance of* {0, 1}-*Costs Surviv-able Network, we are given a node subset $Z \subseteq V$ such that $Z \setminus \partial \mathbb{A} \neq \emptyset$ for every biset $\mathbb{A} \in \mathcal{V}$ with $f_r(\mathbb{A}) > 0$. Then the problem admits ratio $O(|Z| \ln |D|)$. For undirected graphs, the problem also admits ratio $O(|Z| \ln^2 k)$.*

Note that the following choices of $Z$ satisfy the assumption of Theorem 13.7: (i) $Z = \{z\}$ for some $z \in V \setminus Q$; (ii) $Z = \{s\}$ in the case of rooted demands; (iii) any $Z \subseteq V$ with $|Z| = k$. Thus we have

**Corollary 13.4** {0, 1}-*Costs Survivable Network admits ratios $O(\ln |D|)$ if $Q \neq V$ and $O(k \ln |D|)$ if $Q = V$. For undirected graphs the problem also admits ratios $O(\ln^2 k)$ if $Q \neq V$ or if the demands are rooted, and ratio $O(k \ln^2 k)$ otherwise.*

For $Q \neq V$, the ratio $O(\ln |D|)$ is tight as the problem is Set-Cover hard [25]; for directed graphs, this is so even for rooted 0, 1 requirements [26]. For $Q = V$, the ratio $O(k \ln |D|)$ is tight if $k$ is small, but may seem weak if $k$ is large. However, a much better ratio might not exist, see Theorem 13.10.

The proof of Theorem 13.7 follows. An undirected edge set (or a graph) $S$ is a **star** if all its edges are incident to the same node, called the **center** of the star. In the case of directed graphs, $S$ should be either an **outstar**—all edges in $S$ leave the center, or an **instar**—all edges in $S$ enter the center. Let Star Survivable Network be the restriction of Survivable Network to instances when the set of positive cost edges in $\hat{E}$ is a star $S$ with center $z \in V \setminus Q$. Note that here the edges incident to $z$ may have arbitrary costs.

**Lemma 13.9** *If Star Survivable Network admits ratio $\rho$, then any* {0, 1}-*Costs Survivable Network instance $\mathcal{I}$ with $Z$ as in Theorem 13.7 admits ratio $2|Z|\rho$.*

*Proof.* For every $z \in Z$ obtain from $\mathcal{I}$ an instance $\mathcal{I}_z$, in which $z$ is excluded from $Q$, and except the edges in $E$, only edges incident to $z$ are allowed, by cost 1 each. In the undirected case, $\mathcal{I}_z$ is a Star Survivable Network instance; for each $z \in Z$ we compute a $\rho$-approximate solution $I_z$ for $\mathcal{I}_z$ and return $I = \bigcup_{z \in Z} I_z$. In the directed case, we further split the instance $\mathcal{I}_z$ into two Star Survivable Network instances $\mathcal{I}_z^{out}$ and $\mathcal{I}_z^{in}$.

- $\mathcal{I}_z^{out}$ is obtained from $\mathcal{I}_z$ by adding to $E$ zero cost edges from every $v \in V \setminus \{z\}$ to $z$, and allowing only additional edges leaving $z$, by cost 1 each.
- $\mathcal{I}_z^{in}$ is obtained from $\mathcal{I}_z$ by adding to $E$ zero cost edges from $z$ to every $v \in V \setminus \{z\}$, and allowing only additional edges entering $z$, by cost 1 each.

Each of $\mathcal{I}_z^{out}, \mathcal{I}_z^{in}$ is a Star Survivable Network instance; for each $z \in Z$, we compute a $\rho$-approximate solution $I_z^{out}$ and $I_z^{in}$ for $\mathcal{I}_z^{out}$ and $\mathcal{I}_z^{in}$, respectively, set $I_z = I_z^{out} \cup I_z^{in}$, and return $I = \bigcup_{z \in Z} I_z$.

We show that $I$ is a feasible solution to $\mathcal{I}$. We prove this for the directed case, and the proof for the undirected case is similar. Suppose to the contrary that there is $st \in D$ such that $\lambda_{G \cup I}^Q(s, t) < r_{st}$. Then there is an $st$-biset $\mathbb{A}$ such that $|\partial \mathbb{A}| + d_{G \cup I}(\mathbb{A}) < r_{st}$. By the assumption on $Z$, there exists $z \in Z$ such that $z \in A$ or $z \in A^*$. As $I_z^{out}, I_z^{in} \subseteq I$, $|\partial \mathbb{A}| + d_{G \cup I^{out}}(\mathbb{A}) < r_{st}$ and $|\partial \mathbb{A}| + d_{G \cup I^{in}}(\mathbb{A}) < r_{st}$. If $z \in A$, the former inequality contradicts the feasibility of $I_z^{out}$ for $\mathcal{I}_z^{out}$, as adding to $G$ edges entering $z$ does affect the inequality. The contradiction for the case $z \in A^*$ is obtained in a similar way.

Now let $J$ be an optimal solution to $\mathcal{I}$. We will show that $I_z \leq 2\rho|J|$ for every $z \in Z$, and thus $|I| \leq |Z| \cdot 2\rho|J|$. Obtain from $J$ a solution $J_z$ for the instance $\mathcal{I}_z$ by subdividing every edge in $J$ by a new node, and then identifying all new nodes into $z$. It is easy to see that $J_z$ is a feasible solution for $\mathcal{I}_z$, and clearly $|J_z| = 2|J|$. Thus, in the undirected case, our $\rho$-approximate solution $I_z$ satisfies $|I_z| \leq 2\rho|J_z| = 2\rho|J|$. In the directed case, we further split $J_z$ into sets $J_z^{out}$ and $J_z^{in}$ of edges leaving and entering $z$, respectively. Then $|I_z^{out}| \leq \rho|J_z^{out}|$ and $|I_z^{in}| \leq \rho|J_z^{in}|$, implying $|I_z| = |I_z^{out}| + |I_z^{in}| \leq \rho(|J_z^{out}| + |J_z^{in}|) = \rho|J_z| = 2\rho|J|$. This concludes the proof of the lemma. ∎

We prove the following theorem that together with Lemma 13.9 implies Theorem 13.7.

**Theorem 13.8 ([24,30,44])** *Directed Star Survivable Network admits ratio $H(|D|)$, where $H(j) = \sum_{i=1}^{j} 1/i$ is the jth harmonic number. For undirected graphs, the problem admits ratio $H(k) \cdot H(8k^2 - 14k + 7)$, $k \geq 2$.*

## 13.6.1 Ratio $H(|D|)$ for Directed Star Survivable Network

Here we prove the directed part of Theorem 13.8. As it concerns directed graphs only, we may consider only the edge connectivity case by applying Lemma 13.1, as the reduction in the proof of Lemma 13.1 preserves our problem type and does not change the parameter $|D|$. We will also assume that the set $S$ of positive cost edges is an outstar with center $z$.

We use a result due to Wolsey [45] about the performance of a greedy algorithm for the Submodular Cover problem. Here we are given two set functions on subsets of a groundset $S$: a cost-function $c$ and and an integer valued "progress function" $p$. Each of the functions may be given by a value oracle, meaning that there exists an oracle that given a subset of $S$ return the corresponding function value. The goal is to find $J \subseteq S$ of minimum cost such that $p(J) = p(S)$. In the Submodular Cover problem, the function $p$ is submodular and nondecreasing, and the cost of $J \subseteq S$ is $c(J) = \sum_{e \in J} c_e$ for some nonnegative costs $\{c_e : e \in S\}$. Wolsey [45] proved that the greedy algorithm, that starts with $J = \emptyset$ and as long as $p(J) < p(S)$ repeatedly adds to $J$ an element $e \in S \setminus J$ with maximum $\frac{p(J \cup \{e\}) - p(J)}{c_e}$, has ratio $H\left(\max_{e \in S} p(\{e\}) - p(\emptyset)\right)$.

In our case $S$ is an outstar, and for $J \subseteq S$ let

$$p(J) = \sum_{st \in D} \min\{p_{st}(J), r_{st}\} \quad \text{where} \quad p_{st}(J) = \lambda_{G \cup J}(s, t)$$

It is not hard to verify that $p$ is nondecreasing, and that $J$ is a feasible solution to a Survivable Network instance if and only if $p(J) = r(D)$. It is also easy to see that $\lambda_{G \cup \{e\}}(s, t) - \lambda_G(s, t) \leq 1$ for any graph $G$ and any edge $e$, as adding a single edge can increase the $st$-connectivity by at most 1. Thus, for any edge $e$, we have $p_{st}(\{e\}) - p_{st}(\emptyset) \leq 1$, which implies $p(\{e\}) - p(\emptyset) \leq |D|$.

Note that if each function $p_{st}(J) = \lambda_{G \cup J}(s, t)$ is submodular, so is $p$. Indeed, it is known [4] that if $f$ is submodular and nondecreasing, then $\min\{f, \ell\}$ is submodular for any constant $\ell$; thus, $\min\{p_{st}(J), r_{st}\}$ is submodular. As a sum of submodular functions is also submodular, we obtain that $p$ is submodular.

Fix $s, t \in V$ and for $J \subseteq S$ let $f(J) = \lambda_{G \cup J}(s, t)$. We show that if $S$ is a star then $f$ is submodular (this may not be so if $S$ is not a star). We use the following known characterization of submodularity [4]: *A set-function $f$ on a groundset $S$ is submodular iff*

$$f(J_0 \cup \{e\}) + f(J_0 \cup \{e'\}) \geq f(J_0) + f(J_0 \cup \{e, e'\}) \quad \forall J_0 \subseteq S, e, e' \in F \setminus J_0$$

Let $J_0 \subseteq S$. Revising our notation to $G \leftarrow G \cup J_0$, $S \leftarrow S \setminus J_0$, and $f(J) \leftarrow f(J_0 \cup J) - f(J_0)$, we get that the above-mentioned condition, is equivalent to

$$f(\{e\}) + f(\{e'\}) \geq f(\{e, e'\}) \quad \forall e, e' \in S$$

As before, $S$ is an outstar and $f(J) = \lambda_{G \cup J}(s, t) - \lambda_G(s, t)$ is the increase in the $(s, t)$-edge-connectivity as a result of adding $J$ to $G$. We prove the following general statement that implies the earlier; it says that if an augmenting edge set $J$ is an outstar that increases the $st$-edge-connectivity by $h$, then there are $h$ edges in $J$ that cover all $st$-tight sets, and thus each of these edges increases the $st$-edge-connectivity by 1.

**Lemma 13.10** *Let $G = (V, E)$ be a directed graph and $J$ an outstar with center $z$ on $V$. Let $s, t \in V$, and let $h = \lambda_{G \cup J}(s, t) - \lambda_G(s, t)$. Then there is $I \subseteq J$ of size $h$ such that $\lambda_{G \cup \{e\}}(s, t) = \lambda_G(s, t) + 1$ for every $e \in I$.*

*Proof.* Let $\mathcal{F}$ be the family of $st$-tight sets. Then $\mathcal{F}$ is nonempty, and by Lemma 13.6(i) $A \cap B, A \cup B \in \mathcal{F}$ for any $A, B \in \mathcal{F}$. As the intersection of all sets in $\mathcal{F}$ contains $s$ and thus nonempty, $\mathcal{F}$ has a unique inclusion-minimal set $A_{\min}$ and a unique inclusion-maximal set $A_{\max}$, and $A_{\min} \subseteq A_{\max}$.

Let $I = \{zv \in J : a \in A_{\min}, v \in V \setminus A_{\max}\}$ be the set of edges in $J$ that go from $A_{\min}$ to $V \setminus A_{\max}$. Each edge in $I$ covers $\mathcal{F}$; hence, by Menger's Theorem, $\lambda_{G \cup \{e\}}(s, t) = \lambda_G(s, t) + 1$ for every $e \in I$.

It remains to prove that $|I| \geq h$. We claim that as $J$ is an outstar, $\lambda_{G \cup J}(s, t) \leq \lambda_G(s, t) + |I|$, hence $|I| \geq \lambda_{G \cup J}(s, t) - \lambda_G(s, t) = h$. Note that from Menger's Theorem, we have

$$\lambda_{G \cup J}(s, t) \leq \lambda_G(s, t) + |\delta_J(A_{\min})| \qquad \lambda_{G \cup J}(s, t) \leq \lambda_G(s, t) + |\delta_J(A_{\max})|$$

The first inequality implies that if $\delta_J(A_{\min}) = \emptyset$, then $\lambda_{G \cup J}(s, t) = \lambda_G(s, t)$, and then we are done. Else, we must have $z \in A_{\min}$. In this case, $I = \delta_J(A_{\max})$, as $J$ is a star. Then, the second inequality implies $\lambda_{G \cup J}(s, t) \leq \lambda_G(s, t) + |I|$, as claimed. ∎

## 13.6.2 Ratio $O(\ln^2 k)$ for Undirected Star Survivable Network

Here we prove the undirected part of Theorem 13.8. For simplicity of exposition, we consider the case $Q = V \setminus \{z\}$ where $z$ is the center of the star of the available edges.

As an intermediate problem, we consider the augmentation version of the problem, namely

---

**Star Survivable Network Augmentation**

*Input:* A graph $G = (V, E)$, a star $S$ with center $z$ and costs $\{c_e : e \in S\}$, and a set $D$ of demand pairs.

*Output:* A minimum cost edge set $J \subseteq S$ such that $\lambda_{G \cup J}^Q(s, t) \geq \lambda_G^Q(s, t) + 1$ for all $st \in D, Q = V \setminus \{z\}$.

---

Here the family of tight bisets is

$$\{\mathbb{A} \in \mathcal{V} : |\partial \mathbb{A}| + d_G(\mathbb{A}) = \lambda_G^Q(s, t) \text{ for some } st \in D, z \notin \partial \mathbb{A}\}$$

Note that this family is symmetric, as if $\mathbb{A}$ is tight, so is $\mathbb{A}^*$. Let us say that $U$ is a **transversal** (hitting set) of a biset family $\mathcal{C}$ or that $U$ is a $\mathcal{C}$-**transversal** if $U \cap C \neq \emptyset$ for all $\mathbb{C} \in \mathcal{C}$. A **fractional** $\mathcal{C}$-**transversal** is a function $h : V \longrightarrow [0, 1]$ such that $h(C) \geq 1$ for all $\mathbb{C} \in \mathcal{C}$. Given node-weights $\{w_v : v \in V\}$, let $\tau_w(\mathcal{C})$ denote the optimal value of a fractional $\mathcal{C}$-transversal, namely

$$\tau_w(\mathcal{C}) = \min\{\sum_{v \in V} w_v x_v : x(C) \geq 1 \; \forall \mathbb{C} \in \mathcal{C}, \; x_v \geq 0 \; \forall v \in V\}$$

Let $\Delta(\mathcal{C})$ denote the maximum degree in the (multi)hypergraph $\{C : \mathbb{C} \in \mathcal{C}\}$ formed by the inner parts of the members of $\mathcal{C}$. The problem of finding a minimum weight $\mathcal{C}$-transversal is the Hitting Set problem, and it is a particular case of the Submodular Cover problem. In this case, the greedy algorithm computes a solution of weight at most $H(\Delta(\mathcal{C}))\tau_w(\mathcal{C})$. It is easy to see that if $J$ covers a biset family $\mathcal{F}$, then the set of endnodes of $J$ is a $\mathcal{C}(\mathcal{F})$-transversal. The following lemma shows that in our case, the inverse is also true.

**Lemma 13.11** *Let $U$ be a transversal of a symmetric biset family $\mathcal{F}$ and let $J$ be a star on $U$ with center $z$ such that $z \notin \partial \mathbb{A}$ for every $\mathbb{A} \in \mathcal{F}$. Then $J$ covers $\mathcal{F}$.*

*Proof.* Let $\mathbb{A} \in \mathcal{F}$. As $z \notin \partial \mathbb{A}$, $z \in A$ or $z \in V \setminus A^+$. If $z \in V \setminus A^+$, then as $U$ is an $\mathcal{F}$-transversal and as $\mathcal{F}$ is symmetric, there is $u \in U \cap A$. If $z \in A$, then there $u \in U \cap A^*$, by a similar argument. In both cases, the edge $zu$ belongs to the star $J$ and covers $\mathbb{A}$. ∎

Lemma 13.11 implies that Star Survivable Network Augmentation is equivalent to finding a minimum weight $\mathcal{C}$-transversal, where $\mathcal{C}$ is the family of cores of tight bisets and the weights are $w_v = c_{zv}$ if $v \neq z$ and $w_z = 0$. Thus, the problem admits LP-ratio $H(\Delta(\mathcal{C}))$. Hence, if $\Delta_k$ is a bound on $\Delta(\mathcal{C})$ for all instances with maximum demand $k$, then we can apply the backward augmentation method as described in Lemma 13.5 and achieve an LP-ratio $H(k) \cdot H(\Delta_k)$. Consequently, the following theorem finishes the proof of Theorem 13.8.

**Theorem 13.9 (Nutov [30])** *Let $\mathcal{C}$ be the family of cores of the family of tight bisets, and let $\gamma = \max_{\mathbb{A} \in \mathcal{C}} |\partial \mathbb{A} \cap T|$. Then, $\Delta(\mathcal{C}) \le 8\gamma^2 + 2\gamma + 1$; thus, $\Delta(\mathcal{C}) \le 8(k-1)^2 + 2(k-1) + 1 = 8k^2 - 14k + 7$.*

*Proof.* Fix some $v \in V$ and let $\mathcal{C}' = \{\mathbb{C} \in \mathcal{C} : v \in C\}$. For every $\mathbb{C}_i \in \mathcal{C}'$, choose one pair $s_i t_i \in D$ such that $\mathbb{C}_i$ is $s_i t_i$-tight. Let $(T', D')$ be the directed graph induced by the chosen pairs. Also construct an auxiliary directed graph $(\mathcal{C}', \mathcal{J}')$ with edge labels where for every $u \in \{s_i, t_i\} \cap \partial \mathbb{C}_j$ there is an edge $\mathbb{C}_i \mathbb{C}_j \in \mathcal{J}'$ with label $u$. Note that for any $i \neq j$, we have $v \in C_i \cap C_j$, hence by Corollary 13.1 $\{s_i, t_i\} \cap \partial \mathbb{C}_j \neq \emptyset$ or $\{s_j, t_j\} \cap \partial \mathbb{C}_i \neq \emptyset$. Thus, the (simple) underlying graph of $(\mathcal{C}', \mathcal{J})$ is a clique.

**Claim 13.2** *For any $u \in T'$, the number of edges in $D'$ incident to $u$ is at most $2\gamma + 1$. Consequently, for any $st \in D'$, there are at most $4\gamma + 1$ edges in $D'$, that are incident to $s$ or to $t$.*

*Proof.* Let $D'_u$ be the set of edges in $D'$ incident to $u$ and $\mathcal{C}'_u$ the corresponding set of cores. Let $d = |D'_u|$. For every $s_i t_i \in D'_u$, let $v_i = \{s_i, t_i\} \setminus \{u\}$, namely, $v_i = t_i$ if $u = s_i$ and $v_i = s_i$ if $u = t_i$. Consider a pair $s_i t_i, s_j t_j \in D'_u$. As $v \in C_i \cap C_j$, by Lemma 13.6 and the minimality of $\mathbb{C}_i, \mathbb{C}_j$, we must have $v_i \in \partial \mathbb{C}_j$ or $v_j \in \partial \mathbb{C}_i$. Now consider the subgraph $(\mathcal{C}'_u, \mathcal{J}'_u)$ of $(\mathcal{C}', \mathcal{J}')$ induced by $\mathcal{C}'_u$. The underlying graph of this subgraph is a clique, hence $|\mathcal{J}'_u| \geq \frac{d(d-1)}{2}$. Consequently, $(\mathcal{C}'_u, \mathcal{J}'_u)$ has a node $\mathbb{C}$ of indegree at least $\frac{d-1}{2}$. Note that each edge in $\mathcal{J}'_u$ that enters $\mathbb{C}$ contributes the node of its label to $\partial \mathbb{C} \cap T'$, and no two edges entering $\mathbb{C}$ have the same label. Hence $\frac{d-1}{2} \leq |\partial \mathbb{C} \cap T'| \leq \gamma$, implying $d \leq 2\gamma + 1$. ∎

We now finish the proof of Theorem 13.9. Let $\Delta = |\mathcal{C}'|$. The graph $(\mathcal{C}', \mathcal{J}')$ has a node $\mathbb{C}$ of indegree $\geq (\Delta - 1)/2$, as its underlying graph is a clique. Now consider the labels of the arcs entering $\mathbb{C}$. By Claim 13.2, there are at least $(\Delta - 1)/(8\gamma + 2)$ edges that enter $\mathbb{C}$, such that no two edges have the same label. Each one of these edges contributes a node to $\partial \mathbb{C} \cap T$. Consequently, we must have $(\Delta - 1)/(8\gamma + 2) \leq \gamma$, which implies $\Delta \leq 8\gamma^2 + 2\gamma + 1$. ∎

## 13.6.3 Hardness of Approximation of the {0,1}-Costs Case

We will show that Survivable Network with $\{0, 1\}$-costs is almost as hard to approximate as the following variant of the Label-Cover problem:

---

Min-Rep

*Instance:* A bipartite graph $H = (A \cup B, I)$ and partitions $\mathcal{A}$ of $A$ and $\mathcal{B}$ of $B$ into parts of equal size.
*Objective:* Find a minimum size node set $A' \cup B'$, where $A' \subseteq A$ and $B' \subseteq B$, such that for any $A_i \in \mathcal{A}, B_j \in \mathcal{B}$ with $\delta_I(A_i, B_j) \neq \emptyset$ there are $a \in A' \cap A_i, b \in B' \cap B_j$ such that $ab \in I$.

---

By the Parallel Repetition Theorem [46] (see Reference 47 for details) Min-Rep cannot be approximated within $O(2^{\log^{1-\varepsilon} n})$ unless NP has quasipolynomial time algorithms. The Min-Rep problem was defined by Kortsarz [47] and was used later by Dodis and Khanna [20] to show a $2^{\log^{1-\varepsilon} n}$-approximation hardness of the Directed Steiner Forest problem. This was extended to high connectivity undirected Survivable Network problems by Kortsarz, Krauthgamer, and Lee [28], and further generalized and simplified in Reference 39. Here we will describe a slightly more complicated variant that proves the same hardness already for the $\{0, 1\}$-costs case.

**Theorem 13.10 (Kortsarz, Krauthgamer, & Lee [28], Nutov [29])** *Directed* Survivable Network *with* $\{0, 1\}$-*costs cannot be approximated within* $2^{\ln^{1-\varepsilon} n}$ *for any fixed* $\varepsilon > 0$, *unless NP has quasipolynomial time algorithms.*

*Proof.* Given an instance of Min-Rep, construct an instance $G = (V, E), r$ of Survivable Network with $\{0, 1\}$-costs as follows, where edges in $E$ have cost 0. Let $\mathcal{E} = \{ij : A_i \in \mathcal{A}, B_j \in \mathcal{B}, \delta_H(A_i, B_j) \neq \emptyset\}$. Direct the edges of $H$ from $A$ to $B$. Then the graph $G = (V, E)$ is obtained from $H$ as follows.

1. Add to $H$: A set $\{a_1, \ldots, a_{|\mathcal{A}|}, b_1, \ldots, b_{|\mathcal{B}|}\}$ of $|\mathcal{A}| + |\mathcal{B}|$ nodes, and for every $ij \in \mathcal{E}$, a pair of nodes $a_{ij}, b_{ij}$ (so a total number of nodes added to $H$ is $|\mathcal{A}| + |\mathcal{B}| + 2|\mathcal{E}|$). Thus,

$$V = A \cup B \cup \{a_1, \ldots, a_{|\mathcal{A}|}, b_1, \ldots, b_{|\mathcal{B}|}\} \cup \{a_{ij} : ij \in \mathcal{E}\} \cup \{b_{ij} : ij \in \mathcal{E}\}$$

2. For every $ij \in \mathcal{E}$: Connect $a_{ij}$ to every node that is not in $\bar{A}_{ij} = A_i \cup B_j \cup \{b_j, b_{ij}\}$, and connect every node that is not in $\bar{B}_{ij} = A_i \cup B_j \cup \{a_i, a_{ij}\}$ to $b_{ij}$. Thus,

$$E = I \cup \{a_{ij}v : ij \in \mathcal{E}, v \in V \setminus \bar{A}_{ij}\} \cup \{ub_{ij} : ij \in \mathcal{E}, u \in V \setminus \bar{B}_{ij}\}$$

For $ij \in \mathcal{E}$ let $C_{ij} = \{v \in V : a_{ij}v, vb_{ij} \in E\}$. By the construction

$$C_{ij} = V \setminus (\bar{A}_{ij} \cup \bar{B}_{ij}) = V \setminus (A_i \cup B_j \cup \{a_i, b_j, a_{ij}, b_{ij}\})$$

As the sets $A_i$ have the same size and the sets $B_j$ have the same size, the sets $C_{ij}$ are also all of the same size, say $k-1$. Every node in $C_{ij}$ is an internal node of an $a_{ij}b_{ij}$-path of length 2. Let $G_{ij} = G \setminus C_{ij}$ be the subgraph of $G$ induced by $A_i \cup B_j \cup \{a_i, b_j, a_{ij}, b_{ij}\}$. By the construction, $G_{ij}$ has no $a_{ij}b_{ij}$-path. Thus, $\kappa_G(a_{ij}, b_{ij}) = k-1$ for every $ij \in \mathcal{E}$. We will ask to increase the connectivity by 1 between pairs $\{a_{ij}b_{ij} : ij \in \mathcal{E}\}$, so the demands are

$$r(a_{ij}, b_{ij}) = k \text{ for every } ij \in \mathcal{E}$$

For $ij \in \mathcal{E}$, let $F_{ij}$ be the set of edges in $F$ with both endnodes in $G_{ij}$. Note that for every $ij \in \mathcal{E}$

(i) $G_{ij}$ has no $a_{ij}b_{ij}$-path.
(ii) $\kappa_{G \cup F}(a_{ij}, b_{ij}) \geq k$ iff $G_{ij} \cup F_{ij}$ has an $a_{ij}b_{ij}$-path.

Let us say that a new edge is **proper** if it goes from $a_i$ to $A_i$ for some $i$, or if it goes from $B_j$ to $b_j$ for some $j$. Let $F$ be a feasible solution to the constructed Survivable Network instance, and let $e \in F$. Assume that $e$ belongs to $F_{ij}$ for some $ij \in \mathcal{E}$, as otherwise $e$ can be removed. We claim if $e$ is nonproper, then $e$ can be replaced by at most two proper edges $e', e''$ while keeping $F$ feasible, as follows:

- If $e$ goes from $A_i \cup \{a_i, a_{ij}\}$ to $B_j \cup \{b_j, b_{ij}\}$ then $\{e', e''\} = \{a_i a, b b_j\}$ for some $ab \in \delta_I(A_i, B_j)$.
- If $e = a_{ij}a$ for some $a \in A_i$ or $e = bb_{ij}$ for some $b \in B_j$ then $e' = e'' = a_i a$ or $e' = e'' = bb_j$, respectively.
- If $e = a'a''$ for some $a', a'' \in A_i$ or $e = b'b''$ for some $b', b'' \in B_j$, then $\{e', e''\} = \{a_i a', a_i a''\}$ or $\{e', e''\} = \{b'b_j, b''b_j\}$, respectively.

In each one of the cases, it is not hard to verify that $(F \setminus \{e\}) \cup \{e', e''\}$ is a feasible solution as well. This implies that there exists a proper feasible solution $F'$ with $|F'| \leq 2|F|$.

Note that for every $v \in A \cup B$ corresponds a unique proper edge, namely, $a_i v$ if $v \in A_i$ and $v b_j$ if $v \in B_j$. Thus, there is a bijective correspondence between proper edge sets $F'$ and node subsets $A' \cup B'$ of $A \cup B$, where $A' \subseteq A$ and $B' \subseteq B$. Let $F'$ be a proper edge set and let $ij \in \mathcal{E}$. Then there are edges $a_i a, b b_j \in F'_{ij}$ such that $ab \in I$ iff $G_{ij} \cup F'_{ij}$ has an $a_{ij} b_{ij}$-path (the path $(a_{ij}, a_i, a, b, b_j, b_{ij})$ of the lenght 5). Thus, $F'$ is a feasible solution for the Survivable Network instance if and only if the set $A' \cup B'$ of end-nodes in $A \cup B$ of the edges in $F'$ is a feasible solution to the Min-Rep instance. As in the construction $|V| = O(n^2)$, where $n = |A| + |B|$, the theorem follows. ∎

# 13.7 Open Problems

In the current section, we list some open problems in the field.

*Undirected* Survivable Network *with general demands and costs.*

Best known ratio: $O(k^3 \ln |T|)$. Does the problem admit ratio that depends on $k$ only, as in the case of rooted demands?

Rooted Survivable Network *with rooted subset uniform demands.*

Best known ratio: $O(k \ln k)$ for undirected graphs, and $|D|$ for directed graphs. Does the directed variant admit ratio $n^{1-\epsilon}$ for some $\epsilon > 0$? A related (harder) problem is whether the undirected variant admits ratio $k^{1-\epsilon}$.

*Undirected* Rooted Survivable Network *with $\{0, 1\}$-costs.*

Best known ratio: $O(\min\{\ln^2 k, \ln n\})$. The problem also has an $\Omega(\ln k)$ approximation threshold when $k = \Theta(n)$. Does the problem admit ratio $O(\ln k)$? This is open even for Star Survivable Network with rooted demands.

*Undirected* Survivable Network *with metric costs.*

Best known ratio: $O(\ln k)$ for general demands, and 24 for subset uniform and rooted subset uniform demands. Is the general demands case indeed $\Omega(\ln k)$ hard to approximate, or can we get a constant ratio? Can we substantially improve the constant ratio 24 for subset uniform and rooted subset uniform demands?

# References

1. G. Kortsarz and Z. Nutov. Approximating minimum cost connectivity problems. In T. Gonzalez, Ed., *Approximation Algorithms and Metaheuristics*, chapter 58. Chapman & Hall, Boca Raton, FL, 2007.
2. S. Khuller. Approximation algorithms for finding highly connected subgraphs. In D. Hochbaum, Ed., *Approximation Algorithms for NP-hard Problems*, chapter 6, pp. 236–265. PWS, Boston, MA, 1995.
3. A. Frank. *Connections in Combinatorial Optimization*. Oxford University Press, Oxford, UK, 2011.
4. A. Schrijver. *Combinatorial Optimization, Polyhedra and Efficiency*. Springer-Verlag, Berlin, Germany, 2003.
5. A. Frank and É. Tardos. An application of submodular flows. *Linear Algebra Appl.*, 114/115: 329–348, 1989.
6. A. Frank. Rooted $k$-connections in digraphs. *Disc. Appl. Math.*, 157(6):1242–1254, 2009.
7. J. Byrka, F. Grandoni, T. Rothvoß, and L. Sanità. Steiner tree approximation via iterative randomized rounding. *J. ACM*, 60(1):6, 2013.

8. M. Goemans, N. Olver, T. Rothvoß, and R. Zenklusen. Matroids and integrality gaps for hypergraphic steiner tree relaxations. In *STOC*, pp. 1161–1176, 2012.

9. A. Agrawal, P. Klein, and R. Ravi. When trees collide: An approximation algorithm for the generalized steiner problem on networks. *SIAM J. Comput.*, 24(3):440–456, 1995.

10. M. Goemans and D. Williamson. A general approximation technique for constrained forest problems. *SIAM J. Comput.*, 24(2):296–317, 1995.

11. A. Vetta. Approximating the minimum strongly connected subgraph via a matching lower bound. In *SODA*, pp. 417–426, 2001.

12. A. Zelikovsky. A series of approximation algorithms for the acyclic directed steiner tree problem. *Algorithmica*, 18(1):99–110, 1997.

13. G. Kortsarz and D. Peleg. Approximating the weight of shallow Steiner trees. *Disc. Appl. Math.*, 93(2–3):265–285, 1999.

14. M. Charikar, C. Chekuri, T. Cheung, Z. Dai, A. Goel, S. Guha, and M. Li. Approximation algorithms for directed Steiner problems. *J. Algorithms*, 33:73–91, 1999.

15. C. Helvig, G. Robins, and A. Zelikovsky. Improved approximation scheme for the group steiner problem. *Networks*, 37(1):8–20, 2001.

16. E. Halperin and R. Krauthgamer. Polylogarithmic inapproximability. In *STOC*, pp. 585–594, 2003.

17. C. Chekuri, G. Even, A. Gupta, and D. Segev. Set connectivity problems in undirected graphs and the directed steiner network problem. *ACM Trans. Algorithms*, 7(2):18, 2011.

18. P. Berman, A. Bhattacharyya, K. Makarychev, S. Raskhodnikova, and G. Yaroslavtsev. Approximation algorithms for spanner problems and directed steiner forest. *Inf. Comput.*, 222:93–107, 2013.

19. M. Feldman, G. Kortsarz, and Z. Nutov. Improved approximation algorithms for directed steiner forest. *J. Comput. Syst. Sci.*, 78(1):279–292, 2012.

20. Y. Dodis and S. Khanna. Design networks with bounded pairwise distance. In *STOC*, pp. 750–759, 1999.

21. Z. Nutov. The *k*-connected subgraph problem. In T. Gonzalez, Ed., *Approximation Algorithms and Metaheuristics*, chapter 12, Chapman & Hall, 2018.

22. J. Cheriyan and A. Vetta. Approximation algorithms for network design with metric costs. *SIAM J. Disc. Math.*, 21(3):612–636, 2007.

23. A. Frank and T. Kiraly. A survey on covering supermodular functions. In W. Cook, L. Lovsz, and J. Vygen, Eds., *Research Trends in Combinatorial Optimization*, pp. 87–126. Springer, Berlin, Germany, 2009.

24. G. Kortsarz and Z. Nutov. Tight approximation for connectivity augmentation problems. *J. Comput. Syst. Sci.*, 74(5):662–670, 2008.

25. Z. Nutov. Approximating rooted connectivity augmentation problems. *Algorithmica*, 44(3): 213–231, 2006.

26. A. Frank. Augmenting graphs to meet edge-connectivity requirements. *SIAM J. Disc. Math.*, 5(1):25–53, 1992.

27. Z. Nutov. Approximating subset *k*-connectivity problems. *J. Disc. Algorithms*, 17:51–59, 2012.

28. G. Kortsarz, R. Krauthgamer, and J. R. Lee. Hardness of approximation for vertex-connectivity network design problems. *SIAM J. Comput.*, 33(3):704–720, 2004.

29. Z. Nutov. Approximating connectivity augmentation problems. *ACM Trans. Algorithms*, 6(1):11, 2009.

30. Z. Nutov. Approximating node-connectivity augmentation problems. *Algorithmica*, 63(1–2): 398–410, 2012.

31. Z. Nutov. Approximating minimum cost connectivity problems via uncrossable bifamilies. *ACM Trans. Algorithms*, 9(1):1, 2012.

32. J. Cheriyan, B. Laekhanukit, G. Naves, and A. Vetta. Approximating rooted steiner networks. In *SODA*, pp. 1499–1511, 2012.

33. K. Jain. A factor 2 approximation algorithm for the generalized Steiner network problem. *Combinatorica*, 21(1):39–60, 2001.

34. J. Chuzhoy and S. Khanna. An $O(k^3 \log n)$-approximation algorithms for vertex-connectivity survivable network design. *Theor. Comput.*, 8(1):401–413, 2012.

35. B. Laekhanukit. Parameters of two-prover-one-round game and the hardness of connectivity problems. In *SODA*, pp. 1626–1643, 2014.

36. L. Fleischer, K. Jain, and D. Williamson. Iterative rounding 2-approximation algorithms for minimum-cost vertex connectivity problems. *J. Comput. Syst. Sci.*, 72(5):838–867, 2006.

37. Z. Király, B. Cosh, and B. Jackson. Local edge-connectivity augmentation in hypergraphs is NP-complete. *Disc. Appl. Math.*, 158(6):723–727, 2010.

38. J. Cheriyan, T. Jordán, and Z. Nutov. On rooted node-connectivity problems. *Algorithmica*, 30(3):353–375, 2001.

39. Y. Lando and Z. Nutov. Inapproximability of survivable networks. *Theor. Comput. Sci.*, 410(21–23):2122–2125, 2009.

40. A. Frank and T. Jordán. Minimal edge-coverings of pairs of sets. *J. Comb. Theor. B*, 65:73–110, 1995.

41. M. Goemans, A. Goldberg, S. Plotkin, D. Shmoys, E. Tardos, and D. Williamson. Improved approximation algorithms for network design problems. In *SODA*, pp. 223–232, 1994.

42. Z. Nutov. A note on rooted survivable networks. *Inf. Process. Lett.*, 109(19):1114–1119, 2009.

43. T. Fukunaga. Approximating the generalized terminal backup problem via half-integral multiflow relaxation. In *STACS*, pp. 316–328, 2015.

44. G. Kortsarz and Z. Nutov. Approximating source location and star survivable network problems. *Theor. Comput. Sci.*, 674: 32–42, 2017.

45. L. A. Wolsey. An analysis of the greedy algorithm for the submodular set covering problem. *Combinatorica*, 2:385–393, 1982.

46. R. Raz. A parallel repitition theorem. *SIAM J. Comput.*, 27(3):763–803, 1998.

47. G. Kortsarz. On the hardness of approximating spanners. *Algorithmica*, 30(3):432–450, 2001.

# 14

# Optimum Communication Spanning Trees

Bang Ye Wu

Chuan Yi Tang

Kun-Mao Chao

## 14.1 Introduction

A spanning tree for a graph $G$ is a subgraph of $G$ that is a tree and contains all the vertices of $G$. Besides numerous network design applications, spanning trees also play important roles in several newly established research areas, such as biological sequence alignments and evolutionary tree construction. On the other hand, there exist many spanning tree problems that have been proved to be NP-hard. Thus, designing approximation algorithms for those hard spanning-tree problems has become an exciting and important field in theoretical computer science. The current chapter focuses on the approximation algorithms for constructing efficient communication spanning trees.

Let $G = (V, E, w)$ be an undirected graph with nonnegative edge length function $w$ and $\lambda(u, v)$ be the requirements for each pair of vertices. The *optimum communication spanning tree* (OCT) problem is defined as follows. For any spanning tree $T$ of $G$, the communication cost between two vertices is defined to be the requirement multiplied by the path length of the two vertices on $T$, and the communication cost of $T$ is the total communication cost summed over all pairs of vertices. Our goal is to construct a spanning tree with minimum communication cost. That is, we want to find a spanning tree $T$ such that $\sum_{u,v \in V} \lambda(u, v) d_T(u, v)$ is minimized.

The requirements in the OCT problem are arbitrary nonnegative values. By restricting the requirements, several special cases of the problem were defined in the literature. In the following, we compile a list of the problems, in which $r : V \rightarrow Z_0^+$ is a given vertex weight function, and $S \subset V$ is be a set of sources.

- $\lambda(u, v) = 1$ for each $u, v \in V$: This version is called the MINIMUM ROUTING COST SPANNING TREE (MRCT) problem.
- $\lambda(u, v) = r(u)r(v)$ for each $u, v \in V$: This version is called the OPTIMUM PRODUCT-REQUIREMENT COMMUNICATION SPANNING TREE (PROCT) problem.
- $\lambda(u, v) = r(u) + r(v)$ for each $u, v \in V$: This version is called the OPTIMUM SUM-REQUIREMENT COMMUNICATION SPANNING TREE (SROCT) problem.
- $\lambda(u, v) = 0$ if $u \notin S$: This version is called the $p$-SOURCE OCT ($p$-OCT) problem. In other words, the goal is to find a spanning tree minimizing $\sum_{u \in S} \sum_{v \in V} \lambda(u, v)d_T(u, v)$.
- $\lambda(u, v) = 1$ if $u \in S$, and $\lambda(u, v) = 0$ otherwise: This version is called the $p$-Source MRCT ($p$-MRCT) problem. In other words, the goal is to find a spanning tree minimizing $\sum_{u \in S} \sum_{v \in V} d_T(u, v)$.

Figure 14.1 depicts the relationships between the problems, and Table 14.1 gives the currently best approximation ratio of each problem.

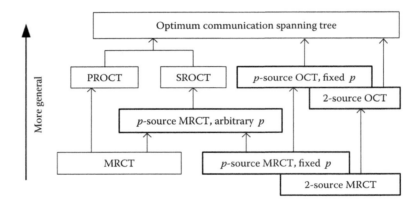

**FIGURE 14.1** The relationships between OCT problems. (From Wu, B.Y. and Chao, K.-M., *Spanning Trees and Optimization Problems*, Chapman & Hall/CRC Press, Boca Raton, FL, 2004. With Permission.)

**TABLE 14.1** The Restrictions and Currently Best Ratios of the OCT Problems

| Problem | Restriction on Requirements | Ratio | Reference |
|---|---|---|---|
| MRCT | $\lambda(u, v) = 1$ | PTAS | [2–4] |
| PROCT | $\lambda(u, v) = r(u)r(v)$ | PTAS | [5,6] |
| SROCT | $\lambda(u, v) = r(u) + r(v)$ | 2 | [5] |
| $p$-MRCT | $\lambda(u, v) = r(u) + r(v)$ | 2 | [5] |
|  | $r(s) = 1$ for $s \in S$, and $r(v) = 0$ otherwise |  |  |
| 2-MRCT | The same as $p$-MRCT but $|S| = 2$ | PTAS | [7] |
| Weighted | The same as 2-MRCT but $r(s_1) \neq r(s_2)$ | 2 (general graphs) | [7] |
| 2-MRCT |  | PTAS (metric graphs) | [7] |
| $p$-OCT | $\lambda(u, v) = 0$ for $u, v \notin S$ and $|S| = p$ is a constant | 2 (metric graphs) | [8] |
| 2-OCT | $\lambda(u, v) = 0$ for $u, v \notin S$, $|S| = 2$ | 3 (general graphs) | [8] |

The rest of the chapter is organized as follows. Section 14.2 presents some approximation schemes for the MRCT problem. We give a polynomial-time approximation scheme (PTAS) for the PROCT problem in Section 14.3, and a 2-approximation algorithm for the SROCT problem in Section 14.4. Sections 14.5 and 14.6 propose some approximation algorithms for the multiple-sources MRCT and OCT problems, respectively. Finally, Section 14.7 concludes the chapter with a few remarks.

## 14.2 Minimum Routing Cost Spanning Tree

In the MRCT problem, the requirements between any pair of vertices are the same. In other words, we want to find the spanning tree minimizing in the all-to-all distance. When there is no ambiguity, we assume that $G = (V, E, w)$ is the given graph, which is simple, connected, and undirected. In this section, $\widehat{T}$ denotes an MRCT of $G$, and $c(T)$ is the routing cost of a spanning tree $T$.

### 14.2.1 Shortest Paths Trees and Solution Decomposition

Let $r$ be the *median* of graph $G$, that is, the vertex with minimum total distance to all vertices, and $Y$ be any shortest paths tree rooted at $r$. By the triangle inequality, we have $d_Y(u, v) \leq d_Y(u, r) + d_Y(v, r)$ for any vertices $u$ and $v$. Summing up over all pairs of vertices, we obtain that $c(Y) \leq 2n \sum_v d_Y(v, r)$. As $r$ is a median, $\sum_v d_G(r, v) \leq \sum_v d_G(u, v)$ for any vertex $u$, it follows $\sum_v d_G(r, v) \leq (1/n) \sum_{u,v} d_G(u, v)$. By the property of a shortest paths tree, $d_Y(r, v) = d_G(r, v)$ for each vertex $v$, and consequently $c(Y) \leq 2n \sum_v d_G(r, v) \leq 2 \sum_{u,v} d_G(u, v)$. As $c(\widehat{T}) \geq \sum_{u,v \in V} d_G(u, v)$, we have that $Y$ is a 2-approximation of an MRCT.

**Theorem 14.1** *A shortest paths tree rooted at the median of a graph is a 2-approximation of an MRCT of the graph [2].*

The median of a graph can be found easily once the distances of all pairs of vertices are known. By Theorem 14.1, we have a 2-approximation algorithm and the time complexity is dominated by that of finding all-pairs shortest path lengths of the input graph.

**Corollary 14.1** *An MRCT of a graph can be approximated with ratio 2 in $O(n^2 \log n + mn)$ time.*

Now we introduce another proof of the approximation ratio of the shortest paths tree. The analysis technique we used is called *solution decomposition*, which is widely used in algorithm design, especially for approximation algorithms. To design an approximation algorithm for an optimization problem, we first suppose that $X$ is an optimal solution. Then we decompose $X$ and construct another feasible solution $Y$. To our aim, $Y$ is designed to be a good approximation of $X$ and belongs to some restricted class of feasible solutions, of which the best solution can be found efficiently. The algorithm is designed to find an optimal solution of the restricted problem, and the approximation ratio is ensured by that of $Y$. It should be noted that $Y$ plays a role only in the analysis of approximation ratio but not in the designed algorithm. In the following, we show how to design a 2-approximation algorithm by this method.

For any tree, we can always cut it at a node $r$ such that each branch contains at most half of the nodes. Such a node is usually called a *centroid* of the tree in the literature. Suppose that $r$ is the centroid of the MRCT $\widehat{T}$. If we construct a shortest paths tree $Y$ rooted at the centroid $r$, the routing cost will be at most twice that of $\widehat{T}$. This can be easily shown as follows. First, if $u$ and $v$ are two nodes not in a same branch, $d_{\widehat{T}}(u, v) = d_{\widehat{T}}(u, r) + d_{\widehat{T}}(v, r)$. Consider the total distance of all pairs of nodes on $\widehat{T}$. For any node $v$, as each branch contains no more than half of the nodes, the term $d_{\widehat{T}}(v, r)$ will be counted in the total distance at least $n$ times, $n/2$ times for $v$ to others and $n/2$ times for others to $v$. Hence, we have $c(\widehat{T}) \geq n \sum_v d_{\widehat{T}}(v, r)$. As in the proof of Theorem 14.1, $c(Y) \leq 2n \sum_v d_G(v, r)$, it follows that $c(Y) \leq 2c(\widehat{T})$. We have decomposed the optimal solution $\widehat{T}$ and construct a 2-approximation $Y$. Of course, there is no way to know what $Y$ is as the optimal $\widehat{T}$ is unknown. But we have the next result.

**Lemma 14.1** *There exists a vertex such that any shortest paths tree rooted at the vertex is a 2-approximation of the MRCT.*

By Lemma 14.1, we can design a 2-approximation algorithm which constructs a shortest paths tree rooted at each vertex and chooses the best of them. As there are only $n$ vertices and a shortest paths tree can be constructed in $O(n \log n + m)$ time, the algorithm runs in $O(n^2 \log n + mn)$ time, which is the same as the result stated in Corollary 14.1.

## 14.2.2 Routing Loads, Separators, and General Stars

We introduce a term, *routing load*, which provides us an alternative formula to compute the routing cost of a tree. For any edge $e \in E(T)$, let $x$ and $y$, $x \le y$, be the number of vertices in the two subtrees resulting by removing $e$. The routing load on $e$, denoted by $l(T, e)$, is $2xy = 2x(n - x)$. Notice that $x \le n/2$, and the routing load increases as $x$ increases. The following property can be easily shown by definition.

**Fact 14.1** *For any edge $e \in E(T)$, if the number of vertices in both sides of $e$ are at least $\delta n$, the routing load on $e$ is at least $2\delta(1 - \delta)n^2$. Furthermore, for any edge of a tree $T$, the routing load is upper bounded by $n^2/2$.*

Intuitively the routing load is the number of paths passing through the edge. To compute the routing load of an edge of a tree, all we need to do is to compute the number of vertices in both sides of the edge. By rooting the tree at an arbitrary vertex and traveling in a post order, we can compute the routing load of every edge in linear time.

**Lemma 14.2** *For a tree $T$ with edge length $w$, $c(T) = \sum_{e \in E(T)} l(T, e)w(e)$. In addition, $c(T)$ can be computed in $O(n)$ time.*

A key point to the 2-approximation in the last section is the existence of the centroid, which separates a tree into sufficiently small components. To generalize the idea, we define the separator as follows. Let $\delta \le 1/2$. Root a tree $T$ at its centroid, and then remove all the vertices of which the number of descendants (including itself) are equal to or less than $\delta n$. The remaining subgraph is defined as a *minimal $\delta$-separator* of $T$. Obviously, the separator is a connected subgraph. A star is a tree with only one internal vertex (center). We define a *general star* as follows.

**Definition 14.1** *Let $R$ be a tree contained in graph $G$. A spanning tree $T$ is a general star with core $R$ if each vertex is connected to $R$ by a shortest path.*

Let $S$ be a connected subgraph of a spanning tree $T$. Let $d_T(v, S)$ denote the minimum distance from $v$ to any vertex of $S$ on the graph $T$. For a graph $G$ and $u, v \in V(G)$, we use $SP_G(u, v)$ to denote a shortest path between $u$ and $v$ in $G$. For the sake of convenience, we define $d_T^S(u, v) = w(SP_T(u, v) \cap S)$. Obviously $d_T(u, v) \le d_T(v, S) + d_T^S(u, v) + d_T(u, S)$, and the equality holds if $v$ and $u$ are in different branches. Summing up the inequality for all pairs of vertices, we have

$$c(T) \le 2n \sum_{v \in V} d_T(v, S) + \sum_{u, v \in V} d_T^S(u, v).$$

By the definition of routing load,

$$\sum_{u, v \in V} d_T^S(u, v) = \sum_{e \in E(S)} l(T, e)w(e).$$

Suppose that $T$ is a general star with core $S$. We can establish an upper bound of the routing cost by observing that $d_T(v, S) = d_G(v, S)$ for any vertex $v$ and $l(T, e) \le \frac{n^2}{2}$ for any edge $e$ (Fact 14.1).

**Lemma 14.3** *If $T$ is a general star with core $S$, $c(T) \le 2n \sum_{v \in V(G)} d_G(v, S) + (n^2/2)w(S)$.*

Let $S$ be a minimal $\delta$-separator of a spanning tree $T$. The following lower bound of the minimum routing cost was established.

**Lemma 14.4** *If S is a minimal $\delta$-separator of $\widehat{T}$, then*

$$c(\widehat{T}) \geq 2(1 - \delta)n \sum_{v \in V} d_{\widehat{T}}(v, S) + 2\delta(1 - \delta)n^2 w(S).$$

## 14.2.3 Approximating by a General Star

As shown previously, a 1/2-separator is used to derive a 2-approximation algorithm. The idea is now generalized to show that a better approximation ratio can be obtained by using a 1/3-separator. Let $r$ be a centroid of $T$. There are at most 2 branches of $r$ with more than $n/3$ vertices. Therefore, there exists a path $P \subset T$ such that $P$ is a 1/3-separator of $T$, and we say that $P$ is a *path separator* of $T$.

Substituting $\delta = 1/3$ in Lemma 14.4, we obtain a lower bound of the minimum routing cost.

**Corollary 14.2** *If P is a path separator of $\widehat{T}$, then*

$$c(\widehat{T}) \geq \frac{4n}{3} \sum_{v \in V} d_{\widehat{T}}(v, P) + \frac{4n^2}{9} w(P).$$

The following result can then be shown by Lemma 14.3 and Corollary 14.2.

**Lemma 14.5** *There exist $r_1, r_2 \in V$ such that if $R = SP_G(r_1, r_2)$ and $T$ is a general star with core $R$, then $c(T) \leq (15/8)c(\widehat{T})$.*

By Lemma 14.5 we have a 15/8-approximation algorithm for the MRCT problem. For every $r_1$ and $r_2$ in $V$, we construct a shortest path $R = SP_G(r_1, r_2)$ and a general star $T$ with core $R$. The one with the minimum routing cost must be a 15/8-approximation of the MRCT. All-pairs shortest paths can be found in $O(n^3)$ time. A direct method takes $O(n \log n + m)$ time for each pair $r_1$ and $r_2$, and therefore $O(n^3 \log n + n^2 m)$ time in total. By avoiding some redundant computations, the time complexity can be reduced to $O(n^3)$, and the following result was obtained.

**Theorem 14.2** *There is a 15/8-approximation algorithm for the MRCT problem with time complexity $O(n^3)$.*

Let $P$ be a path separator of an optimal tree. By Lemma 14.3, if $X$ is a general star with core $P$, then

$$c(X) \leq 2n \sum_{v \in V} d_G(v, P) + (n^2/2)w(P).$$

By Lemma 14.2, it can be shown that $X$ is a 3/2-approximation solution. However, it costs exponential time to try all possible paths. Let $P = (p_1, p_2, \ldots, p_k)$. It is easy to see that a centroid must be in $V(P)$. Let $p_q$ be a centroid of $\widehat{T}$. Construct $R = SP_G(p_1, p_q) \cup SP_G(p_q, p_k)$. Let $T$ be any general star with core $R$. One can show that such $T$ is a 3/2 approximation. For every triple $(r_1, r_0, r_2)$ of vertices, we construct $R = SP_G(r_1, r_0) \cup SP_G(r_0, r_2)$ and find a general star with core $R$. The one with the minimum routing cost is a 3/2-approximation.

**Theorem 14.3** *The MRCT can be approximated with error ratio 3/2 in $O(n^4)$ time.*

## 14.2.4 A Reduction to the Metric Case

Let $S$ be a minimal $\delta$-separator of $\widehat{T}$. The strategy of algorithms shown earlier is to "guess" the structure of $S$ and to construct a general star with the guessed structure as the core. If $T$ is a general star with core $S$, by Lemmas 14.3 and 14.4,

$$c(T) \leq 2n \sum_{v \in V(G)} d_G(v, S) + (n^2/2)w(S)$$

and

$$c(\widehat{T}) \geq 2(1-\delta)n \sum_{v \in V} d_{\widehat{T}}(v, S) + 2\delta(1-\delta)n^2 w(S).$$

The approximation ratio, by comparing the two inequalities, is $\max\{\frac{1}{1-\delta}, \frac{1}{4\delta(1-\delta)}\}$. The ratio achieves its minimum when the two terms coincide, that is, $\delta = 1/4$, and the minimum ratio is $4/3$. In fact, by using a general star and a $(1/4)$-separator, it is possible to approximate an MRCT with ratio $(4/3) + \varepsilon$ for any constant $\varepsilon > 0$ in polynomial time. The additional error $\varepsilon$ is due to the difference between the guessed and the true separators.

By this strategy, the approximation ratio is limited even if $S$ was known exactly. The limit of the approximation ratio may be mostly due to that we consider only general stars. In a general star, the vertices are always connected to their closest vertices of the core. In extreme cases, there are roughly half of the vertices connected to the both sides of a costly edge. This results in the cost $(n^2/2)w(S)$ in the upper bound of a general star. To make a breakthrough, the restriction that each vertex must be connected to the closest vertex of the core needs to be relaxed.

A metric graph is a complete graph with triangle inequality, that is, each edge is a shortest path of its two endpoints. Define $k$-stars to be the trees with at most $k$ internal vertices. Importantly, $k$-stars have no such restriction such as general stars and can be used to approximate an MRCT more precisely. However, $k$-stars work only for metric graphs. So we should first clarify the computational complexity and the approximability of the MRCT problem on metric graphs. The following transformation provides us the answers.

The metric closure of a graph $G = (V, E, w)$ is the complete graph $\bar{G} = (V, V \times V, \bar{w})$ in which $\bar{w}(u, v) = d_G(u, v)$ for all $u, v \in V$. Any edge $(a, b)$ in $\bar{G}$ is called a *bad edge* if $(a, b) \notin E$ or $w(a, b) > \bar{w}(a, b)$. It was shown that given any spanning tree $T$ of $\bar{G}$, in $O(n^3)$ time, we can construct another spanning tree $Y$ without any bad edge such that $c(Y) \leq c(T)$. As $Y$ has no bad edge, it can be thought of as a spanning tree of $G$ with the same routing cost. As a result, we have the next theorem, in which $\Delta$MRCT denotes the MRCT problem with metric inputs. By the transformation, it is straightforward that the $\Delta$MRCT problem is NP-hard.

**Theorem 14.4** *If there is an approximation algorithm for $\Delta$MRCT with time complexity $O(f(n))$, then there is an approximation algorithm for MRCT with the same approximation ratio and time complexity $O(f(n) + n^3)$.*

## 14.2.5 A Polynomial Time Approximation Scheme

By Theorem 14.4, we may focus on metric graphs. The $k$-stars, that is, trees with no more than $k$ internal nodes, are used as a basis of the approximation scheme. The design of the PTAS consists of two parts: the existence of a $k$-star which is a $(k+3)/(k+1)$-approximation and how to compute the best $k$-star in polynomial time.

### 14.2.5.1 Approximation Ratio

Root $\widehat{T}$ at its centroid $r$. For a desired positive $\delta \leq 1/2$, removing all vertices with no more than $\delta n$ descendants, we obtain a minimal $\delta$-separator $S$. We then choose some critical vertices, defined as the *cut and leaf set*, to partition $S$ into some edge-disjoint paths, called as $\delta$-paths, each of which has only few nodes (at most $\delta n/2$) hanging at its internal nodes. It was shown that the number of the necessary critical vertices is at most $2/\delta - 3$. By the following steps, we construct a $k$-star used to argue the upper bound on the routing cost.

1. Replace the $\delta$-paths with the short-cutting edges to construct a tree structure.
2. All vertices in subtrees hanging at the cut and leaf nodes are connected directly to their closest node in the tree.

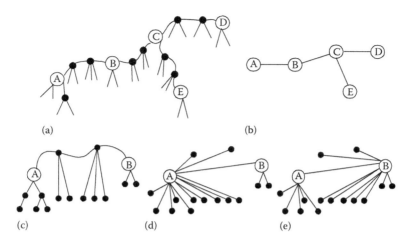

**FIGURE 14.2**  Constructing the $k$-star from an optimal tree: (a) the separators, (b) the skeletal structure, (c) the nodes hanging at a $\delta$-path, (d) connecting to the endpoint $A$, and (e) connecting to the endpoint $B$.

3. Along a $\delta$-path, all the internal nodes and nodes in subtrees hanging at internal nodes are connected to one of the two endpoints of this path (notice that both are in the cut and leaf set) in such a way as to minimize the resulting routing cost.

In Figure 14.2, we illustrate how to construct the desired $k$-star from an optimal tree. Frame (a) is an optimal tree in which the separator is shown and the cut and leaf set is $\{A, B, C, D, E\}$. Frame (b) is the tree spanning the cut and leaf nodes, which has the same skeletal structure as the separator. Frames (c), (d), and (e) illustrate how to connect other nodes to the cut and leaf nodes. Frame (c) exhibits the nodes hanging at a $\delta$-path. These nodes will be connected as in either Frame (d) or (e). The nodes hanging at the endpoints of the path will be connected to the endpoints in either case. All the internal nodes of the path and nodes hanging at the internal nodes will be connected to one of the two endpoints. Notice that they are connected to the same endpoint either as Frame (d) or Frame (e), but not connected to the two endpoints partially.

By establishing a more precise lower bound than Lemma 14.4 and using the properties of $\delta$-separator and $\delta$-paths, it can be shown that the routing cost of such a $k$-star is at most $1/(1 - \delta)$ times the optimal. For any integer $k \geq 1$, we take $\delta = \frac{2}{k+3}$ and obtain the next result.

**Lemma 14.6**  *A $k$-star of minimum routing cost is a $(k + 3)/(k + 1)$ approximation of an MRCT.*

### 14.2.5.2  Finding the Optimal $k$-Star

For a given $k$, to find an optimal $k$-star, we consider all possible subsets $S$ of vertices of size $k$ and for each such choice, find an optimal $k$-star where the remaining vertices have degree one. Any $k$-star can be described by a triple $(S, \tau, \mathcal{L})$, where $S = \{v_1, \ldots, v_k\} \subseteq V$ is the set of $k$ distinguished vertices which may have degree more than one, $\tau$ is a spanning tree topology on $S$, and $\mathcal{L} = (L_1, \ldots, L_k)$, where $L_i \subseteq V - S$ is the set of vertices connected to vertex $v_i \in S$. For any $r \in Z^+$, an $r$-vector is an integer vector with $r$ components. Let $l = (l_1, \ldots, l_k)$ be a nonnegative $k$–vector such that $\sum_{i=1}^{k} l_i = n - k$. We say that a $k$-star $(S, \tau, \mathcal{L})$ has the configuration $(S, \tau, l)$ if $l_i = |L_i|$ for all $1 \leq i \leq k$.

For a fixed $k$, the total number of configurations is $O(n^{2k-1})$ as there are $\binom{n}{k}$ choices for $S$, $k^{k-2}$ possible tree topologies on $k$ vertices, and $\binom{n-1}{k-1}$ possible such $k$-vectors. Notice that any two $k$-stars with the same configuration have the same routing load on their corresponding edges. Any edge cross the cut $(S, V - S)$ connects a leaf to a node in $S$, and therefore has the same routing load $2(n - 1)$. As all these routing loads are the same, the best way of connecting the vertices in $V - S$ to nodes in $S$ can be solved in polynomial time for a given configuration by a straightforward reduction to an instance of minimum-cost perfect matching. The minimum-cost perfect matching problem, also called the *assignment* problem, has been well studied and can be solved in $O(n^3)$ time [9]. Therefore, the overall complexity is $O(n^{2k+2})$ for finding an optimal $k$-star.

In the PTAS, we need to solve many matching problems, each for one configuration. It takes polynomial time to solve these problems individually and this result is sufficient for showing the existence of the PTAS. Although there is no obvious way to reduce the time complexity for one matching problem, the total time complexity can be significantly reduced when considering all these matching problems together. By carefully ordering the matching problems for the configurations and exploiting the common structure of two consecutive problems, it was shown that the best $k$-star for any configuration in this order can be obtained from the optimal solution of the previous configuration in $O(nk)$ time. As a result, an optimal $k$-star of a metric graph can be constructed in $O(n^{2k})$ time. The next theorem concludes the result for the PTAS of the MRCT.

**Theorem 14.5** *There is a PTAS for finding a minimum routing cost tree of a weighted undirected graph. Specifically, we can find a $(1 + \varepsilon)$-approximation solution in time $O(n^{2\lceil \frac{2}{\varepsilon} \rceil - 2})$.*

## 14.3 Product-Requirement Communication Spanning Tree

The *product-requirement communication* (or p.r.c. in abbreviation) cost of a tree $T$ is defined by $c_p(T) = \sum_{u,v} r(u)r(v)d_T(u,v)$, in which $r$ is a nonnegative vertex weight. Recall that the PTAS for the MRCT problem is obtained by showing the following properties:

1. The MRCT problem on general graphs is equivalent to the problem on metric graphs.
2. The best $k$-star is a $((k + 3)/(k + 1))$-approximation solution for the metric MRCT problem.
3. For a fixed $k$, the best $k$-star of a metric graph can be found in polynomial time.

The PROCT problem is a weighted counterpart of the MRCT problem. A vertex with weight $r(v)$ can be regarded as a super node consisting of $r(v)$ nodes of unit weight and connected by edges of zero length. In fact, the first and the second properties remain true for the PROCT problem. They can be obtained by straightforward generalizations of the previous results. However, there is no obvious way to generalize the algorithm for the minimum routing cost $k$-star to that for the minimum p.r.c. cost $k$-star. A straightforward generalization conducts to a pseudo polynomial time algorithm whose time complexity depends on the total weight of all vertices.

For the sake of convenience, we define a *balanced $k$-star* by adding a restriction that the core must be a minimal $(2/(k + 3))$-separator of the spanning tree. The performance ratio of an optimal balanced $k$-star is the same as in Lemma 14.6. Therefore, to approximate a PROCT, we can only focus on the problem of finding an optimal balanced $k$-star on a metric graph. When $k$ is a constant, the number of all possible cores is polynomial, and we may reduce the problem to the following subproblem.

> PROBLEM: Optimal Balanced $k$-Stars with a Given Core
> INSTANCE: A metric graph $G = (V, E, w)$, a tree $S$ in $G$ with $|V(S)| = k$, and a vertex weight
>     $r : V \to Z^+$.
> GOAL: Find an optimal balanced $k$-star with core $S$ if it exists.

In the current section, two approximation algorithms are presented. The first algorithm approximates a PROCT by finding the minimum p.r.c. cost 2-star, and the second one is a PTAS, which employs a PTAS for a minimum p.r.c. cost $k$-star.

For a vertex set $U$, we use $r(U)$ to denote $\sum_{u \in U} r(u)$, and $r(H) = r(V(H))$ for a graph $H$. Let $R = r(G)$ denote the total vertex weight of the input graph. Similar to a centroid of an unweighted graph, we define the $r$-centroid of a tree with vertex weight function $r$ as follows. Let $T$ be a tree with vertex weight function $r$. The *r-centroid* of a tree $T$ is a vertex $m \in V(T)$ such that if we remove $m$, then $r(H) \leq r(T)/2$ for any branch $H$. The p.r.c. routing load on the edge $e$ of a tree $T$ is defined by $l_p(T, e) = 2r(T_u)r(T_v)$, where $T_u$ and $T_v$ are the two subgraphs obtained by removing $e$ from $T$. The p.r.c. routing cost on the edge $e$ is defined to be $l_p(T, e)w(e)$. Similarly, the p.r.c. routing cost of a tree can also be computed by summing up the edge lengths multiplied by their p.r.c. routing load.

**Lemma 14.7** *Let $T$ be any spanning tree of a graph $G = (V, E, w)$ and $r$ be a vertex weight function.* $c_p(T) = \sum_{e \in E(T)} l_p(T, e)w(e)$.

## 14.3.1 Approximating by 2-Stars

The core of a 2-star is an edge and there are $O(n^2)$ possible cores. A 2-star $T$ can be represented by an edge $(x, y)$ and a bipartition $(X, Y)$ of $V$, in which $X$ and $Y$ are the sets of nodes connected to $x$ and $y$, respectively. By definition, its cost can be calculated by the following formula.

$$c_p(T) = 2r(X)r(Y)w(x, y) + 2 \sum_{v \in X} r(v)(R - r(v))w(x, v) + 2 \sum_{v \in Y} r(v)(R - r(v))w(y, v) \qquad (14.1)$$

For each possible edge $(x, y)$, the goal is to find the corresponding bipartition such that the p.r.c. routing cost is minimized. It is not hard to find that such a bipartition can be found by solving a minimum cut problem on an auxiliary graph. As the minimum cut of a graph can be found in $O(n^3)$ [10], the minimum p.r.c. cost 2-star can be found in $O(n^5)$ time. By a result similar to Lemma 14.6, such a 2-star is a $(5/3)$-approximation of a PROCT. In the following, with a more precise analysis, we show that the approximation ratio is 1.577.

Let $T$ be a PROCT of the metric graph $G$ and $p$ the $r$-centroid of $T$. We are going to construct two 2-stars and show that one of them is a 1.577-approximation of $T$. First we establish a lower bound of the optimal cost. Let $1/3 < \delta < 0.5$ be a real number to be determined later. As $\delta > 1/3$, there exists a path $P$, which is a minimal $\delta$-separator of $T$. Let $a$ and $b$ be the two endpoints of $P$. Similar to Lemma 14.4, we have

$$c_p(T) \geq 2(1 - \delta) R \sum_x r(x)d_T(x, P) + 2\delta(1 - \delta)R^2 w(P). \qquad (14.2)$$

Let $V_a$ and $V_b$ be the sets of vertices which are connected to $P$ at $a$ and $b$, respectively. Consider two 2-stars $T^*$ and $T^{**}$ with the same core $(a, b)$ and their corresponding bipartitions are $(V - V_b, V_b)$ and $(V_a, V - V_a)$, respectively. By (14.1) and the triangle inequality, we can show that

$$\min\{c_p(T^*), c_p(T^{**})\} \leq 2R \sum_{v \in V} r(v)d_T(v, P) + (1 - 2\delta^2)R^2 w(P) \qquad (14.3)$$

By (14.2) and (14.3), the approximation ratio is $\max\{1/(1 - \delta), (1 - 2\delta^2)/(2\delta(1 - \delta))\}$, in which $1/3 < \delta < 1/2$. By setting $\delta = (\sqrt{3} - 1)/2 \simeq 0.366$, we get the ratio 1.577. Combining with the time complexity of finding the minimum p.r.c. cost 2-star and the reduction from general to metric graphs, we obtain the next result.

**Theorem 14.6** *A PROCT of a general graph can be approximated with ratio 1.577 in $O(n^5)$ time.*

## 14.3.2 A Polynomial Time Approximation Scheme

We now show that the PROCT problem admits a PTAS. Instead of finding a minimum p.r.c. cost $k$-star exactly, the PTAS finds an approximation of an optimal balanced $k$-star. Let $G = (V, E, w)$ be a metric graph and $S$ a given core consisting of $k$ vertices. Let $U = V - V(S)$. The goal is to connect every vertex in $U$ to the core so as to make the p.r.c. cost as small as possible. For each vertex $v \in U$, we regard $v$ as a super node consisting of $r(v)$ nodes of weight one and connected by zero-length edges. As all these nodes have weight one, by the technique in the PTAS of MRCT, the best leaf connection can be found by solving a series of assignment problems. The time complexity is $O(R^k)$, in which $R = r(V)$ is the total vertex weight. However, the time complexity depends on the total weight of the vertices, that is, it is a pseudo polynomial time algorithm.

To reduce the time complexity, a natural idea is to scale down the weight of each vertex by a common factor. As the algorithm works only on vertices of integer weights, there are rounding errors. To ensure the quality of the solution, some details of the algorithm should be designed carefully and we need to show that the rounding errors on the vertex weights do not affect too much the cost of the solution.

By a selected threshold, we first divide $U$ into a light part and a heavy part according to their weights. Then, for each vertex in the heavy part, we scale down their weights by a scaling factor and round them to integers. When the weights are all integers, the best connection (with respect to the scaled weights) can be determined by a pseudo polynomial time algorithm. Finally, each vertex in the light part is connected to its closest vertex in $S$. It will be shown that the approximation ratio and time complexity are determined by $k$, the scaling factor, and the threshold for dividing the vertices into the light and heavy parts. The PTAS is given in the following.

---

### ALGORITHM 14.1   PTAS_Star

**Input:** A metric graph $G = (V, E, w)$ with vertex weight $r$, a tree $S$ in $G$
with $|V(S)| = k$, a positive number $\lambda < 1$ and a positive integer $q$.

**Output:** A $k$-star with core $S$.
/* assume that $V(S) = \{s_i | 1 \leq i \leq k\}$ and $U = \{1..n - k\}$
in which $r(i) \leq r(i + 1)$ for each $i \in U$. */

1.  Find the maximum $j$ such that $r(\{1 \ldots j\}) \leq \lambda R$.
    Let $V_L = \{1..j\}$ and $V_H = \{j + 1 \ldots n - k\}$ and $\mu = r(j + 1)$.
2.  Let $\bar{r}(v) = \lfloor qr(v)/\mu \rfloor$ for each $v \in V_H$;
    and $\bar{r}(v) = qr(v)/\mu$ for each $v \in V(S)$.
3.  Find an optimal $k$-star $T_1$ with respective to $\bar{r}$.
4.  Construct $T$ from $T_1$ by connecting each vertex in $V_L$
    to the closest vertex in $S$.
5.  Output $T$.

---

The time complexity and approximation ratio are shown in the next theorem. The approximation ratio approaches to 1 as $q$ and $\lambda^{-1}$ go to infinity. Therefore, for any desired approximation ratio $1 + \varepsilon > 1$, we can choose suitable $q$ and $\lambda$, and the time complexity is polynomial when they are fixed.

**Theorem 14.7** *Algorithm PTAS_Star is a PTAS for an optimal balanced k-star with a given core. For any positive integer q and positive number $\lambda < 1$, the time complexity is $O((nq/\lambda)^k)$ and approximation ratio is $((1 + q^{-1})^2 + \lambda(k + 3)^2/(k + 1))$.*

We conclude the section by the next theorem.

**Theorem 14.8** *The PROCT problem on general graphs admits a PTAS.*

## 14.4 Sum-Requirement Communication Spanning Tree

The *sum-requirement communication* (or s.r.c. in abbreviation) cost of a tree $T$ is defined by $c_s(T) = \sum_{u,v}(r(u) + r(v))d_T(u, v)$. Similar to the PROCT problem, the SROCT problem includes the MRCT problem as a special case and is therefore NP-hard. The s.r.c. cost of a tree can also be computed by summing the routing costs of edges. The only difference is the definition of routing load. We define the s.r.c. routing load on the edge $e$ to be $l_s(T, e) = 2(r(T_u)|V(T_v)| + r(T_v)|V(T_u)|)$, where $T_u$ and $T_v$ are the two subgraphs obtained by removing $e$ from $T$.

**Lemma 14.8** *Let $T$ be any spanning tree of a graph $G = (V, E, w)$ and $r$ be a vertex weight function. $c_s(T) = \sum_{e \in E(T)} l_s(T, e) w(e)$.*

For the PROCT problem, it has been shown that an optimal solution for a graph has the same value as the one for its metric closure. In other words, using bad edges cannot lead to a better solution. However, the SROCT problem has no such a property. By giving a counterexample, it was shown that a bad edge may reduce the s.r.c. cost. It is still unknown if any approximation algorithm on metric graphs ensures the same approximation ratio for general graphs.

In the current section, we introduce a 2-approximation algorithm for the SROCT problem on general graphs. For each vertex $v$, the algorithm finds the shortest paths tree rooted at $v$. Then it outputs the shortest paths tree with minimum s.r.c. cost. Obviously, the time complexity of the algorithm is $O(n^2 \log n + mn)$ as it constructs $O(n)$ shortest paths trees and each takes $O(n \log n + m)$ time. Similar to the MRCT problem, we obtain the approximation ratio by showing that there always exists a vertex $x$ such that any shortest paths tree rooted at $x$ is a 2-approximation solution.

We use $\widehat{T}$ to denote an optimal spanning tree of the SROCT problem, and use $x_1$ and $x_2$ to denote a centroid and an $r$-centroid of $\widehat{T}$, respectively. Let $P$ be the path between the two vertices $x_1$ and $x_2$ on the tree. For any edge $e \in E(P)$, let $T_1$ and $T_2$ be the two subtrees resulting by deleting $e$ from $\widehat{T}$. Assume that $x_1 \in V(T_1)$ and $x_2 \in V(T_2)$. By definition, $|V(T_1)| \geq n/2$ and $r(T_2) \geq R/2$, in which $R = r(V)$. Then,

$$l_s(\widehat{T}, e)/2 = |V(T_1)|r(T_2) + |V(T_2)|r(T_1) = 2\left(|V(T_1)| - n/2\right)\left(r(T_2) - R/2\right) + nR/2 \geq nR/2$$

By the inequality, we are able to establish a lower bound of the minimum s.r.c. cost. Recall that $d_{\widehat{T}}(v, P)$ denotes the shortest path length from vertex $v$ to path $P$.

**Lemma 14.9** $c_s(\widehat{T}) \geq \sum_{v \in V} (nr(v) + R) d_{\widehat{T}}(v, P) + nRw(P)$.

For any vertex $v$, let $f_1(v) = d_{\widehat{T}}(v, x_1) - d_{\widehat{T}}(v, P)$ and $f_2(v) = d_{\widehat{T}}(v, x_2) - d_{\widehat{T}}(v, P)$. It should be noted that $f_1(v) + f_2(v) = w(P)$. Let $Y^*$ and $Y^{**}$ be the shortest paths trees rooted at $x_1$ and $x_2$, respectively. By the triangle inequality, we can obtain that $c_s(Y^*) \leq 2 \sum_{v \in V} (nr(v) + R) (d_{\widehat{T}}(v, P) + f_1(v))$ and $c_s(Y^{**}) \leq 2 \sum_{v \in V} (nr(v) + R) (d_{\widehat{T}}(v, P) + f_2(v))$. Taking the mean of the two inequalities, we have

$$\min\{c_s(Y^*), c_s(Y^{**})\} \leq 2 \sum_{v \in V} (nr(v) + R) d_{\widehat{T}}(v, P) + 2nRw(P),$$

which is at most $2c_s(\widehat{T})$ by Lemma 14.9. The next theorem summaries the result in this section.

**Theorem 14.9** *There exists a 2-approximation algorithm with time complexity $O(n^2 \log n + mn)$ for the SROCT problem.*

## 14.5 Multiple Sources Minimum Routing Cost Spanning Tree

In the multiple sources MRCT ($p$-MRCT) problem, we are given a set $S \subset V$ of $p$ sources, and the cost function is defined by $c_m(T) = \sum_{u \in S} \sum_{v \in V} d_T(u, v)$. If there is only one source, the problem is reduced to the shortest paths tree problem, and therefore the 1-MRCT problem is polynomial time solvable. For the other extreme case that all vertices are sources, the problem is reduced to the MRCT problem, and is therefore NP-hard. By a transformation from the well-known *satisfiability problem* [11,12], it was shown that the $p$-MRCT problem is also NP-hard for any fixed $p > 1$ even when the input is a metric graph [8].

Recall that in the SROCT problem, the objective function is $c_s(T) = \sum_{u,v} (r(u) + r(v)) d_T(u, v)$ in which $r$ is the given vertex weight. By setting $r(v) = 1$ for each $v \in S$ and $r(v) = 0$ for other vertices, it is easy to see that the $p$-MRCT problem is just a special case of the SROCT problem, and therefore can be approximated with ratio two.

**Theorem 14.10** *The $p$-MRCT problem admits a 2-approximation algorithm with time complexity $O(n^2 \log n + mn)$.*

## 14.5.1 Approximating the 2-Minimum Routing Cost Spanning Tree

For the 2-MRCT problem, there are only two sources. We shall assume that $s_1$ and $s_2$ are the given sources. Let $T$ be a tree and $P$ the path between the two sources in $T$. For any $v \in V$, $d_T(v, s_1) + d_T(v, s_2) = w(P) + 2d_T(v, P)$. Summing over all vertices in $V$, we obtain that $c_m(T) = nw(P) + 2\sum_{v \in V} d_T(v, P)$. Therefore, once a path $P$ between the two sources has been chosen, it is obvious that the best way to extend $P$ to a spanning tree is to add the shortest paths forest using the vertices of $P$ as multiple roots, that is, the distance from each vertex to the path is made as small as possible. To approximate the 2-MRCT, it is natural to connect the two sources by a shortest path.

---

## ALGORITHM 14.2    2MRCT

1.  Find a shortest path $P$ between $s_1$ and $s_2$ on $G$.
2.  Find the shortest paths forest with multiple roots in $V(P)$.
3.  Output the tree $T$ which is the union of the forest and $P$.

---

We are going to show the performance of the algorithm. First we establish a lower bound of the optimum. Let $\widehat{T}$ be an optimal tree of the 2-MRCT problem. As $d_{\widehat{T}}(v, s_i) \geq d_G(v, s_i)$ for any vertex $v$ and for $i \in \{1, 2\}$, the optimal cost is lower bounded by $\sum_{v \in V} \left( d_G(v, s_1) + d_G(v, s_2) \right)$. By the triangle inequality, we may obtain another lower bound $nd_G(s_1, s_2)$. By taking the mean of the two lower bound, we have

$$c_m(\widehat{T}) \geq \frac{1}{2} \sum_v \left( d_G(v, s_1) + d_G(v, s_2) \right) + \frac{n}{2} d_G(s_1, s_2) \tag{14.4}$$

Let $T$ be the tree constructed by Algorithm 2MRCT and $P$ a shortest path between the two sources. As each vertex is connected to $P$ by a shortest path, for any vertex $v$,

$$d_T(v, P) \leq \min\{d_G(v, s_1), d_G(v, s_2)\} \leq \frac{1}{2}(d_G(v, s_1) + d_G(v, s_2))$$

Therefore $c_m(T) \leq nd_G(s_1, s_2) + \sum_v(d_G(v, s_1) + d_G(v, s_2))$. Comparing with (14.4), we have $c_m(T) \leq 2c_m(\widehat{T})$. As the total time complexity is dominated by the step of finding the shortest paths tree, we have the next result.

**Theorem 14.11** *The 2MRCT algorithm finds a 2-approximation of a 2-MRCT in $O(n \log n + m)$ time.*

The ratio shown in Theorem 14.11 is tight in the sense that there exists an instance such that the spanning tree constructed by the algorithm has a routing cost twice as the optimum. Consider a complete graph in which $w(v, s_1) = w(v, s_2) = 1$ and $w(s_1, s_2) = 2$ for each vertex $v$. The distance between any other pair of vertices is zero. At Step 1, the algorithm may find edge $(s_1, s_2)$ as the path $P$, and then all other vertices are connected to one of the two sources. The routing cost of the constructed tree is $4n - 4$. On an optimal tree, the path between the two sources is a two-edge path, and all other vertices are connected to the middle vertex of the path. The optimal routing cost is therefore $2n$. The increased cost is due to missing the vertex on the path. On the other hand, the existence of the vertex reduces the cost at an amount of $w(P)$ for each vertex.

The worst case instance of the simple algorithm gives us some intuitions to improve the error ratio. To reduce the error, we may try to guess some vertices of the path. Let $r$ be a vertex of the path between the two sources on an optimal tree, and $U$ be the set of vertices connected to the path at $r$. If the path $P$ found in Step 1 of the simple algorithm includes $r$, the distance from any vertex in $U$ to each of the sources will be no more than the corresponding distance on the optimal tree. In addition, the vertex $r$ partitions the path into two subpaths. The maximal increased cost by one of the vertices is the length of the subpath instead of the whole path. In the following, we introduce a PTAS based on this idea. For the sake of convenience, we shall first assume that $G$ is a metric graph, and the generalization to general graphs will be discussed later.

By at most $k$ vertices on the path, $P$ can be partitioned into $k + 1$ subpaths such that the number of vertices hanging at the internal nodes of each subpath is at most $n/(k + 1)$. Here a path or a subpath may consist of only one vertex. Suppose that $(P_0, P_1, \ldots, P_k)$ is such a partition of $P$ and the endpoints of $P_i$ are $m_i$ and $m_{i+1}$ for each $0 \le i \le k$, in which $m_0 = s_1$ and $m_{k+1} = s_2$. Let $U_i$ be the set of vertices hanging at the internal nodes of $P_i$ in $\widehat{T}$, and $U = V - \bigcup_{0 \le i \le k} U_i$ be set of the remaining vertices. Construct a path $X$ from $P$ by replacing each subpath $P_i$ with the short-cut edge $(m_i, m_{i+1})$, and extend $X$ to a spanning tree $T$ by adding the shortest paths forest using the vertices of $X$ as multiple roots. By the construction of $X$, it is obvious that $w(X) \le w(P)$. Similar to the proof of the 2-approximation, for any vertex $v \in U_i$, $0 \le i \le k$, the routing cost of $v$ is increased by at most $w(P_i)$. As $|U_i| \le \frac{n}{k+1}$ and $\sum_i w(P_i) = w(P)$, the total increased cost is upper bounded by $nw(P)/(k + 1)$. As $nw(P)$ is a lower bound of the optimal, $T$ is a $((k + 2)/(k + 1))$-approximation of $\widehat{T}$.

Thus, once we correctly guess the $k$ vertices $m_i$, we can construct an approximation of the 2-MRCT with ratio $(k + 2)/(k + 1)$. By trying all possible $k$-tuples, we can ensure such an error ratio in $O(n^{k+1})$ time as there are $O(n^k)$ possible $k$-tuples and it takes $O(kn)$ time to connect the remaining vertices to their closest vertices in $V(X)$. For any $\varepsilon > 0$, we set $k = \lceil \frac{1}{\varepsilon} - 1 \rceil$, and the approximation ratio is $1 + \varepsilon$.

**Theorem 14.12** *The 2-MRCT problem on metric graphs admits a PTAS. For any constant $\varepsilon > 0$, a $(1 + \varepsilon)$-approximation of a 2-MRCT of a graph $G$ can be found in $O(n^{\lceil 1/\varepsilon \rceil})$ time.*

To generalize the PTAS to general graphs, the only difficulty is how to construct a tree structure playing the same role as $X$ in the above-mentioned PTAS. The following result was shown to overcome the difficulty.

**Lemma 14.10** *Let $m_0, m_1, \ldots, m_{k+1}$ be $k$ vertices in a general graph $G$ and $P$ be a path connecting the consecutive $m_i$. Given the graph $G$ and $m_i$, in $O(kn^2)$ time, we can construct a tree $X$ such that $m_i \in V(X)$ and $d_X(v, m_0) + d_X(v, m_{k+1}) \le w(P)$ for any $v \in V(X)$.*

As a result, the 2-MRCT problem on general graphs also admits a PTAS.

**Theorem 14.13** *The 2-MRCT problem on general graphs admits a PTAS. For any constant $\varepsilon > 0$, a $(1 + \varepsilon)$-approximation of a 2-MRCT of a graph $G$ can be found in $O(n^{\lceil 1/\varepsilon + 1 \rceil})$ time.*

## 14.5.2 The Weighted 2-Minimum Route Cost Spanning Tree

We now turn to a weighted version of the 2-MRCT problem. In such a problem, we want to minimize $\sum_{v \in V} (\beta_1 d_T(s_1, v) + \beta_2 d_T(s_2, v))$, in which $\beta_1$ and $\beta_2$ are given positive real number. Without loss of generality, we define the objective function as $c_m(T, \beta) = \sum_{v \in V} (\beta d_T(s_1, v) + d_T(s_2, v))$, in which $\beta \ge 1$. As in Theorem 14.10, the weighted 2-MRCT admits a 2-approximation algorithm with time complexity $O(n^2 \log n + mn)$. We shall first present a more efficient 2-approximation algorithm, and then show that the problem admits a PTAS if the input is restricted to metric graphs.

### 14.5.2.1 On General Graphs

First, we present the 2-approximation algorithm for general graphs. Basically each vertex is greedily connected to one of the two sources, and then the two sources are connected by a shortest path.

## ALGORITHM 14.3    W2MRCT

1.  Partition $V$ into $(V_1, V_2)$ such that $v \in V_1$ if
    $$(\beta + 1)d_G(v, s_1) + d_G(s_1, s_2) \le (\beta + 1)d_G(v, s_2) + \beta d_G(s_1, s_2);$$
2.  For each $V_i$, with $s_i$ as the root, find a shortest paths tree $T_i$ spanning $V_i$.
3.  Find a shortest path between $s_1$ and $s_2$, and then
    connect $T_1$ and $T_2$ by inserting an edge of the path.
4.  Output the tree obtained in the last step.

Let $\widehat{T}$ be the optimal tree and $P$ be the path from $s_1$ to $s_2$ on $\widehat{T}$. Let $f_1(v) = d_{\widehat{T}}(v, s_1) - d_{\widehat{T}}(v, P)$ and $f_2(v) = d_{\widehat{T}}(v, s_2) - d_{\widehat{T}}(v, P)$ for each vertex $v$. By the definition of routing cost, we have

$$c_m(\widehat{T}, \beta) = \sum_{v \in V}((\beta + 1)d_{\widehat{T}}(v, P) + \beta f_1(v) + f_2(v)) \tag{14.5}$$

Consider the cost in the case that vertex $v$ is connected to $s_1$ by a shortest path. As $d_G(v, s_1) \le d_{\widehat{T}}(v, P) + f_1(v)$ and $d_G(s_1, s_2) \le w(P) = f_1(v) + f_2(v)$, we have

$$(\beta + 1)d_G(v, s_1) + d_G(s_1, s_2) \le (\beta + 1)d_{\widehat{T}}(v, P) + (\beta + 2)f_1(v) + f_2(v)$$
$$= \beta d_{\widehat{T}}(v, s_1) + d_{\widehat{T}}(v, s_2) + 2f_1(v). \tag{14.6}$$

That is, the cost is increased by at most $2f_1(v)$. Similarly, in the case that $v$ is connected to $s_2$ by a shortest path, it can be shown that the cost is increased by at most $2\beta f_2(v)$. Consequently the routing cost of $v$ is increased by at most $\min\{2f_1(v), 2\beta f_2(v)\}$. By taking a weighted mean of the two terms, we have

$$\min\{2f_1(v), 2\beta f_2(v)\} \le \frac{\beta^2}{\beta^2 + 1}(2f_1(v)) + \frac{1}{\beta^2 + 1}(2\beta f_2(v))$$
$$= \frac{2\beta}{\beta^2 + 1}(\beta f_1(v) + f_2(v))$$
$$\le \beta f_1(v) + f_2(v). \tag{14.7}$$

The last step is obtained by $2\beta \le \beta^2 + 1$ as $\beta^2 + 1 - 2\beta = (\beta - 1)^2 \ge 0$. Summing over all vertices and comparing with (14.5), we have that the approximation ratio is 2.

**Theorem 14.14** *For a general graph, a 2-approximation of the weighted 2-MRCT can be found in* $O(n \log n + m)$ *time.*

### 14.5.2.2  On Metric Graphs

The main idea of the PTAS for the weighted case is similar to the unweighted one. We also try to guess $k$ vertices of the path between the two sources on the optimal tree. For each possible $k$-tuple $(m_1, m_2, \ldots, m_k)$ of vertices, we construct a path $X$ starting at $s_1$, passing through the consecutive $m_i$, and ending at $s_2$. The only difference is the way to connect the remaining vertices to the path $X$. In the unweighted case, each remaining vertex is connected to the closest vertex in $X$. In the weighted case, the vertices are also connected to minimize the routing cost. But this time, due to the different cost function, we choose the vertex $m_i$ such that $(\beta + 1)w(v, m_i) + \beta d_X(m_i, s_1) + d_X(m_i, s_2)$ is minimized.

By an analysis similar to the unweighted case, it can be shown that the constructed tree is a $\left(\frac{k+3}{k+1}\right)$-approximation of the weighted 2-MRCT. Therefore it is a PTAS but is less efficient than the one of the unweighted problem.

**Theorem 14.15** *The weighted 2-MRCT problem on metric graphs admits a PTAS. For any* $\varepsilon > 0$, *a* $(1 + \varepsilon)$-*approximation of the optimal can be found in* $O(n^{\lceil 2/\varepsilon \rceil})$ *time.*

# 14.6 Multiple Sources OCT

In the $p$-source optimum communication spanning tree ($p$-OCT) problem, the requirement between a source and a destination is any nonnegative number, whereas there is no requirement between two non source vertices. Let $T$ be a tree and $S = \{s_1, s_2, \ldots, s_p\} \subset V(T)$ the set of given sources. For any vertex $v \in V(T)$, the communication cost of $v$ on $T$ is defined by $D_T(v) = \sum_{i=1}^{p} r_i(v) d_T(v, s_i)$, where $r_i(v)$ is the given nonnegative requirement between $s_i$ and $v$. The communication cost of $T$ is defined by $c(T) = \sum_{v \in V(T)} D_T(v)$.

## 14.6.1 The $p$-OCT on Metric Graphs

### 14.6.1.1 The 2-OCT

To approximate the 2-OCT, the algorithm starts at the edge $(s_1, s_2)$, and then inserts other vertices one by one in an arbitrary order. In each iteration, we greedily connect a vertex $v$ to either $s_1$ or $s_2$ depending on the communication cost. Precisely speaking, for each non source vertex $v$, connect $v$ to $s_1$ if

$$(r_1(v) + r_2(v))w(v, s_1) + r_2(v)w(s_1, s_2) \leq (r_1(v) + r_2(v))w(v, s_2) + r_1(v)w(s_1, s_2);$$

and connect to $s_2$ otherwise. We shall show that the approximation ratio is two.

For the sake of convenience, we define some notations. Let $\widehat{T}$ be the 2-OCT and $P$ be the path between $s_1$ and $s_2$ on $\widehat{T}$. We define $f_1(v) = d_{\widehat{T}}(v, s_1) - d_{\widehat{T}}(v, P)$ and $f_2(v) = d_{\widehat{T}}(v, s_2) - d_{\widehat{T}}(v, P)$ for each vertex $v$. The next formula directly comes from the previous notations and the definition of the communication cost.

$$c(\widehat{T}) = \sum_{v \in V} ((r_1(v) + r_2(v))d_{\widehat{T}}(v, P) + r_1(v)f_1(v) + r_2(v)f_2(v)) \tag{14.8}$$

Let $T$ be the tree delivered by the greedy method. To show that $T$ is a 2-approximation, it suffices to show that $D_T(v) \leq 2D_{\widehat{T}}(v)$ for any vertex $v$. By the triangle inequality, $w(v, s_1) \leq d_{\widehat{T}}(v, P) + f_1(v)$. By a similar analysis to (14.6), we can show that if $v$ is connected to $s_1$, the cost of $v$ is increased by at most $2f_1(v)r_2(v)$, and the cost is increased by at most $2f_2(v)r_1(v)$ if $v$ is connected to $s_2$. As the vertex $v$ is connected to either $s_1$ or $s_2$ by choosing the minimum of the two costs, $D_T(v) \leq D_{\widehat{T}}(v) + \min\{2f_1(v)r_2(v), 2f_2(v)r_1(v)\}$.

As the minimum of two number is no more than their weighted mean, we have

$$\min\{2f_1(v)r_2(v), 2f_2(v)r_1(v)\} \leq \frac{r_1(v)^2}{r_1(v)^2 + r_2(v)^2} 2f_1(v)r_2(v) + \frac{r_2(v)^2}{r_1(v)^2 + r_2(v)^2} 2f_2(v)r_1(v)$$

$$= \frac{2r_1(v)r_2(v)}{r_1(v)^2 + r_2(v)^2} \times (f_1(v)r_1(v) + f_2(v)r_2(v)). \tag{14.9}$$

As $r_1(v)^2 + r_2(v)^2 - 2r_1(v)r_2(v) = (r_1(v) - r_2(v))^2 \geq 0$, we have

$$D_T(v) \leq D_{\widehat{T}}(v) + f_1(v)r_1(v) + f_2(v)r_2(v) \leq 2D_{\widehat{T}}(v). \tag{14.10}$$

**Theorem 14.16** *The greedy method finds a 2-approximation of the 2-OCT of a metric graph.*

### 14.6.1.2 The $p$-OCT

To approximate the $p$-OCT, we define the *reduced skeleton* of a tree as follows. The $S$-skeleton of $T$ is the subgraph obtained by repeatedly removing the non source leaves from $T$ until all the leaves are sources. The reduced skeleton is obtained from the skeleton by eliminating the non source vertices of degree two.

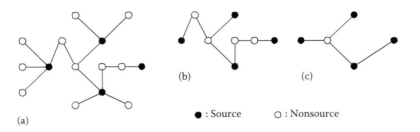

**FIGURE 14.3**   (a) A tree with four sources, (b) the skeleton, and (c) the reduced skeleton.

By eliminating a vertex of degree two, we mean that the two edges incident to the vertex is substituted by the short-cut edge. An example of the $S$-skeleton and reduced $S$-skeleton of a tree is illustrated in Figure 14.3.

The reduced skeleton is a tree spanning $S$ and possibly some other vertices. By the definition and the property of a tree structure, it is not hard to show that the number of vertices of $X$ is bounded by $2|S| - 2$. In other words, there are at most $|S| - 2$ non source vertices in $X$.

By introducing the reduced skeleton, the 2-approximation algorithm for the 2-OCT is generalized to the case of $p$ sources, where $p \geq 2$ is a constant. The approximation algorithm tries to guess the reduced $S$-skeleton $X$ of the OCT, and the other vertices are connected to one of the vertices of $X$ by making the cost as small as possible. By a technique similar to the case of two sources, it was shown that the approximation ratio is two.

The algorithm tries each tree $X$ spanning the $p$ sources and $(p - 2)$ other vertices. The total number of such trees is $\binom{n-p}{p-2}(2p-2)^{2p-4}$. For each $X$ and each $v \in V \setminus V(X)$, it takes $O(p)$ time to determine a vertex $u^* \in V(X)$ and insert edge $(v, u^*)$. The total time complexity is therefore $O(n^{p-1})$ as $p$ is a constant.

**Theorem 14.17** *For a metric graph, a 2-approximation of the p-source OCT can be found in $O(n^{p-1})$ time, where $p \geq 2$ is a constant.*

## 14.6.2 The 2-OCT on General Graphs

In the following, we shall show that the 2MRCT algorithm in Section 14.5.1 is a 3-approximation algorithm of the 2-OCT problem in the case that the input is a general graph. Remember that the algorithm finds a shortest path between the two sources and then constructs a shortest paths forest with all the vertices of the path as the multiple roots. The output tree is the union of the forest and the path. We now show the approximation ratio. Let $\widehat{T}$ be the 2-OCT and $T$ be the spanning tree obtained by the approximation algorithm. Suppose that $v$ is connected to the path $P$ at vertex $x$ of $P$. In other words, among the trees of the shortest paths forest, $x$ is the root of the tree containing $v$. Therefore, $d_T(v, x) = d_G(v, x) \leq d_G(v, s_1)$. As $P$ is a shortest path between $s_1$ and $s_2$, $d_T(s_1, x) = d_G(s_1, x) \leq d_G(s_1, v) + d_G(v, x)$. Therefore,

$$d_T(v, s_1) = d_T(v, x) + d_T(x, s_1) \leq 2d_G(v, x) + d_G(s_1, v) \leq 3d_G(v, s_1).$$

Similarly, $d_T(v, s_2) \leq 3d_G(v, s_2)$. By the definition of the communication cost, it is easy to see that $D_T(v) \leq 3D_{\widehat{T}}(v)$. As $c(T) = \sum_v D_T(v)$, we have that $c(T) \leq 3c(\widehat{T})$.

**Theorem 14.18** *The 2MRCT algorithm finds a 3-approximation of the 2-OCT of a general graph in $O(m + n \log n)$ time.*

## 14.7 Concluding Remarks

The optimum communication spanning tree problem was first discussed in Reference 13, and the NP-hardness in the strong sense of the MRCT problem was shown in Reference 14. The first constant ratio approximation algorithm for the MRCT appeared in Reference 2. More details for the approximation algorithms surveyed in this chapter can be found in Reference 1. Besides approximation algorithms, exact algorithms for the MRCT have also been studied Reference 15.

We close this chapter by mentioning a few open problems. First, it would be interesting to improve the approximation ratio for the weighted 2-MRCT of a general graph. By the previous result for the SROCT problem, the $p$-MRCT admits a 2-approximation algorithm for arbitrary $p$ and for both weighted and unweighted cases. The 2-approximation algorithm in Section 14.5 only improves the time complexity. Although there is a PTAS for metric graphs, we did not find a similar result for general graphs.

Another open problem is how to approximate the $p$-OCT of general graphs. The $O(n^{p-1})$-time algorithm in Section 14.6 only works for metric graphs, and we did not find a similar result for general graphs. It would be nice to find a more efficient algorithm to approximate the $p$-OCT with good ratio. Another interesting issue concerns the development of an approximation scheme, by which one can control the trade-off between the time complexity and the approximation ratio.

## References

1. Wu, B.Y. and Chao, K.-M., *Spanning Trees and Optimization Problems*, Chapman & Hall/CRC Press, Boca Raton, FL, 2004.
2. Wong, R., Worst-case analysis of network design problem heuristics, *SIAM J. Algebra. Discr.*, 1, 51, 1980.
3. Wu, B.Y., Chao, K.-M., and Tang, C.Y., Approximation algorithms for the shortest total path length spanning tree problem, *Disc. Appl. Math.*, 105, 273, 2000.
4. Wu, B.Y., Lancia, G., Bafna, V., Chao, K.-M., Ravi, R., and Tang, C.Y., A polynomial time approximation scheme for minimum routing cost spanning trees, *SIAM J. Comput.*, 29, 761, 1999.
5. Wu, B.Y., Chao, K.-M., and Tang, C.Y., Approximation algorithms for some optimum communication spanning tree problems, *Disc. Appl. Math.*, 102, 245, 2000.
6. Wu, B.Y., Chao, K.-M., and Tang, C.Y., A polynomial time approximation scheme for optimal product-requirement communication spanning trees, *J. Algorithms*, 36, 182, 2000.
7. Wu, B.Y., A polynomial time approximation scheme for the two-source minimum routing cost spanning trees, *J. Algorithms*, 44, 359, 2002.
8. Wu, B.Y., Approximation algorithms for the optimal p-source communication spanning tree, *Disc. Appl. Math.*, 143, 31, 2004.
9. Ahuja, R.K., Magnanti, T.L., and Orlin, J.B., *Network Flows—Theory, Algorithms, and Applications*, Prentice-Hall, Englewood Cliff, NJ, 1993.
10. Cormen, T.H., Leiserson, C.E., and Rivest, R.L., *Introduction to Algorithms*, MIT Press, Cambridge, MA, 1994.
11. Cook, S.A., The complexity of theorem-proving procedures, in *Proceedings of the third annual ACM symposium on theory of computing (STOC)*, pp. 151–158, ACM Press, New York.
12. Garey, M.R. and Johnson, D.S., *Computers and Intractability: A Guide to the Theory of NP-Completeness*, W.H. Freeman and Company, San Francisco, CA, 1979.
13. Hu, T.C., Optimum communication spanning trees, *SIAM J. Comput.*, 3, 188, 1974.
14. Johnson, D.S., Lenstra, J.K., and Rinnooy Kan, A.H.G., The complexity of the network design problem, *Networks*, 8, 279, 1978.
15. Fischetti, M., Lancia, G., and Serafini, P., Exact algorithms for minimum routing cost trees, *Networks*, 39, 161, 2002.

# 15

# Activation Network Design Problems

Zeev Nutov

## 15.1 Introduction

In Network Design problems, the goal is to select a "cheap" graph that satisfies some property $\mathcal{G}$, meaning that the graph belongs to a family $\mathcal{G}$ of subgraphs of a given graph $G$. Many fundamental properties can be characterized by **degree demands** (existence of a given number of edges incident to a node) or pairwise **connectivity demands** (existence of a given number of disjoint paths between node pairs). Traditionally, "cheap" means that the edges of the input graph $G = (V, E)$ have costs $\mathbf{c} = \{c_e : e \in E\}$, and the cost of a subgraph of $G$ is the sum of the costs of its edges. Some classic examples of "low demands" problems are Edge-Cover, $st$-Path, Spanning Tree, Steiner Tree, Steiner Forest, Out-Arborescence, and others. Examples of "high demands" problems are Edge-Multi-Cover, $k$ Disjoint Paths, $k$-Out-Connected Subgraph, $k$-Connected Subgraph, and others. Refer to, for example, References 1 and 2, for polynomial time solvable problems of this type. Here we discuss Activation Network Design problems, in which we seek an assignment $\mathbf{a} = \{a_v \geq 0 : v \in V\}$ to the nodes, such that the activated graph $G_{\mathbf{a}} = (V, E_{\mathbf{a}})$ satisfies a given property, and the value $\mathbf{a}(V) = \sum_{v \in V} a_v$ of the assignment is minimized. We now give three examples of such problems.

*Node-weighted network design.*

Here we have node-weights $\mathbf{w} = \{w_v : v \in V\}$ instead of edge-costs. The goal is to find a node subset $V' \subseteq V$ of minimum total weight $\mathbf{w}(V') = \sum_{v \in V'} w_v$, such that the graph $(V, E')$ satisfies the given property, in which $E'$ is the set of edges with both endnodes in $V'$. This can be formulated as an activation problem, in which the graph $G_{\mathbf{a}} = (V, E_{\mathbf{a}})$ activated by an assignment $\mathbf{a}$ has edge set $E_{\mathbf{a}} = \{uv : a_u \geq w_u, a_v \geq w_v\}$.

*Min-power network design.*

Consider the following scenario with motivation in wireless networks. We are given a set $V$ of nodes (transmitters) and power thresholds $\mathbf{p} = \{p_{uv} : uv \in V\}$, in which $p_{uv}$ is the minimum power (energy level) needed at $u$ to reach $v$. If $u$ can reach $v$, then we can include the directed edge $uv$ in the activated communication graph. The goal is to find an assignment $\mathbf{a} = \{a_v : v \in V\}$ of power levels to the nodes such that the activated directed graph $G_{\mathbf{a}} = (V, E_{\mathbf{a}})$, in which $E_{\mathbf{a}} = \{uv : a_u \geq p_{uv}, u, v \in V\}$ satisfies the given property. Often one is interested in the undirected network in which we have an edge between $u$ and $v$ if each of $u, v$ can reach the other; namely, we have $p_{uv} = p_{vu}$ for all $u, v \in V$, and the activated graph is undirected and has edge set $E_{\mathbf{a}} = \{uv : a_u, a_v \geq p_{uv}, u, v \in V\}$.

*Installation network design.*

Suppose that the installation cost of a wireless network is dominated by the cost of building towers at the nodes for mounting antennas, which in turn is proportional to the height of the towers. An edge $uv$ is activated if the towers at its endpoints $u$ and $v$ are tall enough to overcome obstructions in the middle and establish line of sight between the antennas mounted on the towers. This is modeled as each edge $uv$ has a height-threshold requirement $h_{uv}$, and an edge $uv$ is activated if the scaled heights $s_{uv}a_u, s_{vu}a_v$ at its endpoints sum to at least $h_{uv}$. Namely, the activated graph $G_{\mathbf{a}} = (V, E_{\mathbf{a}})$ is undirected and has edge set $E_{\mathbf{a}} = \{uv : s_{uv}a_u + s_{vu}a_v \geq h_{uv}\}$.

Panigrahi [3] suggested the following common generalization of these and several other problems.

**Definition 15.1 (Panigrahi [3])** *Let $G = (V, E)$ be a graph such that each edge $e = uv \in E$ has an* **activating function** *$f^e = f^{uv}$ from $W^{uv} \subseteq \mathbb{R}^2_+$ to $\{0, 1\}$, in which $f^{uv}(x_u, x_v) = f^{vu}(x_v, x_u)$ if $e$ is an undirected edge. Given a nonnegative* **assignment** *$\mathbf{a} = \{a_v : v \in V\}$ on $V$, we say that an edge $uv \in E$ is* **activated** by *$\mathbf{a}$ if $f^{uv}(a_u, a_v) = 1$. Let $E_{\mathbf{a}} = \{uv \in E : f^{uv}(a_u, a_v) = 1\}$ denote the set of edges activated by $\mathbf{a}$. The* **value** *of an assignment $\mathbf{a}$ is $\mathbf{a}(V) = \sum_{v \in V} a_v$.*

---

**Activation Network Design**

*Input:* A graph $G = (V, E)$, a family $\mathbf{f} = \{f^{uv}(x_u, x_v) : e = uv \in E\}$ of activating functions from $W^{uv} \subseteq \mathbb{R}^2_+$ to $\{0, 1\}$ each, and a graph property $\mathcal{G}$ (namely, a family $\mathcal{G}$ of subgraphs of $G$).

*Output:* An assignment $\mathbf{a} = \{a_v \geq 0 : v \in V\}$ of minimum value $\mathbf{a}(V) = \sum_{v \in V} a_v$ such that the graph $G_{\mathbf{a}} = (V, E_{\mathbf{a}})$ activated by $\mathbf{a}$ satisfies $\mathcal{G}$.

---

We consider degree and connectivity variants of the above-mentioned problem. In general, the input graph $G$ may have parallel edges with distinct activation functions. For simplicity of exposition, we will assume that $G$ is a simple graph, and use $uv$ to denote the edge from $u$ to $v$; note that if $G$ is an undirected graph, then $uv = vu$ and $f^{uv} = f^{vu}$. In what follows, we will make the following assumptions about the activating functions.

**Assumption 15.1 (Monotonicity)** *For every edge $uv \in E$, $f^{uv}$ is monotone nondecreasing, namely, $f^{uv}(x_u, x_v) = 1$ implies $f^{uv}(y_u, y_v) = 1$ whenever $y_u, y_v \in W^{uv}$, $y_u \geq x_u$, and $y_v \geq x_v$.*

**Assumption 15.2 (Polynomial Domain)** *For every $uv \in E$, $W^{uv} = W^u \times W^v$ in which $|W^u|, |W^v|$ are bounded by a polynomial in $n = |V|$.*

For a node $v$ we call $W^v$ the set of **levels of** $v$. Note that the Monotonicity Assumption holds for the above-mentioned three examples, and we are not aware of any practical problem when it does not hold. The Polynomial Domain Assumption also holds in many applications; moreover, making this assumption often incurs only a small loss in the approximation ratio. Assumptions 1 and 2 are the default in this survey, but often we can replace the Polynomial Domain Assumption by the weaker assumption as follows:

**Assumption 15.3 (Polynomial Computability)** *For any $uv \in E$ we can compute in polynomial time $a_u, a_v$ with $f^{uv}(a_u, a_v) = 1$ and $a_u + a_v$ is minimum.*

Let us discuss directed Activation Network Design problems. Then we study the case when each activating function $f^{uv}$ depends only on the assignment at the tail $u$ of the edge $uv$, so it is a function $f^{uv}(x_u) = f^{uv}(x)$ of one variable. Then by the Monotonicity Assumption, each edge $uv$ has a threshold $p_{uv}$ such that $f^{uv}(x) = 1$ iff $x \geq p_{uv}$. This gives the directed min-power variant discussed earlier, in which $p_{uv}$ is the minimum power needed at $u$ to reach to $v$. Consequently, the directed variant can be stated as follows.

---

**Directed Activation Network Design (Directed Min-Power Network Design)**

*Input:* A graph $G = (V, E)$ with power thresholds $\mathbf{p} = \{p_e : e \in E\}$ and a property $\mathcal{G}$.

*Output:* An assignment $\mathbf{a} = \{a_v \geq 0 : v \in V\}$ of minimum value $\mathbf{a}(V) = \displaystyle\sum_{v \in V} a_v$ such that the graph $G_\mathbf{a} = (V, E_\mathbf{a})$ activated by $\mathbf{a}$ satisfies $\mathcal{G}$, where $E_\mathbf{a} = \{uv \in E : a_u \geq p_{uv}, \ u, v \in V\}$.

---

We now specify the degree and connectivity problems to be considered. In both types of problems, we are given a graph $G = (V, E)$ and certain nonnegative integral demands (a.k.a. requirements). In degree problems, we have **degree demands** $\mathbf{r} = \{r_v : v \in V\}$ and in connectivity problems we have **connectivity demands** $\mathbf{r} = \{r_{st} : s, t \in V\}$. In both cases, we use $k$ to denote the maximum demand. In the case of degree demands, we say that a graph $(V, J)$ (or $J$) **satisfies** $\mathbf{r}$ if $\deg_J(v) \geq r_v$ for all $v \in V$, in which $\deg_J(v)$ denotes the degree of $v$ in the graph $(V, J)$. In the case of connectivity demands, we say that $(V, J)$ (or $J$) **satisfies** $\mathbf{r}$ if the graph $(V, J)$ contains $r_{st}$ pairwise disjoint $st$-paths for all $s, t \in V$. In edge-connectivity problems, the path should be edge disjoint, whereas in node-connectivity problems, the paths should be internally disjoint. In the Edge-Multi-Cover problem we need to satisfy degree demands, whereas in the Survivable Network problem, we need to satisfy connectivity demands. Let us state the min-cost versions of these problems formally.

---

**Min-Cost Edge-Multi-Cover**

*Input:* A graph $G = (V, E)$ with edge-costs $\mathbf{c} = \{c_e : e \in E\}$ and degree demands $\mathbf{r} = \{r_v : v \in V\}$.

*Output:* A minimum-cost edge set $J \subseteq E$ that satisfies $\mathbf{r}$.

---

**Min-Cost Survivable Network**

*Input:* A graph $G = (V, E)$ with edge-costs $\mathbf{c} = \{c_e : e \in E\}$ and connectivity demands $\mathbf{r} = \{r_{st} : s, t \in V\}$.

*Output:* A minimum-cost edge set $J \subseteq E$ that satisfies $\mathbf{r}$.

---

In activation version of these problems—Activation Edge-Multi-Cover and Activation Survivable Network, instead of edge-costs we have activating functions $\mathbf{f} = \{f^e : e \in E\}$ and seek an assignment $\mathbf{a} = \{a_v : v \in V\}$ to the nodes with $\mathbf{a}(V) = \displaystyle\sum_{v \in V} a_v$ minimum such that the graph $G_\mathbf{a} = (V, E_\mathbf{a})$ activated by **a** satisfies the demands. In Node-Weighted Edge-Multi-Cover and Node-Weighted Survivable Network, we have node-weights $\mathbf{w} = \{w_v : v \in V\}$ and $E_\mathbf{a} = \{uv : a_u \geq w_u, a_v \geq w_v\}$. In Min-Power Edge-Multi-Cover and Min-Power Survivable Network, we have power thresholds $\mathbf{p} = \{p_{uv} : uv \in E\}$ and $E_\mathbf{a} = \{uv \in E : a_u, a_v \geq p_{uv}\}$. In the directed general case, we also have thresholds $\mathbf{p} = \{p_{uv} : uv \in E\}$ but $E_\mathbf{a} = \{uv \in E : a_u \geq p_{uv}\}$. Moreover, note that in directed Edge-Multi-Cover problems, we may have both outdegree and indegree demands $\{(r_v, r^{in}(v)) : v \in V\}$, and $J \subseteq E$ is a feasible solution if $\deg_J(v) \geq r_v$ and $\deg_J^{in}(v) \geq r^{in}(v)$ for all $v \in V$, in which $\deg_J(v)$ and $\deg_J^{in}(v)$ denote the outdegree and the indegree of $v$ in the graph $(V, J)$, respectively. In what follows, the Edge-Multi-Cover problem with $0, 1$ demands will be called the Edge-Cover problem.

We summarize the best known ratios for undirected problems with $0, 1$ demands in Table 15.1.

**TABLE 15.1** Best Known Approximation Ratios for Low Demands Undirected Activation Problems

| Problem/Activation fn. | General | Node-Weighted | Power | Cost |
|---|---|---|---|---|
| $st$-Path | In P [3] | In P | In P [4] | In P |
| Spanning Tree | $O(\ln n)$ [3] | In P | 1.5 [5] | In P |
| Steiner Tree | $O(\ln n)$ [3] | $O(\ln n)$ [6] | $3\ln 4 - \frac{9}{4} + \epsilon$ [5] | $\ln 4 + \epsilon$ [7] |
| Steiner Forest | $O(\ln n)$ [3] | $O(\ln n)$ [6] | 4 [8] | 2 [9] |
| Edge-Cover | $O(\ln n)$ | $O(\ln n)$ | 1.5 [8] | in P |

The known approximability of installation problems coincides with those known for the general case. The problems in the table that have ratio $O(\ln n)$ are Set-Cover hard [6], and thus have an approximation threshold $\Omega(\ln n)$.

We now describe the high demands connectivity problems that we consider. Each problem can be defined on directed or undirected graphs. Let us state the node-connectivity versions of these problems.

- $k$ Disjoint Paths: Here $r_{st} = k$ for a given pair of nodes $s, t \in V$ and $r_{uv} = 0$ otherwise, namely, the solution graph should contain $k$ internally disjoint $st$-paths. For $k = 1$, we get the $st$-path problem.
- $k$-Out-Connectivity and $k$-In-Connectivity: A graph is $k$-**out-connected from** $s$ if it contains $k$ internally disjoint paths from $s$ to any other node; similarly, a graph is $k$-**in-connected to** $s$ if it contains $k$ internally disjoint paths from every node to $s$ (for undirected graphs these two concepts mean the same). In $k$-Out-Connectivity problems, the activated graph should be $k$-out-connected from $s$, namely, it should satisfy the node-connectivity demands $r_{st} = k$ for all $t \in V \setminus \{s\}$. In $k$-In-Connectivity problems, the activated graph should be $k$-in-connected to $s$. For $k = 1$, we get the Spanning Tree problem in the undirected case, and the problems Out-Arborescence and In-Arborescence in the directed case.
- $k$-Connectivity: Here the graph should be $k$-connected, namely, it should satisfy the node-connectivity demands $r_{st} = k$ for all $s, t \in V$. For $k = 1$, we get the Spanning Tree problem in the undirected case, and the Strong Connectivity problem in the directed case.

The corresponding edge connectivity problems are $k$ Edge-Disjoint Paths, $k$-Edge-Out-Connectivity, $k$-Edge-In-Connectivity, and $k$-Edge-Connectivity (for undirected graphs the later three problems are equivalent). We abbreviate Edge-Connectivity Survivable Network by EC-Survivable Network. We summarize the best known ratios for high demands undirected and directed activation problems in Tables 15.2 and 15.3; for the best known ratios for min-cost connectivity problems see surveys [10,11].

High demands edge-connectivity problems in the last table are $\Omega(2^{\ln^{1-\varepsilon} n})$-hard, assuming NP has no quasi-polynomial time algorithms [13,15]. The corresponding undirected problems are "Densest $k$-Subgraph hard" [15,18], meaning that if the problem admits ratio $\rho$, then the Densest $k$-Subgraph problem admits ratio $O(\rho^2)$. The Densest $k$-Subgraph problem was studied extensively, and the best ratio known for it is $\Omega(n^{1/4+\epsilon})$ [22].

**TABLE 15.2**  Best Known Ratios for High Demands Undirected Activation Problems

| Problem/Activation fn. | General | Node-Weighted | Power |
|---|---|---|---|
| $k$ Disjoint Paths | 2 [12] | In P | 2 [13] |
| $k$-Out-Connectivity | $O(k \ln n)$ [12] | In P | $\min\{k + 1, O(\ln k)\}$ [14–16] |
| $k$-Connectivity | $O(k \ln n)$ [12] | In P | $O\left(\ln k \ln \frac{n}{n-k}\right)$ [16,17] |
| $k$ Edge-Disjoint Paths | $k$ [15] | $k$ [15,18] | $k$ [15] |
| $k$-Edge-Connectivity | $O(k \ln n)$ [12] | In P | $\min\{2k - 1/2, O(\sqrt{n})\}$ [8,13] |
| EC-Survivable Network | $O(k \ln n)$ [12] | $O(k \ln n)$ [18] | $4k$ [8] |
| Edge-Multi-Cover | $O(k \ln n)$ | $O(\ln n)$ | $\min\{k + 1/2, O(\ln k)\}$ [16] |

**TABLE 15.3**  Best Known Ratios for Directed Activation Problems

| Problem | | Problem | Node-Connectivity | Edge-Connectivity |
|---|---|---|---|---|
| $st$-Path | In P | $k$ Disjoint Paths | In P [13] | $k$ [15] |
| In-Arborescence | In P | $k$-In-Connectivity | In P [15] | $k$ [19] |
| Out-Arborescence | $O(\ln n)$ [20,21] | $k$-Out-Connectivity | $O(k \ln n)$ [12] | $O(k \ln n)$ [19] |
| Strong Connectivity | $O(\ln n)$ [20,21] | $k$-Connectivity | $O(k \ln n)$ [12] | $O(k \ln n)$ [19] |

*Notation.*

An edge from $u$ to $v$ is denoted by $uv$. An $st$-**path** is a path from $s$ to $t$. For sets $A, B$ of nodes and edges (or graphs) $A \backslash B$ is the set (or graph) obtained by deleting $B$ from $A$, in which deletion of a node implies deletion of all the edges incident to it; similarly, $A \cup B$ is the set (graph) obtained by adding $B$ to $A$. A set of values given to nodes or edges of a graph is denoted by a bold letter and treated as vector, for example, $\mathbf{w} = \{w_v : v \in V\}$ usually denotes node-weights, and $\mathbf{w} \cdot \mathbf{x} = \sum_{v \in V} w_v x_v$ for another vector $\mathbf{x} = \{x_v : v \in V\}$. For $V' \subseteq V$, let $\mathbf{w}(V') = \sum_{v \in V'} w_v$. Given a directed/undirected graph or an edge set $J$, $\delta_J(v)$ is the set of edges in $J$ leaving $v$ and $\deg_J(v) = |\delta_J(v)|$ the **degree** of $v$ in $J$; $\Delta_J = \max_{v \in V} \deg_J(v)$ denotes the maximum degree of a node w.r.t. $J$. For directed graphs, degree and outdegree means the same, and $\delta_J^{in}(v)$ and $\deg_J^{in}(v)$ denotes the set of edges in $J$ entering $v$ and the in-degree of $v$ in $J$, respectively. For a Network Design problem instance, $k$ denotes the maximum demand and opt the optimal solution value.

*Organization.*

The rest of this survey is organized as follows. In Sections 15.2 through 15.4, we give some approximation and exact algorithms using various simple reductions. In Section 15.5, we consider the undirected Min-Power Edge-Multi-Cover problem. In Section 15.6, we survey the algorithm of Klein and Ravi [6] for the Node-Weighted Steiner Forest problem. In Section 15.7, we consider the directed Min-Power $k$-Edge-Out-Connectivity problem. We conclude in Section 15.8 with some open problems.

# 15.2 Levels Reduction

Here we discuss a method which we call the **Levels Reduction** that converts an Activation Network Design problem into a Node-Weighted Network Design problem. This reduction was designed in Reference 15 to solve certain min-power problems, but the reduction and the analysis extend to several activation problems.

**Definition 15.2** *The **levels graph** of a graph $(V, E)$ with a family $\mathbf{f} = \{f^e : e \in E\}$ of activating functions is a node-weighted graph obtained as follows. For every $v \in V$, take a star with center $v = v(0)$ of weight 0, in which for each level $\ell \in W^v \setminus \{0\}$, the star has a leaf $v(\ell)$ of weight $\ell$; then replace every edge $uv \in E$ by the edge set $\{u(i)v(j) : f^{uv}(i, j) = 1\}$.*

Note that all nodes of the original graph are present in the levels graph and have weight 0. We consider a natural algorithm that given an Activation Network Design problem computes a solution $J'$ to the corresponding node-weighted problem in the levels graph. The algorithm returns the assignment $\mathbf{a}$ defined by the set $V'$ of endnodes of $J'$, namely, $a_v = \max\limits_{v(\ell) \in V'} \ell$ for all $v \in V$, in which the maximum taken over the empty set is assumed to be 0. Clearly, we have $\mathbf{a}(V) \le \mathbf{w}(V')$, but the computed solution may not be feasible. To show that the algorithm is both valid and preserves approximability, we need to show that:

(i) For any feasible solution $J'$ to the node-weighted problem in the levels graph, the assignment $\mathbf{a}$ defined by $J'$ is a feasible solution to the original activation problem.

(ii) For any feasible assignment $\mathbf{a}$ to the original activation problem, the node-weighted problem in the levels graph has a feasible solution of weight at most the value $\mathbf{a}(V)$ of $\mathbf{a}$.

Consider, for example, the Activation $st$-Path problem. Let $P'$ be an $st$-path in the levels graph. If $P'$ contains two leaves of the same star say $v(i)$ and $v(j)$ with $j > i$, then we can "shortcut" $P'$ by replacing the subpath of $P'$ between $v(j)$ and a neighbor $q$ of $v(i)$ on $P$ by the edge $v(j)q$, in which $q$ is the successor of $v(i)$ in $P$ if $v(j)$ precedes $v(i)$, and $q$ is a predecessor of $v(i)$ in $P$ otherwise. This is possible as $j > i$, and thus by the Monotonicity Assumption, if $v(i)q$ is an edge in the level graph, then so is $v(j)q$. Thus, we may assume that $P'$ contains at most one leaf from each star. By the construction, the assignment $\mathbf{a}$ defined by $J'$ activates an $st$-path in the original graph, and property (i) above-mentioned holds. Conversely, if $\mathbf{a}$ is an assignment activating an $st$-path $P$ in the original graph, then the edge set $\{(vv(a_v)) : v \in V\} \cup \{(v(a_v)u(a_u)) : uv \in P\}$ in the levels graph contains an $st$-path of weight $\mathbf{a}(V)$, and property (ii) above-mentioned holds. As the Node-Weighted $st$-Path problem can be solved in polynomial time, we obtain a polynomial time algorithm for the Activation $st$-Path problem.

Consider the Activation Steiner Forest problem. Recall that in this problem we are given an undirected graph $G = (V, E)$ and a set $R$ of demand node pairs, and seek to activate $J \subseteq E$ such that the graph $(V, J)$ has an $st$-path for every demand pair $\{s, t\} \in R$. The node-weighted version of the problem admits ratio ratio $2 \ln |U|$, in which $U$ is the union of the demand pairs, see Section 15.6. Note that the parameter $|U|$ in the levels graph of $G$ is the same as in $G$. Thus by the same analysis as in the $st$-Path case we have:

**Theorem 15.1 (Panigrahi [3])** *If Node-Weighted Steiner Forest admits ratio $\rho(|U|)$, then so is Activation Steiner Forest, in which $U$ is the union of the demand pairs. Thus, Activation Steiner Forest admits ratio $2 \ln |U|$.*

Let us consider a slightly more complicated example of the Levels Reduction. The Activation $k$ Edge-Disjoint Paths Augmentation problem is a particular case of the Activation $k$ Edge-Disjoint Paths problem when the edge set of the input graph $G = (V, E)$ contains an "initial" edge set $E_0 \subseteq E$ activated by the zero assignment, such that $E_0$ is a union of $k - 1$ pairwise edge-disjoint $st$-paths. The goal is to find an assignment that activates an edge set $J \subseteq E \setminus E_0$ such that the graph $(V, E_0 \cup J)$ contains $k$ pairwise edge-disjoint $st$-paths. Clearly, ratio $\rho$ for this augmentation problem implies ratio $k\rho$ for Activation $k$ Edge-Disjoint Paths. Note that for $k = 1$ and $E_0 = \emptyset$, we get the Activation $st$-Path problem.

**Theorem 15.2 (Lando & Nutov [15])** *Undirected Activation $k$ Edge-Disjoint Paths Augmentation admits a polynomial time algorithm. Thus, undirected Activation $k$ Edge-Disjoint Paths admits ratio $k$.*

*Proof.* Let $D_0$ be a set of directed edges obtained by directing $k - 1$ pairwise edge-disjoint $st$-paths in $(V, E_0)$ from $t$ to $s$. A graph that has both directed and undirected edges will be called a **mixed graph**. In the following algorithm, the Activation $k$ Edge-Disjoint Paths Augmentation problem is reduced to

 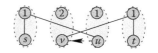

**FIGURE 15.1** Illustration to Algorithm 15.1 for $k = 2$. Edges in $E_0$ are shown by dashed lines, $v$ has level set $W^v = \{0, 2\}$, and any other node has level set $\{0, 1\}$. The edge $su$ is activated if $x_s + x_u \geq 1$ and $vt$ is activated if $x_v + x_t \geq 1$. The minimum weight $st$-path in the levels graph is given by the sequence $s = s(0), s(1), u = u(0), v = v(0), t(1), t = t(0)$, and it defines the assignment $(a_s, a_v, a_u, a_t) = (1, 0, 0, 1)$.

the Node-Weighted $st$-Path problem in a mixed graph. The later problem can be solved in polynomial time by a reduction to the directed Min-Cost $st$-Path problem (with edge-costs) by elementary constructions: replacing every undirected edge by two opposite directed edges and converting node-weights to edge-costs. Formally, the algorithm is as follows (for illustration see Figure 15.1).

---

## ALGORITHM 15.1 Activation $k$ Edge-Disjoint Paths Augmentation $((V,E),E_0,f,\{s,t\},k)$

1 let $G'$ be the mixed graph obtained from the levels graph of $(V, E \setminus E_0)$ by adding the edges in $D_0$
2 in the mixed graph $G'$ compute a minimum weight $st$-path $P'$
3 return the assignment **a** defined by the nodes of $P'$

---

We explain why the algorithm is correct. From the correctness of the Ford-Fulkerson algorithm we have:

*The graph $(V, E_0 \cup J)$ has $k$ edge-disjoint $st$-paths if and only if the mixed graph $(V, D_0 \cup J)$ has an $st$-path.*

Thus, our problem is equivalent to the Activation $st$-Path problem in the mixed graph $(V, (E \setminus E_0) \cup D_0)$, in which the set $D_0$ of directed edges is activated by the zero assignment. The Activation $st$-Path problem in a mixed graph as earlier is equivalent to the Node-Weighted $st$-Path problem in the mixed graph $G'$ obtained by adding to the levels graph of $(E \setminus E_0, V)$ the edges in $D_0$. This follows by the same argument as used for the $st$-Path problem in ordinary undirected graphs. ∎

## 15.3 Min-Cost Reduction

Let $\tau(J) = \tau_f(J)$ denote the optimal value of an assignment activating an edge set $J$; we use $\tau(e)$ instead of $\tau(\{e\})$. For node-weighted and for min-power problems we will sometimes use the notation $\tau_w(J)$ and $\tau_p(J)$, respectively. For these two problems, an optimal assignment **a** activating $J$ can be computed in polynomial time; in node-weighted problems $a_v = w_v$ if $v$ is an endnode of an edge in $J$ and $a_v = 0$ otherwise, and in min-power problems $a_v = \max_{e \in \delta_J(v)} p_e$. In general activation problems, computing $\tau_f(J)$ is NP-hard. However, we can always find in polynomial time an assignment **a** activating $J$ such that $\mathbf{a}(V) \leq \sum_{e \in J} \tau(e)$, as follows. For every $e \in J$, compute an optimal pair $(a_u^e, a_v^e) \in W^u \times W^v$ that activates $e$, so $f^e(a_u^e, a_v^e) = 1$ and $a_u^e + a_v^e = \tau(e)$; this can be done in polynomial time, by the Polynomial Computability Assumption. The assignment **a** defined by $a_v = \max_{e \in \delta_J(v)} a_v^e$ activates $J$ and has value $\mathbf{a}(V) = $

$$\sum_{v \in V} \max_{e \in \delta_J(v)} a_v^e \leq \sum_{e \in J} \tau(e).$$

Assume that we are given an instance of Activation Network Design such that the corresponding min-cost version admits ratio $\rho$. We will analyze the performance of the following natural algorithm.

## ALGORITHM 15.2    Min-Cost Reduction($G=(V,E),f,\mathcal{G}$)

1 let **c** be edge-costs defined by $c_e = \tau(e)$ for every $e \in E$
2 compute a $\rho$-approximate **c**-cost solution $J \in \mathcal{G}$
3 return an assignment **a** that activates $J$ of value $\mathbf{a}(V) \leq \sum_{e \in J} \tau(e)$

The algorithm can be implemented in polynomial time, by the Polynomial Computability Assumption. As we shall see, the algorithm has a good performance when inclusion minimal feasible solutions have small degree, or are (undirected) forests and the activating functions have small "slope." To state this formally, we need some definitions.

**Definition 15.3** *The **slope** $\theta(e)$ (of an activating function) of an undirected edge $e = uv$ is defined by $\theta(e) = \frac{\tau(e)}{\min\{\mu_u^e, \mu_v^e\}}$ in which $\mu_u^e = \min\{x_u \in W^u : f(x_u, \max_{w_v \in W^v} w_v) = 1\}$ is the minimum assignment value needed at $u$ to activate $e$. The slope of $J \subseteq E$ is defined by $\theta_J = \max_{e \in J} \theta(e)$, and we denote $\theta = \theta_E$.*

In min-power problems, the slope of every edge is exactly 2. In node-weighted problems, the slope of an edge $e = uv$ is $\theta(e) = \frac{w_u + w_v}{\min\{w_u, w_v\}} = 1 + \frac{\max\{w_u, w_v\}}{\min\{w_u, w_v\}}$. For example, if $\frac{\max\{w_u, w_v\}}{\min\{w_u, w_v\}} \leq 2$ for every edge $uv \in E$, then the slope of the instance is at most 3. The following statement which particular cases were considered in various papers [12,13,23] will enable us to estimate the approximation ratio of Algorithm 15.2 in terms of various parameters defined.

**Lemma 15.1** *Given an instance of* Activation Network Design, *for any $J \subseteq E$ the following holds:*

(i) $\sum_{e \in J} \tau(e) \leq \Delta_J \cdot \tau(J)$ *if $J$ is directed or undirected.*

(ii) $\sum_{e \in J} \tau(e) \leq \theta_J \cdot \tau(J)$ *if $J$ is a forest.*

(iii) $\sum_{e \in J} \tau(e) \leq \theta_J \sqrt{|J|/2} \cdot \tau(J)$ *if $J$ is undirected and has no parallel edges.*

*Proof.* We prove (i) for the case when $J$ is an an undirected edge set; the proof of the directed case is similar. Let $\mathbf{a}^*$ be an optimal assignment that activates $J$. Note that $\alpha(e) \leq a_u^* + a_v^*$ for every $e = uv \in J$. This implies

$$\sum_{uv \in J} \tau(uv) \leq \sum_{uv \in J}(a_u^* + a_v^*) = \sum_{v \in V} a_v^* \deg_J(v) \leq \Delta_J \sum_{v \in V} a_v^* = \Delta_J \tau(J)$$

It is sufficient to prove (ii) for the case when $J$ is a tree. Root it at some node $s$. Then for each $v \neq s$, $\mu(e(v)) \leq a_v^*$ where $e(v)$ is the parent edge of $v$. This implies

$$\sum_{e \in J} \mu(e) = \sum_{v \in V \setminus \{s\}} \mu(e(v)) \leq \sum_{v \in V} a_v^* = \tau(J)$$

As $\tau(e) \leq \theta_J \mu(e)$ for every $e \in J$, (ii) follows.

We prove (iii). Let **a** be an assignment that activates $J$. As $\tau(e) \leq \theta \min\{a_u, a_v\}$ for any edge $e = uv$, it is sufficient to prove that for any undirected simple graph $(V, J)$ with node-weights $\mathbf{w} = \{w_v : v \in V\}$

$$\sum_{uv \in J} \min\{w_u, w_v\} \leq \sqrt{|J|/2} \cdot \mathbf{w}(V)$$

The proof is by induction on the number of distinct weights in **w**. The induction base is when all nodes have the same weight. Then the above-mentioned inequality reduces to $|J| \leq n^2/2$, which holds for simple graphs. Otherwise, let $q$ be the difference between the maximum and the second maximum node-weight.

Let $V'$ be the set of maximum weight nodes and $J'$ the set of edges in $J$ with both endnodes in $V'$. Let $\mathbf{w}'$ be defined by $w'_v = w_v - q$ if $v \in V'$ and $w'_v = w_v$ otherwise. By the induction hypothesis we have

$$|J'| \leq \sqrt{|J'|/2} \cdot |V'| \quad \text{and} \quad \sum_{uv \in J} \min\{w'_u, w'_v\} \leq \sqrt{|J|/2} \cdot \mathbf{w}'(V)$$

Applying the induction hypothesis, we get

$$\sum_{uv \in J} \min\{w_u, w_v\} = |J'|q + \sum_{uv \in J} \min\{w'_u, w'_v\} \leq \sqrt{|J'|/2} \cdot |V'|q + \sqrt{|J|/2} \cdot \mathbf{w}'(V) \leq \sqrt{|J|/2} \cdot \mathbf{w}(V)$$

as required. ∎

**Corollary 15.1** *For any optimal solution $J^*$, Algorithm 15.2 admits the following approximation ratios.*

(i) *For both directed and undirected graphs, ratio $\rho \Delta_{J^*}$.*
(ii) *For undirected graphs, ratio $\rho \theta_{J^*}$ if $J^*$ is a forest.*
(iii) *For undirected simple graphs, ratio $\rho \theta_{J^*} \sqrt{|J^*|/2}$.*

*Proof.* Note that $c(J) \leq \rho c(J^*)$, as $J$ is a $\rho$-approximate $\mathbf{c}$-cost solution. Thus, we have

$$\mathbf{a}(V) \leq \sum_{e \in J} \tau(e) = c(J) \leq \rho c(J^*) = \rho \sum_{e \in J^*} \tau(e)$$

Now the statement follows by applying Lemma 15.1 on $J^*$. ∎

### 15.3.1 Applications for Directed Graphs

Here we consider some applications of Corollary 15.1(i) to directed graphs, when inclusionwise minimal feasible solution to the problem at hand have low maximum degree.

Recall that in Theorem 15.2, we considered the undirected Activation $k$ Edge-Disjoint Paths Augmentation problem, when we are given an "initial" edge set $E_0 \subseteq E$ of $k - 1$ pairwise edge-disjoint $st$-paths, and seek to activate an edge set $J \subseteq E \setminus E_0$ such that the graph $(V, E_0 \cup J)$ contains $k$ pairwise edge-disjoint $st$-paths. Here we consider the directed variant of this problem. In a similar way, we define the Activation $k$-Edge-In-Connectivity Augmentation problem, in which $(V, E_0)$ is $(k - 1)$-edge-in-connected to the root $s$, and $(V, E_0 \cup J)$ should be $k$-edge-in-connected to $s$. Note that the case $k = 1$ and $E_0 = \emptyset$ is the Activation In-Arborescence problem. Clearly, ratio $\rho$ for the augmentation version implies ratio $k\rho$ for the "non-augmentation" version.

**Corollary 15.2** *The following directed activation problems admit a polynomial time algorithm: $k$ Edge-Disjoint Paths Augmentation, $k$-Edge-In-Connectivity Augmentation, and $k$ Disjoint Paths.*

*Proof.* For $k$ Edge-Disjoint Paths Augmentation and $k$-Edge-In-Connectivity Augmentation, it is known that $\Delta_J \leq 1$ holds for any inclusionwise minimal solution $J$, and the min-cost version of the problem admits a polynomial time algorithm [2]. Hence, we can apply Corollary 15.1(i) with $\Delta = \rho = 1$ and get ratio $\Delta\rho = 1$, namely, a polynomial time algorithm.

Let us consider the $k$ Disjoint Paths problem. We may assume that we know the value $a_s^*$ of some optimal solution $\mathbf{a}^*$; by the Polynomial Domain Assumption, there is a polynomial number of choices. We set $a_s = a_s^*$, meaning that the set of edges activated by this assignment are included in any feasible solution, whereas other edges leaving $s$ are removed. We then apply Algorithm 15.2 on the modified instance, and now any inclusionwise feasible solution has maximum degree 1. The min-cost version admits a polynomial time algorithm, and thus we get ratio $\rho\Delta = 1$, namely, a polynomial time algorithm. ∎

In some cases, we can achieve a good ratio using Corollary 15.1 after adding a certain set of edges and considering the resulting residual problem.

**Theorem 15.3 (Lando & Nutov [15])** *Directed* Activation $k$-In-Connectivity *admits a polynomial time algorithm.*

*Proof.* In Reference 15, the following is proved as follows:

*Let $G'$ be a directed graph with $\deg_{G'}(v) \geq k$ for all $v \in V \setminus \{s\}$ and let $J$ be an inclusionwise minimal augmenting edge set on $V$ such that $G' \cup J$ is $k$-inconnected to $s$. Then $\Delta_J \leq 1$.*

The problem of finding a minimum-cost augmenting edge set $J$ as earlier can be solved in polynomial time, thus by the previous result of Reference 15 and Corollary 15.1(i), the activation version of the problem can also be solved in polynomial time. Now we state the algorithm.

---

## ALGORITHM 15.3    Directed Activation $k$-In-Connectivity$(G=(V,E),s,k,f)$

1 find a minimum value assignment $\mathbf{a}'$ such that in $(V, E_{\mathbf{a}'})$ every node $v \in V \setminus \{s\}$ has out-degree $\geq k$; namely, $a'_v$ is the $k$-th least power of an edge in $\delta_E(v)$ for every $v \in V \setminus \{s\}$, and $a'_s = 0$
2 with power thresholds $p'_{uv} = p_{uv} - a'_u$ for all $uv \in E \setminus E_{\mathbf{a}'}$ find an optimal assignment $\mathbf{a}$ that activates an edge set $J \subseteq E \setminus E_{\mathbf{a}'}$ such that $(V, E_{\mathbf{a}'} \cup J)$ is $k$-inconnected to $s$
3 return $\mathbf{a}' + \mathbf{a}$

---

Clearly, the computed assignment is feasible. We explain why the assignment $\mathbf{a}' + \mathbf{a}$ is optimal. Let $\mathbf{a}^*$ be an optimal assignment. Clearly, $\mathbf{a}' \leq \mathbf{a}^*$, hence $\mathbf{a}^* - \mathbf{a}' \geq \mathbf{0}$. As the assignment $\mathbf{a}$ is optimal, $\mathbf{a}(V) \leq (\mathbf{a}^* - \mathbf{a}')(V)$, hence $(\mathbf{a} + \mathbf{a}')(V) \leq \mathbf{a}^*(V)$, as required. ∎

### 15.3.2 Applications for Undirected Graphs

We consider consequences from Corollary 15.1 for undirected graphs. Among the problems considered in the next Corollary 15.3 is the (undirected) Activation EC-Survivable Network Augmentation problem. In this problem, we are given an "initial" graph $(V, E_0)$ and a set of demand node pairs. The goal is to activate an edge set $J \subseteq E \setminus E_0$ such that for every demand pair $\{s, t\}$ the number of pairwise edge-disjoint $st$-paths in $(V, E_0 \cup J)$ is larger by one than in $(V, E_0)$. Clearly, ratio $\rho$ for this problem implies ratio $k\rho$ for Activation EC-Survivable Network. Moreover, note that for $k = 1$ and $E_0 = \emptyset$, we get the Activation Steiner Forest problem.

**Corollary 15.3** Activation Spanning Tree *admits ratio $\theta$,* Activation Steiner Tree *admits ratio $(\ln 4 + \epsilon)\theta$, and undirected* Activation EC-Survivable Network Augmentation *admits ratio $2\theta$.*

*Proof.* For each one of the problems, any inclusionwise minimal solution is a forest; for Spanning Tree and Steiner Tree this is obvious, whereas for EC-Survivable Network Augmentation this is proved in Reference 24. For the min-cost versions of the problems, the following is known: Spanning Tree admits a polynomial time algorithm, Steiner Tree admits ratio $\ln 4 + \epsilon$ [7], and EC-Survivable Network Augmentation admits ratio 2 [24]. For the activation variants, we get ratios larger by a factor of $\theta$, by Corollary 15.1(ii). ∎

In min-power problems $\theta = 2$, and thus Corollary 15.3 implies that Min-Power Spanning Tree admits ratio 2 and Min-Power Steiner Tree admits ratio $2 \ln 4 + \epsilon$. We have the following improvement over these ratios:

**Theorem 15.4 (Grandoni [5])** Min-Power Spanning Tree *admits ratio 1.5 and* Min-Power Steiner Tree *admits ratio $3 \ln 4 - \frac{9}{4} + \epsilon < 1.909$.*

The proof of Theorem 15.4 relies on a method that is beyond the scope of this survey.
We now continue our list of applications of Corollary 15.1.

**Corollary 15.4** *Undirected* Activation $k$ Disjoint Paths *admits ratio 2.*

*Proof.* We "guess" the values of $a_s^*$ and $a_t^*$ of some optimal solution $\mathbf{a}^*$, as in the proof of Corollary 15.2 for the directed $k$ Disjoint Paths problem. For every edge $sv \in E$, we set the cost of $sv$ to be $c(sv) = \min\{x_v : f^{sv}(a_s^*, x_v) = 1\}$, and for every edge $ut \in E$, we set $c(ut) = \min\{x_u : f^{ut}(a_u, a_t^*) = 1\}$. Now we apply Algorithm 15.2 with these costs and with costs $\tau(e)$ for edges that are not incident to $s$ or to $t$. The min-cost solution $J$ computed can be activated by an assignment $\mathbf{a}$ of value $c(J) + a_s^* + a_t^*$, whereas by an analysis similar to the proof of Lemma 15.1(i), we get that $a_s^* + a_t^* + c(J) \le a_s^* + a_t^* + 2\mathbf{a}^*(V \setminus \{s, t\}) \le 2\mathbf{a}^*(V)$. ∎

**Corollary 15.5 ([13,15])** *For undirected graphs, if* Activation Edge-Multi-Cover *admits ratio* $\alpha(k)$ *then:*

(i)   Activation $k$-Connectivity *admits ratio* $\alpha(k) + O(\theta \ln k \ln \frac{n}{n-k})$ *and ratio* $\alpha(k) + 6\theta$ *if* $n \ge k^3$.
(ii)  Activation $k$-In-Connectivity *admits ratio* $\alpha(k) + 2\theta$.
(iii) Activation $k$-Edge-Connectivity *admits ratio* $\alpha(k) + O(\theta \sqrt{n})$.

*Proof.* We compute an $\alpha(k)$-approximate solution $I$ to Activation Edge-Multi-Cover with demands $r_v = k$ for all $v \in V$. Clearly, $\tau(I) \le \alpha$opt. Then we use Algorithm 15.2 to find an inclusion minimal augmenting edge set $J$ such that $(V, I \cup J)$ satisfies the connectivity demands.

In the case of $k$-Connectivity, $J$ is a forest [25], the min-cost version admits ratio $O(\ln k \ln \frac{n}{n-k})$ [17], and ratio 6 if $n \ge k^3$ [26] (see also Reference 27). In the case of $k$-In-Connectivity, $J$ is a forest [28] and the min-cost version admits ratio 2 [29]. In both cases, we get the stated ratio from Corollary 15.1(ii). In the case of $k$-Edge-Connectivity, $J$ has at most $\frac{kn}{k+1} < n$ edges [30], the min-cost version admit ratio 2 [31], and the statement follows from Corollary 15.1(iii). ∎

As Min-Power Edge-Multi-Cover admits ratio $O(\ln k)$ [16], and in min-power problems $\theta = 2$, we get:

**Corollary 15.6** *For undirected graphs,* Min-Power $k$-Connectivity *admits ratio* $O(\ln k \ln \frac{n}{n-k})$, Min-Power $k$-In-Connectivity *admits ratio* $O(\ln k)$, *and* Min-Power $k$-Edge-Connectivity *admits ratio* $O(\sqrt{n})$.

## 15.4 Bidirection Reduction

Finally, we discuss factors invoked in the approximation ratio when undirected min-power problems are reduced to directed ones. The **bidirected graph** of an undirected graph $(V, J)$ with edge-costs is a directed graph obtained by replacing every undirected edge $e = uv$ of $J$ by two opposite directed edges $uv$ and $vu$ each having the same cost as $e$. Clearly, if $(V, D)$ is a bidirection of $(V, J)$, then $\mathbf{a}$ activates $J$ if and only if $\mathbf{a}$ activates $D$. The **underlying graph** of a directed graph $D$ is obtained from $D$ by ignoring the directions of the edges. We will analyze the performance of the following natural algorithm.

---
**ALGORITHM 15.4   Bidirection Reduction($G=(V,E),p,\mathcal{G}$)**
---
1  let $\mathcal{D}$ be a family of subgraphs of the bidirected graph of $G$ such that the following holds:
   (i) the underlying graph of every $D \in \mathcal{D}$ is in $\mathcal{G}$ (ii) the bidirected graph of every $J \in \mathcal{G}$ is in $\mathcal{D}$
2  compute a $\rho$-approximate min-power subgraph $D \in \mathcal{D}$
3  return the underlying graph $J$ of $D$ and an assignment $\mathbf{a}$ of value $\tau_{\mathbf{p}}(J)$ activating $J$

---

The following statement will enable us to estimate the approximation ratio of Algorithm 15.4.

**Lemma 15.2 ([14])** $\tau_{\mathbf{p}}(J) \le (\Delta_D + 1)\tau_{\mathbf{p}}(D)$ *if* $(V, J)$ *is the underlying graph of a directed graph* $(V, D)$, *in which* $\Delta_D$ *is the maximum out-degree in the graph* $(V, D)$.

*Proof.* By induction on $|D|$. For $|D| = 1$, the statement is obvious. Otherwise, let $v \in V$ be a node in $(V, D)$ of maximum power $p$. Let $D'$ be obtained from $D$ by removing the edges leaving $v$, and let $(V, J')$ be the underlying graph of $(V, D')$. Then $\tau_{\mathbf{p}}(D) = \tau_{\mathbf{p}}(D') + p$ and $\tau_{\mathbf{p}}(J) \leq \tau_{\mathbf{p}}(J') + (\Delta_D + 1)p$. By the induction hypothesis, $\tau_{\mathbf{p}}(J') \leq (\Delta_{D'} + 1)\tau_{\mathbf{p}}(D')$. Thus, we get

$$\tau_{\mathbf{p}}(J) \leq \tau_{\mathbf{p}}(J') + (\Delta_D + 1)p \leq (\Delta_{D'} + 1)\tau_{\mathbf{p}}(D') + (\Delta_D + 1)p \leq (\Delta_D + 1)(\tau_{\mathbf{p}}(D') + p) = (\Delta_D + 1)\tau_{\mathbf{p}}(D)$$

as required.                                                                                                    ∎

**Lemma 15.3 ([14])** *Algorithm 15.4 admits ratio* $\rho(\Delta_D + 1)$.

*Proof.* As $D \in \mathcal{D}$, we have $J \in \mathcal{G}$, by property (i) of $\mathcal{D}$; hence, the computed solution $J$ is feasible. We prove the approximation ratio. Let $J^*$ be an optimal solution to the undirected instance and let $D^*$ be the bidirection of $J^*$. Then $\tau_{\mathbf{p}}(D) \leq \tau_{\mathbf{p}}(D^*)$, by property (ii) of $\mathcal{D}$. Applying Lemma 15.2, we get $\tau_{\mathbf{p}}(J) \leq (\Delta_D + 1)\tau_{\mathbf{p}}(D) \leq (\Delta_D + 1)\tau_{\mathbf{p}}(D^*) = (\Delta_D + 1)\tau_{\mathbf{p}}(J^*)$, as required.                           ∎

**Corollary 15.7** *Undirected* Min-Power Edge-Multi-Cover *admits ratio* $k + 1$.

*Proof.* Let $\mathcal{D}$ be the family of subgraphs of the bidirection of $G$ that are **r**-edge-covers. Then properties (i) and (ii) hold for $\mathcal{D}$, and $\Delta_D = k$ for every inclusionwise minimal member $D \in \mathcal{D}$. Directed Edge-Multi-Cover admits a polynomial time algorithm cf. [1,2]. Thus, we can apply Lemma 15.3 with $\rho = 1$ and $\Delta_D = k$ and get ratio $k + 1$.                                                           ∎

**Corollary 15.8** *Undirected* Min-Power $k$-In-Connectivity *admits ratio* $k + 1$.

*Proof.* Let $\mathcal{D}$ be the family of subgraphs of the bidirection of $G$ that are $k$-in-connected to $s$. Then properties (i) and (ii) hold for $\mathcal{D}$, and $\Delta_D = k$ for every inclusionwise minimal member $D \in \mathcal{D}$. Thus, we can apply Lemma 15.3 with $\rho = 1$ and $\Delta_D = k$ and get ratio $k + 1$.                                                                    ∎

## 15.5 Undirected Min-Power Edge-Multi-Cover Problems

Ratio 2 for Min-Power Edge-Cover follows from Corollary 15.1. We survey the following improvement.

**Theorem 15.5 (Kortsarz & Nutov [8])** Undirected Min-Power Edge-Cover *admits ratio* 3/2.

*Proof.* Given a node subset $S$, we say that an edge set $I$ is an $S$-**edge-cover** if $\delta_I(v) \neq \emptyset$ for all $v \in S$. Let $S = \{v \in V : r_v = 1\}$ be the set of nodes we need to cover. The algorithm is as follows.

---

**ALGORITHM 15.5    Min-Power Edge-Cover($G=(V,E),p,S$) (Ratio 3/2)**

1  for all $u, v \in S$ (possibly $u = v$) compute an optimal $\{u, v\}$-edge-cover $J_{uv}$
2  let $(S, E')$ be a complete graph with all loops and edge-costs $c_{uv} = \tau_{\mathbf{p}}(J_{uv})$ for all $u, v \in S$
3  compute a minimum **c**-cost $S$-edge-cover $J' \subseteq E'$
4  return an optimal assignment activating $J = \bigcup_{uv \in J'} J_{uv}$

---

Step 1 can be implemented in polynomial time as any inclusion minimal $\{u, v\}$-edge-cover has at most two edges. Other steps of the algorithm can also be implemented in polynomial time. For the approximation ratio, we prove that:

*For any $S$-edge-cover $I \subseteq E$, there exists an $S$-edge-cover $I' \subseteq E'$ such that* $\mathbf{c}(I') \leq \frac{3}{2}\tau_{\mathbf{p}}(I)$.

In particular, if $I$ is an optimal edge-cover and $J$ is the edge set computed by the algorithm, then we get

$$\tau_{\mathbf{p}}(J) \leq \sum_{uv \in J'} \tau_{\mathbf{p}}(J_{uv}) = \mathbf{c}(J') \leq \mathbf{c}(I') \leq \frac{3}{2}\tau_{\mathbf{p}}(I)$$

It is sufficient to prove existence of $I'$ as earlier for the case when $I$ is a star with all leaves in $S$, as any inclusion minimal $S$-edge-cover is a union of such node-disjoint stars. Let $v_0$ be the center of the star. Let $e_1 = v_0 v_1, \ldots, e_d = v_0 v_d$ be the edges of $I$ sorted by nonincreasing powers, so $p_1 \leq p_2 \leq \cdots \leq p_d$, in which $p_i = p(v_0 v_i)$ for $i = 1, \ldots, d$. Note that $2p_{i-1} + p_i \leq \frac{3}{2}(p_{i-1} + p_i)$ for all $i$ and that $\tau_{\mathbf{p}}(I) = \sum_{i=1}^{d} p_i + p_d$.

If $d$ is odd, then we let $I' = \{v_0 v_1, v_2 v_3, \ldots, v_{d-1} v_d\}$. Then:

$$\mathbf{c}(I') \leq 2p_1 + p_2 + 2p_3 + p_4 + \cdots + 2p_{d-2} + p_{d-1} + 2p_d \leq \frac{3}{2} \sum_{i=1}^{d} p_i + \frac{1}{2} p_d = \frac{3}{2} \tau_{\mathbf{p}}(I) - p_d$$

If $d$ is even, then we let $I' = \{v_0 v_1, v_0 v_2\} \cup \{v_3 v_4, \ldots, v_{d-1} v_d\}$. Then:

$$\mathbf{c}(I') \leq 2p_1 + 2p_2 + p_3 + \cdots + 2p_{d-2} + p_{d-1} + 2p_d \leq \frac{3}{2} \sum_{i=1}^{d} p_i + \frac{1}{2}(p_1 + p_d) \leq \frac{3}{2} \tau_{\mathbf{p}}(I) - \frac{1}{2} p_d$$

In both cases $\mathbf{c}(I') \leq \frac{3}{2} \tau_{\mathbf{p}}(I) - \frac{1}{2} p_d \leq \frac{3}{2} \tau_{\mathbf{p}}(I)$, as claimed. ∎

By a similar method, we have the following generalization:

**Theorem 15.6 (Cohen & Nutov [16])** Min-Power Edge-Multi-Cover *admits ratio $k + 1/2$.*

For small values of $k$, for example, for $k \leq 6$, the ratio $k + 1/2$ is currently the best known one. Based on an earlier work on the Min-Power Edge-Multi-Cover problem by Hajiaghayi et al. [13] that obtained ratio $O(\ln^4 n)$, and Kortsarz et al. [32] that obtained ratio $O(\ln n)$, the following is proved in Reference 16.

**Theorem 15.7 (Cohen & Nutov [16])** Min-Power Edge-Multi-Cover *admits ratio $O(\ln k)$.*

In the rest of this section, we survey the proof of Theorem 15.7. We start by a standard reduction. Add a copy $V'$ of $V$ and replace every edge $uv$ by the edges $u'v$ and $uv'$, each of the same power as $uv$, in which $v'$ denotes the copy of $v$. It is easy to see that ratio $\rho$ for the obtained instance implies ratio $2\rho$ for the original instance. We thus obtain a Bipartite Min-Power Edge-Multi-Cover instance, when the input graph is bipartite with sides $V$ and $V'$ and the demands are $\mathbf{r} = \{r_v : v \in V\}$ (nodes in $V'$ have no demands).

**Lemma 15.4** *Let $J'$ be an edge set obtained by picking $r_v$ least power edges in $\delta_E(v)$ for every $v \in V$, and let $\mathbf{a}'$ be an optimal assignment activating $J'$. Then $\mathbf{a}'(V) \leq \text{opt}$ and $\mathbf{a}'(V') \leq \sum_{v \in V} a'_v r_v \leq k \cdot \text{opt}$.*

*Proof.* It is clear that $\mathbf{a}'(V) \leq \text{opt}$. Moreover, $\sum_{v \in V} a'_v r_v \leq \mathbf{a}'(V) \cdot \max_{v \in V} r_v \leq k \cdot \text{opt}$. Finally, as no edge joins two nodes in $V'$, we have $\mathbf{a}'(V') \leq \mathbf{p}(J') \leq \sum_{v \in V} a'_v r_v$. ∎

Given a partial solution $J$ to our problem, let $\mathbf{r}^J = \{r_v^J : v \in V\}$ be the residual demands w.r.t. $J$, in which $r_v^J = \max\{r_v - \deg_J(v), 0\}$. Given node-weights $\mathbf{w} = \{w_v : v \in V\}$, we denote by $\mathbf{w} \cdot \mathbf{r}^J = \sum_{v \in V} w_v r_v^J$ the total weighted residual demand, and call $\mathbf{w} \cdot (\mathbf{r} - \mathbf{r}^J)$ the amount of "weighted demand covered" by $J$. The main step of the algorithm is given in the following lemma.

**Lemma 15.5** *There exists a polynomial time algorithm that given a Bipartite Min-Power Edge-Multi-Cover instance with node-weights $\mathbf{w} = \{w_v : v \in V\}$ and a parameter $\gamma > 1$, returns an edge set $I \subseteq E$ such that $\tau_{\mathbf{p}}(I) \leq (\gamma + 1)\text{opt}$ and $\mathbf{w} \cdot \mathbf{r}^I \leq \alpha(\mathbf{w} \cdot \mathbf{r})$, in which $\alpha = 1 - (1 - \frac{1}{\gamma})(1 - \frac{1}{e})$.*

*Proof.* We describe an algorithm that given an integer $\tau$ returns an assignment $\mathbf{a}$ and $I \subseteq E_{\mathbf{a}}$ such that:

(i)  $\mathbf{a}(V) \leq \gamma \tau$ and $\mathbf{a}(V') \leq \tau$.
(ii) If $\tau \geq \text{opt}$ then $\mathbf{w} \cdot \mathbf{r}^I \leq \alpha(\mathbf{w} \cdot \mathbf{r})$, where $\alpha = 1 - (1 - \frac{1}{\gamma})(1 - \frac{1}{e})$.

With such an algorithm, we use binary search to find the least integer $\tau$ for which an edge set $I$ satisfying $\mathbf{w} \cdot \mathbf{r}^I \leq \alpha(\mathbf{w} \cdot \mathbf{r})$ is returned. Then $\tau \leq \text{opt}$, and we have both $\tau_{\mathbf{p}}(I) \leq (\gamma + 1)\text{opt}$ and $\mathbf{w} \cdot \mathbf{r}^I \leq \alpha(\mathbf{w} \cdot \mathbf{r})$.

In Reference 33, it is shown that the following problem admits ratio $1 - 1/e$.

---

**Bipartite Power-Budgeted Maximum Edge-Multi-Coverage**

*Instance:* A bipartite graph $G = (V \cup V', F)$ with edge-powers $\{p_e : e \in F\}$ and node-weights $\{w_v : v \in V\}$, degree bounds $\{r_v : v \in V\}$, and a budget $\tau$.
*Objective:* Find $I \subseteq F$ with $\sum_{v \in V'} \max_{e \in \delta_I(v)} p_e \leq \tau$ that maximizes $\sum_{v \in V} w_v \cdot \min\{\deg_I(v), r_v\}$.

---

The algorithm computes a $\left(1 - \frac{1}{e}\right)$-approximate solution $I \subseteq F$ to the above-mentioned problem with

$$F = \bigcup_{v \in V} \left\{ e \in \delta_E(v) : p_e \leq \frac{w_v r_v}{\mathbf{w} \cdot \mathbf{r}} \cdot \gamma \tau \right\}$$

Let $\mathbf{a}$ be an optimal assignment activating $I$, namely $a_v = \max_{e \in \delta_I(v)} p_e$ for every $v \in V \cup V'$. Clearly,

$\mathbf{a}(V) \leq \sum_{v \in V} \frac{w_v r_v}{\mathbf{w} \cdot \mathbf{r}} \cdot \gamma \tau = \gamma \tau$ and $\mathbf{a}(V') \leq \tau$.

We prove that if $\tau \geq \text{opt}$, then $\mathbf{w} \cdot \mathbf{r}^I \leq \alpha(\mathbf{w} \cdot \mathbf{r})$. Let $J^*$ be an optimal solution to our Bipartite Min-Power Edge-Multi-Cover instance. Let $B = \{v \in V : \delta_{J^* \backslash F}(v) \neq \emptyset\}$. For any assignment $\mathbf{a}$ activating $J^* \backslash F$, we have $a_v \geq \frac{w_v r_v}{\mathbf{w} \cdot \mathbf{r}} \cdot \gamma \tau$ for all $v \in B$. This implies $\tau \geq \tau_{\mathbf{p}}(J^* \cap F) \geq \sum_{v \in B} \frac{w_v r_v}{\mathbf{w} \cdot \mathbf{r}} \cdot \gamma \tau$, namely,

$\sum_{v \in B} w_v r_v \leq \frac{\mathbf{w} \cdot \mathbf{r}}{\gamma}$. As the amount of weighted demand covered by $J^* \backslash F$ is at most $\sum_{v \in B} w_v r_v$, the amount of weighted demand covered by $J^* \cap F$ is at least $(1 - \frac{1}{\gamma})(\mathbf{w} \cdot \mathbf{r})$. As $\tau_{\mathbf{p}}(J^* \cap F) \leq \tau_{\mathbf{p}}(J^*) \leq \tau$, the amount of weighted demand $I$ covers is at least $(1 - \frac{1}{\gamma})(1 - \frac{1}{e})(\mathbf{w} \cdot \mathbf{r})$. Consequently, $\mathbf{w} \cdot \mathbf{r}^I \leq \alpha(\mathbf{w} \cdot \mathbf{r})$. ∎

Theorem 15.7 is deduced from Lemmas 15.4 and 15.5 as follows. We let $\gamma$ to be a constant strictly greater than 1, say $\gamma = 2$. Then $\alpha = 1 - \frac{1}{2}(1 - \frac{1}{e})$. For $v \in V$, we set $w_v$ to be the $r_v$-th least power of an edge in $\delta_E(v)$, and apply the algorithm from Lemma 15.5 iteratively $\lceil \log_{1/\alpha} k \rceil = O(\ln k)$ times. We then extend the partial solution computed to a feasible solution by adding an edge set as in Lemma 15.4.

---

# ALGORITHM 15.6   Min-Power Edge-Multi-Cover($G=(V \cup V', E)$,p,r) (ratio $O(\ln k)$)

1  $J \leftarrow \emptyset$
2  for every $v \in V$ set $w_v$ to be the $r_v$-th least power of an edge in $\delta_E(v)$
3  repeat $\lceil \log_{1/\alpha} k \rceil$ times:
     compute an edge set $I$ as in Lemma 15.5 and update: $\mathbf{r} \leftarrow \mathbf{r}^I, J \leftarrow J \cup I, E \leftarrow E \backslash I$
4  compute an edge set $J'$ as in Lemma 15.4
5  return $J \cup J'$ and an optimal assignment $\mathbf{a}$ activating $J \cup J'$

It is clear that the algorithm computes a feasible solution. We prove the approximation ratio. At each iteration in the loop of step 3, we compute an edge set $I$ with $\tau_{\mathbf{p}}(I) \leq (1 + \gamma)$opt and add $I$ to $J$. We apply this $\lceil \log_{1/\theta} k \rceil$ times, hence $\tau_{\mathbf{p}}(J) \leq \lceil \log_{1/\alpha} k \rceil (1 + \gamma)$opt $= O(\ln k)$opt. We show that $\tau_{\mathbf{p}}(J') \leq 2$opt. Let $\mathbf{a}'$ be an optimal assignment activating $J'$. By Lemma 15.4, $\mathbf{a}'(V) \leq$ opt. We show that $\mathbf{a}'(V') \leq$ opt. Note that $a'_v \leq w_v$ for every $v \in V$. Thus, by Lemma 15.4 we have $\mathbf{a}'(V') \leq \sum_{v \in V} a'_v r^J_v \leq \sum_{v \in V} w_v r^J_v = \mathbf{w} \cdot \mathbf{r}^J$. We claim that $\mathbf{w} \cdot \mathbf{r}^J \leq$ opt. By applying Lemma 15.4 on the initial instance, we have $\mathbf{w} \cdot \mathbf{r} \leq k \cdot$ opt. At each iteration $\mathbf{w} \cdot \mathbf{r}^J$ becomes smaller by a factor of $\alpha$, hence at the end of the step 3 loop we have

$$\mathbf{w} \cdot \mathbf{r}^J \leq (\mathbf{w} \cdot \mathbf{r}) \cdot \alpha^{\lceil \log_{1/\alpha} k \rceil} \leq \frac{\mathbf{w} \cdot \mathbf{r}}{k} \leq \frac{k \cdot \text{opt}}{k} = \text{opt}$$

Consequently, we get that

$$\tau_{\mathbf{p}}(J \cup J') \leq \tau_{\mathbf{p}}(J) + \tau_{\mathbf{p}}(J') \leq \lceil \log_{1/\alpha} k \rceil (1 + \gamma)\text{opt} + 2\text{opt}$$

By choosing $\gamma = 2$, we get $\alpha = 1 - \frac{1}{2}(1 - \frac{1}{e})$, hence $\tau_{\mathbf{p}}(J \cup J') = O(\ln k)$, as required. This concludes the proof of Theorem 15.7.

We also mention that for unit/uniform powers, the problem admits a constant ratio.

**Theorem 15.8 (Cohen & Nutov [16])** Min-Power Edge-Multi-Cover *with uniform powers admits a randomized approximation algorithm with expected approximation ratio $\rho$, in which $\rho < 2.16851$ is the real root of the cubic equation $e(\rho - 1)^3 = 2\rho$.*

## 15.6 The Node-Weighted Steiner Forest Problem

Let us recall the Node-Weighted Steiner Forest problem. We are given an undirected graph $G = (V, E)$ with node-weights $\mathbf{w} = \{w_v : v \in V\}$ and a set $R$ of demand pairs from $V$, and seek $J \subseteq E$ such that the graph $(V, J)$ has an $st$-path for every demand pair $\{s, t\} \in R$. We want to minimize the node-weight of $J$, namely, the weight of the set of endnodes of the edges in $J$. We survey the proof of the following seminal result:

**Theorem 15.9 (Klein & Ravi [6])** Node-Weighted Steiner Forest *admits ratio $2 \ln |U|$, in which $U$ is the union of the demand pairs.*

To prove Theorem 15.9, we use a $\rho$-**Density Algorithm** for the following generic problem:

---

Covering Problem

*Input:* Integral nonnegative set functions $\nu, \tau$ on a groundset $E$, such that $\nu(\emptyset) > \nu(E) \geq 0$ and $\tau(\emptyset) = 0$.
*Output:* $J \subseteq E$ with $\nu(J) = \nu(E)$ and with $\tau(J)$ minimized.

---

A set function $f$ is **increasing** if $f(A) \leq f(B)$ whenever $A \subseteq B$; $f$ is **decreasing** if $-f$ is increasing, and $f$ is **subadditive** if $f(A \cup B) \leq f(A) + f(B)$ for any two subset $A, B$ of the groundset. It is easy to see that if $f$ is subadditive (and nonnegative), then $f$ is increasing.

We call $\nu$ the **deficiency function** and $\tau$ a **payment function**; for a partial solution $J$, $\nu(J)$ measures how far is $J$ from being feasible, whereas the function $\tau$ is our "payment" for $J$. In all our applications, the function $\nu$ is decreasing and $\tau$ is subadditive.

For a subset $S \subseteq E \setminus J$, the quantity $\sigma_J(S) = \dfrac{\tau(J \cup S) - \tau(J)}{v(J) - v(J \cup S)}$ is called the **density of** $S$ w.r.t $J$. Let $\rho \geq v(E) + 1$ be a parameter and let opt be the optimal solution value for an instance of a Covering Problem. The $\rho$-**Density Algorithm** starts with $J = \emptyset$, and as long as $v(J) > v(E)$, it adds to $J$ an augmenting set $S \subseteq E \setminus J$ with $v(J \cup S) \leq v(J) - 1$ that satisfies the $\rho$-**Density Condition**. As at each iteration $v(J \cup S) \leq v(J) - 1$, the algorithm terminates. It is easy to see that if $v$ is decreasing and $\tau$ is subadditive, then for any optimal solution $J^*$, the set $S = J^* \setminus J$ satisfies the $\rho$-Density Condition with $\rho = v(E) + 1$. Thus, if $v(E)$ is small, then a low density set exists, and the problem is to find one in polynomial time. The following is implicitly proved in References 6 and 34.

**Theorem 15.10 (Johnson [34], Klein & Ravi [6])** *The $\rho$-Density Algorithm computes a feasible solution $J$ such that:* $\tau(J) \leq \rho \ln \dfrac{v(\emptyset)}{v(E)} \cdot$ opt *if* $v(E) \geq 1$ *and* $\tau(J) \leq \rho(\ln v(\emptyset) + 1) \cdot$ opt *if* $v(E) = 0$.

*Proof.* Let $\ell$ be the number of the iterations of the algorithm. Let $J_i$ be the partial solution stored in $J$ at the end of iteration $i$, and let $J_0 = \emptyset$. Let $S_i$ be the set added to $J_{i-1}$ at iteration $i$, so $J_i = J_{i-1} \cup S_i$, $i = 1, \ldots, \ell$. As $S_i$ satisfies the $\rho$-Density Condition w.r.t. $J_{i-1}$, we have $\dfrac{\tau(J_i) - \tau(J_{i-1})}{v(J_{i-1}) - v(J_i)} \leq \rho \cdot \dfrac{\text{opt}}{v(J_{i-1})}$. Denote $\tau_i = \tau(J_i)$ and $v_i = v(J_i)$, in which $v_0 = v(\emptyset)$. Then for every $i = 1, \ldots, \ell$

$$v_i \leq v_{i-1}\left(1 - \frac{\tau_i - \tau_{i-1}}{\rho \cdot \text{opt}}\right)$$

Unraveling the last inequality gives that for any $j$ with $v_j > 0$:

$$\frac{v_j}{v_0} \leq \prod_{i=1}^{j}\left(1 - \frac{\tau_i - \tau_{i-1}}{\rho \cdot \text{opt}}\right)$$

Taking natural logarithm from both sides and using the fact that $\ln(1 - x) \leq -x$ for $x < 1$ we obtain:

$$\ln\left(\frac{v_j}{v_0}\right) \leq \sum_{i=1}^{j} \ln\left(1 - \frac{\tau_i - \tau_{i-1}}{\rho \cdot \text{opt}}\right) \leq -\sum_{i=1}^{j} \frac{\tau_i - \tau_{i-1}}{\rho \cdot \text{opt}}$$

Consequently, $\tau_j = \tau_j - \tau_0 = \displaystyle\sum_{i=1}^{j}(\tau_i - \tau_{i-1}) \leq \rho \ln \dfrac{v_0}{v_j} \cdot$ opt.

In the case $v(E) \geq 1$, we apply the last inequality for $j = \ell$ to get $\tau(J) = \tau_\ell \leq \rho \ln \dfrac{v_0}{v_\ell} \cdot$ opt, as required.

In the case $v(E) = 0$, we apply the last inequality for $j = \ell - 1$ to get $\tau_{\ell-1} \leq \rho \ln \dfrac{v_0}{v_{\ell-1}} \cdot$ opt $\leq$ $\rho \ln v_0 \cdot$ opt. Observing that $\tau_\ell - \tau_{\ell-1} \leq \rho \cdot$ opt, we get $\tau(J) = \tau_\ell \leq \tau_{\ell-1} + \rho \cdot$ opt $\leq \rho(\ln v_0 + 1) \cdot$ opt, as required. ∎

If $\tau$ is subadditive, then $\tau(J \cup S) - \tau(J) \leq \tau(S)$, and thus we achieve the same performance as in Theorem 15.10 by replacing in the $\rho$-Density Algorithm the $\rho$-Density Condition by a stronger condition

$$\frac{\tau(S)}{v(J) - v(J \cup S)} \leq \rho \cdot \frac{\text{opt}}{v(J)}$$

In our setting of the Node-Weighted Steiner Forest problem, the groundset is the set $E$ of edges, and for a partial solution $J$, we define $\tau(J)$ to be the node-weight of $J$. It is easy to see that $\tau$ is subadditive. The function $v$ is defined by $v(J) = \min\{|\mathcal{C}^J|, 1\}$, in which $\mathcal{C}^J$ is the family of inclusionwise minimal **deficient sets** (a.k.a. tight sets) uncovered by $J$. More precisely, an undirected edge set $J$ **covers** a node subset $A \subseteq V$ if $J$ has an edge between $A$ and $V \setminus A$, and $A \subseteq V$ is deficient if there exists a demand pair $\{s, t\}$ with $|\{s, t\} \cap A| = 1$. Then $\mathcal{C}^J$ is the family of deficient connected components of the graph $(V, J)$.

Note that $v(\emptyset) = |U|$, as when $J = \emptyset$, the minimal deficient sets are the singletons in $U$. It is easy to see that $v$ is decreasing and that $J$ is a feasible solution iff $v(J) = 1$ (note that $|\mathcal{C}^J| \geq 1$ implies $|\mathcal{C}^J| \geq 2$).

With these definitions, we would like to apply the $\rho$-Density Algorithm with $\rho = 2$ for our problem, namely, to design a polynomial time algorithm that for any partial solution $J$ finds an augmenting edge set $S \subseteq E \setminus J$ of density $\leq 2\dfrac{\text{opt}}{v(J)}$. We start by showing that there exists an augmenting edge set $S$ of good density that has a simple structure.

**Definition 15.4** *A **spider** is a rooted tree such that only its root, called the **head**, may have degree $\geq$ 3. Equivalently, a spider is a union of paths that start at the same node—the head, such that no two paths have other node in common. A **spider decomposition** of a graph with a designated node subset $U$ of terminals is a family $\mathcal{S}$ of node-disjoint spiders in the graph, with at least two terminals each, such that every terminal is a leaf or the head of some $S \in \mathcal{S}$.*

**Lemma 15.6 ([6])** *Any tree $T$ with a set $U$ of at least two terminals has a spider decomposition.*

*Proof.* The proof is by induction on $|U|$. Root $T$ at some node. Let $v$ be a farthest node from the root such that the subtree of $T$ that consists of $v$ and its descendants contains at least two terminals. By the choice of $v$, the paths in the subtree from the terminals to $v$ form a spider $S$ with at least two terminals. Let $T'$ be obtained from $T$ by deleting the subtree rooted at $v$ and let $U'$ be the set of terminals in $T'$. If $U' = \emptyset$, we are done. If $T'$ has a single terminal, then we joint to $S$ the path from this terminal to $v$ and obtain a spider that contains all terminals. Otherwise, by the induction hypothesis the pair $T', U'$ admits a spider decomposition $\mathcal{S}'$. Then $\mathcal{S}' \cup \{S\}$ is a spider decomposition of $T, U$, as required. ∎

Let $J$ be a partial solution for a Node-Weighted Steiner Forest instance $G, R, \mathbf{w}$. We may consider the equivalent instance obtained from $G, R, \mathbf{w}$ by contracting into a single node of weight zero every nontrivial connected component of the graph $(V, J)$, and updating accordingly the set $R$ of demand pairs. Then for any $S \subseteq E \setminus J$, $\tau(J \cup S) - \tau(J)$ equals the weight of the endnodes of the edges in $S$ in the new instance. Thus, in what follows we will assume that $J = \emptyset$. Then the deficient sets are the singletons in $U$. We may also assume that the nodes in $U$ have weight 0. For a subgraph $S$ of $G$, we denote by $S \cap U$ be the set of nodes in $U$ that belong to $S$ but also use the notation $v(S)$ and $\tau(S)$ by considering $S$ as the set of its edges. It is not hard to verify the following.

**Lemma 15.7** *Let $S$ be a connected subgraph of $G$ with $|S \cap U| \geq 2$. Then $v(\emptyset) - v(S) = |S \cap U| - 1 \geq |S \cap U|/2$.*

Let us fix some inclusion minimal optimal solution $F$ to the residual problem, so $\tau(F) \leq$ opt. Then $F$ a forest, and any its nontrivial connected component is a tree with at least two terminals. Thus, by Lemma 15.6, the pair $F, U$ admits a spider decomposition $\mathcal{S}$. In the next two lemmas 15.8 and 15.9, we show that there is $S \in \mathcal{S}$ of low density, and how to find an edge set of density at most the density of any $S \in \mathcal{S}$.

**Lemma 15.8** *There is a spider $S \in \mathcal{S}$ of density $\sigma_\emptyset(S) \leq 2\dfrac{\text{opt}}{v(\emptyset)}$.*

*Proof.* Clearly, $\tau(F) \leq$ opt. As the spiders in $\mathcal{S}$ are node-disjoint $\sum\limits_{S \in \mathcal{S}} \tau(S) \leq$ opt. As the sets $S \cap U$ partition $U$ and from Lemma 15.16 we have

$$\sum_{S \in \mathcal{S}}(v(\emptyset) - v(S)) \geq \sum_{S \in \mathcal{S}} |S \cap U|/2 = |U|/2 = v(\emptyset)/2$$

Consequently, by an averaging argument, there is $S \in \mathcal{S}$ of density $\sigma_\emptyset(S) = \dfrac{\tau(S)}{v(\emptyset) - v(S)} \leq \dfrac{\text{opt}}{v(\emptyset)/2} = 2\dfrac{\text{opt}}{v(\emptyset)}$, as claimed. ∎

Now we show how to find in polynomial time an edge set $S$ such that $\sigma_\emptyset(S) \leq \sigma_\emptyset(S')$ for any spider $S' \in \mathcal{S}$.

**Lemma 15.9** *There exists a polynomial time algorithm that finds a subgraph $S$ such that $\sigma_\emptyset(S) \leq \sigma_\emptyset(S^*)$, in which $S^*$ is a minimum density spider in $\mathcal{S}$.*

*Proof.* We may assume that we know the head $h$ of $S^*$ and the number $\ell = |S^* \cap U|$ (the "guess" of $\ell$ can be avoided, by using a slightly more complicated algorithm). Note that $\ell \geq 2$ and that $h \in U$ may hold. There is a polynomial number of choices, so we can try all choices and return the best outcome. The algorithm computes a set $P_1, \ldots, P_\ell$ of the lightest $\ell$ paths from $h$ to a set of distinct nodes $u_1, \ldots, u_\ell$ in $U$ (one of these nodes may be $h$, if $h \in U$), and returns their union $S$. It is easy to see that the algorithm can be implemented in polynomial time. We show that $\sigma_\emptyset(S) \leq \sigma_\emptyset(S^*)$. Let $P_1^*, \ldots, P_\ell^*$ be the $\ell$ paths from $h$ to the terminals in $S^*$. Then

$$\tau(S) \leq \sum_{i=1}^{\ell} w(P_i) - (\ell - 1)w_h \leq \sum_{i=1}^{\ell} w(P_i^*) - (\ell - 1)w_h = \tau(S^*)$$

By Lemma 15.7, $\nu(\emptyset) - \nu(S) \geq \ell - 1 = \nu(\emptyset) - \nu(S^*)$. Thus, $\sigma_\emptyset(S) = \dfrac{\tau(S)}{\nu(\emptyset) - \nu(S)} \leq \dfrac{\tau(S^*)}{\nu(\emptyset) - \nu(S^*)} = \sigma_\emptyset(S^*)$, as claimed. ∎

Lemma 15.8 implies that the algorithm from Lemma 15.9 finds $S \subseteq E \setminus J$ of density at most $\sigma_J(S) \leq 2\dfrac{\mathrm{opt}}{\nu(J)}$. Thus, we can find in polynomial time an edge set $S$ obeying the $\rho$-Density Condition with $\rho = 2$. As $\nu(\emptyset) = |U|$ and as $\tau = \tau_{\mathbf{w}}$ is subadditive, we get ratio $2 \ln |U|$ from Theorem 15.10.

## 15.7 The Min-Power $k$-Edge-Out-Connectivity Problem

In the current section, we survey an $O(k \ln n)$-approximation algorithm for the directed Activation $k$-Edge-Out-Connectivity problem. For this, we will consider the augmentation variant of the problem, defined as follows.

---

Directed Activation $k$-Edge-Out-Connectivity Augmentation

*Input:* A directed graph $G_0 = (V, E_0)$ that is $(k-1)$-edge-out-connected from a given root node $s$, and an edge set $E$ on $V$ with power thresholds $\mathbf{p} = \{p_e : e \in E\}$.
*Output:* $J \subseteq E$ such that $G_0 \cup J$ is $k$-out-connected from $s$ and $\tau_{\mathbf{p}}(J)$ is minimized.

---

It is easy to see that ratio $\rho$ for the above-mentioned augmentation version implies ratio $k\rho$ for directed Activation $k$-Edge-Out-Connectivity. In the rest of this section, we describe the proof of the following result.

**Theorem 15.11 (Nutov [19])** *The directed* Min-Power $k$-Edge-Out-Connectivity Augmentation *problem admits ratio* $3(\ln(n-1) + 1)$.

**Corollary 15.9** *Directed* Activation $k$-Edge-Connectivity *admits ratio* $O(k \ln n)$.

*Proof.* By Menger's Theorem, a directed graph is $k$-edge-connected iff for a given node $s$, the graph is both $k$-edge-out-connected from $s$ and $k$-edge-in-connected to $s$. From this, we get that ratio $\alpha$ for Activation $k$-Edge-Out-Connectivity and ratio $\beta$ for Activation $k$-Edge-In-Connectivity implies ratio $\alpha + \beta$ for Activation $k$-Edge-Connectivity. From Theorem 15.11, we get $\alpha = 3k(\ln(n-1)+1)$ and from Corollary 15.2, we get $\beta = k$, and the ratio $O(k \ln n)$ follows. ∎

Note that for $k = 1$, we have the directed Activation Out-Arborescence problem in Theorem 15.11 and the Activation Strong Connectivity problem in Corollary 15.9. For this particular case, a slightly better ratio is known.

**Theorem 15.12 (Calinescu et al. [20])** Activation Out-Arborescence *admits ratio* $2(\ln(n-1)+1)$ *and* Activation Strong Connectivity *admits ratio* $2\ln(n-1)+3$.

## 15.7.1 Set-Family Edge-Cover Formulation

Let $\delta_J^{in}(A)$ denote the set of edges in $J$ entering $A$. Let us say that a directed edge set $J$ **covers** $A \subseteq V$ if $J$ has an edge that enters $A$, namely, if $\delta_J^{in}(A) \neq \emptyset$. Given a set-family $\mathcal{F}$, we say that $J$ **covers a set-family** $\mathcal{F}$ or that $J$ is an **edge-cover of** $\mathcal{F}$ if every set in $\mathcal{F}$ is covered by some edge in $J$. The directed Activation $k$-Edge-Out-Connectivity Augmentation problem can be formulated as a particular case of the following problem.

---

Directed Activation Set-Family Edge-Cover

*Input:* A directed graph $G = (V, E)$ with power thresholds $\mathbf{p} = \{p_e : e \in E\}$ and a set-family $\mathcal{F}$ on $V$.

*Output:* An edge-cover $J \subseteq E$ of $\mathcal{F}$ such that $\tau_{\mathbf{p}}(J)$ is minimized.

---

We will assume that $\emptyset, V \notin \mathcal{F}$, as otherwise the problem has no feasible solution. In this problem, the family $\mathcal{F}$ may not be given explicitly, and for a polynomial time implementation of algorithms we just need that some queries related to $\mathcal{F}$ can be answered in polynomial time. The inclusion minimal sets of a set-family $\mathcal{F}$ are called $\mathcal{F}$-**cores**, or just **cores**, if $\mathcal{F}$ is clear from the context. We denote the family of $\mathcal{F}$-cores by $\mathcal{C}(\mathcal{F})$. Given an edge set $J$, let $\mathcal{F}^J$ denote the **residual family of** $\mathcal{F}$ (w.r.t. $J$), that consists of members of $\mathcal{F}$ not covered by $J$. We will assume that for any edge set $J$, the family $\mathcal{C}(\mathcal{F}^J)$ of $\mathcal{F}^J$-cores can be computed in polynomial time. For our problem, this can be done using $n - 1$ min-cut computations.

By Menger's Theorem, $J \subseteq E$ is a feasible solution to our problem iff $J$ covers the family
$$\mathcal{F}_{OC} = \{\emptyset \neq A \subseteq V \setminus \{s\} : |\delta_G^{in}(A)| = k - 1\}$$

The following definition and lemma gives the essential property of the set-family $\mathcal{F}_{OC}$ that we use.

**Definition 15.5** *A set-family $\mathcal{F}$ on $V$ with $\emptyset, V \notin \mathcal{F}$ is an* **intersecting family** *if $A \cap B, A \cup B \in \mathcal{F}$ holds for any $A, B \in \mathcal{F}$ that intersect.*

The following is known, cf. [2].

**Lemma 15.10** $\mathcal{F}_{OC}$ *is an intersecting family.*

It is easy to see that the cores of an intersecting family are pairwise disjoint. Thus, $|\mathcal{C}(\mathcal{F}_{OC})| \leq n - 1$. Hence, to prove Theorem 15.11, it is sufficient to prove the following.

**Theorem 15.13 (Nutov [19])** *Directed* Activation Set-Family Edge-Cover *with intersecting set-family $\mathcal{F}$ admits ratio* $3(\ln|\mathcal{C}(\mathcal{F})|+1)$.

For simplicity of exposition, we give a proof of a slightly worse ratio $\frac{9}{2}(\ln|\mathcal{C}(\mathcal{F})|+1)$. We will again use a $\rho$-Density Algorithm for an appropriate Covering Problem. As before, $E$ is the set of edges and $\tau(J)$ is the optimal value of an assignment activating $J$. The function $\nu$ is defined by $\nu(J) = |\mathcal{C}(\mathcal{F}^J)|$; note that $J$ is a feasible solution iff $\nu(J) = 0$. To define an analogue of spiders, we study in the next section a special simple type of intersecting families.

### 15.7.2 Ring Families

An intersecting set-family that has a unique core is called a **ring family**; equivalently, a set-family $\mathcal{F}$ with $\emptyset, V \notin \mathcal{F}$ is a **ring family** if $A \cap B, A \cup B \in \mathcal{F}$ for any $A, B \in \mathcal{F}$. Ring families often arise from intersecting families as follows.

**Definition 15.6** *Let $\mathcal{F}$ be a set-family on $V$. For an $\mathcal{F}$-core $C \in \mathcal{C}(\mathcal{F})$, let $\mathcal{F}(C)$ denote the family of the sets in $\mathcal{F}$ that contain $C$ and contain no $\mathcal{F}$-core distinct from $C$; for $h \in V$, let $\mathcal{F}(h, C) = \{A \in \mathcal{F}(C) : h \notin A\}$.*

**Lemma 15.11** *Let $\mathcal{F}$ be an intersecting set-family on a groundset $V$.*

(i) *For any $\mathcal{F}$-core $C \in \mathcal{C}(\mathcal{F})$ and $h \in V$, $\mathcal{F}(h, C)$ is a ring family; in particular, $\mathcal{F}(C)$ is a ring family.*
(ii) *For any distinct $\mathcal{F}$-cores $C_i, C_j \in \mathcal{C}(\mathcal{F})$, no $A_i \in \mathcal{F}(C_i)$ and $A_j \in \mathcal{F}(C_j)$ intersect.*

*Proof.* We prove (i). Let $A, B \in \mathcal{F}(h, C)$. Then $A \cap B, A \cup B \in \mathcal{F}, C \subseteq A \cap B \subseteq A \cup B$, and $h \notin A \cup B \supseteq A \cap B$. It remains to prove is that $A \cup B$ contains no $\mathcal{F}$-core $C'$ distinct from $C$. Otherwise, $C'$ and one of $A, B$ intersect, say $C' \cap A \neq \emptyset$. Then $C' \cap A \in \mathcal{F}$, hence by the minimality of $C'$ we must have $C' \subseteq A$. This contradicts that $A \in \mathcal{F}(C)$.

We prove (ii). If $A_i \cap A_j \neq \emptyset$, then $A_i \cap A_j \in \mathcal{F}$, hence $A_i \cap A_j$ contains some $\mathcal{F}$-core $C$. This implies $C_i = C = C_j$, contradicting that $C_i, C_j$ are distinct $\mathcal{F}$-cores. ∎

It is easy to see that if $\mathcal{F}$ is an intersecting or a ring family, then so is the residual family $\mathcal{F}^J$ of $\mathcal{F}$, for any edge set $J$. In the following lemma, we summarize the essential properties of ring families that we use.

**Lemma 15.12** *Let $J$ be an inclusionwise minimal directed edge-cover of a ring family $\mathcal{F}$ with core $C$. Then there is an ordering $e_1, \ldots, e_q$ of $J$ and sets $C_1 \subset \cdots \subset C_q$ in $\mathcal{F}$ in which $C_1 = C$, such that $\delta_J^{in}(C_i) = \{e_i\}$, and if $e_i = v_i u_i$ in which $u_i \in C_i$, then $\{e_1, \ldots, e_i\}$ covers both $\mathcal{F}(v_i, C)$ and $\mathcal{F}(u_{i+1}, C)$. Furthermore, $v_q$ does not belong to any set in $\mathcal{F}$.*

*Proof.* The proof of the main statement is by induction on $q = |J|$. For $q = 1$, the statement is obvious. If $|J| \geq 2$, let $e_1 \in \delta_J^{in}(C)$. Then $J' = J \setminus \{e_1\}$ is an inclusion minimal edge-cover of the residual family $\mathcal{F}' = \mathcal{F}^{\{e_1\}}$ of members of $\mathcal{F}$ not covered by $e_1$. By the induction hypothesis, there is an ordering $e_2, \ldots, e_q$ of $J'$ and sets $C_2 \subset \cdots \subset C_q$ in $\mathcal{F}'$ as in the lemma. As $C_1 = C$ is the unique $\mathcal{F}$-core, $C_1 \subset C_2$. To prove the lemma for $\mathcal{F}$ and $J$, we just need to show that $e_2$ does not cover $C_1$. Suppose to the contrary that $e_2 \in \delta_J^{in}(C_1)$. By the minimality of $J$, there is $A_1 \in \mathcal{F}$ such that $\delta_J(A_1) = \{e_1\}$. There is an edge in $J$ covering $A_1 \cup C_2$, as $A_1 \cup C_2 \in \mathcal{F}$. This edge is one of $e_1, e_2$, as if an edge-covers a union of two sets, then it covers one of the sets. Each of $e_1, e_2$ covers $A_1 \cap C_2$, as $e_1, e_2 \in \delta_J^{in}(C_1)$ and $C_1 \subseteq A_1 \cap C_2$. Thus, one of $e_1, e_2$ covers both $A_1 \cap C_2$ and $A_1 \cup C_2$. However, if an edge-covers both $A \cap B, A \cup B$, then it covers both $A$ and $B$. Hence, one of $e_1, e_2$ covers both $A_1, C_2$. This contradicts our choice of $A_1$.

We show that there is no set $A \in \mathcal{F}$ with $v_q \in A$. Otherwise, all edges in $J$ have both endnodes in $A \cup C_q$, hence $\delta_J^{in}(A \cup C_q) = \emptyset$. As $\mathcal{F}$ is a ring family, $A \cup C_q \in \mathcal{F}$. This contradicts that $J$ covers $\mathcal{F}$. ∎

**Lemma 15.13** *The directed* Activation Set-Family Edge-Cover *problem with ring family $\mathcal{F}$ admits a polynomial time algorithm.*

*Proof.* It is known that computing a minimum-cost edge-cover of a ring family $\mathcal{F}$ can be done in polynomial time. Hence, we can apply Corollary 15.1 with $\rho = 1$. From Lemma 15.12, it follows that if $J$ is an inclusionwise minimal cover of $\mathcal{F}$, then $\Delta_J \leq 1$. Now the lemma follows from Corollary 15.1. ∎

### 15.7.3 Spider Decompositions

To get some intuition, let us first restate some concepts from Section 15.6 in terms of directed graphs. A **(directed) spider** is a union of directed paths that starts at the same node such that no two paths have other node in common. We say that a spider **hits** a node $u$ if $u$ is a leaf or the head of $S$; a family $\mathcal{S}$ of spiders hits a node set $U$ if every $u \in U$ is hit by some $S \in \mathcal{S}$. Recall that Lemma 15.6 states that any (undirected) tree $T$ with a set of terminals has a spider decomposition—a family $\mathcal{S}$ of node-disjoint spiders, such that every $S \in \mathcal{S}$ hits at least two terminals, and such that $\mathcal{S}$ hits all terminals. Here we consider spider decompositions of directed graphs when the spiders should still be node-disjoint, but the other two conditions are relaxed. Spiders that hit just one terminal are allowed, but should satisfy a certain condition. Moreover, spiders may not hit all terminals, but just some fraction of them. For example, by the same proof as in Lemma 15.6, one can prove the following lemma, that allows spiders hitting just one terminal.

**Lemma 15.14** *Any out-arborescence $T$ with a set $U$ of terminals and root $s$ contains a family $\mathcal{S}$ of node-disjoint spiders that hits $U$, such that any $S \in \mathcal{S}$ that contains a single terminal $u$ is the $su$-path in $T$.*

In the current section, we define spiders related to set families, and then state and prove an appropriate spider-decomposition theorem of edge-covers of intersecting families. Let us say that two edge sets $S$ and $S'$ are $V'$-disjoint if no $e \in S$ and $e' \in S'$ have a common endnode in $V'$.

**Definition 15.7** *Let $\mathcal{F}$ be a set-family on $V$. An $\mathcal{F}$-**spider** is a triple $h, C, S$, in which $h \in V$ is the **head** of the $\mathcal{F}$-spider, $C \subseteq \mathcal{C}(\mathcal{F})$ is the set of cores **hit by** the $\mathcal{F}$-spider, and $S$ is an edge set that is a union of (possibly empty) pairwise $(V \setminus \{h\})$-disjoint $\mathcal{F}(h, C)$-covers $\{S_C : C \in C\}$, such that if $C = \{C\}$ then $h$ does not belong to any set in $\mathcal{F}(C)$. We will often denote an $\mathcal{F}$-spider just by $S$, meaning that there exists an appropriate choice of $h$ and $C$.*

The purpose of this section is to prove the following.

**Theorem 15.14 (Nutov [19])** *Any directed cover $J$ of an intersecting family $\mathcal{F}$ with $\ell$ cores contains a family $\mathcal{S}$ of node-disjoint $\mathcal{F}$-spiders that hits at least $\frac{2}{3}\ell$ distinct cores in $\mathcal{C}(\mathcal{F})$.*

Note that the spider decomposition in the theorem differs from the one in Lemma 15.6 in two ways: an $\mathcal{F}$-spider $S \in \mathcal{S}$ may hit just one $\mathcal{F}$-core, and the spiders in $\mathcal{S}$ may not hit all $\mathcal{F}$-cores.

A simple proof of Theorem 15.14 relies on a spider decomposition of families of directed paths.

**Definition 15.8** *Let $\mathcal{P}$ be a family of simple directed paths with a set $U(\mathcal{P})$ of distinct endnodes. We say that a spider $S$ is a $\mathcal{P}$-**spider** if $S$ is a union of internally disjoint $(h, U)$-subpaths (possibly of length 0) of the paths in $\mathcal{P}$, for some $U \subseteq U(\mathcal{P})$ and a node $h$ called the **head of** $S$ (possibly $h \in U$), such that if $|U| = 1$ then $S \in \mathcal{P}$.*

We will need the following spider decomposition lemma that was implicitly proved in Reference 19.

**Lemma 15.15 (Nutov [19])** *Let $\mathcal{P}$ be a family of $\ell$ simple directed paths in a graph $G$ with a set $U(\mathcal{P})$ of distinct endnodes. If $G$ is an out-arborescence then $G$ contains a family of node-disjoint $\mathcal{P}$-spiders that hits $U(\mathcal{P})$. If $G$ has maximum indegree 1, then $G$ contains a family of node-disjoint $\mathcal{P}$-spiders that hits at least $\frac{2}{3}\ell$ nodes in $U(\mathcal{P})$.*

*Proof.* We prove the arborescence case by induction on $\ell$. If $\ell = 1$, then $\mathcal{S} = \mathcal{P}$ is a family as required. Suppose that $\ell \geq 2$. If there is a path $P$ that has no node in common with other paths, then the induction is obvious. Otherwise, let $h$ be a farthest node from the root such that the subtree of $G$ induced by $h$ and its descendants hits at least two nodes in $U(\mathcal{P})$. By the choice of $h$, in the subtree, the paths from $h$ to the nodes in $U(\mathcal{P})$ form a $\mathcal{P}$-spider that hits at least two nodes in $U(\mathcal{P})$. Let $G'$ be obtained from $G$ by removing this subtree and let $\mathcal{P}'$ be obtained by removing from $\mathcal{P}$ all the paths that have a node in this subtree. If $\mathcal{P}' = \emptyset$, then $\mathcal{S} = \{S\}$ is a family of $\mathcal{P}$-spiders as required. Otherwise, by the induction hypothesis, there exists a family $\mathcal{S}'$ of node-disjoint spiders for $\mathcal{P}'$ as in the lemma. Then $\mathcal{S} = \mathcal{S}' \cup \{S\}$ is a family of $\mathcal{P}$-spiders as required. This concludes the proof of the arborescence case.

Suppose that $G$ has maximum indegree 1. Then $G$ is a collection of node-disjoint directed graphs of the following type: each of the graphs is a cycle (that may be a single node) with node-disjoint arborescences (that may be single nodes) attached to the cycle by the roots. As these graphs are node-disjoint, it is sufficient to consider the case when $G$ is such a graph. If $G$ is acyclic, then $G$ is an out-arborescence and we are done. If $\ell \geq 3$, then we arrive at the arborescence case by removing one edge from the cycle of $G$ and removing the path that contains this edge from $\mathcal{P}$ – then we get a family $\mathcal{S}$ of node-disjoint spiders that hits least $\ell - 1 \geq \frac{2}{3}\ell$ nodes in $U(\mathcal{P})$. The remaining case is $\ell = 2$, say $\mathcal{P} = \{P_1, P_2\}$ and $U(\mathcal{P}) = \{u_1, u_2\}$. Let $h$ be a common node of $P_1$ and $P_2$. Then the union of the $hu_1$-subpath of $P_1$ and the $hu_2$-subpath of $P_2$ is a $\mathcal{P}$-spider that hits all nodes in $U(\mathcal{P})$. ∎

We note that Lemma 15.15 was extended by Chuzhoy and Khanna [35] to an arbitrary graph $G$, but the proof of the case when $G$ has maximum indegree 1 is simpler, and Lemma 15.15 suffices for the proof of Theorem 15.14.

Now we use Lemmas 15.15 and 15.12 to prove Theorem 15.14.

### *Proof of Theorem 15.14:*

For every $C \in \mathcal{C}(\mathcal{F})$, fix some inclusionwise-minimal edge-cover $J_C \subseteq J$ of $\mathcal{F}(C)$. By lemma 15.11(i), $\mathcal{F}(C)$ is a ring family. Let $e_1, \ldots, e_q$ be an ordering of $J_C$ and $C_1 \subset \cdots \subset C_q$ sets in $\mathcal{F}(C)$ as in Lemma 15.12, in which $e_i = v_i u_i$ is as in the lemma. Obtain a directed path $P_C$ by adding for every $i = q, \ldots, 2$ the directed edge $u_i v_{i-1}$, if $u_i \neq v_{i-1}$; for example, if $u_i \neq v_{i-1}$ for all $i$, then the node sequence of $P_C$ is $(v_q, u_q, v_{q-1}, u_{q-1}, \ldots, v_1, u_1)$. Denote $M_C = C_q$ and $u_C = u_1$, and note that $u_C \in C \subseteq M_C \in \mathcal{F}(C)$.

Let $\mathcal{P} = \{P_C : C \in \mathcal{C}(\mathcal{F})\}$. Each directed path $P_C \in \mathcal{P}$ has only its starting node outside $M_C$, and the sets $\{M_C : C \in \mathcal{C}(\mathcal{F})\}$ are pairwise disjoint, by Lemma 15.11(ii). Thus, any two paths in $\mathcal{P}$ have distinct endnodes and the union of the paths in $\mathcal{P}$ is a graph of maximum indegree 1. Hence, Lemma 15.15 applies, and there exists a family $\hat{S}$ of node-disjoint $\mathcal{P}$-spiders that hit at least $\frac{2}{3}\ell$ nodes in $U(\mathcal{P}) = \{u_C : C \in \mathcal{C}(\mathcal{F})\}$.

Consider a $\mathcal{P}$-spider $\hat{S} \in \hat{\mathcal{S}}$. Let $h$ be the head of $\hat{S}$ and let $\{u_C : C \in \mathcal{C}\}$ be the set of nodes in $U(\mathcal{P})$ hit by this spider. Let $S = \hat{S} \cap J$ be the edge set obtained from $\hat{S}$ by removing the added edges. To prove the theorem, it is sufficient to show that the triple $h, \mathcal{C}, S$ is an $\mathcal{F}$-spider. For $C \in \mathcal{C}$, let $\hat{S}_C$ be the $hu_C$-path in $\hat{S}$ and let $S_C = \hat{P}_C \cap J$. As $\hat{S}$ is a spider, the edge sets $\{S_C : C \in \mathcal{C}\}$ are pairwise $(V \setminus \{h\})$-disjoint. By Lemma 15.12, each $S_C$ is an $\mathcal{F}(h, C)$-cover. Furthermore, if $\mathcal{C} = \{C\}$, then $\hat{S}_C = P_C$, as $\hat{S}$ is a $\mathcal{P}$-spider. This implies that $S_C = J_C$ and $h = v_q$. By Lemma 15.12, $v_q$ does not belong to any set in $\mathcal{F}(C)$. Thus, the triple $h, \mathcal{C}, S$ is an $\mathcal{F}$-spider, and the proof is complete.

## 15.7.4 Finding a Low-Density $\mathcal{F}$-Spider

Let $J$ be a partial solution for a directed Activation Set-Family Edge-Cover instance $G, \mathbf{p}, \mathcal{F}$. We may consider the equivalent instance obtained by removing $J$ from $G$ and replacing $\mathcal{F}$ by the residual family $\mathcal{F}^J$. Clearly, the optimal solution value of the new instance is at most the optimal solution value of the original instance. Thus, in what follows, we will assume that $J = \emptyset$. In the following lemma, we lower bound the decrease of the deficiency function caused by a union of $\mathcal{F}(h, C)$-covers $\{S_C : C \in \mathcal{C}\}$, and in particular by an $\mathcal{F}$-spider.

**Lemma 15.16** *Let $\mathcal{F}$ be an intersecting set-family on $V$, let $\mathcal{C} \subseteq \mathcal{C}(\mathcal{F})$, let $h \in V$, and let $S$ be a directed edge set that covers $\mathcal{F}(h, C)$ for every $C \in \mathcal{C}$, such that if $\mathcal{C} = \{C\}$ then $h$ does not belong to any set in $\mathcal{F}(C)$. Then $\nu(\emptyset) - \nu(S) \geq |\mathcal{C}|/3$.*

*Proof.* The $\mathcal{F}^S$-cores are pairwise disjoint and each of them contains some $\mathcal{F}$-core. Let $t$ be the number of $\mathcal{F}^S$-cores that contain exactly one $\mathcal{F}$-core. Any other $\mathcal{F}^S$-core contains at least two $\mathcal{F}$-cores. Thus, $\nu(\emptyset) - \nu(S) \geq \lceil (\nu(\emptyset) - t)/2 \rceil$. We upper bound $t$ as follows. By the definition of $S$, any $\mathcal{F}^S$-core $C'$ that contains some $C \in \mathcal{C}$, contains $h$ or contains some $\mathcal{F}$-core distinct from $C$. Furthermore, if $\mathcal{C} = \{C\}$, then the latter must hold. As the $\mathcal{F}^S$-cores are pairwise disjoint, $h$ belongs to at most one of them. Thus, $t \leq \nu(\emptyset) - (|\mathcal{C}| - 1)$ if $|\mathcal{C}| \geq 2$, and $t \leq \nu(\emptyset) - 1$ if $|\mathcal{C}| = 1$. In both cases, we have $\nu(\emptyset) - \nu(S) \geq |\mathcal{C}|/3$. ∎

In Reference 19, a better bound $v(\emptyset) - v(S) \geq |\mathcal{C}|/2$ is established under additional assumptions on $S$. This is why the ratio $\frac{9}{2}(\ln|\mathcal{C}(\mathcal{F})| + 1)$ proved here is worse by a factor $\frac{3}{2}$ than the ratio $3(\ln|\mathcal{C}(\mathcal{F})| + 1)$ in Reference 19.

The following lemma shows that there exists an $\mathcal{F}$-spider of low density.

**Lemma 15.17** *Let $\mathcal{S}$ be a family of $\mathcal{F}$-spiders as in Theorem 15.14 for an optimal directed cover of an intersecting family $\mathcal{F}$. There is an $\mathcal{F}$-spider $S^*, \mathcal{C}^*, h$ in $\mathcal{S}$ of such that $\dfrac{\tau(S^*)}{|\mathcal{C}^*|/3} \leq \dfrac{9}{2} \cdot \dfrac{\text{opt}}{v(\emptyset)}$.*

*Proof.* Let $\mathcal{C}_S$ denote the set of $\mathcal{F}$-cores hit by a spider $S \in \mathcal{S}$. As the spiders in $\mathcal{S}$ are node-disjoint,
$$\sum_{S \in \mathcal{S}} \tau(S) \leq \text{opt}. \text{ As } \mathcal{S} \text{ hits at least } \frac{2}{3}v(\emptyset) \text{ distinct } \mathcal{F}\text{-cores} \sum_{S \in \mathcal{S}} |\mathcal{C}_S| \geq \frac{2}{3}v(\emptyset). \text{ Thus, } \sum_{S \in \mathcal{S}} |\mathcal{C}_S|/3 \geq \frac{2}{9}v(\emptyset).$$
Consequently, by an averaging argument, there is $S^* \in \mathcal{S}$ as required. ∎

**Lemma 15.18** *There exists a polynomial time algorithm that given an instance of directed* Activation Set-Family Edge-Cover *with intersecting $\mathcal{F}$ finds an edge set $S \subseteq E$ of density $\dfrac{\tau(S)}{v(\emptyset) - v(S)} \leq \dfrac{9}{2} \cdot \dfrac{\text{opt}}{v(\emptyset)}$.*

*Proof.* Let $S^*, \mathcal{C}^*, h$ be a spider as in Lemma 15.17. As in the proof of Lemma 15.9, we may assume that we know $h$, the power level $w_h$ of $h$ in $S^*$, and the number $\ell = |\mathcal{C}^*|$. Then the algorithm is as follows.

---

**ALGORITHM 15.7    Low-Density $\mathcal{F}$-Spider$(G = (V, E), \mathbf{p}, \mathcal{F}, h, w_h, \ell)$**

1  for every edge $hv \in E$ do: $p_{hv} \leftarrow 0$ if $p_{hv} \leq w_h$ and $E \leftarrow E \setminus \{hv\}$ otherwise
2  for every core $C \in \mathcal{C}(\mathcal{F})$ compute an optimal $\mathcal{F}(h, C)$-cover $P_C$
3  if $\ell = 1$ then return $S = \arg\min\{\tau(P_C) : C \in \mathcal{C}(\mathcal{F}), v(P_C) \leq v(\emptyset) - 1\}$
4  else return the union $S$ of $\ell$ lowest value sets $P_C$

---

The algorithm can be implemented in polynomial time using the algorithm from Lemma 15.13. We show that $\tau(S) \leq \tau(S^*)$. Let $\tau'(J)$ denote the optimal assignment value activating $J$ with the modified power thresholds after step 1. Then $\tau'(S) \leq \tau'(S^*)$, as $S^*$ is a union of $\ell$ pairwise $(V \setminus \{h\})$-disjoint $\mathcal{F}(h, C)$ covers whereas $S$ is a union of $\ell$ lowest $\tau'$-value $\mathcal{F}(h, C)$-covers. Moreover, $\tau(S) \leq \tau'(S) + w_h$ whereas $\tau(S^*) = \tau'(S^*) + w_h$. Thus, we get $\tau(S) \leq \tau'(S) + w_h \leq \tau'(S^*) + w_h = \tau(S^*)$. Consequently, from Lemma 15.16 and our choice of $S^*$, we will get $\dfrac{\tau(S)}{v(\emptyset) - v(S)} \leq \dfrac{\tau(S^*)}{\ell/3} \leq \dfrac{9}{2} \cdot \dfrac{\text{opt}}{v(\emptyset)}$, as required. ∎

Lemma 15.17 implies that the algorithm from Lemma 15.18 finds an edge set $S \subseteq E \setminus J$ of density $\leq \dfrac{9}{2} \cdot \dfrac{\text{opt}}{v(\emptyset)}$; namely, $J$ satisfies the $\rho$-Density Condition with $\rho = \dfrac{9}{2}$. Thus, we can apply the $\rho$-Density Algorithm with $\rho = \dfrac{9}{2}$. As $v(\emptyset) = |\mathcal{C}(\mathcal{F})|$, we get ratio $\dfrac{9}{2}(\ln|\mathcal{C}(\mathcal{F})| + 1)$ from Theorem 15.10.

## 15.8  Open Problems

In the current section, we list some open problems in the field, most of them for the case of high demands.

***The undirected*** Min-Power Edge-Multi-Cover ***problem.***

The currently best known ratio for the problem is $\min\{O(\ln k), k + 1/2\}$, whereas for unit/uniform powers, a constant ratio is known [16]. A constant ratio for the problem would imply several consequences, via Corollary 15.5. For example, we would get a constant ratio for the Min-Power $k$-Out-Connectivity problem. More importantly, we get that for the $k$-Connectivity problem, the approximability of the min-cost and the min-power versions differs by a constant factor. It is an old open problem whether the Min-Cost

$k$-Connectivity problem admits a constant ratio, and relation between the min-cost and the min-power variants might help to resolve it.

**Problems with linear ratios.**

For several min-power and node-weighted problems, the currently best known ratio is $O(k)$, or even $O(k \ln n)$. The simplest examples are directed/undirected Min-Power $k$ Edge-Disjoint Paths, and directed Min-Power $k$-Connectivity and Min-Power $k$-Edge-Connectivity. As was mentioned in the Introduction, these problems are unlikely to admit polylogarithmic ratios [13,15]. However, this does not exclude ratios sublinear in $k$. The simplest open problem is whether directed or undirected $k$ Edge-Disjoint Paths problems admit ratio $k^{1-\epsilon}$ for some $\epsilon > 0$.

**Problems with undetermined complexity status.**

The best known ratio for the min-power undirected $k$ Disjoint Paths problem is 2. However, the problem is not known to be NP-hard. Does the problem admit a polynomial time algorithm? A related (and probably easier) question is whether the problem of covering a ring set-family by undirected edges admits a polynomial time algorithm for min-power setting. The currently best known ratio for this problem is also 2.

# References

1. A. Schrijver. *Combinatorial Optimization, Polyhedra and Efficiency.* Springer-Verlag, Berlin, Germany, 2003.
2. A. Frank. *Connections in Combinatorial Optimization.* Oxford University Press, Oxford, UK, 2011.
3. D. Panigrahi. Survivable network design problems in wireless networks. In *SODA*, pp. 1014–1027, 2011.
4. E. Althaus, G. Calinescu, I. Mandoiu, S. Prasad, N. Tchervenski, and A. Zelikovsky. Power efficient range assignment for symmetric connectivity in static ad-hoc wireless networks. *Wirel. Networks*, 12(3):287–299, 2006.
5. F. Grandoni. On min-power steiner tree. In *ESA*, pp. 527–538, 2012.
6. P. Klein and R. Ravi. A nearly best-possible approximation algorithm for node-weighted steiner trees. *J. Algorithms*, 19(1):104–115, 1995.
7. J. Byrka, F. Grandoni, T. Rothvoß, and L. Sanitá. Steiner tree approximation via iterative randomized rounding. *J. ACM*, 60(1):6, 2013.
8. G. Kortsarz and Z. Nutov. Approximating minimum-power edge-covers and 2, 3-connectivity. *Disc. Appl. Math.*, 157(8):1840–1847, 2009.
9. A. Agrawal, P. Klein, and R. Ravi. When trees collide: An approximation algorithm for the generalized steiner problem on networks. *SIAM J. Comput.*, 24(3):440–456, 1995.
10. Z. Nutov. Node-connectivity survivable network problems. In T. Gonzalez, Ed., *Approximation Algorithms and Metaheuristics*, chapter 13, Chapman & Hall, 2018.
11. Z. Nutov. The $k$-connected subgraph problem. In T. Gonzalez, Ed., *Approximation Algorithms and Metaheuristics*, chapter 12, Chapman & Hall, 2018.
12. Z. Nutov. Survivable network activation problems. *Theor. Comput. Sci.*, 514:105–115, 2013.
13. M. Hajiaghayi, G. Kortsarz, V. Mirrokni, and Z. Nutov. Power optimization for connectivity problems. *Math. Prog.*, 110(1):195–208, 2007.
14. Z. Nutov. Approximating minimum-power $k$-connectivity. *Ad Hoc Sens. Wirel. Networks*, 9(1–2):129–137, 2010.
15. Y. Lando and Z. Nutov. On minimum power connectivity problems. *J. Dis. Algorithms*, 8(2): 164–173, 2010.
16. N. Cohen and Z. Nutov. Approximating minimum power edge-multi-covers. *J. Comb. Optim.*, 30(3):563–578, 2015.

17. Z. Nutov. Approximating minimum-cost edge-covers of crossing biset-families. *Combinatorica*, 34(1):95–114, 2014.

18. Z. Nutov. Approximating steiner networks with node-weights. *SIAM J. Comput.*, 39(7):3001–3022, 2010.

19. Z. Nutov. Approximating minimum power covers of intersecting families and directed edge-connectivity problems. *Theor. Comput. Sci.*, 411(26–28):2502–2512, 2010.

20. G. Calinescu, S. Kapoor, A. Olshevsky, and A. Zelikovsky. Network lifetime and power assignment in ad hoc wireless networks. In *ESA*, pp. 114–126, 2003.

21. I. Caragiannis, C. Kaklamanis, and P. Kanellopoulos. Energy-efficient wireless network design. *Theor. Comput. Syst.*, 39(5):593–617, 2006. Preliminary version in ISAAC 2003, 585–594.

22. A. Bhaskara, M. Charikar, E. Chlamtac, U. Feige, and A. Vijayaraghavan. Detecting high log-densities: an $O(n^{1/4})$ approximation for densest $k$-subgraph. In *STOC*, pp. 201–210, 2010.

23. W. Chen and N. Huang. The strongly connecting problem on multihop packet radio networks. *IEEE Trans. Commun.*, 37(3):293–295, 1989.

24. M. Goemans, A. Goldberg, S. Plotkin, D. Shmoys, E. Tardos, and D. Williamson. Improved approximation algorithms for network design problems. In *SODA*, pp. 223–232, 1994.

25. W. Mader. Ecken vom grad $n$ in minimalen n-fach zusammenhängenden graphen. *Arch. Math.*, 23:219–224, 1972.

26. J. Cheriyan and L. Végh. Approximating minimum-cost $k$-node connected subgraphs via independence-free graphs. *SIAM J. Comput.*, 43(4):1342–1362, 2014.

27. T. Fukunaga, Z. Nutov, and R. Ravi. Iterative rounding approximation algorithms for degree-bounded node-connectivity network design. *SIAM J. Comput.*, 44(5):1202–1229, 2015.

28. J. Cheriyan, T. Jordán, and Z. Nutov. On rooted node-connectivity problems. *Algorithmica*, 30(3):353–375, 2001.

29. A. Frank and É. Tardos. An application of submodular flows. *Linear Algebra Appl.*, 114/115:329–348, 1989.

30. J. Cheriyan and R. Thurimella. Approximating minimum-size $k$-connected spanning subgraphs via matching. *SIAM J. Comput.*, 30(2):528–560, 2000.

31. S. Khuller and U. Vishkin. Biconnectivity approximations and graph carvings. *J. Assoc. Comput. Mach.*, 41(2):214–235, 1994.

32. G. Kortsarz, V. Mirrokni, Z. Nutov, and E. Tsanko. Approximating minimum-power degree and connectivity problems. *Algorithmica*, 60(4):735–742, 2011.

33. E. Tsanko. Approximating minimum power network design problems. Msc thesis, Department of Computer Science, The Open University of Israel, 2006.

34. D. Johnson. Approximation algorithms for combinatorial problems. *J. Comput. Syst. Sci.*, 9(3):256–278, 1974.

35. J. Chuzhoy and S. Khanna. Algorithms for single-source vertex connectivity. In *FOCS*, pp. 105–114, 2008.

# 16

# Stochastic Local Search Algorithms for the Graph Coloring Problem

Marco Chiarandini

Irina Dumitrescu

Thomas Stützle

## 16.1 Introduction

The (vertex) graph coloring problem (GCP) is a central problem in graph theory [1] and it arises in many real-life applications like register allocation [2], air traffic flow management [3], frequency assignment [4], light wavelengths assignment in optical networks [5], or timetabling [6,7].

In the GCP, one is given an undirected graph $G = (V, E)$, with $V$ being the set of $|V| = n$ vertices and $E$ being the set of edges, we call a *k-coloring* of $G$, a mapping $\varphi : V \mapsto \Gamma$, in which $\Gamma = \{1, 2, \ldots, k\}$ is the set of $|\Gamma| = k$ integers, each one representing one color. We say that a coloring is *feasible* or *legal* if for all $[u, v] \in E$ we have that $\varphi(u) \neq \varphi(v)$; otherwise, we say that the coloring is *infeasible*. If for some $[u, v] \in E$ we have that $\varphi(u) = \varphi(v)$, the vertices $u$ and $v$ are *in conflict*. The *conflict set* $V^c$ is the set of all vertices that are in conflict. A $k$-coloring can also be seen as a partitioning of the set of vertices into $k$ disjoint sets, called *color classes*, and represented as a partitioning of $V$, $C = \{C_1, \ldots, C_k\}$. Finally, if some vertices are assigned to color classes whereas others are not, we say that we have a *partial coloring*.

The GCP can be posed as a decision or an optimization problem. In the decision version, also called the *(vertex) k-coloring problem*, the question to be answered is whether for some given $k$ a feasible $k$-coloring exists. The optimization version of the GCP asks for the smallest number $k$ such that a feasible $k$-coloring exists; for a graph $G$, this number is called the *chromatic number* $\chi_G$. The problem of finding the chromatic number is often approached by solving a sequence of $k$-coloring problems: An initial value of $k$ is considered and each time a feasible $k$-coloring is found, the value of $k$ is decreased by one. The chromatic number is found when for some $k$ the answer to the decision version is "no", that is, a feasible $k$-coloring does not exist. In this case, $\chi_G = k + 1$. If a feasible coloring cannot be found but no proof of its nonexistence is given, as it is typically the case with heuristic algorithms, $k+1$ is an upper bound on the chromatic number.

It is well known that the $k$-coloring problem for general graphs is $\mathcal{NP}$-complete and that the chromatic number problem is $\mathcal{NP}$-hard [8]; only for a few special cases, polynomial time algorithms are known. In some sense, the GCP is among the hardest problems in $\mathcal{NP}$, as approximation ratios of $n^{1-\epsilon}$ cannot be achieved in polynomial time, unless $\mathcal{NP} = \mathcal{ZPP}$ [9]; that is, it is very unlikely to find polynomial time algorithms that guarantee a constant approximation ratio. The best absolute performance guarantee of $\mathcal{O}(n(\log \log n)^2 / \log^3 n)$ is given for an approximation algorithm presented in Reference 10. More details on the approximability of graph coloring can be found in Reference 11.

Due to its importance, many attempts have been made to tackle the GCP algorithmically. Various exact algorithms, including specialized branch–and–bound algorithms [12,13] or approaches based on general integer programming formulations of the GCP have been tested [14–16]. Probably the best known exact algorithm is Brélaz' modification of Randall-Brown's coloring algorithm [12]. Although exact algorithms can be very effective on specific classes of graphs, their performance for many large graphs is often rather poor [16,17]. Therefore, a significant amount of research has focused on stochastic local search (SLS) algorithms for the GCP (see Vol. 1, Chap. 17 for an overview of SLS methods). The available SLS approaches range from rather simple but very fast construction methods over iterative improvement algorithms to sometimes rather complex SLS algorithms, which are currently the best performing approximate algorithms for many classes of GCP instances. In this chapter, we give an overview of available SLS algorithms and present some indicative comparison of the performance of several such algorithms.

## 16.2  Benchmark Instances

The available benchmark instances for the GCP are either randomly generated graphs or derived from some practically relevant application. The most frequently used benchmark instances are available from the web-page *COLOR02/03/04: Graph Coloring and its Generalizations* at http://mat.tepper. cmu.edu/COLOR04/ (A more updated repository with best known results is available at https: //sites.google.com/site/graphcoloring/vertex-coloring) and an earlier center for Discrete Mathematics and Theoretical Computer Science (DIMACS) challenge [18]. Next, we describe some instance classes that will be used in the experimental analysis presented in Section 16.7.

*Uniform random graphs:* This class comprises graphs of a variable number of vertices, $n$, in which each of the $n(n - 1)/2$ possible edges is generated independently at random with probability $p$. The instance has identifiers DSJC$n.p$, with $n \in \{125, 250, 500, 1000\}$ and $p \in \{0.1, 0.5, 0.9\}$. They stem from one of the first thorough experimental studies of SLS algorithms for the GCP [19]. For these graphs, probabilistic estimates of the chromatic number exist [20–22].

*Flat graphs:* These graphs are random graphs generated according to an equi-partitioning of vertices into $k$ sets. Edges are then generated to respect some constraints on the resulting degrees of the vertices until a given edge density is obtained [23]. The value of $k$ is an upper bound on the chromatic number of the graph. These instances are denoted as flat$n\_k$, with $n = 300$; $k \in \{20, 26, 28\}$ and $n = 1000$; $k \in \{50, 60, 76\}$.

*Leighton graphs:* Leighton graphs are random graphs of density below 0.25, which are constructed by first partitioning vertices into $k$ distinct sets representing the color classes and then assigning edges only between vertices that belong to different sets. The chromatic number is guaranteed to be $k$ by implanting cliques of sizes ranging from 2 to $k$ into the graph. The graphs are denoted as le450_$kx$, in which 450 is the number of vertices, $k$ is the chromatic number of the graph,

$x \in \{a, b, c, d\}$ is a letter used to distinguish different graphs with the same characteristics, with $c$ and $d$ graphs having higher edge density than the $a$ and $b$ ones.

*Quasigroup graphs:* A Quasigroup is an algebraic structure on a set with a binary operator. The constraints on this structure define a Latin square, which is a square matrix with elements such that entries in each row and column are distinct. A Latin square of order $n$ has $n^2$ vertices and $n^2(n-1)$ edges, corresponding to $2n$ cliques, a clique per row/column, each of size $n$. The chromatic number of Latin square graphs is $n$. We denote Latin square graphs originated by Quasigroups by `qg.ordern`. Latin square graphs arise also in experimental design for statistical analysis. The instance `latin_square_10`, which has an unknown chromatic number, is one such example with preassigned experiments [24].

*Queens graphs:* The $n$-queens problem asks whether it is possible to place $n$ queens on a $n \times n$ grid such that no pair of queens is in the same row, column, or diagonal. This problem can be posed as a GCP and a feasible solution with $n$ queens exists if, and only if, the resulting queen graphs have a feasible coloring with $n$ colors. Queen graphs are highly structured instances and their edge density decreases with their size; they are denoted by `queenn_n`.

*Wavelength assignment problem (WAP) graphs:* These graphs arise in the design of transparent optical networks [5] and are denoted by `wap0ma`, in which $m = \{1, \dots, 8\}$. They have between 905 and 5231 vertices. All instances have a clique of size 40.

*Jacobian estimation graphs:* These graphs stem from a matrix partitioning problem in the segmented columns approach to determine sparse Jacobian matrices [25]. They range in size from 662 to 1916 vertices and they are identified with the following names `abb313GPIA`, `ash331GPIA`, `ash608GPIA`, `ash958GPIA`, and `will199GPIA`.

The DIMACS benchmark repository contains some other instances that we identified as *easy*, because some combination of preprocessing rules and simple construction heuristics, which are introduced in the next two sections, are enough to find a coloring with the known chromatic number. We identified 45 instances as easy. They are the Mycielski graphs, the graphs from register allocation for variables in compiled code [24], graphs from Knuth's Stanford GraphBase [26], and the almost 3-colorable graphs with embedded four cliques [27]. Two further classes of graphs that are a generalization of Mycielski graphs (the insertions and full insertions graphs due to Caramia and Dell'Olmo, pers. comm.) and graphs for course scheduling [24] are not useful to determine differences among algorithms.

# 16.3 Preprocessing

In the chromatic number problem, preprocessing can be applied to reduce a graph $G$ to a graph $G'$ such that a feasible $k$-coloring for $G$ can be derived by construction rules from any feasible $k$-coloring of $G'$. Next we give the two preprocessing rules presented in Reference 28.

*Rule 1:* Remove all vertices in $G$ that have a degree less than the size of the largest known clique $\widehat{\omega}(G)$. Knowing that the degree of a vertex $u$ is less than $\widehat{\omega}(G)$ guarantees that at least one color that is not used in the set of adjacent vertices can be assigned to $u$ without breaking feasibility. (This rule can be applied when solving the $k$-coloring problem by replacing $\widehat{\omega}(G)$ by $k$.)

*Rule 2:* Remove any vertex $v \in V$ for which there is a $u \in V$, $v \neq u$ and $[u, v] \notin E$, such that $u$ is adjacent to every vertex to which $v$ is adjacent (subsumption). In this case, any color that can be assigned to $u$ can also be assigned to $v$.

These two rules can be applied in any order and if one rule applies, it may make possible further reductions by the other rule. Hence, in the preprocessing stage, the two rules are applied iteratively until no vertex can be removed anymore. The rules are easy to apply. Rule 1 requires $\mathcal{O}(|V|)$ operations once a clique has been found heuristically and the degree of each vertex is known. Rule 2 is more costly and its time-complexity is $\mathcal{O}(|V|^3)$. The overall reduction time is, however, insignificant in practice. Among the challenging instances that we consider, preprocessing is effective only for the WAP instances.

The graph can also be reduced heuristically. One such procedure consists of reducing the graph by removing maximal independent sets [29–31]. Typically, this procedure is accomplished with the companion strategy of minimizing the number of edges in the residual graph. In contrast to the previously mentioned preprocessing rules, this procedure may make it impossible to find the chromatic number of the original graph because it is a priori unknown how the independent sets should be constructed such that an optimal coloring is obtained from a coloring of the residual graph and the coloring for the independent sets. Nevertheless, for some graphs such a heuristic procedure has contributed significantly to the improvement of the solution quality for SLS algorithms [29,30,32,33].

## 16.4 Construction Heuristics

The fastest methods to generate a feasible coloring are construction heuristics. Most of these heuristics for the GCP belong to the class of *sequential* heuristics that start from a set of empty color classes $\{C_1, \ldots, C_k\}$, in which $k = |V|$, and then iteratively add vertices to color classes until a complete coloring is reached. Each iteration of the heuristic consists of two steps: first, the next vertex to be colored is chosen, and second, this vertex is assigned to a color class. The order in which the vertices are colored corresponds to a permutation $\pi$ of the vertex indices that can be determined either *statically* before assigning the colors, or *dynamically* by taking into account the partial coloring for the choice of the next vertex. The choice of the color class at each construction step is typically based on the *greedy heuristic* that adds at iteration $i$ the vertex $\pi(i)$ to the color class with the lowest possible index such that the partial coloring obtained after coloring vertex $\pi(i)$ remains feasible, therefore trying to minimize the number of non-empty color classes.

Clearly, the result of the greedy heuristic depends on the permutation $\pi$. The simplest way to derive $\pi$ in a static way is by using the *random order sequential* (ROS) *heuristic* that simply generates a random permutation. Several other ways for generating $\pi$ statically exist, like largest degree first and smallest degree last. However, static sequential methods are typically dominated by dynamic ones [21]. Probably the most widely known dynamic heuristic is the Degree SATURation (DSATUR) *heuristic* that is derived from the exact DSATUR algorithm of Brélaz [12]. In DSATUR, the vertices are first arranged in decreasing order of their degrees and a vertex of maximal degree is inserted into $C_1$. Then, at each construction step the next vertex to be inserted is chosen according to the *saturation degree*, that is, the number of differently colored adjacent vertices. The vertex with the maximal saturation degree is chosen and inserted according to the greedy heuristic. Ties are broken preferring vertices with the maximal number of adjacent, still uncolored vertices; if further ties remain, they are broken randomly. Other dynamic heuristics for determining $\pi$ were studied in References 34, 35.

A different strategy for generating colorings is to iteratively extract independent sets from the graph. The most famous such heuristic is the *recursive largest first* (RLF) heuristic proposed by Leighton [7]. RLF iteratively constructs a color class $C_i$ by first assigning to it a vertex $v$ with maximal degree from the set $V'$ of still uncolored vertices; initially we have $V' = V$. Next, all the vertices in $V'$ that are adjacent to $v$ are removed from $V'$ and inserted into a set $U$ that is initially empty; $U$ is the set of vertices that cannot be added to color class $C_i$ anymore. Then, while $V'$ is not empty, the vertex $v' \in V'$ that has the largest number of edges $[v', u]$, with $u \in U$, is chosen; $v'$ is added to $C_i$, removed from $V'$, and all vertices in $V'$ adjacent to $v'$ are moved into $U$. Ties are broken, if possible, choosing the vertex that has minimal degree in $V'$, otherwise randomly. These steps are iterated until $V'$ is empty and the same steps are repeated with the residual graph consisting of the vertices in $U$.

We have compared experimentally the three construction heuristics DSATUR, RLF, and ROS in Reference 17. As they are stochastic procedures, due to the random tie breaking employed, we ran each of the algorithms 10 times on all 125 instances of the two previously mentioned benchmark sets as of end 2004. The conclusion was that RLF performs statistically significantly better than DSATUR for most instance classes and both heuristics are by a large margin better than ROS. RLF is not significantly better than DSATUR only on the insertions, full insertions, and course scheduling graphs. With respect to computation time, although RLF and DSATUR have the same time complexity $\mathcal{O}(|V|^3)$, RLF is in practice much more time-consuming and ROS with a complexity of $\mathcal{O}(|V|^2)$ is the fastest. For example, on the largest instance tested (10,000 vertices), RLF takes on average about 18.5 seconds to color the graph with 101 colors, DSATUR 3 seconds to color it with 103 colors, and ROS less than 1 second to produce a coloring with 106 colors (the computer used was a 2 GHz AMD Athlon MP 2400+ Processor with 256 KB cache and 1 GB of RAM memory). A more detailed analysis showed that the computation time of RLF depends not only on the instance size, as it is the case for DSATUR, but it is also affected by the graph density and the final number of colors used [17].

Note that the construction heuristics described earlier can also be used if a fixed number $k < |V|$ of color classes is available. In that case, the usual steps of the construction algorithms are followed as long as the number of used color classes remains below $k$. Then, if a vertex cannot be colored legally, it is added to a color class randomly or according to some other heuristic that tries to minimize the number of arising conflicts. The result of these so modified construction heuristics is typically an infeasible $k$-coloring, which can serve as an initial solution for improvement methods.

## 16.5 Neighborhoods for Local Search

Once an initial (feasible or infeasible) coloring is generated by construction algorithms, one may try to improve it through the application of iterative improvement algorithms. These algorithms may be used within two different approaches to solve the GCP: as a sequence of decision problems or directly as an optimization problem. The first approach corresponds to leaving the numbers of colors *fixed* to some value $k$ at each stage. Subsequently, once a feasible coloring with $k$ colors is found, the number of colors is reduced by one. The second approach allows the number of used colors to *vary* throughout one single trial of the local search algorithm. To apply iterative improvement algorithms, one needs to define a set of candidate solutions, a neighborhood relation, and an evaluation function. Two choices are possible for the set of *candidate solutions*: One may opt for the algorithm to work on *complete colorings*, in which each candidate solution represents a possibly infeasible partitioning of the set of vertices into color classes, or to work on *partial colorings*, in which candidate solutions are also those with a subset of the vertices left without any color assigned. Next, we present several neighborhood relations and evaluation functions of candidate solutions, differentiating among possible choices of how to tackle the GCP.

### 16.5.1 Strategy 1: $k$ Fixed, Complete Colorings

This approach works on a partitioning of $V$ into $k$ color classes: $C = \{C_1, \ldots, C_k\}$. The most widely used evaluation function counts the number of edges that have their end points in the same color class. Formally, the function can be described as $g(C) = \sum_{i=1}^{k} |E_i|$, in which $E_i$ is the set of edges with both end points in $C_i$. A candidate solution with an evaluation function value of zero corresponds to a proper $k$-coloring. An alternative evaluation function would be $g(C) = |V^C|$. However, this function is much less used, possibly because it would lead to a large number of neighbors with the same evaluation function value, therefore inducing large plateaus.

*1-Exchange neighborhood*: The most frequently used neighborhood structure in this setting is the 1-exchange neighborhood, in which two colorings are neighbors if they differ in the color class assignment of exactly one vertex. That is, to obtain a neighbor of $C$, one vertex $u$ is moved from a color class $C_j$,

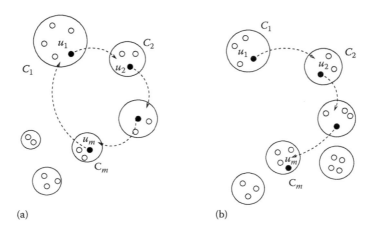

**FIGURE 16.1**   The black vertices are the vertices involved in the cyclic and path exchange: (a) cyclic exchange and (b) path exchange.

$j \in \{1, \ldots, k\}$, into a different color class $C_l, l \neq j$. Often, the 1-exchange neighborhood is further restricted to change only the color class assignment of vertices that are in conflict, as only these modifications can lead to a decrease of the evaluation function; we call this the *restricted 1-exchange neighborhood*.

*Swap neighborhood*:  In the swap neighborhood, exactly one vertex $v \in V^c$ exchanges the color class with another vertex $u \in V$. This neighborhood is of quadratic size and it is used only very rarely.

*Cyclic exchange and path exchange neighborhoods*:  An extension of the 1-exchange and swap neighborhoods are the path and cyclic exchange neighborhoods (Figure 16.1). Both the cyclic and the path exchange are sequences of 1-exchanges. A *cyclic exchange* of length $m$ acts on a sequence of distinctly colored vertices $(u_1, \ldots, u_m)$. For simplicity, we will denote the color class of any $u_i, i = 1, \ldots, m$, by $C_i$. The cyclic exchange moves any $u_i, i = 1, \ldots, m$, from $C_i$ into $C_{i+1}$. We use the convention $C_{m+1} = C_1$. A cyclic exchange does not change the cardinality of the color classes involved in the move. In a *path exchange* instead, $u_m$ remains in $C_m$. Hence, the sequence of exchanges is not closed and the cardinality of $C_1$ and $C_m$ is modified. The cyclic and path exchange neighborhoods are examples of very large-scale neighborhoods (VLSN). The problem of finding a good neighbor within these neighborhoods can be reduced to searching the least cost cycle or cost path in an improvement graph. If these problems can be solved exactly, then the best neighbor is found. We refer to Volume 1, Chapter 18 for more details on how an improvement graph is built and can be searched exactly and heuristically. Details on the implementation of a VLSN for the GCP can be found in Reference 17.

*Other neighborhoods*:  Similar to the path exchange neighborhood is the ejection-chain neighborhood defined in Reference 36, in which the length of the sequences was limited to 3. Other neighborhoods, rather fancy at times, were proposed in the context of a variable neighborhood approach [37]. However, the overall contribution of these neighborhoods is somewhat unclear, especially when they are used inside a more complex SLS algorithm.

## 16.5.2  Strategy 2: $k$ Fixed, Partial Colorings

This approach works on a partitioning of $V$ into $k + 1$ color classes: $C = \{C_1, \ldots, C_k, C_{imp}\}$. The partial coloring $\bar{C} = \{C_1, \ldots, C_k\}$ is usually required to be feasible. The color class $C_{imp}$ is called the *impasse* class [33], and it contains the "uncoloured" vertices. The goal of the local search is to try to empty color class $C_{imp}$, while maintaining the partial coloring $\bar{C}$ feasible.

The most widely used evaluation function in this case is $g(C) = \sum_{v \in C_{imp}} d(v)$, in which $d(v)$ is the degree of vertex $v$. Analogously to complete colorings, an alternative would be to minimize $g(C) = |C_{imp}|$. However, this appears to lead to worse performance [31].

*i-swap neighborhood*: A neighbor of $C$ in the *i*-swap neighborhood is obtained by moving a vertex $v$ from $C_{imp}$ into another color class $C_i$, followed by moving all vertices of $C_i$ that share an edge with $v$ into $C_{imp}$ so that the partial coloring $\bar{C}$ remains feasible [33].

*Other neighborhoods*: The neighborhoods discussed in the section dedicated to Strategy 1 could, at least in principle, also be applied for algorithms using strategy 2; however, it would probably be difficult to maintain the partial coloring feasible and additional penalties for infeasibility may be required in the evaluation function. So far, such an approach appears not to have been tried.

### 16.5.3 Strategy 3: $k$ Variable, Complete Colorings

The final strategy we discuss allows the number of color classes to vary during the local search. In almost all these approaches, the current candidate coloring is forced to be complete and feasible. The local search then tries to minimize the number of color classes. Hence, the simplest evaluation function would be to count the number of colors currently used; however, as moves that remove one color class completely will be certainly very rare, the guidance provided by this evaluation function would be minor. As an alternative, Johnson et al. [19] proposed to use $g(C) = -\sum_{i=1}^{k}(|C_i|)^2$, which biases the search towards a small number of color classes. The most widely used neighborhood structure for this strategy is based on Kempe chains.

*Kempe-chains neighborhood*: A *Kempe chain K* is a set of vertices that form a maximal connected component in the subgraph $G'$ of $G$ induced by the vertices that belong to two (disjoint) color classes $C_i$ and $C_j$ of $C$. A Kempe chain exchange applied to a feasible coloring produces again a feasible coloring (a *Kempe chain neighbor*) by moving all vertices of $C_i$ that belong to the Kempe chain $K$ into the color class $C_j$ and vice versa. An example of a Kempe chain is given in Figure 16.2.

An alternative to enforcing only proper colorings is to enlarge the search space and allow also improper colorings and variable $k$. In that case, the same neighborhoods as when keeping $k$ fixed can be applied, but the evaluation function needs now to also guide the search towards feasible candidate solutions. One such evaluation function is $g'(C) = -\sum_{i=1}^{k} |C_i|^2 + \sum_{i=1}^{k} 2|C_i||E_i|$ in which $E_i$ is the set of vertices with both end points in $C_i$ [19]. The second term in $g'$ is used to penalise conflicts between adjacent vertices.

Finally, note that we are not aware of any approach that leaves $k$ variable and uses partial colorings, although such an approach would be conceivable.

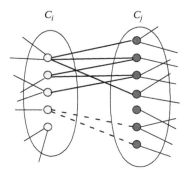

**FIGURE 16.2** Two Kempe chains are available between color classes $C_i$ and $C_j$. They are indicated by the thick and dashed lines, respectively.

### 16.5.4 Neighborhood Examination and Data Structures

The various neighborhoods we have described can be restricted and explored in various ways. It is intuitively clear that neighborhood restrictions are more important for large neighborhoods than for the small ones. Nevertheless, computational results with SLS algorithms suggest that restrictions to only moves that involve vertices from $V^c$ are essential even for the 1-exchange neighborhood.

Regarding the neighborhood search strategy, we can distinguish mainly between best and first-improvement strategies. In best improvement, the best neighboring candidate solution (best w.r.t. neighborhood restrictions, if these are used) is accepted, breaking ties randomly; in first-improvement, the first improving candidate solution found when scanning the neighborhood replaces the current one. Somehow intermediate between these two is the strategy followed in the *min-conflicts heuristic* [38], which searches the restricted 1-exchange neighborhood in a two-stage process. In a first stage, a vertex is chosen uniformly at random from the conflict set; in a second stage, this vertex is moved into a color class such that the number of conflicts is minimized, breaking ties randomly.

Finally, let us note that the local search algorithms for the GCP require the usage of appropriate data structures to support caching and updating strategies to make the evaluation of the neighboring candidate solutions as efficient as possible. Typically, the use of elaborated data structures is more important for best improvement local search algorithms and those making use of large neighborhoods. We refer to Reference 17 for a short discussion of efficient caching and updating strategies.

## 16.6 Stochastic Local Search Algorithms

In the case of the GCP, iterative improvement algorithms have received rather little attention, probably because of their generally poor performance when used as stand-alone algorithms. Much more attention has focused on the application of more complex SLS algorithms. In what follows, we present a selection of some of the best performing SLS algorithms, as well as a few recently developed ones. The selection was biased by the desire to cover different approaches and methods. In which appropriate, we also give a short overview of algorithms that are related to the ones we describe, covering in this way the majority of the SLS algorithms proposed so far for the GCP. If nothing else is said in the text, the evaluation functions used are the standard ones given for the various strategies described in the previous section.

### 16.6.1 Strategy 1: $k$ Fixed, Complete Colorings

*Tabu search with 1-exchange neighborhood*: Tabu search (TS) algorithms based on the 1-exchange neighborhood ($TS_{1\text{-ex}}$) are probably the most frequently applied SLS algorithms for the GCP (see Vol. 1, Chap. 21 for details on tabu search). Such an algorithm was first proposed by Hertz and de Werra [30] and was later improved leading to the most performing tabu search variant to date by Dorne et al. [39,40]. $TS_{1\text{-ex}}$ chooses at each iteration a best non-tabu or tabu but aspired neighboring candidate solution from the restricted 1-exchange neighborhood. If a 1-exchange move puts vertex $v$ from color class $C_i$ into $C_j$, it is forbidden to reassign vertex $v$ to $C_i$ in the next $tt$ steps; the tabu status of a neighboring solution is overruled if it would improve over the best candidate solution found so far (aspiration criterion). If more than one move produces the same effect on the evaluation function, one of those moves is selected uniformly at random. The tabu list length in $TS_{1\text{-ex}}$ is set to $tt = random(A) + \delta \cdot |V^c|$, in which $random(A)$ is an integer uniformly chosen from $\{0, \ldots, A\}$ and $\delta$ and $A$ are parameters. As $tt$ depends on $|V^c|$, the tabu list length varies dynamically with the evaluation function value.

*Tabu search with very large-scale neighbourhood*: In this algorithm, a neighborhood obtained from the composition of the path exchange and cyclic exchange neighborhoods is used, referred to as the *cyclic and path exchange neighborhood.* The embedding of this neighborhood into a tabu search algorithm analogous to $TS_{1-ex}$ results in an algorithm called $TS_{VLSN}$ [17]. Several variants of $TS_{VLSN}$ have been studied and the best performing one of these first selects the best non-tabu move in the restricted 1-exchange neighborhood as in $TS_{1-ex}$. If this move is a plateau-move (i.e., the best neighboring solution has the same evaluation function value as the current one), then the cyclic and path exchange neighborhood is searched. In all other cases, the 1-exchange move is applied and the tabu list is updated. The tabu mechanism is applied to the search for cyclic and path exchanges by discarding any neighboring candidate solution that involves the reassignment of a vertex to some color class that is currently tabu. The tabu list is updated by considering the path or the cyclic exchange as a composition of 1-exchanges and the tabu duration $tt$ for a specific vertex–color class pair $(v, i)$ is chosen using the rule of $TS_{1-ex}$. Yet, contrarily to $TS_{1-ex}$, $TS_{VLSN}$ does *not* use an aspiration criterion.

*Min-conflicts heuristic*: One of the most effective extensions of the basic min-conflicts (MC) heuristic is a tabu search variant of it (MC-$TS_{1-ex}$) [41]. It uses the same two-stage selection process as the min-conflicts heuristic, which was described in the previous section, but in the second stage it only allows to move the vertex into a color class that is not tabu, analogously to $TS_{1-ex}$. If all color classes are tabu, one is chosen randomly. One advantage of this neighborhood examination strategy is that it does not require the usage of sophisticated caching and updating schemes as required, for example, by $TS_{1-ex}$; hence, it allows for an easier implementation. In addition, the chances of cycling are reduced due to the random choices especially in the first stage of the selection process, allowing for shorter tabu lists.

*Guided local search*: Guided local search (GLS) is an SLS method that modifies the evaluation function to escape from local optima [42]. An application of GLS to the GCP was proposed in Reference 17. In this algorithm, GLS uses an augmented evaluation function $g'$ defined as

$$g'(C) = g(C) + \lambda \cdot \sum_{i=1}^{|E|} w_i \cdot I(C, i),$$

where $g(C)$ is the usual evaluation function, $\lambda$ a parameter that determines the influence of the penalties on the augmented cost function, $w_i$ the penalty weight associated to edge $i$, and $I(C, i)$ an indicator function that takes the value 1 if the end points of edge $i$ are in conflict in $C$ and 0 otherwise. The penalties are initialized to 0 and are updated each time an iterative improvement algorithm reaches a local optimum of $g'$. The modification of the penalty weights is done by first computing a utility $u_i$ for each violated edge, $u_i = I(C, i)/(1 + w_i)$, and then incrementing the penalties of all edges with maximal utility by one. The underlying local search is a best improvement algorithm in the restricted 1-exchange neighborhood. Once a local optimum is reached, the search continues for a maximum number of $sw$ plateau moves before the evaluation function $g'$ is updated.

*Iterated local search*: $TS_{1-ex}$ can be used as a local search inside hybrid SLS methods like iterated local search (ILS) [43]. In the ILS algorithm presented in Reference 44, $TS_{1-ex}$ is run until the best solution found does not change for $l_{LS}$ iterations. A perturbation is then applied to the best coloring found so far and $TS_{1-ex}$ is run again. In the perturbation, a number $k_r$, $k_r < k$, of color classes is randomly chosen and the color class membership of all vertices in those color classes is changed. The ROS heuristic bounded by $k$ and with the further strong constraint of avoiding the reinsertion of a vertex into its previous color class is used to accomplish this task. The tabu list of $TS_{1-ex}$ is emptied before applying the perturbation, whereas the exchanges caused by the perturbation are inserted in the tabu list.

Other approaches that are based on the same or similar SLS methods have been studied: A predecessor of the ILS algorithm described earlier is presented in Reference 45, an ILS algorithm that uses a permutation of the color classes in subgraphs as a perturbation is given in Reference 46. Similar in spirit is also the iterated greedy solution reconstruction [47].

*Evolutionary algorithms*: The first evolutionary algorithm (EA) for the GCP is reported in Reference 48. The most successful EAs are hybrid methods that use $TS_{1\text{-ex}}$ to improve candidate solutions [29,32,39, 40,49]. Among them, the best results so far have been reported for the hybrid evolutionary algorithm (HEA) [40]. HEA starts with a population $P$ of candidate solutions, which is initialized by using the DSATUR construction heuristic restricted to $k$ colors, and then iteratively generates new candidate solutions by first recombining two members of the current population that are improved by local search. For the recombination, the greedy partition crossover (GPX) is used [40]. Starting with two candidate partitionings (parents) $C^1 = \{C_1^1, \dots, C_k^1\}$ and $C^2 = \{C_1^2, \dots, C_k^2\}$, GPX generates a candidate solution (offspring) by alternately selecting color classes of each parent. At step $i$ of the crossover operator, $i = 1, \dots, k$, GPX chooses a color class with maximal cardinality from parent $C^1$ (if $i$ is odd) or from parent $C^2$ (if $i$ is even). This color class will become color class $C_i$ of the offspring. Once $C_i$ is chosen, the vertices that belong to it are removed from both parents. The vertices that remain in $C^1$ and $C^2$ after step $k$ are added to a color class of the child, which for each vertex is chosen uniformly at random. The new candidate partitioning returned by GPX is then improved by $TS_{1\text{-ex}}$, run for $l_{LS}$ iterations, and it is inserted in the population $P$ replacing the worse parent. The population is reinitialized if the average distance between colorings in the population falls below a threshold of 20. Responsible for the high performance reported for the algorithms appears to be mainly the GPX operator [46].

The adaptive search algorithm of Galinier et al. [50] also makes use of the GPX, but it does not further improve on the performance of HEA. Another evolutionary algorithm was proposed in Reference 51 and a scatter search algorithm was proposed in Reference 52; however, none of these algorithms appears to reach the performance of HEA.

*Other methods*: A greedy randomized adaptive search procedure was proposed in Reference 53. It uses a randomization of RLF for the candidate solution construction and an iterative improvement algorithm in the 1-exchange neighborhood. The reported results for low-degree graphs appear to be good. Another SLS method, ant colony optimization (ACO), has also been applied to the GCP in Reference 54 (see Vol. 1, Chap. 23 for more details on ACO). In that approach, several ways of defining the heuristic information were studied; the computational results appear to be worse than state-of-the-art, however, no local search was used to improve candidate solutions.

## 16.6.2 Strategy 2: $k$ Fixed, Partial Colorings

*Distributed coloration neighborhood search*: The distributed coloration neighborhood algorithm proposed by Morgenstern [33] can be seen as an ILS algorithm that uses a simulated annealing (SA) algorithm for the local search. The SA algorithm is based on the $i$-swap neighborhood and is run for $I$ iterations or until a certain solution quality threshold is passed. Upon termination of the SA algorithm, a perturbation is applied that is defined by a random *s-chain exchange* that moves from the current coloring configuration to a new one with the same solution quality. An $s$-chain exchange can be seen as a generalization of the Kempe chain exchange. It is defined through a vertex $v$ and an ordered sequence of non-empty color classes $C_1, \dots C_s$, in which we have that $v \in C_1$, all color classes are distinct, and $s \le k$. From this sequence, a directed graph with vertex set $V' = C_1 \cup \dots \cup C_s$ and arc set $A = \{(u, w) \mid (u, w) \in E, u \in C_i$ and $w \in C_{i+1}\}$ is derived (we use the convention that $C_{s+1} = C_1$). In an $s$-chain, each vertex that is reachable from $v$ in the digraph $(V', A)$ is moved from color class $C_i$ to $C_{i+1}$. Note that an $s$-chain, in which all vertices in $V'$ are reachable, would simply correspond to a relabeling of the color classes and, hence, would result in the very same partitioning; therefore, such a *total* $s$-chain is to be avoided and the neighborhood is restricted to non-total $s$-chains. Different ways of combining these two steps and including them into a distributed computational environment were studied in Reference 33.

More recently, a tabu search algorithm based on the $i$-swap neighborhood was proposed [55]. The tabu criterion in this algorithm forbids adding into $C_i$ vertices adjacent to a vertex $v$ that was moved into a color class $C_i$. This interdiction acts for the $tt$ iterations successive to the move of $v$.

### 16.6.3 Strategy 3: *k* Variable, Complete Coloring

*Simulated annealing with Kempe chain neighborhood*: SA was among the first SLS methods applied to the GCP [19,56]. A comprehensive study of three SA algorithms was presented in Reference 19; among these three variants, two work with the number of colors $k$ variable. The more promising one of the two allows only feasible colorings and uses the Kempe chain neighborhood ($SA_{Kempe}$), whereas the other allows infeasible colorings and uses the 1-exchange neighborhood. $SA_{Kempe}$ uses the evaluation function $g(C) = -\sum_{i=1}^{k}(|C_i|)^2$ and starts from an initial partitioning generated by the ROS heuristic. At each iteration of $SA_{Kempe}$, a neighboring solution is generated in three steps; firstly, a non-empty color class $C_i$, a vertex $v \in C_i$, and a non-empty color class $C_j$ are chosen uniformly at random but avoiding that $C_i$ and $C_j$ form a full Kempe chain, which would result simply in a relabeling of the two color classes; secondly, the Kempe chain $K_{ij}$ of color classes $C_i$ and $C_j$ that contains vertex $v$ is determined; thirdly, the Kempe chain exchange is applied. The generated neighboring candidate solution $C'$ is always accepted if it improves over the current candidate solution $C$; otherwise, it is accepted with a probability of $\exp((g(C)-g(C'))/T)$, where $T$ is a parameter called temperature. The parameter $T$ is modified according to a rather standard cooling schedule [19].

## 16.7 Experimental Comparison of Stochastic Local Search Algorithms

In the current section, we give numerical results for the SLS algorithms that were described in detail in Section 16.6. For this purpose, we use the challenging benchmark instances described in Section 16.2, that is, those that are not recognized as easy; the benchmark instances were not treated by the preprocessing rules given in Section 16.3. We compare $TS_{1\text{-ex}}$, $TS_{VLSN}$, $MC\text{-}TS_{1\text{-ex}}$, ILS, GLS, HEA, $SA_{Kempe}$, and XRLF [19], an extension of RLF. All algorithms were implemented under the same environment, sharing the same data structures as much as reasonable. We used a specific experimental setup considering the GCP in its optimization version: each algorithm started from the number of colors $k^{RLF}$ returned by the RLF heuristic. When a feasible $k$-coloring is found, a new coloring with $k - 1$ colors is created by uncoloring the vertices assigned to one selected color and recoloring each of the vertices by randomly choosing any of the remaining colors. Finally, all algorithms were allowed the same maximum computation time $t_{max}$; after preliminary experiments, we decided to use $TS_{1\text{-ex}}$ as the reference algorithm and to set $t_{max}$ to the average time $TS_{1\text{-ex}}$ needs to perform $I_{max} = 10^4 \times |V|$ iterations (averages are taken across 10 trials per instance). For each of the algorithms, we tried to link parameter settings to instance features in which possible and set the remaining ones to some constant value that resulted in good performance across the whole benchmark set. Thus, the results presented in the following give rather an indication of the algorithms' robustness than necessarily their true peak performance.

Each of the algorithms was run 10 times on each instance. For each trial, we measured the minimal number of colors found by the algorithm and we analyzed the resulting data using rank-based statistical methods. In particular, the measured results on each instance are transformed into a rank value in $1, \ldots, 80$ (we compare eight algorithms and each algorithm is run 10 times) and then we aggregated the ranks within the instance classes described in Section 16.2. Note that applying the statistical tests to each of the instance classes separately avoids the incorrect bias towards problem classes that comprise more instances and it may help in distinguishing the particular strength or weakness of the algorithms for the various instance classes.

In Figure 16.3, we report for each instance class the analysis produced by the nonparametric rank-based Friedman test for an all-pairwise comparison [57]. The graphs are obtained by attaching error bars to a scatter plot of the estimated average rank versus algorithm labels. The length of the bars is adjusted so that the average rank of a pair of algorithms can be inferred to be different with statistical significance at the level of $\alpha = 0.05$ if their bars do not overlap. Numerical results on the performance of the algorithms are given in Table 16.1.

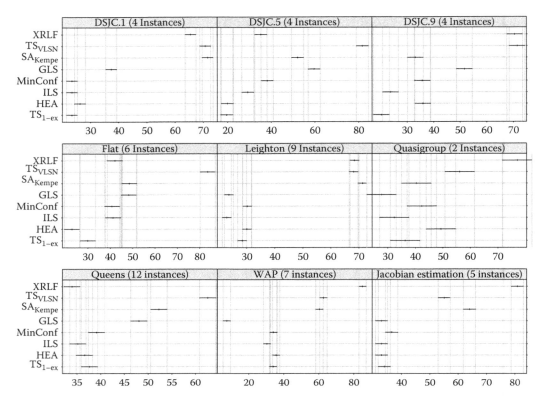

**FIGURE 16.3** All-pairwise comparisons through confidence intervals for the SLS algorithms discussed in the text. The *x*-axis gives the average rank for the SLS algorithms.

From these results, we can draw the following conclusions. Most importantly, there are strong differences in the relative order of the algorithms among the various instance classes and, hence, it is not possible to declare any single algorithm to be the best performing one. On the uniform random graphs, $TS_{1-ex}$, ILS, and HEA are the most competitive algorithms. HEA is the significantly best performing algorithm on the Flat graphs, whereas on the Leighton graphs, ILS and GLS are the best ones. The largest variation in the relative order of the algorithms appears to be due to GLS; GLS is the best or among the best algorithms for Leighton, Jacobian estimation, and WAP graphs, but on the other classes of graphs, its performance is significantly worse than, for example, that of $TS_{1-ex}$. Two further results are significant. First, the exploration of large neighborhoods in $TS_{VLSN}$ does not pay off; in fact, it is among the worst performing algorithms. Further analysis showed that this is mainly due to the higher computational cost per search step [17]. Second, on most instance classes XRLF performs rather poorly and it is among the best algorithms only on the Queens graphs. This contradicts somehow the reputation it gained, which, however, is mainly due to its very good performance on large uniform random graphs with edge density 0.5. Finally, note that the performance of HEA is worse than reported in Reference 40. We verified that this difference is mainly due to our experimental setup and the usage of a single parameter setting; in Reference 40, HEA was tuned for each specific graph and even the value of $k$ when solving the $k$-coloring problem. When we fine-tuned our implementation of HEA, our implementation roughly matched the results presented in Reference 40. Across all the instances, the very good performance of $TS_{1-ex}$ is most noteworthy, because it is also one of the algorithms that are among the most easy ones to implement.

**TABLE 16.1** Numerical Results on the DIMACS Graph Colouring Instances. Given are the Chromatic Number and Best Known Colourings $(\chi, \hat{\chi}^{best})$, the Maximal: Computation Time for the Algorithms in Seconds (Time), the Best and the Median Coloring Found by an Algorithm, and the Median Computation Time for Reaching a Solution with Median (Solution) Quality. The computational experiments were run on a machine with a 2 GHz AMD Athlon MP 2400 + processor, 256 KB cache and 1 GB of RAM memory

| Instance | Bench. $(\chi, \hat{\chi}^{best})$ | time sec. | TS$_{1-ex}$ min | med | sec. | HEA min | med | sec. | ILS min | med | sec. | MC-TS$_{1-ex}$ min | med | sec. | GLS min | med | sec. | SA$_{Kempe}$ min | med | sec. | TS$_{VLSN}$ min | med | sec. | XRLF min | med | sec. |
|---|---|---|---|---|---|---|---|---|---|---|---|---|---|---|---|---|---|---|---|---|---|---|---|---|---|---|
| DSJC125.1 | (−,5) | 10 | 5 | 5 | 0 | 5 | 5 | 0 | 5 | 5 | 0 | 5 | 5 | 0 | 5 | 5 | 0 | 6 | 6 | 0 | 5 | 6 | 0 | 5 | 6 | 0 |
| DSJC250.1 | (−,8) | 30 | 8 | 8 | 0.1 | 8 | 8 | 0 | 8 | 8 | 0.2 | 8 | 8 | 0.3 | 8 | 9 | 0 | 9 | 9 | 0.2 | 9 | 9 | 0 | 9 | 9 | 24.4 |
| DSJC500.1 | (−,12) | 37 | 13 | 13 | 0.1 | 13 | 13 | 0.1 | 13 | 13 | 0.1 | 13 | 13 | 0.2 | 13 | 13 | 0.2 | 14 | 14 | 1.6 | 14 | 14 | 4.5 | 14 | 14 | 29.9 |
| DSJC1000.1 | (−,20) | 174 | 21 | 21 | 2.8 | 21 | 21 | 164.3 | 21 | 21 | 5.9 | 21 | 21 | 10.4 | 21 | 22 | 0.8 | 23 | 23 | 46.7 | 23 | 23 | 90.5 | 22 | 22 | 169.9 |
| DSJC125.5 | (−,17) | 14 | 17 | 17 | 0.5 | 17 | 17 | 1.3 | 17 | 17 | 1.6 | 17 | 17 | 6.6 | 18 | 18 | 0 | 18 | 18 | 0.4 | 18 | 19 | 2.1 | 18 | 18 | 2.5 |
| DSJC250.5 | (−,22) | 47 | 28 | 28 | 22.3 | 28 | 28 | 30.8 | 28 | 28 | 33.6 | 29 | 29 | 2.8 | 29 | 30 | 0.9 | 29 | 30 | 2.7 | 32 | 32 | 6.2 | 29 | 30 | 5.4 |
| DSJC500.5 | (−,48) | 168 | 49 | 50 | 35 | 50 | 50 | 100.3 | 50 | 50 | 105.8 | 50 | 51 | 20.4 | 52 | 52 | 81.4 | 51 | 51 | 47.3 | 55 | 55 | 138.5 | 50 | 50 | 123 |
| DSJC1000.5 | (−,83) | 1102 | 89 | 90 | 309.7 | 89 | 90 | 962.5 | 90 | 91 | 303.5 | 90 | 91 | 496.9 | 93 | 93 | 546.3 | 90 | 91 | 409.7 | 97 | 98 | 981.1 | 86 | 86 | 514.8 |
| DSJC125.9 | (−,30) | 12 | 44 | 44 | 0.1 | 44 | 44 | 0.1 | 44 | 44 | 0.1 | 44 | 44 | 0.2 | 44 | 44 | 0.3 | 44 | 44 | 2 | 44 | 44 | 9.9 | 44 | 45 | 3.4 |
| DSJC250.9 | (−,72) | 60 | 72 | 72 | 3.8 | 72 | 72 | 28.2 | 72 | 72 | 5.6 | 72 | 72 | 26.7 | 72 | 73 | 6.3 | 72 | 72 | 26.6 | 74 | 74 | 39 | 75 | 77 | 12 |
| DSJC500.9 | (−,126) | 398 | 127 | 127 | 234.4 | 128 | 129 | 180.8 | 127 | 128 | 82.3 | 128 | 129 | 127 | 129 | 130 | 154 | 127 | 128 | 377.2 | 134 | 135 | 340.3 | 132 | 132 | 204.4 |
| DSJC1000.9 | (−,224) | 2693 | 226 | 227 | 1983.5 | 230 | 232 | 1869.2 | 227 | 228 | 245 | 230 | 230 | 2382.1 | 233 | 234 | 1621.1 | 226 | 229 | 2401.4 | 245 | 247 | 2143.7 | 232 | 233 | 125.9 |
| flat300_20_0 | (20,20) | 112 | 20 | 20 | 0.3 | 20 | 20 | 0.3 | 20 | 20 | 0.4 | 20 | 20 | 0.6 | 20 | 20 | 0.6 | 20 | 20 | 1.1 | 33 | 34 | 76.9 | 20 | 20 | 2.9 |
| flat300_26_0 | (26,26) | 89 | 26 | 26 | 5.5 | 26 | 26 | 16.1 | 26 | 26 | 16.6 | 26 | 26 | 19.8 | 33 | 33 | 4.3 | 32 | 33 | 4.3 | 35 | 35 | 53.5 | 33 | 34 | 2.9 |
| flat300_28_0 | (28,31) | 61 | 31 | 32 | 3.4 | 31 | 31 | 54.6 | 31 | 32 | 3.3 | 31 | 32 | 7.8 | 33 | 33 | 7.2 | 33 | 33 | 5.1 | 35 | 36 | 17.3 | 33 | 34 | 2.9 |
| flat1000_50_0 | (50,50) | 1076 | 85 | 86 | 957.4 | 50 | 78 | 1004.9 | 88 | 88 | 729.5 | 87 | 88 | 713.3 | 50 | 50 | 636.3 | 86 | 88 | 470.3 | 95 | 96 | 939.4 | 84 | 86 | 359.3 |
| flat1000_60_0 | (60,60) | 1119 | 88 | 89 | 245.2 | 87 | 88 | 918.5 | 89 | 90 | 128.2 | 89 | 90 | 372.3 | 90 | 91 | 719.2 | 88 | 89 | 1014.6 | 96 | 97 | 624.5 | 87 | 87 | 235.4 |
| flat1000_76_0 | (76,83) | 1147 | 88 | 89 | 618.1 | 88 | 89 | 957.6 | 89 | 90 | 188.7 | 90 | 90 | 712 | 92 | 92 | 605 | 89 | 90 | 399 | 96 | 97 | 869 | 87 | 87 | 306 |
| le450_5a | (5,5) | 230 | 5 | 5 | 0.1 | 5 | 5 | 0.1 | 5 | 5 | 0.1 | 5 | 5 | 0.9 | 5 | 5 | 0.2 | 5 | 7 | 0 | 6 | 6 | 0 | 6 | 7 | 47.4 |
| le450_5b | (5,5) | 232 | 5 | 5 | 0.3 | 5 | 5 | 0.5 | 5 | 5 | 0.6 | 5 | 5 | 0.6 | 5 | 5 | 0.3 | 6 | 7 | 0 | 6 | 6 | 59.6 | 7 | 7 | 37.1 |
| le450_5d | (5,5) | 191 | 5 | 5 | 0 | 5 | 5 | 0 | 5 | 5 | 0 | 5 | 5 | 0 | 5 | 5 | 0 | 5 | 5 | 0 | 5 | 5 | 0 | 5 | 6 | 16.4 |
| le450_15a | (15,15) | 68 | 15 | 15 | 0.2 | 15 | 15 | 3.4 | 15 | 15 | 0.1 | 15 | 15 | 15 | 15 | 15 | 2.2 | 16 | 16 | 0 | 16 | 16 | 0 | 16 | 17 | 9.3 |
| le450_15b | (15,15) | 76 | 15 | 15 | 0.1 | 15 | 15 | 0.2 | 15 | 15 | 0.1 | 15 | 15 | 5.8 | 15 | 15 | 0.3 | 16 | 16 | 0 | 16 | 16 | 0 | 16 | 16 | 18.9 |
| le450_15c | (15,15) | 45 | 16 | 16 | 13.5 | 15 | 15 | 19.5 | 15 | 15 | 19.1 | 15 | 16 | 8 | 15 | 15 | 5.9 | 23 | 23 | 0 | 23 | 23 | 0 | 19 | 21 | 216.4 |
| le450_15d | (15,15) | 42 | 16 | 16 | 21.7 | 16 | 16 | 13 | 15 | 15 | 20.3 | 15 | 16 | 7.1 | 15 | 15 | 7.8 | 22 | 23 | 0 | 22 | 23 | 0 | 20 | 21 | 189.6 |
| le450_25c | (25,26) | 56 | 26 | 26 | 0.7 | 26 | 27 | 0 | 26 | 26 | 2 | 26 | 27 | 0.1 | 26 | 26 | 18 | 27 | 28 | 0 | 27 | 28 | 0 | 27 | 28 | 32.3 |
| le450_25d | (25,26) | 59 | 26 | 26 | 0.5 | 26 | 27 | 0 | 26 | 26 | 0.8 | 26 | 27 | 0.2 | 26 | 26 | 4.7 | 28 | 28 | 0 | 27 | 28 | 0 | 27 | 27 | 38.1 |

*(Continued)*

**TABLE 16.1 (*Continued*)**    Numerical Results on the DIMACS Graph Colouring Instances. Given are the Chromatic Number and Best Known Colourings $(\chi, \hat{\chi}^{best})$, the Maximal: Computation Time for the Algorithms in Seconds (Time), the Best and the Median Coloring Found by an Algorithm, and the Median Computation Time for Reaching a Solution with Median (Solution) Quality. The computational experiments were run on a machine with a 2 GHz AMD Athlon MP 2400 + processor, 256 KB cache and 1 GB of RAM memory

| Instance | Bench. $(\chi, \hat{\chi}^{best})$ | time sec. | TS$_{1-ex}$ min | med | sec. | HEA min | med | sec. | ILS min | med | sec. | MC-TS$_{1-ex}$ min | med | sec. | GLS min | med | sec. | SA$_{Kempe}$ min | med | sec. | TS$_{VLSN}$ min | med | sec. | XRLF min | med | sec. |
|---|---|---|---|---|---|---|---|---|---|---|---|---|---|---|---|---|---|---|---|---|---|---|---|---|---|---|
| latin_square_10 | (-,99) | 1242 | 103 | 104 | 617.8 | 106 | 107 | 889.4 | 103 | 104 | 510.4 | 104 | 105 | 458.4 | 102 | 103 | 214.9 | 101 | 102 | 369.9 | 111 | 114 | 798.4 | 117 | 118 | 970.7 |
| qg.order100 | (100,100) | 12102 | 100 | 100 | 17.9 | 100 | 100 | 18.5 | 100 | 100 | 18.3 | 100 | 100 | 19.9 | 100 | 100 | 36.6 | 100 | 101 | 14.8 | 100 | 100 | 875.1 | 100 | 101 | 3971.9 |
| queen6_6 | (7,7) | 2 | 7 | 7 | 0 | 7 | 7 | 0 | 7 | 7 | 0 | 7 | 7 | 0 | 7 | 7 | 0 | 7 | 7 | 0 | 7 | 7 | 1.5 | 7 | 7 | 0 |
| queen7_7 | (7,7) | 4 | 7 | 7 | 0 | 7 | 7 | 0 | 7 | 7 | 0 | 7 | 7 | 0 | 7 | 7 | 0 | 7 | 7 | 0 | 7 | 7 | 2.1 | 7 | 7 | 0 |
| queen8_12 | (12,-) | 7 | 12 | 12 | 0 | 12 | 12 | 0 | 12 | 12 | 0 | 12 | 12 | 0 | 12 | 12 | 0 | 12 | 12 | 0.1 | 12 | 12 | 0.4 | 12 | 12 | 0.9 |
| queen8_8 | (9,9) | 5 | 9 | 9 | 0 | 9 | 9 | 0 | 9 | 9 | 0 | 9 | 9 | 0 | 9 | 9 | 0 | 9 | 9 | 0.1 | 9 | 10 | 0 | 9 | 9 | 0.2 |
| queen9_9 | (10,10) | 6 | 10 | 10 | 0 | 10 | 10 | 0 | 10 | 10 | 0 | 10 | 10 | 0 | 10 | 10 | 0 | 10 | 10 | 0.1 | 10 | 11 | 0 | 10 | 10 | 0.4 |
| queen10_10 | (11,11) | 10 | 11 | 11 | 0.1 | 11 | 11 | 0.1 | 11 | 11 | 0 | 11 | 11 | 0.1 | 11 | 11 | 0.8 | 11 | 12 | 0.1 | 11 | 11 | 0 | 11 | 11 | 0.9 |
| queen11_11 | (11,12) | 14 | 12 | 12 | 0.1 | 12 | 12 | 0.1 | 12 | 12 | 0.2 | 12 | 12 | 0.2 | 12 | 13 | 0 | 12 | 13 | 0.1 | 12 | 12 | 0.1 | 12 | 12 | 2 |
| queen12_12 | (12,12) | 18 | 13 | 13 | 1 | 13 | 13 | 1.4 | 13 | 13 | 0.9 | 13 | 13 | 3.8 | 13 | 14 | 0 | 13 | 13 | 0.2 | 13 | 13 | 0.4 | 13 | 13 | 13.3 |
| queen13_13 | (13,14) | 22 | 14 | 14 | 2.9 | 14 | 14 | 2.3 | 13 | 14 | 1.3 | 14 | 14 | 13.2 | 15 | 15 | 0 | 15 | 14 | 0.3 | 14 | 14 | 1.1 | 14 | 14 | 21.4 |
| queen14_14 | (14,-) | 21 | 15 | 16 | 0 | 15 | 16 | 0 | 15 | 15 | 20 | 15 | 16 | 0 | 16 | 16 | 0 | 16 | 16 | 0.5 | 15 | 15 | 2.7 | 15 | 15 | 32.2 |
| queen15_15 | (15,17) | 24 | 16 | 17 | 0 | 16 | 17 | 0 | 16 | 16 | 23.9 | 16 | 17 | 0 | 17 | 17 | 0 | 17 | 17 | 0.8 | 16 | 16 | 5 | 16 | 17 | 23.9 |
| queen16_16 | (16,18) | 24 | 18 | 18 | 0 | 18 | 18 | 0 | 18 | 18 | 0 | 18 | 18 | 0 | 18 | 18 | 0 | 17 | 18 | 1.2 | 17 | 18 | 5.7 | 17 | 17 | 33.4 |
| wap01a | (-,-) | 412 | 43 | 44 | 1.2 | 43 | 44 | 1.6 | 43 | 44 | 1.5 | 42 | 42 | 217.1 | 42 | 42 | 55 | 44 | 45 | 107.8 | 44 | 46 | 30.8 | 47 | 48 | 131.6 |
| wap02a | (40,-) | 318 | 42 | 43 | 0.7 | 42 | 43 | 0.8 | 42 | 42 | 251.6 | 41 | 42 | 4.9 | 41 | 41 | 159.9 | 43 | 43 | 97.8 | 43 | 44 | 0.5 | 46 | 47 | 150.8 |
| wap03a | (-,-) | 1395 | 46 | 47 | 3.8 | 46 | 47 | 4.5 | 46 | 46 | 365.2 | 45 | 47 | 5.8 | 44 | 44 | 782 | 46 | 47 | 198.9 | 47 | 48 | 339.1 | 50 | 51 | 884.3 |
| wap04a | (40,-) | 2125 | 44 | 44 | 170 | 45 | 45 | 2.8 | 44 | 44 | 484.3 | 43 | 44 | 31.2 | 43 | 43 | 833.6 | 45 | 46 | 1.6 | 45 | 46 | 1.8 | 47 | 49 | 1073.5 |
| wap06a | (40,-) | 138 | 41 | 42 | 0.5 | 42 | 42 | 0.7 | 42 | 42 | 0.5 | 42 | 43 | 0.2 | 40 | 41 | 8.1 | 42 | 44 | 0.2 | 42 | 43 | 4.7 | 44 | 44 | 24.3 |
| wap07a | (-,-) | 341 | 43 | 44 | 0.7 | 43 | 43 | 1.9 | 43 | 44 | 0.7 | 42 | 43 | 30.9 | 42 | 42 | 215.2 | 44 | 45 | 9.1 | 44 | 45 | 39 | 47 | 47 | 80.7 |
| wap08a | (40,-) | 373 | 42 | 43 | 10.3 | 42 | 43 | 1.4 | 43 | 43 | 56.1 | 42 | 44 | 0.7 | 42 | 42 | 41.4 | 45 | 45 | 0.4 | 44 | 45 | 0.4 | 46 | 47 | 89.4 |
| abb313GPIA | (9,10) | 328 | 9 | 9 | 4.1 | 9 | 9 | 26.4 | 9 | 9 | 0.9 | 9 | 9 | 30.7 | 9 | 9 | 1.1 | 11 | 11 | 0 | 11 | 11 | 0.1 | 12 | 13 | 36.4 |
| ash331GPIA | (4,4) | 200 | 4 | 4 | 0 | 4 | 4 | 0 | 4 | 4 | 0 | 4 | 4 | 0 | 4 | 4 | 0 | 4 | 4 | 0 | 4 | 4 | 0 | 5 | 5 | 25.8 |
| ash608GPIA | (4,4) | 633 | 4 | 4 | 0.1 | 4 | 4 | 0.2 | 4 | 4 | 0.1 | 4 | 4 | 0.2 | 4 | 4 | 0.1 | 4 | 5 | 0 | 4 | 4 | 409.3 | 5 | 5 | 27.4 |
| ash958GPIA | (5,4) | 1627 | 4 | 4 | 0.5 | 4 | 4 | 0.5 | 4 | 4 | 0.6 | 4 | 4 | 0.9 | 4 | 4 | 0.4 | 5 | 5 | 0 | 4 | 5 | 0 | 5 | 5 | 121.5 |
| will199GPIA | (7,7) | 31 | 7 | 7 | 0 | 7 | 7 | 0 | 7 | 7 | 0 | 7 | 7 | 0 | 7 | 7 | 0 | 7 | 7 | 0 | 7 | 7 | 0 | 7 | 8 | 43.9 |

## 16.8 Summary

The current chapter gives an overview of the main SLS algorithms described in the literature dedicated to the GCP. Most of these SLS algorithms follow the strategy of keeping the number of colors fixed to a value $k$ and trying to minimize the number of conflicts. The optimization of the GCP is then tackled by solving a series of $k$-coloring problems. Among the algorithms that follow this strategy, a simple tabu search algorithm based on the restricted 1-exchange neighborhood is a very robust and fast approach that achieves competitive results for many instance classes. Other SLS algorithms either show better performance only on a few instance classes (e.g., GLS on the WAP instances or Jacobian estimation graphs) or when very high computation times are available (e.g., HEA when appropriately tuned). On most classes of graphs, SLS algorithms also outperform exact algorithms for the GCP. However, exact algorithms appear to be competitive or even preferable, if the chromatic number is equal or very close to the clique number, the size of the largest clique, of a graph [17].

A promising direction for future research on the GCP appears to be the integration of exact and SLS algorithms, given that they show particular advantages for different instance classes. Another challenge for research on SLS algorithms for the GCP is to get a better understanding of the performance of these algorithms in dependence of instance features. Insights into this relationship may help to increase the robustness of the algorithms across the various instance classes and may finally lead to new developments and possibly even better performing algorithms.

## Acknowledgments

This work was supported in part by the "Metaheuristics Network," a Research Training Network funded by the Improving Human Potential programme of the CEC, grant HPRN-CT-1999-00106, and by a European Community Marie Curie Fellowship, contract HPMF-CT-2001. The information provided is the sole responsibility of the authors and does not reflect the Community's opinion. The Community is not responsible for any use that might be made of data appearing in this publication. Irina Dumitrescu acknowledges support of the Canada Research Chair in distribution management, HEC Montreal, and Thomas Stützle support of the Belgian FNRS, of which he is a research associate.

## References

1. Jensen, T.R. and Toft, B., *Graph Coloring Problems*, John Wiley & Sons, New York, 1994.
2. Allen, M., Kumaran, G., and Liu, T., A combined algorithm for graph-coloring in register allocation, in *Proceedings of the Computational Symposium on Graph Coloring and Its Generalizations*, Ithaca, NY, 2002, p. 100.
3. Barnier, N. and Brisset, P., Graph coloring for air traffic flow management, in *Proceedings of the Fourth International Workshop on Integration of AI and OR Techniques in Constraint Programming for Combinatorial Optimisation Problems*, Le Croisic, France, 2002, p. 133.
4. Gamst, A., Some lower bounds for a class of frequency assignment problems, *IEEE Trans. Veh. Technol.*, 35(1), 8, 1986.
5. Zymolka, A., Koster, A.M.C.A., and Wessäly, R., Transparent optical network design with sparse wavelength conversion, in *Proceedings of the 7th IFIP Working Conference on Optical Network Design & Modelling*, Budapest, Hungary, 2003, p. 61.
6. de Werra, D., An introduction to timetabling, *Eur. J. Oper. Res.*, 19(2), 151, 1985.
7. Leighton, F.T., A graph coloring algorithm for large scheduling problems, *J. Res. Nat. Bur. Stand.*, 84(6), 489, 1979.
8. Karp, R.M., Reducibility among combinatorial problems, in *Complexity of Computer Computations*, Miller, R.E. and Thatcher, J.W., Eds., Plenum Press, New York, 1972, p. 85.

9. Feige, U. and Kilian, J., Zero knowledge and the chromatic number, *JCSS*, 57(2), 187, 1998.

10. Halldórsson, M.M., A still better performance guarantee for approximate graph coloring, *Inf. Proc. Lett.*, 45(1), 19, 1993.

11. Paschos, V.T., Polynomial approximation and graph-coloring, *Computing*, 70(1), 41, 2003.

12. Brélaz, D., New methods to color the vertices of a graph, *CACM*, 22(4), 251, 1979.

13. Caramia, M. and Dell'Olmo, P., Bounding vertex coloring by truncated *multistage* branch and bound, *Networks*, 44(4), 231, 2004.

14. Mendez Diaz, I. and Zabala, P., A branch-and-cut algorithm for graph coloring, in *Proceedings of the Computational Symposium on Graph Coloring and Its Generalizations*, Ithaca, NY, 2002, p. 55.

15. Gomes, C. and Shmoys, D., Completing quasigroups or latin squares: A structured graph coloring problem, in *Proceedings of the Computational Symposium on Graph Coloring and Its Generalizations*, Ithaca, NY, 2002, p. 22.

16. Mehrotra, A. and Trick, M.A., A column generation approach for graph coloring, *INFORMS J. Comput.*, 8(4), 344, 1996.

17. Chiarandini, M., *Stochastic Local Search Methods for Highly Constrained Combinatorial Optimization Problems*, PhD thesis, FG Intellektik, FB Informatik, TU Darmstadt, 2005.

18. Johnson, D.S. and Trick, M.A., Eds., *Cliques, Coloring, and Satisfiability: Second DIMACS Implementation Challenge*, vol. 26 of *DIMACS Series on Discrete Mathematics and Theoretical Computer Science*, AMS, Providence, RI, 1996.

19. Johnson, D.S., Aragon, C.R., McGeoch, L.A., and Schevon, C., Optimization by simulated annealing: An experimental evaluation: Part II, graph coloring and number partitioning, *Oper. Res.*, 39(3), 378, 1991.

20. Achlioptas, D. and Naor, A., The two possible values of the chromatic number of a random graph, *Ann. Math.* 62, 1335–1351, 2005.

21. Johri, A. and Matula, D.W., Probabilistic bounds and heuristic algorithms for coloring large random graphs, Technical Report 82-CSE-6, Southern Methodist University, Dallas, TX, 1982.

22. Luczak, T., The chromatic number of random graphs, *Combinatorica*, 11(1), 45, 1991.

23. Culberson, J., Beacham, A., and Papp, D., Hiding our colors, in *Proceedings of the CP'95 Workshop on Studying and Solving Really Hard Problems*, Cassis, France, 1995, p. 31.

24. Lewandowski, G. and Condon, A., Experiments with parallel graph coloring heuristics and applications of graph coloring, in *Cliques, Coloring, and Satisfiability: Second DIMACS Implementation Challenge*, vol. 26 of *DIMACS Series on Discrete Mathematics and Theoretical Computer Science*, AMS, Providence, RI, 1996, p. 309.

25. Hossain, S. and Steihaug, T., Graph coloring in the estimation of mathematical derivatives, in *Proceedings of the Computational Symposium on Graph Coloring and Its Generalizations*, Ithaca, NY, 2002, p. 9.

26. Knuth, D.E., *The Stanford GraphBase: A Platform for Combinatorial Computing*, ACM Press, New York, 1993.

27. Mizuno, K. and Nishihara, S., Toward ordered generation of exceptionally hard instances for graph 3-colorability, in *Proceedings of the Computational Symposium on Graph Coloring and Its Generalizations*, Ithaca, New York, 2002, p. 1.

28. Cheeseman, P., Kanefsky, B., and Taylor, W.M., Where the really hard problems are, in *Proceedings of IJCAI*, Morgan Kaufmann Publishers, San Francisco, CA, 1991, p. 331.

29. Fleurent, C. and Ferland, J.A., Genetic and hybrid algorithms for graph coloring, *Ann. Oper. Res.*, 63, 437, 1996.

30. Hertz, A. and de Werra, D., Using tabu search techniques for graph coloring, *Computing*, 39, 345, 1987.

31. Morgenstern, C. and Shapiro, H., Coloration neighbourhood structures for general graph coloring, in *Proceedings of SODA*, SIAM, San Francisco, CA, 1990, p. 226.

32. Costa, D., Hertz, A., and Dubois, O., Embedding of a sequential procedure within an evolutionary algorithm for coloring problems in graphs, *J. Heuristics*, 1(1), 105, 1995.

33. Morgenstern, C., Distributed coloration neighbourhood search, in *Cliques, Coloring, and Satisfiability: Second DIMACS Implementation Challenge*, vol. 26 of *DIMACS Series on Discrete Mathematics and Theoretical Computer Science*, AMS, Providence, RI, 1996, p. 335.

34. Chaitin, G., Register allocation and spilling via graph coloring, *SIGPLAN Notices*, 39(4), 66–74, 2004.

35. de Werra, D., Heuristics for graph coloring, *Comput. Suppl.*, 7, 191, 1990.

36. González-Velarde, J.L. and Laguna, M., Tabu search with simple ejection chains for coloring graphs, *Ann. Oper. Res.*, 117(1–4), 165, 2002.

37. Avanthay, C., Hertz, A., and Zufferey, N., A variable neighborhood search for graph coloring, *Eur. J. Oper. Res.*, 151(2), 379, 2003.

38. Minton, S., Johnston, M.D., Philips, A.B., and Laird, P., Minimizing conflicts: A heuristic repair method for constraint satisfaction and scheduling problems, *Artif. Intell.*, 58(1–3), 161, 1992.

39. Dorne, R. and Hao, J.K., A new genetic local search algorithm for graph coloring, in *Proceedings of PPSN-V*, Eiben, A. E., et al., Eds., LNCS, 1498, Springer Verlag, Berlin, Germany, 1998, p. 745.

40. Galinier, P., and Hao, J.K., Hybrid evolutionary algorithms for graph coloring, *J. Comb. Opt.*, 3(4), 379, 1999.

41. Stützle, T., *Local Search Algorithms for Combinatorial Problems: Analysis, Improvements, and New Applications*, DISKI, 220, Infix, Sankt Augustin, Germany, 1999.

42. Voudouris, C., *Guided Local Search for Combinatorial Optimization Problems*, PhD thesis, University of Essex, Department of Computer Science, Colchester, UK, 1997.

43. Lourenço, H.R., Martin, O., and Stützle, T., Iterated local search, in *Handbook of Metaheuristics*, Glover, F. and Kochenberger, G., Eds., Kluwer Academic Publishers, Norwell, MA, 2002, p. 321.

44. Chiarandini, M. and Stützle, T., An application of iterated local search to the graph coloring problem, in *Proceedings of the Computational Symposium on Graph Coloring and Its Generalizations*, Ithaca, NY, 2002, p. 112.

45. Paquete, L. and Stützle, T., An experimental investigation of iterated local search for coloring graphs, in *Applications of Evolutionary Computing*, Cagnoni, S., et al., Eds., LNCS, 2279, Springer Verlag, Berlin, Germany, 2002, p. 122.

46. Glass, C.A. and Prügel-Bennett, A., A polynomially searchable exponential neighbourhood for graph colouring, *J. Oper. Res. Soc.*, 56(3), 324, 2005.

47. Culberson, J.C. and Luo, F., Exploring the *k*-colorable landscape with iterated greedy, in *Cliques, Coloring, and Satisfiability: Second DIMACS Implementation Challenge*, vol. 26 of *DIMACS Series on Discrete Mathematics and Theoretical Computer Science*, AMS, American Mathematical Society Providence, RI, 1996, p. 245.

48. Davis, L., Order-based genetic algorithms and the graph coloring problem, in *Handbook of Genetic Algorithms*, Davis, L., Ed., Van Nostrand Reinhold, New York, 1991, p. 72.

49. Marino, A. and Damper, R.I., Breaking the symmetry of the graph colouring problem with genetic algorithms, in *Late Breaking Papers at the GECCO Conference*, Las Vegas, NV, 2000, p. 240.

50. Galinier, P., Hertz, A., and Zufferey, N., Adaptive memory algorithms for graph colouring, in *Proceedings of the Computational Symposium on Graph Coloring and Its Generalizations*, Ithaca, NY, 2002, p. 75.

51. Eiben, A.E., Hauw, J.K., and Van Hemert, J.I., Graph coloring with adaptive evolutionary algorithms, *J. Heuristics*, 4(1), 25, 1998.

52. Hamiez, J.-P. and Hao, J.-K., Scatter search for graph coloring, in *Artificial Evolution*, Collet, P., et al., Eds., LNCS, 2310, Springer Verlag, Berlin, Germany, 2001, p. 168.

53. Laguna, M. and Martí, R., A GRASP for coloring sparse graphs, *Comput. Optim. Appl.*, 19(2), 165, 2001.

54. Costa, D. and Hertz, A., Ants can colour graphs, *J. Oper. Res. Soc.*, 48(3), 295, 1997.

55. Blöchliger, I. and Zufferey, N., A reactive tabu search using partial solutions for the graph coloring problem, Technical Report 04/03, Ecole Ploytechnique Fédérale de Lausanne, Recherche Opérationelle Sud-Est, Lausanne, Switzerland, 2004.

56. Chams, M., Hertz, A., and De Werra, D., Some experiments with simulated annealing for coloring graphs, *Eur. J. Oper. Res.*, 32(2), 260, 1987.

57. Conover, W.J., *Practical Nonparametric Statistics*, John Wiley & Sons, New York, 3rd Ed., 1999.

# 17

# On Solving the Maximum Disjoint Paths Problem with Ant Colony Optimization

Maria J. Blesa

Christian Blum

## 17.1 Introduction

The efficient use of modern communication networks depends on our capabilities for solving a number of demanding algorithmic problems, some of which are concerned with the allocation of network resources to individual connections. One of the basic operations in communication networks consists in establishing routes for *connection requests* between physically separated network endpointsv that wish to establish a connection for information exchange. Many connection requests occur simultaneously in a network, and it is desirable to establish routes for as many requests as possible. In many situations, either due to technical constraints or just to improve the communication, it is required that no two routes interfere with each other, which implies not to share network resources such as links or switches. This scenario can be modeled as follows. Let $G = (V, E)$ be an edge-weighted undirected graph representing a network in which the nodes represent the hosts and switches, and the edges represent the links. Let $T = \{(s_j, t_j) \mid j = 1, \ldots, \mathbb{T}; \ s_j \neq t_j \in V\}$ be a list (of size $\mathbb{T}$) of *commodities*, that is, pairs of nodes in $G$, representing endpoints demanding to be connected by a path in $G$. $T$ is said to be *realizable* in $G$ if there exist mutually edge-disjoint (respectively vertex-disjoint) paths from $s_j$ to $t_j$ in $G$, for every $j = 1, \ldots, \mathbb{T}$.

The question whether $T$ is realizable was early known to be NP-complete [1] in arbitrary graphs. The problem remains NP-complete for specific types of graphs such as planar graphs [2,3], series-parallel graphs (a.k.a. partial 2-trees) [4], and grid graphs [5,6]. For several types of graphs, this problem belongs to the class of APX-hard problems [7–10]. This fact explains the notorious hardness of the edge-disjoint paths (EDP) problem in terms of approximation, despite the attention and effort that researchers have put on it. Interestingly, for the specific case of complete graphs, we are not aware of any inapproximability results. In particular, it is not even known whether the problem in complete graphs is APX-hard.

The combinatorial optimization version of this problem consists in satisfying as many of the requests as possible, which is equivalent to finding a realizable subset of $T$ of maximum cardinality. A solution $S$ to the combinatorial optimization problem is a set of disjoint paths, in which each path satisfies the connection request for a different commodity. For any solution $S$, the objective function value $f(S)$ is defined as

$$f(S) = |S|. \tag{17.1}$$

In general, the "disjointness" of paths may refer to nodes or to edges. We decided to consider the latter case because it seems of higher importance in practical applications. We henceforth refer to our problem as the maximum EDP problem. The EDP problem is obtained from the more general *unsplittable flow problem* by considering the demands, profits, and capacities to be one. In the extreme case in which the list of commodities is composed by repetitions of the same pair $(s, t)$, the problem is known as *edge-disjoint Menger problem* [11].

The EDP problem is interesting for different research fields such as combinatorial optimization, algorithmic graph theory, and operations research. It has a multitude of applications in areas such as real-time communications, VLSI-design, scheduling, bin packing, load balancing, and it has recently been brought into focus in works discussing applications to routing and admission control in modern networks, namely, large-scale, high-speed, and optical networks [12–15]. Concerning real-time communications, the EDP problem is very much related to survivability and information dissemination. Concerning survivability, having several disjoint paths available may avoid the negative effects of possible failures occurring in the base network. Furthermore, to communicate via multiple disjoint paths can increase the effective bandwidth between pairs of nodes, reduce congestion in the network, and increase the velocity and the probability of receiving the information [16,17]. This becomes especially important nowadays due to the type of information that circulates over networks (e.g., media files), which requires fast, qualified, and reliable connections.

In general, there is a lack of efficient algorithms for tackling the EDP problem. Only some greedy approaches (which we will mention in Section 17.3) and a preliminary ant colony optimization (ACO) approach [18] exist for tackling the problem. The greedy approaches are used as approximation algorithms for theoretical purposes, but the quality of the solutions they obtain are susceptible to improvement. The direct application of a basic ACO scheme to a problem achieves sometimes quite good results. However, the performance of such an algorithm can often be improved by applying some additional features to the search process, especially when a rather unusual problem such as the EDP is tackled. Based on the (basic) approach in Reference 18, we have evolved a more sophisticated ACO algorithm (see Reference 19 for details). We present and evaluate this ACO approach in the remainder of this chapter.

# 17.2  A Greedy Approach

A greedy heuristic is a constructive algorithm that builds a solution step-by-step starting from an empty solution. At each construction step, an element from a finite set of solution components is added to the current partial solution. The element to be added is chosen at each step according to some greedy function, which lends the name to the algorithm. A characteristic feature of the greedy algorithms is that once a decision is made on which element to add, this decision is never reconsidered again. Advantages of greedy heuristics are that they are usually easy to implement and that they are fast in execution. In contrast, the disadvantage is that the quality of the solutions provided by greedy algorithms is often far from being optimal.

Greedy algorithms are often used as *approximation algorithms* to solve optimization problems with a *guaranteed performance*. With this aim, some greedy algorithms were proposed for the EDP problem;

examples are the *simple greedy algorithm* [20], its constrained variant the *bounded-length greedy algorithm* [20–24], and the *greedy path algorithm* [25,26]. Due to its lower time complexity when compared with the other greedy approaches, we decided to implement the simple greedy algorithm (henceforth denoted by SGA) and a multistart version, which we both outline in the following.

The SGA algorithm (Algorithm 17.1) is a natural way of approximating the EDP problem that works as follows. It starts with an empty solution $S$. Then, it proceeds through the commodities in the order that is given as input. For routing each commodity $T_j \in T$, it considers the graph $G$ without the edges that are already in the paths of the solution $S$ under construction. The shortest path (with respect to the number of edges) between $s_j$ and $t_j$ is assigned as path for the commodity $T_j = (s_j, t_j)$. In addition to its simplicity, the SGA algorithm can be naturally considered an on-line algorithm, and then the lower bounds of Reference 27 would imply that it does not achieve a good performance ratio on graphs such as trees and two-dimensional meshes. However, it works well for many other types of graphs.

Observe that the SGA algorithm is deterministic and that the quality of the solutions it provides depends heavily on the order in which the commodities are treated. A simple way of overcoming that dependence on the order is to develop a multi-start version of the SGA by permuting—for each restart—the order of the commodities. This approach is pseudocoded in Algorithm 17.2, in which $N_{perm}$ denotes the number of restarts, $S_i$ denotes the solution under construction in the embedded SGA, and $S_{best}$ denotes the best solution found so far. In the following, we refer to this algorithm as *multi-start greedy algorithm* (MSGA).

---

## ALGORITHM 17.1    Simple Greedy Algorithm (SGA) for the EDP Problem

INPUT: a problem instance $(G, T)$, consisting of a graph $G$ and a commodity list $T$
$S \leftarrow \emptyset, \hat{E} \leftarrow E$
**for** $j = 1, \ldots, |T|$ **do**
    **if** $s_j$ and $t_j$ can be connected by a path in $G = (V, \hat{E})$ **then**
        $P_j \leftarrow$ shortest path from $s_j$ to $t_j$ in $G = (V, \hat{E})$
        $S \leftarrow S \cup P_j, \hat{E} \leftarrow \hat{E} \setminus \{e \mid e \in P_j\}$
    **end if**
**end for**
OUTPUT: the solution $S$

---

## ALGORITHM 17.2    Multi-Start Simple Greedy Algorithm (MSGA) for the EDP Problem

INPUT: a problem instance $(G, T, N_{perm})$, where $N_{perm}$ is the number of restarts
$S_{best} \leftarrow \emptyset, T_{(1)} \leftarrow T$                                    {$T_{(i)}$ denotes the $i$-th permutation of $T$}
**for** $i = 1$ to $N_{perm}$ **do**
    $S_i \leftarrow$ Simple Greedy Algorithm SGA$(G, T^i)$                 {See Algorithm 17.1}
    **if** $f(S_i) > f(S_{best})$ **then** $S_{best} \leftarrow S_i$ **end if**
    **if** $i < N_{perm}$ **then**
        $\pi \leftarrow$ random permutation of size $|T|$
        $T_{(i+1)} \leftarrow (\pi(1), \pi(2), \ldots, \pi(|T| - 1), \pi(|T|))$
    **end if**
**end for**
OUTPUT: $S_{best}$

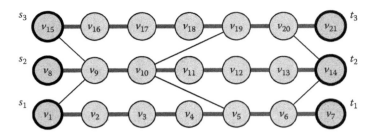

**FIGURE 17.1**  Example of an instance of the EDP problem (with $T = \{(v_1, v_7), (v_8, v_{14}), (v_{15}, v_{21})\}$) for which neither the SGA nor the MSGA greedy algorithm can find the solution of size 3 emphasized with bold font.

## 17.2.1 An Example Where Simple Greedy Algorithm and Multi-Start Greedy Algorithm Fail

As we have commented, the main disadvantage of some existing greedy algorithms that approximate hard problems is the low quality of the solutions they provide. Due to the deterministic greedy decisions that they take during the solution construction, it is sometimes not possible for them to find an existing optimal solution. This is also the case for the SGA and MSGA presented here.

Consider, for example, the instance of the EDP problem depicted in Figure 17.1, which consist in the depicted graph and the set $T = \{(v_1, v_7), (v_8, v_{14}), (v_{15}, v_{21})\}$ of three commodities to join. The optimal solution in which all three commodities are connected is also shown in bold font in Figure 17.1.* Observe, however, that there is no way for any of the greedy algorithms, neither SGA nor MSGA, to find the solution of size greater than two. As these greedies are based on the shortest paths (in terms of the number of edges), the algorithms will tend to connect the commodities through nonconsecutively-numbered vertices. For example, when trying to connect first the commodity $(v_1, v_7)$, the SGA algorithm will establish the path $\{v_1, v_9, v_{10}, v_5, v_6, v_7\}$. This excludes edge $\{v_9, v_{10}\}$ as a possibility for being used in other paths, which makes it impossible to build disjoint paths simultaneously for the remaining two commodities, independently of which one is built next. Analogous situations occur when starting from any of the other two commodities. Thus, as no possible permutation of the commodities would provide a solution of size three, neither the SGA nor the MSGA algorithm will find the optimal solution of size three.

## 17.3  An Ant Colony Optimization Approach

ACO [28,29] is a metaheuristic for solving hard combinatorial optimization problems. Apart from the application to static combinatorial optimization problems (see Reference 30 for an extensive overview), the method has also gained recognition for the applications to adaptive routing in static and dynamic communication networks [31,32]. ACO algorithms are composed by independently operating computational units, namely, *artificial ants*, that generate a global perspective without the necessity of direct interaction. This exclusive use of local information is an advantageous and desirable feature when applications in large-scale environments are concerned in which the computation of global information is often too costly. This property makes ACO algorithms a natural choice for the application to the EDP problem.

---

* This solution is found by our ACO algorithm, which is presented next, in an small amount of time (less than 30 milliseconds).

Ant colony optimization is inspired by the foraging behavior of real ants. While walking from food sources to the nest and vice versa, ants deposit a chemical substance called *pheromone* on the ground. When they decide about a direction to go, they choose probabilistically paths marked by strong pheromone concentrations. This behavior is the basis for a cooperative interaction which leads to the emergence of shortest paths between food sources and their nest. In ACO algorithms, artificial ants incrementally construct a solution by adding appropriately defined solution components to the current partial solution. Each of the construction steps is a probabilistic decision based on local information, which is represented by the *pheromone* information.

In the following we outline our ACO approach, which is based on a decomposition of the EDP problem. Each problem instance $\mathcal{P} = (G, T)$ of the EDP problem can be naturally decomposed into $|T|$ subproblems $\mathcal{P}_j = (G, T_j)$, with $j \in \{1, \ldots, |T|\}$, by regarding the task of finding a path for a commodity $T_j \in T$ as a problem itself. With respect to this problem decomposition, we use a number of $|T|$ ants each of which is assigned to exactly one of the subproblems. Therefore, the construction of a solution consists of each ant building a path $P_j$ between the two endpoints of her commodity $T_j$. Obviously, the subproblems are not independent as the set of $|T|$ paths constructed by the ants should be mutually edge-disjoint.

## 17.3.1 Ant Solutions and Pheromone Model

Our algorithm will deal with solutions that contain a path for each commodity. A solution $S$ constructed by the $|T|$ ants is a set of unnecessarily edge-disjoint paths. We henceforth refer to them as *ant solutions*, in contrast to the EDP solutions, which only consist of disjoint paths. From each ant solution a valid EDP solution can be produced by iteratively removing the path which has most edges in common with the remaining paths, until all remaining paths are mutually edge-disjoint.

The objective function $f(\cdot)$ of the problem (Equation 17.1) is characterized by having many plateaus when it is applied to ant solutions. This is because many ant solutions have the same number of disjoint paths. Thus, a consequence of decomposing the EDP problem is the need to define a more fine-grained objective function $f^a(\cdot)$ for ant solutions. Therefore, referring to $f(S)$ as a *first criterion*, we introduce a *second criterion* $C(S)$, which is defined as follows:

$$C(S) = \sum_{e \in E} \left( \max \left\{ 0, \left( \sum_{P_j \in S} \delta^j(S, e) \right) - 1 \right\} \right), \text{ where } \delta^j(S, e) = \left\{ \begin{array}{lll} 1 & : & e \in P_j \in S \\ 0 & : & \text{otherwise.} \end{array} \right.$$

This second criterion quantifies the degree of nondisjointness of an ant solution. If all the paths in a solution $S$ are edge-disjoint, $C(S)$ is zero. In general, $C(S)$ increases when increasing the number of edges in $S$ which are common to more than one path. Therefore, based on the idea that *the fewer edges are shared in a solution, the closer the solution is to disjointness*, a function $f^a(\cdot)$ that differentiates between ant solutions can be defined as follows. For two ant solutions $S$ and $S'$, it holds that

$$f^a(S) > f^a(S') \Leftrightarrow \underbrace{(f(S) > f(S'))}_{1^{st} \text{ criterion}} \text{ or } \underbrace{((f(S) = f(S') \text{ and } (C(S) < C(S')))}_{2^{nd} \text{ criterion}}. \tag{17.2}$$

The problem decomposition as described earlier requires that we use a pheromone model $\tau^j$ for each subproblem $\mathcal{P}_j$. Each pheromone model $\tau^j$ consists of a pheromone value, that is, a positive numerical value, $\tau^j_e$ for each edge $e \in E$. The set of $|T|$ pheromone models is henceforth denoted by $\tau = \{\tau^1, \ldots, \tau^{|T|}\}$. The pheromone values are bounded between 0 and 1, as our ACO algorithm is implemented in the hypercube framework [33]. Furthermore, to prevent the algorithm from converging to a solution, we borrow an idea from the so-called $\mathcal{MAX}$-$\mathcal{MIN}$ Ant Systems [34] and forbid the extreme pheromone values of 0 or 1 by introducing new pheromone value limits $\tau_{\min} = 0.001$ and $\tau_{\max} = 0.999$.

## 17.3.2 Algorithmic Framework

In the following, we give a high-level description of an ACO algorithm for the EDP problem (see Algorithm 17.3). The main procedures used by the algorithm are explained in detail in the following of the section. First, all the variables are initialized. In particular, the pheromone values are set to their initial value $\tau_{min}$ by the procedure InitializePheromoneValues($\tau$), which initializes all the pheromone values $\tau_e^j \in \tau^j \in \tau$ to the value $\tau_{min}$. Second, $N_{sols}$ ant solutions are constructed per iteration. To construct a solution, each ant applies the function ConstructSolution($G,\pi$) (see the forthcoming Section 17.3.2.1 for details), where $\pi$ is a permutation of $T$. At each iteration, the first of those $N_{sols}$ ant solutions is constructed with the identity permutation, that is, by sending the ants in the order in which the commodities are given in $T$. However, for each further ant solution construction in the same iteration, $\pi$ is randomly generated by the function GenerateRandomPermutation($|T|$) to avoid bias.

Three different ant solutions are kept in the algorithm: $S_{ibest}$ is the *iteration-best* solution, that is, the best ant solution generated in the current iteration, and $S_{gbest}$ is the *best-so-far* solution, that is, the best ant solution found since the start of the algorithm. In addition to them, an ant solution $S_{pbest}$ is also kept, which is the *currently best* solution, that is, the best ant solution generated since the last escape

---

## ALGORITHM 17.3   ACO Algorithm for the EDP Problem

INPUT: a problem instance $(G, T)$
$S_{gbest} \leftarrow \emptyset$, $S_{pbest} \leftarrow \emptyset$, $c_{crit1} \leftarrow 0$, $c_{crit2} \leftarrow 0$, all_update $\leftarrow$ FALSE
InitializePheromoneValues($\tau$)
**while** termination conditions not met **do**
    $\pi \leftarrow (1, 2, \ldots, |T| - 1, |T|)$
    **for** $i = 1$ to $N_{sols}$ **do**
        $S_i \leftarrow$ ConstructSolution($G,\pi$)         {See Algorithm 17.4}
        **if** $i < N_{sols}$ **then** $\pi \leftarrow$ GenerateRandomPermutation($|T|$) **end if**
    **end for**
    Choose $S_{ibest} \in \{S_i \mid i = 1, \ldots, N_{sols}\}$ s.t. $f^a(S_{ibest}) \geq f^a(S)$, $\forall S \in \{S_i \mid i = 1, \ldots, N_{sols}\}$
    **if** $f(S_{ibest}) > f(S_{gbest})$ **then** $S_{gbest} \leftarrow S_{ibest}$ **end if**
    **if** $f^a(S_{ibest}) > f^a(S_{pbest})$ **then**
        $c_{crit1} \leftarrow c_{crit1} + 1$, $c_{crit2} \leftarrow 0$, $S_{psave} \leftarrow S_{pbest}$, $S_{pbest} \leftarrow S_{ibest}$
        **if** $f(S_{ibest}) > f(S_{psave})$ **then**
            $S_{update} \leftarrow$ ExtractDisjointPaths($S_{pbest}$)         {Update for first phase}
            $c_{crit1} \leftarrow 0$, all_update $\leftarrow$ FALSE
        **end if**
        **if** all_update **then** $S_{update} \leftarrow S_{pbest}$ **end if**         {Update for second phase}
    **else** $c_{crit2} \leftarrow c_{crit2} + 1$
    **end if**
    **if** all_update **and** $c_{crit2} > c2_{max}$ **then**
        $S_{pbest} \leftarrow$ DestroyPartially($S_{pbest}$)         {Escape mechanism}
        $S_{update} \leftarrow$ MyDarkBlueExtractDisjointPaths($S_{pbest}$)
        $c_{crit2} \leftarrow 0$, $c_{crit1} \leftarrow 0$
    **else if** not all_update **then** all_update $\leftarrow (c_{crit1} > c1_{max})$
    **end if**
    UpdatePheromoneValues($\tau, S_{update}$)
**end while**
OUTPUT: the EDP solution generated from the best solution $S_{gbest}$

action (see Section 17.3.2.3). The values of these three variables are always kept updated. In addition, there is the $S_{update}$ solution, which is generated from $S_{pbest}$ and which is used for updating the pheromone values.

The search process has two differentiated phases (see Section 17.3.2.2) and two variables, $c_{crit1}$ and $c_{crit2}$, are introduced to control them. The variable $c_{crit1}$ determines the first phase by counting the number of successive iterations without improvement of the first criterion of the objective function. The variable $c_{crit2}$ counts the number of successive iterations without improvement of the second criterion, thus defining the second phase. Limits $c1_{max}$ (for $c_{crit1}$) and $c2_{max}$ (for $c_{crit2}$) are used to determine when the algorithm should change phases.[*] The direct repercussion of the phase distinction is the selection of edges whose pheromone is updated, that is, the construction of $S_{update}$ from $S_{pbest}$. When the algorithm is in the first phase, only the disjoint paths of solution $S_{pbest}$ are used for updating, but when the algorithm is in the second phase, all paths of $S_{pbest}$ are used for updating. In addition, the escape mechanism might be applied by destroying $S_{pbest}$ partially (see Section 17.3.2.3).

Finally, the pheromone values are updated in the method UpdatePheromoneValues($\tau$, $S_{update}$) depending on the edges of the paths included in $S_{update}$. The algorithm is iterated until some opportunely defined termination conditions are satisfied, and it returns the EDP solution generated from the ant solution $S_{gbest}$.

In the forthcoming Sections 17.3.2.1 through 17.3.2.3, we explain in more detail the features concerning the solution construction, the search procedure and its different search phases, and the escape mechanism of our algorithm, respectively.

### 17.3.2.1 Solution Construction

The solution construction is performed in method ConstructSolution($G$,$\pi$), whose high-level description is shown in Algorithm 17.4. That construction is done as follows: at each construction step, each ant moves from the node in which it is currently located to a neighboring node by traversing one of the available edges that is not already in its path $P_{\pi(j)}$ under construction, and that is not labelled forbidden

---

### ALGORITHM 17.4  Method ConstructSolution($G$,$\pi$) of Algorithm 17.3

INPUT: a graph $G$ from a problem instance $(G, T)$, and a permutation $\pi$ of $T$.
$S \leftarrow \emptyset$, $nb\_paths\_finished \leftarrow 0$, $j \leftarrow 0$
**for** $i = 1$ to $|T|$ **do** $P_{\pi(i)} \leftarrow \emptyset$ **end for**
**repeat**
  **if not** isFinishedPath($P_{\pi(j+1)}$) **then**
    $P_{\pi(j+1)} \leftarrow$ ExtendOneStepPath($P_{\pi(j+1)}, \tau^{\pi(j+1)}$)
    **if** isFinishedPath($P_{\pi(j+1)}$) **then**
      $nb\_paths\_finished \leftarrow nb\_paths\_finished + 1$
      $S \leftarrow S \cup \{P_{\pi(j+1)}\}$
    **end if**
  **end if**
  $j \leftarrow (j + 1) \bmod |T|$
**until** ($nb\_paths\_finished = |T|$)
EvaporatePheromone($\tau$, $S$)
OUTPUT: an ant solution $S$

---

[*] After parameter tuning we chose a setting of $c1_{max} = c2_{max} = 20$.

by a backtracking move. Note, that with this strategy the ant will find a path between its source and its destination, if there exists one. Otherwise, the ant returns an empty path. This way of constructing the solution emulates that the ants build concurrently their paths, in contrast to a sequential way in which, for each commodity, a path between its endpoints would be built completely before the next commodity is considered.

The procedures of Algorithm 17.4 are detailed in the following:

- *isFinishedPath*($P_i$):[*] This method returns a boolean value indicating whether the path $P_i$ is finished, that is, whether a path could be established from $s_i$ to $t_i$.
- *ExtendOneStepPath*($P_i, \tau^i$):[†] The path $P_i$ passed as parameter is the path under construction by the $i$-th ant. For constructing a path between the endpoints of the commodity $(s_i, t_i)$, an ant first chooses randomly to start either from the source $s_i$ or from the target $t_i$; this is done when the path $P_i$ is empty. Afterwards, this method either tries to extend the path $P_i$ by adding exactly one edge, or it performs a backtracking step. Backtracking is done in case the ant finds itself in a node in which all the incident edges have been used, or if all the incident edges are labeled forbidden.

Once one of the two endpoints of the commodities is chosen as starting point, the remaining endpoint becomes the so-called goal node and will be denoted by $v_g$. In addition, let us denote by $v_c$ the current node, and by $\mathcal{I}_{v_c}^\star$ the set of allowed edges in $G$, that is, those incident to $v_c$ which are not used yet in the path and not labeled as forbidden. The length of the shortest path between two vertices $u$ and $v$ in $G$ is henceforth denoted by $\sigma(u, v)$, and it is measured in terms of the number of edges. From the set $\mathcal{I}_{v_c}^\star$ of allowed edges, only the two best edges will be actually considered as candidates. This is called a *candidate list strategy* in ACO. The best two edges are those that maximize the value of the following expression:

$$\tau_e^j \cdot \mathbf{p}(D_e) \cdot \mathbf{p}(U_e),$$

where $\mathbf{p}(D_e)$ is a value that determines the influence of the distance from $v_c$ via $u$ to the goal vertex $v_g$, and $\mathbf{p}(U_e)$ is a value that determines the influence of the overall usage of edge $e$, which is the information whether $e$ is already used in the path of another ant for the same solution. The terms $\mathbf{p}(D_e)$ and $\mathbf{p}(U_e)$ are defined as follows:

$$\mathbf{p}(D_{e=\{v_c,u\}}) \leftarrow \frac{(\sigma(u,v_g)+w(e))^{-1}}{\sum\limits_{e'=\{v_c,u'\}\in \mathcal{I}_{v_c}^\star}(\sigma(u',v_g)+w(e'))^{-1}}$$

$$\mathbf{p}(U_e) \leftarrow \frac{U(e)^{-1}}{\sum\limits_{e'\in \mathcal{I}_{v_c}^\star}U(e')^{-1}} \,, \quad \text{in which} \quad U(e) = \begin{cases} 2 : e \text{ already used in } S_i \\ 1 : \text{otherwise} \end{cases}$$

Thus, using this candidate list strategy, we can reduce the set of allowed edges in $\mathcal{I}_{v_c}^\star$ and just consider a new two-cardinality set $\mathbb{I}_{v_c}^\star = \{e_1^*, e_2^*\}$, where $e_1^*$ is the best edge in $\mathcal{I}_{v_c}^\star$, that is,

$$e_1^* = \{v_c, u\} \leftarrow \text{argmax}\{\tau_e^j \cdot \mathbf{p}(D_e) \cdot \mathbf{p}(U_e) \mid e \in \mathcal{I}_{v_c}^\star\},$$

and $e_2^*$ is the second best edge in $\mathcal{I}_{v_c}^\star$.

At each construction step, the choice of where to move to has a certain probability $p$ to be done deterministically, and a certain probability $1 - p$ to be chosen probabilistically among the elements in $\mathbb{I}_{v_c}^\star$. This is a feature that we adopt from a particularly effective ACO variant called Ant Colony

---

[*] For the sake of readability, we substitute $\pi(j + 1)$ in the description of the functions by $i$.

System (ACS [35]). In 75% of the cases, the next edge to join the path $P_{\pi(k)}$ under construction will be $e_1^*$, whereas in the remaining 25% of the cases, the next edge is chosen from $\mathbb{I}_{v_c}^*$ according to the following transition probabilities:

$$\mathbf{p}(e \mid \mathbb{I}_{v_c}^*) = \frac{\tau_e^j \cdot \mathbf{p}(D_e) \cdot \mathbf{p}(U_e)}{\sum\limits_{e' \in \mathbb{I}_{v_c}^*} \tau_{e'}^j \cdot \mathbf{p}(D_{e'}) \cdot \mathbf{p}(U_{e'})} \quad, \forall \, e \in \mathbb{I}_{v_c}^* \tag{17.3}$$

In general, if the probability of doing a deterministic construction step is too high, there is the danger that the algorithm gets stuck in low quality regions of the search space. On the other side, doing deterministic construction steps bears the potential of leading the algorithm quite quickly to good areas of the search space. Concerning the composition of the transition probabilities, the use of the pheromone information $\tau_e^j$ ensures the flexibility of the algorithm, whereas the use of $\mathbf{p}(D_e)$ ensures a bias towards short paths, and $\mathbf{p}(U_e)$ ensures a bias towards disjointness of the $|T|$ paths constituting a solution.

- *EvaporatePheromone*$(\tau, S)$: After every ant has constructed its path and the solution $S$ is completed, we apply another feature of ACS, namely the evaporation of some amount of pheromone from the edges that were used by the ants. Given a solution $S$, the evaporation is done as follows:

$$\tau_e^j \leftarrow \begin{cases} (1-\varepsilon) \cdot \tau_e^j & : \quad e \in P_{\pi(j)} \in S, \ j = 1, \ldots, |T| \\ \tau_e^j & : \quad \text{otherwise.} \end{cases} \tag{17.4}$$

The reason for this pheromone evaporation is the desire to diversify the search in each iteration.[*]

## 17.3.2.2 Search with Distinguished Phases

In general, the pheromone update procedure is an important component of every ACO algorithm. In fact, it determines to a large degree the failure or the success of the algorithm. Most of the existing generic variants of ACO only differ in the pheromone update. In the case of the EDP application, we propose a pheromone updating scheme that is based on the idea that, to maintain a higher degree of freedom for finding also edge-disjoint paths for the commodities that initially prove to be problematic, it might be better not to use the nondisjoint paths for updating the pheromone at the beginning of the search. Therefore, we propose a two-phases search process based on the two criteria of function $f^a(\cdot)$ (Equation 17.2): A first phase of the algorithm in which the algorithm will try to improve the first criterion of $f^a(\cdot)$ (whereas disregarding the second one) and only disjoint paths are used for updating the pheromone values; the first phase is followed by a second phase which is initiated when no improvements of the first criterion can be found over a certain time bounded by $c1_{max}$. In this second phase, the algorithm will try to improve the second criterion of $f^a(\cdot)$ and all the paths are used for updating the pheromone values. Once the second phase leads to an improvement also in terms of the first criterion, the algorithm changes back to the first phase.

In the first phase, the solution $S_{update}$ that is used for updating the pheromone values is obtained by applying function ExtractDisjointPaths$(S_{pbest})$, which implements the process of returning a valid EDP solution from the ant solution $S_{pbest}$ as explained in Section 17.3.1. In the second phase, the solution $S_{update}$ that is used for updating the pheromone values is a copy of the current solution $S_{pbest}$, including possibly nondisjoint paths. If for a number of $c2_{max}$ iterations the second criterion could not be improved

---

[*] After parameter tuning we chose a setting of $\varepsilon = 0.10$.

neither, then some of the paths from the EDP solution that can be produced from $S_{pbest}$ are deleted from $S_{pbest}$. This action can be seen as a mechanism to escape from the current area of the search space and it is explained in Section 17.3.2.3.

After the solution $S_{update}$ is constructed, the pheromone of the edges conforming its paths are updated as follows:

$$\tau_e^j \leftarrow \max\left\{\tau_e^j + \rho \cdot \left(1 - \tau_e^j\right), \tau_{\max}\right\} \quad \forall\, e \in P_j \in S_{update}, \tag{17.5}$$

where $\rho \in (0, 1]$ is a constant value which is called *learning rate* in algorithms that are implemented in the hypercube framework.[*] This pheromone update is performed in function UpdatePheromoneValues($\tau$, $S_{update}$).

### 17.3.2.3 Escape Mechanism

One of the main problems of metaheuristic search procedures is to detect situations in which the search process gets stuck, that is, when some local minimum is reached. Most of the successful applications incorporate algorithm features to escape from these situation once they are detected. In case of our algorithm for the EDP problem, we propose as escape mechanism the partial destruction of the disjoint part of the solution which is used for updating the pheromone values. This escape mechanism is implemented through the function DestroyPartially($S_{pbest}$), whose pseudocode is outlined in Algorithm 17.5. This mechanism is triggered once the algorithm is unable to improve the currently best solution for a number of subsequent applications of first and second phase, as that situation indicates that the search

---

**ALGORITHM 17.5    Method DestroyPartially($S_{pbest}$) of Algorithm 17.3. ExtractDisjointPaths($S_{pbest}$) implements the process of returning a valid EDP solution from an ant solution as explained in Section 17.3.1. The method Cost($S_{temp}$) returns the number of disjoint paths in $S_{temp}$. The method ChooseLongestPath($S_{temp}$) return the longest disjoint path of $S_{temp}$. The method ResetPheromoneModel($\tau^i$) resets to $\tau_{\min}$ all the pheromone values of the pheromone model $\tau^i$, i.e., $\tau_e^i \leftarrow \tau_{\min}$, $\forall e \in E$.**

> INPUT: an ant solution $S_{pbest}$
> $S_{pbest} \leftarrow$ ExtractDisjointPaths($S_{pbest}$)
> $nb_{paths} \leftarrow \left\lceil \dfrac{1}{4} \cdot \text{Cost}(S_{pbest}) \right\rceil$
> **while** ($nb_{paths} > 0$) **do**
>   $P_i \leftarrow$ ChooseLongestPath($S_{pbest}$)
>   $S_{pbest} \leftarrow S_{pbest} \setminus \{P_i\}$
>   $nb_{paths} \leftarrow nb_{paths} - 1$
>   ResetPheromoneModel($\tau^i$)
> **end while**
> OUTPUT: the solution $S_{pbest}$ partially destroyed

---

[*] For all our experiments we have set $\rho$ to 0.1.

process is stuck in a localized area. Similar ideas are applied in backtracking procedures, or in the perturbation mechanism of local search based methods, such as iterated local search or variable neighborhood search [36].

One fourth of the disjoint paths composing solution $S_{pbest}$ are destroyed. The disjoint paths to be destroyed are chosen according to their lengths, giving priority to the longest paths, that is, those paths with the highest number of edges. The idea behind this choice is that the longer a path is, the more restrictions it introduces to assure disjointness of the paths that still conflict with others. Thus, by removing the longest disjoint paths, the number of total edges available is maximized.*

## 17.4 Experiments

We present the experimental evaluation of our ACO approach in comparison to the results obtained by the greedy approaches that we outlined in Section 17.2. As commented before, the ACO algorithm presented here resulted from a detailed algorithm design process that started with a very simple ACO approach [18] that was improved and enriched until the algorithm explained in the present work was obtained. In the same process, the values for those parameters involved in the algorithm were fixed (after a careful tuning) to the values provided here. The experiments to be presented in the following were done with those settings (see Reference 19 for more details)

All the algorithms were implemented in C++ and compiled using GCC 2.95.2 with the -o3 option. The experiments have been run on a PC with Intel(R) Pentium(R) 4 processor at 3.06 GHz and 900 Mb of memory running a Linux operating system. Moreover, our algorithms were all implemented on the same data structures. Information about the shortest paths in the respective graphs is provided to all of them as input. Notice however that, although the greedy approaches need to partially recompute this information after the routing of each commodity, this is not necessary for our ACO algorithm.

### 17.4.1 Problem Instances

In the following, we present the set of benchmark instances that we used to experimentally evaluate our ACO approach. This set of instances includes graphs representing different communication network topologies. Recall that an instance of the EDP problem consists of a graph and a set of commodities.

Concerning the graphs, we adopt graph3 and graph4 from [18], whose structure resembles parts of the communication network of the Deutsche Telekom AG, Germany. In addition, we include graphs which we created with the network topology generator BRITE [37] according to the parameter values specified in Table 17.1. These three generated graphs are named bl-wr2-wht2.10-50, AS-BA.R-Wax.v100e190, and

**TABLE 17.1** Parameters Used for the Generation of Network Topologies with BRITE [37]. In Each Value Tuple $(X_{as}, X_{rou})$, $X_{as}$ is the Value of the Parameter at the AS Level, and $X_{rou}$ is the Value of the Parameter at the Router Level

| Graph | $|V|$ | Model | Node Placement | $m$ |
| --- | --- | --- | --- | --- |
| bl-wr2-wht2.10-50 | (10,50) | (Waxman, Waxman) | (random, heavy-tailed) | (2, 2) |
| AS-BA.R-Wax.v100e190 | (20, 5) | (Barabási-Albert [39], Waxman) | (random, random) | (2, 2) |
| AS-BA.R-Wax.v100e217 | (10,10) | (Barabási-Albert [39], Waxman) | (random, random) | (2, 2) |

Notes: Parameter $m$ specifies the number of links for each new node that is added while constructing the topology. For all the graphs, the growth type (i.e., how nodes join the topology) is incremental. In graph bl-wr2-wht2.10-50, the edge connections between the AS level and the router level are introduced using the Waxman probability model [38] with parameters $\alpha = 0.15$ and $\beta = 0.20$; in graphs AS-BA.R-Wax.v100e190 and AS-BA.R-Wax.v100e217 both levels are interconnected by choosing edges at random.

---

* Other options were tried, both considering different values for the percentage of solution's paths to be destroyed, as well as a random selection of the paths to be destroyed. However, these options were not providing us with better results [19].

**TABLE 17.2**    Main Quantitative Measures of Our Benchmark Graphs

| Graph | $|V|$ | $|E|$ | Degree | | | Diameter | Clustering |
|---|---|---|---|---|---|---|---|
| | | | min. | avg. | max. | | Coefficient |
| graph3 [18] | 164 | 370 | 1 | 4.51 | 13 | 16 | 0.226161 |
| graph4 [18] | 434 | 981 | 1 | 4.52 | 20 | 22 | 0.155547 |
| bl-wr2-wht2.10-50 [18] | 500 | 1020 | 2 | 4.08 | 13 | 23 | 0.102385 |
| AS-BA.R-Wax.v100e190 | 100 | 190 | 2 | 3.80 | 7 | 11 | 0.378524 |
| AS-BA.R-Wax.v100e217 | 100 | 217 | 2 | 4.34 | 8 | 13 | 0.411119 |

AS-BA.R-Wax.v100e217. They consist of a two-level top-down hierarchical topology (autonomous system level plus router level), which are typical for Internet topologies. Table 17.2 summarizes the main features and quantitative measures of all the considered graphs.

For each of the five graphs we have randomly generated different sets of commodities. Hereby, we made the size of the commodity sets dependent on the number of vertices of the graph. For each graph $G = (V, E)$, we generated 20 different instances with $0.10|V|$, $0.25|V|$, and $0.40|V|$ commodities. This makes a sum of 60 instances for each graph and 300 instances altogether.

## 17.4.2 Experiments and Results

We applied the algorithms presented in the current chapter (namely, SGA, MSGA, and ACO) to all 300 instances exactly once. First, we applied MSGA with 50 restarts (i.e., $N_{perm} = 50$) to each of the 300 instances. The computation time of MSGA was used as a maximum CPU time limit for the ACO algorithm. We present the results as averages over the 20 instances of each combination of graph and commodity number in Table 17.3. The layout of this table is explained in its caption.

Concerning the comparison between SGA and MSGA, we observe a clear advantage of MSGA. This means that the order in which the commodities are treated is crucial in achieving a good performance. However, as there is no obvious way of determining a good commodity order beforehand, the only way of exploiting this knowledge is by randomly permuting the commodity list and running MSGA. The price we have to pay for exploiting this knowledge is the increased computation time.

When comparing the SGA and MSGA algorithms with the ACO, we can observe that in 11 out of 15 cases the ACO approach beats the greedy approaches. The ACO approach is on average 4.69% better than MSGA, and in one case (graph4, 173 commodities) it is even 15.07% better. In addition, the ACO approach needs, in general, less computation time than the greedy approaches. This advantage in computation time increases with increasing number of commodities. Exceptions are some of the results for small numbers of commodities, namely, for 10% of the number of nodes. For this combination MSGA has often slight advantages over the ACO approach. Therefore, we recommend to use a greedy approach when easy problem instances are concerned, but to use the ACO approach for instances with a higher number of commodities, since then a clear advantage of the latter is observed in comparison to MSGA both in quality and time.

An additional analysis concerns the run-time behaviour of the algorithms. Figure 17.2 shows that the ACO approach finds relatively good solutions already after a very short computation time. In general, already the first solutions produced by the ACO are quite good, whereas the greedy approaches reach a comparable solution quality only much later in time. This property of our ACO approach is a desirable feature in the context of communication networks as the quality of the solutions that are found after a short execution time might be often sufficient in practice. Also of interest is Figure 17.3 in which a representative example of the usefulness of ACO's escape mechanism is shown. In addition, the evolution of the second criterion as a measure for disjointness can be clearly observed.

TABLE 17.3 Comparison of the Results Obtained by the SGA, the MSGA, and the ACO Algorithm

| Graph | Number of commodities | SGA | | | MSGA | | | ACO | | | |
|---|---|---|---|---|---|---|---|---|---|---|---|
| | | $\bar{q}$ | $\sigma$ | $\bar{t}$ | $\bar{q}$ | $\sigma$ | $\bar{t}$ | $\bar{q}$ | $\sigma$ | $\bar{t}$ | Avg. CPU time |
| graph3 | 16 | 15.30 | 0.781 | 0.566 | 15.70 | 0.557 | 0.960 | 15.70 | 0.557 | 0.457 | 30.582 |
| graph3 | 41 | 29.00 | 2.864 | 1.298 | 32.00 | 2.302 | 25.235 | 31.80 | 1.990 | 27.953 | 79.619 |
| graph3 | 65 | 33.70 | 2.777 | 2.156 | 37.60 | 2.577 | 49.267 | 40.30 | 2.571 | 57.899 | 126.945 |
| graph4 | 43 | 40.50 | 1.628 | 12.121 | 42.05 | 1.024 | 95.744 | 41.45 | 1.284 | 168.871 | 237.520 |
| graph4 | 108 | 58.10 | 4.194 | 31.138 | 64.10 | 3.064 | 697.456 | 68.15 | 2.725 | 730.436 | 1656.475 |
| graph4 | 173 | 66.75 | 4.846 | 49.281 | 73.95 | 3.542 | 974.350 | 85.10 | 3.534 | 1111.982 | 2603.872 |
| bl-wr2-wht2.10-50 | 50 | 19.70 | 2.238 | 17.926 | 22.55 | 2.397 | 318.518 | 24.10 | 1.947 | 155.899 | 971.488 |
| bl-wr2-wht2.10-50 | 125 | 34.15 | 4.464 | 46.387 | 38.10 | 4.369 | 1004.462 | 42.30 | 4.540 | 344.092 | 2425.090 |
| bl-wr2-wht2.10-50 | 200 | 46.70 | 4.961 | 62.158 | 50.85 | 4.892 | 1151.197 | 56.30 | 5.245 | 847.415 | 3124.550 |
| AS-BA.R-Wax.v100e190 | 10 | 8.75 | 0.942 | 0.114 | 9.10 | 0.943 | 0.579 | 8.95 | 0.973 | 0.611 | 6.665 |
| AS-BA.R-Wax.v100e190 | 25 | 12.30 | 1.900 | 0.280 | 14.25 | 1.374 | 4.809 | 14.85 | 1.195 | 3.718 | 16.740 |
| AS-BA.R-Wax.v100e190 | 40 | 15.45 | 2.500 | 0.443 | 17.95 | 1.624 | 7.796 | 19.45 | 1.936 | 4.121 | 26.850 |
| AS-BA.R-Wax.v100e217 | 10 | 7.00 | 1.225 | 0.103 | 8.05 | 0.921 | 0.427 | 7.88 | 0.927 | 0.164 | 6.892 |
| AS-BA.R-Wax.v100e217 | 25 | 11.40 | 1.882 | 0.300 | 13.60 | 1.463 | 4.330 | 13.83 | 1.579 | 1.816 | 17.622 |
| AS-BA.R-Wax.v100e217 | 40 | 14.60 | 1.685 | 0.497 | 17.00 | 1.949 | 9.833 | 17.80 | 1.646 | 2.212 | 28.318 |

The table layout is as follows. The first column gives the name of the graph and the second column, the number of the commodities, which are obtained as the 10%, 25%, and 40% of the number of nodes of the graphs. For each algorithm there are three columns reporting on the average results obtained for the 20 instances of each combination of graph topology and number of commodities. The first of these three columns (headed by $\bar{q}$) shows the average of the values of the best solutions found for the 20 instances. Such an average is in boldface when the result is the best in the comparison. In case of ties, the computation time decides. The second column provides the standard deviation of the 20 values used to compute $\bar{q}$, and the third column (headed by $\bar{t}$) reports on the average time in seconds) needed to find the best solution values for the 20 instances. Finally, the last column shows the average computation time of MSGA.

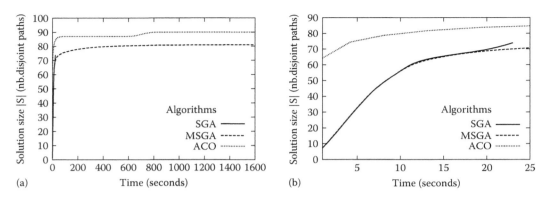

**FIGURE 17.2**  A representative example of the run-time behavior of the algorithms presented in this work. All the curves are smoothed with gnuplots' *sbezier* function.

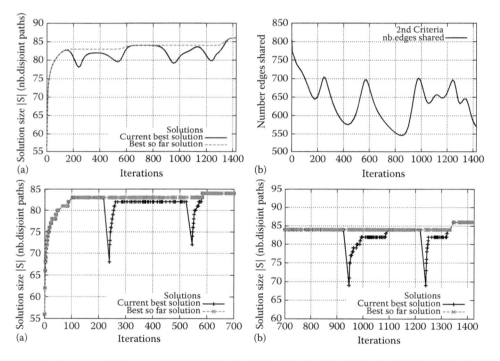

**FIGURE 17.3**  A representative example of the behavior of the ACO algorithm. The effect of the mechanism for the partial destruction of the current best solution can be clearly observed. It is also interesting to observe the evolution of the second criterion as a measure for disjointness. (A) Example of the evolution of the quality of the current best solution $S_{pbest}$ and the best-so-far solution $S_{gbest}$ during the search (a), and the number of shared edges (2nd criterion) of the solution $S_{pbest}$ (b). The behavior shown here corresponds to the application to one of the 20 instances composed by graph4 and a list of 173 commodities. All the curves are smoothed with gnuplots' *sbezier* function. (B) Zoom on the 700 first (a) and the 700 last (b) iterations of part (A) of this figure. On the **left**, the best solution found is quickly improved. At about iteration 250, the algorithm destroys part of the $S_{pbest}$ solution, which produces an instantaneous worsening in the quality (a); another solution destruction takes places around iteration 550, which helps in achieving an improvement soon afterwards (a). Analogous effects can be observed around iteration 950 and 1250 (b). In Figure 17.3(A)(b), we can observe that there exists an (inverted) relation between the number of edges shared and the quality of the solutions obtained. Thus validating our choice of the second criterion as a part of the objective function.

## Acknowledgments

This work is partially supported by EU Programmes under contract numbers IST-2004-15964 (AEOLUS), COST-295 (DYNAMO), and by the Spanish CICYT project TIC-2002-04498-C05-03 (TRACER).

The authors would like to thank Martin Oellrich from the Combinatorial Optimization & Graph Algorithms group of the Technische Universität Berlin for providing them with graph3 and graph4.

## References

1. Karp, R., Reducibility among combinatorial problems, in *Complexity of Computer Computations*, Miller, R.E. and Thatcher, J.W., Eds., 1972, p. 85. Plenum Press, New York.
2. Middendorf, M. and Pfeiffer, F., On the complexity of the disjoint path problem, *Combinatorica*, 13(1), 97, 1993.
3. Vygen, J., NP-completeness of some edge-disjoint paths problems, *Disc. Appl. Math.*, 61(1), 83, 1995.
4. Nishizeki, T., Vygen, J., and Zhou, X., The edge-disjoint paths problem is np-complete for series-parallel graphs, *Disc. Appl. Math.*, 115(1–3), 177, 2001.
5. Kramer, M. and van Leeuwen, J., The complexity of wire-routing and finding minimum area layouts for arbitrary VLSI circuits, in *Advances in Computing Research, Vol. 2: VLSI Theory*, Preparata, F.P. Ed., 1984, p. 129. JAI Press, Reading, MA.
6. Marx, D., Eulerian disjoint paths problem in grid graphs is NP-complete, *Disc. Appl. Math.*, 143(1–3), 336, 2004.
7. Garg, N., Vazirani, V., and Yannakakis, M., Primal-dual approximation algorithms for integral flow and multicut in trees, *Algorithmica*, 18(1), 3, 1997.
8. Ma, B. and Wang, L., On the inapproximability of disjoint paths and minimum steiner forest with bandwidth constraints, *JCSS*, 60(1), 1, 2000.
9. Erlebach, T., Approximation algorithms and complexity results for path problems in trees of rings, in *International Symposium on Mathematical Foundations of Computer Science*, LNCS, 2136, 2001, p. 351. Springer, Berlin, Germany.
10. Guruswami, V., Khanna, S., Rajaraman, R., Shepherd, B., and Yannakakis, M., Near-optimal hardness results and approximation algorithms for edge-disjoint paths and related problems, *JCSS*, 67(3), 473, 2003.
11. Menger, K., Zur allgemeinen kurventheorie, *Fundam. Math.*, 10, 96, 1927.
12. Awerbuch, R., Gawlick, R., Leighton, F.T., and Rabani, Y., On-line admission control and circuit routing for high performance computing and communication, in *Proceedings of FOCS*, 1994, p. 412. IEEE Press, Piscataway, NJ.
13. Raghavan, P. and Upfal, E., Efficient all-optical routing, in *Proceedings of STOC*, 1994, p. 134. ACM Press, New York.
14. Aggarwal, A., Bar-Noy, A., Coppersmith, D., Ramaswami, R., Schieber, B., and Sudan, M., Efficient routing and scheduling algorithms for optical networks, in *Proceedings of SODA*, 1994, p. 412. ACM/SIAM, Philadelphia, PA.
15. Aumann, Y. and Rabani, Y., Improved bounds for all-optical routing, in *Proceedings of SODA*, 1995, p. 567. ACM/SIAM, Philadelphia, PA.
16. Hromkovič, J., Klasing, R., Stöhr, E., and Wagener, H., Gossiping in vertex-disjoing paths mode in *d*-dimensional grids and planar graphs, in *European Symposium on Algorithms*, LNCS, 726, 1993, p. 200. Springer, Berlin, Germany.
17. Sidhu, D., Nair, R., and Abdallah, S., Finding disjoint paths in networks, *ACM SIGCOMM Comp. Comm. Rev.*, 21(4), 43, 1991.

18. Blesa, M. and Blum, C., Ant colony optimization for the maximum edge-disjoint paths problem, in *Proceedings of the European Workshop on Evolutionary Computation in Communications, Networks, and Connected Systems*, LNCS, 3005, 2004, p. 160. Springer, Berlin, Germany.

19. Blesa, M. and Blum, C., Finding edge-disjoint paths with artificial ant colonies, TR LSI-05-13-R, ALBCOM research group, Dept. Llenguatges i Sistemes Informàtics, Universitat Politècnica de Catalunya, 2005. www.lsi.upc.edu/dept/techreps/techreps.html.

20. Kleinberg, J., Approximation algorithms for disjoint paths problems, PhD Thesis, MIT, Cambridge, MA, 1996.

21. Kolman, P. and Scheideler, C., Simple on-line algorithms for the maximum disjoint paths problem, in *Proceedings of SPAA*, 2001, p. 38. ACM Press, New York.

22. Kolman, P. and Scheideler, C., Improved bounds for the unsplittable flow problem, in *Proceedings of SODA*, 2002, p. 184. ACM/SIAM, Philadelphia, PA.

23. Carmi, P., Erlebach, T., and Okamoto, Y., Greedy edge-disjoint paths in complete graphs, in *Graph-Theoretic Concepts in Computer Science*, LNCS, 2880, 2003, p. 143. Springer, Berlin, Germany.

24. Kolman, P., A note on the greedy algorithm for the unsplittable flow problem, *Info. Proc. Lett.*, 88(3), 101, 2003.

25. Kolliopoulos, S. and Stein, C., Approximating disjoint-path problems using packing integer programs, *Math. Prog.*, 99(1), 63, 2004.

26. Chekuri, C. and Khanna, S., Edge disjoint paths revisited, in *Proceedings of SODA*, 2003, p. 628. ACM/SIAM, Philadelphia, PA.

27. Awerbuch, B., Azar, Y., and Plotkin, S., Throughput-competitive online routing, in *Proceedings of FOCS*, 1993, p. 32. IEEE Press, Piscataway, NJ.

28. Dorigo, M., *Ottimizzazione, Apprendimento Automatico, ed Algoritmi basati su Metafora Naturale*, PhD Thesis, DEI, Politecnico di Milano, Milan, Italy, 1992.

29. Dorigo, M., Maniezzo, V., and Colorni, A., Ant System: Optimization by a colony of cooperating agents, *IEEE Trans. Syst. Man Cybern. B*, 26(1), 29, 1996.

30. Dorigo, M. and Stützle, T., *Ant Colony Optimization*, MIT Press, Cambridge, MA, 2004.

31. Di Caro, G. and Dorigo, M., AntNet: Distributed stigmergetic control for communications networks, *J. Artif. Intell. Res.*, 9, 317, 1998.

32. Di Caro, G., Ducatelle, F., and Gambardella, L.M., AntHocNet: An adaptive nature-inspired algorithm for routing in mobile ad hoc networks, *Eur. Trans. Telecom.*, 16(2), 443–455, 2005.

33. Blum, C. and Dorigo, M., The hyper-cube framework for ant colony optimization, *Trans. Syst. Man Cybern. B*, 34(2), 1161, 2004.

34. Stützle, T. and Hoos, H., $\mathcal{MAX}-\mathcal{MIN}$ ant system, *Future Gener. Comp. Sys.*, 16(8), 889, 2000.

35. Dorigo, M. and Gambardella, L., Ant Colony System: A cooperative learning approach to the traveling salesman problem, *IEEE Trans. Evol. Comput.*, 1(1), 53, 1997.

36. Blum, C. and Roli, A., Metaheuristics in combinatorial optimization: Overview and conceptual comparison, *ACM Comput. Surv.*, 35(3), 268, 2003.

37. Medina, A., Lakhina, A., Matta, I., and Byers, J., BRITE: Boston University Representative Internet Topoloy Generator, 2001, http://cs-pub.bu.edu/brite/index.htm.

38. Waxman, B., Routing of multipoint connections, *IEEE J. Sel. Areas Commun.*, 6(9), 1671, 1988.

39. Barabási, A. and Albert, R., Emergence of scaling in random networks, *Science*, 286, 509, 1999.

# Efficient Approximation Algorithms in Random Intersection Graphs

Sotiris E. Nikoletseas

Christoforos
L. Raptopoulos

Paul G. Spirakis

## 18.1 Introduction

A *proper coloring* of a graph $G = (V, E)$ is an assignment of colors to all vertices in $V$ in such a way that no two adjacent vertices have the same color. A *k-coloring* of $G$ is a coloring that uses $k$ colors. The minimum number of colors that can be used to properly color $G$ is the *(vertex) chromatic number* of $G$ and is denoted by $\chi(G)$. Finding the chromatic number of a graph is a fundamental problem in computer science, with various applications related to collision avoidance and message inhibition methods [1], range assignment problems in directional antennas' optimization [2], coordination aspects of MAC access in sensor networks [3], and more.

Deciding whether a given graph admits a $k$-coloring for a given $k \geq 3$ is well known to be NP-complete. In particular, it is NP-hard to compute the chromatic number [4]. The best known approximation algorithm computes a coloring of size at most within a factor $O(\frac{n(\log \log n)^2}{(\log n)^3})$ of the chromatic number [5]. Furthermore, for any constant $\epsilon > 0$, it is NP-hard to approximate the chromatic number within a factor $n^{1-\epsilon}$ [6].

The intractability of the vertex coloring problem for arbitrary graphs lead researchers to the study of the problem on random instances of various random graphs models. Coloring Erdős–Rényi random graphs (where edges appear independently) was considered in Reference 7 and also in Reference 8. As it seems to be implied by these two works, randomness sometimes allows for smaller chromatic number than maximum degree with high probability (whp). For $G_{n,\hat{p}}$, it is shown that whp $\chi(G_{n,\hat{p}}) \sim \frac{d}{\log d}$, where $d$ is the mean degree. We have to point out here that both References 7, 8 prove that there exists a coloring of $G_{n,\hat{p}}$ using around $\frac{d}{\log d}$, but their proof does not lead to polynomial time algorithms.

In this chapter, we consider coloring the vertices of random instances of the *random intersection graphs model*. In this model, there is a universe $\mathcal{M}$ of *labels*, and each one of $n$ vertices selects a random subset of $\mathcal{M}$. Two vertices are connected if and only if their corresponding subsets of labels intersect.

Random intersection graphs may model several real-life applications quite accurately. In fact, there are practical situations in which each communication agent (e.g., a wireless node) gets access only to some ports (statistically) out of a possible set of communication ports. When another agent also selects a communication port, then a communication link is implicitly established, and this gives rise to communication graphs that look like random intersection graphs. RIG modeling is useful in the efficient blind selection of few encryption keys for secure communications over radio channels [9], as well as in $k$-Secret sharing between swarm mobile devices [10]. Furthermore, random intersection graphs are relevant to and capture quite nicely, social networking. Indeed, a social network is a structure made of nodes tied by one or more specific types of interdependency, such as values, visions, financial exchange, friends, conflicts, and web links. Other applications may include oblivious resource sharing in a distributed setting, interactions of mobile agents traversing the web, social networking, and so on. Even epidemiological phenomena (such as spread of disease between individuals with common characteristics in a population) tend to be more accurately captured by this "proximity-sensitive" family of random graphs.

## 18.2 The Model

Random intersection graphs (also referred to as binomial, or uniform random intersection graphs) were introduced by Karoński et al. [11] and K.B. Singer-Cohen [12]. The formal definition of the model is given in the following:

**Definition 18.1 (Uniform/Binomial Random Intersection Graph—$\mathcal{G}_{n,m,p}$ [11,12])** *Consider a universe $\mathcal{M} = \{1, 2, \ldots, m\}$ of elements and a set of $n$ vertices $V$. Assign independently to each vertex $v \in V$ a subset $S_v$ of $\mathcal{M}$, choosing each element $i \in \mathcal{M}$ independently with probability $p$ and draw an edge between two vertices $v \neq u$ if and only if $S_v \cap S_u \neq \emptyset$. The resulting graph is an instance $G_{n,m,p}$ of the random intersection graphs model.*

We will say that a property holds in $\mathcal{G}_{n,m,p}$ *whp* if the probability that a random instance of the $\mathcal{G}_{n,m,p}$ model has the property is at least $1 - o(1)$.

In this model, we also denote by $L_i$ the set of vertices that have chosen label $i \in M$. Given $G_{n,m,p}$, we will refer to $\{L_i, i \in \mathcal{M}\}$ as its *label representation*. It is often convenient to view the label representation as a bipartite graph with vertex set $V \cup \mathcal{M}$ and edge set $\{(v, i) : i \in S_v\} = \{(v, i) : v \in L_i\}$. We refer to this graph as the *bipartite random graph $B_{n,m,p}$ associated to $G_{n,m,p}$*. Notice that the associated bipartite graph is uniquely defined by the label representation.

We note that by selecting the label set of each vertex using a different distribution, we get random intersection graphs models whose statistical behavior can vary considerably from that of $G_{n,m,p}$. Two of these models that have also been considered in the literature are the following: (a) In the **General Random Intersection Graphs Model** $G_{n,m,\vec{p}}$ [13], where $\vec{p} = [p_1, p_2, \ldots, p_m]$, the label set $S_v$ of a vertex $v$ is formed by choosing independently each label $i$ with probability $p_i$. (b) In the **Regular Random Intersection Graphs Model** $G_{n,m,\lambda}$ [14], where $\lambda \in \mathbb{N}$, the label set of a vertex is chosen independently, uniformly at random (i.u.a.r) for the set of all subsets of $\mathcal{M}$ of cardinality $\lambda$.

It is worth mentioning that the edges in $G_{n,m,p}$ are not independent. In particular, there is a strictly positive dependence between the existence of two edges that share an endpoint (i.e., $\Pr(\exists\{u, v\} | \exists\{u, w\}) > \Pr(\exists\{u, v\})$). This dependence is stronger the smaller the number of labels $\mathcal{M}$ includes, whereas it seems to fade away as the number of labels increases. In fact, by using a coupling technique, the authors in Reference 15 prove the equivalence (measured in terms of total variation distance) of uniform random intersection graphs and Erdős–Rényi random graphs, when $m = n^\alpha, \alpha > 6$. This bound on the number of labels was improved in Reference 16, by showing equivalence of sharp threshold functions among the two models for $\alpha \geq 3$. These results show that random intersection graphs are quite general and that known techniques for random graphs can be used in the analysis of uniform random intersection graphs with a large number of labels.

The similarity between uniform random intersection graphs and Erdős–Rényi random graphs vanishes as the number of labels $m$ decreases below the number of vertices $n$ (i.e., $m = n^\alpha$, for $\alpha \leq 1$). This dichotomy was initially pointed out in Reference 12, through the investigation of connectivity of $G_{n,m,p}$. In particular, it was proved that the connectivity threshold for $\alpha > 1$ is $\sqrt{\frac{\ln n}{nm}}$, but it is $\frac{\ln n}{m}$ (i.e., quite larger) for $\alpha \leq 1$. Therefore, the mean number of edges just above connectivity is approximately $\frac{1}{2} n \ln n$ in the first case (which is equal to the mean number of edges just above the connectivity threshold for Erdős–Rényi random graphs), but it is larger by at least a factor of $\ln n$ in the second case. Other dichotomy results of similar flavor were pointed out in the investigation of the (unconditioned) vertex degree distribution by D. Stark [17], through the analysis of a suitable generating function and in the investigation of the distribution of the number of isolated vertices by Shang [18].

## 18.3 Coloring Non-Sparse Random Intersection Graphs

In Reference 19 the authors propose algorithms that whp probability color sparse instances of $G_{n,m,p}$. In particular, for $m = n^\alpha, \alpha > 0$ and $p = o(\sqrt{\frac{1}{nm}})$ they show that $G_{n,m,p}$ can be colored optimally. Moreover, in the case where $m = n^\alpha, \alpha < 1$, and $p = o\left(\frac{1}{m \ln n}\right)$, they show that $\chi(G_{n,m,p}) \sim np$ whp. To do this, they prove that $G_{n,m,p}$ is chordal whp (or equivalently, the associated bipartite graph does not contain cycles), and so a perfect elimination scheme can be used to find a coloring in polynomial time.

In the current section, we present in greater detail the results of Reference 20 for coloring non-sparse random instances of $\mathcal{G}_{n,m,p}$. The range of values we consider here is different than the one needed for the algorithms in Reference 19 to work. We study coloring $G_{n,m,p}$ for the case $m = n^\alpha, \alpha \in (0, 1)$, where random intersection graphs differ the most from Erdős–Rényi random graphs, and in particular for the range $mp \leq (1 - \alpha) \ln n$, as well as the denser range $mp \geq \ln^2 n$. We also have to note that the proof techniques used in Reference 19 cannot be used in the range we consider, as the properties that they examine do not hold in our case.

### 18.3.1 Coloring Almost All Vertices

We are going to consider the case where $m = n^\alpha$, for $\alpha \in (0, 1)$ some fixed constant. As mentioned earlier, the area $mp = o\left(\frac{1}{\ln n}\right)$ gives almost surely instances in which the label graph (i.e., the dual graph where the labels in $\mathcal{M}$ play the role of vertices and the vertices in $V$ play the role of labels) is quite sparse and so $G_{n,m,p}$ can be colored optimally using $\max_{l \in \mathcal{M}} |L_l|$ colors [19]. We will here consider the denser area $mp = \Omega\left(\frac{1}{\ln n}\right)$. In this range of values, it is easy to see that the values of $|L_l|$ are concentrated around $np$. We were able to prove that even for values of the parameters $m, p$ that give quite denser graphs, we can still use $np$ colors to properly color most of the graph.* Our proof technique is inspired by analogous ideas of Frieze [21] (see also in Reference 8). The proof of Theorem 18.1 uses the following Lemma, which first appeared in Reference 13.

**Lemma 18.1** ([13]) *Let $G_{n,m,p}$ be a random instance of the random intersection graphs model. Then the conditional probability that a set of $k$ vertices is an independent set, given that $s$ of them are already an independent set, is equal to $((1 - p)^{k-s} + (k - s)p(1 - p)^{k-s-1}(1 - \frac{sp}{1+(s-1)p}))^m$, where $0 \leq s \leq k$.*

---

* Note, however, that this does not mean that the chromatic number is close to $np$, as the part that is not colored could be a clique in the worst case.

*Proof.* For a vertex set $V'$, let $X_{V'}$ be the indicator random variable that $V'$ is an independent set, that is,

$$X_{V'} = \begin{cases} 1 & \text{if } V' \text{ is an independent set} \\ 0 & \text{otherwise} \end{cases}$$

Let also $V_1'$, $V_2'$ be two sets of $k$ vertices with $s$ vertices in common. Therefore, we need to find the conditional probability $P\{X_{V_1'} = 1 | X_{V_2'} = 1\}$, that is, the probability that $V_1'$ is an independent set given that $V_2'$ is an independent set. The main technical tool of the proof is a vertex contraction technique; in particular, we merge several vertices into one *supervertex* and study its probabilistic behavior.

Towards this goal, fix an element $i$ of $\mathcal{M} = \{1, 2, \ldots, m\}$ and consider two (super)vertices $v_1, v_2$ of $G(n, m, p)$ that choose each label $i$ independently with probability $p^{(1)}$ and $p^{(2)}$, respectively (the exact values of those probabilities will be determined in the following special case that supervertices consist of independent sets). Let also $S_{v_1}, S_{v_2}$ denote the sets of elements of $\mathcal{M}$ assigned to $v_1$ and $v_2$, respectively. Then,

$$P\{i \in S_{v_1} | \nexists(v_1, v_2)\} = P\{i \in S_{v_1}, i \notin S_{v_2} | \nexists(v_1, v_2)\}$$

$$= \frac{P\{i \in S_{v_1}, i \notin S_{v_2}, \nexists(v_1, v_2)\}}{P\{\nexists(v_1, v_2)\}} = \frac{p^{(1)}(1 - p^{(2)})}{1 - p^{(1)}p^{(2)}} \tag{18.1}$$

where $(v_1, v_2)$ is an edge between $v_1$ and $v_2$. From this we get

- Conditional on the fact that $(v_1, v_2)$ does not exist, the probabilistic behavior of vertex $v_1$ is identical to that of a single vertex that chooses element $i$ of $\mathcal{M}$ independently with probability $\frac{p^{(1)}(1-p^{(2)})}{1-p^{(1)}p^{(2)}}$.
- Conditional on the fact that $(v_1, v_2)$ does not exist, the probabilistic behavior of $v_1$ and $v_2$ considered as a unit is identical to that of a single vertex that chooses element $i$ of $\mathcal{M}$ independently with probability

$$P\{i \in S_{v_1} \cup S_{v_2} | \nexists(v_1, v_2)\} = P\{i \in S_{v_1} | \nexists(v_1, v_2)\} + P\{i \in S_{v_2} | \nexists(v_1, v_2)\}$$

$$= \frac{p^{(1)} + p^{(2)} - 2p^{(1)}p^{(2)}}{1 - p^{(1)}p^{(2)}} \tag{18.2}$$

where $i$ is a fixed element of $\mathcal{M}$. The first of the above-mentioned equations follows from the observation that if there is no edge between $v_1$ and $v_2$, then the sets $S_{v_1}$ and $S_{v_2}$ are disjoint, meaning that element $i$ cannot belong to both of them. The second equation follows from symmetry.

Consider now merging one by one the vertices of $G(n, m, p)$ into one supervertex. Let $w_j$ denote a supervertex of $j$ simple vertices that form an independent set. Note that the probabilistic behavior of $w_j$ is not affected by the way the merging is done. If $w_{j_1}, w_{j_2}$ are two supervertices representing two disjoint sets of simple vertices, we say that an edge $(w_{j_1}, w_{j_2})$ exists iff any edge connecting a simple vertex in $w_{j_1}$ and a simple vertex in $w_{j_2}$ exists. Thus, the event $\{\nexists(w_{j_1}, w_{j_2})\}$ is equivalent to the event {the vertices in $w_{j_1}$ together with those in $w_{j_2}$ form an independent set}.

By using Equation 18.2, we can show that $P\{i \in S_{w_2}\} = \frac{2p}{1+p}$, $P\{i \in S_{w_3}\} = \frac{3p}{1+2p}$, and by induction

$$P\{i \in S_{w_j}\} = \frac{jp}{1 + (j-1)p} \tag{18.3}$$

where $i$ is a fixed element of $\mathcal{M}$ and $S_{w_j}$ is the union of all the sets of elements of $\mathcal{M}$ assigned to each simple vertex in $w_j$, that is, $S_{w_j} = \bigcup_{v \in w_j} S_v$, where $v$ is a simple vertex and $S_v$ is the set of elements of $\mathcal{M}$ assigned to $v$. Because of the definition of $w_j$, the subsets $S_v$ in the above-mentioned union are disjoint.

Thus, let $V_1'$ be any set of $k$ (simple) vertices and let $V_2'$ be an independent set of $k$ vertices that has $s$ vertices in common with $V_1'$. As there is no edge between any vertices in $V_2'$, we can treat the $k - s$ vertices of $V_2'$ not belonging to $V_1'$ and the $s$ vertices belonging to both $V_1'$ and $V_2'$ as two seperate supervertices $w_{k-s}$ and $w_s$, respectively, that do not communicate by an edge. Hence, by Equations 18.1, 18.2 and 18.3, the probabilistic behavior of $w_s$ identical to that of a single vertex $w_s'$ that chooses each element of $\mathcal{M}$ independently with probability $p^{(w_s')}$, given by

$$p^{(w_s')} = \frac{p^{(w_s)}(1 - p^{(w_{k-s})})}{1 - p^{(w_s)}p^{(w_{k-s})}} = \frac{sp}{1 + (k-1)p}. \tag{18.4}$$

Let now $V''$ be a set of $k - s$ simple vertices and a vertex identical to $w_s'$. Then, for a fixed element $i$ of $\mathcal{M}$, each of the $k - s$ simple vertices chooses $i$ independently with probability $p$, whereas the supervertex $w_s'$ chooses $i$ independently with probability $p^{(w_s')}$. Therefore, the probability that $V_1'$ is an independent set, given that $V_2'$ is an independent set is the probability that there is no edge between the simple vertices in $V_1' \backslash V_2'$ and the vertex $w_s'$. In particular, this is equal to the probability that these vertices have not selected any elements of $\mathcal{M}$ in common. But this is exactly equal to $((1-p)^{k-s} + (k-s)p(1-p)^{k-s-1}(1 - \frac{sp}{1+(s-1)p}))^m$ as stated in the Lemma. ∎

We are now ready to present our theorem on coloring almost all vertices of $G_{n,m,p}$.

**Theorem 18.1** *When $m = n^\alpha, \alpha < 1$, and $mp \le \beta \ln n$, for any constant $\beta < 1 - \alpha$. Then a random instance of the random intersection graphs model $G_{n,m,p}$ contains a subset of at least $n - o(n)$ vertices that can be colored using $np$ colors, with probability at least $1 - e^{-n^{0.99}}$.*

*Proof.* In what follows, we will denote by $G_{n,m,p}$ an instance of the random intersection graphs model $G_{n,m,p}$. We also denote by $B_{n,m,p}$, the bipartite graph associated to $G_{n,m,p}$. We prove a slightly stronger property than what the lemma requires.

Assume an arbitrary ordering of the vertices $v_1, v_2, \ldots, v_n$. For $i = 1, 2, \ldots, n$, let $B_i$ be the subgraph of $B_{n,m,p}$ induced by $\cup_{j=1}^i v_j \bigcup \mathcal{M}$. We denote by $H_i$ the intersection graph whose bipartite graph has vertex set $V \bigcup \mathcal{M}$ and edge set that is exactly as $B_i$ between $\cup_{j=1}^i v_j$ and $\mathcal{M}$, whereas every other edge (i.e., the ones between $\cup_{j=i}^n v_j$ and $\mathcal{M}$) appears independently with probability $p$.

Set $x = np$. Let $X$ denote the size of the largest $x$-colorable subset of vertices in $G_{n,m,p}$ and let $X_i$ denote the expectation of the largest $x$-colorable subset in $H_i$. Notice that $X_i$ is a random variable depending on the overlap between $G_{n,m,p}$ and $H_i$. Obviously, $X = X_n$ and setting $X_0 = E[X]$, we have $|X_i - X_{i+1}| \le 1$, for all $i = 1, 2, \ldots, n$. It is straightforward to verify that the sequence $X_0, X_1, \ldots, X_n$ is a Doob Martingale (see also Vol. 1, Chap. 9 of [22]). Hence, by applying Azuma's inequality, we have that

$$\Pr(|X - E[X]| \ge t) \le 2e^{-\frac{t^2}{2n}}.$$

Set now $k_0 = \frac{(1-\epsilon^2)n}{x}$, where $\epsilon$ is a positive constant that is arbitrarily close to 0. For $t = \epsilon \frac{k_0 x}{1+\epsilon} = \epsilon(1-\epsilon)n$, Azuma's inequality becomes

$$\Pr\left(|X - E[X]| \ge \epsilon(1-\epsilon)n\right) \le 2e^{-\frac{\epsilon^2 n}{3}}. \tag{18.5}$$

Let now $Y$ denote the number of $x$-colorable subsets of $(1+\epsilon)\frac{xk_0}{1+\epsilon}$ vertices in $G_{n,m,p}$ that can be split in exactly $x$ independent sets (i.e., chromatic classes) of size exactly $k_0$. We can now verify that proving that $\Pr(Y > 0)$ is greater or equal to the right hand side of inequality (18.5), that is, $2e^{-\frac{\epsilon^2 n}{3}}$, then we will have proven that (a) $E[X] \ge \frac{xk_0}{1+\epsilon}$ and (b) that the values of $X$ are concentrated around something greater than $\frac{xk_0}{1+\epsilon}$ whp. More specifically, (a) comes from the observation that the event $\{Y > 0\}$ implies the event $\{X \ge xk_0\}$, hence $\Pr(Y > 0) \le \Pr(X \ge xk_0) = \Pr(X - \frac{xk_0}{1+\epsilon} \ge \frac{\epsilon xk_0}{1+\epsilon}) = \Pr(X - \frac{xk_0}{1+\epsilon} \ge \epsilon(1-\epsilon)n)$. If now $E[X]$ was strictly less than $\frac{xk_0}{1+\epsilon}$, then this would mean that $\Pr(Y > 0) < \Pr(X - E[X] \ge \epsilon(1-\epsilon)n)$ which by (18.5) is less than $2e^{-\frac{\epsilon^2 n}{3}}$. Hence, proving that $\Pr(Y > 0) \ge 2e^{-\frac{\epsilon^2 n}{3}}$ could only mean that $E[X] \ge \frac{xk_0}{1+\epsilon}$. Part (b) then follows as well.

The remarks (a) and (b) described earlier are sufficient to prove the theorem, as $\epsilon$ can be as small as possible. As $Y$ is a nonnegative random variable that takes only integral values, to bound $\Pr(Y > 0)$, we will use the well-known inequality (see also exercise 1 of Vol. 1, Chap. 4 in Reference 23)

$$\Pr(Y > 0) \geq \frac{E^2[Y]}{E[Y^2]}.$$

Since every color class considered in $Y$ must have exactly $k_0$ vertices and obviously different color classes must not overlap, we get that

$$E[Y] = \prod_{i=1}^{x} \binom{n - (i-1)k_0}{k_0} \left( (1-p)^{k_0} + k_0 p(1-p)^{k_0-1} \right)^m$$

where the term $((1-p)^{k_0} + k_0 p(1-p)^{k_0-1})^m \stackrel{def}{=} p_1$ is the probability that a color class is indeed an independent set that is no two vertices in it have a common label. Similarly, we have that

$$E[Y^2] \leq E[Y] \sum_{k_1,\dots,k_x \leq k_0} \prod_{i=1}^{x} \binom{k_0}{k_i} \binom{n - ik_0}{k_0 - k_i} p_2$$

where $p_2$ is the conditional probability that a color class of $k_0$ vertices is an independent set, given that $k_i$ of them are already an independent set. By Lemma 18.1, we have that

$$p_2 \stackrel{def}{=} \left( (1-p)^{k_0-k_i} + (k_0 - k_i)p(1-p)^{k_0-k_i-1} \left( 1 - \frac{k_i p}{1 + (k_i - 1)p} \right) \right)^m.$$

Combining the earlier, we conclude that

$$\frac{1}{\Pr(Y > 0)} \leq \frac{E[Y^2]}{E^2[Y]} \leq \sum_{k_1,\dots,k_x \leq k_0} \prod_{i=1}^{x} \frac{\frac{k_0!}{k_i!(k_0-k_i)!} \frac{(n-ik_0)!}{(k_0-k_i)!(n-(i+1)k_0+k_i)!}}{\frac{(n-(i-1)k_0)!}{k_0!(n-ik_0)!}} \frac{p_2}{p_1}$$

$$\leq \sum_{k_1,\dots,k_x \leq k_0} \prod_{i=1}^{x} \frac{\left( \frac{k_0!}{(k_0-k_i)!} \right)^2}{k_i!(n - ik_0)^{k_i}} \frac{p_2}{p_1} \qquad (18.6)$$

The fraction $\frac{p_2}{p_1}$ can be bounded in a quite straightforward manner as follows

$$\sqrt[m]{\frac{p_2}{p_1}} \leq \frac{(1-p)^{k_0-k_i} + (k_0 - k_i)p(1-p)^{k_0-k_i-1}}{(1-p)^{k_0} + k_0 p(1-p)^{k_0-1}}$$

$$= \frac{1 - p + (k_0 - k_i)p}{1 - p + k_0 p}(1-p)^{-k_i} = \left( 1 - \frac{k_i p}{1 - p + k_0 p} \right)(1-p)^{-k_i}$$

$$\leq e^{-\frac{k_i p}{1-p+k_0 p} + k_i p} = e^{\frac{k_0 k_i p^2 - k_i p^2}{1-p+k_0 p}} \leq e^{k_0 k_i p^2}$$

where the last inequality follows as $k_0 \to \infty$ for $mp = O(\ln n)$ and $m = n^\alpha, \alpha < 1$.

For $i = 1, \dots, x$, let $A_i \stackrel{def}{=} \frac{\left( \frac{k_0!}{(k_0-k_i)!} \right)^2}{k_i!(n-ik_0)^{k_i}} \frac{p_2}{p_1}$, so that $\frac{E[Y^2]}{E^2[Y]} \leq \sum_{k_1,\dots,k_x \leq k_0} \prod_{i=1}^{x} A_i$. When $k_i = 0$, then trivially $A_i = 1$. On the other hand, when $1 \leq k_i \leq k_0$, using the inequalities $\frac{k_0!}{(k_0-k_i)!} \leq k_0^{k_i}, k_i! \geq \left( \frac{k_i}{e} \right)^{k_i}$ and the fact that $xk_0 = (1 - \epsilon^2)n$, we can see that

$$A_i \leq \frac{k_0^{2k_i}}{k_i^{k_i}(n)^{k_i}} e^{mk_0 k_i p^2} = e^{2k_i \ln k_0 - k_i \ln k_i - k_i \ln n + mk_0 k_i p^2 + O(k_i \ln \ln n)} \qquad (18.7)$$

We now distinguish two cases.

(a) $1 \leq k_i \leq \frac{k_0}{\ln^2 n}$. Then $A_i \leq e^{2k_i \ln n + mk_0 k_i p^2} \leq e^{k_i(2\ln n + mp)} = e^{O\left(\frac{k_0}{\ln n}\right)}$, as $mp = O(\ln n)$.

(b) $\frac{k_0}{\ln^2 n} < k_i \leq k_0$. Then $A_i \leq e^{\alpha k_i \ln n - k_i \ln n + mk_0 k_i p^2 + O(k_i \ln \ln n)} \leq e^{(\alpha - 1 + \beta)k_i \ln n + O(k_i \ln \ln n)} = o(1)$, as $\beta < 1 - \alpha$. We should also mention that the $O(\cdot)$ part of the exponent is different than the $O(\cdot)$ part of the exponent in (18.7).

The crucial observation now is that, for all values of $k_i$, $A_i^x \leq e^{O\left(\frac{n}{\ln n}\right)}$. As a final note, the total number of terms in the sum $\sum_{k_1,\ldots,k_x \leq k_0}$ is $(k_0 + 1)^x = e^{x \ln (k_0 + 1)} \leq e^{n^{1-\alpha} \ln^2 n}$.

By (18.6), we then have that

$$\Pr(Y > 0) \geq e^{-n^{1-\alpha} \ln^2 n - O\left(\frac{n}{\ln n}\right)} \geq 2e^{-\frac{\epsilon^2 n}{3}}$$

which concludes the proof. ∎

It is worth noting here that the proof of Theorem 18.1 can also be used similarly to prove that $\Theta(np)$ colors are enough to color $n - o(n)$ vertices even in the case where $mp = \beta \ln n$, for any constant $\beta > 0$. However, finding the exact constant multiplying $np$ is technically more difficult.

## 18.3.2 A Polynomial Time Algorithm for the Case $mp \geq \ln^2 n$

In the following algorithm, every vertex chooses i.u.a.r a preference in colors, denoted by $shade(\cdot)$ and every label $l$ chooses a preference in the colors of the vertices in $L_l$, denoted by $c_l(\cdot)$.

---

## ALGORITHM 18.1   Algorithm CliqueColor:

**Input:** An instance $G_{n,m,p}$ of $\mathcal{G}_{n,m,p}$ and its associated bipartite $B_{n,m,p}$.
**Output:** A proper coloring $G_{n,m,p}$.

1. For every $v \in V$ choose a color denoted by $shade(v)$ i.u.a.r among those in $\mathcal{C}$;
2. For every $l \in \mathcal{M}$ choose a coloring of the vertices in $L_l$ such that for every color in $\{c \in \mathcal{C} : \exists v \in L_l \text{ with } shade(v) = c\}$ there is exactly one vertex in the set $\{u \in L_l : shade(u) = c\}$ having $c_l(u) = c$ while the rest remain uncolored;
3. Set $U = \emptyset$ and $C = \emptyset$;
4. **For** $l = 1$ **to** $m$ **do** {
5. Color every vertex in $L_l \backslash U \cup C$ according to $c_l(\cdot)$ iff there is no collision with the color of a vertex in $L_l \cap C$;
6. Include every vertex in $L_l$ colored that way in $C$ and the rest in $U$;}
7. Let $\mathcal{H}$ denote the (intersection) subgraph of $G_{n,m,p}$ induced by the vertices in $U$;
8. Give a proper coloring of $\mathcal{H}$ using a new set of colors $\mathcal{C}'$;
9. **Output** a coloring of $G_{n,m,p}$ using $|\mathcal{C} \cup \mathcal{C}'|$ colors;

---

The following result concerns the correctness of Algorithm CliqueColor.

**Theorem 18.2 (Correctness)** *Given an instance $G_{n,m,p}$ of the random intersection graphs model, algorithm CliqueColor always finds a proper coloring.*

*Proof.* For the sake of contradiction, suppose that in the coloring proposed by the algorithm there are two vertices $v_1$ and $v_2$ that are connected and have been assigned to the same color $c$. This of course means that these two vertices have at least one label in common. As the sets $\mathcal{C}$ and $\mathcal{C}'$ are disjoint and the coloring of $\mathcal{H}$ provided at step 8 of the algorithm is proper, the only way that such a collision would arise is if both $v_1$ and $v_2$ belong to $C$. This means that both were colored by the first pass of the algorithm and also $shade(v_1) = shade(v_2) = c$. Let $l$ be the smallest indexed label in $|S_{v_1} \cap S_{v_2}|$. It is easy to see then that we come to a contradiction, as label $l$ and step 5 will guarantee that at least one of the two vertices lies in $U$. ∎

The following theorem concerns the efficiency of algorithm CliqueColor, provided that $mp \geq \ln^2 n$ and $p = o(\frac{1}{\sqrt{m}})$. Notice that for $p$ larger than $\frac{1}{\sqrt{m}}$, every instance of the random intersection graphs model $G_{n,m,p}$, with $m = n^\alpha, \alpha < 1$, is complete whp.

**Theorem 18.3 (Efficiency)** *Algorithm CliqueColor succeeds in finding a proper $\Theta(\frac{nmp^2}{\ln n})$-coloring of $G_{n,m,p}$ in polynomial time whp, provided that $mp \geq \ln^2 n, p = o(\frac{1}{\sqrt{m}})$ and $m = n^\alpha, \alpha < 1$.*

*Proof.* For $s \in C$, let $Z_c$ denote the number of vertices $v \in V$ such that $shade(v) = c$. $Z_c$ is a binomial random variable, so by Chernoff bounds we can see that, for any positive constant $\beta_1$ that can be arbitrarily small

$$\Pr\left(\left|Z_c - \frac{n}{|C|}\right| \geq \frac{\beta_1 n}{|C|}\right) \leq 2e^{-\frac{\beta_1^2 n}{3|C|}}.$$

For $|C| = \Theta(\frac{mnp^2}{\ln n})$ and $p = o(\frac{1}{\sqrt{m}})$, we can then use Boole's inequality to see that there is no $c \in C$ such that $|Z_c - \frac{n}{|C|}| \geq \frac{\beta_1 n}{|C|}$, with probability $1 - o(1)$, that is, almost surely.

By using the same type of arguments, we can also verify that for arbitrarily small positive constants $\beta_2$ and $\beta_3$, we have that $\Pr(\exists v \in V : ||S_v| - mp| \geq \beta_2 mp) = o(1)$ and $\Pr(\exists l \in \mathcal{M} : ||L_l| - np| \geq \beta_3 np) = o(1)$ for all $mp = \omega(\ln n)$ and $m = n^\alpha, \alpha < 1$.

We will now prove that the maximum degree of the graph $\mathcal{H}$ is small enough to allow a proper coloring of $\mathcal{H}$ using $C' = \Theta(\frac{nmp^2}{\ln n})$ colors. For a label $l \in \mathcal{M}$, let $Y_l$ denote the number of vertices $v \in L_l$ such that $c_l(v) \neq shade(v)$. In order for a label $l$ not to be able to assign color $shade(v)$ to $v \in L_l$, it should be the case that it has assigned color $shade(v)$ to another vertex $u \in L_l$ with $shade(u) = shade(v)$. Hence, the only way to have a collision is when two or more vertices with the same shade have all chosen label $l$. Notice also that have $Y_l \geq k$, the number of different shades appearing among the vertices that have chosen label $l$ should be at most $|L_l| - k$. This means that $\Pr(Y_l \geq k) \leq \binom{|L_l|}{k}(\frac{|L_l|-k}{|C|})^k$. Given the concentration bound for $|L_l|$, we have that

$$\Pr(\exists l : Y_l \geq k) \leq m\binom{(1+\beta_3)np}{k}\left(\frac{(1+\beta_3)np - k}{|C|}\right)^k + o(1) \leq m\left(\frac{3np}{k}\right)^k\left(\frac{2np}{|C|}\right)^k + o(1).$$

By now setting $k = \frac{np}{\ln n}$ and for $|C| \geq 18\frac{mnp^2}{\ln n}$, we then have that, with probability $1 - o(1)$, there is no label $l \in \mathcal{M}$ such that $Y_l \geq \frac{np}{\ln n}$.

For a label $l \in \mathcal{M}$ now let $W_l$ be the number of vertices $v \in L_l$ such that $shade(v) = c_l(v)$ but they remained uncolored, hence included in $\mathcal{H}$. In order for a vertex $v \in L_l$ to be counted in $W_l$, there should exist a label $j$ prior to $l$ (i.e., a label among $1, \ldots, l - 1$) such that $v \in L_j$ and there is another vertex $u \in L_j$ with $shade(u) = shade(v)$. The probability that this happens is at most $p\left(1 - (1 - p)^{Z_{shade(v)}}\right)(1 + (1 - p) + (1-p)^2 + \cdots) = 1 - (1 - p)^{Z_{shade(v)}}$. The crucial observation now is that because choices of labels by vertices (of the same shade or not) is done independently and because the vertices counted in $W_l$ have (by definition of the coloring $c_l(\cdot)$ in step 2 of the algorithm) different shades, the inclusion in $W_l$ of any vertex $u \in L_l$ with $shade(u) = c_l(u)$ does not affect the inclusion of another $v \in L_l\backslash\{u\}$ with $shade(v) = c_l(v)$. Hence, taking also into account the concentration bound for $Z_{shade(v)}$ and $|L_l|$, we have that

$$\Pr(\exists l : W_l \geq k') \leq m\binom{(1+\beta_3)np}{k'}\left(1 - (1 - p)^{(1+\beta_1)\frac{n}{|C|}}\right)^{k'} + o(1).$$

By now setting $k' = \frac{np}{\ln n}$ and using the relation $(1-x)^y \sim 1 - xy$, valid for all $x, y$ such that $xy = o(1)$, we have that when $|\mathcal{C}| \geq 18\frac{mnp^2}{\ln n}$, there is no label $l$ such that $W_l \geq \frac{np}{\ln n}$, whp.

We have then proved that the number of vertices in $U$ of the algorithm that have chosen a specific label is whp at most $\frac{2np}{\ln n}$. As, for any vertex $v$ in $G_{n,m,p}$ has $|S_v| \leq (1 + \beta_2)mp$, we conclude that the maximum degree in $\mathcal{H}$ satisfies $\max_{v \in \mathcal{H}} degree_{\mathcal{H}}(v) \leq (1+\beta_2)mp\frac{2np}{\ln n}$. It is then evident that we can colour $\mathcal{H}$ greedily, in polynomial time, using $\frac{2.1nmp^2}{\ln n}$ more colors, whp. Hence, we can color $G_{n,m,p}$ in polynomial time, using at most $\frac{20.1nmp^2}{\ln n}$ colors in total. ∎

It is worth noting here that the number of colors used by the algorithm in the case $mp \geq \ln^2 n, p = O(\frac{1}{\sqrt[4]{m}})$ and $m = n^\alpha, \alpha < 1$ is of the correct order of magnitude. Indeed, by the concentration of the values of $|S_v|$ around $mp$ for any vertex $v$ whp, one can use the results of Reference 24 for the uniform random intersection graphs model $G_{n,m,\lambda}$, with $\lambda \sim mp$ to provide a lower bound on the chromatic number. Indeed, it can be easily verified that the independence number of $G_{n,m,\lambda}$, for $\lambda = mp \geq \ln^2 n$ is at most $\Theta(\frac{\ln n}{mp^2})$, which implies that the chromatic number of $G_{n,m,\lambda}$ (and hence of the $G_{n,m,p}$ because of the concentration of the values of $|S_v|$) is at least $\Omega(\frac{nmp^2}{\ln n})$.

### 18.3.3 Coloring Random Hypergraphs

The model of random intersection graphs $\mathcal{G}_{n,m,p}$ could also be though of as generating random hypergraphs. The hypergraphs generated have vertex set $V$ and edge set $\mathcal{M}$. There is a huge amount of literature concerning coloring hypergraphs. However, the question about coloring there seems to be different from the one we answer in this paper. More specifically, a proper coloring of a hypergraph seems to be any assignment of colors to the vertices, so that no monochromatic edge exists. This of course implies that fewer colors than the the the chromatic number (studied in this paper) are needed to achieve this goal.

We would also like to mention that as far as $\mathcal{G}_{n,m,p}$ is concerned, the problem of finding a coloring such that no label is monochromatic seems to be quite easier when $p$ is not too small.

**Theorem 18.4** *Let $G_{n,m,p}$ be a random instance of the model $\mathcal{G}_{n,m,p}$, for $p = \omega(\frac{\ln m}{n})$ and $m = n^\alpha$, for any fixed $\alpha > 0$. Then whp, there is a polynomial time algorithm that finds a $k$-coloring of the vertices such that no label is monochromatic, for any fixed integer $k \geq 2$.*

*Proof.* By Chernoff bounds and Boole's inequality we can easily show that for any constant $\epsilon > 0$ that can be arbitrarily small

$$\Pr(\exists l : ||L_l| - np| \geq \epsilon np) \leq 2me^{-\frac{\epsilon^2 np}{3}} \to 0$$

for any $p = \omega(\frac{\ln m}{n})$.

If we were to choose the color of each vertex i.u.a.r among the available colors, then the mean number of monochromatic edges in $G_{n,m,p}$, would be almost surely (given that the previous concentration bound holds)

$$E[\# \text{ monochromatic edges}] = \sum_{l \in \mathcal{M}} k^{1-|L_l|} \leq mk^{1-(1-\epsilon)np} < 1.$$

Then, using the method of conditional expectations [25,26] we can derive an algorithm that finds the desired coloring in time $O(mn^{k+2})$. Indeed, as $E[\# \text{ monochromatic edges}] < 1$, there must be a vertex $v$ and a color $c$, such that coloring $v$ with $c$ guarantees that $E[\# \text{ monochromatic edges}|color(v) = c] < 1$. This, combined with the fact that, given any coloring $\mathcal{C}_S$ of any subset $S$ of vertices, we can compute $E[\# \text{ monochromatic edges}|\mathcal{C}_S]$ in time $O(nm)$, leads to the desired algorithm. ∎

# 18.4 Other Combinatorial Problems in RIGs

We conclude this chapter by briefly mentioning some works related to the design and average case analysis of efficient approximation algorithms on RIGs for various combinatorial problems. Some of these results, as well as the techniques used for the analysis, highlight and take advantage of the intricacies and special structure of random intersection graphs, whereas others are adapted from the field of Erdős–Rényi random graphs. For further results on various models of random intersection graphs, we refer the reader to the recent review paper [27].

## 18.4.1 Independent Sets

The problem of the existence and efficient construction of large independent sets in general random intersection graphs is considered in Reference 13. Concerning existence, exact formulae are derived for the expectation and variance of the number of independent sets of any size, by using a *vertex contraction technique*. This technique involves the characterization of the statistical behavior of an independent set of any size and highlights an *asymmetry* in the edge appearance rule of random intersection graphs. In particular, it is shown that the probability that any fixed label $i$ is chosen by some vertex in a $k$-size $S$ with no edges is exactly $\frac{kp_i}{1+(k-1)p_i}$. On the other hand, there is no closed formula for the respective probability when there is at least one edge between the $k$ vertices (or even when the set $S$ is complete)! The special structure of random intersection graphs is also used in the design of efficient algorithms for constructing quite large independent sets in uniform random intersection graphs. By analysis, it is proved that the approximation guarantees of algorithms using the label representation of random intersection graphs are superior to that of well-known greedy algorithms for independent sets when applied to instances of $\mathcal{G}_{n,m,p}$.

## 18.4.2 Hamilton Cycles

In Reference 28, the authors investigate the existence and efficient construction of *Hamilton cycles* in uniform random intersection graphs. In particular, for the case $m = n^\alpha, \alpha > 1$ the authors first prove a general result that allows one to apply (with the same probability of success) any algorithm that finds a Hamilton cycle whp in a $G_{n,M}$ random graph (i.e., a graph chosen equiprobably form the space of all graphs with $M$ edges). The proof is done by using a simple coupling argument. A more complex coupling was given in Reference 29, resulting in a more accurate characterization of the threshold function for Hamiltonicity in $G_{n,m,p}$ for the whole range of values of $\alpha$. From an algorithmic perspective, the authors in Reference 28 provide an expected polynomial time algorithm for the case where $m = O(\sqrt{\frac{n}{\ln n}})$ and $p$ is constant. For the more general case where $m = o(\frac{n}{\ln n})$, they propose a *label exposure* greedy algorithm that succeeds in finding a Hamilton cycle in $G_{n,m,p}$ whp, even when the probability of label selection is just above the connectivity threshold.

## 18.4.3 Maximum Cliques

In Reference 30, the authors consider maximum cliques in the uniform random intersection graphs model $\mathcal{G}_{n,m,p}$. It is proved that when the number of labels is not too large, we can use the label choices of the vertices to find a maximum clique in polynomial time (in the number of labels $m$ and vertices $n$ of the graph). Most of the analytical work in the paper is devoted in proving the *Single Label Clique Theorem*. Its proof includes a coupling to a graph model where edges appear independently and in which we can bound the size of the maximum clique by well-known probabilistic techniques. The theorem states that when the number of labels is less than the number of vertices, any large enough clique in a random instance of $\mathcal{G}_{n,m,p}$ is formed by a single label. This statement may seem obvious when $p$ is small, but it is hard to imagine that it still holds for *all* "interesting" values for $p$. Indeed, when $p = o(\sqrt{\frac{1}{nm}})$, by slightly modifying an argument of Reference 19, one can see that $G_{n,m,p}$ almost surely has no cycle of size $k \geq 3$ whose edges are formed by $k$ distinct labels (alternatively, the intersection graph produced by reversing the roles of labels and vertices is a tree). On the other hand, for larger $p$, a random instance of $\mathcal{G}_{n,m,p}$ is far from perfect[*] and the techniques of Reference 19 do not apply. By using the Single Label Clique Theorem, a tight bound on the clique number of $G_{n,m,p}$ is proved, in the case where $m = n^{\alpha}, \alpha < 1$. A lower bound in the special case where $mp^2$ is constant was given in Reference 12. We considerably broaden this range of values to also include vanishing values for $mp^2$ and also provide an asymptotically tight upper bound.

Finally, as yet another consequence of the Single Label Clique Theorem, the authors in Reference 30 prove that the problem of inferring the complete information of label choices for each vertex from the resulting random intersection graph is *solvable* whp, namely, the maximum likelihood estimation method will provide a unique solution (up to permutations of the labels).[*] In particular, given values $m$, $n$, and $p$, such that $m = n^{\alpha}, 0 < \alpha < 1$, and given a random instance of the $\mathcal{G}_{n,m,p}$ model, the label choices for each vertex are uniquely defined.

### 18.4.4 Expansion and Random Walks

The edge expansion and the cover time of uniform random intersection graphs is investigated in Reference 31. In particular, by using first moment arguments, the authors first prove that $G_{n,m,p}$ is an expander whp when the number of labels is less than the number of vertices, even when $p$ is just above the connectivity threshold (i.e., $p = (1 + o(1))\tau_c$, where $\tau_c$ is the connectivity threshold). Second, the authors show that random walks on the vertices of random intersection graphs are whp *rapidly mixing* (in particular, the mixing time is logarithmic on $n$). The proof is based on upper bounding the second eigenvalue of the random walk on $G_{n,m,p}$ through coupling of the original Markov Chain describing the random walk to another Markov Chain on an associated random bipartite graph whose conductance properties are appropriate. Finally, the authors prove that the *cover time* of the random walk on $G_{n,m,p}$, when $m = n^{\alpha}, \alpha < 1$ and $p$ is at least 5 times the connectivity threshold is $\Theta(n \log n)$, which is optimal up to a constant. The proof is based on a general theorem of Cooper and Frieze [32]; the authors prove that the degree and spectrum requirements of the theorem hold whp in the case of uniform random intersection graphs. The authors also claim that their proof also carries over to the case of smaller values for $p$, but the technical difficulty for proving the degree requirements of the theorem of Reference 32 increases.

## References

1. P. Leone, L. Moraru, O. Powell and J.D.P. Rolim. Localization algorithm for wireless ad-hoc sensor networks with traffic overhead minimization by emission inhibition. In *Proceedings of the 2nd International Workshop on Algorithmic Aspects of Wireless Sensor Networks (ALGOSENSORS)*, S. Nikoletseas and J.D.P. Rolim (Eds.), LNCS 4240, Springer-Verlag, Berlin, Germany, pp. 119–129, 2006.
2. I. Caragiannis, C. Kaklamanis, E. Kranakis, D. Krizanc and A. Wiese. Communication in wireless networks with directional antennas. In *Proceedings of the 20th Annual ACM Symposium on Parallelism in Algorithms and Architectures (SPAA)*, ACM, New York, pp. 344–351, 2008.
3. C. Busch, M. Magdon-Ismail, F. Sivrikaya and B. Yener. Contention-free MAC protocols for asynchronous wireless sensor networks. *Distributed Computing* 21(1), 23–42, 2008.

---

[*] A *perfect graph* is a graph in which the chromatic number of every induced subgraph equals the size of the largest clique of that subgraph. Consequently, the clique number of a perfect graph is equal to its chromatic number.

[*] More precisely, if $\mathcal{B}$ is the set of different label choices that can give rise to a graph $G$, then the problem of inferring the complete information of label choices from $G$ is *solvable* if there is some $B^* \in \mathcal{B}$ such that $\Pr(B^*|G) > \Pr(B|G)$, for all $\mathcal{B} \ni B \neq B^*$.

4. M.R. Garey, D.S. Johnson and L. Stockmeyer. Some Simplified NP-Complete Problems. In *Proceedings of the Sixth Annual ACM Symposium on Theory of Computing*, pp. 47–63, 1974. doi:10.1145/800119.803884.

5. M.M. Halldórsson. A still better performance guarantee for approximate graph coloring. *Information Processing Letters* 45, 19–23, 1993. doi:10.1016/0020-0190(93)90246-6.

6. D. Zuckerman. Linear degree extractors and the inapproximability of Max Clique and Chromatic Number. *Theory of Computing* 3, 103–128, 2007. doi:10.4086/toc.2007.v003a006.

7. B. Bollobás. The chromatic number of random graphs. *Combinatorica* 8(1), 49–55, 1988.

8. T. Łuczak: The chromatic number of random graphs. *Combinatorica* 11(1), 45–54, 2005.

9. S. Dolev, S. Gilbert, R. Guerraoui and C.C. Newport. Secure communication over radio channels. In *Proceedings of the ACM Symposium on Principles of Distributed Computing (PODC)*, ACM, New York, pp. 105–114, 2008.

10. S. Dolev, L. Lahiani and M. Yung. Secret swarm unit: Reactive $k$-secret sharing. *Ad Hoc Networks* 10, 1291–1305, 2012.

11. M. Karoński, E.R. Sheinerman and K.B. Singer-Cohen. On random intersection graphs: The subgraph problem. *Combinatorics, Probability and Computing Journal* 8, 131–159, 1999.

12. K.B. Singer-Cohen. Random intersection graphs. PhD thesis, John Hopkins University, 1995.

13. S. Nikoletseas, C. Raptopoulos and P.G. Spirakis. Large independent sets in general random intersection graphs. *Theoretical Computer Science* 406, 215–224, 2008.

14. E. Godehardt and J. Jaworski. Two models of random intersection graphs for classification. In *Studies in Classification, Data Analysis and Knowledge Organization*, O. Opitz and M. Schwaiger (Eds.), Springer Verlag, Berlin, Germany, pp. 67–82, 2002.

15. J.A. Fill, E.R. Sheinerman and K.B. Singer-Cohen. Random intersection graphs when $m = \omega(n)$: an equivalence theorem relating the evolution of the $G(n, m, p)$ and $G(n, p)$ models. *Random Structures & Algorithms* 16(2), 156–176, 2000.

16. K. Rybarczyk. Equivalence of a random intersection graph and $G(n, p)$. *Random Structures and Algorithms* 38(1–2), 205–234, 2011.

17. D. Stark. The vertex degree distribution of random intersection graphs. *Random Structures & Algorithms* 24(3), 249–258, 2004.

18. Y. Shang. On the isolated vertices and connectivity in random intersection graphs. *International Journal of Combinatorics* 2011, 2011, Article ID 872703. doi:10.1155/2011/872703.

19. M. Behrisch, A. Taraz and M. Ueckerdt. Coloring random intersection graphs and complex networks. *SIAM Journal on Discrete Mathematics* 23, 288–299, 2008.

20. S. Nikoletseas, C. Raptopoulos and P. Spirakis. Colouring non-sparse random intersection graphs. In *Proceedings of the 34th International Symposium on Mathematical Foundations of Computer Science (MFCS)*, LNCS 5734, Springer-Verlag, Berlin, Germany, pp. 600–611, 2009.

21. A. Frieze. On the independence number of random graphs. *Discrete Mathematics* 81, 171–175, 1990.

22. S.M. Ross. *Stochastic Processes*, 2nd ed. John Wiley & Sons, New York, 2000.

23. N. Alon and J. Spencer, *The Probabilistic Method*. John Wiley & Sons, New York, 2000.

24. S. Nikoletseas, C. Raptopoulos and P. Spirakis. On the independence number and hamiltonicity of uniform random intersection graphs. *Theoretical Computer Science* 412, 6750–6760, 2011.

25. P. Erdős and J. Selfridge. On a combinatorial game. *Journal of Combinatorial Theory A* 14, 298–301, 1973.

26. M. Molloy and B. Reed. *Graph Colouring and the Probabilistic Method*. Springer, New York, 2002.

27. M. Bloznelis, E. Godehardt, J. Jaworski, V. Kurauskas and K. Rybarczyk. Recent progress in complex network analysis—Models of random intersection graphs. In *Data Science, Learning by Latent Structures, and Knowledge Discovery*, B. Lausen, S. Krolak-Schwerdt, and M. Bhmer (Eds.), Springer, Berlin, Germany, pp. 59–68, 2015.

28. C. Raptopoulos and P. Spirakis. Simple and efficient greedy algorithms for Hamilton cycles in random intersection graphs. In *Proceedings of the 16th International Symposium on Algorithms and Computation (ISAAC)*, X. Deng and D. Du (Eds.), LNCS 3827, Springer-Verlag, Berlin, Germany, pp. 493–504, 2005.

29. C. Efthymiou and P.G. Spirakis. Sharp thresholds for Hamiltonicity in random intersection graphs. *Theoretical Computer Science* 411(40–42), 3714–3730, 2010.

30. S. Nikoletseas, C. Raptopoulos and P.G. Spirakis. Maximum cliques in graphs with small intersection number and random intersection graphs. In *Proceedings of the 37th International Symposium on Mathematical Foundations of Computer Science (MFCS)*, B. Rovan, V. Sassone, and P. Widmayer (Eds.), LNCS 7464, Springer-Verlag, Berlin, Germany, pp. 728–739, 2012.

31. S. Nikoletseas, C. Raptopoulos and P.G. Spirakis. Expander properties and the cover time of random intersection graphs. *Theoretical Computer Science* 410(50), 5261–5272, 2009.

32. C. Cooper and A. Frieze. The cover time of sparse random graphs. *Random Structures and Algorithms* 30, 1–16, 2007.

# Approximation Algorithms for Facility Dispersion

## 19.1 Introduction

Many problems in location theory deal with the placement of facilities on a network so as to minimize some function of the distances between facilities or between facilities and other nodes of the network [1]. Such problems model the placement of "desirable" facilities such as warehouses, hospitals, and fire stations. However, there are situations in which facilities must be located so as to *maximize* a given function of the distances. Such location problems are referred to as **dispersion** problems [2] as they model situations in which proximity of facilities is undesirable. One example of such a situation involves placing "obnoxious" (also called "undesirable") facilities such as nuclear power plants, oil storage tanks, and ammunition dumps [2,3]. These facilities need to be spread out to the greatest possible extent so that an accident at one of the facilities does not damage any of the others. Another example in which dispersion problems arise is in the distribution of business franchises in a city [2,4]. In this case, separation of business units is desirable to minimize the competition for customers among the units. In these examples, the facilities to be dispersed are assumed to be of the same type. Applications involving multiple types of facilities (e.g., incinerators, landfills) have also been considered in the literature [3].

* Supported by NSF Grant CCR-97-34936.
† Supported by NSF Grant CCR-01-05536.

The concept of dispersion is also useful in the context of multi objective decision-making [5]. When the number of nondominated[*] solutions is large, a decision-maker may be interested in selecting a manageable collection of solutions that are dispersed as far as possible with respect to the various objectives. Identifying such a collection is useful in obtaining an understanding of the range of available alternatives. The concept of dispersion has also been used in statistics in the context of experimental design [6,7].

Dispersion problems can be formulated under a variety of maximization objectives. Many of these optimization problems are **NP**-hard. So, it is unlikely that optimal solutions to these problems can be obtained efficiently. The practical importance of these problems motivates the study of efficient approximation algorithms that provide near-optimal solutions. In this chapter, we present some approximation algorithms for several versions of the dispersion problem. This chapter is not a survey on facility dispersion (FD); rather, the focus is on approximation algorithms for which performance guarantee results have been established.

The remainder of this chapter is organized as follows. Section 19.2 provides a general formulation of the dispersion problem and mentions a number of objectives considered in the literature. Section 19.3 presents approximation algorithms for the dispersion problem under several objectives. The discussion also includes a greedy approach that is useful in designing such approximation algorithms. Section 19.4 presents approximation algorithms for capacitated versions of dispersion problems. Section 19.5 discusses several directions for future research.

## 19.2 Analytical Model and Problem Formulation

### 19.2.1 Formulation of Dispersion Problems

Analytical models for the dispersion problem assume that the input consists of a set $V = \{v_1, v_2, \ldots, v_n\}$ of $n$ nodes, a nonnegative distance (also called **edge weight**) between each pair of nodes in $V$, and an integer $p \leq n$. Distances are assumed to be symmetric so that the input can be thought of as an undirected complete graph on $n$ nodes with a nonnegative weight on each edge. The weight of the edge $\{v_i, v_j\}$ $(i \neq j)$ is denoted by $w(v_i, v_j)$. It is assumed that $w(v_i, v_i) = 0$ for $1 \leq i \leq n$. Under these assumptions, a general formulation of the FD problem is as follows.

#### 19.2.1.1 Facility Dispersion

> *Instance:* A complete graph $G(V, E)$, where $V = \{v_1, v_2, \ldots, v_n\}$, a nonnegative distance $w(v_i, v_j)$ for each edge $\{v_i, v_j\}$ in $E$, and an integer $p \leq n$.
>
> *Requirement:* Find a subset $P \subseteq V$, with $|P| = p$, such that a specified objective function $f(P)$ is maximized.

To avoid trivial conditions, we assume throughout this chapter that $p \geq 2$. The chosen placement $P$ induces a complete subgraph of $G$, denoted by $G(P)$. The objective $f(P)$ is a function of the edge distances in $G(P)$. A number of alternatives for $f(P)$ have been considered in the literature (see [3,8–16] and the references cited therein). Some examples of such objective functions are listed in the following.

1. *Max–min dispersion*: Here, the function $f(P)$ is defined to be the smallest edge weight in $G(P)$. Thus, the goal of the corresponding dispersion problem is to maximize the smallest edge weight in $G(P)$. (This is referred to as the **remote edge** problem in [16].)

---

[*] Given two solutions $S_1$ and $S_2$ to a multi objective optimization problem, $S_1$ **dominates** $S_2$ if $S_1$ is no worse than $S_2$ in every objective and $S_1$ is strictly better than $S_2$ in at least one objective. It can be seen that the domination relation is a partial order. Each maximal element of this partial is a **nondominated** solution.

2. *Max–average dispersion*: Here, the function $f(P)$ is defined to be the average edge weight in $G(P)$. As the number of edges in $G(P)$ is $p(p-1)/2$, $f(P)$ is given by

$$f(P) = \frac{2\sum_{\{v_i,v_j\}\in G(P)} w(v_i, v_j)}{p(p-1)}. \tag{19.1}$$

As the quantity $2/[p(p-1)]$ is independent of which nodes are in $P$, maximizing the value of $f(P)$ given by Equation 19.1 is equivalent to maximizing the sum of the edge weights in $G(P)$. (This is referred to as the **remote clique** problem in [16].)

3. *Max-minimum spanning tree (MST) dispersion*: Here, the function $f(P)$ is defined to be the weight of a MST of $G(P)$. A variant of this is the Max-Steiner tree (ST) dispersion measure, where $f(P)$ is the weight of a minimum graph Steiner tree (ST) of $G(P)$. (See [17] for a definition of the graph ST problem.)

4. *Max-traveling salesperson (TSP) dispersion*: Here, the function $f(P)$ is defined to be the weight of a minimum TSP tour of $G(P)$; that is, $f(P)$ is the minimum weight of a simple cycle containing all the nodes in $P$.

5. *Max-matching dispersion*: Here, $f(P)$ is the weight of a minimum perfect matching* in $G(P)$.

6. *Max-star dispersion*: Here, the dispersion objective $f(P)$ is given by

$$f(P) = \min_{v_i \in P} \left\{ \sum_{v_j \in P} w(v_i, v_j) \right\}. \tag{19.2}$$

(This problem is referred to as the **remote star** problem in [16].)

Under all of the above-mentioned objectives, the dispersion problem is known to be **NP**-hard. This motivates the study of approximation algorithms for such problems.

The edge distances specified in an instance of the dispersion problem are said to satisfy the **triangle inequality** when for any three distinct nodes $x$, $y$, and $z$, $w(x, z) \leq w(x, y) + w(y, z)$. We refer to such instances as **metric** instances. Most of the approximation algorithms reported in the literature are for metric instances of dispersion problems. Nonapproximability results are known for some nonmetric instances of dispersion problems [14–16].

Geometric versions of dispersion problems have been considered in the literature under two models, which we refer to as the **discrete placement** and **continuous placement** models, respectively. Under the discrete placement model [13–15,18,19], the node set $V$ consists of $n$ points in an appropriate metric space (e.g., Euclidean space) along with a chosen distance function (e.g., $L_k$ distance metric for some $k \geq 1$), and the placement must be a subset of the points in $V$. Under the continuous placement model, a region in which facilities must be placed is specified geometrically (e.g., a polygonal region possibly containing holes), and each facility may be placed at any point inside or on the boundary of the specified region [20]. In discussing geometric versions, we use the terms "point" and "node" interchangeably.

## 19.2.2 Additional Definitions

The current section contains several definitions that are used throughout this chapter. Definitions of common graph theoretic and algorithmic notions used in this chapter can be found in standard texts such as [21,22].

For a maximization problem, a polynomial time approximation algorithm provides a **performance guarantee** of $\rho \geq 1$, if for each instance of the problem, the solution value produced by the algorithm is at least $1/\rho$ of the optimal solution value. We will also refer to such an algorithm as

---

* See Section 19.2.2 for the definition of perfect matching.

a $\rho$-**approximation algorithm**. A **polynomial time approximation scheme** (PTAS) for a problem is a family of polynomial time algorithms such that for each fixed $\epsilon > 0$, the family contains an algorithm that provides a performance guarantee of $(1 + \epsilon)$.

A **matching** in an undirected graph $G(V, E)$ is a subset $E'$ of edges such that no two edges in $E'$ are incident on the same node. A **maximum matching** is a matching containing the largest number of edges. When the edges of $G$ have weights, a **maximum weight matching** is a matching with the largest total weight. For each $k \geq 1$, a $k$-**matching** is a matching consisting of exactly $k$ edges. When the edges of $G$ have weights, a **maximum weight $k$-matching** (**minimum weight $k$-matching**) is a $k$-matching with the largest (smallest) total weight. A graph $G(V, E)$ has a **perfect matching** if it has a matching of size $\lfloor |V|/2 \rfloor$. When the edges of $G$ have weights, a **maximum weight perfect matching** (**minimum weight perfect matching**) is a perfect matching with the largest (smallest) total weight.

A subset $V'$ of nodes of an undirected graph $G(V, E)$ is an **independent set** if there is no edge between any pair of nodes in $V'$. Given an undirected graph $G(V, E)$ and an integer $J \leq |V|$, the goal of the MAX-IMUM INDEPENDENT SET (MIS) decision problem is to determine whether $G$ has an independent set of size at least $J$. This decision problem is known to be **NP**-complete [17]. When there are nonnegative weights on nodes, one can generalize the problem of finding a MIS to the problem of finding an independent set of maximum weight. Strong nonapproximability results are known for the MIS (optimization) problem for general graphs. In particular, it has been shown [23] that for any $\epsilon > 0$, the MIS problem cannot be approximated to within a factor $O(n^{1-\epsilon})$, unless the complexity classes[*] **NP** and **ZPP** coincide.

In developing approximation algorithms for capacitated dispersion problems (Section 19.4), the class of **unit disk graphs** plays an important role. An undirected graph is a **unit disk graph** if its vertices can be placed in one-to-one correspondence with a set of circles of equal radius in the plane so that two vertices of the graph are joined by an edge if and only if the corresponding circles touch or intersect [25]. The geometric representation for a unit disk graph consists of the radius value and the coordinates of the center of each disk. The recognition problem for unit disk graphs is **NP**-hard [26]. When the geometric representation is not available, the MIS problem for unit disk graphs has a constant factor approximation [27]. However, given the geometric representation, there is a PTAS for the MIS problem for unit disk graphs [28]. As mentioned in [29], this PTAS can be extended in a straightforward manner to the weighted MIS problem for unit disk graphs.

# 19.3 Approximating Dispersion Problems

## 19.3.1 Overview

In the current section, we discuss approximation algorithms with good performance guarantees for dispersion problems under several objectives. We begin with a greedy approach for designing an approximation algorithm for metric instances of the Max–Min dispersion problem. We observe that the greedy approach can also be used to obtain approximation algorithms for other dispersion objectives. Further, we point out that for some objectives, approximation algorithms developed from the greedy framework can be improved using other techniques. Known results for geometric instances of problems are also summarized.

## 19.3.2 Approximation Algorithms for Max–Min Dispersion

### 19.3.2.1 Results for Metric Instances

We abbreviate the Max–Min facility dispersion problem by MMFD. It is shown in [14] that for any $\rho \geq 1$, if there is a polynomial time $\rho$-approximation algorithm for nonmetric instances of the MMFD problem, then the MIS problem can be solved in polynomial time. In other words, for any $\rho \geq 1$, there is no

---

[*] For definitions of complexity classes such as **ZPP** [24].

**Input:** A complete graph $G(V, E)$ with a nonnegative edge weight $w(v_i, v_j)$ for each edge $\{v_i, v_j\} \in E$ and an integer $p \leq |V|$. (The edge weights satisfy the triangle inequality.)

**Output:** A set $P \subseteq V$ such that $|P| = p$. (The goal is to make the smallest edge weight in $G(P)$ close to the optimum value.)

**Algorithm:**

1. Let $P = \emptyset$.

2. Let $v_i$ and $v_j$ be the endpoints of an edge of maximum weight. Add the nodes $v_i$ and $v_j$ to $P$.

3. **while** $(|P| < p)$ **do**

   (a) Find a node $v \in V - P$ such that $\min_{v' \in P}\{w(v, v')\}$ is maximum among the nodes in $V - P$.

   (b) Add $v$ to $P$.

4. Output $P$.

**FIGURE 19.1**   Details of heuristic GMM.

$\rho$-approximation algorithm for nonmetric instances of MMFD, unless $\mathbf{P} = \mathbf{NP}$. Therefore, in the current section, we restrict our attention to metric instances of MMFD, denoted by MMFD:TI, where "TI" is used to indicate that the distances satisfy the triangle inequality.

A greedy heuristic for MMFD:TI, called GMM, is shown in Figure 19.1. This heuristic is similar to "furthest point outside the neighborhood" heuristic, described in [5]. For results regarding the performance guarantee provided by this heuristic [12–14,30]. The presentation in the following is based on [14].

The heuristic begins by initializing $P$ (the set of nodes at which facilities are placed) to contain a pair of nodes in $V$, which are joined by an edge of maximum weight. Subsequently, each iteration of GMM chooses a node $v$ from $V - P$ such that the minimum distance from $v$ to a node in $P$ is the largest among all the nodes in $V - P$. In each step, ties are broken arbitrarily. Heuristic GMM terminates when $|P| = p$. The solution value of the placement $P$ produced by GMM is equal to $\min_{x,y \in P}\{w(x, y)\}$. Using standard techniques, it can be seen that the running time of the heuristic is $O(n^2)$.

Our next theorem shows that GMM provides a performance guarantee of 2. It will also be shown (Theorem 19.2) that unless $\mathbf{P} = \mathbf{NP}$, no polynomial time heuristic can provide a better performance guarantee.

**Theorem 19.1** *Let I be an instance of MMFD:TI. Let OPT(I) and GMM(I) denote respectively the solution values of an optimal placement and that produced by GMM for the instance I. Then OPT(I)/GMM(I) $\leq$ 2.*

*Proof.* Consider the set-valued variable $P$ in the description of heuristic GMM (Figure 19.1). Let $f(P) = \min_{x,y \in P}\{w(x, y)\}$. We will show by induction that the condition

$$f(P) \geq \text{OPT}(I)/2 \tag{19.3}$$

holds after each addition to $P$. As GMM$(I) = f(P)$ after the last addition to $P$, the theorem would follow.

As the first addition inserts two nodes joined by an edge of largest weight into $P$, Condition (19.3) clearly holds after the first addition. So, assume that the condition holds after $k$ additions to $P$, for some $k$, $1 \leq k < p - 1$. We will prove that the condition holds after the $(k + 1)$st addition to $P$ as well.

To that end, let $P^* = \{v_1^*, v_2^*, \ldots, v_p^*\}$ denote an optimal placement. For convenience, we use $\ell^*$ for OPT($I$). The following observation is an immediate consequence of the definition of the solution value corresponding to a placement for an MMFD instance.

**Observation 19.1** *For every pair $v_i^*$, $v_j^*$ of distinct nodes in $P^*$, $w(v_i^*, v_j^*) \geq \ell^*$.* ∎

Let $P_k = \{x_1, x_2, \ldots, x_{k+1}\}$ denote the set $P$ after $k$ additions, $1 \leq k \leq p - 1$. (Note that $|P_k| = k + 1$, as the first addition inserts two nodes into $P$). As GMM adds at least one more node to $P$, the following is a trivial observation.

**Observation 19.2** *For $1 \leq k < p - 1$, $|P_k| = k + 1 < p$.* ∎

For each $v_i^* \in P^*$ ($1 \leq i \leq p$), define $S_i^* = \{u \in V \mid w(v_i^*, u) < \ell^*/2\}$. The following claim provides two useful properties of these sets.

**Claim 19.1** *(a) For $1 \leq i \leq p$, $S_i^*$ is nonempty. (b) For $i \neq j$, $S_i^*$ and $S_j^*$ are disjoint.*

**Proof of Claim:**  Part (a) is obvious, as $v_i^* \in S_i^*$ for $1 \leq i \leq p$. To prove Part (b), suppose $S_i^* \cap S_j^* \neq \emptyset$ for some $i \neq j$. Let $u \in S_i^* \cap S_j^*$. Thus, $w(v_i^*, u) < \ell^*/2$ and $w(v_j^*, u) < \ell^*/2$. By Observation 19.1, $w(v_i^*, v_j^*) \geq \ell^*$. These three inequalities together imply that the triangle inequality does not hold for the three nodes $u$, $v_i^*$, and $v_j^*$. Part (b) follows. ∎

We now continue with the main proof. As for $k < p - 1$, $P_k$ has *less than $p$* nodes (Observation 19.2), and there are $p$ disjoint sets $S_1^*$, $S_2^*$, $\ldots$, $S_p^*$, there must be at least one set, say $S_r^*$ (for some $r$, $1 \leq r \leq p$), such that $P_k \cap S_r^* = \emptyset$. Therefore, by the definition of $S_r^*$, we must have for each $u \in P_k$, $w(v_r^*, u) \geq \ell^*/2$. As $v_r^*$ is available for selection by GMM, and GMM selects a node $v \in V - P_k$ for which $\min\limits_{v' \in P_k} w(v, v')$ is a maximum among the nodes in $V - P_k$, it follows that Condition (19.3) holds after the $(k + 1)$st addition to $P$. This completes the proof of Theorem 19.1. ∎

The next theorem provides a lower bound on the obtainable performance guarantee for MMFD:TI.

**Theorem 19.2** *If $\mathbf{P} \neq \mathbf{NP}$, no polynomial time approximation algorithm can provide a performance guarantee of $(2 - \epsilon)$ for any $\epsilon > 0$ for MMFD:TI.*

*Proof.* We use a reduction from the MIS problem. Given an instance of the MIS problem consisting of graph $G(V, E)$ and an integer $J$, construct an instance of MMFD:TI as follows. The nodes of the MMFD:TI instance are in one-to-one correspondence with the nodes in $V$. For each edge $\{v_i, v_j\}$ in $E$, the weight $w(v_i, v_j)$ is chosen as 1. All the other weights are chosen as 2. Obviously, the resulting distances satisfy the triangle inequality. The number $p$ of facilities to be chosen is set to $J$. If $G$ contains an independent set $V'$ of size at least $J$, then placing the facilities on the nodes of in $V'$ provides a solution value of 2 for the MMFD:TI instance; otherwise, every placement has a solution value of 1. Therefore, any polynomial time approximation algorithm $\mathcal{A}$ with a performance guarantee of $2 - \epsilon$ for some $\epsilon > 0$, for MMFD:TI will output a solution value of 2 if and only if $G$ contains an independent set of size at least $J$. Thus, we obtain a polynomial time algorithm for the MIS problem, contradicting the assumption that $\mathbf{P} \neq \mathbf{NP}$. ∎

Theorems 19.1 and 19.2 indicate that the simple greedy algorithm of Figure 19.1 provides the best performance guarantee for the MMFD:TI problem that one can hope to obtain in polynomial time. A variant of this algorithm which can be used to approximate a more general form of MMFD, where facilities may be placed on nodes or edges, is analyzed in [12].

### 19.3.2.2 Results for Geometric Versions

Geometric versions of the MMFD problem have also been considered in the literature. We will first mention the results under the discrete placement model. The 1-dimensional version of the problem (in which all the points are on a line) can be solved in $O(pn + n \log n)$ time [18]. When the points are in two (or higher)-dimensional Euclidean space, the problem is **NP**-hard [18,31]. As the triangle inequality holds for all geometric instances, the greedy algorithm of Figure 19.1 provides a performance guarantee of 2 for geometric instances of the MMFD problem under the discrete placement model.

Baur and Fekete [20] considered the MMFD problem and several of its variants under the continuous placement model, in which the region is specified by a rectilinear polygon, and the distance metric is $L_1$ or $L_\infty$. If the specified region is not required to be connected, they show that the MMFD problem cannot be approximated to within the factor $2 - \epsilon$ for any $\epsilon > 0$, unless **P** $=$ **NP**. For a more general version of the problem, in which the goal is to maximize the minimum distance between each pair of chosen points as well as the closest distance between a chosen point and the boundary of the region, they present an approximation algorithm with a performance guarantee of $3/2$. They also show that the problem cannot be approximated to within the factor $14/13$, unless **P** $=$ **NP**. Additional approximation results for dispersional packing problems (e.g., finding the largest value of $L$ such that a specified number of $L \times L$ squares can be packed in a given region) are also presented in [20].

## 19.3.3 Approximation Algorithms for Max–Average Dispersion

### 19.3.3.1 Results for Metric Instances

Recall that in the Max–Average dispersion problem, the goal is to maximize the average distance between a pair of chosen facilities. We use Max-Average facility dispersion (MAFD) to denote this problem and MAFD:TI to denote its restriction to metric instances. We discuss approximation results for MAFD:TI in this section. Subsequent sections consider nonmetric and geometric instances of MAFD.

Reference [14] presents a modified version of the greedy approach used in Figure 19.1 to obtain an approximation algorithm for MAFD:TI. The modification is to replace Step 3(a) in that figure by the following: Find a node $v \in V - P$ such that the average distance from $v$ to the nodes in $P$ is a maximum among all the nodes in $V - P$. An inductive argument is used in [14] to show that this modification yields an approximation algorithm with a performance guarantee of 4. Birnbaum and Goldman [32] present a more sophisticated analysis of the same heuristic to show that it actually provides a performance guarantee of 2. This analysis is based on the idea of factor-revealing linear programs introduced in [33]. Hassin et al. [34] present two approximation algorithms for MAFD:TI. Each of these algorithms provides a performance guarantee of 2. We will discuss one of their approximation algorithms in the following as it is also used in [16] for approximating another dispersion problem.

The details of the heuristic from [34], which we call Heuristic-HRT, are shown in Figure 19.2. The key step in the heuristic is the computation of a maximum weight $\lfloor p/2 \rfloor$-matching in $G$ (Step 2). Using the fact that the problem of finding a maximum weighted matching in any graph can be solved in polynomial time [35], it is shown in [34] that the problem of computing a maximum weight $\lfloor p/2 \rfloor$-matching can also be solved efficiently. Thus, Heuristic-HRT runs in polynomial time. A careful implementation of the heuristic which leads to a running time of $O(n^2(p + \log n))$ is presented in [34]. We now prove the performance guarantee provided by the heuristic, closely following the presentation in [34].

The following notation is used in the proof. As before, for any nonempty node subset $V_1$, $G(V_1)$ denotes the complete subgraph of $G$ induced on $V_1$. Further, $W(V_1)$ and $\overline{W}(V_1)$ denote respectively the total weight and the average weight of the edges in the complete subgraph $G(V_1)$. Similarly, for any nonempty subset $E_1$ of edges, $W(E_1)$ and $\overline{W}(E_1) = W(E_1)/|E_1|$ denote respectively the total and average weight of the edges in $E_1$. The following lemma establishes a property of maximum weight $\lfloor p/2 \rfloor$-matchings.

**Input:** A complete graph $G(V, E)$ with a nonnegative edge weight $w(v_i, v_j)$ for each edge $\{v_i, v_j\} \in E$ and an integer $p \le |V|$. (The edge weights satisfy the triangle inequality.)

**Output:** A set $P \subseteq V$ such that $|P| = p$. (The goal is to make the average edge weight in $G(P)$ close to the optimum value.)

**Algorithm:**

1. Let $P = \emptyset$.

2. Compute a $\lfloor p/2 \rfloor$-matching $M^*$ of maximum weight in $G$.

3. Add both end points of the edges in $M^*$ to $P$.

4. If $p$ is odd, add an arbitrary node from $V-P$ to $P$.

5. Output $P$.

**FIGURE 19.2**    Details of heuristic-HRT.

**Lemma 19.1** *Let $V_1$ be a subset with at least $p \ge 2$ nodes and let $M_1^*$ denote a maximum weight $\lfloor p/2 \rfloor$-matching in $G(V_1)$. Then, $\overline{W}(V_1) \le \overline{W}(M_1^*)$.*

*Proof.* Let $q$ denote the number of $\lfloor p/2 \rfloor$-matchings of $G(V_1)$, and let $M_1, M_2, \ldots, M_q$ denote the matchings themselves. Let $A_1 = \sum_{i=1}^{q} W(M_i)/q$ denote the average of the $q$ values $W(M_1), W(M_2), \ldots, W(M_q)$. We will show that $A_1 = \lfloor p/2 \rfloor \overline{W}(V_1)$. As $W(M_1^*) \ge A_1$, the lemma would then follow.

Consider the summation $\sum_{i=1}^{q} W(M_i)/q$. The number of edge weights included in this summation is $q\lfloor p/2 \rfloor$, as $|M_i| = \lfloor p/2 \rfloor$, for $1 \le i \le q$. By symmetry, the weight of each edge of $G(V_1)$ occurs the same number of times, say $t$, in the summation. Therefore, the number of occurrences of edge weights in the summation is also equal to $t|V_1|(|V_1| - 1)/2$. Hence,

$$q \lfloor p/2 \rfloor \;=\; t|V_1|(|V_1| - 1)/2. \tag{19.4}$$

Now,

$$A_1 \;=\; \sum_{i=1}^{q} W(M_i)/q \;=\; t\, W(V_1)/q \;=\; \frac{t\,\overline{W}(V_1)\,|V_1|\,(|V_1| - 1)}{2q}. \tag{19.5}$$

Now, using Equation 19.4, we get $A_1 = \lfloor p/2 \rfloor \overline{W}(V_1)$ as desired.   ■

The next lemma, which relies on the triangle inequality, shows a relationship between the average edge weight of a complete subgraph and the average weight of any $\lfloor p/2 \rfloor$-matching in the subgraph.

**Lemma 19.2** *Let $V_1$ be a subset containing $p \ge 2$ nodes, and let $M$ be any $\lfloor p/2 \rfloor$-matching in $G(V_1)$. Then, $\overline{W}(V_1) > \overline{W}(M)/2$.*

*Proof.* Let $M = \{\{a_i, b_i\} \;:\; 1 \le i \le \lfloor p/2 \rfloor\}$, and let $V_M$ denote the set of nodes that are endpoints of the edges in $M$. Consider each edge $\{a_i, b_i\}$ in $M$, and let $E_i$ denote the set of edges in $G(V_1)$ incident on $a_i$ or $b_i$, except for the edge $\{a_i, b_i\}$ itself. By the triangle inequality, for any node $v \in V_M - \{a_i, b_i\}$, we have

$w(v, a_i) + w(v, b_i) \geq w(a_i, b_i)$. When this inequality is summed up over all the nodes in $V_M - \{a_i, b_i\}$, we get

$$W(E_i) \geq (p - 2) w(a_i, b_i). \tag{19.6}$$

Now, we consider two cases.

*Case 1:* Let $p$ be even. So, $\lfloor p/2 \rfloor = p/2$.

Consider the summation of Inequality (19.6) over all the edge sets $E_i$, $1 \leq i \leq p/2$. On the left side of that summation, each edge of $G(V_1)$, except for those in $M$, appears twice. Therefore, $2[W(V_1) - W(M)] \geq (p - 2) W(M)$. Expressing this inequality in terms of $\overline{W}(V_1)$ and $\overline{W}(M)$, we get $\overline{W}(V_1) \geq p \overline{W}(M)/[2(p - 1)]$; that is, $\overline{W}(V_1) > \overline{W}(M)/2$. This completes the proof for Case 1.

*Case 2:* Let $p$ be odd. So, $\lfloor p/2 \rfloor = (p - 1)/2$.

Let $x$ be the node in $V_1 - V_M$, and let $E_x$ denote the set of edges incident on $x$ in $G(V_1)$. By the triangle inequality, we have

$$W(E_x) \geq W(M). \tag{19.7}$$

Again, consider the summation of Inequality (19.6) over the edge sets $E_i$, $1 \leq i \leq \lfloor p/2 \rfloor$. On the left side of that summation, each edge of $G(V_1)$ appears twice, except that the edges in $M$ don't appear at all and the edges in $E_x$ appear only once. Therefore, $2[W(V_1) - W(M)] - W(E_x) \geq (p - 2) W(M)$. Using Inequality (19.7), we get $2[W(V_1) - W(M)] \geq (p - 1) W(M)$. Again, expressing this inequality in terms of $\overline{W}(V_1)$ and $\overline{W}(M)$, and using the fact that $\lfloor p/2 \rfloor = (p - 1)/2$, we get $\overline{W}(V_1) \geq (p + 1) \overline{W}(M)/(2p)$. Consequently, $\overline{W}(V_1) > \overline{W}(M)/2$. This completes the proof for Case 2 and also that of the lemma. ∎

We are now ready to establish the performance guarantee of Heuristic-HRT.

**Theorem 19.3** *Let $I$ be an instance of MAFD:TI. Let $OPT(I)$ and $HRT(I)$ denote the value of an optimal solution and that of the solution produced by Heuristic-HRT, respectively. Then, $OPT(I)/HRT(I) < 2$.*

*Proof.* Let $P^*$ and $P$ denote respectively the set of nodes in an optimal solution and that in the solution produced by Heuristic-HRT for the instance $I$. By definition, $OPT(I) = \overline{W}(P^*)$ and $HRT(I) = \overline{W}(P)$. Let $M^*$ and $M$ denote respectively a maximum weight $\lfloor p/2 \rfloor$-matching in $P^*$ and $P$. By Lemma 19.1, we have

$$OPT(I) \leq \overline{W}(M^*). \tag{19.8}$$

Further, by Lemma 19.2, we have

$$HRT(I) > \overline{W}(M)/2. \tag{19.9}$$

As Heuristic-HRT computes a maximum weight $\lfloor p/2 \rfloor$-matching for the graph $G$, we have $\overline{W}(M) \geq \overline{W}(M^*)$. This fact in conjunction with Inequalities (19.8) and (19.9) yields the theorem. ∎

A lower bound example is presented in [34] to show that the performance guarantee of Heuristic-HRT can be made arbitrarily close to 2.

### 19.3.3.2 Results for Nonmetric Instances

Here, we briefly mention the known results for nonmetric instances of the MAFD problem. When the weight of every edge in the complete graph $G$ (which is part of the MAFD problem instance) is 0 or 1, the input can be thought of as a graph $G'(V, E')$, where $E'$ contains only the edges of weight 1 in $G$. Then, the MAFD problem corresponds to the following problem: select a subset $V'$ of $p$ nodes such that the number of edges in the subgraph of $G'$ induced on $V'$ is a maximum over all induced subgraphs on $p$ nodes. This problem is called the **Dense $p$-Subgraph** (D$p$S) problem in the literature. Feige et al. [36]

present an approximation algorithm with a performance guarantee of $O(n^\delta)$, where $\delta < 1/3$, for the problem. They also show that the algorithm can be modified to obtain a performance guarantee of $O(n^\delta \log n)$ for a more general version of the problem, where the edges in $G'$ have nonnegative edge weights and the goal is to choose a subset of $p$ nodes so that the weight of all the edges in the chosen induced subgraph is maximized. For any fixed $\epsilon > 0$, Bhaskara et al. [37] present an approximation algorithm that runs in $O(n^{1/\epsilon})$ time and provides a performance guarantee of $O(n^{1/4+\epsilon})$. When $p = \Omega(n)$, Asahiro et al. [38] show that there is an approximation algorithm with a constant performance guarantee for the DpS problem. When the graph $G'$ is dense (i.e., the number of edges in $G' = \Omega(n^2)$) and $p = \Omega(n)$, there is a PTAS for the DpS problem [39]. When the edge weights in the input graph are from the set $\{1, 2, \ldots, K\}$, for some fixed positive integer $K$ and the number of edges in an optimal solution is $\Omega(n^2)$, Czygrinow [40] presents a faster PTAS for the weighted version of the DpS problem. Feige [41] shows the existence of a constant $\rho > 1$ such that if there is a polynomial time $\rho$-approximation algorithm for the DpS problem, then a natural conjecture regarding the average case hardness of the Boolean Satisfiability problem would be violated. Khot [42] rules out the existence of a PTAS for the DpS problem under a different complexity theoretic assumption.

### 19.3.3.3 Geometric Versions

Geometric versions of the MAFD problem seem to have been considered only under the discrete placement model. The approximation algorithm for the MAFD:TI problem presented in Section 19.3.3.1 provides a performance guarantee of 2 for geometric versions. Better performance guarantee results have been obtained by using ideas from computational geometry. In [14,30], it is shown that the 1-dimensional version of the MAFD problem can be solved efficiently. Using this result, an approximation algorithm for the 2-dimensional MAFD problem under Euclidean distances was presented in [14]. The basic idea of the approximation algorithm is to create a polynomial number of instances of the 1-dimensional MAFD problem by projecting the given set of points onto an appropriate set of lines, solve each 1-dimensional problem optimally and choose the best of these solutions. It is shown that this scheme provides an asymptotic performance guarantee* of $\pi/2 \approx 1.571$ for the 2-dimensional MAFD problem.

Fekete and Meijer [19] consider the MAFD problem under rectilinear distances for points in $d$-dimensional space, for any fixed $d \geq 2$. When $p$, the number of points to be selected, is fixed, they show that the MAFD problem can be solved in $O(n)$ time. When $p$ is part of the problem instance, they present a PTAS for the problem. By appropriately modifying this PTAS, they show that an approximation algorithm with an asymptotic performance guarantee of $\sqrt{2}$ can be obtained for the case of Euclidean distances in 2-dimensional space, thus improving the asymptotic bound of $\pi/2$ in [14].

## 19.3.4 Results for Other Dispersion Problems

Approximation results for a number of other dispersion objectives have been reported in the literature [15,16]. In particular, results for objectives Max-MST, Max-ST, and Max-TSP are presented in [15], whereas results for Max-Matching and Max-Star are presented in [16]. In the following, we summarize these results.

Halldórsson et al. [15] consider dispersion problems under three objectives, namely, minimum weight spanning tree, minimum TSP tour length, and minimum ST. Since the corresponding objective must be maximized in each case, we abbreviate these problems as Max-MST, Max-TSP, and Max-ST respectively. Under the assumption that $\mathbf{P} \neq \mathbf{NP}$, they also show lower bounds of 2, 2, and 4/3, respectively, on the performance guarantees obtainable in polynomial time for the three problems. As constructing a near-minimum ST is done using shortest path distances between nodes, the greedy approach provides a

---

* An algorithm provides an **asymptotic** performance guarantee of $\rho$, if for any fixed $\epsilon > 0$, the performance guarantee of the algorithm is at most $\rho + \epsilon$.

performance guarantee of 3 for the Max-ST dispersion problem even for nonmetric instances. However, nonmetric instances of Max-MST and Max-TSP are shown to be at least as hard to approximate as the MIS problem. In [15], geometric versions of dispersion problems under the discrete placement model are also considered. For 2-dimensional versions of Max-MST and Max-ST, in which the distance between two points is the Euclidean distance, they present approximation algorithms with performance guarantees of 2.25 and 2.16, respectively.

In [16], an approximation algorithm with a performance guarantee of $O(\log p)$ for the metric instances of the Max-Matching problem. They obtain the algorithm by suitably modifying the greedy approach (used in Section 19.3.2.1). For metric instances of the Max-Star problem, they show that the HRT-heuristic [34] (discussed in Section 19.3.3.1) provides a performance guarantee of 2. They also discuss other applications of the HRT-heuristic.

Further, Chandra and Halldórsson [16] carry out a more detailed study of the greedy approach for approximating dispersion problems. They show that for many dispersion objectives, the ratio of the optimal solution value to the solution value produced by the greedy approach may be arbitrarily large. They also show that the greedy approach is useful for a more general version of the Max-MST, Max-TSP, and Max-ST dispersion problems. In this general version, an integer $k \leq p$ is also specified as part of the input. After an algorithm chooses a subset of $p$ nodes, an adversary partitions the chosen subset into $k$ subsets, and the value of the objective is the sum of the weights of the corresponding minimum structure (MST, minimum TSP tour or minimum ST) on the $k$ subgraphs. For the generalized version of the Max-MST problem, it is shown that the greedy approach provides a performance guarantee of $4 - 2/(p - k + 1)$. For the generalized versions of Max-TSP and Max-ST problems, the greedy approach is shown to provide a performance guarantee of $\min\{5, 2p/(p - k)\}$.

# 19.4 Approximation Algorithms for Capacitated Versions

## 19.4.1 Motivation and Problem Definition

The model of dispersion considered in previous sections does not explicitly consider the storage capacity of a node. In general, different nodes may have different capacities. This practical aspect adds a new dimension and leads to **capacitated dispersion problems** [29]. These are typically constrained optimization problems in which the constraints or the optimization objectives involve capacities of nodes. One example of such a problem is the following: Choose a subset of nodes so that the minimum distance between any pair of chosen nodes is at least a specified threshold, and the total storage capacity of the chosen nodes is a maximum over all subsets that satisfy the distance constraint. The focus of this section is on approximation algorithms for such capacitated dispersion problems.

The following notation is used throughout this section. For a subset of nodes $V'$, CAP($V'$) denotes the sum of the storage capacities of the nodes in $V'$. For any subset of nodes $V'$ with $|V'| \geq 2$, DIST($V'$) denotes the *minimum* distance between a pair of nodes in $V'$; this is the Max–Min dispersion measure for the set $V'$.

Formulations of some capacitated dispersion problems are given in the following. In naming these problems, we use the following convention. Each problem has a maximization objective and one or more constrained measures. The maximization objective is indicated by the prefix "Max", and each constrained measure is preceded by a slash ("/"). For example, in the Max-Cap/Dist problem, the objective is to maximize the total capacity and the constraint is on the inter node distance. A precise formulation of this problem is as follows.

### 19.4.1.1 Maximizing Capacity under a Distance Constraint (Max-Cap/Dist)

*Instance*: A complete graph $G(V, E)$ with an edge weight $w(v_i, v_j)$ for each edge $\{v_i, v_j\}$ in $E$, a storage capacity $c_i$ (rational number) for each node $v_i \in V$ and a positive rational number $\alpha$.

*Requirement*: Select a subset $V' \subseteq V$ so that $CAP(V')$ is a maximum over all subsets $V'$ that satisfy the constraint $DIST(V') \geq \alpha$.

The dual* version of Max-Cap/Dist, where a required capacity value $B$ is given and the goal is to select a subset $V'$ of nodes so that $DIST(V')$ is a maximum among all subsets satisfying the constraint $CAP(V') \geq B$, is denoted by Max-Dist/Cap. We will use Max-Cap/Dist:TI (Max-Dist/Cap:TI) to denote the restriction of Max-Cap/Dist (Max-Dist/Cap) to metric instances.

Geometric versions of the capacitated dispersion problems seem to have been considered only for the discrete placement model [29]. In our notation, the geometric version of a problem will be specified with the appropriate dimension as the prefix. For example, the one- and two-dimensional versions of the Max-Cap/Dist problem will be denoted by 1D:Max-Cap/Dist and 2D:Max-Cap/Dist, respectively. In all problems involving a capacity constraint, we assume without loss of generality that no single node has a large enough capacity to satisfy the constraint.

## 19.4.2 Approximation Results for Capacitated Dispersion

### 19.4.2.1 A Nonapproximability Result for Maximizing Total Capacity

The Max-Cap/Dist:TI problem is at least as hard to approximate as the MIS problem. This can be seen as follows. Given an instance of the MIS problem represented by an undirected graph $G(V, E)$, construct a Max-Cap/Dist:TI instance as follows. Treat $V$ as the set of nodes, each with a capacity of one unit. For any pair of nodes $u$ and $v$, let $w(u, v) = 1$ if $\{u, v\}$ is an edge in $G$ and let $w(u, v) = 2$ otherwise. Obviously, the resulting distances satisfy the triangle inequality. Let the minimum distance constraint be set to 2. It can be verified that any feasible solution of capacity $\gamma$ to the constructed Max-Cap/Dist:TI instance corresponds to an independent set of size $\gamma$ in $G$ and vice versa. As a consequence, nonapproximability results for the MIS problem (mentioned in Section 19.2.2) carry over to the Max-Cap/Dist:TI problem. The following proposition provides a formal statement of this result.

**Proposition 19.1** *Unless* **NP = ZPP**, *there is no polynomial time approximation algorithm with a perfor-mance guarantee of $O(n^{1-\epsilon})$ for any $\epsilon > 0$ for the Max-Cap/Dist:TI problem.* ∎

### 19.4.2.2 Results for Maximizing Inter Node Distance

The approximation problem for nonmetric instances of Max-Dist/Cap is at least as hard as that for the nonmetric instances of Max–Min facility dispersion (MMFD). To see this, consider an instance of the MMFD problem with $n$ nodes, and let $p$ denote the number of facilities to be placed. Construct an instance of the Max-Dist/Cap problem as follows. The two instances have the set of nodes and inter node distances. The capacity of each node is set to 1, and the capacity requirement is set to $p$. Now, it is straightforward to verify that any solution to the MMFD instance with a Max–Min objective value of $\alpha$ is a feasible solution with the same objective value for the Max-Dist/Cap instance. As a consequence, the known nonapprox-imability result for nonmetric instances of MMFD (mentioned in Section 19.3.2.1) is applicable to the Max-Cap/Dist problem. This result is stated formally in the following.

**Proposition 19.2** *Unless* **P = NP**, *for any $\rho \geq 1$, there is no polynomial time $\rho$-approximation algorithm for nonmetric instances of the Max-Dist/Cap problem.* ∎

We now consider metric instances of the Max-Dist/Cap problem, denoted by Max-Dist/Cap:TI. A fur-ther restricted version of this problem, where all nodes have a capacity of 1, is the MMFD:TI problem considered in Section 19.3.2. Thus, there is a 2-approximation algorithm for this restricted version of

---

* We use the term "dual" in a narrow sense to mean that the maximization objective and constrained measure are interchanged.

**Input:** A complete graph $G(V, E)$ with edge weights satisfying the triangle inequality, a capacity $c_i$ for each node $v_i \in V$ and a capacity requirement $B$.

**Output:** A subset $V'$ of $V$ such that $\text{CAP}(V') \geq B$. (The goal is to make $\text{DIST}(V')$ close to the optimum value.)

**Algorithm:**

1. Sort the nodes in nonincreasing capacity order, and create a list, denoted by `SiteList`.

2. Sort the inter-node distances in nonincreasing order and eliminate duplicate distances. Let the resulting sorted list of distances be stored in the array $D[1 .. t]$ such that $D[1] > D[2] > \cdots > D[t]$.

3. Carry out a binary search over the array $D$ to find the index $i$ such that for $\alpha = D[i]$, the call Greedy($\alpha$, $B$) returns "success" and for $\alpha' = D[i+1]$, the call Greedy($\alpha'$, $B$) returns "failure".

4. Output the inter-node distance $\alpha$ along with the corresponding placement $V'$ found in Step 3.

**procedure** Greedy $(\alpha, \ B)$

(a) Let $L = $ `SiteList` and $V' = \emptyset$.

(b) **while** $L$ is not empty **do**

    (i) Add the first node $v$ from $L$ to $V'$.

    (ii) Remove from $L$ all nodes (including $v$) whose distance to $v$ is strictly less than $\alpha$.

(c) **if** $CAP(V') \geq B$ **then return** "success" and the set $V'$ **else return** "failure".

**FIGURE 19.3** Details of the heuristic for Max-Dist/Cap:TI.

Max-Dist/Cap:TI problem, and this approximation cannot be improved unless **P = NP**. We now show that the more general version of the Max-Dist/Cap:TI problem, in which nodes have arbitrary capacities, can also be approximated to within a factor of 2.

Recall that the specification of the Max-Dist/Cap:TI problem includes a lower bound $B$ on the required capacity. To ensure feasibility, we assume that the sum of the capacities of all the nodes in $G$ is at least $B$. As mentioned earlier, we also assume that no node has a capacity of $B$ or more. Our heuristic for Max-Dist/Cap:TI is shown in Figure 19.3. It is based on a binary search over the inter node distances. For each query distance $\alpha$, the heuristic invokes procedure Greedy (also shown in Figure 19.3) to try to find a subset of nodes $V'$ for which $DIST(V') \geq \alpha$ and $CAP(V') \geq B$. It will be shown (Lemma 19.3) that procedure Greedy returns "success" (and the corresponding placement $V'$) for any inter node distance that is at least half the optimal value.

To establish the performance guarantee provided by the heuristic, we have the following lemma.

**Lemma 19.3** *Let I denote any instance of Max-Dist/Cap:TI problem for which there is a feasible solution. Let $V^*$ denote an optimal set of nodes for the instance I, and let $OPT(I) = DIST(V^*)$. Let $\gamma$ be the smallest inter node distance $\geq OPT(I)/2$. For any inter node distance $\alpha \leq \gamma$, the call Greedy($\alpha$, B) returns "success."*

*Proof.* Let $V^* = \{v_{i_1}, v_{i_2}, \ldots, v_{i_r}\}$ denote the nodes in the optimal solution. Without loss of generality, we assume that $c_{i_1} \geq c_{i_2} \geq \ldots \geq c_{i_r}$. Note that for any two nodes $v_{i_a}$ and $v_{i_b}$ in $V^*$, $w(v_{i_a}, v_{i_b}) \geq OPT(I)$.

Consider the call Greedy($\alpha$, B). Let $v_{j_1}$ be the first node added to $V'$ in this call. As the procedure considers nodes in nonincreasing order of capacities, we have $c_{j_1} \geq c_{i_1}$. We now have the following claim.

**Claim:** $w(v_{j_1}, v_{i_1}) \geq \alpha$ or $w(v_{j_1}, v_{i_2}) \geq \alpha$.

**Proof of Claim:** Suppose the claim is false; that is, $w(v_{j_1}, v_{i_1}) < \alpha$ and $w(v_{j_1}, v_{i_2}) < \alpha$. As $\alpha \leq \gamma$ and $\gamma$ is the smallest inter node distance $\geq OPT(I)/2$, it follows that $w(v_{j_1}, v_{i_1}) < OPT(I)/2$ and $w(v_{j_1}, v_{i_2}) < OPT(I)/2$. Consequently, $w(v_{j_1}, v_{i_1}) + w(v_{j_1}, v_{i_2}) < OPT(I)$. However, as $v_{i_1}$ and $v_{i_2}$ are both in the optimal solution $V^*$, $w(v_{i_1}, v_{i_2}) \geq OPT(I)$. This violates the triangle inequality, and the claim follows.

During the call Greedy($\alpha$, B), after selecting $v_{j_1}$, only those nodes that are at a distance strictly less than $\alpha$ are removed from $L$. Thus, from the above-mentioned claim, at most one of the nodes $v_{i_1}$ and $v_{i_2}$ is eliminated when $v_{j_1}$ is chosen. Consequently, a node of capacity at least $c_{i_2}$ is available for selection during the second iteration.

A straightforward extension of the previous argument shows that for $1 \leq q \leq r$, where $r$ is the number of nodes in the optimal solution $V^*$ under consideration, a node of capacity at least $c_{i_q}$ is available for selection during iteration $q$ of the **while** loop of procedure Greedy. As a consequence, the call Greedy($\alpha$, B) will select a set $V'$ of nodes with $CAP(V') \geq B$ and return "success." ∎

**Theorem 19.4** *Let I denote any instance of Max-Dist/Cap:TI problem for which there is a feasible solution. Let $V^*$ denote an optimal set of nodes for the instance I, and let $OPT(I) = DIST(V^*)$. Let $HEU(I)$ be the distance output by the heuristic shown in Figure 19.3. Then $OPT(I)/HEU(I) \leq 2$.*

*Proof.* Because of the binary search used in the heuristic, the distance value $\alpha$ returned by the heuristic is such that Greedy($\alpha$,B) returns "success," but if $\alpha'$ is the next largest distance after $\alpha$, Greedy($\alpha'$,B) returns "failure." By Lemma 19.3, $\alpha \geq \gamma$, the smallest inter node distance $\geq OPT(I)/2$. ∎

A lower bound example is presented in [29] to show that the bound of 2 established in the above-mentioned theorem is tight. It can be seen that the running time of the heuristic in Figure 19.3 is $O(n^2 \log n)$.

## 19.4.3 Results for Geometric Versions

All the results mentioned in this section are for the discrete placement model. Problems 1D:Max-Cap/Dist and 1D:Max-Dist/Cap can be solved in polynomial time [29] and hence are not discussed further. Using a straightforward reduction from the MIS problem for unit disk graphs, both 2D:Max-Cap/Dist and 2D:Max-Dist/Cap can be shown to be **NP**-complete, even when each node has a capacity of 1.

As mentioned in Section 19.2.2, a PTAS is known for the weighted MIS problem for unit disk graphs when the geometric representation is available as part of the input [28,29]. Using this result, a PTAS for the 2D:Max-Cap/Dist problem can be obtained. This is done by reducing the 2D:Max-Cap/Dist problem to the weighted MIS problem for unit disk graphs in the following manner. Treat each given point in the 2D:Max-Cap/Dist problem as the center of a unit disk with radius = $\alpha/2$, where $\alpha$ is the distance constraint. The weight of each disk is the storage capacity of the corresponding point. Now, it can be seen that any independent set of weight $\gamma$ for the constructed unit disk graph is a feasible solution of capacity $\gamma$ for the 2D:Max-Cap/Dist problem. Therefore, there is a PTAS for the 2D:Max-Cap/Dist problem.

A heuristic with a performance guarantee of 2 for the Max-Dist/Cap:TI was presented in Section 19.4.2. As the inter node distances in any instance of 2D:Max-Dist/Cap satisfy the triangle inequality, it follows that there is a 2-approximation algorithm for the 2D:Max-Dist/Cap problem. The following result summarizes the above-mentioned discussion.

**Proposition 19.3**     *(a) There is a PTAS for the 2D:Max-Cap/Dist problem.*

  *(b) There is a 2-approximation algorithm for the 2D:Max-Dist/Cap problem.*                     ∎

## 19.4.4  Capacitated Dispersion with Storage Costs

In [29], capacitated dispersion problems with storage costs were also considered. Their model assumes that the storage cost at a node is a linear function of the amount of material stored at the node. Such dispersion problems, which involve three measures, namely, capacity, distance, and cost, are formulated by choosing one of these measures as the optimization objective and specifying constraints on the other two measures. The formulation leads to minimization problems when the optimization objective is cost and to maximization problems when the optimization objective is either capacity or distance. Likewise, the constraint on cost specifies an upper bound (budget), whereas constraints on the other two measures specify lower bounds (minimum required value). A solution to such a problem consists of a placement along with the amount of material stored at each node in the placement. In specifying these problems, we use the notation introduced in Section 19.4.1, in which the optimization objective (preceded by "Min" or "Max") is given first, and each constrained measure is preceded by a "/." For example, in the Min-Cost/Cap/Dist problem, the objective is to minimize the storage cost while satisfying constraints on capacity and minimum inter node distance. In the remainder of this section, we will summarize the known results for such problems.

In general, many of the problems involving storage costs are difficult to approximate. For example, strong nonapproximability results are known for metric instances of Min-Cost/Cap/Dist, Max-Dist/Cap/Cost, and Max-Cap/Cost/Dist problems. Further, the 2D:Min-Cost/Cap/Dist problem has no $\rho$-approximation for any $\rho \geq 1$, unless **P** = **NP**. A similar nonapproximability result is also known for the 1D:Max-Dist/Cap/Dist problem. Positive results are known only for geometric versions of some of the problems. For example, there is a PTAS for the 1D:Min-Cost/Cap/Dist problem. This PTAS is obtained by starting with a pseudo polynomial algorithm[*] that solves the problem optimally, and converting the algorithm into a PTAS using standard scaling and rounding techniques [17]. Moreover, a PTAS is known for the 2D:Max-Cap/Dist/Cost problem. This PTAS is obtained by combining the known PTAS for the weighted MIS problem for unit disk graphs [28] with another PTAS for the problem of finding a solution of maximum capacity under a cost constraint. Details regarding the earlier results can be found in [29].

# 19.5  Future Directions

In the previous sections, we discussed approximation algorithms for several versions of dispersion problems. We conclude this chapter by presenting some directions for future research.

For the metric instances of MAFD, the best known approximation algorithm [34] provides a performance guarantee of 2. Whether this performance guarantee can be improved remains an open problem. For the nonmetric instances of MAFD, there is a significant gap between the upper bound of $O(n^{1/4+\epsilon})$ [37] and the lower bound of $O(1)$ [41,42] on the performance guarantee. Narrowing this gap is an interesting research problem. For metric instances of the Max-Matching dispersion problem, an approximation algorithm with a performance guarantee of $O(\log p)$ is known [16], but no lower bounds on the achievable

---

[*] A **pseudo polynomial** algorithm has a running time that is a polynomial function of the size of the input and the maximum value that appears in the input [17].

performance guarantee are known. Many of the known results for the geometric versions of dispersion problem are for the discrete placement model. Under the continuous placement model, results are available for the Max–Min objective [20]. Investigating other dispersion problems under the continuous placement model is a useful research direction. The topic of capacitated dispersion also provides several open questions. For example, all the known results for capacitated dispersion [29] use the Max–Min measure for the inter node distance. It is of interest to consider capacitated dispersion problems for other distance measures (e.g., Max–Average distance). Dispersion problems involving multiple types of facilities are of practical importance. Exploring approximation algorithms for such problems would be a fruitful research endeavor.

Another research direction is offered by the **online** model for facility placement. In this model, introduced in [43], the distance between each pair of nodes is given as part of the input, but $p$, the number of facilities to be placed is not known a priori. Instead, requests for placing facilities arrive one at a time. An algorithm must choose a new node for each request. Previously placed facilities cannot be moved or eliminated. Under this online model, constant factor approximations were presented in [43] for location problems involving desirable facilities. By using the same online model, some results for dispersion problems are presented in [44]. In particular, it is shown that for dispersion objectives satisfying certain properties, approximation algorithms for the offline model (where the value of $p$ is known a priori) can be used to obtain approximation algorithms under the online model, with only an $O(1)$ loss in the performance guarantee. This result also holds for nonmetric instances. Results on **robust subgraphs** developed in [45] are also germane to the context of online facility-placement problems. In general, developing approximation algorithms for FD problems under the online model remains an interesting research direction.

# References

1. P. B. Mirchandani and R. L. Francis. *Discrete Location Theory*. John Wiley and Sons, New York, 1990.
2. E. Erkut and S. Neuman. Analytical models for locating undesirable facilities. *European Journal of Operational Research*, 40(3):275–291, 1989.
3. K. M. Curtin and R. L. Church. A family of models for multiple-type discrete dispersion. Political Economy Working Paper 34/03, School of Social Sciences, University of Texas at Dallas, August 2003.
4. M. J. Kuby. Programming models for facility dispersion: The $p$-dispersion and maximum dispersion problems. *Geographical Analysis*, 19(4):315–329, 1987.
5. R. E. Steuer. *Multiple Criteria Optimization: Theory and Applications*. John Wiley, New York, 1986.
6. T. J. Santner, B. J. Williams, and W. I. Notz. *The Design and Analysis of Computer Experiments*. Springer, New York, 2003.
7. A. Dimnaku, R. Kincaid, and M. W. Trosett. Approximate solutions of continuous dispersion problems. *Annals of Operations Research*, 136(1):65–80, 2005.
8. R. L. Church and R. S. Garfinkel. Locating an obnoxious facility on a network. *Transportation Science*, 12(2):107–118, 1978.
9. R. Chandrasekharan and A. Daughety. Location on tree networks: $p$-Centre and $n$-dispersion problems. *Mathematics of Operations Research*, 6(1):50–57, 1981.
10. I. D. Moon and S. S. Chaudhry. An analysis of network location problems with distance constraints. *Management Science*, 30(3):290–307, 1984.
11. E. Melachrinoudis and T. P. Cullinane. Locating an undesirable facility with a minimax criterion. *European Journal of Operational Research*, 24(2):239–246, 1986.
12. A. Tamir. Obnoxious facility location on graphs. *SIAM Journal on Discrete Mathematics*, 4(4):550–567, 1991.

13. D. J. White. The maximal dispersion problem and the "first point outside the neighborhood" heuristic. *Computers in Operations Research*, 18(1):43–50, 1991.

14. S. S. Ravi, D. J. Rosenkrantz, and G. K. Tayi. Heuristic and special case algorithms for dispersion problems. *Operations Research*, 42(2):299–310, 1994.

15. M. M. Halldórsson, K. Iwano, N. Katoh, and T. Tokuyama. Finding subsets maximizing minimum structures. *SIAM Journal on Discrete Mathematics*, 12(3):342–359, 1999.

16. B. Chandra and M. M. Halldórsson. Facility dispersion and remote subgraphs. *Journal of Algorithms*, 38(2):438–465, 2001.

17. M. R. Garey and D. S. Johnson. *Computers and Intractability: A Guide to the Theory of NP-Completeness*. W. H. Freeman and Co., San Francisco, CA, 1979.

18. D. W. Wong and Y. S. Kuo. A study of two geometric location problems. *Information Processing Letters*, 28(6):281–286, 1988.

19. S. P. Fekete and H. Meijer. Maximum dispersion and geometric maximum weight cliques. *Algorithmica*, 38(3):501–511, 2003.

20. C. Baur and S. P. Fekete. Approximation of geometric dispersion problems. *Algorithmica*, 30(3):451–470, 2001.

21. D. B. West. *Introduction to Graph Theory*, 2nd ed. Prentice-Hall, Inc., Englewood Cliffs, NJ, 2001.

22. T. Cormen, C. E. Leiserson, R. Rivest, and C. Stein. *Introduction to Algorithms*, 2nd ed. MIT Press/McGraw-Hill, Cambridge, MA, 2001.

23. J. Håstad. Clique is hard to approximate within $n^{1-\epsilon}$. *Acta Mathematica*, 182(1):105–142, 1999.

24. C. H. Papadimitriou. *Computational Complexity*. Addison-Wesley Publishing Co., Reading, MA, 1994.

25. B. N. Clark, C. J. Colbourn, and D. S. Johnson. Unit disk graphs. *Discrete Mathematics*, 86(1–3):165–177, 1990.

26. H. Breu and D. G. Kirkpatrick. Unit disk graph recognition is NP-hard. *Computational Geometry*, 9(1–2):3–24, 1998.

27. M. V. Marathe, H. Breu, H. B. Hunt III, S. S. Ravi, and D. J. Rosenkrantz. Simple heuristics for unit disk graphs. *Networks*, 25(2):59–68, 1995.

28. H. B. Hunt III, M. V. Marathe, V. Radhakrishnan, S. S. Ravi, D. J. Rosenkrantz, and R. E. Stearns. NC-approximation schemes for NP- and PSPACE-hard problems for geometric graphs. *Journal of Algorithms*, 26(2):238–274, 1988.

29. D. J. Rosenkrantz, G. K. Tayi, and S. S. Ravi. Facility dispersion problems under capacity and cost constraints. *Journal of Combinatorial Optimization*, 4(1):7–33, 2000.

30. A. Tamir. Comments on the paper: "Heuristic and special Case algorithms for dispersion problems" by S. S. Ravi, D. J. Rosenkrantz and G. K. Tayi. *Operations Research*, 46(1):157–158, 1998.

31. R. J. Fowler, M. Patterson, and S. L. Tanimoto. Optimal packing and covering in the plane are NP-complete. *Information Processing Letters*, 12(3):133–137, 1981.

32. B. Birnbaum and K. J. Goldman. An improved analysis for a greedy remote-clique algorithm using factor-revealing LPs. *Algorithmica*, 55(1):42–59, 2009.

33. K. Jain, M. Mahadian, E. Markakis, A. Saberi, and V. V. Vazirani. Greedy facility location algorithms analyzed using dual fitting with factor-revealing LP. *Journal of the Association for Computing Machinery*, 50(6):795–824, 2003.

34. R. Hassin, S. Rubinstein, and A. Tamir. Approximation algorithms for maximum dispersion. *Operations Research Letters*, 21(3):133–137, 1997.

35. R. K. Ahuja, T. L. Magnanti, and J. B. Orlin. *Network Flows: Theory, Algorithms and Applications*. Prentice-Hall, Upper Saddle River, NJ, 1993.

36. U. Feige, G. Kortsarz, and D. Peleg. The dense *k*-subgraph problem. *Algorithmica*, 29(3):410–421, 2001.

37. A. Bhaskara, M. Charikar, E. Chlamatac, U. Feige, and A. Vijayaraghavan. Detecting high log-densities: An $O(n^{1/4})$ approximation for densest $k$-subgraph. In *Proceedings of the ACM International Conference on Theory of Computing (STOC 2010)*, pp. 201–210, Cambridge, MA, June 2010.

38. Y. Asahirso, K. Iwama, H. Tamaki, and T. Tokuyama. Greedily finding a dense subgraph. *Journal of Algorithms*, 34(2):203–221, 2000.

39. S. Arora, D. Karger, and M. Karpinski. Polynomial time approximation schemes for dense instances of NP-hard problems. *Journal of Computer and System Sciences*, 58(1):193–210, 1999.

40. A. Czygrinow. Maximum dispersion problem in dense graphs. *Operations Research Letters*, 27(5):223–227, 2000.

41. U. Feige. Relations between average case complexity and approximation complexity. In *Proceedings of the ACM Symposium on Theory of Computing (STOC'02)*, pp. 534–543, Montreal, Canada, May 2002.

42. S. Khot. Ruling out PTAS for graph min-bisection, densest subgraph and bipartite clique. In *Proceedings IEEE International Symposium on Foundations of Computer Science (FOCS 2004)*, pp. 136–145, Rome, Italy, October 2004.

43. R. R. Mettu and C. G. Plaxton. The online median problem. *SIAM Journal on Computing*, 32(3):816–832, 2003.

44. D. J. Rosenkrantz, G. K. Tayi, and S. S. Ravi. Obtaining online approximation algorithms for facility dispersion from offline algorithms. *Networks*, 47(4):206–217, 2006.

45. R. Hassin and D. Segev. Robust subgraphs for trees and paths. In *Proceedings of the Scandinavian Workshop on Algorithm Theory (SWAT 2004)*, pp. 51–63, Humlebaek, Denmark, July 2004.

# II

# Large-Scale and Emerging Applications

# 20

# Cost-Efficient Multicast Routing in Ad Hoc and Sensor Networks

Pedro M. Ruiz*

Ivan Stojmenovic†

## 20.1 Introduction

A mobile ad hoc network (MANET) consists of a number of devices equipped with wireless interfaces. Ad hoc nodes are free to move and communicate with each other using their wireless interfaces. Communications among nodes that are not within the same radio range are carried on via multihop routing. That is, some of the intermediate nodes between the source and the destination act as relays to deliver the messages. Hence, these networks can be deployed without any infrastructure, making them specially interesting for dynamic scenarios such as battlefield, rescue operations, and even as flexible extensions of mobile networks for operators.

Wireless sensor network (WSN) follow a similar communication paradigm based on multihop paths. Although wireless sensor nodes are not usually mobile, their limited resources in terms of battery life and computational power pose additional challenges to the routing task. For instance, wireless sensor nodes operate following a duty-cycle, allowing them to save energy while they are sleeping. The different timings for sleep and awake periods across sensors makes the topology change. In addition, these networks are usually densely populated compared with ad hoc networks, requiring very efficient and scalable mechanisms to provide the routing functions. Examples of such techniques gaining momentum nowadays are geographic routing and localized algorithms in general, in which nodes take individual decisions

---

* Work supported by the "Ramon y Cajal" work program from the Spanish Education Ministry.
† Professor Stojmenovic passed away in November 2014.

solely based on the local information about itself and its neighbors. These networks have a lot of potential applications, which is one of the reasons why they are receiving so much attention within the research community.

The most commonly used model for these networks is called "Unit Disk Graph." The network is modeled as an undirected graph $G = (V, E)$ where $V$ is the set of vertices and $E$ is the set of edges. The model assumes that the network is two dimensional (every node $v \in V$ is embedded in the plane), and wireless nodes are represented by vertices of the graph. Each node $v \in V$ has a transmission range $r$. Let $dist(v_1, v_2)$ be the distance between two vertices $v_1, v_2 \in V$. An edge between two nodes $v_1, v_2 \in V$ exists iff $dist(v_1, v_2) \leq r$ (i.e., $v_1$ and $v_2$ are able to communicate directly).

Unicast routing both for MANETs and WSNs can be defined as the process of finding a paths in the network to deliver a message from the originator to the destination. As we mentioned before, in these networks, such paths are formed by a set of nodes acting as relays. The multicast routing task is similar to the unicast routing except that there are a number of destinations instead of a single node. These destinations are often referred as "receivers" in the literature. In this particular case, the set of relay nodes usually forms a tree, commonly known as "multicast tree." In the following, we define more precisely the problem of unicast and multicast routing in these networks.

**Definition 20.1** *Given a graph $G = (V, E)$, a source node $s \in V$ and a destination node $D \in V$, the unicast routing problem can be defined as finding a set of relay nodes $F \subset V$ s.t. $\{s\} \cup F \cup \{D\}$ is connected.*

Similarly, the multicast routing problem can be defined as follows:

**Definition 20.2** *Given a graph $G = (V, E)$, a source node $s \in V$ and a set of destinations $R \subseteq V$, the multicast routing problem, can be defined as finding a set of relay nodes $F \subset V$ s.t. $\{s\} \cup F \cup R$ is connected.*

Of course, routing algorithms are designed to avoid cycles, and usually select paths according to some metric or combination of metrics such as hop count and delays. In fact, most of the existing routing protocols use the hop count as the path selection metric.

The problem of unicast routing is well known, and there are many distributed algorithms such as Dijkstra and Bellman-Ford. For the problem of multicast routing, there are also algorithms to build shortest path tree (SPT), shared trees, and so on. In fact, the problem of the efficient distribution of traffic from a set of senders to a group of receivers in a datagram network was already studied by Deering [1] in the late 80s. Several multicast routing protocols such as DVMRP [2], MOSPF [3], Core Based Trees (CBT) [4], and PIM [5] have been proposed for IP multicast routing in fixed networks. These protocols have not been usually considered in MANET because they do not properly support mobility. In the case of mesh networks, one may believe that they can be a proper solution. However, they were not designed to operate on wireless links, and they lead to suboptimal routing solutions that are not able to take advantage of the broadcast nature of the wireless medium (i.e., sending a single message to forward a multicast message to all the next hops rather than replicating the message for each neighbor). Moreover, their routing metrics do not aim at minimizing the cost of the multicast tree, which limits the overall capacity of the mesh network.

Within the next sections, we describe existing multicast routing protocols for ad hoc and sensor networks, and we analyze the issue of computing minimum cost multicast trees. In fact, we will show the NP-completeness of the problem.

Given that the use of approximation algorithms is fully justified in the multicast routing case, we focus the rest of the chapter on the multicast routing problem, and its approximation algorithms for MANETs and WSNs. The remainder of the chapter is organized as follows: Section 20.2 describes existing multicast routing protocols for MANETs, and their inability to approximate minimum cost multicast trees. Section 20.3 focuses on the issue of computing minimum bandwidth multicast trees in wireless ad hoc networks, shows the NP-completeness of the problem, and offers approximation algorithms that offer better performance than Steiner trees. We focus on the problem of geographic multicast routing in Section 20.4. Finally, we provide some discussion and conclusions in Section 20.5.

# 20.2 Multicast Routing in Ad Hoc Networks

A plethora of protocols have been proposed for multicast routing in MANET. We focus our discussion on the most representative protocols. They can be classified into tree- or mesh-based depending upon the underlying forwarding structure that they use. Tree-based schemes [6–10] construct a multicast tree from each of the sources to all the receivers using generally a SPT, or a shared tree. Mesh-based approaches [11,12] compute several paths among senders and destinations. Thus, when the mobility rate increases, they are able to tolerate link breaks better than tree-based protocols at the expense of a usually higher overhead. Hybrid approaches [13,14] try to combine the robustness of mesh-based ad hoc routing and the low overhead of tree-based protocols. Finally, there are stateless multicast protocols [15] in which there is no need to maintain a forwarding state on the nodes (for instance, if the nodes to traverse are included in the data packets themselves). We will not discuss further about the latter category given the very limited applicability of those variants.

Regarding tree-based protocols, ad hoc multicast routing protocol utilizing increasing id-numberS (AMRIS) [6] builds a shared multicast tree among a set of sources and receivers. There is a root node, which is the one with the smallest ID (Sid). These ID numbers are assigned dynamically within a multicast session, and based on these IDs the multicast tree is built. The numbering process starts at the Sid, and other nodes always select an ID being higher that the one of his upstream node in the tree. Multicast Ad hoc On-Demand Distance Vector Routing (MAODV) [7] is an extension of the well-known Ad hoc On-Demand Distance Vector Routing (AODV) protocol. The route creation is similar to the route request/route reply (RREQ/RREP) process in AODV, except that the source unicasts a Multicast Activation message through the selected paths, which usually form a shortest path tree (SPT) based on the hop count.

Regarding mesh-based multicast routing protocols for ad hoc networks, On-Demand Multicast Routing Protocol (ODMRP) [11] works reactively to build a multicast mesh connecting senders and receivers. All the intermediate nodes taking part in the multicast mesh are said to belong to the forwarding group (FG). When a multicast node has data to send and it does not have a route for that multicast group, it starts a periodic broadcasting of JOIN_QUERY (JQ) packets. These messages are propagated through the entire ad hoc network avoiding duplicates, so that every ad hoc node can learn which of his neighbors is in its shortest path to that source. Upon reception of a nonduplicate JQ message, a receiver broadcasts a JOIN_REPLY (JR) message in which it includes the ID of the neighbor selected as the next hop to reach each of the multicast sources. When a node receives a JR message, it checks out if its own ID is listed as the selected next hop for any of the sources. If that is the case, then it realizes that it is in the SPT to any of the sources, and it adds itself to the FG by activating its FG_FLAG. In addition, the selected node sends out a JR which he fills with the IDs of its selected neighbors to reach those sources for which it was selected in the received JR message. In this way, the FG is populated until the JR messages reach the source. This whole process is repeated periodically to update the FG over time. Once the mesh is built, data forwarding is very simple. Only those nodes whose FG_FLAG is active are allowed to forward data packets generated by the sources. In addition, in the case of receiving the same data packet several times, a node within the FG shall only forward it the first time it is received. Core-Assisted Mesh Protocol (CAMP) [12] was designed as an extension of the CBT [4] protocol. However, unlike CBT, in which there was not link redundancy, CAMP builds a multicast mesh to offer a much better performance and resilience in the case of link breaks. Although in CBT core nodes were used for data forwarding, they are used in CAMP to reduce the overhead for a node to find out a node belonging to the multicast mesh. Data packets are not required to go through core nodes. Thus, given that there can be several core nodes for a multicast group, the tolerance of mobility is increased. CAMP uses a receiver-initiated approach to build the multicast mesh. Using the cores as well-known mesh nodes, CAMP avoids relying on periodic flooding of the network to find multicast routes. However, this comes at the cost of depending upon the unicast routing protocol as well as the need of a mapping service from multicast groups to core nodes. CAMP ensures that the mesh contains all the reverse shortest paths between the source and the recipients. It uses a so-called heartbeat mechanism by which each node periodically monitors its packet cache. If the node finds out that some data packets that it is receiving are not coming from its shortest path neighbor, then it sends a HEARTBEAT message to its successor in the reverse shortest path to the source. The HEARTBEAT triggers a push-join message that (if the successor is not a mesh member) forces the successor and all the nodes in the path to join the mesh.

All these solutions are not aimed at minimizing the cost of multicast trees due to the difficulty of computing such trees. In fact, when the goal is to find multicast trees with minimum edge cost, the problem becomes NP-complete and requires heuristic solutions. Such minimum cost multicast tree is well known as the Steiner tree problem. However, Ruiz et al. [16] showed that ODMRP (and accordingly other similar protocols) can benefit from the use of Steiner trees rather than SPTs and shared trees. Karp [17] demonstrated that the Steiner tree problem is NP-complete even when every link has the same cost using a transformation from the exact cover by 3-sets. There are some heuristic algorithms [18] to compute minimal Steiner trees. For instance, the minimum spanning tree (MST) heuristic ([19,20]) provides a 2-approximation, and Zelikovsky [21] proposed an algorithm that obtains a 11/6-approximation. Recently, Rajagopalan and Vazirani [22] proposed a 3/2-approximation algorithm. Given the complexity of computing this kind of trees in a distributed way, most of the existing multicast routing protocols use shortest path trees or suboptimal shared trees, which can be easily computed in polynomial time.

Recently, Ruiz et al. [23] showed that the Steiner tree is not the best solution for wireless ad hoc networks and provided some approximation algorithms that will be analyzed in this chapter. The problem of minimizing the bandwidth consumption of a multicast tree in an ad hoc network needs to be reformulated in terms of minimizing the number of data transmissions. By assigning a cost to each link of the graph computing the tree which minimizes the sum of the cost of its edges, existing formulations have implicitly assumed that a given node $v$ needs $k$ transmissions to send a multicast data packet to $k$ of its neighbors. However, in a broadcast medium, the transmission of a multicast data packet from a given node $v$ to any number of its neighbors can be done with a single data transmission. Thus, in ad hoc networks, the minimum cost tree is the one which connects sources and receivers by issuing a minimum number of transmissions, rather than having a minimal edge cost. We will discuss further this problem along next sections.

## 20.3 Minimum Bandwidth Consumption Multicast Tree

When nodes are mobile, such as in traditional MANETs, it is not really interesting to approximate optimal multicast trees. The reason is that by the time the tree is computed, it may no longer exist. However, in some new static ad hoc network scenarios being deployed (i.e., wireless mesh networks), these algorithms may become of utmost relevance. In wireless mesh networks, devices are powered. So, the main concern, rather than being the power consumption, is the proper utilization of the bandwidth. Thus, the computation of bandwidth-optimal multicast trees becomes really important.

Given a multicast source $s$ and a set of receivers $R$ in a network represented by a undirected graph, we are interested in finding the multicast tree with the minimal cost in terms of the total amount of bandwidth required to deliver a packet from $s$ to every receiver.

In wired networks, the computation of such minimum bandwidth consumption multicast tree is equivalent to the Steiner tree over a graph $G = (V, E)$ so that $w(e_i) = b_s, \forall i = 1..|E|$, being $b_s$ the rate at which the source $s$ is transmitting, and $w(e_i)$ the cost of the link number $i$. Unitary costs (i.e., $w(e_i) = 1, \forall i = 1..|E|$) are usually assumed for simplicity. The bandwidth consumption of such a Steiner tree $T^* = (V^*, E^*)$ is then proportional to $|E^*|$. This means that the problem is NP-complete in wired networks. As we mentioned before, in wireless multihop networks, a node can send a message to all neighbors with a single transmission. Thus, the bandwidth consumption is different from the edge cost, and the problem requires being reformulated in terms of the number of transmissions rather than the number of traversed links. To account for the excessive bandwidth consumption due to suboptimal trees, we will define a new metric called "data overhead."

## 20.3.1 Problem Formulation

Before going into details, we need some definitions that are used in the sequel.

**Definition 20.3** *Given a graph $G = (V, E)$, a source $s \in V$ and a set of receivers $R \subset V$, we define the set $T$ as the set of the possible multicast trees in $G$ that connect the source $s$ to every receiver $r_i \in R$. We denote by $F_t$, the set of relay nodes in the tree $t \in T$, consisting of every nonleaf node, which relays the message sent out by the multicast source. We define a function $C_t : T \rightarrow \mathbb{Z}^+$ so that given a tree $t \in T$, $C_t(t)$ is the number of transmissions required to deliver a message from the source to every receiver induced by that tree.*

**Lemma 20.1** *Given a tree $t \in T$ as defined earlier, then $C_t(t) = 1 + |F_t|$.*

*Proof.* By definition relay nodes, forward the message sent out by $s$ only once. In addition, leaf nodes do not forward the message. Thus, the total number of transmissions is one from the source, and one from each relay node. Making a total of $1 + |F_t|$. ∎

So, as we can see from Lemma 20.1, to minimize $C_t(t)$, we must somehow reduce the number of forwarding nodes $|F_t|$. Note that some receivers may serve also as relay nodes.

**Definition 20.4** *Under the conditions of definition 20.3, let $t^* \in T$ be the multicast tree such that $C_t(t^*) \leq C_t(t)$ for any possible $t \in T$. We define the data overhead of a tree $t \in T$, as $\omega_d(t) = C_t(t) - C_t(t^*)$. Obviously, with this definition, $\omega_d(t^*) = 0$.*

Based on the previous definitions, the problem can be formulated as follows. Given a graph $G = (V, E)$, a source node $s \in V$, a set of receivers $R \subset V$, and given $V' \subseteq V$ defined as $V' = R \cup \{s\}$, find a tree $T^* = (V^*, E^*) \subset G$ such that the following conditions are satisfied:

(1) $V^* \supseteq V'$.
(2) $C_t(T^*)$ is minimized.

From the condition of $T^*$ being a tree, it is obvious that it is connected, which combined with condition (1) establishes that $T^*$ is a multicast tree. Condition (2) is equivalent to $\omega_d(T^*) = 0$ and establishes the optimality of the tree. As we show in the following, this problem is NP-complete.

## 20.3.2 NP-Completeness

Ruiz et al. [23] demonstrated that this problem is NP-complete. We show in the following a similar demonstration based on the inclusion of the Minimim Common Dominating Set as a particular case of the problem.

**Theorem 20.1** *Given a graph $G = (V, E)$, a multicast source $s \in V$ and a set of receivers $R$, the problem of finding a tree $T^* \supseteq R \cup \{s\}$ so that $C_t(T^*)$ is minimum is NP-complete.*

*Proof.* According to Lemma 20.1, minimizing $C_t(T^*)$ is equivalent to minimize the number of relay nodes $F \subseteq T^*$. So, the problem is finding the smallest set of forwarding nodes $F$ that connects $s$ to every $r \in R$. If we consider the particular case in which $R = V - \{s\}$, the goal is finding the smallest connected $F \subseteq T^*$ which covers the rest of nodes in the graph $(V - \{s\})$. This problem is one of finding a minimum connected dominating set, which is known to be NP-complete [24]. ∎

Ruiz et al. [23] also proved the suboptimality of Steiner trees for this particular problem in ad hoc networks. We give a simple example to prove it.

**Theorem 20.2** *Let $G = (V, E)$ be an undirected graph. Let $s \in V$ be a multicast source and $R \subseteq V$ be the set of receivers. The Steiner multicast tree $T^* \subseteq G$ so that $C_e(T^*)$ is minimal may not be the minimal data-overhead multicast tree.*

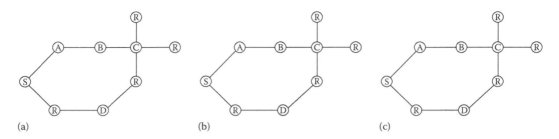

**FIGURE 20.1** Differences in cost for several multicast trees over the same ad hoc network: (a) Shortest path tree: 8 edges and 6 transmissions. (b) Steiner tree: 6 edges and 5 transmissions. (c) Minimum data overhead tree: 7 edges and 4 transmissions.

*Proof.* It is immediate from the examples shown in Figure 20.1 that the Steiner tree may not minimize the bandwidth consumption, leading to suboptimal solutions in MANETs and WSNs. ∎

In general, Steiner tree heuristics try to reduce the cost by minimizing the number of Steiner nodes $|\mathbb{S}^*|$. In Section 20.3.3, we will present our heuristics being able to reduce the bandwidth consumption of multicast trees, by just making receivers be leaf nodes in a cost-effective way.

### 20.3.3 Heuristic Approximations

Given the NP-completeness of the problem, within the next subsections, we describe two heuristic algorithms proposed by Ruiz et al. [23] to approximate minimal data-overhead multicast trees. As we learned from the demonstration of Theorem 20.2, the best approach to reduce the data overhead is reducing the number of forwarding nodes, while increasing the number of leaf nodes. The two heuristics presented in the following try to achieve that trade-off.

#### 20.3.3.1 Greedy-Based Heuristic Algorithm

The first proposed algorithm is suited for centralized wireless mesh networks, in which the topology can be known by a single node, which computes the multicast tree.

Inspired on the results from Theorem 20.2, this algorithm first systematically builds different cost-effective subtrees. The cost-effectiveness refers to the fact that a node $v$ is selected to be a forwarding node only if it covers two or more nodes. That is, if it has two or more multicast receivers as neighbors.

The algorithm shown in Algorithm 20.1 starts by initializing the nodes to cover (aux) to all the sources except those already covered by the source $s$. Initially, the set of forwarding nodes (multicast

---

## ALGORITHM 20.1    Greedy Minimal Data Overhead Algorithm MNT

1:   MF ← ∅ / * mcast − forwarders * /
2:   V ← V - {s}
3:   aux ← R-Cov(s) + {s} / * nodes − to − cover * /
4:   **repeat**
5:      node ← $argmax_{v \in V}(|Cov(v)|)$ s.t. Cov(v)≥2
6:      aux ← aux-Cov(v)+{v}
7:      V ← V-{v}
8:      MF ← MF + {v}
9:   **until** aux = ∅ or node = *null*
10:  **if** V!=∅ **then**
11:     Build Steiner tree among nodes in aux
12:  **end if**

forwarders [MF]) is empty. After the initialization, the algorithm repeats the process of building a cost-effective tree, starting with the node $v$ that covers the largest number of nodes in "aux." Then, $v$ is inserted into the set of forwarding nodes (MF), and it becomes a node to cover. In addition, the receivers covered by $v$ (Cov($v$)) are removed from the list of nodes to cover denoted by "aux." This process is repeated until all the nodes are covered, or it is not possible to find more cost-effective subtrees. In the latter case, the different subtrees are connected by a Steiner tree among their roots, which are in the list "aux" (i.e., among the nodes which are not covered yet). For doing that, one can use any Steiner tree heuristic. In our simulations we use the MST heuristic for simplicity.

### 20.3.3.2 Distributed Variant

The previous algorithm may be useful for some kind of networks. However, in general, a distributed algorithm is preferred for wireless ad hoc networks. Hence, Ruiz et al. [23] proposed the distributed approach described in the following.

The previous protocol consists of two different parts: (i) construction of cost-efficient subtrees and (ii) building a Steiner tree among the roots of the subtrees.

To build a Steiner tree among the roots of the subtrees, Ruiz assumed in the previous protocol the utilization of the MST heuristic. This is a centralized heuristic consisting of two different phases. First, the algorithm builds the metric closure for the receivers on the whole graph, and then, a MST is computed on the metric closure. Finally, each edge in the MST is substituted by the shortest path (in the original graph) between the two nodes connected by that edge. On the contrary, the metric closure of a graph is hard to build in a distributed way. Thus, Ruiz approximated the MST heuristic with the simple, yet powerful, algorithm presented in Algorithm 20.2. The source, or the root of the subtree in which the source is (called source-root), will start flooding a RREQ message. Intermediate nodes, when propagating that message, will increase the hop count. When the RREQ is received by a root of a subtree, it sends a RREP back through the path which reported the lowest hop count. Those nodes in that path are selected as MF. In addition, a root of a subtree, when propagating the RREQ, will reset the hop count field.

---

## ALGORITHM 20.2   Distributed Approximation of MST Heuristic MNT2

```
 1: if thisnode.id = source − root then
 2:     Send RREQ with RREQ.hopcount=0
 3: end if
 4: if rcvd non duplicate RREQ with better hopcount then
 5:     prevhop ← RREQ.sender
 6:     RREP.nexthop ← prevhop
 7:     RREQ.sender ← thisnode.id
 8:     if thisnode.isroot then
 9:        send(RREP)
10:        RREQ.hopcount ← 0
11:     else
12:        RREQ.hopcount++;
13:     end if
14:     send(RREQ)
15: end if
16: if received RREP and RREP.nexthop = thisnode.id then
17:     Activate MF_FLAG
18:     RREP.nexthop ← prevhop
19:     send(RREP)
20: end if
```

This is what makes the process very similar to the computation of the MST on the metric closure. In fact, we achieve the same effect, which is that each root of the subtrees, will add to the Steiner tree the path from itself to the source-root, or the nearest root of a subtree. In the case in which two neighboring nodes are far away from S but at the same hop count, the node-ID is used as a tie-breaker. This avoids a deadlock by preventing each of them from calling the other as its selected next hop. The one with the lowest ID will always select the other. This mechanism and the way in which the algorithm is executed from the source-root to the other nodes guarantees that the obtained tree is connected.

The second part of the algorithm to make distributed is the creation of the cost-effective subtrees. However, this part is much simpler, and it can be done locally with just a few messages. Receivers flood a Subtree_Join (ST_JOIN) message only to its 1-hop neighbors indicating the multicast group to join. These neighbors answer with a Subtree_Join_Ack (ST_ACK) indicating the number of receivers they cover. This information is known locally by counting the number of (ST_JOIN) messages received. Finally, receivers send again a ST_JOIN_Activation message including their selected root, which is the neighbor that covers the greatest number of receivers. This is also known locally from the information in the ST_ACK. Those nodes that are selected by any receiver, repeat the process acting as receivers. Nodes that already selected a root do not answer this time to ST_JOIN messages.

### 20.3.4 Performance Evaluation

Ruiz et al. [23] presented some performance evaluation of their approach. We have reproduced their simulations with the same parameters except that the number of nodes has been fixed at 600, and the area has been ranged from 750 × 750 m to 2250 × 2250 m. The simulated algorithms are the two minimum bandwidth algorithms (Minimum Number of Transmissions [MNT] and MNT2 respectively) as well as the MST heuristic to approximate Steiner trees. In addition, we also simulated the SPT algorithm, which is the one used by most multihop multicast routing protocols proposed to date.

Performance metrics are also the same as Ruiz et al. used in Reference 23, and the reader can refer there for details. The results shown correspond to the 1250 × 1250 m area and 150 receivers. For each combination of simulation parameters, a total of 91 simulation runs with different randomly generated graphs were performed. The error in the graphs shown in the following are obtained using a 95% confidence level.

#### 20.3.4.1 Performance Analysis

In Figure 20.2, we show for a network with an intermediate density (600 nodes in 1250 × 1250 m area) how the number of transmissions required varies with respect to the number of receivers. For a low number of them, the minimum bandwidth schemes do not offer significant differences compared with the Steiner tree heuristic. This is because receivers tend to be very sparse, and it is less likely that cost-effective trees are built. However, as the number of receivers increases, the creation of cost-effective trees is favored, making the MNT and MNT2 algorithms achieve significant reductions in the number of transmissions required. In addition, given that the SPT approach doesn't aim at minimizing the cost of the trees; it shows a lower performance compared with any of the other approaches. The distributed MNT2 algorithm, by not using the metric closure, gets a slightly lower performance compared with the centralized approach. However, both of them have very similar performance, which allow them to offer substantial bandwidth savings compared with the Steiner tree (i.e., MST heuristic).

In Figure 20.3, we represent the mean path length. MNT and MNT2 offer a higher mean path length because grouping paths for several receivers requires deviating from their shortest paths for some of the receivers. As we can see, this metric is much more variable to the number of receivers than the number of transmissions was for the heuristic approaches. This is why the error bars are reporting a larger confidence interval for MST, MNT, and MNT2.

In Figure 20.4, we present the variation of the number of transmissions as the density varies, for 150 receivers. As the figure depicts, the higher the density, the better is the performance of all the approaches.

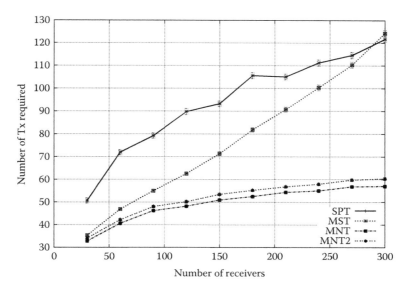

**FIGURE 20.2** Number of Tx at increasing number of receivers.

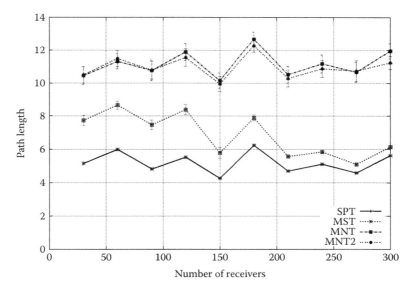

**FIGURE 20.3** Mean path length at increasing number of receivers.

The reason is that higher densities imply shorter path length (note that number of nodes is fixed). So in general, one can reach the receivers with less number of transmissions regardless of the routing scheme. However, if we compare the performance across approaches, we can see that the reduction in the number of transmissions achieved by MNT and MNT2 is higher as the density increases. This can be easily explained by the fact that for higher densities it is more likely that several receivers can be close to the same node, which facilitates the creation of cost-effective subtrees.

We can observe that the number of receivers has small impact on the performance comparison compared with that of the density of the network.

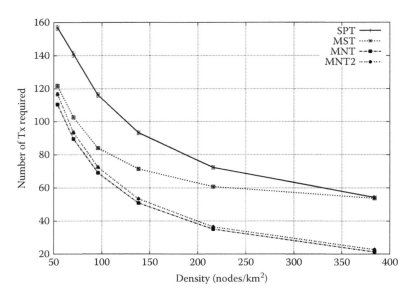

**FIGURE 20.4**   Number of Tx with varying network density for 150 receivers.

# 20.4  Cost-Efficient Geographic Multicast Routing

Routing in sensor networks differ from routing in ad hoc networks. Position information is almost intrinsic to WSNs. In fact, measurements providing from sensor nodes do not usually have proper meaning unless the geographical information regarding that sensor is also reported. Position information, if available to each sensor, also simplifies routing task. This approach is commonly known as "geographic routing." A survey of position based routing schemes is given in Reference 25.

The general idea is very simple. Given a source $s$ and a destination $d$ (the destination is normally a fixed sink whose location is known to all sensors), the source $s$ sends the data packet to be delivered, to the neighbor that is closest to the destination. This neighbor will repeat the process again, until the message is eventually delivered to the destination. For the case in which the algorithm reaches a local minima (i.e., there is no neighbor making progress towards the destination), Bose, Morin, Stojmenovic, and Urrutia proposed a recovery scheme [26] called "Face routing," which guarantees delivery.

Similarly to ad hoc network routing, different metrics can be minimized when finding the route towards a destination using geographic routing. The most used one is the reduction of the hop count. However, there are also proposals to reduce the power consumption [27], delay, and so on. Stojmenovic et al. [28] proposed a general framework to optimize different metrics. The rest of the chapter will be devoted to the explanation of that framework, and its application to the problem of efficient geographic multicast routing.

## 20.4.1  The Cost over Progress Framework

A frequent solution when dealing with multiple metrics in geographic routing in the literature is to introduce additional parameters, not present in the problem formulation, as part of the solution protocol. For instance, the neighbor selection function is changed to something similar to

$$\alpha(metric_1) + \beta(metric_2) + \cdots + \omega(metric_n)$$

where Greek letters are the additional parameters considered, and $metric_i$ are the different metrics (delay, hop count, power consumption, etc.).

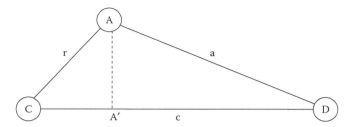

**FIGURE 20.5**  Best neighbor selection in localized routing schemes.

So, the performance of such a protocol often depends on the particular values for the set of parameters. In most cases, the optimal values for these parameters depend on global network conditions, which may be beyond the knowledge available to tiny sensors. In other cases, the computational and communication cost required to obtain such information is higher than the benefits provided by the particular protocol. Another typical approach is to use thresholds for the solutions, which has the effect of eliminating certain options in the protocols that may lead to suboptimal solutions, or to failures.

To avoid those issues, a simple but elegant solution is to use as the neighbor selection function, a cost over progress ratio. The idea is to optimize the ratio of operation costs (in terms of the particular metrics considered in the problem statement) and progress made (e.g., reduction in distance to the destination in the case of routing). For network coverage problems, the same concept can by applied by redefining the cost function and considering an appropriate progress function (e.g., additional area covered). To better understand this idea, we will give a couple of examples of its application.

**Example 20.1** *We consider the case in which we want to provide geographic routing with minimal power consumption. We consider Figure 20.5 as a reference, where C is the source, D the destination, and A is a candidate neighbor and A' is the projection of A on $\overline{CD}$. In addition, we have that $|\overline{CD}| = c$, $|\overline{AD}| = a$ and $|\overline{CA}| = r$.*

*The power needed to send a message from C to A is proportional to $r^{\alpha} + k$, where $\alpha$ is the power attenuation factor ($2 \le \alpha \le 6$), whereas k is a constant ($k > 0$) that accounts for running circuits at the transmitter and receiver nodes. In our framework, this power can be used as the cost measure. The progress can be measured according to Figure 20.5 as $c - a$. Therefore, the neighbor that minimizes $(r^{\alpha} + k)/(c - a)$ is the one to be selected.*

In Section 20.4.2, we discuss how the same framework can be applied to the geographic multicast routing problem.

## 20.4.2 Geographic Multicasting with Cost over Progress

We now focus on the multicasting problem represented in Figure 20.6. A source node $C$ wishes to send a packet to several destinations (sinks) with known positions $(D_1, \ldots, D_n)$. It is assumed that the number of such destinations is small, which is reasonable for the scenario in which a sensor reports to several sinks.

Mauve et al. [29] proposed a geographic multicast protocol that considers the total hop count as the metric to optimize, and distances from neighbors to destinations as part of the criterion to optimize. The impact of each of these aspects in the final neighbor selection is controlled by an external parameter ($\lambda$) whose best value is to be separately determined. We describe in the following how the same problem can be solved with the cost over ratio framework, without the need for such additional parameters.

Let's assume a node $C$, after receiving a multicast message including the positions of the destinations $D_1, D_2, \ldots, D_k$, and lets also assume that $C$ evaluates neighbors $A_1, A_2, \ldots, A_m$ for forwarding. If there is one neighbor that is closer to all destinations than $C$, then it may happen that there is only one next

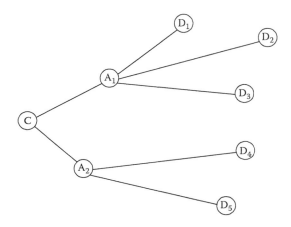

**FIGURE 20.6**  Evaluating candidate forwarding set from neighbors.

hop selected. However, it may also happen that the multicast routing task needs to be split across multiple neighbors, each handling a subset of destinations.

Consider the case in Figure 20.6 as illustration of the general principle. The current total distance to deliver the message from $C$ to all receivers is $T_1 = |\overline{CD_1}| + |\overline{CD_2}| + |\overline{CD_3}| + |\overline{CD_4}| + |\overline{CD_5}|$. If $C$ now considers neighbors $A_1$ and $A_2$ as forwarding nodes covering $D_1, D_2, D_3$, $D_4$, and $D_5$ respectively, the new total remaining distance would be $T_2 = |\overline{A_1D_1}| + |\overline{A_1D_2}| + |\overline{A_1D_3}| + |\overline{A_2D_4}| + |\overline{A_2D_5}|$, and the "progress" made is $T_1 - T_2$. The "cost" is the number of selected neighbors (i.e., 2 in the previous figure). Thus, the forwarding set $\{A_1, A_2\}$ is evaluated as $2/(T_1 - T_2)$. So, among all candidate forwarding sets, the one with optimal value of this expression is selected. Those destinations for which there is no neighbor closer to them will be reached using Greedy-Face-Greedy (GFG) routing [26] directly to them. Note that the number of expressions to evaluate grows with number of neighbors and number of destinations. We provide an enhanced algorithm in the following to reduce the number of evaluations.

### 20.4.2.1 Exhaustive Enumeration by Set Partitioning

Given $k$ destinations, the algorithm can consider all $S_k$ partitions. For each set given in the set partition, check whether there is a node that is closer to the destinations in the set than the current node $C$. If it is not possible to find such a node for a set, that particular partition is ignored. For those being possible, the cost/progress ratio is computed, and the best one is selected. This solution is applicable for small number of destinations (e.g., up to 5). For a larger number, it becomes exponential in $k$, and therefore a faster greedy solution is needed. A fast algorithm for generating set partitions is given in Reference 30.

### 20.4.2.2 Greedy Selection of Set Partitions

The goal of this greedy selection of set partitions is to reduce the number of partitions (destination sets) being evaluated. The destinations for which there is no closer neighbor are served using the GFG protocol [26]. Thus, we start with the set of destinations $\{D_1, D_2, \ldots, D_k\}$, for which there is at least one node closer to them than $C$.

For each of these $D_i$, we choose the best neighbor $A_i$ of $C$ as if it was the only destination. If there are several destinations for which the best neighbor is the same, then we merge those destinations into a single set $\{M_i\}$. At the end of this phase, we have an initial set partition of the destinations $\{M_1, M_2, \ldots M_l\}$, and there is a different neighbor $A_i$ for each subset $M_i$. Each $M_i$ has its cost over progress $1/P_i$, and the whole partition also has its overall cost over progress being equal to $l/(P_1 + P_2 + \cdots + P_l)$. In the example of Figure 20.6, there are two subsets $M_1 = \{D_1, D_2, D_3\}$ and $M_2 = \{D_4, D_5\}$. The cost over progress of $M_1$ is $1/P_1$ being $P_1 = (|\overline{CD_1}| + |\overline{CD_2}| + |\overline{CD_3}|) - (|\overline{A_1D_1}| + |\overline{A_1D_2}| + |\overline{A_1D_3}|)$. Similarly, the cost over progress for $M_2$ is $1/P_2$ being $P_2 = (|\overline{CD_4}| + |\overline{CD_5}|) - (|\overline{A_2D_4}| + |\overline{A_2D_5}|)$. The overall cost over progress ratio of the whole partition is then $2/(P_1 + P_2)$.

After this first iteration, we will repeatedly try to improve the cost over progress ratio, until this was not possible in a given iteration. In each iteration, we try to merge each pair $M_i, M_j$ to see whether they can improve the cost-progress ratio by merging into one set. This merge is done selecting a new $A_j$, being the neighbor of $C$ which is closer than $C$ to all destinations in $M_i \cup M_j$ and provides best ratio. If such $A_j$ does not exist, merge is not possible. From all possible merges that improve the ratio compared with the partition set, select the one with higher ratio, and update the set partition by effectively merging those sets, and decreasing $l$ by 1. The next iteration starts with this new set partition. The process is repeated until no new merging improves the ratio.

It is easy to prove that this algorithm would test $O(k^3)$ cases rather than $k!$.

The described algorithm is based on position information. If it is not available, it can be applied on a version in which distances between nodes are replaced by hop counts between them. Each receiver may flood the network, so that each node may learn hop count distances to all receivers. These distances can be used in the described protocol to determine multicast routes.

## 20.5 Discussion

Multicast routing has proven to be an interesting problem requiring approximation algorithms, and simple yet efficient heuristic solutions. As we have seen, multicast-routing problems are NP-complete even when the metrics to optimize are not very complicated.

In the current chapter, we have shown example of such greedy algorithms and their applicability to the multicast routing problem in two different environments: wireless ad hoc networks and WSN.'

In ad hoc networks, we have shown that the traditional Steiner tree problem does not guarantee optimality regarding the overall bandwidth consumption of the multicast tree, and we have applied a heuristic epidemic algorithm to approximate efficiently those bandwidth-efficient trees.

For sensor network scenarios, we have shown how the general geographic routing concept can be extended to solve the multicast routing task in a localized way. In addition, we have explained how the general cost-progress ratio framework can be also used for such problem. As we showed, using this framework, we avoid adding extra parameters to the problem formulation, which is one of the issues present in most of the existing solutions in the literature.

We believe that the concepts presented here will be useful for solving a number of other network layer problems, or even variants of these ones based on different assumptions or optimality criteria.

## References

1. Deering, S., Multicast routing in a datagram internetwork, PhD Thesis, Electrical Engineering Department of Stanford University, 1991.
2. Deering, S. and Cheriton, D.R., Multicast routing in datagram internetworks and extended LANs, *Trans. Comput. Syst.,* 8(2), 85, 1990.
3. Moy, J., Multicast routing extensions for OSPF, *CACM,* 37(8), 61, 1994.
4. Ballardie, T., Francis, P., and Crowcroft, J., Core Based Trees (CBT)—An architecture for scalable inter-domain multicast routing, in *Proceedings of ACM SIGCOMM*, ACM, 1993, p. 85.
5. Deering, S., Estrin, D., Farinacci, D., Jacobson, V., Liu, C.G., and Wei, L., The PIM architecture for wide-area multicast routing, *IEEE/ACM Trans. Netw.,* 4(2), 153, 1996.
6. Wu, C.W., Tay Y.C., and Toh, C.K., Ad hoc multicast routing protocol utilizing increasing id-numberS (AMRIS) functional specification, Internet-draft, work in progress, draft-ietf-manet-amris-spec-00.txt, 1998.
7. Belding-Royer, E.M. and Perkins, C.E., Multicast operation of the ad-hoc on-demand distance vector routing protocol, in *Proceedings of ACM/IEEE MOBICOM,* ACM, 1999, p. 207.

8. Ji, L. and Corson, S., A lightweight adaptive multicast algorithm, in *Proceedings of IEEE GLOBE-COM*, 1998, p. 1036.

9. Jetcheva, J.G. and Johnson, D.B., Adaptive demand-driven multicast routing in multi-hop wireless ad hoc networks, in *Proceedings of ACM MobiHoc*, ACM, 2001, p. 33.

10. Toh, C.K., Guichala, G., and Bunchua, S., ABAM: On-demand associativity-based multicast routing for mobile ad Hoc networks, in *Proceedings of IEEE VTC*, IEEE, 2000, p. 987.

11. Lee, S.J., Su, W., and Gerla, M., On-demand multicast routing protocol in multihop wireless mobile networks, *ACM/Kluwer Mob. Netw. Appl.*, 7(6), 441, 2002.

12. Garcia-Luna-Aceves, J.J. and Madruga, E.L., The core-assisted mesh protocol, *IEEE J. Sel. Areas Commun.*, 17(8), 1380, 1999.

13. Bommaiah, E., Liu, M., MacAuley, A., and Talpade, R., AMRoute: Ad hoc multicast routing protocol, Internet-draft, work in progress, draft-talpade-manet-amroute-00.txt, 1998.

14. Sinha, P., Sivakumar, R., and Bharghavan, V., MCEDAR: Multicast core-extraction distributed ad hoc routing, in *Proceedings of Wireless Communications and Networking Conference*, IEEE, 1999, p. 1313.

15. Ji, L. and Corson, M.S., Differential destination multicast: A MANET multicast routing protocol of small groups, in *Proceedings of INFOCOM*, 2001, p. 1192.

16. Ruiz, P.M. and Gomez-Skarmeta, A.F., Mobility-aware mesh construction algorithm for low data-overhead multicast ad hoc routing, *J. Commun. Netw.*, 6(4), 331, 2004.

17. Karp, R.M., Reducibility among combinatorial problems, in *Complexity of Computer Computations*, Plenum Press, New York, 1975, p. 85.

18. Waxman, B.M., Routing of multipoint connections, *IEEE J. Sel. Areas Commun.*, 6(9), 1617, 1998.

19. Kou, L., Markowsky, G., and Berman, L., A fast algorithm for Steiner trees, *Acta Inf.*, 2(15), 141, 1981.

20. Plesnik, J., The complexity of designing a network with minimum diameter, *Networks*, 11, 77, 1981.

21. Zelikovsky, A., An 11/6-approximation algorithm for the network Steiner problem, *Algorithmica*, 9, 463, 1993.

22. Rajagopalan, S. and Vazirani, V.V., On the bidirected cut relaxation for the metric Steiner tree problem, in *Proceedings of SODA*, Society for Industrial and Applied Mathematics, 1999, p. 742.

23. Ruiz, P.M. and Gomez-Skarmeta, A.F., Approximating optimal multicast trees in wireless multihop networks, in *Proceedings of Symposium on Computers and Communications*, IEEE, 2005, p. 686.

24. Clark, B.N., Colbourn, C.J., and Johnson, D.S., Unit disk graphs, *Disc. Math.*, 86, 165, 1990.

25. Giordano, S. and Stojmenovic, I., Position based routing in ad hoc networks, a taxonomy, in *Ad Hoc Wireless Networking*, Cheng, X., Huang, X., and Du, D.Z., Eds., Kluwer, Boston, MA, 2003.

26. Bose, P., Morin, P., Stojmenovic, I., and Urrutia, J., Routing with guaranteed delivery in ad hoc wireless networks, *ACM Wirel. Netw.*, 7(6), 609, 2001.

27. Stojmenovic, I. and Lin, X., Power aware localized routing in wireless networks, *IEEE Trans. Parallel Distrib. Syst.*, 12(11), 1122, 2001.

28. Stojmenovic, I., Localized network layer protocols in wireless sensor networks based on optimizing cost over progress ratio, *IEEE Netw.*, 20(1), 21–27, 2006.

29. Mauve, M., Füßler, H., Widmer, J., and Lang, T., Position-based multicast routing for mobile ad-hoc networks, TR-03-004, Department of Computer Science, University of Mannheim, 2003.

30. Djokic, B., Miyakawa, M., Sekiguchi, S., Semba, I., and Stojmenovic, I., A fast iterative algorithm for generating set partitions, *Comput. J.*, 32(3), 281, 1989.

21

# Approximation Algorithm for Clustering in Mobile Ad-Hoc Networks

Lan Wang

Xianping Wang

Stephan Olariu

## 21.1 Introduction

This chapter offers a survey of clustering schemes in the context of providing general-purpose virtual infrastructures for Mobile Ad-hoc Networks (MANET). We also propose a novel clustering scheme based on a number of properties of diameter-2 graphs. The scheme is cluster-centric and works in the presence of node mobility. The resulting virtual infrastructure is more symmetric and stable than those relying on the selection of central nodes, but still light-weight. Extensive simulation results show the effectiveness of our scheme when compared with other clustering schemes proposed in the recent literature.

As flat networks do not scale, it is a time-honored strategy to overlay a virtual infrastructure on a physical network. There are, essentially, two approaches to doing this. The first approach is protocol-driven and involves crafting a virtual infrastructure in support of whatever protocol happens to be of immediate interest. Although the resulting virtual infrastructure is likely to serve the protocol well, more often than not, the infrastructure is not useful for other purposes. This is unfortunate, as its consequence is that a new infrastructure has to be invented and installed from scratch for each individual protocol in a given suite. In bandwidth-constraint MANET, maintaining different virtual infrastructures for different protocols may involve excessive overhead.

The alternate approach is to design a virtual infrastructure with no particular protocol in mind. The challenge, of course, is to design the virtual infrastructure in such a way that it can be leveraged by a *multitude* of different protocols. Such a virtual infrastructure is called general-purpose as opposed to *special-purpose* if it is designed in support of just one protocol. The benefits of a general-purpose virtual infrastructure are obvious.

To the best of our knowledge, research studies addressing MANET have, thus far, taken only the first approach. Indeed, an amazing array of special-purpose virtual infrastructures have been proposed in support of various sorts of protocols, but only a few of them may have the potential of becoming general-purpose. Our point is that the important problem of identifying general-purpose infrastructures that can be leveraged by a multitude of different protocols has not yet been addressed in MANET.

We view the main contribution of this work as the first step in this direction. Specifically, we identify *clustering* as the archetypal candidate for establishing a general-purpose virtual infrastructure for MANET. We begin by a survey of various MANET clustering schemes proposed in the recent literature, showing that most of these schemes are designed for some specific purposes, and the resulting virtual infrastructures may not be reused effectively by the other applications. For example, in clusters predicated on the existence of a centrally placed cluster-head, such a central node can easily become a communication bottleneck and a single point of failure. Consequently, the resulting virtual infrastructure is not suitable for a number of important network control functions including routing [1] and security.

Motivated by the idea that a virtual infrastructure having a decent chance of becoming truly general-purpose should be able to make a large MANET appear *smaller* and *less dynamic*, we propose a novel clustering scheme based on a number of properties of *diameter-2* graphs. Compared with virtual infrastructures based on central nodes, our virtual infrastructure is more symmetric and more stable, but still light-weight. In our clustering scheme, *cluster initialization* naturally blends into *cluster maintenance*, showing the fundamental unity of these two operations. Unlike the cluster maintenance algorithm in [2], our algorithm does not require maintaining complete cluster topology information at each node. We call our algorithm *tree-based* as cluster merge and split operations are performed on the basis of a spanning tree maintained at some specific nodes. Extensive simulation results have shown the effectiveness of our clustering scheme when compared with other schemes proposed in the literature.

The remainder of this chapter is organized as follows: Section 21.2 provides a succinct survey of MANET clustering schemes. Section 21.3 presents our tree-based clustering scheme. Following the background of our work described in Section 21.3.1, Section 21.3.2 presents technicalities that underlie the tree-based clustering scheme; Section 21.3.3 provides the details of the tree-based clustering algorithms; Section 21.3.4 presents our simulation results. Finally, Section 21.7 offers concluding remarks and directions for further work.

## 21.2  Clustering Schemes for MANET: A Quick Review

A significant number of clustering (*cluster initialization* and *cluster maintenance*) schemes for MANET have been proposed in various contexts. For example, at the medium access layer, clustering helps increase system capacity by promoting the spatial reuse of wireless channel [2]; at the network layer, clustering helps broadcast efficiency [3,4], reduce the size of routing tables [5], and strike a balance between reactive and proactive routing control overhead [6,7]. Although, on the surface, these clustering schemes are quite different, they can be broadly classified into two categories—*node-centric* and *cluster-centric*—depending on what is considered to be atomic. In the node-centric schemes, the atomic entities are the nodes, and clustering amounts to identifying special nodes, commonly referred to as *cluster-heads*, that attract neighboring nodes into clusters. By contrast, in the cluster-centric schemes the cluster is atomic: here, clustering amounts to merging and splitting clusters to keep certain properties.

In the remainder of this section, we review various MANET clustering schemes falling into the two major categories mentioned earlier. In each category, we further group schemes according to different *clustering goals*, that is, the desirable properties of the virtual infrastructure that the clustering schemes *generate* and *maintain*. In the node-centric schemes, the clustering goals include *dominating set (DS)*, *maximal independent set (MIS)*, *connected DS (CDS)*, and so on. In the cluster-centric schemes, the clustering goals include $k$-clustering, $(\alpha, t)$-clustering, multimedia support for wireless network system (*MMWN*) clustering, and so on.

In our discussion, we choose to focus more on the general properties of the proposed virtual infrastructures than on the optimizations targeted at specific applications as we believe that such a relatively application-independent discussion can help identify and compare the contributions and limitations of different clustering schemes more fairly and clearly in the broader context of achieving scalability in MANET.

For each clustering goal, we present a representative sample of the various approaches proposed in the literature. In particular, we are more interested in those approaches that exhibit *local* behavior. A *localized algorithm* was originally defined as a distributed computation in which nodes only communicate with nodes within some neighborhood, yet the overall computation achieves a desired global objective. In [8], a *strictly localized protocol* is defined as a localized protocol in which all information processed by a node is either: (a) local in nature (i.e., they are properties of the node itself or of its neighborhood); or (b) global in nature (i.e., they are properties of the network as a whole), but obtainable immediately (in constant time) by querying only the node's neighbors. For example, consider a protocol that builds a *spanning tree* by performing a distributed breadth-first search (BFS) involving only local communications. Such a protocol is localized but *not* strictly localized is it takes time proportional to the *diameter* of the network, and the entire network must be traversed before the spanning tree can be constructed. This definition of a *strictly localized* algorithm better characterizes the capability of a good localized algorithm to perform *independent* and *simultaneous* operations which is especially desirable for MANET.

In this chapter, the *strictly localized* criterion is adopted as an important yardstick for comparing different clustering schemes.

## 21.2.1 Node-Centric Schemes

In node-centric schemes, a subset of the network nodes is selected to perform network control functions. For example, these special nodes can work as local transmission coordinators [9]; they also naturally form a network backbone to achieve efficient broadcasting [4].

Using graph theory terminology, these nodes form a *DS*, *MIS*, or *CDS* of the network. A more precise definition of these structures follows. Consider a graph $G = (V, E)$, a subset $D$ of $V$ is a DS if each node in $V - D$ is adjacent to some node in $D$. If the subgraph induced by $D$ is connected then $D$ is a CDS. In general graphs, the complexity of finding a *minimum DS* (MDS) or a minimum *CDS* (MCDS) is NP-hard. A subset $S$ of $V$ is an *independent set* (IS) if there is no edge between any pair of nodes in $S$. If no proper superset of $S$ is also an IS then $S$ is an MIS. Note that an MIS is a DS in which no two nodes are adjacent.

### 21.2.1.1 Basic Heuristics: LCA and LCA2

Baker and Ephremides propose two basic clustering heuristics—linked cluster algorithm (LCA) [10] and LCA2 [11]. In LCA, a node $x$ becomes a cluster-head if at least *one* of the following conditions is satisfied: (1) $x$ has the highest nodeID among all its 1-hop neighbors; (2) there exists at least one neighboring node $y$ such that $x$ is the highest ID node in $y$'s 1-hop neighborhood. The distributed implementation of the LCA heuristic terminates in $O(1)$ message rounds under the synchronous network model [12]. Amis et al. [13] generalize LCA to $d$ hops (i.e., each node in the cluster is up to $d$ hops away from the cluster-head), and the corresponding *max-min* heuristic terminates in $O(d)$ message rounds.

The LCA heuristic was revised in LCA2 to decrease the number of cluster-heads. In LCA2, a node is said to be *covered* if it is in the 1-hop neighborhood of a node that has declared itself to be a cluster-head. Starting from the lowest ID node to the highest ID node, a node declares itself to be a cluster-head if it has the lowest ID among the *un-covered* nodes in its 1-hop neighborhood. A distributed implementation of the LCA2 heuristic is described in [2]. It terminates in $O(diam)$ message rounds (*diam* is the *diameter*, or strictly speaking, the *blocking diameter* [14], of the network), and each node transmits exactly one message during the execution of the algorithm.

It is interesting to compare the differences between LCA and LCA2: LCA requires the nodeIDs of both 1-hop and 2-hop neighbors, whereas LCA2 only requires the nodeIDs of 1-hop neighbors; on the other hand, LCA is *strictly localized*, whereas LCA2 is not. In addition, the cluster-heads in LCA form a DS, whereas the cluster-heads in LCA2 form a MIS.

Many heuristics are derived from LCA and LCA2, such as the degree-based heuristic described in [9,15]. Other solutions base the election of cluster-heads on degree of connectivity [9], not node ID. Each node broadcasts the number of nodes that it can hear, including itself. A node is elected as a cluster-head if it is the highest connected node in all of the uncovered neighboring nodes. In the case of a tie, the lowest or highest ID may be used. As the network topology changes, this approach can result in a high turnover of cluster-heads [9]. This is undesirable due to the high overhead associated with cluster-head change over.

All these heuristics make the implicit assumption that each node has a *globally unique* ID. MAC address or IP address are examples of such IDs. However, in some form of ad-hoc networks, such a globally unique ID may not be available in advance. The Clubs [16] algorithm tries to do clustering in such a scenario. In Clubs, the nodes compete by choosing random numbers from a fixed integer range $[0, R)$. Then each node counts down from that number silently. If it reaches zero without receiving a message, the node becomes a cluster-head and broadcasts a cluster-head declaration message. A node that hears the cluster-head declaration message before it has had the chance to declare itself a cluster-head becomes a member of the cluster-head node from which it first receives the cluster-head declaration. The Clubs algorithm takes exactly $R$ rounds to terminate. When duplicate random numbers are chosen, neighboring cluster-heads (*leadership conflict*) may happen. The expected number of *leadership conflicts* is proved to be at most $\frac{D_{avg} * N}{2R}$ ($D_{avg}$ is the average node degree, $N$ is the total number of nodes in the network).

The random count-down mechanism described in Clubs is quite similar to the CSMA/CA medium access control mechanism widely used in wireless networks. This suggests the possibility of integrating the clustering algorithm directly into MAC layer [17,18]. Such an approach is efficient as far as the number of control messages is concerned; however, it is very inflexible as its clustering criterion is based solely on channel access.

### 21.2.1.2 Maximal Independent Set

Basagni's distributed and mobility adaptive clustering (DMAC) [14] algorithm further generalizes the LCA2 heuristic by allowing the selection of cluster-heads based on a generic *weight* associated with each node (instead of using nodeID or degree), and the resulting cluster-heads form a *maximal weighted IS*. The dynamically changing weight values are intended to express how suitable a node is for the role of cluster-head. How to calculate the weight is application-dependent and may include factors such as transmission power, mobility, and remaining battery power, among others [19–21].

The author of DMAC also tries to generalize the algorithm, so that it is suitable for *both* cluster initialization *and* maintenance. This is achieved by augmenting a similar implementation as in [2] so that each node reacts not only to the reception of a message from other nodes, but also to the breakage/formation of a link.

At any time, DMAC guarantees that the following properties are satisfied: (1) Every ordinary node has a cluster-head as its neighbor (*dominance property*); (2) Every ordinary node affiliates with the neighboring cluster-head that has the largest weight; (3) No two cluster-heads can be neighbors (*independence property*).

To enforce the above-mentioned properties, DMAC requires that when a cluster-head $v$ becomes the neighbor of an ordinary node $u$ whose current cluster-head has weight smaller than $v$, $u$ has to affiliate with $v$. Furthermore, when two or more cluster-heads become neighbors, those with the smaller weights have to resign and affiliate with the now largest weight neighboring cluster-head. A node $x$ that originally affiliated with the resigning cluster-head tries to affiliate with an existing cluster-head in its neighborhood *with a larger weight*. If such a node does not exist, $x$ becomes a cluster-head itself. This may trigger further violations of the *independence property*. In such a way, resignation of one cluster-head may cause a *rippling effect* such that some nearby cluster-heads may also have to resign. In the worst case, all the clusters in the whole network have to be reformed. Obviously, this is not what we expect from a good cluster maintenance algorithm.

In an attempt to eliminate the global rippling effect exhibited by DMAC, in the least cluster change (LCC) algorithm described in [22], an ordinary node never challenges current cluster-heads even if it has a larger weight. In G-DMAC [23], adjacent cluster-heads are allowed (hence the *independence property* is no longer enforced), and a node does not have to change its cluster even if it moves in the vicinity of a *better* cluster-head. In ARC [24], a cluster-head change occurs only when one cluster becomes a *subset* of another. These solutions greatly improve the cluster stability compared to with [14]. However, a central node is still assumed in each cluster, and the dominance property of cluster-heads is always enforced.

### 21.2.1.3 Connected Dominating Set

The straightforward application of CDS as network backbone (*spine*) has motivated a significant amount of research effort aiming to design efficient heuristics to achieve *small CDS*. Some approaches are based on clustering algorithms [19,25], whereas others [26,27] build CDS from scratch. We include both approaches here for completeness.

The algorithm proposed by Alzoubi et al. [25] consists of two phases to construct a CDS: the first phase is the construction of a MIS; in the second phase, some special nodes (called *connectors*) are selected to connect the MIS nodes together. *The MIS nodes and the connector nodes jointly form the resulting CDS.* The MIS construction algorithm is essentially the same as LCA2. After the MIS construction phase, nodes exchange messages so that a cluster-head knows the nodeIDs of all the cluster-heads that are located in its 3-hop neighborhood. A cluster-head selects a connector node for all the 2-hop and 3-hop cluster-heads with higher nodeID. A selected connector node *c* further selects a second connector to connect its selector *s* to cluster-heads 3-hop away from *s* and with larger nodeID than *s*. The *maintenance* of CDS involves maintaining the MIS first (similar to the maintenance algorithm in LCC), and then maintaining the connection between all MIS nodes within 3-hop distance through connector nodes. Compared with those algorithms that require a separate phase of constructing a global spanning tree as in [28], this maintenance algorithm is strictly localized, hence is more practical for mobile environment. Using the unit-disk graph model, [25] shows that the size of CDS maintained is within a constant factor (192) of the size of the MCDS.

In [19], Bao and Garcia-Luna-Aceves present a similar two-phase algorithm to construct a CDS. In the first phase, a *priority*-based heuristic similar to LCA is used; hence, the result is a DS instead of a MIS. During the second phase, two types of connector nodes are identified: *doorways* and *gateways*. Accordingly, there are two steps in the second phase: in the *first* step, if two cluster-heads in the DS are 3-hop away and there are no other cluster-heads between them, a node with the highest *priority* on the shortest paths between the two cluster-heads is selected as a *doorway*; in the *second* step, if two cluster-heads *or* one cluster-head and one doorway node are only 2-hop away and there are no other cluster-heads between them, the node between them with the largest nodeID becomes a *gateway* to connect the two cluster-heads or the doorway and the cluster-head. After the two steps, the CDS is formed. Unlike [25] in which cluster-heads are responsible for choosing connector nodes, in [19], each node determines itself whether it becomes a connector. However, as each node only relies on 2-hop neighborhood information to make such a decision, the strictly localized algorithms described in [19] are only approximation of the proposed heuristics for determining connector nodes.

In both of the above-mentioned algorithms, the approach is to first construct a basic DS, and then to add some nodes to get a CDS. The strictly localized algorithm proposed by Wu and Li [27] takes an opposite approach. The algorithm first finds a CDS and then prunes certain *redundant* nodes from the CDS. The initial CDS *U* consists of all nodes which have at least two nonadjacent neighbors. Any node in this set is called an *intermediate node*. Two rules are proposed to eliminate redundant nodes: *Rule 1*: An intermediate node *u* is considered as redundant if it has a neighbor in *U* with larger ID which dominates all the neighbors of *u*. After eliminating the redundant nodes according to Rule 1, the nodes left in *U* are called *intergateway* nodes. *Rule 2*: Assume that $u$, $v$, and $w$ are three intergateway nodes that are mutual neighbors with nodeID satisfying: $u < v$ and $u < w$. If $v$ and $w$ together dominate all the neighbors of $u$, then $u$ is considered as redundant. After eliminating the redundant nodes according to Rule 2, the nodes left in $U$ are called *gateway* nodes. These *gateway* nodes form the resulting CDS. In [26], Stojmenovic et al. improve the previous nodeID-based heuristic by using ($degree$, $x$, and $y$) as the key. Detail simulation results comparing different versions of the heuristic, as well as the cluster approach without any optimization for reducing the number of connector nodes (i.e., all the nodes that have neighbors in different clusters are considered as *border nodes*) are also discussed in [26] in the context of achieving efficient network broadcasting.

Besides, both CDS and DS/MIS have been studied extensively in CEDAR [29] and its precursor Spine [30] to support Quality of Service (QoS) routing in MANET. The rationale for preferring DS/MIS to CDS in such a context is that maintaining a good-quality (small) CDS is much more expensive than maintaining a small DS/MIS in MANET [29].

#### 21.2.1.4 Other Node-Centric Schemes

Some other graph theoretic structures are also proposed as virtual infrastructures for MANET, such as *weakly CDS* [30–32], *d-hop CDS* with the shortest path property [33], and *k-Tree core* [34]. [35] proposes a virtual infrastructure that imposes more constraints on a generalized MIS, that is, the network is partitioned into a forest with a small number of trees, and the root of each tree works as cluster-head. These trees also satisfy depth, weight, and some other constraints for QoS guarantees. The algorithms proposed in the above-mentioned work mainly target static ad-hoc networks, hence the question of how to maintain these virtual infrastructures in response to topology changes is left open.

Several *geometric* structures were also proposed for MANET. These algorithms generally assume unit-disk graph model and require node location information provided by GPS. An excellent survey on these structures can be found in [36]. We shall not elaborate on these structures here.

### 21.2.2 Cluster-Centric Schemes

Cluster-centric schemes focus on dividing a large network into manageable subnetworks to form a hierarchical structure over which essential network control functions can be efficiently supported. For example, each cluster can be assigned a unique code to promote spatial reuse of the wireless channel [2]. Each cluster can also naturally act as unit for abstracting and propagating routing state information [5,7,24,37–44]. In the cluster-centric schemes, there is no special node in a cluster, and each node is capable of assuming the role of logical cluster representative if necessary. Such a more *symmetric* cluster has the potential to form a more *stable* and *robust* virtual infrastructure compared with the node-centric schemes.

#### 21.2.2.1 *k*-Clustering

*k*-Clustering has been suggested by several papers [37,39,40]. Fernandess and Malkhi formally define *minimum k-clustering* in [39] as follows: Given $G = (V, E)$ and a positive integer $k$, find the *smallest* value of $l$ such that there is a partition of $V$ into $l$ disjoint subsets, and the diameter of the graph induced by each subset is not larger than $k$. *k-Clustering* is NP-hard for general graphs.

A cluster initialization algorithm forming diameter-$k$ clusters is presented in [39]. The algorithm works in two stages: in the first stage, a spanning tree of the network is constructed using the MCDS approximation algorithm in [28] (which works in two stages itself); in the second stage, the spanning tree is partitioned into subtrees with bounded diameter. How to maintain such a diameter-$k$ cluster in the face of mobility is not discussed in [39]. In [40], forming and maintaining diameter-1 (*clique*) clusters is discussed in the context of MANET routing.

There are also several clustering schemes imposing *implicit* constraints on cluster diameter, such as the ($\alpha$, $t$)-clustering [7,45] and *MMWN* [38,43] discussed in the following.

### 21.2.2.2 $(\alpha, t)$-Clustering

The objective of the $(\alpha, t)$-clustering framework [7] is to maintain an effective virtual infrastructure that adapts to node mobility so that a hybrid routing protocol can be adopted to balance the tradeoff between proactive and reactive routing control overhead according to the temporal and spatial dynamics of the network. Specifically, the $(\alpha, t)$-clustering framework dynamically organizes mobile nodes into clusters in which the probability of path availability $(\alpha)$ can be bounded for a period of time $(t)$. As $\alpha$ establishes a lower bound on the probability that a given cluster path will remain available for time $t$, it controls the cluster's inherent stability. For a given $\alpha$ (stability level), the role of $t$ is to manage the cluster size, which controls the balance between routing optimality and efficiency.

However, the definition of $(\alpha, t)$-cluster needs to be refined for working effectively in a general MANET. Note that the $(\alpha, t)$-*reachable* relation is *not* transitive. This, together with the fact that $(\alpha, t)$-clusters do not overlap, implies that two nodes that are relatively stable with each other are not necessarily affiliated within the same cluster, defeating the ultimate goal of $(\alpha, t)$-clustering. Indeed, the values of $\alpha$ and $t$ are crucial for the effectiveness of the protocol, and the optimum values depend on the mobility pattern of nodes in the network. How to determine such values is not discussed in [7].

Besides, implementing the cluster maintenance algorithm described in [7] is not an easy task. Consider the following scenario, when a node $X$ detects that a cluster member $Y$ is connected within the cluster, but not $(\alpha, t)$-reachable, $X$ will voluntarily leave the cluster. However, it is possible that $Y$ detects the same situation simultaneously and also voluntarily leaves the cluster. Obviously, this is not an optimal behavior. Even worse, the leaving of nodes will further trigger the $(\alpha, t)$-unreachability of the other nodes that still stay in the original cluster. Hence, a series of leaving events may happen, leading to single-node clusters, which further triggers node joining. This example clearly illustrates the potential convergence problem of an $(\alpha, t)$-cluster, especially when considering the mobile nature of MANET.

McDonald and Znati [45] later propose two major modifications to the original $(\alpha, t)$-clustering framework to address the above-mentioned problems: (a) The *pairwise* $(\alpha, t)$-reachability in an $(\alpha, t)$-cluster is considered too restrictive; hence, the cluster definition is revised so that $(\alpha, t)$-reachability is *only* required between a *potential joining* node and the *parent node* of the cluster; (b) A node does not leave a cluster until the cluster becomes *disconnected*.

### 21.2.2.3 Multimedia Support for Wireless Network System (MMWN)

MMWN [43] presents a hierarchical routing scheme designed for multimedia support in large ad-hoc networks. In MMWN, cluster plays a central role in aggregating QoS routing information and limiting the propagation of topology changes.

The centralized cluster initialization algorithm described in [43] uses global link-state information and *recursive* bisection to produce connected clusters within prescribed *size* limit. In each cluster, a single node, the *cluster leader*, performs cluster split and merge to keep clusters within the size bounds as nodes move.

Based on the MMWN framework, [38] proposes a centralized cluster initialization algorithm that can generate clusters with the following desired properties: (a) Each cluster is *connected*; (b) All clusters should have a min and max *size constraint*; (c) A node belongs to a *constant* number of clusters; (d) Two clusters should have *low* overlap; (e) Clusters should be *stable* across node mobility. The distributed implementation of the centralized algorithm involves creating a *BFS* tree and traversing the tree in *post order*.

Cluster maintenance is also considered in [38]: (1) New node *joins* can cause the violation of property (b) and (c). If (b) is violated, the above-mentioned spanning-tree based clustering algorithm is executed on the current *cluster*; if (c) is violated, the clustering algorithm is executed on the *whole network*, hence *not* strictly localized. (2) Existing node *leaves* may cause the violation of (b), hence the nodes in the smaller clusters must *join* some other cluster. (3) A link breakage may split the cluster into disconnected components, hence is equivalent to the scenario where an existing node *leaves*.

## 21.2.3 Comparing Node-Centric and Cluster-Centric Schemes

The previous subsections have shown that there exists a huge variety of clustering schemes in the literature, each with specific properties. Bettstetter and Krausser [20] propose several general performance metrics that can be used to analyze and compare these significantly different schemes. The major metric proposed is the *stability* of the cluster infrastructure. Indeed, a good clustering algorithm should be designed to maintain its cluster infrastructure as stable as possible while the topology changes [2]. Other proposed metrics include control overhead, level of adaptiveness, convergence time, required neighbor knowledge, etc.

It is important to point out that in the series of CDS algorithms for efficient broadcasting we discussed in Section 21.2.1.3, the major performance metric used to compare different algorithms is the *ratio of nodes in the resulting CDS*. Note that this performance metric does not reflect anything about the *stability* of the CDS. Such a discrepancy in the performance evaluation criteria used again reflects the fact that a virtual infrastructure that can be exploited by and optimized for a *specific* purpose does not necessarily mean a good *general-purpose* virtual infrastructure.

Generally speaking, one advantage of the node-centric schemes is that cluster-heads (and connectors) naturally form a network backbone that can be exploited for broadcasting and activity scheduling [4]. However, constraining all traffic to traverse such special nodes may reduce the throughput and is likely to impact the robustness of the network as the cluster-heads can easily become *traffic bottlenecks* and *single points of failure* [1]. On the other hand, the *cluster-centric* scheme organizes the network into clusters that need not contain a cluster-head. In this scenario, each node can potentially be the logical representative of the cluster, and different nodes can work as cluster representatives for different applications. In MANET where topology changes occur frequently, this implies a potentially more stable general-purpose infrastructure that can be leveraged by a multitude of applications without introducing traffic bottlenecks and single points of failure. Some of the most important differences between the various virtual infrastructures as well as the corresponding clustering schemes are summarized in Table 21.1.

**TABLE 21.1**  A Comparison of the Virtual Infrastructures as well as the Corresponding Clustering Schemes

| Virtual Infrastructure | Defining Properties | Symmetry | Adaptivity | Clustering Scheme |
|---|---|---|---|---|
| LCA [10], LCA2 [11], DMAC [14], G-DMAC [23], LCC [22], ARC [24] | Cluster-heads form a DS or MIS | With a central node in each cluster | Max number of neighboring cluster-heads in the network | Node-centric; strictly localized maintenance algorithms requiring only *1-hop* neighborhood info are proposed |
| Max-min *d*-clustering [13,46] | Cluster-heads form a *d*-hop DS | With a node that is within *d* hops away from every other node in the cluster | *d* | Node-centric; a node needs to maintain *2d*-hop neighborhood info; cluster maintenance is done by periodical cluster initialization |
| Network backbone [4,19,25–27] | Cluster-heads and connectors form a small CDS | With a central node in each cluster | Ratio of CDS nodes | Node-centric; strictly localized maintenance algorithms requiring only *2-hop* neighborhood info are proposed; some algorithms need partial 3-hop neighborhood info for smaller CDS |

*(Continued)*

**TABLE 21.1 (*Continued*)**    A Comparison of the Virtual Infrastructures as well as the Corresponding Clustering Schemes

| Virtual Infrastructure | Defining Properties | Symmetry | Adaptivity | Clustering Scheme |
|---|---|---|---|---|
| $k$-Clustering [37,39,40] | Upper bound ($k$) on cluster diameter | Symmetric | $k$ | Cluster-centric; centralized initialization algorithms (and *non*-strictly localized distributed implementations) are proposed |
| ($\alpha$, $t$)-clustering [7,45] | Lower bound on intracluster path availability | Symmetric | $\alpha$ and $t$ | Cluster-centric; pairwise intracluster ($\alpha$, $t$)-reachability is very difficult to maintain |
| *MMWN*-clustering [38,43] | Lower and upper bounds on cluster size; number of hierarchy levels | Symmetric | Cluster size | Cluster-centric; centralized initialization algorithms (and *non*-strictly localized distributed implementations) are proposed; maintenance algorithms assume complete topology of the cluster |

# 21.3  A Tree-Based Clustering Scheme for MANET

## 21.3.1  Motivation

Essentially, a cluster is a subset of the nodes of the underlying network that satisfies a certain property *P*. At the network initialization stage, a *cluster initialization* algorithm is invoked and the network is partitioned into individual clusters each satisfying property *P*. Due to node mobility, new links may form and old ones may break, leading to changes in the network topology and, thus, to possible violations of property *P*. When property *P* is violated, a *cluster maintenance* algorithm must be invoked. It is intuitively clear that the less stringent property *P*, the less frequently is cluster maintenance necessary.

As discussed in Section 21.2, the precise definition of the desirable property *P* of a cluster varies in different contexts. However, there are some general guidelines suggesting instances of *P* that are desirable in all contexts. One of them is that a *consensus* must be reached quickly in a cluster in order for a cluster to work efficiently. As the time complexity of the task of reaching a distributed consensus increases with the diameter of the underlying graph [12], *small-diameter* clusters are generally preferred in MANET [38]. As an illustration, some authors define property *P* such that every node in the cluster is *1-hop* away from every other node, that is, each cluster is a diameter-1 graph [40]. A less restrictive widely adopted definition of *P* is the *dominance property* [9,10,14] which insists on the existence of a central cluster-head adjacent to all the remaining nodes in the cluster. In the presence of a central node, consensus is reached trivially: indeed, the cluster-head dictates the consensus.

Motivated by the fact that a cluster-head may easily become a traffic bottleneck and a single point of failure in the cluster, and inspired by the instability of the virtual infrastructures maintained by the node-centric clustering schemes, in the clustering scheme proposed by Lin and Gerla [2], although the cluster initialization algorithm used is node-centric with the clusters featuring a central cluster-head, once clusters are constructed, [2] eliminates the requirement for a central node, defining the cluster simply as a diameter-2 graph. Only when the cluster is no longer a diameter-2 graph will a cluster change occur. This definition imposes fewer constraints on a cluster and hence may result in significant improvement on the stability of the resulting virtual infrastructure. In addition, Nakano and Olariu [47] have shown that a distributed consensus can be reached fast in a diameter-2 cluster. In the light of these observations, in this work we adopt the *diameter-2 property* as the *defining property of a cluster*.

The basic idea of the *degree*-based *cluster maintenance algorithm* of [2] is the following: when a violation of the diameter-2 property is detected, the highest degree node and its 1-hop neighbors remain in the original cluster and all the other nodes leave the cluster. It is expected that a leaving node will join another cluster or form a new cluster by itself. Unfortunately, the description of the algorithm in [2] is very succinct and many important details are glossed over.

In fact, there are several problems with the above-mentioned degree-based cluster maintenance algorithm as discussed in [2]. To illustrate, consider the cluster topology in Figure 21.1a. When the link (3,4) is broken due to mobility, the diameter-2 property is violated. One problem is that various nodes have a different local view, precluding them from reaching a global consensus as to which node(s) should leave the cluster. To wit, even if the highest degree of nodes in Figure 21.1b is propagated throughout the entire topology, the nodes still do not have sufficient information to decide whether or not they should leave the cluster. For example, node 3 is adjacent to node 2 which has degree two, thus being a highest degree node. Consequently, node 3 decides that it should not leave the cluster. Likewise, node 5 is adjacent to node 4 which also has the highest degree and decides that it should not leave the cluster. The net effect is that no node will leave, invalidating the correctness of the cluster maintenance algorithm.

Notice that the insecurity we just outlined stems, in part, from the fact that in Figure 21.1b there are three highest degree nodes: nodes 1, 2, and 5. The previous problem can be helped somewhat by using the lowest nodeID criterion to break ties. Under this criterion, node 1 and its 1-hop neighbors, nodes 2 and 5, stay in the original cluster, and nodes 3 and 4 leave. Thus, in this case, the original cluster is partitioned into three clusters: {1,2,5}, {3}, and {4}.

Furthermore, if the cluster maintenance algorithm of [2] is to be fully distributed, each node must maintain the whole topology of the cluster; otherwise, the nodes cannot reach a consensus as to which is the *unique* node with the highest degree. Note that maintaining the complete topology of the cluster at each member node requires flooding the formation and breakage of *every* link to *all* the other nodes in the cluster, involving a large overhead.

The cluster maintenance algorithm of [2] tries to minimize the *number of node transitions* between clusters, and this number is used to evaluate the *stability* of the cluster infrastructure. However, there is no guarantee that this algorithm will minimize node transitions. In the example shown in Figure 21.2a, there are $2n+1$ nodes in the cluster, numbered from 1 to $2n+1$. Nodes $1, 2, \ldots, n$ are within transmission range $(R)$ from each other; similarly, the nodes $n+1, n+2, \ldots, 2n-1$ are within transmission range from each other. With the breakage of link between nodes $2n - 1$ and $2n$, the cluster is no longer diameter-2. Nodes $1, 2, \ldots, n$ have degree $n + 2$ and are the highest degree nodes. Assume that node 1 is chosen as the maintenance leader. In this case, according to the degree-based algorithm, $n - 1$ nodes (viz. nodes

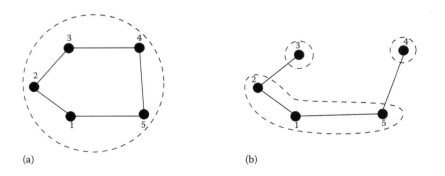

(a)                                                        (b)

**FIGURE 21.1**   An example of the degree-based cluster maintenance algorithm.

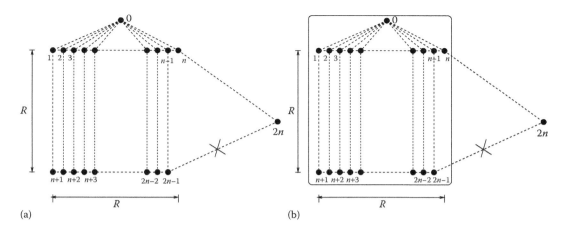

(a)

(b)

**FIGURE 21.2** An example in which the degree-based algorithm generates a lot of leaving nodes.

$n + 2, n + 3, \ldots, 2n$) leave the cluster while, in fact, the minimum number of nodes that have to leave the cluster is just one as shown in Figure 21.2b.

Moreover, using the number of node transitions as the *sole* criterion to assess the goodness of a cluster maintenance algorithm is misleading as: (a) It implicitly assumes that the highest degree node is the same as the logical cluster representative. This assumption is not attractive as during normal operation of a cluster, the highest degree node may change frequently due to link changes. If every highest degree node change results in a migration of the logical cluster representative, a significant amount of overhead will be involved. (b) It assumes that only *leaving* nodes are responsible for the overhead in the cluster maintenance procedure. In reality, during the maintenance procedure, all nodes in the involved clusters participate in computation and message passing for determining the new cluster membership. Consider an example simulation for two clustering schemes 1 and 2. During the simulation, in Scheme 1, a cluster with 100 nodes are split once into two clusters, each with 50 nodes; in Scheme 2, a cluster with 100 nodes decreases its size by one node for 30 times. It is not clear that Scheme 2 is definitely more stable than Scheme 1; (c) In many cases, the degree-based algorithm generates *single-node* clusters. Such a cluster is of little use and must merge with some other existing cluster. This operation should be considered part of the overhead introduced by the cluster maintenance algorithm. Consider the following cluster infrastructure: each node is a single-node cluster, and cluster merge never occurs. In such an infrastructure, the number of node transitions is 0. However, this is a very poor cluster infrastructure and the benefits of clustering are lost. This example clearly illustrates the tradeoff between cluster *stability* and *quality*. We must consider both metrics when evaluating the performance of a cluster maintenance algorithm.

## 21.3.2 Technicalities

The main goal of this subsection is to develop the graph-theoretic machinery that will be used by our clustering algorithms. As customary, we model a multihop ad-hoc network by an undirected graph $G = (V, E)$ in which $V$ is the set of nodes and $E$ is the set of links between nodes. The edge $(u, v) \in E$ exists whenever nodes $u$ and $v$ are 1-hop neighbors. Each node in the network is assigned a unique identifier (nodeID). The distance between two nodes is the length of the *shortest* path between them. The diameter of a graph is the largest distance between any pair of nodes. Our cluster maintenance algorithm relies on the following theorems of diameter-$d$ graphs.

**Theorem 21.1** *Consider a diameter-d graph G and an arbitrary edge e of G. Let $G' = G - e$ be the graph obtained from G by removing edge e. If $G'$ is connected, then there must exist a node in $G'$ whose distance to every other node is at most d. Moreover, the diameter of $G'$ is at most 2d.*

*Proof.* Assume that the edge $e = (u, v)$ is removed. As $G'$ is connected, there must exist a shortest path $P'(u, v) : u = x_1, x_2, \ldots, x_k = v$ joining $u$ and $v$ in $G'$. Consider node $x_{\lceil \frac{k}{2} \rceil}$. Clearly, the distance from $x_{\lceil \frac{k}{2} \rceil}$ to both $u$ and $v$ is unaffected by the removal of the edge $e = (u, v)$. We claim that the distance in $G'$ from $x_{\lceil \frac{k}{2} \rceil}$ to all the remaining nodes is bounded by $d$. To see this, consider an arbitrary node $y$ in $G$ and let $\Pi$ be the shortest path in $G$ joining $x_{\lceil \frac{k}{2} \rceil}$ to $y$. If $\Pi$ does not use the edge $e$, then the removal of $e$ does not affect $\Pi$. Assume, therefore, that $\Pi$ involves the edge $e$. Assume, without loss of generality, that in $\Pi$ node $v$ is closer to $y$ than $u$. However, our choice of $x_{\lceil \frac{k}{2} \rceil}$ guarantees that the path consisting of the nodes $x_{\lceil \frac{k}{2} \rceil}, x_{\lceil \frac{k}{2} \rceil + 1}, \ldots, x_{k-1}, v, \ldots y$ cannot be longer than $\Pi$, completing the first part of the claim.

Consider a BFS tree of $G'$ rooted at $x_{\lceil \frac{k}{2} \rceil}$. We just proved that the depth of this tree is bounded by $d$, confirming that the diameter of $G'$ is, indeed, bounded by $2d$. ∎

Theorem 21.1 has the following important consequence that lies at the heart of our cluster maintenance algorithm.

**Corollary 21.1** *Consider a diameter-2 graph $G$ and an edge $e$ of $G$. Let $G' = G - e$ be the graph obtained from $G$ by removing edge $e$. If $G'$ is connected, then there must exist at least one node in $G'$ whose distance to every other node is at most two. Furthermore, the diameter of $G'$ is at most four.*

**Theorem 21.2** *Let $G$ be a diameter-d graph, and let $x$ and $y$ be a pair of nodes that achieve the diameter of $G$. Then the graph $G' = G - \{x\}$ is connected. Furthermore, in $G'$, any BFS tree rooted at $y$ has depth at most $d$.*

*Proof.* In $G$, $x$ is a level-$d$ (leaf) node of any BFS tree rooted at $y$. Hence, removing node $x$ does not affect the distance from $y$ to any other node. Thus, $G'$ must be connected, and in $G'$, any BFS tree rooted at $y$ has depth at most $d$. ∎

Theorem 21.2 has the following important consequence that will be used in our cluster maintenance algorithm.

**Corollary 21.2** *Let $G$ be a diameter-2 graph, and let $x$ be a node in $G$ such that there exists at least one node $y$ in $G$ that is not adjacent to $x$. In the graph $G' = G - \{x\}$, any BFS tree rooted at $y$ has depth at most two.*

**Theorem 21.3** *Consider a graph $G = (V, E)$, disjoint subsets $V_1, V_2$ of $V$, and let $G'$ be the subgraph of $G$ induced by $V_1 \cup V_2$.*

1. *If the subgraphs of $G$ induced by $V_1$ and $V_2$ are diameter-d graphs, and*
2. *if for every node $x$ of $V_1$, the BFS tree of $G'$ rooted at $x$ has depth at most $d$*

*then $G'$ is a diameter-d graph.*

*Proof.* Consider an arbitrary pair of nodes $u, v$ in $G$. We need to show that $u$ and $v$ are at distance at most $d$ in $G'$. Indeed, if $u, v \in V_1$ (resp. $V_2$), the conclusion is implied by assumption (1). Consequently, we may assume, without loss of generality, that $u \in V_1$ and that $v \in V_2$. By assumption (2), the BFS tree of $G'$ rooted at $u$ has depth at most $d$, implying that the distance between $u$ and $v$ is bounded by $d$. This completes the proof of Theorem 21.3. ∎

## 21.3.3 Our Tree-Based Clustering Algorithm

In MANET, link failures caused by node mobility can be predicted by the gradual weakening of the radio signal strength. In addition, as mechanical mobility and radio transmission occur at vastly different time scales, multiple link failures can be treated as a series of single-link failures. With this in mind, in this work, we adopt the *single-link failure* and *single-node failure* models where either one *link* or one *node* fails at any one time. We also note that the single-node failure model can be used to account for the scenarios where link breakages occur unpredictably.

We make the following two assumptions: (1) a message sent by a node is received correctly by all its neighbors within a finite time, called a *message round*; (2) the cluster split algorithm is atomic in the sense that no new link/node failure occurs during its execution.

### 21.3.3.1 The Tree-Based Cluster Split Algorithm: Single-Link Failure

In this subsection, we discuss the details of our cluster split algorithm in the case where a single-link failure occurs.

When a node detects the formation/breakage of one of its immediate links, it broadcasts a HELLO beaconing message containing its nodeID, clusterID, cluster size, the nodeIDs, and clusterIDs of its 1-hop neighbors, as well as the signal strength of each link to its 1-hop neighbors. By receiving such beaconing messages, each node $u$ maintains a depth-2 BFS tree $T(u)$ rooted at $u$ itself and containing only the nodes belonging to the same cluster as $u$. Clearly, as long as the diameter-2 property holds, the distance between each pair of nodes is at most two, and the tree $T(u)$ contains *all* the nodes in the cluster. Thus, each node knows the number $n$ of nodes in its own cluster.

In our model, each node monitors the signal strength of the links joining it with its 1-hop neighbors. When a generic node $u$ detects that the signal strength of one of its links weakens below a threshold value, it reconstructs $T(u)$. By comparing the size $|T(u)|$ of $T(u)$ with $n$, node $u$ determines whether all the cluster members are still at most two hops away. If it finds that some member cannot be reached in two hops, it broadcasts a VIOLATION message to all of its 1-hop neighbors, identifying the single-link failure causing the violation of diameter-2 property. Each node $v$ receiving a VIOLATION message reconstructs its own tree $T(v)$ and checks whether $|T(v)|$ matches $n$. If there is a mismatch, the node forwards the VIOLATION message to all its neighbors; otherwise, it declares itself a maintenance leader. In other words, a maintenance leader is a node which can reach every other node in at most two hops. By Corollary 21.1, after being forwarded at most once, the VIOLATION message will reach a maintenance leader. Note that there might be multiple maintenance leaders: each of them runs an instance of the cluster split algorithm independently. Finally, *the instance which yields the best quality new clusters is adopted*.

For a generic maintenance leader $x$, the tree $T(x)$ is composed of (1) node $x$ itself—the root of the tree; (2) level-1 nodes, that is, $x$'s 1-hop neighbors in the original cluster; (3) level-2 nodes, all the remaining nodes in the original cluster.

During the split procedure, there can be several different considerations as to how to split the original cluster. Our motivation is to minimize the number of newly generated clusters when splitting. In addition, by considering link stability during a split, the newly generated clusters tend to be more stable.

Specifically, a generic maintenance leader $x$ performs the following steps:

**Step 1:** Node $x$ tries to find the minimum number of level-1 nodes to cover all the level-2 nodes. A level-1 node $y$ can cover a level-2 node $z$ if and only if $x$ can reach $z$ through $y$. This is an instance of the well-known minimum set covering problem and can be solved using the following greedy heuristic [48]:

Initially, all level-2 nodes are marked *uncovered*, and all the level-1 nodes constitute the *total level-1 set*. For each node $y$ in the total level-1 set, $x$ calculates the number $N_y$ of uncovered level-2 nodes that can be covered by $y$. The node $y$ with the largest $N_y$ is deleted from the total level-1 set, added to the *critical level-1 set* and marked as *new leader*. All the $N_y$ level-2 nodes covered by $y$ are marked *covered*. Node $x$ continues the previous process until all the level-2 nodes are marked covered. We call the current total level-1 set as *redundant level-1 set*. For each level-2 node $z$ marked covered, $x$ calculates the stability (i.e., signal strength) of the link $STA_{zw}$ between $z$ and every critical level-1 node $w$. Denote the node $w$ that has the largest $STA_{zw}$ as $p$. Node $x$ marks $w$ as new member of $p$.

**Step 2:** Next, $x$ tries to use the nodes in the critical level-1 set to cover the nodes which are left in the redundant level-1 set. For each node $r$ in the redundant level-1 set, $x$ determines the stability of the link between $r$ and each of the critical level-1 nodes adjacent to $r$. Denote the one that has the most stable link as $w$; $x$ marks $r$ as a new member of $w$.

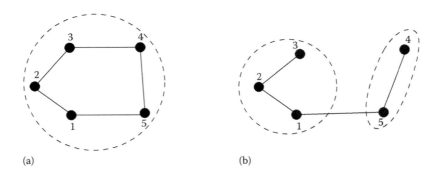

(a)                                                    (b)

**FIGURE 21.3**   An example of the tree-based cluster split algorithm.

**Step 3:** $x$ checks whether there exist nodes in the redundant level-1 set. If so, $x$ marks itself new leader and all the uncovered nodes in the redundant level-1 set as new members of $x$. Otherwise, $x$ finds a new leader $q$ which has the largest link stability value in the critical level-1 set and marks itself as new member of $q$.

At this point, $x$ has reached its cluster split decision. It broadcasts the result through a MAINTENANCE-RESULT message to all its 1-hop neighbors. A node finding itself chosen as a new leader further broadcasts a MEMBER-ENLIST message containing its new cluster membership list. Upon receiving such a message, each node in the original cluster knows its new membership. This completes the split procedure in the case of a single-link failure.

We now illustrate the tree-based split algorithm using an example. There are five nodes in the cluster shown in Figure 21.3a. When the link (3,4) is broken, nodes 3 and 4 detect that the diameter-2 property is violated. Each of them broadcasts a VIOLATION message. Upon receiving the VIOLATION message, nodes 2 and 5 reconstruct their respective BFS trees. As neither of them can work as maintenance leader, they forward the VIOLATION message. When node 1 receives the VIOLATION message, it reconstructs $T(1)$ and finds that $|T(1)| = 5$. At this point, node 1 knows that it is a maintenance leader. In $T(1)$, node 2 covers node 3, and node 5 covers node 4. Hence nodes 2 and 5 are chosen as critical level-1 nodes. Assuming that link (1,2) is more stable than link (1,5), node 1 chooses to be covered by node 2. The result of this split procedure is two new clusters: {1,2,3} and {4,5}, as shown in Figure 21.3b.

### 21.3.3.2  The Tree-Based Cluster Split Algorithm: Single-Node Failure

Our cluster split algorithm for the case when a single-node failure occurs relies, in part, on Corollary 21.2. Indeed, by Corollary 21.2, when a single-node failure occurs in a cluster and the tree maintained by the failed node (just before its failure) has depth two, then the resulting graph is still connected (although it need not be diameter-2), and there must be some node that still maintains a BFS tree with depth at most two. This means that a maintenance leader still exists, and that we can still use our tree-based cluster split algorithm. Specifically, when a node detects the *sudden* breakage of a link to/from a 1-hop neighbor, it assumes a node failure, deletes the failed node from its cluster membership list, and reconstructs the BFS tree. A VIOLATION message is sent out when necessary, identifying the single-node failure causing the violation of diameter-2 property. The remaining steps are the same as those described in Section 21.3.3.1.

However, if the failed node maintains a depth-1 (as opposed to depth-2) tree before its failure, it is possible that none of the remaining nodes can play the role of maintenance leader. To solve this problem, during the cluster's normal operation phase, when a node finds that it is the only node maintaining a depth-1 tree in the cluster, it periodically runs a MDS algorithm (using a greedy algorithm similar to that described in Section 21.3.3.1) on its 1-hop neighbors, and notifies the nodes in the MDS to become candidate maintenance leaders. When the node fails, each candidate maintenance leader detects this failure and immediately broadcasts a MEMBER-ENLIST message containing its new cluster membership list. Upon receiving such a message, each node in the original cluster knows its new membership. This completes the split procedure in the case of a single-node failure.

### 21.3.3.3 Merging Clusters

The previous discussion focused on one aspect of cluster maintenance: the cluster split procedure. Clearly, cluster maintenance cannot rely on cluster splitting only, for otherwise the size of the clusters will continually decrease, and we would end up with many one-node clusters, defeating the purpose of clustering. To prevent this phenomenon from occurring, the other necessary component is a mechanism for merging two clusters. The main goal of this subsection is to discuss a simple tree-based cluster merge procedure.

When the members of two clusters move close so that they can reach each other in two hops, the two clusters may be merged. To better control the cluster merge procedure and to prevent it from being invoked too frequently, we introduce the concept of *desirable size of a cluster*. Specifically, given system parameters—desirable cluster size $k$ and tolerances $\alpha$ and $\beta$, we insist that clusters should have size in the range $[k - \alpha, k + \beta]$. Clusters of size less than $k - \alpha$ are said to be *deficient*. Only deficient clusters are seeking neighboring clusters with which to merge.

For definiteness, consider a deficient cluster $A$ of size $|A| < k - \alpha$. By receiving HELLO beaconing messages described in Section 21.3.3.1, the nodes in $A$ maintain a list of feasible clusters for merging. Among these, the one, say, $B$ such that $|A| \leq |B|$ and $|A| + |B|$ is as close as possible to $k$ but not exceeding $k + \beta$ is selected. *ClusterID* is used to break ties. Upon selecting $B$ as a candidate, the nodes of $A$ that have a 1-hop neighbor in $B$ broadcast a MERGE-REQUEST message. If $B$ is not involved in a merge operation, the nodes of $B$ that have received the MERGE-REQUEST message send back a MERGE-ACK message. At this point, every node in cluster $A$ computes its BFS tree involving nodes in $A \cup B$. A node in $A$ for which the size of the corresponding tree differs from $|A| + |B|$ sends a VIOLATION message to the other nodes in $A$. By virtue of Property 21.3, if no VIOLATION message is received, $A \cup B$ is a diameter-2 graph. In this case, the nodes in cluster $A$ broadcast a MERGE-CONFIRMATION message to cluster $B$ indicating the new cluster membership and the merge procedure terminates. If, however, a VIOLATION message was received, the merge operation is aborted, a MERGE-ABORT message is sent to the nodes of cluster $B$, and a new candidate for merging is examined.

We note that the merge operation takes precedence over split. To explain the intuition behind this design decision refer to Figure 21.4. Here cluster $X$ consisting of nodes $e$ and $f$ attempts to merge with cluster $Y$ consisting of nodes $a$, $b$, $c$, and $d$. Assume that either while the request to merge is issued or just prior to that the edge ($a$ and $d$) broke, invalidating $Y$ as a cluster. Rather than proceeding with the split operation, as would normally be the case, the merge operation is given priority. As illustrated in the figure, all nodes in $X$ and $Y$ detect that $X \cup Y$ has diameter two and is, therefore, a valid cluster. We note, however, that had $X \cup Y$ had diameter larger than two, the merge operation would have failed and the nodes in $Y$ would have proceeded with the split operation.

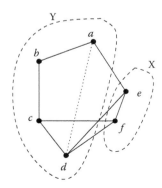

**FIGURE 21.4** Illustrating the priority of merge over split.

### 21.3.3.4 Cluster Initialization

The cluster merge algorithm described in Section 21.3.3.3 is perfectly general and can, in fact, be used for the purpose of cluster initialization. Initially, each *node* is in a *cluster* by itself. The cluster merge algorithm is started as described earlier. The initialization algorithm naturally blends into cluster maintenance as more and more clusters reach desirable size.

It is worth noting that our cluster initialization algorithm has a number of advantages over the nodeID-based algorithms. First, our algorithm is cluster-centric, as opposed to node-centric. Second, the natural blend of cluster initialization and cluster maintenance shows the unity between these two operations. This is certainly not the case in the vast majority of clustering papers in the literature. Third, our cluster initialization algorithm (just as the cluster merge) can be performed in the presence of node mobility.

Last, our initialization algorithm results in better quality clusters than the nodeID-based algorithms. To see this, consider the subnetwork in Figure 21.5a and assume that the desirable cluster size ($k$) is seven with tolerances $\alpha = \beta = 2$. It is not hard to see that our initialization algorithm actually returns the entire sub-network as a single cluster – for this graph is diameter-2. On the other hand, the nodeID-based algorithm results in many deficient clusters, as illustrated in Figure 21.5b.

## 21.3.4 Performance Analysis and Simulation Results

In this subsection, we use simulation to demonstrate the effectiveness of our tree-based clustering scheme compared with other clustering schemes in the literature. We choose LCC [22] as a representative of the node-centric clustering schemes as it avoids the global rippling effect and greatly reduces cluster changes compared with the other nodeID-based algorithms. In addition, it is shown in [49] that in the unit-disk graph model, LCC is asymptotically optimal with respect to the number of clusters maintained in the system.

### 21.3.4.1 Performance Metrics

As discussed in Section 21.3.1, we need to consider both cluster *quality* and cluster *stability* in our comparison. The number of clusters in the system is generally considered as a good indication of the quality of a cluster infrastructure [13,49]. A clustering scheme that generates and maintains fewer clusters is potentially able to accommodate more nodes in a cluster, hence providing better load balancing among clusters. In our simulation, we count the number of clusters in the system once every second of simulation time. We calculate the sum of these numbers divided by the total simulation time, and we use this *average number of clusters* maintained in the system to characterize cluster *quality*. We note, however, that the number of clusters maintained does not tell the whole story. Given two clustering algorithms that

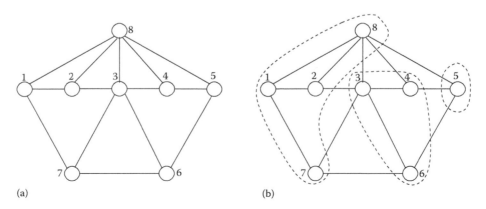

(a)                                                                    (b)

**FIGURE 21.5**   An example of the cluster initialization algorithm.

maintain, essentially, the same number of clusters, we prefer the one that generates clusters of roughly equal size to the one that generates a mix of very large and very small clusters. Indeed, clustering schemes that generate very small clusters have to rely on frequent cluster merges to keep cluster quality, clearly an undesirable situation.

To evaluate cluster *stability*, we assume that each cluster chooses one of its member as cluster leader and takes its nodeID as the clusterID. When a node is no longer in the same cluster as its latest cluster leader, this node is considered as a node *changing cluster*. Note that the cluster leader defined here serves only as a reference point that allows us to count and compare the number of *node transitions* in different clustering schemes. In LCC, the central node of a cluster is always the cluster leader. In the diameter-2 schemes, each node initially chooses its nodeID as the clusterID of the single-node cluster. When two clusters merge, the clusterID of the cluster with larger size is used as the new clusterID. When a cluster split happens, among the new clusters, the one which contains the original cluster leader still keeps the original clusterID, and all the other clusters choose the minimum nodeID of its members as the new clusterID. Further, we need to clearly identify the events that can cause cluster changes. In LCC, there are two types of events that can cause nodes to change clusters: (1) a nonleader node is no longer adjacent to its leader; in this case, the node joins another leader, or becomes itself a new leader; (2) when two cluster leaders become neighbors, the one with larger nodeID gives up its role, and all the nodes in its cluster either join a new cluster, or become new leaders by themselves. In the diameter-2 schemes, the two types of events that can cause nodes to change clusters are (1) a cluster is no longer diameter-2 and is split to several subclusters; (2) a cluster merges with another cluster.

With the above-mentioned assumptions in mind, we define two measurements to evaluate cluster *stability*: (1) *total number of nodes changing clusters*; (2) *average cluster lifetime*. Specifically, we compare the snapshots of the system taken exactly before and after the execution of the maintenance algorithm triggered by either of the earlier events. If node $x$'s clusterID after the event is different from its clusterID before the event, then it is counted as a *node changing its cluster*. If a node $x$ is a cluster leader before the event, but no longer a leader after the event, then the cluster is considered as disappearing and we stop increasing its lifetime. If a node $x$ is not a cluster leader before the event, but becomes one after the event, then a new cluster is considered generated, and we start increasing its lifetime. The *average cluster lifetime* is calculated as the sum of all the cluster lifetimes divided by the number of clusters generated in the simulation.

### 21.3.4.2 Simulation Results

We simulate a MANET by placing $N$ nodes within a bounded region of area $A$. The nodes move according to the random way-point model [50] with zero pause time and constant node speed $V$. All the nodes have uniform transmission range, which varies from 30 m to 210 m in different simulations. For each simulation, we examine the first 300 seconds of simulation time. All the simulation results presented here are an average of 10 different simulation runs. We also plot 95% confidence intervals for the means. The small confidence intervals show that our simulation results precisely represent the unknown means.

A set of representative simulation results ($N = 100$, $A = 500$ m $\times$ 500 m, $V = 5$ m/s) are shown in Figures 21.6–21.9. For the tree-based algorithm, we implement a *baseline* version which does not consider link stability during cluster split. Also, as the tree-based algorithm allows for controlling cluster merging frequency and LCC and the degree-based algorithm do not, we have set the *desirable size of a cluster* to $\infty$.

(A) *Comparing the node-centric LCC and the cluster-centric diameter-2 schemes*
Figure 21.6 indicates that the average number of clusters in the system maintained by the diameter-2 clustering schemes is about *half* of that maintained by LCC. Figure 21.7 shows that the number of nodes changing clusters in LCC is significantly larger than in either of the diameter-2 schemes. This is hardly a surprise as LCC is node-centric, and it is obvious that clusters predicated on the existence of a central node (the cluster-head) are more brittle than regular diameter-2 clusters. This is further confirmed by Figure 21.8 that illustrates that the average lifetime of clusters generated by LCC is shorter than the

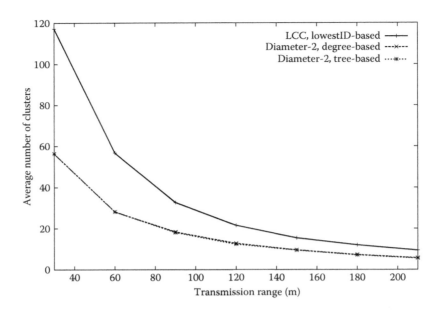

**FIGURE 21.6**   Comparing the performance of different clustering schemes—average number of clusters.

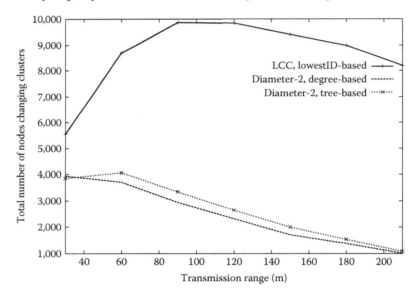

**FIGURE 21.7**   Comparing the performance of different clustering schemes—total number of nodes changing clusters.

lifetime of clusters generated by either of the diameter-2 schemes. These results demonstrate that by removing the central-node constraint, the diameter-2 cluster is a much more *stable* structure and can provide *better quality* clusters, especially in MANET applications where central node is not necessary, such as [2,7,37,43].

(B) *Comparing the tree-based algorithm and the degree-based algorithm*
In terms of the average number of clusters maintained in the system, the tree-based algorithm is slightly better than the degree-based algorithm as shown in Figure 21.6. Figure 21.8 shows that the average cluster

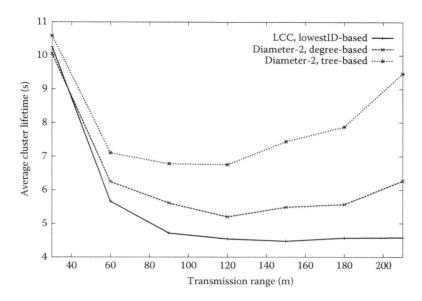

**FIGURE 21.8** Comparing the performance of different clustering schemes—average cluster lifetime.

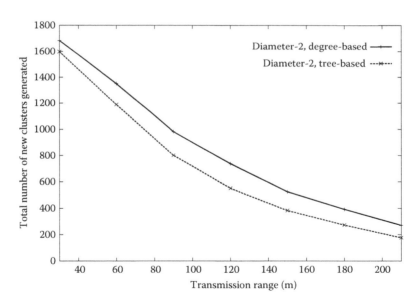

**FIGURE 21.9** Comparing the performance of different clustering schemes—total number of clusters generated during simulation.

lifetime in the tree-based algorithm is longer than in the degree-based algorithm. From Figure 21.9, we can see that the degree-based algorithm generates many more new clusters than the tree-based algorithm. On the other hand, Figure 21.7 shows that the total number of nodes changing clusters is significantly larger in the tree-based algorithm than in the degree-based algorithm. The explanation is simple: the degree-based algorithm tends to generate single-node clusters during cluster split, while the clusters generated by the tree-based algorithm are much more *balanced*. The net effect is that when a cluster split/merge happens, a larger number of nodes change clusters in the tree-based algorithm than in the degree-based algorithm.

This result shows that the number of nodes changing clusters is not always indicative of the quality of the cluster maintenance algorithm. Note that the single-node clusters generated in the degree-based algorithm are *short-living* and will be merged with other clusters soon; hence, they do not significantly influence the *average number of clusters* maintained in the system shown in Figure 21.6.

It is important to realize that what really distinguishes the tree-based algorithm and the degree-based algorithm is the *cluster maintenance overhead*. As the degree of a node is a rather unstable parameter, in the degree-based algorithm, *every* link change (formation and breakage) has to be forwarded to *all* the cluster members. This is certainly not the case in the tree-based algorithm where, as long as the cluster is still diameter-2, link formation and link breakage are propagated in the HELLO beaconing message as described in Section 21.3.3 and will not be forwarded by the other nodes.

To take this point one step further, we count the total number of intracluster link changes during the simulation. We call a link change between nodes $A$ and $B$ in the same cluster *benign* if after the change nodes $A$ and $B$ remain in the original diameter-2 cluster, and $A$ and $B$ have a common 2-hop neighbor. For example, in the cluster shown in Figure 21.5a, the breakage of link (6,7) is benign as the resultant graph is still diameter-2, and nodes 6 and 7 have a common 2-hop neighbor (node 8). However, the breakage of link (3,8) is not benign as nodes 3 and 8 do not have a common 2-hop neighbor. We note that, trivially, the tree-based algorithm saves *at least* one message forwarding per benign link change over the degree-based algorithm. We count the number and ratio of benign link changes, and the corresponding simulation results are shown in Figure 21.10. As the simulation result shows, the ratio of benign link changes is quite significant, and as the node density becomes higher, the savings become more and more significant.

Our simulation results have revealed an interesting piece of evidence that speaks for the robustness of our tree-based algorithm: even when multiple link failures occur in a cluster, the probability of the existence of a maintenance leader is still very high. Theoretically, when multiple edges are removed from a diameter-2 graph, there may no longer exist a maintenance leader in the resulting graph. There are two approaches that can be employed by the tree-based algorithm to deal with this situation. The first approach is to predict link failure ahead of time whenever possible. Thus, when multiple link failures occur at the same time, all these links are actually still there, and the maintenance leader can arbitrarily choose one link as the only broken link. Essentially, this prevents real link failures from occurring in the first place. The second approach is to simply let multiple link failures occur. By Corollary 21.1, if a maintenance leader exists, each node will know the maintenance result in at most four message rounds. A node sets a 4-message round long timer when violation is detected. Upon time-out, each node uses the cluster initialization algorithm described in Section 21.3.3.4 as the last resort for cluster maintenance.

# 21.4 Application

In this section, we discuss *topology control* [51] in mobile ad-hoc networks as a sample application of the cluster-based general-purpose infrastructure we have proposed.

Cluster-based infrastructure provides a natural framework for designing topology control algorithms. In such a framework, no node maintains the global topology. Instead, the framework relies on clustering where nodes autonomously form groups. In each cluster, a centralized topology control algorithm is executed by a cluster-head, or a distributed topology control algorithm is executed by all the nodes, so that a some desirable topology properties are achieved in the cluster. The desired topology properties between clusters are achieved by exchanging information between adjacent clusters.

Motivated by the above-mentioned idea, we propose a cluster-based algorithm to construct an approximate Minimum Spanning Tree (MST). The algorithm has three phases: (1) Phase 1: Cluster formation. A distributed clustering algorithm is used to generate and maintain clusters in the network. In this work, our focus is the diameter-2 clustering we have proposed; (2) Phase 2: Forming intracluster MST. In our infrastructure, as each cluster is diameter-2, a distributed MST algorithm exists that finishes very quickly [12]. Alternatively, a leader for topology control can be elected in each cluster, which is responsible for running a centralized MST algorithm (such as Kruskal's algorithm [48]). Note that this

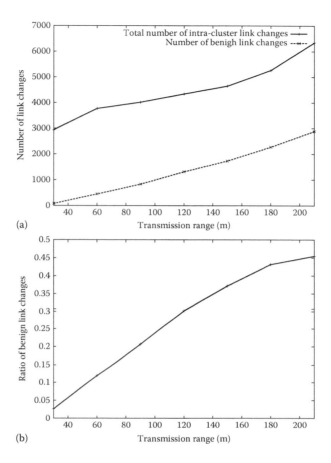

**FIGURE 21.10** Comparing the maintenance overhead of tree-based and degree-based algorithms: (a) Number of benign intracluster link changes and (b) Ratio of benign link changes.

leader is a *logical* leader for the topology control application only; (3) Phase 3: Connecting clusters. In this phase, connectivity between adjacent clusters is considered. Each cluster runs the following algorithm: by exchanging information with neighboring clusters only, a cluster knows its *shortest* link to each of its adjacent clusters, as well as the shortest links between each pair of its adjacent clusters. Based on this information, a cluster constructs a Localized MST (LMST) [52]. Note that each node in the LMST is a cluster, and each edge is the LMST is an actual link between two nodes. When running the LMST to establish connections between two adjacent clusters, the power assigned to the involved nodes is increased only. The collections of all edges in the LMSTs constructed by all nodes, as well as the links selected in Phase 2, form the resulting structure.

We have conducted a simulation study to determine the effectiveness of our cluster-based MST algorithm. In this study, 100–500 nodes were distributed uniformly at random in an area of $1000 \times 1000\,\mathrm{m}^2$. When operating at full transmission power, each node has a transmission range of 250 m. In the simulation, for a specific number of nodes, we generate 50 different topologies. And the result is the average of these 50 simulation runs. Also, in this simulation, we consider static topology only.

We consider the following metrics in the simulation: (1) The two most important metrics, average link length and number of links in the resulting topology, consider only the bidirectional links in the resulting connected structures. For a connected network with $N$ nodes, its MST has $N - 1$ links. The average link length is calculated as the sum of the length of each link divided by the number of links;

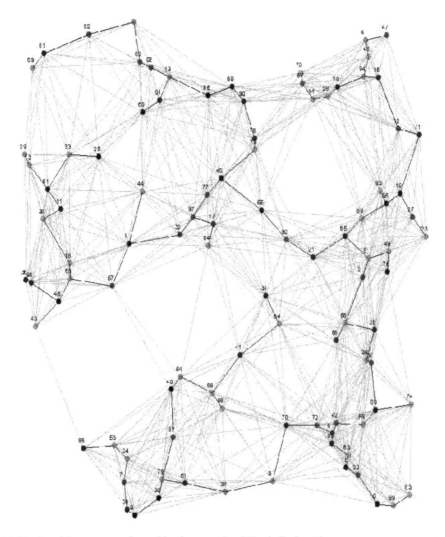

**FIGURE 21.11**   Resulting structure formed by the centralized Kruskal's algorithm.

(2) The degree of the node is counted in the following way: for a node $u$ with transmission power $P_u$, and a node $v$ with transmission power $P_v$, if the distance between $u$ and $v$ is not larger than $P_v$, then node $v$ is considered as a neighbor of $u$. Note that this relationship is not symmetric; (3) Average node power is calculated as the sum of the powers assigned to each node divided by the total number of nodes in the network; (4) Max node power: It is the maximum value among the powers assigned to the nodes in the network.

A sample topology and the resulting structures generated by the three different topology control algorithms are shown in Figures 21.11, through 21.13. From Figures 21.12 and 21.13, we can see that for this specific topology, seven clusters are generated by the diameter-2 clustering scheme (clusters 0, 1, 2, 3, 4, 9, and 55), whereas eleven clusters are generated by the lowestID clustering scheme (clusters 0, 1, 2, 3, 4, 9, 13, 43, 54, 55, and 74). More generally, for any given topology, the diameter-2 clustering scheme can potentially generate/maintain fewer (or equal) number of clusters than any central-node-based clustering scheme; hence, there are more topology information available for making *intra*-cluster decisions (as there are *more* nodes in a cluster) and for making *inter*-cluster decisions (as there are *fewer* clusters in the network), leading to a better quality global structure.

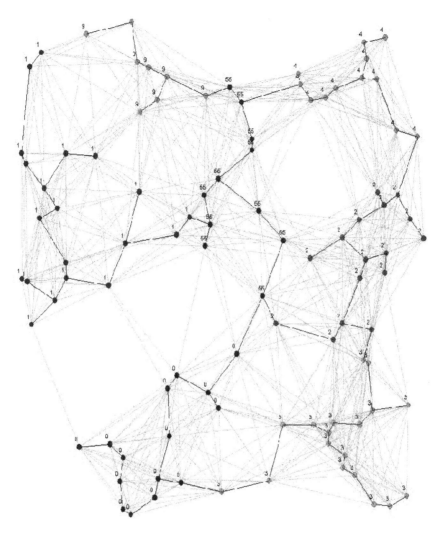

**FIGURE 21.12** Resulting structure formed by the cluster-based MST algorithm. The clustering algorithm used is diameter-2 clustering.

More detail simulation results are shown in Table 21.2. In the table, *MST* is the result using a centralized Kruskal's algorithm; *Diameter-2* and *LowestID* are the results of diameter-2 clustering and the lowestID clustering, respectively.

From the simulation result, it is evident that resulting topology constructed by our cluster-based MST algorithm approximates the MST effectively in terms of all the four performance metrics used. Specifically, (1) The average link length of the resulting structure is very close to the optimal value (the approximation ratio is 1.06, 1.06, 1.05, 1.08, and 1.05 as the number of nodes increases from 100 to 500); the number of links in the resulting structure is about three more than the optimal value, regardless of the number of nodes in the networks (the approximation ratio is 1.03, 1.02, 1.01, 1.01, and 1.01 as the number of nodes increases from 100 to 500); (2) The average node degree keeps stable when the number of nodes increases; (the approximation ratio is 1.15, 1.16, 1.14, 1.12, and 1.16 as the number of nodes increases from 100 to 500); (3) The average node power is very close to the optimal value (as the number of nodes increases from 100 to 500; the approximation ratio of the average node power is 1.09, 1.08, 1.07,

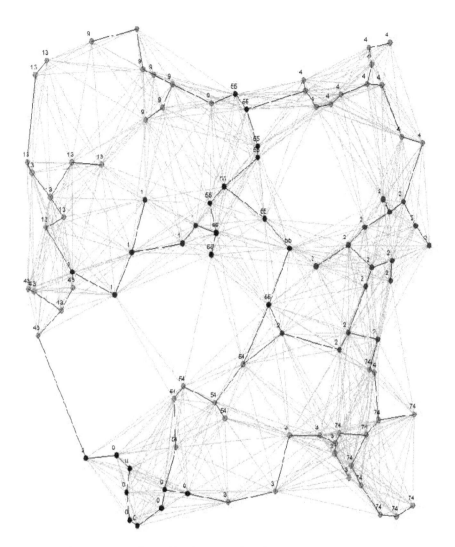

**FIGURE 21.13**   Resulting structure formed by the cluster-based MST algorithm. The clustering algorithm used is the lowestID clustering.

1.07, and 1.07); (4) The approximation ratio of the max node power is a little high (1.16, 1.27, 1.25, 1.27, and 1.27 as the number of nodes increases from 100 to 500). This is expected as max node power is determined by the critical part of a network. In fact, it is proved in [51] that it is impossible for any localized algorithm to construct a connected structure such that the max node power based on this structure is within a constant factor of that based on MST.

In the simulation result, the diameter-2 clustering consistently generates better quality structures than the lowestID clustering in terms of all the performance metrics used; however, the difference between the two is small. The reason is that the simulation is conducted on static topologies only, and under static topologies, the difference between these two clustering schemes is not as dramatic as the difference in face of mobility (Figure 21.6). The advantage of diameter-2 clustering scheme is expected to be more obvious in face of node mobility.

Finally, it is worth emphasizing that in this section, we use MST construction as an illustration of the application of our proposed general-purpose infrastructure, but in fact, the cluster-based infrastructure

**TABLE 21.2** Performance Comparison of the Three Topology Control Algorithms

| Number of Nodes | Algorithm | MST | Diameter-2 | LowestID |
|---|---|---|---|---|
| 100 | Max node power | 164.44 | 190.32 | 192.23 |
| 100 | Average node power | 82.19 | 89.89 | 90.89 |
| 100 | Average node degree | 2.51 | 2.89 | 2.93 |
| 100 | Average link length | 68.06 | 72.14 | 72.77 |
| 100 | Number of links | 99 | 102.42 | 103.20 |
| 200 | Max node power | 116.74 | 147.80 | 149.71 |
| 200 | Average node power | 57.73 | 62.51 | 62.88 |
| 200 | Average node degree | 2.51 | 2.90 | 2.92 |
| 200 | Average link length | 47.42 | 50.32 | 50.50 |
| 200 | Number of links | 199 | 202.58 | 202.94 |
| 300 | Max node power | 99.44 | 124.19 | 125.79 |
| 300 | Average node power | 46.97 | 50.41 | 50.53 |
| 300 | Average node degree | 2.50 | 2.85 | 2.86 |
| 300 | Average link length | 38.66 | 40.70 | 40.79 |
| 300 | Number of links | 299 | 302.78 | 303.06 |
| 400 | Max node power | 86.70 | 110.51 | 113.82 |
| 400 | Average node power | 40.28 | 42.94 | 43.15 |
| 400 | Average node degree | 2.51 | 2.81 | 2.83 |
| 400 | Average link length | 33.19 | 35.74 | 34.85 |
| 400 | Number of links | 399 | 402.92 | 403.52 |
| 500 | Max node power | 78.44 | 99.75 | 100.42 |
| 500 | Average node power | 36.00 | 38.36 | 38.48 |
| 500 | Average node degree | 2.51 | 2.80 | 2.81 |
| 500 | Average link length | 29.67 | 31.09 | 31.15 |
| 500 | Number of links | 499 | 503.58 | 504.04 |

provides a powerful general framework, and similar approaches can be used to establish many other global structures such as *strongly connected* graphs [53].

## 21.5 Proof of Some Properties of Diameter-2 Graphs

In this section, we prove some properties of diameter-2 graphs.

Let $T$ be a set of nodes (with transmission range $D$) on the plane with the following property $P1$:

**P1 (diameter-2 property)**: For every two nodes $p, q \in T$, there exists a node $r \in T$ such that $|pr| \leq D$ and $|qr| \leq D$. If $|pq| \leq D$, we can take $r$ to be either $p$ or $q$.

Let $T_d \subseteq T$ be a subset of $T$ with the following property $P2$.

**P2 (dominating property)**: For every point $x \in T$, there exists a point $y \in T_d$ such that $|xy| \leq D$.

**Lemma 21.1** *Let $V$ be a circle. The chord $\overline{pq}$ divides $V$ into two parts $V = V_+ \cup V_-$. Let $U_r$ be the circle of centered at $r$. If $U_r$ covers both point $p$ and point $q$, then $U_r$ covers either $V_+$ or $V_-$.*

*Proof.* We prove by contradiction. In the following, we assume that chord $\overline{pq}$ divides $V$ into a *left* part and a *right* part.

Case 1: Assume that both $p$ and $q$ are on the boundary of the circle $U_r$. As $U_r$ can cover neither $V_+$ nor $V_-$, then if there is a third intersecting point between circle $U_r$ and circle $V$, $U_r$ is same as $V$. So, $p, q$ must be the only two intersecting points between $U_r$ and $V$. Consider the arc of $U_r$ on the right side of $\overline{pq}$, the center of $U_r$ must be to the left of the center of $V$. On the other hand, consider the arc of $U_r$ on the left side of $\overline{pq}$, the center of $U_r$ must be to the right of the center of $V$. This is a contradiction.

Case 2: Assume that only $p$ is on the boundary of circle $U_r$. If $U_r$ cannot cover either $V_+$ or $V_-$, circle $U_r$ and circle $V$ must have at least an intersecting point on the left side of $\overline{pq}$ and at least an intersecting point on the right side of $\overline{pq}$. This contradicts the fact that three points determine a circle.

Case 3: Assume that neither $p$ nor $q$ is on the boundary of circle $U_r$. If $U_r$ can cover neither $V_+$ nor $V_-$, circle $U_r$ and circle $V$ must have two intersecting points on the left side of $\overline{pq}$ and two intersecting point on the right side of $\overline{pq}$. This contradicts the fact three points determine a circle.                                                                  ■

**Lemma 21.2** *Let $V$ be a circle that contains $T$. There are two nodes $p, q \in T$ lying on the boundary of $V$. The chord $\overline{pq}$ divides $V$ into two parts $V = V_+ \cup V_-$, where $V_+$ is the larger part in terms of area. Let $r$ be a node in $T$ such that $|pr| \leq D$ and $|qr| \leq D$. Let $U_r$ be the circle of radius $D$ centered at $r$. Suppose that there is a node $r \in T$ such that $V_+ \subset U_r$, then there exists a subset $T_d$ of $T$ with property P2 and $|T_d| \leq 3$.*

*Proof.* Let $S$ be the set of nodes of $T$ that lie in $V_-$, but not on the chord $\overline{pq}$. If $S$ is empty, then we are done. Otherwise, we prove by **induction on the number of nodes in $S$**.

Let $w \in S$ be the node such that the angle $\angle pwq$ achieves the *minimum* for any $w \in S$ (see Figure 21.14). Let $U_{pwq}$ be the circle that passes through $p, q$, and $w$. Our choice of $w$ implies that $T \subset U_{pwq}$.

Let $a \in T$ be the node such that $|pa| \leq D$ and $|wa| \leq D$. Also let $U_a$ be the circle of radius $D$ centered at $a$. Let $\widehat{pw}$ and $\widehat{wqp}$ be two arcs of the boundary of $U_{pwq}$ divided by the chord $\overline{pw}$. According to Lemma 21.1, we have either (a) $\widehat{pw} \subset U_a$, or (b) $\widehat{wqp} \subset U_a$.

If (b) is true then we can replace $V$ with $U_{pwq}$, replace node $r$ with node $a$, and replace $p, q$ with $p, w$. We have reduced the number of nodes in $S$ by *one* (point $w$), and hence, it follows from induction hypothesis that Lemma 21.2 holds.

Similarly, let $b \in T$ be the node such that $|wb| \leq D$ and $|qb| \leq D$. Then either $\widehat{wq} \subset U_b$ or $\widehat{qpw} \subset U_b$. Again, if $\widehat{qpw} \subset U_b$ is true, it follows from induction hypothesis and we are done.

So, the remaining case is that both $\widehat{pw} \subset U_a$ and $\widehat{wq} \subset U_b$. In the remaining of this proof, we are going to prove that in this case, $T \subset U_a \cup U_b \cup U_r$.

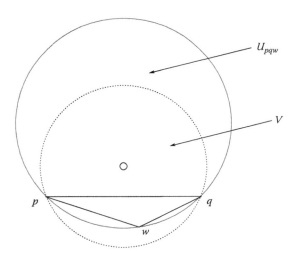

**FIGURE 21.14**   Proof of Lemma 21.2: $w \in S$ achieves the minimum angle, and $U_{pwq}$ is the circle passing through $p, q$, and $w$.

Let $V_{pwq} \subset U_{pwq}$ be the region bounded by $\widehat{pwq}$ and chord $\overline{pq}$. Then $T \subset V_{pwq} \cup V_{+}$. As we assume that $V_{+} \subset U_r$, it is enough to show that $V_{pwq} \subset U_a \cup U_b$. Let $x$ be the midpoint of the chord $\overline{pq}$. If we can show that $x \in U_a$ and $x \in U_b$ then the convex region bounded by $\widehat{pw}$, $\overline{xp}$, and $\overline{wx}$ lies in $U_a$, and the convex region bounded by $\widehat{wq}$, $\overline{qx}$, and $\overline{xw}$ lies in $U_b$; hence, $V_{pwq} \subset U_a \cup U_b$. So it is enough to show that $|ax| \leq 2$ and $|bx| \leq 2$.

We first prove that $|ax| \leq 2$. Note that $a$ can be chosen to be any point in $T$ satisfying $|pa| \leq 2$ and $|wa| \leq 2$. So if $|pw| \leq 2$, we may choose $a = p$. Similarly, if $|wq| \leq 2$, we choose $b = q$.

As $|pq| \leq 4$, so for every point $y \in V_{-}$ we have $|xy| \leq 2$. So if $a \in V_{-}$, then $|ax| \leq 2$, and we are done. Also, if $|pw| \leq 2$, then $a = p$ and $|ax| \leq 2$, and we are done.

The remaining case is that $a \in V_{+}$ and $|pw| > 2$. There are two subcases here based on whether $x$ is located inside $\triangle paw$.

Case (1): $x$ is located outside $\triangle paw$ (see Figure 21.15). Assume $|ax| > 2$. Then we have $|ax| + |pw| > 4$. Consider the quadrilateral $apwx$, we have $|px| + |aw| > |ax| + |pw|$. Hence, $|px| + |aw| > 4$. However, this is impossible as $|px| \leq 2$ and $|aw| \leq 2$.

Case (2): $x$ is located inside $\triangle paw$ (see Figure 21.16). Assume $|ax| > 2$. In $\triangle pax$, we have $|px| \leq 2$, $|ap| \leq 2$, so $\angle apx > \pi/3$. In $\triangle wax$, we have $|wx| \leq 2$, $|aw| \leq 2$, so $\angle awx > \pi/3$. So, In $\triangle paw$, $\angle paw < \pi/3$. However, as $|pw| > 2$, $|ap| \leq 2$, $|aw| \leq 2$, we have $\angle paw > \pi/3$. This is a contradiction.

Hence $|ax| \leq 2$. Similarly, we can prove that $|bx| \leq 2$. ∎

**Lemma 21.3** *Let $V$ be the smallest circle that contains $T$, and there are three nodes $p, q$, and $r \in T$ lying on the boundary of $V$. Among the chord $|pq|$, $|qr|$, and $|pr|$, if at least two are $\leq D$, then $V$ is covered by at most three nodes in $T$.*

*Proof.* Without loss of generality, we assume that $|pr| \leq D$ and $|qr| \leq D$. Also we assume that $|pr| \leq |qr|$. Now we draw a circle with $U_r$ with $r$ as center and $|qr|$ as radius. The area covered by $U_r$ includes the following three parts: (1) the area bounded by $\widehat{pr}$ and $\overline{pr}$; (2) the area bounded by $\widehat{qr}$ and $\overline{qr}$; (3) $\triangle pqr$.

According to Jung's Theorem [54], the center of circle $V$ must be located inside $\triangle pqr$. This means that $U_r$ covers the *larger* part of $V$. Based on Lemma 21.2, $V$ can be covered by at most three nodes in $T$. ∎

**Theorem 21.4** *Let $T$ be set of nodes with property P1, then there exists a subset $T_d$ of $T$ with property P2 and $|T_d| \leq 3$.*

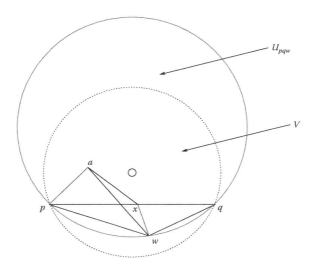

**FIGURE 21.15** Proof of Lemma 21.2 case (1): $x$ is located outside $\triangle paw$.

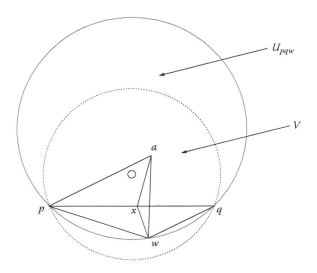

**FIGURE 21.16**   Proof of Lemma 21.2 case (2): $x$ is located inside $\triangle paw$.

*Proof.* Let $V$ be the *smallest* circle that contains $T$. By Jung's Theorem [54], we know that one of the following holds:

    **Case (1):** There are *two* nodes $p$ and $q \in T$ lying on the boundary of $V$, and $|pq|$ is the diameter of $V$;

    **Case (2):** There are *three* nodes $p, q$, and $r \in T$ lying on the boundary of $V$, and the center of $V$ lies inside the triangle $\triangle pqr$.

    Let $R$ be the radius of $V$. In case (1), it is obvious that $R = d(T)/2 \leq 2$. In case (2), it can be shown that $R \leq d(T)/\sqrt{3} \leq 4/\sqrt{3}$. (The equality holds when $|pq| = |qr| = |rp| = d(T) = 4$.)

For the case (1) in Theorem 21.4, it follows directly from the Lemma 21.2 as node $r$ always exists.

For the case (2) in Theorem 21.4, there are three nodes $p, q$, and $r \in T$ lying on the boundary of $V$. Let $x, y$, and $z \in T$ be the nodes such that $|pz|, |qz|, |qx|, |rx|, |ry|$, and $|py| \leq D$. Let $U_x, U_y$, and $U_z$ be the circles of radius $D$ centered at $x, y$, and $z$, respectively. (Figure 21.17).

    Let $\widehat{pq}, \widehat{qr}$, and $\widehat{rp}$ be the arcs of the boundary of $V$ such that $\widehat{pq} \cup \widehat{qrp} = \widehat{qr} \cup \widehat{rpq} = \widehat{rp} \cup \widehat{pqr} = boundary(V)$.

    As $q, r \in U_x$, we have either (a) $\widehat{qr} \subset U_x$ or (b) $\widehat{rpq} \subset U_x$; similarly, either (a) $\widehat{rp} \subset U_y$ or (b) $\widehat{pqr} \subset U_y$; either (a) $\widehat{pq} \subset U_z$ or (b) $\widehat{qrp} \subset U_z$ .

    If any one of the above-mentioned three (b)s is true, Theorem 21.4 follows directly from Lemma 21.2.

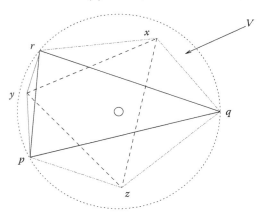

**FIGURE 21.17**   Proof of Theorem 21.4: $x, y, z \in T$ are three points such that $|pz|, |qz|, |qx|, |rx|, |ry|, |py| \leq 2$.

So, in the following, we assume $\widehat{qr} \subset U_x$, $\widehat{rp} \subset U_y$, and $\widehat{pq} \subset U_z$.

Further, if $x$ and $y$ are the same node, then $x$ covers $\triangle pqr$. As the center of $V$ is located inside $\triangle pqr$, $x$ covers the *larger* part of $V$. According to Lemma 21.2, we are done.

Hence in the following, we assume that $x, y$, and $z$ are three *unique* nodes.

It remains to prove that $\triangle pqr \subset U_x \cup U_y \cup U_z$.

To show that $\triangle pqr \subset U_x \cup U_y \cup U_z$, it suffices to show that there does *not* exists a point $w$ inside $\triangle pqr$ such that $|wx| > 2, |wy| > 2$, and $|wz| > 2$.

**We prove by contradiction. Assume that there exists a point $w$ inside $\triangle pqr$ such that $|wx| > 2$, $|wy| > 2$, and $|wz| > 2$.**

We have three cases based on the length of edges of $\triangle pqr$: (1) at least two of them $\leq 2$; (2) exactly two edges $> 2$; (3) all three edges $> 2$.

For case (1), by Lemma 21.3, we are done.

For case (2), assume $|pr| > 2, |qr| > 2, |pq| \leq 2$ (see Figure 21.18). We choose $p$ to be $z$. We have $\angle pqy > \angle pwy, \angle xry > \angle xwy$, and $\angle xpw > \angle xwp$. So, $\angle pqy + \angle xry + \angle xpw > \angle pwy + \angle xwy + \angle xwp = 2 \times \Pi$. In pentagon $pqyrx$, we have $\angle pqy + \angle xry + \angle xpw + \angle pxy + \angle qyr + \angle wpq = 3 \times \Pi$. So, $\angle pxy + \angle qyr + \angle wpq < \Pi$.

On the other hand, on the boundary of $V$, we have $\angle pxy \geq \angle px'r = \widehat{pqr}/2 \geq \Pi/2$, $\angle qyr \geq \angle qy'r = \widehat{rpq}/2 \geq \Pi/2$. So, $\angle pxy + \angle qyr \geq \Pi$. We have a contradiction.

For case (3), see Figure 21.19. We have $\angle ypz + \angle xqz + \angle xyr > 2 \times \Pi$. In hexagon $pzqxyr$, we have $\angle ypz + \angle xqz + \angle xyr + \angle pzq + \angle qxr + \angle ryp = 4 \times \Pi$. So, $\angle pzq + \angle qxr + \angle ryp < 2 \times \Pi$. However, $\angle pzq + \angle qxr + \angle ryp \geq \angle pz'q + \angle qx'r + \angle ry'p = 2 \times \Pi$. This is a contradiction. ∎

**Theorem 21.5** *Consider a graph G and an arbitrary edge e of G. Let $G' = G - e$ be the graph obtained from G by removing edge e. If $G'$ is connected then $|MDS(G')| = |MDS(G)|$ or $|MDS(G')| = |MDS(G)| + 1$.*

*Proof.* Assume that the edge $e = (u, v)$ is removed.

First, if neither $u$ nor $v$ is in the *MDS* then the removal of $e$ does not affect the dominance property of the *MDS*, and any *MDS* in G is still a *MDS* in $G'$, so we have $|MDS(G')| = |MDS(G)|$.

Second, if both $u$ and $v$ are in the *MDS*, then it is obvious that $|MDS(G')| = |MDS(G)|$.

Third, if exactly one of $u$ and $v$ is in the *MDS* of G. Let us assume that $u$ is in *MDS* of G. Then, if $v$ is not dominated by $u$ in G, it is obvious that $|MDS(G')| = |MDS(G)|$. If $v$ is dominated by $u$ in G, and if $v$

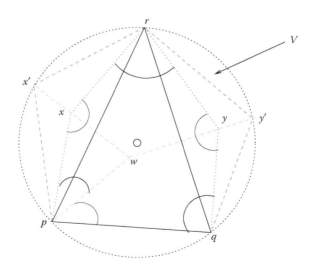

**FIGURE 21.18** Proof of Theorem 21.4 case (2): among the three edges of $\triangle pqr$, exactly two edges ($|pr|, |qr|$) are longer than two.

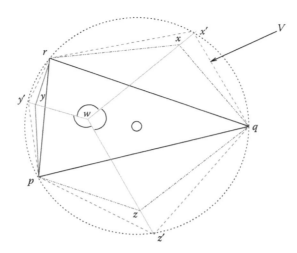

**FIGURE 21.19**  Proof of Theorem 21.4 case (3): all three edges of $\triangle pqr$ are longer than two.

can also be dominated by another node in *MDS*, then we have $|MDS(G')| = |MDS(G)|$. If $v$ is dominated by $u$ in $G$, but it cannot be dominated by another node in *MDS*, then we need to add $v$ into the *MDS*, and $MDS \cup v$ is a DS in $G'$. So, $|MDS(G')| = |MDS(G)|$ or $|MDS(G')| = |MDS(G)| + 1$.                                ∎

**Theorem 21.6**  *Consider any diameter-2 graph G and an arbitrary edge e of G. Let $G' = G - e$ be the graph obtained from G by removing edge e. If $G'$ is connected, then $|MDS(G')| \leq 4$.*

*Proof.* This theorem follows immediately from Theorem 21.4 and Theorem 21.5.                                ∎

**Theorem 21.7**  *There exists a unit-disk diameter-2 graph G such that $|MDS(G)| = 3$.*

*Proof.* We have written a Java program to generate random unit-disk diameter-2 graphs, and in the one million instances of graphs that was generated by the program, all can be dominated by two nodes. This suggests that the probability that a unit-disk diameter-2 graph is dominated by two nodes is very high.

On the other hand, we have been able to construct a counter example that cannot be dominated by two nodes. Consider the unit-disk diameter-2 graph shown in Figure 21.20. In the figure, $D = 10000$ unit, and the coordinates of nodes are shown in Table 21.3. This counterexample is inspired by Figure 11 in [54]. It can be verified by hand or program that this *unit-disk* graph is *diameter-2* but cannot be dominated by *two* nodes.                                ∎

From Theorems 21.4 and 21.7, we know that any *unit-disk* diameter-2 graph can be dominated by at most *three* nodes. Note that this is not true for a *general* diameter-2 graph. For a *general* diameter-2 graph, there is no proved constant upper bound on the size of its *MDS*. In the examples shown in [55], there is a *general* diameter-2 graph with 198 nodes and max node degree 16, so its $|MDS| \geq 12$.

## 21.6 Further Researches

As an updated version, we included further researches in this section.

To mitigate the impact of node mobility on ad-hoc clustering, Ghosh et al. [56] generalized a typical clustering protocol for MANET—DMAC [57] to decrease the number of cluster updates, clustering overhead imposed by mobility, and the corresponding maintenance cost. They verified the effectiveness of the generalized DMAC by simulation, especially in situations where the nodes move according to three different mobility models: the random way-point model, the Brownian motion and the Manhattan mobility model.

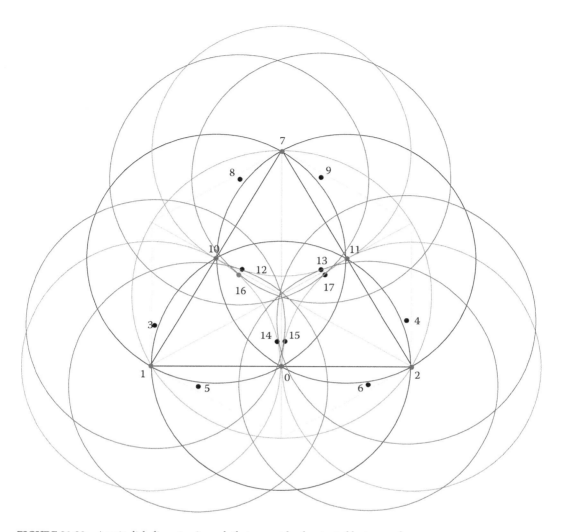

**FIGURE 21.20** A unit-disk diameter-2 graph that cannot be dominated by two nodes.

Deval et al. [58] argued that our tree-based unifying clustering algorithm that combines cluster initialization and cluster maintenance results in significant storage overhead as it requires each node to maintain a depth-2 BFS tree that roots at itself. They developed an algorithm that integrates initial clustering and backbone construction with clustering and backbone maintenance. By keeping the density of leader nodes (DS) constant, the information about gateway nodes leading to neighboring leader nodes is limited to a constant amount of storage irrespective of the network density or total number of nodes in the network. They claim that their protocol is the first one to directly exploit physical carrier sense for the network layer task of distributed CDS construction.

With prediction of mobility, the impact on clustering of ad-hoc networks can be mitigated greatly. Dekar et al. [59] use our algorithm to construct the topology of MANETs and develop a topology maintenance algorithm to predict mobility of nodes. They use the evidence theory of Dempster-Shafer to predict the future location of the mobile by basing itself on relevant criteria. As the future movement of the mobile can be predicted in advance, so the MANETs can reserve resources and react in advance to provide a better QoS.

**TABLE 21.3** Coordinates of
the Nodes in Figure 21.20

| Node | Coordinates $(x, y)$ |
|------|----------------------|
| 0 | $(0, 0)$ |
| 1 | $(-10000, 0)$ |
| 2 | $(10000, 0)$ |
| 3 | $(-8710, -4943)$ |
| 4 | $(8710, -4943)$ |
| 5 | $(-5080, 1360)$ |
| 6 | $(5080, 1360)$ |
| 7 | $(0, -17320)$ |
| 8 | $(-3630, -13737)$ |
| 9 | $(3630, 13737)$ |
| 10 | $(-5000, -8660)$ |
| 11 | $(5000, -8660)$ |
| 12 | $(-2540, -7652)$ |
| 13 | $(2540, -7652)$ |
| 14 | $(-360, -2629)$ |
| 15 | $(360, -2629)$ |
| 16 | $(-2900, -7034)$ |
| 17 | $(2900, -7034)$ |

## 21.7 Concluding Remarks

A large number of clustering schemes for MANET have been proposed in the recent literature. In general, we believe that a clustering scheme that can generate a more *stable* and *symmetric* virtual infrastructure is especially suitable for MANET, and such a virtual infrastructure can be leveraged by a number of MANET applications without introducing traffic bottlenecks and single points of failure.

To illustrate the feasibility of this concept, we have proposed a tree-based cluster initialization/maintenance algorithm for MANET based on a number of properties of diameter-2 graphs. The resulting algorithm is cluster-centric and works in the presence of node mobility. Simulation results demonstrated the effectiveness of our algorithm when compared with other clustering schemes in the literature.

The tree-based clustering algorithm proposed in this chapter can be further generalized to achieve ($d1$, $d2$)-clustering in which two clusters merge when the diameter of the resulting cluster is not larger than $d1$, and a cluster is split into several diameter-$d1$ clusters if its diameter is larger than $d2$. By adaptively changing the values of $d1$ and $d2$, a stable and symmetric general-purpose virtual infrastructure can be achieved efficiently in large-scale MANET.

## References

1. M. Steenstrup, Cluster-based networks, *Ad Hoc Networking*, Addison-Wesley, Boston, MA, 2001.
2. C. R. Lin and M. Gerla, Adaptive clustering for mobile wireless networks, *IEEE Journal on Selected Areas in Communications*, 15(7), 1997, 1265–1275.
3. D. Bhagavathi, P. J. Looges, S. Olariu, and J. L. Schwing. A fast selection algorithms on meshes with multiple broadcasting, *IEEE Transactions on Parallel and Distributed Systems*, 5(7), 1994, 772–778.
4. I. Stojmenovic and J. Wu, Broadcasting and activity-scheduling in ad hoc networks, *Ad Hoc Networking*, IEEE Press, Eds. S. Basagni, M. Conti, S. Giordano and I. Stojmenovic, 2004.

5. A. Iwata, C. Chiang, G. Pei, M. Gerla, and T. Chen, Scalable routing strategies for ad hoc wireless networks, *IEEE Journal on Selected Areas in Communications, Special Issue on Ad-Hoc Networks*, 17(8), 1999, 1369–1379.

6. S. Olariu, M. Eltoweissy, and M. Younis. ANSWER: Autonomous networked sensor system, *Journal of Parallel and Distributed Computing*, 2007, 114–126.

7. A. B. McDonald and T. Znati, A mobility based framework for adaptive clustering in wireless ad-hoc networks, *IEEE Journal on Selected Areas in Communications, Special Issue on Ad-Hoc Networks*, 17(8), 1999, 1466–1487.

8. H. Chan and A. Perrig, ACE: An emergent algorithm for highly uniform cluster formation, *Proceedings of the 2004 European Workshop on Sensor Networks*, Richardson, TX, pp. 154–171.

9. M. Gerla and J. Tsai, Multicluster, Mobile, Multimedia radio network, *Wireless Networks*, 1(3), 1995, 255–265.

10. D. J. Baker and A. Ephremides, The architectural organization of a mobile radio network via a distributed algorithm, *IEEE Transactions on Communications*, 29(11), 1981, 1694–1701.

11. A. Ephremides, J. E. Wieselthier, and D. Baker, A design concept for reliable mobile radio networks with frequency hopping signaling, *Proceedings of the of IEEE*, 75(1), 1987, 56.

12. N. Lynch, *Distributed Algorithms*, Morgan Kaufmann Publishers, Burlington, MA, 1996.

13. A. D. Amis, R. Prakash, D. Huynh, and T. Vuong, Max-min d-cluster formation in wireless ad hoc networks, *Proceedings of the INFOCOM 2000*, Vol. 1, pp. 32–41, 2000.

14. S. Basagni, Distributed clustering for ad hoc networks, *Proceedings of the IEEE International Symposium on Parallel Architectures, Algorithms, and Networks* (*I-SPAN*), Perth, Western Australia, June 1999, pp. 310–315.

15. F. G. Nocetti, J. S. Gonzalez, and I. Stojmenovic, Connectivity based k-hop Clustering in wireless networks, *Telecommunication System*, 22(1–4), 2003, 205–220.

16. R. Nagpal and D. Coore, An algorithm for group formation in an amorphous computer, *Proceedings of the 10th International Conference on Parallel and Distributed Computing Systems* (*PDCS'98*), Las Vegas, NV, October 1998.

17. T. C. Hou and T. J. Tsai, An access-based clustering protocol for multi-hop wireless ad hoc networks, *IEEE Journal on Selected Areas in Communications*, 19(7), 2001, 1201–1210.

18. T. J. Kwon and M. Gerla, Efficient flooding with passive clustering (PC) in ad hoc networks, *ACM SIGCOMM Computer Communication Review*, 32(1), 2002, 44–56.

19. L. Bao and J. J. Garcia-Luna-Aceves, Topology management in ad hoc networks, *Proceedings of the ACM MobiHoc 2003*, Annapolis, MD, June 2003.

20. C. Bettstetter and R. Krausser, Scenario-based stability analysis of the distributed mobility-adaptive clustering (DMAC) algorithm, *Proceedings of the MobiHoc 2001*, Long Beach, CA, October 2001, pp. 232–241.

21. M. Chatterjee, S. Das, and D. Turgut, WCA: A weighted clustering algorithm for mobile ad hoc networks, *Cluster Computing*, 5(2), 2002, 193–204.

22. C. C. Chiang, H. K. Wu, W. Liu, and M. Gerla, Routing in clustered multi-hop, mobile wireless networks with fading channel, *Proceedings of the IEEE Singapore International Conference on Networks* (*SICON*), pp. 197–211, April 1997.

23. S. Basagni, Distributed and mobility-adaptive clustering for multimedia support in multi-hop wireless networks, *Proceedings of the IEEE 50th International Vehicular Technology Conference*, Vol. 2, pp. 889–893, Amsterdam, the Netherlands, September 1999.

24. E. M. Belding-Royer, Hierarchical routing in ad hoc mobile networks, *Wireless Communication & Mobile Computing*, 2(5), 2002, 515–532.

25. K. M. Alzoubi, P.-J. Wan, and O. Frieder, Message-optimal connected-dominating-set construction for routing in mobile ad hoc networks, *Proceedings of the ACM MobiHoc 2002*, Las Vegas, NV, June 2002.

26. I. Stojmenovic, M. Seddigh, and J. Zunic, Dominating sets and neighbor elimination based broadcasting algorithms in wireless networks, *IEEE Transactions on Parallel and Distributed Systems*, 13(1), 2002, 14–25.

27. J. Wu and H. Li, On calculating connected dominating set for efficient routing in ad hoc wireless networks, *Proceedings of the 3rd International Workshop on Discrete Algorithms and Methods for Mobile Computing and Communications (DIALM'99)*, August 1999, pp. 7–14.

28. K. M. Alzoubi, P.-J. Wan, and O. Frieder, New distributed algorithm for connected dominating set in wireless ad hoc networks, *Proceedings of the IEEE HICSS35*, Hawaii, January 2002.

29. R. Sivakumar, P. Sinha, and V. Bharghavan, CEDAR: A core-extraction distributed ad hoc routing algorithm, *IEEE Journal on Selected Areas in Communications, Special Issue on Wireless Ad Hoc Networks*, 17(8), 1999, 1454–1465.

30. R. Sivakumar, B. Das, and V. Bharghavan, Spine routing in ad hoc networks, *ACM/Baltzer Publications Cluster Computing Journal, Special Issue on Mobile Computing*, 1(2), 1998, 237–248.

31. K. M. Alzoubi, P.-J. Wan, O. Frieder, Weakly connected dominating sets and sparse spanners for wireless ad hoc networks, *Proceedings of the 23rd International Conference on Distributed Computing Systems (ICDCS)*, Atlanta, GA, 2003.

32. Y. Chen and A. Liestman, A zonal algorithm for clustering ad hoc networks, *International Journal of Foundation of Computer Science*, 14(2), 2003, 305–322.

33. M. Q. Rieck, S. Pai, and S. Dhar, Distributed routing algorithms for wireless ad hoc networks using d-hop connected dominating sets, *Proceedings of the 6th International Conference on High Performance Computing: Asia Pacific Region (HPC Asia 2002)*, ACM, New York, Vol 2, pp. 443–450, 2002.

34. S. Srivastava and R. K. Ghosh, Cluster based routing using a k-tree core backbone for mobile ad hoc networks, *Proceedings of the 6th International Workshop on Discrete Algorithms and Methods for Mobile Computing and Communications (DIALM'02)*, Atlanta, GA, September 28, 2002.

35. Y. Bejerano, Efficient integration of multi-hop wireless and wired networks with QoS constraints, *Proceedings of the MobiCom 2002*, Atlanta, GA, September 2002, pp. 215–226.

36. X. Y. Li, Topology control in wireless ad hoc networks, *Ad Hoc Networking*, IEEE Press, Eds. S. Basagni, M. Conti, S. Giordano and I. Stojmenovic, 2003.

37. B. An and S. Papavassiliou, A mobility-based clustering approach to support mobility management and multicast routing in mobile ad-hoc wireless networks, *International Journal of Network Management*, 11(6), 2001, 387–395.

38. S. Banerjee and S. Khuller, A clustering scheme for hierarchical control in multi-hop Wireless Networks, *Proceedings of the INFOCOM*, Anchorage, AK, April 2001.

39. Y. Fernandess and D. Malkhi, K-clustering in wireless ad-hoc networks, *Proceedings of the ACM Workshop on Principles of Mobile Computing (POMC 2002)*, Toulouse, France, 2002, pp. 31–37.

40. P. Krishna, M. Chatterjee, N. Vaidya, and D. Pradhan, A cluster-based approach for routing in ad-hoc networks, *ACM SIGCOMM Computer Communication Review*, 27(2), 1997, 49–64.

41. K. Nakano and S. Olariu. Randomized leader election protocols in radio networks with no collision detection, *Algorithms and Computation*, 2000, 362–373.

42. S. Olariu, J. L. Schwing, and J. Zhang. Fast computer vision algorithms for reconfigurable meshes, *Image and Vision Computing*, 10(9), 1992, 610–616.

43. S. Ramanathan and M. Steenstrup, Hierarchically-organized, Multihop mobile networks for multimedia support, *ACM/Baltzer Mobile Networks and Applications*, 3(1), 1998, 101–119.

44. A. Zomaya, M. Clements, and S. Olariu. A framework for reinforcement-based scheduling in parallel processor systems, *IEEE Transactions on Parallel and Distributed Systems*, 9(3), 1998, 249–260.

45. A. B. McDonald and T. Znati, Design and performance of a distributed dynamic clustering algorithm for ad hoc networks, *Proceedings of the 34th Annual IEEE/ACM Simulation Symposium (AISS)*, Seattle WA, April 2001, 27–35.

46. A.D. Amis and R. Prakash. Load-balancing clusters in wireless ad hoc networks, *Proceedings of the ASSET 2000*, Richardson, TX, March 2000.

47. K. Nakano and S. Olariu, Randomized leader election protocols in ad-hoc networks, *Proceedings of the 7th International Symposium on Structural Information and Communication Complexity*, L'Aquila, Italy, June 2000, 253–268.

48. T. H. Cormen, C. E. Leiserson, and R. L. Rivest, *Introduction to Algorithms*, MIT Press and McGraw-Hill, Cambridge, MA, 1992.

49. H. Huang, A. W. Richa, and M. Segal, Approximation algorithms for the mobile piercing set problem with applications to clustering in ad-hoc networks, *Proceedings of the International Workshop on Discrete Algorithms and Methods for Mobile Computing and Communications (DIAL-M 2002)*, Atlanta, GA, 2002, pp. 61–62.

50. T. Camp, J. Boleng, and V. Davies, A survey of mobility models for ad hoc network research, *Wireless Communication & Mobile Computing: Special issue on Mobile Ad Hoc Networking: Research, Trends and Applications*, 2(5), 2002, 483–502.

51. X.-Y. Li, Topology control in wireless ad hoc networks, *Ad Hoc Networking*, IEEE Press, Eds. S. Basagni, M. Conti, S. Giordano, and I. Stojmenovic, 2003.

52. N. Li, J. Hou, and L. Sha, Design and analysis of an MST-based topology control algorithm, *Proceedings of the INFOCOM Conference 2003*, April 2003.

53. C. Shen, C. Srisathapornphat, R. Liu, Z. Huang, C. Jaikaeo, and E. Lloyd, CLTC: A cluster-based topology control framework for ad hoc networks, *IEEE Transactions on Mobile Computing*, 3(1), 2004, 1–15.

54. H. Hadwiger, H. Debrunner, and V. Klee, *Combinatorial Geometry in the Plane*, Holt, Rinehart and Winston, New York, 1964.

55. L. Wang, Clustering and Hybrid Routing in Mobile Ad-hoc Networks, Doctoral Dissertation, Old Dominion University, Norfolk, VA, 2005.

56. R. Ghosh and S. Basagni. Mitigating the impact of node mobility on ad hoc clustering, *Wireless Communications and Mobile Computing*, 8(3), 2008, 295–308.

57. S. Basagni. Distributed clustering for ad hoc networks, *Parallel Architectures, Algorithms, and Networks, 1999.(I-SPAN'99) Proceedings Fourth International Symposium on*, pp. 310–315. IEEE, 1999.

58. S. Deval, L. Ritchie, M. Reisslein, and A. W. Richa. Evaluation of physical carrier sense based backbone maintenance in mobile ad hoc networks, *International Journal of Vehicular Technology*, 2009, 2009, pp. 1–13.

59. L. Dekar and H. Kheddouci. A cluster based mobility prediction scheme for ad hoc networks, *Ad Hoc Networks*, 6(2), 2008, 168–194.

# Topology Control Problems for Wireless Ad Hoc Networks

Gruia Călinescu

Errol L. Lloyd

S. S. Ravi

## 22.1 Introduction

Ad hoc networks are formed by collections of nodes which communicate with each other through radio propagation. Topology control problems in such networks deal with the assignment of power values to the nodes so that the power assignment leads to a graph topology satisfying some specified properties. In such applications, it is important to minimize a specified function of the powers assigned to the nodes. Several versions of this minimization problem are **NP**-complete. This chapter discusses some known approximation algorithms for these problems. The focus is on approximation algorithms with proven performance guarantees.

An ad hoc network consists of a collection of transceivers (nodes). All communication among these nodes is based on radio propagation. The term "ad hoc" is used to indicate that the nodes organize themselves into a network without any pre existing infrastructure. Wireless ad hoc networks are used in a number of applications including military operations, cellular phones and emergency search-and-rescue operations [1,2]. In such networks, battery power is a precious resource. This chapter presents approximation algorithms for several problems that arise in the context of wireless ad hoc networks, in which the main goal is to minimize the power used by the nodes.

To develop precise formulations of the problems considered in the current chapter, we need to introduce some basic concepts and terminology regarding the nodes used in the ad hoc network. For each ordered pair $(u, v)$ of nodes, there is a **transmission power threshold**, denoted by $p(u, v)$, with the following significance: A signal transmitted by the node $u$ can be received by $v$ only when the transmission power of $u$ is at least $p(u, v)$. The transmission power threshold for a pair of nodes depends on a number of factors including the distance between the nodes, the antenna gain at the sender and the receiver, interference, noise, and so on [3].

Each assignment of powers to the nodes of an ad hoc network induces a directed graph in the following manner. The nodes of this directed graph are in one-to-one correspondence with the nodes of the ad hoc network. A directed edge $(u, v)$ is in this graph if and only if the transmission power of $u$ is at least the transmission power threshold $p(u, v)$. The main goal of **topology control** is to assign transmission powers to nodes so that the resulting directed graph satisfies some specified properties. As the battery power of each node is an expensive resource, it is important to achieve the goal while minimizing a given function of the transmission powers assigned to the nodes. Examples of desirable graph properties are connectivity, small diameter, and so on. The primary minimization objectives considered in the literature are the maximum power assigned to a node and the total power assigned to all nodes. (The latter objective is equivalent to minimizing the average power assigned to a node.) Most of the results presented in this chapter deal with the minimum total power objective.

Unless otherwise mentioned, the power threshold values are assumed to be *symmetric*; that is, for any pair of nodes $u$ and $v$, $p(u, v) = p(v, u)$. Under this assumption, the power threshold values can be represented by an undirected complete graph which we call the **threshold graph**. The weight of each edge $\{u, v\}$ in this graph is the transmission power threshold $p(u, v)$.

As stated earlier, the main motivation for studying topology control problems is to make efficient use of available power at each node. In addition, using a minimum amount of power at each node to achieve a desired topological objective is also likely to decrease the interference in the medium access (MAC) layer between adjacent radios. We refer the reader to References 3–5 for a thorough discussion of the power control issues in ad hoc networks. A book [6] and several surveys on topology control [7–11] have appeared in the literature. The emphasis of Reference 7 is on how topology control can facilitate the design of routing protocols for ad hoc networks. References 6, 8–11 provide comprehensive coverage of the practical aspects of topology control.

Precise formulations of the optimization problems considered in this chapter are provided in Section 22.2. Many of these problems are **NP**-hard. The practical importance of these problems motivates the study of approximation algorithms for them. The focus of this chapter is on approximation algorithms with proven performance guarantees. Our goal is to discuss some known techniques for developing such approximation algorithms rather than to provide a survey on the topic of topology control.

The remainder of this chapter is organized as follows. Section 22.2 provides the necessary definitions and develops a notation for specifying topology control problems. Section 22.3 presents approximation algorithms for topology control problems in which the goal is to minimize the total power assigned to the nodes of the network. Both centralized and distributed approximation algorithms are discussed in that section. Section 22.4 summarizes known approximation results for other objectives. Finally, some directions for future research are presented in Section 22.5.

# 22.2 Model and Problem Formulation

## 22.2.1 Graph Models for Topology Control

Topology control problems have been studied under two graph models. The discussion in Section 22.1 corresponds to the **directed graph model** studied in Reference 3. In the **undirected graph model** considered in Reference 12, each power assignment induces an undirected graph in the following manner: An undirected edge $\{u, v\}$ is in this graph if and only if the transmission powers of $u$ and $v$ are both at least the transmission power threshold $p(u, v)$. Under this model, the goal of a topology control problem is to assign transmission powers to nodes such that the resulting undirected graph has a specified property and a specified function of the powers assigned to nodes is minimized. Note that the directed graph model allows two-way communication between some pairs of nodes and one-way communication between other pairs of nodes. In contrast, every edge in the undirected graph model corresponds to a two-way communication.

## 22.2.2 Notation for Topology Control Problems

In general, a topology control problem can be specified by a triple of the form $\langle \mathbb{M}, \mathbb{P}, \mathbb{O} \rangle$. In such a specification, $\mathbb{M} \in \{\text{DIR, UNDIR}\}$ represents the graph model, $\mathbb{P}$ represents the desired graph property, and $\mathbb{O}$ represents the minimization objective. For the problems considered in this chapter, $\mathbb{O} \in \{\text{MAXP, TOTALP}\}$ (where MAXP and TOTALP are abbreviations of Max Power and Total Power, respectively). For example, in the $\langle \text{DIR, STRONGLY CONNECTED, MAXP} \rangle$ problem, powers must be assigned to nodes so that the resulting directed graph (induced by the assigned powers in relation to the edges of the threshold graph) is strongly connected and the maximum power assigned to a node is minimized. Similarly, the $\langle \text{UNDIR},$ 2-NODE-CONNECTED, TOTALP$\rangle$ problem seeks to assign powers to the nodes so that the resulting undirected graph has a node connectivity* of (at least) 2 and the sum of the powers assigned to all nodes is minimized. Throughout this chapter, an instance of an $\langle \mathbb{M}, \mathbb{P}, \mathbb{O} \rangle$ problem is specified by a threshold graph.

## 22.2.3 Additional Definitions

The current section collects together the definitions of some graph theoretic and algorithmic terms used throughout this chapter.

Given an undirected graph $G(V, E)$, an **edge subgraph** $G'(V, E')$ of $G$ has all of the nodes of $G$ and the edge set $E'$ is a subset of $E$. Further, if $G$ is an edge weighted graph, then the weight of each edge in $G'$ is the same as its weight in $G$.

The **node connectivity** of an undirected graph is the smallest number of nodes that must be deleted from the graph so that the resulting graph is disconnected. The **edge connectivity** of an undirected graph is the smallest number of edges that must be deleted from the graph so that the resulting graph is disconnected. For example, a tree has node and edge connectivities equal to 1, whereas a simple cycle has node and edge connectivities equal to 2. When the node (edge) connectivity of a graph is greater than or equal to $k$, the graph is said to be $k$-**node connected** ($k$-**edge connected**). Given an undirected graph, polynomial algorithms are known for finding its node and edge connectivities [13].

Some of the results presented in this chapter use the following definition.

**Definition 22.1** *A property $\mathbb{P}$ of the (directed or undirected) graph induced by a power assignment is* **monotone** *if the property continues to hold even when the powers assigned to some nodes are increased while the powers assigned to the other nodes remain unchanged.*

**Example:** For any $k \geq 1$, the property $k$-NODE-CONNECTED for undirected graphs is monotone as increasing the powers of some nodes while keeping the powers of other nodes unchanged may only add edges to the graph. However, properties such as ACYCLIC or BIPARTITE are not monotone.

We say that an approximation algorithm provides a **performance guarantee** of $\rho$ if for every instance of the problem, the solution produced by the approximation algorithm is within the multiplicative factor of $\rho$ of the optimal solution.

# 22.3 Approximation Algorithms for Minimizing Total Power

## 22.3.1 A General Framework

Topology control problems in which the minimization objective is total power tend to be computationally intractable. For example, the problem is **NP**-hard even for the (simple) property 1-NODE-CONNECTED [12]. In this section, we first discuss a general framework developed in Reference 14 for approximation algorithms for topology control problems of the form $\langle \text{UNDIR}, \mathbb{P}, \text{TOTALP} \rangle$. We then discuss how approximation algorithms for several graph properties can be derived from this framework.

---

* This concept is defined in Section 22.2.3.

**Input:** An instance $I$ of $\langle$UNDIR, $\mathbb{P}$, TOTALP$\rangle$ where the property $\mathbb{P}$ is monotone and polynomial time testable.

**Output:** A power value $\pi(u)$ for each node $u$ such that the undirected graph induced by the power assignment satisfies property $\mathbb{P}$ and the total power assigned to all nodes is as small as possible.

**Steps:**

1. Let $G_c(V, E_c)$ denote the threshold graph for the given problem instance.

2. Construct an edge subgraph $G'(V, E')$ of $G_c$ such that $G'$ satisfies property $\mathbb{P}$ and the total weight of the edges in $E'$ is minimum among all edge subgraphs of $G_c$ satisfying property $\mathbb{P}$.

3. For each node $u$, assign a power value $\pi(u)$ equal to the weight of the largest edge in $E'$ incident on $u$.

**FIGURE 22.1**    Heuristic GEN-TOTAL-POWER: A general framework for approximating total power.

The framework assumes that the property $\mathbb{P}$ to be satisfied by the graph is *monotone* and that it can be tested in polynomial time. It also assumes *symmetric* power thresholds as in References 12, 15, 16; that is, for any pair of nodes $u$ and $v$, the power thresholds $p(u, v)$ and $p(v, u)$ are equal.

The general approximation framework (called Heuristic GEN-TOTAL-POWER) is shown in Figure 22.1. Note that Steps 1 and 3 of the heuristic can be implemented in polynomial time. The time complexity of Step 2 depends crucially on the property $\mathbb{P}$. For some properties such as 1-NODE CONNECTED, Step 2 can be done in polynomial time. For other properties such as 2-NODE CONNECTED, Step 2 cannot be done in polynomial time, unless $\mathbf{P} = \mathbf{NP}$ [17]. In such cases, an efficient algorithm that produces an approximately minimum solution can be used in Step 2. The following theorem proves the correctness of the general approach and establishes its performance guarantee as a function of some parameters that depend on the property $\mathbb{P}$ and the approximation algorithm used in Step 2 of the framework.

**Theorem 22.1** *Let $I$ be an instance of $\langle$UNDIR, $\mathbb{P}$, TOTALP$\rangle$ where $\mathbb{P}$ is a monotone property. Let $OPT(I)$ and $GTP(I)$ denote respectively the total power assigned to the nodes in an optimal solution and in a solution produced by Heuristic GEN-TOTAL-POWER for the instance $I$. Then, the following results hold.*

(i) *The graph $G_\pi$ induced by the power assignment produced by the heuristic (i.e., Step 3) satisfies property $\mathbb{P}$.*

(ii) *Let $H(V, E_H)$ be an edge subgraph of the threshold graph $G_c$ such that $H$ has the minimum total edge weight among all edge subgraphs of $G_c$ satisfying property $\mathbb{P}$. Let $W(H)$ denote the total edge weight of $H$. Let Step 2 of the heuristic produce an edge subgraph $G'(V, E')$ of $G$ with total edge weight $W(G')$. Suppose there are quantities $\alpha > 0$ and $\beta > 0$ such that (a) $W(H) \leq \alpha\, OPT(I)$ and (b) $W(G') \leq \beta\, W(H)$. Then, $GTP(I) \leq 2\alpha\beta\, OPT(I)$. That is, Heuristic GEN-TOTAL-POWER provides a performance guarantee of $2\alpha\beta$.*

*Proof.* **Part (i):** The edge subgraph $G'(V, E')$ constructed in Step 2 of the heuristic satisfies property $\mathbb{P}$. We show that every edge in $E'$ is also in the subgraph $G_\pi$ induced by the power assignment $\pi$ produced in Step 3. Then, even if $G_\pi$ has other edges, the monotonicity of $\mathbb{P}$ allows us to conclude that $G_\pi$ satisfies $\mathbb{P}$.

Consider an edge $\{u, v\}$ with weight $p(u, v)$ in $E'$. Recall that $p(u, v)$ is the minimum power threshold for the existence of edge $\{u, v\}$ and that the power thresholds are symmetric. As Step 3 assigns to each node the maximum of the weights of edges incident on that node, we have $\pi(u) \geq p(u, v)$ and $\pi(v) \geq p(u, v)$. Therefore, the graph $G_\pi$ induced by the power assignment also contains the edge $\{u, v\}$ and this completes the proof of Part (i).

**Part (ii):** By conditions (a) and (b) in the statement of the theorem, we have $W(G') \leq \alpha\beta \, OPT(I)$. We observe that $GTP(I) \leq 2 \, W(G')$. This is because Step 3 of the heuristic may assign the weight of any edge to at most two nodes (namely, the endpoints of the edge). Combining the two inequalities, we get $GTP(I) \leq 2\alpha\beta \, OPT(I)$. ∎

## 22.3.2 Applications of Theorem 22.1

The current section presents several examples of approximation algorithms derived from the general framework outlined in Figure 22.1.

### 22.3.2.1 An Approximation Algorithm for ⟨UNDIR, 1-NODE-CONNECTED, TOTALP⟩

As a first example, we observe that the 2-approximation algorithm presented in References 12, 18 for the ⟨UNDIR, 1-NODE-CONNECTED, TOTALP⟩ problem* can be derived from the above-mentioned general framework. In Step 2 of the framework, the algorithm in Reference 18 constructs a minimum spanning tree of $G_c$. It is also shown that the total power assigned by any optimal solution is at least the weight of a minimum spanning tree of $G_c$ (see Lemma 22.9 in Reference 18). Thus, using the notation of Theorem 22.1, $\alpha = \beta = 1$ for their approximation algorithm. As 1-NODE-CONNECTED is a monotone property, it follows from Theorem 22.1 that the performance guarantee of their algorithm is 2. The best known approximation algorithm for this problem provides a performance guarantee of $3/2 + \epsilon$, for any fixed $\epsilon > 0$ [19]. It is not a practical algorithm as it is based on the framework for the Steiner Tree algorithm presented in Reference 20. Althaus et al. [21] provide practical algorithms with performance guarantees better than 2, also using ideas which were first developed in context of the Steiner tree problem [22,23].

### 22.3.2.2 An Approximation Algorithm for ⟨UNDIR, 2-NODE-CONNECTED, TOTALP⟩

The approximation algorithm discussed in this section is from Reference 14. The **NP**-hardness of the ⟨UNDIR, 2-NODE-CONNECTED, TOTALP⟩ problem was established in Reference 24. We note that the property 2-NODE-CONNECTED is monotone. The following notation is used throughout this section. $I$ denotes the given instance of ⟨UNDIR, 2-NODE-CONNECTED, TOTALP⟩ with $n$ nodes. For each node $u$, $\pi^*(u)$ denotes the power assigned to $u$ in an optimal solution. Further, $OPT(I)$ denotes the sum of the powers assigned to the nodes in an optimal solution.

The approximation algorithm in Reference 14 for the ⟨UNDIR, 2-NODE-CONNECTED, TOTALP⟩ problem is obtained from the framework of Figure 22.1 by using an approximation algorithm from Reference 25 for the minimum weight 2-NODE-CONNECTED subgraph problem in Step 2. This approximation algorithm provides a performance guarantee of $(2 + 1/n)$. Thus, using the notation of Theorem 22.1, we have $\beta \leq (2 + 1/n)$.

Lemma 22.1 in the following shows that the threshold graph $G_c(V, E_c)$ of the instance $I$ contains an edge subgraph $G_1(V, E_1)$ such that $G_1$ is 2-NODE-CONNECTED and the total weight $W(G_1)$ of the edges in $G_1$ is at most $(2 - 2/n) \, OPT(I)$. Again, using the notation of Theorem 22.1, this result implies that $\alpha \leq (2 - 2/n)$.

Thus, once Lemma 22.1 is established, it would follow from Theorem 22.1 that the performance guarantee of the resulting approximation algorithm is $2(2 - 2/n)(2 + 1/n)$, which approaches 8 asymptotically from the following. The remainder of this section is devoted to the formal statement and proof of Lemma 22.1.

---

* Actually, both References 12, 18 discuss their results for ⟨DIR, STRONGLY CONNECTED, TOTALP⟩; however, the result discussed in this paragraph is implicit in those references.

**Lemma 22.1** *Let I denote an instance of the* ⟨UNDIR, 2-NODE-CONNECTED, TOTALP⟩ *problem with n nodes. Let $G_c(V, E_c)$ denote the threshold graph for the instance I. Let OPT(I) denote the total power assigned to the nodes in an optimal solution to I. There is an edge subgraph $G_1(V, E_1)$ of $G_c$ such that $G_1$ is 2-NODE-CONNECTED, and the total weight $W(G_1)$ of the edges in $G_1$ is at most $(2 - 2/n)$ OPT(I).*

Our proof of Lemma 22.1 begins with an optimal power assignment $\pi^*$ to instance $I$ and constructs a graph $G_1$ satisfying the properties mentioned in the above-mentioned statement. This construction relies on several definitions and known results from graph theory.

**Definition 22.2** *Let $G(V, E)$ be an undirected graph. Suppose the node sequence ⟨$v_1$, $v_2$, $v_3$, ..., $v_k$, $v_1$⟩ forms a simple cycle C of length at least 4 in G. Any edge $\{v_i, v_j\}$ of G ($1 \le i \ne j \le k$) which is not in C is a* **chord**.

**Definition 22.3** *An undirected graph $G(V, E)$ is* **critically** *2-NODE-CONNECTED if it satisfies both of the following conditions: (i) G is 2-NODE-CONNECTED. (ii) For every edge $e \in E$, the subgraph of G obtained by deleting the edge e is not 2-NODE-CONNECTED.*

For example, a simple cycle on three or more nodes is critically 2-NODE-CONNECTED. This is because such a cycle is 2-NODE-CONNECTED, and deleting any edge of the cycle yields a simple path which is not 2-NODE-CONNECTED. A number of properties of critically 2-NODE-CONNECTED graphs have been established in the literature [26–28]. In proving Lemma 22.1, we use the following property established[*] in References 26, 27.

**Proposition 22.1** *If a graph G is critically 2-NODE-CONNECTED then no cycle of G has a chord.* ∎

We also need some terminology associated with **Depth-First-Search** (DFS) [29]. When DFS is carried out on a connected undirected graph $G(V, E)$, a spanning tree $T(V, E_T)$ is produced. Each edge in $T$ is called a **tree edge**. Each tree edge joins a child to its parent. An **ancestor** of a node $u$ in $T$ is a node which is not the parent of $u$ but which is encountered in the path from $u$ to the root of $T$. Each edge in $E - E_T$ is called a **back edge**. Each back edge joins a node $u$ to an ancestor of $u$ in $T$. The following lemma establishes a simple property of back edges that arise when DFS is carried out on a critically 2-NODE-CONNECTED graph.

**Lemma 22.2** *Let $G(V, E)$ be a critically 2-NODE-CONNECTED graph, and let $T(V, E_T)$ be a spanning tree for G produced using DFS. For any node u, there is at most one back edge from u to an ancestor of u in T.*

*Proof.* The proof is by contradiction. Suppose a node $u$ has two or more back edges. Let $v$ and $w$ be two ancestors of $u$ in $T$ such that both $\{u, v\}$ and $\{u, w\}$ are back edges. Note that these two edges are in $G$. Without loss of generality, let $w$ be encountered before $v$ in the path in $T$ from the root to $u$. The path from $w$ to $u$ in $T$ together with the edge $\{u, w\}$ forms a cycle in $G$. By our choice of $w$, this cycle also includes the node $v$. Therefore, the edge $\{u, v\}$ is a chord in the cycle. This contradicts the assumption that $G$ is critically 2-NODE-CONNECTED as by Proposition 22.1, no cycle in $G$ can have a chord. The lemma follows. ∎

We now prove several additional lemmas that are used in our proof of Lemma 22.1. Consider the given instance $I$ of the ⟨UNDIR, 2-NODE-CONNECTED, TOTALP⟩ problem and let $V$ denote the set of nodes. For the chosen optimal power assignment $\pi^*$, let $p^*$ denote the maximum power value assigned to a node. Let the chosen optimal power assignment induce the graph $G_{\pi^*}$. Note that $G_{\pi^*}$ is 2-NODE-CONNECTED. Let $G_1^*(V, E_1^*)$ be an edge subgraph of $G_{\pi^*}$ such that $G_1^*$ is critically 2-NODE-CONNECTED. Such a subgraph can be obtained by starting with $G_{\pi^*}$ and repeatedly removing edges until no further edge deletion is possible without violating the property. For each edge $\{u, v\}$ of $G_1^*$, we assign a weight $w_1(u, v)$ as follows.

---

[*] It should be noted that the graph theoretic terminology used in References 26, 27 is different from that used in this chapter. The statement of Proposition 22.1 given earlier is from Reference 28.

1. Let $r$ be a node such that $\pi^*(r) = p^*$. By using $r$ as the root, perform a DFS of $G_1^*$. Let $T(V, E_T)$ be the resulting spanning tree. Thus, each edge of $G_1^*$ is either a tree edge or a back edge.
2. For each tree edge $\{u, v\}$ where $v$ is the parent of $u$, let $w_1(u, v) = \pi^*(u)$.
3. For each back edge $\{u, v\}$ where $v$ is an ancestor of $u$, let $w_1(u, v) = \pi^*(u)$.

We can now bound the total weight $W_1(G_1^*)$ of the edges in $G_1^*$ under the edge weight function $w_1$:

**Lemma 22.3** $W_1(G_1^*) \leq (2 - 2/n)\, OPT(I)$.

*Proof.* As mentioned earlier, each edge of $G_1^*$ is either a tree edge or a back edge. Consider the tree edges first. For each tree edge $\{u, v\}$, where $v$ is the parent of $u$, $w_1(u, v) = \pi^*(u)$. Thus, the weight $\pi^*(u)$ is assigned to at most one tree edge (namely, the edge that joins $u$ to the parent of $u$ if any in $T$). The power value of the root $r$ in the optimal solution, namely $p^*$, is not assigned to any tree edge (since the root has no parent). Thus, the total weight of all of the tree edges under the weight function $w_1$ is bounded by $OPT(I) - p^*$.

Now consider the back edges. For each back edge $\{u, v\}$, where $v$ is an ancestor of $u$, $w_1(u, v) = \pi^*(u)$. As $G_1^*$ is critically 2-NODE-CONNECTED, by Lemma 22.2, each node has at most one back edge to an ancestor. Thus, the weight $\pi^*(u)$ is assigned to at most one back edge. Again, the power value $p^*$ of the root $r$ in the optimal solution is not assigned to any back edge. Thus, the total weight of all of the back edges under the weight function $w_1$ is also bounded by $OPT(I) - p^*$.

Therefore, the total weight $W_1(G_1^*)$ of all of the edges in $G_1^*$ under the edge weight function $w_1$ is at most $2\, OPT(I) - 2\, p^*$. As $p^*$ is the largest power value assigned to a node in the optimal solution, $p^*$ is at least $OPT(I)/n$. Hence, $W_1(G_1^*)$ is bounded by $(2 - 2/n)\, OPT(I)$ as required. ∎

The following lemma relates the weight $w_1(u, v)$ of an edge $\{u, v\}$ to the power threshold $p(u, v)$ needed for the existence of the edge.

**Lemma 22.4** *For any edge $\{u, v\}$ in $G_1^*$, $p(u, v) \leq w_1(u, v)$.*

*Proof.* Consider any edge $\{u, v\}$ in $G_1^*$. As $G_1^*$ is an edge subgraph of $G_{\pi^*}$ (the graph induced by the chosen optimal power assignment), $\{u, v\}$ is also an edge in $G_{\pi^*}$. Moreover, recall that the minimum power threshold values are symmetric. Therefore, $\pi^*(u) \geq p(u, v)$ and $\pi^*(v) \geq p(u, v)$. Hence $\min\{\pi^*(u), \pi^*(v)\} \geq p(u, v)$. The weight assigned to the edge $\{u, v\}$ by the edge weight function $w_1$ is either $\pi^*(u)$ or $\pi^*(v)$. Therefore, $w_1(u, v) \geq \min\{\pi^*(u), \pi^*(v)\}$. It follows that $w_1(u, v) \geq p(u, v)$. ∎

We are now ready to complete the proof of Lemma 22.1.

*Proof of Lemma 22.1.* Starting from the optimal power assignment $\pi^*$, construct the graph $G_1^*(V, E_1^*)$ as described previously. As the threshold graph $G_c$ is complete, every edge in $G_1^*$ is also in $G_c$. Consider the edge subgraph $G_1(V, E_1)$ of $G_c$ where $E_1 = E_1^*$. As $G_1^*$ is 2-NODE-CONNECTED, so is $G_1$. By Lemma 22.4, for each edge $\{u, v\}$ in $E_1$, $p(u, v) \leq w_1(u, v)$. Therefore, the total weight $W(G_1)$ of all of the edges in $G_1$ under the edge weight function $p$ is at most $W_1(G_1^*)$. By Lemma 22.3, $W_1(G_1^*)$ is bounded by $(2 - 2/n)\, OPT(I)$. Therefore, $W(G_1)$ is also bounded by $(2 - 2/n)\, OPT(I)$. In other words, the edge subgraph $G_1(V, E_1)$ is 2-NODE-CONNECTED and the total weight of all its edges is at most $(2 - 2/n)\, OPT(I)$. This completes the proof of Lemma 22.1. ∎

The following is a direct consequence of the above-mentioned discussion.

**Theorem 22.2** *There is a polynomial time approximation algorithm with a performance guarantee of $2\,(2 - 2/n)\,(2 + 1/n)$ (which approaches 8 asymptotically from the following) for the ⟨UNDIR, 2-NODE-CONNECTED, TOTALP⟩ problem.* ∎

The best known performance guarantee for the ⟨UNDIR, 2-NODE-CONNECTED, TOTALP⟩ problem is 3; the corresponding algorithm appears in Reference 30.

### 22.3.2.3 An Approximation Algorithm for ⟨UNDIR, 2-EDGE-CONNECTED, TOTALP⟩

A result analogous to Theorem 22.2 has also been obtained in Reference 14 for the the ⟨UNDIR, 2-EDGE-CONNECTED, TOTALP⟩ problem, in which the goal is to induce a graph that is 2-EDGE-CONNECTED. This problem has also been shown to be NP-complete in Reference 24. To obtain an approximation algorithm for this problem from the general framework, an approximation algorithm of Khuller and Vishkin [31] is used. Their approximation algorithm produces a 2-EDGE-CONNECTED subgraph whose cost is at most twice that of a minimum 2-EDGE-CONNECTED subgraph. In the notation of Theorem 22.1, we have $\beta \leq 2$. Again using the notation of Theorem 22.1, it is also possible to show that $\alpha \leq (2 - 1/n)$. The proof of this result is almost identical to that for the 2-NODE-CONNECTED case, except that we need an analog of Proposition 22.1. Before stating this analog, we have the following definition (which is analogous to Definition 22.3).

**Definition 22.4** *An undirected graph* $G(V, E)$ *is* **critically** 2-EDGE-CONNECTED *if it satisfies both of the following conditions: (i) G is* 2-EDGE-CONNECTED. *(ii) For every edge* $e \in E$, *the subgraph of G obtained by deleting the edge e is* not 2-EDGE-CONNECTED.

The following is the analog of Proposition 22.1 for critically 2-EDGE-CONNECTED graphs.

**Proposition 22.2** *If a graph G is critically* 2-EDGE-CONNECTED *then no cycle of G has a chord.*

*Proof.* The proof is by contradiction. Suppose $G$ is critically 2-EDGE-CONNECTED but there is a cycle $C = \langle v_1, v_2, \ldots, v_r \rangle$, with $r \geq 4$, with a chord $\{v_i, v_j\}$. Consider the graph $G'$ obtained from $G$ by deleting the chord $\{v_i, v_j\}$. We will show that $G'$ is 2-EDGE-CONNECTED, thus contradicting the assumption that $G$ is critically 2-EDGE-CONNECTED.

To show that $G'$ is 2-EDGE-CONNECTED, it suffices to show that $G'$ cannot be disconnected by deleting any single edge. Consider any edge $\{x, y\}$ of $G'$, and let $G''$ denote the graph created by deleting $\{x, y\}$ from $G'$. As we deleted only one edge from $G'$, all the nodes of the cycle $C$ are in the same connected component of $G''$. Thus, if we create the graph $G_1$ by adding the chord $\{v_i, v_j\}$ to $G''$, the two graphs $G_1$ and $G''$ have the same number of connected components. However, $G_1$ is also the graph obtained by deleting the edge $\{x, y\}$ from $G$. As $G$ is 2-EDGE-CONNECTED, $G_1$ is connected. Thus, $G''$ is also connected. We therefore conclude that $G'$ is 2-EDGE-CONNECTED, and this contradiction completes the proof of Proposition 22.2. ∎

The remainder of the proof to show that $\alpha \leq (2 - 2/n)$ is identical to that for the 2-NODE-CONNECTED case. With $\alpha \leq (2 - 2/n)$ and $\beta \leq 2$, the following theorem is a direct consequence of Theorem 22.1.

**Theorem 22.3** *There is a polynomial time approximation algorithm with a performance guarantee of* $8(1 - 1/n)$ *(which approaches* 8 *asymptotically from below) for the* ⟨UNDIR, 2-EDGE-CONNECTED, TOTALP⟩ *problem.* ∎

### 22.3.2.4 Improvement by Calinescu and Wan

The approximation algorithms of Sections 22.3.2.2 and 22.3.2.3 were shown to provide a performance guarantee of at most 8 using the general bound given in Theorem 22.1. By a more careful analysis of the two algorithms, Calinescu and Wan [24] have shown that the algorithms actually provide a performance guarantee of at most 4. In particular, they show that the approximation algorithm of Section 22.3.2.3 for the ⟨UNDIR, 2-EDGE-CONNECTED, TOTALP⟩ problem can be generalized to obtain an approximation algorithm with a performance guarantee of $2k$ for the ⟨UNDIR, $k$-EDGE-CONNECTED, TOTALP⟩ problem, for any $k \geq 2$. Here, we discuss this general result.

We need to introduce some notation. Given an undirected graph $G$, the directed graph obtained by replacing each undirected edge $\{u, v\}$ by the pair of directed edges $(u, v)$ and $(v, u)$ is denoted by $\overrightarrow{G}$. When each edge $e = \{u, v\}$ of the graph $G$ has an edge weight $w(e)$, the weights of the directed edges $(u, v)$ and $(v, u)$ in $\overrightarrow{G}$ are also set to $w(e)$. For a directed graph $D(V, A)$, we use $\overline{D}$ to denote the undirected graph obtained from $D$ by erasing the directions on all the edges and combining multi-edges into a single edge. For an edge weighted undirected graph $G$ (directed graph $D$), the total weight of all the edges is denoted by $W(G)$ ($W(D)$).

Suppose $G(V, E)$ is the undirected graph induced by assigning powers to the nodes of an ad hoc network. For any node $u \in V$, the **power of $u$ with respect to** $G$, denoted by $p_G(u)$, is given by $p_G(u) =$ $\max\{p(u, v) : \{u, v\} \in E\}$. The **power for inducing** $G$, denoted by $P(G)$, is given by $P(G) = \sum_{u \in V} p_G(u)$.

We will also need the directed graph model (Section 22.1) of graphs induced by power assignments. Recall that in that model, a directed edge $(u, v)$ is in the induced directed graph if and only if the power assigned to $u$ is at least $p(u, v)$. The definitions of power of a node and the power for inducing a graph given earlier for undirected graphs can be readily extended to directed graphs. For any node $u$ of a directed graph $D(V, A)$, we let $p_D(u) = \max\{p(u, v) : (u, v) \in A\}$ and $P(D) = \sum_{u \in V} p_D(u)$.

A directed graph is STRONGLY $k$-EDGE-CONNECTED if for each pair of nodes $u$ and $v$, there are at least $k$ edge disjoint directed paths from $u$ to $v$. A directed graph $D(V, A)$ is an **Inward Branching** rooted at a node $s \in V$ if $|A| = |V| - 1$ and there is a directed path from each node in $V - \{s\}$ to $s$. Thus, in an inward branching rooted at $s$, the outdegree of each node except $s$ is 1 and the outdegree of $s$ is 0. The following three-part observation is a simple consequence of the above-mentioned definitions.

**Observation 22.1** *Let $G$ be an undirected graph and $D$ be a directed graph, both having edge weights. Then, the following properties hold: (i) $P(G) \leq 2W(G)$. (ii) $W(\overline{D}) \leq W(D)$. (iii) If $D$ is an inward branching, then $P(D) = W(D)$.*

Given a directed graph $G(V, A)$, a vertex $s \in V$ and an integer $k \geq 1$, we say that $D$ is $k$-EDGE-INCONNECTED to $s$ if there are at least $k$ edge-disjoint paths from each vertex in $V - \{s\}$ to $s$. The following known results about $k$-EDGE-INCONNECTED graphs are used in proving the main result of this section.

**Theorem 22.4**  (a) *Suppose $D(V, A)$ is $k$-EDGE-INCONNECTED to a vertex $s \in V$. Then $A$ contains $k$ pairwise disjoint subsets $A_1, A_2, \ldots, A_k$ such that for each $i$, $1 \leq i \leq k$, the directed graph $B_i(V, A_i)$ is an inward branching rooted at $s$ [32].*

(b) *Given a directed graph $D(V, A)$ with a nonnegative weight for each edge, an integer $k \geq 2$ and a vertex $s \in V$, the problem of obtaining a subgraph $D_1(V, A_1)$ of minimum total weight such that $D_1$ is $k$-EDGE-INCONNECTED to $s$ can be solved in polynomial time [33,34].* ∎

Recall that the approximation algorithm of Section 22.3.2.3 was obtained using the Khuller-Vishkin Algorithm (KV-Algorithm) [31] in Step 2 of the general framework. Relying on Part (b) of Theorem 22.4, the KV-Algorithm actually provides a performance guarantee of 2 for the following problem: Given an integer $k \geq 2$ and an undirected and $k$-EDGE-CONNECTED graph $G$ with nonnegative edge weights, find a subgraph $H(V, E_H)$ of minimum weight such that $H$ is also $k$-EDGE-CONNECTED. We give the steps of the KV-Algorithm in the following as they are used in the proof of the main result.

1. From the graph $G$, construct $\overrightarrow{G}$.
2. Choose any vertex $s$ of $G$ and construct a subgraph $D$ of $\overrightarrow{G}$ such that $D$ is $k$-EDGE-INCONNECTED to $s$ and has the smallest total weight among all such subgraphs. (By Part (b) of Theorem 22.4, this step can be done in polynomial time.)
3. Construct and output $\overline{D}$.

It is shown in Reference 31 that $\overline{D}$ is $k$-EDGE-CONNECTED and that $W(D)$ is at most twice the weight of an optimal solution.

We are now ready to prove the main result of this section. The following proof follows the presentation in Reference 24.

**Theorem 22.5** *For any $k \geq 2$, the approximation algorithm obtained by using the KV-Algorithm in Step 2 of the general framework (Figure 22.1) provides a performance guarantee of $2k$ for the ⟨UNDIR, $k$-EDGE-CONNECTED, TOTALP⟩ problem.*

*Proof.* Let $I$ denote the given instance of the $\langle$UNDIR, $k$-EDGE-CONNECTED, TOTALP$\rangle$ problem. Let $OPT(I)$ and $HEU(I)$ denote respectively the total power of an optimal assignment $\pi^*$ and that of the power assignment $\pi$ produced by the heuristic referred to in the statement of the theorem. Note that $\pi^*$ is an optimal power assignment under the undirected graph model. Our goal is to show that $HEU(I) \leq 2k\,OPT(I)$.

Consider an optimal power assignment $\pi_d^*$ for the instance $I$ under the *directed* graph model. Let $\pi_d^*$ induce the directed graph $D^*(V, A^*)$. Note that $D^*$ is STRONGLY $k$-EDGE-CONNECTED. As the power threshold values are symmetric, it follows that $P(D^*) \leq OPT(I)$.

Consider any vertex $s \in V$. As $D^*(V, A^*)$ is STRONGLY $k$-EDGE-CONNECTED, $D^*$ is also $k$-EDGE-INCONNECTED to $s$. By Part (a) of Theorem 22.4, $A^*$ contains $k$ pairwise disjoint subsets $A_1, A_2, \ldots, A_k$ such that each directed graph $B_i(V, A_i)$, $1 \leq i \leq k$, is an inward branching with $s$ as the root. Thus, the directed graph $D'(V, \cup_{i=1}^k A_i)$ is $k$-EDGE-INCONNECTED to $s$. Therefore, the weight of the solution $D$ found in Step 2 of the KV-Algorithm is at most $W(D')$. By using this observation, the following claim provides a bound on $W(D)$.    ∎

**Claim 22.1** $W(D) \leq k\,OPT(I)$.

*Proof.* As mentioned earlier, $W(D) \leq W(D')$. As $W(D') = \sum_{i=1}^k W(B_i)$, we have

$$
\begin{aligned}
W(D) &\leq \sum_{i=1}^k W(B_i) \\
&= \sum_{i=1}^k P(B_i) \quad \text{(Part (iii) of Observation 22.1)} \\
&\leq \sum_{i=1}^k P(D^*) \\
&= k\,P(D^*) \\
&\leq k\,OPT(I) \quad \text{(since } p(D^*) \leq OPT(I))
\end{aligned}
$$

as indicated in the statement of the claim.

The power assignment $\pi$ output by the approximation algorithm is obtained from the graph $\overline{D}$. Therefore,

$$
\begin{aligned}
HEU(I) &= P(\overline{D}) \\
&\leq 2\,W(\overline{D}) \quad \text{(Part (i) of Observation 22.1)} \\
&\leq 2\,W(D) \quad \text{(Part (ii) of Observation 22.1)} \\
&\leq 2k\,OPT(I) \quad \text{(Claim 22.1)}.
\end{aligned}
$$

This completes the proof of Theorem 22.5.    ∎

### 22.3.2.5 Bicriteria Approximation for Diameter and Total Power

We now present an application of Theorem 22.1 to the topology control problem in which the goal is to induce a graph whose diameter is bounded by a given value. For an undirected graph $G(V, E)$, recall that the **diameter** is given by $\max\{d(u,v) : u,v \in V\}$, where $d(u,v)$ denotes the number of edges in a shortest path between nodes $u$ and $v$. Graphs with small diameters are desirable in the context of ad hoc networks as the diameter determines the largest end-to-end delay in such a network. We assume that a diameter bound $\gamma$ is given as part of the problem instance and the goal is to find a power assignment such that the diameter of the induced graph is at most $\gamma$ and the total power assigned to the nodes is minimized. For this problem, denoted by $\langle$UNDIR, DIAMETER, TOTALP$\rangle$, approximation algorithms with similar performance guarantees were presented in References 35, 36. We discuss the algorithm from Reference 35.

The $\langle$UNDIR, DIAMETER, TOTALP$\rangle$ problem involves two minimization objectives, namely, the diameter and the total power. Such **bicriteria** minimization problems are typically handled by designating one of the objectives as a budget constraint and the other as the minimization objective [37]. Following

Reference 35, we will designate the diameter bound as a constraint and total power as the minimization objective. It is shown in Reference 35 that if the diameter constraint must be satisfied, the total power cannot be approximated to within a factor $\lambda \, \log n$, for some $\lambda$, $0 < \lambda < 1$, unless $\mathbf{P} = \mathbf{NP}$. So, the approximation algorithm presented in Reference 35 is a **bicriteria approximation**, in which the diameter constraint is violated by a certain factor and the total power is approximated to within another factor. A formal definition of bicriteria approximation is as follows.

**Definition 22.5** *Suppose a problem* $\Pi$ *with two minimization objectives A and B is posed in the following manner: Given a budget constraint on objective A, find a solution which minimizes the value of objective B among all solutions satisfying the budget constraint. A* $(\rho_1, \rho_2)$-***approximation algorithm** for problem* $\Pi$ *is a polynomial time algorithm that provides for every instance of* $\Pi$ *a solution satisfying the following two conditions.*

1. *The solution violates the budget constraint on objective A by a factor of at most* $\rho_1$.
2. *The value of objective B in the solution is within a factor of at most* $\rho_2$ *of the minimum possible value satisfying the budget constraint.*

To obtain bicriteria approximation algorithms for the ⟨UNDIR, DIAMETER, TOTALP⟩ problem, we rely on known approximation results for another problem, called the **Minimum Cost Tree with a Diameter Constraint** (MCTDC), also involving two minimization objectives. A formal definition of this problem is as follows.

**Minimum Cost Tree with a Diameter Constraint**
*Instance:* A connected undirected graph $G(V, E)$, a nonnegative weight $w(e)$ for each edge $e \in E$, an integer $\gamma \leq n - 1$.

*Requirement:* Find a spanning tree $T(V, E_T)$ of $G$ such that $\text{DIA}(T) \leq \gamma$ and the total edge weight of $T$ is minimum among all the trees satisfying the diameter constraint.

MCTDC is known to be NP-hard [37]. Bicriteria approximations for this problem have been presented in References 37–39.

The bicriteria approximation algorithm for ⟨UNDIR, DIAMETER, TOTALP⟩ in Reference 35 is based on the general framework (Figure 22.1). The steps of the approximation algorithm, which we call GEN-DIAMETER-TOTAL-POWER, are shown in Figure 22.2. In Step 2 of Figure 22.2, any approximation algorithm $\mathcal{A}$ for the MCTDC problem can be used. As long as $\mathcal{A}$ runs in polynomial time,

---

**Input:** An instance $I$ of the ⟨UNDIR, DIAMETER, TOTALP⟩ problem, with diameter bound $\gamma$.

**Output:** A power value $\pi(u)$ for each node $u$ such that the diameter of the undirected graph induced by the power assignment is close to $\gamma$ and the total power assigned to all nodes is as small as possible.

1. Let $G_c(V, E_c)$ denote the threshold graph for the given instance of the ⟨UNDIR, DIAMETER, TOTALP⟩ problem.

2. Use any approximation algorithm $\mathcal{A}$ for the MCTDC problem on $G_c(V, E_c)$ with diameter bound $2\gamma$, and obtain a spanning tree $T(V, E_T)$ of $G_c$.

3. For each node $u$, assign a power value $\pi(u)$ equal to the weight of the largest edge incident on $u$ in $T$.

**FIGURE 22.2** Outline of heuristic GEN-DIAMETER-TOTAL-POWER.

GEN-DIAMETER-TOTAL-POWER also runs in polynomial time. The performance guarantee provided by GEN-DIAMETER-TOTAL-POWER is a function of the performance guarantee provided by Algorithm $\mathcal{A}$.

The solution produced by Heuristic GEN-DIAMETER-TOTAL-POWER is approximate in terms of both diameter and total power. So, we cannot directly rely on Theorem 22.1 to derive the performance guarantee provided by the heuristic.

For the remainder of this section, we use the following notation. Let $I$ denote the given instance of the $\langle$UNDIR, DIAMETER, TOTALP$\rangle$ problem with $n$ nodes and diameter bound $\gamma \geq 1$. Let $\pi^*$ denote an optimal power assignment such that the graph $G_{\pi^*}$ induced by $\pi^*$ has diameter at most $\gamma$, and let $OPT(I) = \sum_{v \in V} \pi^*(v)$. Let $\pi$ denote the power assignment produced by the heuristic and let $G_\pi$ denote the graph induced by $\pi$. Let $DTP(I) = \sum_{v \in V} \pi(v)$, the total power assigned by the heuristic for the instance $I$. The bicriteria approximation result proved in this section is as follows.

**Theorem 22.6** *Suppose Algorithm $\mathcal{A}$ used in Step 2 of Heuristic* GEN-DIAMETER-TOTAL-POWER *is a $(\rho_1, \rho_2)$-approximation algorithm for the* MCTDC *problem. For any instance I of the* $\langle$UNDIR, DIAMETER, TOTALP$\rangle$ *problem, Heuristic* GEN-DIAMETER-TOTAL-POWER *produces a power assignment $\pi$ satisfying the following two properties: (1)* DIA$(G_\pi) \leq 2\rho_1 \gamma$, *(2)* $DTP(I) \leq 2\rho_2 (1 - 1/n) OPT(I)$. *In other words, Heuristic* GEN-DIAMETER-TOTAL-POWER *provides a $(2\rho_1, 2\rho_2(1-1/n))$ bicriteria approximation for the* $\langle$UNDIR, DIAMETER, TOTALP$\rangle$ *problem.*

Several lemmas are needed to prove Theorem 22.6. We begin with a simple lemma about spanning trees generated by carrying out a **breadth-first-search** (BFS) on a connected graph [29].

**Lemma 22.5** *Let G be a connected graph with diameter $\delta$. Let T be any spanning tree for G generated by BFS. Then* DIA$(T) \leq 2\delta$.

*Proof.* Suppose $T$ was generated by carrying out BFS starting with node $v$ (as the root). For any node $x$, the distance $d(v, x)$ between the root $v$ and $x$ in $T$ is the length of a shortest path between them in $G$ [29]. Thus, for any node $x$, $d(v, x) \leq \delta$ in $T$. It follows that the maximum distance between any pair of nodes in $T$, that is DIA$(T)$, is at most $2\delta$. ∎

The next lemma indicates why in Step 2 of Heuristic GEN-DIAMETER-TOTAL-POWER, the diameter bound of $2\gamma$ is used.

**Lemma 22.6** *Consider the threshold graph $G_c(V, E_c)$ of the $\langle$UNDIR, DIAMETER, TOTALP$\rangle$ instance I. There is a spanning tree $T_1(V, E_{T_1})$ of $G_c$ satisfying the following two properties: (a)* DIA$(T_1) \leq 2\gamma$, *(b) Let $W(E_{T_1})$ denote the total edge weight of $T_1$. Then, $W(E_{T_1}) \leq (1 - 1/n) OPT(I)$.*

*Proof.* **Part (a):** Consider the graph $G_{\pi^*}$ induced by the optimal power assignment $\pi^*$. Note that DIA$(G_{\pi^*}) \leq \gamma$. Let $v$ be a node such that $\pi^*(v)$ has the largest value among all the nodes in $V$. Let $T_1(V, E_{T_1})$ be a spanning tree of $G_{\pi^*}$ generated by carrying out a BFS on $G_{\pi^*}$ with $v$ as the root. Then, from Lemma 22.5, we have DIA$(T_1) \leq 2\gamma$.

**Part (b):** Consider another assignment $w$ of weights to the edges of $T_1$ as indicated in the following. Consider each edge $\{x, y\}$ in $T_1$, where $y$ is the parent of $x$. Let $w(x, y) = \pi^*(x)$. Thus, the power value assigned by the optimal solution to each node except the root becomes the weight of exactly one edge of $T_1$. The power value $\pi^*(v)$ of the root is not assigned to any edge. Therefore,

$$\sum_{\{x,y\} \in E_{T_1}} w(x, y) = OPT(I) - \pi^*(v).$$

As $v$ has the maximum power value under $\pi^*$ among all the nodes, we have $\pi^*(v) \geq OPT(I)/n$. Therefore,

$$\sum_{\{x,y\} \in E_{T_1}} w(x, y) \leq (1 - 1/n) OPT(I).$$

The following claim relates the weight $w(x, y)$ to the power threshold value $p(x, y)$.

**Claim 22.2** *For each edge* $\{x, y\} \in E_{T_1}$, $w(x, y) \geq p(x, y)$.

*Proof of Claim.* Consider edge $\{x, y\} \in E_{T_1}$, where $y$ is the parent of $x$. Then, by definition, $w(x, y) = \pi^*(x)$. As $T_1$ is a spanning tree of $G_{\pi^*}$, the edge $\{x, y\}$ is also in $G_{\pi^*}$. This fact, in conjunction with the assumption that the power threshold values are symmetric, implies that $\pi^*(x) \geq p(x, y)$. The claim follows.

As a simple consequence of the above-mentioned claim, we have

$$W(E_{T_1}) \leq \sum_{\{x,y\} \in E_{T_1}} w(x, y) \leq (1 - 1/n) \, OPT(I),$$

and this completes the proof of Part (b) of the lemma. ∎

The next lemma uses the performance guarantee provided by the approximation algorithm $\mathcal{A}$ used in Step 2 of the heuristic.

**Lemma 22.7** *Let $T(V, E_T)$ denote the tree produced by $\mathcal{A}$ at the end of Step 2 of Heuristic* GEN-DIAMETER-TOTAL-POWER. *Let $W(E_T)$ denote the total weight of the edges in $T$. Let $(\rho_1, \rho_2)$ denote the performance guarantee provided by $\mathcal{A}$ for the* MCTDC *problem. Then the following inequalities hold: (a)* $\mathrm{DIA}(T) \leq 2 \, \rho_1 \, \gamma$ *and (b)* $W(E_T) \leq \rho_2 \, (1 - 1/n) \, OPT(I)$.

*Proof.* By Lemma 22.6, the edge weighted graph $G_c(V, E_c)$ has a spanning tree of diameter at most $2\gamma$ and total edge weight at most $(1 - 1/n) \, OPT(I)$. Now, Lemma 22.7 is a simple consequence of the performance guarantee provided by Algorithm $\mathcal{A}$. ∎

We are now ready to prove Theorem 22.6.

*Proof of Theorem 22.6.* Consider the spanning tree $T(V, E_T)$ produced in Step 2 of the heuristic. It is straightforward to verify that every edge $\{x, y\} \in E_T$ is also in $G_\pi (V, E_\pi)$, the graph induced by the power assignment constructed in Step 3 of the heuristic. As $\mathrm{DIA}(T) \leq 2 \, \rho_1 \, \gamma$, and the addition of edges cannot increase the diameter, it follows that $\mathrm{DIA}(G_\pi) \leq 2 \, \rho_1 \, \gamma$.

To bound $DTP(I)$, we note from Lemma 22.7 that $W(E_T) \leq \rho_2 \, (1 - 1/n) \, OPT(I)$. In the power assignment constructed in Step 3, the weight of any edge can be assigned to at most two nodes (namely, the end points of that edge). Thus, the total power assigned to all the nodes is at most $2 \, W(E_T)$. In other words, $DTP(I) \leq 2 \, \rho_2 \, (1 - 1/n) \, OPT(I)$. ∎

We now briefly indicate how several bicriteria approximation algorithms for the $\langle$UNDIR, DIAMETER, TOTALP$\rangle$ problem can be obtained using GEN-DIAMETER-TOTAL-POWER in conjunction with known bicriteria approximation results for the MCTDC problem.

1. For any $\epsilon > 0$, a $(2 \lceil \log_2 n \rceil, (1 + \epsilon) \lceil \log_2 n \rceil)$-approximation algorithm is presented in Reference 37 for the MCTDC problem. By using this algorithm and setting $\epsilon < 1/n$, we can obtain a $(4 \lceil \log_2 n \rceil, 2 \lceil \log_2 n \rceil)$-approximation algorithm for the $\langle$UNDIR, DIAMETER, TOTALP$\rangle$ problem.

2. For any *fixed* $\gamma \geq 1$, a $(1, O(\gamma \log n))$-approximation algorithm for the MCTDC problem is presented in Reference 39. Thus, for any fixed $\gamma \geq 1$, we can obtain a $(2, O(\gamma \log n))$-approximation algorithm for the $\langle$UNDIR, DIAMETER, TOTALP$\rangle$ problem.

3. For any $\gamma$ and any fixed $\epsilon > 0$, a $(1, O(n^\epsilon \log n))$-approximation algorithm for the MCTDC problem is presented in Reference 38. Thus, for this case, we can obtain a $(2, O(n^\epsilon \log n))$-approximation algorithm for the $\langle$UNDIR, DIAMETER, TOTALP$\rangle$ problem.

### 22.3.3 Some Extensions

The general framework for minimizing total power can also be used to obtain polynomial time approximation algorithms for topology control problems wherein the connectivity requirements are specified using $\{0,1\}$-*proper functions* [40]. To obtain this result, the general method outlined in References 40, 41 is used as the algorithm in Step 2 of the framework. The method of References 40, 41 gives a 2-approximation algorithm for network design problems specified using $\{0,1\}$-proper functions. Thus, using the notation of Theorem 22.1, we have $\beta = 2$. It is also straightforward to show that the threshold graph contains an appropriate subgraph of weight at most the optimal solution value. In other words, $\alpha \le 1$. Thus, we obtain a 4-approximation algorithm for the general class of problems defined in References 40, 41. An important example of a problem in this class is the Steiner variant of connectivity, in which the goal is to assign power levels so as to induce a graph which connects a specified subset of nodes, called **terminals**; the induced graph may use nodes which are not terminals. A better approximation algorithm for this problem appears in Reference 19. This algorithm uses the iterative randomized rounding framework of Reference 20 and provides a performance guarantee of $3 \ln 4 - \frac{9}{4} + \epsilon$, for any fixed $\epsilon > 0$. By choosing an appropriate value of $\epsilon$, its performance guarantee can be made less than 1.91.

The bicriteria results for minimizing the diameter and total power can also be extended to the Steiner version, in which only the terminals need to be connected together into a graph of bounded diameter. Letting $\eta$ denote the number of terminals, Reference 37 presents an $(O(\log \eta), O(\log \eta))$-approximation algorithm for the Steiner version of the MCTDC problem. By using this approximation algorithm in Step 2 of Figure 22.2, we obtain an $(O(\log \eta), O(\log \eta))$-approximation algorithm for the Steiner version of the ⟨UNDIR, DIAMETER, TOTALP⟩ problem.

## 22.3.4 Distributed Approximation Algorithms

Sections 22.3.1 to 22.3.3 considered centralized approximation algorithms for minimizing total power used by the nodes. In the current section we discuss *distributed* topology control algorithms for minimizing total power. These algorithms also have provably good performance guarantees.

### 22.3.4.1 Preliminaries

The distributed algorithms that we discuss all utilize the *geometric model*. Typically, in geometric instances of topology control problems, nodes are points in a metric space, and the transmission power threshold $p(u, v)$ is determined by the distance $d(u, v)$ between $u$ and $v$. Specifically, the power threshold $p(u, v)$ is $d(u, v)^\alpha$, where $\alpha$ is the *attenuation constant* associated with path loss. The *path loss* is the ratio of the received power to the transmitted power of the signal [42]. The value of $\alpha$ is typically between 2 and 4.

For consistency with the notation used in previous sections, we assume that instances of topology control problems under the geometric model are also specified through threshold graphs, in which the threshold values are determined from the distances as discussed previously. The following result providing a *weak triangle inequality* between the threshold values under the geometric model is established in Reference 43:

**Lemma 22.8** *For nodes $u$, $v$, and $w$, $p(u, v) \le 2^{\alpha-1}(p(u, w) + p(w, v))$.*  ∎

It is shown in Reference 24 that:

**Theorem 22.7** *The* ⟨GEOMETRIC, 2-NODE-CONNECTED, TOTAL POWER⟩ *problem is NP-hard.*  ∎

### 22.3.4.2 Distributed Algorithms for ⟨GEOMETRIC, 2-NODE-CONNECTED, TOTAL POWER⟩

A framework for distributed topology control algorithms for the ⟨GEOMETRIC, 2-NODE-CONNECTED, TOTAL POWER⟩ problem is given in Figure 22.3. The framework consists of first finding a minimum spanning tree of the threshold graph, and then, for each non leaf node $u$ in that tree, adding edges to connect all of the neighbors of $u$.

The framework given in Figure 22.3 is easy to implement in a distributed fashion: Step 1 utilizes the distributed minimum spanning tree algorithm given by Gallager et al. [44]; the computation in Step 2 is handled by each node $u$, which then distributes the relevant results to its neighbors; finally, each node computes its own power value in Step 3.

**Input:** A threshold graph $G_t$ under the geometric model.

**Output:** A power value $\pi(u)$ for each node $u$ such that the undirected graph induced by that power assignment is 2-NODE-CONNECTED.

**Steps:**

1. Compute a minimum spanning tree $T$ of $G_t$ and let $G' = T$.

2. For each nonleaf node $u$ in $T$, let $V_u$ consist of the neighbors of $u$ in $T$. Then, from $G_t$, add edges $E_u$ to $G'$ such that the graph $G_u = (V_u, E_u)$ is connected.

3. For each node $u$, assign a power value $\pi(u)$ equal to the largest threshold among the edges incident on $u$ in $G'$.

**FIGURE 22.3** A distributed framework for ⟨GEOMETRIC, 2-NODE-CONNECTED, TOTAL POWER⟩.

**Theorem 22.8** *The framework in Figure 22.3 produces a* 2-NODE-CONNECTED *network.*

*Proof.* As the resulting network contains a minimum spanning tree of $G_t$, the network is at least 1-NODE-CONNECTED. Thus, by way of contradiction, assume that $G'$ is not 2-NODE-CONNECTED. This means that some node $u$ is an articulation point of $G'$, hence $u$ must be a node of degree two or more in the minimum spanning tree $T$ that was placed into $G'$. Let $v$ and $w$ be neighbors of $u$ that are in different connected components if $u$ is removed from $G'$. This is not possible as Step 2 of the framework adds edges to $G'$ to connect all of the neighbors of $u$ without using any of the edges in $T$. ∎

Although the framework does produce a network that is 2-NODE-CONNECTED, the specific method used to connect the neighbors of node $u$ in Step 2 is left open. Two natural methods for connecting the neighbors of $u$ are ones in which

- Each graph $G_u$ is a simple path [43].
- Each graph $G_u$ is a minimum spanning tree of the restriction of $G_t$ to $V_u$ [24].

The quality of the solutions produced by algorithms based on the framework in Figure 22.3 depends on both the path loss exponent ($\alpha$) and on the particular method utilized in Step 2 to connect the neighbors of node $u$. To state this dependence, we use the following notation. For any instance $I$ of ⟨GEOMETRIC, 2-NODE-CONNECTED, TOTAL POWER⟩, let $OPT(I)$ and $DF(I)$ denote the total power assigned by an optimal solution and that produced by the above-mentioned framework, respectively. We begin with the following result which was originally shown in Reference 43.

**Theorem 22.9** *Let $I$ be an instance of* ⟨GEOMETRIC, 2-NODE-CONNECTED, TOTAL POWER⟩. *By using the distributed framework and finding a simple path in Step 2, $DF(I) \leq (2 + 2^{\alpha+2}) OPT(I)$.*

To prove this theorem we begin by letting $C(G)$ denote the sum of the threshold values associated with the edges in a graph $G$. Note that this is different from the total power value needed to induce $G$. We have the following lemma whose proof also appears in Reference 12.

**Lemma 22.9** *For $G_t$, let $\pi'$ be an optimal power assignment such that the induced graph $G_{\pi'}$ is 1-NODE-CONNECTED. Then $C(T) \leq \sum_{v \in G_t} \pi'(v)$, where $T$ is the minimum spanning tree computed in Step 1 of the framework.*

*Proof.* Consider any spanning tree $T'$ of $G_{\pi'}$, and any leaf $x$ in $T'$. The threshold value of the edge between $x$ and its parent $y$, is no more than the power value of $x$ in $G_{\pi'}$. Now remove node $x$ and edge $\{x,y\}$ from the tree. By recursively applying this node removal, it is seen that the weight of each edge in $T'$ is bounded from above by the power value assigned to some node, and each node is utilized at most once in this fashion. Thus, $C(T') \leq \sum_{v \in G_t} \pi'(v)$. The lemma follows as $T'$ is also a spanning tree of $G_t$, and $T$ is a minimum spanning tree of $G_t$. Hence $C(T) \leq C(T')$. ∎

*Proof of Theorem* 22.9. Consider the graph $G'$ computed by the framework. As each edge in $G'$ determines the power assignment to at most two nodes, it follows that

$$DF(I) \leq 2C(G'). \tag{22.1}$$

Note that $G'$ consists of the minimum spanning tree $T$ computed in Step 1 along with the edges added in Step 2. Let $G''$ be a graph over the nodes of $G'$ that contains all of the edges added in Step 2. As the edges of $T$ and $G''$ are disjoint,

$$C(G') = C(T) + C(G''). \tag{22.2}$$

Consider any edge $\{u,v\}$ in $G''$. Recall that $u$ and $v$ are both neighbors of some $w$, such that edges $\{w,u\}$ and $\{w,v\}$ are both in $T$. From Lemma 22.8, $p(u,v) \leq 2^{\alpha-1}(p(u,w)+p(w,v))$. Note that for any edge $\{s,t\}$ in $T$, there are at most four edges of $G''$ that are adjacent to it. Thus,

$$C(G'') \leq 4 * 2^{\alpha-1}C(T) = 2^{\alpha+1}C(T). \tag{22.3}$$

Combining Equations 22.1 through 22.3, we have that

$$DF(I) \leq 2(C(T) + 2^{\alpha+1}C(T)) = (2 + 2^{\alpha+2})C(T).$$

By applying Lemma 22.9, it follows that $DF(I) \leq (2 + 2^{\alpha+2})\sum_{v \in G_t} \pi'(v)$. Finally, the theorem follows by noting that $\sum_{v \in G_t} \pi'(v)$ cannot exceed the optimal power $OPT(I)$ needed to produce a 2-NODE-CONNECTED network. ∎

Recall from earlier that finding a minimum spanning tree in Step 2 is an alternative to finding a simple path. For that case, a somewhat better performance guarantee was shown in Reference 24.

**Theorem 22.10** *Let $I$ be an instance of* ⟨GEOMETRIC, 2-NODE-CONNECTED, TOTAL POWER⟩. *By using the distributed framework and finding a minimum spanning tree in Step 2, $DF(I) \leq 8\,OPT(I)$ when $\alpha = 2$ and $DF(I) \leq (3.2 * 2^\alpha)\,OPT(I)$ for any $\alpha > 2$.* ∎

The above-mentioned discussion focused on distributed algorithms for ⟨GEOMETRIC, 2-NODE-CONNECTED, TOTAL POWER⟩. By using that framework, one can also obtain the following result for ⟨GEOMETRIC, 1-NODE-CONNECTED, TOTAL POWER⟩:

**Theorem 22.11** *Let $I$ be an instance of* ⟨GEOMETRIC, 1-NODE-CONNECTED, TOTAL POWER⟩. *By using a distributed algorithm consisting of Steps 1 and 3 of the framework, $DF(I) \leq 2\,OPT(I)$.* ∎

That is, the algorithm* finds a minimum spanning tree of $G_t$, and the result follows from the proof of Theorem 22.9.

## 22.4 A Summary of Results for Other Topology Control Problems

The literature contains approximation results for a variety of topology control problems. In the current section we provide a brief overview of some of these results.

---

* In fact, this is the algorithm given in Reference 12 for the ⟨UNDIR, 1-NODE-CONNECTED, TOTALP⟩ problem.

Results similar to those presented in Section 22.3.2 have been obtained for higher node connectivities. Reference 30 provides approximation algorithms for ⟨UNDIR, $k$-NODE-CONNECTED, TOTALP⟩ for $k = 3$ (with a performance guarantee of 4), and for $k \in \{4, 5\}$ (with a performance guarantee of $k + 3$). For arbitrary $k$, the best known performance guarantee is $O(\log k \log(\frac{n}{n-k}))$; the corresponding algorithm appears in Reference 45. To within a constant factor, this performance guarantee matches the best known performance guarantee for the Minimum Cost $k$-Node-Connected Subgraph problem. For earlier but weaker results, see Reference 46.

For ⟨UNDIR, $k$-EDGE-CONNECTED, TOTALP⟩, an algorithm with a performance guarantee of $O(\sqrt{n})$ is given in Reference 47; this provides an improvement over the bound of $2k$ presented in Theorem 22.5 only for very large values of $k$.

Reference 35 considers the topology control problem in which the goal is to compute a power assignment $\pi$ such that the undirected graph $G_\pi$ is connected, the degree of each node in $G_\pi$ is at least a specified value $\Delta$, and the total power is minimized. It is shown there that the problem is **NP**-complete for every fixed integer $\Delta \geq 2$. Moreover, an approximation algorithm based on the general framework in Figure 22.1 with a performance guarantee of $2(\Delta + 1)(1 - 1/n)$ is presented for the problem. An improved performance guarantee of $O(\log(\Delta))$ for this problem follows from the results in Reference 45.

Topology control problems in which the power threshold values may be *asymmetric* were considered in References 35, 48, 49. In this model, there may be pairs of nodes $u$ and $v$ such that $p(u, v) \neq p(v, u)$. It was shown in References 35, 48 that the ⟨UNDIR, 1-NODE-CONNECTED, TOTALP⟩ problem under this model can be approximated to within $O(\log n)$ and that this result cannot be improved beyond a constant factor, unless **P** = **NP**. This in sharp contrast to the symmetric threshold case, where there is a $(3/2 + \epsilon)$-approximation algorithm (Section 22.3.2.1). Reference 49 presents other positive and negative approximation results for asymmetric power thresholds.

As mentioned in our initial formulation of topology control problems (Section 22.2.2), the objective of minimizing the maximum power assigned to any node has also been studied [3,14]. For any monotone and efficiently testable property $\mathbb{P}$, it was shown in Reference 14 that there is a polynomial time algorithm for minimizing the maximum power. This algorithm makes $O(\log n)$ calls to an algorithm for testing whether a graph $G$ has the property $\mathbb{P}$. It was also shown in Reference 14 that for any given $\epsilon > 0$, it is possible to get a $(1 + \epsilon)$-approximation for minimizing the maximum power using only $O(\log \log (p_{max}/p_{min}))$ calls to the algorithm for testing $\mathbb{P}$, where $p_{max}$ and $p_{min}$ are the largest and the smallest power thresholds, respectively. This approximation algorithm is particularly useful when the ratio $p_{max}/p_{min}$ is small. For ⟨UNDIR, 1-NODE CONNECTED, MAXP⟩, it is known that the bottleneck spanning tree achieves the optimum, and it can be computed in linear time (as in Problem 23-3 of Reference 29).

The polynomial algorithm for minimizing maximum power given in Reference 14 may assign the maximum power value to a larger than necessary number of nodes. The problem of minimizing the number of nodes to which the maximum power is assigned was considered in Reference 50. It was shown that even for simple properties such as 1-NODE-CONNECTED, the problem is **NP**-complete. Further, it was shown that for any property, the problem can be reduced in an approximation-preserving manner to the problem of minimizing the total power. As a consequence, the $(5/3 + \epsilon)$-approximation algorithm for ⟨UNDIR, 1-NODE-CONNECTED, TOTALP⟩ given in Reference 21 provides the same performance guarantee for minimizing the number of nodes assigned the maximum power. However, as pointed out in Reference 21, the algorithm is not practical as it involves solving large linear programs. In Reference 50, a 5/3-approximation algorithm was presented for the problem. This algorithm does not involve solving linear programs, and its running time is $O(n^3 \alpha(n))$, where $\alpha(n)$ is the functional inverse of the Ackermann function [29]. The running time of the 5/3-approximation has been reduced to $O(|E|\alpha(n))$ in Reference 51. The performance guarantee has been improved to 3/2 in Reference 52.

Other topology control problems which address issues such as maximizing network lifetime, power assignment using game theoretic considerations, and handling unreliable nodes are considered in References 48, 53–55.

## 22.5 Directions for Future Research

We start by mentioning some open problems in the context of minimizing total power. First, it is of interest to investigate whether there are approximation algorithms with better performance guarantees for inducing graphs with various node and edge connectivity requirements. A second open problem is whether there is a better bicriteria approximation algorithm for the problem of inducing a graph whose diameter is bounded by a specified value. A more general research issue is to identify other graph topologies that are useful in the context of ad hoc networks and to study the approximation issues for inducing such graph topologies.

Most of the results mentioned in this chapter are for symmetric power thresholds. Investigating approximation algorithms for inducing various graph topologies under the asymmetric power thresholds is also of interest. All of the known theoretical results on topology control are for ad hoc networks in which nodes of the network are stationary. Extending these results to the case in which the nodes are mobile is a challenging research direction. In practice, centralized algorithms are likely to be of limited value. So, the development of distributed approximation algorithms for inducing graphs with various topologies is necessary if such algorithms are to be deployed in actual networks.

## Disclaimer

The views and conclusions contained in this document are those of the authors and should not be interpreted as representing the official policies, either expressed or implied, of the Army Research Laboratory or the U.S. Government.

## Acknowledgments

We thank Professor Madhav V. Marathe (Biocomplexity Institute and Virginia Tech) and Dr. Ram Ramanathan (BBN) for their valuable suggestions.

## References

1. C. E. Perkins, (Ed.). *Ad Hoc Networking*. Addison-Wesley, Boston, MA, 2001.
2. I. Stojmenović, (Ed.). *Handbook of Wireless Networks and Mobile Computing*. John Wiley & Sons, Inc., New York, 2002.
3. R. Ramanathan and R. Rosales-Hain. Topology control of multi-hop wireless networks using transmit power adjustment. In *Proceedings of the IEEE International Conference on Computer Communications (INFOCOM 2000)*, pp. 404–413, March 2000.
4. E. M. Royer, P. Melliar-Smith, and L. Moser. An analysis of the optimum node density for ad hoc mobile networks. In *Proceedings of the IEEE International Conference on Communication (ICC'01)*, pp. 857–861, June 2001.
5. V. Radoplu and T. H. Meng. Minimum energy mobile wireless networks. *IEEE Journal on Selected Areas in Communications*, 17(8):1333–1344, 1999.
6. P. Santi. *Topology Control in Wireless Ad Hoc and Sensor Networks*. John Wiley & Sons, New York, 2005.
7. R. Rajaraman. Topology control and routing in ad hoc networks: A survey. *SIGACT News*, 33(2): 60–73, 2002.
8. R. Ramanathan. Antenna beam forming and power control for ad hoc networks. In S. Basagni, M. Conti, S. Giordano, and I. Stojmenović, (Eds.), *Mobile Ad Hoc Networking*, pp. 139–173. IEEE Press and Wiley, New York, 2004.
9. P. Santi. Topology control in wireless ad hoc and sensor networks. *ACM Computing Surveys*, 37(2):164–194, 2005.

10. M. Li, Z. Li, and A. V. Vasilakos. A survey on topology control in wireless sensor networks: Taxonomy, comparative studies and open issues. *Proceedings of IEEE*, 101(12):1.1–1.20, 2013.

11. A. A. Aziz, Y. A. Şekercioğlu, P. Fitzpatrick, and M. Ivanovich. A survey on distributed topology control techniques for extending the lifetime of battery powered wireless sensor networks. *IEEE Communicatios Surveys & Tutorials*, 15(1):2538–2557, 2013.

12. L. M. Kirousis, E. Kranakis, D. Krizanc, and A. Pelc. Power consumption in packet radio networks. *Theoretical Computer Science*, 243(1–2):289–305, 2000.

13. J. van Leeuwen. Graph algorithms. In J. van Leeuwen, (Ed.), *Handbook of Theoretical Computer Science, Volume A*, chapter 10. MIT Press and Elsevier, Cambridge, MA, 1990.

14. E. L. Lloyd, R. Liu, M. V. Marathe, R. Ramanathan, and S. S. Ravi. Algorithmic aspects of topology control problems for ad hoc networks. *Mobile Networks and Applications (MONET)*, 10:19–34, 2005.

15. A. E. F. Clementi, P. Penna, and R. Silvestri. Hardness results for the power range assignment problem in packet radio networks. In *Proceedings of the International Workshop on Randomization and Approximation in Computer Science (APPROX 1999)*, volume 1671 of *Lecture Notes in Computer Science*, pp. 195–208. Springer Verlag, Berlin, Germany, July 1999.

16. A. E. F. Clementi, P. Penna, and R. Silvestri. The power range assignment problem in packet radio networks in the plane. In *Proceedings of the Annual Symposium on Theoretical Aspects of Computer Science (STACS 2000)*, volume 1770 of *Lecture Notes in Computer Science*, pp. 651–660. Springer Verlag, Berlin, Germany, June 2000.

17. M. R. Garey and D. S. Johnson. *Computers and Intractability: A Guide to the Theory of NP-Completeness*. W. H. Freeman and Co., San Francisco, CA, 1979.

18. W. Chen and N. Huang. The strongly connecting problem on multi-hop packet radio networks. *IEEE Transactions on Communications*, 37(3):293–295, 1989.

19. F. Grandoni. On min-power Steiner tree. In *Proceedings of the European Symposium on Algorithms (ESA 2012)*, volume 7501 of *Lecture Notes in Computer Science*, pp. 527–538. Springer, 2012.

20. J. Byrka, F. Grandoni, T. Rothvoss, and L. Sanità. Steiner tree approximation via iterative randomized rounding. *Journal of the ACM*, 60(1):6:1–6:33, 2013.

21. E. Althaus, G. Călinescu, I. Mandoiu, S. Prasad, N. Tchervenski, and A. Zelikovksy. Power efficient range assignment for symmetric connectivity in static ad hoc wireless networks. *Wireless Networks*, 12(3):287–299, 2006.

22. A. Z. Zelikovsky. An 11/6-approximation algorithm for the network Steiner problem. *Algorithmica*, 9(5):463–470, 1993.

23. H. J. Prömel and A. Steger. A new approximation algorithm for the Steiner tree problem with performance ratio 5/3. *Journal of Algorithms*, 36(1):89–101, 2000.

24. G. Călinescu and P-J. Wan. Symmetric high connectivity with minimum total power consumption in multi-hop packet radio networks. In *Proceedings of the International Conference on Ad hoc and Wireless Networks (ADHOC-NOW'03)*, volume 2865 of *Lecture Notes in Computer Science*, pp. 235–246. Springer Verlag, October 2003.

25. S. Khuller and B. Raghavachari. Improved approximation algorithms for uniform connectivity problems. *Journal of Algorithms*, 21(2):434–450, 1996.

26. G. A. Dirac. Minimally 2-connected graphs. *Journal für Reine und Angewandte Mathematik*, 228: 204–216, 1967.

27. M. D. Plummer. On minimal blocks. *Transactions of the American Mathematical Society*, 134: 85–94, 1968.

28. D. B. West. *Introduction to Graph Theory*. Prentice-Hall, Englewood Cliffs, NJ, 2nd ed., 2001.

29. T. Cormen, C. Leiserson, R. Rivest, and C. Stein. *Introduction to Algorithms*. MIT Press and McGraw-Hill, Cambridge, MA, 2nd ed., 2001.

30. Z. Nutov. Approximating minimum-power $k$-connectivity. *Ad Hoc & Sensor Wireless Networks*, 9(1–2):129–137, 2010.

31. S. Khuller and U. Vishkin. Biconnectivity approximations and graph carvings. *Journal of the ACM*, 41(2):214–235, 1994.

32. J. Edmonds. Edge-disjoint branchings. In R. Rustin, (Ed.), *Combinatorial Algorithms*, pp. 91–96. Algorithmic Press, New York, 1972.

33. J. Edmonds. Matroid intersection. *Annals of Discrete Mathematics*, 4:185–204, 1979.

34. H. N. Gabow. A matroid approach to finding edge connectivity and packing arborescences. In *Proceedings of the ACM Symposium on Theory of Computing (STOC'91)*, pp. 112–122. ACM Media, New York, May 1991.

35. S. O. Krumke, R. Liu, E. L. Lloyd, M. V. Marathe, R. Ramanathan, and S. S. Ravi. Topology control problems under symmetric and asymmetric power thresholds. In *Proceedings International Conference on Ad hoc and Wireless Networks (ADHOC-NOW'03)*, volume 2865 of *Lecture Notes in Computer Science*, pp. 187–198. Springer Verlag, Berlin Heidelberg, New York, October 2003.

36. G. Călinescu, S. Kapoor, and M. Sarwat. Bounded-hops power assignment in ad hoc wireless networks. *Discrete Applied Mathematics*, 154(9):1358–1371, 2006.

37. M. V. Marathe, R. Ravi, R. Sundaram, S. S. Ravi, D. J. Rosenkrantz, and H. B. Hunt III. Bicriteria network design problems. *Journal of Algorithms*, 28(1):142–171, 1998.

38. G. Kortsarz and D. Peleg. Approximating the weight of shallow light trees. *Discrete Applied Mathematics*, 93(2–3):265–285, 1999.

39. M. Charikar, C. Chekuri, T. Cheung, Z. Dai, A. Goel, S. Guha, and M. Li. Approximation algorithms for directed Steiner problems. *Journal of Algorithms*, 33(1):73–91, 1999.

40. M. Goemans and D. P. Williamson. A general approximation technique for constrained forest problems. *SIAM Journal on Computing*, 24(2):296–317, 1995.

41. A. Agrawal, P. Klein, and R. Ravi. When trees collide: An approximation algorithm for the generalized Steiner problem on networks. *SIAM Journal on Computing*, 24(3):440–456, 1995.

42. T. S. Rappaport. *Wireless Communications: Principles and Practice*. Prentice-Hall, Englewood Cliffs, NJ, 1996.

43. M. T. Hajiaghayi, N. Immorlica, and V. Mirrokni. Power optimization in fault-tolerant topology control algorithms for wireless multi-hop networks. *IEEE-ACM Transactions on Networking*, 15(6):1345–1358, 2007.

44. R. Gallager, P. Humblet, and P. Spira. A distributed algorithm for minimum-weight spanning trees. *ACM Transactions on Programming Languages and Systems*, 5(1):66–77, 1983.

45. N. Cohen and Z. Nutov. Approximating minimum power edge-multi-covers. *Journal of Combinatorial Optimization*, 30(3):563–578, 2015.

46. X. Jia, D. Kim, S. Makki, P. Wan, and C. Yi. Power assignment for $k$-connectivity in wireless ad hoc networks. *Journal of Combinatorial Optimization*, 9(2):213–222, 2005.

47. M. T. Hajiaghayi, G. Kortsarz, V. S. Mirrokni, and Z. Nutov. Power optimization for connectivity problems. *Mathematical Programming*, 110(1):195–208, 2007.

48. G. Călinescu, S. Kapoor, A.Olshevsky, and A. Zelikovsky. Network lifetime and power assignment in ad-hoc wireless networks. In *Proceedings of the European Symposium on Algorithms (ESA 2003)*, volume 2832 of *Lecture Notes in Computer Science*, pp. 114–126. Springer Verlag, September 2003.

49. I. Caragiannis, C. Kaklamanis, and P. Kanellopoulos. Energy-efficient wireless network design. *Theory of Computing Systems*, 39(5):593–617, 2006.

50. E. L. Lloyd, R. Liu, and S. S. Ravi. Approximating the minimum number of maximum power users in ad hoc networks. *Mobile Networks and Applications (MONET)*, 11(2):129–142, 2006.

51. B. Grimmer and K. Qiao. Near linear time 5/3-approximation algorithms for two-level power assignment problems. In *Proceedings of the 10th ACM International Workshop on Foundations of Mobile Computing (FOMC 2014)*, pp. 29–38. ACM, 2014.

52. Z. Nutov and A. Yaroshevitch. Wireless network design via 3-decompositions. *Information Processing Letters*, 109(19):1136–1140, 2009.

53. V. S. Anil Kumar, S. Eidenbenz, and S. Zust. Equilibria in topology control games for ad hoc networks. In *Proceedings of the DIALM-POMC Workshop*, pp. 2–11, September 2003.

54. M. Khan, V. S. Anil Kumar, M. V. Marathe, G. Pandurangan, and S. S. Ravi. Bi-criteria approximation algorithms for power-efficient and low-interference topology control in unreliable ad hoc networks. In *Proceedings of the 28th IEEE International Conference on Computer Communications (INFOCOM 2009)*, pp. 370–378, 2009.

55. P. Floreen, P. Kaski, J. Kohonen, and P. Orponen. Multicast time maximization in energy-constrained wireless networks. *IEEE Journal on Selected Areas in Communications*, 23(1):117–126, 2005.

# 23

# QoS Multimedia Multicast Routing

Ion Mandoiu

Alex Olshevsky

Alex Zelikovsky

## 23.1 Introduction

Recent progress in audio, video, and data storage technologies has given rise to a host of high-bandwidth real-time applications such as video conferencing. These applications require Quality of Service (QoS) guarantees from the underlying networks. Thus, multicast routing algorithms that manage network resources efficiently and satisfy the QoS requirements have come under increased scrutiny in recent years [1]. The focus on multimedia data transfer capability in networks is expected to further increase as video conferencing applications gain popularity.

It is becoming apparent that new network mechanisms will be required to provide differentiated quality guarantees to network users. Of particular importance is the problem of optimal multimedia distribution from a source to a collection of users with heterogeneous demands. Multimedia distribution is usually done via multicast trees. There are two main reasons for using trees in multicast routing: (a) the data can be transmitted concurrently to destinations along the branches of the tree and (b) only a minimum number of copies of the data must be transmitted as information replication is limited to the branching points of the tree [2]. The bandwidth savings obtained from the use of multicast trees can be maximized by using optimal or nearly optimal multicast tree algorithms, and future networks are expected to integrate such algorithms into their operation [3].

Several versions of the QoS multicast problem have been studied in the literature. These versions seek routing tree cost minimization subject to (1) end-to-end delay, (2) delay variation, and/or (3) minimum bandwidth constraints [3–5]. This chapter deals with the case of minimum bandwidth constraints, that is, the problem of finding an optimal multicast tree when each terminal possesses a different rate of receiving information. This problem is a generalization of the classical Steiner tree problem and therefore

NP-hard [6]. Formally, given a graph $G = (V, E)$, a source $s$, a set of terminals $S$, and two functions: $length : E \to R^+$ representing the length of each edge and $rate : S \to R^+$ representing the rate of each terminal, a *multicast tree* $T$ is a tree in $G$ spanning $s$ and $S$. The *rate* of an edge $e$ in a multicast tree $T$, denoted by $rate(e, T)$, is the maximum rate of a downstream terminal, that is, of a terminal in the connected component of $T - e$ which does not contain $s$. The *cost* of a multicast tree $T$ is defined as

$$cost(T) = \sum_{e \in T} length(e) \cdot rate(e)$$

QUALITY OF SERVICE MULTICAST TREE (QoSMT) PROBLEM: Given a network $G = (V, E, length, rate)$ with source $s \in V$ and set of terminals $S \subseteq V$, find a minimum cost multicast tree in $G$.

Without loss of generality, in this chapter, we further assume that the rates belong to a given discrete set of possible rates: $0 = r_0 < r_1 < \cdots < r_N$. The QoSMT problem is equivalent to the Grade of Service Steiner Tree problem [7], which has a slightly different formulation: In the latter, the network has no source node, and rates $r_e$ must be assigned to edges so that the minimum edge rate on the tree path from a terminal with rate $r_i$ to a terminal with rate $r_j$ is at least $min(r_i, r_j)$, and such that the total tree cost is minimized. A more general *QoSMT with Priorities* was considered by Charikar et al. [6]. In this version of the problem, the cost of an edge $e$ is given arbitrarily instead of being equal to the length times the rate. In other words, edge costs in QoSMT with Priorities are not required to be proportional to edge rates. This generalization seems more difficult—the best known approximation ratio is logarithmic (which also holds for multiple multicast groups) [6].

The QoSMT problem was introduced in the context of multimedia distribution over communication networks by Maxemchuk [5]. Maxemchuk suggested a low-complexity heuristic that can be used to build reliable multicast trees in many practical applications. Following Maxemchuk, Charikar et al. [6] gave a useful approximation algorithm that finds a solution within $e\alpha$ of the optimal, where $\alpha < 1.550$ is the best approximation ratio of an algorithm for the Steiner tree problem. This is the first known algorithm with a constant approximation ratio for this problem. Recently, an approximation ratio of 3.802 based on accurate estimation of Steiner tree length has been achieved in Reference 8.

We note that the problem QoSMT problem was also considered previously (under different names) in the operations research literature. A number of results for particular instances of the problem were obtained: Current et al. [9] gave an integer programming formulation for the problem and proposed a heuristic algorithm for its solution. Some results for the case of few rates were obtained by Balakrishnan et al. [10,11]. Specifically, Reference 11 (see also Reference 7) suggested an algorithm for the case of two nonzero rates with approximation ratio of $\frac{4}{3}\alpha < 2.066$. An improved approximation algorithm with a ratio of 1.960 was proposed in Reference 8. For the case of three nonzero rates, Mirchandani [12] gave an 1.522-approximation algorithm.

The current chapter is organized as follows. First, we describe centralized algorithms for the QoSMT problem, spending the bulk of the time on the algorithms in Reference 8, which have best approximation factors to date. Although these algorithms have superior quality, they cannot be easily adjusted for operation in a distributed environment. Thus, we then describe a more practical primal–dual approach to the QoSMT problem following Reference 13. This approach yields algorithms that have natural distributed implementations, and work well even when the multimedia source does not have exact knowledge of network topology. We conclude with an experimental comparison showing the advantage of the primal–dual approach over practical heuristics proposed in the literature.

## 23.2 Centralized Approximation Algorithms

Tables 23.1 and 23.2 summarize the approximation ratios of known centralized algorithms for the QoSMT problem, for the cases of two nonzero rates and unbounded number of nonzero rates, respectively. In this table, we present the approximation ratios achievable using various Steiner tree approximation algorithms

**TABLE 23.1** QoSMT Problem with Two Rates. Runtime and Approximation Ratios of Previously Known Algorithms and of the Algorithms Given in This Chapter. In the Runtime, $n$ and $m$ Denote the Number of Nodes and Edges in the Original Graph $G = (V, E)$, Respectively. Approximation Ratios Associated with Polynomial-Time Approximation Schemes Are Accompanied by a $+ \epsilon$ to Indicate That They Approach the Quoted Value from Above and Do Not Reach This Value in Polynomial Time

| Steiner Tree Algorithm | LCA [14] | LCA +RNS [15] | BR [18,16] | MST [17] |
|---|---|---|---|---|
| Runtime | Polynomial | Polynomial | $O(n^3)$ [23] | $O(n \log n + m)$ [24] |
| Approximation ratio | $\frac{4}{3}\frac{1+\ln 3}{2} + \epsilon$ | $\frac{20}{9} + \epsilon$ | $\frac{22}{9}$ | $\frac{8}{3}$ |
| in References 7, 11 | $< 2.066 + \epsilon$ | $< 2.223 + \epsilon$ | $< 2.445$ | $< 2.667$ |
| Improved ratio [8] | $-$ | $1.960 + \epsilon$ | $2.237$ | $2.414$ |

**TABLE 23.2** Approximation Ratios for QoSMT Problem with an Arbitrary Number of Rates

| Steiner Tree Algorithm | LCA [14] | RNS [15] | BR [18,16] | MST [17] |
|---|---|---|---|---|
| Approximation | $e\frac{1+\ln 3}{2} + \epsilon$ | $e\frac{5}{3} + \epsilon$ | $e\frac{11}{6}$ | $2e$ |
| ratio in Reference 15 | $< 4.212 + \epsilon$ | $< 4.531 + \epsilon$ | $< 4.984$ | $< 5.44$ |
| Improved ratio [8] | $-$ | $3.802 + \epsilon$ | $4.059$ | $4.311$ |

as a subroutine. Note that along with the best approximation ratios resulting from the use of the loss-contracting Steiner tree algorithm in Reference 14, we also give approximation ratios resulting from the use of the more practical Steiner tree algorithms from References 15–18. In this section, we briefly discuss Maxemchuk's [5] and Charikar et al. [6] methods and then give a detailed description of best-to-date $\beta$-convex approximation algorithms of Karpinski et al. [8].

## 23.2.1 Maxemchuk's Algorithm

Maxemchuk [5] introduced the QoSMT problem and proposed the first heuristic to solve this problem. His algorithm is a modification of the MST heuristic for Steiner Trees [17] (Figure 23.1).

The extensive experiments given in Reference 5 demonstrate that this method works well in practice. Nevertheless, the following example shows that the method may produce arbitrarily large error (linear in

---

**Input:** A graph $G = (V, E, length, rate)$ with a source $s$ in $V$ and a collection of terminals $S \subseteq V$.

**Output:** A QoSMT spanning the source and the terminals.

---

1. Initialize the current tree to $\{s\}$.

2. Find a non-reached terminal $t$ of highest rate with the shortest distance to the current tree.

3. Add $t$ to the current tree along with a shortest path connecting it to the current tree.

4. Repeat until all terminals are spanned.

**FIGURE 23.1** Maxemchuk's Algorithm for the QoSMT problem.

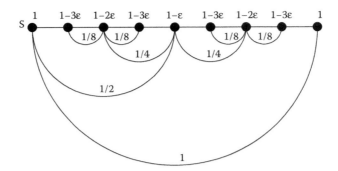

**FIGURE 23.2**  A bad example for Maxemchuk's algorithm, with $k = 4$ rates. In the figure, $\varepsilon = 1/2^{2k-1}$. The rate of each node is given above the node. The edge lengths are given on the thin curved arcs, while on the solid horizontal line each segment has length $1/2^{k-1} + \varepsilon$. The optimum, of total cost $1 + 2^{k-1}\varepsilon = 1 + 2^{k-1}(1/2^{2k-1}) = 1 + 1/2^k$, uses the solid horizontal line at rate 1. Maxemchuk's algorithm picks the thin curved arcs at a cost of $1 + (1/2)(1 - \varepsilon) + 2(1/4)(1 - 2\varepsilon) + 4(1/8)(1 - 3\varepsilon) \geq ((k+1)/2)(1 - 1/2^k)$.

the number of rates) compared with the optimal tree. Consider the natural generalization of the example in Figure 23.2 with an arbitrary number $k$ of distinct rates. Its optimal solution has a cost of about 1, whereas Maxemchuk's method returns a solution of cost about $(k+1)/2$. As there are $2^{k-1} + 1$ nodes, this cost can also be written as $1 + \frac{1}{2}\log_2(n - 1)$, where $n$ is the number of nodes in the graph. We conclude that the approximation ratio of Maxemchuk's algorithm is no better than linear in the number of rates and no better than logarithmic in the number of nodes in the graph.

### 23.2.2 The Charikar–Naor–Schieber Algorithms

Charikar et al. [6] gave the first constant-factor approximation algorithms for the QoS Steiner tree problem. The simplest version is a *binary rounding* algorithm. In its first step, all rates are rounded to the closest power of two to produce the rounded up instance of this problem (clearly, this at most doubles the cost of an optimal solution). In its second step, Steiner trees are computed separately for each rate (within some approximation ratio $\alpha$). The union of these trees is the final solution.

Consider the network obtained by replacing each edge of rate $2^i$ in an optimal solution by $i + 1$ parallel edges of rates $2^0, 2^1, \ldots, 2^{i-1}, 2^i$, respectively. In the new network, edges of a specific rate form a Steiner tree spanning all terminals of the respective rate. As the optimal cost in this new network is no more than twice the cost of the rounded up instance, taking the union of all the computed Steiner trees introduces another factor of two to the approximation ratio. Thus, the final approximation factor is $2 \cdot \alpha \cdot 2 = 4\alpha$.

Using a randomization technique, Charikar, Naor, and Schieber [6] reduce the approximation ratio to $e\alpha \approx 4.21$, where $e \approx 2.71$ is the Euler constant and $\alpha \approx 1.55$ is the currently best approximation ratio for the Steiner Tree problem.

### 23.2.3 $\beta$-Convex Steiner Tree Approximation Algorithms

In the current section, we introduce the notion of $\beta$-convex Steiner tree approximation algorithms and show tighter upper bounds on their output when applied to the QoSMT problem.

We begin by reviewing some Steiner tree definitions. A Steiner tree is a minimum-length tree connecting a subset of the graph's nodes. The nodes in a subset are usually referred to as *terminal* nodes. A Steiner tree is called *full* if every terminal is a leaf. A Steiner tree can be decomposed into components that are full by breaking the tree up at the non leaf terminals. A Steiner tree is called *k-restricted* if every full component has at most $k$ terminals. Let us denote the length of the optimum $k$-restricted Steiner tree as $opt_k$ and the length of the optimum unrestricted Steiner tree as $opt$. Let the $k$-restricted Steiner ratio $\rho_k$ be $\rho_k = sup \frac{opt_k}{opt}$, where the supremum is taken over all instances of the Steiner tree problem. It has been shown in Reference 19 that $\rho_k = \frac{(r+1)2^r+s}{r2^r+s}$, where $r$ and $s$ are obtained from the decomposition $k = 2^r + s, 0 \leq s < 2^r$. A slightly tighter bound on the length of the optimal $k$-restricted Steiner tree has been established in Reference 8.

**Theorem 23.1** [8] *For every Steiner tree T partitioned into edge-disjoint full components $T^i$,*

$$opt_k \leq \sum_i \left(\rho_k(l(T^i) - D(T^i)) + D(T^i)\right),$$

*where $l(T^i)$ is the length of the full component $T^i$, and $D(T^i)$ is the length of the longest path in $T^i$.*

$\beta$-convexity of Steiner tree approximation has been introduced in Reference 8. A Steiner tree heuristic $A$ is called a $\beta$-*convex* $\alpha$-*approximation Steiner tree algorithm* if there exist an integer $m$ and non negative real numbers $\lambda_i, i = 2, \ldots, m$, with $\beta = \sum_{i=2}^{m} \lambda_i$ and $\alpha = \sum_{i=2}^{m} \lambda_i \rho_i$ such that the length of the tree computed by $A$, $l(A)$, is upper bounded by

$$l(A) \leq \sum_{i=2}^{m} \lambda_i opt_i,$$

where $opt_i$ is the length of the optimal $i$-restricted Steiner tree.

The MST-algorithm [17] is 1-convex 2-approximation as its output is the optimal 2-restricted Steiner tree of length $opt_2$. Every $k$-restricted approximation algorithm from Reference 16 is 1-convex – the sum of coefficients in the approximation ratio always equals to 1, for example, for $k = 3$, it is 1-convex 11/6-approximation algorithm as the output tree is bounded by $\frac{1}{2}opt_2 + \frac{1}{2}opt_3$. The output tree for PTAS [15] converges to the optimal 3-restricted Steiner tree and has length $(1+\epsilon)opt_3$; therefore, it is $(1+\epsilon)$-convex $\frac{5}{3}(1+\epsilon)$-approximation algorithm. The currently best approximation ratio of $1 + \frac{\sqrt{3}}{2}$ is achieved by heuristic from Reference 14 which is not known to be $\beta$-convex for any value of $\beta$.

Given a $\beta$-convex $\alpha$-approximation algorithm $A$, it follows from Theorem 23.1 that

$$l(A) \leq \sum_i \lambda_i opt_i \leq \sum_i \lambda_i \rho_i(opt - D) + \beta D = \alpha(opt - D) + \beta D \qquad (23.1)$$

Let $OPT$ be the optimum cost QoSMT tree $T$, and let $t_i$ be the length of rate $r_i$ edges in $T$. Then,

$$cost(OPT) = \sum_{i=1}^{N} r_i t_i$$

Let $OPT_k$ be the subtree of the optimal QoSMT $OPT$ induced by edges of rate $r_i, i \geq k$. The tree $OPT_k$ spans the source $s$ and all nodes of rate $r_k$ and, therefore, an optimal Steiner tree connecting $s$ and rate-$r_k$ nodes cannot be longer than

$$l(OPT_k) = \sum_{i=k}^{N} t_i$$

The main idea of the QoSMT algorithms in Reference 8 is to reuse connections for the higher rate nodes when connecting lower rate nodes. When connecting nodes of rate $r_k$, we collapse nodes of rate strictly higher than $r_k$ into the source $s$, thus allowing to reuse higher rate connections for free. Let $T_k$ be an approximate Steiner tree connecting the source $s$, and all nodes of rate $r_k$ after collapsing all nodes of rate strictly higher than $r_k$ into the source $s$ and treating all nodes of rate lower than $r_k$ as Steiner points. If we apply an $\alpha$-approximation Steiner tree algorithm for finding $T_k$, then the resulted length can be bounded as follows

$$l(T_k) \leq \alpha l(OPT_k) = \alpha t_k + \alpha t_{k+1} + \cdots + \alpha t_N$$

The following lemma shows that if the tree $T_k$ is obtained using $\beta$-convex $\alpha$-approximation Steiner tree algorithm, then a tighter upper bound on the length of $T_k$ holds.

**Lemma 23.1** [8] *Given an instance of the QoSMT problem, the cost of the tree $T_k$ computed by a $\beta$-convex $\alpha$-approximation Steiner tree algorithm is at most*

$$cost(T_k) \leq \alpha r_k t_k + \beta(r_k t_{k+1} + r_k t_{k+2} + \cdots + r_k t_N)$$

*Proof.* Let $OPT_k$ be the subtree of the optimal QoSMT $OPT$ induced by edges of rate $r_i$, $i \geq k$. By duplicating nodes and introducing zero length edges, it can be assumed that $OPT_{k+1}$ is a complete binary tree with the set of leaves consisting of the source $s$ and all nodes of rate at least $r_{k+1}$. The edges of rate $r_k$ form subtrees attached to the tree $OPT_{k+1}$ connecting rate $r_k$ nodes to $OPT_{k+1}$ (Figure 23.3a).

Note that edges of any binary tree $T$ can be partitioned into the edge-disjoint paths connecting internal nodes with leaves as follows. Each internal node $v$ (including the degree-2 root) is split into two nodes $v_1$ and $v_2$ such that $v_1$ becomes a leaf incident to one of the downstream edges and $v_2$ becomes a degree-2 node (or a leaf if $v$ is the root) incident to an edge connecting $v$ to its parent (if $v$ is not the root) and another downstream edge. As each node is incident to a downstream edge, each resulted connected component will be a path containing exactly one leaf of $T$ connected to an internal node of $T$.

The binary tree $OPT_{k+1}$ broken into edge-disjoint paths described earlier along with all nodes of rate $r_k$ that attached to them is shown in Figure 23.3b. A resulted connected component $OPT_k^i$ consisting of a path $D_k^i = OPT_k^i \cap OPT_{k+1}$ and attached Steiner trees with edges of rate $r_k$ is shown in Figure 23.3c. Note that the total length of the paths $D_k^i$ is $l(OPT_{k+1}) = t_{k+1} + t_{k+2} + \cdots + t_N$. By Theorem 23.1, decomposing the tree $T_k$ along these full components $OPT_k^i$ results in the following upper bound:

$$l(T_k) \leq \sum_i \left[ \alpha(l(OPT_k^i) - D_k^i) + \beta D_k^i \right]$$
$$= \alpha t_k + \beta(t_{k+1} + t_{k+2} + \cdots + t_N)$$

The lemma follows by multiplying the last inequality by $r_k$. ∎

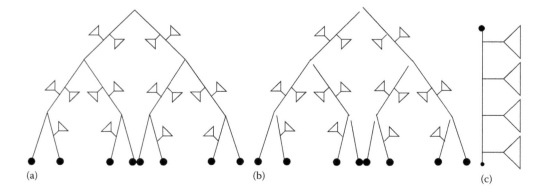

**FIGURE 23.3** (a) The subtree $OPT_k$ of the optimal QoSMT $OPT$ induced by edges of rate $r_i$, $i \geq k$. Edges of rate greater than $r_k$ (shown as solid lines) form a Steiner tree for $s \cup S_{k+1} \cup \ldots S_N$ (filled circles); attached triangles represent edges of rate $r_k$. (b) Partition of $OPT_k$ into edge-disjoint connected components $OPT_k^i$ each containing a single terminal of rate $r_i$, $i > k$. (c) A connected component $OPT_k^i$ that consists of a path $D_k^i$ containing all edges of rate $r_i$, $i > k$, and attached Steiner trees containing edges of rate $r_k$.

## 23.2.4 $\beta$-Convex Approximation for QoSMT with Two Rates

In practice, it is often the case that only few distinct rates are requested by the terminals. This is why the QoS problem with two or three rates has a long history [7,10–12]. The previously-best results of References 7, 12 have produced algorithms with approximation factor equal to 2.667 (provided that the MST heuristic is used to compute Steiner trees).

In the current section, approximation factors for the QoSMT problem with two non-zero rates are derived for the balancing algorithm based on $\beta$-convex Steiner tree approximation (Figure 23.4) [8].

Recall that an edge $e$ has rate $r_i$ if the largest node rate in the component of $T - \{e\}$ that does not contain the source is $r_i$. Let the optimal Steiner tree in $G$ have cost $opt = r_1t_1 + r_2t_2$, with $t_1$ being the total length of the edges of rate $r_1$ and $t_2$ being the total length of the edges of rate $r_2$. The algorithm in Figure 23.4 uses as subroutines two Steiner tree algorithms: an algorithm $A_1$ with an approximation ratio of $\alpha_1$, and a $\beta$-convex algorithm $A_2$ with an approximation ratio of $\alpha_2$. It outputs the minimum cost Steiner tree between the tree $ST1$ obtained by running $A_1$ with a set of terminals containing the source and the nodes with both high and low nonzero rate, and the tree $ST2$ obtained by running $A_1$ with a set of terminals containing the source and all high rate nodes, contracting the resulting tree into the source, and running $A_2$ with a set of terminals containing the contracted source and the low rate nodes.

**Theorem 23.2** [8] *The algorithm in Figure 23.4 has an approximation ratio of*

$$\max\left\{\alpha_2, \ \max_r \ \alpha_1\frac{\alpha_1 - \alpha_2r + \beta r}{\alpha_1 - \alpha_2r + \beta r^2}\right\}$$

*Proof.* The cost of ST1 is bounded by $cost(ST1) \le \alpha_1r_2(t_1 + t_2)$. To obtain a bound on the cost of ST2 note that $cost(T_2) \le \alpha_1r_2t_2$, and that, by Lemma 23.1, $cost(T_1) \le \alpha_2r_1t_1 + \beta r_1t_2$.

Thus, the following two bounds for the costs of $ST1$ and $ST2$ follow

$$cost(ST1) \le \alpha_1r_2t_1 + \alpha_1r_2t_2$$
$$cost(ST2) \le \alpha_1r_2t_2 + \alpha_2r_1t_1 + \beta r_1t_2$$

---

**Input:** Graph $G = (V, E, l)$ with two nonzero rates $r_1 < r_2$, source $s$, terminal sets $S_1$ of rate $r_1$ and $S_2$ of

rate $r_2$, Steiner tree $\alpha_1$-approximation algorithm $A_1$ and a $\beta$-convex $\alpha_2$-approximation algorithm $A_2$

**Output:** Low cost QoSMT spanning all terminals

---

1. Compute an approximate Steiner tree ST1 for $s \bigcup S_1 \bigcup S_2$ using algorithm $A_1$

2. Compute an approximate Steiner tree $T_2$ for $s \bigcup S_2$ (treating all other points as Steiner points) using

   algorithm $A_1$. Next, contract $T_2$ into the source $s$ and compute the approximate Steiner tree $T_1$ for $s$ and

   remaining rate $r_1$ points using algorithm $A_2$. Let ST2 be $T_1 \bigcup T_2$

3. Output the minimum cost tree among ST1 and ST2

**FIGURE 23.4** QoSMT approximation algorithm for two non-zero rates.

Let us distinguish the following two cases:

**Case 1:** Let $\beta r_1 \leq (\alpha_2 - \alpha_1)r_2$. Then,

$$
\begin{aligned}
cost(ST2) &\leq \alpha_1 r_2 t_2 + \alpha_2 r_1 t_1 + \beta r_1 t_2 \\
&\leq \alpha_1 r_2 t_2 + \alpha_2 r_1 t_1 + (\alpha_2 - \alpha_1)r_2 t_2 \\
&\leq \alpha_2(r_2 t_2 + r_1 t_1) \\
&= \alpha_2 opt
\end{aligned}
$$

**Case 2:** Let $\beta r_1 > (\alpha_2 - \alpha_1)r_2$. Then the following two values are positive

$$
\begin{aligned}
x_1 &= \frac{r_1}{\alpha_1 r_2}(\beta r_1 - (\alpha_2 - \alpha_1)r_2) \\
x_2 &= r_2 - r_1
\end{aligned}
$$

The following linear combination will be bounded

$$
\begin{aligned}
x_1 cost(ST1) + x_2 cost(ST2) &= \frac{r_1(\beta r_1 - (\alpha_2 - \alpha_1)r_2)}{\alpha_1 r_2}cost(ST1) + (r_2 - r_1)cost(ST2) \\
&\leq r_1(\beta r_1 - (\alpha_2 - \alpha_1)r_2)(t_1 + t_2) \\
&\quad + (r_2 - r_1)(\alpha_1 r_2 t_2 + \alpha_2 r_1 t_1 + \beta r_1 t_2) \\
&= ((\beta - \alpha_2)r_1^2 + r_1 r_2 \alpha_1)t_1 + ((\beta - \alpha_2)r_1 r_2 + r_2^2 \alpha_1)t_2 \\
&= ((\beta - \alpha_2)r_1 + r_2 \alpha_1)(r_1 t_1 + r_2 t_2) \\
&\leq (\beta r_1 + \alpha_1 r_2 - \alpha_2 r_1)opt \quad\quad\quad\quad\quad (23.2)
\end{aligned}
$$

Let *Approx* be the cost of the tree produced by the approximation algorithm. The inequality (23.2) implies that

$$
\begin{aligned}
Approx &= \min\{cost(ST1), cost(ST2)\} \\
&= \frac{x_1 \min\{cost(ST1), cost(ST2)\} + x_2 \min\{cost(ST1), cost(ST2)\}}{x_1 + x_2} \\
&\leq \frac{x_1 cost(ST1) + x_2 cost(ST2)}{x_1 + x_2} \\
&\leq \frac{\beta r_1 + \alpha_1 r_2 - \alpha_2 r_1}{\frac{r_1}{\alpha_1 r_2}(\beta r_1 - (\alpha_2 - \alpha_1)r_2) + r_2 - r_1}opt \\
&\leq \alpha_1 \frac{\beta r_1 r_2 + \alpha_1 r_2^2 - \alpha_2 r_1 r_2}{\beta r_1^2 - (\alpha_2 - \alpha_1)r_2 r_1 + \alpha_1 r_2^2 - \alpha_1 r_1 r_2}opt \\
&\leq \alpha_1 \frac{\alpha_1 - \alpha_2 r + \beta r}{\alpha_1 - \alpha_2 r + \beta r^2}opt
\end{aligned}
$$

where $r = \frac{r_1}{r_2}$.

Summarizing the two cases, we obtain that *Approx* is at most the maximum of two values – $\alpha_2 opt$ and $\alpha_1 \frac{\alpha_1 - \alpha_2 r + \beta r}{\alpha_1 - \alpha_2 r + \beta r^2}opt$ – which proves the theorem.

Theorem 23.2 implies numerical bounds on the approximation ratios. Using that $\alpha_1 = 1 + \ln 3/2 + \epsilon$ for the algorithm from Reference 14, $\alpha_2 = 5/3 + \epsilon$ for the algorithm from Reference 15, $\alpha_1 = \alpha_2 = 11/6$ for the algorithm from Reference 16, and $\alpha_1 = \alpha_2 = 2$ for the MST heuristic, and $\beta \to 1$ for all of the above-mentioned algorithms (except for the algorithm from Reference 14), we maximize the expression in Theorem 23.2 to obtain the following theorem. ∎

**Theorem 23.3** [8] *If the algorithm from Reference 14 is used as $A_1$ and the algorithm from Reference 15 is used as $A_2$, then the approximation ratio of the QoSMT algorithm in Figure 23.4 is $1.960 + \epsilon$. If the algorithm from Reference 15 is used in place of both $A_1$ and $A_2$, then the approximation ratio is $2.059 + \epsilon$. If the algorithm from Reference 16 is used in place of both $A_1$ and $A_2$, then the ratio is 2.237. If the MST heuristic is used in place of both $A_1$ and $A_2$, then the ratio is 2.414.*

### 23.2.5 $\beta$-Convex Approximation for QoSMT with Unbounded Number of Rates

In the current section, we describe and prove the performance ratios of $\beta$-convex approximation algorithms for the case of the QoSMT problem with arbitrarily many nonzero rates $r_1 < r_2 < \cdots < r_N$ [8]. The algorithm (Figure 23.5) is a modification of the algorithm in Reference 6. As in Reference 6, node rates are rounded up to the closest power of some number $a$ starting with $a^y$, where $y$ is picked uniformly at random between 0 and 1. In other words, the given rates are replaced with numbers from the set $\{a^y, a^{y+1}, a^{y+2}, \ldots\}$. The major difference is that each approximate Steiner tree, $T_k$, constructed over nodes of rounded rate $a^{y+k}$ is contracted in increasing order of $k$ instead of simply taking union of $T_k$'s according to Reference 6. This allows contracted edges to be reused at zero cost by Steiner trees connecting lower rate nodes. The following analysis from Reference 8 of this improvement shows that it decreases the approximation ratio from 4.211 to 3.802.

Let $T_{opt}$ be the optimal QoSMT, and let $t_i$ be the total length of the edges of $T_{opt}$ with rates rounded to $a^{y+i}$. First, we prove the following "randomized doubling" lemma corresponding to Lemma 4 from Reference 6.

---

**Input:** Graph $G = (V, E, l)$, source $s$, sets $S_i$ of terminals with rate $r_i$, positive number $a$, and

$\alpha$-approximation $\beta$-convex Steiner tree algorithm

**Output:** Low cost QoSMT spanning all terminals

---

1. Pick $y$ uniformly at random between 0 and 1. Round up each rate to the closest power of some number $a$ starting with $a^y$, i.e., round up to numbers in the set $\{a^y, a^{y+1}, a^{y+2}, \ldots\}$. Form new terminal sets $S_i'$ which are unions of terminal sets with rates rounded to the same number $r_i'$

2. $T \leftarrow \emptyset$

3. For each non-zero rounded rate $r_i'$, in decreasing order, do:

   Find an $\alpha$-approximate Steiner tree $T_i$ spanning $s \bigcup S_i'$

   $T \leftarrow T \cup T_i$

   Contract $T_i$ into source $s$

4. Output $T$

---

**FIGURE 23.5** Approximation algorithm for multirate QoSMT.

**Lemma 23.2** [8] *Let $S$ be the cost of $T_{opt}$ after rounding node rates as in Figure 23.5, that is,*
$S = \sum_{i=0}^{n} t_i a^{y+i}$. *Then,*

$$S \leq \frac{a-1}{\ln(a)} cost(T_{opt})$$

*Proof.* First, note that an edge $e$ used at rate $r$ in $T_{opt}$ will be used at the rate $a^{y+m}$, where $m$ is the smallest integer $i$ such that $a^{y+i}$ is no less than $r$. Indeed, $e$ is used at rate $r$ in $T_{opt}$ if and only if the maximum rate of a node connecting to the source via $e$ is $r$, and every such node will be rounded to $a^{y+m}$. Next, let $r = a^{x+m}$. If $x \leq y$, then the rounded up cost is $a^{y-x}$ times the original cost; otherwise, if $x > y$, is $a^{y+1-x}$ times the original cost. Hence, the expected factor by which the cost of each edge increases is

$$\int_0^x a^{y+x-1} dy + \int_x^1 a^{y-x} dy = \frac{a-1}{\ln a}$$

By linearity of expectation, the expected cost after rounding of $T_{opt}$ is

$$S \leq \frac{a-1}{\ln a} cost(T_{opt}) \qquad \blacksquare$$

**Theorem 23.4** [8] *The algorithm given in Figure 23.5 has an approximation ratio of*

$$\min_a \left( \alpha \frac{a}{\ln a} - (\alpha - \beta) \frac{1}{\ln a} \right)$$

*Proof.* (Sketch) Let *Approx* be the cost of the tree returned by the algorithm in Figure 23.5, and *Approx_k* be the cost of the tree $T_k$ constructed by the algorithm when considering rate $r_k$. Then, by Lemma 23.1,

$$Approx_k \leq \alpha a^{y+k} t_k + \beta a^{y+k+1} t_{k+1} + \beta a^{y+k+2} t_{k+2} + \cdots + \beta a^{y+n} t_n$$

Summing up all the *Approx_k*'s (we omit the details), we get an upper bound of

$$(\alpha - \beta)S + \beta S \left( 1 + \frac{1}{a} + \frac{1}{a^2} + \cdots \right) \leq (\alpha - \beta) \frac{a-1}{\ln a} cost(T_{opt}) + \beta \frac{a}{\ln a} cost(T_{opt})$$

$$= \left( \alpha \frac{a}{\ln a} - (\alpha - \beta) \frac{1}{\ln a} \right) cost(T_{opt})$$

where the last inequality follows from Lemma 23.2.

Note that the corresponding approximation ratio in Reference 6 is larger and equals $\alpha \frac{a}{\ln a}$ attaining minimum for $a = e$. The minimum of the approximation ratio in Theorem 23.4 can be obtained numerically—it is equal to 3.802, 4.059, 4.311, respectively, when the $\beta$-convex $\alpha$-approximation Steiner tree algorithm used in Figure 23.4 is the algorithm in References 15, 16, respectively, the MST heuristic. Finally, the algorithm in Figure 23.5 can be derandomized using the same techniques as in Reference 6. $\qquad \blacksquare$

# 23.3 Primal–Dual Approach to the Quality of Service Multicast Tree Problem

In the current section, we discuss several primal–dual heuristics for the QoSMT problem due to Calinescu et al. [13]. A simpler integer linear program and two primal-dual algorithms based on it are discussed in Sections 23.3.1 and 23.3.2. A tighter Integer Linear Program (ILP) and an associated 4.311-approximation primal–dual algorithm are then described in Section 23.3.3.

### 23.3.1 A Simpler ILP Formulation

The QoSMT problem can be formulated as an integer program as follows. Consider a network $G = (V, E, length, rate)$ with a source node $s$ and a set of terminal nodes. Let $r_1 < r_2 < \cdots < r_N$ be all rate values assigned to the terminals. To simplify our notation, we assume that every node has an extra rate $r_0 = 0$ (i.e., assign rate $r_0$ to each non terminal node). As before, the source $s$ has the highest rate. Construct a new network $G' = (V, E', cost, rate)$ by replacing each edge $e$ of $G$ with $k$ edges $(e, r_1), (e, r_2), \ldots, (e, r_k)$ and setting $cost((e, r_i)) = r_i \cdot length(e)$.

Let $x_{(e,r)}$ be a boolean variable denoting whether edge $e$ is used at rate $r$ in an optimum tree. The QoS Steiner tree problem can be formulated as

$$\min \quad \sum_{(e,r)\in E'} x_{(e,r)} \cdot r \cdot length(e) \tag{23.3}$$

$$\text{s.t.} \quad \sum_{\substack{(e,r)\in\delta(C) \\ r\geq r_C}} x_{(e,r)} \geq 1, \quad \forall C \subseteq V \setminus \{s\} \tag{23.4}$$

$$x_{(e,r)} \in \{0,1\} \tag{23.5}$$

where $\delta(C)$ denotes the set of edges with exactly one endpoint in $C$ and $r_C$ denotes the maximum rate of a node in $C$. Note that (23.3) gives the cost of an optimal solution, while (23.4) guarantees that each terminal is connected to the source through a collection of edges of rate no less than its rate.

After relaxing the integrality constraints (23.5), the dual linear program can be written as follows. For each $(e, r)$, $C^*(e, r)$ is defined as $\{C \in V \setminus \{s\} : (e, r) \in \delta(C), r \geq r_C\}$. In words, $C^*(e, r)$ is the set of subsets $C$ of $V \setminus \{s\}$ such that $(e, r)$ has at least one endpoint in $C$ and $r$ is at least as large as $r_C$. Using this definition, the dual is as follows:

$$\max \quad \sum_C y_C$$

$$\text{s.t.} \quad \sum_{C\in C^*(e,r)} y_C \leq r \cdot length(e), \quad \forall(e, r)$$

$$y_C \geq 0$$

### 23.3.2 Two Primal–Dual Methods for the Simple ILP Formulation

The primal–dual framework applied to network design problems usually grows uniformly the dual variables associated to the "active" components of the current forest [20]. This approach fails to take into account the different rates of different nodes in the QoSMT problem. The *Naive Primal–Dual algorithm* [13] (Figure 23.6) takes into account different rates by varying the speed at which each component grows. Although the simulations in the ensuing sections, show that this is a good method in practice, the solution it produces on some graphs may be very large compared with the optimal solution, as shown by the following example with two rates.

Consider two nodes of rate 1 connected by an edge of length 1 (Figure 23.7). There is an arc between these two nodes, and on this arc, there is a chain of nodes of rate $\epsilon$. Each two consecutive nodes in the chain are at a distance $\delta$ from each other, where $\delta < 1$. Each extreme node in the chain is at a distance $\delta/2$ of its neighboring rate-1 node.

The Naive Primal–Dual applied to this graph connects the rate-$\epsilon$ nodes first, as $\frac{\delta}{2} < \frac{1}{2}$. So, the algorithm connects the rate-1 nodes via the rate-$\epsilon$ nodes, and not via the direct edge connecting them. Thus, the Naive Primal–Dual can make arbitrarily large errors (just take an arbitrarily long chain).

An improved *Restarting Primal–Dual algorithm* [13] is given in Figure 23.8. One can easily see that this is a primal–dual algorithm. Indeed, each addition of an edge to the current solution is the result of growing dual variables. Moreover, as the feasibility requirement for edge $a$ is $\Sigma_{a\in\delta(C)}y_C \leq r \cdot length(a)$,

---

**Input:** A graph $G = (V, E, length, rate)$ with a source $s$ in $V$ and a collection of terminals $S \subseteq V$.

**Output:** A QoSMT spanning the source and the terminals.

---

1. Start from the spanning forest of $G$ with no edges.

2. Grow $y_C$ with speed $r_C$ for each "active" component $C$ of the current forest. (A component $C$ is *inactive* if it contains $s$ and all vertices of rate $r_C$.)

3. Stop growing once the dual inequality for a pair $(e, r)$ becomes tight, with $e$ connecting two distinct components of the forest.

4. Add $e$ to the forest, collapsing the two components.

5. Terminate when there is no active component left.

6. Keep an edge of the resulting tree at the minimum needed rate.

---

**FIGURE 23.6**   The Naive Primal–Dual algorithm for the QoSMT problem.

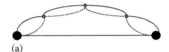 

(a)                              (b)

**FIGURE 23.7**   The Restarting Primal–Dual avoids the mistake of the Naive Primal–Dual. Part (a) shows duplication of the edges. Part (b) shows the components growing along the respective edges.

---

**Input:** A Graph $G' = (V, E, cost, rate)$ with source $s$, and a collection of terminals $S$.

**Output:** A QoSMT spanning the source and the terminal.

---

1. Grow each active $C_{r_i}$ with speed $r_i$ along incident edges $(e, r_j)$, $j \leq i$, picking edges which become tight.

2. Continue this process until there is no active component of rate $r_k$.

3. Remove all edges which are not necessary for maintaining connectivity of nodes of rate $r_k$.

4. Accept (keep in the solution) and contract all edges of $C_{r_k}$ (i.e., set their length/cost to 0)

5. Restart the algorithm with the new graph

---

**FIGURE 23.8**   The Restarting Primal–Dual algorithm for the QoSMT problem.

---

this addition preserves the feasibility of the dual solution. The algorithm maintains forests $F^{r_i}$ given by the edges picked at rate $r_i$, and the connected components of $F^{r_i}$, seen as sets of vertices, are denoted in the algorithm by $C_{r_i}$. Such a component is *active* if $r_{C_{r_i}} = r_i$, and $C_{r_i}$ is disjoint from components of higher rate.

The Restarting Primal–Dual algorithm avoids the mistake made by the Naive Primal–Dual algorithm on the frame example in Figure 23.7a. Then, at time $\frac{\delta}{2}$, the rate-$\epsilon$ nodes become connected. This means that $\delta(1 - \epsilon)$ of each rate-1 edge between the $\epsilon$-rate nodes is not covered. Meanwhile, the rate-1 nodes are growing on the respective edges as shown in Figure 23.7b.

Let us assume that the Restarting Primal–Dual algorithm uses the chain of rate-$\epsilon$ nodes to connect the two rate-1 nodes instead of the direct edge. This would imply that it takes less time to cover the chain, that is, $\frac{1}{2}\delta(1 - \epsilon)n \le \frac{1}{2} - \frac{\delta}{2}$, where $n$ is the number of rate-$\epsilon$ nodes. When $\epsilon$ is small, then $n\delta \le 1$, so if the Restarting Primal–Dual algorithm uses the chain then it is correct to do so.

### 23.3.3 Primal–Dual 4.311-Approximation Algorithm

A constant–factor primal–dual approximation algorithm is obtained in Reference 13 based on an enhanced integer linear programming formulation of the QoSMT problem. The enhanced formulation takes into account the fact that if a set $C \subset V \setminus \{s\}$ is connected to the source with edges of rate $r' > r_C$, then there should be at least **two** edges of rate $r'$ with exactly one endpoint in $C$. The integer program is

$$\min \quad \sum_{(e,r)\in E'} x_{(e,r)} \cdot r \cdot length(e)$$

$$\text{s.t.} \quad \sum_{\substack{e\in\delta(C) \\ r=r_C}} x_{(e,r)} + \frac{1}{2} \sum_{\substack{e\in\delta(C) \\ r>r_C}} x_{(e,r)} \ge 1, \quad \forall C \subseteq V \setminus \{s\}$$

$$x_{(e,r)} \in \{0, 1\}$$

The corresponding dual of the LP relaxation is

$$\max \quad \sum_{C\subseteq V\setminus\{s\}} y_C$$

$$\text{s.t.} \quad \sum_{\substack{C\,:e\in\delta(C) \\ r_C=r}} y_C + \frac{1}{2} \sum_{\substack{C\,:e\in\delta(C) \\ r_C<r}} y_C \le r \cdot length(e) \qquad (23.6)$$

$$y_C \ge 0$$

The core algorithm presented in Figure 23.9 is preprocessed with random bucketing of rates as in Section 23.2.5 (see also Step 1 in Figure 23.5). Let $a$ be a real (to be picked later) and $y$ be a real picked uniformly at random from the interval $[0 \cdots 1]$. Every node of rate $r$ is replaced by a node of rate $a^{\gamma+j}$, where $j$ is the integer satisfying $a^{\gamma+j-1} < r \le a^{\gamma+j}$.

The primal–dual part follows the classical framework [20] and works in stages starting from the lower rate to the highest. During the execution of the algorithm, edges are picked at a certain rate (in other words, $x_{(e,r)}$ is set to 1) one by one. Before executing step 3 at rate $r$ for the $i$th time, the set of edges picked at rate $r$ by the algorithm forms a forest $F_i^r$. (An edge can be picked at several rates, but it is kept in at most one such rate in the final solution because of the reverse delete step.) A component $C$ of $F_i^r$ is called an $r$-component if $r_C = r$.

Using Constraint (23.6), it follows by induction on $j$ that, for an edge $e$ and a rate $a^{\gamma+j}$, we have

$$\sum_{\substack{C\,:e\in\delta(C) \\ r_C\le a^{\gamma+j}}} y_C \le length(e)a^{\gamma+j} \sum_{i=0}^{j} \left(\frac{1}{2a}\right)^i$$

$$\le length(e)a^{\gamma+j}\frac{2a}{2a - 1}.$$

---

**Input:** A graph $G = (V, E, length, rate)$ with source $s$ in $V$ and a collection of terminals $S \subseteq V$.

**Output:** A QoSMT spanning the source and the terminal.

---

1. For each $r = r_1, r_2, \ldots, r_k$, execute steps 2-6.

2. Start from the spanning forest $F^r$ of $G$ with no edges.

3. Grow $y_C$ uniformly for each $r$-component $C$ of the current forest $F^r$.

4. Stop growing once the dual inequality for a pair $(e, r)$ becomes tight, with $e$ connecting two distinct

   components of $F^r$.

5. Add $(e, r)$ to $F^r$, collapsing two of its components.

6. Terminate when there is no $r$-component of $F^r$ left.

7. Traversing the list of picked edges in reverse order, remove an edge $(e, r)$ from $F^r$ if after $(e, r)$'s removal

   the set of edges picked form a feasible tree.

---

**FIGURE 23.9** The 4.311-approximation algorithm for QoSMT problem.

For an edge picked by the algorithm at rate $r$, Constraint (23.6) is tight and therefore

$$\sum_{\substack{C \, : \, e \in \delta(C) \\ r_C \leq a^{\gamma+j}}} y_C \geq length(e) \frac{2a-2}{2a-1} a^{\gamma+j}. \tag{23.7}$$

Exactly as in Reference 20, the number of edges of rate $r$ in the final solution that cross the active $r$-components at some moment (an edge being counted twice if it crosses two $r$-components) is at most twice the number of active $r$-components. Equation (23.7) and exactly the same argument as in Theorem 4.2 of Reference 20 imply that the cost of the solution of the algorithm is bounded by $(2(2a-1)/(2a-2)) \sum y_C \leq ((2a-1)/(a-1))$ opt, as any feasible solution for the dual linear program has value at most the value of any feasible solution of the primal.

The same argument as in Section 23.2.5 shows that the approximation ratio of the above-mentioned algorithm is $(2a-1)/\ln a$—numerically picking the same best value for $a$ as in Section 23.2.5.

**Theorem 23.5 [13]** *The output cost of the algorithm on Figure 23.9 is at most* 4.311 *times the optimum cost.*

## 23.4 Experimental Study

In the current section, we report experimental results with several QoSMT heuristics: Maxemchuk's [5], binary rounding [6], naive primal–dual, and restarting primal–dual algorithms. The heuristics were implemented in C++ and compiled using gpp with -O2 optimization, and run on a Sun workstation Ultra-60. The experiments were run on random testcases generated using GT-ITM generator [21] which is used for modelling internet networks [22]. Table 23.3 gives a comparison of the performance of of the aforementioned algorithms. The experiments were conducted in the presence of no Steiner nodes, respectively 50% Steiner nodes. Moreover, both arithmetic and geometric distributions of rates were tested.

**TABLE 23.3**   Cost Improvement Over Maxemchuck's Algorithm (%) and CPU Seconds for Binary Rounding and Two Primal–Dual Algorithms (Averages Over 10 Testcases)

| | | 50% Steiner Nodes, Geometric Progression Rates | | | | | | |
|---|---|---|---|---|---|---|---|---|
| R | N | Maxemchuk's | Binary Rounding | | Naive-PD | | Restart-PD | |
| | | CPU | %G | CPU | %G | CPU | %G | CPU |
| 1 | 200 | 0.017 | 0.00 | 0.017 | −0.01 | 0.544 | −0.01 | 0.325 |
| 1 | 300 | 0.050 | 0.00 | 0.052 | 0.04 | 1.372 | 0.04 | 0.946 |
| 2 | 200 | 0.027 | 0.00 | 0.026 | 0.43 | 1.271 | 1.03 | 1.125 |
| 2 | 300 | 0.070 | 0.00 | 0.072 | 0.93 | 4.573 | 2.17 | 3.747 |
| 5 | 200 | 0.044 | 0.00 | 0.044 | −2.13 | 1.490 | 1.30 | 5.321 |
| 5 | 300 | 0.123 | 0.00 | 0.120 | −0.91 | 5.221 | 1.10 | 16.798 |
| 10 | 200 | 0.065 | 0.00 | 0.068 | −2.53 | 1.636 | 0.66 | 17.848 |
| 10 | 300 | 0.180 | 0.00 | 0.176 | −2.61 | 6.582 | 0.24 | 107.125 |
| | | 50% Steiner Nodes, Arithmetic Progression Rates | | | | | | |
| 1 | 200 | 0.016 | 0.00 | 0.017 | −0.01 | 0.541 | −0.01 | 0.327 |
| 1 | 300 | 0.052 | 0.00 | 0.051 | 0.04 | 1.370 | 0.04 | 0.946 |
| 2 | 200 | 0.027 | 0.00 | 0.023 | −0.69 | 1.373 | −0.00 | 1.136 |
| 2 | 300 | 0.071 | 0.00 | 0.070 | −0.32 | 4.491 | 0.24 | 3.773 |
| 5 | 200 | 0.043 | −0.01 | 0.040 | 1.70 | 1.564 | 2.66 | 5.256 |
| 5 | 300 | 0.123 | −0.10 | 0.107 | 1.92 | 5.392 | 4.19 | 17.271 |
| 10 | 200 | 0.067 | 1.79 | 0.043 | 4.25 | 1.556 | 6.11 | 16.856 |
| 10 | 300 | 0.181 | 2.36 | 0.126 | 3.38 | 5.444 | 5.73 | 92.575 |
| | | 0% Steiner Nodes, Geometric Progression Rates | | | | | | |
| 1 | 100 | 0.002 | 0.00 | 0.002 | 0.00 | 0.052 | 0.00 | 0.077 |
| 1 | 200 | 0.028 | 0.00 | 0.028 | 0.00 | 0.251 | 0.00 | 0.465 |
| 2 | 100 | 0.007 | 0.00 | 0.007 | 1.21 | 0.088 | 1.69 | 0.185 |
| 2 | 200 | 0.038 | 0.00 | 0.033 | 2.14 | 0.698 | 2.31 | 1.517 |
| 5 | 100 | 0.012 | 0.00 | 0.013 | 1.24 | 0.120 | 2.82 | 0.665 |
| 5 | 200 | 0.059 | 0.00 | 0.056 | −0.25 | 1.296 | 1.70 | 6.314 |
| 10 | 100 | 0.019 | 0.00 | 0.018 | −0.68 | 0.133 | 1.63 | 1.953 |
| 10 | 200 | 0.090 | 0.00 | 0.091 | −1.97 | 1.466 | 0.73 | 20.525 |
| | | 0% Steiner Nodes, Arithmetic Progression Rates | | | | | | |
| 1 | 100 | 0.005 | 0.00 | 0.005 | 0.00 | 0.054 | 0.00 | 0.078 |
| 1 | 200 | 0.026 | 0.00 | 0.026 | 0.00 | 0.247 | 0.00 | 0.457 |
| 2 | 100 | 0.005 | 0.00 | 0.006 | −0.11 | 0.111 | −0.04 | 0.187 |
| 2 | 200 | 0.036 | 0.00 | 0.034 | −0.02 | 1.078 | 0.30 | 1.570 |
| 5 | 100 | 0.011 | −0.17 | 0.011 | 3.70 | 0.114 | 4.60 | 0.656 |
| 5 | 200 | 0.059 | −0.15 | 0.052 | 3.13 | 1.235 | 3.85 | 5.952 |
| 10 | 100 | 0.019 | 2.62 | 0.012 | 6.65 | 0.113 | 7.12 | 1.922 |
| 10 | 200 | 0.091 | 2.67 | 0.058 | 5.83 | 1.203 | 6.38 | 17.689 |

Table 23.3 gives the results for instances generated using several sets of parameters. The relative solution quality of various heuristics is fairly independent on the class of instances. We note that the Naive Primal–Dual and the Charikar–Naor–Schieber algorithms most often produce comparable results which are slight improvements over the results produced by Maxemchuk's algorithm. The Restarting Primal–Dual

typically produces solutions of best quality, typically 0.25%−6% better than solutions produced by Max-emchuk's algorithm; this, however, occurs at the expense of greater CPU time. We also note that the difference between algorithms increases as the number of rates increases. Figures 23.10 and 23.11 illustrate this observation in a graphical form.

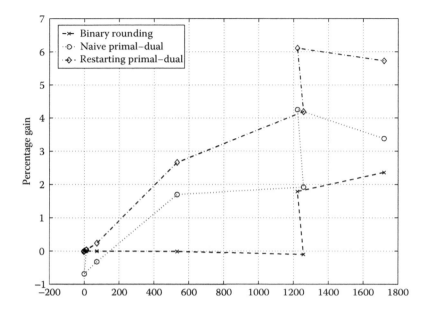

**FIGURE 23.10**   The gain of several algorithms versus Maxemchuk's algorithm, 50% Steiner nodes.

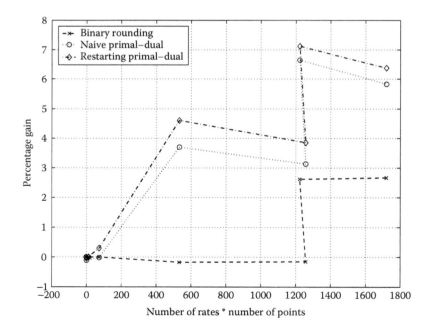

**FIGURE 23.11**   The gain of several algorithms versus Maxemchuk's algorithm, 0% Steiner nodes.

# References

1. Salama, H.F., Reeves, D.S., and Viniotis, Y., Evaluation of multicast routing algorithms for real-time communication on high-speed networks, *IEEE J. Sel. Areas Commun.*, 3, 332, 1997.

2. Ural, H., and Zhu, K., An efficient distributed QoS based multicast routing algorithm, in *Proceedings of the International Performance Computing and Communications Conference*, 2002, p. 27.

3. Bajaj, R., Ravikumar, C.P., and Chandra, S., Distributed delay constrained multicast path setup high speed networks, in *Proceedings of the International Conference on High Performance Computing*, 1997, p. 438.

4. Rouskas, R.N. and Baldine, I., Multicast routing with end-to-end delay and delay variation constraints, *IEEE J. Sel. Areas Commun.*, 15, 346, 1997.

5. Maxemchuk, N., Video distribution on multicast networks, *IEEE J. Sel. Areas Commun.*, 15, 357, 1997.

6. Charikar, M., Naor, J., and Schieber, B., Resource optimization in QoS multicast routing of real-time multimedia, *IEEE/ACM Trans. Networking*, 12, 340, 2004.

7. Xue, G., Lin, G.-H., and Du, D.-Z., Grade of service Steiner minimum trees in the Euclidean plane, *Algorithmica*, 31, 479, 2001.

8. Karpinski, M., Măndoiu, I., Olshevsky, A., and Zelikovsky, A., Improved approximation algorithms for the quality of service Steiner tree problem, *Algorithmica*, 42, 109, 2005.

9. Current, J.R., Revelle, C.S., and Cohon, J.L., The hierarchical network design problem, *Eur. J. Oper. Res.*, 27, 57, 1986.

10. Balakrishnan, A., Magnanti, T.L., and Mirchandani, P., Modeling and heuristic worst-case performance analysis of the two-level network design problem, *Manage. Sci.*, 40, 846, 1994.

11. Balakrishnan, A., Magnanti, T.L., and Mirchandani, P., Heuristics, LPs, and trees on trees: Network design analyses, *Oper. Res.*, 44, 478, 1996.

12. Mirchandani, P., The multi-tier tree problem, *INFORMS J. Comput.*, 8, 202, 1996.

13. Calinescu, G., Fernandes, C., Mandoiu, I., Olshevsky, A., Yang, K., and Zelikovsky, A., Primal-dual algorithms for QoS multimedia multicast, in *Proceedings of IEEE GLOBECOM*, 2003, p. 3631.

14. Robins, G. and Zelikovsky, A., Tighter bounds for graph Steiner tree approximation, *SIAM J. Disc. Math.*, 19, 122, 2005.

15. Promel, H. and Steger, A., A new approximation algorithm for the Steiner tree problem with performance ratio $\frac{5}{3}$, *J. Algorithms*, 36, 89, 2000.

16. Berman, P. and Ramaiyer, V., Improved Approximations for the Steiner tree problem, *J. Algorithms*, 17, 381, 1994.

17. Takahashi, H. and Matsuyama, A., An approximate solution for the Steiner problem in graphs, *Math. Jpn.*, 6, 573, 1980.

18. Zelikovsky, A., An 11/6-approximation algorithm for the network Steiner problem, *Algorithmica*, 9, 463, 1993.

19. Borchers, A. and Du, D.Z., The $k$-Steiner ratio in graphs, *SIAM J. Comput.*, 26, 1997, 857.

20. Goemans, M. and Williamson, D., The primal–dual method for approximation algorithms and its application to network design problems, in *Approximation Algorithms*, Hochbaum, D., Ed., 1997, p. 144. Boston, MA:PWS Publishing Company.

21. http://www.cc.gatech.edu/fac/Ellen.Zegura/gt-itm/gt-itm.tar.gz

22. Zegura, E.W., Calvert, K., and Bhattacharjee, S., How to model an Internetwork?, in *Proceedings of INFOCOM*, 1996, p. 594.

23. Zelikovsky, A., A faster approximation algorithm for the Steiner tree problem in graphs, *Inf. Proc. Lett.*, 46, 79, 1993.
24. Mehlhorn, K., A faster approximation algorithm for the Steiner problem in graphs, *Inf. Proc. Lett.*, 27, 125, 1988.
25. Colbourn, C.J. and Xue, G.L., Grade of service Steiner trees in series-parallel networks, in *Advances in Steiner Trees*, Du, D.Z., Smith, J.M., and Rubinstein, J.H., Eds., 2000, p. 163. Dordrecht, the Netherlands: Kluwer Academic Publishers.

# Overlay Networks for Peer-to-Peer Networks

Andréa W. Richa

Christian Scheideler

Stefan Schmid

## 24.1 Introduction

At the heart of any distributed system lies some kind of logical interconnecting structure, also called *overlay network*, which supports the exchange of information between the different sites. With an increasing scale, distributed systems are likely to become more dynamic and have to deal with sites continuously entering and leaving the system. Reasons for a dynamic membership include, for example, site failures, sites which have to be updated and replaced by new sites, or the addition of new sites or resources which are required to preserve the functionality of the system. Hence, any large-scale distributed system needs an overlay network that supports joining, leaving, and routing between the sites. Without a scalable implementation of such a network, it is impossible to build large high-performance distributed systems.

Scalability is especially critical for peer-to-peer systems. Peer-to-peer systems are self-organizing systems whose members, the *peers*, cooperate without relying on any central server. A key characteristic of peer-to-peer systems is that they typically support an open membership: peers can join and leave at will. Peer-to-peer systems do not require an investment in additional high-performance hardware, and are hence low-cost. In particular, peer-to-peer systems can leverage the tremendous amount of resources (such as computation and storage) available at its constituent parts, the peers: although not used by their owners, these resources may sit idle on the individual computers.

A truly scalable peer-to-peer system must support efficient operations for joining, leaving, and routing between the sites: ideally, the work required for such operations should be at most polylogarithmic in the system size. In particular, the maximum degree and the diameter of (and route lengths in) the overlay network should be at most polylogarithmic in $n$, where $n$ is the total number of peers in the system. The overlay network should also be well-connected, and be robust against faulty peers. The well-connectedness of a graph is usually measured by its *expansion*, which we will formally define later in this chapter. Another important parameter is the *stretch factor* of an overlay network, which measures by how much the length of a shortest route between two nodes $v$ and $w$ *in the overlay network* is off from a shortest route from $v$ to $w$ *when using the underlying physical network*.

To summarize, a scalable peer-to-peer system must offer efficient JOIN, LEAVE (assuming graceful departures), and ROUTE operations such that for any sequence of join, leave, and route request

- The *work* of executing these requests is as small as possible.
- The *degree, diameter, and stretch factor* of the resulting network are as small as possible.
- The *expansion* of the resulting network is as large as possible.

In other words, we are dealing with multiobjective optimization problems.

In addition, we want our systems to be fault-tolerant: our system should not only provide leave operations allowing peers to leave gracefully, but it should also tolerate peer failures. In particular, we aim to design self-stabilizing peer-to-peer systems: peer-to-peer networks which automatically recover from any configuration.

To address these problems, we first introduce some basic notation and techniques for constructing overlay networks (Section 24.2). Afterwards, we discuss supervised overlay network designs (i.e., the topology is maintained by a supervisor but routing is done on a peer-to-peer basis), and then we present various decentralized overlay network designs (i.e., the topology is maintained by the peers themselves). Finally, in Section 24.5, we identify reliable connectivity primitives and study the design of *self-stabilizing* overlay networks.

## 24.2 Basic Notation and Techniques

We start with some basic notation. A graph $G = (V, E)$ consists of a node set $V$ and an edge set $E \subseteq V \times V$. We will only consider directed graphs. The *in-degree* of a node is the number of incoming edges, the *out-degree* of a node is the number of outgoing edges, and the *degree* of a node is the number of incoming and outgoing edges. Given two nodes $v$ and $w$, let $d(v, w)$ denote the length of a shortest directed path from $v$ to $w$ in $G$. $G$ is strongly connected if $d(v, w)$ is finite for every pair $v, w \in V$. In this case,

$$D = \max_{v, w \in V} d(v, w)$$

is the *diameter* of $G$. The *expansion* of $G$ is defined as

$$\alpha = \min_{S \subseteq V, \, |S| \leq |V|/2} \frac{|\Gamma(S)|}{|S|}.$$

where $\Gamma(S) = \{v \in V \setminus S \mid \exists u \in S : (u, v) \in E\}$ is the neighbor set of $S$. The following relationship between the expansion and diameter of a graph is easy to show.

**Fact 24.1** *For any graph $G$ with expansion $\alpha$, the diameter of $G$ is in $O(\alpha^{-1} \log n)$.*

The vast majority of overlay networks for peer-to-peer systems suggested in the literature is based on the concept of virtual space . That is, every site is associated with a point in some space $U$ and connections between sites are established based on rules how to interconnect points in that space. In this case, the following operations need to be implemented:

- JOIN($p$): Add new peer $p$ to the network by choosing a point in $U$ for it.
- LEAVE($p$): Remove peer $p$ from the network.
- ROUTE($m, x$): Route message $m$ to point $x$ in $U$.

Several virtual space approaches are known. The most influential techniques are the hierarchical decomposition technique, the continuous-discrete technique, and the prefix technique. We will give a general outline of each technique in this section. At the end of this section, we present two important families of graphs that we will use later in this chapter to construct dynamic overlay networks.

## 24.2.1 The Hierarchical Decomposition Technique

Consider the space $U = [0,1]^d$ for some fixed $d \geq 1$. The *decomposition tree* $T(U)$ of $U$ is an infinite binary tree, whose root represents $U$. In general, every node $v$ in the tree represents a subcube $U'$ in $U$, and the children of $v$ represent two subcubes $U''$ and $U'''$: $U''$ and $U'''$ are the result of cutting $U'$ in the middle at the smallest dimension in which $U'$ has a maximum side length. The subcubes $U''$ and $U'''$ are closed, that is, their intersection defines the cut. Let every edge to a left child in $T(U)$ be labeled with 0 and every edge to a right child in $T(U)$ be labeled with 1. Then the label of a node $v$, $\ell(v)$, is the sequence of all edge labels encountered when moving along the unique path from the root of $T(U)$ downwards to $v$. For $d = 2$, the result of this decomposition is shown in Figure 24.1.

The goal is to map the peers to nodes in $T(U)$ so that the following conditions are met:

**Condition 24.1**

1. *The interiors of the subcubes associated with the (nodes assigned to the) peers are disjoint.*
2. *The union of the subcubes of the peers gives the entire set $U$.*
3. *Every peer $p$ with subcube $U_p$ is connected to all peers $p'$ with subcubes $U_{p'}$ that are adjacent to $U_p$ (i.e., $U_p \cap U_{p'}$ is a $d-1$-dimensional subcube).*

In the 2-dimensional case, for example, condition (3) means that $p$ and $p'$ share a part of the cut line through their first common ancestor in $T(U)$. It is not difficult to see that the following result is true.

**Fact 24.2** *Consider the space $U = [0,1]^d$ for some fixed $d$ and suppose we have $n$ peers. If the peers are associated with nodes that are within $k$ levels of $T(U)$ and Condition 24.1 is satisfied, then the maximum degree of a peer is at most $(2d) \cdot 2^{k-1}$ and the diameter of the graph is at most $d \cdot n^{1/d} + 2(k-1)$.*

The diameter of the graph can be as large as $d \cdot n^{1/d}$ and therefore too large for a scalable graph if $d$ is fixed, but its degree is small as long as $k = O(\log \log n)$.

An example of a peer-to-peer system based on hierarchical decomposition technique is CAN [1]. In the original CAN construction, a small degree is achieved by giving each peer $p$ a label $\ell(p)$ consisting of a (sufficiently long) random bit string when it joins the system. This bit string is used to route $p$ to the

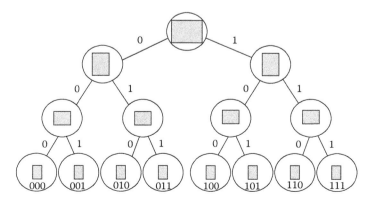

**FIGURE 24.1** The decomposition tree for $d = 2$.

unique peer $p'$ that is reached when traversing the tree $T(U)$ according to $\ell(p)$ (a 0 bit means "go left" and a 1 bit means "go right") until a node $v$ is reached that is associated with a peer. (Such a node must always exist if there is at least one peer in the system and the two rules of assigning peers to nodes in Condition 24.1 are satisfied.) One of $p$ and $p'$ is then placed in the left child of $v$ and the other in the right child of $v$. Leave operations basically reverse join operations so that Condition 24.1 is maintained. Due to the use of a random bit sequence, one can show that the number of levels the peers are apart is indeed $O(\log \log n)$, as desired. However, it can also be as bad as that. Strategies that achieve a more even level balancing were subsequently proposed in several papers (Reference 2 and the references therein).

## 24.2.2 The Continuous-Discrete Technique

The basic idea underlying the continuous-discrete approach [3] is to define a continuous model of graphs and to apply this continuous model to the discrete setting of a finite set of peers. A well-known peer-to-peer system that uses an approach closely related to the continuous-discrete approach is Chord [4].

Consider the $d$-dimensional space $U = [0, 1)^d$, and suppose that we have a set $F$ of continuous functions $f_i : U \rightarrow U$. Then we define $E_F$ as the set of all pairs $(x, y) \in U^2$ with $y = f_i(x)$ for some $i$. Given any subset $S \subseteq U$, let $\Gamma(S) = \{y \in U \setminus S \mid \exists x \in S : (x, y) \in E_F\}$. If $\Gamma(S) \neq \emptyset$ for every $S \subset U$, then $F$ is said to be *mixing*. If $F$ does not mix, then there are disconnected areas in $U$.

Consider now any set of peers $V$, and let $S(v)$ be the subset in $U$ that has been assigned to peer $v$. Then the following conditions have to be met:

**Condition 24.2**

1. $\cup_v S(v) = U$.
2. *For every pair of peers $v$ and $w$, it holds that $v$ is connected to $w$ if and only if there are two points $x, y \in U$ with $x \in S(v)$, $y \in S(w)$ and $(x, y) \in E_F$.*

Let $G_F(V)$ be the graph resulting from the above-mentioned conditions. Then the following fact is easy to see.

**Fact 24.3** *If $F$ is mixing and $\cup_v S(v) = U$, then $G_F(V)$ is strongly connected.*

To bound the diameter of $G_F(V)$, we introduce some further notation. For any point $x$ and any $\epsilon \in [0, 1)$, let $B(x, \epsilon)$ denote the $d$-dimensional ball of volume $\epsilon$ centered at $x$. For any two points $x$ and $y$ in $U$, let $d_n(x, y)$ denote the shortest sequence $(s_1 s_2 s_3 \cdots s_k) \in \mathbb{N}^k$ so that there are two points $x' \in B(x, 1/n)$ and $y' \in B(y, 1/n)$ with $f_{s_1} \circ f_{s_2} \circ \ldots f_{s_k}(x') = y'$. Then we define the diameter of $F$ as

$$D(n) = \max_{x, y \in U} d_n(x, y)$$

Using this definition, it holds:

**Fact 24.4** *If $\cup_v S(v) = U$ and every $S(v)$ contains a ball of volume at least $1/n$, then $G_F(V)$ has a diameter of at most $D(n)$.*

Moreover, the expansion of $G_F(V)$ can be bounded with a suitable parameter for $F$. However, it is easier to consider explicit examples here, and therefore we defer a further discussion to Section 24.4.1.

## 24.2.3 The Prefix Technique

The prefix technique was first presented in References 5, 6 and first used in the peer-to-peer world by Pastry [7] and Tapestry [8]. Given a label $\ell = (\ell_1 \ell_2 \ell_3 \cdots)$, let $\text{prefix}_i(\ell) = (\ell_1 \ell_2 \cdots \ell_i)$ for all $i \geq 1$ and $\text{prefix}_0(\ell) = \epsilon$, the empty label.

In the prefix technique, every peer node $v$ is associated with a unique label $\ell(v) = \ell(v)_1 \ldots \ell(v)_k$, where each $\ell(v)_i \in \{0, \ldots, b-1\}$, for some constant $b \geq 2$ and sufficiently large $k$. The following condition has to be met concerning connections between the nodes.

**Condition 24.3** *For every peer $v$, every digit $\alpha \in \{0, \ldots, b-1\}$, and every $i \geq 0$, $v$ has a link to a peer node $w$ with $\mathrm{prefix}_i(\ell(v)) = \mathrm{prefix}_i(\ell(w))$ and $\ell(w)_{i+1} = \alpha$, if such a node $w$ exists.*

As for some values of $i$ and $\alpha$ there can be many nodes $w$ satisfying the above-mentioned condition, a rule has to be specified which of these nodes $w$ to pick. For example, a peer may connect to the geographically closest peer $w$, or a peer may connect to a peer $w$ it has the best connection to. The following fact is easy to show:

**Fact 24.5** *If the maximum length of a node label is $L$ and the node labels are unique, then any rule of choosing a node $w$ as in Condition 24.3 guarantees strong connectivity. Moreover, the maximum out-degree of a node and the diameter of the network are at most $L$.*

However, the in-degree, that is, the number of incoming connections, can be quite high, depending on the rule. A simple strategy guaranteeing polylogarithmic in- and out-degree and logarithmic diameter is that every node chooses a random binary sequence as its label, and a node $v$ connects to the node $w$ among the eligible candidates with the closest distance to $v$, that is, $|\ell(v) - \ell(w)|$ is minimized. This rule also achieves a good expansion but not a good stretch factor. To address the stretch factor, other rules are necessary. We will discuss them in Section 24.4.2.

### 24.2.4 Basic Classes of Graphs

We will apply our previous basic techniques to two important classes of graphs: the hypercube and the de Bruijn graph. They are defined as follows.

**Definition 24.1** *For any $d \in \mathbb{N}$, the $d$-dimensional hypercube is an undirected graph $G = (V, E)$ with $V = \{0, 1\}^d$ and $E = \{\{v, w\} \mid H(v, w) = 1\}$ where $H(v, w)$ is the Hamming distance between $v$ and $w$.*

**Definition 24.2** *For any $d \in \mathbb{N}$, the $d$-dimensional de Bruijn graph is an undirected graph $G = (V, E)$ with node set $V = \{v \in \{0, 1\}^d\}$ and edge set $E$. It contains all edges $\{v, w\}$ with the property that $w \in \{(x, v_{d-1}, \ldots, v_1) : x \in \{0, 1\}\}$, where $v = (v_{d-1}, \ldots, v_0)$.*

## 24.3 Supervised Overlay Networks

A *supervised overlay network* is a network formed by a supervisor but in which all other activities can be performed in a peer-to-peer fashion involving the supervisor. Supervised overlays therefore lie between server-based overlay networks and pure peer-to-peer overlay networks. In order for a supervised network to be highly scalable, two central requirements have to be satisfied:

1. The supervisor needs to store at most a polylogarithmic amount of information about the system at any time. For example, if there are $n$ peers in the system, storing contact information about $O(\log^2 n)$ of these peers would be fine.
2. The supervisor needs at most a constant number of messages to include a new peer into or exclude an old peer from the network.

The second condition makes sure that the work of the supervisor to include or exclude peers from the system is kept at a minimum. First, we present a general strategy of constructing supervised overlay networks, which combines the hierarchical decomposition technique with the continuous-discrete technique and the recursive labeling technique in the following. Subsequently, we give some explicit examples that achieve near-optimal results for the cost of the join, leave, and route operations as well as the degree, diameter, and expansion of the network.

## 24.3.1 The Recursive Labeling Technique

In the recursive labeling approach, the supervisor assigns a *label* to every peer that wants to join the system. The labels are represented as binary strings and are generated in the following order:

$$0, 1, 01, 11, 001, 011, 101, 111, 0001, 0011, 0101, 0111, 1001, 1011, \ldots$$

Basically, when stripping off the least significant bit, then the supervisor first creates all binary numbers of length 0, then length 1, then length 2, and so on. More formally, consider the mapping $\ell : \mathbb{N}_0 \to \{0, 1\}^*$ with the property that for every $x \in \mathbb{N}_0$ with binary representation $(x_d \cdots x_0)_2$ (where $d$ is minimum possible),

$$\ell(x) = (x_{d-1} \cdots x_0 x_d).$$

Then $\ell$ generates the sequence of labels displayed earlier. In the following, it will also be helpful to view labels as real numbers in $[0, 1)$. Let the function $r : \{0, 1\}^* \to [0, 1)$ be defined so that for every label $\ell = (\ell_1 \ell_2 \ldots \ell_d) \in \{0, 1\}^*$, $r(\ell) = \sum_{i=1}^{d} \frac{\ell_i}{2^i}$. Then the above-mentioned sequence of labels translates into

$$0, \ 1/2, \ 1/4, \ 3/4, \ 1/8, \ 3/8, \ 5/8, \ 7/8, \ 1/16, \ 3/16, \ 5/16, \ 7/16, \ 9/16, \ \ldots$$

Thus, the more labels are used, the more densely the $[0, 1)$ interval will be populated. When employing the recursive approach, the supervisor aims to maintain the following condition at any time:

**Condition 24.4** *The set of labels used by the peers is $\{\ell(0), \ell(1), \ldots, \ell(n-1)\}$, where $n$ is the current number of peers in the system.*

This condition is preserved with the following simple strategy:

- Whenever a new peer $v$ joins the system and the current number of peers is $n$, the supervisor assigns the label $\ell(n)$ to $v$ and increases $n$ by 1.
- Whenever a peer $w$ with label $\ell$ wants to leave the system, the supervisor asks the peer with currently highest label $\ell(n-1)$ to take over the role of $w$ (and thereby change its label to $\ell$) and reduces $n$ by 1.

## 24.3.2 Putting the Pieces Together

We assume that we have a single supervisor for maintaining the overlay network. In the following, the label assigned to some peer $v$ will be denoted by $\ell_v$. Given $n$ peers with unique labels, we define the *predecessor* pred($v$) of peer $v$ as the peer $w$ for which $r(\ell_w)$ is closest from below to $r(\ell_v)$, and we define the *successor* succ($v$) of peer $v$ as the peer $w$ for which $r(\ell_w)$ is closest from above to $r(\ell_v)$ (viewing $[0, 1)$ as a ring in both cases). Given two peers $v$ and $w$, we define their *distance* as

$$\delta(v, w) = \min\{(1 + r(\ell_v) - r(\ell_w)) \bmod 1, \ (1 + r(\ell_w) - r(\ell_v)) \bmod 1\}.$$

To maintain a doubly linked cycle among the peers, we simply have to maintain the following condition:

**Condition 24.5** *Every peer $v$ in the system is connected to pred($v$) and succ($v$).*

Now, suppose that the labels of the peers are generated via the recursive strategy earlier. Then we have the following properties:

**Lemma 24.1** *Let $n$ be the current number of peers in the system, and let $\bar{n} = 2^{\lfloor \log n \rfloor}$. Then for every peer $v \in V$, $|\ell_v| \leq \lceil \log n \rceil$ and $\delta(v, \text{pred}(v)) \in \{1/(2\bar{n}), 1/\bar{n}\}$.*

So the peers are approximately evenly distributed in $[0, 1)$, and the number of bits for storing a label is almost as low as it can be without violating the uniqueness requirement.

Recall the hierarchical decomposition approach. The supervisor will assign every peer $p$ to the unique node $v$ in $T(U)$ at level $\log(1/\delta(p, \text{pred}(p)))$ with $\ell_v$ being equal to $\ell_p$ (padded with 0's to the right so that $|\ell_v| = |\ell_p|$). As an example, if we currently have 4 peers in the system, then the mapping of peer labels to node labels is

$$0 \to 00,\ 1 \to 10,\ 01 \to 01,\ 11 \to 11$$

With this strategy, it follows from Lemma 24.1 that Fact 24.2 applies with $k = 2$.

Consider now any family $F$ of functions acting on some space $U = [0,1)^d$ and let $C(p)$ be the subcube of the node in $T(U)$ that $p$ has been assigned to. Then the goal of the supervisor is to maintain the following condition at any time.

**Condition 24.6** *For the current set $V$ of peers in the system it holds that*

1. *The set of labels used by the peers is $\{\ell(0), \ell(1), \ldots, \ell(n-1)\}$, where $n = |V|$.*
2. *Every peer $v$ in the system is connected to $\text{pred}(v)$ and $\text{succ}(v)$.*
3. *There is an edge $(v, w)$ for every pair of peers $v$ and $w$ for which there is an edge $(x, y) \in E_F$ with $x \in C(v)$ and $y \in C(w)$.*

### 24.3.3 Maintaining Condition 24.6

Next we describe the actions that the supervisor has to perform to maintain Condition 24.6 during a join or leave operation. We start with the following important fact.

**Fact 24.6** *Whenever a new peer $v$ enters the system, then $\text{pred}(v)$ has all the connectivity information $v$ needs to satisfy Condition 24.6(3). Moreover, to maintain Condition 24.6(3), whenever an old peer $w$ leaves the system, it suffices that $w$ transfers all of its connectivity information to $\text{pred}(w)$.*

The first part of the fact follows from the observation that when $v$ enters the system, then the subcube of $\text{pred}(v)$ splits into two subcubes where one resides at $\text{pred}(v)$ and the other is taken over by $v$. Hence, if $\text{pred}(v)$ passes all of its connectivity information to $v$, then $v$ can establish all edges relevant for it according to the continuous-discrete approach. The second part of the fact follows from the observation that the departure of a peer is the reverse of the insertion of a peer.

Thus, if the peers take care of the connections in Condition 24.6(3), the only part that the supervisor has to take care of is maintaining the cycle. For this we require the following condition.

**Condition 24.7** *At any time, the supervisor stores the contact information of $\text{pred}(v)$, $v$, $\text{succ}(v)$, and $\text{succ}(\text{succ}(v))$ where $v$ is the peer with label $\ell(n-1)$.*

To satisfy Condition 24.7, the supervisor performs the following actions. If a new peer $w$ joins, then the supervisor

- Informs $w$ that $\ell(n)$ is its label, $\text{succ}(v)$ is its predecessor, and $\text{succ}(\text{succ}(v))$ is its successor
- Informs $\text{succ}(v)$ that $w$ is its new successor
- Informs $\text{succ}(\text{succ}(v))$ that $w$ is its new predecessor
- Asks $\text{succ}(\text{succ}(v))$ to send its successor information to the supervisor
- Sets $n = n + 1$

If an old node $w$ leaves and reports $\ell_w$, $\text{pred}(w)$, and $\text{succ}(w)$ to the supervisor (recall that we are assuming graceful departures), then the supervisor

- Informs $v$ (the node with label $\ell(n-1)$) that $\ell_w$ is its new label, $\text{pred}(w)$ is its new predecessor, and $\text{succ}(w)$ is its new successor
- Informs $\text{pred}(w)$ that its new successor is $v$ and $\text{succ}(w)$ that its new predecessor is $v$
- Informs $\text{pred}(v)$ that $\text{succ}(v)$ is its new successor and $\text{succ}(v)$ that $\text{pred}(v)$ is its new predecessor

- Asks pred($v$) to send its predecessor information to the supervisor and to ask pred(pred($v$)) to send its predecessor information to the supervisor
- Sets $n = n - 1$

Thus, the supervisor only needs to handle a small constant number of messages for each arrival or departure of a peer, as desired. Next we look at two examples resulting in scalable supervised overlay networks.

### 24.3.4 Examples

For a supervised hypercubic network, simply select $F$ as the family of functions on $[0, 1)$ with $f_i(x) = x + 1/2^i$ (mod 1) for every $i \geq 1$. By using our framework, this gives an overlay network with degree $O(\log n)$, diameter $O(\log n)$, and expansion $O(1/\sqrt{\log n})$, which matches the properties of ordinary hypercubes.

For a supervised de Bruijn network, simply select $F$ as the family of functions on $[0, 1)$ with $f_0(x) = x/2$ and $f_1(x) = (1 + x)/2$. Using our framework, this gives an overlay network with degree $O(1)$, diameter $O(\log n)$, and expansion $O(1/\log n)$. This matches the properties of ordinary de Bruijn graphs.

In both networks, routing with logarithmic work can be achieved by using the bit adjustment strategy.

## 24.4 Decentralized Overlay Networks

Next we show that scalable overlay networks can also be maintained without involving a supervisor. As the hierarchical decomposition technique cannot yield networks of polylogarithmic diameter, in the following, we will only discuss examples for the latter two basic techniques in Section 24.2.

### 24.4.1 Overlay Networks Based on the Continuous-Discrete Approach

Similar to the supervised approach, we first show how to maintain a hypercubic overlay network. Subsequently, we show how to maintain a de Bruijn-based overlay network.

#### Maintaining a Dynamic Hypercube

Let $U = [0, 1)$ and consider the family $F$ of functions on $[0, 1)$ with $f_i(x) = x + 1/2^i$ (mod 1) for every $i \geq 1$. Given a set of points $V \subset [0, 1)$, we define the region $S(v)$ associated with point $v$ as the interval (pred($v$), $v$) where pred($v$) is the closest predecessor of $v$ in $V$ and $U$ is seen as a ring. The following result follows from Reference 9:

**Theorem 24.1** *If every peer is given a random point in $[0, 1)$, then the graph $G_F(V)$ with $|V| = n$ resulting from the continuous-discrete approach has a degree of $O(\log^2 n)$, a diameter of $O(\log n)$, and an expansion of $\Omega(1/\log n)$, with high probability.*

Suppose that Condition 24.2 is satisfied for our family of hypercubic functions. Then it is fairly easy to route a message from any point $x \in [0, 1)$ to any point $y \in [0, 1)$ along edges in $G_F(V)$:

Consider the path $P$ in the continuous space that results from using a bit adjustment strategy to get from $x$ to $y$. That is, given that $x_1 x_2 x_3 \cdots$ is the bit sequence for $x$ and $y_1 y_2 y_3 \cdots$ is the bit sequence for $y$, $P$ is the sequence of points $z_0 = x_1 x_2 x_3 \ldots, z_1 = y_1 x_2 x_3 \ldots, z_2 = y_1 y_2 x_3 \ldots, \ldots, y_1 y_2 y_3 \ldots = y$. Of course, $P$ may have an infinite length, but simulating $P$ in $G_F(V)$ only requires traversing a finite sequence of edges:

We start with the region $S(v)$ containing $x = z_0$. Then we move along the edge $(v, w)$ in $G_F(V)$ to the region $S(w)$ containing $z_1$. This edge must exist because we assume that Condition 24.2 is satisfied. Then we move along the edge $(w, w')$ simulating $(z_1, z_2)$, and so on, until we reach the node whose region contains $y$. By using this strategy, it holds:

**Theorem 24.2** *Given a random node set $V \subset [0, 1)$ with $|V| = n$, it takes at most $O(\log n)$ hops, with high probability, to route in $G_F(V)$ from any node $v \in V$ to any node $w \in V$.*

Next we explain how nodes can join and leave. Suppose that a new node $v$ contacts some node already in the system to join the system. Then $v$'s request is first sent to the node $u$ in $V$ with $u = \text{succ}(v)$, which only takes $O(\log n)$ hops according to Theorem 24.2. Node $u$ forwards information about all of its incoming and outgoing edges to $v$, deletes all edges that it does not need any more, and informs the corresponding endpoints about this. As $S(v) \subseteq S(u)$ for the old $S(u)$, the edges reported to $v$ are a superset of the edges that it needs to establish. Node $v$ checks which of the edges are relevant for it, informs the other endpoint for each relevant edge, and removes the others.

If a node $v$ wants to leave the network, it simply forwards all of its incoming and outgoing edges to $\text{succ}(v)$. Node $\text{succ}(v)$ will then merge these edges with its existing edges and notifies the endpoints of these edges about the changes.

Combining Theorems 24.1 and 24.2 we obtain:

**Theorem 24.3** *It takes a routing effort of $O(\log n)$ hops and an update work of $O(\log^2 n)$ messages that can be processed in $O(\log n)$ communication rounds to execute a join or leave operation.*

### Maintaining a Dynamic deBruijn Graph

Next we show how to dynamically maintain a deBruijn graph (see Figure 24.2 for an illustration). Let $U = [0, 1)$ and $F$ consist of two functions, $f_0$ and $f_1$, where $f_i(x) = (i + x)/2$ for each $i \in \{0, 1\}$. Then one can show the following result:

**Theorem 24.4** *If the peers are mapped to random points in $[0, 1)$, then the graph $G_F(V)$ resulting from the continuous-discrete approach has a degree of $O(\log n)$, diameter of $O(\log n)$, and node expansion of $\Omega(1/\log n)$, with high probability.*

Next we show how to route in the de Bruijn network. Suppose that Condition 24.2 is satisfied. Then we use the following trick to route a message from any point $x \in [0, 1)$ to any point $y \in [0, 1)$ along edges in $G_F(V)$.

Let $z$ be a randomly chosen point in $[0, 1)$. Let $x_1 x_2 x_3 \ldots$ be the binary representation of $x$, let $y_1 y_2 y_3 \ldots$ be the binary representation of $y$, and let $z_1 z_2 z_3$ be the binary representation of $z$. Let $P$ be the path

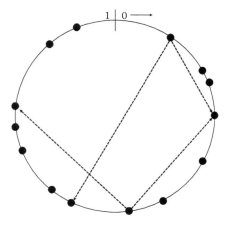

**FIGURE 24.2** An example of a dynamic de Bruijn network (only some short-cut pointers for two nodes are given).

along the points $x = x_1x_2x_3\ldots, z_1x_1x_2\ldots, z_2z_1x_1\ldots,\ldots$ and let $P'$ be the path along the points $y = y_1y_2y_3\ldots, z_1y_1y_2\ldots, z_2z_1y_1\ldots,\ldots$ Then we simulate moving along the points in $P$ by moving along the corresponding edges in $G_F(V)$. We stop when we hit a node $w \in V$ with the property that $S(w)$ contains a point in $P$ and a point in $P'$. At that point, we follow the points in $P'$ backwards until we arrive at the node $w' \in W$ that contains $y$ in $S(w')$. By using this strategy, it holds:

**Theorem 24.5** *Given a random node set $V \subset [0, 1)$ with $|V| = n$, it takes at most $O(\log n)$ hops, with high probability, to route in $G_F(V)$ from any point $x \in [0, 1)$ to any point $y \in [0, 1)$.*

Joining and leaving the network is done in basically the same way as in the hypercube, giving the following result:

**Theorem 24.6** *It takes a routing effort of $O(\log n)$ hops and an update work of $O(\log n)$ messages that can be processed in a logarithmic number of communication rounds to execute a join or leave operation in the dynamic de Bruijn graph.*

## 24.4.2 Overlay Networks Based on Prefix Connections

The efficiency of routing on an overlay network is quite often measured in terms of the number of hops (neighbor links) followed by a message. Although this measure indicates the latency of a message in the overlay network, it fails to convey the complexity of the given operation with respect to the original underlying network. In other words, although in the overlay network all overlay links may have the same cost, this is not true when those overlay links are translated back into paths in the underlying network. Hence, in practice, to evaluate the routing performance of an overlay network, one should not only study the cost of following a path in the overlay network, but also take into account the different internode communication costs *in the underlying network*. For example, a hop from a peer in the USA to a peer in Europe costs a lot more (in terms of reliability, speed, cost of deploying and maintaining the link, etc.) than a hop going between two peers in a local area network. In brief, keeping routing local is important: It may make sense to route a message originated in Phoenix for a destination peer in San Francisco through Los Angeles, but not through a peer in Europe.

In the current section, we present peer-to-peer overlay network design schemes which take locality into account and which are able to achieve constant stretch factors, while keeping polylogarithmic degree and diameter, and polylogarithmic complexity for join and leave operations. All of the work in this section assumes that the underlying peer-to-peer system is a growth-bounded network, which we define a few paragraphs later.

Peers communicate with one another by means of messages; each message consists of at least one word. We assume that the underlying network provides reliable communication. We define the cost of communication by a function $c : V^2 \rightarrow \Re$. This function $c$ is assumed to reflect the combined effect of the relevant network parameter values, such as latency, throughput, congestion, and so on. In other words, for any two peers $u$ and $v$ in $V$, $c(u, v)$ is the cost of transmitting a single-word message from $u$ to $v$. We assume that $c$ is symmetric and satisfies the triangle inequality. The cost of transmitting a message of length $l$ from peer $u$ to peer $v$ is given by $f(l)c(u, v)$, where $f : \mathbf{N} \rightarrow \Re^+$ is any nondecreasing function such that $f(1) = 1$.

A *growth-bounded* network satisfies the following property: Given any $u$ in $V$ and any real $r$, let $B(u, r)$ denote the set of peers $v$ such that $c(u, v) \leq r$. We refer to $B(u, r)$ as the *ball* of *radius* $r$ around $u$. We assume that there exist a real constant $\Delta$ such that for any peer $u$ in $V$ and any real $r \geq 1$, we have

$$|B(u, 2r)| \leq \Delta|B(u, r)|. \tag{24.1}$$

In other words, the number of peers within radius $r$ from $u$ grows polynomially with $r$. This network model has been validated by both theoreticians and practitioners as to model well existing internetworking topologies [7,8,10,11].

Plaxton, Rajaraman, and Richa (PRR) in References 5, 6 pioneered the work on locality-aware routing schemes in dynamic environments. Their work actually addresses a more general problem—namely, the *object location problem*—than that of designing efficient overlay networks with respect to the parameters outlined in Section 24.1. In that early work, Plaxton et al. formalize the problem of object location in a peer-to-peer environment, pinpointing the issue of locality and developing a formal framework under which object location schemes have been rigorously analyzed. In the object location problem, peers seek to find objects in a dynamic and fully distributed environment, in which multiple (identical) copies of an object may exist in the network: The main goal is to be able to locate and find a copy of an object within cost that is proportional to the cost of retrieving the closest copy of the object to the requesting peer, while being able to efficiently support these operations in a dynamic peer-to-peer environment in which copies of the objects are continuously inserted and removed from the network.

Our overlay network design problem can be viewed as a subset of the object location problem addressed by PRR, in which each object is a peer (and hence there exists a single copy of each object in the network). Hence the results in the PRR scheme and other object location schemes to follow, in particular, the LAND scheme to be addressed later in this section, directly apply to our overlay network design problem. Both PRR and LAND assume a growth-bounded network model. For these networks, the LAND scheme provides the best currently known overlay design scheme with $1 + \epsilon$ stretch, and polylogarithmic bounds on diameter and degree, for any fixed constant $\epsilon > 0$. Combining the LAND scheme with a technique by Hildrum et al. [12] to find nearest neighbors enable us to also attain polylogarithmic work for JOIN and LEAVE operations. As the LAND scheme heavily relies on the PRR scheme, we will present the latter in more detail and then highlight the changes introduced by the LAND scheme. We will address both schemes in the light of overlay routing, rather than the object location problem originally addressed by these schemes.

In the PRR and related schemes, each peer $p$ will have two attributes in addition to its exact location in the network (which is unknown to other peers): a virtual location $x$ in $U$, and a *label* $\ell(x)$ *generated independently and uniformly at random*. We call the virtual location $x$ of $p$ the *peer identifier* $x$, or simply, the *ID* $x$. In the remainder of this section, we will *indistinctly use the ID $x$ to denote the peer $p$ itself.*

### 24.4.2.1 The Plaxton, Rajaraman, and Richa Scheme

The original PRR scheme assumes that we have a growth-bounded network with the extra assumption of also having a lower bound on the rate of growth. Later work that evolved from the PRR scheme (e.g., the LAND scheme) showed that this assumption could be dropped by slightly modifying the PRR scheme. The PRR scheme and the vast majority of provably efficient object location schemes rely on a basic yet powerful technique called *prefix routing*, which we outlined in Section 24.4.2. In what follows, we re-visit prefix routing in the context of the PRR scheme.

**Theorem 24.7** *The PRR scheme, when combined with a technique by Hildrum et al. for finding nearest neighbors, achieves an overlay peer-to-peer network with the following properties: expected constant stretch for* ROUTE *operations, $O(\log n)$ diameter, and, with high probability, $O(\log^2 n)$ degree and work for* JOIN, LEAVE, *and* ROUTE *operations.*

### Prefix Routing

The basic idea behind prefix routing (Section 24.4.2) is that the path followed in a routing operation will be guided solely by the ID $y$ we are seeking for: Every time a ROUTE$(m, y)$ request is forwarded from a peer $u$ to a peer $v$ in the overlay path, the prefix of $\ell(v)$ that (maximally) matches a prefix of the ID $y$ is strictly larger than that of $\ell(u)$.

We now sketch the PRR prefix routing scheme (for more details, see Reference 5). Each peer $x$ in the network is assigned a $(\log_b n)$-digit label[*] $\ell(x) = \ell(x)_1 \ldots \ell(x)_{\log_b n}$, where each $\ell(x)_i \in \{0, \ldots, b - 1\}$,

---

[*] Without loss of generality, assume that $n$ is a power of $b$.

uniformly at random, for some large enough constant $b \geq 2$. Recall that each peer also has a unique $(\log_b n)$-digit ID which is independent of this label. We denote the ID of peer $x$ by $x_1 \ldots x_{\log_b n}$, where each $x_i \in \{0, \ldots, b - 1\}$.

The random labels are used to construct a *neighbor table* at each peer. For a base $b$ sequence of digits $\gamma = \gamma_1 \ldots \gamma_m$, $\text{prefix}_i(\gamma)$ denotes the first $i$ digits, $\gamma_1 \ldots \gamma_i$, of $\gamma$. For each peer $x$, each integer $i$ between 1 and $\log_b n$, and each digit $\alpha$ between 0 and $b - 1$, the neighbor table at peer $x$ stores the $(\log_b n)$-digit ID of the *closest* peer $y$ to $x$—that is, the peer with minimum $c(x, y)$—such that $\text{prefix}_{i-1}(\ell(x)) = \text{prefix}_{i-1}(\ell(y))$ and $\ell(y)_i = \alpha$. We call $y$ the $(i, \alpha)$-*neighbor* of peer $x$. There exists an edge between any pair of neighbor peers in the overlay network.

The degree of a peer in the overlay network is given by the size (number of entries) of its neighbor table. The size of the neighbor table at each peer, as constructed earlier, is $O(\log n)$. In the final PRR scheme (and other follow-up schemes) the neighbor table at a peer will have size polylogarithmic in $n$ (namely, $O(\log^2 n)$ in the PRR scheme), as a set of "auxiliary" neighbors at each peer will need to be maintained in order for the scheme to function efficiently. Each peer $x$ will also need to maintain a set of "pointers" to the location of the subset of peers that were published at $x$ by JOIN operations. In the case of routing, each peer will maintain $O(\log^2 n)$ such pointers with high probability[*]. We describe those pointers in more detail while addressing the JOIN operation in PRR.

The sequence of peers visited in the overlay network during a routing operation for ID $y$ initiated by peer $x$ will consist of the sequence $x = x^0, x^1, \ldots, x^q$, where $x^i$ is the $(i, y_i)$-neighbor of peer $x^{i-1}$, for $1 \leq i \leq q$, and $x^p$ is a peer that holds a pointer to $y$ in the network ($p \leq \log_b n$). We call this sequence the *neighbor sequence* of peer $x$ for (peer ID) $y$. In the following, we explain in more detail the ROUTE, JOIN, and LEAVE operations according to PRR.

### Route, Join, and Leave

Before we describe a routing operation in the PRR scheme, we need to understand how JOIN and LEAVE operations are processed. As we are interested in keeping low stretch, the implementation of such operations will be different in the PRR scheme than in the two other overlay network design techniques presented in this chapter. In a ROUTE$(m, x)$ operation, we will, as in the hierarchical decomposition and continuous-discrete approaches, start by routing towards the virtual location $x$; however, as we route toward this virtual location, as soon as we find some information on the network regarding the actual location of the peer $p$ corresponding to $x$, we will redirect our route operation to reach $p$. This way, we will be able to show that the total stretch of a route operation is low. (If we were to route all the way to the virtual location $x$, to find information about the actual location of $p$, the total incurred stretch might be too large.)

There are two main components in a JOIN$(p)$ operation: First, information about the location of peer $p$ joining the peer-to-peer system needs to be published in the network, such that subsequent ROUTE operations can indeed locate peer $p$. Second, peer $p$ needs to build its own neighbor table, and other peers in the network may need to update their neighbor tables given the presence of peer $x$ in the network. Similarly, there are two main components in a LEAVE$(p)$ operation: unpublishing any information about $p$'s location in the network and removing any entries containing peer $p$ in the neighbor tables of other peers. We will address these two issues separately. For the moment, we will only be concerned with how information about $p$ is published or unpublished in the network, as this is what we need in order to guarantee the success of a ROUTE operation. We will assume that the respective routing table entries are updated correctly upon the addition or removal of $p$ from the system. Later, we will explain how the neighbor tables can be efficiently updated.

Whenever a peer $p$ with ID $x$ decides to join the peer-to-peer system, we place (*publish*) a pointer leading to the actual location of $p$ in the network, which we call an *x-pointer*, at up to $\log_b n$ peers of the

---

[*] With probability at least $1 - 1/p(n)$, where $p(n)$ is a polynomial function on $n$.

network. For convenience of notation, in the remainder of this section, we will *always use x to indistinctly denote both the ID of peer p and the peer p itself.* Let $x^0 = x, x^1, \ldots, x^{\log_b n - 1}$ be the neighbor sequence of peer $x$ for node ID $x$. We place an $x$-pointer to $x^{i-1}$ at *each* peer $x^i$ in this neighbor sequence. Thus, whenever we find a peer with an $x$-pointer (at a peer $x^j$, $1 \le j \le \log_b n$) during a ROUTE$(m, x)$ operation, we can forward the message all the way "down" the reverse neighbor sequence $x^j, \ldots, x^0 = x$ to the actual location of peer $x$.

The LEAVE$(p)$ operation is the reverse of a JOIN operation: We simply *unpublish* (remove) all the $x$-pointers from the peers $x^0, \ldots, x^{\log_b n}$.

There is an implicit search tree associated with each peer ID $y$ in prefix routing. Assume for a moment that there is only one peer $r$ matching a prefix of the ID $y$ in the largest number of digits in the network. At the end of this section, we address the case when this assumption does not hold. Let the search tree $T(y)$ for peer $y$ be defined by the network edges in the neighbor sequences for peer ID $y$ for each peer $x$ in the network, where, for each edge of the type $(x^{i-1}, x^i)$ in the neighbor sequence of $x$ for $y$, we view $x^i$ as the parent of $x^{i-1}$ in the tree. The above-mentioned implementation of the publish and unpublish operations trivially maintain the following invariant. Let $T_x(y)$ be the subtree rooted at $x$ in $T(y)$.

**Invariant 24.1** *If peer $y$ belongs to $T_x(y)$, then peer $x$ has a $y$-pointer.*

We now describe how a ROUTE$(m, y)$ operation initiated at peer $x$ proceeds. Let $x^0 = x, x^1, \ldots, x^{\log_b n - 1}$ be the neighbor sequence of peer $x$ for $y$. Starting with $i = 0$, peer $x^i$ first checks whether it has a $y$-pointer. If it does then $x^i$ will forward the message $m$ using its $y$-pointer down the neighbor sequence that was used when publishing information about peer $y$ during a JOIN$(y)$ operation. More specifically, let $j$ be the maximum index such that $\text{prefix}_j(\ell(x^i)) = \text{prefix}_j(y)$ (note that $j \ge i$). Then $x^i$ must be equal to $y^j$, where $y = y^0, y^1, \ldots$ is the neighbor sequence of peer $y$ for the ID $y$, used during JOIN$(y)$. Thus message $m$ will be forwarded using the $y$-pointers at $y^j, \ldots, y^0 = y$ (if $y^j$ has a $y$-pointer, then so does $y^k$ for all $1 \le k \le j$) all the way down to peer $y$. If $x^i$ does not have a $y$-pointer, it will simply forward the message $m$ to $x^{i+1}$.

Given Invariant 24.1, peer $x$ will locate peer $y$ in the network if peer $y$ is indeed part of the peer-to-peer system. The cost of routing to peer $y$ can be bounded by:

**Fact 24.7** *A message from peer $x$ to peer $y$ will be routed through a path with cost $O(\sum_{k=1}^{j}[c(x^{k-1}, x^k) + c(y^{k-1}, y^k)])$.*

The main challenge in the analysis of the PRR scheme is to show that this summation is indeed $O(c(x, y))$ in expectation.

A deficiency of this scheme, as described, is that there is a chance that we may fail to locate a pointer to $y$ at $x^1$ through $x^{\log_b n}$. In this case, we must have more than one "root peer" in $T(y)$ (a root peer is a peer such that the length of its maximal prefix matching the ID of $y$ is maximum among all peers in the network), and hence there is no guarantee that there exists a peer $r$ which will have a global view of the network with respect to the ID of peer $y$. Fortunately, this deficiency may be easily rectified by a slight modification of the algorithm, as shown in Reference 5.

The probability that the $k$-digit prefix of the label of an arbitrary peer matches a particular $k$-digit string $\gamma = \text{prefix}_k(y)$, for some peer $y$, is $b^{-k}$. Consider a ball $B$ around peer $x^{k-1}$ containing exactly $b^k$ peers. Note that there is a constant probability (approximately $1/e$) that no peer in $B$ matches $\gamma$. Thus the radius of $B$ is a lower bound (up to a constant factor) on the expected distance from $x^{k-1}$ to its $(k, y_k)$-neighbor. Is this radius also an upper bound? Not for an arbitrary metric, as (e.g.) the diameter of the smallest ball around a peer $z$ containing $b^k + 1$ peers can be arbitrarily larger than the diameter of the smallest ball around $z$ containing $b^k$ peers. However, it can be shown that the radius of $B$ provides a tight bound on the expected distance from $x^{k-1}$ to its $(k, y_k)$-neighbor. Furthermore, Equation 24.1 implies that the expected distance from $x^{k-1}$ to its $(k, y_k)$-neighbor is geometrically increasing in $k$. The latter observation is crucial as it implies that the expected total distance from $x^{k-1}$ to its $(k, y_k)$-neighbor, summed over all $k$ such that

$1 \le k \le j$, is dominated by (i.e., within a constant factor of) the expected communication cost from $x^{j-1}$ to its $(j, y_j)$-neighbor $x^j$.

The key challenge of the complexity analysis of PRR is to show that, $c(x, y) = O(E[c(x^{j-1}, x^j)])$, and hence that the routing stretch factor in the PRR scheme is constant in expectation. This proof is technically rather involved and we refer the reader to Reference 5. In the next section, we will show how the PRR scheme can be elegantly modified to yield deterministic constant stretch. More specifically, the LAND scheme achieves deterministic stretch $1 + \epsilon$, for any fixed $\epsilon > 0$.

### Updating the Neighbor Tables

To be able to have JOIN and LEAVE operations with low work complexity, while still enforcing low stretch ROUTE operations and polylogarithmic degree, one needs to devise an efficient way for updating the neighbor tables upon the arrival or departure of a peer from the system. The PRR scheme alone does not provide such means. Luckily, we can combine the work by Hildrum et al. [12], which provides an efficient way for finding nearest neighbors in a dynamic and fully distributed environment, with the PRR scheme to be able to efficiently handle the insertion or removal of a peer from the system. Namely, the work by Hildrum et al. presents an algorithm which can build the neighbor table of a peer $p$ joining the system and update the other peers' neighbor tables to account for the new peer in the system with total work $O(\log^2 n)$.

### 24.4.2.2 The LAND Scheme

The LAND scheme proposed by Abraham, Malkhi, and Dobzinski in Reference 10 was the first peer-to-peer overlay network design scheme to achieve constant deterministic stretch for routing, while maintaining a polylogarithmic diameter, degree, and work (for ROUTE, JOIN, and LEAVE) for growth-bounded metrics. Note that LAND does not require a lower bound on the growth as PRR does. Like the PRR scheme, the LAND scheme was also designed for the more general problem of object location in peer-to-peer systems.

Namely, the main results of the LAND scheme are summarized in the following theorem:

**Theorem 24.8** *The LAND scheme, when combined with a technique by Hildrum et al. for finding nearest neighbors, achieves an overlay peer-to-peer network with the following properties: deterministic $(1 + \epsilon)$ stretch for ROUTE operations, for any fixed $\epsilon > 0$, $O(\log n)$ diameter, and expected $O(\log n)$ degree and work for JOIN, LEAVE, and ROUTE operations.*

The LAND scheme is a variant of the PRR scheme. The implementation of ROUTE, JOIN, and LEAVE operations in this later scheme are basically the same as in PRR. The basic difference between the two schemes is on how the peer labels are assigned to the peers during the neighbor table construction phase. A peer may hold more than one label, some of which may not have been assigned in a fully random and independent way.

In a nutshell, the basic idea behind the LAND scheme is that instead of letting the distance between a peer $x$ and its $(i, \alpha)$-neighbor be arbitrarily large, it will enforce that this distance be always at most some constant $\beta$ times $b^i$ by letting peer $x$ emulate a virtual peer with label $\gamma$ such that $\text{prefix}_i(\gamma) = x_1 \dots x_{i-1}\alpha$ if no peer has $x_1 \dots x_{i-1}\alpha$ as a prefix of its label in a ball centered at $x$ with $O(b^i)$ peers in it. Another difference in the LAND scheme which is crucial to guarantee a deterministic bound on stretch is that the set of "auxiliary" neighbors it maintains are only used during join/leave operations, rather than during routing operations such as in the PRR scheme.

The analysis of the LAND scheme is elegant, consisting of short and intuitive proofs. Thus, this is also a main contribution of this scheme, given that the analysis of the PRR scheme is rather lengthy and involved.

# 24.5 Self-Stabilizing Overlay Networks

## 24.5.1 Introduction: From Fault-Tolerance to Self-Stabilization

Besides efficiency, fault-tolerance is arguably one of the most important requirements of large-scale overlay networks. Indeed, at large scale and when operating for long time periods, the probability of even unlikely failure events to happen can become substantial. In particular, the assumption that all peers leave the network gracefully, executing a predefined LEAVE protocol, seems unrealistic. Rather, many peers are likely to leave unexpectedly (e.g., crash). The situation of course becomes worse if the peer-to-peer system is under attack. Intuitively, the larger and hence more popular the peer-to-peer system, the more attractive it also becomes for attackers. For example, Denial-of-Service attacks or partitions of the underlying physical network may push the overlay network into an undesired state. In addition, other kinds of unexpected and uncooperative behaviors may emerge in large-scale networks with open membership, such as selfish peers aiming to obtain an unfair share of the resources.

It is hence difficult in practice to rely on certain invariants and assumptions on what can and what cannot happen during the (possibly very long) lifetime of a peer-to-peer system. Accordingly, it is important that a distributed overlay network be able to recover from unexpected or even *arbitrary* situations. This recovery should also be quick: once in an illegal state, the overlay network may be more vulnerable to further changes or attacks. This motivates the study of self-stabilizing peer-to-peer systems.

Self-stabilzation is a very powerful concept in fault-tolerance. A self-stabilizing algorithm guarantees to "eventually" converge to a desirable system state *from any initial configuration*. Indeed, a self-stabilizing system allows to survive arbitrary failures, beyond Byzantine failures, including for instance a total wipe out of volatile memory at all nodes. Once the external or even adversarial changes stop, the system will simply "self-heal" and converge to a correct state. The idea of self-stabilization in distributed computing first appeared in a classical paper by E.W. Dijkstra in 1974 [13], which considered the problem of designing a self-stabilizing token ring. As Dijkstra's paper, self-stabilization has been studied in many contexts, including communication protocols, graph theory problems, termination detection, clock synchronization, and fault containment [14].

In general, the design of self-stabilizing algorithms is fairly well-understood today. In particular, already in the late 1980s, very powerful results have been obtained on how any synchronous, not fault-tolerant local network algorithm can be transformed into a very robust, self-stabilizing algorithm which performs well both in synchronous and asynchronous environments [15–17]. These transformations rely on synchronizers and on the (continuous) emulation of one-shot local network algorithms.

Although these transformations are attractive to strengthen the robustness of local algorithms *on a given network topology*, for example, for designing self-stabilizing spanning trees, they are not applicable, or only applicable at high costs, in overlay peer-to-peer networks whose topology is subject to change and optimization itself.

Indeed, many decentralized overlay networks (including very well-known examples like Chord) are not self-stabilizing, in the sense that the proposed protocols only manage to recover the network from a restricted class of illegal states [18–20]. Informally, the self-stabilizing overlay network design problem is the following:

1. An adversary can manipulate the peers' neighborhood (and hence topology) information arbitrarily and continuously. In particular, it can remove and add arbitrary nodes and links.
2. As soon as the adversary stops manipulating the overlay topology, say at some unknown time $t_0$, the self-stabilization protocols will ensure that eventually, and in the absence of further adversarial changes, a desired topology is reached.

A topological self-stabilizing mechanism must guarantee *convergence* and *closure* properties: By local neighborhood changes (i.e., by creating, forwarding and deleting links with neighboring nodes), the nodes will eventually form an overlay topology with desirable properties (e.g., polylogarithmic degree and diameter) from any initial topology. The system will also stay in a desirable configuration provided that no further external topological changes occur.

In order for a distributed self-stabilizing algorithm to recover any connected topology, the initial topology must at least be *weakly connected*.

**Condition 24.8**

- *At $t_0$, the peers are weakly connected to each other.*

Designing a topological self-stabilizing algorithm is nontrivial for several reasons. First, as the time $t_0$ is not known to the self-stabilizing algorithm, the self-stabilizing procedure must run continuously, respectively, *local detectability* is required: *At least one peer should notice (i.e., locally detect) an inconsistency*, if it exists. This peer can then trigger (local or global) convergence. Moreover, a key insight is that a topologically self-stabilizing algorithm can never remove links: it may happen that this link is the only link connecting two otherwise disconnected components. Clearly, once disconnected, connectivity can never be established again. Now one may wonder how a self-stabilizing algorithm starting (at $t_0$) from a supergraph of the target topology will ever be able to reach the desired topology without edge deletion. The answer is simple: while an edge cannot be removed, it can be *moved* resp. *delegated* (e.g., one or both of its endpoints can be forwarded to neighboring peers) and *merged* (with parallel edges).

Another fundamental implication of the self-stabilization concept is that self-stabilizing algorithms cannot cycle through multiple stages during their execution. For instance, it may be tempting to try to design an algorithm which, given the initial weakly connected topology, first converts the topology into a clique graph, a "full-mesh"; once in the clique state, any desirable target topology could be achieved efficiently, simply be removing (i.e., merging) unnecessary links. The problem with this strategy is twofold:

1. Self-stabilizing algorithms can never assume a given stage of the stabilization has been reached, and based on this assumption, move into a different mode of stabilization. The problem is that this assumption can be violated anytime by the adversary, ruining the invariant and hence correctness of the stabilization algorithm.
2. The *transient* properties of the dynamic topology, that is, the topological properties during convergence, may be undesirable: Even though the initial and the final peer degrees may be low, the intermediate clique topology has a linear degree and hence does not scale.

Given these intuitions, we will next identify fundamental connectivity primitives and topological operations which are sufficient and necessary to transform any initial network into any other network. Subsequently, we will discuss how these primitives can be exploited systematically for the design of distributed algorithms. Finally, we present two case studies for topological self-stabilization: self-stabilizing linearization and the self-stabilizing construction of skip graphs.

## 24.5.2 Universal Primitives for Reliable Connectivity

The current section identifies *universal connectivity primitives*: local graph operations which allow us to transform any topology into any other topology. Although we in this section focus on feasibility, we will later use these primitives to design topologically self-stabilizing algorithms.

Let us first define the notion of links $(u, v)$. Links can either be explicit or implicit. An explicit link $(u, v)$ (in the following depicted as solid line) means that $u$ knows $v$, that is, $u$ stores a reference of $v$ (e.g., $v$'s IP address). An implicit link $(u, v)$ (depicted as dashed line) means that a message including $v$'s reference is currently in transit to $u$ (from some arbitrary sender). We are often interested in the union of the two kinds of links.

As discussed previously, a first most fundamental principle in the design of distributed self-stabilizing algorithms is that links can never be deleted:

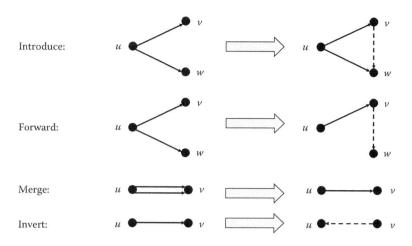

**FIGURE 24.3** Basic connectivity primitives.

**Rule 24.1** *During the execution of a topologically self-stabilizing algorithm, weak connectivity must always be preserved. In particular, a pointer (i.e., information about a peer) can never be deleted.*

We next identify four basic primitives which preserve connectivity (cf Figure 24.3): INTRODUCE, FORWARD, MERGE, INVERT.

1. INTRODUCE: Assume node $u$ has a pointer to nodes $v$ and $w$: there are two directed links $(u, v)$ and $(u, w)$. Then, $u$ can introduce $w$ to $v$ by sending the pointer to $w$ to $v$.
2. FORWARD: Assume node $u$ has a pointer to nodes $v$ and $w$, that is, $(u, v)$ and $(u, w)$. Then, $u$ can forward the reference to $w$ to $v$.
3. MERGE: If $u$ has two pointers to $v$, that is, $(u, v)$ and $(u, v)$, then $u$ can merge the two.
4. INVERT: If $u$ is connected to $v$, it can invert the link $(u, v)$ to $(v, u)$ by forwarding a pointer to itself to $v$, and delete the reference to $v$.

It is easy to see that these primitives indeed preserve weak connectivity. In fact, the INTRODUCE, FORWARD, and MERGE operations even preserve strong connectivity. We also note that we need a compare operation to implement the merge operation: Namely, we need to be able to test whether two references point to the same node.

These operations turn out to be very powerful. In the following, we first show that three of them are sufficient to transform any weakly connected graph into any strongly connected graph. In other words, they are *weakly universal*. Subsequently, we show that all four of them together are even *universal*: they are sufficient to transform any weakly connected graph into any weakly connected graph. Finally, we prove that these primitives are also necessary.

**Theorem 24.9** *The three primitives* INTRODUCE, FORWARD, *and* MERGE *are weakly universal: They are sufficient to turn any weakly connected graph $G = (V, E)$ into any strongly connected graph $G' = (V, E')$.*

Let us provide some intuition why this theorem is true. Note that we only need to prove feasibility, that is, that a transformation exists; how to devise a distributed algorithm that finds such a transformation is only discussed later in this chapter.

The proof proceeds in two stages, from $G = (V, E)$ to the clique, and from the clique to $G' = (V, E')$. From the definition of weak connectivity, it follows that for any two nodes $v$ and $w$, there is a path from $v$

to $w$ (ignoring link directions). It is easy to see that if in each communication round, each node introduces its neighbors to each other as well as itself to its neighbors, we reach a complete network (the clique) after $O(\log n)$ communication rounds.

So now assume $G = (V, E)$ is a clique. Then using FORWARD and MERGE operations, we can transform $G$ into $G'$ using the following steps (without removing edges in $G'$):

1. Let $(u, w)$ be an arbitrary edge which needs to be removed, that is, $(u, w) \notin E'$. Since $G' = (V, E')$ is strongly connected, there is a shortest directed path from $u$ to $w$ in $G'$. Let $v$ be the next node along this path.
2. Node $u$ forwards ("delegates") $(u, w)$ to $v$, that is, $(u, w)$ becomes $(v, w)$. This reduces the distance between an unused node pair in $G'$ by 1.
3. As the maximal distance is $n - 1$, the distance of a superfluous edge can be reduced at most $n - 1$ many times before it merges with an edge in $G'$. Thus, we eventually obtain $G'$.

**Theorem 24.10** *The four primitives* INTRODUCE, FORWARD, MERGE, *and* INVERT *are universal: they are sufficient to turn any weakly connected graph* $G = (V, E)$ *into any weakly connected graph* $G' = (V, E')$.

We again provide some intuition for this theorem:

1. Let $G'' = (V, E'')$ be the graph in which for each edge $(u, v) \in E'$, both edges $(u, v)$ and $(v, u)$ are in $E''$. Note that $G''$ is strongly connected.
2. According to Theorem 24.9, it is possible to transform any $G$ to $G''$.
3. To transform $G''$ to $G'$, we need the INVERT primitive, to remove undesired edges: We invert any undesired edge $(u, v)$ to $(v, u)$ and then merge it with $(v, u)$.

Significantly, the primitives are not only sufficient but also necessary.

**Theorem 24.11** *The four primitives* INTRODUCE, FORWARD, MERGE, *and* INVERT *are also necessary.*

The reason is that INTRODUCE is the only primitive which generates an edge, FORWARD is the only primitive which separates a node pair, MERGE is the only primitive which removes an edge, and INVERSION is the only primitive rendering a node unreachable.

## 24.5.3 Distributed Algorithms for Self-Stabilization

In Section 24.5.2, we have presented universal primitives that allow to transform any weakly-connected graph $G$ into any weakly-connected graph $G'$. However, the mere *existence* or *feasibility* of such transformations is often not interesting in practice, if there do not exist efficient distributed algorithms to find a transformation.

In the following, we will show how to devise distributed algorithms exploiting our primitives to render systems truly self-stabilizing. We first need to introduce some terminology. We first differentiate between the following notions of *state*.

1. *State of a process:* The state of a process includes all variable information stored at the process, excluding the messages in transit.
2. *State of the network:* The network states includes all messages currently in transit.
3. *State of the system:* The system state is the state of all proceses plus the network state.

Reformulating our objective accordingly, we aim to transition from any initial network state $S$ to a legal network state $S'(S)$. For the algorithm design, we usually assume node actions to be *locally atomic*: At each moment in time, a process executes a single action. Multiple processes, however, may execute multiple actions simultaneously: actions are not globally atomic. Moreover, it is usually assumed that the scheduler is fair: every enabled node action will eventually be executed.

Concretely, we consider the following action types:

1. *Name(Object-List)* → *Commands*:
   a. A local call of action *A* is executed immediately.
   b. *Incoming Request*: The correspnding action will eventually be scheduled (the request never expires).
2. *Name: Predicate* → *Commands*: The rule will only be executed in finite time if the the predicate is always enabled, for example, the rule does not time out.

Independently of the initial states as well as possible messages in transit, a self-stabilizing system should fulfill the following crteria:

**Definition 24.3** *A system is self-stabilizing with respect to a given network problem P, if the following requirements are fulfilled when no failure or external error occurs and if nodes are static:*
Convergence: *For all initial states S and all fair executions, the system eventually reaches a state S' with S' ∈ L(S): a legal state.*
Closure: *For all legal initial states S, also each subsequent state is legal.*

A central requirement in topologically self-stabilizing system is the *monotonicity of reachability*: if $v$ is reachable from $u$ at time $t$, using explicit or implicit edges, then, if no further failures or errors occur and given a static node set, $v$ is also reachable from $u$ at any time $t' > t$.

The following theorem can easily be proved using induction, as long as there are no references to non-existent nodes in the system.

**Theorem 24.12** *The* INTRODUCE, FORWARD, *and* MERGE *operations fulfill monotonic reachability.*

Remarks:

1. One particularly annoying challenge in the design of self-stabilizing algorithms is due to the fact that there may still be corrupt messages in transit in the system. Such messages can threaten the correctness of an algorithm later.
2. In particular, corrupted message may violate the closure property: Although initially in a legal state, the system may move to an illegal state.
3. The set of legal states is hence only a subset of the "correct states."

In general, the following performance metrics are most relevant in topological self-stabilization:

1. *Convergence Time:* Assuming a synchronous environment (or assuming an upper bound on the message transmission per link), the distributed convergence time measures how many (parallel) communication rounds are required until the final topology is reached.
2. *Work:* The work measures how many edges are inserted, changed, or removed in total, during the covergence process.
3. *Locality:* Although a self-stabilizing algorithm by definition will reestablish a desired property from *any* initial configuration, it is desirable that the parallel convergence time as well as the overall work is proportional to "how far" the initial topology is from the desired one. In particular, if there are only one or two links missing, it should be possible to handle these situations more efficiently than performing a complete stabilization. Similarly, single peer JOIN and LEAVE operations should be efficient (in terms of time and work).
4. *Transient Behavior:* Although the initial and the final network topologies are given, it is desirable that during convergence, only efficient topologies transiently emerge. For example, it may be desirable that no topology during convergence will have a higher degree or diameter than the initial or the final topology.

Usually, we do not assume synchronous executions or that nodes process requests at the same speed. Rather, requests may be processed asynchronously. To measure time in asynchronous executions, we use the notion of a round: In a round, each node which has to process one or more requests, completed at least one of these requests. The time complexity is measured in terms of number of rounds (usually in the worst-case).

### 24.5.3.1 Case Study Linearization

Linearization is a most simple example for topological self-stabilization [21–23]: Essentially, we are looking for a local-control strategy for converting an arbitrary connected graph (with unique node IDs) into a sorted list. To provide some intuition and for the sake of simplicity, let us assume an undirected network topology, in which peers have unique identifiers. To sort and linearize this overlay network in a distributed manner, two basic rules are sufficient, defined over node triples (cf Figure 24.4):

1. *Linearize right:* Any node $u$ which currently has two neighbors $v, w$ with larger IDs than its own ID, that is, $u < v < w$, introduces these two nodes to each other, essentially forwarding the edge. That is, the edge $\{u, w\}$ is forwarded from $u$ to $v$ and becomes edge $\{v, w\}$.
2. *Linearize left:* Any node $w$ which currently has two neighbors $u, v$ with lower IDs than its own ID, that is, $u < v < w$, introduces these two nodes to each other, essentially forwarding the edge. That is, the edge $\{u, w\}$ is forwarded from $w$ to $v$ and becomes edge $\{u, v\}$.

Note that the algorithm indeed does not remove any edges, but only forwards and merges them. It is fairly easy to see that connectivity is preserved: If there is a path between $x$ and $y$ at time $t$, then there also exists a path between the two nodes after the linearization step. To prove that the algorithm will eventually converge, we can use a potential function argument. First, however, we note that if the network is in a configuration in which it does not constitute the linear chain graph yet, then there must exist a node having at least two left neighbors or at least two right neighbors. Accordingly, the linearize left or linearize right rule is enabled and will continue to change the topology in this step. Now, to show eventual convergence to the unique legal configuration (the linear graph), we will prove that after any execution of the linearize left or linearize right rule, the topology will come closer to the linearized configuration, in the following sense: We can define a potential function whose value is (strictly) monotonically decreased with each executed rule. Consider the potential function that sums up the lengths (differences) of all existing links with respect to the linear ordering of the nodes. As the initial configuration is connected, this sum is at least $n - 1$ ($n - 1$ links of length 1). Whenever an action is executed (in our case, a linearization step is performed), the potential is reduced by at least the length of the shorter edge in the linearization triple. Thus, we will eventually reach a topology of minimal potential, in which all actions are disabled.

### 24.5.3.2 Case Study Skip Graphs

The first self-stabilizing and scalable overlay network is SKIP+ [24], a self-stabilizing variant of the skip graph family [25,26]. Similarly to the original skip graphs, SKIP+ features a polylogarithmic degree and diameter. However, in contrast to the original skip graph versions, SKIP+ contains additional edges which enable *local detectability*: only with these edges it can be ensured that at least one peer will always notice, locally, if the overall network is not in the desired state yet.

SKIP+ distinguishes between stable edges and temporary edges. Similar to the linearization example earlier, temporary edges will travel through the topology (i.e., they are forwarded), and eventually merge

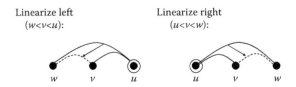

**FIGURE 24.4** Linearize operations.

or stabilize. Node $v$ considers an edge $(v, w)$ to be temporary if from $v$'s point of view $(v, w)$ does not belong to SKIP+ and so $v$ will try to forward it to some of its neighbors for which the edge would be more relevant. Otherwise, $v$ considers $(v, w)$ to be a stable edge and will make sure that the connection is bidirected, that is, it will propose $(w, v)$ to $w$.

As many self-stabilizing algorithms, the self-stabilizing protocol for SKIP+ is very simple: the peers in SKIP+ continuously must execute three rules:

1. *Rule 1: Create Reverse Edges and Introduce Stable Edges.* This rule makes sure that a directed edge becomes a bidirected edge, introducing the peers to each other. Moreover, stable edges are created where needed.
2. *Rule 2: Forward Temporary Edges.* This rule is used for forwarding temporary edges to neighboring nodes. Eventually, the edges will stabilize or merge.
3. *Rule 3: Introduce All and Linearize.* The rule has two parts. It performs some kind of local transitive closure, in which peers introduce all their neighbors to each other. Moreover, the rule is responsible for sorting neighboring nodes according to their identifiers. (In a skip graph, nodes are ordered on each level, facilitating search operations.)

The three rules are continuously checked and executed in parallel by all nodes. However, while the algorithm itself is simple, its analysis is nontrivial. In a nutshell, the stabilization proof is based on the observation that the execution of the algorithm can be divided into phases in which certain properties (milestones) are achieved. In particular, the execution can be thought of being divided into a bottom-up and a top-down phase. The bottom-up phase (i.e., from skip graph level 0 upwards), connected components for increasingly larger prefixes are formed in the identifier space. This will be accomplished by Rules 1 (where new nodes in the range of a node are discovered and where ranges may be refined) and Rules 3 (where an efficient variation of a local transitive closure is performed). Once the connected components are formed, in the second phase of the algorithm (recall that the division into phases is a purely analytical one) will form a sorted list out of each prefix component. This is accomplished in a top-down fashion by merging the two already sorted subcomponents into a sorted larger component until all nodes in the bottom level form a sorted list.

In summary, Jacob et al. [24] show the following result.

**Theorem 24.13** *SKIP+ converges, for any initial state in which the nodes are weakly connected, in $O(\log^2 n)$ rounds. A single join event (i.e., a new node connects to an arbitrary node in the system) or leave event (i.e., a node just leaves without prior notice) can be handled with polylogarithmic work.*

One drawback of this approach, however, is its transient behavior: It may happen that node degrees can increase significantly during covergence (namely, due to the Introduce All in Rule 3), even though the initial and the final topology have low degrees, that is, are scalable.

We conclude this section by emphasizing that the field of topological self-stabilization is relatively young and many algorithmic techniques and limitations still wait to be discovered.

## 24.6 Other Related Work

There is a wealth of literature on peer-to-peer systems, and papers on this subject can be found in every major computer science conference. Peer-to-peer overlay networks can roughly be classified into three categories: social networks, random networks, and structured networks.

Examples of social networks are Gnutella and KaZaA. Their basic idea of interconnecting peers is that connections follow the principle of highest benefit: A peer preferably connects to peers with similar interests by maintaining direct links to those peers that can successfully answer queries.

An example for random networks is JXTA, a Java library developed by SUN to facilitate the development of peer-to-peer systems. The basic idea behind the JXTA core is to maintain a random-looking network between the peers. In this way, peers are very likely to stay in a single connected component because random graphs are known to be robust against even massive failures or departures of nodes. Theory work on random peer-to-peer networks can be found, for example, in References 27, 28.

Most of the scientific work on peer-to-peer networks has focused on structured overlay networks, that is, networks with a regular structure that makes it easy to route in them with low overhead. The vast majority of these networks is based on the concept of virtual space. The most prominent among these are Chord [4], CAN [1], Pastry [7], and Tapestry [8]. The virtual space approach has the problem that it requires node labels to be evenly distributed in the space to obtain a scalable overlay network. Alternative approaches that yield scalable overlay networks for arbitrary distinct node labels are skip graphs [25], skip nets [26], and the hyperring [29].

As we have already seen previously, structured overlay networks are also known that can take locality into account. All of this work is based on the results in References 6, 5. The first peer-to-peer systems were Tapestry [8] and Pastry [7]. Follow-up schemes addressed some of the shortcomings of the PRR scheme. In particular, the LAND scheme [10] improves the results in PRR as seen in Section 24.4.2 algorithms for efficiently handling peer arrival and departures are presented in References 12, 30; a simplified scheme with provable bounds for ring networks is given in Reference 31; a fault-tolerant extension is given in Reference 32; a scheme that addresses general networks, at the expense of an $O(\log n)$ stretch bound, is given in Reference 33; a scheme that considers object location under more realistic networks is given in Reference 34; and a first attempt at designing overlay networks in peer-to-peer systems consisting of mobile peers is presented in Reference 35 (no formal bounds are proven in Reference 35 for any relevant network distribution though).

Finally, overlay networks have not only been designed for wired networks but also for wireless networks. See, for example, Reference 36 and the references therein for more results in this area. We also refer the reader to the various comprehensive surveys and books on peer-to-peer computing [37,38].

# References

1. S. Ratnasamy, P. Francis, M. Handley, R. Karp, and S. Shenker. A scalable content-addressable network. In *Proceedings of the ACM SIGCOMM '01*, 2001.
2. X. Wang, Y. Zhang, X. Li, and D. Loguinov. On zone-balancing of peer-to-peer networks: Analysis of random node join. In *Proceedings of ACM SIGMETRICS*, 2004.
3. M. Naor and U. Wieder. Novel architectures for P2P applications: The continuous-discrete approach. In *Proceedings of the 15th ACM Symposium on Parallel Algorithms and Architectures (SPAA)*, 2003.
4. I. Stoica, R. Morris, D. Karger, M.F. Kaashoek, and H. Balakrishnan. Chord: A scalable peer-to-peer lookup service for Internet applications. In *Proceedings of the ACM SIGCOMM '01*, 2001.
5. C. Plaxton, R. Rajaraman, and A. Richa. Accessing nearby copies of replicated objects in a distributed environment. *Theory of Computing Systems*, 32:241–280, 1999. A preliminary version of this paper appeared in *Proceedings of the 9th Annual ACM Symposium on Parallel Algorithms and Architectures (SPAA)*, pp. 311–320, June 1997.
6. C. G. Plaxton, R. Rajaraman, and A. W. Richa. Accessing nearby copies of replicated objects in a distributed environment. In *Proceedings of ACM SPAA*, Springer, New York, pp. 311–320, 1997.
7. P. Druschel and A. Rowstron. Pastry: Scalable, distributed object location and routing for large-scale peer-to-peer systems. In *Proceedings of the 18th IFIP/ACM International Conference on Distributed Systems Platforms (Middleware 2001)*, 2001.
8. B.Y. Zhao, J. Kubiatowicz, and A. Joseph. Tapestry: An infrastructure for fault-tolerant wide-area location and routing. Technical report, University of California at Berkeley, Computer Science Department, 2001.
9. B. Awerbuch and C. Scheideler. *Chord++: Low-congestion routing in chord.* Unpublished manuscript, Johns Hopkins University, June 2003. Retrieved from http://www.in.tum.de/personen/scheideler/index.html.en.

10. I. Abraham, D. Malkhi, and O. Dobzinski. LAND: Stretch $(1+\epsilon)$ locality aware networks for DHTs. In *Proceedings of the 9th ACM-SIAM Symposium on Discrete Algorithms*, 2004.

11. D. R. Karger and M. Ruhl. Finding nearest neighbors in growth-restricted metrics. In *Proceedings of ACM STOC*, ACM, New York, pp. 741–750, 2002.

12. K. Hildrum, J. Kubiatowicz, S. Rao, and B. Y. Zhao. Distributed object location in a dynamic network. In *Proceedings of ACM SPAA*, ACM, Winnipeg, Canada, pp. 41–52, 2002.

13. E. W. Dijkstra. Self-stabilizing systems in spite of distributed control. *Communications of the ACM*, 17(11):643–644, 1974.

14. S. Dolev. *Self-stabilization*. MIT Press, Cambridge, MA, 2000.

15. B. Awerbuch and M. Sipser. Dynamic networks are as fast as static networks. In *Proceedings of the 29th Annual Symposium on Foundations of Computer Science (SFCS)*, IEEE, pp. 206–219, 1988.

16. B. Awerbuch and G. Varghese. Distributed program checking: A paradigm for building self-stabilizing distributed protocols. In *Proceedings of the Annual Symposium on Foundations of Computer Science (FOCS)*, IEEE, pp. 258–267, 1991.

17. C. Lenzen, J. Suomela, and R. Wattenhofer. Local algorithms: Self-stabilization on speed. In *Proceedings of the 11th International Symposium on Stabilization, Safety, and Security of Distributed Systems (SSS)*, November 2009.

18. D. Angluin, J. Aspnes, J. Chen, Y. Wu, and Y. Yin. Fast construction of overlay networks. In *Proceedings of the 17th Annual ACM Symposium on Parallelism in Algorithms and Architectures (SPAA)*, 2005.

19. J. Aspnes and Y. Wu. O (logn)-time overlay network construction from graphs with out-degree 1. In *Proceedings of the International Conference on Principles of Distributed Systems*, Springer, Berlin, Germany, pp. 286–300, 2007.

20. I. Stoica, R. Morris, D. Karger, M. F. Kaashoek, and H. Balakrishnan. Chord: A scalable peer-to-peer lookup service for internet applications. In *Proceedings of the Technical Report MIT-LCS-TR-819*, 2001.

21. D. Gall, R. Jacob, A. Richa, C. Scheideler, S. Schmid, and H. Taeubig. A note on the parallel runtime of self-stabilizing graph linearization. *Journal Theory of Computing Systems (TOCS)*, 55(1):110–135, 2014.

22. R. Jacob, S. Ritscher, C. Scheideler, and S. Schmid. Towards higher-dimensional topological self-stabilization: A distributed algorithm for delaunay graphs. *Elsevier Theoretical Computer Science (TCS)*, 457:137–148, 2012.

23. C. Rickmann, C. Wagner, U. Nestmann, and S. Schmid. Topological self-stabilization with name-passing process calculi. In *Proceedings of the 27th International Conference on Concurrency Theory (CONCUR)*, 2016.

24. R. Jacob, A. Richa, C. Scheideler, S. Schmid, and H. Taeubig. Skip+: A self-stabilizing skip graph. *Journal of the ACM (JACM)*, 61(6):36, 2014.

25. J. Aspnes and G. Shah. Skip graphs. In *Proceedings of the 14th ACM Symposium on Discrete Algorithms (SODA)*, Society for Industrial and Applied Mathematics, Philadelphia, PA, pp. 384–393, 2003.

26. N. J. Harvey, M. B. Jones, S. Saroiu, M. Theimer, and A. Wolman. Skipnet: A scalable overlay network with practical locality properties. In *Proceedings of the 4th USENIX Symposium on Internet Technologies and Systems (USITS '03)*, 2003.

27. C. Cooper, M. Dyer, and C. Greenhill. Sampling regular graphs and a peer-to-peer network. In *Proceedings of the 16th ACM Symposium on Discrete Algorithms (SODA)*, 2005.

28. P. Mahlmann and C. Schindelhauer. Peer-to-peer networks based on random transformations of connected regular undirected graphs. In *Proceedings of the 17th ACM Symposium on Parallel Algorithms and Architectures (SPAA)*, pp. 155–164, 2005.

29. B. Awerbuch and C. Scheideler. The Hyperring: A low-congestion deterministic data structure for distributed environments. In *Proceedings of the 15th ACM Symposium on Discrete Algorithms (SODA)*, 2004.

30. K. Hildrum, J. Kubiatowicz, S. Ma, and S. Rao. A note on the nearest neighbor in growth-restricted metrics. In *Proceedings of ACM Symposium on Discrete Algorithms (SODA)*, Society for Industrial and Applied Mathematics, Philadelphia, PA, pp. 560–561, 2004.

31. X. Li and C. G. Plaxton. On name resolution in peer-to-peer networks. In *Proceedings of the 2nd ACM Worskhop on Principles of Mobile Commerce (POMC)*, ACM, New York, pp. 82–89, October 2002.

32. K. Hildrum and J. Kubiatowicz. Asymptotically efficient approaches to fault-tolerance in peer-to-peer networks. In *Proceedings of DISC*, Springer, Berlin, Germany, pp. 321–336, 2003.

33. R. Rajaraman, A. W. Richa, B. Vöcking, and G. Vuppuluri. A data tracking scheme for general networks. In *Proceedings of the 12th ACM Symposium on Parallel Algorithms and Architectures*, ACM, New York, pp. 247–254, July 2001.

34. K. Hildrum, R. Krauthgamer, and J. Kubiatowicz. Object location in realistic networks. In *Proceedings of ACM SPAA*, ACM, New York, pp. 25–35, 2004.

35. I. Abraham, D. Dolev, and D. Malkhi. LLS: A locality aware location service for mobile ad hoc networks. In *Proceedings of DIALM-POMC*, 2004.

36. C. Scheideler. Overlay networks for wireless ad hoc networks. In *Proceedings of the IMA Workshop on Wireless Communication, to appear*, 2005. Retrieved from http://www.in.tum.de/personen/scheideler/index.html.en.

37. R. Steinmetz and K. Wehrle. Peer-to-peer-networking &-computing. *Informatik-Spektrum*, 27(1):51–54, 2004.

38. X. S. Shen, H. Yu, J. Buford, and M. Akon. *Handbook of peer-to-peer networking*, volume 34. Springer Science & Business Media, Berlin, Germany, 2010.

# 25

# Data Broadcasts on Multiple Wireless Channels: Exact and Time-Optimal Solutions for Uniform Data and Heuristics for Nonuniform Data[*][†]

Alan A. Bertossi

Cristina M. Pinotti

Romeo Rizzi

## 25.1 Introduction

In wireless asymmetric communication, broadcasting is an efficient way of simultaneously disseminating data to a large number of clients. Consider data services on cellular networks, such as stock quotes, weather infos, traffic news, where data are continuously broadcast to clients that may desire them at any instant of time. In this scenario, a server at the base-station repeatedly transmits data items from a given set over a wireless channel, whereas clients passively listen to the shared channel waiting for their desired item. The server follows a broadcast schedule for deciding which item of the set has to be transmitted at

[*] Portions of this chapter reprinted, with permission, from [1] *IEEECS Log Number TC-0238-0704*, © 2005, IEEE.

[†] This work has been partially supported by the Italian project "RISE", Fondazione Cassa Risparmio Perugia, code 2016.0104.021.

any time instant. An efficient broadcast schedule minimizes the client expected delay, that is, the average amount of time spent by a client before receiving the item he needs. The client expected delay increases with the size of the set of the data items to be transmitted by the server. Indeed, the client has to wait for many unwanted data before receiving his own data. The efficiency can be improved by augmenting the server bandwidth, for example, allowing the server to transmit over multiple disjoint physical channels and therefore defining a shorter schedule for each single channel. In a multichannel environment, in addition to a broadcast schedule for each single channel, an allocation strategy has to be pursued so as to assign data items to channels. Moreover, each client can access either only a single channel or any available channel at a time. In the former case, if the client can access only one prefixed channel and can potentially retrieve any available data, then all data items must be replicated over all channels. Otherwise, data can be partitioned among the channels, thus assigning each item to only one channel. In this latter case, the efficiency can be improved by adding an index that informs the client at which time and on which channel the desired item will be transmitted. In this way, the mobile client can save battery energy and reduce the tuning time because, after reading the index info, it can sleep and wake up on the proper channel just before the transmission of the desired item.

Several variants for the problem of data allocation and broadcast scheduling have been proposed in the literature, which depend on the perspectives faced by the research communities [2–12].

Specifically, the networking community faces a version of the problem, known as the *broadcast problem*, whose goal is to find an infinite schedule on a single channel [4,6,7,10]. Such a problem was first introduced in the teletext systems by [2,3]. Although it is widely studied (e.g., it can be modeled as a special case of the maintenance scheduling problem and the multi-item replenishment problem [4,6]), its tractability is still under consideration. Therefore, the emphasis is on finding near optimal schedules for a single channel. Almost all the proposed solutions follow the *square root rule* [3]. The aim of such a rule is to produce a broadcast schedule in which each data item appears with equally spaced replicas, whose frequency is proportional to the square root of its popularity and inversely proportional to the square root of its length. The multichannel schedule is obtained by distributing in a round robin fashion the schedule for a single channel [10]. As each item appears in multiple replicas which, in practice, are not equally spaced, these solutions make indexing techniques not effective. Briefly, the main results known in the literature for the broadcast problem can be summarized as follows. For *uniform* lengths, namely, all items of the same length, it is still unknown whether the problem can be solved in polynomial time or not. For a constant number of channels, the best algorithm proposed so far is the Polynomial Time Approximation Scheme (PTAS) devised in Reference 7. In contrast, for *nonuniform* lengths, the problem has been shown to be strong *NP*-hard even for a single channel, a 3-approximation algorithm was devised for one channel, and a heuristic has been proposed for multiple channels [6].

On the other hand, the database community seeks for a periodic broadcast scheduling that should be easily indexed [5]. For the single channel, the obvious schedule that admits index is the *flat* one. It consists in selecting an order among the data items, and then transmitting them once at a time, in a round-robin fashion [13], producing an infinite periodic schedule. In a flat schedule, indexing is trivial, as each item will appear once, and exactly at the same relative time, within each period. Although indexing allows the client to sleep and save battery energy, the client expected delay is half of the schedule period and can become infeasible for a large period. To decrease the client expected delay, still preserving indexing, flat schedules on multiple channels can be adopted [8,9,12]. However, in such a case, the allocation of data to channels becomes critical. For example, allocating items in a balanced way simply scales the expected delay by a factor equal to the number of channels. To overcome this drawback, *skewed* allocations have been proposed in which items are partitioned according to their popularities so that the most requested items appear in a channel with shorter period [8,12]. Hence, the resulting problem is slightly different from the broadcast problem as, to minimize the client expected delay, it assumes skewed allocation and flat scheduling. This variant of the problem is easier than the broadcast problem. Indeed, as proved in Reference 12, the optimal solution for uniform lengths can be found in polynomial time. In contrast, the problem becomes computationally intractable for nonuniform lengths [1]. For this latter case, several heuristics have been developed in References 12, 14, which have been tested on some benchmarks whose popularities follow Zipf distributions. Such distributions are used to characterize the popularity of one element among a set of similar data, similar to web page in a web site [15].

The present chapter reviews the work of References 1, 12, 14, 16 on the broadcasting problem of $N$ data items over $K$ wireless channels, under the assumptions of skewed data allocation to channels and flat data scheduling per channel. Both the uniform and nonuniform length cases are surveyed showing their exact and heuristic solutions, respectively.

For the case of data items with uniform lengths, four exact polynomial time algorithms are presented, all based on dynamic programming. The first algorithm, called *DP* and originally proposed in Reference 12, takes $O(N^2K)$ time, whereas the second algorithm, called *Dichotomic* and proposed later in Reference 1, is faster as it runs in $O(NK \log N)$ time. The third algorithm, presented in Reference 14, is designed for the specific case of $K = 2$. Although it requires $O(N \log N)$ time, and hence it is asymptotically not faster than Dichotomic, it exploits a specific characterization of the optimal solution that we also use in developping the heuristics for the nonuniform case. Finally, the fourth algorithm *Smawk-AED* has been very recently presented in Reference 16, and it takes $O(NK)$ time. Note that this algorithm matches the trivial lower bound for the time complexity of any dynamic programming algorithm for the $K$-uniform allocation problem that solves all the $NK$ subproblems of allocating a prefix of the items $\{1, \ldots, n\}$ to $k$ channels, $1 \le n \le N$, $1 \le k \le K$.

For the case of data items with nonuniform lengths, the problem is *NP*-hard when $K = 2$, and *strong NP*-hard for arbitrary $K$. In this latter case, the *Optimal* algorithm presented in Reference 1 is reviewed. It requires $O(KN^{2z})$ time, where $z$ is the maximum data length, and reduces to the DP algorithm when $z = 1$. As algorithm Optimal can solve only small instances in a reasonable time, three heuristics are described, all having an $O(N(K + \log N))$ time complexity.

The first heuristic, called *Greedy*, has been proposed in Reference 12. Fixed $N$, Greedy starts with all data items assigned to one channel, and then proceeds by splitting the items of one channel between two channels, thus adding a new channel, until $K$ channels are reached. The other two heuristics, both presented in Reference 14, pretend that the characterization of the optimal solution of the problem for $K = 2$ and uniform lengths holds also for the general case of arbitrary $K$ and nonuniform lengths. One heuristic is called *Greedy+* as it combines such solution characterization with the Greedy approach, whereas the second heuristic is called *Dlinear* and combines the same characterization with the dynamic programming relation proposed in Reference 12.

All the three heuristics are then tested on benchmarks whose popularities are characterized by Zipf distributions. The experimental tests reveal that Dlinear finds optimal solutions almost always, requiring reasonable running times. Although Greedy remains the fastest heuristic, it gives the worst sub optimal solutions. Both the running times and the quality of the solutions of Greedy+ are intermediate between those of Dlinear and Greedy. However, Greedy and Greedy+ have the feature to scale well with respect to the parameter changes.

The rest of this chapter is so organized. Section 25.2 gives notations, definitions, and the problem statement. Section 25.3 illustrates the DP and Dichotomic algorithms for items of uniform lengths, as well as the solution characterization for the particular case of two channels. Section 25.3.4 describes in details the new Smawk-AED algorithm. In contrast, Section 25.4 studies the nonuniform length case. It first recalls the strong *NP*-hardness for an arbitrary number of channels, and then presents the exponential time Optimal algorithm. Then, Section 25.5 gives the Greedy, Greedy+, and Dlinear heuristics. Section 25.6 reports the experimental tests on some benchmarks, whose popularities follow Zipf distributions. Finally, conclusions are offered in Section 25.7.

## 25.2 Problem Formulation

Consider a set of $K$ identical channels, and a set $D = \{d_1, d_2, \ldots, d_N\}$ of $N$ data items. Each item $d_i$ is characterized by a *probability* $p_i$ and a *length* $z_i$, with $1 \leq i \leq N$. The probability $p_i$ represents the popularity of item $d_i$, namely its probability to be requested by the clients, and it does not vary along the time. Clearly, $\sum_{i=1}^{N} p_i = 1$. The length $z_i$ is an integer number, counting how many time units are required to transmit item $d_i$ on any channel. When all data lengths are the same, that is, $z_i = z$ for $1 \leq i \leq N$, the lengths are called *uniform* and are assumed to be unit, that is, $z = 1$. When the data lengths are not the same, the lengths are said *nonuniform*.

The items have to be partitioned into $K$ groups $G_1, \ldots, G_K$. Group $G_j$ collects the data items assigned to channel $j$, with $1 \leq j \leq K$. The cardinality of $G_j$ is denoted by $N_j$, the sum of its item lengths is denoted by $Z_j$, that is, $Z_j = \sum_{d_i \in G_j} z_i$, and the sum of its probabilities is denoted by $P_j$, that is, $P_j = \sum_{d_i \in G_j} p_i$. Note that as the items in $G_j$ are cyclically broadcast according to a flat schedule, $Z_j$ is the schedule period on channel $j$. Clearly, in the uniform case $Z_j = N_j$, for $1 \leq j \leq K$. If item $d_i$ is assigned to channel $j$, and assuming that clients can start to listen at any instant of time with the same probability, the *client expected delay* for receiving item $d_i$ is half of the period, namely, $\frac{Z_j}{2}$. Assuming that indexing allows clients to know in advance the content of the channels [12], the *average expected delay* (AED) over all channels is

$$\text{AED} = \frac{1}{2} \sum_{j=1}^{K} Z_j P_j \tag{25.1}$$

Given $K$ channels, a set $D$ of $N$ items, where each data item $d_i$ comes along with its probability $p_i$ and its integer length $z_i$, the *$K$-nonuniform allocation problem* consists in partitioning $D$ into $K$ groups $G_1, \ldots, G_K$, so as to minimize the objective function AED given in Equation 25.1. In the special case of equal lengths, the above-mentioned problem is called *$K$-uniform allocation problem* and the corresponding objective function is derived replacing $Z_j$ with $N_j$ in Equation 25.1.

As an example, consider a set of $N = 6$ items with uniform lengths and $K = 3$ channels. Let the demand probabilities be $p_1 = 0.37$, $p_2 = 0.25$, $p_3 = 0.18$, $p_4 = 0.11$, $p_5 = 0.05$, and $p_6 = 0.04$. The optimal solution assigns item $d_1$ to the first channel, items $d_2$ and $d_3$ to the second channel, and the remaining items to the third channel. The corresponding AED is $\frac{1}{2}(0.37 + 2(0.25 + 0.18) + 3(0.11 + 0.05 + 0.04)) = 0.915$.

Consider the sequence $d_1, \ldots, d_N$ of items ordered by their indices, and assume that each channel contains items with consecutive indices. Then, the *single-channel cost* $C_{i,j}$ is the cost of assigning to a single channel the consecutive items $d_i$ to $d_j$, and it is defined as $C_{i,j} = \frac{1}{2} \left( \sum_{h=i}^{j} p_h \right) \left( \sum_{h=i}^{j} z_h \right)$. Letting $P_{i,j} = \sum_{h=i}^{j} p_h$ and $Z_{i,j} = \sum_{h=i}^{j} z_h$, one notes that all the $P_{1,n}$ and $Z_{1,n}$, for $1 \leq n \leq N$, can be computed in $O(N)$ time by two prefix sum computations. Hence, a single $C(i,j)$ can be computed on the fly in constant time as $C(i,j) = \frac{1}{2}(P_{1,j} - P_{1,i-1})(Z_{1,j} - Z_{1,i-1})$. From now on, to simplify the presentation, $C(i,j)$ is defined to be 0 whenever $i > j$. Note that, for uniform lengths, the formula of $C_{i,j}$ simplifies as $C(i,j) = \frac{1}{2}(j - i + 1) \sum_{h=i}^{j} p_h$.

Moreover, a *segmentation* is a partition of the ordered sequence $d_1, \ldots, d_N$ into $G_1, \ldots, G_K$, such that if $d_i \in G_k$ and $d_j \in G_k$ then $d_h \in G_k$ whenever $i \leq h \leq j$. A segmentation

$$\underbrace{d_1, \ldots, d_{B_1}}_{G_1}, \underbrace{d_{B_1+1}, \ldots, d_{B_2}}_{G_2}, \ldots, \underbrace{d_{B_{K-1}+1}, \ldots, d_N}_{G_K}$$

is compactly denoted by the $(K-1)$-tuple

$$(B_1, B_2, \ldots, B_{K-1})$$

of its *right borders*, where border $B_k$ is the index of the last item that belongs to group $G_k$. Notice that it is not necessary to specify $B_K$, the index of the last item of the last group, because its value will be $N$ for any solution. From now on, $B_{K-1}$ will be referred to as the *final border* of the solution. The cardinality of $G_k$, that is the number $N_k$ of items in the group, is $N_k = B_k - B_{k-1}$, where $B_0 = 0$ and $B_K = N$ are assumed.

# 25.3 Uniform Lengths

The current section is devoted to take a look to the dynamic programming algorithms proposed in References 1, 12, 14, 16 for the $K$-uniform allocation problem. The following results show that the optimal solutions for the $K$-uniform allocation problem can be sought within the class of segmentations.

**Lemma 25.1** [12] *Let $G_h$ and $G_j$ be two groups in an optimal solution for the K-uniform allocation problem. Let $d_i$ and $d_k$ be items with $d_i \in G_h$ and $d_k \in G_j$. If $N_h < N_j$, then $p_i \geq p_k$. Similarly, if $p_i > p_k$, then $N_h \leq N_j$.*

In other words, the most popular items are allocated to less loaded channels so that they appear more frequently. As a consequence, if the items are sorted by nonincreasing probabilities, then the group sizes are nondecreasing.

**Corollary 25.1** [12] *Let $d_1, d_2, \ldots, d_N$ be $N$ uniform length items with $p_i \geq p_k$ whenever $i < k$. Then, there exists an optimal solution for partitioning them into $K$ groups $G_1, \ldots, G_K$, where each group is made of consecutive elements.*

Thus, assuming the items sorted by nonincreasing probabilities, any sought solution $S$ will be a segmentation. Moreover, the sought segmentations $S = (B_1, B_2, \ldots, B_{K-1})$ for the uniform case can be restricted to those verifying $N_1 \leq N_2 \leq \ldots \leq N_K$, which will be called *feasible* segmentations.

## 25.3.1 The DP Algorithm

To describe the DP algorithm [12], let $OPT(k, n)$ denote an optimal solution for grouping items $d_1, \ldots d_n$ into $k$ groups, and let $opt(k, n)$ be its corresponding cost, for any $n \leq N$ and $k \leq K$. As $d_1, d_2, \ldots, d_N$ are sorted by nonincreasing probabilities, one has

$$opt(k, n) = \begin{cases} C(1, n) & \text{if } k = 1 \\ \min_{1 \leq \ell \leq n-1}\{opt(k - 1, \ell) + C(\ell + 1, n)\} & \text{if } k > 1 \end{cases} \tag{25.2}$$

The DP algorithm is a dynamic programming implementation of Recurrence 25.2. Indeed, to find $OPT(k, n)$, consider the $K \times N$ matrix $M$ with $M(k, n) = opt(k, n)$. The entries of $M$ are computed row by row applying Recurrence 25.2. Clearly, $M(K, N)$ contains the cost of an optimal solution for the $K$-uniform allocation problem. To actually construct an optimal partition, a second matrix $F$ is employed to keep track of the final borders of segmentations corresponding to entries of $M$. In Recurrence 25.2, the value of $\ell$ which minimizes the right-hand-side is the *final border* for the solution $OPT(k, n)$ and is stored in $F(k, n)$.

$$F(k, n) = \begin{cases} 1 & \text{if } k = 1 \\ \arg\min_{1 \leq \ell \leq n-1}\{opt(k - 1, \ell) + C(\ell + 1, n)\} & \text{if } k > 1 \end{cases} \tag{25.3}$$

Hence, the optimal segmentation is given by $OPT(K, N) = (B_1, B_2, \ldots, B_{K-1})$ where, starting from $B_K = N$, the value of $B_k$ is equal to $F(k + 1, B_{k+1})$, for $k = K - 1, \ldots, 1$.

To evaluate the time complexity of the DP algorithm, observe that $O(n)$ comparisons are required to fill the entry $M_{k,n}$, which implies that $O(N^2)$ comparisons are required to fill a row. As there are $K$ rows, the complexity of the DP algorithm is $O(N^2 K)$.

## 25.3.2 The DICHOTOMIC Algorithm

To improve on the time complexity of the DP algorithm for the $K$-uniform allocation problem, the properties of optimal solutions have to be further exploited.

**Definition 25.1** *Let $d_1, d_2, \ldots, d_N$ be uniform length items sorted by nonincreasing probabilities. An optimal solution $OPT(K, N) = (B_1, B_2, \ldots, B_{K-1})$ is called* left-most optimal *and denoted by $LMO(K, N)$ if, for any other optimal solution $(B'_1, B'_2, \ldots, B'_{K-1})$, it holds $B_{K-1} \leq B'_{K-1}$.*

Clearly, as the problem always admits an optimal solution, there is always a LMO. Although the LMO solutions do not need to be unique, it is easy to check that there exists a unique $(B_1, B_2, \ldots, B_{K-1})$ such that $(B_1, B_2, \ldots, B_i)$ is a LMO solution for partitioning into $i + 1$ groups the items $d_1, d_2, \ldots, d_{B_{i+1}}$, for every $i < K$.

**Definition 25.2** *A LMO solution $(B_1, B_2, \ldots, B_{K-1})$ for the $K$-uniform allocation problem is called* strict LMO solution *and denoted by $SLMO(K, N)$, if $(B_1, B_2, \ldots, B_i)$ is a $LMO(i + 1, B_{i+1})$, for every $i < K$.*

The Dichotomic algorithm computes a LMO solution for every $i < K$, and thus it finds the unique strict LMO solution.

**Lemma 25.2** [1] *Let $d_1, d_2, \ldots, d_N$ be uniform length items sorted by nonincreasing probabilities. Let $LMO(K, N - 1) = (B_1, B_2, \ldots, B_{K-1})$ and $OPT(K, N) = (B'_1, B'_2, \ldots, B'_{K-1})$. Then, $B'_{K-1} \geq B_{K-1}$.*

In words, Lemma 25.2 implies that, given the items sorted by nonincreasing probabilities, if one builds an optimal solution for $N$ items from an optimal solution for $N - 1$ items, then the final border $B_{K-1}$ can only move to the right. Such a property can be easily generalized as follows to problems of increasing sizes. From now on, let $B_h^c$ denote the $h$-th border of $LMO(k, c)$, with $k > h \geq 1$.

**Corollary 25.2** [1] *Let $d_1, d_2, \ldots, d_N$ be uniform length items sorted by nonincreasing probabilities, and let $l < j < r \leq N$. Then, $B_{K-1}^l \leq B_{K-1}^j \leq B_{K-1}^r$.*

Corollary 25.2 plays a fundamental role in speeding up the DP algorithm. Indeed, assume that $LMO(k - 1, n)$ has been found for every $1 \leq n \leq N$. If the $LMO(k, l)$ and $LMO(k, r)$ solutions are also known for some $1 \leq l \leq r \leq N$, then one knows that $B_{k-1}^j$ is between $B_{k-1}^l$ and $B_{k-1}^r$, for any $l \leq j \leq r$. Thus, Recurrence 25.2 can be rewritten as

$$opt(k, j) = \min_{B_{k-1}^l \leq \ell \leq B_{k-1}^r} \{opt(k - 1, \ell) + C(\ell + 1, j)\} \tag{25.4}$$

As the name suggests, the $O(KN \log N)$ time Dichotomic algorithm is derived by choosing $j = \lceil \frac{l+r}{2} \rceil$ in Recurrence 25.4, thus obtaining

$$opt\left(k, \left\lceil \frac{l+r}{2} \right\rceil\right) = \min_{B_{k-1}^l \leq \ell \leq B_{k-1}^r} \left\{opt(k - 1, \ell) + C\left(\ell + 1, \left\lceil \frac{l+r}{2} \right\rceil\right)\right\} \tag{25.5}$$

where $B_{k-1}^l$ and $B_{k-1}^r$ are, respectively, the final borders of $LMO(k, l)$ and $LMO(k, r)$. Such a recurrence is iteratively solved within three nested loops (Algorithm 25.1).

As for DP algorithm, the Dichotomic algorithm uses the two matrices $M$ and $F$, whose entries are filled up row by row. A generic row $k$ is filled in $\log N$ stages which overall perform $O(N)$ comparisons. As there are $K$ rows, the time complexity of the Dichotomic algorithm is $O(NK \log N)$.

In details, the Dichotomic algorithm is shown in Algorithm 25.1. It uses the two matrices $M$ and $F$, whose entries are again filled up row by row (Loop 1). A generic row $k$ is filled in stages (Loop 2). Each stage corresponds to a particular value of the variable $t$ (Loop 3). The variable $j$ corresponds to the index of the entry that is currently being filled in stage $t$. The variables $l$ (left) and $r$ (right) correspond to the indices of the entries nearest to $j$ which have been already filled, with $l < j < r$.

Input: $N$ items sorted by non-increasing probabilities, and $K$ groups;
Initialize: **for** $i$ **from** 1 **to** $N$ **do**
   **for** $k$ **from** 1 **to** $K$ **do**
      **if** $k = 1$ **then** $M_{k,i} \leftarrow C_{k,i}$ **else** $M_{k,i} \leftarrow \infty$;
Loop 1: **for** $k$ **from** 2 **to** $K$ **do**
   $F_{k,0} \leftarrow F_{k,1} \leftarrow 1$; $F_{k,N+1} \leftarrow N$;
Loop 2:    **for** $t$ **from** 1 **to** $\lceil \log N \rceil$ **do**
Loop 3:       **for** $i$ **from** 1 **to** $2^{t-1}$ **do**
         $j \leftarrow \lceil \frac{2i-1}{2^t}(N+1) \rceil$; $l \leftarrow \lceil \frac{i-1}{2^{t-1}}(N+1) \rceil$; $r \leftarrow \lceil \frac{i}{2^{t-1}}(N+1) \rceil$;
         **if** $M_{k,j} = \infty$ **then**
Loop 4:          **for** $\ell$ **from** $F_{k,l}$ **to** $F_{k,r}$ **do**
            **if** $M_{k-1,\ell} + C_{\ell+1,j} < M_{k,j}$ **then**
               $M_{k,j} \leftarrow M_{k-1,\ell} + C_{\ell+1,j}$;
               $F_{k,j} \leftarrow \ell$;

**ALGORITHM 25.1** The Dichotomic algorithm for the $K$-uniform allocation problem. (From Ardizzoni, E. et al., *IEEE Trans. Comput.*, 54, 558–572, 2005. ©2005 IEEE. With Permission.)

If no entry before $j$ has been already filled, then $l = 1$, and therefore the final border $F_{k,1}$ is initialized to 1. If no entry after $j$ has been filled, then $r = N$, and thus the final border $F_{k,N+1}$ is initialized to $N$. To compute the entry $j$, the variable $\ell$ takes all values between $F_{k,l}$ and $F_{k,r}$. The index $\ell$ which minimizes the recurrence in Loop 4 is assigned to $F_{k,j}$, whereas the corresponding minimum value is assigned to $M_{k,j}$.

To show the correctness, consider how a generic row $k$ is filled up. In the first stage (i.e., $t = 1$), the entry $M_{k,\lceil \frac{N+1}{2} \rceil}$ is filled and $\ell$ ranges over all values $1, \ldots, N$. By Corollary 25.2, observe that to fill an entry $M_{k,l}$ where $l < \lceil \frac{N+1}{2} \rceil$, one needs to consider only the entries $M_{k-1,\ell}$ where $\ell \leq F_{k,\lceil \frac{N+1}{2} \rceil}$. Similarly, to fill an entry $M_{k,l}$ where $l > \lceil \frac{N+1}{2} \rceil$, one needs to consider only the entries $M_{k-1,\ell}$ where $\ell \geq F_{k,\lceil \frac{N+1}{2} \rceil}$. In general, one can show that in stage $t$, to compute the entries $M_{k,j}$ with $j = \lceil \frac{2i-1}{2^t}(N+1) \rceil$ and $1 \leq i \leq 2^{t-1}$, only the entries $M_{k-1,\ell}$ must be considered, where $F_{k,l} \leq \ell \leq F_{k,r}$ and $l$ and $r$ are $\lceil \frac{i-1}{2^{t-1}}(N + 1) \rceil$ and $\lceil \frac{i}{2^{t-1}}(N + 1) \rceil$, respectively. Notice that these entries have been computed in earlier stages. The above-mentioned process repeats for every row of the matrix. The algorithm proceeds till the last entry $M_{K,N}$, the required optimal cost, is computed. The strict LMO solution $SLMO_{N,K} = (B_1, B_2, \ldots, B_{K-1})$ is obtained, where $B_{k-1} = F_{k,B_k}$ for $1 < k \leq K$ and $B_K = N$.

As regard to the time complexity, first note that the total number of comparisons involved in a stage of the Dichotomic algorithm is $O(N)$ as it is equal to the sum of the number of values the variable $\ell$ takes in Loop 3, that is

$$\sum_{i=1}^{2^{t-1}} \left( F_{k,\lceil \frac{i}{2^{t-1}}(N+1) \rceil} - F_{k,\lceil \frac{i-1}{2^{t-1}}(N+1) \rceil} + 1 \right) = F_{k,N+1} - F_{k,0} + 2^{t-1} = N - 1 + 2^{t-1} = O(N)$$

As Loop 2 runs $\lceil \log N \rceil$ times and Loop 1 is repeated $K$ times, the overall time complexity is $O(NK \log N)$.

### 25.3.3 Two Channels

The current section exploits the structure of the optimal solution in the special case in which the item lengths are uniform, and there are only two channels. Indeed, as shown later, the values assumed varying $\ell$ in the right hand side of Recurrence 25.2 for $k = 2$ form a *unimodal* sequence. That is, there is a particular index $\ell$ such that the values on its left are in nonincreasing order, whereas those on its right are in increasing order. By this fact, one can search the minimum of Recurrence 25.2 in a very effective way, improving on the overall running time.

---

## ALGORITHM 25.2    BinSearch

**Input**: Vector $f$, Start-index $i$, End-index $j$
**Output**: Min-index $m$
$m \leftarrow \lfloor \frac{i+j}{2} \rfloor$;
**if** $i = j$ **then**
  |   **return** $m$
**else**
     **if** $f(m) \geq f(m+1)$ **then**
       |   BinSearch $(m+1, j)$
     **else**
       └  BinSearch $(i, m)$

---

Formally, the 2-uniform allocation problem consists in finding a partition $S$ into two groups $G_1$ and $G_2$ such that $AED_S = \frac{1}{2}(N_1 P_1 + N_2 P_2)$ is minimized. Clearly, $N = N_1 + N_2$, and by Lemma 25.1, $N_1 \leq N_2$ holds for any optimal solution. Moreover, recall that any feasible segmentation $S$ for $K = 2$ can be denoted by the single border $B_1$, which coincides with $N_1$.

**Lemma 25.3** [14] *Consider the uniform length items $d_1, d_2, \ldots, d_N$ sorted by nonincreasing probabilities, and $K = 2$ channels. Let $S = (N_1)$ be a feasible segmentation such that $P_1 \leq P_2$. If the segmentation $S' = (N_1 + 1)$ is feasible, then $AED_{S'} \leq AED_S$.*

Although Lemma 25.1 gives the upper bound $N_1 \leq \lfloor \frac{N}{2} \rfloor$ on the cardinality of group $G_1$, Lemma 25.3 provides a lower bound $b$ on $N_1$. Indeed, it guarantees that any optimal solution contains at least the first $b$ items $d_1, \ldots, d_b$, where $b$ is the largest index for which $P_1 = \sum_{h=1}^{b} p_h \leq P_2 = \sum_{h=b+1}^{N} p_h$. Formally, Recurrence 25.2 for $K = 2$ can be rewritten as follows:

$$opt_{N,2} = \min_{b \leq \ell \leq \lfloor \frac{N}{2} \rfloor} \{ C_{1,\ell} + C_{\ell+1,N} \} \tag{25.6}$$

where

$$b = \max_{1 \leq s \leq \lfloor \frac{N}{2} \rfloor} \left\{ s : \sum_{h=1}^{s} p_h \leq \sum_{h=s+1}^{N} p_h \right\}.$$

The following lemma improves on the upper bound of $N_1$ given by Lemma 25.1 and shows that the values of the feasible segmentations assumed in the right-hand side of Equation 25.6 form a unimodal sequence.

**Lemma 25.4** [14] *Consider the uniform length items $d_1, d_2, \ldots, d_N$ sorted by nonincreasing probabilities, and $K = 2$ channels. Let $S = (N_1)$ be a feasible solution such that $P_1 > P_2$. Consider the solutions $S' = (N_1 + 1)$ and $S'' = (N_1 + 2)$. If $AED_{S'} > AED_S$, then $AED_{S''} > AED_{S'}$.*

In practice, one can scan the feasible solutions of Equation 25.6 by moving the border $\ell$ rightwards, one position at a time, starting from the lower bound $b$ obtained applying Lemma 25.3. The scan continues even when the AED of the current solution does not increase but stops as soon as the AED starts to increase. Indeed, by Lemma 25.4, further moving the border $\ell$ to the right can only increase the cost of the solutions. Hence, the border $m$ that minimizes Equation 25.6, that is the optimal solution of the problem, is given by

$$opt_{N,2} = C_{1,m} + C_{m+1,N} \tag{25.7}$$

where

$$m = \min_{b \leq \ell \leq \lfloor \frac{N}{2} \rfloor} \left\{ \ell : C_{1,\ell} + C_{\ell+1,N} < C_{1,\ell+1} + C_{\ell+2,N} \right\}.$$

Due to the unimodal property of the sequence of values on the right-hand side of Equation 25.7, the search of $m$ can be done in $O(\log N)$ time by a suitable modified binary search. Let $f(\ell) = C_{1,\ell} + C_{\ell+1,N} = \frac{\ell}{2} \sum_{h=1}^{\ell} p_h + \frac{N-\ell}{2} \sum_{h=\ell+1}^{N} p_h$. Then, the unimodal sequence consists of the values $f(b), f(b+1), \ldots, f(\lfloor \frac{N}{2} \rfloor)$. As said, solving Equation 25.7 is equivalent to find the index $m$ such that $f(b) \geq \ldots \geq f(m) < f(m+1) < \ldots < f(\lfloor \frac{N}{2} \rfloor)$. This can be done by invoking the recursive procedure *BinSearch*, given in Algorithm 25.1, with parameters $i = b$ and $j = \lfloor \frac{N}{2} \rfloor$. The BinSearch procedure first computes the middle point $m = \lfloor \frac{i+j}{2} \rfloor$. Then, the values $f(m)$ and $f(m+1)$ are compared in the light of the unimodal sequence definition. If $f(m) \geq f(m+1)$, the minimum must belong to the right half, otherwise it must be in the left half. Procedure BinSearch proceeds recursively on the proper half until the minimum is reached.

## 25.3.4 The New Optimal Algorithm

In the current section, we present a fast implementation of the DP algorithm that exploits the main property of the concave Monge's matrices: the $m$ row minima in an $m \times n$ Monge's matrix can be found in $O(m + n)$ time (e.g., see [17–20]).

### 25.3.4.1 Monge Matrices and the SMAWK Algorithm

**Definition 25.3** *A $2 \times 2$ matrix $\begin{bmatrix} a & b \\ c & d \end{bmatrix}$ is* concave Monge *if $a + d \leq b + c$. An $m \times n$ matrix $\mathbb{A}$ is* concave Monge *if every $2 \times 2$ submatrix is concave Monge. That is, for all $1 \leq i < m$ and $1 \leq j < n$,*

$$A[i,j] + A[i+1,j+1] \leq A[i+1,j] + A[i,j+1]$$

*or, equivalently*

$$A[i,j+1] - A[i,j] \geq A[i+1,j+1] - A[i+1,j].$$

**Definition 25.4** *A $2 \times 2$ matrix is* monotone *if the minimum of the upper row is not to the right of the minimum of the lower row. More formally, $\begin{bmatrix} a & b \\ c & d \end{bmatrix}$ is* monotone *if $b < a$ implies that $d < c$ and $b = a$ implies that $d \leq c$. An $m \times n$ matrix $\mathbb{A}$ is* totally monotone *if every $2 \times 2$ submatrix of $\mathbb{A}$ is monotone.*

It is well known [17,21,22]:

**Lemma 25.5** *Every concave Monge matrix is totally monotone.*

For a totally monotone matrix, we mention the following important result:

**Lemma 25.6** [17] *The minimum in each row of a totally monotone matrix of size $m \times n$ can be computed in $O(m + n)$ time by applying the SMAWK algorithm proposed in Reference 17 and reported in Algorithm 25.3.*

Let $A$ be a totally monotone matrix of size $m \times n$, with $m \leq n$, and assume that we can compute or access each $A[i,j]$ in constant time, for any $i, j$.

The row-minima of $A$ can be computed in $O(n+m)$ time by invoking the SMAWK algorithm. An example of execution is illustrated in Figure 25.1. The main block of SMAWK is the subroutine REDUCE. It takes as input matrix $A$ of size $m \times n$, with $m \leq n$, and returns an $m \times m$ submatrix $G \subset A$ with only the $m$ columns of $A$ which contain the row-minima.

## ALGORITHM 25.3   SMAWK

**Input:** Matrix $A$, Rows $m$, Columns $n$
**Output:** $m$ minima, one for each row of $A$
$G = \text{REDUCE}(A)$;
**if** $n = 1$ **then**
 |  **return** output the minimum
**else**
 |  $H \leftarrow G\left[1, 3, \ldots 2\lfloor \frac{m}{2}\rfloor + 1\right]$, i.e., select the odd rows;
 |  $\text{SMAWK}(H, 2\lfloor \frac{m}{2}\rfloor + 1, n)$;
 |  **return** the minima of the even rows.

## ALGORITHM 25.4   REDUCE

**Input:** $A$, Rows $m$, Columns $n$, with $m \leq n$
**Output:** $G$, Rows $m$, Columns $m$
$G \leftarrow A; k \leftarrow 1$;
**while** $n > m$ **do**
 |  **if** $G[k, k] < G[k, k + 1]$ **then**
 |  |  **if** $k < m$ **then**
 |  |  |  $k \leftarrow k + 1$
 |  |  **else**
 |  |  |  Delete column $k + 1$
 |  **else**
 |  |  **if** $G[k, k] \geq G[k, k + 1]$ **then**
 |  |  |  Delete column $k$;
 |  |  |  **if** $k > 1$ **then**
 |  |  |  |  $k \leftarrow k + 1$

**return** $G$;

An element of $A$ in row $i$ is *dead* if, using the results of any comparisons made so far and by the total monotonicity of $A$, it can be shown that it is not the minimum of row $i$. A column is dead if all its elements are dead. Only dead columns can be deleted. Let the *frontier* of $A$ be the largest index $k$ such that for all $1 \leq i < k$ and $1 \leq j \leq k$, element $A[i, j]$ is dead. Note that every matrix has frontier at least 1.

The REDUCE algorithm is given in Algorithm 25.4. Initially $k = 1$. During the REDUCE algorithm, $k$ is always the frontier of $G$. As $k$ is the frontier, $G[1 : j - 1, j]$ cannot contain row-minima, for all $1 \leq j \leq k$. If $G[k, k] < G[k, k + 1]$ then $G[1 : k, k + 1]$ cannot contain row-minima by the total monotonicity property. Therefore, if $k < m$, $k$ is increased by 1. If $k = m$, column $k + 1$ is dead, and it is deleted. Thus, $k$ is unchanged. If $G[k, k] \geq G[k, k + 1]$ then $G[k : m, k]$ is dead by the total monotonicity property. As also $G[1 : k - 1, k]$ consists of dead elements, column $k$ is dead and hence it is deleted. Thus, if $k > 1$ then $k$ is set to $k - 1$.

The row-minima of an $m \times n$ totally monotone matrix $A$ with $m \leq n$ are then computed as follows. First, the SMAWK algorithm invokes REDUCE on $A$ to return an $m \times m$ matrix $G$ whose columns contain the row-minima of $A$, and then recursively the row-minima of the submatrix $H$ of $G$, which consists of the odd rows of $G$, are found. The recursion continues until $H$ has just 1 column. Then, the minima in

Matrix $A$, $3 \times 6$; $G \leftarrow A$;

$$\begin{pmatrix} \mathbf{2} & \mathbf{6} & 5 & 7 & 18 \\ \mathbf{10} & \mathbf{11} & 9 & 3 & 6 \\ 50 & 30 & 19 & 8 & 10 \end{pmatrix}$$

Reduce $(G)$, $k = 1$

$$\begin{pmatrix} 2 & 6 & 5 & 7 & 18 \\ 10 & 11 & 9 & 3 & 6 \\ 50 & 30 & 19 & 8 & 10 \end{pmatrix}$$

Reduce $(G)$, $k = 2$

$$\begin{pmatrix} 2 & 6 & 5 & 7 & 18 \\ 10 & 11 & 9 & 3 & 6 \\ 50 & 30 & 19 & 8 & 10 \end{pmatrix}$$

Reduce $(G)$, $k = 1$

$$\begin{pmatrix} 2 & 6 & 5 & 7 & 18 \\ 10 & 9 & 11 & 3 & 6 \\ 50 & 30 & 19 & 8 & 10 \end{pmatrix}$$

Reduce $(G)$, $k = 2$

$$\begin{pmatrix} 2 & 6 & 5 & 7 & 18 \\ 10 & 9 & 11 & 3 & 6 \\ 50 & 30 & 19 & 8 & 10 \end{pmatrix}$$

Reduce $(G)$, $k = 2$

$$\begin{pmatrix} 2 & 6 & 5 & 7 & 18 \\ 10 & 9 & 11 & 3 & 6 \\ 50 & 30 & 19 & 8 & 10 \end{pmatrix}$$

Matrix $G$, $3 \times 3$

$$\begin{pmatrix} 2 & 7 & 18 \\ 10 & 3 & 6 \\ 50 & 8 & 10 \end{pmatrix}$$

$H$ = Odd rows of $G$, $2 \times 3$

$$\begin{pmatrix} 2 & 7 & 18 \\ 50 & 8 & 10 \end{pmatrix}$$

$G \leftarrow H$; Reduce $(G)$, $k = 1$

$$\begin{pmatrix} 2 & 7 & 18 \\ 50 & 8 & 10 \end{pmatrix}$$

Reduce $(G)$, $k = 2$

$$\begin{pmatrix} 2 & 7 & 18 \\ 50 & 8 & 10 \end{pmatrix}$$

Reduce $(G)$, $k = 2$

$$\begin{pmatrix} 2 & 7 & 18 \\ 50 & 8 & 10 \end{pmatrix}$$

Matrix $G$, $2 \times 2$

$$\begin{pmatrix} \mathbf{2} & 7 \\ 50 & \mathbf{8} \end{pmatrix}$$

$H$ = Odd rows of $G$, $1 \times 2$

$$(\ 2 \quad 7\ )$$

$G \leftarrow H$, Reduce $(G)$, $k = 1$

$$(\ 2 \quad 7\ )$$

Computing minima $(G)$, $2 \times 2$

$$\begin{pmatrix} \mathbf{2} & 7 \\ 50 & 8 \end{pmatrix}$$

Minima $G$, $2 \times 2$; $H \leftarrow G$;

$$\begin{pmatrix} \mathbf{2} & 7 \\ 50 & \mathbf{8} \end{pmatrix}$$

Computing minima $G$, $1 \times 3$

$$\begin{pmatrix} \mathbf{2} & 7 & 18 \\ 10 & 3 & 6 \\ 50 & 8 & 10 \end{pmatrix}$$

Minima:

$$\begin{pmatrix} \mathbf{2} & 7 & 18 \\ 10 & \mathbf{3} & 6 \\ 50 & \mathbf{8} & 10 \end{pmatrix}$$

**FIGURE 25.1** Example of how to compute the row minima by the SMAWK algorithm.

the even rows of $G$ are found, starting from the known positions of the minima in the odd rows of $G$, by closing the recursive calls in reverse order.

### 25.3.4.2 The SMAWK-AED Algorithm

The rest of this section explains how to apply the SMAWK algorithm to optimally solve the $K$-uniform allocation problem in $O(NK)$ time.

Define the upper triangular single-channel cost matrix $\mathbb{C}$ of size $N \times N$ as the matrix of the costs of the single-channel, i.e. $C(i, j) = \frac{j - i + 1}{2} \sum_{k=i}^{j} P_k$

**Lemma 25.7** *The single-channel cost (upper triangular) matrix $\mathbb{C}$ is a concave Monge matrix, for $0 \leq \ell < n < N$:*

$$C(\ell + 1, n) + C(\ell + 2, n + 1) \leq C(\ell + 1, n + 1) + C(\ell + 2, n).$$

*Proof.* The simplest way to prove this result is to observe that for $0 \leq \ell < n \leq N$:

$$C(\ell + 1, n + 1) - C(\ell + 1, n) = \frac{n - \ell + 1}{2} p_{n+1} + \frac{1}{2} \sum_{q=\ell+1}^{n} p_q$$

$$\geq \frac{n - \ell}{2} p_{n+1} + \frac{1}{2} \sum_{q=\ell+2}^{n} p_q = C(\ell + 2, n + 1) - C(\ell + 2, n). \quad \blacksquare$$

It is worthy to point out that the Monge property for the single-channel costs holds whatever is the order in which the items $p_1, p_2, \ldots, p_N$ are considered, and not only if the $p$'s are sorted in increasing/decreasing order.

As the upper triangular single-channel cost matrix $\mathbb{C}$ is a concave Monge matrix, if we fill the lower triangular matrix $\mathbb{C}$ with $+\infty$ values, the matrix $\mathbb{C}$ is totally monotone.

From now on, let us denote $S_{\ell,n}^k = opt(k - 1, \ell) + C(\ell + 1, n)$. Thus, $S_{\ell,n}^k$ can be seen as a single entry of a three-dimensional matrix $\mathbb{S}$, with the first dimension $k$ varying on the set of channels $[1, \ldots, K]$, the second dimension $\ell$ on the possible positions of the final border of the group $G_{k-1}$ (i.e., $k-1 \leq \ell \leq N-1$), and the third dimension $n$ on the set of data items $[1, \ldots, N]$. After having fixed the first dimension of $\mathbb{S}$ to the value $k$, we may extract from $\mathbb{S}$ the two-dimensional matrix $\mathbb{S}^k = [S_{\ell,n}^k]$, $k - 1 \leq \ell \leq N - 1, k \leq n \leq N$. Thus, the three dimensional matrix $\mathbb{S} = \cup_{k=1}^{K} \mathbb{S}^k$ can be decomposed into $K$ two-dimensional matrices, one for each value of $k$. According to the new notation, for $k \geq 2$, we can rewrite Equations 25.2 and 25.3 as

$$opt(k, n) = \min_{k-1 \leq \ell \leq n-1} \{S_{\ell,n}^k\}; \tag{25.8}$$

$$F[k, n] = \arg \min_{k-1 \leq \ell \leq n-1} \{S_{\ell,n}^k\} \tag{25.9}$$

Then, as $S_{\ell,n}^k = +\infty$ when $1 \leq \ell \leq k - 1$, $opt(k, n)$ is the minimum in row $n$ of $\mathbb{S}^k$.

Now we prove that each matrix $\mathbb{S}^k$, for $1 \leq k \leq K$, is totally monotone. Namely, if we add the same value $opt(k - 1, \ell) + opt(k - 1, \ell + 1)$ to both sides of the concave Monge condition for the single-channel cost matrix $\mathbb{C}$, we obtain the concave Monge condition for matrix $\mathbb{S}^k$:

$$\underbrace{opt(k - 1, \ell) + C(\ell + 1, n)}_{S_{\ell,n}^k} + \underbrace{opt(k - 1, \ell + 1) + C(\ell + 2, n + 1)}_{S_{\ell+1,n+1}^k}$$
$$\leq \underbrace{opt(k - 1, \ell) + C(\ell + 1, n + 1)}_{S_{\ell,n+1}^k} + \underbrace{opt(k - 1, \ell + 1) + C(\ell + 2, n)}_{S_{\ell+1,n}^k} \tag{25.10}$$

Hence, to compute the row $k$ of the matrix $\mathbb{M}$ of the classical dynamic programming algorithm, it is sufficient to apply the SMAWK algorithm to matrix $\mathbb{S}^k$. Specifically, by applying Lemma 25.6, it holds

**Lemma 25.8** *Fixed any $k \geq 2$, the SMAWK algorithm can compute the values $opt(k, n)$ for $k \leq n \leq N$ in $O(N)$ time, if the values $opt(k - 1, n)$ for $k - 1 \leq n \leq N$ are known (that is, if the row $k - 1$ of the matrix $\mathbb{M}$ is known).*

*Proof.* To apply the SMAWK algorithm to the totally monotone matrix $\mathbb{S}^k$, it is required that each entry $S_{\ell,n}^k$ can be computed in constant time. This trivially holds as we assume to know $opt(k - 1, n)$ for $1 \leq n \leq N$ and as we have shown that each entry of the single-channel cost matrix $\mathbb{C}$ can be computed in constant time. $\quad \blacksquare$

## ALGORITHM 25.5   SMAWK-AED

**Input**: $\mathbb{C}, K, N$
**Output**: $opt(k, n), 1 \leq k \leq K, 1 \leq n \leq N$
$opt(1, n) \leftarrow C(1, n), 1 \leq n \leq N;$
**for** $k$ **from** 2 **to** $K$ **do**
> $opt(k, n) \leftarrow +\infty$ for $1 \leq n \leq k - 1;$
> $opt(k, k), \ldots, opt(k, N) \leftarrow \text{SMAWK}(\mathbb{S}^k, N, N)$

Finally:

**Theorem 25.1** [16] *The K-uniform allocation problem can be solved in $O(NK)$ time by applying $K - 1$ times the SMAWK algorithm as illustrated in Algorithm* SMAWK-AED*. The $O(NK)$ time complexity of such an algorithm is optimal.*

Clearly, the SMAWK-AED algorithm is always faster than the Dichotomic algorithm. Moreover, the time complexity of this algorithm cannot be improved as far as we apply the dynamic programming technique which solves $NK$ subproblems to find the optimal data input segmentation.

Lastly, assume that the $K$-uniform allocation problem has already been calculated for $N$ items, and the new items $d_{N+1}$ is given with minimum $p_{N+1}$. We then want to calculate the values $M[k, N + 1] = opt(k, N + 1)$, with $1 \leq k \leq K$, or equivalently, add column $(N + 1)$ to matrix $\mathbb{M}$.

This online version of the problem is quite useful in a dynamic context. We now prove that such online problem can be solved by applying the SMAWK algorithm to a suitable decomposition of $\mathbb{S}$.

Namely, fix $n$ in $1 \ldots N$ and extract from $\mathbb{S}$ the two-dimensional matrix $\mathbb{S}_n = [S^k_{\ell,n}]$ for $1 \leq k \leq K, 1 \leq \ell \leq N - 1$. Thus, $\mathbb{S} = \cup^N_{n=1} \mathbb{S}_n$, that is, $\mathbb{S}$ can be decomposed into $N$ two-dimensional matrices, one for each value of $n$.

Adding the rightmost item $d_{N+1}$ to previously existing items $d_1, d_2, \ldots, d_N$ and computing the values $opt(k, N + 1)$ with $1 \leq k \leq K$ using Equation 25.8 may require up to $O(NK)$ comparisons. In fact, for each $k$ in $1 \ldots K$, $M[k, N + 1] = opt(k, N + 1) = \min_{k-1 \leq \ell \leq N} S^k_{\ell,N+1}$ that is, $opt(k, N + 1)$ corresponds to the minimum in row $k$ of $\mathbb{S}_{N+1}$.

We now prove that the previous online problem can be solved in $O(N)$ time because $\mathbb{S}_{N+1}$ is totally monotone. We first state that the matrix $\mathbb{M}$ of the classical dynamic programming, that is, $M[k, n] = opt(k, n)$, satisfies the concave Monge property.

**Lemma 25.9** [16] *Matrix $\mathbb{M}$ is a concave Monge matrix.* ∎

**Corollary 25.3** *Given $1 \leq n \leq N + 1$, matrix $\mathbb{S}_n$ satisfies the concave Monge property, and therefore it is totally monotone.*

*Proof.* $\mathbb{S}_n$ is a concave Monge matrix if and only if $\mathbb{M}$ is a concave Monge matrix. Namely,

$$\underbrace{opt(k - 1, \ell) + w(\ell, n)}_{S^k_{\ell,n}} + \underbrace{opt(k, \ell + 1) + w(\ell + 1, n)}_{S^{k+1}_{\ell+1,n}}$$
$$\leq \underbrace{opt(k, \ell) + w(\ell, n)}_{S^{k+1}_{\ell,n}} + \underbrace{opt(k - 1, \ell + 1) + w(\ell + 1, n)}_{S^k_{\ell+1,n}}$$

or equivalently, if and only if $opt(k - 1, \ell) + opt(k, \ell + 1) \leq opt(k, \ell) + opt(k - 1, \ell + 1)$. ∎

Notice that the both DP and DICHOTOMIC can accommodate the introduction of a new data item, but this would give suboptimal $O(NK)$ time.

It is worth to point out that all the matrices $\mathbb{C}$, $\mathbb{M}$, $\mathbb{S}^k$, and $\mathbb{S}_n$ are concave Monge matrices whatever is the order of the data items. Therefore, the online algorithm computes the best $K$-partition for any given order of the data items. However, this corresponds to an optimal solution for the $K$-uniform allocation problem only if the items are sorted. If the increasing or decreasing order of the items is not preserved by the new inserted item $d_{N+1}$, the online version can be considered a fast heuristic for the allocation problem in the dynamic context.

Similar to the online problem that adds a new item, we can define the online problem that fills another row of $\mathbb{M}$ when a new channel is released. It is obvious that this problem can be solved in $O(N)$ time because indeed the SMAWK-AED algorithm fills row by row matrix $\mathbb{M}$. Note that the same problem can be solved in $O(N^2)$ and $O(N \log N)$ time, respectively, by the DP and DICHOTOMIC algorithms. In conclusion:

**Theorem 25.2** [16] *Adding a new item $d_{N+1}$ to the set of data items, the K-uniform allocation problem can be solved in $O(N)$ time by applying the SMAWK algorithm to the two-dimensional matrix $\mathbb{S}_{N+1}$. Similarly, adding a new channel, the online AED problem can be solved in $O(N)$ time by applying the SMAWK algorithm to matrix $\mathbb{S}^{K+1}$.*

## 25.4 Nonuniform Lengths

Consider now the $K$-nonuniform allocation problem for an arbitrary number $K$ of channels. In contrast to the uniform case, introducing items with different lengths makes the problem computationally intractable.

**Theorem 25.3** [1] *The K-nonuniform allocation problem is NP-hard for $K = 2$, and strong NP-hard for an arbitrary K.*

As a consequence of the above-mentioned result, there is no pseudopolynomial time optimal algorithm or Fully PTAS for solving the $K$-nonuniform allocation problem (unless P=NP). However, when the maximum item length $z$ is bounded by a constant, a polynomial time optimal algorithm can be derived where $z$ appears in the exponent. When $z = 1$, this algorithm reduces to the DP algorithm. The following result generalizes Lemma 25.1.

**Lemma 25.10** [1] *Let $G_h$ and $G_j$ be two groups in an optimal solution for the K-nonuniform allocation problem. Let $d_i$ and $d_k$ be items with $z_i = z_k$ and $d_i \in G_h$, $d_k \in G_j$. If $Z_h < Z_j$, then $p_i \geq p_k$. Similarly, if $p_i > p_k$, then $Z_h \leq Z_j$.*

Based on the previous lemma, some additional notations are introduced. The set $D$ of items can be viewed as a union of disjoint subsets $D_i = \{d_1^i, d_2^i, \ldots, d_{L_i}^i\}$, $1 \leq i \leq z$, where $D_i$ is the set of items with length $i$, $L_i$ is the cardinality of $D_i$, and $z$ is the maximum item length. Let $p_j^i$ represent the probability of item $d_j^i$, for $1 \leq j \leq L_i$.

The following corollary generalizes Corollary 25.1.

**Corollary 25.4** [1] *Let $d_1^i, d_2^i, \ldots, d_{L_i}^i$ be the $L_i$ items of length $i$ with $p_m^i \geq p_n^i$ whenever $m < n$, for $i = 1, \ldots, z$. There is an optimal solution for partitioning the items of $D$ into $K$ groups $G_1, \ldots, G_K$, such that if $a < b < c$ and $d_a^i, d_c^i \in G_j$, then $d_b^i \in G_j$.*

In the rest of this section, the items in each $D_i$ are assumed to be sorted by nonincreasing probabilities, and optimal solutions will be sought of the form:

$$\underbrace{d_1^1,\ldots,d_{B_1^{(1)}}^1}_{G_1},\underbrace{d_{B_1^{(1)}+1}^1,\ldots,d_{B_2^{(1)}}^1}_{G_2},\ldots,\underbrace{d_{B_{K-1}^{(1)}+1}^1,\ldots,d_{N_1}^1}_{G_K}$$

$$\underbrace{d_1^2,\ldots,d_{B_1^{(2)}}^2}_{G_1},\underbrace{d_{B_1^{(2)}+1}^2,\ldots,d_{B_2^{(2)}}^2}_{G_2},\ldots,\underbrace{d_{B_{K-1}^{(2)}+1}^2,\ldots,d_{N_2}^2}_{G_K}$$

$$\vdots$$

$$\underbrace{d_1^z,\ldots,d_{B_1^{(z)}}^z}_{G_1},\underbrace{d_{B_1^{(z)}+1}^z,\ldots,d_{B_2^{(z)}}^z}_{G_2},\ldots,\underbrace{d_{B_{K-1}^{(z)}+1}^z,\ldots,d_{N_z}^z}_{G_K}$$

where $B_j^{(i)}$ is the highest index among all items of length $i$ in group $G_j$. The solution will be represented as $(\bar{B}_1,\bar{B}_2,\ldots,\bar{B}_{K-1})$, where each $\bar{B}_j$ is the $z$-tuple $(B_j^{(1)},B_j^{(2)},\ldots,B_j^{(z)})$ for $1 \leq j \leq K-1$. From now on, $B_{K-1}^{(i)}$ will be referred to as the *final border for length $i$* and $\bar{B}_{K-1}$ as the *final border vector*.

Let $OPT_{n_1,\ldots,n_z,k}$ denote the optimal solution for grouping the $\sum_{i=1}^z n_i$ items $d_1^i,d_2^i,\ldots,d_{n_i}^i,1 \leq i \leq z$, into $k$ groups and let $opt(k,n_1,\ldots,n_z)$ be its corresponding cost. Let $C(l_1,n_1,\ldots,l_z,n_z)$ be the cost of putting items $l_i$ through $n_i$, for all $i = 1,2,\ldots,z$, into one group, that is,

$$C(l_1,n_1,\ldots,l_z,n_z) = \frac{1}{2}\left(\sum_{i=1}^z i(n_i-l_i+1)\right)\left(\sum_{i=1}^z \sum_{j=l_i}^{n_i} p_j^i\right)$$

Now, consider the recurrence:

$$opt(k,n_1,\ldots,n_z) = \min_{\substack{\bar{\ell}=(\ell_1,\ldots,\ell_z)\\ 0\leq\ell_i\leq n_i, 1\leq i\leq z}}\left\{opt(k-1,\ell_1,\ldots,\ell_z)+C(\ell_1+1,n_1,\ldots,\ell_z+1,n_z)\right\} \qquad (25.11)$$

To solve this recurrence by using dynamic programming, consider a $(z+1)$-dimensional matrix $M$, made of $K$ rows in the first dimension and $L_i$ columns in dimension $i+1$ for $i = 1,\ldots,z$. Each entry is represented by a $(z+1)$-tuple $M(k,n_1,\ldots,n_z)$, where $k$ corresponds to the row index, and $n_i$ corresponds to the index of the column in dimension $i+1$. The entry $M(k,n_1,\ldots,n_z)$ represents the optimal cost for partitioning items $d_1^i$ through $d_{n_i}^i$, for $i = 1,2,\ldots z$, into $k$ groups. There is also a similar matrix $F$ where the entry $F(k,n_1,\ldots,n_z)$ corresponds to the final border vector of the solution whose cost is $M(k,n_1,\ldots,n_z)$. The matrix entries are filled row by row. The optimal solution is given by $OPT(K,L_1,\ldots,L_z) = (\bar{B}_1,\bar{B}_2,\ldots,\bar{B}_{K-1})$ where, starting from $\bar{B}_K = (L_1,L_2,\ldots,L_z)$, the value of $\bar{B}_k$ is obtained from the value of $\bar{B}_{k+1}$ and by $F$ as $\bar{B}_k = F(k+1,\bar{B}_{k+1})$, for $k = 1,\ldots,K-1$. The Optimal algorithm derives directly from Recurrence 25.11. Since the computation of every entry $M(k,n_1,\ldots,n_z)$ and $F(k,n_1,\ldots,n_z)$ requires $\prod_{i=1}^z(n_i+1) \leq \prod_{i=1}^z(L_i+1)$ comparisons, and every row has $\prod_{i=1}^z L_i$ entries, the overall time complexity is $O(K\prod_{i=1}^z(L_i+1)^2) = O(KN^{2z})$.

## 25.5 Heuristics

As the $K$-nonuniform allocation problem is strong NP-hard, it results to be computationally intractable (unless P=NP). In practice, this implies that one is forced to abandon the search for efficient algorithms which find optimal solutions. Therefore, one can devise fast and simple heuristics that provide solutions which are not necessarily optimal but usually fairly close. This strategy is followed in this section, where the main heuristics are reviewed. All heuristics assume that the items are sorted by nonincreasing $\frac{p_i}{z_i}$ ratios, which can be done in $O(N \log N)$ time during a preprocessing step.

### 25.5.1 The Greedy Algorithm

The Greedy heuristic [11,12] initially assigns all the $N$ data items to a single group. Then, for $K - 1$ times, one of the groups is split in two groups, that will be assigned to two different channels. To find which group to split along with its actual split point, all the possible points of all groups are considered as split point candidates, and the one that decreases AED the most is selected. In details, assume that the channel to be split contains the items from $d_i$ to $d_j$, with $1 \le i < j \le N$, and let $cost_{i,j,2}$ denote the cost of a feasible solution for assigning such items to two channels. Then, the split point is the index $m$ that satisfies

$$cost_{i,j,2} = C(i, m) + C(m + 1, j) = \min_{i \le \ell \le j-1} \{C(i, \ell) + C(\ell + 1, j)\} \tag{25.12}$$

An efficient implementation takes advantage from the fact that, between two subsequent splits, it is sufficient to recompute the costs for the split point candidates of the last group that has been actually split. The time complexity of the Greedy heuristic is $O(N(K + \log N))$ and $O(N \log N)$ in the worst and average cases, respectively [14].

Note that Greedy scales well when changes occur on the number of channels, on the number of items, on item probabilities, as well as on item lengths. Indeed, adding or removing a channel simply requires doing a new split or removing the last introduced split, respectively. Adding a new item first requires to insert such an item in the sorted item sequence. Assume the new item is added to group $G_j$, then the border of the two-channel subproblem including items of $G_j$ and $G_{j+1}$ is recomputed by applying Equation 25.12. Similarly, deleting an item that belongs to group $G_j$ requires to solve again the two-channel subproblem including items of $G_j$ and $G_{j+1}$. Finally, a change in the probability/length of an item is equivalent to first removing that item and then adding the same properly modified item.

### 25.5.2 The Greedy+ Algorithm

The Greedy+ heuristic [14] is a refinement of the Greedy heuristic and consists of two phases. In the first phase, it behaves as Greedy, except the way the split point is determined. In the second phase, the solution provided by the first phase is refined by working on pairs of consecutive channels. Specifically, in the first phase, Greedy+ uses an approach similar to that of Equation 25.7 to determine the split point. This is because splitting one channel is the same as solving the problem for two channels. In details, the split point $m$ is given by

$$cost_{i,j,2} = C_{i,m} + C_{m+1,j} \tag{25.13}$$

where

$$m = \min_{i \le \ell \le j-1} \{\ell : C_{i,\ell} + C_{\ell+1,j} < C_{i,\ell+1} + C_{\ell+2,j}\}.$$

Note that, as the item lengths are not the same, the sequence of values $C_{i,\ell} + C_{\ell+1,j}$, for $i \le \ell \le j - 1$, is not unimodal. However, Greedy+ behaves as such a sequence were unimodal. Instead of trying all the possible values of $\ell$ between $i$ and $j$, as done by Greedy, Greedy+ performs a left-to-right scan starting from $i$ and stopping as soon the AED increases. In this way, a sub optimal solution $S = (B_1, B_2, \ldots . B_{K-1})$ is found.

The second phase is performed only when $K \ge 3$ and consists in refining the solution $S$ by recomputing its borders. The phase consists in a sequence of odd steps, followed by a sequence of even steps. During the $t$th odd step, $1 \leq t \leq \lfloor \frac{K}{2} \rfloor$, the two-channel subproblem including the items assigned to groups $G_{2t-1}$ and $G_{2t}$ is solved. Specifically, Equation 25.13 is applied choosing $i = B_{2t-2} + 1$ and $j = B_{2t}$, thus recomputing the border $B_{2t-1}$ of $S$. Similarly, during the $t$th even step, $1 \leq t \leq \lfloor \frac{K-1}{2} \rfloor$, the two-channel subproblem including the items assigned to groups $G_{2t}$ and $G_{2t+1}$ is solved by applying Equation 25.13 with $i = B_{2t-1} + 1$ and $j = B_{2t+1}$, recomputing the border $B_{2t}$ of $S$.

The initial sorting requires $O(N \log N)$ time. As each split runs in $O(N)$ time, and $K$ splits are computed, the first phase of Greedy+ takes $O(NK)$ time. The second phase of Greedy+ requires $O(N)$ time as each item is considered as a candidate split point at most in a single split computation among all the odd steps, and in a single split computation among the even steps. Therefore, the overall time required in the worst case by the Greedy+ heuristic is $O(N(K + \log N))$. Clearly, Greedy+ maintains the same scaling features as Greedy.

## 25.5.3 The Dlinear Algorithm

The Dlinear heuristic [14] follows a dynamic programming approach similar to that provided by Recurrence 25.2. Fixed $k$ and $n$, Dlinear computes a solution for $n$ items from the previously computed solution for $n - 1$ items and $k$ channels, exploiting the characteristics of the optimal solutions for two channels and uniform lengths.

Let $M_{k,n}$ and $F_{k,n}$ be defined as in Section 25.3.1. Dlinear selects the feasible solutions that satisfy the following Recurrence:

$$M_{k,n} = \begin{cases} C_{1,n} & \text{if } k = 1 \\ M_{k-1,m} + C_{m+1,n} & \text{if } k > 1 \end{cases} \tag{25.14}$$

where

$$m = \min_{F_{k,n-1} \leq \ell \leq n-1} \{\ell : M_{k-1,\ell} + C_{\ell+1,n} < M_{k-1,\ell+1} + C_{\ell+2,n}\}.$$

In practice, Dlinear pretends to adapt Equation 25.7, that holds for the 2-uniform allocation problem also to the $K$-nonuniform allocation problem. In particular, the choice of the lower bound $F_{k,n-1}$ in the formula of $m$ is suggested by Lemma 25.2 that says that the border of channel $k - 1$ can only move right when a new item with the smallest probability is added. Moreover, $m$ is determined as in Equation 25.7 pretending that the sequence $M_{k-1,\ell} + C_{\ell+1,n}$, obtained for $F_{k,n-1} \leq \ell \leq n - 1$, be unimodal. Therefore, the solution provided by Dlinear is a suboptimal one.

As regard to the time complexity, computing $M_{k,n}$ requires $O(F_{k,n} - F_{k,n-1})$ time. Hence, row $k$ of $M$ is filled in $\sum_{n=1}^{N} O(F_{k,n} - F_{k,n-1}) = O(F_{k,N} - F_{k,1}) = O(N)$ time. As $M$ has $K$ rows and the sorting step takes $O(N \log N)$ time, the overall time complexity of the Dlinear algorithm is $O(N(K + \log N))$.

## 25.6 Experimental Tests

In the current section, the behavior of the Greedy, Greedy+, and Dlinear heuristics is tested. The algorithms are written in C, and the experiments are run on an AMD Athlon XP 2500+, 1.84 GHz, with 1 GB RAM.

The heuristics are executed on the following nonuniform length instances. Given the number $N$ of items and a real number $0 \leq \theta \leq 1$, the item probabilities are generated according to a Zipf distribution whose skew is $\theta$, namely,

$$p_i = \frac{(1/i)^{\theta}}{\sum_{i=1}^{N} (1/i)^{\theta}} \qquad 1 \leq i \leq N$$

In the above-mentioned formula, $\theta = 0$ stands for a uniform distribution with $p_i = \frac{1}{N}$, whereas $\theta = 1$ implies a high skew, namely, the range of $p_i$ values becomes larger. The item lengths $z_i$ are integers randomly generated according to a uniform distribution in the range $1 \le z_i \le z$. The items are sorted by non increasing $\frac{p_i}{z_i}$ ratios. The parameters $N, K, z,$ and $\theta$ vary, respectively, in the ranges: $500 \le N \le 2500$, $10 \le K \le 500, 3 \le z \le 10,$ and $0.5 \le \theta \le 1$.

As the Optimal algorithm can find the exact solutions in a reasonable time only for small instances, a *lower bound* on AED is used for large values of $N, K,$ and $z$. The lower bound for a nonuniform instance is obtained by transforming it into a uniform instance as follows. Each item $d_i$ of probability $p_i$ and length $z_i$ is decomposed in $z_i$ items of probability $\frac{p_i}{z_i}$ and length 1. As more freedom has been introduced, it is clear that the optimal AED for the so transformed problem is a lower bound on the AED of the original problem. As the transformed problem has uniform lengths, its optimal AED is obtained by running the Dichotomic algorithm.

The simulation results are exhibited in Tables 25.1 through 25.4. The tables report the time (measured in microseconds), the AED, and the percentage of error, which is computed as

$$\left( \frac{\text{AED}_{\text{heuristic}} - \text{AED}_{\text{lowerbound}}}{\text{AED}_{\text{lowerbound}}} \right) 100$$

The running times reported in the tables do not include the time for sorting.

By observing the tables, one notes that Greedy+ and Dlinear always outperform Greedy in terms of solution quality. In particular, Greedy+ at least halves the error of Greedy, producing solutions whose errors is at most 5.7%. Moreover, Dlinear reaches the optimum almost in all cases, and its maximum error

**TABLE 25.1**   Experimental Results When $K = 20, \theta = 0.8,$ and $z = 3$

| $N/K/\theta/z$ | Algorithm | AED | % Error | Time |
|---|---|---|---|---|
| 500/20/0.8/3 | Greedy | 18.72 | 7.1 | 102 |
| | Greedy+ | 17.58 | 0.6 | 3514 |
| | Dlinear | 17.47 | | 2106 |
| | Lower bound | 17.47 | | |
| 1500/20/0.8/3 | Greedy | 53.85 | 7.9 | 283 |
| | Greedy+ | 51.71 | 3.6 | 21240 |
| | Dlinear | 49.90 | | 6519 |
| | Lower bound | 49.90 | | |
| 1750/20/0.8/3 | Greedy | 62.64 | 7.9 | 326 |
| | Greedy+ | 58.92 | 1.5 | 31137 |
| | Dlinear | 58.04 | | 7488 |
| | Lower bound | 58.04 | | |
| 2000/20/0.8/3 | Greedy | 71.24 | 7.9 | 373 |
| | Greedy+ | 66.93 | 1.4 | 38570 |
| | Dlinear | 65.98 | | 8602 |
| | Lower bound | 65.98 | | |
| 2250/20/0.8/3 | Greedy | 79.70 | 7.8 | 457 |
| | Greedy+ | 75.06 | 1.6 | 45170 |
| | Dlinear | 73.87 | | 9749 |
| | Lower bound | 73.87 | | |
| 2500/20/0.8/3 | Greedy | 88.40 | 7.8 | 474 |
| | Greedy+ | 82.51 | 0.7 | 62376 |
| | Dlinear | 81.93 | | 10920 |
| | Lower bound | 81.93 | | |

**TABLE 25.2** Experimental Results When $N = 2500$, $\theta = 0.8$, and $z = 3$

| $N/K/\theta/z$ | Algorithm | AED | % Error | Time |
|---|---|---|---|---|
| 2500/10/0.8/3 | Greedy | 179.16 | 7.8 | 381 |
| | Greedy+ | 167.86 | 1.0 | 97356 |
| | Dlinear | 166.14 | | 4919 |
| | Lower bound | 166.14 | | |
| 2500/40/0.8/3 | Greedy | 44.04 | 7.9 | 562 |
| | Greedy+ | 41.58 | 1.9 | 34147 |
| | Dlinear | 40.79 | | 22771 |
| | Lower bound | 40.79 | | |
| 2500/80/0.8/3 | Greedy | 21.98 | 7.9 | 685 |
| | Greedy+ | 20.72 | 1.7 | 19179 |
| | Dlinear | 20.37 | | 46545 |
| | Lower bound | 20.37 | | |
| 2500/100/0.8/3 | Greedy | 17.14 | 5.2 | 740 |
| | Greedy+ | 16.75 | 2.8 | 27452 |
| | Dlinear | 16.29 | | 57906 |
| | Lower bound | 16.29 | | |
| 2500/200/0.8/3 | Greedy | 8.56 | 5.1 | 1009 |
| | Greedy+ | 8.37 | 2.8 | 12974 |
| | Dlinear | 8.15 | 0.1 | 116265 |
| | Lower bound | 8.14 | | |
| 2500/500/0.8/3 | Greedy | 3.4 | 4.2 | 2313 |
| | Greedy+ | 3.35 | 2.7 | 21430 |
| | Dlinear | 3.32 | 1.8 | 273048 |
| | Lower bound | 3.26 | | |

**TABLE 25.3** Experimental Results When $N = 2500$, $K = 50$, and $z = 3$

| $N/K/\theta/z$ | Algorithm | AED | % Error | Time |
|---|---|---|---|---|
| 2500/50/0.5/3 | Greedy | 47.74 | 9.7 | 595 |
| | Greedy+ | 46.02 | 5.7 | 23175 |
| | Dlinear | 43.52 | 0.02 | 29075 |
| | Lower bound | 43.51 | | |
| 2500/50/0.7/3 | Greedy | 39.59 | 6.8 | 600 |
| | Greedy+ | 38.47 | 3.8 | 23606 |
| | Dlinear | 37.05 | 0.02 | 29132 |
| | Lower bound | 37.04 | | |
| 2500/50/0.8/3 | Greedy | 34.33 | 5.2 | 603 |
| | Greedy+ | 33.49 | 2.6 | 24227 |
| | Dlinear | 32.61 | | 29121 |
| | Lower bound | 32.61 | | |
| 2500/50/1/3 | Greedy | 23.10 | 3.2 | 609 |
| | Greedy+ | 22.53 | 0.6 | 27566 |
| | Dlinear | 22.38 | | 28693 |
| | Lower bound | 22.38 | | |

**TABLE 25.4** Experimental Results When $N = 500$, $K = 50$, and $\theta = 0.8$

| $N/K/\theta/z$ | Algorithm | AED | % Error | Time |
|---|---|---|---|---|
| 500/50/0.8/3 | Greedy | 7.34 | 5.3 | 147 |
| | Greedy+ | 7.19 | 3.1 | 2517 |
| | Dlinear | 6.98 | 0.1 | 5423 |
| | Lower bound | 6.97 | | |
| 500/50/0.8/5 | Greedy | 10.78 | 5.3 | 147 |
| | Greedy+ | 10.52 | 2.8 | 2938 |
| | Dlinear | 10.25 | 0.1 | 5490 |
| | Lower bound | 10.23 | | |
| 500/50/0.8/7 | Greedy | 14.50 | 4.9 | 146 |
| | Greedy+ | 14.16 | 2.4 | 3329 |
| | Dlinear | 13.85 | 0.2 | 5499 |
| | Lower bound | 13.82 | | |
| 500/50/0.8/10 | Greedy | 19.48 | 5.1 | 145 |
| | Greedy+ | 18.97 | 2.3 | 3899 |
| | Dlinear | 18.58 | 0.2 | 5507 |
| | Lower bound | 18.53 | | |

is as high as 1.8% only in one instance. As regard to the running times, although all the three heuristics have the same asymptotic worst case time, Greedy is the fastest in practice. Although Greedy+ and Dlinear are slower than Greedy, their running times are always less than one tenth of second. The experiments show that Greedy+ and Dlinear behave well when the item probabilities follow a Zipf distribution. This suggests that, in most cases, the AED achieved in correspondence of the leftmost value of $\ell$ satisfying Recurrences 25.13 and 25.14 is the optimal AED or it is very close to the optimal AED. In other words, the sequence of values obtained by varying $\ell$ is almost unimodal.

In summary, the experimental tests show that the Dlinear heuristic finds optimal solutions almost always. In contrast, Greedy is the fastest heuristic but produces the worst solutions. Finally, Greedy+ presents running times and suboptimal solutions which are both intermediate between those of Greedy and Dlinear. Therefore, the choice among the heuristics depends on the goal to be pursued. If one is interested in finding the best suboptimal solutions, then Dlinear should be adopted. Instead, if the running time is the main concern, then Greedy should be chosen, whereas if adaptability to parameter changes is the priority, then either Greedy or Greedy+ could be applied. In this scenario, Greedy+ represents a good compromise as it is scalable and produces fairly good solutions.

## 25.7 Conclusion

In the current chapter, the problem of data broadcasting over multiple channels, with the objective of minimizing the AED of the clients, was considered under the assumptions of skewed allocation to multiple channels and flat scheduling per channel. Both the uniform and nonuniform length problems were solved to the optimum, illustrating exact algorithms based on dynamic programming. The new algorithm SMAWK-AED, which solves the $NK$ prefix subproblems of the $K$-uniform allocation problem, matches the time-complexity lower bound, and for this, it cannot be further improved. It can also be adapted to compute online in $O(N)$ time the solution for the $K$-uniform allocation problem when a new item or a new channel are added. Moreover, effective heuristics for nonuniform lengths have also been shown. All the results reviewed in this chapter are summarized in Table 25.5.

**TABLE 25.5** Results for Broadcasting $N$ Data Items on $K$ Channels with Skewed Allocation and Flat Scheduling

| Item Lengths | Complexity | Solution | Algorithm | Time | Time Optimality | References |
|---|---|---|---|---|---|---|
| Uniform | $P$ | Optimal | DP | $O(KN^2)$ | No | [12] |
| | | Optimal | Dichotomic | $O(KN \log N)$ | No | [1] |
| | | Optimal | Smawk-AED | $O(KN)$ | Yes | [16] |
| Nonuniform | Strong | Optimal | Optimal | $O(KN^{2z})$ | – | [1] |
| | $NP$-hard | Heuristic | Greedy, Greedy+, Dlinear | $O(N(K + \log N))$ | – | [11,14] |

In the current chapter, the client delay has been defined as the overall time elapsed from the moment the client desires a data item to the moment the item download starts. Such a definition assumes that indexing is already available to the client. Hence, the client delay does not include the tuning time spent by the client for actively retrieving the index information and the data item. Thus, after reading the index, the client can be turned into a power saving mode until the data item appears on the proper channel. Therefore, our solution minimizes the AED and keeps as low as possible the tuning time provided that an efficient index strategy is adopted on one or more separate channels. In our solution, the index can be readily derived from the $(K-1)$-tuple $(B_1, B_2, \ldots, B_{K-1})$, which compactly represents the data allocation. However, this tuple is enough for indexing only if all the clients know, as a global information, the relative position of each data item within the set of all data items sorted by probabilities. To overcome this assumption, new solutions can be sought that, without using global information on data items, either mix index and data items within the same channels or optimize the index broadcasting on dedicated channels [23,24].

# References

1. E. Ardizzoni, A.A. Bertossi, M.C. Pinotti, S. Ramaprasad, R. Rizzi, and M.V.S. Shashanka. Optimal skewed data allocation on multiple channels with flat broadcast per channel. *IEEE Transactions on Computers*, 54(5):558–572, 2005.
2. M.H. Ammar and J.W. Wong. The design of teletext broadcast cycles. *Performance Evaluation*, 5(4):235–242, 1985.
3. M.H. Ammar and J.W. Wong. On the optimality of cyclic transmission in teletext systems. *IEEE Transactions on Communications*, 35(11):1159–1170, 1987.
4. A. Bar-Noy, R. Bhatia, J.S. Naor, and B. Schieber. Minimizing service and operation costs of periodic scheduling. In *Proceedings of the Ninth ACM-SIAM Symposium Discrete Algorithms (SODA)*, pp. 11–20, SIAM, Philadelphia, PA, 1998.
5. T. Imielinski, S. Viswanathan, and B.R. Badrinath. Energy efficient indexing on air. In *Proceedings of the SIGMOD*, Minneapolis, MN, May 1994.
6. C. Kenyon and N. Schabanel. The data broadcast problem with non-uniform transmission time. In *Proceedings of the Tenth ACM-SIAM Symposium on Discrete Algorithms (SODA)*, pp. 547–556, SIAM, Philadelphia, PA, 1999.
7. C. Kenyon, N. Schabanel, and N. Young. Polynomial time approximation scheme for data broadcast. In *Proceedings of the ACM Symposium on Theory of Computing (STOC)*, pp. 659–666, ACM, New York, 2000.
8. W.C. Peng and M.S. Chen. Efficient channel allocation tree generation for data broadcasting in a mobile computing environment. *Wireless Networks*, 9(2):117–129, 2003.
9. K.A. Prabhakara, K.A. Hua, and J. Oh. Multi-level multi-channel air cache designs for broadcasting in a mobile environment. In *Proceedings of the Sixteenth IEEE International Conference on Data Engineering (ICDE)*, February 2000.

10. N. Vaidya and S. Hameed. Log time algorithms for scheduling single and multiple channel data broadcast. In *Proceedings of the Third ACM-IEEE Conference on Mobile Computing and Networking* (*MOBICOM*), September 1997.

11. W.G. Yee. Efficient data allocation for broadcast disk arrays. Technical Report, GIT-CC-02-20, Georgia Institute of Technology, Atlanta, GA, 2001.

12. W.G. Yee, S. Navathe, E. Omiecinski, and C. Jermaine. Efficient data allocation over multiple channels at broadcast servers. *IEEE Transactions on Computers*, 51(10):1231–1236, 2002.

13. S. Acharya, R. Alonso, M. Franklin, and S. Zdonik. Broadcast disks: Data management for asymmetric communication environments. In *Proceedings of the SIGMOD*, San Jose, CA, May 1995.

14. S. Anticaglia, F. Barsi, A.A. Bertossi, L. Iamele, and M.C. Pinotti. Efficient heuristics for data broadcasting on multiple channels. Technical Report, 2005/5, Department of Mathematics and Computer Science, University of Perugia, 2005.

15. L. Breslau, P. Cao, L. Fan, G. Phillips, and S. Shenker. Web caching and Zipf-like distributions: Evidence and implications. In *Proceedings of the IEEE INFOCOM*, 1999.

16. G. Audrito, D. Diodati, and C.M. Pinotti. Optimal skewed allocation on multiple channels for broadcast in smart cities. In *2nd IEEE SMARTCOMP 2016*, pp. 1–8, St. Louis, MO, May 18–20, 2016.

17. A. Aggarwal, M. Klawe, S. Moran, P. Shor, and R. Wilber. Geometric applications of a matrix-searching algorithm. *Algorithmica*, 2(1–4):195–208, 1987.

18. A. Aggarwal, B. Schieber, and T. Tokuyama. Finding a minimum-weight $k$-link path in graphs with the concave monge property and applications. *Discrete & Computational Geometry*, 12:263–280, 1994.

19. A. Aggarwal, B. Schieber, and T. Tokuyama. Finding a minimum weight $k$-link path in graphs with monge property and applications. In *Proceedings of the Ninth Annual Symposium on Computational Geometry*, SCG '93, pp. 189–197, ACM, New York, 1993.

20. B. Schieber. Computing a minimum weight $k$-link path in graphs with the concave monge property. *Journal of Algorithms*, 29(2):204–222, 1998.

21. W. Bein, M.J. Golin, L.L. Larmore, and Y. Zhang. The knuth-yao quadrangle-inequality speedup is a consequence of total monotonicity. *ACM Transactions on Algorithms*, 6(1):17:1–17:22, 2009.

22. R.E. Burkard, B. Klinz, and R. Rudolf. Perspectives of Monge properties in optimization. *Discrete Applied Mathematics*, 70(2):95–161, 1996.

23. S.-C. Lo and A.L.P. Chen. Optimal index and data allocation in multiple broadcast channels. In *Proceedings of the Sixteenth IEEE International Conference on Data Engineering* (*ICDE*), February 2000.

24. I. Stojmenovic (Ed.). *Handbook of Wireless Networks and Mobile Computing*. Wiley, Chichester, UK, 2002.

# Strategies for Aggregating Time-Discounted Information in Sensor Networks

Xianping Wang

Stephan Olariu

## 26.1 Introduction and Motivating Scenarios

Sensor networks are deployed to monitor a seemingly endless list of events in a multitude of application domains. Through data collection and aggregation, many static and dynamic patterns can be found by sensor networks. The aggregation problem is complicated by the fact that the value of data collected by the sensors deteriorates, often dramatically, over time.

In the current chapter, we present formal algebraic models for the time-discounted value of information and for the interaction between data aggregation and time-discounting. These models are applied to emergency response with natural thresholding strategies and verified by extensive simulation. Due to the limited resources in each sensor, we approximate the exponential discount function by low-degree polynomials and capture the major characteristics through Primary Component Analysis (PCA).

Consider a sensor network deployed with the specific mission of protecting a power plant. The information about a possible attack on the power plant will be more valuable the less it is delayed. It will lose value continuously as there will not be adequate time to prepare. Thus, there are powerful incentives for reporting an attack as soon as possible. However, the cost of a false alarm is considered to be prohibitive in terms of the amount of human attention it requires. Thus, there are powerful incentives for aggregating individual sensor data before reporting.

To begin, imagine that the power plant is threatened by an intruder who intends to sabotage the turbines. Again, the less delay in the information about the intrusion event, the better. The sensors that have detected the event need to decide whether to report an intrusion (and risk triggering a false alarm) or wait until several other sensors have corroborated the intrusion. With each moment of delay in notification of the intrusion, the ability to find the intruder decreases, as the intruder may be moving, and the area to search increased quadratically in the time since detection.

Next, imagine a foreign hacker who launches an attack on the network equipment controlling the power plant. The earlier the cyber-attack is detected, the higher the chance of thwarting the intruder. But as time goes on, the worse the attack gets. This type of cyber-attack may well double or triple the malicious network traffic with each time increment. Thus, the value of the information to the decision-maker will deteriorate rapidly, as it becomes harder and harder to fight the attack as the network becomes overwhelmed.

The common characteristic of the above-mentioned scenarios is that getting information quickly has value. On the other hand, there are costs associated with obtaining information.

## 26.2 A Quick Literature Survey

Information is a good that has value and, consequently, can be exchanged or traded. There are many aspects of information that may increase or decrease its value: Timeliness is an important one; accuracy is another [1]. Assessing the value of information and understanding the dynamics of its change over time has been a topic of research in economics [2–6], information systems [7,8], psychology [9,10], social and political sciences [11–13], among many others.

It is intuitively clear that information is more valuable new than old. For example, real-time stock quotes are more valuable than quotes that are delayed by 30 min. We see evidence of this in the tiered pricing offered by exchanges for pricing information based on the delay: The less delay, the more expensive the service. In everyday situations, we recognize that today's newspaper is more important than yesterday's. This deterioration of information value is discussed in a general way in the economics literature [2,6].

In psychology, information aggregation [9,10] studies how humans handle data. New information flows into a human mind and is merged into already processed information. Our ability to perceptually aggregate information is an indicator of our overall thinking abilities [14]. Research shows that some forms of human memory decay exponentially. As humans, we also discount past information in favor of present information [15].

At the social level, economists discuss the half-life of an academic field or even of a scientific paper. Recently, the Web has provided empirical evidence of information decay in academia. For example, researchers in physics post their articles online in standard venues: The citation rates decay exponentially starting from the time of posting [16].

The common denominator of the research summarized earlier is that a number of entities are associated with particular pieces of information. This information has value that is deteriorating over time. Each entity must decide to aggregate or not with others. There are powerful incentives for aggregation. For example, in sensor networks, individual sensors may report an event based solely on the data they hold, but they risk reporting a false positive and triggering a false alarm. In many systems, the cost of a false alarm is considered to be prohibitive and must be avoided to the largest extent possible.

On the other hand, aggregation involves costs, too. Chief among these costs is interentity coordination and communication that, in wireless sensor networks, may be high. Moreover, aggregation takes time, and during this time, the value of the information continues to decay.

Deutsch [11] modeled aggregation in the context of politics, investigating the way coalitions form and dissolve. This kind of merger when states form coalitions is analogous, at a higher scale, to the aggregation we describe between nodes in a sensor network. Work on preferential attachment [17] has suggested that a tendency to attach to the node in a network that already has more connections will create realistic networks. Work in this area has focused on the purely formal properties of the network: Once a vertex has many edges, more vertices will attach to it.

Lawrence and Lorsch [12] proposed a model of aggregation in which individual ambassadors carry information back and forth between culturally distinct parts of the corporation. Porter's [13] ideas on how organizational units specialize and, as a result, create boundaries underlying the justifications offered for most aggregation technologies [18].

In work on coordination science [19–21], patterns of behavior are classified with respect to producers and consumers of resources. Aggregation, then, can be seen as a linking between places.

In a series of papers, Jones et al. [22–26] have investigated a particular instance of information aggregation in sensor networks, namely that of reaching a consensus. Although their model of a sensor network is similar to ours, their approach is not explicitly probabilistic.

In previous work, Nickerson and Olariu [27] have focused on latency as a key aspect of modern communication networks and have investigated its implications in the context of sensor networks [28,29]. In References 30, 31, they have proposed metrics for measuring aggregation. They have introduced the idea of *couriers* moving entities that visit sensors and aggregate the information between sensors or sensor clusters [32]. This is similar to aggregating information by using mobile data collectors [33].

Nickerson and Olariu [31] proposed a model of aggregation and introduced a discount function [4] effectively reducing the value of a network when latency delays communication.

The size of data to aggregate will affect the time to aggregate. In fully electronic networks, the throughput of a link and the message size together affect overall latency (see Reference 27 for a discussion on this topic related to mobile computing). Thus, in a more complex model, the size of the message and the electronic throughput may be important to include.

In their decision model presented in Reference 1, Nickerson and Olariu looked at the aggregation of a particular event an intrusion across a set of sensors, provided an incentive-based probabilistic model for pairwise aggregation and showed that the model finds applications to sensor networks, social networks, and business applications such as merging between companies and departments. They did not model the continuous monitoring of a stream of events. In such cases, the decay of information becomes important: It will not make sense to aggregate old, lower value information in comparison with newer, higher value information. When there is an event stream, it becomes possible for the sensors to learn something about the environment and about their fellow sensors' behaviors. Therefore, over time, the sensors might make more effective decisions about who to aggregate with, based on past experiences, using reinforcement learning techniques [34–36].

## 26.3 Modeling the Process of Time-Discounting

We assume that the time-discounting process occurs in continuous time. Consider an arbitrary sensor and let $X$ be the random variable that describes some characteristic of a sensed attribute. To specify that the sensor has collected the data at time $r$, we shall write $X(r)$ and refer to it as the *value* of the data at time $r$. To avoid trivialities, we shall assume that $X(r) \neq 0$.

The value of the information collected by the sensor decays over time. In its most general form, for $t \geq r$, the *discounted* value, $X(t)$, of $X(r)$ at time $t$ is given by

$$X(t) = X(r)g(r, t) \tag{26.1}$$

where $g : \mathbb{R}^+ \cup \{0\} \times \mathbb{R}^+ \cup \{0\} \rightarrow [0, 1]$ is referred to as a *discount function*.

We assume that $g(r, t) = 1$ when if $t = r$, in other words, that there is no discount on intervals of measure 0.

In a variety of practical applications, the discount function in Equation 26.1 is, actually, a function of the difference $t - r$ only, that is, a function of the difference between the time of data collection and the current time. With this in mind, in this chapter we are interested in discount functions satisfying the condition

$$X(t) = X(r)\delta(t - r) \tag{26.2}$$

with $\delta : \mathbb{R}^+ \cup \{0\} \longrightarrow [0, 1]$.

Equation 26.2 tells us that the *penalty* of waiting for $t - r$ time is that the value of the information collected by the sensor decreases from $X(r)$ to $X(t)$. Refer to Figure 26.1 for an illustration.

It is worth noting at this point that one can specialize the discount function $g(\cdot, \cdot)$ in Equation 26.1 in myriad ways, each relevant to some practical application [35,37–39]. By the same token, one can specialize the discount function $\delta(t - r)$ in Equation 26.2 in various ways suggested by practical scenarios [40–42].

Obviously, $X(r) = X(r)\delta(0)$ implies

$$\delta(0) = 1. \tag{26.3}$$

Further, we assume that after a very long time, the value of information vanishes. Formally, we assume that

$$\lim_{x \to \infty} \delta(x) = 0. \tag{26.4}$$

We begin by proving the following useful result that will be instrumental in obtaining a closed form for $\delta$.

**Lemma 26.1** *If $X(r) \neq 0$, then for all $r$, $s$, $t$ with $0 \leq r \leq s \leq t$*

$$\delta(t - r) = \delta(s - r)\delta(t - s). \tag{26.5}$$

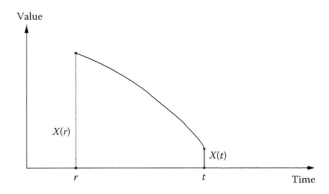

**FIGURE 26.1**   Illustrating the time discounting of information.

*Proof.* Applying Equation 26.2 to the pairs $(r, s)$, $(s, t)$, $(r, t)$ we obtain $X(s) = X(r)\delta(s - r)$, $X(t) = X(s)\delta(t - s)$ and $X(t) = X(r)\delta(t - r)$ which, combined, yield

$$X(r)\delta(t - r) = X(r)\delta(s - r)\delta(t - s).$$

As $X(r) \neq 0$, the conclusion follows. ∎

Observe that by virtue of Equation 26.4, $\delta$ cannot be identically 1 on $\mathbb{R}^+ \cup \{0\}$. Our next result shows that, in fact, $\delta$ takes on the value 1 if and only if $x = 0$.

**Lemma 26.2** $\delta(x) = 1$ *if and only if* $x = 0$.

*Proof.* Recall that by Equation 26.3, if $x_0 = 0$, then $\delta(x_0) = 1$. To prove the converse, let $x_0$ be the *largest* nonnegative real for which $\delta(x_0) = 1$. It suffices to show that $x_0 = 0$. Suppose not and consider $\delta(2x_0)$. We can write

$$\delta(2x_0) = \delta(2x_0 - 0)$$
$$= \delta(2x_0 - x_0)\delta(x_0 - 0) \quad \text{[by Equation 26.5]}$$
$$= \delta(x_0)\delta(x_0)$$
$$= 1, \quad [\text{since } \delta(x_0) = 1]$$

contradicting the maximality of $x_0$. Thus, $x_0 = 0$, and the proof of the lemma is complete. ∎

**Corollary 26.1** *For all* $x > 0$, $0 < \delta(x) < 1$.

*Proof.* Follows immediately from Equation 26.3 and Lemma 26.2, combined. ∎

For all $r$, $s$, $t$ with $0 \leq r \leq s \leq t$, let $x$ and $y$ stand for $s - r$ and $t - s$, respectively. In this notation, $t - r = x + y$ and Equation 26.5 can be written in the equivalent form

$$\delta(x + y) = \delta(x)\delta(y) \tag{26.6}$$

with both $x$ and $y$ nonnegative. It is well known [43] that the functional Equation 26.6 has a *unique* solution that we discuss next.

**Theorem 26.1** *If the function*

$$f : [0, \infty) \longrightarrow \mathbb{R}$$

*satisfies the functional equation* $f(x + y) = f(x)f(y)$ *and is not identically zero then there exists a constant a such that*

$$f(x) = e^{ax} \tag{26.7}$$

We are now in a position to show that the discount function $\delta$ is, in fact, an exponential. The details are spelled out by the following theorem.

**Theorem 26.2** *For all* $r$ *and* $t$ *with* $0 \leq r \leq t$,

$$\delta(t - r) = e^{-\mu(t-r)}$$

*where*

$$\mu = -\ln \delta(1) > 0$$

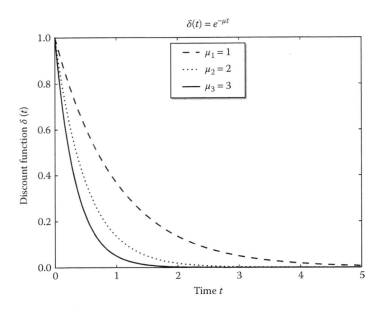

**FIGURE 26.2**   Three exponential time-discount functions with increasing discount constant.

*Proof.* Recall that by Equation 26.6 the discount function $\delta$ satisfies the conditions of Theorem 26.1. Moreover, by Corollary 26.1 $0 < \delta(1) < 1$ and so $\ln \delta(1) < 0$. Thus, with

$$\mu = -\ln \delta(1) > 0$$

the expression of $\delta(t - r)$ becomes $\delta(t - r) = e^{-\mu(t-r)}$, as claimed.  ∎

Theorem 26.2 shows that, under mild assumptions, the discount function is an exponential. We note that a similar result was derived by References 1, 31 in the case of discrete time.

Figure 26.2 features three exponential time-discount functions with increasing *discount rate* $\mu$ in which $r = 0$.

Thus, assuming a discount rate $\mu$, the residual value $X(t)$ at time $t$, $(t \geq r)$, of the initial value $X(r)$, can be determined by equation Equation 26.8.

$$X(t) = X(r)e^{-\mu(t-r)}. \qquad (t \geq r) \tag{26.8}$$

Equation 26.8 tells us that the value of information decays exponentially with time. Consequently, it is extremely important to find aggregation strategies to offset this decaying effect.

## 26.4 Aggregation: Counteracting Discounting

For information, the value of which is subject to time-discounted, we use aggregation to counteract the effect of time discounting. Consider two sensors that have collected data about an event at times $r$ and $s$, respectively. Let $X(r)$ and $Y(s)$ be, respectively, the values of the information collected by the two sensors. At some later time $\tau$, the two sensors decide to aggregate their information.

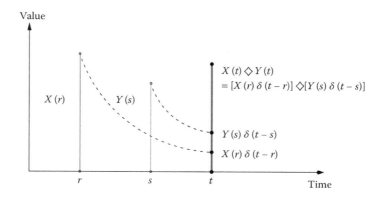

**FIGURE 26.3** Illustrating the aggregation at time $t$ of $X(r)$ and $Y(s)$.

By definition, the aggregated value of $X(r)$ and $Y(s)$ at time $t$, with $0 \leq r \leq t$ and $0 \leq s \leq t$, is $X(t)\diamond Y(t)$. By Equation 26.2, it follows that

$$X(t)\diamond Y(t) = [X(r)\delta(t-r)] \diamond [Y(s)\delta(t-s)]. \tag{26.9}$$

Thus, what is being aggregated at time $t$ are the discounted values $X(r)\delta(t-r)$ and $Y(s)\delta(t-s)$. The aggregator can be extended, in the obvious way, to an arbitrary number of values that need to be aggregated. Refer to Figure 26.3 for a pictorial representation of the aggregation process of $X(t)$ and $Y(t)$.

Useful instances of $\diamond$ include $+$, max, min, XOR, OR, among many others. It is worth observing that the type of aggregator that should be applied is application-dependent. In fact, an aggregator that makes sense in one application may be totally irrelevant in a different context.

Most of the aggregation operators $\diamond$ of practical relevance have the following fundamental properties as follows:

*Commutativity:* $X(t)\diamond Y(t) = Y(t)\diamond X(t)$ for all $t \geq 0$. In other words, the result of the aggregation does not depend on the order in which the values are aggregated.

*Associativity:* $[X(t)\diamond Y(t)]\diamond Z(t) = X(t)\diamond[Y(t)\diamond Z(t)]$ for all $t \geq 0$. If several values are aggregated in groups, the value of the aggregated information does not depend on the order in which groups are formed. We shall follow established practice to write $X(t)\diamond Y(t)\diamond Z(t)$ instead of the cumbersome parenthesized expressions. A straightforward inductive argument shows that if $\diamond$ is associative, then for an arbitrary collection of $n$ values $X_1(t)$, $X_2(t)$, $\cdots$, $X_n(t)$, we have

$$[X_1(t)\diamond X_2(t)\diamond \cdots \diamond X_{n-1}(t)]\diamond X_n(t)$$
$$= X_1(t)\diamond[X_2(t)\diamond X_3(t)\diamond\cdots\diamond X_n(t)]$$
$$= X_1(t)\diamond X_2(t)\diamond X_3(t)\diamond\cdots\diamond X_n(t).$$

To simplify notation, when aggregating several values $X_1,(t)\, X_2(t),\, \cdots,\, X_n(t)$, we shall write $\diamond_{i=1}^{n}X_i(t)$ instead of $X_1(t)\diamond X_2(t)\diamond\cdots\diamond X_n(t)$.

*Idempotency:* If $Y(t) = 0$, then $X(t)\diamond Y(t) = X(t)$. In other words, aggregation with data of value 0 has no effect. It may be worth noting that the idempotency property may be extended to read $X\diamond Y = X$ whenever $Y \leq X$, giving expression to our intuitive idea that one does not stand to gain by aggregating with information of lesser value [35].

### 26.4.1 A Taxonomy of Aggregation Operators

Consider, again, two sensors that have collected, at time $t$, data about some attribute of an event they have witnessed and let $X(t)$ and $Y(t)$ be, respectively, the values of the information collected. It is important for $X(t)$ and $Y(t)$ to be aggregated to obtain a more reliable and, perhaps, more relevant information about the event at hand.

Suppose, further, that aggregation takes time and that the aggregated information is available at time $\tau > t$. Given that the value of the data decays over time exponentially according to Equation 26.2, and given an aggregator $\diamond$, what strategy maximizes the value of the information at time $\tau$?

The answer to this question depends on the type of aggregator used. Indeed, the aggregator $\diamond$ can be one of three distinct types defined as follows:

- *Type 1:* For all $t$, $\tau$ with $0 \leq t \leq \tau$, $[X(t)\diamond Y(t)]\,\delta(\tau - t) < X(\tau)\diamond Y(\tau)$. In other words, in the case of a Type 1 aggregator, it is best to defer aggregation as long as the semantics of the application permit;
- *Type 2:* For all $t$, $\tau$ with $0 \leq t \leq \tau$, $[X(t)\diamond Y(t)]\,\delta(\tau - t) = X(\tau)\diamond Y(\tau)$. As it is apparent, in the case of a Type 2 aggregator, the order between aggregation and discount does not matter. In reality, we need to aggregate as soon as possible because the value may decay quickly; and
- *Type 3:* For all $t$, $\tau$ with $0 \leq t \leq \tau$, $[X(t)\diamond Y(t)]\,\delta(\tau - t) > X(\tau)\diamond Y(\tau)$. Thus, in the case of a Type 3 aggregator, the best strategy is to aggregate as early as the data are available and/or the semantics of the application permit.

Observe that for a Type 2 aggregator, the discount function distributes over $\diamond$. Consequently, for such an aggregator it does not matter whether we first aggregate and then discount the aggregated information or vice versa. In Sections 26.5 and 26.6, we take a closer look at aggregators of Types 1 and 2, respectively.

## 26.5 Aggregators of Type 1

Let $\diamond$ be an *arbitrary* Type 1 aggregator and assume that $n$, $(n \geq 2)$, sensors have collected data about an event at times $t_1, t_2, \ldots, t_n$. Let $X_1(t_1), X_2(t_2), \ldots, X_n(t_n)$ be, respectively, the values collected by the sensors. At some later time $\tau \geq \max_{1 \leq i \leq n} t_i$, the sensors decide to aggregate their information. We take note of the following relevant result.

**Lemma 26.3** *Assume an associative Type 1 aggregator $\diamond$. For all $t$, $\tau$ with $\max_{1 \leq i \leq n}\{t_i\} \leq t \leq \tau$ we have*

$$\left[\diamond_{i=1}^n X_i(t)\right]\delta(\tau - t) < \diamond_{i=1}^n X_i(\tau).$$

*Proof.* The proof is by induction on $n$. For $n = 2$, the conclusion follows from the definition of a Type 1 aggregator. For the inductive step, let $n \geq 3$ be arbitrary and assume that the property holds for $n - 1$ aggregated values. We write

$$\begin{aligned}
\left[\diamond_{i=1}^n X_i(t)\right]\delta(\tau - t) \\
= \left[\left(\diamond_{i=1}^{n-1} X_i(t)\right)\diamond X_n(t)\right]\delta(\tau - t) \\
< \left[\diamond_{i=1}^{n-1} X_i(t)\delta(\tau - t)\right]\diamond[X_n(t)\delta(\tau - t)] \\
= \diamond_{i=1}^n [X_i(t)\delta(\tau - t)] \\
= \diamond_{i=1}^n X_i(\tau),
\end{aligned}$$

completing the proof of Lemma 26.3. ∎

The left-hand side of the above-mentioned inequality is the discounted value of $\diamondsuit_{i=1}^n X_i(t)$ at time $\tau$, whereas the right hand is the aggregated value of the discounted values of $X_i(t_i)$ at time $\tau$. Lemma 26.3 asserts that the defining inequality of the Type 1 aggregator holds when an arbitrary number, $n$, of values are being aggregated.

Consider an event witnessed by $n$, $(n \geq 2)$, sensors and let the sensed values collected, respectively, at times $t_1, t_2, \ldots, t_n$ be denoted by $X_1(t_1), X_2(t_2), \ldots, X_n(t_n)$. Assume, further, that various groups of sensors have aggregated their data before time $t$ and that, finally, at time $t$ the aggregation has been completed. We are interested in evaluating the time-discounted value of the information collected by the sensors at time $t$, in which $t \geq \max\{t_1, t_2, \ldots, t_n\}$. The answer to this natural question is provided by the following fundamental result.

**Theorem 26.3** *Assuming that the Type 1 aggregator $\diamondsuit$ is associative and commutative, the discounted value of the aggregated information at time t is upper-bounded by $\diamondsuit_{i=1}^n X_i(t)$, regardless of the order in which the values were aggregated.*

*Proof.* The proof is by induction on $n$. For $n = 2$, the conclusion follows at once from definition. Now, let $n \geq 3$ be arbitrary and assume the statement true for all $m$, $(m < n)$. We assume, without loss of generality, that the last aggregation takes place at time $t$. This aggregation must have involved a number of disjoint groups $G_1, G_2, \ldots, G_p$, each of them is the result of a previous aggregation at times, respectively, $u_1, u_2, \ldots, u_p$. Observe that we can always relabel the groups in such a way that their aggregation times are ordered as $u_1 < u_2 < \cdots < u_p$.

Let us look at group $G_k$. By the induction hypothesis, the value of information in group $G_k$ aggregated at time $u_k$ is upper-bounded by $\diamondsuit_{j=1}^{n_k} X_{k_j}(u_k)$ where, of course, we assume that group $G_k$ involves $n_k$ sensors whose values were aggregated.

Assuming $t \geq u_k$, Lemma 26.3 guarantees that the discounted value at time $t$ is upper-bounded by

$$\left[\diamondsuit_{j=1}^{n_k} X_{k_j}(u_k)\right] \delta(t - u_k) < \diamondsuit_{j=1}^{n_k} X_{k_j}(t),$$

which is an upper bound on the value of information collected by sensors in group $G_k$, had it been aggregated at time t. As $G_k$ was arbitrary, the conclusion follows. ∎

Theorem 26.3, in effect, says that the maximum value of the aggregated information that can be attained is independent of the order in which the values are aggregated. In practical terms, Theorem 26.4 gives the algorithm designer the freedom to schedule aggregation in a random manner, much in line with the stochastic nature of wireless communication and sensor data aggregation.

Recall that for Type 1 aggregators, aggregation can be delayed as long as the semantics of the application permit. In the next subsection, we look at a thresholding mechanism that defines how long it is feasible to delay aggregation.

## 26.5.1 A Special Type 1 Aggregator

We begin by taking note of a nontrivial Type 1 aggregation operator that turns out to have interesting applications. Imagine that the data collected by sensors take on values in the range $[0, 1]$ and consider the aggregator $\diamondsuit$ defined as

$$X(t)\diamondsuit Y(t) = X(t) + Y(t) - X(t)Y(t). \tag{26.10}$$

It is straightforward to verify that $\diamondsuit$ satisfies the associativity, commutativity, and idempotency properties defined earlier. To prove that $\diamondsuit$ is a Type 1 aggregator, consider an arbitrary $\tau$ with $0 \leq t \leq \tau$ and write

$$[X(t)\Diamond Y(t)]\,\delta(\tau - t) = [X(t) + Y(t) - X(t)Y(t)]\,\delta(\tau - t)$$
$$= X(t)\delta(\tau - t) + Y(t)\delta(\tau - t) - X(t)Y(t)\delta(\tau - t)$$
$$= X(\tau) + Y(\tau) - X(\tau)Y(t) \quad [\text{by Equation 26.2}]$$
$$\leq X(\tau) + Y(\tau) - X(\tau)Y(\tau) \quad [\text{since } Y(\tau) \leq Y(t)]$$
$$= [X(t)\delta(\tau - t)] \Diamond [Y(t)\delta(\tau - t)]$$
$$= X(\tau)\Diamond Y(\tau),$$

confirming that the aggregator defined in Equation 26.10 is of Type 1.

For later reference, we take note of the following useful property of the aggregator in Equation 26.10, established in Reference 35.

**Lemma 26.4** *Consider values $X_1, X_2, \cdots, X_n$ in the range $[0,1]$ acted upon by the aggregator $\Diamond$ defined in Equation 26.10. Then the aggregated value $\Diamond_{i=1}^n X_i$ satisfies the condition*

$$\Diamond_{i=1}^n X_i = 1 - \Pi_{i=1}^n (1 - X_i). \tag{26.11}$$

*Proof.* The proof is by induction. To settle the basis, consider $n = 2$ and assume that both $X_1$ and $X_2$ are reals in the interval $[0,1]$. Observe that $1 - (X_1\Diamond X_2) = 1 - (X_1 + X_2 - X_1X_2) = (1 - X_1)(1 - X_2)$.

For the inductive step, let $n$ be arbitrary and assume that the statement of the lemma true for the chosen value of $n$. With this in hand, we need to show that $1 - \Diamond_{i=1}^{n+1} X_i = \Pi_{i=1}^{n+1}(1 - X_i)$. Write $Y = 1 - \Pi_{i=1}^n (1 - X_i)$. In this notation

$$1 - \Diamond_{i=1}^{n+1} X_i = 1 - \left(\Diamond_{i=1}^n X_i\right)\Diamond X_{n+1}$$
$$= 1 - Y\Diamond X_{n+1}$$
$$= 1 - Y - X_{n+1} + YX_{n+1}$$
$$= \Pi_{i=1}^n (1 - X_i) - X_{n+1}\Pi_{i=1}^n (1 - X_i)$$
$$= \Pi_{i=1}^{n+1}(1 - X_i)$$

and the proof of the lemma is complete.                    ∎

## 26.6 Aggregators of Type 2

For a Type 2 aggregator defined in Section 26.4.1, the order between aggregation and discount does not matter; however, we need to aggregate as soon as possible because the value will decay quickly in reality.

With the same notations, assumptions, and reasoning, we have similar *lemma* and *theorem* for Type 2 aggregator as for Type 1.

**Lemma 26.5** *Assume an associative Type 2 aggregator $\Diamond$. For all $t$, $\tau$ with $\max_{1\leq i\leq n}\{t_i\} \leq t \leq \tau$ we have*

$$\left[\Diamond_{i=1}^n X_i(t)\right]\delta(\tau - t) = \Diamond_{i=1}^n X_i(\tau).$$

**Theorem 26.4** *Assuming that the Type 2 aggregator $\Diamond$ is associative and commutative, the discounted value of the aggregated information at time $t$ is $\Diamond_{i=1}^n X_i(t)$, regardless of the order in which the values were aggregated.*

These *lemma* and *theorem* have similar implications as Type 1 aggregator and can be proved in a similar way as in Section 26.5.

To be able to understand how time discounting affects aggregated values, we shall find it convenient to assume that the discount operator — multiplication '·' distributes over the aggregator $\Diamond$.

*Distributivity:* For all $t$, $\tau$ with $0 \leq t \leq \tau$, we can write $[X(t)\diamond Y(t)] \cdot \delta(\tau - t) = [X(t) \cdot \delta(\tau - t)] \diamond [Y(t) \cdot \delta(\tau - t)]$. The discounted value at time $\tau - t$ of the information $X(t)\diamond Y(t)$ aggregated at time $t$ matches the aggregated value at time $\tau$ of $X(t) \cdot \delta(\tau - t)$ and $Y(t) \cdot \delta(\tau - t)$. In other words, it does not matter whether we first aggregate and then discount the aggregated information or vice versa.

The distributivity property is fundamental in understanding the interplay between time discounting and aggregation. We mention in passing that, in general the distributivity property need not be verified. However, for exponentially time-discounted information, we look specifically at aggregation operators in which distributivity holds.

**Lemma 26.6** *Assuming the distributivity property, for all $0 \leq r \leq s \leq t \leq \tau$, we have*

$$[X(t)\diamond Y(t)] \cdot \delta(\tau - t) = [X(r) \cdot \delta(\tau - r)] \diamond [Y(s) \cdot \delta(\tau - s)]. \qquad (26.12)$$

*Proof.* By using distributivity, we write

$$[X(t)\diamond Y(t)] \cdot \delta(\tau - t)$$
$$= [X(t) \cdot \delta(t - r)\diamond Y(t) \cdot \delta(t - s)] \cdot \delta(\tau - t) \ \text{[by (26.2)]}$$
$$= [X(t) \cdot \delta(t - r) \cdot \delta(\tau - t)] \diamond [Y(t) \cdot \delta(t - s) \cdot \delta(\tau - t)]$$
$$= X(r) \cdot \delta(\tau - r)\diamond Y(t) \cdot \delta(\tau - s). \ \text{[by Lemma 26.1]} \qquad \blacksquare$$

The left-hand side of Equation 26.12 is the discounted value of $X(t)\diamond Y(t)$ at time $\tau$, whereas the right hand is the aggregated value of the discounted values of $X(r)$ and $Y(s)$ at time $\tau$.

## 26.6.1 Some Special Classes of Type 2 Aggregators

There are some special classes of Type 2 aggregation operators such as $+, \min, \max$ that turn out to have wide applications. Assume that the data collected by sensors take on values in $\mathbb{R}$, these aggregation operators $\diamond$ can be defined as

$$+ : \ X(t)\diamond Y(t) = X(t) + Y(t)$$
$$\max : \ X(t)\diamond Y(t) = \max\{X(t), Y(t)\}$$
$$\min : \ X(t)\diamond Y(t) = \min\{X(t), Y(t)\}$$

It is easy to verify that $\diamond$ satisfies the associativity and commutativity properties defined earlier. For idempotency, $+$, max, and min satisfy obvious relationships, which we omit at this time.

The proof that these $\diamond$s are Type 2 aggregators is immediate. We only prove the statement for $+$. Consider an arbitrary $\tau$ with $0 \leq t \leq \tau$ and write

$$[X(t)\diamond Y(t)] \, \delta(\tau - t)$$
$$= [X(t) + Y(t)] \, \delta(\tau - t)$$
$$= X(t)\delta(\tau - t) + Y(t)\delta(\tau - t)$$
$$= X(\tau) + Y(\tau) \ \text{[by Equation 26.2]}$$
$$= X(\tau)\diamond Y(\tau)$$

confirming that the aggregator $+$ defined in Equation 26.13 is of Type 2.

## 26.6.2 Exponentially-Discounted Value of Aggregated Information

Consider an event witnessed by $n$, $(n \geq 2)$, sensors and let the sensed values collected, respectively, at times $t_1, t_2, \cdots, t_n$ be denoted by $X_1(t_1), X_2(T_2), \cdots, X_n(T_n)$. Assume, further, that various groups of sensors have aggregated their information before time $t$ and that, finally, at time $t$ the aggregation has been completed. We are interested in evaluating the time-discounted value of the information collected by the sensors at time $t$, in which $t \geq \max\{t_1, t_2, \cdots, t_n\}$. The answer to this natural question is provided by the following fundamental result.

**Theorem 26.5** *Assuming distributivity of the discount operator $\cdot$ over the aggregator $\diamond$, the discounted value $V(t)$ of the aggregated information at time $t$ is*

$$V(t) = \diamond_{i=1}^n X_i(t_i) \cdot \delta(t - t_i), \tag{26.13}$$

*regardless of the order in which the values were aggregated.*

*Proof.* The proof is by induction on $n$. For $n = 2$, the conclusion follows at once from Lemma 26.6. Now, let $n \geq 2$, be arbitrary and assume the statement true for all $m$, $(m < n)$. We assume, without loss of generality, that the last aggregation takes place at time $t$. This aggregation must have involved a number of disjoint groups $G_1, G_2, \ldots, G_p$, each of them is the result of a previous aggregation at times, respectively, $u_1, u_2, \ldots, u_p$.

Observe that we can always relabel the groups in such a way that their aggregation times are ordered as $u_1 < u_2 < \cdots < u_p$.

Let us look at group $G_k$. By the induction hypothesis, the value of information in group $G_k$ aggregated at time $u_k$ was

$$V(u_k) = \diamond_{j=1}^{n_k} X_{k_j} \cdot \delta(u_k - t_{k_j})$$

where, of course, we assume that group $G_k$ involves $n_k$ sensors whose values were aggregated.

Assuming $t \geq u_k$, the discounted value of $V(u_k)$ at time $t$ is

$$
\begin{aligned}
V_k(t) &= \left[ \diamond_{j=1}^{n_k} X_{k_j} \cdot \delta(u_k - t_{k_j}) \right] \cdot \delta(t - u_k) \\
&= \diamond_{j=1}^{n_k} X_{k_j} \cdot \delta(u_k - t_{k_j}) \cdot \delta(t - u_k) \quad \text{[by distributivity]} \\
&= \diamond_{j=1}^{n_k} X_{k_j} \cdot \delta(t - t_{k_j}) \quad \text{[by Lemma 26.1].}
\end{aligned}
$$

which is exactly the discounted value of information collected by sensors in group $G_k$, had it been aggregated at time $t$. As $G_k$ was arbitrary, the conclusion follows. ∎

Theorem 26.5, in effect, says that the order in which the values are aggregated does not matter as long as each is aggregated only once. In practical terms, Theorem 26.5 gives the algorithm designer the freedom to schedule aggregation in a random manner, much in line with the stochastic nature of wireless communication and sensor data aggregation [42].

# 26.7 Approximating the Exponential Discount Function

## 26.7.1 The Approximation Problem

As the sensors are generally viewed as computationally challenged [42], we are interested in replacing the task of evaluating an exponential function by a suitable approximation thereof. One of the most natural such approximations is offered by the well-known Taylor expansion of the exponential function. However, some of these approximations may be converging too slowly to be efficient. As an alternative, we propose to approximate the exponential function by a different polynomial that we discuss in the following. However, before we do so, we remind the reader that a well-known approximation theorem of

Weierstrass [44] guarantees that every continuous function defined on a closed interval $[a, b]$ can be uniformly approximated as closely as desired by a polynomial. As the exponential function is continuous on any closed interval, Weierstrass' theorem guarantees that it can be approximated by a polynomial. One of the most natural instances of such a polynomial is offered by the well-known Taylor expansion of the exponential function. As an alternative, we propose to approximate the exponential function by a different polynomial that we discuss in the following.

For simplicity, suppose a variable $X$ has a value $X(0)$ collected at time 0, then at any late time $t$, its exponentially discounted value $X(t)$ at rate $\mu$ is

$$X(t) = X(0)e^{-\mu t} \qquad (t \geq 0). \qquad (26.14)$$

Equation 26.14 can be rewritten as

$$X(t) = X(0)e^{-\frac{t}{\tau}} \qquad (t \geq 0), \qquad (26.15)$$

where $\tau = 1/\mu$.

Let us take a few special time points to get a feel for this exponential discount function $\delta(t) = e^{-\mu t}$:

- $t = 0$, $\delta(0\tau) = 1$;
- $t = \tau$, $\delta(1\tau) = e^{-1} \approx 0.3679$;
- $t = 5\tau$, $\delta(5\tau) = e^{-5} \approx 0.0067 < 1\%$. This is considered as "vanished"; and
- Half-life $t_{1/2}$: $\delta(t_{1/2}) = \frac{1}{2} \Rightarrow t_{1/2} = \tau \ln 2 \approx 0.693\tau$, that is, after half-life $t_{1/2}$, half of the original value is left.

A time table covering the range from $t = 0$ to $t = 5\tau$ is suitable as a value is considered as "vanished" after $5\tau$.

To find the impact of approximation on aggregation, suppose $\hat{\delta}(t)$ is the approximation function for the exponential discount function $\delta(t)$ with error $\epsilon$, that is,

$$|\hat{\delta}(t) - \delta(t)| \leq \epsilon \quad \forall t \in [0, 5\tau]. \qquad (26.16)$$

Meanwhile, $\hat{\delta}(t)$ satisfies $0 \leq \hat{\delta}(t) \leq 1$. As the type of an aggregator is determined by the order of $[X(t) \diamond Y(t)] \delta(r - t)$ and $X(r) \diamond Y(r)$, for all $t$, $r$ with $0 \leq t \leq r$, so we require $\hat{\delta}(t)$ keeping the order.

In our application, we don't need a range as wide as $[0, 5\tau]$. In the application of Type 1 aggregator, we discard those values below 1/2, in which we only need approximate values within $[0, t_{1/2}]$. In the application of Type 2 aggregator, we discard values below the reporting threshold, the lowest ratio of decayed value to its initial value is about 0.1, in which the range is about $[0, 2.3\tau]$. With this in mind, in the following approximation, we discuss these two ranges. We set up *two indexes* to measure the approximating quality: One is the *maximum absolute error* $E_{am}$,

$$E_{am} = \max_{t \in [0, T]} |\hat{\delta}(t) - \delta(t)|, \qquad (26.17)$$

the other is the *expected absolute error* $E_{ae}$,

$$E_{ae} = \int_0^T |\delta(x) - \hat{\delta}(x)| dF(x), \qquad (26.18)$$

when all the approximated values within the range are used under a distribution $P\{X \leq x\} = F(x)$. Here, we take uniform distribution, the expected absolute error becomes the mean absolute error.

For $x = \mu t$, the exponential discount function can be rewritten as

$$\delta(x) = e^{-x} \qquad x \in [0, T]. \qquad (26.19)$$

Then $t \in [0, t_{1/2}]$ corresponds to $x \in [0, \ln 2]$ and $t \in [0, 2.3\tau]$ to $x \in [0, 2.3]$.

## 26.7.2 Approximation by Primary Component Analysis

To find the major characteristics of exponential time-discounting effect, we use PCA to decompose $e^{-x}$ into a series of components on the polynomial function space. Instead of using an infinite number of components, we use only the first four components that project in the orthogonal polynomial function space constructed through *Gram–Schmidt* process. With *Gram–Schmidt* process, we construct a set of orthogonal polynomial bases $\{p_0, p_1, p_2, p_3\}$ from the set of $\{1, x, x^2, x^3\}$ on the interval $[0, T]$ as follows:

$$p_0(x) = 1$$

$$p_1(x) = x - \frac{T}{2}$$

$$p_2(x) = x^2 - Tx + \frac{T^2}{6}$$

$$p_3(x) = x^3 - \frac{3}{2}Tx^2 + \frac{3}{5}T^2x + \frac{T^3}{5}$$

Accordingly, $e^{-x}$ is expanded in this space and represented as a linear combination of the above-mentioned bases as follows:

$$e^{-x} \approx P_3(x) = \sum_{k=0}^{3} a_k p_k(x)$$

$$a_k = \frac{\int_0^T e^{-x} p_k(x) dx}{\int_0^T p_k^2(x) dx},$$

where $a_k$ is the $k$th component projected on the $k$th base.
Through definite integral, $a_k$ is determined as

$$a_0 = \frac{1 - e^{-T}}{T}$$

$$a_1 = \frac{6\left(2 - T - e^{-T}(T + 2)\right)}{T^3}$$

$$a_2 = \frac{30\left[T^2 + 12 - 6T - e^{-T}(T^2 + 6T + 12)\right]}{T^5}$$

$$a_3 = \frac{140\left[12T^2 + 120 - T^3 - 60T - e^{-T}(T^3 + 12T^2 + 60T + 120)\right]}{T^7}$$

When $T = 0.693$ and $T = 2.3$, these components and polynomial bases are in Tables 26.1 and 26.2. The smaller $k$, the more information $p_k$ contains, as illustrated in Figure 26.4.

**TABLE 26.1**  Components

| $T$ | $a_0$ | $a_1$ | $a_2$ | $a_3$ |
|-----|-------|-------|-------|-------|
| 0.693 | 0.7214 | −0.7157 | 0.3566 | −0.1186 |
| 2.3 | 0.3912 | −0.3605 | 0.1738 | −0.0568 |

**TABLE 26.2**    Orthogonal Polynomial Bases

| $T$ | 0.693 | 2.3 |
|---|---|---|
| $p_0(x)$ | 1 | 1 |
| $p_1(x)$ | $x - 0.3465$ | $x - 1.15$ |
| $p_2(x)$ | $x^2 - 0.693x + 0.08$ | $x^2 - 2.3x + 0.8817$ |
| $p_3(x)$ | $x^3 - 1.0395x^2 + 0.2881x + 0.0666$ | $x^3 - 3.45x^2 + 3.174x + 2.4334$ |

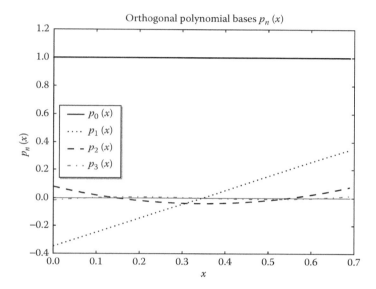

**FIGURE 26.4**    The amplitudes of four bases when $T = 0.693$.

Hence, the first four approximating polynomials are: when $T = 0.693$,

$$P_0(x) = \sum_{k=0}^{0} a_k p_k(x) = 0.7214$$

$$P_1(x) = \sum_{k=0}^{1} a_k p_k(x) = -0.7157x + 0.9694$$

$$P_2(x) = \sum_{k=0}^{2} a_k p_k(x) = 0.3566x^2 - 0.9628x + 0.9979$$

$$P_3(x) = \sum_{k=0}^{3} a_k p_k(x) = -0.1186x^3 + 0.4799x^2 - 0.9970x + 0.9999$$

The corresponding curves are shown in Figure 26.5 and approximating qualities in Table 26.3.

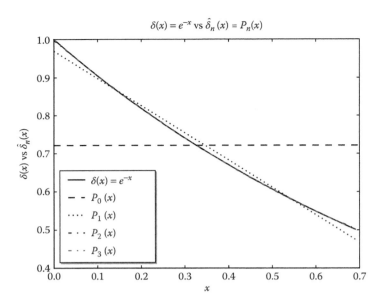

**FIGURE 26.5**   Polynomial approximations of $e^{-x}$ when $T = 0.693$.

**TABLE 26.3**   The Approximating Quality of $P_n(x)$ When $T = 0.693$

| $P_n(x)$ | $P_0(x)$ | $P_1(x)$ | $P_2(x)$ | $P_3(x)$ |
|---|---|---|---|---|
| $E_{am}$ | 0.2213 | 0.02666 | $9 \times 10^{-4}$ | $4.08 \times 10^{-5}$ |
| $E_{ae}$ | 0.12416 | 0.01099 | $6.4 \times 10^{-4}$ | $2.8 \times 10^{-5}$ |

**TABLE 26.4**   The Approximating Quality of $P_n(x)$ When $T = 2.3$

| $P_n(x)$ | $P_0(x)$ | $P_1(x)$ | $P_2(x)$ | $P_3(x)$ |
|---|---|---|---|---|
| $E_{am}$ | 0.6088 | 0.12368 | 0.01665 | $2.212 \times 10^{-3}$ |
| $E_{ae}$ | 0.2101 | 0.0593 | 0.011255 | $1.6 \times 10^{-3}$ |

when $T = 2.3$,

$$P_0(x) = \sum_{k=0}^{0} a_k p_k(x) = 0.3912$$

$$P_1(x) = \sum_{k=0}^{1} a_k p_k(x) = -0.3605x + 0.8058$$

$$P_2(x) = \sum_{k=0}^{2} a_k p_k(x) = 0.1738x^2 - 0.7604x + 0.9591$$

$$P_3(x) = \sum_{k=0}^{3} a_k p_k(x) = -0.0568x^3 + 0.3697x^2 - 0.9405x + 0.9936$$

The corresponding curves are shown in Figure 26.6 and approximating qualities in Table 26.4.

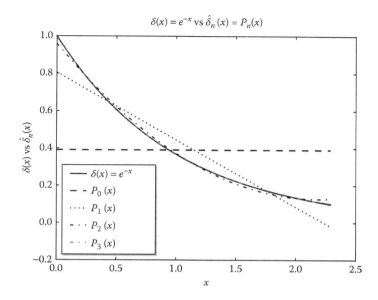

**FIGURE 26.6** Polynomial approximations of $e^{-x}$ when $T = 2.3$.

# 26.8 Applications of Data Aggregation

## 26.8.1 Effect of the Aggregation Method

Consider a scenario in which the sensors collect information and report it to a central sink [41,42]. We want to evaluate the effect of wireless communication and aggregation methods on the *average time to aggregate*. Collisions in the wireless channel typically garble messages beyond recognition. Thus, messages need to be retransmitted, which will increase the time to receive all pieces of information to be aggregated. Clearly, the denser the traffic, the more collisions and the more retransmissions causing delays in aggregation [45]. A nonnegligible side effect of all this is that due to collision-caused time delays, the value of individual pieces of information decays and what is being aggregated has lesser value.

Furthermore, the aggregation method could affect the time to achieve the defined aggregation. This latency is due, to a large extent, to the logic behind the transfer of information for aggregation. For example, the aggregation method may use one of the following strategies:

1. Wait until at least $k$ nodes report.
2. Wait until the aggregated value exceeds a threshold.
3. Wait until there is enough spatial diversity in the reported information.

In all these examples, the sensor nodes can decide about reporting in a way that the only the effective ones reports. For example, in the former method, it would be enough if only one node out of several neighbor nodes transmits the message. This can be performed through many methods such as cluster head or a probabilistic transmission policy. Independent of policy, transmitting less number of messages would reduce the traffic and faster reception of required information by the aggregation method.

## 26.8.2 Applications of Type 1 Aggregators

### 26.8.2.1 Thresholding

Sensor networks deployed in support of emergency response applications must provide timely and accurate reports of detected events. Aggregation of sensor data is required to accomplish this in an efficient manner.

The aggregation problem is complicated by the fact that the perceived value of the data collected by the sensors deteriorates, often dramatically, over time. Individual sensors must determine whether to report a perceived event immediately or to defer reporting until the confidence has increased after aggregating data with neighboring nodes. However, aggregation takes time and the longer the sensors wait, the lower the value of the aggregated information.

As already mentioned, we assume that reporting a false positive involves a huge overhead and is considered prohibitively expensive. Mindful of this state of affairs, and having aggregated, at time $t$, the information collected by the various sensors, it is important to decide whether this information warrants reporting.

### 26.8.2.2 A Fixed Aggregation Strategy

One of the natural strategies employed is thresholding. Specifically, a policy is followed by first setting up an application-dependent threshold $\Delta$ and then reporting an event only if the aggregated information exceeds $\Delta$. Refer to Figure 26.7 for an illustration.

Assume that $n$, ($n \geq 2$), sensors have collected data about an event at times $t_1$, $t_2$, ..., $t_n$ and let $t = \max\{t_1, t_2, \ldots, t_n\}$. Further, let $X_1(t_1), X_2(t_2), \ldots, X_n(t_n)$ be the values of the data collected by the sensors. Assuming that $\Diamond_{i=1}^{n} X_i(t) > \Delta$, the time at which the aggregation is performed is critical. We have seen that for a Type 1 operator, aggregation may be delayed as long as the semantics of the application permit. It is, however, intuitively clear that if aggregation is delayed too much, the aggregated value might not exceed the threshold and a relevant event would go unreported. Thus, the question is to determine the *time window* during which the sensors need to aggregate their values in order for the aggregated value to exceed the threshold $\Delta$.

Let $\tau$ be the *latest* time at which aggregation should be performed. As we are interested in Type 1 operators, it is natural to insist that $\left[\Diamond_{i=1}^{n} X_i(t)\right] \delta(\tau - t) > \Delta$. Recalling that by, Theorem 26.2, $\delta(\tau - t) = e^{-\mu(\tau-t)}$, the above-mentioned inequality becomes $e^{-\mu(\tau-t)} > \frac{\Delta}{\Diamond_{i=1}^{n} X_i(t)}$ which, upon taking logarithms on both sides, yields

$$-\mu(\tau - t) > \ln \frac{\Delta}{\Diamond_{i=1}^{n} X_i(t)}.$$

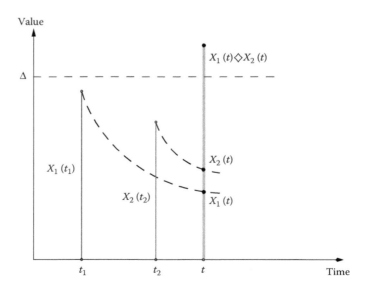

**FIGURE 26.7**  Illustrating how the aggregation of two sensor readings $X_1(t_1)$ and $X_2(t_2)$ can exceed a threshold.

Upon solving for $\tau$ we obtain

$$\tau < t + \frac{1}{\mu} \ln \frac{\diamond_{i=1}^n X_i(t)}{\Delta}. \tag{26.20}$$

As the last data were collected at time $t = \max\{t_1, t_2, \ldots, t_n\}$, Equation 26.20 specifies that, past $t$, there is a time window of size $\frac{1}{\mu} \ln \frac{\diamond_{i=1}^n X_i(t)}{\Delta}$ during which aggregation must occur. This result gives the system designer a handle on the types of aggregation protocols to use. In Equation 26.20, to get a certain length $W$ of time window, that is, the following inequality needs to be satisfied:

$$W \le \frac{1}{\mu} \ln \frac{\diamond_{i=1}^n X_i(t)}{\Delta} \tag{26.21}$$

As well, the following inequality should satisfy to make sense for inequalities Equations 26.20 and 26.21.

$$\diamond_{i=1}^n X_i(t) > \Delta \tag{26.22}$$

Inequality (26.22) is intuitively satisfying because the aggregated value should be larger than the alarm threshold to trigger an alarm and larger enough to have a time window for emergency response.

From Lemma 26.4, it is easy to get an estimation of the length of time window in fixed strategy. Suppose $X_1, X_2, \ldots, X_n$ are all 0.9, then $1 - X_i = 0.1 = 10^{-1}$, substitute into Equation 26.11,

$$\diamond_{i=1}^n X_i = 1 - \Pi_{i=1}^n (1 - X_i)$$
$$= 1 - \Pi_{i=1}^n 0.1$$
$$= 1 - 10^{-n}$$
$$= 0.\underbrace{9 \cdots 9}_{n \text{ nines}}$$

Substitute back into inequality (26.21) and take equal sign, get

$$W = \frac{1}{\mu} \ln \frac{1 - 10^{-n}}{\Delta} \tag{26.23}$$

Treat $W$ in Equation 26.23 as a function of $n$, shown in Figure 26.8. Actually,

$$\lim_{n \to \infty} W(n) = \lim_{n \to \infty} \frac{1}{\mu} \ln \frac{1 - 10^{-n}}{\Delta}$$
$$= \frac{1}{\mu} \ln \frac{1}{\Delta}$$
$$= -\frac{1}{\mu} \ln \Delta$$
$$\approx \frac{0.01005}{\mu} |_{\Delta = 0.99}$$
$$\approx \frac{0.01}{\mu}$$

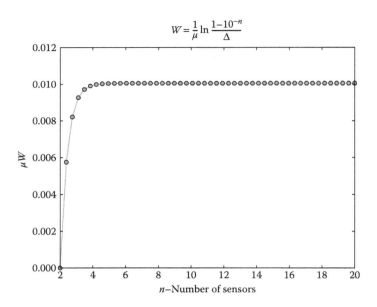

**FIGURE 26.8** Ratio of the length of time window to the mean life time of value versus number of sensors in fixed aggregation strategy.

The limit of $W$ says there is a upper bound, which is roughly 1% of the mean lifetime of values when $\Delta = 0.99$ that is a relatively high threshold. The good news is that *three to four* sensors suffice to raise the alert provided that each sensor's value is no less than 0.9.

### 26.8.2.3 An Adaptive Aggregation Strategy

As before, sensor readings about an event were collected and the resulting values $X_1, X_2, \cdots$ are reals in $[0, 1]$. Assume that one of the network sensors is in charge of the aggregation process and that the operator $\diamond$ defined in Equation 26.4 is employed in conjunction with a threshold $\Delta > 0$.

We now state and prove a technical result that will motivate our adaptive aggregation strategy.

**Theorem 26.6** *If* $X_{i_1}, X_{i_2}, \ldots, X_{i_m}$, $m > 1$, *satisfy* $X_{i_j} > 1 - \sqrt[m]{1 - \Delta}$, $j = 1, 2, \ldots, m$, *then* $\diamond_{j=1}^{m} X_{i_j} > \Delta$.

*Proof.* By Lemma 26.4,

$$\diamond_{j=1}^{m} X_{i_j} = 1 - \Pi_{j=1}^{m}(1 - X_{i_j})$$
$$> 1 - \Pi_{j=1}^{m}\left(1 - (1 - \sqrt[m]{1 - \Delta})\right)$$
$$= 1 - \Pi_{j=1}^{m} \sqrt[m]{1 - \Delta}$$
$$= 1 - (1 - \Delta)$$
$$= \Delta.$$

Notice what Theorem 26.6 says: if there are two sensors whose individual values exceed $1 - \sqrt{1 - \Delta}$, then the two should aggregate their values and, having exceeded $\Delta$, should report the event. Similarly, if there are three sensors whose individual values exceed $1 - \sqrt[3]{1 - \Delta}$, then the result of their aggregated data exceeds $\Delta$, and so on.

In turn, this observation suggests the following *adaptive aggregation* strategy: In the first aggregation round, the aggregator will announce the target $1 - \sqrt{1 - \Delta}$. If at least two sensors (including the aggregator) hold values in excess of the target $1 - \sqrt{1 - \Delta}$, then the event will be reported. If the first round of aggregation suffices, all is well. If, however, there is an insufficient number of sensors holding suitable values, the second round begins. In this round, the aggregator announces the target $1 - \sqrt[3]{1 - \Delta}$. If three or more sensors can be identified that exceed this target, then by Theorem 26.6, the aggregated value must exceed $\Delta$ and so the event is reported. This aggregation strategy is continued, as described, until either an event is reported, or else the results are inconclusive and no event is reported.

For the sake of illustration, and to fix ideas, consider that a fire just broke out on a ship. There are seven sensors on the ship, of which five, namely $A$, $B$, $C$, $E$, and $G$, are in close proximity of the location of the fire. As the fire spreads, these sensors will detect abnormal temperatures at times $t_1 < t_2 < t_3 < t_4 < t_5$. Further, let $X_1(t_1), X_2(t_2), X_3(t_3), X_4(t_4), X_5(t_5)$ be the values thus collected, normalized to $[0, 1]$. Given the layout of the sensors, it is reasonable to assume that $X_1(t_1) \geq X_2(t_2) \geq X_3(t_3) \geq X_4(t_4) \geq X_5(t_5)^*$. Assuming that sensor $B$ is closest to the fire, it will be the first one to sense high temperature and, thus, will become an aggregator. Sensor $B$ will wait for other sensors to report a reading over $1 - \sqrt{1 - \Delta}$ and will attempt aggregation with such sensors.

The first two rounds of our adaptive aggregation strategy do not yield a sufficient number of sensors to effect an aggregation. In the third round, $B$ announces the target $1 - \sqrt[4]{1 - \Delta}$. This third rounds yields four sensors whose individual values exceed the announced target and, consequently, the fire event will be reported.

Notice that with each round, the value of the information decays. When the aggregator and other sensors wait for more sensors to confirm, their value will decay. In order that their values are all above the final threshold, their time interval should satisfy the following constraint.

For any $i$ and $k$ that $1 \leq i < k \leq n$, value $X_i$ collected at time $t_i$ is larger than value $X_k$ collected at time $t_k$ in which $X_i > X_k$, $t_i < t_k$, $X_i \geq 1 - \sqrt[i]{1 - \Delta}$ and $X_k \geq 1 - \sqrt[k]{1 - \Delta}$, after $X_i$ decreased from $t_i$ to $t_k$, it should still be larger than the $k$th threshold $1 - \sqrt[k]{1 - \Delta}$, so the constraint is what described in inequality (26.24)

$$t_k - t_i < \frac{1}{\mu} \ln \frac{1 - \sqrt[i]{1 - \Delta}}{1 - \sqrt[k]{1 - \Delta}}, \qquad 1 \leq i < k \leq n, \qquad (26.24)$$

whereas $X_1(t_1) > X_1(t_2) > \cdots > X_n(t_n)$. ∎

*Proof.* For $\forall i, \forall k$ such that $1 \leq i < k \leq n$, we have

$$X_i > X_k$$

$$t_i < t_k$$

$$X_i > 1 - \sqrt[i]{1 - \Delta}$$

$$X_k > 1 - \sqrt[k]{1 - \Delta}$$

The value of $X_i$ collected at time $t_i$ decrease to time $t_k$ is

$$X_i(t_i)e^{-\mu(t_k - t_i)}$$

---

* This ordering is assumed for illustration purposes but is not really necessary.

then,

$$X_i(t_i)e^{-\mu(t_k-t_i)} > (1 - \sqrt[i]{1-\Delta})e^{-\mu(t_k-t_i)}$$

$$> 1 - \sqrt[k]{1-\Delta}$$

$$\Rightarrow$$

$$(1 - \sqrt[i]{1-\Delta})e^{-\mu(t_k-t_i)} > 1 - \sqrt[k]{1-\Delta}$$

$$e^{-\mu(t_k-t_i)} > \frac{1 - \sqrt[k]{1-\Delta}}{1 - \sqrt[i]{1-\Delta}} \qquad (\because 0 < \Delta < 1, \text{ so } 0 < 1 - \sqrt[i]{1-\Delta} < 1)$$

$$-\mu(t_k - t_i) > \ln \frac{1 - \sqrt[k]{1-\Delta}}{1 - \sqrt[i]{1-\Delta}}$$

$$t_k - t_i < \frac{1}{\mu} \ln \frac{1 - \sqrt[i]{1-\Delta}}{1 - \sqrt[k]{1-\Delta}}. \qquad\qquad \blacksquare$$

In reality, we need to respond to emergency in a timely manner. It is practical to aggregate values before their flip-flop of confidence, that is,

$$1 - \sqrt[k]{1-\Delta} > \frac{1}{2}.$$

From which, we get

$$k < -\log_2(1 - \Delta). \tag{26.25}$$

Equation 26.25 gives engineers a design guidance to determine the practical number of sensors for aggregating values in an emergency for alert. Substitute $\Delta = 0.99$ into it and get $k < 7$, that is, five or six sensors is a practical choice for aggregation in emergency. Figure 26.9 shows the practical number of sensors for different final thresholds. It shows that the higher threshold, the more threshold-levels, or sensors for aggregation are needed. We will choose $\Delta = 0.99$, a relatively high threshold for following simulations.

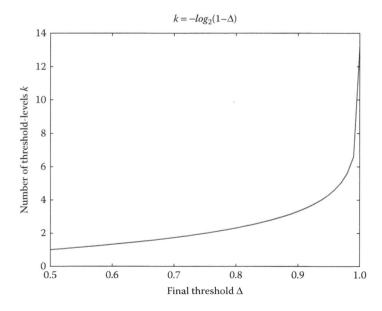

**FIGURE 26.9** Practical number of sensors for final threshold.

### 26.8.2.4 Simulation Results

Imagine that fire just broke out on a ship instrumented by a set of relevant sensors, including temperature and humidity sensors and light and smoke detectors [46]. To keep the aggregator simple, in what follows we assume the existence of temperature sensors only. Aggregating data across sets of sensors detecting different attributes of a fire event would proceed along similar lines but is not covered here.

Imagine a *critical* temperature range $K = [100; 1000]°C$. The various sensors take temperature readings $T_1, T_2, \cdots$. The value $X_i$ associated with temperature $T_i$ is defined to be

$$X_i = \Pr[T_i \in K|F]$$

that is, the *conditional probability* of temperature $T_i$ being recorded by a sensor, given the event $F$ that fire is present. To stamp out noise, $X_i$ is taken to be 0.9 if $T_i$ is in the critical range $K$ and 0 otherwise.

$$X_i = \begin{cases} 0.9 & T_i \in K \\ 0 & T_i \notin K. \end{cases} \tag{26.26}$$

We define the aggregator $\diamond$ as follows

$$X_i \diamond X_j = \Pr[\{T_i \in K\} \cup \{T_j \in K\}|F]. \tag{26.27}$$

In other words, aggregating two values is tantamount to computing the conditional probability of a union of events. Assuming that the sensors act independently, Equation 26.27 implies that

$$\begin{aligned} X_i \diamond X_j &= P[\{T_i \in K\} \cup \{T_j \in K\}|F] \\ &= \Pr[T_i \in K|F] + P[T_j \in K|F] \\ &\quad - \Pr[\{T_i \in K\} \cap \{T_j \in K\}|F] \\ &= \Pr[T_i \in K|F] + P[T_j \in K|F] \\ &\quad - \Pr[T_i \in K|F] \, P[T_j \in K|F] \\ &= X_i + X_j - X_i X_j, \end{aligned}$$

confirming that the aggregator $\diamond$ is the one defined in Equation 26.13. One can also see that $\diamond$ has the property that the value of the aggregated information increases with the number of sensor readings in the critical temperature range.

A group of $n$ sensors, once their temperature reading is available, can evaluate the corresponding $X_i$ by a simple table lookup. The aggregated value of the various $X_i$s is described in Equation 26.28.

$$\begin{aligned} \diamond_{i=1}^n X_i &= \Pr\{T_1 \cup T_2 \cup \cdots \cup T_n|F\} \tag{26.28} \\ &= \sum_{m=1}^n (-1)^{m+1} \sum_{1 \leq i_1 < i_2 < \cdots < i_m \leq n} \Pr\{T_{i_1} \cap T_{i_2} \cap \cdots \cap T_{i_m}|F\} \\ &= \sum_{m=1}^n (-1)^{m+1} \sum_{1 \leq i_1 < i_2 < \cdots < i_m \leq n} X_{i_1} X_{i_2} \cdots X_{i_m}. \end{aligned}$$

By Lemma 26.4, Equation 26.28 can be rewritten as Equation 26.29, which gives us a simple way to calculate the aggregation of arbitrary number of values.

$$\diamond_{i=1}^n X_i = Pr\{T_1 \cup T_2 \cup \cdots \cup T_n | F\} \qquad (26.29)$$

$$= \sum_{m=1}^n (-1)^{m+1} \sum_{1 \le i_1 < i_2 < \cdots < i_m \le n} Pr\{T_{i_1} \cap T_{i_2} \cap \cdots \cap T_{i_m} | F\}$$

$$= \sum_{m=1}^n (-1)^{m+1} \sum_{1 \le i_1 < i_2 < \cdots < i_m \le n} X_{i_1} X_{i_2} \cdots X_{i_m}$$

$$= 1 - \prod_{i=1}^n (1 - X_i)$$

$$= 1 - \prod_{i=1}^n (1 - Pr\{T_i | F\}).$$

In the following simulations using MATLAB®, the distribution of temperature in the fire is approximated by a linear model [47] with a plateau temperature of 1000°C and an ambient temperature of 20°C. The critical temperature range is [100; 1000]°C, and the value discount constant $\mu$ is $0.1^{s-1}$ [35]. The fire propagation model is approximated by a dot source spreading out at the same rate in all directions at a speed of 1 m/s [47]. The fire front has a temperature of 1000°C that decreases to the ambient temperature within 3 m. The temperature sensors are deployed in rectangular lattices of size $3 \times 2$ in a plane with every side of 3 m. The fire source location is randomly generated in one of the rectangles. Due to the small distances involved, wireless communication delays are ignored. We assume a sensor sampling time of 2 s, and that the sensors are asynchronous, that is, a new temperature reading will be reported every 2 s.

As the fire spreads, every sensor will sense a rising temperature and will compute the corresponding value. Figure 26.10 illustrates this process for all six sensors. The location for the fire center is (3.758, 4.32). As the location of sensors with respect to the center of the fire is, in ascending order, $B < C < A < E < F < D$, it follows that sensor B takes value of 0.9 first, then sensor C, and so on.

Figure 26.11 illustrates fixed aggregation strategy for six sensors. Sensor B, the first sensor got value 0.9 and sent this value to the base station. This single value is below the fixed threshold 0.99, so no alarm is triggered at this time. Then, the second sensor C got its value 0.9 and send it to the base station, however, the first value decayed a lot, the aggregation of these two values are still below the threshold. During the time sensor A got its value 0.9 and send to the base station, even though the first and the second values have decayed a certain amount, finally, the aggregation of these three values are above the final threshold 0.99, so an alarm is triggered. With time going by, as long as the alarm is not handled, the sensor network will keep triggering alarm as the aggregated value keeps rushing above the threshold. Through this automatic periodic triggering mechanism, the alarm is reported timely and reliably.

Figure 26.12 illustrates our adaptive aggregation strategy for six sensors. In order for a pairwise comparison, the fire scenario here is the same as that for fixed aggregation strategy. In adaptive aggregation, there is no fixed aggregator, every sensor can work as an aggregator if applicable. When sensor B got value of 0.9 first, it broadcasts this value to all other sensors. Only sensors in a certain distance from B will accept this value; here all six sensors are not far away from sensor B, so the accident happened at B very possible affects them as well. Sensor C got value of 0.9 second, it broadcast this value again as sensor B and aggregated this fresh value with the decayed value it received from B a moment ago, sensor B also did the same aggregation as sensor C. As the aggregated values are not above the final threshold 0.99, so no alarm is triggered, and these values continue decaying. When sensor A got value of 0.9, it broadcast this value and aggregated it with those decayed values from sensors B and C, sensors B and C also did the same aggregation. This time, these three sensors, sensors B, C, and A all had aggregated value above 0.99, they consequently all triggered alarms. All other sensors do not have values of 0.9, they will keep passive

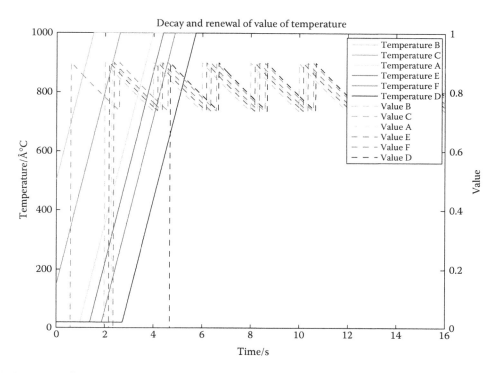

**FIGURE 26.10** Illustrating renewal and decay of values at six sensors.

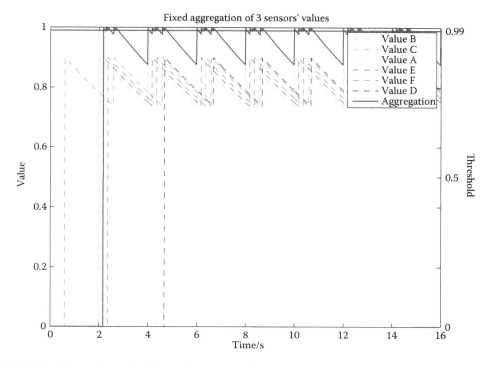

**FIGURE 26.11** Illustrating our fixed aggregation strategy for six sensors.

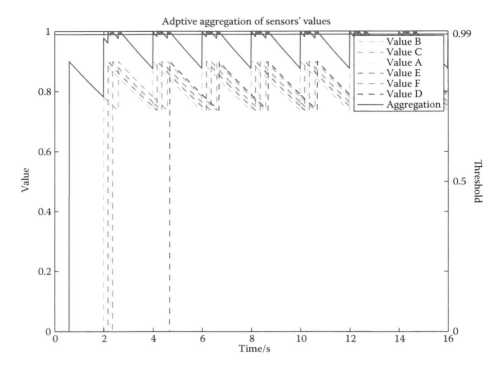

**FIGURE 26.12**  Illustrating our adaptive aggregation strategy for six sensors.

and do not trigger alarm to save energy. From this adaptive aggregation strategy and alarm procedure, it is clear that adaptive aggregation strategy works more timely and reliably than fixed aggregation strategy that depends heavily on the base station.

Figure 26.13 illustrates this process for a single sensor to show more details. The dots are sensed temperatures, sampled every 2 s. The solid line is the corresponding value, which jumps from 0 to 0.9 when the sensed temperature enters the critical range. It then decays until the next sample and the value is renewed. The process repeats to form the shape of a sawtooth.

With fixed thresholding, the fire alarm is set if the aggregated value of three sensors is above a threshold of 0.99. In Figure 26.14, the top six lines that oscillate between 0.7 and 0.9 are the values of six sensors' temperature. The middle panel shows the aggregated value with fixed thresholding. The alarm is triggered repeatedly every time the aggregated value rises above the threshold $\Delta$.

With fixed thresholding, a base station is required. We can achieve the same triggering of the alarm without a base station by using adaptive thresholding. Here, the initial threshold $\Delta_0$ is 0.8, and the final threshold $\Delta$ is 0.99. In this case, the maximum number of sensors needed for reporting a fire event is five. The bottom panel in Figure 26.14 shows the adaptive aggregated value. Once a single sensor's value rises above the initial threshold, it becomes the aggregator (in this case, sensor $B$). As new readings are shared, the aggregated value rises above the 0.99 threshold, and the alarm is triggered. As can be seen, the pattern of alarms in the bottom panel (adaptive thresholding) is similar to that in the middle panel (fixed thresholding), without the need for a base station performing the aggregation.

### 26.8.3 Application of Type 2 Aggregators

The main goal of this subsection is to show how the theoretical concepts developed for Type 2 aggregators apply to a practically relevant scenario.

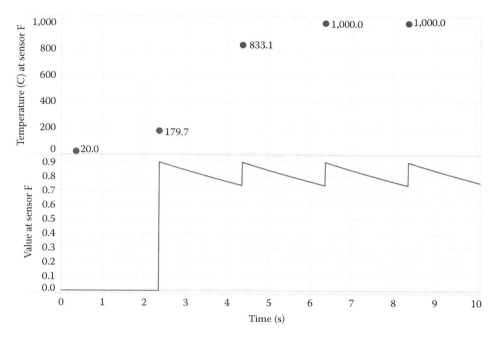

**FIGURE 26.13** Illustrating the decay of temperature values.

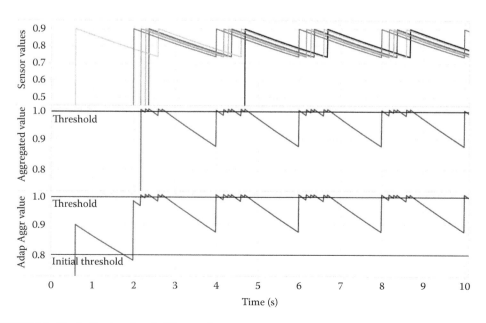

**FIGURE 26.14** Illustrating two thresholding modes.

### 26.8.3.1 The Scenario

Consider a fire event witnessed by a number of sensors deployed in a given area. For simplicity, assume that each sensor has collected a temperature value. Let $X_1(t_1), X_2(t_2), \ldots$ be, respectively, the sensed temperature values collected by the various sensors at times $t_1, t_2, \ldots$. As the sensors have witnessed the same event, it is natural to assume that the random variables $X_1(t_1), X_2(t_2), \cdots$ come from the same underlying distribution $X$ with finite expectation $E[X] < \infty$.

We assume that the $X_i$s are independent and, moreover, they are independent of the times $t_1, t_2, \cdots$ at which the data were collected. These assumptions can be justified by the spatial diversity of the sensors.

For a generic sensor that has collected data at time $t_i$, we let $X_i(t_i)$ denote the value of this information when it was collected. By Equation 26.2, the discounted value of this information at a later time $t$ is

$$X_i(t) = X_i(t_i)e^{-\mu(t-t_i)}. \tag{26.30}$$

Further, as a QoS parameter intended to avoid reporting a false positive, we need a minimum of $k$ individual temperatures to be aggregated. Given an expected temperature reported of $100°C$, this requirement is tantamount to insisting on accumulating a total of $\Delta = k \times 100$ "temperature points" as a result of aggregation. In turn, this suggests $\diamond = $ "+" as a suitable aggregator. It is easy to confirm that for the chosen $\diamond$, the distributivity property holds and so the results of Section 26.4 apply.

In this context, we are interested in evaluating the *expected* time-discounted value, $V(t)$, at time $t$, of the information collected by the sensors in which $t \geq \max\{t_1, t_2, \ldots\}$. To answer this natural question, we make the simplifying assumption that $t_1, t_2, \ldots$ are the times of a Poisson process with parameter $\lambda > 0$. In other words, $\lambda$ is the rate at which the sensors that witnessed an event are ready to report their sensory data.

**Theorem 26.7** *The expected time-discounted value, $E[V(t)]$, of the information collected by sensors at times $t_1, t_2, \ldots$ is*

$$E[V(t)] = \frac{\lambda}{\mu}E[X]\left[1 - e^{-\mu t}\right] \tag{26.31}$$

*where $\lambda > 0$ is the rate at which the sensors collect their data and $E[X]$ is the common expectation of $X_1, X_2, \ldots$.*

*Proof.* Recall, we assumed that the sensors collected their data at the times of a Poisson process with parameter $\lambda$. By the Law of Total Expectation,

$$E[V(t)] = \sum_{n \geq 1} E[V(t)|\{N = n\}]P[\{N = n\}], \tag{26.32}$$

where $N$ is the random variable that counts the number of sensors that have data ready for aggregation by time $t$. By Equation 26.30, Theorem 26.1, the conditional expectation, $E[V(t)|N = n]$, can be written as

$$E[V(t)|\{N = n\}] = E[\sum_{i=1}^{N} X_i(t_i)e^{-\mu(t-t_i)}|\{N = n\}]$$

$$= \sum_{i=1}^{n} E[X_i(t_i)e^{-\mu(t-t_i)}]$$

It is well known [43] that, given that $n$ Poisson events were recorded in $(0, t]$, their conditional distribution is uniform. Thus,

$$E[V(t)|\{N = n\}]$$

$$= \sum_{i=1}^{n} E[X_i(t_i)e^{-\mu(t-t_i)}]$$

$$= \sum_{i=1}^{n} E[X_i(t_i)e^{-\mu(t-U_i)}]$$

[where the $U_i$s are uniform in $(0, t]$]

$$= \sum_{i=1}^{n} E[X_i(t_i)]E[e^{-\mu(t-U_i)}]$$

[because the $X_i$s and $U_i$s are independent]

$$= \sum_{i=1}^{n} E[X]E[e^{-\mu(t-U_i)}]$$

[recall, $X$ is the common distribution of the $X_i$s]

$$= E[X]e^{-\mu t} \sum_{i=1}^{n} E[e^{\mu(U_i)}]$$

$$= e^{-\mu t}E[X] \sum_{i=1}^{n} \int_0^t e^{\mu u} \frac{du}{t}$$

$$= e^{-\mu t}E[X] \sum_{i=1}^{n} \frac{e^{\mu t} - 1}{\mu t} = \frac{E[X]}{\mu t} \sum_{i=1}^{n}[1 - e^{-\mu t}]$$

$$= \frac{nE[X]}{\mu t}[1 - e^{-\mu t}]. \tag{26.33}$$

On plugging Equation 26.33 back into Equation 26.32, we obtain

$$E[V(t)] = \sum_{n \geq 1} \frac{nE[X]}{\mu t}[1 - e^{-\mu t}]P[\{N = n\}]$$

$$= \sum_{n \geq 1} \frac{nE[X]}{\mu t}[1 - e^{-\mu t}]\frac{(\lambda t)^n}{n!}e^{-\lambda t}$$

$$= \frac{e^{-\lambda t}E[X][1 - e^{-\mu t}]}{\mu t} \sum_{n \geq 1} \frac{(\lambda t)^{(n-1)}}{(n-1)!}$$

$$= \frac{e^{-\lambda t}E[X][1 - e^{-\mu t}]\lambda t}{\mu t}e^{\lambda t}$$

$$= \frac{\lambda}{\mu}E[X][1 - e^{-\mu t}]. \qquad \blacksquare$$

There are a number of interesting things to note here as follows:

- The actual distribution of the $X_i$s does not appear explicitly in Theorem 26.7. This is telling us that two quite different distributions with the same expectation are equivalent as far as Theorem 26.7 is concerned;

- $E[V(t)] = \frac{\lambda}{\mu}E[X][1 - e^{-\mu t}]$ is an increasing function of time and

$$\lim_{t \to \infty} E[V(t)] = \frac{\lambda}{\mu}E[X].$$

Thus, for every application-dependent threshold $\Delta$, there exists an earliest time when $\Delta$ is exceeded.

Note that, as mentioned earlier, Theorem 26.7 allows us to evaluate the earliest time $t$ at which the expected discounted value of the information collected by the sensors exceeds an application-dependent threshold $\Delta$. Thus, at time $t$, $E[V(t)] \geq \Delta$, or equivalently,

$$\frac{\lambda}{\mu}E[X][1 - e^{-\mu t}] \geq \Delta.$$

Solving for $t$, we obtain

$$t \geq \frac{1}{\mu} \ln \frac{\lambda E[X]}{\lambda E[X] - \Delta\mu} \tag{26.34}$$

In fact, Equation 26.34 states that a value of $t$ exists only if $\lambda E[X] > \Delta\mu$ or, equivalently,

$$\Delta < \frac{\lambda}{\mu}E[X]. \tag{26.35}$$

Inequality (26.35) can be rewritten as

$$\Delta < \frac{\frac{1}{\mu}}{\frac{1}{\lambda}}E[X]. \tag{26.36}$$

where $\frac{1}{\mu}$ is the mean life-time of values, and $\frac{1}{\lambda}$ is the mean length of reporting interval. If $\frac{1}{\mu} \leq \frac{1}{\lambda}$, that is, collected values vanished because of the untimely reporting, from inequality (26.36), we get $E(X) > \Delta$; however, in practice it is the case that $E[X] < \Delta$, for otherwise there are no incentives for aggregation. This implies $\frac{1}{\mu} > \frac{1}{\lambda}$, the reporting interval should be shorter than the mean lifetime of values for a timely aggregation and reporting. However, if the reporting interval is too short, the probability of collisions will rise and lots of valuable resource especially wireless transmission resource will be wasted, so an application-dependent trade-off is needed.

### 26.8.3.2 Time for the Expectation of Aggregated Value to Exceed a Threshold

Recall that because of information value decay, the longer the time to aggregate, the lower the value of the aggregated information. The delay due to waiting for arriving pieces of information as well as collision in the reception of those pieces of information will result in reducing the value of information. From Equation 26.8, the decayed value $X_d$ of information with an initial value $X_0$ can be deduced as follows:

$$X_d = X_0 - X(t) \tag{26.37}$$
$$= X_0 - X_0 e^{-\mu t}$$
$$= X_0(1 - e^{-\mu t})$$
$$= X_0(1 - e^{-\frac{t}{\tau}}), \tag{26.38}$$

where $\tau = \frac{1}{\mu}$, which is the mean lifetime of values. Hence, the ratio $R_d$ of decayed value at time $t$ to the initial value is

$$R_d = \frac{X(t)}{X_0} = 1 - e^{-\frac{t}{\tau}}. \tag{26.39}$$

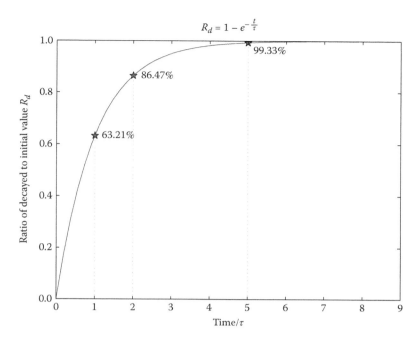

**FIGURE 26.15** Ratios of decayed value to its initial value.

Figure 26.15 illustrates the ratio of decayed value to its initial value. It shows that

- After 1 mean lifetime, 63.21% value of initial information decayed
- After 2 mean lifetimes, 86.47% value of initial information decayed
- After 5 mean lifetimes, 99.33% value of initial information decayed, which can be considered the value is vanished

This delay may be due to waiting for pieces of information to be generated by different sources and arriving at the aggregator, as well as delay due to the wireless channel conditions.

In the following simulation, we investigate how the rate of information generation in relation to decay should be designed to achieve the value of information as soon as possible, so that a decision-maker can make a proper determination. Here, we evaluate the applicability of the defined threshold time in Equation 26.34. Recall that Equation 26.34 shows the minimum time at which the expected value of information will exceed the threshold. This can be useful, for example, to plan for capturing an intruder. If we assume that an intruder staying time in the monitored area is an exponential random variable $S$ with expected value $E[S] < \infty$ and the reaction time of the security personnel to arrive at location after receiving the alarm is a random variable $R$ with expected value $E[R]$, then the expected detection time $D$ can be written as $D \leq E[S] - E[R]$. In other words, the design of $\lambda$, that is, sensor reporting periods, should not let $t$ in Equation 26.34 exceed $D$. In the following, we simulate this scenario to show how $\lambda$ and $\mu$ can affect the effectiveness of the monitoring application.

Assume that $E[S] = 300$ s and $E[R] = 100$ s. This means that the detection time should be less than 200 s. In this simulation, we assume that $V$ is selected uniformly from $\mathbb{R}[V_{min}, V_{max}]$, which are set to 0 and 100, respectively. This can be a valid setting as it could mean a direct functional mapping from the percentage of confidence that a sensor has in detecting the intruder. Then, we could set the threshold $\Delta$ to 196, which means that we need to have a confidence equal to two 98% confidence in detection of intrusion before reporting. Letting $VoI(t)$ stand for the random variable that keeps track of the value of information at time $t$, Table 26.5 shows the moments in time when the expected value, $E[VoI(t)]$, of $VoI(t)$

**TABLE 26.5**   Time When $E[VoI(t)]$ Exceeds a Threshold

| Configuration No. | $\lambda$ | $\mu$ | Theoretical Result | Simulated Result |
|---|---|---|---|---|
| 1 | 0.01 | $2.50 \times 10^{-3}$ | 1475 s | 1700 s |
| 2 | 0.01 | $1.25 \times 10^{-3}$ | 534 s | 557 s |
| 3 | 0.02 | $2.50 \times 10^{-3}$ | 267 s | 280 s |
| 4 | 0.02 | $1.25 \times 10^{-3}$ | 223 s | 247 s |
| 5 | 0.10 | $2.50 \times 10^{-3}$ | 41 s | 47 s |
| 6 | 0.10 | $1.25 \times 10^{-3}$ | 40 s | 46 s |

will be above $\Delta$ for theory and simulation. We show the average time over 10,000 trials for various values of $\lambda$ and $\mu$. As it turns out, the simulation results match the theoretical predictions well. In our defined scenario, we required that $E[VoI(t)]$ exceed $\Delta$ within 200 s. Table 26.5 shows that only configurations 5 and 6 meet that requirement, due to the arrival rate of 0.1, which has sensors reporting every 10 s.

### 26.8.3.3 Simulation Results

As for Type 1 aggregator, we use similar situation to verify our theoretical conclusions. A wireless sensor network is deployed for monitoring wildfires through measuring ambient temperature. It is supposed to aggregate the value of temperature at a central node; for simplicity, the value of temperature is taken as the number value of temperature, for example, the value of 100°C is dimensionless 100, and "+" is taken for aggregation, as proved in Section 26.6.1 which is a Type 2 aggregator. A temperature is reported to the central base station only when it is in the critical temperature range [100, 500]°C. The temperature is modeled as a uniform random variable in the range [10, 450]°C. An alarm is triggered when the aggregated value is above 400. Here, $\Delta = 400$ and $E(X) = \frac{10+450}{2} = 230 < \Delta$, so aggregation is needed for reliable emergency report.

The wireless sensor network covers 100 km$^2$ area of forest with a density of 1 sensor per 8000 m$^2$, the area is evenly divided into 100 patches, each patch is installed with a central base station for information aggregation. All sensors in this wireless sensor network work asynchronously; when an emergency happens, usually at a single location, only those sensors close to this emergency locale can sense abnormal temperature and report it to their base station, due to the asynchronous working mode, transmission delay and collision, the base station will see the incoming reports as a Poisson process [39], the reporting rate is supposed to be $0.2^{s-1}$, or once every 5 s. It is further assumed that the decay rate $\mu = 5 \times 10^{-3s-1}$, that is, the mean lifetime of value is 200 s, or roughly 3 min. The reporting time is far shorter than the mean life time of value.

In Figure 26.16, 50 points of temperature are collected, there are 10 points below the report threshold, two points are above alert threshold, all others are between the two thresholds. Only the decays of those temperature in the critical temperature range are shown. If the only two points above the alert threshold are missed, this emergency will be missed; for a complement, those values include decayed values above reporting threshold are aggregated to trigger more alerts.

In Figure 26.17, the number of reporting temperatures that are above alert threshold without aggregation is 4 in 350 s, about 86 s, more than 1 min for each alert averagely. Significantly, with aggregation, there are 13 alerts from aggregation, total 17 alerts, about 20 s for each alert. This greatly improves the reliability of alerts. However, some aggregated values go to above 1000, in some case, we may not need so frequent alert.

In Figure 26.18, the reporting rate is decreased to $0.01^{s-1}$, once every 100 s. In 100 min, there are three alerts without aggregation, 23 alerts from aggregation, for a total of 26 alerts, roughly 4 min. for every alert. This way benefits both the wireless sensor network and the alert reaction time, less frequent alerts avoid lots of wireless transmission to save energy, also make the alert period reasonably longer for related departments to intervene.

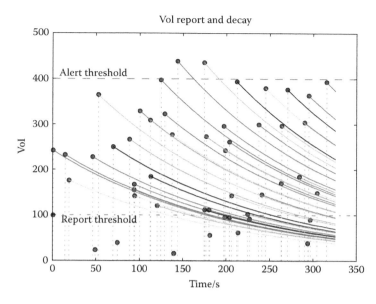

**FIGURE 26.16** Reporting and decaying of the value of temperature.

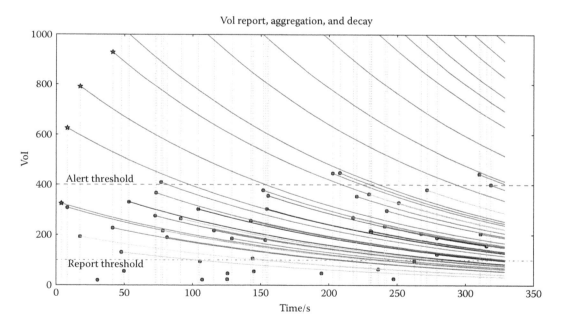

**FIGURE 26.17** Reporting, aggregation, and decaying of the value of temperature with reporting rate $0.2^{s-1}$

## 26.9 Concluding Remarks

The current chapter has reviewed formal ways of looking at aggregation of information in networks in which individual sensors possess information whose value decays over time. We offered a formal model for the valuation of time-discounted information and of the algebra of its aggregation.

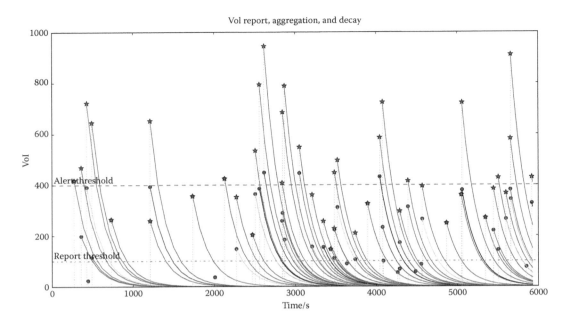

**FIGURE 26.18**   Reporting, aggregation, and decaying of the value of temperature with reporting rate $0.01^{s-1}$

We allowed aggregation of time-discounted information to proceed in an arbitrary, not necessarily pairwise, manner. We have shown that the resulting value of the aggregate does not depend on the order in which aggregation of individual values takes place.

Our results suggest natural thresholding strategies for the aggregation of the information collected by sets of network actors. Our theoretical predictions were confirmed by extensive simulation.

# References

1. S. Olariu and J. V. Nickerson. A probabilistic model of integration. *Decision Support Systems*, 45(4):746–763, 2008.
2. K. J. Arrow. The value of and the demand for information. In C. B. McGuire and R. Radner (Eds.), *Decision and Organization. A Volume in Honor of Jacob Marschak*, pp. 131–140. North-Holland Publishing, Amsterdam, the Netherlands, 1972.
3. P. Bolton, G. Roland, and E. Spolaore. Economic theories of the break-up and integration of nations. *European Economic Review*, 40:697–705, 1996.
4. S. Frederick, G. Loewenstein, and T. O'Donoghue. Time discounting and time preference: A critical review. *Journal of Economic Literature*, XL:351–401, 2002.
5. J. Marschak. Economics of information systems. In M. D. Intriligator (Ed.), *Frontiers of Quantitative Economics: Papers Invited for Presentation at the Econometric Society Winter Meetings, New York, 1969*, pp. 32–108. North Holland Publishing, Amsterdam, the Netherlands, New York, USA, 1971.
6. J. Marschak and R. Radner. *Economic Theory of Teams*. Yale University Press, New Haven, CT, 1972.
7. N. Ahituv. A systematic approach toward assessing the value of an information system. *Management Information Systems Quarterly*, 4(4):61–75, 1980.

8. D. R. Raban and S. Rafaeli. The effect of source nature and status on the subjective value of information. *Journal of the American Society for Information Science and Technology*, 57(3):321–329, 2006.

9. N. H. Anderson. *Foundations of Information Integration Theory*. Academic Press, New York, 1981.

10. N. H. Anderson. *Contributions To Information Integration Theory: Volume 1: Cognition*. Psychology Press, New York, 2014.

11. K. Deutsch. Communication theory and political integration. In P. E. Jacob and J. V. Toscano (Eds.), *The Integration of Political Communities*. Lippincott, Philadelphia, PA, 1964.

12. P. R. Lawrence and J. W. Lorsch. *Organization and Environment; Managing Differentiation and Integration*. Harvard University, Boston, MA, 1967.

13. M. E. Porter. *Competitive Strategy: Techniques for Analyzing Industries and Competitors*. Free Press, New York, 1980.

14. P. Kozma-Wiebe, S. M. Silverstein, A. Feher, I. Kovacs, P. Ulhaas, and S. M. Wilkniss. Development of a world-wide web based contour integration test. *Computers in Human Behavior*, 22(6):971–980, 2006.

15. J. L. Lu, S. J. Williams, and L. Kaufman. Behavioral lifetime of human auditory sensory memory predicted by physiological measures. *Science*, 258:1668–1670, 1992.

16. T. Brody, S. Harnad, and L. Carr. Earlier web usage statistics as predictors of later citation impact. *Journal of the American Society for Information Science and Technology*, 57(8):1060–1072, 2006.

17. A. L. Barabási and R. Albert. Emergence of scaling in random networks. *Science*, 286(5439): 509–512, 1999.

18. E. A. Stohr and J. V. Nickerson. Intra enterprise integration. In J. Luftman (Ed.), *Competing in the Information Age: Align in the Sand*, pp. 227–251. Oxford University Press, New York, 2002.

19. K. Crowston. A taxonomy of organizational dependencies and coordination mechanisms. Center for Coordination Science, Alfred P. Sloan School of Management, Massachusetts Institute of Technology, Cambridge, MA, 1994.

20. T. W. Malone and K. Crowston. Interdisciplinary study of coordination. *ACM Computing Surveys*, 26(1):87–119, 1994.

21. T. W. Malone, K. Crowston, J. Lee, B. Pentland, C. Dellarocas, G. Wyner, J. Quimby et al. Tools for inventing organizations: Toward a handbook of organizational processes. *Management Science*, 45(3):425–443, 1999.

22. K. H. Jones, K. N. Lodding, S. Olariu, L. Wilson, and C. Xin. Biology-inspired distributed consensus in massively-deployed sensor networks. In *Ad-Hoc, Mobile, and Wireless Networks*, pp. 99–112. Springer, Berlin, Germany, 2005.

23. K. H. Jones, K. N. Lodding, A. Wadaa, S. Olariu, L. Wilson, and M. Eltoweissy. Biomimetic models for sensor networks: Towards a social sensor network. *Handbook of Bio-Inspired Algorithmic Techniques*, CRC Press, Boca Raton, FL, 2005.

24. K. H. Jones, K. N. Lodding, S. Olariu, L. Wilson, and C. Xin. Energy usage in biomimetic models for massively-deployed sensor networks. In *Parallel and Distributed Processing and Applications-ISPA 2005 Workshops*, pp. 434–443. Springer, Berlin, Germany, 2005.

25. K. H. Jones, K. N. Lodding, S. Olariu, L. Wilson, and C. Xin. Sensor networks for situation management: A biomimetic model. In *Military Communications Conference, 2005. MILCOM 2005. IEEE*, pp. 1787–1793. IEEE, 2005.

26. K. H. Jones, K. N. Lodding, S. Olariu, L. Wilson, and C. Xin. Communal cooperation in sensor networks for situation management. In *Information Fusion, 2006 9th International Conference on*, pp. 1–8. IEEE, 2006.

27. J. V. Nickerson. A concept of communication distance and its application to six situations in mobile environments. *IEEE Transactions on Mobile Computing*, 5(4):409–419, 2005.

28. J. V. Nickerson and S. Olariu. Protecting with sensor networks: Attention and response. In *Proceedings of the 40th Annual Hawai'i International Conference on System Sciences*, IEEE, 2007.

29. S. Olariu and J. V. Nickerson. Protecting with sensor networks: Perimeters and axes. In *MILCOM*, IEEE, 2005.

30. J. V. Nickerson. Flying sinks: Heuristics for movement in sensor networks. In *Proceedings of the 39th Annual Hawaii International Conference on System Sciences*, IEEE, 2006.

31. J. V. Nickerson and S. Olariu. A measure for integration and its application to sensor networks. In *Workshop on Information Technology and Systems (WITS)*, Dallas, TX, 2005.

32. J. V. Nickerson and S. Olariu. Courier assignment in social networks. In *Proceedings of the 40th Annual Hawaii International Conference on System Sciences*, HICSS '07, pp. 46, Washington, DC, IEEE Computer Society, 2007.

33. M. Kamat, A. S. Ismail, and S. Olariu. Efficient in-network aggregation for wireless sensor networks with fine grain location-based clustering. In *Proceedings of the 9th ACM International Conference on Advances in Mobile Computing and Multimedia*, MoMM'11, pp. 66–71, December 5–8, 2011.

34. L. P. Kaelbling, M. L. Littman, and A. W. Moore. Reinforcement learning: A survey. *Journal of Artificial Intelligence Research*, 4:237–285, 1996.

35. S. Olariu, S. Mokhrekesh, and M. C. Weigle. Toward aggregating time discounted information. In *MiSeNet'2013*, pp. 57–66. Miami, FL, September 2013.

36. A. Zomaya, M. Clements, and S. Olariu. A framework for reinforcement-based scheduling in parallel processor systems. *IEEE Transactions on Parallel and Distributed Systems*, 9(3):249–260, 1998.

37. D. Bhagavathi, P. J. Looges, S. Olariu, and J. L. Schwing. A fast selection algorithms on meshes with multiple broadcasting. *IEEE Transactions on Parallel and Distributed Systems*, 5(7):772–778, 1994.

38. S. Olariu, J. L. Schwing, and J. Zhang. Fast computer vision algorithms for reconfigurable meshes. *Image and Vision Computing*, 10(9):610–616, 1992.

39. S. Olariu, S. Mohrehkesh, X. Wang, and M. C. Weigle. On aggregating information in actor networks. *SIGMOBILE Mobile Computing and Communications Review*, 18(1):85–96, 2014.

40. H. S. AbdelSalam and S. Olariu. Toward adaptive sleep schedules for balancing energy consumption in wireless sensor networks. *IEEE Transactions on Computers*, 61(10):1443–1458, 2012.

41. S. Olariu, A. Wadaa, L. Wilson, and M. Eltoweissy. Wireless sensor networks: Leveraging the virtual infrastructure. *IEEE Network*, 18(4):51–56, 2004.

42. S. Olariu, M. Eltoweissy, and M. Younis. ANSWER: Autonomous networked sensor system. *Journal of Parallel and Distributed Computing*, 67:114–126, 2007.

43. J. Aczel. *Lectures on functional equations and their applications*. Academic Press, New York, 1966.

44. J. E. Marsden. *Elementary classical analysis*. W. H. Freeman and Company, New York, 1974.

45. K. Nakano and S. Olariu. Randomized leader election protocols in radio networks with no collision detection. *Algorithms and Computation*, pp. 362–373, 2000.

46. Z. Li, Y. Wang, and Y. Song. Wireless sensor network design for wildfire monitoring. *Proceedings of the Sixth World Congress on Intelligent Control and Automation WCICA 2006*, 1, pp. 109–113, 2006.

47. E. Manolakos and G. Xanthopoulos. Temperature field modelling and simulation of wireless sensor network behaviour during a spreading wildfire. *Proceedings of the European Signal Processing Conference (EUSIPCO 2008)*, IEEE, 2008.

# 27

# Approximation and Exact Algorithms for Optimally Placing a Limited Number of Storage Nodes in a Wireless Sensor Network

Gianlorenzo D'Angelo

Alfredo Navarra

Cristina M. Pinotti

## 27.1 Introduction

Networks of sensor nodes are usually employed to monitor large areas, collecting data with regular frequency. This large volume of data has to be stored somewhere for answering to external user queries [1]. There are usually two main ways to store data. Source nodes, which are responsible for collecting data, can either locally store the data or transmit them to the *sink*, a powerful node connected to the external world. Both solutions present some disadvantages. If data are locally stored, several problems may arise: (1) data cannot be accumulated for long periods because nonsink nodes are equipped with only limited memory space, (2) stored data are lost once the energy of a source node—battery operated—is depleted, and (3) searching data for serving query demand results in network-wide communications. If data are forwarded from the source nodes to the sink node, the network might become congested, especially if *raw* (i.e., uncompressed) data are transmitted. Limitations to the number of packets a sensor can transmit to the sink per time unit must be also considered [2,3]. Nonetheless, sensors around the sink are generally highly used and exhausted easily; thus, the network may be partitioned rapidly due to the so-called *sink-hole problem* [4]. This problem has an increasing impact, for example, in visual sensor networks, a special class of wireless sensor network with smart-cameras, that handle larger set of data and that are an effective tool for many large area surveillance, environmental monitoring, and objects tracking applications [5].

Recently, hybrid solutions have been proposed for sensor networks that make use of a limited number of "special" sensors, more powerful than standard ones in terms of storage, energy, and computational capabilities [6–10]. Under this model, source nodes may forward their raw data to such special nodes, referred to as *storage nodes*. Here, raw data are stored and *compressed*, that is, reduced in size, to be transmitted to the sink at the time a query demand from external users is submitted. With this two-tier model, if the number of storage nodes is kept limited, the network becomes less congested at the price of a moderate increase of the sensor cost of the network.

Indeed, the integration of storage nodes in the tiered architecture for sensor networks is made possible by the new storage-enriched hardware [11–13] and considered to be very practical [14]. The introduction of the storage nodes helps to alleviate the transmission bandwidth problem by distributing the local data transmission to the storage nodes. This hierarchical structure has been instantiated by the popular stargate device [13] and the memory-enhanced sensor nodes by UC Riverside [12]. Those special powerful nodes take advantage of their high transmission, storage, and even computational capabilities to alleviate the bandwidth limitation, and also provide auxiliary support for surrounding vulnerable sensors for data back-up. The introduction of the storage nodes is also supported by the concept of *data-centric storage* [15].

In References 6, 8, 9, the problem of selecting a subset of storage nodes so as the overall communication cost is minimized is called *optimal storage placement* problem. When the number of storage nodes is limited by an integer $k$, we talk about the *minimum k-storage problem* (MSP).

The problem of suitably placing storage nodes to mitigate the energy consumption requirement is a main challenging problem in the field of sensor networks. Moreover, different storage nodes might be used to collect different type of data. In so doing, when external users have to retrieve specific data, not all storage nodes must be involved. Actually, data query is a very important service in networking applications. Thus, the MSP appears also in other contexts, such as web caching, peer-to-peer, and database systems. Security aspects have been considered in Reference 16, whereas large-scale sensor networks have been addressed in Reference 17.

There has been a lot of prior research on data collection in sensor networks. Initially, no in-network storage was considered: the request for data was routed from the sink to every sensor by flooding messages. The data were sent to the sink by following the same path but in the reverse direction [18]. To reduce the communication cost toward the sink, clustering routing protocols have been then considered [19]. Later, the data-centric in-network model has been introduced [20], which stores different data types in different places in the network to facilitate the archiving and retrieving process. Due to the large storage capabilities required by the in-network model, in Reference 21, it has been proposed to store data in a degrading model: fresh-data are stored raw, whereas long-term data are kept, but compressed. More recently, a two-tier model has been proposed [9] to ameliorate the problem of communication congestion. The authors formulate the problem as an integer programming problem and propose a 10-approximation rounding algorithm. Differently from us, they assume that (1) raw data have size independent from the source node and (2) the energy spent for transmitting one unit data between any pair of sensors is proportional to their Euclidean distance. For us, instead, different source nodes may generate data of different size as sensors can monitor different environment aspects. Moreover, we assume that communications follow an underlying network represented by a graph. Each edge of the graph has its own weight that measures the energy required to traverse it.

In References 6, 8, the problem is solved assuming that the communication network topology is a directed tree, rooted at the sink, and a path, respectively. The arcs directed toward the sink are used to collect the data, and the arcs directed away from the sink are used to broadcast the query. When a sensor $s$ sends one unit data upwards to the sink, the energy cost is fixed, whereas when a sensor $s$ sends one unit data downwards, the energy cost can be high because it is proportional to the number of children of $s$.

In our research and in Reference 9, instead, storage nodes simply send query replies in a proactive manner with a predefined query frequency, and hence the query cost is null. Here we consider as communication network topology both undirected and general directed graphs, whereas the restriction to specific classes of graphs is dictated only by the requirement to provide polynomial time optimal algorithms.

Finally, the MSP is strongly related to the well-known $k$-median problem [22–24] defined in the following. Let $G = (V, E)$ be a complete directed weighted graph, $k$ be an integer, and $dist(u, v) \in \mathbb{N}$ be the weight associated with the arc $(u, v) \in E$. A $k$-median set for $G$ is a subset $V' \subseteq V$ with $|V'| \leq k$. The *minimum $k$-median problem* (briefly, $\text{M}^3\text{P}$) consists in finding a $k$-median set $V'$ that minimizes the sum of the weights from each vertex to its nearest median, that is, $\sum_{u \in V} \min_{v \in V'} dist(u, v)$.

In general, the minimum $k$-median problem is not in *APX* [25]. In the minimum *metric $k$-median* problem, it is assumed that the weight function is symmetric and satisfies the triangle inequality, that is, it is a distance function.

Alternatively, the minimum metric $k$-median problem can be defined on a general weighted *undirected* graph instead of a complete directed graph with metric weight function. We observe that these two definitions are equivalent as a general graph can be reduced to a complete directed graph with metric weight function in polynomial time by computing the distances between each pair of nodes. Moreover, reducing a complete directed graph to a general graph is straightforward if the weight function is symmetric.

A harder variant of the minimum metric $k$-median problem in which the medians must be selected from a specific set of nodes different from the set of nodes they must serve has been shown to be not approximable within a factor of $1 + \frac{2}{e}$ [24]. Note that such a bound does not hold in the metric $k$-median problem defined earlier.

In the MSP, if data compressed are assumed to be of negligible size (i.e., there is no cost in sending data from storage nodes to the sink), so as for the energy spent for the queries, and if the sink is assumed to be one of the $k$ selected nodes, then MSP coincides with the minimum $k$-median problem. Under these assumptions, $k$-storage minimizes only the energy cost required to send raw data to the storage nodes. This is the classical $k$-median problem, except that the sink is a special, predefined median. Clearly, the relation with the $k$-median also suggests affinities with facility location problems [22–24]. However, to the best of our knowledge, there is no variant of facility location problems that coincides with our $k$-storage.

## 27.1.1 Results in This Chapter

The results contained in this chapter summarize what has been obtained in References 26, 27 in terms of approximation and exact algorithms for optimally solving the MSP. As the MSP is similar to the well-known metric $k$-median problem, then it is easy to show that also in this case, the problem is *NP*-hard. However, when dealing with directed graphs, we show that in general, the MSP is not in *APX*, unless $P = NP$. The result actually holds for *directed acyclic graphs (DAG)*.

Still, we are able to show that on the restricted case in which the topology is a tree rooted at the sink and all the arcs are directed towards the sink, the problem can be solved in polynomial time by a dynamic programming algorithm. Our algorithm requires $O(\min\{kn^2, k^2P\})$ time, where $n$ and $P$ are the number of nodes and the *path length* of the tree, respectively. The path length is defined as the sum over the whole tree of the number of arcs on the path from each tree node to the root. For a balanced binary tree with $n$ nodes $P = \Theta(n \log n)$, for random general trees $P = \Theta(n\sqrt{n})$, and in the worst case $P = O(n^2)$ [28].

In the case of undirected graphs, we first prove that it is *NP*-hard to approximate the metric $k$-median problem within a factor of $1 + \frac{1}{e}$. The obtained result is then extended to the MSP by means of a polynomial time reduction that preserves approximation. A local search algorithm that guarantees a constant approximation ratio is then proposed.

Moreover, we are able to show that for graphs with bounded *treewidth*, the problem is optimally solvable in polynomial time. Intuitively, the treewidth of a graph represents a sort of distance of the graph topology from being a tree, see Section 27.5.2 for a formal definition. We note that the obtained result for graphs of bounded teewidth also holds for the metric $k$-median problem that is interesting by itself.

### 27.1.2  Structure of the Presentation

First, in Section 27.2, we give the necessary notation and define the problem. In Section 27.3, we provide hardness results for both directed and undirected graphs. Section 27.4 is devoted to the design of a constant factor approximation algorithm for undirected graphs. Section 27.5 is instead devoted to the design of resolution algorithms for particular graph classes that are: directed trees and undirected graphs with bounded treewidth. Finally, Section 27.6 concludes the chapter.

## 27.2  Notation and Problem Statement

Let $G = (V, E)$ be a weighted graph of $n$ nodes representing a sensor network. If $G$ is a directed graph, then $G$ is assumed to be weakly connected, and the energy cost propagation of a message over the edge $(u, v) \in E$ is denoted by the weight $w(u, v)$. If $G$ is an undirected graph, then $G$ is assumed to be connected and the energy cost propagation of a message over the edge $\{u, v\} \in E$ is indiscriminately denoted by the weight $w(u, v)$ or $w(v, u)$. Let $d(u, v)$ be the minimum energy cost for propagating a message from $u$ to $v$ which is given by the shortest path distance from $u$ to $v$ in $G$.

Each node $v \in V$ can be selected to serve as a *storage* node. A storage node is a special node with higher storage capacity than regular nodes and with enough computational capability and energy for compressing data. As storage nodes have higher cost than regular nodes, it is reasonable to bound their number in the network, as done in References 8, 9.

A feasible solution for the MSP is a set $S \subseteq V$ of storage nodes such that $|S| \le k$, for some $k \in \mathbb{N}$. Given $S$, each node $v$ in $V$ is associated with a storage node in $S$, denoted as $\sigma(v, S)$ or $\sigma(v)$ when $S$ is clear by the context. Clearly, if $v \in S$, then $\sigma(v, S) = v$.

Given the set $S$ of storage nodes, we associate each node $v \in V$ with the storage node $\sigma(v, S)$ such that

$$\sigma(v, S) = \arg\min_{s \in S} s_d(v)(d(v, s) + \alpha d(s, r)).$$

The total cost per time unit for a set $S$ of storage nodes is then given by

$$cost(S) = \sum_{v \in V} s_d(v)(d(v, \sigma(v, S)) + \alpha d(\sigma(v, S), r)).$$

In conclusion, the *MSP* consists in finding a subset $S \subseteq V$, with $|S| \le k$ that minimizes $cost(S)$.

We observe that any solution $S$ to *MSP* has size equal to $k$ because reducing the number of storage nodes does not decrease the cost.

## 27.3  Hardness of Approximation

In the current section, we give hardness of approximation results for the *MSP* problem in both the directed and the undirected case. In particular, in Section 27.3.1, we show that in the directed case, the problem does not admit an algorithm that guarantees any *constant* approximation factor, unless $P = NP$. In other words, the problem does not belong to *APX*. The proof used for the directed case holds even in the case of DAG but does not hold for the undirected graphs. Indeed, in Section 27.4, we will give a constant factor approximation algorithm for this latter case. However, in Section 27.3.2, we show that, in this case, the problem cannot be approximated in polynomial time within a factor smaller than $1 + \frac{1}{e}$, unless $P = NP$ which implies that, under the same hypothesis, the problem does not admit a Polynomial Time Approximation Scheme (PTAS). To show such result, we show that the same lower bound on the approximation holds for the minimum metric $k$-median problem. All the above-mentioned inapproximability results hold also in the case of unitary weights.

## 27.3.1 Directed Graphs

To show that, unless $P = NP$, the *MSP* problem on DAGs cannot be approximated within any constant, we provide an approximation factor preserving reduction, similar to the technique used in Reference 23, from the Set-Cover problem defined as follows. Let $X = \{x_1, \ldots, x_n\}$ be a universe and $\mathcal{C} = \{C_1, C_2, \ldots, C_m\}$ be a collection of subsets of $X$. The *Set-Cover problem* consists in finding the minimum collection of subsets that covers each element of $X$. Formally, the Set Cover requires to find a set $\mathcal{C}' \subseteq \mathcal{C}$ such that $\bigcup_{C_j \in \mathcal{C}'} C_j = X$ and $|\mathcal{C}'|$ is minimum.

Given an instance $\{X, \mathcal{C}\}$ of Set Cover, we build an instance of *MSP* on a DAG $G = (V, E)$, as follows (Figure 27.1 for a visualization). The set $V$ consists of the sink node $r$, a node for each subset $C_i \in \mathcal{C}$, and a pair of nodes $x'$ and $x''$, for each $x \in X$. Formally, $V = \{r\} \cup Q \cup X' \cup X''$ where $Q = \{q_i \mid C_i \in \mathcal{C}\}$, $X' = \{x_i' \mid x_i \in X\}$, and $X'' = \{x_i'' \mid x_i \in X\}$. From now on, $x_i'$ and $x_i''$ are called twins. The set of edges $E$ is defined as follows. The twin nodes $x_j'$ and $x_j''$ are connected to node $q_i$ if $x_j \in C_i$, and the arcs $(x_j', q_i)$ and $(x_j'', q_i)$ have unit weight. Finally, each node $q_i \in Q$ is connected to the sink $r$ by the arc $(q_i, r)$ of weight $\rho > 1$. The so constructed graph is clearly a DAG. Values of $s_d(v)$ are unitary for all the nodes in $X' \cup X''$, whereas they are equal to zero for the nodes in $Q$ and for $r$. The value of $\alpha$ is set to zero. The value of $k$ will be specified later.

Given a solution $S$ to *MSP*, we denote by $SC(S)$ the subsets of $\mathcal{C}$ corresponding to nodes in $S \cap Q$.

**Lemma 27.1** *Let $S$ be a set of storage nodes in graph $G$. If $S \cap \{X' \cup X''\} \neq \emptyset$, then starting from $S$, we can generate a solution $S'$ such that $cost(S') \leq cost(S)$ and $S' \subseteq Q$, or $SC(S)$ corresponds to a set cover for $\{X, \mathcal{C}\}$.*

*Proof.* Given a set $S$ of storage nodes, let $In(v) = \{x \mid (x, v) \in E\}$ be the set of nodes that have an arc incoming in $v$ and let $\Delta(S) = \bigcup_{v \in S} In(v)$. If $v \neq r$ is a storage node, then $In(v)$ are the nodes associated with $v$. Moreover, a node $z$ is a *candidate storage* if $z \in Q \setminus S$ and $In(z) \setminus \{S \cup \Delta(S)\} \neq \emptyset$.

Let $\mathcal{P} = \{z \mid z \in Q \setminus S \text{ and } In(z) \setminus (S \cup \Delta(S)) \neq \emptyset\}$ be the set of candidate storage nodes in $G$ given $S$. Note that if $\mathcal{P} = \emptyset$, $SC(S)$ is a set cover for $X$.

Four possible distinct cases may arise (Figure 27.2).

a. $x' \in S$, $x'' \notin S$, and $q \notin S$, for each $q \in Q$ such that $(x', q) \in E$.
b. $x' \in S$, $x'' \notin S$, and there exists a node $q \in Q$ such that $(x', q) \in E$ and $q \in S$.
c. $x', x'' \in S$ and $q \notin S$, for each $q \in Q$ such that $(x', q) \in E$.
d. $x', x'' \in S$ and there exists a node $q \in Q$ such that $(x', q) \in E$ and $q \in S$.

For the sake of simplicity, we do not consider the symmetric cases of cases (a) and (b) in which $x''$ belongs to $S$ instead of $x'$.

To refine $S$ so as it contains only nodes of $Q$, we swap the storage nodes in $S \cap \{X' \cup X''\}$ with some node in $\mathcal{P}$. In detail, the following operations are repeatedly applied.

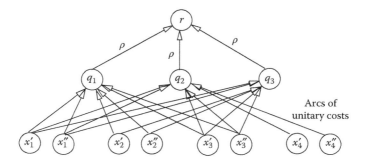

**FIGURE 27.1** Directed graph obtained from instance $\{X, \mathcal{C}\}$ of Set Cover, with $X = \{x_1, x_2, x_3, x_4\}$ and $\mathcal{C} = \{C_1, C_2, C_3\} = \{\{x_1, x_2\}, \{x_1, x_3, x_4\}, \{x_1, x_2, x_3\}\}$.

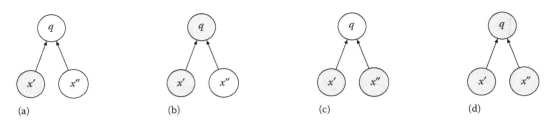

**FIGURE 27.2**  Four cases of the refining stage of Lemma 27.1, storage nodes are in gray. Node $q \in In(r)$; $x'$ and $x''$ are twins.

In case $(a)$, $x'$ can be swapped with $q$, which is a candidate storage node, obtaining a set $S'$ where $q$ is a storage node, whereas $x'$ is not. In case $(b)$, $x'$ can be swapped with a candidate storage node $z$ obtaining $S' = S \setminus x' \cup z$. In cases $(c)$ and $(d)$, we first swap $x''$ with a candidate storage node, and then we are left with case $(a)$ or $(b)$, respectively.

To prove that $cost(S') \leq cost(S)$, let define $gain(u, v)$ as the difference between the cost before and after the swap of $u$ with $v$. In the case $(a)$, we move the storage from $x'$ to $q$, loosing one unit from $x'$ and gaining at least $\rho > 1$ from $x''$, then $gain(x', q) \geq \rho - 1 > 0$. In case $(b)$, we move the storage $x'$ to a node $v \in \mathcal{P}$, loosing one unit from $x'$ and gaining $2\rho$ from each couple of nodes in $In(v) \setminus S \cup \Delta(S)$, then $gain(x', v) \geq 2\rho - 1$. Cases $(c)$ and $(d)$ are similar to case $(b)$ but for the choice of the storage to be removed that is $x''$, then $gain(x'', v) \geq 2\rho - 1$.

Note that, as soon as there are no nodes in $\mathcal{P}$ to be swapped with the storage nodes in $S \cap \{X' \cup X''\}$, a solution for the Set Cover has been found.  ∎

**Theorem 27.1**  *For DAGs, MSP does not belong to APX, unless $P = NP$.*

*Proof.* We prove the statement by contradiction, that is we assume there exists an approximation algorithm $\gamma$-*MSP* for *MSP* guaranteeing a constant approximation factor $\gamma$. We devise an approximation algorithm $\mathcal{B}$ for the set cover problem on $\{X, \mathcal{C}\}$ which applies $\gamma$-*MSP* on $G$. The pseudocode of $\mathcal{B}$ is given in Algorithm 27.1. Let us suppose for the moment that the size $k$ of an optimal set cover is known. Algorithm $\mathcal{B}$ selects a set cover $\mathcal{C}'$ by repeatedly applying Algorithm $\gamma$-*MSP* with parameter $k$ on subgraphs of $G$.

Initially, $\mathcal{C}' = \emptyset$ and $G_1 = (V_1, E_1)$ with $V_1 = V$ and $E_1 = E$. At each iteration $p \geq 1$, the $\gamma$-*MSP* algorithm selects a set $S \subseteq V \setminus \{r\}$ of $k$-storage nodes. Then, it refines solution $S$, obtaining the set $S'$, with the postprocessing stage described in the proof of Lemma 27.1. It follows that $S'$ is an approximate solution for the *MSP* instance with a better performance guarantee than $S$, that is $cost(S') \leq \beta OPT_{MSP}$, with $\beta \leq \gamma$.

At each iteration $p \geq 1$, algorithm $\mathcal{B}$ selects the subsets $SC(S')$ and add them to $\mathcal{C}'$. Then, a smaller instance $G_{p+1} = (V_{p+1}, E_{p+1})$ is created, where $V_{p+1} = V_p \setminus \{S' \cup \Delta(S')\}$ and $E_{p+1} = \{(u, v) \in E_p | u, v \in V_{p+1}\}$, and $\gamma$-*MSP* is performed on it. The algorithm ends when all nodes in $X$ are covered by $\mathcal{C}'$.

To evaluate the approximation provided for the Set-Cover problem, observe that if $\mathcal{B}$ terminates after $\lambda$ iterations, then $\lambda$ is the approximation factor guaranteed by $\mathcal{B}$ for Set Cover. In fact, an optimal algorithm for Set Cover would select exactly $k$ storage nodes at the first iteration as by hypothesis, $k$ is the size of an optimal set cover. The proposed algorithm, instead, selects $k$ storage nodes at each iterations, hence approximating the optimal solution of a factor of $\lambda$, the number of iterations.

After the $p$th iteration, let $d_p = |\Delta(S')|$ be the number of neighbors of any storage nodes, and $i_p = |X' \cup X''| - d_p$ be the number of remaining nodes. In other words, $d_p$ are the nodes that pay a cost of one in the $p$th instance of *MSP*, and $i_p$ are those that pay a cost of $1 + \rho$. Let $\lambda$ be the last iteration.

After the first iteration, by definition $d_1 + i_1 = |X_1' \cup X_1''| = 2n$. As $k$ is the size of an optimal set cover, then there exists a solution for *MSP* on $G_1$ of cost $2n$ obtained by selecting the storage nodes in $Q$ corresponding to the optimal set cover. After applying algorithm $\gamma - MSP$ and the procedure in Lemma 27.1,

---

## ALGORITHM 27.1 Approximation Algorithm $\mathcal{B}$ for Set Cover Used in the Proof of Theorem 27.1

**Input**: An instance of set cover, $\{X, \mathcal{C} = \{C_1, \ldots, C_m\}\}$
**Output**: An approximated solution for set cover
**begin**

Let $G = (V, E)$ be the DAG obtained by $\{X, \mathcal{C}\}$;
$w_{\min} = +\infty$;
**foreach** $1 \leq k \leq m$ **do**
$\quad \mathcal{C}' = \emptyset$;
$\quad p = 1$;
$\quad G_1 = (V_1, E_1) = (V, E)$;
$\quad$ **while** $V_p \neq \emptyset$ **do**
$\quad\quad$ Run $\gamma$-MSP on graph $G_p$;
$\quad\quad$ Let $S \subseteq V \setminus \{r\}$ the storages selected by $\gamma$-MSP;
$\quad\quad$ Obtain $S'$ by applying the procedure in Lemma 27.1;
$\quad\quad V_{p+1} = V_p \setminus \{S' \cup \Delta(S')\}$;
$\quad\quad E_{p+1} = \{(u, v) \in E_p | u, v \in V_{p+1}\}$;
$\quad\quad G_{p+1} = (V_{p+1}, E_{p+1})$;
$\quad\quad \mathcal{C}' = \mathcal{C}' \cup SC(S')$;
$\quad\quad p = p + 1$;
$\quad$ **if** $|\mathcal{C}'| < w_{\min}$ **then** $w_{\min} = |\mathcal{C}'|$;

---

$d_1 + (1 + \rho)i_1 \leq \beta OPT_{MSP} \leq \beta(2n) \leq \gamma(2n)$, and thus $i_1 \leq 2n(\frac{\gamma-1}{\rho})$ as $d_1 + i_1 = 2n$. After the $p$th iteration, $i_p + d_p = i_{p-1}$, and $d_p + (1 + \rho)i_p \leq \beta(i_{p-1}) \leq \gamma(i_{p-1})$, it holds

$$i_p \leq 2n \left(\frac{\gamma - 1}{\rho}\right)^p.$$

At the beginning of the last iteration $\lambda$, we have at most $2n \left(\frac{\gamma-1}{\rho}\right)^{\lambda-1} = \eta$ uncovered elements with $1 \leq \eta \leq 2n$, and then

$$\left(\frac{\gamma - 1}{\rho}\right)^{\lambda-1} = \frac{\eta}{2n}$$

For $n$ sufficiently large, we have that

$$\lambda - 1 = \log_{\frac{\gamma-1}{\rho}} \frac{\eta}{2n} \leq \log_{\frac{\rho}{\gamma-1}} 2n - \log_{\frac{\rho}{\gamma-1}} \eta < \log_{\frac{\rho}{\gamma-1}} 2n \leq \frac{\ln 2n}{\ln \frac{\rho}{\gamma-1}} \leq \frac{(1+\epsilon)\ln n}{\ln \frac{\rho}{\gamma-1}}, \qquad (27.1)$$

for each $\epsilon > 0$.

As there is no $(c \ln n)$-approximation algorithm for the Set-Cover problem for each $c < 1$, unless $P = NP$ [29], and as $c' \ln n \leq c \ln n - 1$ for each $0 < c' < c < 1$, it follows that

$$\lambda > c \ln n \geq c' \ln n + 1$$

From Inequality 27.1,

$$\lambda \le \frac{(1+\epsilon)\ln n}{\ln \frac{\rho}{\gamma-1}} + 1.$$

Therefore,

$$\frac{1+\epsilon}{\ln \frac{\rho}{\gamma-1}} > 1 \quad \Rightarrow \quad \frac{\rho}{\gamma-1} < e^{1+\epsilon} \quad \Rightarrow \quad \gamma > 1 + \frac{\rho}{e^{1+\epsilon}}.$$

As we do not know the size $k$ of the optimal Set Cover, we repeat the approximation algorithm based on $\gamma$-MSP for all the values of $k$, $1 \le k < m$, where $m$ is the size of $\mathcal{C}$, and we return the Set Cover of minimum size. Hence, there is no approximation algorithm for the *MSP* with approximation factor $\gamma < 1 + \frac{\rho}{e^{1+\epsilon}}$ unless $P = NP$. As the above-mentioned proof can be repeated for any constant $\rho$ the theorem holds. ∎

Note that the above-mentioned proof holds also for unitary arc weights. In fact, for any constant $\rho$, graph $G$ can be modified by substituting, for each $q \in Q$, the arcs from $q$ to $r$ with a path made of $c$ arcs, where $c = \lceil \rho \rceil$. On the other hand, the above-mentioned reduction is strictly based on the fact that the graph is directed and does not hold in an undirected graph.

## 27.3.2 Undirected Graphs

To show that it is *NP*-hard to approximate *MSP* within a factor of $1 + \frac{1}{e}$, we provide reduction from the minimum metric $k$-median problem ($M^3P$) that preserves the approximation factor. Therefore, in the next theorem, we show that it is *NP*-hard to approximate $M^3P$ within a factor of $1 + \frac{1}{e}$. This result is of its own interest and holds even if the distance function assumes only values in $\{1, 2\}$.

**Theorem 27.2** *It is NP-hard to approximate* $M^3P$ *within a factor* $\gamma < 1 + \frac{1}{e}$.

*Proof.* Our proof is based on a reduction that preserves approximation from minimum dominating set to $M^3P$. The minimum dominating set problem is defined as follows. Let $G = (V, E)$ be an undirected graph, a dominating set for $G$ is a subset $V' \subseteq V$ such that for each $u \in V \setminus V'$ there is a $v \in V'$ for which $\{u, v\} \in E$. The *minimum dominating set problem* consists in finding the minimum cardinality dominating set.

The proof uses the same technique used in Theorem 27.1. We show that, if there exists an algorithm $\gamma$-$M^3P$ with approximation factor $\gamma < 1 + \frac{1}{e}$ for $M^3P$, then there exists a $(c \ln n)$-approximation algorithm for the minimum dominating set for some $c < 1$. This implies that it is *NP*-hard to approximate $M^3P$ within a factor $\gamma$ as it has been shown that it is *NP*-hard to approximate the minimum dominating set within a factor $c \ln n$ for any $c < 1$ [29,30].

Given an instance $G = (V, E)$ of the minimum dominating set problem, we define the following distance function for each $(u, v) \in V \times V$.

$$dist(u, v) = \begin{cases} 1 & \text{if } \{u, v\} \in E \\ 2 & \text{otherwise.} \end{cases} \tag{27.2}$$

We give an approximation algorithm $\mathcal{A}$ for the minimum dominating set problem on $G$ that exploits the algorithm $\gamma$-$M^3P$. The pseudocode of $\mathcal{A}$ is given in Algorithm 27.2.

Let us suppose for a while that the size $k$ of an optimal dominating set is known. Fixed such a value of $k$, algorithm $\mathcal{A}$ selects a dominating set $V'$ by repeatedly applying algorithm $\gamma$-$M^3P$ by using parameter $k$ for the instance of $M^3P$.

## ALGORITHM 27.2   Approximation Algorithm $\mathcal{A}$ for Minimum Dominating Set Problem Used in Theorem 27.2

**Input**: An instance of Minimum Dominating Set, $G = (V, E)$
**Output**: An approximated solution for Minimum Dominating Set
$V_{\min} = V$;
**foreach** $1 \leq k \leq |V|$ **do**
     $V' = \emptyset; p = 1; V_1 = V$;
     **while** $|V_p| > 0$ **do**
         $G_p = (V_p, E_p)$ where $E_p = V_p \times V_p$;
         Define $dist : V_p \times V_p \to \{1, 2\}$ according to Equation 27.2;
         Run $\gamma\text{-M}^3\text{P}$ on graph $G_p$ with distance function $dist$ and parameter $k$;
         Let $S$ be the set of medians selected by $\gamma\text{-M}^3\text{P}$;
         $\Delta(S) = \bigcup_{v \in S} Adj(v)$;
         $V' = V' \bigcup S$;
         $V_{p+1} = V_p \setminus \{S \cup \Delta(S)\}$;
         $p = p + 1$;
     **if** $|V'| < |V_{\min}|$ **then** $V_{\min} = V'$;

     **return** $V_{\min}$;

Initially, $V' = \emptyset$ and $G_1 = (V_1, E_1)$ with $V_1 = V$ and $E_1 = V_1 \times V_1$. At each iteration $p \geq 1$, the $\gamma\text{-M}^3\text{P}$ algorithm selects the set $S$ of the $k$-medians that will be added to the dominating set $V'$, then $\mathcal{A}$ computes the set $\Delta(S) = \bigcup_{v \in S} Adj(v)$, where $Adj(v)$ is the set of neighbors of $v$ in $G$. Then, it creates a smaller instance $G_{p+1}$ for $\text{M}^3\text{P}$ with $V_{p+1} = V_p \setminus (S \cup \Delta(S))$. The algorithm ends when all nodes are covered.

To evaluate the approximation provided for the Minimum Dominating Set problem, observe that if $\mathcal{A}$ terminates after $\lambda$ iterations, the returned dominating set $V'$ has size at most $\lambda \cdot k$. Then, as $k$ is the size of an optimal dominating set, $\lambda$ is the approximation factor guaranteed by $\mathcal{A}$ for minimum dominating set.

We now show an upper bound on $\lambda$. After the $p$th iteration of the while loop, let $d_p = |\Delta(S)|$ be the number of neighbors in $G$ of the medians in $S$ and $i_p = |V_p| - d_p - k$ be the number of remaining nodes. In other words, according to Equation 27.2, $d_p$ are the nodes that pay a cost of 1 in the $p$th instance of $\text{M}^3\text{P}$ and $i_p$ are those that pay a cost of 2. After the first stage of the while loop, by definition $d_1 + i_1 + k = |V_1| = n$. As $k$ is the size of an optimal dominating set on $G$, then there exists a solution for $\text{M}^3\text{P}$ on $G_1$ of cost $n - k$. Therefore, the value of an optimal solution for $\text{M}^3\text{P}$ on $G_1$ is $OPT \leq n - k$. Algorithm $\gamma\text{-M}^3\text{P}$ outputs a solution of cost $d_1 + 2i_1 \leq \gamma OPT \leq \gamma(n - k)$. By substitution, $n - k + i_1 \leq \gamma(n - k)$ and finally $i_1 \leq (n - k)(\gamma - 1) \leq n(\gamma - 1)$.

After the $p$th stage of the while loop, $d_p + i_p + k = i_{p-1}$, and $d_p + 2i_p \leq \gamma(i_{p-1} - k)$ and then $i_p \leq n(\gamma - 1)^p$.

At the beginning of the last iteration $\lambda$, we have at most $n(\gamma - 1)^{\lambda - 1} = \eta$ uncovered nodes, for some $1 \leq \eta \leq n$, and then, $\lambda - 1 = \log_{(\gamma - 1)} \frac{\eta}{n} \leq \log_{(\gamma - 1)} \frac{1}{n} = \frac{\ln n}{\ln \frac{1}{\gamma - 1}}$, where the inequality holds because $\gamma - 1 < \frac{1}{e}$.

We know that it is *NP*-hard to approximate the Minimum Dominating Set problem within a factor $(c \ln n)$-approximation, for each $c < 1$ [29,30]. Moreover, for each $c'$ such that $0 < c' < c < 1$, we have $c' \ln n \leq c \ln n - 1$, for $n$ sufficiently large. It follows that for each $c' < 1$,

$$c' \ln n + 1 \leq c \ln n < \lambda \leq \frac{\ln n}{\ln \frac{1}{\gamma - 1}} + 1.$$

Therefore, $\frac{1}{\ln \frac{1}{\gamma - 1}} \geq 1$ which implies $\frac{1}{\gamma - 1} \leq e$, and hence $\gamma \geq 1 + \frac{1}{e}$.

As we do not know the size $k$ of the optimal dominating set, we repeat the approximation algorithm based on $\gamma$-$M^3P$ for all the values of $k$, $1 \leq k \leq |V|$ and we return the dominating set of minimum size. ∎

In Theorem 27.3, we will show that it is *NP*-hard to approximate *MSP* within a factor $\gamma < 1 + 1/e$. The proof is based on a reduction that preserves approximation from $M^3P$ to *MSP*.

Given an instance of $M^3P$ made of a graph $G = (V, E)$ and an integer $k$, where $dist(u, v) \in \{1, 2\}$ with $\{u, v\} \in E$, we build a graph $G' = (V', E')$, where $V' = V \cup I$. An *intermediate* node $i_{\{u,v\}}$ is inserted into $I$ for each edge $\{u, v\} \in E$ such that $dist(u, v) = 2$. The set $E'$ consists of one edge for each edge of $E$ with unit distance and of two edges for each edge of $E$ with distance two. In detail, for each $\{u, v\} \in E$:

- $\{u, v\} \in E'$ if $dist(u, v) = 1$
- $\{u, i_{\{u,v\}}\}, \{i_{\{u,v\}}, v\} \in E'$ if $dist(u, v) = 2$

The size of raw data is $s_d(i_{\{u,v\}}) = 0$, for each $i_{\{u,v\}} \in I$, and $s_d(v) = 1$, for each $v \in V$. For each $\{u, v\} \in E'$, $w(u, v) = 1$.

Let $\gamma$-*MSP* be the $\gamma-$approximation algorithm for *MSP* that will be applied to find the approximation algorithm for $M^3P$. Before showing how to iteratively apply the $\gamma$-*MSP* algorithm to approximate the $M^3P$ problem, in the next lemma, we describe how to refine the solution $S$ returned by the $\gamma$-*MSP* algorithm to obtain a new solution $S'$ with no intermediate nodes and smaller cost than $S$.

**Lemma 27.2** *Let $S$ be the set of storages returned by $\gamma$-MSP executed on graph $G'$. If $S \cap I \neq \emptyset$, starting from $S$, we can generate a solution $S'$ such that $cost(S') \leq cost(S)$ by iteratively swapping each storage $i_{\{u,v\}} \in S \cap I$ with $u$ or $v$.*

*Proof.* Let $N(u)$ and $N(v)$ be the sets of nodes that send data to $i_{\{u,v\}}$, across $u$ and $v$, respectively. If $|N(u)| = |N(v)|$, we can swap the storage in $i_{\{u,v\}}$ with either $u$ or $v$ without changing the cost of the solution. If $|N(u)| \neq |N(v)|$, we can assume without loss of generality that $|N(u)| > |N(v)|$. Hence, we swap the storage in $i_{\{u,v\}}$ with $u$ decreasing the cost of the solution of $|N(u)| - |N(v)| > 0$. ∎

**Theorem 27.3** *MSP is at least as hard to approximate as $M^3P$.*

*Proof.* To find an approximation algorithm for $M^3P$ given $\gamma$-*MSP*, we proceed as described in Algorithm 27.3. Specifically, for each $r \in V$, we run $\gamma$-*MSP* on an instance of *MSP* given by the graph

---

## ALGORITHM 27.3 Approximation Algorithm for $M^3P$ Used in Theorem 27.3

**Input**: An instance of $M^3P$, $G = (V, E)$, $k$, and $dist : V \times V \rightarrow \{1, 2\}$
**Output**: An approximated solution for $M^3P$
Let $G' = (V', E')$ be the graph obtained by $G$, where $V' = V \cup I$;
**foreach** $r \in V$ **do**
  Run $\gamma$-*MSP* on graph $G'$, with $r$ being the sink node;
  Let $S(r)$ be the set of storages selected by $\gamma$-*MSP*;
  $S'(r) = S(r)$;
  **foreach** $i_{\{u,v\}} \in S'(r) \cap I$ **do**
    Let $N(u)$ and $N(v)$ be the sets of nodes that send data to $i_{\{u,v\}}$, across $u$ and $v$, respectively;
    **if** $|N(u)| \geq |N(v)|$ **then** $S'(r) = \{S'(r) \cup \{u\}\} \setminus \{i_{\{u,v\}}\}$ ;
    **else** $S'(r) = \{S'(r) \cup \{v\}\} \setminus \{i_{\{u,v\}}\}$;

**return** $S^*$ *such that* $cost(S^*) = \min_{r \in V} cost(S'(r))$;

$G' = (V', E')$ defined earlier, with sink $r$ and $\alpha = 0$. Let $S(r)$ be the set of storages selected by $\gamma$-*MSP*, and $OPT_{MSP}(r)$ be the cost of an optimal solution for *MSP* with sink $r$. We refine solution $S(r)$, obtaining the set $S'(r)$, with the postprocessing stage described in the proof of Lemma 27.2. Then, $S'(r)$ is an approximate solution for the *MSP* instance with a better performance guarantee than $S(r)$, that is $cost(S'(r)) \leq \beta OPT_{MSP}(r)$, with $\beta \leq \gamma$.

Let $S^*$ be the subset of medians generated by the algorithm, that is, $cost(S^*) = \min_{r \in V} cost(S'(r))$. Let $OPT_{M^3P}$ be the cost of an optimal solution for $M^3P$ on $G$, and let $v$ be a node in an optimal solution for $M^3P$. As a solution for *MSP* with sink $v$ is a feasible solution for the instance of $M^3P$, $OPT_{MSP}(v) \geq OPT_{M^3P}$. However, $OPT_{MSP}(v) \leq OPT_{M^3P}$ as there exists an optimal solution for $M^3P$ that contains $v$ and then such a solution is feasible for the instance of *MSP* with sink $v$. Hence, $OPT_{MSP}(v) = OPT_{M^3P}$ and $cost(S^*) \leq cost(S'(v)) \leq \beta OPT_{MSP}(v) = \beta OPT_{M^3P} \leq \gamma OPT_{M^3P}$, as $cost(S^*) = \min_{r \in V} cost(S'(r)) \leq cost(S'(v))$ and $S'(v)$ is a $\beta$-approximate solution for *MSP* with $\beta \leq \gamma$. Finally, observe that as in the reduction $\alpha = 0$, then $cost(S^*)$ is the cost of the set of medians $S^*$ in the instance of $M^3P$. Hence, from Theorem 27.2, $\beta$ cannot be smaller than $1 + \frac{1}{e}$. ∎

In conclusion:

**Corollary 27.1** *It is NP-hard to approximate MSP within a factor $\gamma < 1 + \frac{1}{e}$.*

It is worth to note that the above-mentioned lower bound holds even with unitary edge weights. The reduction can be modified by shortcutting each intermediate node with an edge of weight 2. In this way, we can prove that the above-mentioned theorem and corollary hold with unitary $s_d$ and weights in $\{1, 2\}$.

## 27.4 Local Search Algorithm for Arbitrary Undirected Graphs

In the current section, we propose a local search algorithm for solving *MSP* and show that it guarantees a constant approximation ratio. The algorithm is denoted by $\mathcal{L}$ and is reported in Algorithm 27.4.

---

**ALGORITHM 27.4    Algorithm $\mathcal{L}$ When $t = 1$ and $\epsilon = 0$**

**Input**: $G = (V, E)$: an instance of *MSP*
**Output**: A set $S$ of storage nodes and the associated cost
Select an initial solution $S \subseteq V$ of size $k$;

local-minimum=false;
**while** $\neg$ *local-minimum* **do**
    Generate the list $L$ of all the pairs $(s, s')$ such that $s \in S$ and $s' \in V \setminus S$;
    swap = false;
    **while** $(\exists (s, s') \in L \ not \ yet \ examined) \wedge (\neg swap)$ **do**
        $S' = S \cup \{s'\} \setminus \{s\}$;
        Compute $\sigma(v, S')$ for each $v \in V$;
        **if** $cost(S') < cost(S)$ **then**
            $cost(S) = cost(S')$;
            $S = S'$;
            swap = true;
    **if** $\neg$ *swap* **then**
        local-minimum=true;
        **return** $S$ and $cost(S)$

---

Each solution is specified by a subset $S \subseteq V$ of exactly $k$ nodes. To move from one feasible solution $S$ to a neighboring one $S'$, we define a *swap* operation between two nodes $s \in S$ and $s' \in V \setminus S$, which consists in adding $s'$ and removing $s$, that is $S' = S \cup \{s'\} \setminus \{s\}$. In our local search algorithm, we repeatedly check whether any swap move yields a solution of lower cost. In the affirmative case, we apply to the current solution any swap move that improves the solution cost, and the resulting solution is set to be the new current solution. This is repeated until, from the current solution, no swap operation decreases the cost, that is, the current solution represents a local optimum.

Let us define the following three functions:

- $f : (0,1) \to \mathbb{R}, f(\alpha) = 2/\alpha$
- $g : [0, \frac{1}{2}) \to \mathbb{R}, g(\alpha) = \frac{12\alpha}{1-2\alpha}$
- $h : [0,1) \to \mathbb{R}$

$$
h(\alpha) = \begin{cases} g(\alpha) & \text{if } \alpha = 0 \\ \min\{f(\alpha), g(\alpha)\} & \text{if } \alpha \in (0, \frac{1}{2}) \\ f(\alpha) & \text{if } \alpha \in [\frac{1}{2}, 1). \end{cases}
$$

**Theorem 27.4** *The local search algorithm $\mathcal{L}$ for MSP exhibits a locality gap of at most $5 + h(\alpha)$.*

*Proof.* The proof follows the scheme of Reference 22, in which it is shown that a local search algorithm for *MMP* provides a 5-approximation. We consider, for any input instance of *MSP*, the optimal solution, denoted by $S^*$, and the one provided by $\mathcal{L}$, denoted by $S$. Clearly, sink $r$ belongs to both $S$ and $S^*$. For both solutions, each node $v$ is assigned to a storage node according to the definition in Section 27.2. These assignments are denoted as $\sigma(v) = \sigma(v, S)$ and $\sigma^*(v) = \sigma(v, S^*)$, respectively. Similarly, $C$ and $C^*$ denote the total cost of $S$ and $S^*$, respectively. By definition, $S$ represents a local minimum, that is, it cannot be improved anymore by means of one further step of algorithm $\mathcal{L}$.

The relation between $S$ and $S^*$ is obtained by considering $k$ ideal swaps, called *crucial swaps*, between the nodes in $S$ and the ones in $S^*$. As $S$ is locally optimal, each swap, singularly taken, does not improve the objective function of the resulting solution. Each crucial swap consists in swapping one node $s^* \in S^*$ into solution $S$ and swapping out one node $s \in S$. The main property for each swap is that any element $s^* \in S^*$ will participate in exactly one of these $k$ crucial swaps, and each $s \in S$ will participate in at most two of these $k$ swaps. The case where $s^* \equiv s$ is possible, but still it does not improve the current solution. We remark that, in the following, each crucial swap is considered individually, that is we consider solutions that differ from $S$ only by one node which is identified by the crucial swap.

The assignment $\sigma$ can be used to categorize the nodes in $S$ as follows:

- Let $A \subseteq S$ be the set of nodes $s \in S$ that have exactly one node $s^* \in S^*$ with $\sigma(s^*) = s$.
- Let $B \subseteq S$ be the set of nodes $s \in S$ for which none of the nodes $s^* \in S^*$ have $\sigma(s^*) = s$.
- Let $C \subseteq S$ be the set of nodes $s \in S$ such that $s$ has at least two nodes $s_1^*, s_2^*$ in $S^*$ assigned to it in the current solution i.e. $\sigma(s_1^*) = \sigma(s_2^*) = s$.

The assignment $\sigma$ provides a matching between a subset $A^* \subseteq S^*$ and set $A \subseteq S$. Hence, if $\ell$ denotes the number of nodes in $R^* = S^* \setminus A^*$, then $|B \cup C| = \ell$ as $|S^*| = |S| = k$. This implies that $|B| \geq \frac{\ell}{2}$ as $|C| \leq \frac{\ell}{2}$, by definition of $C$.

We can now construct the crucial swaps as follows: for each node $s^* \in A^*$, we swap it with $\sigma(s^*)$. As $r$ belongs to both $S$ and $S^*$, either it is in $A$, or it is in $C$. In both cases, it is swapped with itself, and then other at most $\ell$ swaps remain to be defined. Each of such swaps moves into the solution a distinct node in $R^*$, and moves out a node from $B$, so that each node in $B$ appears in at most two swaps. For those swaps involving nodes in $R^*$ and $B$, we are free to choose any assignment provided that each element of $R^*$ is swapped in exactly once, and each element of $B$ is swapped out once or twice.

For each crucial swap $cs(s, s^*)$ between $s \in S$ and $s^* \in S^*$, let $S'$ be the solution obtained by applying $cs(s, s^*)$, that is, $S' = (S \backslash \{s\}) \bigcup \{s^*\}$. We set the associations of nodes to storage nodes in $S'$ as follows:

- For any $v$ such that $\sigma^*(v) = s^*$, $v$ is associated with $s^*$, since $s^* \in S'$.
- For any $v$ such that $\sigma^*(v) \neq s^*$ and $\sigma(v) = s$, $v$ is associated with $\sigma(\sigma^*(v))$.
- Any other $v$ remains associated with $\sigma(v)$.

To proceed with the proof, we need the following proposition that proves that $\sigma(\sigma^*(v)) \in S'$.

**Proposition 27.1** *If $\sigma^*(v) \neq s^*$, then $\sigma(\sigma^*(v)) \neq s$.*

*Proof.* By contradiction, let $\sigma(\sigma^*(v)) = s$. We know that $s \in A$. In fact, each node swapped out by a crucial swap is either in $B$ or in $A$. However, the former case is not possible as by the definition of $B$, the nodes in $B$ have no associated nodes of $S^*$. As $s \in A$, a single element of $A^*$ is mapped to $s$ by function $\sigma$, and we build a crucial swap by swapping $s$ with that one element. Hence, $\sigma^*(v) = s^*$ which contradicts the hypothesis. ∎

We can now evaluate the cost of the solution $S'$ which is obtained after a crucial swap. By the previous discussion, $S'$ differs from $S$ only for elements $v$ such that $\sigma^*(v) = s^*$, and those for which $\sigma^*(v) \neq s^*$ but $\sigma(v) = s$. Moreover, $cost(S) \leq cost(S')$ as $S$ is locally optimal. Hence,

$$0 \leq cost(S') - cost(S)$$

$$\leq \sum_{v : \sigma^*(v) = s^*} s_d(v) \cdot [(d(v, \sigma^*(v)) + \alpha d(\sigma^*(v), r)) - (d(v, \sigma(v)) + \alpha d(\sigma(v), r))] \qquad (27.3)$$

$$+ \sum_{\substack{v : \sigma^*(v) \neq s^* \\ \sigma(v) = s}} s_d(v) \cdot [(d(v, \sigma(\sigma^*(v))) + \alpha d(\sigma(\sigma^*(v)), r)) - (d(v, \sigma(v)) + \alpha d(\sigma(v), r))] \quad (27.4)$$

By summing over all crucial swaps $cs(s, s^*)$, the term Equation 27.3 of the inequality becomes

$$\sum_{cs(s, s^*)} \sum_{v : \sigma^*(v) = s^*} s_d(v) \cdot [(d(v, \sigma^*(v)) + \alpha d(\sigma^*(v), r)) - (d(v, \sigma(v)) + \alpha d(\sigma(v), r))]$$

$$= \sum_{v \in V} s_d(v) \cdot [(d(v, \sigma^*(v)) + \alpha d(\sigma^*(v), r)) - (d(v, \sigma(v)) + \alpha d(\sigma(v), r))] = C^* - C.$$

In fact, $\bigcup_{s^* \in S^*} \{v : \sigma^*(v) = s^*\} = V$, and, for any $s_1^*, s_2^* \in S^*$, $\{v : \sigma^*(v) = s_1^*\} \cap \{v : \sigma^*(v) = s_2^*\} = \emptyset$, implies $\sum_{cs(s, s^*)} \sum_{v : \sigma^*(v) = s^*} m(v) = \sum_{s^* \in S^*} \sum_{v : \sigma^*(v) = s^*} m(v) = \sum_{v \in V} m(v)$, for any function $m$.

From the term Equation 27.4 of the inequality, we first consider $d(v, \sigma(\sigma^*(v))) + \alpha d(\sigma(\sigma^*(v)), r)$, which can be upper bounded by

$$d(v, \sigma^*(v)) + \alpha d(\sigma^*(v), r) + (1 + \alpha) d(\sigma^*(v), \sigma(\sigma^*(v))),$$

as, by the triangle inequality, $d(v, \sigma(\sigma^*(v))) \leq d(v, \sigma^*(v)) + d(\sigma^*(v), \sigma(\sigma^*(v)))$ and $d(\sigma(\sigma^*(v)), r) \leq d(\sigma(\sigma^*(v)), \sigma^*(v)) + d(\sigma^*(v), r)$, and, by the symmetric property, $d(\sigma^*(v), \sigma(\sigma^*(v))) = d(\sigma(\sigma^*(v)), \sigma^*(v))$. See Figure 27.3 for a visualization of these bounds.

We now give two different upper bounds on $d(\sigma^*(v), \sigma(\sigma^*(v)))$, for different values of $\alpha$ which imply two different upper bounds on the term Equation 27.4 of the inequality.

**Bound 1.** Clearly, $d(\sigma^*(v), \sigma(\sigma^*(v))) \leq d(\sigma^*(v), r)$ as $r \in S$ and $\sigma^*(v)$ is associated with $\sigma(\sigma^*(v))$ in $S$, hence for $\alpha > 0$:

$$d(v, \sigma^*(v)) + \alpha d(\sigma^*(v), r) + (1 + \alpha) d(\sigma^*(v), r) \leq (x + 1)(d(v, \sigma^*(v)) + \alpha d(\sigma^*(v), r)),$$

with $x = \frac{1+\alpha}{\alpha}$, that is, $(1 + \alpha) d(\sigma^*(v), r) = x \alpha d(\sigma^*(v), r) \leq x d(\sigma^*(v), r)$.

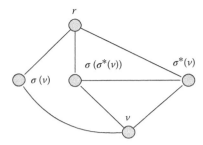

**FIGURE 27.3**  Upper bounds on distance $d(\sigma^*(v), \sigma(\sigma^*(v)))$.

From the earlier bounding, by summing up over all the crucial swaps:

$$\sum_{cs(s,s^*)} \sum_{\substack{v \,:\, \sigma^*(v) \neq s^* \\ \sigma(v) = s}} s_d(v) \cdot [(d(v, \sigma(\sigma^*(v))) + \alpha d(\sigma(\sigma^*(v)), r)) - (d(v, \sigma(v)) + \alpha d(\sigma(v), r))]$$

$$\leq \sum_{cs(s,s^*)} \sum_{\substack{v \,:\, \sigma^*(v) \neq s^* \\ \sigma(v) = s}} s_d(v) \cdot (x + 1)(d(v, \sigma^*(v)) + \alpha d(\sigma^*(v), r))$$

$$\leq 2(x + 1) \sum_{v \in V} s_d(v) \cdot (d(v, \sigma^*(v)) + \alpha d(\sigma^*(v), r)) = 2(x + 1)C^*,$$

where the last inequality is due to the fact that each $s \in S$ appears in at most two crucial swaps.

Hence, for any $\alpha > 0$, $0 \leq cost(S') - cost(S) \leq C^* - C + 2(x + 1)C^* = C^*(2x + 3) - C$, that is, $\frac{C}{C^*} \leq 2x + 3 = 5 + \frac{2}{\alpha}$.

**Bound 2.** As solution $S$ contains both $\sigma(\sigma^*(v))$ and $\sigma(v)$, then (Figure 27.3)

$$d(\sigma^*(v), \sigma(\sigma^*(v))) + \alpha d(\sigma(\sigma^*(v)), r) \leq d(\sigma^*(v), \sigma(v)) + \alpha d(\sigma(v), r),$$

hence,

$$d(\sigma^*(v), \sigma(\sigma^*(v))) \leq d(\sigma^*(v), \sigma(v)) + \alpha d(\sigma(v), r) \leq d(\sigma^*(v), v) + d(v, \sigma(v)) + \alpha d(\sigma(v), r)$$
$$\leq d(\sigma^*(v), v) + \alpha d(\sigma^*(v), r) + d(v, \sigma(v)) + \alpha d(\sigma(v), r).$$

The penultimate inequality holds by the triangle inequality (Figure 27.3), and in the last one we added $\alpha d(\sigma^*(v), r)$.

By summing over all crucial swaps,

$$\sum_{cs(s,s^*)} \sum_{\substack{v \,:\, \sigma^*(v) \neq s^* \\ \sigma(v) = s}} s_d(v) \cdot [d(\sigma^*(v), v) + \alpha d(\sigma^*(v), r) + (1 + \alpha)(d(\sigma^*(v), v) + \alpha d(\sigma^*(v), r)$$

$$+ d(v, \sigma(v)) + \alpha d(\sigma(v), r)) - (d(v, \sigma(v)) + \alpha d(\sigma(v), r))]$$

$$= \sum_{cs(s,s^*)} \sum_{\substack{v \,:\, \sigma^*(v) \neq s^* \\ \sigma(v) = s}} s_d(v) \cdot [(2 + \alpha)(d(\sigma^*(v), v) + \alpha d(\sigma^*(v), r)) + \alpha(d(v, \sigma(v)) + \alpha d(\sigma(v), r))]$$

$$\leq 2 \sum_{v \in V} s_d(v) \cdot [(2 + \alpha)(d(\sigma^*(v), v) + \alpha d(\sigma^*(v), r)) + \alpha(d(v, \sigma(v)) + \alpha d(\sigma(v), r))]$$

$$= 2(2 + \alpha)C^* + 2\alpha C$$

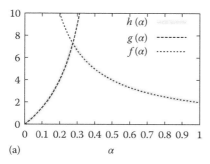
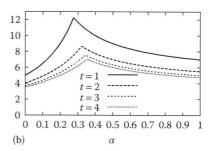

**FIGURE 27.4** (a) The two upper bound functions to the locality gap. (b) Function $h'$ by varying on the number $t$ of simultaneous swaps.

Hence obtaining, $0 \leq cost(S') - cost(S) \leq C^* - C + 2(2 + \alpha)C^* + 2\alpha C = (5 + 2\alpha)C^* + (2\alpha - 1)C$. If $\alpha < \frac{1}{2}, 2\alpha - 1 < 0$, and $(1 - 2\alpha)C \leq (5 + 2\alpha)C^*$, that is $\frac{C}{C^*} \leq 5 + \frac{12\alpha}{1 - 2\alpha}$. ∎

Theorem 27.4 provides two upper bounds to the locality gap given by $5 + f(\alpha)$ and $5 + g(\alpha)$. Functions $f, g$, and $h$ are plotted in Figure 27.4a. Function $f$ is monotonic decreasing, whereas $g$ is monotonic increasing, in their intervals of definition. We have that $f(\alpha) = g(\alpha)$ for $\alpha = \frac{1}{6}(\sqrt{7} - 1) \approx 0.274$ where $f(\alpha) = g(\alpha) < 7.3$. For all the other values of $\alpha$, one of the two functions is always below such a threshold, that is, the approximation ratio is always below 12.3.

Actually, algorithm $\mathcal{L}$ is not yet an approximation algorithm, as the number of iterations needed to find a local optimum solution might be superpolynomial. To fix this problem, as in Reference 22, we can change the stopping condition of $\mathcal{L}$ so it finishes as soon as it finds an approximate local optimum solution, i.e., when the solution $S$ is such that every neighboring solution $S'$ of $S$ has $cost(S') > (1 - \epsilon)$ $cost(S)$, for some $\epsilon \in (0, 1)$. This leads to at most $\frac{\log(\frac{cost(S_0)}{cost(S^*)})}{\log(\frac{1}{1 - \epsilon})}$ iterations, where $S_0$ is the initial solution selected in the first iteration of $\mathcal{L}$. This is polynomial in the size of the input.

**Corollary 27.2** *There exists an $\frac{1}{1 - \epsilon}\left(5 + h(\alpha)\right)$-approximation algorithm for MSP for any $\epsilon \in (0, 1)$.*

Finally, by following the arguments in Reference 22, the algorithm can be improved by allowing $t$ simultaneous swaps. For such algorithm, the analysis given in Theorem 27.4 can be extended by defining the crucial swaps in a way that each storage in the local optimal solution appears in at most $\frac{t+1}{t}$ swaps. This leads to a locality gap of $h'(\alpha)$, where

- $h' : [0, 1) \to \mathbb{R}$

$$h'(\alpha) = \begin{cases} g'(\alpha) & \text{if } \alpha = 0 \\ \min\{f'(\alpha), g'(\alpha)\} & \text{if } \alpha \in (0, \frac{t}{t+1}) \\ f'(\alpha) & \text{if } \alpha \in [\frac{t}{t+1}, 1), \end{cases}$$

- $f' : (0, 1) \to \mathbb{R}, f'(\alpha) = 1 + \frac{t+1}{t}\frac{1+2\alpha}{\alpha}$
- $g' : [0, \frac{t}{t+1}) \to \mathbb{R}, g' = \frac{(3+\alpha)t+2+\alpha}{(1-\alpha)t-\alpha}$

Function $h'$ is plotted in Figure 27.4b for $t = 1, 2, 3, 4$. To give an idea on the improvement provided by this method, we computed the maximum value of the upper bounds on the approximation ratio for $t = 2, 3, 4$, which is less than 8.67, 7.78, and 7.05, respectively. It follows that for $t \geq 2$ and any value of $\alpha$, our algorithm improves over the 10-approximation algorithm provided in Reference 9.

## 27.5 Polynomial Time Solvable Cases

In the current section, we consider specific graph classes where we are able to devise polynomial time optimal algorithms. In particular, we deal with the case of directed trees and undirected graphs with bounded treewidth.

### 27.5.1 Directed Trees

When the input graph is a tree rooted at the sink and all the arcs are directed towards the sink, we show that the complexity bound of the *MSP* on rooted trees can be reduced to $O(\min\{kn^2, k^2P\})$, where $P$ is the *path length* of the tree. The path length is defined as the sum over the whole tree of the number of arcs on the path from each tree node to the root. For a balanced binary tree with $n$ nodes $P = \Theta(n \log n)$, for random general trees $P = \Theta(n\sqrt{n})$, and in the worst case $P = O(n^2)$ [28].

We are given a generic weighted and rooted tree $T = (V, E)$ where the root is given by the sink $r$ and all the arcs are directed towards it. Let us denote by $c_i$ the number of children of node $i \in V$. We transform $T$ into a weighted rooted binary tree $B$ by adding *dummy* nodes as follows:

1. Let the root $r$ of $T$ be the root of $B$.
2. For each nonleaf node $i$ such that $c_i = 1$, introduce a new dummy node $u_i$, as the right child of $i$. Note that $u_i$ will be a leaf in $B$. Assign a weight of zero to the arc $(u_i, i)$.
3. For each node $i$ such that $t = c_i \geq 3$, let $i_1, i_2, \ldots i_t$ be the children of $i$. Add $t - 2$ dummy nodes $u_{i_2}, u_{i_3}, \ldots, u_{i_{t-2}}, u_{i_{t-1}}$ linked in the path $(u_{i_2}, i), (u_{i_3}, u_{i_2}), \ldots, (u_{i_{t-1}}, u_{i_{t-2}})$. For each child $i_s$ of $i$ with $2 \leq s \leq t - 1$, delete arc $(i_s, i)$ and add arc $(i_s, u_{i_s})$ with weight $w(i_s, i)$.
   Moreover, let $i_s$ and $u_{i_{s+1}}$ be the left and right child of $u_{i_s}$, for $2 \leq s \leq t - 1$, respectively. Replace the arc $(i_t, i)$ with the right arc $(i_t, u_{i_{t-1}})$ of weight $w(i_t, i)$. Finally, consider $i_1$ as the left child of $i$ and give the weight of $w(i_1, i)$ to the arc $(i_1, i)$. We assign weight zero to arcs $(u_{i_{s+1}}, u_{i_s})$, for $2 \leq s \leq t - 1$, and to arc $(u_{i_2}, i)$.

Nondummy nodes are called *original* nodes.

The new tree $B$ has at most $2n - 3$ nodes, out of which $n$ are the *original* nodes of $T$ and at most $n - 3$ are dummy nodes. Let the distance $d_B(i, j)$ between two nodes $i$ and $j$ in $B$ be the sum of the weights on the path from $i$ to $j$ in $B$. Let $\ell_B(i)$ and $\ell_T(i)$ be the depth of $i$ in $B$ and in $T$, respectively. Since each dummy node is linked to its father in $B$ by an arc of weight zero while each original node by an arc of weight equal to the weight of the corresponding arc in $T$, each original node $i$ yields $d_B(i, r) = d_T(i, r)$. Whereas, for each node $i \in T$, the dummy node $u_{i_s}$ associated to the $s$-th child of $i$, with $2 \leq s \leq c_i - 1$, yields $d_B(u_{i_s}, r) = d_T(i, r)$. Hence, for each node $i \in B$, $d_B(i, r) = d_T(\hat{i}, r)$, where $\hat{i} \in B$ is the closest ancestor of $i$ which is an original node (possibly $\hat{i} = i$).

Furthermore, for each node $i \in B$, let $B_i$ denote the subtree of $B$ rooted at $i$ and $T(B_i)$ be the subtree of $T$ induced by $B_i$ (note that $i$ belongs to $T(B_i)$ only if $i$ is an original node). Actually, when $i$ is a dummy node, $T(B_i)$ is a forest.

The size $\varsigma(i)$ of $T(B_i)$ is the number of original nodes in $B_i$. Clearly, for each node $i$, values $\ell_B(i)$, $\varsigma(i)$, and $d_B(i, r)$ can be computed in $B$ by a preprocessing step in $O(|B|) = O(n)$ time.

In the light of the above-mentioned transformation, *MSP* on the tree $T$ is solved on the transformed binary tree $B$. Each nonleaf node $i \in B$ has exactly two children denoted as $L(i)$ (left child) and $R(i)$ (right child). Each node is either dummy or original. We impose that dummy nodes cannot be selected as storage nodes.

Following the dynamic-programming algorithm proposed in Reference 8, we solve *MSP* starting from the leaves of $B$. For each node $i$, let $D(i)$ be the set of distance values from a node $x$ to the sink $r$, for each $x \neq r$ in the path from $i$ to $r$. Formally,

$$D(i) = \{d_B(x, i) | x \neq i \text{ is in the path from } i \text{ to } r\}.$$

We observe that $|D(i)| = \ell_T(i)$.

We define $E_i(m,j)$ as the cost of a solution with $m$ storages in the subtree rooted in $i$, given that the closest storage node not in such subtree is at distance $j$ from $i$. In the following we compute values $E_i^d$, $E_i^s$, and $E_i^f$ corresponding to $E_i$ when we assume that $i$ is a dummy, storage, or forward (original but not storage) node, respectively.

For each leaf $i \in B$, we solve *MSP* assuming that there are 0 or 1 storage nodes in $T(B_i)$ and that the deepest storage node $a_i$, not including $i$, on the path from $i$ to $r$ is at distance $j$ from $i$, that is, $d_B(i, a_i) = j$, with $j \in D(i)$. If the leaf $i$ is a dummy node, as $i$ cannot be a storage node, it must be $m = 0$ and hence we define $E_i^d(1,j) = +\infty$ for each $j \in D(i)$. Moreover, as $i$ does not collect data, $E_i^d(0,j) = 0$ for each $j \in D(i)$. If the leaf $i \in B$ is an original node, we solve two subproblems. If we assume that $i$ is a storage node, then we define $E_i^s(1,j) = \alpha s_d(i) d_B(i,r)$ and $E_i^s(0,j) = +\infty$, for each $j \in D(i)$. If we assume that $i$ is a forward node (i.e., $m = 0$), then we define $E_i^f(0,j) = s_d(i)(j + \alpha(d_B(i,r) - j))$ and $E_i^f(1,j) = +\infty$, for each $j \in D(i)$. In conclusion, for each leaf in $B$, $0 \le m \le 1$, and $j \in D(i)$

$$E_i(m,j) = \begin{cases} E_i^d(m,j) & \text{if } i \text{ is dummy and } m = 0 \text{ or } m = 1 \\ E_i^s(1,j) & \text{if } i \text{ is original and } m = 1 \\ E_i^f(0,j) & \text{if } i \text{ is original and } m = 0 \end{cases}$$

Proceeding the visit of $B$ in post order, for each node $i \in B$, we solve *MSP* assuming that there are $0 \le m \le \min\{\varsigma(i), k\}$ storage nodes in $B_i$ and assuming that the deepest ancestor storage node $a_i$ yields $d_B(i, a_i) = j$, with $j \in D(i)$. If $i$ is a *dummy* node, for $0 \le m \le \min\{\varsigma(i), k\}$ and $j \in D(i)$, we compute $E_i^d(m,j)$ as follows:

$$E_i^d(m,j) = \min_{\substack{m_1 + m_2 = m \\ m_1 \le \min\{m, \varsigma(L(i))\} \\ m_2 \le \min\{m, \varsigma(R(i))\}}} \left\{ E_{L(i)}(m_1, j + w(i, L(i))) + E_{R(i)}(m_2, j + w(i, R(i))) \right\}.$$

If $i$ is an original node, we assume that $i$ can be either a forwarding or a storage node. Assuming $i$ to be a *forwarding* node, we compute, $E_i^f(m,j)$ for each $0 \le m \le \min\{\varsigma(i) - 1, k\}$ and $j \in D(i)$, as follows:

$$E_i^f(m,j) = \min_{\substack{m_1 + m_2 = m \\ m_1 \le \min\{m, \varsigma(L(i))\} \\ m_2 \le \min\{m, \varsigma(R(i))\}}} \left\{ E_{L(i)}(m_1, j + w(i, L(i))) + E_{R(i)}(m_2, j + w(i, R(i))) \right\}$$
$$+ s_d(i)(j + \alpha(d_B(i,r) - j)).$$

Furthermore, assuming $i$ to be a *storage* node, we compute, $E_i^s(m,j)$ for each $1 \le m \le \min\{\varsigma(i), k\}$ and $j \in D(i)$, as follows:

$$E_i^s(m,j) = \min_{\substack{m_1 + m_2 = m - 1 \\ m_1 \le \min\{m - 1, \varsigma(L(i))\} \\ m_2 \le \min\{m - 1, \varsigma(R(i))\}}} \left\{ E_{L(i)}(m_1, w(i, L(i))) + E_{R(i)}(m_2, w(i, R(i))) \right\}$$
$$+ s_d(i) \alpha d_B(i,r).$$

If $i$ is a storage node, it must hold $m > 0$ and hence we initialize $E_i^s(0,j) = +\infty$ for $j \in D(i)$.

In conclusion, for each node $i \in B \setminus \{r\}$, we maintain a two dimensional matrix $E_i$ such that, for $0 \le m \le \min\{k, \varsigma(i)\}$ and $j \in D(i)$, it holds

$$E_i(m,j) = \begin{cases} E_i^d(m,j) & \text{if } i \text{ is dummy} \\ \min\{E_i^s(m,j), E_i^f(m,j)\} & \text{if } i \text{ is original} \end{cases}$$

As the root has to be a storage node and it has no ancestors, only the subproblem $E_r^s(k, 0)$ is solved. The optimal value of the problem will be given by $E_r^s(k, 0)$. The correctness follow by simple cut-and-paste arguments.

From the recursive equations and recalling that for $i \in B$, $\ell_T(i) = |D(i)|$, it follows directly that the total effort to fill the matrix $E_i$ for the node $i \in B$ is $O(k^2 \ell_T(i))$ because for each entry $E_i(m, j)$ of the matrix the minimum among at most $k$ values has to be computed. We assume that we keep a table that maps each value $j$ of $D(i)$ with a node $x$ such that $d(i, x) = j$. Overall, the time complexity can be bounded from above by $O(k^2 n^2)$ observing that $\ell_T(i) \leq n$. Such a bound can be reduced to $O(kn^2)$ by a more careful analysis, based on the shape of $T$, reported in Reference 31. On the other hand, the time complexity can be bounded from above by $O(k^2 P)$ noting that $\sum_{i \in B} |D(i)| = \sum_{i \in B} \ell_T(i) = O(P)$, where $P$ is the path length of $T$. In conclusion, the time complexity is $O(k^2 P)$, which can be expressed as $O(kn^2)$ if $P \in \Omega(\frac{n^2}{k})$; while the space complexity is $O(kn)$ since the recurrence to compute $E_i$ only depends on the values stored in $E_{L(i)}$ and $E_{R(i)}$.

As last remark, it is worth noting that the proposed algorithm can be also applied to the case of rooted trees with multiple sinks in which each node is assigned to its closest ancestor which is a sink. Let $s_i$, $i = 1, 2, \ldots, n_s$ be the set of sinks. To solve the $k$-MSP, it is enough to remove nodes $s_i$, for each $i = 1, 2, \ldots, n_s$, add one sink node $s$ and for each arc $(u, s_i)$ with $i = 1, 2, \ldots, n_s$ add a new arc $(u, s)$ of null weight. Then, on the so built tree, we solve the *MSP* problem selecting overall $k - n_s$ storage nodes.

## 27.5.2 Undirected Graphs with Bounded Treewidth

In the current section, we focus on a special class of undirected graphs, namely, graphs with bounded treewidth. We first give some notations and preliminary results on decomposition of graphs with bounded treewidth. Then, we propose a polynomial time algorithm for the *MSP* problem on such graphs.

### 27.5.2.1 Decomposition of Graphs with Bounded Treewidth

The treewidth of a graph has been introduced in Reference 32. It is based on the concept of tree-decomposition, defined as follows.

**Definition 27.1** *A tree-decomposition of a graph $G = (V, E)$ is a pair $(\mathcal{X}, T)$, where $T = (I, F)$ is a tree and $\mathcal{X} = \{X_i | i \in I\}$ is a family of subsets of $V$ such that*

- $\bigcup_{i \in I} X_i = V$ *(Node coverage)*
- *For all edges $\{u, v\} \in E$ there is an $i \in I$ with $u, v \in X_i$ (Edge coverage)*
- *For all $h, i, j \in I$: if $i$ is on the path from $h$ to $j$ in $T$ then $X_h \cap X_j \subseteq X_i$ (Coherence)*

Given a node $i \in I$, the set of children of $i$ in $T$ is denoted by $\varphi(i)$. The value $|I|$ is called the *size* of the tree-decomposition $(\mathcal{X}, T)$.

The width of a tree-decomposition $(\mathcal{X}, T)$ is $\max_{i \in I} |X_i| - 1$. The treewidth $w$ of $G$ is the minimum width over all the tree-decompositions of $G$. Unfortunately, finding the value of $w$ is NP-hard. The following theorem gives us an algorithm to compute the tree-decomposition of a graph with bounded treewidth.

**Theorem 27.5** ([33]) *Given a graph $G = (V, E)$, there exists a polynomial $p(w)$ and an algorithm that computes a tree-decomposition of $G$ of width $w$ in time at most $2^{p(w)} \cdot n$, where $w$ is the threewidth of $G$.*

From each decomposition, it has been proven in Reference 34 that a small tree-decomposition for which the size $|I| \leq |V| = n$ holds can be produced in linear time. Moreover, if $T$ is rooted, we say that $(\mathcal{X}, T)$ is a rooted tree-decomposition.

**Definition 27.2** *A rooted tree-decomposition is called* nice *if every node $i \in I$ is of one of the following types:*

- **Leaf:** *$i$ is a leaf of $T$, and has $|X_i| = 1$*
- **Introduce:** *$i$ has one child $j$, with $X_i = X_j \cup \{v\}$ for some node $v \in V \setminus \{X_j\}$.*

- **Forget:** $i$ has one child $j$, with $X_j = X_i \cup \{v\}$ for some node $v \in V \setminus \{X_i\}$.
- **Join:** $i$ has two children $j_1, j_2$ with $X_i = X_{j_1} = X_{j_2}$.

**Theorem 27.6** ([35]) *For constant $w$, given a tree-decomposition of a graph $G$ of width $w$ and $O(n)$ nodes, where $n$ is the number of nodes of $G$, one can find a nice tree-decomposition of $G$ of width $w$ and with at most $4n$ nodes in $O(n)$ time.*

From Theorems 27.5 and 27.6, the next corollary follows.

**Corollary 27.3** *Given a graph $G = (V, E)$ with fixed treewidth $w$, we can a nice tree-decomposition $(\mathcal{X}, T)$ of width $w$ and size $O(n)$, in $O(2^{p(w)} \cdot n)$ time.*

Given a node $i \in I$, let $V_i = \bigcup_{j \in T_i} X_j$ be a subset of $V$ induced by the subtree $T_i$ rooted in $i$. The next lemma shows a property of the paths of $G$ related to the tree-decomposition $(\mathcal{X}, T)$.

**Lemma 27.3** *Given a tree-decomposition $(\mathcal{X}, T)$, with $T = (I, F)$, of graph $G = (V, E)$, a node $i \in I$, a node $u \in V_i$, and a node $v \in V \setminus V_i$, each path $p \in G$ from $u$ to $v$ contains a node in $X_i$.*

*Proof.* Let $I(u) = \{i \in I | u \in X_i\}$ be the subset of nodes of $I$ induced by $u \in V$. By the coherence property, the graph induced by $I(u)$ in $T$ is connected. If the path $p$ contains only the edge $\{u, v\} \in E$ then, by the edge coverage property, there is $j \in I$ s.t. $u, v \in X_j$; therefore, $I(u) \cap I(v) \neq \emptyset$ and hence $v \in X_i$ or $u \in X_i$. Similarly, if $p = \{x_1, \cdots, x_m\}$ with $u = x_1$ and $v = x_m$, for each edge $\{x_i, x_{i+1}\} \in p$ there is $j \in I$ s.t. $x_i, x_{i+1} \in X_j$, therefore $I(x_i) \cap I(x_{i+1}) \neq \emptyset$. As $x_1 = u \in V_i = \bigcup_{h \in T_i} X_h$ and $x_m = v \in V \setminus V_i$ and $\bigcup_{j \in \{1, \cdots, m\}} I(x_j)$ is connected, there exists a node of $p$ in the set $X_i$. ∎

### 27.5.2.2 Algorithm for Graphs of Bounded Treewidth

In the current section, we present a dynamic programming algorithm that, given a nice tree-decomposition $(\mathcal{X}, T)$ of graph $G = (V, E)$, optimally solves *MSP* on $G$. When the tree-decomposition has bounded width (i.e., the graph has bounded treewidth), the algorithm runs in polynomial time. An exhaustive algorithm would look for a set of $k$ storage nodes among all $n$ nodes of $G$, hence obtaining an asymptotic complexity of $O\binom{n}{k}$. Exploiting the tree-decomposition, we are able to decrease the complexity to $O(w \cdot k \cdot n^{w+3})$ with $w$ being the treewidth of $G$. This clearly provides a polynomial time algorithm for graphs $G$ of bounded treewidth. To limit the scope of research for possible solutions of size $k$, the basic idea of our algorithm works in two main steps. Starting from the leaves of the nice tree-decomposition, it maintains trace of the possible solutions obtainable when considering different storage nodes for the nodes of $G$ contained in the current $x \in \mathcal{X}$ of the tree. When $G$ has treewidth bounded by a constant $w$, then the size of each subset in $\mathcal{X}$ is at most $w$, hence the size of the subset of nodes of $G$ that one has to consider as storage nodes associated with the nodes contained in each subset of $\mathcal{X}$ limited by $w$ as well. Once reached the root (that contains the sink), one of the entries providing the minimum cost can be used to visit the tree from the root down to the leaves to build the corresponding solution, hence deciding where to place storage devices.

Given a node $v \in V$ and a set of storage nodes $S \subseteq V$, let us denote as $d'(v, S)$ the minimum cost of assigning $v$ to a storage in $S$, that is $d'(v, S) = \min_{z \in S} \{s_d(v) \cdot (d(v, z) + \alpha d(z, r))\}$.

Given a set of nodes $V' \subseteq V$ and a set of storage nodes $S \subseteq V$, we denote as $cost_{V'}(S)$ the component of $cost(S)$ induced by the nodes in $V'$, that is $cost_{V'}(S) = \sum_{v \in V'} d'(v, S)$.

For each node $i$ of $T$, integer $q \leq \min\{|V_i|, k\}$, and set of nodes $C_i \subseteq V$, we define $F(i, q, C_i)$ as the minimum cost for assigning a set $C_i \cup D_i$ of storage nodes in $V_i$ with the following constraints:

- $C_i$ storage nodes are fixed.
- $|D_i \cup (C_i \cap V_i)| = q$.
- For each $s$ in $D_i$, the cost of assigning any node $x$ in $X_i$ to $s$ is greater than or equal to the minimal cost of assigning $x$ to a node in $C_i$.

Note that, for any set $D_i$, the above-mentioned third condition ensures that in the solution given by $C_i \cup D_i$ the nodes in $X_i$ are assigned to the storage nodes in $C_i$. Note also that the storage nodes in $C_i$ might belong to $V \setminus V_i$. Informally, $F(i, q, C_i)$ is the cost of assigning $q$ storage nodes to $V_i$, given that $C_i \subseteq V$ storage nodes are fixed and the nodes in $X_i$ are all constrained to be assigned to the storage nodes in $C_i$. Formally, $F(i, q, C_i) =$

$$\min_{D_i \subseteq V_i \setminus C_i} \left\{ \sum_{x \in X_i} d'(x, C_i) + \sum_{y \in V_i \setminus X_i} d'(y, C_i \cup D_i) \right\}$$

subject to: $\quad |D_i \cup (C_i \cap V_i)| = q$

$$\forall x \in X_i, d'(x, D_i) \geq d'(x, C_i).$$

Our algorithm computes $F(i, q, C_i)$ for each $X_i \in \mathcal{X}$, $q \leq \min\{|V_i|, k\}$, and $C_i \subseteq V$ such that all the following constraints are satisfied.

$$1 \leq |C_i| \leq \min\{|X_i|, k\};$$

For each $z \in C_i, \exists x \in X_i$ such that $\sigma(x, C_i) = z;$ $\qquad\qquad (27.5)$

$$|C_i \cap V_i| \leq q.$$

Let us consider the solution $S$ induced by $F(i, q, C_i)$ in $G$, and the solution $S_i \subseteq S$ induced by the same $F(i, q, C_i)$ in $V_i$, $S_i = S \cap V_i$. We denote $D_i = S_i \setminus C_i$ and for each child $j$ of $i$, we introduce the following definitions.

- $S_j$ is the solution induced by $S_i$ in the subproblem defined in $V_j$, $S_j = S_i \cap V_j$.
- $q_j = |S_j|$.
- $A_j$ is the set of storage nodes of $S_i$ that the nodes in $X_j \setminus X_i$ are associated with, $A_j = \{\sigma(x, S_i)\}_{x \in X_j \setminus X_i}$.
- $B_j$ is the set of storage nodes of $S_i$ which the nodes in $X_j \cap X_i$ are associated with, $B_j = \{\sigma(x, S_i)\}_{x \in X_j \cap X_i}$.
- $C_j = A_j \cup B_j$.
- $D_j = S_j \setminus C_j$.

The following lemma proves that the suboptimal structure of function $F$, that is, that optimal solution $F(i, q_i, C_i)$ is based on optimal solution $F(j, q_j, C_j)$, where $X_j$ is a child of node $X_i$.

**Lemma 27.4** $cost_{V_j}(S_j \cup C_j) = F(j, q_j, C_j)$.

*Proof.* We now show that $S_j \cup C_j$ satisfies the constraints of $F(j, q_j, C_j)$, that is, we show the following two statements.

- $|D_j \cup (C_j \cap V_j)| = q_j$. By definition of $D_j$, $|D_j \cup (C_j \cap V_j)| = |(S_j \setminus C_j) \cup (C_j \cap V_j)|$. As $S_j$ is a subset of $V_j$, $|(S_j \setminus C_j) \cup (C_j \cap V_j)| = |(S_j \setminus (C_j \cap V_j)) \cup (C_j \cap V_j)| = |S_j| = q_j$.
- For each $x \in X_j$, $d'(x, D_j) \geq d'(x, C_j)$. We note that $d'(x, D_j) \geq d'(x, D_i)$ as $D_j \subseteq D_i$. Two cases arise.
  - If $x \in X_i$, then $d'(x, D_i) \geq d'(x, C_i)$. Moreover, since $x \in X_j \cap X_i$, by definition of $B_j$, $d'(x, C_i) = d'(x, C_j)$. It follows that $d'(x, D_j) \geq d'(x, D_i) \geq d'(x, C_i) = d'(x, C_j)$.
  - If $x \notin X_i$, then $x \in X_j \setminus X_i$ and, by definition of $A_j$, $\sigma(x, S_i) \in A_j$. We have that, $d'(x, A_j) = d'(x, S_i) \leq d'(x, D_i)$ because $D_i \subseteq S_i$ and $d'(x, A_j) = d'(x, C_j)$. It follows that $d'(x, D_j) \geq d'(x, D_i) \geq d'(x, A_j) = d'(x, C_j)$.

We now show that the cost of $S_j \cup C_j$ is minimum among the solutions that satisfy the above-mentioned constraints. By contradiction, let us assume that there exists a solution $\bar{S}_j \subseteq V_j$ such that

- $|\bar{D}_j \cup (C_j \cap V_j)| = q_j$
- For each $x \in X_j$, $d'(x, \bar{D}_j) \geq d'(x, C_j)$
- $cost_{V_j}(\bar{S}_j \cup C_j) < cost_{V_j}(S_j \cup C_j)$

where $\bar{D}_j = \bar{S}_j \setminus C_j$. Let us define the set $\bar{S}_i = (S_i \setminus S_j) \cup \bar{S}_j = (S_i \setminus D_j) \cup \bar{D}_j$, and we show that

- $\bar{S}_i$ satisfies the constraints of $F(i, q, C_i)$, that is, $|\bar{D}_i \cup (C_i \cap V_i)| = q$ and for each $x \in X_i$, $d'(x, \bar{D}_i) \geq d'(x, C_i)$
- $cost_{V_i}(\bar{S}_i \cup C_i) < cost_{V_i}(S_i \cup C_i)$

where $\bar{D}_i = \bar{S}_i \setminus C_i$.

To show that $|\bar{D}_i \cup (C_i \cap V_i)| = q$, we observe that $|\bar{D}_j| = |D_j| = q_j - |C_j \cap V_j|$ because $D_j$ and $\bar{D}_j$ are disjoint from $C_j$. Moreover, $D_i \setminus V_j = \bar{D}_i \setminus V_j$. Therefore, $|D_i| = |\bar{D}_i|$ and $|\bar{D}_i \cup (C_i \cap V_i)| = |D_i \cup (C_i \cap V_i)| = q$.

We now show that for each $x \in X_i$ $d'(x, \bar{D}_i) \geq d'(x, C_i)$. Two cases arise:

- If $\sigma(x, \bar{D}_i) \in V_i \setminus V_j$, then $x$ is associated with a node not in $\bar{D}_j$ and therefore, $d'(x, \bar{D}_i) = d'(x, D_i) \geq d'(x, C_i)$.
- If $\sigma(x, \bar{D}_i) \in V_j$, then $d'(x, \bar{D}_i) = d'(x, \bar{D}_j) \geq d'(x, C_j)$. From Lemma 27.3, the path from $x$ to $\sigma(x, \bar{D}_i)$ must contain a node $y$ such that $y \in X_j \cap X_i$ (possibly, $y \equiv x$). By definition, $y$ is associated with a node in $B_j$, that is, $\sigma(y, C_j) \in B_j$. Therefore, $d'(x, C_j) = d'(x, C_i)$.

To conclude the proof, we observe that $cost_{V_i}(\bar{S}_i \cup C_i)$ is made of three components: the one induced by the nodes in $X_i$, the one induced by the nodes in $V_j \setminus X_i$, and the one induced by nodes in $V_i \setminus (V_j \cup X_i)$.

The nodes in $X_i$ are associated with the storage nodes in $C_i$ in both $\bar{S}_i$ and $S_i$. By Lemma 27.3, the nodes in $V_i \setminus (V_j \cup X_i)$ are associated with the same node both in $S_i$ and in $\bar{S}_i$. In solution $\bar{S}_i$, the nodes in $V_j \setminus X_i$ are associated with the storage in $\bar{S}_j$ and hence have a cost $cost_{V_j}(\bar{S}_j \cup C_j)$, whereas, in solution $S_i$, they are associated with nodes in $S_j$ and hence have a cost $cost_{V_j}(S_j \cup C_j) > cost_{V_j}(\bar{S}_j \cup C_j)$. Therefore, $cost_{V_i}(\bar{S}_i \cup C_i) < cost_{V_i}(S_i \cup C_i)$ a contradiction. ∎

Now that the suboptimality property has been proved, we give the recurrences for the *MSP* problem.

If $i$ is a leaf node of $T$, then $F(i, q, C_i) = \sum_{x \in X} d'(x, C_i)$.

By exploiting Lemma 27.4, we can compute the value of $F(i, q, C_i)$ by summing up the values of $F(j, q_j, C_j)$ for each $j \in \varphi(i)$ and the cost paid by the nodes in $X_i \setminus \bigcup_{j \in \varphi(i)} X_j$ and by subtracting the cost paid by the nodes that belong to more than one child. Formally,

$$F(i, q, C_i) = \sum_{j \in \varphi(i)} F(j, q_j, C_j) + \sum_{x \in X_i \setminus \bigcup_{j \in \varphi(i)} X_j} d'(x, C_i)$$

$$F(i, q, C_i) = \begin{cases} F(j, q-1, C_i) + d'(v, C_i) & \text{if } v \in C_i \text{ and } \exists u \in X_j \text{ s.t. } \sigma(u, C_i) = v \\ F(j, q-1, C_i \setminus \{v\}) + d'(v, C_i) & \text{if } v \in C_i \text{ and } \nexists u \in X_j \text{ s.t. } \sigma(u, C_i) = v \\ F(j, q, C_i) + d'(v, C_i) & \text{if } v \notin C_i \text{ and } \exists u \in X_j \text{ s.t. } \sigma(u, C_i) = \sigma(v, C_i) \\ F(j, q, C_i \setminus \{\sigma(v, C_i)\}) + d'(v, C_i) & \text{if } v \notin C_i \text{ and } \nexists u \in X_j \text{ s.t. } \sigma(u, C_i) = \sigma(v, C_i) \end{cases}$$

(27.6)

$$- \sum_{x \in \bigcup_{j \in \varphi(i)} X_j} d'(x, C_i) \left( |\{j \mid x \in X_j\}| - 1 \right)$$

$$= \sum_{j \in \varphi(i)} F(j, q_j, C_j)$$

$$- \sum_{x \in X_i} d'(x, C_i)(|\{j \mid x \in X_j \wedge j \in \varphi(i)\}| - 1).$$

To derive a recurrence relation, we minimize over the possible values of $q_j$ and $A_j$, for each $j \in \varphi(i)$. The value of $q_j$ cannot be smaller than $|C_j \cap V_j|$. Moreover, the values of $q_j$ have to sum up to $q$ minus the number of nodes of $C_i$ that do not belong to any $X_j$, that is, $\sum_{j \in \varphi(i)} q_j = q - |C_i \cap (X_i \setminus \bigcup_{j \in \varphi(i)} X_j)|$.

Finally, $A_j$ must be a subset of $V_j$ such that $|A_j| \leq |X_j \setminus X_i|$ and $d'(x, A_j) \geq d'(x, C_i)$, for each $x \in X_i$. We obtain the following recurrence. $F(i, q, C_i) =$

$$\min \quad \left\{ \begin{array}{l} \sum_{j \in \varphi(i)} F(j, q_j, C_j) \\ - \sum_{x \in X_i} d'(x, C_i) \left( |\{j \mid x \in X_j \wedge j \in \varphi(i)\}| - 1 \right) \end{array} \right\}$$

$$\text{s.t.:} \quad |C_j \cap V_j| \leq q_j;$$

$$\sum_{j \in \varphi(i)} q_j = q - |C_i \cap (X_i \setminus \cup_{j \in \varphi(i)} X_j)|;$$

$$C_j = B_j \cup A_j, A_j \subseteq V_j, |A_j| \leq |X_j \setminus X_i|,$$

$$\forall x \in X_i, d'(x, A_j) \geq d'(x, C_i).$$

(27.7)

The previous recurrence equation cannot be computed in polynomial time for any minimal tree-decomposition. In fact, if the number $|\varphi(i)|$ of children of $i$ in $T$ is not bounded, the number of sets $\{q_j\}_{j \in \varphi(i)}$ satisfying $\sum_{j \in \varphi(i)} q_j = q - |C_i \cap (X_i \setminus \cup_{j \in \varphi(i)} X_j)|$ is exponential. Therefore, we exploit the concept of nice tree-decomposition and obtain the following recurrence relations. Recall that in a nice tree-decomposition, the introduce or forget nodes have just one child, whereas the join nodes have two children. So for each node $i \in I$, $\varphi(i) = \{1, 2\}$.

- If $i$ is a leaf node where $X_i = \{v\}$, then $q \in \{0, 1\}$, $C_i = \{u\}$ for each $u \in V$, and

$$F(i, q, C_i) = d'(v, C_i).$$

Clearly, if $C_i = X_i = \{v\}$, it must be $q = 1$, and thus $F(i, 0, v)$ is set to $+\infty$.

- If $i$ is an introduce node where $j$ is the child of $i$ in $T$, $X_j = X_i \setminus \{v\}$, $0 \leq q \leq |V_i|$, then $|A_j| \leq |X_j \setminus X_i| = 0$ and $\varphi(i) = 1$.
  - If $v$ is a storage node, that is, $v \in C_i$, $|C_i \cap (X_i \setminus \cup_{j \in \varphi(i)} X_j)| = |C_i \cap (X_i \setminus X_j)| = 1$ and two cases arise: If $\exists u \in X_j$ s.t. $\sigma(u, C_i) = v$, then the nodes of $X_j$ and $X_i$ are associated with the same storage nodes. That is, $B_j = \{\sigma(x, S_i)\}_{x \in X_i \cap X_j} = \{\sigma(x, S_i)\}_{x \in X_j} = C_i$. Otherwise, i.e., $v \in C_i$ but it was not associated with any node in $X_j$, $B_j = C_i \setminus \{v\}$.
  - If $v$ is not a storage node, that is, $v \notin C_i$, then $|C_i \cap (X_i \setminus \cup_{j \in \varphi(i)} X_j)| = |C_i \cap v| = 0$ and two cases arise: If $v$ is associated with a storage node $\sigma(v, C_i)$ that has already been selected for $X_j$, that is, $\exists u \in X_j$ s.t. $\sigma(u, C_i) = \sigma(v, C_i)$, then $B_j = C_i$. Otherwise, $B_j = C_i \setminus \{\sigma(v, C_i)\}$.

  These cases can be summarized in Equation 27.6 which is obtained by plugging the above-mentioned values into Equation 27.7. Precisely, any node $x \in X_i \cap X_j$ yields $|\{j \mid x \in X_j \wedge j \in \varphi(i)\}| - 1 = 0$ because $x$ is contained in just one child of $i$, whereas the new node introduced in $X_i$ yields $|\{j \mid x \in X_j \wedge j \in \varphi(i)\}| - 1 = -1$ because $|\{j \mid x \in X_j \wedge j \in \varphi(i)\}| = 0$. Moreover, note that in the first two cases, as $v \in C_i$, then $d'(v, C_i) = s_d(v)\alpha d(v, r)$.

- If $i$ is a forget node where $j$ is the child of $i$ in $T$, $X_i = X_j \setminus \{v\}$, $0 \leq q \leq |V_i|$, then for each $x \in X_i$ it holds $|\{j \mid x \in X_j \wedge j \in \varphi(i)\}| = 1$ and $|C_i \cap (X_i \setminus X_j)| = 0$. Thus, observed that $|A_j| \leq 1$ and $B_j = C_i$, it yields $C_j = C_i$ or $C_j = C_i \cup \{z\}$ with $z$ that cannot be the storage node associated with any $u \in X_i$. Therefore, $C_j$ can contain any node $z \in V_i \setminus C_i$ such that $d'(u, z) \geq d'(u, C_i)$ for each $u \in X_i$. By plugging into Equation 27.7, we obtain $F(i, q, C_i) =$

$$\min \left\{ F(j, q, C_i), \min_{\substack{z \in V_i \setminus C_i : \forall u \in X_i, \\ d'(u,z) \geq d'(u,C_i)}} \{F(j, q, C_i \cup \{z\})\} \right\}.$$

(27.8)

- If $i$ is a join node where $j_1$ and $j_2$ are the children of $i$ and $X_i = X_{j_1} = X_{j_2}$, for each $x \in X_i$ it yields $|\{j \mid x \in X_j \wedge j \in \varphi(i)\}| = 2$. Moreover, $|C_i \cap (X_i \setminus \cup_{j \in \varphi(i)} X_j)| = 0$, $|A_{j_1}| = |A_{j_2}| = 0$, $B_{j_1} = B_{j_2} = C_i$. Then, we obtain the solution $F(i, q, C_i)$ with $q$ storage nodes by combining the solutions for $X_{j_1}$ and for $X_{j_2}$ such that $|C_i \cap V_{j_1}| \leq q_1$, $|C_i \cap V_{j_2}| \leq q_2$ and $q = q_1 + q_2 - |C_i \cap (V_{j_1} \cap V_{j_2})|$. As each node $x \in X_i$ belongs to both $X_{j_1}$ and $X_{j_2}$, and as by the coherence property each node in $V_{j_1} \cap V_{j_2}$

belongs to $X_i$, it holds $q = q_1 + q_2 - |C_i \cap (V_{j_1} \cap V_{j_2})| = q_1 + q_2 - |C_i \cap X_i|$. Thus, Equation 27.7 can be rewritten as $F(i, q, C_i) =$

$$
\min_{q=q_1+q_2-|C_i \cap X_i|} \{F(j_1, q_1, C_i) + F(j_2, q_2, C_i)
$$
$$
- \sum_{x \in X_i} d'(x, C_i) \ \Big| \ |C_i \cap V_{j_1}| \le q_1, |C_i \cap V_{j_2}| \le q_2 \}.
\tag{27.9}
$$

We can now state the following theorem.

**Theorem 27.7** *Given a tree-decomposition of size $w$, there exists an algorithm that optimally solves MSP in $O(w \cdot k \cdot n^{w+3})$ time.*

*Proof.* The algorithm works as follows: First, starting from the given tree-decomposition, it computes a nice tree-decomposition $(\mathcal{X}, T)$ rooted at a node containing $r$. Let $i_r$ be the root of $T$. Then, recursively, starting from the leaves computes the values of $F(i, q, C_i)$, for each $X_i \in \mathcal{X}$, $q \le \min\{|X_i|, k\}$, $C_i \subseteq V$ satisfying conditions 27.5. Finally, the optimal value is given by

$$
\min_{C: r \in C} F(i_r, k, C).
\tag{27.10}
$$

The correctness of the algorithm follows from Lemma 27.4 and the earlier discussion.

By Theorem 27.6, computing a nice tree-decomposition requires linear time. As $|C_i| \le w + 1$ for each $i$, the number of values of $F(i, q, C_i)$ to be computed is at most $|T| \cdot k \cdot n^{w+1}$. Computing each $F(i, q, C_i)$ requires at most $O(1)$, $O(|C_i|)$, $O(n)$, or $O(k \cdot w)$, if $i$ is a leaf, an introduce, a forget, or a join node, respectively. As $|T| = O(n)$, $|C_i| \le n$, and $k \le n$, the overall cost is given by $O((|T| \cdot k \cdot n^{w+1}) \cdot (n + |C_i| + k \cdot w)) = O(w \cdot k \cdot n^{w+3})$. ∎

It is worth noting that the algorithm can be exploited to solve the metric $k$-median problem (see definition in Section 27.1) with the following minor modifications. For the *MMP* problem, the above-mentioned algorithm can be applied by setting $\alpha = 0$ and $s_d(v) = 1$, for each $v \in V$, and rewriting Equation 27.10 as $\min_C F(i_r, k, C)$.

**Corollary 27.4** *Given a tree-decomposition of size $w$, there exists an algorithm that optimally solves MMP in $O(w \cdot k \cdot n^{w+3})$ time.*

To the best of our knowledge, no results were known about the metric $k$-median problem on graphs of bounded treewidth.

## 27.6 Conclusion

We have investigated on the MSP. The obtained results show that the problem is in general not in *APX*, whereas it has been proven to be not approximable within a factor of $1 + \frac{1}{e}$ in the case of undirected graphs. On the other hand, it admits an optimal polynomial-time algorithm for directed trees and undirected graphs of bounded treewidth, and a constant-factor polynomial-time approximation algorithm for general undirected graphs. Finally, the $k$-storage problem generalizes the well-known metric $k$-median problem, and then the above-mentioned results also hold for this case.

As future work, it is worth to consider variants of the studied problem. For example, one may consider the cost of diffusing a query from the sink toward the selected storage nodes or one may introduce the additional constraint of having a bounded number of nodes associated with each storage node. Moreover, it would be interesting to study the $k$-MSP problem under more realistic assumptions. For example, one may conduct stochastic analysis of the cost of the algorithm assuming edge weight distributions suitable to model the traffic cost.

# Acknowledgment

The work has been partially supported by the European project "Geospatial based Environment for Optimisation Systems Addressing Fire Emergencies", contract no. H2020-691161, and by the Italian project "RISE: un nuovo framework distribuito per data collection, monitoraggio e comunicazioni in contesti di emergency response," Fondazione Cassa Risparmio Perugia, code 2016.0104.021.

# References

1. J. Gehrke and S. Madden. Query processing in sensor networks. *IEEE Pervasive Computing*, 3(1):46–55, 2004.
2. E. J. Duarte-Melo and M. Liu. Data-gathering wireless sensor networks: Organization and capacity. *Computer Networks (COMNET)*, 43(4):519–537, 2003.
3. P. Gupta and P. R. Kumar. The capacity of wireless networks. *IEEE Transactions on Information Theory*, 46(2):388–404, 2000.
4. S. K. Das, A. Navarra, and M. C. Pinotti. Dense, concentric and non-uniform multi-hop sensor networks. In S. Nikoletseas and J. D. P. Rolim (Eds.), *Theoretical Aspects of Distributed Computing in Sensor Networks*, Monographs in Theoretical Computer Science, an EATCS Series, Part 5, pp. 515–551. Springer, Berlin, Germany, 2011.
5. Y. Charfi, N. Wakamiya, and M. Murata. Challenging issues in visual sensor networks. *Wireless Communications, IEEE*, 16(2):44–49, 2009.
6. A. A. Bertossi, D. Diodati, and C. M. Pinotti. Storage placement in path networks. *IEEE Transactions on Computers*, 64(4):1201–1207, 2015.
7. P. Kulkarni, D. Ganesan, P. J. Shenoy, and Q. Lu. Senseye: A multi-tier camera sensor network. In *Proceedings of the 13th Annual ACM International Conference on Multimedia*, Singapore, November 6–11, pp. 229–238. ACM, New York, 2005.
8. B. Sheng, Q. Li, and W. Mao. Optimize storage placement in sensor networks. *IEEE Transactions on Mobile Computing*, 9(10):1437–1450, 2010.
9. B. Sheng, C. C. Tan, Q. Li, and W. Mao. An approximation algorithm for data storage placement in sensor networks. In *Proceedings of the 2nd International Conference on Wireless Algorithms, Systems and Applications (WASA)*, pp. 71–78. IEEE, 2007.
10. H. Li, V. Pandit, and D. P. Agrawal. Deployment optimization strategy for a two-tier wireless visual sensor network. *Wireless Sensor Network*, 4(4), 2012.
11. G. Mathur, P. Desnoyers, P. Chukiu, D. Ganesan, and P. Shenoy. Ultra-low power data storage for sensor networks. *ACM Transactions on Sensor Networks*, 5(4):33:1–33:34, 2009.
12. Rise project. http://www.cs.ucr.edu/~rise, 2014. (accessed April 1, 2016).
13. Stargate gateway (spb400). http://www.xbow.com, 2014. (accessed April 1, 2016).
14. J. Paek, B. Greenstein, O. Gnawali, K. Jang, A. Joki, M. Vieira, J. Hicks, D. Estrin, R. Govindan, and E. Kohler. The tenet architecture for tiered sensor networks. *ACM Transactions on Sensor Networks*, 6(4):34:1–34:44, 2010.
15. S. Ratnasamy, B. Karp, S. Shenker, D. Estrin, R. Govindan, L. Yin, and F. Yu. Data-centric storage in sensornets with ght, a geographic hash table. *Mobile Networks and Applications*, 8(4):427–442, 2003.
16. B. Sheng and Q. Li. Verifiable privacy-preserving sensor network storage for range query. *IEEE Transactions on Mobile Computing*, 10(9):1312–1326, 2011.
17. L. Xie, S. Lu, Y. Cao, and D. Chen. Towards energy-efficient storage placement in large scale sensor networks. *Frontiers of Computer Science*, 8(3):409–425, 2014.
18. S. Madden, M. J. Franklin, J. M. Hellerstein, and W. Hong. The design of an acquisitional query processor for sensor networks. In *Proceedings of the 2003 ACM SIGMOD International Conference on Management of Data, SIGMOD '03*, San Diego, CA, pp. 491–502. ACM, New York, 2003.

19. S. Tilak, N. B. Abu-Ghazaleh, and W. R. Heinzelman. A taxonomy of wireless micro-sensor network models. *Mobile Computing and Communication Review*, 6(2):28–36, 2002.

20. S. Shenker, S. Ratnasamy, B. Karp, R. Govindan, and D. Estrin. Data-centric storage in sensornets. *Computer Communication Review*, 33(1):137–142, 2003.

21. D. Ganesan, B. Greenstein, D. Estrin, J. S. Heidemann, and R. Govindan. Multiresolution storage and search in sensor networks. *ACM Transactions on Storage*, 1(3):277–315, 2005.

22. V. Arya, N. Garg, R. Khandekar, A. Meyerson, K. Munagala, and V. Pandit. Local search heuristics for k-median and facility location problems. *SIAM Journal on Computing*, 33(3):544–562, 2004.

23. S. Guha and S. Khuller. Greedy strikes back: Improved facility location algorithms. *Journal of Algorithms*, 31(1):228–248, 1999.

24. K. Jain, M. Mahdian, and A. Saberi. A new greedy approach for facility location problems. In *Proceedings of the 34th Annual ACM Symposium on Theory of Computing (STOC)*, Montreal, Canada, pp. 731–740. ACM, New York, 2002.

25. J. H. Lin and J. S. Vitter. $\epsilon$-approximations with minimum packing constraint violation (extended abstract). In *Proceedings of the 24th Annual ACM Symposium on Theory of Computing (STOC)*, Victoria, Canada, pp. 771–782. ACM, New York, 1992.

26. G. D'Angelo, D. Diodati, A. Navarra, and Cristina M. Pinotti. The minimum k-storage problem on directed graphs. *Theoretical Computer Science*, 596:102–108, 2015.

27. G. D'Angelo, D. Diodati, A. Navarra, and C. M. Pinotti. The minimum k-storage problem: Complexity, approximation, and experimental analysis. *IEEE Transactions on Mobile Computing*, 15(7):1797–1811, 2016.

28. R. Sedgewick and P. Flajolet. *An Introduction to the Analysis of Algorithms*. Addison-Wesley, Upper Saddle River, NJ, 1996.

29. I. Dinur and D. Steurer. Analytical approach to parallel repetition. In *Proceedings of the 46th Annual ACM Symposium on Theory of computing (STOC)*, pp. 624–633. ACM, New York, 2014.

30. R. Bar-Yehuda and S. Moran. On approximation problems related to the independent set and vertex cover problems. *Discrete Applied Mathematics*, 9(1):1–10, 1984.

31. A. Tamir. An $O(pn^2)$ algorithm for the p-median and related problems on tree graphs. *Operations Research Letters*, 19(2):59–64, 1996.

32. N. Robertson and P. D. Seymour. Graph minors. II. algorithmic aspects of tree-width. *Journal of Algorithms*, 7(3):309–322, 1986.

33. J. Flum and M. Grohe. *Parameterized Complexity Theory*, volume 3. Springer, New York, 2006.

34. T. Kloks. *Treewidth, Computations and Approximations*, volume 842 of Lecture Notes in Computer Science. Springer, Berlin, Germany, 1994.

35. H. L. Bodlaender and T. Kloks. Efficient and constructive algorithms for the pathwidth and treewidth of graphs. *Journal of Algorithms*, 21(2):358–402, 1996.

# 28

# Approximation Algorithms for the Primer Selection, Planted Motif Search, and Related Problems[*]

Sudha Balla

Jaime Davila

Marius Nicolae

Sanguthevar Rajasekaran

## 28.1 Primer Selection Problem

In the current chapter, we consider two problems from computational biology, namely, primer selection and planted motif search (PMS). The closest string problems (CSPs) and the closest substring problems (CSSPs) are closely related to the PMS problem. All of these problems have been proven to be NP-hard. We survey some representative approximation algorithms that have been proposed for these problems.

The problem of selecting primers for Polymerase Chain Reaction (PCR) and Multiplex-PCR (MP-PCR) experiments is important in computational biology and has drawn the attention of numerous researchers in the recent past. This is a minimization problem that seeks the minimum set of primers required for a given set of DNA sequences as the input. The primers selected for the input set could be of two different categories, namely, nondegenerate primers and degenerate primers. The latter method of designing degenerate primers for a given input set gives rise to a variant of the primer selection problem (PSP) called the degenerate primer selection problem (DPSP). These two variants have been proven to be NP-Complete in the literature and also intractable to approximation within a constant to the optimal

---

[*] This work has been supported in part by the NSF grant ITR-0326155.

solution [1,2]. Thus, a number of heuristics have been proposed in the literature to select primers and in this chapter, we discuss them in detail. The primers can be viewed as motifs occurring in the input set and hence this problem is related to the problem of identifying motifs in DNA sequence data.

## 28.1.1 Background Information

PCR is a molecular biological method for amplifying, that is, creating multiple copies of, DNA sequences. In its basic form, PCR requires a pair of synthetic DNA sequences, called forward and reverse primers, which are short single-stranded DNA strings, typically 15–20 nucleotides in length, that exactly match the beginning and end of the DNA fragment to be amplified.

MP-PCR is a variant of PCR, which enables simultaneous amplification of multiple DNA fragments of interest in one reaction by using a mixture of multiple primers [3]. This method has been applied in many areas of DNA testing, including analyses of deletions, mutations, and polymorphisms, and, more recently, in genotyping applications requiring simultaneous analysis of up to thousands of markers. A set of nondegenerate primers is selected on each end of the regions to be amplified for the given input set of DNA sequences. This is the basic version of the problem and is called the *PSP* in the literature. We discuss some of the salient algorithms proposed for this problem in some detail in this chapter.

The presence of multiple primers in MP-PCR can lead to severe problems, such as unintended amplification products caused by mis-priming or lack of amplification due to primer cross-hybridization. To minimize these problems, it is critical to minimize the number of primers involved in a single MP-PCR reaction, particularly when the number of DNA sequences to be amplified is large. This can be achieved by selecting primers that would simultaneously act as forward and/or reverse primers for several of the DNA sequences in the input set. A recent technique that enables higher degrees of *primer reuse* is to allow more than one nucleotide at some of the positions of the primer. Remarkably, such primers, called degenerate primers [4], are as easy to synthesize as regular primers as their synthesis requires the same number of biochemical steps (the only difference is that one must add multiple nucleotides in some of the synthesis steps). The degeneracy of a degenerate primer is the number of distinct nondegenerate primers that could be formed out of it. For example, if the degenerate primer $p_d = A\{CT\}GC\{ACG\}T\{GA\}$, it has degeneracy 12; the distinct nondegenerate primers represented in $p_d$ are ACGCATG, ACGCATA, ACGCCTG, ACGCCTA, ACGCGTG, ACGCGTA, ATGCATG, ATGCATA, ATGCCTG, ATGCCTA, ATGCGTG, and ATGCGTA. As highly degenerate primers may give excessive mis-priming, a bound on the degeneracy of a primer is typically imposed, leading to a variant of the PSP called the *DPSP*, discussed in detail in the following sections of this chapter.

## 28.1.2 Polymerase Chain Reaction

The PCR experiment is conducted in a series of cycles, typically 30–40 in number, each cycle divided into three major steps. It requires several components (called the reaction mixture) that are placed in tubes, called the reaction tubes, which are repeatedly heated and cooled in an automated equipment called the thermal cycler. The components that make up the reaction mixture are a DNA template or the sequence that needs to be amplified, two primers that define the start and the end of the region to be amplified, nucleotides from which the new DNA is built by the DNA-polymerase, and a suitable chemical environment provided by the buffer.

The three steps that constitute each cycle of the PCR are as follows:

**Step 1:** The double-stranded DNA is heated to around 94°C–96°C to break the hydrogen bonds that connect the two DNA strands and separate them. This step is called denaturing.

**Step 2:** In this step, called annealing, the temperature is lowered so the primers can attach themselves to the single DNA strands. The temperature of this stage is usually 5°C below melting temperature of the primers (45°C–60°C).

**Step 3:** The DNA-Polymerase fills in the missing strands in this step called elongation. It starts at the annealed primer and works its way along the DNA strand and typically takes place at a temperature of around 72°C.

The total time for a single cycle of the PCR is 3–5 minutes. Because both strands of the DNA sequence are copied during PCR, there is an exponential increase of the number of copies of the sequence. If there is one copy of the sequence before the cycles start, after one cycle, there will be two copies, after two cycles, there will be four copies, three cycles will result in eight copies, and so on.

The quality of the amplifications depends very largely on the primers used in the experiment, thus making the primer selection a very important process. The melting temperature of a primer is the temperature at which at least half of the primer binding sites are occupied in the above-mentioned Step 1 and increases with the increase in the length of the primer. On the other hand, very short primers, although they have low melting temperatures, would result in binding to many locations in the DNA sequence leading to mis-priming. Thus, there arise some experimental constraints for the selection of PCR primers, referred to as the biological constraints:

1. The Guanine-Cytosine (GC) content (the number of G's and C's in the primer) of the primers should be around 40%–60% of its length.
2. The length of the primers should be chosen in such a way that they do not bind themselves to several positions of the DNA sequence.
3. There should not be any complementarity in the primers, that is, they should not be self-complementary, for example, the primer 5′-GCGGTATACCGC-3′ is self-complementary, and they should not be complementary to one another, for example, the primers 5′-CGAAATGCCG-3′ and 5′-CGGCATTTCG-3′ are complementary to each other.
4. The melting temperatures of both primers should not differ by more than 5°C and the melting temperature of the DNA sequence should not differ from that of the primers by more than 10°C.

## 28.1.3 Terminology

In the current section, the terminology adopted to explain the problems under discussion and the algorithms proposed for the same is given in detail.

Let $S = \{S_1, S_2, \ldots, S_n\}$ be the set of input sequences defined over the DNA alphabet $\Sigma = \{A, C, G, T\}$. Let $\Sigma^*$ denote the set of all finite strings defined over the alphabet $\Sigma$. Let $l_i$ be the length of the sequence $S_i, 1 \le i \le n$. Let $k$ be the length of the primer designed for the input set. The number of $k$-mers (a $k$-mer is a substring of length $k$) possible from each input string $S_i$ is $(l_i - k + 1), 1 \le i \le n$. Let $P$ be the set of all $k$-mers of the input set $S$. A primer $p \in P$ of length $k$ is said to cover a subset $S'$ of the set of input sequences $S$, iff $p$ is a substring of every sequence in $S'$. The primer that is designed to bind at the 5′ end of the sequences is called a forward primer and the one designed to bind at the 3′ end is called a reverse primer. The PSP is defined as follows:

**Definition 28.1 Primer Selection Problem (PSP):** *The PSP is to minimize the size of the subset $P'$ of $P$ such that the primers of $P'$ collectively cover the input set $S$, and every sequence $S_i, 1 \le i \le n$, has at least one* forward *and one* reverse *primer in $P'$.*

An *optimal cover* for $S$ is defined as the set $P'$ of minimum size.

A degenerate primer $p_d$ is a primer of length $k$ with one or more symbols of $\Sigma$ occurring in each position. The *degeneracy* $d$ of the degenerate primer $p_d$ is the product of the number of symbols in each position of the primer, that is, $d(p_d) = \Pi_{i=1}^{k} |p_d[i]|$.

The *degeneracy* of a degenerate primer is also the number of distinct nondegenerate primers that could be formed out of it. For example, the degenerate primer $p_d$ = A{CT}GC{ACG}T{GA} has a degeneracy of 12; the distinct nondegenerate primers represented in $p_d$ are ACGCATG, ACGCATA, ACGCCTG, ACGCCTA, ACGCGTG, ACGCGTA, ATGCATG, ATGCATA, ATGCCTG, ATGCCTA, ATGCGTG, and

ATGCGTA. The degenerate primer $p_d$ is said to cover an input sequence $S_i$ iff one of the nondegenerate primers represented in $p_d$ is a substring of $S_i$. If a degenerate primer of length $k$ covers $m$ of the given $n$ input sequences, it is said to have a coverage of size $m$.

The decision version of Degenerate Primer Design Problem (known as DPD [1]) is to find if there exists a degenerate primer of length $k$ and degeneracy at most $d$ that has a coverage of $m$ for a given input set of $n$ sequences. As the length of the primer $k$ is decided beforehand, the algorithms that have been designed for this problem try to optimize either the degeneracy $d$ or the coverage $m$ and hence there are two variants of the DPD problem. The former is called the *Minimum Degeneracy Degenerate Primer Design Problem* and attempts to find a degenerate primer of length $k$ and minimum degeneracy $d_{min}$ that covers all the $n$ input sequences. The latter is called the *Maximum Coverage Degenerate Primer Design (MC-DPD) Problem* that identifies a primer of length $k$ and degeneracy at most $d$ that covers a maximum number of the given $n$ input strings. Linhart and Shamir formulated the above-mentioned versions of the problem in Reference 1. The formulation of above-mentioned MC-DPD is for identifying one degenerate primer and can be extended to find a set of degenerate primers to cover a given set of input sequences as follows:

**Definition 28.2 Degenerate Primer Selection Problem (DPSP):** *Given a set S of n input sequences (DNA sequences) and integers k and d, find a set of degenerate primers $P_d$ such that each primer in $P_d$ has a degeneracy of at most d, the set $P_d$ covers all the input strings S, that is, every sequence $S_i, 1 \leq i \leq n$, has at least one forward and one reverse primer in $P_d$.*

In the following sections, we discuss some salient algorithms that have been proposed for PSP and DPSP. Both these variants have been proven to be NP-Complete in the literature. These proofs are described next.

## 28.1.4 NP-Completeness of Primer Selection Problem and Degenerate Primer Selection Problem

Pearson et al. [2] formulated the problem of finding an optimal cover for the PSP version of primer selection as follows:

*Optimal Primer Cover Problem (OPCP):* Given an input set $S$ of DNA sequences and integer $k$, find an optimal cover of $S$, the primer length being $k$.

They proved the NP-Completeness of OPCP by transforming the Minimum Set Cover problem (MSCP) to OPCP.

*Minimum Set Cover Problem (MSCP):* Let $\mathcal{F} = \{F_j\}$ be a finite family of sets. Let $\mathcal{F}'$ be a subset of $\mathcal{F}$. $\mathcal{F}'$ is a cover of $\mathcal{F}$ iff

$$U = \cup_{F \in \mathcal{F}'} F = \cup_{F \in \mathcal{F}} F.$$

The decision version of MSCP is, for a given family of sets $\mathcal{F}$ and an integer $f$, to determine if $\mathcal{F}$ has a cover $\mathcal{F}'$ such that $|\mathcal{F}'| \leq f$.

Let the number of primers in an optimal cover solution for OPCP be $q$. Let $(S, P, q)$ denote an instance of OPCP and $(\mathcal{F}, f)$ denote an instance of MSCP. An arbitrary instance $(\mathcal{F}, f)$ of MSCP is transformed into an instance $(S, P, q)$ of OPCP such that $(S, P, q)$ has a solution iff $(\mathcal{F}, f)$ has a solution, as given in the following. Let $U = \cup_{F \in \mathcal{F}} F$.

Let $q = f$;

$\Sigma = \{0, 1, b_1, b_2, b_3, \ldots, b_{|U|}\}$. Here the $b_i$'s $(1 \leq i \leq |U|)$ are unique symbols used as separators as explained next.

$k = \log_2 |\mathcal{F}|$;

Construct the set $S$ over the alphabet $\Sigma$ as follows:

Every $F_j \in \mathcal{F}$ is encoded by a unique string $v_j$, of length $k$, over the alphabet $\Sigma' = \{0, 1\}$, $\Sigma'$ being a subset of $\Sigma$. Every $S_i \in S$ represents an unique element $u_i \in U$, $S_i$ encoding details about the subsets $F_j \in \mathcal{F}$ in which element $u_i$ is present. This is achieved by concatenating the string $v_j b_i$ to $S_i$ of all $F_j \in \mathcal{F}$ in which $u_i$ is present. Note that $b_i$ acts as a unique string separator in $S_i$.

In the above-mentioned transformation of the MSCP instance to the OPCP instance, the size of the alphabet $\Sigma$ varies with the size of $U$. But for input sets that are DNA sequences, the alphabet $\Sigma$ is fixed, that is, $\Sigma = \{A, C, G, T\}$. To transform an arbitrary MSCP instance to an OPCP instance using the fixed alphabet, it is sufficient to represent $\{0, 1\}$ above using $\{a, c\}$ and $\{b_1, b_2, b_3, \ldots, b_{|U|}\}$ using $\{g, t\}$. Thus, the OPCP for the DNA alphabet is NP-Complete. ∎

Hence, we get the following theorem:

**Theorem 28.1** *PSP is NP-Complete.*

Linhart and Shamir [1] prove the NP-Completeness of DPSP by proving that the Minimum Primers Degenerate Primer Design problem (MP-DPD), a special case of DPSP in which every input string is of length $k$, is NP-Complete for $|\Sigma| \geq 2$. The proof is based on a reduction from the Minimum Bin Packing problem (MBPP).

*Minimum Bin Packing Problem (MBPP)*: Given are $q$ positive integers or items $a_1, a_2, a_3, \ldots, a_q$, two integers $b$ (the number of bins), and $c$ (the capacity). The goal is to find if the $q$ items can be packed into the $b$ bins such that the total sum of the items in each bin is at most $c$.

Given an instance of MBPP, the instance for DPSP can be constructed as follows: Let $A = \sum_{i=1}^{q} a_i$; $\Sigma = \{0, 1\}$; $k = A$; $d = 2^c$; and $|P_d| = b$. The set $S$ is constructed as follows: each string $S_i \in S$ is of length $A$ and is over the alphabet $\Sigma$. $S_i = s_i^1, s_i^2, \ldots, s_i^A$, in which $s_i^j = 1$ when $A_i \leq j \leq A_i + a_i$ and $s_i^j = 0$ otherwise. Here $A_i = \sum_{x=1}^{i-1} a_x$.

The size of the set $P_d$ is set to $b$ and so the goal is to find if there is a $P_d$ of size $b$, with primers of length $A$, and degeneracy $2^c$ that cover all the $q$ input strings. This polynomial reduction of MBPP to DPSP proves that a solution to MBPP exists iff a solution to MP-DPD exists and hence the following theorem arises.

**Theorem 28.2** *DPSP is NP-Complete.*

Given that PSP and DPSP are NP-hard, researchers have devised several approximation algorithms for these problems. Quality bounds have been proven for some of these algorithms. For the other algorithms, the quality has been measured only empirically. We describe some of the approximation algorithms that have been proposed for PSP and DPSP next.

## 28.1.5 Algorithms for Primer Selection Problem

In the current section, we will survey some of the salient algorithms from the literature that have been proposed for the PSP.

Pearson et al. [2] have proposed a simple greedy algorithm for MSCP called PSP-Greedy. The output of this algorithm is guaranteed to be within an $O(\log n)$ factor of the optimal. They have also proposed an exact branch and bound algorithm, which is not discussed here.

Algorithm **PSP-Greedy** can be used to select forward and the reverse primers for a given set of DNA sequences in two separate steps, namely, by considering the first, say, $r$ nucleotides of the sequences to select the set of forward primers and then the last $r$ nucleotides to select the reverse primers. Another approach is to consider the first $r$ nucleotides and the complement of the last $r$ nucleotides of each sequence, thus building an input dataset of $2n$ sequences of length $r$ each. The latter approach was described by Souvenir et al. [5] to design degenerate primers. Note that the value $r$ must be chosen carefully such that $(l_i - 2r) > 0$ for $1 \leq i \leq n$ to assure that every DNA sequence in the input would have an amplified product of length strictly $> 0$.

Algorithm **PSP-Greedy** {
        Let $P$ be the collection of primers selected; initially, $P := \emptyset$;
        Let $R$ be the set of remaining (uncovered) sequences; initially, $R := \{1, 2, \ldots, n\}$.
        Let $C$ be the collection of $k$-mers from all the $n$ input sequences; Note that each
        input sequence $S_i, 1 \le i \le n$, will have $(r - k + 1)$ $k$-mers, $r$ being the
        length of $S_i$. Each element of $C$ is a tuple of the form $< k - mer, i >$, $i$ being the
        sequence to which the $k$-mer belongs.

        Sort the collection $C$ such that its $k$-mers are in lexicographic order. Scan
        through $C$ and identify unique $k$-mers and the list of sequences in which they
        are present (denoted by the second value in each tuple), called their coverage.
        Let $L$ be the list of all such unique $k$-mers and their coverage.
        While ($R$ is not empty) do {
                Pick $p$ from $L$, $p$ being the $k$-mer that has coverage of maximum
                cardinality among all the elements of $L$;
                $L := L - \{p\}; P := P \cup \{p\}$;
                $R := R - \{\text{coverage of } p\}$;
                For each $q \in L$ do {
                        $\{\text{coverage of } q\} := \{\text{coverage of } q\} - \{\text{coverage of } p\}$;
                }
        }
        Output $P$;
}

It is obvious from its description that PSP-Greedy does not consider the biological constraints seen in the earlier section to select primers. The first efforts in this direction came from Doi and Imai [6–8], who proposed another greedy heuristic, essentially a modification of PSP-Greedy that considered biological constraints such as *GC-content* and *complementarity* in selecting primers. Their algorithm also considered length constraint of the amplified product, namely, the minimum length constraint that ensures that the minimum length of the amplified products is at least of a prespecified length $l_{min}$ (instead of 0 as described earlier). Algorithms in Reference 9 also consider length constraints in designing primers for MP-PCR.

## 28.1.6 Algorithms for Degenerate Primer Selection Problem

The current section discusses some of the known algorithms for the DPSP. Rose et al. [10] proposed algorithm Consensus-Degenerate Hybrid Oligonucleotide Primer (CODEHOP) that designs hybrid primers with nondegenerate consensus clamp at the $5'$ region and a degenerate $3'$ core region. In an effort to identify genes belonging to the same family, Fuchs et al. [11] devised a two phase algorithm called DEFOG. In its first phase, DEFOG introduces degeneracy into a set of nondegenerate primer candidates selected due to their best entropy score. Linhart and Shamir [1] proposed an algorithm called HighlY DEgeNerate (HYDEN) for the first phase of DEFOG. Wei et al. [12] contributed an algorithm based on clustering called Degenerate Primer Design via Clustering (DePiCt) that designs primers of low degeneracy and high coverage for a given set of aligned amino acid sequences. Souvenir et al. [5] proposed the Multiple Iterative Primer Selector (MIPS) algorithm for a variation of DPSP, discussed in their paper as the Partial Threshold Multiple Degenerate Primer Design. Algorithms HYDEN, DePiCt, and MIPS are explained in some detail in this section.

### 28.1.6.1 Algorithm HighlY DEgeNerate primers

The HYDEN algorithm [1] performs the first phase of DEFOG [11]. For a given set of input sequences, HYDEN is run separately on the datasets of the first $r$ residues from every sequence to select the forward degenerate primers (say, the set $P_f$) and the last $r$ residues from each sequence to select the reverse degenerate primers (say, the set $P_r$). Then, the desired set $P_d = P_f \cup P_r$. For a given run, HYDEN designs a degenerate primer that covers the maximum number of the sequences that are yet to be covered (initially, all sequences are yet to be covered), adds the primer to the output set, removes the sequences that it had covered, and repeats the same procedure until all sequences are covered. To select one degenerate primer for the set of sequences alive at a given time of its execution, HYDEN employs a three-phased approach described as follows:

**Phase 1**: Named HYDEN-Align, this phase identifies highly conserved regions of the given set of DNA sequences by locating ungapped local alignments that contribute towards a low entropy score. It enumerates all substrings of length $k$ ($k$-mers) in the input set, generates alignment score for each substring by finding best matches to it with respect to its Hamming distance with substrings of other strings in the input. Let $L = \sum_{i=1}^{n} l_i$. As the number of $k$-mers possible in the input is $O(L)$, HYDEN-Align considers all such possibilities and obtains $O(L)$ alignments. Therefore, the runtime of HYDEN-Align is $O(L^2 k)$. A subset of these alignments, determined by another input parameter (say $a$, i.e., the $a$ best alignments), that have a low entropy score is considered for the next phase. The authors also give a simple heuristic that will speed up this phase, namely, initially each such alignment is generated only on a subset of the input strings (say $\epsilon n, 0 < \epsilon < 1$), a subset of size $a'$ ($a' > a$) of these alignments that have the best entropy scores are selected; for each partial alignment selected, a full alignment is generated, the entropy scores calculated, and the best $a$ alignments are selected for the next phase based on the complete entropy scores. This reduces the runtime of this phase to $O(kL(\epsilon L + a'))$.

The entropy scores are calculated as follows. Let $A$ be a given ungapped alignment. $A$ consists of $n$ $k$-mers one from each input sequence. Let $D_A$ denote a column distribution matrix of $A$. The column distribution matrix $D_A$ is a two-dimensional matrix of size $|\Sigma| \times k$, in which $D_A[\sigma, j], 1 \leq \sigma \leq |\Sigma|$ and $1 \leq j \leq k$, has the value equal to the count or the number of occurrences of the symbol $\sigma (\sigma \in \Sigma)$ in column $j$ of the alignment $A$. The entropy score $E_A$ of alignment $A$ is given as:

$$E_A = -\sum_{j=1}^{k} \sum_{\sigma \in \Sigma} [(D_A[\sigma, j]/n) * \log_2(D_A[\sigma, j]/n)].$$

The lower the entropy score, the less will be the variation of symbols in the alignment $A$. Thus, greater are the chances of finding a $k$-mer that would cover many of the input sequences.

**Phase 2**: In this phase two procedures, namely HYDEN-Contraction and HYDEN-Expansion, are run on the set of alignments from the first phase. HYDEN-Contraction starts with a complete degenerate primer ($p_c$) of degeneracy $4^k$, proceeds by removing the symbol from a position at which it has occurred the minimum number of times in the given alignment until the target degeneracy $d$ is achieved. On the other hand, HYDEN-Expansion starts from a non-degenerate primer ($p_e$), that is, the $k$-mer from which the alignment was obtained, adds to it symbols one at a time at positions in which the symbol added has occurred the maximum number of times in the alignment, increasing the degeneracy until the target degeneracy is achieved. Both the procedures use $D_A$ of each alignment $A$ to eliminate and add symbols in the primers $p_c$ and $p_e$, respectively. Two such primers are designed for every alignment in the set, and a subset of these primers that have maximum coverage is considered for the third phase (the size of the subset is given as an input parameter, say $b$, that is, the $b$ best primers of the $2a$ primers designed). Each run of HYDEN-Contraction or HYDEN-Expansion takes $O(kL)$ time and there are $a$ alignments for which the primers are designed, thus the runtime of phase 2 is $O(akL)$.

**Phase 3**: In this phase, called the HYDEN-Greedy, attempt is made to improve the primers selected from Phase 2 using a greedy hill climbing approach, trying to exclude symbols in some positions of a given primer, and include symbols in other positions to increase coverage.

Although there is no theoretical guarantee on the performance of algorithm HYDEN, the authors have reported good practical performance in experiments on real biological data.

### 28.1.6.2 Algorithm Degenerate Primer Design via Clustering

This algorithm proposed by Wei et al. [12] designs degenerate primers of low degeneracy and high coverage from a given multiple alignment of amino acid (or protein) sequences. It adopts clustering techniques to group the set of input sequences, thus ensuring that sequences that belong to a given cluster would have regions significantly conserved in them that would enable the design of a pair of degenerate primers for them. Conserved regions of a cluster are determined using a novel scoring technique called the Block-Similarity scoring. The degenerate primers are then obtained by reverse translation of the amino acids in the conserved regions to corresponding nucleotides.

The *Genetic Code* consists of triplets of nucleotides, called codons, each such codon encoding one of the 20 amino acids that are used in the synthesis of proteins in organisms. As the DNA alphabet has four symbols, there are 64 triplets in the genetic code, leading to some redundancy that many of the amino acids are encoded by more than one codon. For example, the amino acid Proline (P) is encoded by four codons, namely, CCT, CCG, CCC, and CCA. It is obvious to see that many amino acids have very similar codons too, although they may be very different in their physical and chemical properties. For example, another amino acid Alanine (A) is encoded by the codons GCT, GCG, GCC, and GCA. In calculating the similarity between the input sequences, algorithm DePiCt considers Proline and Alanine in our example to be *similar* as they differ only by one nucleotide in the first position of the codons that encode them.

DePiCt adopts *Hierarchical Clustering* to cluster the set of input sequences into groups that have conserved regions. Initially, there are $n$ groups, each group consisting of one input sequence. A series of iterations are performed to regroup the sequences in the groups based on their similarities. In each iteration, sequences of two groups that have the highest similarity score are grouped together into one, if the resultant group is a *valid* cluster. A *valid* cluster is one that has at least one conserved block of length greater than or equal to the minimum required product length or two blocks separated by a length in the range of the minimum required product length and the maximum required product length. These minimum and the maximum product lengths are specified as input. The iteration stops when no more groups can be combined into one. The similarity scores used are the BlockSimilarity scores calculated as follows:

A multiple alignment $M$ of the sequences to be grouped is obtained. Conserved regions are located in $M$ based on the amino acids that appear in all the sequences. If the amino acids are identical or if they are *similar* according to the explanation given earlier for *similarity* considered by DePiCt in any given column of $M$, then that column is considered as *conserved*. Consecutive conserved columns of $M$ give rise to conserved regions or blocks. The BlockSimilarity score of a block is simply the number of columns in it. If the BlockSimilarity score of a block is $< \lceil k/3 \rceil$ (as the sequences are protein sequences and each amino acid corresponds to three nucleotides in the primer), then it is assigned a score of 0. The BlockSimilarity score of the alignment $M$ is the sum of the BlockSimilarity scores of all the blocks in it.

Two degenerate primers $p_f$ and $p_r$ are designed for each cluster, $p_f$ being the forward primer and $p_r$ the reverse primer, by reverse translating the conserved blocks of the cluster into nucleotide sequences that correspond to the codons of the amino acids in the conserved blocks. For the reverse primers, the complement of the nucleotides is considered. The set of all such primers designed is the desired set $P_d$.

### 28.1.6.3 Algorithm Multiple Iterative Primer Selector

Proposed by Souvenir et al., algorithm MIPS [5] follows an iterative beam search technique to design degenerate primers. It starts with a set of primers that cover two sequences from an input of $n$ sequences. To bring down the time complexity, the 2-primers are formed only by merging a $k$-mer with those $k$-mers that are returned by a technique similar to a FASTA lookup table. Then it extends the coverage of the primers in the candidate set by one additional sequence, introducing degeneracy in the primers if necessary, retains a subset of these primers (the number determined by an input parameter called beam size $b$)

for the next iterative step until none of the primers can be extended further without crossing the target degeneracy. At this point, the primer with the lowest degeneracy is selected and the sequences that it covers (let the number be $q$) are removed from the input set and the procedure is repeated until all the sequences are covered.

The input dataset for the algorithm is generated as follows: Each input sequence has two sequences representing it in the dataset, one sequence is the first $r$ nucleotides and the other the complement of the last $r$ nucleotides of the sequence itself. Thus, the input set will consist of $N = 2n$ sequences of length $r$ each. MIPS has an overall time complexity of $O(bN^3rp)$, in which $b$ is the beam size, $N$ is the number of input sequences, $r$ is the sequence length, and $p$ is the cardinality of the final set of selected degenerate primers ($P_d$). The pseudocode of algorithm MIPS is hereunder:

Algorithm **MIPS** {

        Let $P$ be the list of selected primers, initially, $P$ is empty;

        Let $Q$ be a priority queue of size $b$ that holds the primer candidates;
        (candidates in $Q$ are ordered with respect to their degeneracy).
        Initially all the $N$ input sequences are alive;
        While (# of sequences alive > 0) {
                Let $p$ be the selected primer for the current iteration; Initially $p$ = null;
                Let $C$ be the collection of all substrings of length $k$ ($k$-mers) in the
                sequences alive;
                For each element $k$-mer $u \in C$ {
                        Let $C'$ be the collection of $k$-mers that are obtained from the
                        FASTA lookup of $u$;
                        For each element $k$-mer $v \in C'$ {
                              Form the primer $u' = u \cup v$;
                              Add $u'$ to $Q$ (if the degeneracy of $u'$ is at most
                              the target degeneracy $d$);
                        }
                }
                While ($Q$ is not empty) {
                      Let $Q'$ be a priority queue of the next generation candidates;
                      For each element $q \in Q$ {
                          For each sequence $S_i$ alive and not covered by $q$ {
                              For each $k$-mer $v$ of $S_i$ {
                                  Form the primer $q' = q \cup v$;
                                  Add $q'$ to $Q'$ if the degeneracy of $q'$ is at most the
                                target degeneracy $d$;
                              }
                        }
                    }
                    $p_d$ = the primer of lowest degeneracy in $Q'$;
                    $Q = Q'$;
                }
                Add $p_d$ to $P$;
                Set the sequences not covered by $p_d$ as the sequences alive;
        }
        Output $P$;

}

Let us analyze the time taken by the loop that processes the elements of the priority queue $Q$ to generate primers of higher coverage. Forming each $q'$ takes $O(N + |\Sigma|k)$ time. As each candidate can form at most $O(Nr)$ such $q'$, the time complexity of creating $Q'$ is $O(bNr(N + |\Sigma|k))$. Therefore, the time required for one iteration of the loop is $O(bN^2r)$.

Now, let us look into the number of iterations the algorithm will perform to design one primer of degeneracy at most $d$. The algorithm constructs $i$-primers in each iteration from $(i-1)$-primer candidates of the previous iteration whose degeneracy either remains the same or increases. Thus, the number of iterations performed by algorithm MIPS to identify one primer of the output set is $O(N)$, leading to a runtime of $O(bN^3r)$. If there are $p$ primers in the output, then, the overall time complexity of algorithm MIPS is $O(bN^3rp)$.

### 28.1.6.4 Algorithm DPS

An algorithm called Degenerate Primer Search (DPS) has been given in Reference 13. DPS has been shown to have a better runtime than that of MIPS in the worst case. It employs a new strategy of ranking the primers in every iteration as defined in the following.

**Definition 28.3** *The coverage-efficiency $e(P)$ of a degenerate primer $P$ is the ratio of the number of sequences it amplifies or covers ($c(P)$) to its degeneracy ($d(P)$), that is, $e(P) = c(P)/d(P)$.*

Let $P_1$ and $P_2$ be two degenerate primers in the priority queue of candidate primers and let $e(P_1) > e(P_2)$. Then the priority of $P_1$ is higher than that of $P_2$. If $e(P_1) = e(P_2)$, then the primers are ranked in the nondecreasing order of their degeneracy.

In every iteration, the new algorithm performs additional processing of primer candidates before selecting the $b$ best primers for the next iteration. Instead of adding each $q'$ directly to the priority queue $Q'$, the candidates are collected in a collection $B$, sorted in their lexicographic order and unique primer candidates are identified by scanning the sorted collection $B$, obtaining their coverage by merging the coverage of the duplicates. Each such unique primer candidate is added to the priority queue $Q'$, in which the priority of the candidates are as explained previously. This ensures that the degeneracy of the candidates generated for $(i + 1)$-th iteration from a candidate of $i$-th iteration is strictly greater than that of their predecessor. As the number of symbols that can be added to a nondegenerate primer to create a degenerate primer of degeneracy at most $d$ lies in the range $[\lfloor \log_2 d \rfloor : (|\Sigma| - 1) * \lceil \log_{|\Sigma|} d \rceil]$, the number of iterations the new algorithm performs to identify a single primer of the output set $P$ is $O(|\Sigma| \log_{|\Sigma|} d)$. If $|P| = p$, then the overall time complexity of the algorithm is $O(|\Sigma| \log_{|\Sigma|} dbN^2rp)$, an improvement of the worst case time complexity $O(bN^3rp)$ of algorithm MIPS.

## 28.2 The Planted Motif Search Problem

Motif searching is an important step in the detection of rare events occurring in a set of DNA or protein sequences. The PMS problem, also known as the $(l, d)$-motif problem, has been introduced in Reference 14 with the aim of detecting motifs and significant conserved regions in a set of DNA or protein sequences. PMS receives as input $n$ biological sequences and two integers $\ell$ and $d$. It returns all possible biological sequences $M$ of length $\ell$ such that $M$ occurs in each of the input strings, and each occurrence differs from $M$ in at most $d$ positions. Any such $M$ is called a motif. Given two $\ell$-mers, the number of positions in which they differ is called their Hamming distance. A more general formulation of the problem is called quorum PMS (qPMS). In qPMS, we are interested in motifs that appear in at least $q$ percent of the $n$ input strings. Therefore, when $q = 100\%$ the qPMS problem is the same as PMS.

Buhler and Tompa [15] have employed PMS algorithms to find known transcriptional regulatory elements upstream of several eukaryotic genes. In particular, they have used orthologous sequences from different organisms upstream of four different genes: preproinsulin, dihydrofolate reductase (DHFR), metallothioneins, and c-fos. These sequences are known to contain binding sites for specific transcription factors. Their algorithm successfully identified the experimentally determined transcription factor binding sites. They have also employed their algorithm to solve the ribosome binding site problem for various prokaryotes. Eskin and Pevzner [16] used PMS algorithms to find composite regulatory patterns using their PMS algorithm called MIsmatch TRee (MITRA). They have employed the upstream regions involved in purine metabolism from three *Pyrococcus* genomes. They have also tested their algorithm on four sets of *S.cerevisiae* genes that are regulated by two transcription factors such that the transcription factor binding sites occur near each other. Price et al. [17] have employed their PatternBranching PMS technique to find motifs on a sample containing cAMP receptor protein (CRP) binding sites in *E.coli*, upstream regions of many organisms of the eukaryotic genes: preproinsulin, DHFR, metallothionein, & c-fos, and a sample of yeast promoter regions.

A simple algorithm can be devised for the solution of this problem. Consider every possible $l$-mer one at a time and check if this $l$-mer is the correct motif $M$. There are $4^l$ possible $l$-mers. Let $M'$ be one such $l$-mer. We can check if $M' = M$ as follows. Let the input sequences be $S_1, S_2, \ldots, S_n$. The length of each sequence is $m$. Form all possible $l$-mers from out of these sequences. The total number of $l$-mers is $\leq nm$. Call this collection of $l$-mers $C$. Compute the Hamming distance between $u$ and $M'$ for every $u \in C$. As a result, we can check if $M'$ occurs in each input sequence (at a Hamming distance of $d$). Thus, we can identify all the motifs of interest in a total of $O(nml4^l)$ time. This algorithm becomes impractical even for moderately large values of $l$. Numerous efficient algorithms have been proposed in the literature.

Algorithms for PMS can be broadly classified into *exact* and *approximate* algorithms. An exact algorithm always outputs the planted motif from a given input of sequences. On the other hand, an approximate algorithm may not always output the correct planted motif. Note that this notion of an approximate algorithm is different from the traditional concept of approximation algorithms. The random projection algorithm of Buhler and Tompa [15] is an example of an approximate algorithm, and the PMS algorithms given in [18] are exact.

Furthermore, algorithms for PMS can be categorized into two depending on the basic approach employed, namely, *profile-based algorithms* and *pattern-based algorithms*. Profile-based algorithms predict the starting positions of the occurrences of the motif in each sequence and pattern-based algorithms predict the motif itself.

Examples of pattern-based algorithms include PROJECTION [15], MULTIPROFILER [19], MITRA [16], and PatternBranching [17]. Examples of profile-based algorithms include CONSENSUS [20], Gibbs-DNA [21], Multiple EM for Motif Elicitation (MEME) [22], and ProfileBranching [17]. The performances of profile-based algorithms are specified with a measure called "performance coefficient." The performance coefficient gives an indication of how many positions (for the motif occurrences) have been predicted correctly. These algorithms have been shown to perform well in practice for $l \leq 18$ and $d \leq 6$. A profile-based algorithm could either be approximate or exact. Likewise, a pattern-based algorithm may either be exact or approximate.

Profile-based algorithms such as CONSENSUS, GibbsDNA, MEME, and ProfileBranching tend to take less time than the pattern-based ones. However, these algorithms fall under the approximate category and may not always output the correct answer.

Some of the pattern-based algorithms (such as PROJECTION, MULTIPROFILER, and PatternBranching) also take less time [17]. However, these are approximate as well (though the success rates are close to 100%).

There are many pattern-based exact PMS algorithms in the literature. Most of the exact PMS algorithms use a combination of two fundamental techniques. One technique is sample driven and the other technique is pattern driven. In the sample-driven stage, the algorithm selects a tuple of $\ell$-mers coming from distinct input strings. Then, in the pattern-driven stage, the algorithm generates the common $d$-neighborhood of the $\ell$-mers in the tuple. Each neighbor becomes a motif candidate. The size of the tuple is usually fixed to a value such as 1 [18,23,24], 2 [25], 3 [26–29], or $n$ (see, e.g., [14,30]). The PMS8 [31] and qPMS9 [32] utilize a variable tuple size, which adapts to the problem instance under consideration.

Some of the earlier PMS algorithms include [18,33–39]. As pointed out in Reference 15, these algorithms "become impractical for the sizes involved in the challenge problem." Challenge problems are instances of the PMS problem that have been found to be difficult and were proposed by Pevzner [14]. Essentially, a challenging instance can be defined as an $(\ell, d)$ instance in which $d$ is the largest integer for which the expected number of motifs of length $\ell$ that would occur in the input by random chance does not exceed a constant (e.g., References 31, 32 use the constant 500).

The MITRA algorithm MITRA [16] solves, for example, the $(15, 4)$ instance in 5 minutes using 100 MB of memory [16]. This algorithm is based on the WINNOWER algorithm [14] and uses pairwise similarity information. A new pruning technique enables MITRA to be more efficient than WINNOWER. MITRA uses a mismatch tree data structure and splits the space of all possible patterns into disjoint subspaces that start with a given prefix.

More recent algorithms can solve much larger challenging instances within reasonable amount of time. The algorithm in Reference 40 can solve instances with relatively large $l$ (up to 48) provided that $d$ is at most $l/4$. However, most of the well-known challenging instances have $d > l/4$. PairMotif [25] can solve instances with larger $l$, such as $(27, 9)$ or $(30, 9)$, but these are significantly less challenging than $(23, 9)$. The first algorithm to solve the challenging DNA instance $(23, 9)$ has been the PMS5 algorithm [27] followed by the qPMS7 algorithm [29]. qPMS7 can solve $(23, 9)$ within 24 hours. The first algorithm to solve $(25, 10)$ in a reasonable amount of time (about 17 hours) has been the TraverStringRef algorithm [26]. TraverStringRef also solves $(23, 9)$ in about 4 hours. TraverStringRef [26] is an algorithm for the qPMS problem, which is based on the earlier qPMS7 [29] algorithm. PMS8 [31] can solve DNA instances $(25,10)$ on a single-core machine and $(26,11)$ on a multi-core machine. Its successor, qPMS9 [32], can solve $(28, 12)$ and $(30, 13)$ on a single-core machine. qPMS9 solves $(23, 9)$ in about 2 hours and $(25, 10)$ in about 6 hours. qPMS9 is currently the fastest exact PMS/qPMS algorithm.

## 28.2.1 The WINNOWER Algorithm

The algorithm of Pevzner and Sze [14] (called *WINNOWER*) works as follows: If $A$ and $B$ are two instances (i.e., occurrences) of the motif, then the Hamming distance between $A$ and $B$ is at most $2d$. The algorithm constructs a collection $C$ of all possible $l$-mers in the input. A Graph $G(V, E)$ is then constructed. Each $l$-mer in $C$ will correspond to a node in $G$. Two nodes $u$ and $v$ in $G$ are connected by an edge if and only if the Hamming distance between the two $l$-mers is at most $2d$ and these $l$-mers come from two different sequences.

Clearly, the $n$ instances of the motif $M$ form a clique of size $n$ in $G$. Thus, the problem of finding $M$ reduces to that of finding large cliques in $G$. On the contrary, there will be numerous "spurious" edges (i.e., edges that do not connect instances of $M$) in $G$ and also finding cliques is $\mathcal{NP}$-hard. Pevzner and Sze [14] employ a clever technique to prune spurious edges. More details can be found in Reference 14.

## 28.2.2 Random Projection Algorithm

The algorithm of Buhler and Tompa [15] is based on random projections. Let the motif $M$ of interest be an $l$-mer. Collect all the $l$-mers from all the $n$ input sequences and let $C$ be this collection. Project these $l$-mers along $k$ randomly chosen positions (for some appropriate value of $k$). In other words, for every $l$-mer $u \in C$, generate a $k$-mer $u'$ that is a subsequence of $u$ corresponding to the $k$ random positions chosen. (The random positions are the same for all the $l$-mers.) We can think of each $k$-mer thus generated as an integer. We group the $k$-mers according to their integer values. (i.e. we hash all the $l$-mers using the $k$-mer of any $l$-mer as its hash value.)

If a hashed group has at least a threshold number $s$ of $l$-mers in it, then there is a good chance that $M$ will have its $k$-mer equal to the $k$-mer of this group. (An appropriate value for $s$ is obtained using a probabilistic analysis.) We collect all the $k$-mers (and the corresponding $l$-mers) that pass the threshold and these are processed further to arrive at the final answer $M$. Processing is done using the expectation maximization technique of Lawrence and Reilly [41].

### 28.2.3 Algorithms PMS1 to qPMS7

A simple algorithm called PMS1 has been given in Reference 18. Even this simple algorithm has been shown to solve some of the challenge problems efficiently. Steps involved in this algorithm are as follows:

1. Generate all possible $l$-mers from out of each of the $n$ input sequences. Let $C_i$ be the collection of $l$-mers from out of $S_i$ for $1 \leq i \leq n$.
2. For all $1 \leq i \leq n$ and for all $u \in C_i$, generate all $l$-mers $v$ such that $u$ and $v$ are at a Hamming distance of $d$. Let the collection of $l$-mers corresponding to $C_i$ be $C'_i$, for $1 \leq i \leq n$. The total number of patterns in any $C'_i$ is $O\left(m\binom{l}{d}3^d\right)$.
3. Sort all the $l$-mers in every $C'_i, 1 \leq i \leq n$ and eliminate duplicates in every $C'_i$. Let $L_i$ be the resultant sorted list corresponding to $C'_i$.
4. Merge all the $L_i$s $(1 \leq i \leq n)$ and output the generated (in step 2) $l$-mer that occurs in all the $L_i$s.

The runtime of the above-mentioned algorithm is $O\left(nm\binom{l}{d}3^d\frac{l}{w}\right)$ in which $w$ is the word length of the computer.

Two other exact algorithms called PMS2 and PMS3 have also been proposed in [18]. These algorithms are competitive in practice with other exact algorithms. For a survey on motif search algorithms see Reference 42.

The PMSPrune algorithm [23] generates neighborhoods of individual $l$-mers from the first string, in a branch and bound manner, in which the search space is represented as a tree. During the search procedure, the algorithm keeps track of whether the $l$-mer being generated is found in the remaining strings. This allows for pruning not only of the current $l$-mer but also of its entire subtree, which leads to much better running time than previous algorithms.

The PMS5 algorithm [27] generates common neighborhoods of three $l$-mers at a time. To compute such neighborhoods, the algorithm uses the solutions of an Integer Linear Program (ILP). These solutions are precomputed and loaded into a table every time the algorithm starts. The PMS6 algorithm [28] improves the running time of PMS5 by detecting equivalence classes among the possible triplets of $l$-mers and thus avoiding some redundant neighborhood computation. The qPMS7 [29] algorithm further improves on PMS5 and PMS6 by extending the PMS5 3 $l$-mer idea to which it adds pruning techniques similar to the ones in PMSPrune [23]. qPMS7 can solve the more general version of PMS called qPMS.

The TraverStringRef [26] algorithm extends the qPMS7 algorithm with a clever string reordering heuristic. This heuristic reduces the size of the generated neighborhoods and thus increases efficiency.

### 28.2.4 Algorithms PMS8 and qPMS9

The PMS8 [31] and qPMS9 [32] algorithms make use of new necessary and sufficient conditions for 3 $l$-mers to have a common neighbor. These conditions remove the need for an ILP table such as in PMS5. Furthermore, PMS8 and qPMS9 generate common neighborhoods of $k$ $l$-mers instead of just 3. The value of $k$ is chosen depending on the size of the instance.

As many other motif algorithms, PMS8 and qPMS9 combine a sample-driven approach with a pattern-driven approach. In the sample-driven part, tuples of $\ell$-mers $(t_1, t_2, \ldots, t_k)$ are generated, in which $t_i$ is an $\ell$-mer in $S_i$. Then, in the pattern-driven part, for each tuple, its common $d$-neighborhood is generated. Every $\ell$-mer in the neighborhood is a candidate motif. In PMS8 and qPMS9, the tuple size $k$ is variable. By default, a good value for $k$ is estimated heuristically [31] based on the input parameters, or $k$ can be user specified.

#### 28.2.4.1 Generating Tuples of $\ell$-mers

In the sample-driven part, PMS8 and qPMS9 generate tuples $T = (t_1, t_2, \ldots, t_k)$, in which $t_i$ is an $\ell$-mer from string $s_i$, $\forall i = 1..k$, based on the following principles. First, if $T$ has a common $d$-neighbor, then every subset of $T$ has a common $d$-neighbor. Second, there has to be at least one $\ell$-mer $u$ in each of the remaining strings $s_{k+1}, s_{k+2}, \ldots, s_n$ such that $T \cup \{u\}$ has a common $d$-neighbor. We call such $\ell$-mers $u$ "alive" with respect to tuple $T$.

We now define the consensus $\ell$-mer and the consensus total distance for a tuple of $\ell$-mers. Given a tuple of $\ell$-mers $T = (t_1, \ldots, t_k)$, the **consensus $\ell$-mer** of $T$ is an $\ell$-mer $u$ in which $u[i]$ is the most common character among $(t_1[i], t_2[i], \ldots, t_k[i])$ for each $1 \le i \le \ell$. If the consensus $\ell$-mer for $T$ is $p$, then the **consensus total distance** of $T$ is defined as $Cd(T) = \sum_{u \in T} Hd(u, p)$. Although the consensus string is generally not a motif, the consensus total distance provides a lower bound on the total distance between any motif and a tuple of $\ell$-mers, as proven in the PMS8 [31] paper.

As $\ell$-mers are added to $T$, we update the alive $\ell$-mers in the remaining strings. Based on the number of alive $\ell$-mers, PMS8 reorders the remaining strings increasingly. This is a heuristic that speeds up the search because the first $\ell$-mers in the tuple are the most expensive so we want as few combinations of them as possible. However, qPMS9 uses the following more efficient string reorder heuristic. Let $u$ be an alive $\ell$-mer with respect to $T$. If $u$ is added to $T$, then the consensus total distance of $T$ increases. We compute this additional distance $Cd(T \cup \{u\}) - Cd(T)$. For each of the remaining strings, we compute the minimum additional distance generated by any alive $\ell$-mer in that string. Then we sort the strings decreasingly by the minimum additional distance. Therefore, we give priority to the string with the largest minimum additional distance. The intuition is that larger minimum additional distance could indicate more "diversity" among the $\ell$-mers in the tuple, which means smaller common $d$-neighborhoods. If two strings have the same minimum additional distance, priority is given to the string with fewer alive $\ell$-mers.

The tuple generation is described in algorithm 28.1. We invoke the algorithm as $GenTuples(\{\}, k, R)$ in which $k$ is the desired size of the tuples and $R$ is a matrix that contains all the $\ell$-mers in all the input strings, grouped as one row per string. This matrix is used to keep track of alive $\ell$-mers. To exclude tuples that cannot have a common neighbor, we employ the pruning techniques in Section 28.2.4.3.

---

## ALGORITHM 28.1    GenerateTuples($T, k, R$)

**Input**: $T = (t_1, t_2, \ldots, t_i)$, current tuple of $\ell$-mers;
       $k$, desired size of the tuple;
       $R$, array of $n - i$ rows, in which $R_j$ contains all alive $\ell$-mers from string $s_{i+j}$;
**Result**: Generates tuples of size $k$, containing $\ell$-mers, that have common neighbors, then passes
       these tuples to the `GenerateNeighborhood` function;
**begin**
    **if** $|T| == k$ **then**
        `GenerateNeighborhood`($T, d$);
        **return**;
    outerLoop: **for** $u \in R_1$ **do**
        $T' := T \cup \{u\}$;
        **for** $j \leftarrow 1$ **to** $n - i - 1$ **do**
            $R'_j = \{v \in R_{j+1} | \exists$ common $d$-neighborhood for $T' \cup \{v\}\}$;
            **if** $|R'_j| == 0$ **then**
                `continue` outerLoop;
            $minAdd := \min_{v \in R'_j} Cd(T' \cup \{v\}) - Cd(T')$;
            $aliveLmers := |s_{i+j+1}| - |R'_j|$;
            $sortKey[j] := (minAdd, -aliveLmers)$;
        sort $R'$ decreasingly by $sortKey$;
        `GenerateTuples` $(T', k, R')$;

### 28.2.4.2 Generating Common Neighborhoods

For every tuple that algorithm 28.1 generates, we want to generate a common neighborhood. Namely, given a tuple $T = (t_1, t_2, \ldots, t_k)$ of $\ell$-mers, we want to generate all $\ell$-mers $M$ such that $Hd(t_i, M) \leq d, \forall i = 1..k$. To do this, we traverse the tree of all possible $\ell$-mers, starting with an empty string and adding one character at a time. A node at depth $r$, which represents an $r$-mer, is pruned if certain conditions are met (Section 28.2.4.3). The pseudocode for neighborhood generation is given in algorithm 28.2.

### 28.2.4.3 Pruning Conditions

In the current section, we address the following question. Given a tuple $T = (t_1, t_2, \ldots, t_k)$ of $\ell$-mers and a tuple $D = (d_1, d_2, \ldots, d_k)$ of distances, is there an $\ell$-mer $M$ such that $Hd(M, t_i) \leq d_i, \forall i = 1..k$? This question appears in algorithm 28.1 in which $T$ is the current tuple and $D$ is an array with all values set to $d$. The same question appears in algorithm 28.2 in which $T$ is a tuple of suffixes and $D$ is an array of remaining distances.

Two $\ell$-mers $a$ and $b$ have a common neighbor $M$ such that $Hd(a, M) \leq d_a$ and $Hd(b, M) \leq d_b$ if and only if $Hd(a, b) \leq d_a + d_b$. For 3 $l$-mers, the paper in Reference 31 proves the following simple necessary and sufficient conditions for 3 $\ell$-mers to have a common neighbor. These conditions are also necessary (but not sufficient) for 4 or more $\ell$-mers.

---

## ALGORITHM 28.2 GenerateNeighborhood(*T,d*)

**Input:** $T = (t_1, t_2, \ldots, t_k)$, tuple of $\ell$-mers;
  $d$, maximum distance for a common neighbor;
**Result:** Generates all common $d$-neighbors of the $\ell$-mers in $T$;
**begin**
  **for** $i \leftarrow 1$ **to** $|T|$ **do**
    $r[i] = d$
  GenerateLMers $(x, 0, T, r)$;

**Procedure** GenerateLMers $(x, p, T, r)$
  **Input:** $x$, the current $\ell$-mer being generated;
    $p$, the current length of the $\ell$-mer being generated;
    $T = (t_1, t_2, \ldots, t_k)$, tuple of $(\ell - p)$-mers;
    $r[i]$, maximum distance between the (yet to be generated) suffix of $x$ and $t_i, \forall i = 1..k$;
  **Result:** Generates all possible suffixes of $x$ starting at position $p$ such that the distance between
    the suffix of $x$ and $t_i$ does not exceed $r[i], \forall i = 1..k$;
  **if** $p == \ell$ **then**
    report $\ell$-mer $x$;
  **else**
    **if not** prune $(T, r)$ **then**
      **for** $\alpha \in \Sigma$ **do**
        $x_p = \alpha$;
        **for** $i \leftarrow 1$ **to** $|T|$ **do**
          **if** $t_i[0] == \alpha$ **then**
            $r'[i] = r[i]$;
          **else**
            $r'[i] = r[i] - 1$;
          $t'_i = t_i[1..|t_i|]$;
        GenerateLMers $(x, p + 1, T', r')$;

**Theorem 28.3** *Let $T = (t_1, t_2, t_3)$ be a tuple of three l-mers and $D = (d_1, d_2, d_3)$ be a tuple of three nonnegative integers. There exists an l-mer M such that $Hd(M, T_i) \leq d_i, \forall i, 1 \leq i \leq 3$ if and only if the following conditions hold:*

(i) $Cd(t_i, t_j) \leq d_i + d_j, \forall i, j, 1 \leq i < j \leq 3$
(ii) $Cd(T) \leq d_1 + d_2 + d_3$

#### 28.2.4.4 Adding Quorum Support

In the current section, we extend the above-mentioned techniques to solve the qPMS problem. In the qPMS problem, when we generate tuples of $\ell$-mer, we may "skip" some of the strings. This translates to the implementation as follows: In the PMS version, we successively try every alive $\ell$-mer in a given string by adding it to the tuple $T$ and recursively calling the algorithm for the remaining strings. For the qPMS version, we have an additional step in which, if the value of $q$ permits, we skip the current string and try $\ell$-mers from the next string. At all times we keep track of how many strings we have skipped. The pseudocode is given in algorithm 28.3. We invoke the algorithm as $QGenerateTuples(n - Q + 1, \{\}, 0, k, R)$ in which $Q = \lfloor \frac{qn}{100} \rfloor$ and $R$ contains all the $\ell$-mers in all the strings.

---

## ALGORITHM 28.3   QGenerateTuples(*qTolerance,T,k,R*)

**Input**: *qTolerance*, number of strings we can afford to skip;
$\quad\quad$ $T = (t_1, t_2, \ldots, t_i)$, current tuple of $\ell$-mers;
$\quad\quad$ *i*, last string processed;
$\quad\quad$ *k*, desired size of the tuple;
$\quad\quad$ $R = (R_1, \ldots R_{n-i})$, in which $R_j$ contains all alive $\ell$-mers in $s_{i+j}$;
**Result**: Generates tuples of size *k*, containing $\ell$-mers, that have common neighbors, then passes
$\quad\quad\quad$ these tuples to the `GenerateNeighborhood` function;
**begin**
$\quad$ **if** $|T| == k$ **then**
$\quad\quad$ `GenerateNeighborhood`$(T, d)$;
$\quad\quad$ **return**;

$\quad$ outerLoop: **for** $u \in R_1$ **do**
$\quad\quad$ $T' := T \cup \{u\}$;
$\quad\quad$ *incompat* := 0;
$\quad\quad$ **for** $j \leftarrow 1$ **to** $n - i - 1$ **do**
$\quad\quad\quad$ $R'_j = \{v \in R_{j+1} | \exists$ common $d$-neighborhood for $T' \cup \{v\}\}$;
$\quad\quad\quad$ **if** $|R'_j| == 0$ **then**
$\quad\quad\quad\quad$ **if** *incompat* $\geq$ *qTolerance* **then**
$\quad\quad\quad\quad\quad$ `continue` outerLoop;
$\quad\quad\quad\quad$ *incompat* + +;
$\quad\quad\quad$ *minAdd* := $\min_{v \in R'_j} Cd(T' \cup \{v\}) - Cd(T')$;
$\quad\quad\quad$ *aliveLmers* := $|s_{i+j+1}| - |R'_j|$;
$\quad\quad\quad$ *sortKey*[*j*] := (*minAdd*, −*aliveLmers*)
$\quad\quad$ sort $R'$ decreasingly by *sortKey*;
$\quad\quad$ `QGenerateTuples` $(qTolerance - incompat, T', k, R')$;

$\quad$ **if** *qTolerance* > 0 **then**
$\quad\quad$ `QGenerateTuples` $(qTolerance - 1, T, k, R \setminus R_1)$;

### 28.2.4.5 Parallel Algorithm

PMS8 and qPMS9 are parallel algorithms. Processor 0 acts as both a master and a worker, the other processors are workers. Each worker requests a subproblem from the master, solves it, then repeats until all subproblems have been solved. Communication between processors is done using the Message Passing Interface.

In PMS8, the subproblems are generated as follows: The search space is split into $m = |s_1| - \ell + 1$ independent subproblems $P_1, P_2, \ldots, P_m$, in which $P_i$ explores the $d$-neighborhood of $\ell$-mer $s_1[i..i+\ell-1]$.

In qPMS9, the previous idea is extended to the $q$ version. The problem is split into subproblems $P_{1,1}, P_{1,2}, \ldots, P_{1,|s_1|-\ell+1}, P_{2,1}, P_{2,2}, \ldots, P_{2,|s_2|-\ell+1}, \ldots, P_{r,1}, P_{r,2}, \ldots, P_{r,|s_r|-\ell+1}$ in which $r = n - Q + 1$ and $Q = \lfloor \frac{qn}{100} \rfloor$. Problem $P_{i,j}$ explores the $d$-neighborhood of the $j$-th $\ell$-mer in string $s_i$ and searches for $\ell$-mers $M$ such that there are $Q - 1$ instances of $M$ in strings $s_{i+1}, \ldots, s_n$. Notice that $Q$ is fixed, therefore subproblems $P_{i,j}$ get progressively easier as $i$ increases.

### 28.2.4.6 Speedup Techniques

*Speedup Hamming Distance calculation by packing $\ell$-mers*: By packing $\ell$-mers in advance we can speedup Hamming distance operations. For example, we can pack eight DNA characters in a 16-bit integer. To compute the Hamming distance between two $l$-mers, we first perform an exclusive or of their packed representations. Equal characters produce groups of zero bits, different characters produce nonzero groups of bits. For every possible 16-bit integer $i$, we precompute the number of nonzero groups of bits in $i$ and store it in a table. Therefore, one table look up provides the Hamming distance for eight DNA characters. The same technique applies to any alphabet $\Sigma$ besides DNA.

For an alphabet $\Sigma$, let $b = \lceil \log |\Sigma| \rceil$ be the number of bits required to encode one character. Then, one compressed $\ell$-mer requires $\ell * b$ bits of storage. However, due to the overlapping nature of the $\ell$-mers in our input strings, we can employ the following trick. In a 16-bit integer, we can pack $p = \lfloor 16/b \rfloor$ characters. For every $\ell$-mer, we only store the bit representation of its first $p$ characters. The bit representation of the next $p$ characters is the same as the bit representation of the first $p$ characters of the $l$-mer $p$ positions to the right of the current one. Therefore, the table of compressed $l$-mers requires constant memory per $\ell$-mer for a total of $O(n(m - l + 1))$ words of memory.

*Preprocess Hamming distances for all pairs of input $\ell$-mers*: The filtering step tests many times if two $l$-mers have a distance of no more than $2d$. Thus, for every pair of $l$-mers, this Boolean information is preprocessed, provided the required storage memory is not too high.

*Find motifs for a subset of the strings*: PMS8 and qPMS9 also use the speedup technique described in Reference 24: compute the motifs for $n' < n$ of the input strings, then test each motif to see in which it appears in the remaining $n - n'$ strings.

*Cache locality*: $R$ can be updated in an efficient manner as follows: Every row in the updated matrix $R'$ is a subset of the corresponding row in the current matrix $R$ because some elements will be filtered out. Therefore, we can store $R'$ in the same memory locations as $R$. To do this, in each row, we move the elements belonging to $R'$ at the beginning of the row and keep track of how many elements belong to $R'$. To go from $R'$ back to $R$, we just have to restore the row sizes to their previous values. The row elements will be the same even if they have been permuted within the row. The same process can be repeated at every step of the recursion, therefore the whole "stack" of $R$ matrices is stored in a single matrix. This reduces the memory requirement and improves cache locality. The cache locality is improved because at every step of the recursion, in each row, we access a subset of the elements we accessed in the previous step, and those elements are in contiguous locations of memory.

### 28.2.4.7 Memory

As PMS8 and qPMS9 store all matrices $R$ in the space of a single matrix, they only require $O(n(m - l + 1))$ words of memory. To this we add $O(n^2)$ words to store row sizes for the at most $n$ matrices that share the same space. The bits of information for $l$-mer pairs that have Hamming distance no more than $2d$ require $O((n(m-l+1))^2/w)$ words, in which $w$ is the number of bits in a machine word. The table of compressed $l$-mers takes $O(n(m-l+1))$ words. Therefore, the total memory used by the PMS8 and qPMS9 algorithms is $O(n(n+m-l+1)+(n(m-l+1))^2/w)$.

# 28.3 Closest String and Closest Substring Problems

Two problems that are closely related to the PMS are the CSP and the CSSP. The CSP takes as input $n$ sequences of length $m$ each and the problem is to identify a string $s$ of length $m$ that is the closest to all the input strings. In other words, the maximum distance of $s$ to any input sequence should be minimum. On the other hand, the CSSP takes as input $n$ sequences of length $m$ each. The problem is to identify a string $\bar{s}$ of length $l(< m)$ such that $\bar{s}$ is the closest to some substrings (each of length $l$ and picked one from each input sequence) of the input sequences.

Algorithms that have been proposed for the PMS problem have typically been tested on random inputs. Specifically, the input sequences will be generated randomly such that each symbol in each sequence is uniformly randomly picked from $\Sigma$. A motif $M$ will also be generated randomly in a similar fashion. This motif will then be planted in the input sequences starting from random locations. What is planted in each sequence will be a random neighbor of $M$ that is at a Hamming distance of $\leq d$ from $M$. If $l$ is small enough in relation to $d$, then there could be spurious motifs occurring in the input sequences by random chance [15]. A probabilistic analysis can be performed to figure out values of $l$ and $d$ for which the probability of a spurious motif occurring by random chance is very low. For these values of $l$ and $d$, in fact the planted motif will correspond to the closest substring. If there are spurious motifs in the input sequences, then the closest substring may not be the same as the planted motif. On the other hand, in this case it may not be possible to identify the planted motif using any other algorithm also unless additional information is given for the planted motif. For example, the exact algorithms of Reference 18 will identify all the motifs present in the input (including the planted motif). But the algorithm will not be able to isolate the planted motif.

Both CSP and CSSP have been proven to be NP-hard. In the current section, we present some of the approximation algorithms that have been devised for CSP and CSSP, as well as some simple polynomial-time algorithms for a fixed set of parameters.

## 28.3.1 The Closest String Problem

In the current section, we address the CSP. A formal definition of the CSP follows.

**Definition 28.4** *Given strings $s_1, s_2, \ldots, s_n$ (of length $m$ each) over the alphabet $\Sigma$, the CSP is to find a string $s$ of length $m$ over $\Sigma$ that minimizes $\max\limits_{i=1}^{n} d(s, s_i)$, in which $d$ is the Hamming distance. Let*

$$d_{min} := \max_{i=1}^{n} d(s, s_i).$$

From hereon whenever we refer to $s_1, \ldots, s_n$, we assume that they are strings of length $m$ over the alphabet $\Sigma$.

The first result on the complexity of the problem was obtained in Reference 43 under the context of coding theory and the so-called *minimum radius problem* and states the following.

**Theorem 28.4** *If $\Sigma = \{0, 1\}$, the CSP is NP-Complete.*

After this negative result, two different approaches were used to tackle this problem. The first approach had the goal of finding polynomial approximation schemes with a prescribed accuracy [44–46]. The other approach sought to find exact solutions that take polynomial time for a fixed set of parameters (such as $n$ or $d_{min}$). Examples include References 47, 48.

In the next sections, we will describe some algorithms that show both approaches.

### 28.3.1.1 Simple Approximation Algorithms

Given any instance of the *CSP*, one of the easiest strategies is to output any of the given strings. This gives rise to the following theorem due to Reference 44.

**Theorem 28.5** *Fix $1 \leq i \leq n$. The algorithm that outputs $s_i$ is a 2 approximation algorithm for the CSP.*

*Proof.* Let us call $s$ the optimal solution to the CSP and let $d_{min} := \max\limits_{i=1}^{n} d(s, s_i)$. Given $1 \leq j \leq n$ we have that

$$d(s_i, s_j) \leq d(s_i, s) + d(s, s_j) \leq 2d_{min}.$$

Hence, we have $\max\limits_{j=1}^{n} d(s_i, s_j) \leq 2d_{min}$ and the result follows. ∎

The following result from Reference 46 will be used later on and allows us to find an exact solution when $m \leq c \log n$ for a fixed $c$.

**Theorem 28.6** *There is a polynomial-time algorithm that solves the CSP when $m \leq c \log n$*

*Proof.* We proceed by enumerating all of the strings of length $m$ and picking the one that minimizes the desired distance. As $|\Sigma|^m \leq |\Sigma|^{c \log n} = n^{c'}$ (for some constant $c'$) we have that it takes polynomial time in $n$. ∎

### 28.3.1.2 An Approximate Solution Using Integer Programming

indexinteger programming

A useful and widely used strategy in approximation algorithms is the so-called method of randomized rounding [49]. This method models the given problem as a linear integer program, relaxes the integrality constraints, solves the resultant problem by a polynomial-time linear programming solver, and rounds the possible real solution to an integer solution based on the values obtained. This strategy was first used in Reference 44.

We could use the above-mentioned strategy to solve the CSP as well.

**Definition 28.5** *Given a string $p = p[1] \ldots p[m]$ over an alphabet $\Sigma$, we define the following binary variables, for $\sigma \in \Sigma$ and $i = 1, \ldots, m$:* $p_i^\sigma = \begin{cases} 1 & \text{if } p[i] = \sigma \\ 0 & \text{if not} \end{cases}$

Notice that given a set of binary variables $\{s_i^\sigma\}$ in which $i = 1, \ldots, m$ and $\sigma \in \Sigma$, they represent a string of length $m$ if for every $i = 1, \ldots, m$ there is exactly one 1 in the sequence $\{s_i^\sigma\}_{\sigma \in \Sigma}$ or equivalently if $\sum\limits_{\sigma \in \Sigma} s_i^\sigma = 1$.

We would like to find a formula that will allow us to calculate the Hamming distance between two strings $x$ and $y$ by using the binary variables $x_i^\sigma$ and $y_i^\sigma$. To do that, we introduce the following definition and lemma.

**Definition 28.6** *Given two strings $p = p[1] \ldots p[m]$ and $q = q[1] \ldots q[m]$, for any $i$ $(1 \leq i \leq m)$ we define $\delta_i(p, q) = \begin{cases} 1 & \text{if } p[i] \neq q[i] \\ 0 & \text{if } p[i] = q[i] \end{cases}$*

**Lemma 28.1** *Given strings $p = p[1] \ldots p[m]$ and $q = q[1] \ldots q[m]$ we have*

$$d(p, q) = \sum_{i=1}^{m} \delta_i(p, q) = \sum_{i=1}^{m} \left( 1 - \sum_{\sigma \in \Sigma} p_i^\sigma q_i^\sigma \right).$$

*Proof.* This follows easily from the fact that $p_i^\sigma q_i^\sigma = \begin{cases} 1 & \text{if } p[i] = q[i] = \sigma \\ 0 & \text{otherwise} \end{cases}$. ∎

Notice that if one of the sequences is fixed, the equation obtained in Lemma 28.1 is linear.

To state the CSP as a minimization problem, suppose that $s_i = s_i[1] \ldots, s_i[m]$ for $i = 1, \ldots, n$. The problem can be stated as:

$$
\begin{aligned}
&\min \ \max_{i=1}^{n} \ d(r, s_i) \\
&\sum_{\sigma \in \Sigma} r_i^\sigma = 1 &&& i = 1, \ldots, n \\
&r_i^\sigma \in \{0, 1\} &&& i = 1, \ldots, n \text{ and } \sigma \in \Sigma
\end{aligned}
\tag{28.1}
$$

Let $d$ represent the maximum distance of $r$ to any $s_i$. By using Lemma 28.1, we get the following linear integer program:

$$
\begin{aligned}
&\min \ d \\
&\sum_{j=1}^{m} (1 - \sum_{\sigma \in \Sigma} s_{i,j}^\sigma r_j^\sigma) \le d && i = 1, \ldots, n \\
&\sum_{\sigma \in \Sigma} r_i^\sigma = 1 && i = 1, \ldots, n \\
&r_i^\sigma \in \{0, 1\} && i = 1, \ldots, n \text{ and } \sigma \in \Sigma
\end{aligned}
\tag{28.2}
$$

By allowing the variables to take values in the interval $[0, 1]$, we get the following linear program

$$
\begin{aligned}
&\min \ d \\
&\sum_{j=1}^{m} (1 - \sum_{\sigma \in \Sigma} s_{i,j}^\sigma r_j^\sigma) \le d && i = 1, \ldots, n \\
&\sum_{\sigma \in \Sigma} r_i^\sigma = 1 && i = 1, \ldots, n \\
&0 \le r_i^\sigma \le 1 && i = 1, \ldots, n \text{ and } \sigma \in \Sigma
\end{aligned}
\tag{28.3}
$$

We will call $\hat{d}_{min}$ the solution to Equation 28.3 and we will call $\tilde{r}_i^\sigma$ the values that the variables $r_i^\sigma$ take for $i = 1, \ldots, m$ and $\sigma \in \Sigma$. It is clear that $\hat{d}_{min} \le d_{min}$.

### Definition 28.7

1. Given $\tilde{r}_i^\sigma$ with $i = 1, \ldots, m$ and $\sigma \in \Sigma$, which are solutions to Equation 28.3, we define random variables $x_i$ independently for $i = 1, \ldots, m$ by satisfying the equation $\Pr(\{x_i = \sigma\}) = \tilde{r}_i^\sigma$ for $\sigma \in \Sigma$. Notice that $\sum_{\sigma \in \Sigma} \Pr(\{x_i = \sigma\}) = \sum_{\sigma \in \Sigma} \tilde{r}_i^\sigma = 1$
2. Let $x$ be the string obtained by concatenating the $\{x_i\}_{i=1}^{m}$, that is, $x = x_1 x_2 \ldots x_m$.
3. We call $d_x = \max_{i=1}^{m} d(x, s_i)$

It is clear that one can obtain $x$ by a polynomial-time algorithm by solving the linear programming problem Equation 28.3 and then do the randomized rounding described in Definition (28.7). We will prove now that $x$ is a good approximation to the solution.

**Lemma 28.2** $E[d(x, s_i)] \le d_{min}$ for $i = 1, \ldots, n$.

*Proof.* By Lemma 28.1, we have that $d(x, s_i) = \sum_{j=1}^{m} \delta_j(x, s_i)$. Furthermore, for fixed $i$ (in the range $[1, m]$) and $j$ (in the range $[1, n]$), $\delta_i(x_i, s_j)$ is a Bernoulli trial with

$$\mathrm{E}[\delta_j(x, s_i)] = 1 - \sum_{\sigma \in \Sigma} s_{i,j}^{\sigma} \mathrm{E}[x_j^{\sigma}] = 1 - \sum_{\sigma \in \Sigma} s_{i,j}^{\sigma} \Pr(\{x_j = \sigma\}) = 1 - \sum_{\sigma \in \Sigma} s_{i,j}^{\sigma} \tilde{r}_j^{\sigma}.$$

Hence, by linearity of expectations, we get that $\mathrm{E}[d(x, s_i)] = \sum_{j=1}^{m}(1 - \sum_{\sigma \in \Sigma} s_{i,j}^{\sigma} \tilde{r}_j^{\sigma}))$ and as $r_j^{\sigma}$ is a solution of Equation 28.3, we have that

$$\mathrm{E}[d(x, s_i)] = \sum_{j=1}^{m}(1 - \sum_{\sigma \in \Sigma} s_{i,j}^{\sigma} \tilde{r}_j^{\sigma}) \leq \hat{d}_{min} \leq d_{min} \text{ for } i = 1, \ldots, n$$

∎

We can employ Chernoff bounds to prove a stronger result.

**Lemma 28.3** *Let $Y_1, Y_2, \ldots, Y_n$ be independent Bernoulli trials with $\mathrm{E}(Y_i) = p_i$. If $Y = \sum_{i=1}^{n} Y_i$ and $\mu = \mathrm{E}(Y) = \sum_{i=1}^{n} p_i$ and $0 < \epsilon \leq 1$, we have that*

$$\Pr(Y \geq (1 + \epsilon)\mu) \leq e^{-\frac{1}{3}\mu\epsilon^2} \qquad \Pr(Y \geq \mu + \epsilon n) \leq e^{-\frac{1}{3}n\epsilon^2}$$

$$\Pr(Y \leq (1 + \epsilon)\mu) \leq e^{-\frac{1}{2}\mu\epsilon^2} \qquad \Pr(Y \leq \mu - \epsilon n) \leq e^{-\frac{1}{2}n\epsilon^2}.$$

**Theorem 28.7** *The algorithm that outputs $x$ is a randomized polynomial-time $(1 + \epsilon)$ approximation algorithm for the CSP when $d_{min} \geq \frac{6 \log n}{\epsilon^2}$.*

*Proof.* We have that $d(x, s_i)$ is the sum of independent Bernoulli trials, that is, $d(x, s_i) = \sum_{j=1}^{m} \delta_j(x, s_i)$ and by Lemma 28.2, we have that $\mu = E(d(X, s_i)) \leq d_{min}$. By applying Lemma 28.3, we have that

$$\Pr\left(d(x, s_i) > (1 + \epsilon)d_{min}\right) \leq e^{-\frac{1}{3}\epsilon^2\mu}.$$

Furthermore, we have that

$$\Pr(d_x > (1 + \epsilon)d_{min}) = \Pr(\{\forall i = 1, \ldots, n : d(X, s_i) > (1 + \epsilon)d_{min}\}) \leq n e^{-\frac{1}{3}\epsilon^2\mu}.$$

And as $\mu \geq d_{min} \geq \frac{6 \log n}{\epsilon^2}$, we have that

$$\Pr\left(d_x > (1 + \epsilon)d_{min}\right) \leq n e^{-2 \log n} \leq \frac{1}{n}$$

∎

By using the method of conditional probabilities [49], it is possible to derandomize the previous algorithm. This is done explicitly in References 45, 46.

### 28.3.1.3 A $(1 + \epsilon)$ Polynomial Approximate Scheme

In the current section, we describe the approximation scheme of References 45, 46. Consider the following strategy to solve the *CSP*: Fix $0 \leq k < n$ and align any $k$ strings out of the $n$. In this alignment, there will be *clean* columns and *dirty* columns. A column is clean if a single character occurs in the entire column; it is dirty otherwise. If $j$ is a clean column and $c$ is the character in this column, then in the output string, column $j$ will be set to $c$. Luckily, the number of dirty columns will be relatively small and can be dealt with using the methods introduced in the previous section.

In the remainder of this section, we assume a fixed $k$, such that $1 < k < n$, and we make the previous ideas rigorous in the following way.

**Definition 28.8** *Let $i_1, \ldots, i_k$ be a subset of indices of $\{1, \ldots n\}$. We define*

1. $Q_{i_1,\ldots,i_k} := \{j : s_{i_1}[j] = s_{i_2}[j] = \ldots s_{i_k}[j]\}$, *that is, the set of positions in which $s_{i_1}, \ldots, s_{i_k}$ agree.*
2. $P_{i_1,\ldots,i_k} := \{1, \ldots, m\} \setminus Q_{i_1,\ldots,i_k}$, *that is, the set of positions in which $s_{i_1}, \ldots, s_{i_k}$ have two or more characters.*
3. *Given a string $r$ of length $m$ over $\Sigma$, we define $r|_{\{i_1 \ldots i_k\}} = r[i_1] \ldots r[i_k]$.*

The following lemma gives us a good estimate for the number of dirty columns.

**Lemma 28.4** *Given $1 \le i_1 \le \cdots \le i_k \le n$ we have that*

$$|P_{i_1 \ldots i_k}| \le k d_{min}.$$

*Proof.* Let $j$ be a position (i.e., a column) in which $s_{i_1}, \ldots, s_{i_k}$ have two or more characters. Then, there is an $l$ ($1 \le l \le k$) such that $s[j] \ne s_{i_l}[j]$, in which $s$ is the closest to all the $n$ input strings. By definition, $d(s, s_{i_l}) \le d_{min}$ and hence, every $s_{i_l}$ contributes at most $d_{min}$ dirty columns to $P_{i_1,\ldots,i_k}$ hence, $|P_{i_1 \ldots i_k}| \le k\, d_{min}$. ∎

Theorem 28.8 gives us an effective way to find an approximate solution to the CSP of $s_1, \ldots, s_n$ when we restrict every string to the positions $P_{i_1,\ldots,i_k}$ (i.e., the dirty columns).

**Theorem 28.8** *Let $P_{i_1,\ldots,i_k}$ form a subset of indices from $\{1, \ldots, n\}$ ("dirty columns") and let $0 < \epsilon < 1$. There is a polynomial-time algorithm that produces a string $s'$ of length $|P_{i_1,\ldots,i_k}|$ such that*

$$d(s', s_l|_{P_{i_1,\ldots,i_k}}) \le (1 + \epsilon)d_{min} - d(s_{i_1}|_{Q_{i_1 \ldots i_k}}, s_l|_{Q_{i_1 \ldots i_k}}) \text{ for } l = 1, \ldots, n$$

*Proof.* (Sketch) Consider $\tilde{s}_1, \ldots, \tilde{s}_n$, in which $\tilde{s}_l[j] := \begin{cases} s_l[j] & \text{when } j \in P_{i_1,\ldots,i_k} \\ s_{i_1}[j] & \text{when } j \in Q_{i_1 \ldots i_k} \end{cases}$. Let $\tilde{s}$ be a solution to the CSP over these strings, and define as before $\tilde{d}_{min}$ and $\tilde{P}_{i_1,\ldots,i_k}$. It is simple to notice that $\tilde{d}_{min} \le d_{min}$ and that $\tilde{P}_{i_1,\ldots,i_k} = P_{i_1,\ldots,i_k}$. Hence, by using Lemma 28.4, we infer that $|P_{i_1 \ldots i_k}| = |\tilde{P}_{i_1 \ldots i_k}| \le k\tilde{d}_{min}$.

We can find a solution to the closest string $s'$ if $|P_{i_1 \ldots i_k}| > \frac{6k \log n}{\epsilon^2}$ by using Theorem 28.7 or if $|P_{i_1 \ldots i_k}| \le \frac{6k \log n}{\epsilon^2}$ the result follows by using Theorem 28.6.

If we define $\tilde{s}$ by $\tilde{s}[j] := \begin{cases} s'[j] & \text{when } j \in P_{i_1,\ldots,i_k} \\ s_{i_1}[j] & \text{when } j \in Q_{i_1,\ldots,i_k} \end{cases}$ it is clear that this is a solution to the CSP for $\tilde{s}_1, \ldots, \tilde{s}_n$. Moreover,

$$d(\tilde{s}, \tilde{s}_l) = d(s', s_l|_{P_{i_1,\ldots,i_k}}) + d(s_{i_1}|_{Q_{i_1 \ldots i_k}}, s_l|_{Q_{i_1 \ldots i_k}}) \le \tilde{d}_{min} \le d_{min}.$$ ∎

For the remaining part of this section, we will be interested in knowing how close is $s_{i_1}$ to $s$ when we restrict the problem to the positions $Q_{i_1,\ldots,i_k}$ (i.e., the clean columns). That is, we want to estimate

$$d(s_l|_{Q_{i_1 \ldots i_k}}, s_{i_1}|_{Q_{i_1 \ldots i_k}}) - d(s_l|_{Q_{i_1 \ldots i_k}}, s|_{Q_{i_1 \ldots i_k}}) \text{ for } l = 1, \ldots, n.$$

The following definition and lemma will be a step in that direction. Notice that we will state a collection of lemmas without proof. The interested reader can find those in Reference 46.

**Definition 28.9** *Let $i_1, \ldots, i_k$ be a subset of indices of $\{1, \ldots n\}$ and $1 \le l \le n$, we define*

$$J(l) = \{j \in Q_{i_1 \ldots i_k} : s_{i_1}[j] \ne s_l[j] \wedge s_{i_1}[j] \ne s[j]\}.$$

**Lemma 28.5** *Let* $i_1, \ldots, i_k$ *be a subset of indices of* $\{1, \ldots n\}$ *and* $1 \leq i \leq n$, *then*

$$d(s_l|_{Q_{i_1 \ldots i_k}}, s_{i_1}|_{Q_{i_1 \ldots i_k}}) - d(s_l|_{Q_{i_1 \ldots i_k}}, s|_{Q_{i_1 \ldots i_k}}) \leq J(l). \tag{28.4}$$

Notice that for a fixed $k$, $J(l)$ depends on the set of indices $i_1, \ldots, i_k$ that we choose. For an arbitrary set of indices $i_1, \ldots, i_k$, we cannot bound $J(l)$ but we will prove in the following lemmas that there exists a set of indices $i_1, \ldots, i_k$ in which $J(l)$ is small. By using Lemma 28.5, we know that $s_{i_1}$ is an *approximate solution* for the restriction of the problem to the positions $Q_{i_1, \ldots, i_k}$. To be more precise, we introduce the following.

**Definition 28.10** *Let* $i_1, \ldots, i_k$ *be a subset of indices from* $\{1, \ldots, n\}$ *and* $0 \leq l \leq n$, *we define*

1. $p_{i_1, \ldots, i_k} := d(s_{i_1}|_{Q_{i_1, \ldots, i_k}}, s|_{Q_{i_1, \ldots, i_k}})$ *that is, the number of mismatches between* $s_{i_1}$ *and* $s$ *at positions in* $Q_{i_1, \ldots, i_k}$.

2. $\rho_0 := \max\limits_{1 \leq i,j \leq n} \dfrac{d(s_i, s_j)}{d_{min}}$ *and* $\rho_k := \min\limits_{1 \leq i_1 \leq \ldots i_k \leq n} \dfrac{p_{i_1 \ldots i_k}}{d_{min}}$ *for* $k = 1, \ldots, n$.

**Lemma 28.6** *For any* $2 \leq k' \leq k$, *there are indices* $1 \leq i_1 \leq \cdots \leq i_k \leq n$ *such that for any* $1 \leq l \leq n$

$$J(l) \leq (\rho_{k'} - \rho_{k'+1})d_{min}.$$

**Lemma 28.7** *For* $2 \leq k < n$

$$\min\{\rho_0 - 1, \rho_2 - \rho_3, \ldots, \rho_k - \rho_{k+1}\} \leq \frac{1}{2k - 1}.$$

**Theorem 28.9** *There exists a set of indices* $1 \leq i_1 \leq \cdots \leq i_k \leq n$ *such that*

$$d(s_l|_{Q_{i_1 \ldots i_k}}, s_{i_1}|_{Q_{i_1 \ldots i_k}}) - d(s_l|_{Q_{i_1 \ldots i_k}}, s|_{Q_{i_1 \ldots i_k}}) \leq \frac{1}{2k - 1}d_{min} \text{ for } 1 \leq l \leq n.$$

*Proof.* It is clear by using Lemmas 28.5 through 28.7 in consecutive order. ∎

Based on the ideas presented consider the following algorithm. Notice that $k$ is a fixed parameter with $\frac{1}{2k-1} + \epsilon \leq \delta$.

### Algorithm Closest_String

1. **for** every set of indices $\{i_1, \ldots, i_k\}$ **do**
   a. Let $\hat{s}(i_1, \ldots, i_k)$ be the solution to the problem as in Theorem 28.8.
   b. Define $s(i_1, \ldots, i_k)$ by making $s(i_1, \ldots, i_k)|_{Q_{i_1, \ldots, i_k}} := s_{i_1}|_{Q_{i_1, \ldots, i_k}}$ and
      $s(i_1, \ldots, i_k)|_{P_{i_1, \ldots, i_k}} := \hat{s}(i_1, \ldots, i_k)$.
   c. Let $cost(i_1, \ldots, i_k) := \max\limits_{j=1}^{k} d(s_{i_j}, s(i_1, \ldots, i_k))$.
2. Let $s' := s(i_1, \ldots, i_k)$ be the string that minimizes $cost(i_1, \ldots, i_k)$.
3. **for** i$= 1, \ldots, n$ **do** calculate $cost(i) := \max\limits_{j=1}^{n} d(s_j, s_i)$.
4. Select the string of minimum cost from the two previous steps.

**Theorem 28.10** *Let* $0 < \delta < 1$. *The algorithm* **Closest_String** *is a* $(1 + \delta)$ *polynomial approximation algorithm for the CSP.*

*Proof.* Choose $1 < k < n$ and $0 < \epsilon < 1$ such that $\frac{1}{2k-1} + \epsilon \leq \delta$.
If $\rho_0 - 1 \leq \frac{1}{2k-1}$ then in step (2) we find a solution such that

$$\rho_0 d_{min} \leq \left(1 + \frac{1}{2k - 1}\right) d_{min}.$$

which by definition of $\rho_0$ implies that

$$d(s', s_l) \leq \max_{1 \leq i,j \leq n} d(s_i, s_j) \leq \frac{1}{2k-1} d_{min} \leq (1+\delta)d_{min}.$$

In case $\rho_0 - 1 > \frac{1}{2k-1} d_{min}$, let us first observe that in step 2 we find a set of indices $1 \leq i_1 \leq \ldots \leq i_k \leq n$ such that

$$d(s', s_l) = d(s_{i_1}|_{Q_{i_1,\ldots,i_k}}, s_l|_{Q_{i_1,\ldots,i_k}}) + d(\hat{s}(i_1,\ldots,i_k), s_l|_{P_{i_1,\ldots,i_k}}).$$

By using Theorems 28.8 and 28.9, we note that

$$d(s', s_l) \leq \left(1 + \frac{1}{2k-1}\right) d_{min} + d(s_{i_1}|_{Q_{i_1,\ldots,i_k}}, s_l|_{Q_{i_1,\ldots,i_k}}) +$$

$$+ \epsilon d_{min} - d(s_{i_1}|_{Q_{i_1,\ldots,i_k}}, s_l|_{Q_{i_1,\ldots,i_k}}) \leq (1+\delta)d_{min}.$$

∎

Notice that the time complexity of this algorithm is $O((nm)^k n^{O\left(\log|\Sigma| \times \frac{k^2}{\epsilon^2}\right)}) = O(mn^{O(\epsilon^{-5})})$.

A simpler and more efficient approach is described in Reference 50 by using the linear programming approach from Theorem 28.7 to solve the case $d = \Omega(\frac{\log n}{\epsilon^2})$ and a fixed parameter like the one described in the next section (Section 28.3.1.4) for $d = O(\frac{\log n}{\epsilon^2})$. Such approach is a Polynomial-Time Approximation Scheme (PTAS) with approximation ratio $1 + \epsilon$ and a time complexity of $O(mn^{O(\epsilon^{-2})})$.

### 28.3.1.4 Exact Fixed-Parameter Polynomial Algorithms

In many practical applications, the value of $d$ is small, hence it is useful to study whether there are algorithms that solve the problem in polynomial time for a fixed $d$. The first such solution was described in Reference 47, which achives a time complexity of $(O(nm + nd(d+1)^d)$. The solution is simple and elegant and will be described next. The following algorithm receives as input the set of strings $s_1, \ldots, s_n$, an integer $d$ (the fixed distance), a candidate string $s$ and parameter $h$, which is used recursively. The algorithm is recursive and its initial call is **CSd**($s_1$,$d$).

#### Function CSd(s,h)

1. **If** $(h < 0)$ or $d(s, s_i) > d + h$ for some $i$ return "not found."
2. **If** $d(s, s_i) \leq d$ for all $i$, return $s$.
3. Select $i$ such that $d(s, s_i) > d$. Let $P'$ be a subset of $d + 1$ positions in which $s$ and $s_i$ differ.
4. **for** all $p \in P'$ **do**
   a. Modify $s$ in the $p$-th position to take the value of $s_i[p]$ and call the result $s'$.
   b. Return **CSd**($s'$,$h - 1$).

**Theorem 28.11** *The algorithm* **CSd** *solves the CSP problem for a fixed $d$ in time* $(O(nm + nd(d+1)^d)$.

*Proof.* It is clear that **CSd** generates a recursive tree in which each node generates $d+1$ leafs. Furthermore, it is clear that tree has $d$ levels, so the size of the tree is bounded by $(d+1)^d$. Each step in a particular node can be done in $nd$ time, hence the time complexity follows.

To prove the correctness, let us focus on the recursive call. Let $\tilde{s}$ be the solution to CSP problem. The positions in which $s$ and $s_i$ differ can be divided in two groups, the ones in which additionally $s_i$ is equal to $\tilde{s}$ (call it $P_1$) and the ones in which $s_i$ is different to *tildes* (call it $P_2$). As $d(\tilde{s}, s_i) \leq d$, it implies that $|P_2| \leq d$.

This implies that any subset $P'$ of $d + 1$ elements has element of $P_1$, which is step in the right direction as $s_i$ is equal to $\tilde{s}$. Hence, the result follows. ∎

Numerous algorithms [50,51] obtained improvements on the time complexity of the previous algorithm. In particular, the algorithm in Reference 51 achieves a complexity of $O(nm + nd^3 \cdot 6.74^d)$ for $|\Sigma| = 2$. Their approaches and proofs are not trivial and can be found in the references.

Significantly, a simple randomized approach to the problem was proposed in Reference 52, which we will describe for the case of $|\Sigma| = 2$. The interested reader can find in Reference 52 a more detailed presentation and proofs.

### Function RandomCSd

    i. **Repeat** $r$ times. Set $s = s_1$.
       a. **Repeat** $d$ times.
          i. **if** $d(s, s_i) \leq d$ for all $i$, return $s$ and exit.
          ii. Select $i$ such that $d(s, s_i) > d$. Pick a position here $s$ and $s_i$ differ uniformly at random and modify $s$ in the $p$-th position to take the value of $s_i[p]$.
     1. **if** no solution is found return "No solution."

The previous algorithm is MonteCarlo ramdomized solution with one-sided error, which means that the algorithm sometimes reports "No solution" when one solutions exists but in case it reports a solution, such solution is correct.

One of the key observations is to bound the probability that a solution is found in the inner loop (which is repeated $d$ times). Such probability can be shown to be at least $\frac{d!}{(2d)^d}$ [52]. This implies that the value of $r$ can be set to $c\frac{(2e)^d}{\sqrt{2\pi d}}$ when $|\Sigma| = 2$, which yields an algorithm that can run in $O(nm + nd \cdot 5.43^d)$.

Further extensions to nonbinary alphabets and improvements in the running time can be found in Reference 52.

## 28.3.2 Closest Substring Problem

The *CSSP* takes as input $n$ sequences of length $m$ each. The goal is to find a substring (of length $l$) that is the closest to some As substrings (of length $l$ each) picked one from each input sequence. The substring of interest is also known as the "motif." The notion of a motif is defined rigorously next.

**Definition 28.11** *If $s$ and $s'$ are strings over the alphabet $\Sigma$ and $l$ is such that $0 < l < |s|$, we define:*

    1. $s' \lhd_l s$ *if $s'$ is a substring of length $l$ of $s$.*
    2. $\bar{d}(s', s) := \min_{r \lhd_l s} d(s', r)$.

Notice that if $s' \lhd_l s$, then $\bar{d}(s', s) = 0$.

**Definition 28.12** *Given strings $s_1, s_2, \ldots, s_n$ of length $m$ each over the alphabet $\Sigma$ and an $l$ ($0 < l \leq m$), the CSSP is to find a string $\bar{s}$ of length $l$ over $\Sigma$ that minimizes $\max\limits_{i=1}^{n} \bar{d}(s, s_i)$. We denote by $\bar{t}_i \lhd_l s_i$ the string such that $\bar{d}(\bar{s}, s_i) = d(\bar{s}, \bar{t}_i)$ for $i = 1, \ldots, n$. We denote by $\bar{d}_{min} := \max\limits_{i=1}^{n} \bar{d}(\bar{s}, s_i)$.*

In the remaining part of this section, we will assume that we are given $s_1, \ldots, s_n$ and $l$ that satisfy the conditions of Definition 28.12. $\bar{d}_{min}, \bar{t}_i$, and $\bar{s}$ will satisfy the conditions stated in Definition 28.12.

We now present approximation strategies that have been proposed in References 46, 53, 54. Some of these strategies will be based on the results that were discussed in Section 28.3.

### 28.3.2.1 Simple Approximation Schemes

We start by presenting a simple strategy that obtains a 2 approximation polynomial-time algorithm as it is described in Reference 53.

Algorithm Simple_Closest_Substring

1. **for** every $s' \lhd_l s_1$ **do**
   a. **for** $i := 2, \ldots, n$ **do**

   Let $t_i(s') \lhd_l s_i$ be such that $d(s', t_i(s')) = \bar{d}(s', s_i)$.

   b. Let $cost(s') := \max\limits_{i=1}^{n} \bar{d}(s', s_i)$.

2. Pick the string $\tilde{s} \lhd_l s_1$ that minimizes $cost(\tilde{s})$. Let $\tilde{t}_j \lhd_l s_j$ be such that
   $d(\tilde{s}, \tilde{t}_j) = \bar{d}(\tilde{s}, s_j)$ for $j = 1, \ldots, n$.

**Theorem 28.12** *Algorithm* **Simple_Closest_Substring** *is 2-approximate for CSSP.*

*Proof.* Let $r_1, \ldots, r_n$, where $r_i \lhd_l s_i$, be the strings such that $\bar{d}(\tilde{s}, s_i) = d(\tilde{s}, r_i)$. Then, $d(\tilde{s}, r_i) \leq \bar{d}_{min}$ and hence

$$d(r_1, r_j) \leq d(\tilde{s}, r_1) + d(\tilde{s}, r_j) \leq 2\bar{d}_{min} \text{ for } j = 1, \ldots, n.$$

We conclude by noticing that

$$\bar{d}(\tilde{s}, s_i) = d(\tilde{t}_1, \tilde{t}_i) = \min_{s' \lhd_l s_1} \max_{j=1}^{n} d(s', t_j(s')) \leq \min_{s' \lhd_l s_1} \max_{j=1}^{n} d(s', r_j) = \max_{j=1}^{n} d(r_1, r_j) \leq 2\bar{d}_{min}.$$

∎

Our aim is to describe an approximation algorithm that follows the ideas of algorithm **Closest_String**. To do so we will describe some definitions and theorems that generalize the ones done in Section 28.3.1.3.

**Definition 28.13** *Given a set of indices $1 \leq i_1 \leq \cdots \leq i_k \leq n$ and a set of substrings $T = \{t_{i_1}, \ldots, t_{i_k}\}$ in which $t_{i_j} \lhd_l s_{i_j}$ for $j = 1, \ldots, k$. We define*

1. $Q^T_{i_1, \ldots, i_k} := \{j : t_{i_1}[j] = t_{i_2}[j] = \ldots t_{i_k}[j]\}$, *that is, the set of positions in which $t_{i_1}, \ldots, t_{i_k}$ agree.*

2. $P^T_{i_1, \ldots, i_k} := \{1, \ldots, m\} - Q^T_{i_1, \ldots, i_k}$, *that is, the set of positions in which $t_{i_1}, \ldots, t_{i_k}$ differ.*

The following theorems extend naturally Lemma 28.4 and Theorem 28.9.

**Theorem 28.13** *For $1 \leq i_1 \leq \cdots \leq i_k \leq n$ and $T := \{t_{i_1}, \ldots, t_{i_k}\}$ in which $t_{i_j} \lhd s_j$,*

$$|P^T_{i_1 \ldots i_k}| \leq k\bar{d}_{min}.$$

*Proof.* It follows directly from Lemma 28.4. ∎

**Theorem 28.14** *There exists a set of indices $1 \leq i_1 \leq \cdots \leq i_k \leq n$ such that for $T := \{\bar{t}_{i_1}, \ldots, \bar{t}_{i_k}\}$,*

$$d(\bar{t}_l|_{Q^T_{i_1 \ldots i_k}}, \bar{t}_{i_1}|_{Q^T_{i_1 \ldots i_k}}) - d(\bar{t}_l|_{Q^T_{i_1 \ldots i_k}}, \tilde{s}|_{Q^T_{i_1 \ldots i_k}}) \leq \frac{1}{2k-1}\bar{d}_{min} \text{ for } 1 \leq l \leq n.$$

*Proof.* It follows from Theorem 28.9 ∎

## Algorithm Closest_Small_Substring

1. **for** every set of substrings $T = \{t_{i_1}, \ldots, t_{i_k}\}$ in which $t_{i_j} \lhd_l s_{i_j} \; j = 1, \ldots, k$ **do**

    a. **for** every $t \in \Sigma^p$ in which $p = |P^T_{i_1,\ldots,i_k}|$ **do**

    Build the string $x$ by $x|_{P^T_{i_1,\ldots,i_k}} = t$ and $x|_{Q^T_{i_1,\ldots,i_k}} = s_{i_1}|_{Q^T_{i_1,\ldots,i_k}}$.

    b. Let $x^T_{i_1,\ldots,i_k}$ be the string that minimizes $\max\limits_{i=1}^{n} \bar{d}(x^T_{i_1,\ldots,i_k}, s_i)$ and call

    $cost^T(i_1, \ldots, i_k) := \max\limits_{i=1}^{n} \bar{d}(x^T_{i_1,\ldots,i_k}, s_i)$.

2. Let $x'$ be the string that minimizes $cost^T(i_1, \ldots, i_k)$.

3. **for** i=:1, ..., n and **for** every $r \lhd_l s_i$ **do** calculate $cost^r(i) := \max\limits_{i=1}^{n} d(r, s_i)$.

4. Select the string of minimum cost from the two previous steps.

**Theorem 28.15** *Let* $1 \le k < n$. *The algorithm* **Closest_Small_Substring** *is a* $(1 + \frac{1}{2k-1})$ *polynomial-time approximation algorithm for the CSSP when* $d_{min} \le O(\log(nm))$.

*Proof.* To argue that **Closest_Small_Substring** takes polynomial time we notice by using Theorem 28.13 that $\left|P^T_{i_1,\ldots,i_k}\right| = O(k \log(nm))$ in step 1. This means that the inner loop of step 1 takes $|\Sigma|^{O(k \log(nm))} mnl = O((nm)^{O(\log |\Sigma| k)})$ time and that step 1 takes $O((nm)^{O(\log |\Sigma| k)}) O((nm)^k) = O((nm)^{O(\log |\Sigma| k)})$ time. It is clear that steps 2 and 3 take less time.

Applying Theorems 28.13 and 28.14 in a similar way to the one in the proof of Theorem 28.10, we get a $(1 + \frac{1}{2k-1})$ approximate solution. ∎

### 28.3.2.2 A $(1 + \epsilon)$ Polynomial Approximation Scheme

The following ideas were first described in Reference 54 and later in Reference 46.

We would like to extend the algorithm **Closest_Small_Substring** for the general case. Let us first notice that by trying all possibilities we get a set of indices $0 \le i_1 \le \ldots i_k \le n$ and substrings $T = \{t_{i_1}, \ldots, t_{i_k}\}$, which satisfy the conditions of Theorem 28.14 and hence, we know that if we set $s'|_{Q^T_{i_1,\ldots,i_k}} = s_{i_1}|_{Q^T_{i_1,\ldots,i_k}}$ we get a "good" solution when we restrict ourselves to the positions $Q^T_{i_1,\ldots,i_k}$.

To calculate $s'|_{P^T_{i_1,\ldots,i_k}}$ we would like to use Theorem 28.8 applied to the strings $\tilde{t}_1, \ldots, \tilde{t}_n$ in which $\tilde{t}_j = \bar{t}_j|_{P^T_{i_1,\ldots,i_k}}$. One major difficulty in doing so is that we don't know $\bar{t}_1, \ldots, \bar{t}_n$.

To do this, we want for $t_j \lhd_l s_j$ to estimate $d(t_j, \bar{s})$. To accomplish that, we fix a small set $R \subset P^T_{i_1,\ldots,i_k}$ and we calculate $\bar{s}|_R$ by brute force. Based on the values of $\bar{s}|_R$ and $t_j|_R$, we define $f^R(t_j) \approx d(t_j, \bar{s})$. By choosing $t'_j$ such that $f^R(t'_j) = \min\limits_{t \lhd_l s_j} f^R(t)$, we hope to get a good approximation of $\bar{t}_j$ for $j = 1, \ldots, n$.

To make the preceding discussion more formal, we introduce the following definitions and theorem. We omit the proofs, which can be found in

**Definition 28.14** *Let* $1 \le i_1 \le \cdots \le i_k \le n$, $T := \{\bar{t}_{i_1}, \ldots, \bar{t}_{i_k}\}$, $R$ *be a multiset of positions from* $P^T_{i_1,\ldots,i_k}$, *and let us call* $\rho := \frac{|P^T_{i_1,\ldots,i_k}|}{|R|}$.

1. *For* $t$ *a string of length* $l$, *we define*

$$f^R(t) = \rho \, d(t|_R, \bar{s}|_R) + d\left(t|_{Q^T_{i_1,\ldots,i_k}}, \bar{t}_{i_1}|_{Q^T_{i_1,\ldots,i_k}}\right).$$

2. *For $j = 1, \ldots, n$, let $t'_j$ be such that*

$$f^R(t'_j) := \min_{t \lhd_l s_j} f^R(t).$$

The following lemma implies that the $t'_i$ referred to in Definition 28.14 is a "good" approximation of the $\bar{t}_i$, for $i = 1, \ldots, n$.

**Theorem 28.16** *Suppose the conditions of Definition 28.14 are met and let $s^*$ be a string with $s^*|_{P^T_{i_1,\ldots,i_k}} = \bar{s}|_{P^T_{i_1,\ldots,i_k}}$ and $s^*|_{Q^T_{i_1,\ldots,i_k}} = \bar{t}_{i_1}|_{Q^T_{i_1,\ldots,i_k}}$. Furthermore, let us assume $|R| = \frac{4}{\epsilon^2} \log nm$. Then,*

$$\Pr(\{\forall i = 1, \ldots, n : d(s^*, t'_i) \leq d(s^*, \bar{t}_i) + 2\epsilon |P^T_{i_1,\ldots,i_k}|\}) \leq 2(nm)^{-\frac{1}{3}}.$$

## Algorithm Closest_Substring

1. **for** every set of substrings $T = \{t_{i_1}, \ldots, t_{i_k}\}$ in which $t_{i_j} \lhd_l s_{i_j}$ $j = 1, \ldots, k$ **do**
   a. Let $R$ be a multiset containing $\frac{4}{\epsilon^2} \log nm$ uniformly random positions from $P^T_{i_1,\ldots,i_k}$.
   b. By enumerating all strings in $\Sigma^R$ find $\bar{s}|_R$.
   c. **for** $j = 1, \ldots, n$ **do** find $t'_j \lhd_l s_j$ that satisfies Definition 28.14.
   d. By using Theorem 28.8, find a solution $s'$ to the CSP for
   $t'_1|_{P^T_{i_1,\ldots,i_k}}, \ldots, t'_n|_{P^T_{i_1,\ldots,i_k}}$.
   e. Define $x$ so that $x|_{Q_{i_1,\ldots,i_k}} = s_{i_1}|_{Q^T_{i_1,\ldots,i_k}}$ and $x|_{P^T_{i_1,\ldots,i_k}} = s'$.
   f. Call $cost^T(i_1, \ldots, i_k) := \max_{i=1}^{n} \bar{d}(x, s_i)$.
2. Let $x'$ be the string that minimizes $cost^T(i_1, \ldots, i_k)$.
3. **for** $i =: 1, \ldots, n$ and **for** every $r \lhd_l s_i$ **do** calculate $cost^r(i) := \max_{i=1}^{n} d(r, s_i)$.
4. Select the string with the minimum cost from the two previous steps.

**Theorem 28.17** *Let $0 < \delta < 1$. The algorithm* **Closest_Substring** *is a $(1 + \delta)$ polynomial approximation algorithm for the CSP.*

*Proof.* Let $s'$ be the output of algorithm **Closest_Substring**. By using Theorem 28.16, we note that

$$d(s^*, t'_i) \leq d(s^*, \bar{t}_i) + 2\epsilon |P^T_{i_1,\ldots,i_k}| \text{ with high probability.}$$

An application of Theorem 28.13 implies that $|P^T_{i_1,\ldots,i_k}| \leq k\bar{d}_{min}$. Combining this with Theorem 28.14,

$$d(s^*, t'_i) \leq \left(1 + \frac{1}{2k-1} + 2\epsilon k\right) \bar{d}_{min} \text{ for } i = 1, \ldots, n \text{ with high probability.}$$

Employing Theorem 28.8 with $\epsilon' = \epsilon |P^T_{i_1,\ldots,i_k}| \leq \epsilon k\bar{d}_{min}$,

$$d(s', t'_i) \leq \left(1 + \frac{1}{2k-1} + 3\epsilon k\right) \bar{d}_{min} \text{ for } i = 1, \ldots, n, \text{ with high probability.}$$

Then if we choose $\delta$ such that $\left(\frac{1}{2k-1} + 3\epsilon k\right) \leq \delta$, the result holds. $\blacksquare$

By using the method of conditional probabilities [49], it is possible to eliminate the randomness in algorithm **Closest_Substring**. More details can be found in Reference 46.

# 28.4 Conclusion

In the current chapter, we have considered the problem of selecting primers for a given set of DNA sequences that will be employed in PCR and MP-PCR experiments. The basic version of the problem, namely, the PSP and its variant the DPSP were discussed in detail. Some of the salient algorithms found in the literature for these versions of primer selection were surveyed. An elaborate discussion of other variants of DPD can be found in Reference 1. We have also addressed the PMS problem and some of the algorithms to solve this problem. Furthermore, we consider the related minimization problems: the CSP, and the CSSPs. For the CSP and the CSSP, we presented a survey of the existing polynomial approximation algorithms.

# References

1. C. Linhart and R. Shamir, The degenerate primer design problem—Theory and applications, *Journal of Computational Biology*, Mary Ann Liebert, Inc., publishers, NY, USA, 12(4), 2005, 431–456.
2. W. R. Pearson, G. Robins, D. E. Wrege and T. Zhang, On the primer selection problem in polymerase chain reaction experiments, *Discrete Applied Mathematics* 71, 1996, 231–246.
3. J. S. Chamberlain, R. A. Gibbs, J. E. Rainer, P. N. Nguyen, C. T. Casey, Deletion screening of the Duchenne muscular dystrophy locus via multiplex DNA amplification, *Nucleic Acids Research* 16, 1988, 11141–11156.
4. S. Kwok, S. Y. Chang, J. J. Sninsky and A. Wang, A guide to the design and use of mismatched and degenerate primers, *PCR Methods and Applications*, Cold Spring Harbor Laboratory Press, NY, USA, 3, 1994, S39–S47.
5. R. Souvenir, J. Buhler, G. Stormo and W. Zhang, Selecting degenerate multiplex PCR primers, *Proceedings of the 3rd International Workshop on Algorithms in Bioinformatics (WABI)*, Universal Academy Press, Inc., Japan, 2003, pp. 512–526.
6. K. Doi and H. Imai, Greedy algorithms for finding a small set of primers satisfying cover length resolution conditions in PCR experiments, *Proceedings of the 8th Workshop on Genome Informatics (GIW)*, Universal Academy Press, Japan, 1997, pp. 43–52.
7. K. Doi and H. Imai, A Greedy algorithm for minimizing the number of primers in multiple PCR experiments, *Genome Informatics*, 10, 1999, 73–82.
8. K. Doi and H. Imai, Complexity properties of the primer selection problem for PCR experiments, *Proceedings of the 5th Japan-Korea Joint Workshop on Algorithms and Computation*, 2000, pp. 152–159.
9. K. M. Konwar, I. I. Mandoiu, A. C. Russell, and A. A. Shvartsman, Improved algorithms for multiplex PCR primer set selection with amplification length constraints, *Proceedings of the 3rd Asia Pacific Bioinformatics Conference (APBC)*, Imperial College Press, London pp. 41–50, 2005.
10. T. M. Rose, E. R. Schultz, J. G. Henikoff, S. Pietrokovski, C. M. McCallum and S. Henikoff, Consensus-degenerate hybrid oligonucleotide primers for amplification of distantly related sequences, *Nucleic Acids Research* 26(7), 1998, 1628–1635.
11. T. Fuchs, B. Malecova, C. Linhart, R. Sharan, M. Khen, R. Herwig, D. Shmulevich et al, DEFOG: A practical scheme for deciphering families of genes, *Genomics* 80(3), 2002, 295–302.
12. X. Wei, D. N. Kuhn and G. Narasimhan, Degenerate primer design via clustering, *Proceedings of the 2003 IEEE Bioinformatics Conference (CSB)*, 2003, pp. 75–83.
13. S. Balla, S. Rajasekaran and I. I. Mandoiu, Efficient algorithms for degenerate primer search, UConn BECAT/CSE Technical Report, BECAT/CSE-TR-05-2, October 2005.
14. P. Pevzner and S.-H. Sze, Combinatorial approaches to finding subtle signals in DNA sequences, *Proceedings of the 8th International Conference on Intelligent Systems for Molecular Biology*, Association for the Advancement of Artificial Intelligence (AAAI Press), California, CA, 2000, pp. 269–278.

15. J. Buhler and M. Tompa, Finding motifs using random projections, *Proceedings of the 5th Annual International Conference on Computational Molecular Biology (RECOMB)*, April 2001.

16. E. Eskin and P. Pevzner, Finding composite regulatory patterns in DNA sequences, *Bioinformatics* 18(Suppl-1), Oxford University Press, UK, 2002, 354–363.

17. A. Price, S. Ramabhadran and P. A. Pevzner, Finding subtle motifs by branching from sample strings, *Bioinformatics* 1(1), 2003, 1–7.

18. S. Rajasekaran, S. Balla, and C.-H. Huang, Exact algorithms for the planted motif problem, *Journal of Computational Biology*, 12, 2005, 1117–1128.

19. U. Keich and P. Pevzner, Finding motifs in the twilight zone, *Bioinformatics* 18, 2002, 1374–1381.

20. G. Hertz and G. Stormo, Identifying DNA and protein patterns with statistically significant alignments of multiple sequences, *Bioinformatics* 15, 1999, 563–577.

21. C. E. Lawrence, S. F. Altschul, M. S. Boguski, J. S. Liu, A. F. Neuwald, and J. C. Wootton, Detecting subtle sequence signals: A Gibbs sampling strategy for multiple alignment, *Science* 262, 1993, 208–214.

22. T. L. Bailey and C. Elkan, Fitting a mixture model by expectation maximization to discover motifs in biopolymers, *Proceedings of the 2nd International Conference on Intelligent Systems for Molecular Biology*, Association for the Advancement of Artificial Intelligence (AAAI Press), Palo Alto, CA, 1994, pp. 28–36.

23. J. Davila, S. Balla, and S. Rajasekaran, Fast and practical algorithms for planted (l, d) motif search, *IEEE/ACM Transactions on Computational Biology and Bioinformatics* 4(4), 2007, 544–552.

24. S. Rajasekaran and H. Dinh, A speedup technique for (l, d)-motif finding algorithms, *BMC Research Notes* 4(1), 2011, 54.

25. Q. Yu, H. Huo, Y. Zhang, and H. Guo, Pairmotif: A new pattern-driven algorithm for planted $(l, d)$ DNA motif search, *PLoS ONE* 7(10), 2012, e48442.

26. S. Tanaka, Improved exact enumerative algorithms for the planted (l, d)-motif search problem, *IEEE/ACM Transactions on Computational Biology and Bioinformatics* 11(2), 2014, 361–374.

27. H. Dinh, S. Rajasekaran, and V. Kundeti, Pms5: An efficient exact algorithm for the $(l, d)$-motif finding problem, *BMC Bioinformatics* 12(1), 2011, 410.

28. S. Bandyopadhyay, S. Sahni, and S. Rajasekaran, Pms6: A fast algorithm for motif discovery, *IEEE 2nd International Conference on Computational Advances in Bio and Medical Sciences, ICCABS 2012, Las Vegas, NV, February 23–25, 2012*, IEEE, pp. 1–6.

29. H. Dinh, S. Rajasekaran, and J. Davila, qpms7: A fast algorithm for finding $(l, d)$-motifs in DNA and protein sequences, *PLoS ONE* 7(7), 2012, e41425.

30. I. Roy and S. Aluru, Finding motifs in biological sequences using the micron automata processor, *2014 IEEE 28th International Parallel and Distributed Processing Symposium, IPDPS'14, Washington, DC*, IEEE, pp. 415–424.

31. M. Nicolae and S. Rajasekaran, Efficient sequential and parallel algorithms for planted motif search, *BMC Bioinformatics*, BioMed Central Ltd., London, UK, 15(1), 2014, 34.

32. M. Nicolae and S. Rajasekaran, qPMS9: An efficient algorithm for quorum planted motif search, *Scientific Reports* 5, 2015, Article 7813, doi:10.1038/srep07813.

33. H. M. Martinez, An efficient method for finding repeats in molecular sequences, *Nucleic Acids Research* 11(13), 1983, 4629–4634.

34. A. Brazma, I. Jonassen, J. Vilo, and E. Ukkonen, Predicting gene regulatory elements in silico on a genomic scale, *Genome Research* 15, 1998, 1202–1215.

35. D. J. Galas, M. Eggert, and M. S. Waterman, Rigorous pattern-recognition methods for DNA sequences: Analysis of promoter sequences from *Escherichia coli*, *Journal of Molecular Biology*, Elsevier Science (USA), 186(1), 1985, 117–128.

36. S. Sinha and M. Tompa, A statistical method for finding transcription factor binding sites, *Proceedings of the 8th International Conference on Intelligent Systems for Molecular Biology*, 2000, pp. 344–354.

37. R. Staden, Methods for discovering novel motifs in nucleic acid sequences, *Computer Applications in the Biosciences* 5(4), 1989, 293–298.

38. M. Tompa, An exact method for finding short motifs in sequences, with application to the ribosome binding site problem, *Proceedings of the 7th International Conference on Intelligent Systems for Molecular Biology*, 1999, pp. 262–271.

39. J. van Helden, B. Andre, and J. Collado-Vides, Extracting regulatory sites from the upstream region of yeast genes by computational analysis of oligonucleotide frequencies, *Journal of Molecular Biology* 281(5), 1998, 827–842.

40. S. Desaraju and R. Mukkamala, Multiprocessor implementation of modeling method for planted motif problem, *2011 World Congress on Information and Communication Technologies (WICT)*, December 11–14, 2011, Mumbai, India, IEEE, pp. 524–529.

41. C. E. Lawrence and A. A. Reilly, An expectation maximization (EM) algorithm for the identification and characterization of common sites in unaligned biopolymer sequences, *Proteins: Structure, Function, and Genetics* 7, 1990, 41–51.

42. S. Rajasekaran, Motif search algorithms, in *Handbook of Computational Molecular Biology*, CRC Press, Boca Raton, FL, 2005.

43. M. Frances and A. Litman, On covering problems of codes, *Theory of Computing Systems*, 30, 1997, 113–119.

44. A. Ben-Dor, G. Lancia, J. Perone, and R. Ravi, Banishing bias from consensus sequences, *Proceedings of the 8th Annual Symposium on Combinatorial Pattern Matching Theory Computer Systems*, Lecture Notes In Computer Science; Vol. 1264, 1997, pp. 247–261.

45. M. Li, B. Ma, and L. Wang, Finding similar regions in many strings, *Proceedings of the 31st Anual ACM Symposium on Theory of Computing*, 1999, pp. 473–482.

46. M. Li, B. Ma, and L. Wang, On the closest string and substring problems, *Journal of the ACM* 49(2), 2002, 157–171.

47. J. Gramm, R. Niedermeier, and P. Rossmanith, Exact solutions for closest string and related problems, *Proceedings of the 12th Annual Symposium on Algorithms and Computation (ISAAC 2001)*, Lecture Notes in Computer Science; Vol. 2223, 2001, pp. 441–453.

48. J. Gramm, R. Niedermeier, and P. Rossmanith, Fixed-parameter algorithms for closest string and related problems, *Algorithmica* 37(1), 2003, 25–42.

49. R. Motwani and P. Raghavan, *Randomized Algorithms*, Cambridge University Press, New York, 1995.

50. B. Ma, X. Sun, More efficient algorithms for closest string and substring problems. *SIAM Journal on Computing*, Society for Industrial and Applied Mathematics, PA, USA, 39(4), 2009, 1432–1443.

51. Z. Chen, B. Ma, L. Wang, A three-string approach to the closest string problem, *Journal of Computer and System Sciences* 78(1), 2012, 164–178.

52. Z. Chen, B. Ma, L. Wang, Randomized and parameterized algorithms for the closest string problem. *CPM* 2014, pp. 100–109.

53. K. Lanctot, M. Li, B. Ma, S. Wang, and L. Zhang, Distinguishing string selection problems, *Proceedings of the 10th Annual ACM-SIAM Symposium on Discrete Algorithms (SODA)*, Elsevier Science, pp. 633–642.

54. B. Ma, A polynomial time approximation scheme for the closest substring problem, *Proceedings of the 11th Annual Combinatorial Pattern Matching (CPM) Symposium*, Lecture Notes in Computer Science; Vol. 1848, 2000, pp. 99–107.

# 29

# Dynamic and Fractional Programming-Based Approximation Algorithms for Sequence Alignment with Constraints

Abdullah N. Arslan

Ömer Eğecioğlu

## 29.1 Introduction

There are interesting algorithmic issues that arise when length constraints are taken into account in the formulation of a variety of problems on string similarity, particularly, in the problems related to local alignment. These types of problems have their roots and most striking applications in computational biology. In fact, because of the applications in biological sequence analysis, detection of local similarities in two given strings has become an increasingly important computational problem. When there are additional constraints that need to be satisfied as a part of the search criteria, it is natural to consider approximation algorithms for the resulting computational problems for large parameters.

Given two strings $X$ and $Y$, the classical dynamic programming solution to the local alignment problem searches for two substrings $I \subseteq X$ and $J \subseteq Y$ with maximum similarity score under a given scoring scheme, where $\subseteq$ indicates the substring relation. This classical definition of similarity has certain anomalies mainly because the lengths of the segments $I$ and $J$ are not taken into account. To cope with the possible anomalies of mosaic and shadow effects, many variations of the local alignment problem have been suggested. Mosaic effect is observed when an unrelated segment is sandwiched between two very similar segments. Shadow effect is observed when a biologically important short alignment is not detected because it overlaps with a longer yet biologically inadequate alignment with only a slightly higher score.

The variations suggested either define new objective functions, or include a length constraint on the substrings $I$ and $J$ for optimal alignments sought. This constraint can be driven by practical considerations for various objective functions (e.g., the maximization of *length-normalized* scores) and can be explicitly given such as requiring $|I| + |J| \geq t$ or $|J| \leq T$ for given parameters $t$ and $T$. In addition, in some local alignment problems the constraint may also be implicit, as it happens in the case of cyclic sequence comparison. In Table 29.1, we give a list of local alignment problems, their objectives, and computational results for them. The function $s(I, J)$ denotes the similarity score between $I$ and $J$. The optimizations are over all possible substrings $I$ of $X$, and $J$ of $Y$. In the table, we use "nor. score" as a shorthand for length normalized score. For any optimization problem $\mathcal{P}$, we denote by $\mathcal{P}^*$ its optimum value, and sometimes drop the parameters from the notation when they are obvious from the context. An optimization problem $\mathcal{P}$ is called *feasible* if it has a solution with the given parameters.

In most cases under consideration, there are simple dynamic programming formulations for the solution of the exact version of a given alignment problem with a length constraint. However, the resulting algorithms require cubic or higher time complexity, which is unacceptably high for practical purposes as the sequence lengths can be in the order of millions. To cope with such high complexity, approximations are considered in both definitions of similarity, and in the resulting computations.

There have been approximation algorithms proposed for various alignment problems with constraints, involving applications of techniques from fractional programming, and dynamic programming. In the current chapter, we present a survey of the most interesting of the approximation algorithms for variations of local alignment problems. Our focus is on fractional programming algorithms and algorithms returning results that meet the length constraint only partially but guaranteed to be within a given tolerance. These algorithms can be organized into three main categories:

1. *Fractional programming algorithms:* Application of fractional programming on *adjusted normalized local alignment* (the *ANLA* problem in Table 29.1) is of interest. The local alignment is normally defined as a graph problem. The fractional programming technique offers an iterative solution such that at each iteration an ordinary local alignment problem with modified weights is solved. This mimics the action of manually changing the weights until the results are found satisfactory. Fundamental theorems of fractional programming guarantee an optimal solution at the conclusion of these iterations. The termination properties of the iterative scheme are not obvious at all without referring to the results established for fractional programming.

2. *Approximation algorithms for partial constraint satisfaction:* Another noteworthy feature of some constrained local alignment approximation algorithms is their unusual performance measure. Ordinarily, performance of an approximation algorithm is measured by comparing the returned results against optimum value with respect to the objective function. In some approximation results regarding the length-constrained local alignment problems, such as the problem of finding a sufficiently long alignment with high score (the *LAt* problem in Table 29.1), the alignment returned is assured to have at least the score obtainable with respect to the given constraint, but the length constraint is satisfied to only within a prescribed tolerance from the required length value.

3. *Fractional programming approximation algorithms:* There are fractional programming approximation algorithms for the *normalized local alignment* (*NLAt*) problem with length constraint (the *NLAt* problem in Table 29.1). These algorithms iteratively invoke an approximation algorithm to solve a length-constrained local alignment problem (*LAt*) such that the length constrained is guaranteed to be satisfied within a given tolerance. This length-guarantee carries over for the final result for the normalized local alignment problem. That is, the fractional programming algorithm returns an approximate result for which the guarantee on the satisfaction of the length-constraint within some tolerance is due to the approximation algorithm used at each iteration, and the criteria is preserved over the iterations.

**TABLE 29.1** Variations of Local Alignment Problems. (From Arslan, A.N. and Eğecioğlu, Ö., *INFORMS J. Comput.*, 16, 441, 2004.)

| Alignment Problem | Objective | Algorithm | Time | Space | Returned Alignment Satisfies |
|---|---|---|---|---|---|
| $LA$ | Maximize $s(I,J)$ | Smith-Waterman | $O(nm)$ | $O(m)$ | score $= LA^*$ |
|  |  | Dinkelbach | $O(nm)$ (experimental) | $O(m)$ | score $= ANLA^*$ |
| $ANLA$ | Maximize $\frac{s(I,J)}{|I|+|J|+L}$ for parameter $L \geq 0$ | RationalANLA, Dinkelbach | $O(nm \log n)$ | $O(m)$ | score $= ANLA^*$ |
| $LRLA$ | Maximize $s(I,J)$ such that $|J| \leq T$ | HALF | $O(nm)$ | $O(m)$ | score $\geq \frac{1}{2} LRLA^*$ |
| $CLA$ | $LRLA$ with parameters $X$, $YY$, and $T = |Y|$ | APX-LRLA | $O(nmT/\Delta)$ | $O(mT/\Delta)$ | score $\geq LRLA^* - 2\Delta$ |
|  | The same $LRLA$ algorithms, complexity, and results |  |  |  |  |
| $LAt$ | Maximize $s(I,J)$ such that $|I| + |J| \geq t$ | APX-LAt | $O(rnm)$ | $O(rm)$ | score $\geq LAt^*$, length $\geq (1 - \frac{1}{r})t$ |
| $Qt$ | Find $(I,J)$ such that $\frac{s(I,J)}{|I|+|J|} > \lambda$, and $|I| + |J| \geq t$, for parameter $\lambda > 0$ | APX-LAt | $O(rnm)$ | $O(rm)$ | nor. score $> \lambda$, length $\geq (1 - \frac{1}{r})t$ |
| $NLAt$ | Maximize $\frac{s(I,J)}{|I|+|J|}$ such that $|I| + |J| \geq t$ | Dinkelbach | $O(rnm)$ (experimental) | $O(rm)$ | nor. score $\geq NLAt^*$, length $\geq (1 - \frac{1}{r})t$ |
|  |  | RationalNLAt, Dinkelbach | $O(rnm \log n)$ | $O(rm)$ | nor. score $\geq NLAt^*$, length $\geq (1 - \frac{1}{r})t$ |

In the current chapter, we start with the basic frame-work for local alignment in Section 29.2. We present the details of the topics enumerated earlier in three sections. In Section 29.3, we describe the fractional programming algorithms for the *ANLA* problem. In Section 29.4 we describe an approximation algorithm that uses decomposition of the alignment graph into slabs to find a sufficiently long alignment with high score (*LAt*). This algorithm is used to obtain a fractional programming approximation algorithm for the *NLAt* problem. This is done in such a way that the length constraint is met within a given tolerance, as described in detail in Section 29.5. The main results and the algorithm descriptions given in the subsequent sections of this chapter are a compilation and a reorganization of the results that appear in References 1–4.

## 29.2 Framework for Pairwise Sequence Comparison

Given two strings $X = x_1 x_2 \ldots x_n$ and $Y = y_1 y_2 \ldots y_m$ with $n \geq m$, we use the *alignment graph* $G_{X,Y}$ to analyze *alignments* between all substrings of $X$, and $Y$. The alignment graph is a directed acyclic graph having $(n+1)(m+1)$ lattice points $(u, v)$ as vertices for $0 \leq u \leq n$, and $0 \leq v \leq m$. Figure 29.1 shows an alignment graph for $x_i \cdots x_k = ATTGT$ and $y_j \cdots y_l = AGGACAT$. Matching diagonal arcs are drawn as solid lines, whereas mismatching diagonal arcs are shown by dashed lines. Dotted lines are used for horizontal and vertical arcs. An example alignment path is shown in Figure 29.1. Labels of the arcs on this path are the corresponding edit operations where $\epsilon$ denotes the null string. An *alignment path* for substrings $x_i \cdots x_k$, and $y_j \cdots y_l$ is a directed path from the vertex $(i-1, j-1)$ to $(k, l)$ in $G_{X,Y}$ where $i \leq k$ and $j \leq l$. To each vertex there is an incoming arc from each neighbor if it exists. Horizontal and vertical arcs correspond to insert and delete operations, respectively. We sometimes use *indel* to refer to an insert or a delete operation. The diagonal arcs correspond to substitutions which are either matching (if the corresponding symbols are the same) or mismatching (otherwise). If we trace the arcs of an alignment path for substrings $I$ and $J$ and perform the indicated edit operations in the given order on $I$, we obtain $J$.

Blocks of insertions and deletions are also referred to as *gaps*. The alignment in Figure 29.1 includes two gaps with sizes 1 and 3. We will use the terms alignment and alignment path interchangeably.

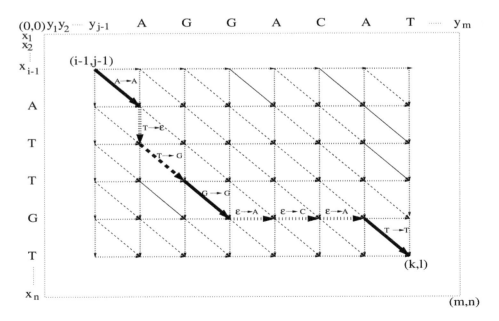

**FIGURE 29.1**    Alignment graph $G_{X,Y}$ where $x_i \cdots x_k = ATTGT$ and $y_j \cdots y_l = AGGACAT$.

The objective of sequence alignment is to quantify the similarity between $X$ and $Y$ under a given *scoring scheme*. In the *simple scoring scheme*, the arcs of $G_{X,Y}$ are assigned weights determined by nonnegative reals $\delta$ (*mismatch penalty*) and $\mu$ (*indel or gap penalty*). We assume that $s(x_i, y_j)$ is the similarity score between the symbols $x_i$ and $y_j$ which is normally 1 for a match ($x_i = y_j$) and $-\delta$ for a mismatch ($x_i \neq y_j$).

Given two strings $X$ and $Y$ the *local alignment* (LA) problem seeks substrings $I \subseteq X$ and $J \subseteq Y$ with the highest similarity score. The optimum value $LA^*(X, Y)$ for this problem is given by

$$LA^*(X, Y) = \max\{s(I, J) \mid I \subseteq X, J \subseteq Y\} \tag{29.1}$$

where $s(I, J)$ is the best alignment score between $I$ and $J$. Alignments have positive scores, or otherwise, they do not exist, that is, $s(I, J) = 0$ iff there is no alignment between $I$ and $J$.

The following is the classical dynamic programming formulation [5] to compute the maximum LA score $\mathcal{S}_{i,j}$ achieved by an optimal LA ending at each vertex $(i, j)$:

$$\mathcal{S}_{i,j} = \max\{0, \mathcal{S}_{i-1,j} - \mu, \mathcal{S}_{i-1,j-1} + s(x_i, y_j), \mathcal{S}_{i,j-1} - \mu\} \tag{29.2}$$

for $1 \leq i \leq n, 1 \leq j \leq m$, with the boundary conditions $\mathcal{S}_{i,j} = 0$ whenever $i = 0$ or $j = 0$. Then

$$LA^*(X, Y) = \max_{i,j} \mathcal{S}_{i,j} \tag{29.3}$$

$LA^*$ can be computed using the Smith–Waterman algorithm [6] in time $O(nm)$. The space complexity is $O(m)$ because only $O(m)$ entries of the dynamic programming matrix need to be stored at any given time.

The simple scoring scheme can be extended such that the scores can vary depending on the individual symbols within the same edit operation type. This leads to arbitrary scoring matrices. In this case, there is a dynamic programming formulation similar to Equation 29.2.

Affine gap penalties is another common scoring scheme in which the total penalty for a gap of size $k$, that is, a block of $k$ insertions (or deletions), is $\alpha + (k - 1)\mu$ where $\alpha$ is the gap open penalty, and $\mu$ is called the gap extension penalty. The dynamic programming formulation for this case can be described as follows [5]: Let $\mathcal{E}_{i,j} = \mathcal{F}_{i,j} = \mathcal{S}_{i,j} = 0$ when $i$ or $j$ is 0, and define

$$\mathcal{E}_{i,j} = \max\{\mathcal{S}_{i,j-1} - \alpha, \mathcal{E}_{i,j-1} - \mu\},$$
$$\mathcal{F}_{i,j} = \max\{\mathcal{S}_{i-1,j} - \alpha, \mathcal{F}_{i-1,j} - \mu\},$$
$$\mathcal{S}_{i,j} = \max\{0, \mathcal{S}_{i-1,j-1} + s(x_i, y_j), \mathcal{E}_{i,j}, \mathcal{F}_{i,j}\} \tag{29.4}$$

By virtue of this formulation, consideration of affine gap penalties does not increase the asymptotic complexity of the LA problem.

We can also express the alignment problems as optimization problems that involve linear functions. In Sections 29.(3-5), we will describe fractional programming algorithms based on these expressions. We define an *alignment vector* as the vector of edit operation frequencies such that the scores, and the lengths of alignments can be expressed as linear functions over alignment vectors. For example, under the basic scoring scheme, we say that $(x, y, z)$ is an alignment vector if there is an alignment path between substrings $I \subseteq X$ and $J \subseteq Y$ with $x$ matches, $y$ mismatches, and $z$ indels. In Figure 29.1, $(3, 1, 4)$ is an alignment vector corresponding to the path shown in the figure. Let $AV$, under a given scoring scheme, denote the set of alignment vectors. Then $s(I, J)$ can be expressed as a linear function $SCORE$ over $AV$ for the scoring schemes we study: the basic scoring scheme, arbitrary scoring matrices, and affine gap penalties. For example when simple scoring is used

$$SCORE(a) = x - \delta y - \mu z \text{ for } a = (x, y, z) \in AV$$

where $x, y$, and $z$ of alignment vector $a$ represent the number of matches, mismatches, and indels, respectively. We can easily verify that also for affine gap penalties, and arbitrary scoring matrices, *SCORE* can be expressed as a linear function.

The LA problem can be rewritten as follows:

$$LA \quad : \quad maximize \; SCORE(a) \quad \text{s.t. } a \in AV$$

## 29.3 Fractional Programming Adjusted Normalized Local Alignment Algorithms

Using length-normalized scores in LA is suggested by Arslan et al. [4] to cope with the mosaic and shadow effects. The objective of the *NLAt* problem [3] is

$$NLAt^*(X, Y) = \max\{s(I, J)/(|I| + |J|) \mid I \subseteq X, J \subseteq Y, |I| + |J| \geq t\}. \tag{29.5}$$

To solve the *NLAt* problem we can extend the dynamic programming formulation for the scoring schemes that we address in this chapter by adding another dimension. At each entry of the dynamic programming matrix we can store optimum scores for all possible alignment lengths up to $m + n$. This increases the time and space complexity to $O(n^2 m)$ and $O(nm)$, respectively. These are unacceptably high because in practice, the values of both $n$ and $m$ may be in the order of millions.

The length of an alignment can appropriately be defined as the sum of the lengths of the substrings involved in the alignment. For an alignment vector $a \in AV$, the length of the corresponding alignment can be expressed as a linear function *LENGTH*. For example when the simple scoring scheme is used

$$LENGTH(a) = 2x + 2y + z \text{ for } a = (x, y, z) \in AV$$

where $x, y, z$ represent the number of matches, mismatches, and indels, respectively. We can easily see that for affine gap penalties, and arbitrary scoring matrices *LENGTH* can be expressed as a linear function. We assume that only the matches have nonnegative scores therefore on any alignment the score cannot exceed the length.

The objective of *NLAt* may be achieved by a reformulation. In *ANLA* problem, we can modify the maximization ratio function in such a way that we drop the length constraint, yet achieve a similar objective: to obtain sufficiently long alignments with a high degree of similarity. The adjusted length normalized score of an alignment is computed by adding some parameter $L \geq 0$ to the denominator in the calculation of the quotient of ordinary scores by the length. Thus the *ANLA* problem [3] is a variant of the normalized local alignment problem in which the length constraint is dropped, and the optimization function is modified by adding a parameter $L$ to the denominator:

$$ANLA^*(X, Y) = \max\{s(I, J)/(|I| + |J| + L) \mid I \subseteq X, J \subseteq Y, L \geq 0\} \tag{29.6}$$

The *ANLA* problem can be rewritten as follows:

$$ANLA \quad : \quad maximize \; \frac{SCORE(a)}{LENGTH(a) + L} \quad \text{s.t. } a \in AV$$

For *ANLA* faster algorithms are possible using *fractional programming* technique. The provable time complexity of the *ANLA* problem for rational weights is $O(nm \log n)$, as we discuss later. Test results of a fractional programming based approach suggests that the time complexity is $O(nm)$, although this result is empirical. Compared with $O(n^2 m)$ time complexity of a naive dynamic programming algorithm for the *NLAt* problem, the *ANLA* problem can be solved much faster.

Fractional programming *ANLA* algorithms [3] use the *parametric method.* They iteratively solve a so-called *parametric problem* $LA_\lambda$ which is the following optimization problem: for a given $\lambda$

$$LA_\lambda^*(X, Y) = \max\{s(I, J) - \lambda(|I| + |J| + L) \mid I \subseteq X, J \subseteq Y\} \quad (29.7)$$

$LA_\lambda(X, Y)$ can also be written as

$$LA(\lambda) \quad : \quad maximize\ SCORE(a) - \lambda\ LENGTH(a) - \lambda L \quad \text{s.t. } a \in AV$$

**Proposition 29.1** [3] *For $\lambda < \frac{1}{2}$, the optimum value $LA^*(\lambda)$ of the parametric LA problem can be formulated in terms of the optimum value $LA^*$ of an LA problem.*

*Proof.* Under the simple scoring scheme the optimum value of the parametric problem, when $\lambda < \frac{1}{2}$, is

$$LA_{\delta,\mu}^*(\lambda) = (1 - 2\lambda)LA_{\delta',\mu'}^* - \lambda L \text{ where } \delta' = \frac{\delta + 2\lambda}{1 - 2\lambda}, \ \mu' = \frac{\mu + \lambda}{1 - 2\lambda}. \quad (29.8)$$

We can easily verify that a similar relation exists in the case of arbitrary scoring matrices, and affine gap penalties. Thus, computing $LA^*(\lambda)$ involves solving the local alignment problem $LA$, and performing some simple arithmetic afterward. $\blacksquare$

We assume without loss of generality that for any alignment the score does not exceed the number of matches. Therefore for any alignment, its normalized score $\lambda \le \frac{1}{2}$. We consider $\lambda = \frac{1}{2}$ as a special case as it can only happen when the alignment is composed of matches only, and $L = 0$.

The thesis of the parametric method of fractional programming is that the optimum solution to the original problem that involves a ratio of two functions can be obtained via optimal solutions of the parametric problem. In this case, an optimal solution to a ratio optimization problem *ANLA* can be achieved via a series of optimal solutions of the parametric problem $LA(\lambda)$ with different parameters $\lambda$. In fact, $\lambda = ANLA^*$ iff $LA^*(\lambda) = 0$. That is, an alignment vector $v \in AV$ has the optimum adjusted normalized score $\lambda$ iff $v$ is an optimal alignment vector for the parametric problem $LA(\lambda)$ with optimum value zero. (See Reference 3 for more details, also see References 7, 8 for many interesting properties of fractional programming). The *Dinkelbach* algorithm for the *ANLA* problem is shown in Figure 29.2. Solutions of the parametric problems through the iterations yield improved (higher) values to $\lambda$ except for the last iteration in which $\lambda$ remains the same, and becomes the optimum value. In fractional programming algorithms, convergence to an optimal result is guaranteed: In infinite sets the convergence to optimum is superlinear. In finite sets the termination is guaranteed. In the case of *ANLA Dinkelbach* algorithm, when the algorithm terminates, the final alignment is optimal with respect to both the ordinary scores used at that iteration, and the adjusted length normalized scoring with the original scores. This mimics manually changing the scores until the result is satisfactory.

As reported by Arslan et al. [3], experiments suggest that the number of iterations in the algorithm is a small constant: 3–5 on average. However, a theoretical tight bound is yet to be established. If we assume

```
Algorithm Dinkelbach
Pick an arbitrary alignment, and let λ* be the adjusted length normalized
score of this alignment
Repeat
    λ ← λ*
    Solve LA(λ) and let λ* be the adjusted length normalized score of
an optimal alignment
Until λ* = λ
Return(λ*)
```

**FIGURE 29.2** Dinkelbach algorithm for *ANLA*. (From Arslan, A.N. et al., *Bioinformatics*, 17, 327, 2001. With Permission.)

```
Algorithm RationalANLA
Let σ be the smallest gap between two adjusted length normalized scores
Initialize [e, f] ← [0, ½σ⁻¹]
While (e + 1 < f) do
    k ← ⌊(e + f)/2⌋
    If LA*(kσ) > 0  then e ← k else f ← k
End {while}
Return(eσ)
```

**FIGURE 29.3**   *ANLA* algorithm `RationalANLA` for rational scores. (From Arslan, A.N. et al., *Bioinformatics*, 17, 327, 2001. With Permission.)

that the sequences involved in alignments are fixed (for example consider the normalized global alignment), and the simple scoring scheme is used then the number of iterations is bounded by the size of the convex hull of lattice points whose diameter is bounded by the length of the strings. In this case, each parametric problem is optimized at one of the extreme points of the convex hull, and each extreme point is visited at most once during the iterations. It is known that the size of a convex hull of diameter $N$ is $O(N^{2/3})$ [3]. Even this rough estimate shows that the algorithm in the worst case is better than the straightforward dynamic programming extension for *ANLA*. Furthermore, it can be shown that problem *ANLA* can be reduced to a *0-1 fractional programming problem* for which an upper bound on the number of iterations of parametric problem is established [9]. The $0-1$ fractional programming problem optimizes the fractional objective function $(c_1x_1 + c_2x_2 + \cdots + c_qx_q)/(d_1x_1 + d_2x_2 + \cdots + d_qx_q)$ under the condition that $(x_1, x_2, \ldots, x_q) \subseteq \{0,1\}^*$ is in the set of feasible solutions where $c_i, d_i$ denote real coefficients for all $i = 1, 2, \ldots, q$. It is shown in Reference 9 that Dinkelbach's algorithm for the $0-1$ fractional problem solves at most $O(\log(qM))$ subproblems in the worst case, where $M = \max\{\max_{i=1,2,\ldots,q} |c_i|, \max_{i=1,2,\ldots,q} |d_i|, 1\}$.

Problem *ANLA* can be reduced to a $0-1$ fractional programming problem by defining the score and length functions for alignments over a larger vector space in which each element is in $\{0,1\}$. For example by associating each arc in the edit graph $G_{X,Y}$ by a separate $x_i$, both score and length can be written as linear functions (with coefficients bounded by constants) over vectors of size $(x_1, x_2, \ldots, x_q)$ where $q = O(nm)$ and each $x_i \in \{0,1\}$. There exists a similar formulation for the case of affine gap penalties, too. The implication of the complexity result in Reference 9 is that the Dinkelbach algorithm for *ANLA* takes $O(\log n)$ iterations, and therefore, its total time complexity is $O(nm \log n)$.

In practice scores are rational, and in the case of rational scores there is another provably good result [3], which is achieved by Algorithm *RationalANLA* given in Figure 29.3. The algorithm uses Megiddo's technique [10] to perform a binary search for optimum adjusted normalized score over an interval of integers. The search is based on the sign of the optimum value of the parametric problem. In this case, if $LA^*(\lambda) = 0$, then $\lambda = ANLA^*$, and an optimal alignment vector of $LA(\lambda)$ is also an optimal solution of *ANLA*. On the other hand, if $LA^*(\lambda) > 0$, then a larger $\lambda$, and if $LA^*(\lambda) < 0$, then a smaller $\lambda$ should be tested (i.e., Problem $LA(\lambda)$ should be solved with a different value of $\lambda$). When the scores are rational numbers the effective search space includes $O(n^2)$ integers because the gap between any two distinct length normalized score is $\Omega(1/n^2)$. The algorithm solves $O(\log n)$ parametric problems. Therefore the resulting time complexity is $O(nm \log n)$, and the space complexity is $O(m)$.

## 29.4 Approximation Algorithms for Partial Constraint Satisfaction

In Table 29.1, we list several local alignment problems with length constraint. For these problems there are approximation algorithms that guarantee the satisfaction of the constraints partially, that is, they return alignments whose lengths are within a given tolerance of the required length.

These algorithms decompose the alignment graph into slabs. The length-restricted local alignment (*LRLA*) problem [4] is suggested to find alignments with optimal score over the alignments that involve substrings of up to a given length. The length limit is only on the substrings of one of the strings. The approximation algorithms for this problem imagine that the alignment graph is partitioned into vertical slabs. The results for this problem are summarized in Table 29.1. The cyclic local alignment *CLA* [4] is a special case of the *LRLA* problem. In the *CLA* problem the length constraint is implicit as shown in the table. The *LRLA* algorithms and results are applicable to the *CLA* problem, too.

We omit the details of *LRLA* approximation algorithms. Instead we describe another algorithm which is also based on the decomposition of the alignment graph into slabs. This algorithm is for the length-constrained local alignment problem *LAt* [1] (Table 29.1).

For a given $t$, we define the *local alignment with length threshold* score between $X$ and $Y$ as

$$LAt^*(X, Y) = \max\{s(I, J) \mid I \subseteq X, J \subseteq Y, \text{ and } |I| + |J| \geq t\} \tag{29.9}$$

Equivalently

$$LAt \quad : \quad maximize\ SCORE(a) \qquad \text{s.t. } a \in AVt$$

where $AVt \subseteq AV$ be a set of alignment vectors corresponding to alignments with length $\geq t$.

Although the problem itself is not very interesting, an algorithm for the problem can be used to find a long alignment with length normalized score $> \lambda$ for a given positive $\lambda$, which is a practical query problem *Qt* included in Table 29.1. We also show that the algorithm for the local alignment with length threshold leads to improved approximation algorithms for the *NLAt* problem (Section 29.5).

To solve *LAt* we can extend the dynamic programming formulation in Equation 29.2 by adding another dimension. At each entry of the dynamic programming matrix we store optimum scores for all possible lengths up to $m + n$, increasing the time and space complexity to $O(n^2m)$ and $O(nm)$, respectively.

We describe an approximation algorithm *APX-LAt* [1] which computes a local alignment whose score is at least $LAt^*$, and whose length is at least $(1 - \frac{1}{r})t$ provided that the *LAt* problem is feasible, that is, the algorithm finds two substrings $\widehat{I} \subseteq X$, and $\widehat{J} \subseteq Y$ such that $s(\widehat{I}, \widehat{J}) \geq LAt^*$ and $|\widehat{I}| + |\widehat{J}| \geq (1 - \frac{1}{r})t$. The algorithm runs in time $O(rnm)$ using $O(rm)$ space. For simplicity, we assume the simple scoring scheme. Instead of a single score, we maintain at each node $(i, j)$ of $G_{X,Y}$, a list of alignments with the property that for positive $s$ where $s$ is the optimum score achievable over the set of alignments with length $\geq t$ and ending at $(i, j)$, at least one element of the list achieves score $s$ and length $t - \Delta$ where $\Delta$ is a positive integral parameter. We show that the dynamic programming formulation can be extended to preserve this property through the nodes. In particular, an alignment with score $\geq LAt^*$, and length $\geq t - \Delta$ will be observed in one of the nodes $(i, j)$ during the computations. We imagine the vertices of $G_{X,Y}$ as grouped into $\lfloor (n + m)/\Delta \rfloor$ diagonal slabs at distance $\Delta$ from each other as shown in Figure 29.4.

As we define the length of an alignment as the sum of the lengths of the substrings involved in the alignment, on a given alignment, the contribution of each diagonal arc to the alignment length is 2 (each match, or mismatch involves two symbols, one from each sequence), whereas that of each horizontal or vertical arc is 1 (each indel involves one symbol from one of the sequences). Equivalently, we say that the length of a diagonal arc is 2, and the length of each horizontal, or vertical arc is 1. The length of an alignment $a$ is the total length of the arcs on $a$. Each slab consists of $\lfloor \Delta /2 \rfloor + 1$ diagonals. Two consecutive slabs share a diagonal that we call a *boundary*. The *left* and the *right boundaries* of slab $b$ are, respectively, the boundaries shared by the left and right neighboring slabs of $b$. As a subgraph, a slab contains all the edges in $G_{X,Y}$ incident to the vertices in the slab except for the horizontal and vertical edges incident to the vertices on the left boundary (which belong to the preceding slab), and the diagonal edges incident to the vertices on the first diagonal following the left boundary.

Now to a given diagonal $d$ in $G_{X,Y}$, we associate a number of slabs as follows. Let *slab 0 with respect to diagonal d* be the slab that contains the diagonal $d$ itself. The slabs to the left of *slab 0* are then ordered consecutively as *slab 1, slab 2, . . .* with respect to $d$. In other words, *slab k* with respect to diagonal $d$ is the

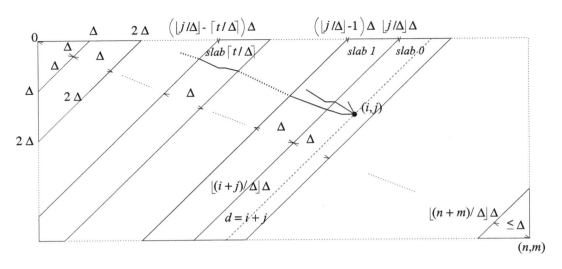

**FIGURE 29.4**   Slabs with respect to diagonal $d$ and alignments ending at node $(i,j)$ starting at different slabs.

subgraph of $G_{X,Y}$ composed of vertices placed inclusively between diagonals $\lfloor d/\Delta \rfloor$ and $d$ if $k = 0$, and between diagonal $(\lfloor d/\Delta \rfloor - k)\Delta$ and $(\lfloor d/\Delta \rfloor - k+1)\Delta$, otherwise. Figure 29.4 includes sample slabs with respect to diagonal $d$, and alignments ending at some node $(i,j)$ on this diagonal.

Let $\mathcal{S}_{i,j,k}$ represent the optimum score achievable at $(i,j)$ by any alignment starting at slab $k$ with respect to diagonal $i + j$ for $0 \leq k < \lceil t/\Delta \rceil$. For $k = \lceil t/\Delta \rceil$, $\mathcal{S}_{i,j,k}$ is slightly different: It is the maximum of all achievable scores by an alignment starting in or before slab $k$. Moreover, let $\mathcal{L}_{i,j,k}$ be the length of an optimal alignment starting at slab $k$ and achieving score $\mathcal{S}_{i,j,k}$. A single slab can contribute at most $\Delta$ to the length of any alignment. We store at each node $(i,j)$ $\lceil t/\Delta \rceil + 1$ score-length pairs $(\mathcal{S}_{i,j,k}, \mathcal{L}_{i,j,k})$ for $0 \leq k \leq \lceil t/\Delta \rceil$ corresponding to $\lceil t/\Delta \rceil + 1$ optimal alignments that end $(i,j)$. Figure 29.5 shows the steps of the algorithm *APX-LAt*. The processing is done row-by-row starting with the top row ($i = 0$) of $G_{X,Y}$.

Step 1 of the algorithm performs the initialization of the lists of the nodes in the top row ($i = 0$). Step 2 implements computation of scores as dictated by the dynamic programming formulation in Equation 29.2. Let maxp of a list of score-length pairs be a pair with the maximum score in the list. We obtain an optimal alignment with score $\mathcal{S}_{i,j,k}$ by extending an optimal alignment from one of the nodes $(i - 1,j)$, $(i - 1,j - 1)$, or $(i,j - 1)$. We note that extending an alignment at $(i,j)$ from node $(i - 1,j - 1)$ increases the length by 2 and the score by $s(x_i, y_j)$, whereas from nodes $(i - 1,j)$ or $(i,j - 1)$ adds 1 to the length and $-\mu$ to the score of the resulting alignment. There are two cases:

**(Case 1)** If the current node $(i,j)$ is not on the first diagonal after a boundary then nodes $(i - 1,j)$, $(i - 1,j - 1)$, and $(i,j - 1)$ share the same slabs with node $(i,j)$. In this case, $(\mathcal{S}_{i,j,k}, \mathcal{L}_{i,j,k})$ is calculated by using $(\mathcal{S}_{i-1,j,k}, \mathcal{L}_{i-1,j,k})$, $(\mathcal{S}_{i-1,j-1,k}, \mathcal{L}_{i-1,j-1,k})$, and $(\mathcal{S}_{i,j-1,k}, \mathcal{L}_{i,j-1,k})$ as shown in Step 2.*b* where $(\mathcal{S}_{i-1,j-1,k}, \mathcal{L}_{i-1,j-1,k}) \oplus (s(x_i, y_j), 2) = (\mathcal{S}_{i-1,j-1,k} + s(x_i, y_j), \mathcal{L}_{i-1,j-1,k} + 2)$ if $\mathcal{S}_{i-1,j-1,k} > 0$ or $k = 0$; and $(0, 0)$ otherwise. This is because, by definition, every local alignment has a positive score, and it is either a single match, or it is an extension of an alignment whose score is positive. Therefore, we do not let an alignment with no score be extended unless the resulting alignment is a single match in the current slab.

**(Case 2)** If the current node is on the first diagonal following a boundary (i.e., $i + j \bmod \Delta = 1$) then the slabs for the nodes involved in the computations for node $(i,j)$ differ as shown in Figure 29.6. In this case, slab $k$ for node $(i,j)$ is slab $k - 1$ for nodes $(i - 1,j)$, $(i - 1,j - 1)$, and $(i,j - 1)$. Moreover, any alignment ending at $(i,j)$ starting at slab 0 for $(i,j)$ can only include one of the edges $((i - 1,j),(i,j))$ or $((i - 1,j - 1),(i,j))$ both of which have negative weight $-\mu$. Therefore, $(\mathcal{S}_{i,j,0}, \mathcal{L}_{i,j,0})$ is set to $(0, 0)$. Steps 2.*a*.1 and 2.*a*.2 show the calculation of $(\mathcal{S}_{i,j,k}, \mathcal{L}_{i,j,k})$, respectively, for $0 < k < \lceil t/\Delta \rceil$ and for $k = \lceil t/\Delta \rceil$.

Algorithm *APX-LAt*$(\delta, \mu)$

1. Initialization:  set $\widehat{LAt} = 0$;  and  $(\mathcal{S}_{0,j,k}, \mathcal{L}_{0,j,k}) = (0,0)$ for all $j, k$,  $0 \leq j \leq m$,  and  $0 \leq k \leq \lceil t/\Delta \rceil$

2. Main computations:

  for $i = 1$ to $n$ do {

    set $(\mathcal{S}_{i,0,k}, \mathcal{L}_{i,0,k}) = (0,0)$ for all $k$, $0 \leq k \leq \lceil t/\Delta \rceil$

    for $j = 1$ to $m$ do {

      if $(i + j \bmod \Delta = 1)$ then {

        set $(\mathcal{S}_{i,j,0}, \mathcal{L}_{i,j,0}) = (0,0)$

        for $k = 1$ to $\lceil t/\Delta \rceil - 1$ do

2.a.1          set $(\mathcal{S}_{i,j,k}, \mathcal{L}_{i,j,k}) = \text{maxp}\{\ (0,0),\ (\mathcal{S}_{i-1,j,k-1}, \mathcal{L}_{i-1,j,k-1}) + (-\mu, 1),$
$(\mathcal{S}_{i-1,j-1,k-1}, \mathcal{L}_{i-1,j-1,k-1}) \oplus (s(x_i, y_j), 2),$
$(\mathcal{S}_{i,j-1,k-1}, \mathcal{L}_{i,j-1,k-1}) + (-\mu, 1)\ \}$

        for $k = \lceil t/\Delta \rceil$

2.a.2          set $(\mathcal{S}_{i,j,k}, \mathcal{L}_{i,j,k}) = \text{maxp}\{\ (0,0),\ (\mathcal{S}_{i-1,j,k-1}, \mathcal{L}_{i-1,j,k-1}) + (-\mu, 1),$
$(\mathcal{S}_{i-1,j-1,k-1}, \mathcal{L}_{i-1,j-1,k-1}) \oplus (s(x_i, y_j), 2),$
$(\mathcal{S}_{i,j-1,k-1}, \mathcal{L}_{i,j-1,k-1}) + (-\mu, 1),\ (\mathcal{S}_{i-1,j,k}, \mathcal{L}_{i-1,j,k}) + (-\mu, 1),$
$(\mathcal{S}_{i-1,j-1,k}, \mathcal{L}_{i-1,j-1,k}) \oplus (s(x_i, y_j), 2),\ (\mathcal{S}_{i,j-1,k}, \mathcal{L}_{i,j-1,k}) + (-\mu, 1)\ \}$

      } else {

        for $k = 0$ to $\lceil t/\Delta \rceil$ do

2.b          set $(\mathcal{S}_{i,j,k}, \mathcal{L}_{i,j,k}) = \text{maxp}\{\ (0,0),\ (\mathcal{S}_{i-1,j,k}, \mathcal{L}_{i-1,j,k}) + (-\mu, 1),$
$(\mathcal{S}_{i-1,j-1,k}, \mathcal{L}_{i-1,j-1,k}) \oplus (s(x_i, y_j), 2),\ (\mathcal{S}_{i,j-1,k}, \mathcal{L}_{i,j-1,k}) + (-\mu, 1)\ \}$

      }

    for $k = \lceil t/\Delta \rceil - 1$ if $\mathcal{L}_{i,j,k} \geq t - \Delta$ then set $\widehat{LAt} = \max\{\widehat{LAt}, \mathcal{S}_{i,j,k}\}$

    for $k = \lceil t/\Delta \rceil$ set $\widehat{LAt} = \max\{\widehat{LAt}, \mathcal{S}_{i,j,k}\}$

  } }

3. Return $\widehat{LAt}$

**FIGURE 29.5**  Algorithm *APX-LAt*. (From Arslan, A.N. and Eğecioğlu, Ö., *INFORMS J. Comput.*, 16, 441, 2004. With Permission.)

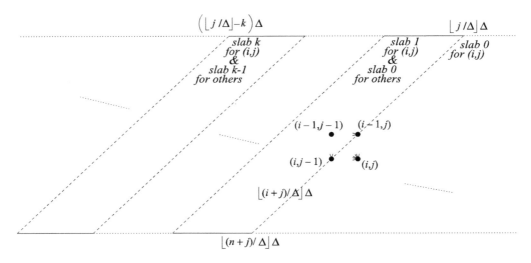

**FIGURE 29.6** Relative numbering of the slabs with respect to $(i,j)$, $(i-1,j)$, $(i-1,j-1)$, and $(i,j-1)$ when node $(i,j)$ is on the first diagonal following boundary $\lfloor (i+j)/\Delta \rfloor$.

The running maximum score $\widehat{LAt}$ is updated whenever a newly computed score for an alignment with length $\geq t - \Delta$ is larger than the current maximum which can only happen with alignments starting in or before slab $\lceil t/\Delta \rceil - 1$. The final value $\widehat{LAt}$ is returned in Step 3. The alignment position achieving this score may also be desired. This can be done by maintaining for each optimal alignment a start and end position information besides its score and length. In this case, in addition to the running maximum score, the start and end positions of a maximal alignment should be stored and updated.

We first show that $S_{i,j,k}$ calculated by the algorithm is the optimum score achievable and $\mathcal{L}_{i,j,k}$ is the length of an alignment achieving this score over the set of all alignments ending at node $(i,j)$ and starting with respect to diagonal $i+j$: 1) at slab $k$ for $0 \leq k < \lceil t/\Delta \rceil$, 2) in or before slab $k$ for $k = \lceil t/\Delta \rceil$. This claim can be proved by induction. If we assume that the claim is true for nodes $(i-1,j)$, $(i-1,j-1)$ and $(i,j-1)$, and for their slabs, then we can easily see by following Step 2 of the algorithm that the claim holds for node $(i,j)$ and its slabs.

Let optimum score $LAt^*$ for the alignments of length $\geq t$ be achieved at node $(i,j)$. Consider the calculations of the algorithm at $(i,j)$ at which an optimal alignment ends. There are two possible orientations of an optimal alignment as shown in Figure 29.7: (1) It starts at some node $(i',j')$ of slab $k = \lceil t/\Delta \rceil - 1$. By a previous claim an alignment starting at slab $k$ with score $S_{i,j,k} \geq LAt^*$ is captured in Step 2. The length of this alignment $\mathcal{L}_{i,j,k}$ is at least $t - \Delta$ as the length of the optimal alignment is $\geq t$, and both start at the same slab and end at $(i,j)$. (2) It starts at some node $(i'',j'')$ in or before slab $k = \lceil t/\Delta \rceil$. Again by the previous claim an alignment starting in or before slab $k$ with score $S_{i,j,k} \geq LAt^*$ is captured in Step 2. The length of this alignment $\mathcal{L}_{i,j,k}$ is at least $t - \Delta$ as slab $k$ is at distance $\geq t - \Delta$ from $(i,j)$. Therefore, the final value $\widehat{LAt}$ returned in Step 3 is $\geq LAt^*$ and it is achieved by an alignment whose length is $\geq t - \Delta$. We summarize these results in the following theorem.

**Theorem 29.1** [1] *For a feasible LAt problem, Algorithm APX-LAt returns an alignment $(\widehat{I},\widehat{J})$ such that $s(\widehat{I},\widehat{J}) \geq LAt^*$ and $|\widehat{I}| + |\widehat{J}| \geq (1 - \frac{1}{r})t$ for any $r > 1$. The algorithm's complexity is $O(rnm)$ time and $O(rm)$ space.*

*Proof.* Algorithm *APX-LAt* is similar to the Smith-Waterman algorithm except that at each node instead of a single score, $\lceil t/\Delta \rceil + 1$ entries for score-length pairs are stored and manipulated. Therefore the resulting complexity exceeds that of the Smith-Waterman algorithm by a factor of $\lceil t/\Delta \rceil + 1$. That is, the time complexity of *APX-LAt* is $O(nmt/\Delta)$. The algorithm requires $O(mt/\Delta)$ space as the computations

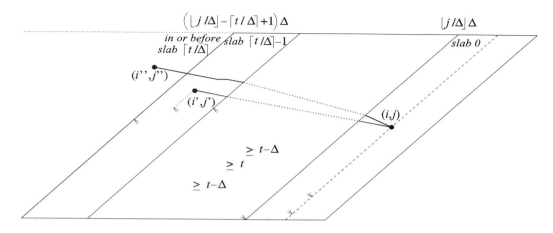

**FIGURE 29.7** Two possible orientations of an optimal alignment of length $\geq t$ ending at $(i, j)$: It starts either at some $(i', j')$ at slab $\lceil t/\Delta \rceil - 1$, or $(i'', j'')$ in or before slab $\lceil t/\Delta \rceil$.

proceed row by row, and we need the entries in the previous and the current row to calculate the entries in the current row. When the *LAt* problem is feasible, it is guaranteed that Algorithm *APX-LAt* returns an alignment $(\widehat{I}, \widehat{J})$ such that $s(\widehat{I}, \widehat{J}) \geq LAt^* > 0$ and $|\widehat{I}| + |\widehat{J}| \geq t - \Delta$ for any positive $\Delta$. Therefore setting $\Delta = \max\{2, \lfloor t/r \rfloor\}$ for a choice of $r$, $1 < r \leq t$, and using Algorithm *APX-LAt* we can achieve the approximation and complexity results expressed in the theorem. We also note that for $\Delta = 2$ the algorithm becomes a dynamic programming algorithm extending the dimension by storing all possible alignment lengths. ∎

A variant of *APX-LAt* for arbitrary scoring matrices can be obtained by simple modifications: At each entry of the dynamic programming matrix, instead of a single score a number of scores (and lengths) are maintained and manipulated as dictated by the underlying dynamic programming formulation (e.g., Equation 29.4).

An application of the *LAt* problem is on problem *Qt* which is defined as the problem of finding two subsequences with normalized score higher than $\lambda$, and total length at least $t$. More formally

$$Qt : \text{ find } (I, J) \text{ such that } I \subseteq X, J \subseteq Y, \frac{s(I, J)}{|I| + |J|} > \lambda \text{ and } |I| + |J| \geq t \qquad (29.10)$$

The following simple query explains the motivation for this problem: "Do two sequences share a (sufficiently long) fragment with more than 70% of similarity?"

The problem is feasible for given thresholds $t$, and $\lambda > 0$, if the answer to this query is not empty, that is, there exists a pair of subsequences $I$ and $J$ with total length $|I| + |J| \geq t$, and normalized score $s(I, J)/(|I| + |J|) > \lambda$. Note that *Qt* is feasible iff $NLAt^* > \lambda$. We describe an algorithm which returns for a feasible problem finds two subsequences $\widehat{I} \subseteq X$, and $\widehat{J} \subseteq Y$ with normalized score higher than $\lambda$, and total length $|\widehat{I}| + |\widehat{J}| \geq (1 - \frac{1}{r})t$. The approximation ratio is controlled by parameter $r$. The computations take $O(rnm)$ time and $O(rm)$ space.

For a given $\lambda$, we define *the parametric local alignment with length threshold problem* $LAt(\lambda)$

$$LAt(\lambda) \quad : \quad maximize \ SCORE(a) - \lambda \ LENGTH(a) \qquad \text{s.t. } a \in AVt$$

**Proposition 29.2 [1]** *For $\lambda < \frac{1}{2}$, the optimum value $LAt^*(\lambda)$ of the parametric LAt problem can be formulated in terms of the optimum value $LAt^*$ of an LAt problem.*

*Proof.* The proof is very similar to that of Proposition 29.1. Under the simple scoring scheme the optimum value of the parametric problem, when $\lambda < \frac{1}{2}$, is

$$LAt^*_{\delta,\mu}(\lambda) = (1 - 2\lambda)LAt^*_{\delta',\mu'} \text{ where } \delta' = \frac{\delta + 2\lambda}{1 - 2\lambda}, \ \mu' = \frac{\mu + \lambda}{1 - 2\lambda}. \tag{29.11}$$

We can easily see that a similar relation exists in the case of arbitrary scoring matrices, and affine gap penalties. Computing $LAt^*(\lambda)$ involves solving the local alignment with length threshold problem $LAt$, and performing some simple arithmetic afterward. ∎

Under the scoring schemes we study we assume without loss of generality that for any alignment, its normalized score is $\leq \frac{1}{2}$. We consider $\lambda = \frac{1}{2}$ as a special case which can only happen when the alignment is composed of matches only.

**Proposition 29.3** [1] *When solving $LAt(\lambda)$, the underlying algorithm for $LAt$ returns an alignment $(\widehat{I},\widehat{J})$ with normalized score higher than $\lambda$, and $|\widehat{I}| + |\widehat{J}| \geq (1 - \frac{1}{r})t$ if Problem Qt is feasible.*

*Proof.* Assume that Problem $Qt$ is feasible. Then $LAt^*(\lambda) > 0$ which implies that the algorithm which solves the corresponding $LAt$ problem (of Proposition 29.3) returns an alignment $(\widehat{I},\widehat{J})$ such that its score is positive (i.e., $s(\widehat{I},\widehat{J}) - \lambda(|\widehat{I}| + |\widehat{J}|) > 0$) and $|\widehat{I}| + |\widehat{J}| \geq (1 - \frac{1}{r})t$ by the approximation results of Algorithm *APX-LAt*. ∎

Thus solving $Qt$ requires a single application of Algorithm *APX-LAt*.

## 29.5 Normalized Local Alignment

Need for a length constraint is clear when length-normalized scores are used because shorter alignments may have high normalized scores but they may not be biologically significant. The definition of the *NLAt* problem contains a length constraint as described in Section 29.3.

Let $AVt \subseteq AV$ be a set of alignment vectors corresponding to alignments with length $\geq t$. The *NLAt* problem can be rewritten as follows:

$$NLAt \quad : \quad maximize \ \frac{SCORE(a)}{LENGTH(a)} \quad \text{s.t. } a \in AVt$$

We present approximation algorithms for the *NLAt* problem that apply fractional programming and use Algorithm *APX-LAt* as a subroutine. The approximation is in the sense that the length constraint is partially satisfied. These algorithms are the *Dinkelbach* algorithm for *NLAt*, and Algorithm *RationalNLAt*. Both algorithms obtain an alignment whose score is no smaller than the optimum score *NLAt** of the original *NLAt* problem, and whose length is at least $(1 - \frac{1}{r})t$ for a given $r$ provided that the original *NLAt* problem is feasible (Theorem 29.2). Algorithm *APX-RationalNLAt* (Figure 29.8) and the Dinkelbach algorithm for *RationalNLAt* (Figure 29.9) are similar to the corresponding *ANLA* algorithms except that they iteratively solve *LAt* problems presented in Section 29.4 instead of *LA* problems. The approximation algorithm *APX-LAt* can be applied to solving the parametric problems that arise in computing *NLAt**.

In both resulting algorithms the space complexity is $O(rm)$. The observed time complexity of the *Dinkebach* algorithm for *NLAt* is $O(rnm)$ (in tests [2], it performs always smaller than 10, and on average 3–5 invocations to Algorithm *APX-LAt*). Algorithm *RationalNLAt* has proven time complexity $O(rnm \log n)$ as in this algorithm $O(\log n)$ invocations of *APX-LAt* is sufficient to solve the *NLAt* problem. The total time complexity of the Dinkelbach algorithm for the *NLAt* problem is also $O(rnm \log n)$ by using the fact that *NLAt* can be formulated as a 0-1 fractional programming problem, and by using the complexity result in Reference 9.

We reiterate the definitions of the local alignment with length threshold *LAt*, *NLAt*, and the parametric local alignment *LAt(λ)* problems as the following optimization problems defined in terms of *SCORE* and

```
Algorithm APX-RationalNLAt
If there is an exact match of size (1 − 1/r)t then return(1/2) and exit
Let σ be the smallest gap between two length normalized scores
[e, f] ← [0, 1/2 σ⁻¹]
λ* ← 0
While (e + 1 < f) do
  k ← ⌈(e + f)/2⌉
  APX-LAt*(kσ) > 0 then {
    e ← k
    λ* ← the normalized score of an optimal alignment obtained
  } else f ← k
End {while}
Return(λ*)
```

**FIGURE 29.8** Algorithm *APX-RationalNLAt* for rational scores. (From Arslan, A.N. and Eğecioğlu, Ö., *INFORMS J. Comput.*, 16, 441, 2004. With Permission.)

```
Algorithm Dinkelbach
If APX-LAt*(0) > 0 then set λ* to the length normalized score of an
optimal alignment else exit
Repeat
  λ ← λ*
  If APX-LAt*(λ) > 0 then set λ* to the length normalized score of an
optimal alignment
Until λ* ≤ λ
Return(λ*)
```

**FIGURE 29.9** Dinkelbach algorithm for *NLAt*. (From Arslan, A.N. and Eğecioğlu, Ö., *INFORMS J. Comput.*, 16, 441, 2004. With Permission.)

*LENGTH* functions that are linear over *AVt* under the scoring schemes we study:

$$
\begin{aligned}
LAt \quad &: \quad maximize \; SCORE(a) &&\text{s.t. } a \in AVt \\
NLAt \quad &: \quad maximize \; \frac{SCORE(a)}{LENGTH(a)} &&\text{s.t. } a \in AVt \\
LAt(\lambda) \quad &: \quad maximize \; SCORE(a) - \lambda \, LENGTH(a) &&\text{s.t. } a \in AVt
\end{aligned}
$$

If we apply the fractional programming to the *NLAt* computation then we can obtain an optimal solution to *NLAt* via a series of optimal solutions of the parametric problem with different parameters $LAt(\lambda)$ such that $\lambda = NLAt^*$ iff $LAt^*(\lambda) = 0$.

**Theorem 29.2** [1] *If NLAt\* > 0 then an alignment with normalized score at least NLAt\*, and total length at least* $(1 - \frac{1}{r})t$ *can be computed for any* $r > 1$ *in time* $O(rnm \log n)$ *and space* $O(rm)$.

*Proof.* Algorithm *RationalNLAt* given in Figure 29.8 accomplishes this. The algorithm is based on a binary search for optimum normalized score over an interval of integers. This takes $O(\log n)$ parametric problems to solve. The algorithm is similar to the *RationalANLA* algorithm in Figure 29.3, and the results are derived similarly. It first determines if there is an exact match of size $(1 - \frac{1}{r})t$, which can easily be done by using the Smith-Waterman algorithm. If the answer is yes, then the algorithm returns the maximum possible normalized score and exits. The skeleton of the rest of the algorithm is the same as Algorithm *RationalNLAt* in Figure 29.3, based on Megiddo's search technique [10]. The difference is that the parametric alignment problems now have a length constraint. The algorithm computes the smallest possible gap $\sigma$ between any two distinct possible normalized scores, which is $\Omega(1/(n+m)^2)$ [3]. It maintains an interval $[e, f]$, on which a binary search is done to find the largest $\lambda$ for which $LAt^*(\lambda)$ is positive where $e$, and $f$ are integer variables. Initially $e$ is set to zero, and $f$ is set to $\frac{1}{2}\sigma^{-1}$ as $NLAt^*$ is in $[0, \frac{1}{2}]$. A parametric $LAt$ problem with parameter $k\sigma$ is iteratively solved, where $k$ is the median of integers in $[e, f]$. At each iteration, the interval is updated according to the sign of the value of the parametric problem. The effective search space is the integers in $[e, f]$ and each iteration reduces this space by half. The iterations end whenever there remains no integer between $e$ and $f$. By Theorem 29.1 and Proposition 29.3 in Section 29.4 for every $k\sigma < NLAt^*$, Algorithm *APX-LAt* returns an alignment with a positive score, and length at least $(1 - \frac{1}{r})t$ as a solution to the parametric problem. After the search ends, $\lambda^* \geq NLAt^*$, and $\lambda^*$ is achieved by an alignment whose length is at least $(1 - \frac{1}{r})t$ for $NLAt$ feasible. Note that if $NLAt^* = 0$ then the algorithm returns 0.

The asymptotic space requirement is the same as that of Algorithm *APX-LAt*, and the loop iterates $O(\log n)$ times. Therefore the complexity results are as described in the theorem. ∎

If $NLAt^* > 0$ then we can also achieve the same approximation guarantee by using a Dinkelbach algorithm given by Arslan et al. [3] as the template. The details of the resulting algorithm appear in Figure 29.9. At each iteration, except for the last, Algorithm *APX-LAt* returns an alignment with a positive score, and length at least $(1 - \frac{1}{r})t$ as a solution to the parametric problem by Theorem 29.1 and Proposition 29.2 in Section 29.4 as $\lambda < NLAt^*$. Solutions of the parametric problems through the iterations yield improved (higher) values to $\lambda$ except for the last iteration. The resulting algorithm performs no more than 3–5 iterations on average, and never more than 9 in the worst case in tests [1]. When the algorithm terminates the optimal alignment whose length normalized score is $\lambda^*$ has the total length at least $(1 - \frac{1}{r})t$, and $\lambda^* \geq NLAt^*$.

## 29.6 Discussion

We would like to point out the relation between the *NLAt*, and a problem known as *parametric sequence alignment* [11] (which is different from the parametric local alignment problem we discuss in this chapter) in the literature. The fractional programming-based *ANLA* and *NLAt* algorithms iteratively and systematically change the four parameters (i.e., match score, mismatch, gap open, and gap extension penalties) until the resulting alignment is satisfactory (i.e., optimal both with respect to ordinary scores at the last iteration and with respect to length-normalized scores with the original scores). It has been known that sequence alignment is sensitive to the choice of these parameters as they change the optimality of the alignments. Parametric sequence alignment studies the relation between the parameter settings and optimal alignments. The goal is to partition the parameter space into convex polygons such that the same alignment is optimal at every point in the same polygon. Clearly a point in one of the polygons computed yields an optimal length-normalized alignment. The following results are summarized by Gusfield [12]: A polygonal decomposition requires $O(nm)$ time per polygon when scores are uniform (i.e., not dependent on individual symbols). When only two parameters are chosen to be variable, then the polygonal decomposition can contain at most $O(nm)$ polygons. When all the four parameters are variables, then there is no known reasonable upper bound on the number of polygons. When the alignment is global, and no scoring matrices are used the number of polygons is bounded from above by $O(n^{2/3})$ [13].

We also remark that to find long regions with high degree of similarity we may also formulate an objective with which we aim to minimize a length-normalized weighted edit distance for substrings, and include a length threshold as a lower bound for the desired length. For solving this problem, Karp's $O(|V||E|)$-time minimum mean-weight cycle algorithm [14] seems a natural candidate. This solution requires adding extra edges to cause cycles of minimum certain length determined by the given length threshold. For an alignment graph for a pair of strings of length $n$ each, the number of vertices $|V|$ and number of edges $|E|$ (excluding the additional edges) are both $O(n^2)$. This is not more efficient than the naive dynamic programming.

We conclude by stating an open problem for further study:

*Are there (provably) faster exact, or better approximation algorithms for the NLAt, LRLA, LAt, or Qt problems?*

# References

1. Arslan, A.N. and Eğecioğlu, Ö., Dynamic programming based approximation algorithms for local alignment with length constraints, *INFORMS Journal on Computing, Special Issue on Computational Molecular Biology/Bioinformatics*, 16(4), 441, 2004.

2. Arslan, A.N. and Eğecioğlu, Ö., An improved upper bound on the size of planar convex-hulls, in *Proceedings of COCOON*, LNCS, 2108, 2001, p. 111, Springer, Berlin, Germany.

3. Arslan, A.N. and Eğecioğlu, Ö and Pevzner, P.A., A new approach to sequence comparison: Normalized local alignment, *Bioinformatics*, 17(4), 327, 2001.

4. Arslan, A.N. and Eğecioğlu, Ö., Approximation algorithms for local alignment with length constraints, *International Journal of Foundations of Computer Science*, 13(5), 751, 2002.

5. Waterman, M.S., *Introduction to Computational Biology*, Chapman & Hall, Boca Raton, FL, 1995.

6. Smith, T.F. and Waterman, M.S., The identification of common molecular subsequences, *Journal of Molecular Biology*, 147, 195, 1981.

7. Craven, B.D., *Fractional Programming*, Helderman Verlag, Berlin, Germany, 1988.

8. Sniedovich, M., *Dynamic Programming*, Marcel Dekker, New York, 1992.

9. Matsui, T., Saruwatari, Y., and Shigeno, M., An analysis of Dinkelbach's algorithm for 0-1 fractional programming problems, Technical Report METR92-14, Department of Mathematical Engineering and Information Physics, Unviersity of Tokyo, Japan, 1992.

10. Megiddo, N., Combinatorial optimization with rational objective functions, *Mathematics of Operations Research*, 4, 414, 1979.

11. Fitch, W.M. and Smith, T.F., Optimal sequence alignments, *Proceedings of the National Academy of Sciences of the United States of America*, 80, 1382, 1983.

12. Gusfield, D., *Algorithm on Strings, Trees, and Sequences: Computer Science and Computational Biology*, The Press Syndicate of The University of Cambridge, New York, 1997.

13. Gusfield, D., Balasubramanian, K. and Naor, D., Parametric optimization of sequence alignment, *Algorithmica*, 12, 312, 1994.

14. Cormen, T.H., Leiserson, C.E., Rivest, R.L. and Stein, C., *Introduction to Algorithms*. 2nd edition, The MIT Press, Cambridge, MA, 2001.

# 30

# Approximation Algorithms for the Selection of Robust Tag SNPs

Yao-Ting Huang

Kui Zhang

Ting Chen

Kun-Mao Chao

## 30.1 Introduction

Each individual has a unique set of genetic blueprints stored in a long and spiral-shaped molecule called *deoxyribonucleic acid*. The genetic blueprints are composed of linked subunits called *nucleotides*. Each nucleotide carries one of the four genetic codes: adenine (A), cytosine (C), guanine (G), and thymine (T). The variations of genetic codes from individual to individual (e.g., insertions, deletions, and mutations) have a major impact on genetic diseases and phenotypic differences. Therefore, correlating genetic variations with diseases or traits is the next important step in human genomics. In the following, we first introduce the related biological background to understand the problem studied in the current chapter. Then we describe a frequently encountered problem in the current experimental environment, which is the main focus of this chapter.

### 30.1.1 Single Nucleotide Polymorphisms and Haplotypes

Among various genetic variations, *Single Nucleotide Polymorphisms* (SNPs) are generally considered to be the most frequent form which has fundamental importance for genetic disease association and drug design. A SNP is a genetic variation when a single nucleotide (i.e., A, C, G, or T) in the genomic sequence is altered and kept through heredity thereafter.[*] It has been shown that approximately 90% of genetic variations are made up of SNPs. Up to the present, millions of SNPs have been identified and these data are now publicly available for researchers [1,2]. The SNPs can be further divided into the following types depending on their region and function to the amino acid sequence.

*Coding SNP (cSNP):* A SNP in the coding region that involves in the regulation of amino acid substitution.

*Synonymous SNP:* A cSNP that synonymously change the codon of an amino acid, which does not alter the amino acid sequence.

*Nonsynonymous SNP:* A cSNP that non-synonymously changes the codon of an amino acid, which alters the amino acid sequence.

Despite there are many types of SNPs, almost all SNPs observed have only two variants called *alleles*. Very few SNPs (about 0.1% ) in the human population have been found to have more than two different nucleotides. In fact, these third type nucleotides are often resulted from possible experimental errors [3]. Consequently, most studies usually assume that the value of a SNP is binary. A SNP is referred to as the *major allele* if it is the wild type and is called the *minor allele* otherwise (i.e., mutant type). In this chapter, each type of SNP is equally treated as a binary variable.

A set of linked SNPs on one genomic sequence is referred to as a *haplotype*. As the value of a SNP is usually assumed to be binary, a haplotype can be simply considered to be a binary string. *Linkage Disequilibrium* (LD), which refers to the nonrandom association of alleles at different loci in haplotypes, plays an important role in genome-wide association studies for identifying genetic variations responsible for common diseases. A number of studies have shown that using haplotypes instead of individual SNP as the basic units for LD analysis can greatly reduce the noise [4]. Recently, the International HapMap Project [1], formed in 2002, aimed to characterize the patterns of LD across the human genome such that the information can be used for large-scale genetic association studies.

## 30.1.2 Haplotype Blocks and Tag Single Nucleotide Polymorphisms

The LD analysis of haplotypes is greatly affected the frequency of past *recombination* events. Recombination (or cross over) is a process during meiosis that the two homologous chromosomes (inside the cells that produce sperms or eggs) break and swap portion with each other. After these two homologous chromosomes glue themselves back, each of them obtains new alleles from the other. Therefore, the result of recombination can produce new chromosomes for the offspring. On the other hand, the nonrandom association of alleles at different loci is broken up by the recombination occurred in between. Consequently, recombination can also reduce the LD observed in the population.

In recent years, the patterns of LD observed in the human population show a block like structure [4–7]. The entire chromosome can be partitioned into high LD regions interspersed by low LD regions. The chromosome recombination almost only takes place at those low LD regions called recombination hotspots. The high LD region between recombination hotspots is often referred to as a "haplotype block." Within a haplotype block, there is little or even no recombination occurred, and the SNPs in the block tend to be inherited together. Due to the low haplotype diversity within a block, the information carried by these SNPs is highly redundant. Thus, a small subset of SNPs, called "tag SNPs," is sufficient to distinguish each pair of patterns in the block [5,7–12]. Haplotype blocks with corresponding tag SNPs are quite useful and cost-effective for association studies as it does not require genotyping all SNPs.

Many studies have tried to find the minimum set of tag SNPs. These studies can be classified into the following categories.

---

[*] The genetic variation is considered to be a SNP only if it is observed with frequency at least 1% in the population. Otherwise, it is considered to be a mutation.

*LD-bins based model:* These methods try to identify minimum bins of SNPs such that all SNPs in the same bin are in high LD (e.g., $r^2 \geq 0.8$) with each other (e.g., Carlson et al. [9]). Most of them solve some variants of the minimum clique cover problem [13].

*Blocks based model:* These methods assume that the block partition is available as input, and try to find a minimum set of SNPs which is able to distinguish each pair of haplotypes in a block [11]. Most of them solve some variants of the minimum test set problem [13].

*Block-free based model:* These methods define an informative measure or prediction accuracy to evaluate a set of tag SNPs. Most of them try to find a minimum set of SNPs which maximizes their criterion [8].

In the current chapter, we study the tag SNP selection problem following the blocks based model. In a large-scale study of human Chromosome 21, Patil et al. [5] developed a greedy algorithm to partition the haplotypes into 4,135 blocks with 4,563 tag SNPs. Zhang et al. [7,11,12] used a dynamic programming approach to reduce the numbers of blocks and tag SNPs to 2575 and 3562, respectively. Bafna et al. [8] showed that the general version of this problem is NP-hard and gave efficient algorithms for special cases of this problem. In the following, we show that the previous studies do not consider the influence of missing data in the current SNP detection environment.

### 30.1.3 The Problem of Missing Data

In reality, a SNP may not be genotyped and is considered to be missing data (i.e., we fail to obtain the allele configuration of the SNP) if it does not pass the threshold of data quality [5,7,14]. The missing rates of SNPs can be up to 10% under the current genotyping experiment. In practice, there could be two kinds of missing data: completely and partially missing data. In this chapter, partially missing data are handled in analogy to completely missing data. These missing data may cause ambiguity when using the minimum set of tag SNPs to distinguish an unknown haplotype sample. As a consequence, the power of using tag SNPs for association study is reduced by missing data.

Figure 30.1 illustrates the influence of missing data when using the minimum set of tag SNPs to identify haplotype samples. In this figure, a haplotype block (Figure 30.1a) defined by 12 SNPs and 4 haplotype patterns is presented (from the haplotype database of human Chromosome 21 by Patil et al. [5]). We follow the same assumption as Patil et al. [5] and Bafna et al. [8] that all SNPs are biallelic (i.e., taking on only two values). Suppose we select SNPs $S_1$ and $S_{12}$ as tag SNPs. The haplotype sample $h_1$ is identified

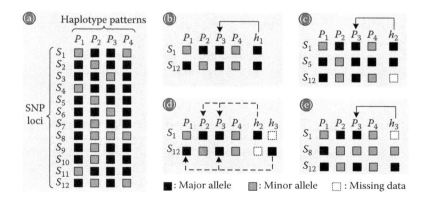

**FIGURE 30.1** The influence of missing data and auxiliary tag SNPs. (a) A haplotype block defined by 12 SNPs and 4 haplotype patterns. Each column represents a haplotype pattern and each row represents a SNP locus. The black and grey boxes stand for the major and minor alleles at each SNP locus, respectively. (b) Tag SNPs genotyped without missing data. (c) Tag SNPs genotyped with missing data. (d) The auxiliary tag SNP $S_5$ for $h_2$. (e) The auxiliary tag SNP $S_8$ for $h_3$.

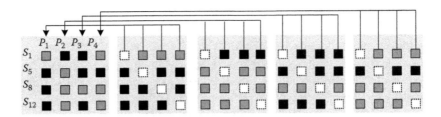

**FIGURE 30.2**    A set of robust tag SNPs for tolerating one missing tag SNP.

as haplotype pattern $P_3$ unambiguously (Figure 30.1b). Consider haplotype samples $h_2$ and $h_3$ with one tag SNP genotyped as missing data (Figure 30.1c). $h_2$ can be identified as haplotype patterns $P_2$ or $P_3$, and $h_3$ can be identified as $P_1$ or $P_3$. As a result, these missing tag SNPs result in ambiguity when identifying haplotype samples.

Although we cannot avoid the occurrence of missing data, the remaining SNPs within the haplotype block may provide abundant information to resolve the ambiguity. For example, if we regenotype an additional SNP $S_5$ for $h_2$ (Figure 30.1d), $h_2$ is identified as haplotype pattern $P_3$ unambiguously. On the other hand, if SNP $S_8$ is regenotyped (Figure 30.1e), $h_3$ is also identified unambiguously. These additional SNPs are referred to as "auxiliary tag SNPs," which can be found from the remaining SNPs in the block and are able to resolve the ambiguity caused by missing data.

Alternatively, instead of regenotyping auxiliary tag SNPs whenever encountering missing data, we work on a set of SNPs which is not affected by the the occurrence of missing data. Figure 30.2 illustrates a set of SNPs which can tolerate one missing SNP. Suppose we select SNPs $S_1$, $S_5$, $S_8$, and $S_{12}$ to be genotyped. Note that no matter which SNP is missing, each pair of patterns can still be distinguished by the remaining three SNPs. Therefore, all haplotype samples with one missing SNP can still be identified unambiguously. We refer to these SNPs as "robust tag SNPs," which are able to tolerate a certain number of missing data. The important feature of robust tag SNPs is that although they consume more SNPs than the "tag SNPs" defined in previous studies, they guarantee that all haplotype patterns with a certain number of missing data can be distinguished unambiguously. When the occurrence of missing data is frequent, the cost of regenotyping processes can be reduced by robust tag SNPs.

This chapter studies the problems of finding robust and auxiliary tag SNPs. Our study indicates that auxiliary tag SNPs can be found efficiently when robust tag SNPs have been computed in advance. This chapter is organized as follows. In Section 30.2, we show that the problem of finding minimum robust tag SNPs (MRTS) is NP-hard, and propose two greedy and one iterative linear programming (LP)-relaxation algorithms which find solutions of $(m + 1) \ln(\frac{K(K-1)}{2})$, $\ln((m + 1)\frac{K(K-1)}{2})$, and $O(m \ln K)$ approximation, respectively. Section 30.3 describes an efficient algorithm to find auxiliary tag SNPs when robust tag SNPs have been computed in advance. Section 30.4 presents the experimental results of our algorithms tested on a variety of simulated and biological data. Finally, concluding remarks are given in Section 30.5.

# 30.2 Finding Robust Tag Single Nucleotide Polymorphisms

Assume we are given a haplotype block consisting of $N$ SNPs and $K$ haplotype patterns. This block is denoted by an $N \times K$ binary matrix $M_h$ (Figure 30.3a). Define $M_h[i, j] \in \{1, 2\}$ for each $i \in [1, N]$ and $j \in [1, K]$, where 1 and 2 represent the major and minor alleles, respectively[*]. Define $C$ as the set of SNPs (i.e., rows) in $M_h$. The set of robust tag SNPs $C' \subseteq C$ (which allows up to $m$ missing SNPs) must satisfy the following two properties: (1) an unknown haplotype sample can be identified (as one of the $K$

---

[*] In reality, the haplotype block may also contain missing data. This formulation can be easily extended to handle missing data by treating them as "don't care" symbols. To simplify the presentation, we will assume no missing data in the block.

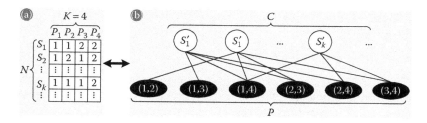

**FIGURE 30.3** (a) The haplotype matrix $M_h$ containing $N$ SNPs and $K$ haplotype patterns. (b) The bipartite graph corresponding to $M_h$.

patterns) by SNPs in $C'$ unambiguously; (2) when at most $m$ SNPs in $C'$ are genotyped as missing data, (1) still holds. Note that to identify a sample unambiguously, each pair of patterns must be distinguished by at least one SNP in $C'$. For example (Figure 30.3a), patterns $P_1$ and $P_2$ can be distinguished by SNP $S_2$ as $M_h[2, 1] \neq M_h[2, 2]$. A formal definition of this problem is given in the following.

Problem:   MRTS

   Input:   An $N \times K$ matrix $M_h$ and an integer $m$.

  Output:   The minimum subset of SNPs $C' \subseteq C$ which satisfies
             (1) For each pair of patterns $P_i$ and $P_j$, there is a SNP $S_k \in C'$ such that $M_h[k, i] \neq M_h[k, j]$.
             (2) When at most $m$ SNPs are discarded from $C'$ arbitrarily, (1) still holds.

We then reformulate MRTS to a variant of the *set covering problem* [13]. Each SNP $S_k \in C$ (i.e., the $k$th row in $M_h$) is reformulated to a set $S'_k = \{(i, j) \mid M[k, i] \neq M[k, j] \text{ and } i < j\}$. For example, suppose the $k$th row in $M_h$ is {1,1,1,2}. The corresponding set $S'_k = \{(1, 4), (2, 4), (3, 4)\}$. In other words, $S'_k$ stores pairs of patterns distinguished by SNP $S_k$. Define $P$ as the set that contains all pairs of patterns (i.e., $P = \{(i, j) \mid 1 \leq i < j \leq K\} = \{(1, 2), (1, 3), ..., (K - 1, K)\}$).

Consider each element in $P$ and each reformulated set of $C$ as nodes in an undirected bipartite graph (Figure 30.3b). If SNP $S_k$ can distinguish patterns $P_i$ and $P_j$ (i.e., $(i, j) \in S'_k$), there is an edge connecting the nodes $(i, j)$ and $S'_k$. The following lemma implies that each pair of patterns must be distinguished by at least $(m + 1)$ SNPs to allow $m$ SNPs genotyped as missing data.

**Lemma 30.1** *$C' \subseteq C$ is the set of robust tag SNPs which allows at most $m$ SNPs genotyped as missing data iff each node in $P$ has at least $(m + 1)$ edges connecting to each node in $C'$.*

*Proof.* Let $C'$ be the set of robust tag SNPs which allows $m$ SNPs genotyped as missing data. Suppose patterns $P_i$ and $P_j$ are distinguished by only $m$ SNPs in $C'$ (i.e., $(i, j)$ has only $m$ edges connecting to nodes in $C'$). However, if these $m$ SNPs are genotyped as missing data, no SNPs in $C'$ are able to distinguish patterns $P_i$ and $P_j$, which is a contradiction. Thus, each pair of patterns must be distinguished by at least $(m + 1)$ SNPs, which implies that each node in $P$ must have at least $(m + 1)$ edges connecting to nodes in $C'$. The proof of the other direction is similar. ∎

In the following, we give a lower bound on the minimum number of robust tag SNPs required.

**Lemma 30.2** *Given $K$ haplotype patterns, the minimum number of robust tag SNPs for tolerating $m$ missing SNPs is at least $m + \log K$.*

*Proof.* Recall that the value of a SNP is binary. The maximum number of distinct haplotypes which can be distinguished by $N$ SNPs is at most $2^N$. As a result, to distinguish $K$ distinct haplotype patterns, at least $\log K$ SNPs are required as $2^{\log K} = K$. In addition, there could be up to $m$ missing SNPs. Therefore, the minimum number of robust tag SNPs required is at least $m + \log K$. ∎

Now we show the NP-hardness of the MRTS problem, which implies there is no polynomial time algorithm to find the optimal solution of MRTS.

**Theorem 30.1** *The MRTS problem is NP-hard.*

*Proof.* When $m = 0$, MRTS is the same as the original problem of finding minimum number of tag SNPs, which is known as the *minimum test set* problem [11,13]. As the minimum test set problem is NP-hard and can be reduced to a special case of MRTS, MRTS is NP-hard.                                      ∎

## 30.2.1 The First Greedy Algorithm

To solve MRTS efficiently, we propose a greedy algorithm which returns a solution with a number of SNPs that is not to extremely far from optimal. By Lemma 30.1, to tolerate $m$ missing tag SNPs, we need to find a subset of SNPs $C' \subseteq C$ such that each pair of patterns in $P$ is distinguished by at least $(m+1)$ SNPs in $C'$. Assume that the SNPs selected by this algorithm are stored in a $(m+1) \times |P|$ table (Figure 30.4a). Initially, each grid in the table is empty. Once a SNP $S_k$ (that can distinguish patterns $P_i$ and $P_j$) is selected, one grid of the column $(i, j)$ is filled in with $S_k$, and we say that this grid is *covered* by $S_k$.

This greedy algorithm works by covering the grids from the first row to the $(m+1)$-th row and greedily selects a SNP which covers most uncovered grids in the $i$th row at each iteration. In other words, while working on the $i$th row, a SNP is selected if its reformulated set $S'$ maximizes $|S' \cap R_i|$, where $R_i$ is the set of uncovered grids at the $i$th row.

Figure 30.4 illustrates an example for this algorithm to tolerate one missing tag SNP (i.e., $m = 1$). The SNPs $S_1$, $S_4$, $S_2$, and $S_3$ are selected in order. When all grids in this table are covered, each pair of patterns is distinguished by $(m+1)$ SNPs in the corresponding column. Thus, the SNPs in this table are the robust tag SNPs which allows at most $m$ SNPs genotyped as missing data. The pseudocode of this greedy algorithm is given in the following.

The time complexity of this algorithm is analyzed as follows. At Line 4, the number of iterations of the intermediate loop is bounded by $|R_i| \leq |P|$. Within the loop body (Lines 5–13), Line 5 takes $O(|C||P|)$ because we need to check all SNPs in $C$ and examine the uncovered grids of $R_i$. The inner loop (Lines 8–13) takes only $O(|S'|)$. Thus, the entire program runs in $O(m|C||P|^2)$.

We now show the number of SNPs in the solution $C'$ returned by the first greedy algorithm is not extremely large compared with the ones in the optimal solution $C^*$. Suppose the algorithm selects the $k$th SNP when working on the $i$th row. Let $|S_k^c|$ be the number of grids in the $i$th row covered by the $k$th selected SNP (i.e., $|S_k^c| = |S' \cap R_i|$; see Line 5 in FIRST-GREEDY-ALGORITHM). For example (see Figure 30.4), $S_2^c = 2$ as the second selected SNP (i.e., $S_4$) covers two grids in the first row. We incur 1 unit of cost to each selected SNP and spread this cost among the grids in $S_k^c$ [15]. In other words, each grid at the $i$th row and $j$th column is assigned a cost $C_j^i$ (Figure 30.5), where

$$C_j^i = \begin{cases} \frac{1}{|S_k^c|} & \text{if the algorithm selects the } k\text{th SNP when covering the } i\text{th row;} \\ 0 & \text{otherwise.} \end{cases}$$

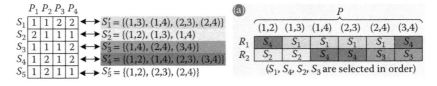

**FIGURE 30.4** The SNPs $S_1$, $S_4$, $S_2$, and $S_3$ are selected by the first greedy algorithm. (a) The table that stores each selected SNP.

$C'$:
- $S'_1 = \{(1,3), (1,4), (2,3), (2,4)\}$
- $S'_4 = \{(1,2), (1,4), (2,3), (3,4)\}$
- $S'_2 = \{(1,2), (1,3), (1,4)\}$
- $S'_3 = \{(1,4), (2,4), (3,4)\}$

$P$:

| (1,2) | (1,3) | (1,4) | (2,3) | (2,4) | (3,4) |
|---|---|---|---|---|---|
| 1/2 | 1/4 | 1/4 | 1/4 | 1/4 | 1/2 |
| 1/2 | 1/2 | 0 | 0 | 1/2 | 1/2 |

**FIGURE 30.5** The cost $C_j^i$ of each grid for the first greedy algorithm.

## ALGORITHM First-Greedy-Algorithm(C,P,m)

$R_i \leftarrow P, \forall i \in [1, m+1]$
$C' \leftarrow \phi$
**for** $i = 1$ to $m + 1$ **do**
  **while** $R_i \neq \phi$ **do**
    select and remove a SNP $S$ from $C$ that maximizes $|S' \cap R_i|$
    $C' \leftarrow C' \cup S$
    $j \leftarrow i$
    **while** $S' \neq \phi$ **and** $j \leq m+1$ **do**
      $S_{tmp} \leftarrow S' \cap R_j$
      $R_j \leftarrow R_j - S_{tmp}$
      $S' \leftarrow S' - S_{tmp}$
      $j \leftarrow j + 1$
    **endwhile**
  **endwhile**
**endfor**
**return** $C'$

As each selected SNP is assigned 1 unit of cost, the sum of $C_j^i$ for each grid in the table is equal to $|C'|$, that is,

$$|C'| = \sum_{i=1}^{m+1} \sum_{j=1}^{\frac{K(K-1)}{2}} C_j^i. \tag{30.1}$$

Let $R_k^i$ be the number of uncovered grids in the $i$th row before the $k$th iteration (i.e., $(k-1)$ SNPs have been selected by the algorithm). For example (Figure 30.5), $R_2^1 = 2$ as two grids in the first row are still uncovered before the second SNP is selected. Define $C_i'$ as the set of iterations used by the algorithm when working on the $i$th row. For example (Figure 30.5), $C_2' = \{3, 4\}$ as this algorithm works on the second row in the third and fourth iterations. We can rewrite Equation 30.1 as

$$\sum_{i=1}^{m+1} \sum_{j=1}^{\frac{K(K-1)}{2}} C_j^i = \sum_{i=1}^{m+1} \sum_{k \in C_i'} (R_{k-1}^i - R_k^i) \frac{1}{|S_k^c|}. \tag{30.2}$$

**Lemma 30.3** *The kth selected SNP has* $|S_k^c| \geq \frac{R_{k-1}^i}{|C^*|}$.

*Proof.* Suppose the algorithm is working on the $i$th row at the beginning of the $k$th iteration. Let $C_k^*$ be the set of SNPs in $C^*$ (the optimal solution) that has been selected by the algorithm before the $k$th iteration and the set of remaining SNPs in $C^*$ be $C_{\bar{k}}^*$. We claim that there exists a SNP in $C_{\bar{k}}^*$ which can cover at least

$\frac{R_k^i}{|C_{\bar{k}}^*|}$ grids in the $i$th row. Otherwise (i.e., each SNP in $C_{\bar{k}}^*$ covers less than $\frac{R_k^i}{|C_{\bar{k}}^*|}$ grids), all SNPs in $C_{\bar{k}}^*$ will cover less than ($\frac{R_k^i}{|C_{\bar{k}}^*|} \times |C_{\bar{k}}^*| = R_k^i$) grids in the $i$th row. But as $C_k^* \cup C_{\bar{k}}^* = C^*$, this implies that $C^*$ cannot cover all grids in $R_k^i$, which is a contradiction. As all SNPs in $C_{\bar{k}}^*$ are candidates to the greedy algorithm, the $k$th selected SNP must cover at least $\frac{R_k^i}{|C_{\bar{k}}^*|}$ grids in the $i$th row, which implies $|S_k^c| \geq \frac{R_{k-1}^i}{|C^*|}$ as $|C^*| \geq |C_{\bar{k}}^*|$ and $|R_k^i| \leq |R_{k-1}^i|$. ∎

**Theorem 30.2** *The first greedy algorithm gives a solution of* $(m+1) \ln \frac{K(K-1)}{2}$ *approximation.*

*Proof.* Define the $d$th harmonic number as $H(d) = \sum_{i=1}^{d} \frac{1}{i}$ and $H(0) = 0$. By Equation 30.2 and Lemma 30.3,

$$\sum_{i=1}^{m+1} \sum_{j=1}^{\frac{K(K-1)}{2}} C_j^i = \sum_{i=1}^{m+1} \sum_{k \in C_i'} (R_{k-1}^i - R_k^i) \frac{1}{|S_k^c|} \leq \sum_{i=1}^{m+1} \sum_{k \in C_i'} (R_{k-1}^i - R_k^i) \frac{|C^*|}{R_{k-1}^i}$$

$$= \sum_{i=1}^{m+1} \sum_{k \in C_i'} \left( \sum_{l=R_k^i+1}^{R_{k-1}^i} \frac{|C^*|}{R_{k-1}^i} \right)$$

$$\leq |C^*| \sum_{i=1}^{m+1} \sum_{k \in C_i'} \sum_{l=R_k^i+1}^{R_{k-1}^i} \frac{1}{l} \qquad (l \leq R_{k-1}^i)$$

$$= |C^*| \sum_{i=1}^{m+1} \sum_{k \in C_i'} \left( \sum_{l=1}^{R_{k-1}^i} \frac{1}{l} - \sum_{l=1}^{R_k^i} \frac{1}{l} \right)$$

$$\leq |C^*| \sum_{i=1}^{m+1} \sum_{k \in C_i'} (H(R_{k-1}^i) - H(R_k^i))$$

$$\leq |C^*| \sum_{i=1}^{m+1} (H(R_0^i) - H(R_{|C_i'|}^i))$$

$$\leq |C^*|(m+1) \max\{H(R_0^i)\} \qquad (R_{|C_i'|}^i = 0 \text{ and } H(0) = 0)$$

$$\leq |C^*|(m+1) \ln |P| \qquad (H(R_0^i) \leq H(|P|)) \qquad (30.3)$$

By Equations 30.1 and 30.3, we get

$$\frac{|C'|}{|C^*|} \leq (m+1) \ln |P| = (m+1) \ln \frac{K(K-1)}{2}.$$

∎

## 30.2.2 The Second Greedy Algorithm

The current section describes the second greedy algorithm which returns a better solution than that the one the first greedy algorithm generates. Let $R_i$ be the set of uncovered grids at the $i$th row. Unlike the row-by-row manner of the first greedy algorithm, this algorithm greedily selects a SNP that covers most uncovered grids in the table (i.e., its reformulated set $S'$ maximizing $|S' \cap (R_1 \cup \cdots \cup R_{m+1})|$). Let $T$ be the collection of $R_i$ (i.e., $T$ is the set of all uncovered grids in the table). If the grids in the $i$th row are all covered (i.e., $R_i = \phi$), $R_i$ is removed from $T$. This algorithm runs until $T = \phi$ (i.e., all grids in the table are covered).

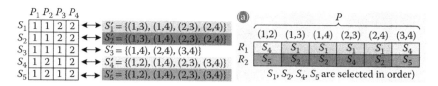

**FIGURE 30.6** The SNPs $S_1$, $S_2$, $S_4$, and $S_5$ are selected by the second greedy algorithm. (a) The table that stores each selected SNP.

Figure 30.6 illustrates an example for this algorithm with $m$ set to 1. The SNPs $S_1$, $S_2$, $S_4$, and $S_5$ are selected in order. As this algorithm runs until all grids are covered, the set of SNPs in this table is able to tolerate $m$ missing tag SNPs. The pseudocode of this algorithm is given in the following.

---

## ALGORITHM   Second-Greedy-Algorithm($C$,$P$,$m$)

```
1    R_i ← P, ∀i ∈ [1, m + 1]
2    T ← {R_1, R_2, . . . , R_{m+1}}
3    C' ← φ
4    while T ≠ φ do
5        select and remove a SNP S from C that maximizes |S' ∩ (R_1 ∪ · · · ∪ R_{m+1})|
6        C' ← C' ∪ S
7        for each R_i ∈ T and S' ≠ φ do
8            S_{tmp} ← S' ∩ R_i
9            R_i ← R_i − S_{tmp}
10           S' ← S' − S_{tmp}
11           if R_i = φ then T ← T − R_i
12       endfor
13   endwhile
14   return C'
```

---

The time complexity of this algorithm is analyzed as follows. At Line 4, the number of iterations of the loop is bounded by $O(|T|)=O(m|P|)$. Within the loop, Line 5 takes $O(|C||P|)$ time because we need to check each SNP in $C$ and examine if it can cover any uncovered grid in each column. The inner loop (Lines 7–12) is bounded by $O(|S'|) < O(|P|)$. Thus, the running time of this program is $O(m|C||P|^2)$.

We now evaluate the solution returned by the second greedy algorithm. Let $C'$ and $C^*$ be the set of SNPs selected by this algorithm and the optimal solution, respectively. Let $|S_k^c|$ be the number of grids in the table covered by the $k$th selected SNP. For example (Figure 30.6), $|S_2^c| = 4$ as the second selected SNP (i.e., $S_2$) covers four grids in the table. Define $T_k$ as the number of uncovered grids in the table before the $k$th iteration. We have the following lemma similar to Lemma 30.3.

**Lemma 30.4** *The $k$th selected SNP has $|S_k^c| \geq \frac{T_{k-1}}{|C^*|}$.*

*Proof.* The proof is similar to that of Lemma 30.3. Let $C_{\bar{k}}^*$ be the set of remaining SNPs in $C^*$ which has not been selected before the $k$th iteration. We claim that there exists a SNP in $C_{\bar{k}}^*$ which can cover at least $\frac{T_k}{|C_{\bar{k}}^*|}$ grids in the table. Otherwise, we can get the same contradiction (i.e., $C^*$ fails to cover all grids) as in Lemma 30.3. As $|C^*| \geq |C_{\bar{k}}^*|$ and $T_{k-1} \leq T_k$, we have $|S_k^c| \geq \frac{T_{k-1}}{|C^*|}$.  ∎

**Theorem 30.3** *The second greedy algorithm gives a solution of $\ln((m + 1)\frac{K(K-1)}{2})$ approximation.*

$$C\begin{cases} S'_1 = \{(1,3),\ (1,4),\ (2,3),\ (2,4)\} \\ S'_3 = \{(1,3),\ (1,4),\ (2,3),\ (2,4)\} \\ S'_4 = \{(1,2),\ (1,4),\ (2,3),\ (3,4)\} \\ S'_5 = \{(1,2),\ (1,4),\ (2,3),\ (3,4)\} \end{cases}$$

| | $P$ | | | | | |
|---|---|---|---|---|---|---|
| | (1,2) | (1,3) | (1,4) | (2,3) | (2,4) | (3,4) |
| | 1/2 | 1/4 | 1/4 | 1/4 | 1/4 | 1/2 |
| | 1/2 | 1/4 | 1/4 | 1/4 | 1/4 | 1/2 |

**FIGURE 30.7**    The cost $C_j^i$ of each grid for the second greedy algorithm.

*Proof.* Each grid at the $i$th row and $j$th column is assigned a cost $C_j^i = \frac{1}{|S_k^c|}$ (Figure 30.7) if it is covered by the $k$th selected SNP. The sum of $C_j^i$ for each grid is

$$|C'| = \sum_{i=1}^{m+1} \sum_{j=1}^{\frac{K(K-1)}{2}} C_j^i = \sum_{k=1}^{|C'|} (T_{k-1} - T_k) \frac{1}{|S_k^c|} \qquad \text{(see Equations 30.1 and 30.2)}$$

$$\le \sum_{k=1}^{|C'|} (T_{k-1} - T_k) \frac{|C^*|}{T_{k-1}} \qquad \text{(by Lemma 30.4)}$$

$$\le |C^*|(H(T_0) - H(T_{|C'|})) \qquad \text{(see the proof in Theorem 30.2)}$$

$$\le |C^*| \ln((m+1)|P|). \tag{30.4}$$

By Equation 30.4, we have

$$\frac{|C'|}{|C^*|} \le \ln((m+1)|P|) = \ln\left((m+1)\frac{K(K-1)}{2}\right). \qquad \blacksquare$$

### 30.2.3 The Iterative Linear Programming-Relaxation Algorithm

In practice, a probabilistic approach is sometimes more useful as the randomization can explore different solutions. In this section, we reformulate the MRTS problem to an *Integer Programming* (IP) problem. Based on the IP problem, we propose an iterative LP-relaxation algorithm. The iterative LP-relaxation algorithm is described in the following.

Step 1:   Given a haplotype block containing $N$ SNPs and $K$ haplotype patterns. Let $\{x_1, x_2, ..., x_N\}$ be the set of integer variables for the $N$ SNPs, where $x_k = 1$ if the SNP $S_k$ is selected and $x_k = 0$ otherwise. Define $D(P_i, P_j)$ as the set of SNPs which are able to distinguish $P_i$ and $P_j$ patterns. By Lemma 30.1, to allow at most $m$ SNPs genotyped as missing data, each pair of patterns must be distinguished by at least $(m+1)$ SNPs. Therefore, for each set $D(P_i, P_j)$, at least $(m+1)$ SNPs have to be selected to distinguish $P_i$ and $P_j$ patterns. As a consequence, the MRTS problem can be formulated as the following IP problem:

$$\textbf{Minimize} \ \sum_{k=1}^{N} x_k$$

$$\textbf{Subject to} \ \sum_{k \in D(P_i, P_j)} x_k \ge m+1, \quad \text{for all } 1 \le i < j \le K, \tag{30.5}$$

$$x_k = 0 \text{ or } 1.$$

Step 2:   As solving the IP problem is NP-hard [13], we relax the integer constraint of $x_k$, and the IP problem becomes a LP problem defined as follows:

$$\textbf{Minimize } \sum_{k=1}^{N} y_k$$

$$\textbf{Subject to } \sum_{k \in D(P_i, P_j)} y_k \geq m + 1, \quad \text{for all } 1 \leq i < j \leq K, \tag{30.6}$$

$$0 \leq y_k \leq 1.$$

The above-mentioned LP problem can be solved by efficient algorithms such as the interior point method [16,17].

Step 3:   Let $\{y_1, y_2, ..., y_N\}$ be the set of linear solutions obtained from Equation 30.6, where $0 \leq y_k \leq 1$. We assign 0 or 1 to $x_k$ by the following randomized rounding method

$$\text{Assign } \begin{cases} x_k = 1 \text{ with probability } y_k, \\ x_k = 0 \text{ with probability } 1 - y_k. \end{cases}$$

Step 4:   The randomized rounding method may invalidate some of the inequalities in Equation 30.5. Thus, we repeat Steps 1–3 for those unsatisfied inequalities until all of them are satisfied. Finally, when all inequalities in Equation 30.5 are satisfied, we construct a final solution by the following rule:

$$\text{Assign } \begin{cases} x_k = 1 & \text{if } x_k \text{ is assigned to 1 in any one of the iterations;} \\ x_k = 0 & \text{otherwise.} \end{cases}$$

We now evaluate the solution returned by the iterative LP-relaxation algorithm. The selection of each SNP is considered as a *Bernoulli* random variable $x_k$ taking values 1 (or 0) with probability $y_k$ (or $1 - y_k$). Let $X_{i,j}$ be the sum of random variables in one inequality of Equation 30.5, that is,

$$X_{i,j} = \sum_{k \in D\{P_i, P_j\}} x_k.$$

By Equation 30.6, the expected value of $X_{i,j}$ (after randomized rounding) is

$$E[X_{i,j}] = \sum_{k \in D\{P_i, P_j\}} E[x_k] = \sum_{k \in D\{P_i, P_j\}} y_k$$

$$\geq m + 1. \tag{30.7}$$

**Lemma 30.5** *The probability that an inequality in Equation 30.5 is not satisfied after randomized rounding is less than* $e^{-\frac{1}{2(m+1)}}$.

*Proof.* The probability that an inequality in Equation 30.5 is not satisfied is $P[X_{i,j} < m+1] = P[X_{i,j} \leq m]$. By the *Chernoff* bound (i.e., $P[X \leq (1 - \theta)E[X]] \leq e^{-\frac{\theta^2 E[X]}{2}}$), we have

$$P[X_{i,j} \leq m] \leq e^{-\frac{(E[X_{i,j}]-m)^2}{2E[X_{i,j}]}}. \tag{30.8}$$

By Equation 30.7, we know $E[X_{i,j}] \geq m + 1$. As the right-hand side of Equation 30.8 decreases when $E[X_{i,j}] > m$, we can replace $E[X_{i,j}]$ with $(m + 1)$ to obtain an upper bound, that is,

$$P[X_{i,j} \leq m] \leq e^{-\frac{(E[X_{i,j}]-m)^2}{2E[X_{i,j}]}} \leq e^{-\frac{(m+1-m)^2}{2(m+1)}}$$

$$\leq e^{-\frac{1}{2(m+1)}}.$$ ∎

**Theorem 30.4** *The iterative LP-relaxation algorithm gives a solution of $O(m \ln K)$ approximation.*

*Proof.* Suppose this algorithm runs for $t$ iterations. The probability that all $\frac{K(K-1)}{2}$ inequalities in Equation 30.5 are satisfied after $t$ iterations is

$$(1 - (e^{-1/2(m+1)})^t)^{\frac{K(K-1)}{2}} = (1 - e^{-t/2(m+1)})^{\frac{K(K-1)}{2}}$$

$$\approx e^{-\frac{K(K-1)}{2}e^{-t/2(m+1)}}.$$

When $t = 2(m + 1) \ln \frac{K(K-1)}{2}$, the algorithm stops and returns a solution with probability $e^{-1}$. Define $OPT(IP)$ and $OPT(LP)$ as the optimal solutions of the IP problem and the LP problem, respectively. As the solution space of LP includes that of IP,

$$OPT(LP) \leq OPT(IP).$$

Let the set of solutions returned in $t$ iterations be $\{Z_1, Z_2, ..., Z_t\}$.

$$E[Z_1] = E[\sum_{k=1}^{N} x_k] = \sum_{k=1}^{N} y_k = OPT(LP).$$

Note that we repeat this algorithm only for those unsatisfied inequalities. Thus, $E[Z_1] \geq E[Z_2] \geq ... \geq E[Z_t]$. Let $x_p$ denote the final solution obtained in Step 4. The expected final solution is

$$E[\sum_{p=1}^{N} x_p] \leq E[\sum_{p=1}^{t} Z_p]$$

$$\leq t \times E[Z_1]$$

$$\leq t \times OPT(LP)$$

$$\leq 2(m + 1) \ln \frac{K(K - 1)}{2} \times OPT(IP)$$

$$= O(m \ln K) \times OPT(IP).$$

With a high probability, the iterative LP-relaxation algorithm stops after $O(m \ln K)$ iterations and finds a solution of $O(m \ln K)$ approximation. ∎

## 30.3 Finding Auxiliary Tag Single Nucleotide Polymorphisms

This section describes an algorithm for finding auxiliary tag SNPs assuming robust tag SNPs have been computed in advance. Given a haplotype block $M_h$ containing $N$ SNPs and $K$ haplotypes, we define $C_{tag} \subseteq C$ as the set of tag SNPs obtained from a haplotype sample and some SNPs in $C_{tag}$ are missing. This haplotype sample may be identified ambiguously due to the lack of missing SNPs. We wish to find the minimum number of auxiliary tag SNPs from the remaining SNPs to resolve the ambiguity. A formal definition of this problem is given in the following.

Problem:   Minimum Auxiliary Tag SNPs (MATS)

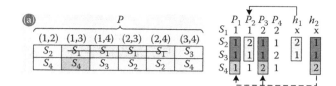

**FIGURE 30.8** An example to find the auxiliary tag SNPs. The SNP $S_1$ is genotyped as missing data and SNP $S_4$ is the auxiliary tag SNP for $h_2$. (a) The table that stores the set of robust tag SNPs.

Input: An $N \times K$ matrix $M_h$, and a set of SNPs $C_{tag}$ genotyped from a sample with missing data.

Output: The minimum subset of SNPs $C_{aux} \subseteq C - C_{tag}$ such that each pair of ambiguous patterns can be distinguished by SNPs in $C_{aux}$.

The following theorem shows the NP-hardness of the MATS problem.

**Theorem 30.5** *The MATS problem is NP-hard.*

*Proof.* Consider that all SNPs in $C_{tag}$ are genotyped as missing data. This special case of the MATS problem is just like finding another set of tag SNPs from $C - C_{tag}$ to distinguish those $K$ patterns, which is already known as NP-hard [11]. ∎

Although the MATS problem is NP-hard, we show that auxiliary tag SNPs can be found efficiently when robust tag SNPs have been computed in advance. Without loss of generality, assume that these robust tag SNPs are stored in an $(m + 1) \times |P|$ table $T_r$ (Figure 30.8a).

Step 1: The patterns that match the haplotype sample are stored into a set $A$. For example (Figure 30.8), if we genotype SNPs $S_1$, $S_2$, and $S_3$ for the sample $h_2$ and the SNP $S_1$ is missing, patterns $P_1$ and $P_3$ both match $h_2$. Thus, $A = \{P_1, P_3\}$

Step 2: If $|A|=1$, the sample is identified unambiguously and we are done (e.g., $h_1$ in Figure 30.8). If $|A| > 1$ (e.g., $h_2$), for each pair of ambiguous patterns in $A$ (e.g., $P_1$ and $P_3$), traverse the corresponding column in $T_r$, find the next unused SNP (e.g., $S_4$), and add the SNP to $C_{aux}$. As a result, the SNPs in $C_{aux}$ can distinguish each pair of ambiguous patterns, which are the auxiliary tag SNPs for the haplotype sample.

The worst case of this algorithm is that all SNPs in $C_{tag}$ are genotyped as missing data, and we need to traverse each column in $T_r$. Thus, the running time of this algorithm is $O(|T_r|) = O(m|P|)$.

## 30.4 Experimental Results

We have implemented the first and second greedy algorithms in JAVA. The LP-relaxation algorithm has been implemented in Perl, in which the LP problem is solved via a program called "lp_solve." [16] The LP-relaxation algorithm is a randomized method. Thus, this program is repeated for 10 times to explore different solutions and the best solution among them is chosen as the output. To compare the solutions (and efficiency) returned by our algorithms with the optimal solution, we also implement a brute force program in JAVA (referred to as "OPT") which enumerates all possible solutions to find the optimal solution. The proposed algorithms along with the brute force program are tested on a variety of simulated and biological data.

### 30.4.1 Results on Simulated Data

We first generate 100 datasets containing short haplotypes. Each dataset consists of 10 haplotypes with 20 SNPs. These haplotypes are created by randomly assigning the major or minor alleles at each SNP locus.

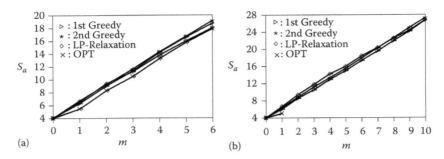

**FIGURE 30.9** Experimental results on random data. (a) Results from datasets containing 10 haplotypes and 20 SNPs. (b) Results from datasets containing 10 haplotypes and 40 SNPs.

Let $m$ be the number of missing SNPs allowed and $S_a$ be the average number of robust tag SNPs over 100 datasets. Figure 30.9a plots $S_a$ with respect to $m$ (roughly corresponding to SNP missing rates from 0% to 33%). When $m = 0$, all programs find the same number of SNPs as the optimal solution. The iterative LP-relaxation algorithm slightly outperforms others as $m$ increases. When $m > 6$, more than 20 SNPs are required to tolerate missing data. Thus, no datasets contain enough SNPs for solutions.

We then generate 100 datasets containing long haplotypes. Each dataset is composed of 10 haplotypes with 40 SNPs. Figure 30.9b illustrates the experimental results on these long datasets (corresponding to SNP missing rates from 0% to 37%). The optimal solutions for $m > 1$ cannot be computed in one day and are not shown in this figure. It is because the number of possible solutions in long datasets is too large to enumerate. On the other hand, both greedy and iterative LP-relaxation algorithms run in polynomial time and always output a solution efficiently. In this experiment, both greedy algorithms slightly outperforms the iterative LP-relaxation algorithm. In addition, the number of SNPs allowed for missing data is larger than those in short datasets. For example, when $m = 10$, all programs output less than 28 SNPs. The remaining SNPs in each dataset are still enough to tolerate more missing SNPs.

Hudson [18] provides a program which can simulate a set of haplotypes under the assumption of neutral evolution and uniformly distributed recombination rate using the coalescent model. We use Hudson's program to generate 100 short datasets with 10 haplotypes and 20 SNPs and 100 long datasets with 10 haplotypes and 40 SNPs. Figure 30.10a shows the experimental results on Hudson's short datasets (corresponding to SNP missing rates from 0% to 23%). The number of missing SNPs allowed are less than that of random data. It is because Hudson's program generates coalescent haplotypes which are similar to each other. As a result, many SNPs cannot be used to distinguish those haplotypes and the amount of tag SNPs is inadequate to tolerate larger missing SNPs. In this experiment, we observe that the iterative LP-relaxation algorithm finds solutions quite close to the optimal solutions and slightly outperforms the other two algorithms.

Figure 30.10b illustrates the experimental results on long datasets generated by Hudson's program (corresponding to SNP missing rates from 0% to 29%). The optimal solutions for $m > 1$ again cannot be computed in one day. In this experiment, the performance of the first greedy and iterative LP-relaxation algorithms are similar, and they slightly outperform the second greedy algorithm as $m$ becomes large.

## 30.4.2 Results on Biological Data

We test these programs on public haplotype data of human Chromosome 21 released by Patil et al. (2001). Patil's data include 20 haplotypes of 24,047 SNPs spanning over about 32.4 MB. Based on the 4,135 haplotype blocks partitioned by Patil et al., we apply all programs to find the robust tag SNPs in each block. Figure 30.11a shows the experimental results on these 4,135 blocks. As there are many long blocks in Patil's data (e.g., more than one hundred SNPs), the optimal solution for $m > 1$ cannot be computed in one day. On the other hand, some short blocks may not have solutions for larger $m$ due to insufficient number of

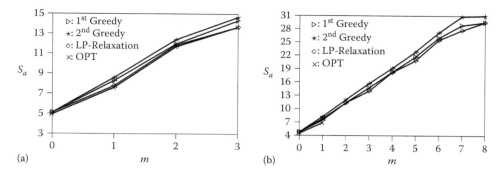

**FIGURE 30.10** Experimental results on Hudson's data. (a) Results from datasets containing 10 haplotypes and 20 SNPs. (b) Results from datasets containing 10 haplotypes and 40 SNPs.

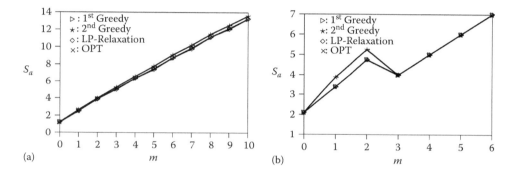

**FIGURE 30.11** Experimental results on biological data. (a) Results from Patil's Chromosome 21 data. (b) Results from Daly's Chromosome 5q31 data.

SNPs in the block. As a consequence, $S_a$ here stands for the average number of robust tag SNPs over those blocks containing solutions. In this experiment, all algorithms find similar number of robust tag SNPs. We observe that the number of robust tag SNPs required in Patil's data is less than those in simulated data. For example, when $m = 8$, all algorithms find less than 12 SNPs in Patil's data, and they find about 28 SNPs in random data and 31 SNPs in Hudson's data. This result implies that genotyping additional tag SNPs to tolerate missing data is more cost-effective on biological data than on simulated data.

Daly et al. [4] studied a 500 kb region on human Chromosome 5q31 which may contain a genetic variant related to the Crohn disease. By genotyping 103 SNPs with minor allele frequency at least 5%, they partition this chromosomal region into 11 haplotype blocks. Figure 30.11b illustrates the experimental results on these 11 blocks. As the blocks partitioned by Daly et al. are very short (e.g., most blocks contains less than 12 SNPs), the optimal solution is still computable. The solutions found by each algorithm is almost the same as optimal solutions. Note that the number of blocks (containing solutions) decreases as $m$ increases. When $m$ increases to 3, some blocks that do not have enough SNPs for a solution are discarded. The remaining blocks requires only 4 SNPs and $S_a$ thus drops down to 4.

### 30.4.3 Discussion

In terms of efficiency, the first and second greedy algorithms are faster than the LP-relaxation algorithm. The greedy algorithms usually returns a solution in seconds and the LP-relaxation algorithm requires about half minute for a solution. It is because the running time of LP-relaxation algorithm is bounded

**TABLE 30.1** The Number of Total Tag SNPs Found by Each Algorithm. The Ratio of Tag SNPs to Total SNPs Is Shown in Parentheses

|  | Random Data | | Hudson's Data | | Patil's Data | Daly's Data |
|---|---|---|---|---|---|---|
| Total blocks | 100 | 100 | 100 | 100 | 4135 | 11 |
| Total SNPs | 2000 | 4000 | 2000 | 4000 | 24047 | 103 |
| 1st Greedy | 400 (20%) | 400 (10%) | 509 (25.5%) | 472 (11.8%) | 4610 (19.2%) | 23 (22.3%) |
| 2nd Greedy | 400 (20%) | 400 (10%) | 509 (25.5%) | 472 (11.8%) | 4610 (19.2%) | 23 (22.3%) |
| LP-relaxation | 400 (20%) | 400 (10%) | 509 (25.5%) | 471 (11.8%) | 4657 (19.4%) | 23 (22.3%) |
| OPT | 400 (20%) | 400 (10%) | 492 (24.6%) | 443 (11.1%) | 4595 (19.1%) | 23 (22.3%) |

by the time of solving the linear programming problem. Furthermore, this LP-relaxation algorithm is repeated and rounded for 10 times to explore 10 different solutions. The brute force program for searching the optimal solution is apparently slower than the others (e.g., taking hours for a solution). The optimal solution usually cannot be found in 24 hours if the size of the block becomes large. When $m$ increases, each block requires more tag SNPs to tolerate more missing SNPs, and the brute force program requires longer execution time to find the optimal solution.

Assuming no missing data (i.e., $m = 0$), we now compare the solutions found by each algorithm with the optimal solution. Table 30.1 lists the numbers of total tag SNPs found by each algorithm in previous experiments. In the experiments on random and Daly's data, the solution found by each algorithm is as good as the optimal solution. In the experiments on Hudson's and Patil's data, these algorithms still find solutions quite close to the optimal solution. For example, the approximation ratios of these algorithms are only $\frac{472}{443} \approx 1.07$ and $\frac{4657}{4595} \approx 1.01$, respectively.

We then analyze the genotyping cost that can be saved by using tag SNPs. In Table 30.1, the ratio of tag SNPs to total SNPs in each dataset is shown in parentheses. The experimental results indicate that the cost of genotyping tag SNPs is much lower than that of genotyping all SNPs in a block. For example, in Patil's data, we only need to genotype about 19% of tag SNPs in each block, which saves about 81% genotyping cost. The genotyping cost saved by using tag SNPs is especially significant in long haplotype blocks. For example, in random and Hudson's long datasets, the saved genotyping cost can be as high as 90%.

Finally, we compute the cost of genotyping extra tag SNPs for tolerating missing data. The biological datasets (i.e., Patil's and Daly's data) have blocks in different sizes. Some of them may not have solutions for larger values of $m$. Therefore, we only consider random and Hudson's 100 datasets in the same size. Each dataset contains 10 haplotypes with 40 SNPs and has solutions for $m$ from 1 to 5. Table 30.2 lists the number of extra tag SNPs used by each algorithm. When $m$ increases to 5, the extra genotyping cost is less than 30% for all algorithms on random data. On the other hand, the extra genotyping cost is higher on

**TABLE 30.2** The Number of Extra Tag SNPs Required to Tolerate Missing Data. The Ratio of Extra Tag SNPs to Total SNPs Is Shown in Parentheses

|  | $m$ | 1 | 2 | 3 | 4 | 5 |
|---|---|---|---|---|---|---|
| Random | 1st Greedy | 200 (5.0%) | 451 (11.3%) | 647 (16.2%) | 889 (22.2%) | 1092 (27.3%) |
| data | 2nd Greedy | 237 (5.9%) | 477 (11.9%) | 714 (17.9%) | 930 (23.3%) | 1144 (28.6%) |
| (4000 SNPs) | LP-relaxation | 262 (6.6%) | 535 (13.4%) | 774 (19.4%) | 1018 (25.5%) | 1194 (29.9%) |
| Hudson's | 1st Greedy | 299 (7.5%) | 656 (16.4%) | 995 (24.9%) | 1351 (33.8%) | 1695 (42.4%) |
| data | 2nd Greedy | 347 (8.7%) | 723 (18.1%) | 1091 (27.3%) | 1439 (36.0%) | 1806 (45.2%) |
| (4000 SNPs) | LP-relaxation | 269 (6.7%) | 657 (16.4%) | 921 (23.0%) | 1344 (33.6%) | 1609 (40.2%) |

Hudson's data due to the coalescent haplotypes. However, in comparison with genotyping all SNPs, the extra genotyping cost is still less than 50% and is thus cost-effective.

# 30.5 Concluding Remarks

In this chapter, we show these exists a set of robust tag SNPs which is able to tolerate a certain number of missing data. Our study indicates that robust tag SNPs is more practical than the minimum tag SNPs if we cannot avoid the occurrence of missing data. We describe two greedy and one LP-relaxation approximation algorithms for finding robust tag SNPs. Our experimental results and theoretical analysis show that these algorithms are not only efficient but the solutions found are also close to the optimal solution. In terms of genotyping cost, we observe that the genotyping cost saved by using tag SNPs can be as high as 90%, and genotyping extra tag SNPs to tolerate missing data is still cost-effective. One future direction is to assign weights to different types of SNPs (e.g., SNPs in coding or non-coding regions), and design algorithms for the selection of weighted tag SNPs.

# References

1. Helmuth, L., Genome research: Map of the human genome 3.0, *Science*, 293(5530), 583, 2001.
2. Hinds, D.A., Stuve, L.L., Nilsen, G.B., Halperin, E., Eskin, E., Ballinger, D.G., Frazer, K.A., and Cox, D.R., Whole-genome patterns of common DNA variation in three human populations, *Science*, 307, 1072, 2005.
3. Bafna, V. and Bansal, V., Improved recombination lower bounds for haplotype data, in *Proceedings of the RECOMB*, 2005.
4. Daly, M.J., Rioux, J.D., Schaffner, S.F., Hudson, T.J., and Lander, E.S., High-resolution haplotype structure in the human genome, *Nat. Genet.*, 29(2), 229, 2001.
5. Patil, N., Berno, A.J., Hinds, D.A., Barrett, W.A., Doshi, J.M., Hacker, C.R., Kautzer, C.R. et al., Blocks of limited haplotype diversity revealed by high-resolution scanning of human chromosome 21, *Science*, 294, 1719, 2001.
6. Halperin, E. and Eskin, E., Haplotype reconstruction from genotype data using imperfect phylogeny, *Bioinformatics*, 20, 1842–1849, 2004.
7. Zhang, K., Qin, Z.S., Liu, J.S., Chen, T., Waterman, M.S., and Sun, F., Haplotype block partition and tag SNP selection using genotype data and their applications to association studies, *Genome Res.*, 14, 908, 2004.
8. Bafna, V., Halldórsson, B.V., Schwartz, R., Clark, A.G., and Istrail, S, Haplotypes and informative SNP selection algorithms: Don't block out information, in *Proceedings of the RECOMB*, 2003, p. 19, Springer, Berlin, Germany.
9. Carlson, C.S., Eberle, M.A., Rieder, M.J., Yi, Q., Kruglyak, L., and Nickerson, D.A, Selecting a maximally informative set of single-nucleotide polymorphisms for association analyses using linkage disequilibrium, *Am. J. Hum. Genet.*, 74, 106, 2004.
10. Halldórsson, B.V., Bafna, V., Lippert, R., Schwartz, R., Vega, F.M., Clark, A.G., and Istrail, S., Optimal haplotype block-free selection of tagging SNPs for genome-wide association studies, *Genome Res.*, 14, 1633, 2004.
11. Zhang, K., Deng, M., Chen, T., Waterman, M.S., and Sun, F., A dynamic programming algorithm for haplotype partitioning, *Proc. Nat. Acad. Sci. U.S.A.*, 99(11), 7335, 2002.
12. Zhang, K., Sun, F., Waterman, M.S., and Chen, T., Haplotype block partition with limited resources and applications to human chromosome 21 haplotype data, *Am. J. Hum. Genet.*, 73, 63, 2003.
13. Garey, M.R. and Johnson, D.S., *Computers and Intractability*, Freeman, New York, 1979.
14. Zhao, J.H., Lissarrague, S., Essioux, L., and Sham, P.C., GENECOUNTING: Haplotype analysis with missing genotypes, *Bioinformatics*, 18, 1694, 2002.

15. Cormen T.H., Leiserson, C.E., Rivest, R.L., and Stein, C, *Introduction to Algorithms*, The MIT Press, Cambridge, MA, 2001.
16. SourceForget. https://sourceforge.net/projects/lpsolve/.
17. Forsgren, A., Gill, P.E., and Wright, M.H., Interior methods for nonlinear optimization, *SIAM Rev.*, 44, 525, 2002.
18. Hudson, R.R., Generating samples under a Wright-Fisher neutral model of genetic variation, *Bioinformatics*, 18, 337, 2002.

# 31

# Large-Scale Global Placement[*]

Jason Cong

Joseph R. Shinnerl

Nearly six decades of steady exponential improvement in the design and manufacture of very large-scale integrated circuits (VLSI) have produced some of the largest combinatorial optimization problems ever considered. *Placement*—arranging the elements of a circuit in the plane—is one of the most difficult of these. As of 2017, mixed integer nonconvex nonlinear-programming (NLP) formulations of placement with more than 100 million variables and constraints are not unusual, and problem sizes continue to grow with Moore's Law. Realistic objectives and constraints for placement incorporate complex models of signal timing, power consumption, wiring routability, manufacturability, noise, temperature, and so on. A popular and very useful simplification is to minimize a standard estimate of total wirelength subject only to pairwise nonoverlap constraints. Although this abstract model problem cannot fully express the scope of the placement challenge, evidence suggests that it does capture a critical part of the core mathematical difficulty [1–4].

Sahni and Gonzalez [5] showed that, unless $P = NP$, no deterministic polynomial-time approximation algorithms for placement exist. In practice, problem sizes and available computing resources prohibit any order of run time beyond approximately $N \log N$, where $N$ is the number of movable objects. Despite these obstacles, progress continues to push back the achievable limits of practical algorithms.

VLSI placement traditionally consists of two stages: (i) global and (ii) detailed. In *global* placement, approximate locations of all modules are computed under relaxed formulations of design constraints. Immediately following global placement, a strictly legal configuration is computed prior to or as part of *detailed placement*, which must maintain strict feasibility of all cell locations. This chapter presents an overview of metaheuristics for global placement. Section 31.1 contains a brief description of the global placement's context and typical abstract formulation. In Section 31.2, dominant placement metaheuristics

---

[*] Partial support for this work has been provided by Semiconductor Research Consortium Contract 2003-TJ-1091 and National Science Foundation Contract CCF 0430077. This chapter is derived from the article Large-scale circuit placement, *ACM Transactions on Design Automation of Electronic Systems*, 10(2), ISSN# 1084-4309 (April) © ACM, 2005. http://doi.acm.org/10.1145/1059876.1059886

are reviewed. In Section 31.3, brief overviews are given of practical formulations modeling signal timing and routability, the identification and placement of regular subcircuits, and placement in three spatial dimensions. Conclusions are drawn in Section 31.4.

# 31.1 Background

Integrated-circuit (IC) design consists of three main stages: behavioral, logical, and physical. The task of *physical design* is to compute a *layout*, that is, spatial positions for all circuit elements and their interconnecting wires, consistent with the logical design. VLSI physical design [6] is usually divided further into several separate but interdependent stages, traditionally performed in the following sequence: placement, clock-network synthesis, routing, and timing-closure optimization. During *placement*, the circuit's movable modules—called *cells*—are arranged without overlap within a two-dimensional rectilinear region of prescribed dimensions. After placement, required connections among modules are explicitly constructed during *routing* as nonintersecting wiring paths in parallel planes over the modules. Wire segments in the same plane are generally oriented in the same direction along one of two possible orthogonal axes, either $x$ or $y$. Successive wire segments along a path are orthogonal, lie in different planes, and must be connected by a *via* in the $z$ direction. Most circuits are *synchronous*, which means that every signal must begin and end each leg of its journey at some state element—register, latch, flip-flop, or primary input or output (PI or PO)—within one clock period of the clock or clocks controlling that leg's local spatial domain. The design of synchronous circuits also includes a clock-network design stage after or jointly with the placement stage.

Wires of synchronous circuits connect placeable cells to four distinct global networks: signal, clock, power-and-ground, and test. The focus of placement is the signal network. Traditionally, the remaining three networks are planned at separate design stages; however, routability of the placement's signal network, that is, the ability to specify physical wire connections between all logically designated pairs of terminals without short or open circuits or design-rule violations, may ultimately depend on routing resource consumption by all of the circuit's networks.

Multiple passes of timing-closure transformations, including module sizing and net buffering—insertion of amplifying signal repeaters along wires—are typically interleaved with the other stages as the placement and routing converge. Deploying larger module sizes and adding net buffers increases the total area and power requirements of the final circuit but are essential for meeting tight delay constraints along timing-critical signal paths. As many as 40% or more of the nodes (or 10% or more by area) in a sub-20 nm design are likely to be buffers. Thus, a successful physical-design flow depends both on the efficiency and quality of results (QoR) of each stage and the ability of each stage to approximate measures from the other stages. In placement, signal delay and routability are the most important approximate measures. Estimated total wirelength and cell-area density ("spread"), as described in the following, serve as simple, robust, and well-correlated global proxies for timing and routability at the earlier stages of placement.

The exponential growth of on-chip complexity has dramatically increased the demand for scalable optimization algorithms for large-scale physical design. Although complex logic functions are usually composed hierarchically, studies [7] show the importance of building a good physical hierarchy from a flattened or nearly flattened logical netlist for performance optimization. When a logical hierarchy is conceived with little or no consideration of the layout and interconnect information, it may not map well to a two-dimensional layout. Therefore, large-scale global placement on a nearly flattened netlist is needed for physical hierarchy generation to achieve the best performance.

This approach is even more important in today's nanometer designs, where the interconnect has become the performance bottleneck. Interconnect delay does not scale linearly unless the maximum signal-propagation distance between cells is limited by insertion of buffers. As module sizes have decreased, their internal signal delays have become small compared with the delays of the wires joining them, especially when the modules are far apart. Placement determines the interconnect structure and, hence, the performance of the resulting circuit is more than any other step in the VLSI design sequence. Thus, the continued exponential decrease of circuit element sizes has increased the relative importance of placement in the VLSI design flow.

## 31.1.1 Mathematical Formulation

An instance of the VLSI placement problem is specified as a hypergraph netlist $\mathcal{H} = (\mathcal{V}, \mathcal{E})$ which is the output of logic synthesis. The terminology is illustrated in Figure 31.1. The vertices $v_i \in \mathcal{V}$ of $\mathcal{H}$ are rectangular modules of prescribed functionality. Each hyperedge or *net* $e \in \mathcal{E}$ is essentially a subset of the vertices, $e \subset \mathcal{V}$. More precisely, because vertices have size and shape, a net is defined as a set of pins—electrical connection points on vertices—one pin per vertex per net.

During global placement, it is customary to project the modules $v \in \mathcal{V}$ into two dimensions ("2-D") and consider only their *widths* along the $x$-direction and *heights* along the orthogonal $y$-direction. Most modules are *standard cells* selected from a given library during *technology mapping*. Such cells consist of up to several dozen logic gates[*] each and have fixed dimensions. Cell widths are diverse integer multiples of unit length, but cells in the same connected subregion have either the same, fixed, uniform height, or some small multiple of that reference height, called "row height." The placement region $\mathcal{R}$ has fixed rectilinear shape ($\mathcal{R}$ is a finite union of axis-aligned rectangular regions) and is partitioned into regions of rows of uniform height. The height of each row equals the standard-cell row height for the region. Increased design complexity has brought ever greater reuse of larger intellectual-property (IP) blocks called *macros*, which may be thousands of times as large as the standard cells and thus span dozens to thousands of standard rows. The multiplicity and diversity of macro sizes in large-scale mixed-size placement presents a particular challenge to the development of robust heuristics.

Let $(x_i, y_i)$ denote the coordinates of the lower-left corner of module $i$, and let $(w_i, h_i)$ denote its width and height. Let $x$ and $y$ denote corresponding vectors of the $x_i$ and $y_i$. An accurate calculation of routed wirelength requires both detailed modeling of 3-D routing topologies and an actual routing solution; projected 2-D examples are illustrated in Figure 31.3c and d. For simplicity, the following *nonconstructive* 2-D bounding-box half-perimeter wirelength (HPWL) approximation depicted in Figure 31.3a is used as a substitute. The bounding-box wirelength of a single net $e \in \mathcal{E}$ is simply half the perimeter of its smallest circumscribing rectangle,

$$w(e) = \max_{v_j \in e} \left(x_j + w_j\right) - \min_{v_k \in e} x_k + \max_{v_m \in e} \left(y_m + h_m\right) - \min_{v_n \in e} y_n. \tag{31.1}$$

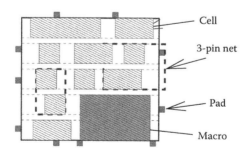

**FIGURE 31.1** A sample placement illustrating basic terminology. Cells and macro (lower right) are shaded. Two nets are shown as dashed lines enclosing subsets of cells and pads.

---

[*] For example, NAND, XOR, or NOT gates, often with different driving strengths.

The corresponding weighted HPWL for a given placement is thus

$$f(x, y) = \sum_{e \in E} \gamma(e)w(e), \tag{31.2}$$

where the weights $\gamma(e)$ may be chosen adaptively for various purposes, for example, dynamically prioritizing nets by timing criticality and/or switching activity (Section 31.3).

It is emphasized that HPWL serves only as a crude approximation of total wirelength, effective primarily for early-stage global placement. As the placement matures, more accurate approximations become necessary, such as the projected 2-D rectilinear steiner models depicted in Figure 31.3c and d. Such *constructive* models generally require execution of a routing algorithm and add complexity associated with alternative possible routing topologies, as shown.

In general, several of the modules' locations will be fixed a priori, but the number of fixed modules is a small fraction (typically, order $1/\sqrt{|\mathcal{V}|}$) of the total.

Given the shapes of the modules and a specification of their hypergraph netlist $\mathcal{H}$, a precise formulation of placement constrains all modules to lie aligned with standard cell row boundaries (macros span multiple rows) with no two modules overlapping; this view amounts to a mixed integer NLP problem (NLP). What distinguishes global placement from general and detailed placement is its approximation of the nonoverlap constraints. Typically, this approximation is expressed simply as a collection of simple constant upper bounds $u_{ij}$ of module areas in subregions ("bins") $B_{ij}$ defined by a regular $m \times n$ rectangular grid $G$. Thus, the global placement model problem may be expressed as a nonconvex, NLP problem,

$$\begin{aligned} \min_{x,y} \quad & \sum_{e \in E} \gamma(e)w(e) \\ \text{subject to} \quad & \sum_{v \in \mathcal{V}} \text{area}\,(v \cap B_{ij}) \;\le\; u_{ij} \quad \text{for all } i \in \{1, \ldots m\},\, j \in \{1, \ldots n\}, \end{aligned} \tag{31.3}$$

where in this formula $v$ and $B_{ij}$ are viewed as rectangles in the plane. The resolution of the grid is usually determined by some empirical estimate of the capabilities and limitations of the legalization and detailed placement steps that follow. Illustrations of a global placement and a corresponding detailed placement, both produced by mPL5 [8] on a synthetic benchmark with over 1 million modules and nets, are shown in Figure 31.2.

While there is general agreement on formulation (31.3) among active researchers, the actual formulation of global placement used, if it is formally stated at all, is often tailored adhoc to suit a given algorithm.

Center-to-center HPWL = 440715913.
Pin-to-pin HPWL = 301919591.

Center-to-center HPWL = 420180624.
Pin-to-pin HPWL = 28251408.

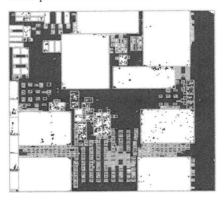
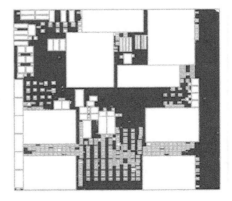

**FIGURE 31.2**  A global placement (left) with a corresponding detailed placement on a synthetic benchmark with over 1M movable objects and hundreds of spatially fixed macros. For simplicity, nets are not shown.

New variations continue to be investigated. For example, (31.3) is often viewed as a discretization of continuous area-density function $d(\bar{x}, \bar{y})$ defined at every point $(\bar{x}, \bar{y}) \in \mathcal{R}$. Extensions or alternatives that incorporate more detailed modeling of complex objectives and constraints, such as routability, signal propagation times, maximum temperature, noise, and so on, are crucial in practice.

## 31.2 Overview of Dominant Metaheuristics

Heuristics for placement may be broadly classified as either hierarchical or flat. Prior to the 1980s, most research focused on flat heuristics for instances up to at most a few thousand modules and nets [9]. With the explosion in instance sizes due to Moore's Law, much research in the 1990s and early 2000s shifted to hierarchical formulations as a means to fast and scalable implementations [10]. Since then, attention has largely returned to fast and scalable flat representations, but in practice, the flat and hierarchical views are combined in various ways. Flat heuristics typically play an enabling role at each level of a hierarchical algorithm. Conversely, flat formulations may rely on hierarchical numerical schemes to accelerate their internal calculations, for example, by linear-system preconditioning or iterative constraint refinement (Section 31.2.2.3).

### 31.2.1 Flat Improvement Heuristics

In their survey article of 1972, Hanan and Kurtzberg [9] divided placement techniques into three categories: (i) constructive initial placement, (ii) iterative placement improvement, (iii) branch and bound. Constructive initial placement incrementally selects and places unplaced movable modules according to the strength of their connectivity to already fixed modules (most circuits have at least some small subset of terminals fixed a priori). The process is simple and fast, but neglecting connections among movable modules diminishes the quality of the final placement. Branch and bound is far more accurate and also constructive but is affordable only for subproblems of approximately 15–20 modules or fewer.

Of these three early kinds of placement techniques, iterative improvement is the only one still widely used. In this approach, a given placement is repeatedly modified as long as sufficient reduction in the objectives is obtained. Early usage of the term is usually restricted to sequences of strictly *feasible*, that is, overlap-free, placements. Although many iterative heuristics today also generate sequences of strictly feasible placements, other *infeasible* methods attain legality only approximately or asymptotically. Most feasible heuristics are either discrete or linear. Dominant infeasible heuristics include nonlinear, analytical formulations such as force-directed methods. Modification strategies may be randomized or deterministic, localized or global.

#### 31.2.1.1 Iterative Improvement over Feasible Placements

Perhaps the simplest but least efficient placement procedure is the global *Monte Carlo* strategy attempted in early work [9]. In this approach, all modules are randomly assigned positions according to a given probability distribution. The resulting placement is retained if and only if it produces lower cost than previously obtained placements. Although the probability distribution can be dynamically adapted to push modules toward subregions likeliest to produce lower cost, results are not generally competitive. Currently, the most successful randomized algorithms employ sequences of local moves guided by *simulated annealing* [11–16]. A given placement is endowed with a neighborhood structure. A pair of neighboring modules is randomly selected, and the change in cost $\Delta C$ associated with exchanging the two modules' positions is computed. If $\Delta C < 0$, then the modules' positions are swapped—the move is accepted. If $\Delta C \geq 0$, then the move is accepted with probability proportional to $e^{-\Delta C/T}$, where $T$ is the parameter simulating temperature. Initially, $T$ is set large, so that *hill-climbing* moves are accepted with relatively high probability. Eventually, $T$ is decreased far enough that such uphill moves are essentially excluded. When $T$ is decreased sufficiently slowly, certain theoretical guarantees exist for asymptotic convergence of the process to a global optimum. In practice, a far more rapid decrease in $T$ must be used to keep run times acceptable. The main drawback of simulated annealing (SA) is its inherent lack of scalability. Genetic algorithms and simulated evolution have also been used in placement [17], but to our knowledge, are not directly used by leading tools.

Early deterministic heuristics include techniques based on linear assignment, network flows, and force equilibration. In a typical assignment-based scheme, a subset of movable vertices is selected. If all movable vertices have the same dimensions, and no movable vertex shares a hyperedge with any other, then linear assignment (bipartite matching) can be used to determine an optimal permutation of the movable vertices over their set of locations. Construction of multiple subsets of nonadjacent vertices by iterative deletion is simple and fast on hypergraphs of bounded degree.

A generalization of this approach, called relaxation-based local search, is introduced by Hur and Lillis [18,19]. In this scheme, the optimal locations of all vertices in a given movable subset are simultaneously determined without regard to overlap via solution of two separate rectilinear distance facility location subproblems, one for each coordinate direction. In the $x$-direction, the problem may be written as follows. Let $M \subset E$ denote the set of nets containing movable vertices; $M$ typically also contains many other, fixed vertices. Variables $r_i$ and $l_i$ are introduced to represent the right and left boundaries of the $i$th net $e_i \in M$.

$$\text{min} \quad \sum_{\{e_i \in M\}} r_i - l_i$$
$$\text{s.t.} \quad l_i \leq x_j \leq r_i \quad \text{for all } v_j \in e_i$$

This problem can be solved efficiently by either a sequence of related network flows or by a single network flow applied to its dual. Once the subproblem has been solved and the optimal locations are determined, cell swapping along monotone chains of bins ("ripple-move") is used to restore area-density legality. For each overfull bin $s$, a nearest underfull bin $t$ is selected, and a chain of cell swaps between neighboring bins $(a_i, a_{i+1})$ leading from $s = a_1$ to $t$ is computed. The chain is monotone in the sense that the Manhattan distance between $a_i$ and $t$ strictly decreases with $i$; this property and memoization reduce computational overhead. Network flows have also been used extensively in legalization and detailed placement algorithms [20,21].

### 31.2.1.2 Iterative Improvement over Infeasible Placements

Contemporary approaches to iterative improvement fall mostly among the so-called analytical methods based on mathematical programming or the equilibration of simulated forces. Generally, these methods all use continuous approximation and seek to satisfy a set of computationally verifiable optimality conditions either repeatedly for sequences of subproblems or asymptotically for the entire circuit. In contrast to the methods described in the previous subsection, they do *not* normally terminate at overlap-free configurations, and they therefore require post processing by a legalization engine.

The idea of modeling an integrated circuit as a system of springs and masses dates back at least as far as 1967 [22]. The mass of a vertex is taken in proportion to its area. The force on a mass $i$ due to mass $j$ is defined by Hooke's law: $F_{ij} = k_{ij} s_{ij}$. Vector $s_{ij}$ is the displacement from the position of $i$ to that of $j$. Spring constant $k_{ij}$ is proportional to the total relative strength of all hyperedges containing both $i$ and $j$. The precise form of $k_{ij}$ amounts to a prescription for approximating netlist $H$ by a graph $G$; for example, $k_{ij} = \sum_{i,j \in e} w(e)/(|e| - 1)$. One simple form of iteration attempts to move vertices one by one to the available location nearest where the sum of the forces on them is zero. Later work of Quinn and Breuer [23] applies a Newton-based algorithm to a simultaneous systems of nonlinear equations for force equilibrium.

Other abstractions of force simulations have been proposed for placement. For example, Cheng and Kuh [24] proceed by an analogy to the minimization of power dissipation in an electrical network. More generally, any explicit modeling of physical forces can be abandoned in favor of a simple mathematical model in which a quadratic, graph-based wirelength approximation is minimized without regard to overlap constraints. This simple unconstrained quadratic placement is widely used to generate both initial placements and subsequent refinements, both local and global. Fixed terminals at various locations, whether native to the given netlist or artificially introduced by the placement algorithm, tend to spread cells enough that cell centers rarely coincide, thus enabling subsequent spreading based on relative-order heuristics [25,26]. Judicious iterative addition and adjustment of fixed pseudo terminals is used by FastPlace [27] and mFAR [28,29] both for accelerating convergence and improving quality.

#### 31.2.1.2.1 Example: Kraftwerk

Seminal work by Eisenmann and Johannes [30] formulates force-directed placement as a sequence of unconstrained quadratic minimizations. The perturbed quadratic-wirelength objective function

$$q(x,y) = \frac{1}{2}(x^T Q x + y^T Q y) + b_x^T x + b_y^T y + f_x^T x + f_y^T y$$

captures both netlist connectivity and area congestion by a graph approximation and force-field calculation, as follows. Each net of 3 or more vertices is decomposed into a collection of simple 2-vertex graph edges by means of a quadratic star-wirelength net model [25] (Figure 31.3b). Cell-to-cell connections determine the off-diagonal entries and part of the diagonal entries in the fixed graph Laplacian matrix $Q$. Cell-to-pad connections contribute to the diagonal elements of $Q$, rendering it positive definite, and determine the linear-term coefficients in the right-hand-side vector $b = (b_x, b_y)$. Viewing this vector $b$ as external spring-like forces following Hooke's law, the circuit connectivity is represented by the (constant) symmetric-positive-definite matrix $Q$ and the vector $b$. The perturbation vector $f = (f_x, f_y)$ represents global area-density distribution forces subject to the following assumptions.

1. For a given placement, the density force acting on a particular cell depends only on the coordinates of that cell.
2. High-density regions are sources of density force. Low-density regions are sinks of density force.
3. Lines of the density force do not form closed loops.
4. The density force goes to zero at infinity.

Under these assumptions, these forces are expressed as negative gradient $f = -\nabla \psi$ of a potential $\psi$ satisfying Poisson's equation*,

$$\nabla \cdot \nabla \psi(\bar{x}, \bar{y}) = -\rho(\bar{x}, \bar{y})$$
$$\int \int_R \psi(\bar{x}, \bar{y}) d\bar{x} d\bar{y} = \int \int_R \rho(\bar{x}, \bar{y}) d\bar{x} d\bar{y}. \tag{31.4}$$

This insight is a key advance, incorporating non local or *far-field* cell-area balancing terms into the spreading force. Thus, at each iteration, vector $f$ is recalculated from the current cell positions by means of a fast Poisson-equation solver. As $Q$ does not change from one iteration to the next unless nets are reweighted, a hierarchical set of approximations to $Q$ can typically be reused over several iterations. We refer to this approach as *Poisson-based* quadratic placement.

Significant improvements to this formulation are introduced in Kraftwerk2 [1,31], including the bound-to-bound (B2B) wirelength model, large-macro density scaling, and a monotonic spreading schedule derived from target displacement measure $\mu_T$ and the "hold force" wirelength gradient bias. The B2B model is illustrated in Figure 31.3e and f. As with HPWL, it decouples in $x$ and $y$. We describe only the $x$-component of the measure; the $y$-component is analogous. For a given net $e$ in a current placement, let $j \in e$ denote the index of a pin of least $x$ and $k \in e$ the index of a pin of greatest $x$ over all pins in $P$-pin net $e$, that is,

$$x_j \le x_i \le x_k \quad \text{for all} \quad i \in e.$$

---

* As $(x, y)$ represents the vectors of module placement coordinates, we use $(\bar{x}, \bar{y})$ to represent an arbitrary point in $R$.

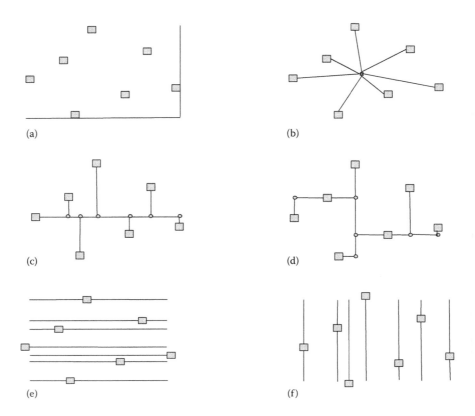

**FIGURE 31.3** Alternative 2-dimensional wirelength models for identical placements of the same 7-pin net: (a) half-perimeter (b) euclidean star (c) rectilinear steiner (d) rectilinear steiner with a different topology (e) bound-to-bound in $x$-direction (f) bound-to-bound in $y$-direction.

The B2B wirelength model in $x$ is

$$B2B_x(e) = w_{kj}^x(x_k - x_j)^2 + \sum_{i \in e | i \neq j, i \neq k} w_{ij}^x(x_i - x_j)^2 + w_{ik}^x(x_k - x_i)^2, \qquad (31.5)$$

where each weight $w_{pq}^x = 1/((P-1)|x_p - x_q|)$ is chosen so that the total $B2B_x(e) = HPWL_x(e) = x_k - x_j$ at the current placement where the model is formed. Error in the B2B model relative to HPWL is introduced at each placement step, but recalibrating the model—resetting weights $w_{ij}^x, w_{ij}^y$ to match HPWL at the current placement—keeps this error well below all previously published quadratic approximations to HPWL [31].

Kraftwerk2 terminates when module overlap area decreases below 20% of total module area; this typically happens after about 25 iterations. Appropriate scaling and modulation of forces in Kraftwerk presents a major challenge to implementors [32] and has been viewed by some as an impediment to consistent QoR [33].

### 31.2.1.2.2 Example: ePlace

ePlace [34,35] directly solves the NLP form of placement (31.3) using Nesterov's method [36] to compute placement iterates from penalty functions [37,38] of the form

$$f_k(x, y) = W(x, y) + \lambda_k N(x, y). \qquad (31.6)$$

Here $W(x, y) = W(x) + W(y)$ denotes the *weighted-average* approximation [39] to HPWL, which in the $x$-direction for one net $e$ is

$$W_{x,y}(e) = \left( \frac{\sum_{i \in e} x_i \exp(x_i/\gamma)}{\sum_{i \in e} \exp(x_i/\gamma)} - \frac{\sum_{i \in e} x_i \exp(-x_i/\gamma)}{\sum_{i \in e} \exp(-x_i/\gamma)} \right), \tag{31.7}$$

and $N(x, y) \equiv \frac{1}{2} \sum_i q_i \psi_i(x, y)$ denotes Poisson-based area-density potential summed over all placeable objects, $q_i$ the area of cell $i$. In contrast to prior academic NLP-based methods—mPL, APlace, NTU-place [1,8,40,41]—ePlace is flat; that is, it does not use any clustering or grid coarsening (Section 31.2.3) whatsoever. Like Kraftwerk and mPL, ePlace uses a Poisson-based model of cell-area density (31.4), but with some important differences. First, global area supply and demand are not only equated but also normalized to zero:

$$\int \int_R \psi(\bar{x}, \bar{y}) d\bar{x} d\bar{y} = \int \int_R \rho(\bar{x}, \bar{y}) d\bar{x} d\bar{y} = 0. \tag{31.8}$$

Normalizing to zero annihilates constant solutions, ensuring uniqueness and eliminating the need for any Helmholtz $\epsilon$-regularization as in mPL (Section 31.2.3). Second, as in mPL, explicit use of Neumann boundary conditions is made:

$$\hat{\mathbf{n}} \cdot \nabla \psi(\bar{x}, \bar{y}) = 0, \quad (\bar{x}, \bar{y}) \in \delta R, \tag{31.9}$$

where $\delta R$ is the boundary of $R$, and $\hat{\mathbf{n}}$ is the outward unit normal vector to $\delta R$. Third, to support fast numerical solution by fast fourier transform (FFT)-based spectral methods, the cell-area density $\rho(\bar{x}, \bar{y})$ is extended periodically from its original domain $[0, m-1] \times [0, m-1]$ to the entire plane $[-\infty, +\infty] \times [-\infty, +\infty]$.

As in References 8, 42, wirelength degradation due to overspreading is prevented by introduction of unconnected "filler cells" whose total area $A_{fc}$ equals the total available white space in the design,

$$A_{fc} = \rho_t A_{ws} - A_m,$$

where $A_m$ is the total area of all placeable objects in the netlist, $\rho_t \in (0, 1)$ is the target placeable-cell area utilization[*], and $A_{ws}$ is the total area of $R$ available for placement. Filler cells are uniformly sized to either (a) the average of the 80% of movable cells obtained by excluding the largest 10% and smallest 10%, or (b) the size of a density bin, if the native design utilization $A_m/A_{ws}$ is small. These cells are placed but are not in the netlist. They are analogous to slack variables in mathematical programming, in effect converting the bin-density area constraints (31.3) to equality constraints. They support fast convergence to spatially nonuniform placement solutions.

ePlace uses a single, fixed-resolution $m \times m$ grid, with $m = \lceil \log_2 \sqrt{n'} \rceil$, where $n'$ is the total number of objects placed, including artificial filler cells. Its initial placement minimizes HPWL without regard to cell area, using iterated, unconstrained minimization of the quadratic B2B approximation (31.5). Penalty-function-based iterations by Nesterov's method then proceed on (31.6) until $\tau \leq 0.1$, where $\tau$ is total excess occupied bin area divided by total movable cell area. At each placement iteration, ePlace does not actually minimize penalty function (31.6). Rather, it uses a single step of Nesterov's method Equation [36] (31.11) with diagonal preconditioning to compute the next placement iterate, then it updates its internal parameters $\lambda$ and $\gamma$ as follows.

---

[*] $\rho_t$ is typically set below 1 to support subsequent routing and timing-closure design stages.

Penalty parameter $\lambda_k$ (31.6) is initially set to balance the relative contributions of wirelength and density in the penalty-function gradient, $\lambda_0 = ||\nabla W||_1/||\nabla N||_1$. After each iteration, $\lambda_k$ is updated by the safeguarded formula $\lambda_{k+1} = \mu_k \lambda_k$, where

$$\mu_k = \max\left(0.75, \min(1.1, \mu_0^{1.0 - \frac{\Delta HPWL_k}{\Delta HPWL_{ref}}})\right),$$

where $\Delta HPWL_k = HPWL(v_k) - HPWL(v_{k-1})$ is the change in HPWL between successive reference iterates in Nesterov's method, and $\Delta HPWL_{ref}$ is expected wirelength increase per iteration ($\mu_0$ is set to 1.1, and $\Delta HPWL_{ref}$ is approximately 0.1% of total final HPWL).

Wirelength smoothness parameter $\gamma$ (31.7) is set to depend linearly on bin width $w_b$ but exponentially on bin-area overflow ratio $\tau$ as follows:

$$\gamma(\tau) = 0.8w_b \times 10^{k\tau + b}, \tag{31.10}$$

where $k \equiv 20/9$ and $b \equiv -11/9$ ensure that initially, $\gamma(1.0) = 80w_b$ and finally, $\gamma(0.1) = 0.8w_b$, which are found empirically to produce the best results.

Each iteration of Nesterov's method [36] makes the following updates

$$\begin{aligned}
u_{k+1} &= v_k - \alpha_k \nabla f_k \\
a_{k+1} &= \left(1 + \sqrt{4a_k^2 + 1}\right)/2 \\
v_{k+1} &= u_{k+1} + (a_k - 1)(u_{k+1} - u_k)/a_{k+1}
\end{aligned} \tag{31.11}$$

to proposed solution $u_{k+1}$ from "reference solution" $v_k$ and penalty gradient $\nabla f_k \equiv \nabla f(u_k)$, with "steplength" $\alpha_k$ satisfying sufficient decrease criterion

$$f(u_k) - f(u_{k+1}) = f_k - f(v_k - \alpha_k \nabla f_k) \geq 0.5\alpha_k ||\nabla f_k||^2, \tag{31.12}$$

where $f_k \equiv f(u_k)$ is the penalty objective (31.6) evaluated at $u_k$. To formally ensure convergence, it is necessary to find such $\alpha_k$ efficiently by *linesearch*. In ePlace, however, strict adherence to this rule is abandoned in favor of safeguarded "linesearch prediction," with $\alpha_k$ *provisionally* set to an estimated inverse Lipschitz constant,

$$\alpha_k = \frac{||v_k - v_{k-1}||}{||\nabla f(v_k) - \nabla f(v_{k-1})||},$$

but forced to remain between strict lower and upper bounds, which are also dynamically updated. By this rule, monotonic decrease in $f$ may be sacrificed, but run time is dramatically reduced, as the "linesearch bottleneck," that is, long sequences of tiny steps forced by monotonic decrease requirements, is avoided.

### 31.2.1.2.3 Hierarchical Methods

Dominant hierarchical metaheuristics used in placement are (1) recursive partitioning, discussed next, and (2) multilevel methods, a.k.a. multiscale methods, discussed in Section 31.2.3.

## 31.2.2 Recursive Partitioning

Top-down methods rely on variants of recursive circuit partitioning. Seminal work on partitioning-based placement was done by Breuer [43] and Dunlop and Kernighan [44]. Subsequent methods have exploited further advances in fast multiscale algorithms for cutsize-driven hypergraph partitioning [45–47] or displacement-minimizing partitioning [26,33,48] to push these frameworks beyond their original capabilities. Fast, high-quality $\mathcal{O}(N)$ partitioning algorithms give top-down partitioning attractive $\mathcal{O}(N \log N)$ scalability overall.

### 31.2.2.1 Cutsize Minimization

At a given level of the top-down hierarchy, each rectangular subregion $S$ and the modules assigned to it are bipartitioned, that is, $S$ is split by a horizontal or vertical *cutline* into two disjoint rectangular subregions $S_1$ and $S_2$, and each module assigned to $S$ is assigned to either $S_1$ or $S_2$. Most partitioning-based top-down placers employ variations of multilevel [45–47] Fiduccia-Matheysses (FM) style [49] iterations to separate the modules. Given some initial partition, subsets of cells are moved across its cutline in a way that reduces the total weight of hyperedges cut without violating a given area-balance constraint (a hyperedge is *cut* if it contains modules in both subsets of the partition). Academic tools based on recursive cutsize-driven partitioning include Capo [50,51] and Feng Shui [52,53]. Spatial cutlines for subregions, either horizontal or vertical, can be carefully chosen, for example, by dynamic programming [54], such that subregion aspect ratios remain bounded. As the recursion proceeds, cell subsets become smaller, and the cell-area distribution over the placement region becomes more uniform. Base cases of the bipartitioning recursion are reached when cell subsets become small enough that special end-case placers can be applied [55]. A small example is illustrated after three levels of bipartitioning in Figure 31.4.

Netlist bipartitioning can be enhanced in a few important ways to support the ultimate goal of wirelength-driven circuit placement. Key considerations include (1) terminal propagation, (2) subproblem ordering, (3) cutline placement, (4) handling small balance tolerances and/or highly nonuniform module areas.

Connections between subregions can be modeled by *terminal propagation* [44,56], in which the usual cutsize objective is augmented by terms incorporating the effect of connections to external subregions. At early stages, when the best positions of these connection points are not clear, several iterations may be used to incorporate feedback [52,57]. Other techniques for organizing local partitioning subproblems use Rent's rule to relate cutsize to wirelength estimation [54,58].

Careful consideration of the order and manner in which subregions are selected for partitioning can be significant. In the *multiway partitioning* framework, intermediate results from the partitioning of each subregion are used to influence the final partitioning of others. Explicit use of multiway partitioning at each stage can in some cases bring the configuration closer to a global optimum than is possible by

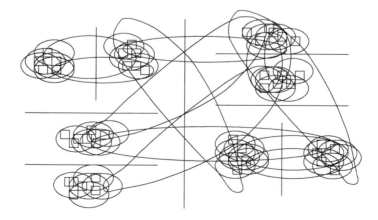

**FIGURE 31.4** Cutsize-driven partitioning-based placement. Rectangles represent movable cells, line segments represent cutlines, and ellipses and other closed curves represent nets. The recursive bipartitioning attempts to minimize the number of nets containing cells in more than one subregion.

recursive bisection alone [52]. Cell replication and iterative deletion have been used for this purpose [59]. Rather than attempting to find the best subregion in which to place a cell, one can replicate the cell enough times to place it in once in every subregion, then iteratively delete only the worst choices. These iterations may continue until only one choice remains, or they may be terminated earlier, allowing a small pool of candidates to be propagated to and replicated at finer levels. By postponing further deletion decisions until better information becomes available, spurious effects from locally optimal subregion partitions can be diminished and the global result improved.

In Capo, horizontal cuts are constrained to lie between uniform-height rows of the standard-cell layout. Respecting standard cell row boundaries in this fashion greatly facilitates legalization of the final global placement. However, one study shows [60] that this restriction occasionally overconstrains end cases and increases wirelength. The authors of this study show that Feng Shui's "fractional-cut" relaxation of row boundaries during the partitioning can considerably improve results, when it is followed by careful displacement-minimizing legalization, such as dynamic-programming based row-assignment.

Much of CAPO's performance derives from its placement-driven enhancements to its core FM partitioner [61,62] to support nonuniform module sizes and tight area-balance constraints. Given any initial partition, FM considers sequences of single, maximum-gain cell moves from one partition block to the other. It maintains a list of "buckets" for each partition block, where the $k$th bucket in each list holds the vertices which, when moved to the opposite block, will reduce the total number of nets cut by $k$. However, a cell will not be moved if the move violates the vertex-weight (area) balance constraint. A large module in an FM gain bucket must not prevent other modules of equal gain from being considered for movement. Capo starts each bipartitioning subproblem with a relaxed area-balance constraint and gradually tightens the constraint as partitioning iterations proceed. As the balance tolerance decreases below the area of any cell, that cell is locked in its current partition block. The final subproblem balance tolerance is selected so that, given an initial whitespace budget, enough relative whitespace in endcase subproblems is ensured that overlap-free configurations can typically be found.

### 31.2.2.2 Incorporating Advances in Floorplanning

Fixed-outline *floorplanning* may be viewed as a generalization of placement, in which each module to be placed has fixed, prescribed area, but some *soft* modules have unspecified rectangular shape. Incorporation of techniques for fixed-outline floorplanning into placement algorithms has improved (a) handling of large macro blocks [63], (b) final legalization of end-case subproblems, and (c) the placement and shaping of clusters of cells in a multiscale algorithm [64]. Alternative approaches to (a) and (b) are taken by Capo [50,51] Patoma/PolarBear [65,66], and Scampi [67].

#### 31.2.2.2.1 Example: Capo

In Capo [1,50,51], min-cut placement proceeds as described earlier until certain adhoc tests suggest that legalization of a subset of macro blocks and cells within their assigned subregion may be difficult. At that point, the cells in that subregion are aggregated into soft clusters, and annealing-based fixed outline floorplanning is applied to the given subproblem [68]. If it succeeds, the macro locations in its solution are fixed. If it fails, it must be merged with its sibling subproblem, and the merged parent subproblem must then be floorplanned. This step therefore initiates a recursive backtracking through ever larger ancestor subproblems. The backtracking terminates when one of these ancestor subproblems is successfully floorplanned. The adhoc tests are chosen to prevent long backtracking sequences on most cases.

#### 31.2.2.2.2 Examples: Patoma and PolarBear

Patoma and PolarBear [66,69] ensure the legalizability of subproblems within min-cut-partitioning-based floorplanning and placement, respectively. Beginning with the given instance itself, PolarBear employs fast and scalable area-minimizing floorplanning before cutsize-driven partitioning to confirm that the problem can be legalized as given. This area-driven "prelegalization" ignores wirelength but serves as a guarantor of the legalizability of subsequent steps. Given the guarantor legalization at a given level, cutsize-driven partitioning proceeds at that level. The flow then proceeds recursively on the subproblems generated by the cutsize-driven partitioning, each subproblem being legalized before it is solved. When prelegalization fails, the failed subproblem is merged with its sibling, and the previously computed legal guarantor solution to this parent subproblem is improved to reduce wirelength. The flow thus guarantees the computation of a legal placement or floorplan, under the modest assumption that the initial attempt to prelegalize the given instance succeeds.

#### 31.2.2.2.3 Example: Scampi

Ng et al. observe that neither Capo nor Patoma performs adequately on the most difficult floorplanning instances [67]. Capo's fixed-outline floorplanner may time out when it falls back to larger subproblems. Patoma's earliest look-ahead guarantor solutions essentially ignore wirelength altogether and thus may exhibit unacceptably high wirelength. Ng et al. demonstrate however that the min-cut placement with look-ahead floorplanning heuristic can still produce high-quality legal placements on the same test cases where Capo and Patoma fail to find acceptable solutions. They reduce the run time of annealing-based look-ahead floorplanning by clustering both standard cells and macros. The largest macros are marked fixed after successful floorplanning steps, and a B*-tree representation [70] is used to prevent overlap with them at subsequent steps. Top-down white space redistribution in effect rebalances the difficulty of child subproblems.

### 31.2.2.3 Partitions Guided by Analytical Placements

An oft-cited disadvantage of cutsize-driven recursive bisection is its tendency to ignore global wirelength as it pursues locally optimal partitions. Approximating wirelength by cutsize in the objective may also degrade the quality of the final placement. A radically different approach, first introduced in Proud [24,71] and subsequently refined by Gordian [25,72], BonnPlace [26,48], and Warp [73], is to use continuous, iteratively-constrained, quadratic star-model wirelength minimization over the entire circuit to guide partitioning decisions. A quadratic-wirelength objective helps avoid long wires and facilitates the construction of efficient numerical linear-system solvers for the optimality conditions, for example, preconditioned conjugate gradients. Fixed I/O terminals prevent modules from simply collapsing to a single point. Linear wirelength can still be asymptotically approximated by iterative adjustments to the net weights [72].

Following this "analytical" placement, each region is then quadrisected, and cells are assigned to subregions to further reduce overlap and area congestion. In Gordian, carefully chosen cutlines and FM-based cutsize-driven partitioning and repartitioning are used. Cell-to-subregion assignments are loosely enforced by imposing and maintaining a single center-of-mass equality constraint for each subregion. As constraints accumulate geometrically, degrees of freedom in cell movement are eliminated, and the quadratic minimization at each step moves cells less and less. In BonnPlace, modules are quadrisected in a manner that essentially minimizes the sum of their rectilinear displacements from their starting positions [74]. BonnPlace does not explicitly impose equality constraints into the subsequent analytical minimization to preserve these partitioning assignments, as Gordian does. Instead, it directly alters the quadratic-wirelength objective to minimize the sum of all cells' squared Manhattan displacements from their assigned subregions.

#### 31.2.2.3.1 Example: Grid Warping

The novelty of grid warping [73] is that, rather than directly move modules based on their spatial distribution, it uses the modules to deform or *warp* the region in which they lie, in an analogy with gravity as described by Einstein's general relativity. The inverse of the deformation is then used to carry modules from their original locations to a more uniform distribution. Figure 31.5 illustrates the approach. As shown, oblique grid lines are used, and although a slicing structure with alternating cutline directions and quadrilateral bins is maintained, gridlines not necessary to the slicing pattern are broken at points in which they intersect other gridlines. This weakening of the grid structure allows close neighbors

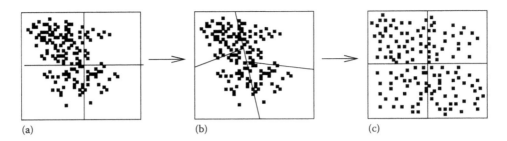

(a)                              (b)                              (c)

**FIGURE 31.5**  Warping. Each bin of a uniform bin grid (a) is mapped to a corresponding quadrilateral in an oblique but slicing bin structure (b) so as to capture roughly equal numbers of cells in each quadrilateral. The inverse bin maps are applied to the cells in order to spread them out (c).

in the original unconstrained placement to be separated a relatively large distance by the warping. The grid points of the warped grid are determined simultaneously by a derivative-free method of nonlinear optimization of Brent and Powell [75]. The required top-down slicing grid structure is maintained by (a) fixing the alternating cutline direction order a priori, by deciding whether to orient the first cut from top to bottom or from side to side and (b) expressing each cutline after the first in terms of 2 variables, one for where it intersects its parent cutline, and another for where it intersects the opposite boundary or cutline. A penalty function $f$ is used as the objective:

$$f = \text{wirelength} + \rho \cdot \sum_{\text{bins}} \beta_{ij},$$

where $\beta_{ij}$ is approximately the square of the difference between the total cell area in bin $(i, j)$, and the target cell area $\kappa = \kappa(i, j)$ for each bin. The wirelength is the total weighted HPWL obtained after the inverse warp. Although evaluating the objective is fairly costly, the number of variables in the optimization is low—only 6 for a $2 \times 2$ grid or 30 for a $4 \times 4$ grid—and convergence is fast.

An alternative formulation of cell spreading by continuous grid deformation has subsequently been developed by Chong and Szegedy [76] for incremental placement following localized netlist changes during physical synthesis.

#### 31.2.2.3.2  Iterative Refinement

Following the initial partitioning at a given level, various means of further improving the result at that level can be used. In BonnPlace (Section 31.2.2.3), unconstrained quadratic wirelength minimization over $2 \times 2$ windows of subregions is followed by a repartitioning of the cells in these windows. Windows can be selected based on routing-congestion estimates. Capo [50] greedily selects cell orientations to reduce wirelength and improve routability. Feng Shui [52] follows $k$-way partitioning by localized repartitioning of each subregion. Some partitioning-based placers also employ time-limited branch-and-bound-based enumeration at the finest levels [55].

In Dragon [15,58], an initial cutsize-minimizing quadrisection is followed by a bin-swapping-based refinement, in which entire partition blocks at that level are interchanged in an effort to reduce total wirelength. At all levels except the last, low-temperature simulated annealing is used; at the finest level, a more detailed and greedy strategy is employed. As the refinement is performed on aggregates of cells rather than on cells from the original netlist, Dragon may also be grouped with the multilevel methods discussed in Section 31.2.3.

#### 31.2.2.3.3  Example: SimPL

SimPL [33] alternates B2B-based quadratic wirelength minimization with detailed, displacement-minimizing or "geometric" partitioning used as look-ahead legalization (LAL). Cells are not actually moved to their LAL locations. Instead, each cell is attached by a 2-pin pseudonet to a fixed pseudoterminal at its LAL location. These pseudonets are in turn added to the B2B wirelength model (31.5) to gradually spread overlapping cells apart, the weights of the pseudonets gradually increasing over the placement iterations. Following initial unconstrained HPWL-minimizing placement, cell locations are highly concentrated, and LAL partitions are essentially top-down. As iterations proceed, however, cell overlap becomes more scattered, and each connected clump of overfilled bins in a cell-area density grid is legalized separately. For each connected clump of overfilled bins, a minimum-area bounding circumscribing rectangle is defined, over which total cell utilization is approximately 100%. Within this rectangle, cells are partitioned across cutlines, and then spread in order-preserving fashion separately in the two coordinate directions by linear scaling. Iterations terminate when the gap between the lower bound HPWL of the quadratic placement and the upper bound HPWL of its successor LAL placement is sufficiently small.

## 31.2.3 Multiscale Methods

Placement algorithms in the multilevel paradigm [77] developed rapidly in the early 2000s [8,15,16,29, 40,58,78–83]. These methods construct a hierarchy of problem approximations, perform optimizations on aggregated variables and data functions at each level, and transfer solutions between levels to obtain a final placement of the given netlist. The following terminology is standard.

1. *Coarsening:* Hierarchies are built recursively, by bottom-up aggregation or top-down partitioning.
2. *Relaxation:* Iterative optimization improves an approximate solution at a given aggregation level.
3. *Interpolation:* Each final approximate placement at a given level is used as the initial placement at its neighboring finer level. (A good placement for the coarsest level can be obtained directly.)

The order in which the various problems at the various levels are solved can also be important. The simplest and most common approach is simply to proceed top down, from the coarsest to the finest level, once the aggregation hierarchy has been constructed [15,40,78,79]. However, studies show that considerable improvement is possible by repeated traversals and reconstructions of the hierarchy in various orderings [81,84], as in traditional multiscale methods for partial differential equation (PDEs) [85]. We refer to this organization of traversals as *iteration flow.*

The scalability of the multilevel approach is obvious. Provided relaxation at each level has order linear in the number $N_a$ of aggregates at that level, and the number of aggregates per level decreases by factor $r < 1$ at each level of coarsening, say $N_a(i) = r^i N$ at level $i$, the total order of a multilevel method is at most $cN(1 + r + r^2 + \cdots) = cN/(1 - r)$. Higher order (nonlinear) relaxations can still be used in a limited way [8,29,40]. For example, a hard limit on the number of global nonlinear relaxation steps can be imposed. Although not strictly scalable, global relaxations often produce better solutions than their localized counterparts and can be tuned to limit run time. Alternatively, relaxation can be applied only to subsets of bounded size, for example, by sweeps over overlapping windows of contiguous clusters at the current aggregation level.

### 31.2.3.1 Coarsening

Traditional multiscale algorithms form their hierarchies by recursive clustering or generalizations thereof. However, the importance of limiting cutsize makes partitioning attractive in the placement context [15]. Typically, clustering algorithms merge tightly connected cells in a way that eliminates as many nets at the adjacent coarser level as possible while respecting some area-balance constraints. Experiments to date suggest that relatively simple, graph-based greedy strategies such as First-Choice vertex matching [86,87] produce fairly good results. More sophisticated ideas like edge-separability clustering [88], wirelength-prediction-based clustering [89], and "best-choice" clustering [87] have also been attempted.

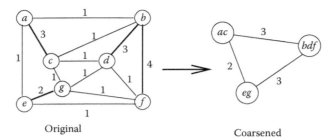

**FIGURE 31.6** First-Choice Clustering on an affinity graph. Darkened edges in the original graph are of maximal weight for at least one of their vertices. Note that vertex $d$ has maximal affinity for vertex $b$, but vertex $b$ has maximal affinity for vertex $f$.

In general, how best to define coarse-level hyperedges without explosive growth in the number and degree of coarsened hyperedges relative to coarsened vertices remains an important open question [90].

First-choice clustering is illustrated in Figure 31.6. A graph is defined on the netlist vertices with each edge weighted by the "affinity" of the given two vertices. The affinity may represent some weighted combination of complex objectives, such as hypergraph connectivity, spatial proximity, timing delay, area balance, coarse-level hyperedge elimination, and so on. Each vertex is paired with some other vertex for which it has its highest affinity. This maximum-affinity pairing is not symmetric and is independent of the order in which vertices are considered (Figure 31.5). The corresponding maximum-affinity edges are marked and define a subgraph of the affinity graph; connected components of this subgraph are clustered and thus define vertices at the next coarser level.

### 31.2.3.2 Initial Placement at Coarsest Level

A placement at the coarsest aggregate level may be derived in various ways. As the initial placement may have a large influence at subsequent iterations, and because the coarsest level problem is relatively small, the placement at this level is typically performed with great care, to the highest quality possible, by an enhanced version of the relaxation engine. How to judge the coarse-level placement quality is not necesssarily obvious, however, as the coarse-level objective may not correlate strictly with the ultimate fine-level objectives.

### 31.2.3.3 Relaxations

The core of a multiscale algorithm is the means by which it improves its approximate solution at a given aggregation level. Almost any algorithm for intralevel optimization can be used, provided that it can support (i) incorporation of complex constraints (ii) restriction to subsets of movable objects. Leading contemporary algorithms build on the iterative improvement heuristics employed by their predecessors. Relaxation in mPG [16], Dragon [15], and Ultrafast Versatile Place and Route (VPR) [78] is by fast annealing. Both APlace and mPL5 use nonlinear programming with the following log-sum-exp smoothing of the HPWL of each net $t = \{(x_i, y_i) \mid i = 1, \ldots, \deg(t)\}$,

$$\ell_{\exp}(t) = \alpha \cdot \left( \ln(\sum e^{x_i/\alpha}) + \ln(\sum e^{-x_i/\alpha}) + \ln(\sum e^{y_i/\alpha}) + \ln(\sum e^{-y_i/\alpha}) \right); \qquad (31.13)$$

$\alpha$ is the smoothing parameter. The formulations used by APlace and mPL5 for nonoverlap constraints are quite different, however, as are the optimization engines used to find solutions. These are reviewed next.

### 31.2.3.3.1 Example: APlace

In APlace [1,91,92], the scalar potential field $\phi(x, y)$ used to generate area-density-balancing forces is defined as a sum over cells and bins as follows. For a single cell $v$ at position $(x_v, y_v)$ overlapping with a single bin $b$ centered at $(x_b, y_b)$, the potential is the bell-shaped function

$$\phi_v(b) = \alpha(v)p(|x_v - x_b|)p(|y_v - y_b|),$$

where $\alpha(v)$ is selected so that $\sum_{b \in G} \phi_v(b) = \text{area}(v)$, and

$$p(d) \equiv \begin{cases} 1 - 2d^2/r^2 & \text{if } 0 \le d \le r/2 \\ 2(d - r)^2/r^2 & \text{if } r/2 \le d \le r \end{cases}, \tag{31.14}$$

and $r$ is the *radius* of the potential. The potential $\phi$ at any bin $b$ is then defined as the sum of the potentials $\phi_v(b)$ for the individual cells overlapping with that bin. Let $(X, Y)$ denote all positions of all cells in the placement region $R$. Let $|G|$ denote the total number of bins in grid $G$. Then the target potential for each bin is simply $\bar{\phi} = \sum_{v \in V} \text{area}(v)/|G|$, and the area-density penalty term for a current placement $(X, Y)$ on grid $G$ is defined as

$$\psi_G(X, Y) = \sum_{b \in G} \left( \phi(b) - \bar{\phi} \right)^2.$$

For the given area density grid $G$, APlace then formulates placement as the unconstrained minimization problem

$$\min_{v \in V} \rho_\ell \left( \sum_{e \in E} \ell_{\exp}(e) \right) + \rho_\psi \psi_G(X, Y)$$

for appropriate, grid-dependent scalar weights $\rho_\ell$ and $\rho_\psi$. This formulation has been successfully augmented in APlace to model routing congestion, movable I/O pads, and symmetry constraints on placed objects.

Optimization in APlace proceeds by the Polak-Ribiere variant of nonlinear conjugate gradients [93] with Golden-Section linesearch [94]. A hard iteration limit of 100 is imposed. The grid size $|G|$, objective weights $\rho_\ell$ and $\rho_\psi$, wirelength smoothing parameter $\alpha$ (31.13), and area-density potential radius $r$ (31.14) are selected and adjusted at each level to guide the convergence. Bin size and $\alpha$ are taken proportional to the average aggregate size at the current level. The potential radius $r$ is set to 2 on most grids but is increased to 4 at the finest grid to prevent oscillations in the maximum cell-area density of any bin. The potential weight $\rho_\psi$ is fixed at one. The wirelength weight $\rho_\ell$ is initially set rather large and is subsequently decreased by 0.5 to escape from local minima with too much overlap. As iterations proceed, the relative weight of the area-density penalty increases, and a relatively uniform cell-area distribution is obtained.

### 31.2.3.3.2 Example: mPL5

mPL5 generalizes the Kraftwerk framework to a more rigorous mathematical formulation suitable for a multilevel implementation. Recall that $x$ and $y$ denote vectors of module coordinates; hence, we let $(\bar{x}, \bar{y})$ denote an arbitrary point in $\mathcal{R}$. Letting $D_{ij}$ denote the cell-area density of bin $B_{ij}$ and $K$ the total cell area divided by the total placement area, the area-density constraints are initially expressed simply as $D_{ij} = K$ over all bins $B_{ij}$. Viewing the $D_{ij}$ as a discretization of the smooth density function $d(\bar{x}, \bar{y})$, these constraints are smoothed by approximating $d$ by the solution $\psi$ to the Helmholtz equation

$$\Delta \psi(\bar{x}, \bar{y}) - \epsilon \psi(\bar{x}, \bar{y}) = d(\bar{x}, \bar{y}), \quad (\bar{x}, \bar{y}) \in \mathcal{R}$$
$$\frac{\partial \psi}{\partial \nu} = 0, \quad (\bar{x}, \bar{y}) \in \partial \mathcal{R} \tag{31.15}$$

where $\epsilon > 0$, $\nu$ is the outer unit normal, $\partial \mathcal{R}$ is the boundary of the placement region $\mathcal{R}$, $d(\bar{x}, \bar{y})$ is the continuous density function at a point $(\bar{x}, \bar{y}) \in \mathcal{R}$, and $\Delta$ is the Laplacian operator $\Delta \equiv \frac{\partial^2}{\partial x^2} + \frac{\partial^2}{\partial y^2}$. The smoothing operator $\Delta_\epsilon^{-1} d(\bar{x}, \bar{y})$ defined by solving (31.15) is well defined, because (31.15) has a unique solution for any $\epsilon > 0$. As the solution of (31.15) has two more derivatives [95] than $d(\bar{x}, \bar{y})$, $\psi$ is a smoothed version of $d$. Discretized versions of (31.15) can be solved rapidly by fast numerical multilevel methods. Recasting the density constraints as a discretization of $\psi$ gives the nonlinear programming problem

$$
\begin{aligned}
\min \quad & W(x, y) \\
s.t. \quad & \psi_{ij} = -K/\epsilon, \quad 1 \le i \le m, 1 \le j \le n,
\end{aligned}
\tag{31.16}
$$

where the $\psi_{ij}$ are obtained by solving (31.15) with the discretization defined by the given bin grid. Interpolation from the adjacent coarser level defines a starting point. This NLP problem is solved by the Uzawa iterative algorithm [96], which does not require second derivatives or large linear-system solves:

$$
\begin{aligned}
& \nabla W(x^{k+1}, y^{k+1}) + \sum_{i,j} \lambda_{ij}^k \nabla \psi_{ij} = 0 \\
& \lambda_{ij}^{k+1} = \lambda_{ij}^k + \alpha(\psi_{ij} + \bar{K}/\epsilon)
\end{aligned}
\tag{31.17}
$$

where $\lambda$ is the Lagrange multiplier, $\lambda^0 = 0$, $\alpha$ is a parameter to control the rate of convergence, and gradients of $\psi_{ij}$ are approximated by simple forward finite differences $\nabla_{\bar{x}_k} \psi_{ij} = \frac{\psi_{i,j+1} - \psi_{i,j}}{h_{\bar{x}}}$, $\nabla_{\bar{y}_k} \psi_{ij} = \frac{\psi_{i+1,j} - \psi_{i,j}}{h_{\bar{y}}}$ when the center of cell $v_k$ is inside $B_{ij}$ and are set to zero otherwise. The nonlinear equation for $(x^{k+1}, y^{k+1})$ is recast as an ordinary differential equation and solved by an explicit Euler method [97].

### 31.2.3.3.3 Comparison to APlace

mPL5 and APlace both employ multilevel adaptations of globalized, analytical, iterative, formulations for placement. The primary difference between their formulations is the manner in which they model the nonoverlap constraints. APlace uses *local* smoothing of area densities derived from symmetric bell-shaped functions, whereas mPL5 uses a *global* smoothing derived from the Helmholtz equation. Although mPL5 specifically targets first-order *constrained* optimality conditions with explicit Lagrange-multiplier updates derived from fast ordinary differential equation (ODE) solves, APlace minimizes each member of a sequence of unconstrained penalty functions. The convergence theory for these sequential unconstrained methods [37] shows that they implicitly maintain Lagrange-multiplier estimates as well. Both methods rely on (1) empirical density estimates for termination criteria and (2) empirically tuned parameters for control of the convergence rate.

Several teams have subsequently developed variations of these techniques, including combinations of global and local smoothing [98], Gaussian [99] and Huber [98] variations of bell-function models, low-pass smoothing filters applied to density potentials [41,99–101], and so on. Analysis [102] based on Green's theorem draws an equivalence between Poisson-based density models like mPL's and bell-function overlap models like APlace's in the placement context, and empirical results have generally borne out the viability of both the Poisson-based and bell-function-based families of density models since their first uses.

### 31.2.3.4 Interpolation

Simple declustering and linear assignment can be effective [79]. With this approach, each component cluster is initially placed at the center of its parent's location. If an overlap-free configuration is needed, a uniform bin grid can be laid down, and clusters can be assigned to nearby bins or sets of bins. The complexity of this assignment can be reduced by first partitioning clusters into smaller windows, for example, of 500 clusters each. If clusters can be assumed to have uniform size, then fast linear assignment can be used. Otherwise, approximation heuristics are needed.

Under algebraic-multigrid-style weighted disaggregation, each finer-level cluster is initially placed at the weighted average of the positions of all coarser-level clusters with which its connection is sufficiently strong [80]. Finer level connections can also be used: Once a finer-level cluster is placed, it can be treated as a fixed, coarser-level cluster for the purpose of placing subsequent finer level clusters.

## 31.3 Extensions in Practice

The models described so far are adapted significantly in practice. Realistic constraints for VLSI placement require limits on maximum signal propagation times and the anticipated routability of wires connecting modules. Automatic identification and carefully aligned placement of regularly structured subcircuits can dramatically improve timing and routability properties while also reducing solution space complexity. Effective extension to 3-D placement requires thermal modeling and additional area-resource modeling and budgeting for interlayer vias (ILVs). The formulation of these constraints and their incorporation within the generic wirelength-driven model problem are described briefly in this section.

### 31.3.1 Timing

A timing-accurate view of signal propagation in an integrated circuit is graph-based rather than hypergraph-based. In each net of a given IC, one vertex serves as signal source, and the other vertices serves as sinks. The directed-graph edge connecting source to one sink is a *timing arc*. Any sequence of timing arcs connecting state elements[*] is known as a timing *path*.

Propagation of a signal from a spatially fixed PI through various functional elements to a spatially fixed PO typically requires multiple clock cycles, often across multiple clock domains. Within any particular clock domain, the signal must typically be captured in a state element a sufficient length of time before the end of the current clock cycle and held there a sufficient length of time after the start of the next clock cycle. These intermediate state elements (registers or latches) are typically movable. Thus, in this jargon, each signal traversal from PI to PO follows a *sequence* of paths, and the timing end points of any particular path are most often movable cells.

Signal delay is typically incorporated into placement in the form of time-of-arrival bound constraints $a_i < r_i$ at the IC's POs $i$ ($a_i$ is called the (actual) arrival time; $r_i$, the required arrival time). The timing *slack* at any such PO is simply $r_i - a_i$. The problem of satisfying these constraints is often generalized to the problem of maximizing the minimum slack at any PO. Both forms are thus referred to collectively as *timing-driven placement*, even when the delay appears only in the constraints, rather than in the objective.

Timing driven placement algorithms fall into two categories: path-based and net-based. The performance of a circuit is determined by the longest delay of any of its signal paths. Path delay is extremely complex, however, as the number of paths grows exponentially with circuit size.

Two *path-based* formulations of timing-driven placement appear frequently. In the first, the maximum signal-propagation time along any path is used directly as the objective to be minimized. In the second, constraints on propagation times are imposed, and the minimum slack along any path is maximized. In either formulation, auxiliary variables representing *arrival times* at circuit nodes are explicitly introduced. In terms of arrival time $a(i)$ at pin $i$, timing constraints may be expressed as follows,

$$a(j) \geq a(i) + d(i,j) \quad \forall (i,j) \in G; \qquad a(j) \leq T \quad \forall j \in PO; \qquad a(i) = 0 \quad \forall i \in PI,$$

where $G$ denotes the timing graph, $d(i,j)$ denotes the delay of timing arc $(i,j)$ either as a constant for cell-internal delay or as a function of cell locations, and $T$ denotes the target longest path-delay. Here we

---

[*] A state or *sequential* element stores a value from one clock period to the next, but a *combinational* element does not.

assume that the arrival time at all *PI* pins is zero, and that all *PO* pins have the same delay targets. Simple changes can be made to the formula to accomodate more complex situations.

The advantage of path-based algorithms is their accurate timing view during the optimization procedure. However, they usually require substantial computation resources, and, in certain placement frameworks such as top-down partitioning, it is very difficult or infeasible to maintain an accurate view of global timing.

*Net-based* algorithms [30,103–105], in contrast, enforce path-based constraints only indirectly by means of net-length constraints or net weights. This information is fed to a weighted-wirelength-minimization-based placement engine to obtain a new placement with better timing. This new placement is then analyzed by a static analyzer, thus generating a new set of timing information to guide the next placement iteration. Usually this process must be repeated for a few iterations until no improvement can be made or until a certain iteration limit has been reached. Net-weighting-based approaches assign weights to nets based on (a) their timing criticality and (b) the number of paths sharing a net. The PATH algorithm proposed by Kong [106] can properly scale the impact of all paths by their relative timing criticalities as measured by their slacks. Under certain conditions, this method is equivalent to enumerating all the paths in the circuit, counting their weights, and then distributing the weights to all edges in the circuit.

Chan et al. [107] have examined properties sufficient to prove convergence of net-based algorithms formulated as penalty methods. Applying their analysis, they demonstrate improved results of Kong's PATH algorithm [106] as well as net-based timing-driven algorithms VPR [108] and APlace [109].

## 31.3.2 Routability

As most wire routes go over the modules in parallel planes known as *routing layers*, during placement it is not strictly necessary to reserve space for wires alongside the modules. However, a tightly packed placement may be difficult or impossible to route. Therefore, quantitative models of estimated routing congestion are incorporated into the objective or constraints of practical placement algorithms.

There are two major categories of routability modeling: topology-free (TP-free), where no explicit routing is done, and topology-based (TP-based), where rectilinear routing trees are explicitly constructed on some routing grid. TP-free modeling is faster in general. In bounding-box modeling [110], for example, the routing supply for each bin in the routing grid structure is modeled according to how the existing wiring of power or clock nets, regular cells, and macros is placed, and the routing demand of a net is modeled by its weighted bounding-box length. Pin densities and stochastic models for 2-pin nets have also been used to compute expected horizontal and vertical track usage with consideration of routing blockages [111,112]. In TP-based modeling, for each net, a Steiner tree is generated on the given routing grid. If a TP-based modeling method uses a topology similar to what the after-placement-router does, the fidelity of the model can be guaranteed. However, topology generation is often of high complexity; therefore, most research focuses mainly on efficiency. In one approach [113], a precomputed Steiner tree topology on a few grid structures is used for wiring-demand estimation. In another approach [16], two algorithms of logarithmic complexity are proposed: A fast congestion-avoidance two-bend routing algorithm, LZ-router, for two-pin nets, and an IncA-tree algorithm, which can support incremental updates for building a rectilinear Steiner arborescence tree (A-tree) for a multipin net.

The results of routability modeling can be applied to placement optimization by net weighting, cell weighting (cell inflation), or white-space allocation. Net weighting directly incorporates a congestion picture into the weighted-wirelength placement objective. Cell weighting (a.k.a. cell inflation) incorporates a congestion picture into nonoverlap constraints by inflating cell sizes based on congestion estimation, so that cells in congested bins can be moved out of the bins after being inflated. White-space allocation can be applied hierarchically to ease routing congestion in hot spots with low perturbation to a given layout.

As minimum feature sizes of manufacturable ICs approach the atomic scale at 20 nm and below, design rules for detailed routing have increased dramatically in number and complexity. Over time, moreover, previous IC designs are mapped into large, fixed macro components of later designs. With this increased design reuse, the floorplans over which circuits must be placed have become increasingly convoluted, full of narrow channels in which cells must be placed and large obstacles around which signal routes must be planned, as illustrated in Figure 31.2. Accurate routability estimation in placement has correspondingly become more challenging, drawing the attention of several studies [114–124] contests [125–131], and proposed solutions [1,132–147].

### 31.3.3 Identification and Placement of Regular Subcircuits

In large VLSI circuits, the manipulation and storage of data can be mostly physically separated from the control logic coordinating the operation of the surrounding system [148]. The so-called datapath circuitry is characterized by repeated bitwise-parallel operations across the width of represented data values [149]. Figure 31.7 shows an example datapath subcircuit with six parallel bit *slices* arranged in horizontal rows and five computation *stages* in vertical columns. Typically, all cells in the same stage have the same type, as suggested by the cell shapes in the figure. Each net containing cells in the subcircuit is typically confined to cells in only one slice or only one stage. Aligning the elements in the rectangular arrangement as shown thus produces a vastly more compact, routable, timing-efficient, and manufacturable [150] placement compared with pattern-oblivious alternatives. The challenge is to identify such regular subcircuits in the netlist, construct a set of efficient candidate placements for them, and incorporate selection from these candidates within the larger placement algorithm for the entire circuit.

Nijssen and van Eijk [149] describe a general methodology for (a) the quantified assessment of datapath subcircuit regularity and (b) partitioning a circuit's netlist into separate regular subcircuits and unstructured control logic. Each pin of the netlist is associated with a *regularity signature* of attributes such as net degree, cell degree, cell type, net function, signal flow direction, and so on. For any triple $(i, j, k)$, define $x(i, j, k)$ to be the total number of pins attached to cells or nets of bit slice $i$ at stage $k$ having regularity signature $j$. Hence, $x(i, j, k)$ is a nonnegative integer, possibly greater than 1. Similarly, for signature $j$ and stage $k$, let $X \equiv X(j, k)$ denote the vector whose $i$th element is $x(i, j, k) \equiv X(j, k)[i]$. The more uniform the distribution of $x(i, j, k)$, the more regular is the subcircuit under consideration. With $Z(X)$ denoting the number of zero components of $X$, and $L(X) \equiv \max_i X_i - \min_i X_i$, and $X_{avg}$ the average of $\{X[i]\}$, Nijssen and van Eijk propose the following regularity metric:

$$\rho(j, k) = c_Z Z(X(j, k)) + c_L L(X(j, k)) + c_{avg}(X_{avg} - 1)$$

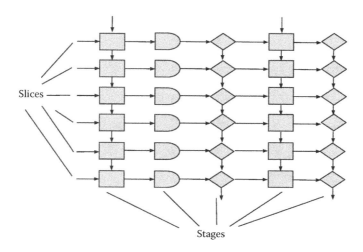

**FIGURE 31.7** A simplified datapath regular subcircuit illustrating bit slices and logic stages.

with suggested values $c_Z = 10000$, $c_L = 100$, and $c_{avg} = 1$. This metric increases monotonically as regularity decreases. To extract regular subcircuits, search waves are propagated stage by stage $k = 1, 2, \ldots$ from specially designated seed stages, each wave continuing until $\min_j \rho(j, k)$ exceeds a specified limit. Wave collision detection prevents duplicate searches over identical subcircuits; hence, the algorithm is $O(|P|)$, where $P$ is the set of all pins in the circuit.

For each regular subcircuit, alignment constraints can be introduced to confine elements of the same bit slice to the same row and elements of the same stage to the same column, or vice versa. Ward et al. [151] describe detailed techniques for incorporating and enforcing such techniques in the context of quadratic placement with a B2B net model (31.5 in Section 31.2.1.2). These include $xy$-skewed net weighting, ordering constraints, step-size scheduling, bit-slice aligned cell swapping, and min-cut alignment-group repartitioning. Chou et al. [152] cluster datapath subcircuits into atomic objects within analytical mixed-size placement. After declustering, control logic is held fixed, whereas slices and stages of datapath blocks are spread by means of a specialized high-accuracy sigmoid overlap smoothing. Similarly, legalization after global placement first legalizes datapath cells before legalizing the remaining logic.

Other methods have been proposed to capture and exploit both rectangular and nonrectangular subcircuit patterns. Chowdhary et al. [153] generate regular subcircuit templates and then compute a covering of a given circuit by the templates. Xiang et al. [154] have observed that the start and end stages of regular subcircuits are typically easy to identify in practice. They therefore express bit-slice identification as bipartite matching of start bits to end bits followed by two-way breadth-first search for the bit-slice paths. The bit matching problem is expressed as a min-cost max-flow problem in such a way that the number of cells in the optimal flow is maximal. Separately, Xiang et al. [155,156] have also developed algorithms for the identification, restructuring, and placement of symmetric-function fanin trees—fanout free cones of logic whose inputs can be freely reordered. Compared with structure-blind methods, their algorithm can reduce the Steiner wirelength of such trees by 80% while also reducing logical depth (number of stages per tree) by 50% or more.

Ward et al. [157] employ support vector machines and neural networks to identify regular subcircuits by supervised learning via offline training, calibration, and validation on test cases from the ISPD 2011 Datapath Benchmark Suite [158]. The resulting method of regularity extraction is incorporated into placement such that intermediate placement information can be used along with connectivity to identify regular subcircuits. Seed-based connectivity clustering [159] is extended to form the initial candidate datapath subcircuits. The number of graph symmetries or automorphisms of the candidates is used with their intermediate, internal placement densities to train the learning algorithms. The bit slices of each candidate classified as regular are then selected by integer linear programming. The authors report a 12% reduction in Steiner wirelength compared with SimPL over a separate suite of untrained industrial benchmarks.

### 31.3.4 Toward 3-D Placement

As chip designs continue to grow in scale and complexity, so grows the challenge of satisfying their signal-propagation delay constraints. For placement algorithms, this challenge amounts to bounding the lengths of increasingly many timing-critical paths. Confining placement solutions to the plane imposes hard limits on the number of objects placeable within any particular distance of some specified location. Vastly tighter integration is possible, if objects can be freely placed in all three spatial dimensions, rather than just two [160–163]. Significantly reduced average spacing between netlist elements presents the opportunity to significantly reduce total wirelength and maximum path delay [164].

Basic notation for a simplified *stacked-die* configuration is illustrated in Figure 31.8. As in previous sections, $x$ and $y$ label coordinates along the usual 2-D coordinate axes; $z$ is now used for the dimension along which multiple *tiers* of the 3-D design are placed. Each tier includes a 2-D device layer and associated metal routing layers. Intertier vias connect adjacent tiers. If a square, $L \times L$, 2-D placement region is replaced by $K$ square, parallel, vertically aligned, $M \times M$, 2-D *tiers* such that total planar area

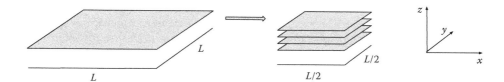

**FIGURE 31.8** Comparing a square 2-dimensional floorplan to an area-equivalent stack of $K$ tiers, with $K = 4$. The spacing between tiers is greatly exaggerated!

is conserved, then it is easily shown that $M = L/\sqrt{K}$. Under the additional assumption* that number of tiers $K$, tier thickness $\tau$, and intertier spacing $\sigma$ are all negligible compared with die width $M$ (i.e., $K(\tau + \sigma) \ll M < L$), the longest possible distance between any two placed design elements is reduced by factor $1/\sqrt{K}$, from $2L$ to $2M \equiv 2L/\sqrt{K}$—a dramatic decrease, even for $K$ as small as 2.

Although this simplistic calculation does illustrate the promise of 3-D integration, it neglects area resources consumed by intertier connections. A more realistic analysis by Mak and Chu [165] concludes that, for a 32nm manufacturing process, under the optimistic assumption of intertier vias 1 row-height square, average wirelength of a 4-tier 3-D design can be reduced by no more than 18% below its single-tier 2-D alternative. Under the less optimistic assumptions of intertier vias 5 row-heights square, average wirelength of the 4-tier design actually *exceeds* the 2-D single tier average wirelength.

Thus, as the physical properties of intertier connections may vary by orders of magnitude across different manufacturing technologies, the manner in which they are modeled leads to various problem formulations and corresponding families of algorithms. Ultimately, the primary challenges [166] associated with 3-D physical design [162,167] concern (a) granularity and discreteness of heterogeneous manufacturing constraints and (b) limits on peak temperatures.

### 31.3.4.1 Granularity

The extremely limited extent of 3-D designs along the $z$ axis—typically just 2–4 tiers—makes the 3-D placement problem akin to 2-D placement combined with partitioning for $z$-layer assignment, regardless of the algorithm used to solve it. Available connectivity in $z$ may also be dramatically less than that in $x$ and $y$, depending on manufacturing technology. Granularity of placement in $z$ depends heavily on $z$-connectivity density; hence, arise the following variants of 3-D integration, from coarsest to finest: core-level, block-level, gate-level, and transistor level. Core-level integration partitions processor cores and large memory units across tiers [168]. Block-level integration partitions large functional sub-circuits such as IP blocks across tiers [169,170]. Gate-level integration partitions logic gates and standard cells across tiers [171,172], with all transistors for the same gate remaining in the same tier. Transistor-level integration partitions the transistors in each logic gate across multiple layers in $z$; the resulting standard-cell placement problem may still be 2-D [173]. Placement formulations and solutions developed for one level of integration typically do not transfer to other levels.

### 31.3.4.2 Temperature

The heat generated by an IC during its operation must not corrupt its functionality. As 3-D design ratchets up the computational power of VLSI circuits, it also endows them with radically higher power densities. Thus, temperature control becomes paramount, especially in placement, which largely determines the distribution of power consumption over the volume of the chip [174,175].

Mathematical work on thermal placement for 2-D designs includes its formulation by Chu and Wong [176] as a *matrix synthesis* problem: Given $mn$ positive numbers representing cell power densities and a positive integer $t \leq \min\{m, n\}$, arrange the numbers in an $m \times n$ matrix $M$ such that the

---

* Typically valid for realistic manufacturing technology and costs as of year 2016.

maximum sum over all $t \times t$ submatrices of $M$ is minimized. Chu and Wong show the problem is NP-complete and derive $O(mn \log(mn))$ approximation algorithms that produce matrices $M$ with maximum submatrix sums no more than $2\times$ optimal when $t$ is a power of 2, and no more than $5\times$ optimal otherwise.

Thermal modeling in 3-D placement [177] is typically accomplished by either finite element analysis (FEA) [178,179] or compact thermal resistive network modeling (CTM) [98,180–183]. At room temperature, the thermal conductivity of the dielectric between device layers is typically less than $0.0005\times$ that of the silicon device layer or the copper used for wiring [184]. Consequently, heat flows out of a 3-D circuit laterally through the silicon device layer and/or power-delivery network and vertically through ILVs [185]. ILVs may be signal vias or dummy thermal vias inserted solely for temperature control. Via size and the granularity of legal via placement locations vary dramatically across different fabrication technologies.

As the run time of high-accuracy thermal analysis generally prohibits its use in the inner loop of physical design, researchers have sought approximations with controllable accuracy/run-time trade-offs. Some have investigated faster thermal profiling by boundary-element-based Green's function evaluation [186] or multigrid [187]. Samal et al. have shown that fast and accurate thermal modeling is possible by non-linear regression per each $100 \times 100\,\mu m$ tile of a monolithic 3-D design [188]. They first precompute FEA thermal evaluation of 17 full-chip test cases with random power distributions. The full-chip results are divided into $100 \times 100\,\mu m$ reference tiles used by the regression to compute a thermal model. Each tile $T$ is parametrized by (a) sums of lumped power densities in tiles directly above and below tile $T$, (b) sums of lumped power densities in other tiles not directly above and below tile $T$, and (c) distances from tile $T$ to chip boundaries. The temperature of a tile in a target design can then be computed from the regression-based model orders of magnitude faster than FEA alone, with average error below 5%.

### 31.3.4.3 Technology Constraints

Different families of 3-D IC architectures and fabrication technologies present different granularities of design alternatives along the $z$ direction [189]. Detailed overviews of these technologies are beyond the scope of this chapter, but realistic problem formulations require some understanding of basic properties and limits. Formulations and algorithms appropriate for one technology may not transfer to another.

In *parallel 3-D* integration [190–193], device layers are manufactured separately and then vertically aligned [194] and bonded together to assemble the final 3-D IC. Parallel 3-D fabrication processes in year 2016 produce device layer thicknesses of approximately $10\,\mu m^*$, with intertier spacings of $10$–$50\,\mu m$. Parallel 3-D relies on *through-silicon vias* (TSVs), which are ILVs passing completely through a device layer. TSVs are quite large, with diameters of $1$–$10\,\mu m$ or more, the rough equivalent of $10$–$50$ standard cells. They are typically placed at densities of $800/mm^2$ or roughly 1200 per tier on a typical parallel-3-D IC [98]. Due to their large size, TSVs are significant obstacles to placement and routing. Hence, placement formulations for parallel 3-D typically attempt to minimize total TSVs, either by count or by area, while distributing them so as to reduce peak temperature. Mechanical stress and electromigration [195] associated with TSVs are also significant and have also been considered in placement [196,197].

In *sequential* or *monolithic* 3-D integration (MI) [189,198,199], each layer is manufactured directly on top of the layer below it. Layer thicknesses are of order $100\,nm \equiv 0.1\,\mu m$, and ILV dimensions are similar to those of traditional vias between routing layers of a conventional 2-D IC. This design style results in ILV densities of order $1000\times$ greater than those of TSVs, enabling vastly finer granularity of design compared with TSV-based parallel 3-D. There are two broad categories of monolithic 3-D IC: gate-level integration (MI-G) and transistor-level (MI-TR) integration. In MI-G, individual standard cells in the same functional circuit block may be assigned to different tiers.

Liu and Lim [189] compare (a) parallel 3-D (a.k.a. TSV-3-D) versus MI-G and (b) 2-D versus MI-TR in two separate design flows over three different designs, using 2 tiers for all 3-D designs. Their MI-G solutions exhibit modestly reduced area, wirelength, power consumption, and total-negative-slack (TNS)

---

$^*$ 1 micron $= 1\,\mu m = 10^{-6}m = 1000\,nm$.

timing-constraint violations compared with TSV-3-D. Their MI-TR solutions exhibit 9%–17% wirelength reduction with 7%–35% reduction in TNS, compared with 2-D.

Estimated 3-D wirelength has been extended from 2-D HPWL (31.1) in different ways. The trivial extension of HPWL to 3-D bounding-box wirelength of a single net $e = \{v_1, \ldots, v_n\}$ is simply

$$w(e) = \max_{v_j \in e} \left(x_j + w_j\right) - \min_{v_k \in e} x_k + \max_{v_m \in e} \left(y_m + h_m\right) - \min_{v_n \in e} y_n. + \max_{v_m \in e} \left(z_m\right) - \min_{v_n \in e} z_n. \tag{31.18}$$

A more accurate *net-splitting* estimate, however, computes the 2-D HPWL of $e$ separately on each individual tier, then adds the sum of these terms to the total $z$ span [200]. Detailed 3-D WL estimation [201] requires careful analysis of area-resource consumption by ILVs.

### 31.3.4.4 Placement Algorithms for 3-D ICs

Algorithms for 3-D placement generally adapt known techniques from 2-D placement under various assumptions of technology and problem granularity. Given the extremely limited number of tiers $K$ available to date, existing algorithms on 3-D placement may be broadly grouped by (a) granularity of objects being placed (large blocks or small cells), (b) type of temperature modeling, if any, and (c) the manner in which $z$-layer assignment of placeable objects is achieved.

Researchers have examined the problem of mapping 2-D placements to 3-D placements [170,202]. Cong et al. [202] describe mappings based on folding and stacking. Under folding transformations, nets cut by fold lines require intertier vias. Under stacking, cells retain their dimensions, whereas their locations on chip are shrunk in $x$ and $y$ by factor $1/\sqrt{K}$ for a $K$-tier solution. The shrink operation initially overcrowds the 2-D placement on the base tier by a factor of $K$; subsequent layer assignment of the cells by Tetris-style *stacking* balances cell displacement against TSV count and temperature estimates from a resistive thermal model. This initial layer assignment is refined using relaxed conflict-net graph algorithm [203] based on linear-time construction of sequences of maximally induced subtrees. Nodes in the graph are cells or vias of the circuit. Edges in the graph model nets in the circuit or, with infinite weight, overlap between cells assigned to the same layer.

Several papers focus on the effective modeling of TSV placement under parallel 3-D IC. Kim et al. [200] begin with a cutsize-driven partitioning of cells across the $K$ layers. They compare two alternative forms of a Kraftwerk-style force-directed algorithm. In the first, TSVs are preplaced at regular intervals and then assigned to nets after standard-cell placement. In the second, TSVs are treated as placeable objects and placed jointly with standard cells. Solutions are fully routed in both cases by a commercial 2-D router. They find solutions by the coplacement approach to have roughly 10% shorter routed wirelengths than solutions using uniformly preplaced TSVs.

Knechtel et al. [170] have examined granularity and placement of TSV "islands" in the context of block-level integration. Starting from a 3-D floorplan [204] with 10% added deadspace, they examine the intersection graph of the (axis-aligned) bounding boxes of 3-D nets. Each bounding box is a vertex; overlapping boxes are connected by a graph edge. Cliques in this graph indicate candidate locations for TSVs. A theorem of Imai and Asano [205] can be used to find a maximal clique in order $n \log n$ time for $n$ vertices, but associating large cliques with TSV islands may not budget sufficient deadspace for the TSVs. Hence, Knechtel et al. partition the edges of the bounding-box intersection graph into cliques of appropriate size by a deadspace-aware clique covering algorithm. Floorplan constraint graphs [206] are used to shift blocks and iteratively redistribute deadspace, as TSVs are associated with net-bounding-box cliques, and nets are assigned to TSVs.

Luo et al. [98,207] present a detailed thermal analysis of an algorithm for TSV and standard-cell co placement. They prove that peak temperature is minimized when TSV area in each bin is proportional to the lumped power consumption in that bin and all bins in all tiers directly above it. By incorporating this result into their analytical 3-D placement formulation, they demonstrate reductions in peak temperature over 4× that of the previously dominant methodology targeting a uniform power distribution.

## 31.4 Conclusion

The investigation of both synthetic benchmarks with known optimality properties [208–211] and lower bounds for globally optimal solutions to real instances [212] have found significant suboptimality in the solutions produced by leading academic placement solutions. For standard-cell benchmarks with known optima [209], leading tools may produce solutions with wirelengths as much as 2× or more above the optimal. For large mixed-size benchmarks with known optima [213], observed wirelengths are often 5× or more above the optimal. Moreover, this gap is observed to increase 20% or more with the size of the parametrized benchmarks. One study suggests that a large gap may result even when almost every module in a global placement is within a very small distance—one or two cell widths—of its optimal location [210]. Such observations have drawn increased attention to both (a) the discrete problem of legalizing a global placement [214–217] and (b) improved estimates of legalizability within global placement [33,66,67,218]. These studies and the techniques they describe have been used to identify algorithm weaknesses that may be difficult to isolate on real instances.

The increased importance of interconnect delay on the performance of VLSI circuits has spurred great advances in algorithms for large-scale global placement. Newer algorithms often generalize earlier heuristics within a hierarchical framework—top-down recursive partitioning, multiscale optimization, or FFT-based numerical acceleration. Heuristics proven successful in one manufacturing technology are soon superseded in subsequent technologies if they cannot readily be adapted to address new constraints and design styles. As the scale and complexity of integrated circuits continue to grow, IC designers demand continued improvement in the extensibility, robustness, QoR, and run time of algorithms for global placement.

## References

1. G.-J. Nam and J. Cong. *Modern Circuit Placement.* Springer, New York, 2007.
2. C.J. Alpert, D.P. Mehta, and S.S. Sapatnekar. *Handbook of Algorithms for Physical Design Automation.* Auerbach Publications, Boston, MA, 1st ed., 2008.
3. Y.-W. Chang, Z.-W. Jiang, and T.-C. Chen. Essential issues in analytical placement algorithms. *Information and Media Technologies*, 4(4):815–836, 2009.
4. I.L. Markov, J. Hu, and M.-C. Kim. Progress and challenges in VLSI placement research. In *Proceedings of the International Conference on Computer-Aided Design*, pp. 275–282. IEEE, 2012.
5. S. Sahni and T. Gonzalez. P-complete approximation problems. *Journal of the ACM*, 23(3): 555–565, 1976.
6. A. B. Kahng, J. Lienig, I.L. Markov, and J. Hu. *VLSI Physical Design: From Graph Partitioning to Timing Closure.* 1st ed., Springer Publishing Company, New York, 2011.
7. J. Cong. An interconnect-centric design flow for nanometer technologies. *Proceedings of the IEEE*, 89(4):505–527, 2001.
8. T.F. Chan, J. Cong, and K. Sze. Multilevel generalized force-directed method for circuit placement. In *Proceedings of the International Symposium on Physical Design*, 2005.
9. M.A. Breuer, Ed. *Design Automation of Digital Systems: Theory and Techniques*, volume 1, Chapter 5, pp. 213–282. Prentice-Hall, Engelwood Cliffs, NJ, 1972.
10. J. Cong, J. R. Shinnerl, M. Xie, T. Kong, and X. Yuan. Large-scale circuit placement. *ACM Transactions on Design Automation of Electronic Systems*, 10(2):389–430, 2005.
11. S. Kirkpatrick, C.D. Gelatt Jr., and M.P. Vecci. Optimization by simulated annealing. *Science*, 220:671ff, 1983.
12. C. Sechen. *VLSI Placement and Global Routing Using Simulated Annealing.* Kluwer Academic Publishers, Boston, MA, 1988.

13. W.-J. Sun and C. Sechen. Efficient and effective placement for very large circuits. *IEEE Transactions on Computer-Aided Design*, pp. 349–359, March 1995.

14. V. Betz and J. Rose. VPR: A new packing, placement, and routing tool for FPGA research. In *Proceedings of the International Workshop on FPL*, New York, pp. 213–222, 1997.

15. M. Sarrafzadeh, M. Wang, and X. Yang. *Modern Placement Techiques*. Kluwer, Boston, MA, 2002.

16. C.-C. Chang, J. Cong, D. Pan, and X. Yuan. Multilevel global placement with congestion control. *IEEE Transactions on Computer-Aided Design of Integrated and Systems*, 22(4):395–409, 2003.

17. N. Sherwani. *Algorithms for VLSI Physical Design Automation*. Kluwer Academic Publishers, Boston, MA, 3rd ed., 1999.

18. S.-W. Hur and J. Lillis. Relaxation and clustering in a local search framework: Application to linear placement. In *Proceedings of the Design Automation Conference*, pp. 360–366, New Orleans, LA, 1999.

19. S.-W. Hur and J. Lillis. Mongrel: Hybrid techniques for standard-cell placement. In *Proceedings of the International Conference on Computer-Aided Design*, pp. 165–170, San Jose, CA, November 2000.

20. K. Doll, F.M. Johannes, and K.J. Antreich. Iterative placement improvement by network flow methods. *IEEE Transactions on Computer-Aided Design*, 13(10):1189–1200, 1994.

21. U. Brenner, A. Pauli, and J. Vygen. Almost optimum placement legalization by minimum cost flow and dynamic programming. In *Proceedings of the International Symposium on Physical Design*, New York, pp. 2–8, 2004.

22. C.J. Fisk, D.L. Caskey, and L.L. West. Accel: Automated circuit card etching layout. *Proceedings of the IEEE*, 55(11):1971–1982, 1967.

23. N. Quinn and M. Breuer. A force-directed component placement procedure for printed circuit boards. *IEEE Transactions on Circuits and Systems CAS*, CAS-26:377–388, 1979.

24. C.K. Cheng and E.S. Kuh. Module placement based on resistive network optimization. *IEEE Transactions on Computer-Aided Design*, CAD-3(3):218–225, 1984.

25. J.M. Kleinhans, G. Sigl, F.M. Johannes, and K.J. Antreich. GORDIAN: VLSI placement by quadratic programming and slicing optimization. *IEEE Transactions on Computer-Aided Design*, 10:356–365, 1991.

26. J. Vygen. Algorithms for large-scale flat placement. In *Proceedings of the Design Automation Conference*, New York, pp. 746–751, 1997.

27. C. Chu and N. Viswanathan. FastPlace: Efficient analytical placement using cell shifting, iterative local refinement, and a hybrid net model. In *Proceedings of the International Symposium on Physical Design*, New York, pp. 26–33, April 2004.

28. B. Hu and M. Marek-Sadowska. FAR: fixed-points addition & relaxation based placement. In *Proceedings of the International Symposium on Physical Design*, pp. 161–166, New York, ACM Press, 2002.

29. B. Hu, Y. Zeng, and M. Marek-Sadowska. mFAR: Fixed-points-addition-based VLSI placement algorithm. In *Proceedings of the International Symposium on Physical Design*, pp. 239–241, New York, April 2005.

30. H. Eisenmann and F.M. Johannes. Generic global placement and floorplanning. In *Proceedings of the Design Automation Conference*, New York, pp. 269–274, 1998.

31. P. Spindler, U. Schlichtmann, and F. M. Johannes. Kraftwerk2—A fast force-directed quadratic placement approach using an accurate net model. *IEEE Transactions on Computer-Aided Design of Integrated Circuits and Systems*, 27(8):1398–1411, 2008.

32. K. Vorwerk, A. Kennings, and A. Vannelli. Engineering details of a stable force-directed placer. In *Proceedings of the International Conference on Computer-Aided Design*, pp. 573–580, November 2004.

33. M.C. Kim, D.J. Lee, and I.L. Markov. SimPL: An effective placement algorithm. *IEEE Transactions on Computer-Aided Design of Integrated Circuits and Systems*, 31(1):50–60, 2012.

34. J. Lu, P. Chen, C.-C. Chang, L. Sha, D.J.-H. Huang, C.-C. Teng, and C.-K. Cheng. ePlace: Electrostatics-based placement using fast fourier transform and nesterov's method. *ACM Transactions on Design Automation of Electronic Systems*, 20(2):17:1–17:34, 2015.

35. J. Lu, H. Zhuang, P. Chen, H. Chang, C.C. Chang, Y.C. Wong, L. Sha, D. Huang, Y. Luo, C.C. Teng, and C.K. Cheng. ePlace-ms: Electrostatics-based placement for mixed-size circuits. *IEEE Transactions on Computer-Aided Design of Integrated Circuits and Systems*, 34(5):685–698, 2015.

36. Y. Nesterov. A method of solving a convex programming problem with convergence rate o (1/k2). *Soviet Mathematics Doklady*, 27(2):372–376, 1983.

37. A.V. Fiacco and G.P. McCormick. *Nonlinear Programming: Sequential Unconstrained Minimization Techniques*. Classics in Applied Mathematics. SIAM, Philadelphia, PA, 1990.

38. A.P. Ruszczyński. *Nonlinear Optimization*, volume 13. Princeton University Press, Princeton, NJ, 2006.

39. M.-K. Hsu, Y.-W. Chang, and V. Balabanov. TSV-aware analytical placement for 3D IC designs. In *Proceedings of the Design Automation Conference*, pp. 664–669. IEEE, 2011.

40. A.B. Kahng and Q. Wang. Implementation and extensibility of an analytic placer. *IEEE Transactions on Computer-Aided Design of Integrated and Systems*, 24(5):734–747, 2005.

41. T.-C. Chen, Z.-W. Jiang, T.-C. Hsu, H.-C. Chen, and Y.-W. Chang. Ntuplace3: An analytical placer for large-scale mixed-size designs with preplaced blocks and density constraints. *IEEE Transactions on Computer-Aided Design of Integrated Circuits and Systems*, 27(7):1228–1240, 2008.

42. S.N. Adya, I.L. Markov, and P.G. Villarrubia. On whitespace and stability in mixed-size placement and physical synthesis. In *Proceedings of the International Conference on Computer-Aided Design*, ICCAD '03, pp. 311–318, Washington, DC, IEEE Computer Society, 2003.

43. M.A. Breuer. Min-cut placement. *ACM Transactions on Design Automation of Electronic Systems*, 1(4):343–362, 1977.

44. A.E. Dunlop and B.W. Kernighan. A procedure for placement of standard-cell VLSI circuits. *IEEE Transactions on Computer-Aided Design*, CAD-4(1):92–98, 1985.

45. J. Cong and M. Smith. A parallel bottom-up clustering algorithm with applications to circuit partitioning in VLSI designs. In *Proceedings of the Design Automation Conference*, pp. 755–760, New York, June 1993.

46. G. Karypis, R. Aggarwal, V. Kumar, and S. Shekhar. Multilevel hypergraph partitioning: Application in VLSI domain. In *Proceedings of the Design Automation Conference*, pp. 526–529, New York, 1997.

47. A.E. Caldwell, A.B.Kahng, and I.L. Markov. Improved algorithms for hypergraph partitioning. In *Asia South Pacific Design Automation Conference*, 2000.

48. U. Brenner and A. Rohe. An effective congestion-driven placement framework. In *Proceedings of the International Symposium on Physical Design*, April 2002.

49. C.M. Fiduccia and R.M. Mattheyses. A linear-time heuristic for improving network partitions. In *Proceedings of the Design Automation Conference*, pp. 175–181, New York, 1982.

50. A.E. Caldwell, A.B. Kahng, and I.L. Markov. Can recursive bisection produce routable placements? In *Proceedings of the Design Automation Conference*, pp. 477–482, New York, 2000.

51. J.A. Roy, D.A. Papa, S.N. Adya, H.H. Chan, A.N. Ng, J.F. Lu, and I.L. Markov. Capo: Robust and scalable open-source min-cut floorplacer. In *Proceedings of the International Symposium on Physical Design*, pp. 224–226, New York, 2005.

52. M.C. Yildiz and P.H. Madden. Global objectives for standard cell placement. In *Eleventh Great-Lakes Symposium on VLSI*, pp. 68–72, New York, 2001.

53. A.R. Agnihotri, S. Ono, and P. Madden. Recursive bisection placement: Feng shui 5.0 implementation details. In *Proceedings of the International Symposium on Physical Design*, pp. 230–232, New York, April 2005.

54. M.C. Yildiz and P.H. Madden. Improved cut sequences for partitioning-based placement. In *Proceedings of the Design Automation Conference,* pp. 776–779, 2001.

55. A.E. Caldwell, A.B. Kahng, and I.L. Markov. Optimal partitioners and end-case placers for standard-cell layout. *IEEE Transactions on CAD,* 19(11):1304–1314, 2000.

56. P. Villarrubia, G. Nusbaum, R. Masleid, and E.T. Patel. IBM RISC chip design methodology. In *ICCD,* pp. 143–147, New York, 1989.

57. A.B. Kahng and S. Reda. Placement feedback: A concept and method for better min-cut placements. In *Proceedings of the Design Automation Conference,* pp. 357–362, June 2004.

58. M. Wang, X. Yang, and M. Sarrafzadeh. Dragon2000: Standard-cell placement tool for large circuits. *Proceedings of the International Conference on Computer-Aided Design,* pp. 260–263, New York, April 2000.

59. P.H. Madden. Partitioning by iterative deletion. In *Proceedings of the International Symposium on Physical Design,* pp. 83–89, New York, ACM Press, 1999.

60. A.R. Agnihotri, M.C. Yildiz, A. Khatkhate, A. Mathur, S. Ono, and P.H. Madden. Fractional cut: Improved recursive bisection placement. In *Proceedings of the International Conference on Computer-Aided Design,* pp. 307–310, New York, 2003.

61. L. Hagen, J.H. Huang, and A.B. Kahng. On implementation choices for iterative improvement partitioning algorithms. *IEEE Transactions on Computer-Aided Design of Integrated and Systems,* 16(10):1199–1205, 1997.

62. A.E. Caldwell, A.B. Kahng, and I.L. Markov. Iterative partitioning with varying node weights. *VLSI Design,* pp. 249–258, New York, 2000.

63. S.N. Adya, S. Chaturvedi, J.A. Roy, D.A. Papa, and I.L. Markov. Unification of partitioning, placement and floorplanning. In *Proceedings of the International Conference on Computer-Aided Design,* pp. 12–17, New York, 2004.

64. J.Z. Yan, N. Viswanathan, and C. Chu. Handling complexities in modern large-scale mixed-size placement. In *Proceedings of the Design Automation Conference,* pp. 436–441. ACM, New York, 2009.

65. J. Cong, M. Romesis, and J.R. Shinnerl. Fast floorplanning by look-ahead enabled recursive bipartitioning. *IEEE Transactions on Computer-Aided Design of Integrated Circuits and Systems,* 25(9):1719–1732, 2006.

66. J. Cong, M. Romesis, and J. Shinnerl. Robust mixed-size placement under tight white-space constraints. In *Proceedings of the International Conference on Computer-Aided Design,* pp. 165–172, New York, November 2005.

67. J.A. Roy, A.N. Ng, R. Aggarwal, V. Ramachandran, and I.L. Markov. Solving modern mixed-size placement instances. *INTEGRATION, the VLSI Journal,* 42(2):262–275, 2009.

68. S.N. Adya and I.L. Markov. Consistent placement of macro-blocks using floorplanning and standard-cell placement. In *Proceedings of the International Symposium on Physical Design,* pp. 12–17, April 2002.

69. J. Cong, M. Romesis, and J.R. Shinnerl. Fast floorplanning by look-ahead enabled recursive bipartitioning. In *Asia South Pacific Design Automation Conference,* 2005.

70. Y.-C. Chang, Y.-W. Chang, G.-M. Wu, and S.-W. Wu. B*-trees: A new representation for non-slicing floorplans. In *Proceedings of the Design Automation Conference,* pp. 458–463. ACM, New York, 2000.

71. R.S. Tsay, E.S. Kuh, and C.P. Hsu. Proud: A fast sea-of-gates placement algorithm. *IEEE Design and Test of Computers,* 5:44–56, 1988.

72. G. Sigl, K. Doll, and F.M. Johannes. Analytical placement: A linear or a quadratic objective function? In *Proceedings of the Design Automation Conference,* pp. 427–432, New York, 1991.

73. Z. Xiu, J.D. Ma, S.M. Fowler, and R.A. Rutenbar. Large-scale placement by grid warping. In *Proceedings of the Design Automation Conference,* pp. 351–356, New York, June 2004.

74. J. Vygen. Four-way partitioning of two-dimensional sets. Report 00900-OR, Research Institute for Discrete Mathematics, University of Bonn, Bonn, Germany, 2000.

75. W.H. Press, S.A. Teukolsky, W.T. Vetterling, and B.P. Flannery. *Numerical Recipes in C: The Art of Scientific Computing.* Cambridge University Press, New Delhi, India, 2nd ed., January 1993.

76. Philip Chong and Christian Szegedy. A morphing approach to address placement stability. In *Proceedings of the International Symposium on Physical Design*, pp. 95–102. ACM, 2007.

77. J. Cong and J.R. Shinnerl, Eds. *Multilevel Optimization in VLSICAD.* Kluwer Academic Publishers, Boston, MA, 2003.

78. Y. Sankar and J. Rose. Trading quality for compile time: Ultra-fast placement for FPGAs. In *FPGA '99, ACM Symposium on FPGAs*, pp. 157–166, New York, 1999.

79. T.F. Chan, J. Cong, T. Kong, and J. Shinnerl. Multilevel optimization for large-scale circuit placement. In *Proceedings of the International Conference on Computer-Aided Design*, pp. 171–176, IEEE/ACM (Association for Computing Machinery), San Jose, CA, November 2000.

80. T.F. Chan, J. Cong, T. Kong, J. Shinnerl, and K. Sze. An enhanced multilevel algorithm for circuit placement. In *Proceedings of the International Conference on Computer-Aided Design*, IEEE/ACM (Association for Computing Machinery), San Jose, CA, November 2003.

81. T.F. Chan, J. Cong, T. Kong, and J.R. Shinnerl. Multilevel circuit placement. In J. Cong and J.R. Shinnerl, Eds., *Multilevel Optimization in VLSICAD.* Kluwer Academic Publishers, Boston, MA, 2003.

82. A.B. Kahng, S. Reda, and Q. Wang. Architecture and details of a high quality, large-scale analytical placer. In *Proceedings of the International Conference on Computer-Aided Design*, November 2005.

83. K. Vorwerk and A.A. Kennings. An improved multi-level framework for force-directed placement. In *DATE*, pp. 902–907, New York, 2005.

84. A. Brandt and D. Ron. *Multigrid Solvers and Multilevel Optimization Strategies*, Chapter 1 of *Multilevel Optimization and VLSICAD.* Kluwer Academic Publishers, Boston, MA, 2002.

85. W.L. Briggs, V.E. Henson, and S.F. McCormick. *A Multigrid Tutorial.* SIAM, Philadelphia, PA, 2nd ed, 2000.

86. G. Karypis. Multilevel hypergraph partitioning. In J. Cong and J.R. Shinnerl, Eds., *Multilevel Optimization and VLSICAD.* Kluwer Academic Publishers, Boston, MA, 2002.

87. C. Alpert, A.B. Kahng G.-J. Nam, S. Reda, and P. Villarrubia. A semi-persistent clustering technique for VLSI circuit placement. In *Proceedings of the International Symposium on Physical Design*, pp. 200–207, Apr 2005.

88. J. Cong and S.K. Lim. Edge separability based circuit clustering with application to circuit partitioning. In *Asia South Pacific Design Automation Conference*, pp. 429–434, 2000.

89. B. Hu and M. Marek-Sadowska. Wire length prediction based clustering and its application in placement. In *Proceedings of the Design Automation Conference*, June 2003.

90. B. Hu and M. Marek-Sadowska. Fine granularity clustering based placement. *IEEE Transactions on Computer-Aided Design of Integrated and Systems*, 23:527–536, 2004.

91. A.B. Kahng, S. Reda, and Q. Wang. Aplace: A general analytic placement framework. In *Proceedings of the International Symposium on Physical Design*, pp. 233–235, April 2005.

92. A.B. Kahng and Q. Wang. A faster implementation of APlace. In *Proceedings of the International Symposium on Physical Design*, ISPD '06, pp. 218–220, New York, USA, 2006. ACM.

93. S.G. Nash and A. Sofer. *Linear and Nonlinear Programming.* McGraw Hill, New York, 1996.

94. P.E. Gill, W. Murray, and M.H. Wright. *Practical Optimization.* Academic Press, London and New York, 1981.

95. L. C. Evans. *Partial Differential Equations.* American Mathematical Society, Providence, RI, 2002.

96. K. Arrow, L. Huriwicz, and H. Uzawa. *Studies in Nonlinear Programming.* Stanford University Press, Stanford, CA, 1958.

97. K.W. Morton and D.F. Mayers. *Numerical Solution of Partial Differential Equations.* Cambridge, Cambridge University Press, 1994.

98. G. Luo, Y. Shi, and J. Cong. An analytical placement framework for 3-D ICs and its extension on thermal awareness. *IEEE Transactions on Computer-Aided Design of Integrated Circuits and Systems*, 32(4):510–523, 2013.

99. Z.-W. Jiang, H.-C. Chen, T.-C. Chen, and Y.-W. Chang. Challenges and solutions in modern VLSI placement. In *VLSI Design, Automation and Test, 2007. VLSI-DAT 2007. International Symposium on*, pp. 1–5. IEEE, 2007.

100. N. Viswanathan, M. Pan, and C. Chu. Fastplace 3.0: A fast multilevel quadratic placement algorithm with placement congestion control. In *Asia South Pacific Design Automation Conference*, ASP-DAC '07, pp. 135–140, Washington, DC, 2007. IEEE Computer Society.

101. N. Viswanathan, G.-J. Nam, C.J. Alpert, P. Villarrubia, H. Ren, and C. Chu. Rql: Global placement via relaxed quadratic spreading and linearization. In *Proceedings of the Design Automation Conference*, DAC '07, pp. 453–458, ACM, New York, 2007.

102. J. Cong, G. Luo, and E. Radke. Highly efficient gradient computation for density-constrained analytical placement. *IEEE Transactions on Computer-Aided Design of Integrated Circuits and Systems*, 27(12):2133–2144, 2008.

103. A.E. Dunlop, V.D. Agrawal, D.N. Deutsch, M.F. Jukl, P. Kozak, and M. Wiesel. Chip layout optimization using critical path weighting. In *Proceedings of the Design Automation Conference*, pp. 133–136, New York, 1984.

104. R. Nair, C.L. Berman, P.S. Hauge, and E.J. Yoffa. Generation of performance constraints for layout. *IEEE Transactions on Computer-Aided Design of Integrated and Systems*, 8(8):860–874, 1989.

105. R.S. Tsay and J. Koehl. An analytic net weighting approach for performance optimization in circuit placement. In *Proceedings of the Design Automation Conference*, pp. 620–625, New York, 1991.

106. T. Kong. A novel net weighting algorithm for timing-driven placement. In *Proceedings of the International Conference on Computer-Aided Design*, pp. 172–176, November 2002.

107. T.F. Chan, J. Cong, and E. Radke. A rigorous framework for convergent net weighting schemes in timing-driven placement. In *Proceedings of the International Conference on Computer-Aided Design*, ICCAD '09, pp. 288–294, New York, 2009. ACM.

108. A. Marquardt, V. Betz, and J. Rose. Timing-driven placement for FPGAs. In *Proceedings of the 2000 ACM/SIGDA Eighth International Symposium on Field Programmable Gate Arrays*, FPGA '00, pp. 203–213, New York, 2000. ACM.

109. A.B. Kahng and Q. Wang. Implementation and extensibility of an analytic placer. *IEEE Transactions on Computer-Aided Design of Integrated Circuits and Systems*, 24(5):734–747, 2005.

110. C.-L.E. Cheng. RISA: accurate and efficient placement routability modeling. In *Proceedings of the International Conference on Computer-Aided Design*, pp. 690–695, New York, November 1994.

111. J. Lou, S. Thakur, S. Krishnamoorthy, and H. Sheng. Estimating routing congestion using probabilistic analysis. *IEEE Transactions on Computer-Aided Design of Integrated and Systems*, 21(1):32–41, 2002.

112. U. Brenner and A. Rohe. An effective congestion-driven placement framework. *IEEE Transactions on Computer-Aided Design of Integrated and Systems*, 22(4):387–394, 2003.

113. S. Mayrhofer and U. Lauther. Congestion-driven placement using a new multi-partitioning heuristic. In *Proceedings of the International Conference on Computer-Aided Design*, pp. 332–335, New York, 1990.

114. A.B. Kahng, S. Mantik, and D. Stroobandt. Toward accurate models of achievable routing. *IEEE Transactions on CAD of Integrated Circuits and Systems*, 20:648–659, 2001.

115. P. Saxena, R.S. Shelar, and S. Sapatnekar. *Routing Congestion in VLSI Circuits: Estimation and Optimization.* Springer Science & Business Media, New York, 2007.

116. C.J. Alpert, Z. Li, M.D. Moffitt, G.-J. Nam, J.A. Roy, and G. Tellez. What makes a design difficult to route. In *Proceedings of the International Symposium on Physical Design*, pp. 7–12. ACM, New York, 2010.

117. H. Shojaei, A. Davoodi, and J.T. Linderoth. Congestion analysis for global routing via integer programming. In *Proceedings of the International Conference on Computer-Aided Design*, pp. 256–262. IEEE Press, 2011.

118. W.-H. Liu, C.-K. Koh, and Y.-L. Li. Case study for placement solutions in ISPD11 and DAC12 routability-driven placement contests. In *Proceedings of the International Symposium on Physical Design*, pp. 114–119. ACM, New York, 2013.

119. J. Hu, M.-C. Kim, and I.L. Markov. Taming the complexity of coordinated place and route. In *Proceedings of the Design Automation Conference*, pp. 1–7. IEEE, 2013.

120. H. Shojaei, A. Davoodi, and J. Linderoth. Planning for local net congestion in global routing. In *Proceedings of the International Symposium on Physical Design*, pp. 85–92. ACM, New York, 2013.

121. M. Sarrafzadeh, M. Wang, and X. Yang. *Modern Placement Techniques*. Springer Science & Business Media, New York, 2013.

122. Y. Wei, C. Sze, N. Viswanathan, Z. Li, C.J. Alpert, L. Reddy, A.D. Huber, G.E. Tellez, D. Keller, and S.S. Sapatnekar. Techniques for scalable and effective routability evaluation. *ACM Transactions on Design Automation of Electronic Systems (TODAES)*, 19(2):17, 2014.

123. W.-H. Liu, T.-K. Chien, and T.-C. Wang. A study on unroutable placement recognition. In *Proceedings of the International Symposium on Physical Design*, pp. 19–26. ACM, New York, 2014.

124. Z. Qi, Y. Cai, and Q. Zhou. Accurate prediction of detailed routing congestion using supervised data learning. In *Proceedings of the International Conference on Computer Design*, pp. 97–103. IEEE, 2014.

125. G.-J. Nam. ISPD 2006 placement contest: Benchmark suite and results. In *Proceedings of the International Symposium on Physical Design*, pp. 167–167. ACM, New York, 2006.

126. N. Viswanathan, C.J. Alpert, C. Sze, Z. Li, G.-J. Nam, and J.A. Roy. The ISPD-2011 routability-driven placement contest and benchmark suite. In *Proceedings of the International Symposium on Physical Design*, pp. 141–146. ACM, New York, 2011.

127. N. Viswanathan, C. Alpert, C. Sze, Z. Li, and Y. Wei. Iccad-2012 cad contest in design hierarchy aware routability-driven placement and benchmark suite. In *Proceedings of the International Conference on Computer-Aided Design*, pp. 345–348. ACM, New York, 2012.

128. N. Viswanathan, C. Alpert, C. Sze, Z. Li, and Y. Wei. The dac 2012 routability-driven placement contest and benchmark suite. In *Proceedings of the Design Automation Conference*, pp. 774–782. ACM, New York, 2012.

129. V. Yutsis, I.S. Bustany, D. Chinnery, J.R. Shinnerl, and W.-H. Liu. ISPD 2014 benchmarks with sub-45nm technology rules for detailed-routing-driven placement. In *Proceedings of the International Symposium on Physical Design*, pp. 161–168. ACM, New York, 2014.

130. I.S. Bustany, D. Chinnery, J.R. Shinnerl, and V. Yutsis. ISPD 2015 benchmarks with fence regions and routing blockages for detailed-routing-driven placement. In *Proceedings of the International Symposium on Physical Design*, pp. 157–164. ACM, New York, 2015.

131. S. Yang, A. Gayasen, C. Mulpuri, S. Reddy, and R. Aggarwal. Routability-driven FPGA placement contest. In *Proceedings of the International Symposium on Physical Design*, pp. 139–143. ACM, New York, 2016.

132. A.B. Kahng and X. Xu. Accurate pseudo-constructive wirelength and Congestion estimation. In *Proceedings of the 2003 Int'l Workshop on System-level Interconnect Prediction*, pp. 61–68. ACM, New York, 2003.

133. J.A. Roy and I.L. Markov. Seeing the forest and the trees: Steiner wirelength optimization in placement. *IEEE Transactions on Computer-Aided Design of Integrated and Systems*, 26(4):632–644, 2007.

134. J.A. Roy, N. Viswanathan, G.J. Nam, C.J. Alpert, and I.L. Markov. Crisp: Congestion reduction by iterated spreading during placement. In *Proceedings of the International Conference on Computer-Aided Design*, pp. 357–362. ACM, 2009.

135. Y. Zhang and C. Chu. Crop: Fast and effective congestion refinement of placement. In *Proceedings of the International Conference on Computer-Aided Design*, pp. 344–350. IEEE, 2009.

136. M.C. Kim, J. Hu, D.J. Lee, and I.L. Markov. A simPLR method for routability-driven placement. In *Proceedings of the International Conference on Computer-Aided Design*, pp. 67–73. IEEE, 2011.

137. X. He, W.K. Chow, and E.F.Y. Young. SRP: Simultaneous routing and placement for congestion refinement. In *Proceedings of the International Symposium on Physical Design*, pp. 108–113. ACM, New York, 2013.

138. J. Cong, G. Luo, K. Tsota, and B. Xiao. Optimizing routability in large-scale mixed-size placement. In *Asia South Pacific Design Automation Conference*, pp. 441–446. IEEE, 2013.

139. T. Lin and C. Chu. Polar 2.0: An effective routability-driven placer. In *Proceedings of the Design Automation Conference*, pp. 1–6. IEEE, 2014.

140. Y.F. Chen, C.C. Huang, C.H. Chiou, Y.-W. Chang, and C.J. Wang. Routability-driven blockage-aware macro placement. In *Proceedings of the Design Automation Conference*, pp. 1–6. IEEE, 2014.

141. M.K. Hsu, Y.F. Chen, C.C. Huang, S. Chou, T.H. Lin, T.C. Chen, and Y.W. Chang. Ntuplace4h: A novel routability-driven placement algorithm for hierarchical mixed-size circuit designs. *IEEE Transactions on Computer-Aided Design of Integrated Circuits and Systems*, 33(12):1914–1927, 2014.

142. C.C. Huang, H.Y. Lee, B.Q. Lin, S.W. Yang, C.H. Chang, S.T. Chen, and Y.W. Chang. Detailed-routability-driven analytical placement for mixed-size designs with technology and region constraints. In *Computer-Aided Design (ICCAD), 2015 IEEE/ACM International Conference on*, pp. 508–513. IEEE, 2015.

143. N.K. Darav, A. Kennings, D. Westwick, and L. Behjat. High performance global placement and legalization accounting for fence regions. In *Proceedings of the International Conference on Computer-Aided Design*, pp. 514–519. IEEE Press, 2015.

144. N.K. Darav, A. Kennings, A.F. Tabrizi, D. Westwick, and L. Behjat. Eh? placer: A high-performance modern technology-driven placer. *ACM Transactions on Design Automation of Electronic Systems (TODAES)*, 21(3):37, 2016.

145. X. He, Y. Wang, Y. Guo, and E.F.Y. Young. Ripple 2.0: Improved movement of cells in routability-driven placement. *ACM Transactions on Design Automation of Electronic Systems (TODAES)*, 22(1):10, 2016.

146. W. Li, S. Dhah, and D. Z. Pan. UTPlaceF: A routability-driven FPGA placer with physical and congestion aware packing. In *Proceedings of the International Conference on Computer-Aided Design*. IEEE/ACM, 2016.

147. W.T.J. Chan, Y. Du, A.B. Kahng, S. Nath, and K. Samadi. Beol stack-aware routability prediction from placement using data mining techniques. In *Proceedings of the International Conference on Computer Design*, 2016.

148. T. Marshburn, I. Lui, R. Brown, D. Cheung, G. Lum, and P. Cheng. Datapath: A cmos data path silicon assembler. In *Proceedings of the Design Automation Conference*, DAC '86, pp. 722–729, Piscataway, NJ, 1986. IEEE Press.

149. R.X.T. Nijssen and C.A.J. Van Eijk. Greyhound: A methodology for utilizing datapath regularity in standard design flows. *INTEGRATION, the VLSI Journal*, 25(2):111–135, 1998.

150. M. Cho, M. Choudhury, R. Puri, H. Ren, H. Xiang, G.J. Nam, F. Mo, and R.K. Brayton. Structured digital design. In L. Lavagno, I.L. Markov, G. Martin, and L.K. Scheffer, Eds., *Electronic Design Automation for IC Implementation, Circuit Design, and Process Technology: Circuit Design, and Process Technology*, Chapter 7, pp. 155–182. CRC Press, Boca Raton, FL, 2016.

151. S.I. Ward, M.C. Kim, N. Viswanathan, Z. Li, C.J. Alpert, E.E. Swartzlander, and D.Z. Pan. Structure-aware placement techniques for designs with datapaths. *IEEE Transactions on Computer-Aided Design of Integrated Circuits and Systems*, 32(2):228–241, 2013.

152. S. Chou, M.K. Hsu, and Y.W. Chang. Structure-aware placement for datapath-intensive circuit designs. In *Proceedings of the Design Automation Conference*, pp. 762–767. ACM, New York, 2012.

153. A. Chowdhary, S. Kale, P.K. Saripella, N.K. Sehgal, and R.K. Gupta. Extraction of functional regularity in datapath circuits. *IEEE Transactions on Computer-Aided Design of Integrated Circuits and Systems*, 18(9):1279–1296, 1999.

154. H. Xiang, M. Cho, H. Ren, M. Ziegler, and R. Puri. Network flow based datapath bit slicing. In *Proceedings of the International Symposium on Physical Design*, pp. 139–146. ACM, New York, 2013.

155. H. Xiang, H. Ren, L. Trevillyan, L. Reddy, R. Puri, and M. Cho. Logical and physical restructuring of fan-in trees. In *Proceedings of the International Symposium on Physical Design*, pp. 67–74. ACM, New York, 2010.

156. H. Xiang, L. Reddy, L. Trevillyan, and R. Puri. Depth controlled symmetric function fanin tree restructure. In *Proceedings of the International Conference on Computer-Aided Design*, pp. 585–591. IEEE, 2013.

157. S. Ward, D. Ding, and D.Z. Pan. Pade: A high-performance placer with automatic datapath extraction and evaluation through high dimensional data learning. In *Proceedings of the Design Automation Conference*, pp. 756–761. ACM, New York, 2012.

158. S.I. Ward, D.A. Papa, Z. Li, C.N. Sze, C.J. Alpert, and E. Swartzlander. Quantifying academic placer performance on custom designs. In *Proceedings of the International Symposium on Physical Design*, pp. 91–98. ACM, New York, 2011.

159. Q. Liu and M. Marek-Sadowska. Pre-layout physical connectivity prediction with application in clustering-based placement. In *Proceedings of the International Conference on Computer Design*, pp. 31–37. IEEE, 2005.

160. A. Papanikolaou, D. Soudris, and R. Radojcic. *Three Dimensional System Integration: IC Stacking Process and Design*. Springer Science & Business Media, New York, 2010.

161. V.F. Pavlidis and E.G. Friedman. *Three-Dimensional Integrated Circuit Design*. Morgan Kaufmann, Cambridge, MA, 2010.

162. Y. Xie, J. Cong, and S.S. Sapatnekar. *Three-Dimensional Integrated Circuit Design*. Springer, Dordrecht, the Netherlands, 2010.

163. S.K. Lim. *Design for High Performance, Low Power, and Reliable 3D Integrated Circuits*. Springer Science & Business Media, New York, 2012.

164. J. Cong and G. Luo. Advances and challenges in 3D physical design. *Information and Media Technologies*, 5(2):321–337, 2010.

165. W.K. Mak and C. Chu. Rethinking the wirelength benefit of 3-D integration. *IEEE Transactions on Very Large Scale Integration (VLSI) Systems*, 20(12):2346–2351, 2012.

166. J. Knechtel and J. Lienig. Physical design automation for 3D chip stacks: Challenges and solutions. In *Proceedings of the International Symposium on Physical Design*, ISPD '16, pp. 3–10, New York, 2016. ACM.

167. S.K. Lim. Physical design for 3D ICs. In L. Lavagno, I.L. Markov, G. Martin, and L.K. Scheffer, Eds., *Electronic Design Automation for IC Implementation, Circuit Design, and Process Technology: Circuit Design, and Process Technology*, Chapter 9, pp. 217–243. CRC Press, Boca Raton, FL, 2016.

168. M.B. Healy, K. Athikulwongse, R. Goel, M.M. Hossain, D.H. Kim, Y.J. Lee, D.L. Lewis, et al. Design and analysis of 3D-maps: A many-core 3D processor with stacked memory. In *Custom Integrated Circuits Conference*, pp. 1–4, IEEE, 2010.

169. D.H. Kim, R.O. Topaloglu, and S.K. Lim. Block-level 3D IC design with through-silicon-via planning. In *Asia South Pacific Design Automation Conference*, pp. 335–340. IEEE, 2012.

170. J. Knechtel, I.L. Markov, and J. Lienig. Assembling 2-d blocks into 3-D chips. *IEEE Transactions on Computer-Aided Design of Integrated Circuits and Systems*, 31(2):228–241, 2012.

171. S.A. Panth, K. Samadi, Y. Du, and S.K. Lim. Design and cad methodologies for low power gate-level monolithic 3D ICs. In *Proceedings of the 2014 International Symposium on Low Power Electronics and Design*, pp. 171–176. ACM, New York, 2014.

172. D.H. Kim, K. Athikulwongse, and S.K. Lim. Study of through-silicon-via impact on the 3-D stacked IC layout. *IEEE Transactions on Very Large Scale Integration (VLSI) Systems*, 21(5): 862–874, 2013.

173. Y.J. Lee, D. Limbrick, and S.K. Lim. Power benefit study for ultra-high density transistor-level monolithic 3D ICs. In *Proceedings of the Design Automation Conference*, p. 104. ACM, New York, 2013.

174. Y. Zhan, S.V. Kumar, and S.S. Sapatnekar. *Thermally-Aware Design*. Now Publishers, Boston, MA, 2008.

175. J. Cong and G. Luo. Thermal-aware 3D placement. In Visileios F Pavlidis and Eby G Friedman, Eds., *Three Dimensional Integrated Circuit Design*, Chapter 5, pp. 103–144. Springer, Dordrecht, the netherlands, 2010.

176. C.C.N. Chu and D.F. Wong. A matrix synthesis approach to thermal placement. *IEEE Transactions on Computer-aided design of Integrated Circuits and Systems*, 17(11):1166–1174, 1998.

177. P. Wilkerson, A. Raman, and M. Turowski. Fast, automated thermal simulation of three-dimensional integrated circuits. In *Thermal and Thermomechanical Phenomena in Electronic Systems, 2004. ITHERM'04. The Ninth Intersociety Conference on*, pp. 706–713. IEEE, 2004.

178. B. Goplen and S. Sapatnekar. Efficient thermal placement of standard cells in 3D ICs using a force directed approach. In *Proceedings of the International Conference on Computer-Aided Design*, pp. 86. IEEE Computer Society, 2003.

179. B. Goplen and S.S. Sapatnekar. Placement of thermal vias in 3-D ICs using various thermal objectives. *IEEE Transactions on Computer-Aided Design of Integrated Circuits and Systems*, 25(4):692–709, 2006.

180. C.H. Tsai and S.M. Kang. Cell-level placement for improving substrate thermal distribution. *IEEE Transactions on Computer-aided design of Integrated Circuits and Systems*, 19(2):253–266, 2000.

181. G. Chen and S. Sapatnekar. Partition-driven standard cell thermal placement. In *Proceedings of the International Symposium on Physical Design*, pp. 75–80. ACM, New York, 2003.

182. J. Cong, J. Wei, and Y. Zhang. A thermal-driven floorplanning algorithm for 3D ICs. In *Computer Aided Design, 2004. ICCAD-2004. IEEE/ACM International Conference on*, pp. 306–313. IEEE, 2004.

183. W. Huang, S. Ghosh, S. Velusamy, K. Sankaranarayanan, K. Skadron, and M.R. Stan. Hotspot: A compact thermal modeling methodology for early-stage VLSI design. *IEEE Transactions on Very Large Scale Integration (VLSI) Systems*, 14(5):501–513, 2006.

184. P. Wilkerson, M. Furmanczyk, and M. Turowski. Compact thermal modeling analysis for 3D integrated circuits. In *11th International Conference Mixed Design of Integrated Circuits and Systems*, vol. 159, 2004.

185. H. Wei, T.F. Wu, D. Sekar, B. Cronquist, R.F. Pease, and S. Mitra. Cooling three-dimensional integrated circuits using power delivery networks. In *Electron Devices Meeting (IEDM), 2012 IEEE International*, pp. 14–2. IEEE, 2012.

186. Y. Zhan and S.S. Sapatnekar. High-efficiency green function-based thermal simulation algorithms. *IEEE Transactions on Computer-Aided Design of Integrated Circuits and Systems*, 26(9):1661–1675, 2007.

187. P. Li, L.T. Pileggi, M. Asheghi, and R. Chandra. Efficient full-chip thermal modeling and analysis. In *Computer Aided Design, 2004. ICCAD-2004. IEEE/ACM International Conference on*, pp. 319–326. IEEE, 2004.

188. S.K. Samal, S. Panth, K. Samadi, M. Saedi, Y. Du, and S.K. Lim. Fast and accurate thermal modeling and optimization for monolithic 3D ICs. In *Proceedings of the Design Automation Conference*, DAC '14, pp. 206:1–206:6, New York, 2014. ACM.

189. C. Liu and S.K. Lim. A design tradeoff study with monolithic 3D integration. In *Thirteenth International Symposium on Quality Electronic Design (ISQED)*, pp. 529–536. IEEE, 2012.

190. A. Todri-Sanial and C.S. Tan. *Physical Design for 3D Integrated Circuits*, volume 55. CRC Press, Boca Raton, FL, 2016.

191. S.K. Lim. *Design for High Performance, Low Power, and Reliable 3D Integrated Circuits*. Springer Science & Business Media, New York, 2013.

192. J.U. Knickerbocker, P.S. Andry, B. Dang, R.R. Horton, M.J. Interrante, C.S. Patel, R.J. Polastre et al. Three-dimensional silicon integration. *IBM Journal of Research and Development*, 52(6):553–569, 2008.

193. A.W. Topol, D.C. La Tulipe, L. Shi, D.J. Frank, K. Bernstein, S.E. Steen, A. Kumar et al. Three-dimensional integrated circuits. *IBM Journal of Research and Development*, 50(4.5):491–506, 2006.

194. S.H. Lee, K.N. Chen, and J.J.Q. Lu. Wafer-to-wafer alignment for three-dimensional integration: A review. *Journal of Microelectromechanical Systems*, 20(4):885–898, 2011.

195. J. Lienig. Introduction to electromigration-aware physical design. In *Proceedings of the International Symposium on Physical Design*, pp. 39–46. ACM, New York, 2006.

196. K. Athikulwongse, A. Chakraborty, J.S. Yang, D.Z. Pan, and S.K. Lim. Stress-driven 3D-IC placement with TSV keep-out zone and regularity study. In *Proceedings of the International Conference on Computer-Aided Design*, ICCAD '10, pp. 669–674, Piscataway, NJ, 2010. IEEE Press.

197. T. Lu, Z. Yang, and A. Srivastava. Electromigration-aware placement for 3DICs. In *2016 17th International Symposium on Quality Electronic Design (ISQED)*, pp. 35–40. IEEE, 2016.

198. P. Batude, T. Ernst, J. Arcamone, G. Arndt, P. Coudrain, and P.E. Gaillardon. 3-d sequential integration: A key enabling technology for heterogeneous co-integration of new function with cmos. *IEEE Journal on Emerging and Selected Topics in Circuits and Systems*, 2(4):714–722, 2012.

199. M.M. Shulaker, T.F. Wu, M.M. Sabry, H. Wei, H.-S.P. Wong, and S. Mitra. Monolithic 3D integration: A path from concept to reality. In *Proceedings of the Design Automation Conference*, DATE '15, pp. 1197–1202, San Jose, CA, 2015. EDA Consortium.

200. D.H. Kim, K. Athikulwongse, and S.K. Lim. A study of through-silicon-via impact on the 3D stacked IC layout. In *Proceedings of the International Conference on Computer-Aided Design*, pp. 674–680. ACM, New York, 2009.

201. D.H. Kim, S. Mukhopadhyay, and S.K. Lim. Through-silicon-via aware interconnect prediction and optimization for 3D stacked ICs. In *Proceedings of the 11th International Workshop on System Level Interconnect Prediction*, pp. 85–92. ACM, New York, 2009.

202. J. Cong, G. Luo, J. Wei, and Y. Zhang. Thermal-aware 3D IC placement via transformation. In *Asia South Pacific Design Automation Conference*, pp. 780–785. IEEE, 2007.

203. C.C. Chang and J.J.S. Cong. An efficient approach to multilayer layer assignment with an application to via minimization. *IEEE Transactions on Computer-Aided Design of Integrated Circuits and Systems*, 18(5):608–620, 1999.

204. P. Zhou, Y. Ma, Z. Li, R.P. Dick, L. Shang, H. Zhou, X. Hong, and Q. Zhou. 3D-STAF: Scalable temperature and leakage aware floorplanning for three-dimensional integrated circuits. In *Proceedings of the International Conference on Computer-Aided Design*, pp. 590–597. IEEE, 2007.

205. H. Imai and T. Asano. Finding the connected components and a maximum clique of an intersection graph of rectangles in the plane. *Journal of Algorithms*, 4(4):310–323, 1983.

206. M.D. Moffitt, A.N. Ng, I.L. Markov, and M.E. Pollack. Constraint-driven floorplan repair. In *Proceedings of the Design Automation Conference*, pp. 1103–1108. ACM, New York, 2006.

207. J. Cong, G. Luo, and Y. Shi. Thermal-aware cell and through-silicon-via co-placement for 3D ICs. In *Proceedings of the Design Automation Conference*, pp. 670–675. ACM, New York, 2011.

208. L.W. Hagen, D.J.H. Huang, and A.B. Kahng. Quantified suboptimality of VLSI layout heuristics. In *Proceedings of the Design Automation Conference*, pp. 216–221, New York, 1995.

209. C. Chang, J. Cong, M. Romesis, and M. Xie. Optimality and scalability study of existing placement algorithms. *IEEE Transactions on Computer-Aided Design of Integrated Circuits and Systems*, pp. 537–549, 2004.

210. S. Ono and P.H. Madden. On structure and suboptimality in placement. In *Asia South Pacific Design Automation Conference*, January 2005.

211. A.B. Kahng and S. Reda. Evaluation of placer suboptimality via zero-change netlist transformations. In *Proceedings of the International Symposium on Physical Design*, pp. 208–215, New York, April 2005.

212. Q. Wang, D. Jariwala, and J. Lillis. A study of tighter lower bounds in LP relaxation based placement. In *ACM Great Lakes Symposium on VLSI*, pp. 498–502, New York, 2005.

213. J. Cong, J. Shinnerl, and M. Xie. A set of large-scale mixed-size placement examples with known optima. Report, Computer Science Department, University of California, Los Angeles, CA, 2005.

214. M. Pan, N. Viswanathan, and C. Chu. An efficient and effective detailed placement algorithm. In *Proceedings of the International Conference on Computer-Aided Design*, pp. 48–55. IEEE, 2005.

215. P. Spindler, U. Schlichtmann, and F.M. Johannes. Abacus: Fast legalization of standard cell circuits with minimal movement. In *Proceedings of the International Symposium on Physical Design*, pp. 47–53. ACM, 2008.

216. U. Brenner. Bonnplace legalization: Minimizing movement by iterative augmentation. *IEEE Transactions on Computer-Aided Design of Integrated Circuits and Systems*, 32(8):1215–1227, 2013.

217. G. Wu and C. Chu. Detailed placement algorithm for VLSI design with double-row height standard cells. *IEEE Transactions on Computer-Aided Design of Integrated Circuits and Systems*, 35(9):1569–1573, 2016.

218. U. Brenner, A. Hermann, N. Hoppmann, and P. Ochsendorf. Bonnplace: A self-stabilizing placement framework. In *Proceedings of the International Symposium on Physical Design*, pp. 9–16. ACM, New York, 2015.

# 32

# Histograms, Wavelets, Streams, and Approximation

Sudipto Guha

## 32.1 Introduction

Over the last decade, the size of data seen by a computational problem have grown immensely. There appears to be more web pages than human beings, and we have successfully indexed the pages. Routers generate huge traffic logs, in the order of terabytes, in a short time. The same explosion of data is felt in observational sciences because our capabilities of measurement have grown significantly. In comparison, computational resources have not increased at the same rate. In particular, it has been found that the ability of random access to data itself is a resource. In settings where each individual input item is not so significant by itself, consider monitoring a network, estimating costs of query plans, and so on, but the quantity we are interested in is the aggregate picture that emerges from the data. In several scenarios, such as network monitoring, some data are never stored but merely used to infer aggregate health of the network. At the same time, in several data-intensive computations, making passes over the data has been found to be significantly more efficient. This has brought the data stream model to the fore. The model consists of an algorithm with a small random access memory, typically sublinear in input size, operating in passes over the input. Any input item not explicitly stored is inaccessible to the algorithm in the same pass. The model captures the essence of a monitoring process that is allowed to observe a system unobtrusively using some small "extra" space. Of particular interest is the one pass model, in which the input may simply not be stored at all. The data stream model poses several challenges. Intriguingly, even simple problems, which were thought to be fully understood at small scales, have been found to be ominous at the current scale of data and have required reexamination. As mentioned earlier, the aggregate picture that emerges from the the data, or the synopsis, is often the desiderata.

The idea of synopses is not new. We can view the data as a function over a suitable domain. Expressing a function accurately as a combination of few simpler functions has been at the heart of approximation theory in mathematics and dates back centuries starting with polynomial approximations and the work of Fourier. Histograms and bar charts have been in use since the middle ages. The Haar system was proposed as early as 1910. Initially, the thrust of synopsis construction had been to project the data on to a fixed space. The benefit of such schemes is that the sum of two synopses is the synopsis of the sum of the original functions. This had led to the bulk of work in mathematical approximation theory, known as linear approximation theories. The theorems proved about these were largely extremal, namely, projections which work for all data: what is the maximum error (again considering all data) given a fixed projection strategy, and so on. Schmidt, in 1909, was one of the earliest to consider nonlinear theories in which the image space is dependent on data; which in modern terms can be viewed as a data-dependent or data-driven synopsis. This immediately raised the question of approximating the given data in the best possible way. This question is closer to common optimization problems—and as the end goal is approximation, it is only natural to consider approximation algorithms for these problems. Most synopses techniques used currently are in the nonlinear category.

In context of databases synopses, date back to Selinger et al. [1] in the late 1970s. They proposed estimating the cost of various operators using synopses of data to decide between alternative query plans. The first synopsis structure proposed for this purpose was a simple division of the domain in equal sizes. Over time, it was recognized that piecewise constant approximations of the data, or serial histograms, are significantly more accurate descriptions of distributions. Subsequently, it was demonstrated that the $\ell_2$ distance between the representation and the data was an accurate estimate, which brought histograms closer to the mathematical definition of approximating functions. Since the late 1980s, Wavelets have become popular as a tool in image processing. Their success was primarily due to the existence of fast algorithms for transforms and their multiresolution nature. They were introduced in databases in the context of "data cubes" to describe the data hierarchically. Wavelets and histograms are, by no means, the only synopses structures used. Quantiles and other estimates have been used as well. We will only consider histograms and wavelet approximations of data in this chapter.

*Our goal in the current chapter* will not be to catalog the problems and the best results known. Our main aim is to introduce the reader to these problems and demonstrate what style of analysis is used. To that end, we will consider the simpler versions of the problems and restrict ourselves to one dimension mostly. The problems that we will focus on will be illustrative and will not be exhaustive. The notes at the end of the chapter contain pointers to the more commonly known variants of these problems and the respective references. We will only refer to works on histograms and wavelets in the main body of the chapter. The sources of the algorithms discussed can be found in the notes. We will focus on those problems that are simplest to state and yet nontrivial to solve in massive dataset context and therefore are the basic problems in this area. We subsequently discuss histograms and then wavelets. We conclude with notes on the literature in this area.

## 32.2 Definitions and Problems

**Definition 32.1 (Data Streams)** *A data stream is a model of computation that treats random access as a resource.*

In particular, given a set of $m$ objects $Y = Y(1), \ldots, Y(m)$, we want to compute a function $f(Y)$ using small space and one pass over the data under the following restrictions: (i) The items $Y(j)$ are inspected in an increasing order of $j$. (ii) Any item not explicitly stored is "forgotten"—we do not know its value any longer (in the same pass). The computation proceeds in passes over the data. Unless otherwise mentioned, a data stream algorithm will refer to a one pass streaming algorithm. The streaming models differ in the *semantics* of the stream items $Y(j)$. There are a multitude of models, but we will discuss the two most common ones: (1) We can have $Y(j) = X(j)$ and $m = n$, where we are considering a function $X(i)$ defined on integers which is specified in an increasing order of ($i$). This is a **Time series model**. (2) We can have $Y(j) = \langle i, \delta_j \rangle$, where $i \in [0, n]$ for some $n$. Each element $Y(j)$ implies "set $X(i) = X(i) + \delta_j$" ($\delta_j$ can be negative) and thus specifies a function $X$ as a sequence of updates. This is the **Update model**; also known as the cash-register/dynamic model.

In the current chapter, we mainly focus on the time series model of data streams. This model is the simplest stream model and relevant in the context of sensor data, stocks, and so on, where the order in which the data arrive has a natural meaning. A good sublinear space algorithm for this model implies a good algorithm with small "extra" space—as is typical in the monitoring setting. Concrete examples of such systems are the "self-tuning" systems [2] in which the system executes queries, and a monitoring component is gathering information about various parameters to optimize/maintain the system performance. In essence, the input to the monitoring process is "free" because that input to the monitoring system is the result of some computation that was necessary anyways. An example in this context is the work of Bruno et al. [3], who consider learning histograms from observing answers to database queries. Any resource allocated to the monitor implies less resource for the actual system, and it is naturally desired that the monitor has a small footprint.

Although the above is true for the update model of streams, the maintenance of relevant information under updates is often a bigger challenge than solving the original problem from the maintained information. These algorithms typically find nontrivial ways of capturing similar computation as in a time series model stream. The techniques used in these algorithms are exciting, but orthogonal to the question of histogram construction.

In summary, the time series models captures the problem at an abstract level that is restricted compared with offline computation but rich enough to allow us to design interesting algorithms. Very often, these algorithms are the stepping stones to the most general results on update streams. In the interest of space, we omit discussing the results in the update stream model, but the notes contain references to them. In this chapter, we focus on the following problems:

**Problem 32.1 (Histograms)** *Given a set of numbers $X = X(0), \ldots, X(n-1)$ in a (time series) stream, find a piecewise constant function $H$ with at most $B$ pieces to minimize some suitable function of the error $X - H$, for example, $||X - H||_2$ and $||X - H||_\infty$. Each "piece" corresponds to a subinterval $[a, b]$ of $[0, n-1]$ and is represented by one number. Unless otherwise specified, we would assume that the $B$ pieces induce a partition of the interval $[0, n-1]$.*

Each of the pieces is defined as a "bucket." Given an $i$, we find the bucket to which $i$ belongs to and return the representative number, say $v$, for the bucket as an estimate of $X(i)$. The error introduced by this process is $X(i) - v$ that can be viewed as $X(i) - H(i)$. A natural goal of any accurate description would be to minimize a suitable function of $X - H$. One of the most natural measures is the $\ell_2$ norm (or its square) of the error vector $X - H$; however, $\ell_1, \ell_\infty$ measures are common as well. We can also consider weighted variants where given a weight vector $\{\pi_i\}$ we seek to minimize a suitable function of the terms $\pi_i(X(i) - H(i))$. The weighted variants are sometimes termed "workload optimal." Several questions arise immediately—are the intervals allowed to overlap, should they cover the entire $[0, n-1]$, and so on. In the case of overlapping intervals, it is unclear how to define the value of a point that belongs to two buckets. However, under any definition, we can easily see that $B$ overlapping buckets define at most $2B-1$ nonoverlapping buckets over any interval. A natural extension of the above is to piecewise polynomials. These have a rich history in numerical estimation algorithms. A priori, it is not clear why we should be able to achieve near optimal solutions for these problems in near linear time, which brings us back to motivation of studying time series models.

**Problem 32.2 (Piecewise polynomials)** *Solve the earlier problem of expressing a function using $B$ nonoverlapping pieces in which the pieces are small degree polynomials.*

We can pose a problem about wavelets analogously and about the connections:

**Problem 32.3 (Wavelets)** *Given a wavelet basis $\{\psi_i\}$ and a set of numbers $X = X(0), \ldots, X(n-1)$ in a data stream, find a set of values $Z(i)$ with at most $B$ nonzero values such that a suitable error of $X - \sum_i Z(i)\psi_i$ is minimized.*

**Problem 32.4 (Connections between Synopses)** *What are the connections between the various synopses and can we leverage them to devise better algorithms?*

The above-mentioned problems do not make an explicit assumption on the dimension of the dataset. For wavelets, there are very few changes in the results in the offline setting. Histograms, on the other hand, turn out to be NP-hard, primarily due to two-dimensional partitioning. The issue of overlap of buckets becomes critical, as the number of nonoverlapping buckets required to express $B$ overlapping buckets is exponential in the dimension.

Further, two or more dimensional streaming is tricky to define except in update streams. The time series model is implicitly one-dimensional, in the special dimension that corresponds to the semantics of time. In this chapter, we will focus on the one-dimensional case and point the reader to the specific papers for the higher dimensional case.

*The above, by no means, is an exhaustive list of interesting problems.* However, the above are indeed *basic* in the sense that they are simple to pose and not always easy to solve in sublinear space.

## 32.3 Histograms

As mentioned earlier, histograms are piecewise constant approximations of data. Recall that the histogram problem is defined as: Given a set of numbers $X = X(0), \ldots, X(n-1)$ in a streaming fashion, find a piecewise constant function $H$ with at most $B$ pieces to minimize some suitable function of the error $X - H$, for example, $||X - H||_2$, $||X - H||_\infty$, and so on. We will assume that $X(i)$ are polynomially bounded integers, as the histograms are most often used to approximate frequency. The discussion will extend to reals provided the minimum nonzero error of estimation using a histogram can be bounded from below—which is also a finite precision assumption. We will focus on the $\ell_2^2$ measure. By using this as an example, we will see how to construct faster approximation algorithms. Subsequently, we will see how to extend the result to measures similar to $\ell_\infty$. All the discussion extends to weighted variants using standard techniques.

### 32.3.1 The Vopt or the $\ell_2^2$ Measure

The measure is a popular measure in databases and is also interesting mathematically. In this problem, the interval $[0, n-1]$ is partitioned into $B$ pieces.

**Observation 32.1** *Due to the partitioning, we can express the $\ell_2^2$ error as a sum of bucket errors. In each bucket, the best representative is the mean/average of the numbers.*

Let TERR$[i, k]$ be the minimum $\ell_2^2$ error of approximating $[0, i]$ using at most $k$ buckets. TERR$[i, k]$ is computed for the points in $[0, i]$ only. A natural dynamic program (DP) that tries all possible guesses of the last interval $[j + 1, i]$ is immediate. If the $\ell_2^2$ error of approximating the interval $[j + 1, i]$ by its mean is SQERROR$(j + 1, i) = \sum_{r=j+1}^i X(r)^2 - \left(\sum_{r=j+1}^i X(r)\right)^2 / (i - j)$, we have TERR$[i, k] = \min_j \{$TERR$[j, k - 1] +$ SQERROR$(j + 1, i)\}$.

The final solution is given by TERR$[n, B]$. Maintaining the prefix sums $\sum_{r=0}^j X(r), \sum_{r=0}^j X(r)^2$, the values of SQERROR$(j + 1, i)$ can be computed in $O(1)$ time. Immediately we arrive at an $O(n^2 B)$ algorithm

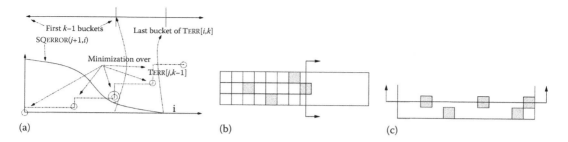

(a)                  (b)                  (c)

**FIGURE 32.1** (a) Approximating $\text{TERR}[i, k - 1]$ by a histogram, (b) front moving left to right, and (c) front moving bottom up.

using $O(nB)$ space. The space requirement can be reduced to $O(n)$, but a natural question arises: "since the primary role of histograms is in approximating data, can we develop linear time algorithms that are near optimal approximations of the best histogram?" In what follows, we show how to achieve such an algorithm. The starting point is the following:

**Observation 32.2** $\text{TERR}[j, \cdot]$ *is nondecreasing and* $\text{SQERROR}(j, \cdot)$ *(and therefore* $\text{SQERROR}(j + 1, \cdot)$*) is nonincreasing.*

It may appear that we can immediately use the above-mentioned properties to get faster algorithms, but that is not the case. Consider $v_1, \ldots, v_n$ where each $v_i \geq 0$. Let $f(i) = \sum_{r=1}^{i} v_r$ and $g(i) = f(n) - f(i - 1)$. The function $f(i)$ is nondecreasing, and $g(i)$ is nonincreasing. But finding the minimum of $f(i) + g(i)$ amounts to minimizing $f(n) + v_i$, or in other words minimizing $v_i$. Note that this does not rule out that over $B$ levels, the cost of the searching can be amortized—but no such analysis exists to date. The interesting aspect of the example is that picking any $i$ gives us a 2 approximation (as $f(n) + v_i \leq 2f(n)$ and the minimum is no smaller than $f(n)$). In essence, the searching can be reduced if we are willing to settle for an approximation.

The central idea is that instead of storing the entire function $\text{TERR}[j, k-1]$, we approximate the function as shown in Figure 32.1. The interval $[1, i]$ is broken down into $\tau$ intervals $(a_u, b_u)$ to approximately represent the function with a "staircase." We have $a_1 = 1$, $a_{u+1} = b_u + 1$, and $b_\tau = n$. Furthermore, the intervals are created such that the value of the function at the right-hand boundary of an interval is at most a factor $(1 + \delta)$ times the value of the function at the left-hand boundary. The number of maximum such intervals is $O(\frac{1}{\delta} \log n)$.

Now for any $a_u \leq j \leq b_u$, we have $(1 + \delta)\text{TERR}[a_u, k - 1] + \text{SQERROR}(a_u + 1, i) \leq (1 + \delta)\text{TERR}[j, k-1] + \text{SQERROR}(j+1, i)$ and which in turn is at most $\text{TERR}[b_u, k-1] + \text{SQERROR}(b_u+1, i)$ this rewrites to $(1+\delta)\text{TERR}[i, k] \leq \text{TERR}[b_u, k-1] + \text{SQERROR}(b_u+1, i)$. This implies that evaluating the sum at $b_u$ gives us a $(1+\delta)$ approximation. However, there is a caveat—we cannot simultaneously approximate $\text{TERR}[i, k]$ and assume that we know $\text{TERR}[j, k - 1]$ exactly for all $j < i, k > 2$. The solution is to employ the "ostrich algorithm," that is, ignore the issue and simply use the approximation $\text{APXERR}[j, k - 1]$ (of $\text{TERR}[j, k - 1]$) to compute $\text{APXERR}[i, k]$. We can show by induction that the ratio of $\text{APXERR}[i, k]$ to $\text{TERR}[i, k]$ is at most $(1 + \delta)^{k-1}$. Setting $\delta = \frac{\epsilon}{2B}$ gives us a $(1 + \epsilon)$ approximation as $(1 + \frac{\epsilon}{2B})^B \leq 1 + \epsilon$ for $\epsilon \leq 1$. The benefit of the algorithm is that we evaluate the sum at $O(\frac{1}{\delta} \log n)$ points assuming the input integers are polynomially bounded. The entire algorithm runs in time $O(\frac{nB}{\delta} \log n)$ time. The algorithm proceeds left-to-right, and as more data arrive, the function $\text{TERR}[i, k]$ does not change, and the staircase we have constructed remains valid. This, along with the fact that we only need to store $\sum_{r=1}^{b_u} X(r), \sum_{r=1}^{b_u} X(r)^2$ for the points $b_u$ allow the algorithm to be a $O(\frac{B^2}{\epsilon} \log n)$ space streaming algorithm. Therefore,

**Theorem 32.1** *We can compute a* $(1 + \epsilon)$ *approximation of the optimal histogram under* $\ell_2^2$ *error over a data stream in* $O(\frac{nB^2}{\epsilon} \log n)$ *time and* $O(\frac{B^2}{\epsilon} \log n)$ *space.*

*The algorithm in retrospect*: A metaphoric view of the algorithm could be the following: Consider the DP table generated by the optimum algorithm with $n$ columns and $B$ rows, the bottommost row corresponding to TERR$[i, 1]$. This new algorithm maintains a "front" that moves from left to right and creates (approximately) the same table as the optimal algorithm, but only chooses to remember a few "highlights" (Figure 32.1). The highlights correspond to boundary points that are sufficient to construct an approximate histogram. We can view the optimum algorithm as using $n$ buckets to represent the nondecreasing error function TERR$[i, k - 1]$ exactly. But we need at most $O(\frac{1}{\delta} \log n)$ buckets for polynomially bounded input if we approximate the function in the above-mentioned geometrically growing fashion. This geometrically growing staircase has been subsequently used in several problems of interest in time windowed data streams [4].

*Improving the above-mentioned algorithm*: The algorithm mentioned earlier still computed all the $\theta(nB)$ table entries, though each entry was computed faster. We were forced to evaluate all entries in the table in the absence of any indication if that value will not be relevant later in a streaming setting.

To improve the algorithm, we begin by ignoring the streaming aspect and develop on an offline algorithm with $O(n)$ memory. We subsequently show how to adapt the improved algorithm to a stream setting. One way of viewing the new offline algorithm is that we want to create a similar dynamic table as the optimal, but we only want to compute the APXERR$[j, k - 1]$ entries that are useful for some APXERR$[i, k]$ entry. *In a sense we want the front to move from bottom to top and only evaluate the necessary values* (Figure 32.1), and this requires the offline setting.

Note, we immediately have a problem that APXERR$[i, k + 1]$ may depend on APXERR$[i - 1, k]$ and APXERR$[i - 1, k]$ was later replaced by some APXERR$[i', k]$ where $i' > i$. If we are only computing the values which are necessary, we will be computing different values as APXERR$[i, k + 1]$ now has to use APXERR$[i', k]$. This is where the induction in the proof of earlier algorithm fails. However, the reassuring aspect is that APXERR$[i', k]$ must have been within a $(1 + \delta)$ factor of APXERR$[i - 1, k]$ and a more subtle induction goes through. This idea in itself gives an algorithm with running time $O(n + \frac{B^3 \log n}{\epsilon^{-2}})$. But we will improve the algorithm even more.

Note that we are interested in the entry APXERR$[B, n - 1]$. This is the top right-hand corner of the table. Now, the elements in the bottom right-hand corner of the table are likely to contain very large values, because they correspond to the approximation by very few buckets. Likewise, the elements in the top left-hand corner are likely to contain very small answers which correspond to approximating very small amount of data with a large number of buckets. Either of these sets of values are unlikely to influence the optimum solution. However, we need to quantify "large" and "small" in this discussion. The idea would be to first find the *scale of the optimum solution* and subsequently search in that scale to find the (near) best solution.

Assume that we were guaranteed that the optimum solution is less than $2\Delta$ for some $\Delta$. We find the largest $i$ s.t., APXERR$[i, 1] = $ SQERROR$(1, i) \leq (1 + \epsilon)2\Delta$. This we set to be $b_u$. We proceed backward to determine the smallest number $a_u$ such that APXERR$[a_u, 1] + \frac{\epsilon\Delta}{B-1} \leq$ APXERR$[b_u, 1]$. This defines the last interval $(a_u, b_u)$. We set $b_{u-1} = a_u - 1$ and proceed (backward). After we have created the list of intervals corresponding to 1 bucket (or $k$ buckets) we will proceed to the list of intervals corresponding to 2 (or $k + 1$) buckets. The size of each list will be $O(\frac{B}{\epsilon})$. Finding the smallest $a_u$ would involve a binary search and each evaluation of APXERR$[i, k]$ would involve using $O(\frac{B}{\epsilon})$ values from the list corresponding to $b_u$ for $k - 1$ buckets. The running time over all the $B$ lists can be shown to be $O(\frac{B^3}{\epsilon^2} \log n)$. Observe that the algorithm incurs approximation error additively and over the entire algorithm the total error can be shown to be an additive $\epsilon \Delta$.

Suppose we start from the smallest possible nonzero value as $\Delta$ and get a solution whose error is more than $2\Delta(1 + \epsilon')$. Then we know that the optimum solution is above $2\Delta$ and we can double the estimate of $\Delta$. This way, after at most $O(\log n)$ rounds, we will get to a point where we get a solution whose error is at most $2\Delta(1 + \epsilon')$, but recursively we have maintained the invariant that the optimum is at least $\Delta$. At this point set $\Delta' = \Delta(1 + \epsilon')$. Moreover, by virtue of the existence of some solution of cost $2\Delta(1 + \epsilon')$, we are guaranteed that the optimum is within $2\Delta'$. We choose a $\epsilon''$ such that $\epsilon'' = \epsilon \Delta / \Delta'$ and apply the above-mentioned algorithm. The final solution has additive error $\epsilon \Delta$ which is a $(1+\epsilon)$ approximation as $\Delta$ is less than the optimum solution. The running time of the algorithm is $O(\frac{B^3 \log n}{\epsilon^2} + \frac{B^3 \log n}{\epsilon'^2} \log n)$ considering all $O(\log n)$ rounds. Now observe that we can set $\epsilon' = 1$, that is, try to get a fast 4 approximation. We compute the sums $\sum_i X(i)$ and so on, in $O(n)$ additional time. In summary,

**Theorem 32.2** *We can find a $(1+\epsilon)$ approximation to the optimal histogram in $O(n + B^3 \log^2 n + \frac{B^3}{\epsilon^2} \log n)$ time.*

*A return to streams*: The above-mentioned algorithm appears to be hopelessly offline. In our metaphor of "fronts," we are proceeding row-by-row upward, and the entire data are required to be present to allow us to compute SQERROR(). The idea we would now use is to read in *a block of data* of size $M$ in left-to-right order, but use the bottom-to-top approach to construct the staircases for the new data (using staircases of the old data). We would have to maintain the geometric approximation as in the first approximation algorithm (as across different blocks we cannot proceed backwards as in the second algorithm). The number of items we would consider for each list would be $O(\frac{B}{\epsilon} \log n)$ as in the first algorithm, and $\frac{n}{M}$, corresponding to the endpoints of the blocks as in each block we will proceed backward and always evaluate the endpoint. To find the smallest $a_u$, however, would only require $O(\log M)$ evaluations as we would be searching over the new elements only. Thus, over all the lists of the algorithm will use $O(B(\frac{n}{M} + \frac{B}{\epsilon} \log n) \log M)$ evaluations of some APXERR$[i, \cdot]$. We will use a better algorithm to compute APXERR$[i, k]$. Along with the lists $Q[k]$ of intervals, where APXERR$[\cdot, k]$ increase geometrically in powers of $(1 + \frac{\epsilon}{2B})$, we would keep track of a sublist $SubQ[k]$ inside this list of intervals where the APXERR$[\cdot, k]$ increase in powers of 2. Thus, the sublist will be of size $O(\log n)$. We will use $SubQ[k - 1]$ to compute a 4 approximation (say $A$) of TERR$[i, k]$ of APXERR$[i, k]$. Then we would proceed backward inside the list $Q[k - 1]$ and only consider the APXERR$[j, k - 1]$ elements which are separated by $A/(cB)$ for some constant $c$ and *use these new elements* to compute a better approximation of TERR$[i, k]$. This approximation can be shown to be $(1 + \delta)^k$, the proof is detailed and is omitted. The upshot of this more complicated algorithm is that the time to compute each APXERR$[i, k]$ is $O(\log n + \frac{B}{\epsilon} \log \tau)$ (where $\tau = \frac{B \log n}{\epsilon}$, the same as earlier)—the extra log term arises from the backward binary search inside $Q[k - 1]$. Thus, the total running time (we add the $O(n)$ time to compute the sums for all the elements) is

$$n + \left( \log n + \frac{B}{\epsilon} \log \tau \right) B \left( \frac{n}{M} + \frac{B}{\epsilon} \log n \right) \log M$$

We can now set $M$ to get the coefficient of $n$ to be a true constant, and thus

**Theorem 32.3** *Let $\tau = \frac{B}{\epsilon} \log n$ and $M = O((\frac{B}{\epsilon} \log \tau + \log n) B \log \tau)$. We can construct a $(1 + \epsilon)$ approximation data stream algorithm for computing the optimal histogram that runs in time $O(n + M\tau)$ and uses space $O(B\tau + M)$.*

For fixed $B, \epsilon$, and any $\gamma > 0$, using $O(\gamma \log n)$ space we get a $(1+\epsilon)$ approximation in $O(n + \frac{n \log \log n}{\gamma})$ time – which is a nice tradeoff. A natural question that would arises at this point—do these approximations help? Note that the approximations were motivated from very common sense "pruning strategies" or heuristics that should be a part of a good code. It is gratifying that in this case we can analyze the pruning strategies and in fact prove their correctness as well as improved performance. For an implementation, the dependence on $\epsilon, B$ matters and getting a better theoretical algorithm does allow us to have better algorithms for practice.

## 32.3.2 Beyond $\ell_2^2$ Error: Workloads, Piecewise Polynomials

If we inspect the algorithms in the previous section, the following ideas were used in the DP and the approximation algorithm(s) respectively:

1. OPT: The error SQERROR($i,j$) of a bucket depends only on the values in the bucket and the endpoints $i,j$. We can maintain small information for each element s.t. given any $i,j$ the value of SQERROR($i + 1,j$) can be computed efficiently. The overall error is the sum of the errors of the buckets.
2. APX: The error is *interval monotone*, that is, a subinterval has error no more than the whole interval. The minimum nonzero error and the maximum error are lower and upper bounded respectively by polynomials in $n$.

We can now revisit the proof in the previous section (with $P = Q = T = O(1)$), and the next theorem follows:

**Theorem 32.4** *Suppose we are given a histogram construction problem where the error $E_T[\cdot,\cdot]$ satisfies the above-mentioned conditions. Suppose the error of a single bucket $E_B(i + 1,j)$ can be computed in time $O(Q)$ from the records INFO[$i$] and INFO[$j$] each requiring $O(P)$ space. Assume that the time to create the $O(P)$ structure is $O(T)$ then by changing the function that computes the error given the endpoints we achieve the following:*

 (i) *We can find the optimum histogram in time $O(nT + n^2(B + Q))$ time and $O(n(P + B))$ space.*
 (ii) *In $O(nT + QB^3(\log n + \epsilon^{-2})\log n)$ time and $O(nP)$ space, we can find a $(1 + \epsilon)$ approximation to the optimum histogram.*
 (iii) *In $O(nT + M_Q\tau)$ time and $O(PB\tau + M_Q)$ space we can find a $(1 + \epsilon)$ approximation to the optimum histogram over a data stream where $M_Q = B(\frac{QB}{\epsilon} + Q\log n + \frac{B}{\epsilon}\log\tau)\log(Q\tau)$ and $\tau = B\epsilon^{-1}\log n$.*

**Example 32.1 (Workloads)**  *Workloads are weighted $\ell_p$ norms. Typically the workload is specified as a $k$-bucket histogram as well (as specifying $n$ weights requires a lot of space) and each $E_B()$ can be computed in time $Q = O(k)$ time. The space requirement to store the lists increases by an additive $O(k)$ as it is simpler to add the endpoints of the workload histogram to all the queues. $T = P = O(1)$ in this case.*

**Example 32.2 (Piecewise Polynomials)**  *For polynomials of degree $d$ we need to store prefix sums such as $\sum_r X(r)^m$ for $0 \le 2d + 2$. To find the best representative we need to solve a $O(d) \times O(d)$ matrix which makes $Q = O(d^3)$. $P = T = O(d)$ in this case.*

**Example 32.3 ($\ell_1$ Error)**  *In this case, the representative of a bucket is the median. In an offline setting, we can preprocess the data in $O(n\log n)$ time and space to achieve $Q = O(\log^2 n)$. In the stream setting, we need to prove that an approximate median of rank within $\frac{n}{2} \pm \epsilon n$ increases the error of a bucket by $(1 + \epsilon)$ factor. Approximate medians can be found using the algorithms of Manku et al. [5] or Greenwald and Khanna [6] using $O(\frac{\log n}{\epsilon})$ space. However, this needs to be done for each of the $B\tau$ endpoints (each of which could potentially form a bucket with the current data we are seeing). Thus we can apply Theorem 32.4(iii) with $T = B\tau\log\frac{\log n}{\epsilon}$, $Q = P = \log\frac{\log n}{\epsilon}$.*

**Example 32.4 ($\chi^2$ Error and Information Distances)**  *In this case, the error of representing a set of numbers $v_1,\dots,v_m$ numbers by $h$ is given by $\sum_r \frac{(v_r - h)^2}{h}$. By using prefix sums similar to $\ell_2^2$ we can show $Q = T = P = O(1)$. This is interesting as one of the objectives of histograms to represent distributions and information theoretic metrics are obviously more suited for comparing distributions.*

**Example 32.5 (Relative Error)**  *One of the issues with $\ell_p$ error is that approximating 1000 by 1010 counts as much towards the error as approximating 1 by 11. One way of ameliorating the problem is to define relative error measure which computes a function ($\ell_2,\ell_1$ norm of the vector) $\frac{|X(i)-\hat{X}(i)|}{\max\{|X(i)|,c\}}$ where $\hat{X}(i)$ is the estimate of $X(i)$ constructed from the synopsis. $c$ is a constant to avoid division by 0 and the effect of arbitrarily small numbers. The different measures lead to different settings of $P, Q, T$. For example, for relative $\ell_2$, we need to compute the harmonic mean, but that can be achieved with $P = Q = T = O(1)$.*

### 32.3.3 $\ell_\infty$ and Variants

The histogram algorithms can be significantly simplified for $\ell_\infty$ variants (workload, relative error, etc.). Note that

**Observation 32.3** *For most reasonable error metrics based on $\ell_\infty$ (relative $\ell_\infty$), the error of a bucket depends on the (suitably weighted) maximum and minimum values in the bucket.*

Thus as long as the maximum and minimum values are fixed, the error of a bucket does not change. If we were told that the error of an interval using $k$ buckets is $\tau$, we can verify the claim by "eating up" maximal subintervals from the left of error $\tau$ and see if we can "cover" the entire interval. Assuming we can compute the error of any interval in $O(Q)$ time, the running time of such an algorithm is $O(kQ\log n)$ using binary search. We can easily maintain a tree using $O(n)$ preprocessing and $O(n)$ space which gives gives $Q = \log n$. Let the error of a single bucket defined by the interval $[a, b]$ be $E_{B,\infty}(a, b)$.

The above gives a simple algorithm in which we guess the *first* bucket. If the first bucket is defined by the interval $[0, i]$, we can check if $B - 1$ buckets cover the interval $[i + 1, n]$ using $\tau = E_{B,\infty}(0, i)$. If yes, we need to try lower values of $i$ (or $i$ may be the correct answer). Otherwise, we know that the error of the optimum solution is larger than $\tau$. Thus either (a) we need to increase $i$ which increases $\tau$ or (b) we need to increase $\tau$ but $[1, i]$ is the first bucket. So we find an $i$ such that if the error of the optimum solution is $\tau$ then it satisfies $E_{B,\infty}(0, i) < \tau \le E_{B,\infty}(0, i + 1)$. This is found in time $O(B\log^3 n)$ using binary search. Now if the optimum error of representing the interval $[i + 1, n]$ using $B - 1$ buckets is $\tau^*$ (which we will find recursively), the final error is $\min\{E_{B,\infty}(0, i + 1), \tau^*\}$. More explicitly, if $\tau^* < E_{B,\infty}(0, i + 1)$, the optimum solution is the solution of $[i + 1, n]$ (found recursively) along with the first bucket defined by the interval $[0, i]$. Otherwise, the solution is the result of taking out intervals whose error is less or equal to $E_{B,\infty}(0, i + 1)$. Note that this sets up a recursion $f(B) = B\log^3 n + f(B - 1)$ and thus we conclude:

**Theorem 32.5** *For variants of $\ell_\infty$ error (weighted, workload, relative error) we can compute the optimal histogram in time $O(n + B^2\log^3 n)$ and $O(n)$ space.*

However, in a streaming scenario where we cannot afford linear space, then we can use an algorithm similar to *(iii)* in Theorem 32.4. We can begin by writing a *worse* algorithm, which is $O(n^2 B)$ time, but computes the optimum solution in a fashion similar to $\ell_2^2$, but uses $E_{T,\infty}[i, k] = \min_j \max\{E_{T,\infty}[j, k - 1], E_{B,\infty}(j + 1, i)\}$. But then,

**Observation 32.4** *We can compute the minimum of $\max\{f(j), g(j)\}$ in $O(\log n)$ evaluations of $g()$ if $f(j) = E_{T,\infty}[j, k - 1]$ is nondecreasing and $g(j) = E_{B,\infty}(j + 1, i)$ is nonincreasing.*

Therefore, we immediately improve the worse optimum algorithm to run in time $O(nB\log^2 n)$ time using the $O(n)$ preprocessing to answer $E_{B,\infty}()$ in $O(\log n)$ time. Now consider maintaining a staircase approximating $E_{T,\infty}[j, k - 1]$ using $\tau = O(\frac{B\log n}{\epsilon})$ endpoints. Now the block-by-block algorithm performs $\frac{n}{M} + \tau$ insertions into each interval list. Each requires a binary search of $\log M$ and over $B$ lists we evaluate $\text{ApxE}_\infty[]$ at most $O(B(\frac{n}{M} + \tau)\log M)$ times each requiring $O(\log \tau)$ time. There is one complication, namely, in evaluating $E_{B,\infty}(b_u, i)$ if $b_u$ was in some block $r - 2$ or before and $i$ was in block $r$, the answer depends on the maximum and minimum values in block $r - 1$. So as we process one block and move to the next, we may have to update the $O(B\tau)$ entries in the list to take care of the above-mentioned issue. This adds $O(\frac{n}{M}B\tau)$ to the running time. The overall running time is thus

$$B(\frac{n}{M} + \tau)(\log M)(\log \tau) + \frac{n}{M}B\tau + n$$

We can set $M = O(B\tau)$ to make the coefficient of $n$ to be $O(1)$ and thus

**Theorem 32.6** *We can compute a $(1 + \epsilon)$ approximation of the $\ell_\infty$ variants (workloads, weighted, etc.) in $O(n + \frac{B^2\log n}{\epsilon}\log^2\frac{B\log n}{\epsilon})$ time and $O(\frac{B^2\log n}{\epsilon})$ space over a data stream.*

## 32.4 Wavelet Synopses

We begin with a brief review of wavelets and their main properties. One of the most important reasons for the popularity of wavelets is captured in Proposition 32.1, which states that (for compact wavelets) at most $O(\log n)$ basis vectors are relevant to a point. Moreover, the fact that the basis set is orthonormal and the existence of fast forward and inverse transforms has been a strong attraction for their wide use.

### 32.4.1 A Compact Primer on Compact Wavelets

**Definition 32.2** *The support of any vector $\Psi$ is the set $\mathrm{SUPP}(\Psi) = \{t | \Psi(t) \neq 0\}$.*

**Definition 32.3** *Let $h[], g[]$ be two arrays defined on $\{0, 1, \ldots, 2q - 1\}$, s.t., $g[k] = (-1)^k h[2q - 1 - k]$. Assume that $\sum_k h[k] = \sqrt{2}$ and $\sum_k g[k] = 0$, (along with few other properties, see [7,8]). Let $\phi_{0,s}(t) = \delta_{st} \in \mathcal{R}^n$, that is, the vector which is 1 at $s$ and 0 everywhere else. Define $\phi_{j+1,s} = \sum_t h[t - 2s]\phi_{j,t}$ and $\psi_{j+1,s} = \sum_t g[t - 2s]\phi_{j,t}$.*

The set of wavelet vectors $\{\psi_{j,s}\}_{(j,s)\in\mathbb{Z}^2}$ define an orthonormal basis for $\mathcal{R}^n$. For ease of notation, we will use both $\psi_i$ and $\psi_{j,s}$ depending on the context and assume there is a consistent map between them. The function $\psi_{j,s}$ is said to be centered at $2^j s$ and of scale $j$ and is defined on at most $(2q - 1)2^j$ points. It can be shown that $\phi_{j,s} = \sum_t h[s - 2t]\phi_{j+1,t} + \sum_t g[s - 2t]\psi_{j+1,t}$, [7]. Further, $\phi_{j,s}(x), \psi_{j,s}(x)$ when scaled to the (continuous) domain $[0, 2q - 1]$ converge to $2^{-j/2}\phi\left(\frac{x-2^j s}{2^j}\right), 2^{-j/2}\psi\left(\frac{x-2^j s}{2^j}\right)$; that is, the vectors look similar, but are shifted and scaled.

**Example 32.6 (Haar Wavelets)** *In this case $q = 1$ and $h[] = \{\frac{1}{\sqrt{2}}, \frac{1}{\sqrt{2}}\}$. Thus $g[] = \{\frac{1}{\sqrt{2}}, -\frac{1}{\sqrt{2}}\}$. Given X, the algorithm to compute the transform finds the "difference" coefficients $d_1[i] = \frac{X(2i)-X(2i+1)}{\sqrt{2}}$. The "averages" $\frac{X(2i)+X(2i+1)}{\sqrt{2}}$, corresponds to $a_1[i]$, and the entire process is repeated on these $a_1[i]$ but with $n := n/2$ as we have halved the number of values. In the inverse transform, we get for example $a_0[0] = (a_1[0] + d_1[0])/\sqrt{2} = X(0)$ as expected. The coefficients naturally define a coefficient tree where the root is $a_{\log n+1}[0]$ (the overall average scaled by $\sqrt{n}$) with a single child $d_{\log n}[0]$ (the scaled differences of the averages of the left and right halves). Underneath $d_{\log n}[0]$ lies a complete binary tree as shown in Figure 32.2c. Note that in this case the support of the basis vectors define a hierarchical structure and each wavelet coefficient "affects" the values in its subtree only.*

*Note that $\psi$ is discontinuous, that is, if a wavelet basis vector with a large support is mapped to the continuous interval $[0, 1]$, the transition from positive to negative values remain and the gap is $2^{j/2}$ at scale $j$. Thus the synopses using Haar wavelets are better suited to handle "jumps" or discontinuities in data. This simple wavelet proposed in 1910 is still useful as it is excellent in concentrating the energy of the transformed signal (sum of squares of coefficients). A natural question arises if "smooth" wavelets exist, that is, when wavelet vector with a large support is mapped to the continuous interval $[0, 1]$, the values get significantly closer. The seminal work of Daubechies gives us several examples which we discuss next, [8].*

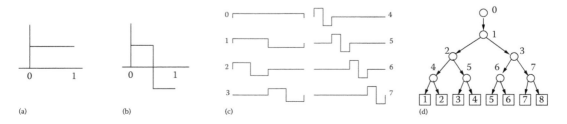

**FIGURE 32.2** (a) $\phi$, (b) $\psi$, (c) the set of Haar basis vectors at scale 3, and (d) the tree defined by the coefficients.

**Example 32.7 (Daubechies Wavelets $D_2$)** *In this case $q = 2$ and $h[] = \left\{ \frac{1+\sqrt{3}}{4\sqrt{2}}, \frac{3+\sqrt{3}}{4\sqrt{2}}, \frac{3-\sqrt{3}}{4\sqrt{2}}, \frac{1-\sqrt{3}}{4\sqrt{2}} \right\}$. Thus $g[] = \{h[3], -h[2], h[1], -h[0]\}$. The $\phi$ and the $\psi$ functions are shown in the following (normalized to the domain $[0, 1]$) and they converge quite rapidly. The coefficients now form a graph rather than a tree, which is given in Figure 32.2. The $D_q$ wavelets have compact support ($q$ is a fixed integer) but are unfortunately asymmetric. It turns out that Haar wavelets are the unique real symmetric compactly supported wavelets [8]. Moving to the complex domain, one can define symmetric biorthogonal wavelets.*

**Proposition 32.1** *For a compactly supported wavelet, there are $O(q \log n)$ basis vectors with a nonzero value at any point $t$. Further, given $t$, $\psi_{j,s}(t)$ and $\phi_{j,s}(t)$ can be computed in $O(q \log n)$ time.*

*The Cascade algorithm for $\langle X, \psi_{j,s} \rangle$, $\langle X, \phi_{j,s} \rangle$:* To compute the **forward transform**: Given a function $X$, set $a_0[i] = X(i)$. Repeatedly compute, $a_{j+1}[t] = \sum_s h[s - 2t]a_j[s]$ and $d_{j+1}[t] = \sum_s g[s - 2t]a_j[s]$. It is easy to see that $a_j[t] = \langle X, \phi_{j,t} \rangle$ and $d_j[t] = \langle X, \psi_{j,t} \rangle$. To compute the **inverse transform**, we compute $a_j[t] = \sum_s h[t - 2s]a_{j+1}[s] + \sum_s g[t - 2s]d_{j+1}[s]$.

**Definition 32.4** *Let $\mathcal{W}(X)$ denote the wavelet transform, that is, $\mathcal{W}(X)(t) = \langle X, \psi_t \rangle$, and let $\mathcal{W}^{-1}(Z) = \sum_i Z(i)\psi_i$ denote the inverse transform.*

Recall that the synopsis problem is: Given a wavelet basis $\{\psi_i\}$ and $X = X(0), \ldots, X(n-1)$ in a data stream, find a set of values $\{Z(i)\}$ with at most $B$ nonzero values minimizing a suitable function of $X - \sum_i Z(i)\psi_i$.

## 32.4.2 Wavelet Synopses and $\ell_2$ Theory

Suppose that we were interested in minimizing $\|X - \mathcal{W}^{-1}(Z)\|_2$. We can use the result of Parseval which states that "lengths are preserved under rotations." As an orthonormal transformation defines a rotation, and the wavelet basis vectors define an orthonormal basis, $\|X - \mathcal{W}^{-1}(Z)\|_2 = \|\mathcal{W}(X - \mathcal{W}^{-1}(Z))\|_2$. As the transformation is linear we get $\|X - \mathcal{W}^{-1}(Z)\|_2 = \|\mathcal{W}(X) - \mathcal{W}(\mathcal{W}^{-1}(Z))\|_2$, which is equivalent to minimizing $\sum_i (Z(i) - \mathcal{W}(X)(i))^2$. The constraint is that at most $B$ of the $Z(i)$'s can be nonzero. The solution is clearly choosing the largest $|\mathcal{W}(X)(i)| = |\langle \psi_i, X \rangle|$ and set $Z(i) = \mathcal{W}(X)(i) = \langle \psi_i, X \rangle$. *Observe that the fact that we retain some of the coefficients was a consequence of the proof and not a constraint.*

The synopsis construction problem for $\ell_2$ error reduces to choosing the largest (ignoring sign) wavelet coefficients of the data. It is not to difficult too see that this can be computed over a data stream. The simplest way of viewing the computation is a "level-by-level" construction of running several algorithms in parallel, each corresponding to a level. The basic insight of the paradigm is reduce-merge [9], and for streaming algorithms, this idea was first used in the context of clustering [10]. We need to implement the cascade algorithm in a similar format.

We describe an algorithm that reads a stream of values $X(0), \ldots, X(n - 1)$ and outputs the set of coefficients $\langle \psi_i, X \rangle$ (in some order). This algorithm uses $O(q \log n)$ space. We can feed the output of this algorithm that maintains the largest (ignoring signs) $B$ values seen in the stream. This can be achieved using $O(B)$ space in $O(n)$ time.

In the lowest level, the algorithm sees the data and computes $d_1[q]$ for the first $2q$ values. For the Haar basis, this is $\frac{X(0)-X(1)}{\sqrt{2}}$. The value $a_1[]$ (for Haar, $a_1[0] = \frac{X(0)+X(1)}{\sqrt{2}}$) is passed to the algorithm in the next (higher) level.

The algorithm in the lowest level now proceeds to read two new values and output $d_1[q + 1]$. For non-Haar basis there is an issue of wrap-around and the first $2q - 2$ data values are useful for a coefficient that depends on data that arrive at the end; thus they need to be stored. For Haar, the values $X(0), X(1)$ can be discarded. It is clear that in each level $j$, we need to store $O(q)$ information and output the coefficients of scale $j$. The total space of the algorithm across all levels is $O(q \log n)$.

**Theorem 32.7** *We can compute the optimal wavelet synopsis under $\ell_2$ error using $O(n)$ space and $O(B + q \log n)$ space, for any compact wavelet basis.*

### 32.4.3 Wavelet Synopses under Non-$\ell_2$ Error

Suppose that we were interested in minimizing $\|X - \mathcal{W}^{-1}(Z)\|_\infty$. We cannot use the result of Parseval as the $\ell_\infty$ norm is not preserved under rotations. In fact, we can easily see that storing any subset Wavelet coefficient is suboptimal. Consider $B = 1$, the Haar basis and $X = \{2, 2, 2, 0\}$, then $\mathcal{W}(X) = \{3, 1, 0, \sqrt{2}\}$. The best solution restricted to storing the coefficients is $Z = \{3, 0, 0, 0\}$ and $\mathcal{W}^{-1}(Z) = \{\frac{3}{2}, \frac{3}{2}, \frac{3}{2}, \frac{3}{2}\}$ with $\|X - \mathcal{W}^{-1}(Z)\|_\infty = 1.5$. It is easy to see that $Z = \{2, 0, 0, 0\}$ gives $\|X - \mathcal{W}^{-1}(Z)\|_\infty = 1$. It is not difficult to construct a similar example for $\ell_1$ error as well. We will prove a polynomial time approximation scheme for the Haar basis which extends to a quasipolynomial time scheme for a general compact basis. We begin by computing a lower bound for any $\ell_p$ error.

Let the minimum possible value of $\|X - \mathcal{W}^{-1}(Z)\|_p = \tau_{opt}$ be achieved by the solution $Z^*$. For all $j$, we have $-\tau_{opt} \leq X(j) - \mathcal{W}^{-1}(Z^*)(j) \leq \tau_{opt}$. Multiplying the equation by $\psi_i(j)$ and summing over $j$, we get $-\|\psi_i\|_1|\tau_{opt}| \leq \langle X, \psi_i \rangle - \langle \psi_i, \mathcal{W}^{-1}(Z^*) \rangle \leq \|\psi_i\|_1|\tau_{opt}|$. But $\langle \psi_i, \mathcal{W}^{-1}(Z^*) \rangle = z_i^*$. Thus we can write a system of equations

$$\min \tau \qquad \text{s.t.}$$
$$-\tau\|\psi_1\|_1 \leq \langle X, \psi_i \rangle - z_1^* \leq \tau\|\psi_1\|_1 \text{ for all } i \qquad (32.1)$$
At most $B$ of the $z_i^*$ are nonzero

The constraints are satisfied by $\tau_{opt}$. Let the optimum solution of the above-mentioned system be $\tau^*$. Thus $\tau^* \leq \tau_{opt}$. The system of equations is *nonlinear*. However, the above system (32.1) can be solved optimally, and the minimum solution is the $(B+1)$th largest (ignoring signs) value of $\langle X, \psi_i \rangle/\|\psi_i\|_1$. We can also derive the following:

$$-\|\psi_i\|_1|\tau_{opt}| \leq \langle X, \phi_i \rangle - \langle \phi_i, \mathcal{W}^{-1}(Z^*) \rangle \leq \|\psi_i\|_1|\tau_{opt}|$$

The previous equation shows the effect of the coefficients whose support contains the entire support of $\psi_i$. In effect, given the optimum error, we have a handle on how the rest of the input must behave in relation to the $\{X(j)|j \in \text{SUPP}(\psi_i)\}$. We are interested in keeping track of all possible scenarios of $\langle \phi_i, \mathcal{W}^{-1}(Z^*) \rangle$.

#### 32.4.3.1 Streaming PTAS for Haar Systems

We solve the problem in a scaled bases and translate the solution to the original basis. We begin with the following:

**Proposition 32.2** *Define $\psi_{j,s}^\mathcal{P} = 2^{-j/2}\psi_{j,s}$ and $\psi_{j,s}^\mathcal{D} = 2^{j/2}\psi_{j,s}$. Likewise, define $\phi_i^\mathcal{P}, \phi_i^\mathcal{D}$. The Cascade algorithm used with $\frac{1}{\sqrt{2}}h[]$ computes $\langle X, \psi_i^\mathcal{P} \rangle$ and $\langle X, \phi_i^\mathcal{P} \rangle$. The problem of finding a synopsis $Z$ with basis $\{\psi_i\}$ is equivalent to finding a synopsis $Y$ using the basis $\{\psi_i^\mathcal{D}\}$ for the inverse transform, that is, we are seeking to minimize a function of $X - \sum_i Y(i)\psi_i^\mathcal{D}$. The correspondence is $Y(i) = 2^{-j/2}Z(i)$ where $i = (j, s)$.*

**Lemma 32.1** *Let $Y^*$ be the optimal solution using the basis set $\{\psi_i^\mathcal{D}\}$ for the reconstruction, that is, $\hat{X} = \sum_i Y^*(i)\psi_i^\mathcal{D}$ and $\|X - \hat{X}\|_p = \tau_{opt}$. Let $Y^\rho$ be the vector where each $Y^*(i)$ is rounded to the nearest multiple of $\rho$. If $X^\rho = \sum_i Y^\rho(i)\psi_i^\mathcal{D}$, $\|X - X^\rho\|_p \leq \tau_{opt} + O(qn^{1/p}\rho \log n)$.*

The proof follows from standard accounting of the error at each point and the triangle inequality. If we can find $Y^\rho$ for $\rho = \frac{\epsilon\tau_{opt}}{qn^{1/p}\log n}$ we have a $(1 + \epsilon)$-approximation. Note that $q = 1$.

---

* The last constraint can be expressed as a linear constraint, but it is not clear how to "round" the fractional solution we would obtain.

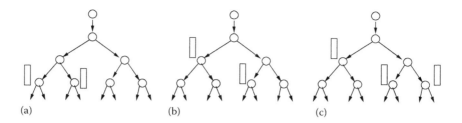

**FIGURE 32.3** The dynamic programming along the stream.

*Haar and the art of streaming*: We can find $Y^*$ using a DP. At each node $i$, we will compute a table $\text{ERR}_i[v, b]$ which corresponds to the contribution of the subtree rooted at $i$ to the error of the minimum solution assuming that the combined effect (signed sum) of all coefficients corresponding to ancestors of $i$ is $v$. In other words, this is the interaction between the subtree rooted at $i$ and the rest of the tree. Note that this value will lie in the range $\langle X, \phi_i^{\mathcal{P}} \rangle \pm \tau^*$. The size of the table is $O(\frac{\tau_{opt} B^2}{\rho})$, the extra $B$ term arises from keeping track of the solution.

We can compute the DP table at $i$, given the tables for its children. As we read data from left to right, we can keep generating the tables in postorder fashion. Figure 32.3a shows the state when we have read 4 values and have created 2 tables for two sibling nodes. We can discard the tables for the children once we have computed the table at their parent. This is shown in Figure 32.3b, along with the evolution of the state corresponding to more data being read. When we reach the configuration in Figure 32.3c, we would recursively collapse the tables and propagate them upwards and finish the computation.

It is immediate that the space taken would be $\log n$ times the space of a table. For each entry, we have to decide on the value of $Z(i)$, and the allocation of $b$ coefficients to the subtrees rooted at its children. This corresponds to $O(\frac{\tau_{opt} B}{\rho})$ choices, but for nodes with very few descendants, this can be reduced. However, there is small complication: we cannot solve the system (32.1) as the order of the coefficients cannot be determined until we have seen all the data. To avoid this, we will guess the value of $\tau_{opt}$ and verify that no more than $B$ coefficients exceed the required limit. This would involve a further $O(\log n)$ term (assuming polynomially bounded input). Thus we can conclude with:

**Theorem 32.8** *We can compute the optimal wavelet synopsis under $\ell_p$ error using $O(n^{1+\frac{2}{p}} B^2 \frac{\log^3 n}{\epsilon^2})$ time and $O(\frac{B^2 n^{\frac{1}{p}} \log^3 n}{\epsilon})$ space, for Haar wavelets.*

The running time can be improved slightly as at the lower part of the tree, the number of descendants are less than $B$. The earlier results extend to multiple dimensions. In case of **Non-Haar compact systems**, for example, Daubechies-4, and so on, we can achieve a quasipolynomial time approximation scheme.

### 32.4.4 Restricted Optimizations

A question arises in this context—"if we are only interested in a synopsis such that $Z(i) = 0$ or $Z(i) = \langle \psi_i, X \rangle$, what is the best possible reconstruction?" One motivation of the question can be that we may be interested in multiple uses of the synopsis and retaining the wavelet coefficients presses some advantage in some other direction. The above-mentioned problem can be solved optimally in $O(n^2 \log B)$ time and $O(n)$ space for a variety of error measures (any weighted $\ell_p$, which subsumes all "workload" versions, for $\ell_\infty$ the time is $O(n^2)$) for the Haar system.

**Problem 32.5** *What is the relationship between the minimum error of the restricted and the original problem?*

We show that a modified greedy heuristic which retains $B$ coefficients of the data gives a $O(n^{\frac{1}{p}} \log n)$ approximation algorithm in $O(n)$ time and $O(B + \log n)$ space for any compact wavelet. Recall the equation system (32.1). Consider the set of values $\{\frac{|\langle X, \psi_{i_j} \rangle|}{\|\psi_{i_j}\|_1}\}$ in the decreasing order. Set $Z(i_j) = \langle X, \psi_{i_j} \rangle$ for $1 \leq j \leq B$ and $Z(i) = 0$ otherwise. The resulting $Z$ is an optimal solution of the system (32.1), but not of the original synopsis construction problem. To analyze the error we would require the next lemma which follows from the properties of compact wavelets.

**Lemma 32.2** *For all basis vectors $\psi_i$, there exists a constant $C$ (depends on $q$) s.t. $\|\psi_i\|_\infty \|\psi_i\|_1 \leq C$.*

The previous proposition allows us to conclude $\|X - W^{-1}(Z)\|_p = O(\tau n^{\frac{1}{p}} \log n)$ where $\tau \leq E$. Thus

**Theorem 32.9** *The algorithm of retaining the $B$ coefficients with largest $\frac{|\langle X, \psi_i \rangle|}{\|\psi_i\|_1}$ is a $O(n^{1/p} \log n)$ approximation for the synopsis construction problem under the $\ell_p$ norm. The algorithm can be implemented over a stream to run in $O(n)$ time and $O(B + \log n)$ space. This is a simultaneous approximation of all $\ell_p$ norms.*

## 32.5 From (Haar) Wavelets Form Histograms

The first idea that comes to one's mind is that—can we simply represent $X$ using a number (to be chosen later) of large projections onto the wavelet basis, that is, using a few large wavelet coefficients? The idea does work but requires many wavelet coefficients. Stepping back, we can reason that what we want to achieve from the process is the discovery of the places where $X$ "jumps" significantly and use that information. This points us to the Haar basis, as the Haar basis is best suitable for functions with discontinuities (jumps). The idea would be to find a few large coefficients and store information relevant to the *boundaries* defined by the function. Note that storing information about the boundaries is tantamount to storing more coefficients, except that we use less space. As mentioned, we will focus on the Haar system. Define $\text{PROJ}(X, \mathcal{V})$, to be the projection of $X$ onto the basis vectors in set $\mathcal{V}$.

**Definition 32.5** *Given a $B$-bucket histogram $H$, define $\text{POSS}(H)$ the set of wavelet vectors $\psi_i$ such that the support $\text{SUPP}(\psi_i)$ is **not** completely contained inside one of the buckets. Observe that if $\psi_i \notin \text{POSS}(H)$, $\psi_i \cdot H = 0$.*

Note that $|\text{POSS}(H)| \leq 2B \log n$ and $\text{POSS}(H)$ is hierarchically closed (upward) with respect to the coefficient tree, as the coefficients form a tree defined by the set $\text{SUPP}()$.

**Lemma 32.3** *Given a $B$-bucket histogram $H$, let $\mathcal{V}_0$ be the set of $2B(1 + \epsilon^{-2}) \log n$ basis vectors which have the largest (unsigned) projection with $X$. Let $\mathcal{V}$ be the hierarchical upward closure of $\mathcal{V}_0$. Then $\|\text{PROJ}(X, \text{POSS}(H) \setminus \mathcal{V})\|_2^2 \leq \epsilon^2 \|X - H\|_2^2$.*

The above follows from the fact that $|\text{POSS}(H) \setminus \mathcal{V}| \leq 2B \log n$. Consider $\mathcal{V}_0$ and $\mathcal{W}(X - H)$; as $H$ can change at most $2B \log n$ coefficients, $\mathcal{W}(X - H)$ has at least $2B\epsilon^{-2} \log n$ coefficients of $\mathcal{V}_0$, each of which are larger than any of the $2B \log n$ coefficient in $\text{PROJ}(X, \text{POSS}(H) \setminus \mathcal{V})$ (as the $\setminus$ operation took out the potentially large coefficients). Therefore $\|\mathcal{W}(X - H)\|_2^2 \leq \epsilon^{-2} \|\text{PROJ}(X, \text{POSS}(H) \setminus \mathcal{V})\|_2^2$. Now by Parseval $\|\mathcal{W}(Y')\|_2 = \|Y'\|_2$ for any $Y'$, and thus the lemma follows.

**Definition 32.6** *Given a $B$-bucket histogram $H$, and a set of hierarchically closed set of basis vectors $\mathcal{V}$, define $\text{FLAT}(X, \mathcal{V}, H) = X - \text{PROJ}(X, \text{POSS}(H) \setminus \mathcal{V})$, that is, all the places where $H$ differs from the projection defined on $\mathcal{V}$ are **flattened**.*

It is immediate if $Y = X - \text{FLAT}(X, \mathcal{V}, H)$, from Lemma 32.3 $\|Y\|_2 \leq \epsilon \|X - H\|_2$. Now from the triangle inequality we have $\|X - H\|_2 + \|Y\|_2 \geq \|\text{FLAT}(X, \mathcal{V}, H) - H\|_2 \geq \|X - H\|_2 - \|Y\|_2$. Summarizing,

**Lemma 32.4** *If $H_f$ minimizes $\|\text{FLAT}(X, \mathcal{V}, H) - H\|_2$, $\|X - H_f\|_2$ is a $\frac{1+\epsilon}{1-\epsilon}$-approximation of the optimum error $\|X - H^*\|_2$.*

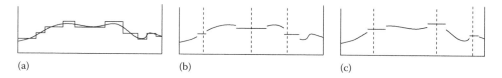

(a)          (b)          (c)

**FIGURE 32.4** The use of the flattened function.

We first show that if we "flatten" the signal at places in which the coefficients are not large, the error of a histogram in estimating the original signal $X$ remains approximately the same if we use the same histogram to estimate the "flattened" signal.

**Finding** $\min_H \|\text{FLAT}(X, \mathcal{V}, H) - H\|$ We would first find the large Haar wavelet coefficients corresponding to $\mathcal{V}_0$ in a streaming fashion. For each of the *four endpoints* (including the middle two values corresponding to the jump) of the wavelets, we store SUM, SQSUM. Then instead of using the function $X$ in the optimization, we would use the function $\text{FLAT}(X, \mathcal{V}, H)$. The illustration is in Figure 32.4, part (a) shows the approximation by the large wavelet functions. Part (b) shows what exactly we are computing, given a set of boundaries for $H$. Note that we are not approximating by wavelets, we are merely using the flat part from the wavelets corresponding to where the boundaries of $H$ lie.

By definition, $\text{FLAT}(X, \mathcal{V}, H) = X - \text{PROJ}(X, \text{POSS}(H) \setminus \mathcal{V})$, that is, when we decide on a particular boundary $u$ of $H$ – $\text{FLAT}(X, \mathcal{V}, H)$ looks *flat* between the two boundary points $v_1, v_2$ defined by $\mathcal{V}$ between which the boundary $u$ falls. Note that this means for different $H, H'$ the $\text{FLAT}(X, \mathcal{V}, H), \text{FLAT}(X, \mathcal{V}, H')$ look different (cf Figure 32.4b and c). As $\text{FLAT}(X, \mathcal{V}, H)$ looks flat, let that flat height between $v_1, v_2$ be $h$, we know this $h$ as we know $\text{PROJ}(X, \mathcal{V})$. Now we also know that $\text{SUM}'[u] = \text{SUM}[v_1] + h(u - v_1)$, where $\text{SUM}'$ refers to $\text{FLAT}(X, \mathcal{V}, H)$. Likewise, $\text{SQSUM}'[u] = \text{SQSUM}[v_1] + h^2(u - v_1)$. At this point we have all the pieces of the algorithm, we run the offline algorithm mentioned in Section 32.3. As the algorithm is described, our space requirement appears to be $O(n)$, but we can avoid that and use $O(B\epsilon^{-2} \log n)$ space corresponding to the size of $\text{PROJ}(X, \mathcal{V})$. We omit the details in the interest of space. The running time is $O(n + B^3(\log n + \frac{1}{\epsilon^2}) \log^2 n)$, the extra $\log n$ appears from the fact that given $u$, we need to find $v_1, v_2, h$ which takes $O(\log n)$ time. This immediately allows us to conclude that:

**Theorem 32.10** *We can compute a $1 + \epsilon$ approximation for the optimal B-bucket histogram under $\ell_2^2$ error in a single pass in time $O(n + B^3(\log n + \frac{1}{\epsilon^2}) \log^2 n)$ and space $O(B\epsilon^{-2} \log n)$.*

## Notes

Equiwidth histograms were introduced by Kooi [11]. Muralikrishna and Dewitt [12] considered equidepth/equiheight histograms which are essentially quantiles. The piecewise constant definition arises from the work of Ioannidis and Christosoudakis [13] and led to the work of Ioannidis [14]. Ioannidis and Poosala [15] proposed several different measures and the $\ell_2^2$/V-Optimal measure was introduced by Poosala et al. [16]. For a history of histograms, see Reference 17.

The $O(n^2 B)$ time $O(nB)$ space dynamic programming algorithm was given by Jagadish et al. [18]. They also showed that a $(B + \ell)$-bucket histogram which has the same error as the optimal $B$-bucket histogram can be constructed in $O(n^2 B/\ell)$ time. The running time remained quadratic even if the approximation algorithm was allowed a constant factor larger resources. Guha et al. [19] gave the first FPTAS running in $O(\frac{nB^2}{\epsilon} \log n)$ time while preserving the number of buckets. Guha and Koudas [20] considered the question of constructing histograms over sliding window data streams and showed that a data structure can be maintained in $O(1)$ time, such that a $(1 + \epsilon)$ approximation of the optimal histogram of the last $n$ values can be constructed on demand in $O(\frac{B^3}{\epsilon^2} \log^3 n)$ time. The net result is a $O(n + poly(B, \epsilon, \log n))$ algorithm for histogram construction. Gilbert et al. [21] gave a polytime approximation scheme for the stronger dynamic/update model. The result holds for all $\ell_p$ norms with $0 < p \leq 2$. Guha et al. [22]

gave the first $\tilde{O}(B)$ space histogram algorithm for the weaker time series streaming model. This is interesting as $B$ is not always a "small" constant and space is the premium in a streaming model; however, this algorithm only works for the $\ell_2$ error. Guha et al. [23], considered the relative error and introduced the block-by-block algorithm and the amortized analysis. The journal version of Reference 19, in Reference 24, subsumed and improved several of these algorithms and provided the experimental validation of the approximation algorithms. Guha [25] improved Reference 18 and gave a $O(n^2 B)$ time $O(n)$ space algorithm as well as improved the space complexity of References 22, 24. The result concerning the $\ell_\infty$ histograms in Section 32.3.3 are from Reference 26. The discussion on piecewise polynomials and other extensions in Section 32.3.2 is from References 19, 24. Donkerjovic et al. [27] considered the $\chi^2$ measure.

The above-mentioned discussion applied to point queries. Range queries form an important class of queries and pose significantly more complicated optimizations. Several early papers only consider the restricted version in which we store the mean of the values in a bucket, which is suboptimal. Koudas et al. [28], considered hierarchical queries and gave an algorithm that runs in time $O(|T|n^6 B^2)$ for a hierarchy of size $|T|$. Gilbert et al. [29] gave a pseudopolynomial time algorithm for the case when all $\binom{n}{2}$ ranges are present. Guha et al. [30] gave a sparse set system based algorithm which improved the running time of the hierarchical case to $O(n + |T|B^2 n^\gamma)$ but used $12B/\gamma$ buckets, whereas preserving the optimum guarantee with respect to $B$ buckets. However all these algorithms considered the restricted model where we are restricted to store the mean of a set of values as a representative. Muthukrishnan and Strauss [31] considered the unrestricted version when all $\binom{n}{2}$ ranges are present and gave an $O(n^2 B)$ algorithm which uses $2B$ buckets and guarantees less or equal error compared with the best $B$-bucket histogram. They also gave an $O(B^3 N^4/\epsilon^2)$ time approximation scheme for this case. The space requirement of most of these algorithms were improved in Reference 25.

**Wavelets** have a rich history dating back to the work of Haar (1910). However, they became significantly more popular in early 1990s primarily due to the application in image analysis [7,32], and the seminal work of Daubechies [8]. In context of database systems, wavelets were introduced by Matias et al. [33]. Their paper proposed greedy algorithms for optimum wavelet selection for several error measures including $\ell_1$. Chakraborty et al. [34], consider using wavelets for modeling time series data. Gilbert et al. [35] gave the streaming algorithm for the $\ell_2$ case for Haar wavelets. Gilbert et al. [21] extend the $\ell_2$ result to the stronger model of update streams. Gibbons and Garofalakis [36] considered constructing synopses based on randomized rounding techniques. Garofalakis and Kumar [37] considered the restricted version (of retaining the coefficients) for $\ell_\infty$ (and similar measures) and gave a $O(n^2 B)$ time and space algorithm. This was improved to $O(n^2)$ time and $O(n)$ space [25], for a broad range of error measures, including workloads. All the above are in the context of Haar wavelets.

Matias and Urieli [38] considered the problem of designing a basis for weighted $\ell_2$ measures such that greedy coefficient selection was optimum for that basis. Guha and Harb [39] gave the first approximation schemes for the original (unrestricted) optimization problem which is discussed in Section 32.4.3. For $\ell_p$ error measures with $p > 2$ (e.g., $\ell_\infty$), the algorithm can be implemented as a small space streaming algorithm for Haar wavelets. They also proved the upper bound on the gap between the optimum of the restricted version and the unrestricted optimum which is discussed in Section 32.4.4. In case of range queries using wavelets, Matias and Urieli [40] show that a scaled greedy strategy is provably optimal when all $\binom{n}{2}$ ranges are equally likely. Gilbert et al. [29] also discusses a similar scenario. Guha et al. [41] consider the hierarchical case in the restricted model.

The algorithm in Section 32.5 is based primarily on the ideas in Reference 22. The improved space bounds follows from Reference 25.

# References

1. Selinger, P.G., Astrahan, M.M., Chamberlin, D.D., Lorie, R.A., and Price, T.G., Access path selection in a relational database management system, in *Proceedings of ACM SIGMOD*, 1979, p. 23.

2. Aboulnaga, A. and Chaudhuri, S., Self tuning histograms: Building histograms without looking at data, in *Proceedings of ACM SIGMOD*, ACM, New York, 1999, p. 181.

3. Bruno, N., Gravano, L., and Chaudhuri, S., STHoles: A workload aware multidimensional histogram, in *Proceedings of ACM SIGMOD*, ACM, New York, 2001.

4. Datar, M., Gionis, A., Indyk, P., and Motwani, R., Maintaining stream statistics over sliding windows, in *Proceedings of SODA*, Society of Industrial and Applied Mathematics, Philadelphia, PA, 2002, p. 635.

5. Manku, G.S., Rajagopalan, S., and Lindsay, B., Approximate medians and other quantiles in one pass and with limited memory, in *Proceedings of ACM SIGMOD*, ACM, New York, 1998, p. 426.

6. Greenwald, M. and Khanna, S., Space-efficient online computation of quantile summaries, in *Proceedings of ACM SIGMOD*, 2001.

7. Mallat, S., *A Wavelet Tour of Signal Processing*, Academic Press, Burlington, MA, 1999.

8. Daubechies, I., *Ten Lectures on Wavelets*, Society of Industrial and Applied Mathematics, Philadelphia, PA, 1992.

9. Bentley, J.L., Multidimensional divide-and-conquer, *CACM*, 23(4), 214, 1980.

10. Guha, S., Mishra, N., Motwani, R., and O'Callaghan, L., Clustering data streams, in *Proceedings of FOCS*, IEEE, New York, 2000, p. 359.

11. Kooi, R., *The optimization of queries in relational databases*, Case Western Reserve University, 1980.

12. Muralikrishna, M. and DeWitt, D.J., Equi-depth histograms for estimating selectivity factors for multidimensional queries, in *Proceedings of ACM SIGMOD*, ACM, New York, 1988, p. 28.

13. Ioannidis, Y. and Christodoulakis, S., Optimal histograms for limiting worst-case error propagation in the size of join results, *ACM Transactions on Database Systems*, 18(4), 709, 1993.

14. Ioannidis, Y.E., Universality of serial histograms, in *Proceedings of the VLDB Conference*, VLDB Endowment, San Jose, CA, 1993, p. 256.

15. Ioannidis, Y. and Poosala, V., Balancing histogram optimality and practicality for query result size estimation, in *Proceedings of ACM SIGMOD*, ACM, New York, 1995, p. 233.

16. Poosala, V., Ioannidis, Y., Haas, P., and Shekita, E., Improved histograms for selectivity estimation of range predicates, in *Proceedings of ACM SIGMOD*, ACM, New York, 1996, p. 294.

17. Ioannidis, Y.E., The history of histograms (abridged), in *Proceedings of VLDB Conference*, VLDB Endowment, San Jose, CA, 2003, p. 19.

18. Jagadish, H.V., Koudas, N., Muthukrishnan, S, Poosala, V., Sevcik, K.C., and Suel, T., Optimal histograms with quality guarantees, in *Proceedings of the VLDB Conference*, VLDB Endowment, San Jose, CA, 1998, p. 275.

19. Guha, S., Koudas, N., and Shim, K., Data streams and histograms, in *Proceedings of STOC*, ACM, New York, 2001, p. 471.

20. Guha S. and Koudas, N., Approximating a data stream for querying and estimation: Algorithms and performance evaluation, in *Proceedings of ICDE*, IEEE Press, New York, 2002, p. 567.

21. Gilbert, A.C., Guha, S., Indyk, P., Kotidis, Y., Muthukrishnan, S., and Strauss, M., Fast, small-space algorithms for approximate histogram maintenance, in *Proceedings of ACM STOC*, ACM Press, New York, 2002, p. 389.

22. Guha, S., Indyk, P., Muthukrishnan, S., and Strauss, M., Histogramming data streams with fast per-item processing, in *Proceedings of ICALP*, Dagstuhl Publishing, Saarbrücken/Wadern, 2002, p. 681.

23. Guha, S., Shim, K., and Woo, J., REHIST: Relative error histogram construction algorithms, in *Proceedings VLDB Conference*, VLDB Endowment, San Jose, CA, 2004, p. 300.

24. Guha, S., Koudas, N., and Shim, K., Approximation and streaming algorithms for histogram construction problems, in *ACM TODS*.

25. Guha, S., Space efficiency in synopsis construction problems, in *Proceedings of VLDB Conference*, VLDB Endowment, San Jose, CA, 2005, p. 409.

26. Guha, S. and Shim, K., $\ell_\infty$ histograms, Technical Report, Department of Computer and Information Science, University of Pennsylvania, Philadelphia, PA, 2005.

27. Donjerkovic, D., Ioannidis, Y.E., and Ramakrishnan, R., Dynamic histograms: Capturing evolving data sets, CS-TR 99-1396, University of Wisconsin, Madison, WI, 1999.

28. Koudas, N., Muthukrishnan, S., and Srivastava, D., Optimal histograms for hierarchical range queries, in *Proceedings of ACM PODS*, ACM, New York, 2000, p. 196.

29. Gilbert, A.C., Kotidis, Y., Muthukrishnan, S., and Strauss, M., Optimal and approximate computation of summary statistics for range aggregates, in *Proceedings of ACM PODS*, ACM, New York, 2001.

30. Guha, S., Koudas, N., and Srivastava, D., Fast algorithms for hierarchical range histogram construction, in *Proceedings of ACM PODS*, ACM Press, New York, 2002, p. 180.

31. Muthukrishnan, S. and Strauss, M., Rangesum histograms, in *Proceedings of SODA*, Society of Industrial and Applied Mathematics, Philadelphia, PA, 2003, p. 233.

32. Jacobs, C.E., Finkelstein, A., and Salesin, D.H., Fast multiresolution image querying, in *Proceedings of the Computer Graphics Conference*, ACM SIGGRAPH, New York, 1995, p. 277.

33. Matias, Y., Vitter, J.S., and Wang, M., Wavelet-based histograms for selectivity estimation, in *Proceedings of ACM SIGMOD*, ACM, New York, 1998, p. 448.

34. Chakrabarti, K., Garofalakis, M.N., Rastogi, R., and Shim, K., Approximate query processing using wavelets, in *Proceedings of VLDB Conference*, VLDB Endowment, San Jose, CA, 2000, p. 111.

35. Gilbert, A., Kotadis, Y., Muthukrishnan, S., and Strauss, M., Surfing wavelets on streams: One pass summaries for approximate aggregate queries, in *Proceedings of VLDB Conference*, VLDB Endowment, San Jose, CA, 2001, p. 79.

36. Garofalakis, M.N. and Gibbons, P.B., Probabilistic wavelet synopses, *ACM TODS*, 29, 43, 2004.

37. Garofalakis, M. and Kumar, A., Deterministic wavelet thresholding for maximum error metric, in *Proceedings of PODS*, ACM, New York, 2004, p. 166.

38. Matias, Y. and Urieli, D., Optimal workload-based weighted wavelet synopses, in *Proceedings of ICDT*, LNCS, Springer-Verlag, Berlin, Germany, 2005, p. 368.

39. Guha S. and Harb, B., Approximation algorithms for wavelet transform coding of data streams, in *Proceedings of SODA*, Society of Industrial and Applied Mathematics, Philadelphia, PA, 2006.

40. Matias, Y. and Urieli, D., On the optimality of the greedy heuristic in wavelet synopses for range queries, Technical Report, Tel Aviv University, Tel Aviv, Israel, 2005.

41. Guha, S., Park, H., and Shim, K., Wavelet synopsis for hierarchical range queries with workloads, *The VLDB Journal*, 17(5), pp. 1079–1099, 2008.

# 33

# Color Quantization

Zhigang Xiang

## 33.1 Introduction

Quantization is originally defined as the digitization of a continuous signal. The signal may represent a monochromatic tone varying between black and white, and may be measured by its intensity. The task of quantization is to map the signal's intensity values into a series of gray levels (e.g., 256 levels from black to white along the gray axis). The signal may also carry chromatic information with multiple attributes that characterize the signal within a multidimensional color space (e.g., $c = (r, g, b)$ for the red, green, and blue [RGB] space). In this case, quantization amounts to mapping color points onto a grid that discretizes the color space (e.g., 256 levels in each dimension).

The advent of digital image processing gives rise to the need to reduce the number of colors that are present in an image (e.g., a 24-bit image with tens of thousands of colors from a $256 \times 256 \times 256$ grid) by reassigning pixels to a smaller set of grid values (e.g., 256 for the 8-bit lookup table representation). Thus the problem of color quantization or color image quantization evolves to become the problem of remapping the already quantized image colors. This serves a variety of purposes because quantized images have lower memory consumption, allow speedier transmission, and place lesser demand on processing hardware.

Let $G$ be the set of grid values (i.e., possible colors) and $C = \{c_1, c_2, \ldots, c_n\} \subseteq G$ be the set of $n$ original colors in an image. The problem of quantizing the image into $k$ final colors, which are now referred to as quantized colors and denoted by $Q = \{q_1, q_2, \ldots, q_k\} \subset G$, can be stated as finding $Q$ along with a mapping from $C$ to $Q$, where $Q$ is generally not a subset of $C$ and is often called the color map color mapor color palette (or codebook, where each quantized color is a code word) for the image. On the other hand, considering that this mapping from $C$ to $Q$ implies that pixels with the same original color are destined to have the same quantized color, which is usually the case but rather restrictive, we may adopt a more general characterization by defining the quantization task as finding $Q$ along with an assignment of a quantized color to each pixel.

Quantization inevitably introduces distortion. An ideal quantization algorithm should make the quantized image look as close to the original image and should exhibit as few objectionable artifacts as possible. Aside from the apparent difficulty in quantifying this objective due to its subjective nature, we also face a couple of complicating factors that are relatively immune to variations in individual judgment. Together they make color quantization a good candidate for approximation algorithms and heuristics.

The first complicating factor stems from the fact that the sensation of color is a psychophysical phenomenon, which results from physical stimulus in the form of visible light entering our visual system. Ideally, distances in a color space where the physical stimulus is measured are proportional to the perceived differences by the viewer—such a color space is often referred to as being perceptually uniform. We can then equate the task of minimizing the distances between original and quantized colors (quantization errors) with the task of minimizing the visible discrepancies between those colors. However, the prevailing RGB color model for image representation is far from being perceptually uniform, and the visualization of an RGB image is affected by the physical characteristics of the display or printing device [1,2]. A numerical displacement to an original color may result in a minute color change that is hardly visible when the displacement occurs in some parts of the RGB color space but manifests as a clearly noticeable color shift when it takes place in other regions. Although we may map RGB colors into another color space that is device independent and significantly more uniform to carry out the quantization task [3,4], it is still work-in-progress to find a color model that is truly perceptually uniform.

The second complicating factor comes from the context-dependent nature of the quantization errors that are considered visually offensive by an average observer. An image is not a simple collection of isolated individual data points in the eyes of the viewer, its quality needs to be judged with all pixels taken as a sophisticated whole. Even if we conduct quantization in a color space that is perfectly perceptually uniform (and we have perfectly calibrated display and printing devices), the resulting distribution of quantization errors (minimized one way or another without regard to context) does not necessarily guarantee the minimization of their visual offensiveness. For instance, consider the impact of a certain visible shift of the colors that are referred to as the skin tones in a quantized image. The shift may very well be acceptable when the colors depict ordinary objects such as a piece of fabric or flowers (the objects are quite likely to look just fine in the shifted colors); however, the same shift can be rather objectionable when the colors happen to portray a human face (proper skin tones are indicative of the subject's well-being). Furthermore, there are factors such as the spatial averaging effect of color stimuli from adjacent pixels through our visual pathway—much like the spatial averaging of subpixels that enables RGB display devices to work, and the phenomenon of simultaneous contrast [5] that affects our perception of color when differently colored patches are viewed in each other's presence—the two small disks shown in Figure 33.1 have exactly the same color but the one on the right looks brighter because it is surrounded by a dark ring. There are also many other aspects (ranging from psychophysical to cognitive) of visual information processing that are not yet fully understood.

Given the complexity of the color quantization problem, it should come as no surprise that existing color quantization methods tend to focus on limited aspects of the problem and have relatively confined goals. Based on the scope of the information that is used in the construction of the color map, we may divide these methods into three broad categories (Figure 33.2): (1) image-independent methods that determine a universal set of quantized colors without regard to any specific image; (2) image-dependent, context-free methods that take into account the actual colors that appear in the input image as well as the frequency (i.e., number of occurrences or pixel population) of each of those colors, and typically focus on the minimization of the numerical discrepancies between original and quantized colors based on certain statistical criteria; and (3) image-dependent, context-sensitive method that make use of additional contextual information beyond original colors and their frequencies, which can be derived from the input image (e.g., spatial relationship between the pixels) as heuristics to help to restrain visible quantization artifacts better.

**FIGURE 33.1** Demonstration of simultaneous contrast.

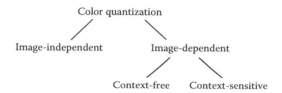

**FIGURE 33.2** Categorization of color quantization methods.

Many algorithms are essentially color space-neutral by design, leaving it a separate issue for the pixel colors to be represented in an appropriate color space. However, the RGB color space is frequently used in experimental implementation.

Three common options exist for the mapping of original colors to quantized colors, or more broadly, the assignment of quantized colors to pixels: (1) Replace each original color with the closest counterpart in the color map (mostly image-independent methods), (2) replace each original color with a specific quantized color whose association with the original color has already been determined during the construction of the color map (mostly image-dependent methods), and (3) make use of such techniques as error diffusion [6–9] to select the appropriate quantized color for each pixel. The latter option helps to smooth out some of the visible quantization artifacts, with the trade-off being that it may also degrade sharp edges and fine details.[*]

## 33.2 Color Spaces for Quantization

The CIELUV and CIELAB are two prominent candidate spaces that are derivatives of the CIE 1931 XYZ color model, which is device independent but nonuniform in terms of perceived color differences [10]. Both are defined as nonlinear transformations of XYZ with respect to a reference white point, which may be the standard illuminant D50 for reflective reproduction with XYZ coordinates being (0.9642, 1.0, 0.8249), or D65 for emissive display with XYZ coordinates being (0.9504, 1.0, 1.0889).

---

[*] Edge detection has been used to suppress error diffusion across edges to preserve image sharpness [28], and edge enhancement may be incorporated into the error diffusion process [54].

The CIELUV or CIE 1976 $L^*u^*v^*$ color space is defined by:

$$L^* = \begin{cases} 116 \left(\dfrac{Y}{Y_w}\right)^{1/3} - 16, & \left(\dfrac{Y}{Y_w}\right) > 0.008856 \\ 903.3 \left(\dfrac{Y}{Y_w}\right), & \text{otherwise} \end{cases}$$

$$u^* = 13L * (u' - u'_w)$$
$$v^* = 13L * (v' - v'_w)$$

where

$$u' = \frac{4X}{X + 15Y + 3Z} \qquad u'_w = \frac{4X_w}{X_w + 15Y_w + 3Z_w}$$
$$v' = \frac{9Y}{X + 15Y + 3Z} \qquad v'_w = \frac{9Y_w}{X_w + 15Y_w + 3Z_w}$$

$X_w$, $Y_w$, and $Z_w$ are determined from the reference white point.

The CIELAB or CIE 1976 $L^*a^*b^*$ color space is defined by:

$$L^* = \begin{cases} 116 \left(\frac{Y}{Y_w}\right)^{1/3} - 16, & \left(\frac{Y}{Y_w}\right) > 0.008856 \\ 903.3 \left(\frac{Y}{Y_w}\right), & \text{otherwise} \end{cases}$$

$$a^* = 500 \left( f\left(\frac{X}{X_w}\right) - f\left(\frac{Y}{Y_w}\right) \right)$$
$$b^* = 200 \left( f\left(\frac{Y}{Y_w}\right) - f\left(\frac{Z}{Z_w}\right) \right)$$

where

$$f(t) = \begin{cases} t^{1/3} & t > 0.008856 \\ 7.787t + \frac{16}{116} & \text{otherwise} \end{cases}$$

$X_w$, $Y_w$, and $Z_w$ are determined from the reference white point.

The $L^*$ component in both trivariant models is designed to carry luminance, and the remaining two specify chrominance. Although often referred to as being perceptually uniform, these two color spaces still depart from uniformity over the visible gamut with variations that may reach as high as 6:1 when color differences are measured by Euclidian distances $\Delta E_{uv} = \sqrt{\Delta L^{*2} + \Delta u^{*2} + \Delta v^{*2}}$ and $\Delta E_{ab} = \sqrt{\Delta L^{*2} + \Delta a^{*2} + \Delta b^{*2}}$ [11,12]. An improved color difference formula was introduced in 1994 [11]:

$$\Delta E^*_{94} = \sqrt{\left(\frac{\Delta L^*}{k_L S_L}\right)^2 + \left(\frac{\Delta C^*_{ab}}{k_C S_C}\right)^2 + \left(\frac{\Delta H^*_{ab}}{k_H S_H}\right)^2}$$

which is based on using polar coordinates to address color points in the CIELAB space in terms of perceived lightness $L^*$, chroma $C^*_{ab} = \sqrt{a^{*2} + b^{*2}}$, and hue angle $H^*_{ab} = \tan^{-1}(\frac{b^*}{a^*})$. Standard reference values for the formula are $k_L = k_C = k_H = 1$, $S_L = 1$, $S_C = 1 + 0.045\, C^*_{ab}$, and $S_H = 1 + 0.015\, C^*_{ab}$.

The conversion between RGB (assumed to be linear without $\gamma$ correction for cathode ray tubes) and XYZ may be carried out by the following standard transformation, with white illuminant D65:

$$\begin{pmatrix} R \\ G \\ B \end{pmatrix} = \begin{pmatrix} 3.0651 & -1.3942 & -0.4761 \\ -0.9690 & 1.8755 & 0.0415 \\ 0.0679 & -0.2290 & 1.0698 \end{pmatrix} \begin{pmatrix} X \\ Y \\ Z \end{pmatrix}$$

# 33.3 Image-Independent Quantization

In contrast with the image-dependent methods that choose quantized colors in an image-specific fashion, quantization may also be carried out on an image-independent basis, with a fixed/universal palette for all images. These image-independent methods enjoy high-computational efficiency as they avoid the need to analyze each original image to determine the quantized colors for that image. Moreover, there is no overhead for the storage and transmission of the individualized color map for each image. The trade-off is that a set of quantized colors that are specifically tailored to the distribution of the original colors in a given image tends to do a better job in approximating those original colors and lowering quantization errors.

## 33.3.1 Uniform Quantization

In this approach, we preselect $k$ colors that are uniformly distributed in the chosen color space (preferably perceptually uniform). A ready example would be the $6 \times 6 \times 6$ browser/web-safe palette with integer values 0, 51, 102, 153, 204, and 255 for each primary for a total of 216 RGB colors [13]. The quantization of an image now entails mapping each pixel to a preselected color (e.g., one that is the closest to the pixel's original color).

An easy and fast implementation of uniform quantization involves the truncation of a few least-significant bits from each component of an original color, rounding the original color down to a quantized color. For example, we may truncate 3 bits from each component of a 24-bit RGB color to arrive at its counterpart in a set of $32 \times 32 \times 32$ quantized colors. On the contrary, by aiming to better preserve luminance (a key ingredient that conveys details) and taking hint from the standard formula for computing luminance from RGB values: $Y = 0.299R + 0.587G + 0.114B$, we may truncate 3 bits from the red component, 2 bits from the green component, and 4 bits from the blue component to partially compensate for the nonuniform nature of the RGB color space.

This bit-cutting technique effectively places all quantized colors below the maximum intensity level in each dimension of the color space and causes a downward shift in intensity (as well as hue shift) across the entire image. These are often unacceptable when a relatively high number of bits are truncated.

## 33.3.2 Trellis-Coded Quantization

Consider the case of uniform quantization using one byte for the direct encoding of pixel colors, for example, 3-3-2 for red-green-blue, we would have a rather coarse grid of $8 \times 8 \times 4$ quantized colors. Now if we can "extend" the capacity of the limited number of bits used for each primary to specify intensity values at a higher resolution, we will be able to approach the effect of uniform quantization within a finer grid. This is made possible by the application of the Viterbi algorithm [14,15] in trellis-coded quantization [16–18].

Take for example the encoding of one of the primaries with $x = 3$ bits, which normally yields $2^x = 2^3 = 8$ intensity levels. On the other hand, we may have two color maps (Figure 33.3), each of which consists of 8 equally spaced intensity levels. The values in one map can be obtained by offsetting the values in the other map by half the distance between two adjacent levels. We further partition the intensity values in each map into two subsets. Given a specific map, only $x = 3$ bits are necessary to identity one of the two subsets (1 bit needed) and the particular intensity level within the chosen subset ($x - 1 = 2$ bits needed). Operating as a finite state machine, described by a trellis, the algorithm uses the bit that identifies the subset within the current map to determine the next state of the trellis, which in turn determines the choice of color map for the next input bit-string. This approximates the effect of quantizing with $2^{x+1} = 2^4 = 16$ intensity levels.

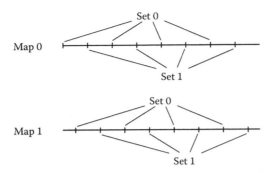

**FIGURE 33.3** Trellis-coded quantization.

### 33.3.3 Sampling by Fibonacci Lattice

Unlike scalar values on the gray axis, points in a multidimensional color space do not lend themselves to easy manipulation and ordering. In this variation of uniform quantization [19], the universal color palette is constructed by sampling within a series of cross planes along the luminance axis of the CIELAB color space. Each cross plane is a complex plane centered at the luminance axis, and the sample points $z_j$ in the plane (with equal luminance) are determined by the Fibonacci spiral lattice (Figure 33.4):

$$z_j = j^\delta e^{i\theta}, \theta = 2\pi j\tau + \alpha_0$$

where:

parameter $\delta$ controls the radial distribution of the points (higher value produces greater dispersion)

$\tau$ determines the overall pattern of distribution (a Markoff irrational number yields the most uniform distribution)

$\alpha_0$ denotes an initial angle of rotation to better align the sample points with the boundaries of the color space

The values $\delta = 0.5$ and $\tau = \frac{\sqrt{5}-1}{2}$ (the golden mean) are used in implementation, along with an additional scaling factor to help adjust the sample points' coverage of the color space within each cross plane.

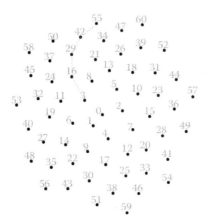

**FIGURE 33.4** Points on the Fibonacci spiral lattice.

To produce a universal palette of a certain size, the number of luminance levels and the number of sample points per level need to be carefully chosen, and the set of luminance values need to be determined through image-dependent quantization of luminance values from a large set of training images.

A unique aspect of this sampling method comes from the Fibonacci lattice. Each sample point $z_j$ in the spiral lattice is uniquely determined by its scalar index $j$, and two neighboring points are always some Fibonacci number apart in their indices. These plus other useful properties of the Fibonacci lattice make the resulting color palette amenable to fast quantization and ordered dither. In addition, a number of gray-scale image processing operations such as gradient-based edge detection can be readily applied to the quantized color images.

## 33.4 Image-Dependent, Context-Free Quantization

Image-dependent, context-free quantization methods select quantized colors based solely on original colors and their frequencies, without regard to the spatial relationship of the pixels and the context of the visual information that the image conveys. A basic strategy shared by numerous quantization algorithms in this category is to proceed in two steps. The first step partitions the $n$ original image colors into $k$ disjoint clusters $S_1, S_2, \ldots, S_k$ based on a certain numerical criterion—this makes color quantization a part of the broader area of data clustering [20], and the second step computes a representative (i.e., a quantized color) for each cluster. The quantized image may then be constructed by recoloring each pixel with the representative of the cluster that contains the pixel's original color or with the application of techniques such as error diffusion using the resultant color map.

Intuitively, these methods differ in how to balance two interrelated and competing objectives: the preservation of popular colors versus the minimization of maximum quantization error (Figure 33.5). The former may be characterized as achieving an error-free mapping for the $k$ most popular original colors, whereas the latter is the minimization of the upper bound for all $d(c, q_i)$, $1 \leq i \leq k$, where $c$ is an original color in the $i$th cluster $S_i$, $q_i$ is the representative of $S_i$, and $d(c, q_i)$ is the nonnegative quantization error, typically the Euclidean distance between $c$ and $q_i$.

A classic approach to strike a balance between the two objectives is to minimize the sum of squared errors $\sum_{1 \leq i \leq k} \sum_{c \in S_i} P(c) d^2(c, q_i)$ across the entire image, where $P(c)$ is the frequency (pixel population) of color $c$. As an alternative, we may try to minimize the total quantization errors $\sum_{1 \leq i \leq k} \sum_{c \in S_i} P(c) d(c, q_i)$, which represent a lesser bias toward capping the maximum quantization error. Such statistical criteria can trace their origin to the quantization of a continuous-tone black-and-white signal [21,22] and have been proven to be NP-complete [23–26].

Regardless of the operational principle (e.g., limiting the spatial extent of each cluster) for the clustering step of an approximation algorithm, the frequency-weighed mean of the original colors in each resulting cluster is almost always used as the cluster's representative, often called the cluster's centroid but sometimes referred to as the center of gravity (the two notions are equivalent in this context). This reflects a common consensus on minimizing intracluster variance.

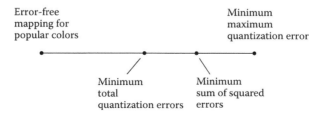

**FIGURE 33.5**  An intuitive scale for comparing statistical criteria.

### 33.4.1 The Popularity Algorithm

This early quantization method aims at the preservation of popular colors [27]. It creates a histogram of the colors in the original image and selects the $k$ most popular ones as quantized colors. Pixels in other colors are simply mapped to the closest quantized colors, respectively. The advantage here is that relatively large and similarly colored image areas are kept little changed after quantization; however, smaller areas, some of which may carry crucial information (e.g., a uniquely colored signal light), can take on significant distortion as a result of being mapped to popular colors.

An implementation technique that may alleviate this problem preprocesses the 24-bit original colors by truncating a few least-significant bits from each color component (i.e., performing a uniform quantization), effectively combining several popular colors that are very similar to each other into a single quantized color, thus allowing some of the less popular colors to be selected as quantized colors. This preprocessing step (e.g., 3-3-3 or 3-2-4 bit-cutting for red-green-blue) can also be used to achieve color reduction and to alter color granularity for other algorithms. The downside here is that bit-cutting itself can cause false contours to appear on smoothly shaded surfaces.

One may also avoid having several popular colors that are neighbors of each other as quantized colors by choosing one quantized color at a time and by artificially reducing the pixel count of the remaining colors in the vicinity of the chosen color (currently the most popular), with the reduction being based on a spherically symmetric exponential function in the form of $1 - e^{Kr^2}$ (Note that this sets the pixel count of the chosen color to 0, so it will never be selected again in subsequent iterations), where $r$ is the radius of the sphere that is centered at the chosen color and $K$ is an experimentally determined constant [28].

### 33.4.2 Detecting Peaks in Histogram

Instead of choosing the $k$ most popular original colors for the color map, we may find peaks in the histogram and use colors at the peaks as quantized colors. The peaks may be identified by a multiscale clustering scheme based on discrete wavelet transform (DWT), where computational efficiency comes from carrying out three-dimensional DWT as a series of independent one-dimensional transforms followed by downsampling [29]. The quantizer can determine the value of $k$ from the number of detected peaks, or it may be adjusted to produce a preset number of quantized colors.

### 33.4.3 Peano Scan

In another color technique to lessen the difficulty associated with multidimensional data processing, a recursively defined space-filling curve, referred to as a Peano curve, is used to traverse the space (e.g., the RGB color cube), thus creating a one-to-one mapping between points in space and their counterparts along the curve. Being subjected to the spatial relationships that are preserved by the mapping, certain spatially oriented operations may now be carried out along a single dimension. For example, as points close on the Peano curve are also close in space, given a specific color, we may easily find some of its neighbors by searching along the curve [30,31]. The shortfall of this approach comes from the fact that points close in space are only likely, but not necessarily, to be close on the curve.

### 33.4.4 The Median-Cut Algorithm

This two-step algorithm conducts a hierarchical subdivision of clusters that have high pixel populations, attempting to achieve an even distribution of pixels among the quantized colors [27]. We first fit a rectangular box over an initial cluster containing all original colors and then split the box into two with a plane that is orthogonal to its longest dimension to bisect the cluster in such a way that each new cluster is now responsible for half of the pixel population in the original cluster. The new clusters are then treated the same way repeatedly until we have $k$ clusters. The criterion for selecting the next cluster to split is based on pixel count. By splitting the most popular cluster in each step, the algorithm will eventually produce $k$ clusters, each of which is responsible for roughly $1/k$ of the image's pixel population.

In comparison with the popularity algorithm, this alternative for resource distribution often brings about better quantization results. However, having the same number of pixels mapped to each quantized color does not necessarily lead to effective control of quantization errors.

### 33.4.5 The Center-Cut Algorithm

As a variation to the median-cut algorithm, this method bisects a cluster at the midpoint of the longest dimension of its bounding box without regard to pixel population [32] and it ranks candidate clusters for subdivision based on the longest dimension of their bounding boxes—the longest one is split first. These changes put more emphasis on restraining the spatial extent of the clusters and do a better job in keeping grossly distinct colors from being grouped into the same cluster and mapped to the same quantized color.

Both the median-cut and the center-cut algorithms take a top–down approach to partitioning a single cluster into $k$ clusters. On the other hand, we may follow a bottom–up strategy that merges the $n$ original colors into the desired number of clusters. To this end the octree data structure [33] can be used to provide a predetermined hierarchical subdivision of the RGB color space for merging clusters.

### 33.4.6 Octree Quantization

With the entire color cube represented by the root node and each octant of the color cube by a child node descending from the root, an individual 24-bit RGB color corresponds to a leaf node at depth 8 of the octree [34]. Conceptually, once we populate an octree with pixel colors from an input image, we may start from the bottom of the octree (greatest depth) and recursively merge leaf nodes that have the same parent into the parent node, thereby transforming the parent node into a leaf node at the next level, until we reduce the number of leaf nodes from $n$ to $k$. Each remaining leaf node now represents a cluster of original colors that inhabit the spatial extent of the node.

In an actual implementation, only an octree structure with no more than $k$ leaf nodes needs to be maintained, where each leaf node has a color accumulator and a pixel counter for the eventual calculation of its centroid. As we scan an original image, the color of each pixel is processed as follows. If the color falls within the spatial extent of an existing leaf node, then add it to the node's color accumulator and increase the node's pixel count by 1. Otherwise, use the color to initialize the color accumulator of a new leaf node and set the node's pixel counter to 1. If this increases the number of leaf nodes to $k + 1$, merge some of the existing leaf nodes (leaves with greatest depth first) into their parent node, which becomes a new leaf node whose color accumulator takes on the sum of the accumulated color values from the children and whose pixel counter gets the total of the children's pixel count.

Note that each splitting operation in median-cut or center-cut is performed either at the median or the midpoint of the longest dimension of a bounding box, and the merging operation in the octree algorithm is along predetermined spatial boundaries. This leaves the possibility of separating color points that are close to each other in a naturally forming cluster into different clusters.

### 33.4.7 Agglomerative Clustering

There are other bottom–up approaches where we start with $n$ clusters each of which contains one original color and merge clusters without regard to preset spatial boundaries. In a method that relies on a three-dimensional representation of all 24-bit original colors and the clusters to which they belong to [35], we begin with $n$ initial clusters that have the smallest bounding boxes. By gradually increasing the size limit on bounding boxes (with increments 2, 1, and 4 for red, green, and blue, respectively, to partially compensate for the nonuniform nature of the RGB color space), we merge neighboring clusters into larger ones to reduce the number of clusters. For a given size limit, we search the vicinity of each existing cluster $S$ in the three-dimensional data structure to find candidates to merge, that is, clusters that can fit into a new bounding box for $S$ that satisfies the size limit. The process terminates when $k$ clusters remain.

In addition to limiting the size of bounding boxes, the criterion for merging clusters may also be based on variance or distance between centroids (see below).

## 33.4.8 Variance-Based Methods

The two-step top–down or bottom–up methods we have discussed so far decouple the formation of clusters and the computation of a representative for each cluster (the centroid) in the sense that the two steps are designed to achieve different numerical objectives: evenly distributed pixel population or size-restricted bounding boxes for clustering and minimum variance for selecting cluster representatives after clustering. Several approximation algorithms are devised with variance-based criteria for the clustering step as well.

A $K$-means algorithm starts with an initial selection of quantized colors $q_1, q_2, \ldots, q_k$, which may simply be evenly spaced points in the color space or the result of some other algorithm. It then partitions the $n$ original colors into $k$ clusters $S_1, S_2, \ldots, S_k$ such that $c \in S_i$ if $d(c, q_i) \leq d(c, q_j)$ for all $j$, $1 \leq j \leq k$. After the partition, it calculates the centroid of each cluster $S_i$ as the cluster's new representative $q_i'$. The algorithm terminates when the relative reduction in overall quantization error from the previous choice of quantized colors is below a preset threshold. This relative reduction may be defined as $\frac{E - E'}{E'}$, where previous overall quantization error $E = \sum_{1 \leq i \leq k} \sum_{c \in S_i} P(c) d^2(c, q_i)$ and current overall quantization error $E' = \sum_{1 \leq i \leq k} \sum_{c \in S_i} P(c) d^2(c, q_i')$. Otherwise, the algorithm reiterates the partitioning step (followed by the recalculation of centroids) using the newly selected quantized colors. This quantization method is rather time-consuming (with $O(nk)$ for each iteration), and its convergence at best leads to a locally optimal solution that is influenced by the initial selection of quantized colors [36–38].

In one of the bottom–up approaches, we merge clusters under the notion of pairwise nearest neighbors [39]. Each iteration of the algorithm entails searching among current clusters to find two candidates, viz., $S_i$ and $S_j$ that are the closest neighbors, that is, two that when merged together into $S_{ij} = S_i \cup S_j$, will result in minimum sum of squared errors for $S_{ij}$: $\sum_{c \in S_{ij}} P(c) d^2(c, \mu_{ij})$, where $\mu_{ij}$ is the centroid of $S_{ij}$. A full implementation of this method is rather time-consuming as it would take at least $O(n \log n)$ just for the first iteration. To this end, a $k$-$d$ tree in which existing clusters (each cluster is spatially located at its centroid) are grouped into buckets (roughly equal number of clusters in each buckets) is used to restrict the search for pairwise nearest neighbors within each bucket (one pair per bucket). The pair that will result in the lowest sum of squared errors squared errors is merged first, then the pair in another bucket that yields the second lowest error sum is merged, and so on. The tree is rebalanced to account for the merged clusters when a certain percentage (e.g., 50%) of the identified pairs has been merged.

Another bottom–up method [40] randomly samples the input image for original colors and their frequencies; sorts the list of sampled original colors based on their frequencies in ascending order; and merges each color $c_i$, starting from the top of the list (i.e., low frequency first), with its nearest neighbor $c_j$, chosen based on a weighted squared Euclidean distance $\frac{P(c_i) P(c_j)}{P(c_i) + P(c_j)} d^2(c_i, c_j)$ to favor the merging of pairs of low-frequency colors. Each pair of merged colors is removed from the current list and replaced by $c_{ij} = \frac{P(c_i) c_i + P(c_j) c_j}{P(c_i) + P(c_j)}$, with $P(c_{ij}) = P(c_i) + P(c_j)$, which will be handled as an ordinary color during the next iteration of sorting and pairwise merging. The algorithm terminates when $k$ colors remain on the list, which are used as quantized colors.

In a couple of approaches that follow the strategy of hierarchical subdivision, we start with a single cluster containing all original colors and repeatedly partition the cluster $S$ whose sum of squared errors

$\sum_{c \in S} P(c) d^2(c, \mu)$, also termed weighted variance, with $\mu$ being the centroid of $S$, is the highest. We move an orthogonal cutting plane along each of the three dimensions of the RGB color cube to search for a position that divides the chosen cluster into two. One way to determine the orientation and position of the cutting plane is to project color points in the cluster, bounded by $r_1 \leq r \leq r_2$, $g_1 \leq g \leq g_2$, and $b_1 \leq b \leq b_2$, onto each of the three color axis; find the threshold that minimizes the weighted sum of projected variances of the two intervals adjoining at the threshold for each axis; and run the cutting plane perpendicular to and through the threshold on the axis that gives the minimum sum of projected variances [41]. More specifically, the frequency of a projected point on the $r$-axis is $P(r, 0, 0) = \sum_{g_1 \leq g \leq g_2} \sum_{b_1 \leq b \leq b_2} P(r, g, b)$. Similarly, we have $P(0, g, 0) = \sum_{r_1 \leq r \leq r_2} \sum_{b_1 \leq b \leq b_2} P(r, g, b)$ for the $g$-axis and $P(0, 0, b) = \sum_{r_1 \leq r \leq r_2} \sum_{g_1 \leq g \leq g_2} P(r, g, b)$ for the $b$-axis. Given an axis along with a series of projected points between $l$ and $m$, a threshold $l < t \leq m$ partitions the points into two intervals $[l, t-1]$ and $[t, m]$, with the resulting weighted sum of projected variances being $E_t = \sum_{l \leq i \leq t-1} P_i(i - \mu_1)^2 + \sum_{t \leq i \leq m} P_i(i - \mu_2)^2$, where $\mu_1$ and $\mu_2$ are the means of the two intervals, respectively, and $P_i = P(i, 0, 0)$, $P(0, i, 0)$, or $P(0, 0, i)$. The optimal threshold value that minimizes $E_t$ is in the range of $[\frac{l+\mu}{2}, \frac{\mu+m}{2}]$ and maximizes $E_t$ is in the range of $\frac{w_1}{w_2}(\mu - \mu_1)^2$, where $\mu$ is the mean of the projected points in $[l, m]$, and $w_1 = \sum_{l \leq i \leq t-1} P_i$ and $w_2 = \sum_{t \leq i \leq m} P_i$ are the weights for the two respective intervals [42].

Another way to determine the cutting plane is to minimize the sum of weighted variances (without projecting points onto the three color axes) on both sides of the plane [43]. A rectangular bounding box is now defined by $r_1 < r \leq r_2$, $g_1 < g \leq g_2$ and $b_1 < b \leq b_2$; and it is denoted by $\Omega(c_l, c_m]$, where $c_l = (r_1, g_1, b_1)$ and $c_m = (r_2, g_2, b_2)$. And we define $M_d(c_t) = \sum_{c \in \Omega(o, c_t]} c^d P(c)$, with $d = 0, 1, 2$; $c^0 = 1$; $c^2 = cc^T$, and $o$ being a reference point such that $\sum_{c \in \Omega(-\infty, o]} P(c) = 0$. We precompute and store $M_d(c)$, $d = 0, 1, 2$, for each grid point in the RGB space to facilitate efficient computation of the pixel population $w(c_l, c_m]$, mean $\mu(c_l, c_m]$, and weighted variance $E(c_l, c_m]$ of any cluster of image colors bounded by $\Omega(c_l, c_m]$:

$$w(c_l, c_m] = \sum_{c \in \Omega(c_l, c_m]} P(c)$$

$$\mu(c_l, c_m] = \frac{\sum_{c \in \Omega(c_l, c_m]} c P(c)}{w(c_l, c_m]}$$

$$E(c_l, c_m] = \sum_{c \in \Omega(c_l, c_m]} P(c) d^2(c, \mu(c_l, c_m]) = \sum_{c \in \Omega(c_l, c_m]} c^2 P(c) - \frac{\left(\sum_{c \in \Omega(c_l, c_m]} c P(c)\right)^2}{w(c_l, c_m]}$$

The evaluation of these items in $O(1)$ time is made possible by designating the remaining six corners of the bounding box as:

$$c_a = (r_2, g_1, b_1) \quad c_b = (r_1, g_2, b_1) \quad c_c = (r_1, g_1, b_2)$$
$$c_d = (r_1, g_2, b_2) \quad c_e = (r_2, g_1, b_2) \quad c_f = (r_2, g_2, b_1)$$

and applying the rule of inclusion–exclusion to obtain:

$$\sum_{c \in \Omega(c_l, c_m]} f(c) P(c) =$$

$$\left( \sum_{c \in \Omega(o, c_m]} + \sum_{c \in \Omega(o, c_a]} + \sum_{c \in \Omega(o, c_b]} + \sum_{c \in \Omega(o, c_c]} - \sum_{c \in \Omega(o, c_d]} - \sum_{c \in \Omega(o, c_e]} - \sum_{c \in \Omega(o, c_f]} - \sum_{c \in \Omega(o, c_l]} \right) f(c) P(c)$$

where $f(c)$ may be $1$, $c$, or $c^2$. Furthermore, to determine a cutting plane for $\Omega(c_l, c_m]$, we need to minimize $E(c_l, c_t] + E(c_t, c_m]$, with $c_t = (r, g_2, b_2)|_{r_1 < r \le r_2}$ or $(r_2, g, b_2)|_{g_1 < g \le g_2}$ or $(r_2, g_2, b)|_{b_1 < b \le b_2}$. As

$$E(c_l, c_t] + E(c_t, c_m] = \sum_{c \in \Omega(c_l, c_m]} c^2 P(c) - \frac{\left( \sum_{c \in \Omega(c_l, c_t]} cP(c) \right)^2}{w(c_l, c_t]} - \frac{\left( \sum_{c \in \Omega(c_t, c_m]} cP(c) \right)^2}{w(c_t, c_m]},$$

minimizing $E(c_l, c_t] + E(c_t, c_m]$ is equivalent to maximizing

$$\frac{\left( \sum_{c \in \Omega(c_l, c_t]} cP(c) \right)^2}{w(c_l, c_t]} + \frac{\left( \sum_{c \in \Omega(c_t, c_m]} cP(c) \right)^2}{w(c_t, c_m]} = \frac{\left( \sum_{c \in \Omega(c_l, c_t]} cP(c) \right)^2}{w(c_l, c_t]} + \frac{\left( \sum_{c \in \Omega(c_l, c_m]} cP(c) - \sum_{c \in \Omega(c_l, c_t]} cP(c) \right)^2}{w(c_l, c_m] - w(c_l, c_t]}$$

where $w(c_l, c_m]$ and $\sum_{c \in \Omega(c_l, c_m]} cP(c)$ are constants.

In addition to exploring cutting planes that are perpendicular to axes of the color coordinate system, we may also look into other orientations as well. For example, we may take a random sample of the cluster $S$ that has the highest weighted variance, and subdivide it into two in the following way [44]. For every linearly separable 2-clustering of the sample set $T$ into $T_1$ and $T_2$, compute their centroids $t_1$ and $t_2$, divide $S$ by the perpendicular bisector of $\overline{t_1 t_2}$, then compute the centroids $s_1$ and $s_2$ of the two resulting subsets, and divide $S$ again by the perpendicular bisector of $\overline{s_1 s_2}$. Finally, choose among all second bisectors the one that yields the minimum sum of weighted variances of the two subsets of $S$ to divide $S$.

Alternatively, we may place cutting planes orthogonally to each cluster's principal axis, which is found from the largest eigenvalue and the corresponding principal eigenvector of the cluster's covariance matrix [45]. During each iteration of the subdivision process, the algorithm finds a cluster whose principal eigenvalue is the highest among all existing clusters and bisects the found cluster with a plane that is perpendicular to the corresponding eigenvector and through the cluster's mean. The two resulting clusters become independent candidates for further partitioning in subsequent iterations.

### 33.4.9 Minimizing Total Quantization Errors

Principal analysis is also the basis for an approximation method for the minimization of total quantization errors [46]. The algorithm is inspired by the observation that colors in any given image tend to form a cluster that spreads out more in terms of differences in luminance than in chromaticity variations. It first finds the principal axis for the entire set of original colors. It then introduces parallel cutting planes that are perpendicular to the principal axis to minimize $\sum_{1 \le i \le \kappa} \sum_{c \in S_i} P(c) d(c, \mu_i)$, where $\mu_i$ is the centroid of $S_i$, and $\kappa$ is increased as cutting planes are introduced one by one by way of dynamic programming until none of the resulting clusters has a strongly biased orientation in the principal direction of the original set. Now if $\kappa = k$ the algorithm terminates with $k$ clusters; otherwise, it continues to subdivide the $\kappa$ existing clusters, either by using one of the hierarchical methods or by splitting each chosen cluster with a cutting plane that is perpendicular to the cluster's principal axis and minimizes the total quantization errors of the two resulting subsets.

### 33.4.10 Minimizing Maximum Intercluster Distance

Being unique among quantization algorithms that are based on numerical criteria, and unlike hierarchical methods that partition or merge clusters based on local information (i.e., color points inside a restricted spatial extent), the following method achieves proven tight approximation to global optimality for its clustering operation. The method attempts to minimize the maximum quantization error across the entire image by partitioning original colors into tight clusters under a formal notion of minimizing the maximum intercluster* distance: finding a partition of $n$ points in an $m$-dimensional Euclidean space into $k$ disjoint clusters $S_1, S_2, \ldots, S_k$ such that $\max(M_1, M_2, \ldots, M_k)$, where $M_i$ is the maximum distance between two points in cluster $S_i$, is minimized [47]. This minimization problem is polynomial solvable for $m = 1$ [23] and NP-hard for $m = 2$ [48]. When $m = 3$, which is typically the case in color image quantization, even finding a partition with maximum intercluster distance less than two times the optimal solution value, referred to as the $(2 - \varepsilon)$-approximation problem, is NP-hard for all $\varepsilon > 0$ [48,49]. Hence we make use of an efficient 2-approximation algorithm that has worst case time complexity $O(nk)$ [48]:

$S_1 = \{c_1, c_2, \ldots, c_n\}$; // Start with a single cluster containing all original colors
$h_1 = c_1$; // Each cluster has a designated point as the head of the cluster
for $(x = 1; x < k; x++)$ {

    $d = \max\{d(c_i, h_j) \,|\, c_i \in S_j, 1 = i = n,$ and $1 = j = x\}$;
    $c = $ one of the points whose distance to its respective cluster head is $d$;
    move $c$ to $S_{x+1}$;
    $h_{x+1} = c$;
    for each $c' \in (S_1 \cup S_2 \cup \ldots \cup S_x)$ {
        let $j$ be such that $c' \in S_j$;
        if $(d(c', h_j) = d(c', c))$ move $c'$ from $S_j$ to $S_{x+1}$;
    }

}

## 33.5 Image-Dependent, Context-Sensitive Quantization

Context-sensitive quantization methods work with not only original colors and their frequencies but also the image's context. The latter spans from the adjacency relationship between pixels to the primitive elements of visual information above the pixel level (e.g., edges and boundaries) and to the overall meaning of the visual information that the image and various parts of the image convey. Effective use of contextual information should bring about a better balance for the allocation of resources (i.e., selection of quantized colors along with the mapping of colors) in terms of moderating visually offensive distortion than what we may achieve with context-free quantization.

### 33.5.1 Dithered Quantization

This color method takes into consideration the impact of neighboring pixels on the viewer's perception of each individual pixel in both the original and the quantized images [50]. Let $c_{x,y}$ be the color of pixel $(x, y)$ in the original image and $c'_{x,y}$ be the perceived color at $(x, y)$, calculated by a linear blurring operation that convolutes the original image with a localized kernel to account for the phenomenon of spatial averaging in human vision (e.g., a Gaussian kernel of identical standard deviation for all color components, with choice of neighborhood size from $3 \times 3$ to $11 \times 11$). Similarly, let $q_{x,y}$ be the color of pixel $(x, y)$ in the quantized image and $q'_{x,y}$ be the perceived color at $(x, y)$, calculated by convoluting the quantized image with the same kernel. The goal of color quantization is now defined as the minimization of $\sum_{1 \le x \le w} \sum_{1 \le y \le h} d^2(c'_{x,y}, q'_{x,y})$, where $w$ and $h$ are the width and height of the image, respectively. Hence the task of quantization becomes finding the right set $Q$ of quantized colors along with a proper assignment $A$ of each pixel to a quantized color—simultaneous quantization and dithering.

---

* As we are trying to minimize the maximum distance between color points in each cluster, it might be more appropriate to use the word intracluster. However, if we view each color point as a singleton cluster we are indeed minimizing the maximum intercluster distance. We adopt this second view to be consistent with the existing literature on clustering.

A two-fold minimization scheme—first optimize $A$ for a fixed $Q$ and then optimize $Q$ for a fixed $A$—is iterated to achieve convergence to a local minimum of the stated goal. The optimization of $A$ is solved by a local iterative conditional mode (ICM) algorithm in [50], whereas more accurate results can be produced with deterministic annealing [51].

## 33.5.2 Feedback-Based Quantization

Being short of being able to reliably predict the type, severity, and location of eye-catching artifacts in the quantized image, we may try to develop techniques to detect the artifacts and may use the findings as feedback to modify the behavior of the quantizer to alleviate the distortion. A couple of studies have looked into a commonly occurring type of visible artifacts, viz., the appearance of false contours in areas that have gradual shadings before quantization. These false contours are the direct result of mapping a series of smoothly changing colors into a low number of quantized colors that make up a staircase-like profile across a troubled area—the effect of simultaneous contrast can make the steps look more profound than they really are.

In an extension to an aforementioned variance-based quantization method [45], a weighting mechanism is activated after the number of clusters have reached a preset threshold (e.g., $\frac{2}{3}k$) to adjust the ranking of existing clusters for further subdivision. Each candidate cluster is now given a weight that represents the size (number of interior pixels) of a continuous region that is colored by the cluster's representative in the quantized image. The candidate cluster whose principal eigenvalue multiplied by its weight is the highest is chosen for splitting. Doing so helps to eliminate large and uniformly colored regions. However, these regions are only necessary, rather than sufficient conditions for false contours. Even when they do border false contours, we need more than their sizes to distinguish a severe false contour from a minor one.

Another investigation involves an iterative process where findings from the last complete quantized image are used to requantize the original for better results [52,53]. An agglomerative clustering quantizer [35] that operates on 24-bit RGB colors is adapted to be the embedded quantization mechanism, as the usual 3-3-3 or 3-2-4 bit-cutting color-reduction technique itself causes false contours to appear when the original is a computer-synthesized or high-quality photographic image that depicts smooth/glossy surfaces.*

The system detects false contours in the quantized image by convoluting both the original (Figure 33.6a) and the quantized (Figure 33.6b)[†] images with a set of $5 \times 5$ directional edge detectors. The detectors are first applied independently to each primary component. A magnitude value in Euclidean color distance is then calculated for each detector (the corresponding results in the red, green, and blue directions are first scaled by 2, 4, and 1, respectively, to make the measurement more indicative of the change in luminance), and the highest magnitude is recorded as the magnitude of the edge element in an edge map. Next, we construct a mask (Figure 33.6c) based on the edges in the original image's edge map and use the mask to suppress their counterparts (i.e., the true edges) in the quantized image's edge map (Figure 33.6d). After further elimination of relatively insignificant edge elements by thresholding, the resulting edge map for the quantized image becomes a false contour map (Figure 33.6e), which identifies areas in the quantized image that border false contours. Each of these areas in turn identifies a set of original colors that have been mapped to the same quantized color and need to be better preserved during requantization to alleviate the artifacts (Figure 33.6f—no error diffusion). The latter is accomplished by increasing the importance factor (initially all original colors have equal importance) of each affected

---

* The introduction of a random perturbation to slightly degrade the original before quantization tends to inhibit the occurrence of false contours in the quantized image.
† The effect of simultaneous contrast is visible when the image is reproduced with good fidelity: Each uniformly shaded band on the spherical surface in the foreground looks nonuniform—darker on the side that is adjacent to a brighter band and brighter on the side that is adjacent to a darker band.

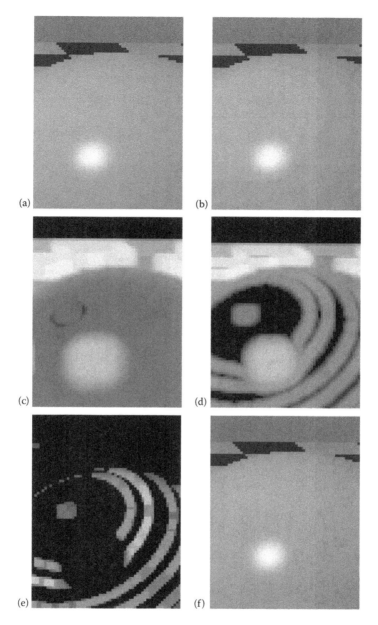

**FIGURE 33.6** Detecting and reducing false contours: (a) Original image, (b) Quantized image, (c) Mask constructed from original image's edge map, (d) Edge map for the quantized image, (e) False contour map, and (f) Re-quantized image.

original color and by having the quantizer restrict the growth of each cluster based on the highest importance factor of its constituents. The increment in importance for colors in an area that needs to reduce quantization errors is proportional to:

$$maxedge^2 \times (1 + tcd/maxtcd) \times (1 + pp/maxpp)$$

where:

    *maxedge* is the magnitude of the highest edge element from the corresponding region of the false contour map

    *tcd* is the total discrepancies between the quantized color and the colors in the area

    *pp* is the area's pixel population

    *maxtcd* and *maxpp* are the maximum *tcd* and *pp*, respectively, of all areas in the image that are identified by the false contour detection process.

Figure 33.7 shows a gray-scale reproduction of a ray-traced image with photoedited background (blue sky, white cloud, and brown mountain), where the five model cars are in silver, green, gold, orange, and cyan, respectively. Figure 33.8 is the gray-scale reproduction of the result of quantizing Figure 33.7 to

**FIGURE 33.7**   Gray-scale reproduction of a ray-traced color image with added background.

**FIGURE 33.8**   Context-free quantization.

**FIGURE 33.9**  Context-sensitive quantization.

256 colors using the original agglomerative quantizer, where false contours are clearly visible on four of the cars. Figure 33.9 is the gray-scale reproduction of what is produced by the feedback-based system at the end of the 47th iteration (without error diffusion), where false contours are greatly reduced with no clear degradation elsewhere.

# References

1. Lindbloom, B. J., Accurate color reproduction for computer graphics applications, *Computer Graphics*, 23(3), 117, 1989.
2. Stone, M. C., Cowan, W. B., and Beatty, J. C., Color gamut mapping and the printing of digital color images, *ACM Trans. Graph.*, 7(4), 249, 1988.
3. Gentile, R. S., Allebach, J. P., and Walowit, E., Quantization of color images based on uniform color spaces, *J. Imaging Technol.*, 16(1), 11, 1990.
4. Kurz, B. J., Optimal color quantization for color displays, in *IEEE Proceedings of Computer Vision and Pattern Recognition*, 1983, p. 217.
5. Itten, J., *The Elements of Color*, John Wiley & Sons, New York, 2003.
6. Floyd, R. and Steinberg, L., An adaptive algorithm for spatial gray scale, *International Symposium Digest of Technical Papers, Society for Information Display*, Campbell, CA, 1975, 36.
7. Knuth, D. E., Digital halftones by dot diffusion, *ACM Trans. Graph.*, 6, 245, 1987.
8. Ostromoukhov, V., A simple and efficient error-diffusion algorithm, in *Proceedings of Conference on Computer Graphics and Interactive Techniques*, ACM, New York, 2001, p. 567.
9. Zhang, Y. and Webber, R. E., Space diffusion: An improved parallel halftoning technique using space-filling curves, in *Proceedings of Conference on Computer Graphics and Interactive Techniques*, ACM, New York, 1993, p. 305.
10. Wyszecki, G. and Stiles, W. S., *Color Science: Concepts and Methods, Quantitative Data and Formulae*, 2nd ed., John Wiley & Sons, New York, 1982.
11. Hill, B., Roger, T., and Vorhagen, F. W., Comparative analysis of the quantization of color spaces on the basis of the CIELAB color-difference formula, *ACM Trans. Graph.*, 16(2), 109, 1997.
12. Kasson, J. M. and Plouffe, W., An analysis of selected computer interchange color spaces, *ACM Trans. Graph.*, 11(4), 373, 1992.

13. Weinman, L. and Heavin, B., *Coloring Web Graphics*, 2nd ed., New Riders, Indianapolis, IN, 1997.

14. Forney Jr., G. D., The Viterbi algorithm, *Proc. IEEE*, 61, 169, 1984.

15. Viterbi, A. J., Error bounds for convolutional codes and an asymptotically optimum decoding algorithm, *IEEE Trans. Inf. Theor.*, 13, 260, 1967.

16. Cheng, S. S., Xiong, Z., and Wu, X., Fast trellis-coded color quantization of images, *Real-Time Imag.*, 8, 265, 2002.

17. Marcellin, M. W. and Fischer, T. R., Trellis coded quantization of memoryless and Gaussian-Markov sources, *IEEE Trans. Commun.*, 38, 82, 1990.

18. Ungerboeck, G., Channel coding with multilevel/phase signals, *IEEE Trans. Inf. Theor.*, 28, 55, 1982.

19. Mojsilović, A. and Soljanin, E., Color quantization and processing by Fibonacci lattices, *IEEE Trans. Image Process.*, 10(11), 1712, 2001.

20. Jain, A. K., Murty, M. N., and Flynn, P. J., Data clustering: A review, *ACM Computing Surveys*, 31(3), 264, 1999.

21. Lloyd, S. P., Least squares quantization in PCM, Unpublished Bell Laboratories memorandum, 1957; also *IEEE Trans. Inf. Theor.*, IT-28, 129, 1982.

22. Max, J., Quantizing for minimum distortion, *IRE Trans. Inf. Theor.*, IT-6, 7, 1960.

23. Brucker, P., On the complexity of clustering problems, in *Optimization and Operations Research*, Henn, R., Korte, B., and Oettli, W. (Eds.), Springer-Verlag, Berlin, Germany, 1978, p. 45.

24. Garey, M. R. and Johnson, D. S., *Computers and Intractability: A Guide to the Theory of NP-Completeness*, W. H. Freeman and Company, New York, 1979, Problem MS9.

25. Garey, M. R., Johnson, D. S., and Witsenhausen, H. S., The complexity of the generalized Lloyd-Max problem, *IEEE Trans. Inf. Theor.*, IT-28(2), 255, 1982.

26. Megiddo, N. and Supowit, K. J., On the complexity of some common geometric location problems, *SIAM J. Comput.*, 13, 182, 1984.

27. Heckbert, P., Color image quantization for frame buffer display, *Comput. Graph.*, 16(3), 297, 1982.

28. Braudaway, G. W., A procedure for optimum choice of a small number of colors from a large color palette for color imaging, in *Electronic Imaging*, 1987, p. 71.

29. Kim, N. and Kehtarnavaz, N., DWT-based scene-adaptive color quantization, *Real-Time Imag.*, 11(5–6), 443, 2005.

30. Lehar, A. F. and Stevens, R. J., High-speed manipulation of the color chromaticity of digital images, *IEEE Comput. Graph. Appl.*, 4, 34, 1984.

31. Stevens, R. J., Lehar, A. F., and Preston, F. H., Manipulation and presentation of multidimensional image data using the Peano scan, *IEEE Trans. Pattern Anal. Mach. Intell.*, 5, 520, 1983.

32. Joy, G. and Xiang, Z., Center-cut for color-image quantization, *Vis. Comput.*, 10(1), 62, 1993.

33. Samet, H., *Applications of Spatial Data Structures*, Addison-Wesley, Reading, MA, 1990.

34. Gervautz, M. and Purgathofer, W., A simple method for color quantization: Octree quantization, in *New Trends in Computer Graphics*, Magnenat-Thalmann, N. and Thalmann, D. (Eds.), Springer-Verlag, Berlin, Germany, 1988, 219.

35. Xiang, Z. and Joy, G., Color image quantization by agglomerative clustering, *IEEE Comput. Graph. Appl.*, 14(3), 44, 1994.

36. Gray, R. M., Kieffer, J. C., and Linde, Y., Locally optimal block quantizer design, *Inf. Contr.*, 45, 178, 1980.

37. Linde, Y., Buzo, A., and Gray, R. M., An algorithm for vector quantizer design, *IEEE Trans. Commun.*, 28(1), 84, 1980.

38. Selim, S. Z. and Ismail, M. A., K-means-type algorithms: A generalization convergence theorem and characterization of local optimality, *IEEE Trans. Pattern Anal. Mach. Intel.*, PAMI-6(1), 81, 1984.

39. Equitz, W. H., A new vector quantization clustering algorithm, *IEEE Trans. Acoustics, Speech, Signal Process.*, 37(10), 1568, 1989.

40. Dixit, S. S., Quantization of color images for display/printing on limited color output devices, *Comput. Graph.*, 15(4), 561, 1991.

41. Wan, S. J., Prusinkiewicz, P., and Wong, S. K. M., Variance-based color image quantization for frame buffer display, *Color Res. Appl.*, 15(1), 52, 1990.

42. Wong, S. K. M., Wan, S. J., and Prusinkiewicz, P., Monochrome image quantization, in *Proceedings of Canadian Conference on Electrical and Computer Engineering*, Canadian Society for Electrical and Computer Engineering, Montreal, Canada, 1989, p. 28.

43. Wu, X., Efficient statistical computations for optimal color quantization, *Graphics Gems II*, Arvo, J. (Ed.), Academic Press, Boston, MA, 126, 1991.

44. Inaba, M., Imai, H., Nakade, M., and Sekiguchi, T., Application of an effective geometric clustering method to the color quantization problem, in *Proceedings of Symposium on Computational Geometry*, ACM, New York, 1997, p. 477.

45. Orchard, M. T. and Bouman, C. A., Color quantization of images, *IEEE Trans. Signal Process.*, 39(12), 2677, 1991.

46. Wu, X., Color quantization by dynamic programming and principle analysis, *ACM Trans. Graph.*, 11(4), 348, 1992.

47. Xiang, Z., Color image quantization by minimizing the maximum intercluster distance, *ACM Trans. Graph.*, 16(3), 260, 1997.

48. Gonzalez, T. F., Clustering to minimize the maximum intercluster distance, *Theor. Comput. Sci.*, 38(2–3), 293, 1985.

49. Sahni, S. and Gonzalez, T., P-complete approximation problems, *JACM*, 23(3), 555, 1976.

50. Buhmann, J. M., Fellner, D. W., Held, M., Ketterer, J., and Puzicha, J., Dithered color quantization, *Comput. Graph. Forum*, 17(3), 219, 1998.

51. Ketterer, J., Puzicha, J., Held, M., Fischer, M., Buhmann, J. M., and Fellner, D., On spatial quantization of color images, in *Proceedings of European Conference on Computer Vision*, Springer-Verlag, Berlin, Germany, 1998, p. 563.

52. Joy, G. and Xiang, Z., Reducing false contours in quantized color images, *Comput. Graph.*, 20(2), 231, 1996.

53. Xiang, Z. and Joy, G., Feedback-base d quantization of color images, in *Proceedings of SPIE 2182: Image and Video Processing II*, SPIE, Bellingham, WA, 1994, p. 34.

54. Eschbach, R. and Knox, K. T., Error-diffusion algorithm with edge enhancement, *J. Opt. Soc. Am. A*, 8(12), 1844, 1991.

# A GSO-Based Swarm Algorithm for Odor Source Localization in Turbulent Environments

Joseph Thomas

Debasish Ghose

## 34.1 Introduction

Olfaction is a long-distance sense and is used widely in the animal kingdom for performing many basic tasks such as finding food, marking paths, and so on. An artificial olfaction system could possibly replace canines in various search and rescue operations. The problem of odor source localization has received attention recently due to its potential applications in areas such as detection of sources of toxic gas leaks [1,2], fire-origins of forest fires [3], leak point determination in pressurized systems [4], chemical discharge in water bodies [5], and detection of mines and explosives [6]. From the point of view of security, an artificial olfaction system would be handy in detecting explosives, locating harmful gas leaks, and so on.

Two approaches, namely chemotaxis and anemotaxis, have been used for odor source localization. The chemotaxis approach utilizes the difference in concentration near the source, to navigate toward the source. This is inspired by bacteria that utilize local concentration gradients to guide themselves toward nutrient deposits [7]. The chemotactic algorithm has been shown to work in experimental settings in which the concentration field is smooth. However, due to turbulence, the flowing medium breaks up into random and disconnected patches thus rendering the gradient information unreliable [6].

Anemotaxis is a popular approach inspired by the zigzagging of moths in the upwind direction [8]. An anemotactic robot moves upwind within the plume to reach the source. This strategy is successful in problems in which the flow has no large-scale turbulence. However, large turbulent and circulatory eddies create a region where simple upwind travel will result in a cycle, causing the anemotaxis technique to fail [9].

One of the first works in the field of odor source localization was carried out by Russell [2] using a trail-following robot. Several odor source locating systems followed subsequently. Ishida et al. [10] used an autonomous mobile system to locate an ethanol source by moving upwind and along the gas-concentration gradient. One of the common issues with the detection of chemicals using mobile robots has been based on experimental setups, in which the distance between the source and the sensor following the odor trail has been minimized to limit the influence of turbulent transport [11,12]. Recently, algorithms utilizing multiple robots for odor source localization have been proposed. Such an approach offers advantages such as averaging of measurement errors, tolerance to unpredictable failures, and the possibility of replacing a single complex (and potentially expensive) robot with many low-cost simple robots [13]. Lytridis et al. [14] utilized a strategy combining the chemotaxis and biased random walk (BRW) for a multi-robot team, their approach was motivated by the observation that chemotaxis is efficient in stable fields, but unreliable in noisy fields, whereas BRW was robust even in turbulent fields but is inefficient. Hayes et al. [15] describe a spiral-surge algorithm in which a collection of autonomous mobile robots use spiral plume finding, surge, and spiral casting behaviors to find the source of an odor plume. Jatmiko et al. [16] and Marques et al. [44] used a particle swarm optimization (PSO) algorithm for locating odor sources. They used a filament-based approach to model the odor plume, which is valid for small-scale eddies that shred an odor plume into filaments. However, large-scale eddies result in random, disconnected odor patches [17]. Few algorithms have been shown to work in such a patchy environment. Vergassola et al. [6] have proposed an infotaxis algorithm that utilizes a strategy of locally maximizing the expected rate of information gain to locate the odor source. This single robot algorithm is shown to work well in a patchy environment. However, the convergence time for single robots can be rather high. Moreover, single robot algorithms have standing issues of fault tolerance and are prone to inaccurate measurements.

The motivation behind this work was to design an efficient swarm robotic algorithm that is capable of locating odor sources in a natural environment. The advantages of the swarm robotic approach include scalability from a few to several units, flexibility as units can be dynamically added or removed without explicit reorganization, and increased system robustness, not only through unit redundancy but also through the design of minimalist units [15]. The agents in the proposed algorithm utilize three modes of behavior, namely the chemotactic, anemotactic, and spiraling modes. The Glowworm Swarm Optimization (GSO) algorithm, which has been used for multimodal function optimization [18,19], is chosen as the basic algorithm for the chemotactic mode. GSO has been used for a solving a wide variety of problems including sensor deployment [20], sensor placement [21], crop planning [22], space suit puncture repair [23], web services design [24], clustering analysis [25], load allocation [26], fixed charge transportation problem [27], and various optimization problems [28–36].

The GSO algorithm models agents as glowworms, which communicate with each other to locate the multiple optima of a multimodal function simultaneously. As opposed to earlier approaches, which utilize instantaneous concentration measurements for chemotaxis [16], we utilize the maximum concentration measured in the recent past. This is shown to significantly improve the convergence time. The algorithm is later modified so as to counter the deterioration in performance, observed after the inclusion of obstacle avoidance behavior. In the absence of wind information, obstacle avoidance behavior is modified and utilized to locate the odor source. The capability of the proposed algorithm to locate an odor source in turbulent environments is verified on data obtained from a dye mixing experiment [37]. An ethanol source localization experiment was then carried out using a set of gas sensors. These experiments show the validity of our approach for gas source localization in the presence and absence of wind.

The paper is organized as follows. The odor propagation model developed for generating odor profiles is presented in Section 34.2. Section 34.3 explains the proposed algorithm and the different modes of robot behavior used. In Section 34.4, the performance of the algorithm under varied conditions is studied. In Section 34.5, a gyroscopic obstacle avoidance strategy is introduced among robots and the algorithm is modified and shown to work efficiently in such environments. In Section 34.6, a variant of the algorithm is proposed, which is capable of locating odor sources even in the absence of wind information. In Section 34.7, the proposed approaches are shown to work well on experimental dye mixing data. In Section 34.8, the suitability of the proposed algorithm under real-world conditions is studied by localizing an ethanol source using a set of gas sensors. In Section 34.9, the algorithm is compared with earlier proposed algorithms. Finally, we provide some concluding remarks in Section 34.10.

# 34.2 Odor Propagation Model

The concentration of a substance advected by a turbulent flow exhibits a complex, chaotically evolving structure [38]. The formation of eddies due to turbulence leads to disconnected patches. If one measures the concentration of an odor patch, the maximum amplitude of the concentration within the patch decreases as one moves away from the source, and the average time between encountering two successive patches increases [17]. A patch-based model is developed here to generate concentration profiles similar to the ones observed in a turbulent environment.

The patch-based model assumes that fluid flows in turbulence in the form of patches, which are transported by wind and turbulence. Although the wind has the same effect on all patches, turbulence works differently on each patch. The intensity of a patch is modeled as

$$f(r) = C\left(ae^{-\sqrt{a}\,r}\right) \tag{34.1}$$

where $f(r)$ is the patch intensity function, $r$ is the distance from the center of the patch, $C$ is used to scale concentration levels to realistic values, and $a$ is a parameter used to control patch decay with time. The parameter $a$ decreases exponentially with time as

$$a = a_0 e^{-t/\tau} \tag{34.2}$$

where $a_0$ is the initial value of the parameter $a$, $t$ is time in seconds, and $\tau$ is the decay rate of the patch. Figure 34.1 shows patch decay with time. Initially, the patch has a sharp concentration peak at its center, with time this peak goes down and the patch spreads out. The decrease in peak value is a result of a decrease in the value of the parameter $a$ with time, as can be seen from Equation 34.2. A decrease in the value of $a$ also results in the patch spreading out, as can be seen from Equation 34.1. The volume of gas contained in a patch is given by:

$$\int_0^{2\pi} \int_0^{\infty} f(r)dr\, rd\phi = C \int_0^{2\pi} \int_0^{\infty} are^{-\sqrt{a}\,r} dr d\phi = 2\pi C \tag{34.3}$$

where $\phi$ represents the angle. The volume is independent of $a$ and hence invariant with time. Thus, as the patch spreads out, the amount of gas contained in a patch remains the same.

The odor source is assumed to emit a fixed number of $N$ patches per second. These patches are transported around the field by turbulence and wind. As the patches move around the field, they also spread out. Patch concentrations are summed up to obtain the dynamic concentration profiles. In realistic odor profiles, a concentration peak is generally observed at the odor source location. This is modeled by assuming that $N$ patches are present at the source location always, that is, there are initially $N$ patches at the source and when a group of $N$ patches is released, they are immediately replaced by a new set of $N$ patches.

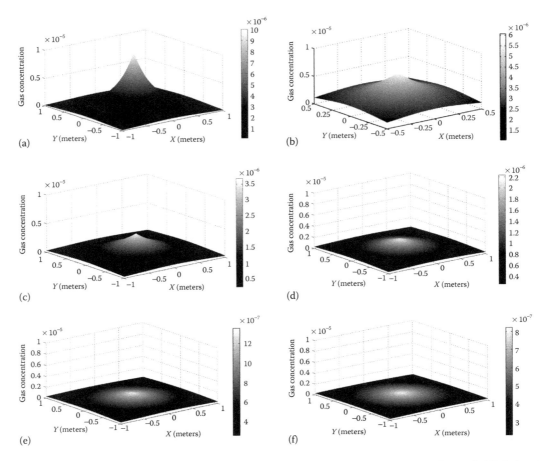

**FIGURE 34.1** Patch decay with time. After 25 seconds, the maximum gas concentration of a patch falls below 1 ppm ($10^{-6}$). The darker portions in the periphery signify low or zero gas concentration. Note that at 25 seconds, the patch becomes almost completely dark. The parameter values assumed to generate these profiles are $a_0 = 10$, $\tau = 10$, and $C = 10^{-6}$: (a) time $= 0$ second, (b) time $= 5$ seconds, (c) time $= 10$ seconds, (d) time $= 15$ seconds, (e) time $= 20$ seconds, and (f) time $= 25$ seconds.

The movement of each patch is assumed to be governed by two velocity components, the turbulent transport velocity and the wind velocity. These velocities are updated after every second. The turbulent transport velocity assumes a direction that is chosen randomly (with a uniform distribution) between 0 and $2\pi$ initially and then, in subsequent steps, the direction of turbulent transport velocity is chosen uniformly within $\pm \pi/4$ of the previous direction. Turbulent transport of a single patch can be represented as

$$\vec{x}(i) = \vec{x}_r(i) + \vec{w}, \ \forall \, i \geq 0 \tag{34.4}$$

where $\vec{x}(i)$ is the resultant patch velocity at the $i$th second, $\vec{w}$ is the wind velocity vector, and $\vec{x}_r(i)$ is the turbulent transport velocity vector after the $i$th second, which is given by

$$\vec{x}_r(i) \in \left\{ \vec{x} \mid \vec{x} = \eta \vec{x}_1 + (1 - \eta)\vec{x}_2, \ \vec{x}_1 = R(\pi/4)\,\vec{x}_r(i - 1), \ \vec{x}_2 = R(-\pi/4)\,\vec{x}_r(i - 1) \right\} \tag{34.5}$$

where $\eta$ is a parameter uniformly distributed between 0 and 1 and $R(\theta)$ is the rotation matrix. As can be seen from Figure 34.2, the new turbulent transport velocity vector $\vec{x}_r(i)$ chooses a direction uniformly

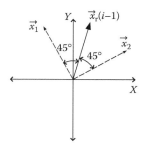

**FIGURE 34.2** Turbulent transport direction.

between $\vec{x}_1$ and $\vec{x}_2$, in which, $\vec{x}_1$ and $\vec{x}_2$ are obtained by rotating $\vec{x}_r(i-1)$ in the anticlockwise and clockwise direction by $\pi/4$, respectively.

Figure 34.3 shows the trajectory of a patch, originating from the source location S. Here, the initial turbulent transport vector $\vec{x}_r(0)$ chooses a direction uniformly between 0 and $2\pi$. The wind velocity vector $\vec{w}$ is then added to $\vec{x}_r(0)$ to obtain the resultant patch velocity vector $\vec{x}(0)$ at time $i = 0$. The resultant patch velocity $\vec{x}(0)$ takes the patch to A. The new turbulent transport velocity vector direction is chosen uniformly within $\pm\pi/4$ of the previous direction, as given by Equation 34.5. The new turbulent transport vector $\vec{x}_r(1)$ is added to the wind velocity vector $\vec{w}$ so as to compute the resultant patch velocity $\vec{x}(1)$, which takes the patch to B. The above-mentioned process is repeated at every subsequent step.

To generate the concentration profiles, we assume that $N$ patches are released simultaneously by the source after every second. Each one of these patches is transported by turbulence and wind as explained earlier. The simulation is seen to generate complex structures as observed in an actual gas leak. Figure 34.4 shows snapshots of the concentration profile generated. The source is placed at the center of the field and the wind is assumed to blow constantly along the positive $x$-direction. As in turbulent environments, the profile generated consists of disconnected patches that constantly evolve with time. Due to the wind in the positive $x$-direction, all the odor patches are transported downwind, resulting in a zero concentration measurement behind the odor source in the upwind direction.

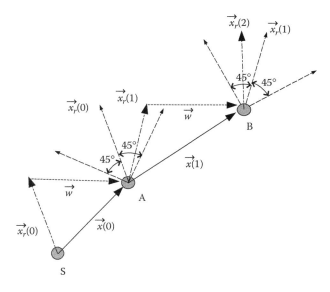

**FIGURE 34.3** Patch movement.

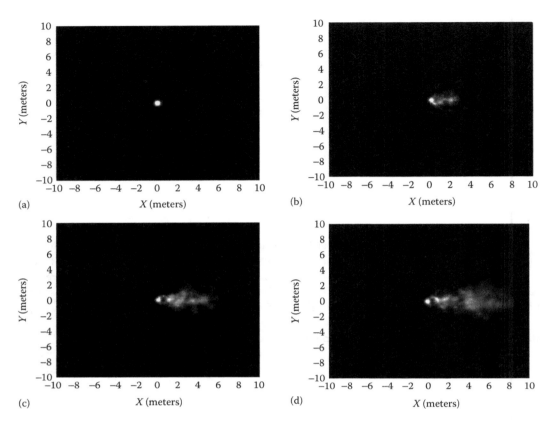

**FIGURE 34.4** Odor concentration profiles generated by the model. The snapshots show the propagation of odor patches with time. As before, the peripheral dark portion represents low or zero gas concentration. The parameter values used are $a_0 = 10$, $\tau = 10$, $C = 10^{-6}$, Turbulent transport speed = 0.5 m/sec, Wind speed = 0.5 m/sec, and $N = 20$: (a) time = 1 second, (b) time = 5 seconds, (c) time = 10 seconds, and (d) time = 15 seconds.

## 34.3 The Modified Glowworm Swarm Optimization Algorithm

In this paper, we modify the GSO algorithm to solve the odor source localization problem. The glowworms have three different modes of behavior, namely the chemotactic, anemotactic, and spiraling modes.

The basic GSO algorithm, as introduced in Reference 39, and with different variations discussed in References 18, 40, is an optimization algorithm inspired by the grouping behavior of glowworms. Analytical results related to the convergence of GSO algorithm are available in Reference 19. In the GSO algorithm, the agents are initially deployed randomly in the objective function space. The agents in the GSO algorithm are thought of as glowworms carry a luminescence quantity called luciferin along with them. The luciferin level of an agent is related to the function profile value that the glowworm has encountered in its path. Agents emit a light whose intensity of luminescence is proportional to the associated luciferin. Each glowworm uses its luciferin to (indirectly) communicate the function-profile information at its current location to the neighbors. To obtain a direction for movement, each glowworm selects a neighbor that has a luciferin value more than its own, using a probabilistic mechanism, and moves toward it. The glowworms have a dynamic decision domain radius that is updated at regular intervals.

*Chemotactic mode:* The GSO algorithm is utilized here by the glowworms in the chemotactic mode to move closer to the source. In turbulent flows, the peak concentration value within a patch and the frequency of encountering a patch increases as the glowworm gets closer to the source [17]. However,

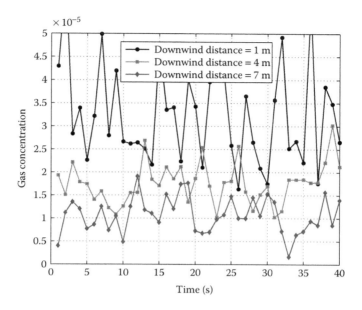

**FIGURE 34.5** Concentration measured by glowworms placed directly downwind to the source at different times.

the instantaneous concentration value might mislead the movement decision of the glowworms. Figure 34.5 shows the variation in concentration measured by glowworms placed directly downwind to the source. The glowworms measure concentration once every second. It can be seen that even for locations separated by 3 meters, the instantaneous concentration value can be higher at locations farther from the source, whereas the maximum concentration encountered in an interval is well separated. Hence, we define the luciferin value to be the maximum gas concentration value encountered by a glowworm in the last $N_{mem}$ seconds in its trajectory, as given by Equation 34.6. We assume a value of 1 ppm ($10^{-6}$) for the gas sensor threshold. Hence, gas concentration measurements below this value are taken to be zero.

The various steps of the GSO algorithm are given in the inset box, where $\ell_i(t)$ represents the luciferin level associated with Glowworm $i$ at time $t$, $C(x_i(t))$ represents the concentration value measured at Glowworm $i$'s location at time $t$, $x_i(t)$ is the location of Glowworm $i$ at time $t$, $N_i(t)$ is the neighborhood of Glowworm $i$ at time $t$, $d_{i,j}(t)$ represents the Euclidian distance between Glowworms $i$ and $j$ at time $t$, $r_d^i(t)$ represents the variable local-decision range associated with Glowworm $i$ at time $t$, $s$ ($>0$) is the step-size, $w$ is the wind direction, $r_s$ represents communication range, the Glowworm $i$ selects to move toward a Glowworm $j \in N_i(t)$ with probability $p_j(t)$, $n_t$ represents the desired number of neighbors, and $\beta$ is a parameter to control the number of neighbors.

Glowworms update their luciferin value after every step. Each glowworm defines its set of neighbors, $N_i(t)$ to be the set of glowworms within a distance of $r_d^i(t)$ from it and having a higher luciferin value, as given by Equation 34.7. The dynamic decision domain radius, $r_d^i(t)$ is useful while localizing multiple sources simultaneously [18]. Each glowworm then probabilistically selects a neighbor and moves toward it. This probability $p_j(t)$ depends on the difference in the luciferin value between the glowworm and its neighbor. Thus, neighbors with high luciferin value will have a higher probability of selection as seen in Equation 34.8. After selecting a neighbor to move toward, the glowworm takes a step toward that neighbor. In the absence of a neighbor, the glowworm switches its behavioral mode either to anemotactic mode, if it has a nonzero luciferin value, or to the spiraling mode in case of zero luciferin. In the original GSO algorithm [18], the luciferin value update contained information about the glowworm's earlier luciferin value with a temporal decay term associated with it. The idea was to retain information from earlier measurements made by the glowworm. Although this works well in case of a stable time-invariant environment, the odor profiles are subject to rapidly varying odor concentrations, which is likely to make past information unreliable. This is the reason why we do not use this term in the modified GSO algorithm.

THE MODIFIED GLOWWORM SWARM OPTIMIZATION (GSO) ALGORITHM

*deploy_glowworms*;
$\forall\, i$, set $\ell_i(0) = 0$
$\forall\, i$, set $r_d^i(0) = r_s$

For each glowworm *i*,

$$\ell_i(t) \;\leftarrow\; \max\{C(x_i(t - N_{mem} + 1)), \ldots, C(x_i(t))\} \tag{34.6}$$

For each glowworm *i*,

$$N_i(t) = \{j : d_{ij}(t) < r_d^i(t); \ell_i(t) < \ell_j(t)\} \tag{34.7}$$

*If* $(N_i(t) \neq 0)$ (chemotactic mode)
    For each glowworm $j \in N_i(t)$
$$p_j(t) = \frac{\ell_j(t) - \ell_i(t)}{\sum_{k \in N_i(t)} \ell_k(t) - \ell_i(t)} \tag{34.8}$$

$j =$ select glowworm  (using $p_j(t)$)

$$x_i(t) \;\leftarrow\; x_i(t) + s\left(\frac{x_j(t) - x_i(t)}{\|x_j(t) - x_i(t)\|}\right) \tag{34.9}$$

        *end*
    *end*

*If* $(N_i(t) = 0)$ *and* $(\ell_i(t) \neq 0)$(anemotactic mode)
    *If* $(C(x_i(t)) \neq 0)$
        $x_i(t) \;\leftarrow\; x_i(t) - sw \tag{34.10}$
        *else*
        $x_i(t) \;\leftarrow\; x_i(t) \tag{34.11}$
        *end*
    *end*

*If* $(N_i(t) = 0)$ *and* $(\ell_i(t) = 0)$
    Execute spiraling motion
    *end*

$$r_d^i(t) \;\leftarrow\; \min\{r_s, \max\{0, r_d^i(t) + \beta(n_t - |N_i(t)|)\}\} \tag{34.12}$$

*end*
    $t \leftarrow t + 1$

*Anemotactic mode:* A glowworm enters the anemotactic mode when it does not have any neighbors to move toward but has a nonzero luciferin value. A glowworm in this mode is within the gas plume but does not have a neighbor, that is, it does not have another glowworm within its local-decision domain, which has a higher luciferin value than itself. The anemotactic mode is the exploratory mode, in which glowworms move upwind searching for the source. The glowworm with the highest luciferin value will have no neighbors, hence it switches to the anemotactic mode and moves upwind. A glowworm in this mode takes a step in the upwind direction, as given by Equation 34.10, when the measured concentration is above a threshold. This behavior prevents a glowworm from leaving the plume and proceeding upwind away from the source. In case the concentration measured at its current position is below the threshold value, the glowworm stays at its current position.

*Spiraling mode:* A glowworm enters the spiraling mode when it is completely lost, that is, it has no neighbor to move toward nor it is within the plume so that it could move upwind and get to the source. The spiral motion serves as a heuristic to explore the field to find a neighbor or to capture the plume. A glowworm switches to spiraling mode once its luciferin value becomes zero, that is, if the glowworm has measured zero concentration for the last $N_{mem}$ seconds. Square spiraling is used for ease in robotic implementation. During spiraling, the glowworm takes $w$ steps along a straight line, in which $w$ is defined as the spiraling width. After which it performs a 90° turn and moves $w$ steps followed by another 90° turn and movement by $2w$ steps, and so on. The length of the spiral is incremented by $w$ steps after every two turns.

The mode-switching behavior explained previously is encapsulated in Figure 34.6. The figure is seen to be a fully connected graph. The conditions leading to a mode transition are also mentioned. A glowworm selects one of the three modes which best suits its environment.

As described in Section 34.2, odor concentration profile for a turbulent environment is generated using the patch-based model. It is assumed that the glowworms are deployed in this environment after 15 seconds, so that by this time the odor has already spread and has reached a reasonable distance downwind from the odor source. The glowworms are initialized with zero luciferin value. They adopt one of the three search modes depending on the concentration values measured. A glowworm that is unable to detect a patch and is without neighbors will adopt the spiraling mode. It will continue spiraling until it either finds a neighbor or detects a patch.

The chemotactic mode of the algorithm tries to maximize the luciferin level of the agents. The glowworm with the highest luciferin value attracts other agents toward it. High concentration spots are generally found on locations directly downwind to the source as can be seen from Figure 34.4. The luciferin value of a glowworm is the maximum concentration measured by the glowworm in the past $N_{mem}$ seconds and hence the chemotaxis essentially takes the glowworms toward locations where high gas concentrations have been measured. This results in glowworms bunching at locations directly downwind of the source.

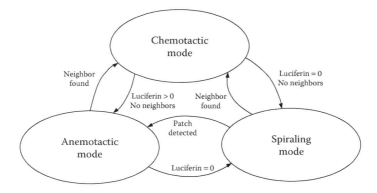

**FIGURE 34.6** Glowworm mode transition diagram.

A glowworm with no neighbors and a nonzero luciferin value enters the anemotactic mode. A glowworm in this mode takes a step in the upwind direction every time it detects a patch. Due to the property of the chemotactic mode as explained earlier, the glowworms tend to bunch together at locations directly downwind to the source location. An upwind motion from this point is likely to take them to the source. This is the intended function of the anemotactic mode. A zigzagging motion, as observed in moth flight [8], is not required as the bunching of agents generally takes place directly downwind of the source.

## 34.4 Effect of Different Key Parameters

In the current section, the performance of the algorithm under different conditions is evaluated. The set of parameters in Table 34.1 are assumed for the simulations unless mentioned otherwise. The simulations are run 100 times each for statistical evaluation.

### 34.4.1 Glowworm Convergence

In Figure 34.7, the glowworms are initially placed in a line. A glowworm close to the $X$-axis, that is directly downwind to the source location, acquires the highest luciferin value. This glowworm attracts all others toward it, resulting in a glowworm cluster near the $X$-axis. As the glowworm with the highest luciferin value does not have any neighbors, it enters the anemotactic mode and moves upwind. Other glowworms try to follow it and are thus carried upwind. This results in convergence of all the agents to the source location. The glowworms are considered to have reached the source when they get within 0.5 meters of the source location.

### 34.4.2 Effect of Communication Range

The communication range of the glowworms is the upper bound to the dynamic decision domain radius of a glowworm [18]. This parameter is used to incorporate the limitations of the communication equipment into the algorithm. The communication range affects the convergence of the glowworms. If it is too small, then it tends to isolate the glowworms thus forcing them into anemotactic or spiraling mode. When the communication range is large, as in Figure 34.8a, the glowworms find neighbors easily and hence converge quickly. In Figure 34.8a, complete convergence occurred in 244 seconds. For a small communication range limit, the glowworms are forced into spiraling behavior due to the absence of neighbors as shown in Figure 34.8b. The isolated glowworm keeps spiraling until it finds a neighbor or detects an odor patch. This results in delayed convergence, as shown in Figure 34.8b, in which complete convergence took 636 seconds. Figure 34.8c shows that the improvement in convergence time is rapid when communication

**TABLE 34.1**   Parameter Values Assumed for Simulations

| Parameter | Value |
| --- | --- |
| Number of agents | 9 |
| Gas sensor threshold | 1 ppm ($10^{-6}$) |
| Step size | 0.05 meters |
| $\beta$ | 0.25 |
| $n_t$ | 3 |
| Spiraling width | 10 |
| $a_0$ | 10 |
| Communication range ($r_s$) | 6 meters |
| $N_{mem}$ | 20 |
| Source strength | 20 patches/second |
| Field size | $20 \times 20$ meters |

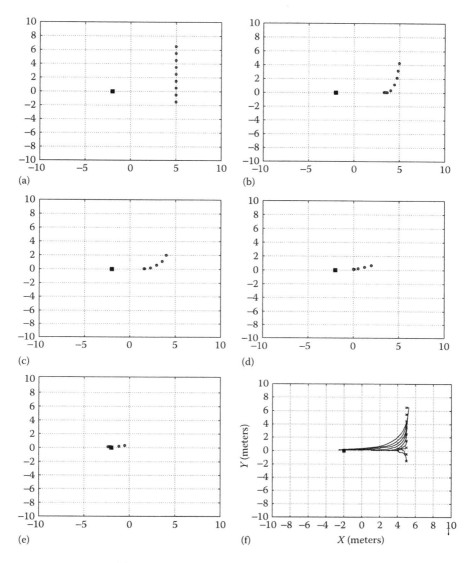

**FIGURE 34.7** Convergence of glowworms toward the source location: (a) time = 0 second, (b) time = 50 seconds, (c) time = 100 seconds, (d) time = 150 seconds, (e) time = 200 seconds, and (f) glowworm trajectory.

range is small. However, once the communication range exceeds a certain limit, further increase has little effect on the algorithm performance. In the given case, this happens at 6-meter communication range. This value of communication range ensures that the glowworms are not forced into anemotactic or spiraling mode due to communication range limitations, and further increase in communication range does not benefit convergence.

## 34.4.3 Effect of Source Location

To study algorithmic performance, the location of the source with respect to the glowworms is varied by locating them at grid points. Then the time required for convergence of the glowworms to the source for different source locations is simulated. For all simulations, the glowworms are initially deployed as shown

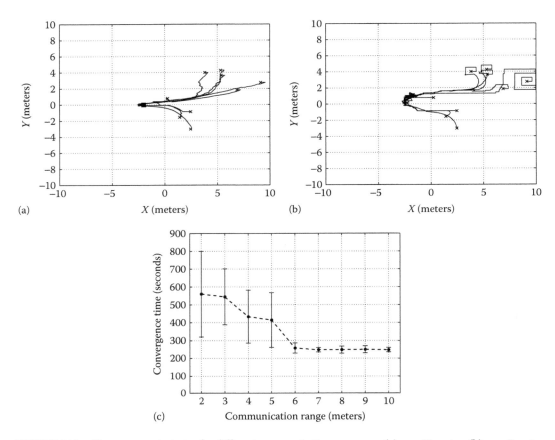

**FIGURE 34.8**  Glowworm trajectories for different communication ranges, $r_s$: (a) $r_s$ = 10 meter. (b) $r_s$ = 2 meter. (c) Algorithm behavior with variation in communication range.

in Figure 34.9a. The source location is varied and the convergence time for various source locations is plotted against the source coordinates as shown in Figure 34.9b and c. The convergence time is seen to increase rapidly if the source is close to an edge of the field. When the source is placed close to the edge of the field, no glowworm will be within the plume initially, hence they all enter the spiraling mode. This is shown in Figure 34.9d, whereas the source is far away from the glowworms, the gas concentration measured by the glowworms will be below the threshold level. This forces all glowworms into the spiraling mode. They spiral until one of them gets to the plume and measures a nonzero gas concentration. This glowworm attracts other glowworms within its communication range and with the help of anemotaxis gets to the source. As the spiraling behavior is inefficient, the convergence time is high. In Figure 34.9d, convergence took 1824 seconds. Convergence here is defined as the time taken for at least seven out of nine glowworms to get within 0.5 meters of the source location. The simulations are run 10 times each.

### 34.4.4 Effect of Sensor Noise

To study the performance of the algorithm in the presence of noise, zero mean Gaussian noise is added to the gas sensor readings as

$$C(x_i(t)) = F(x_i(t)) + \gamma(0, \sigma)(t) \tag{34.13}$$

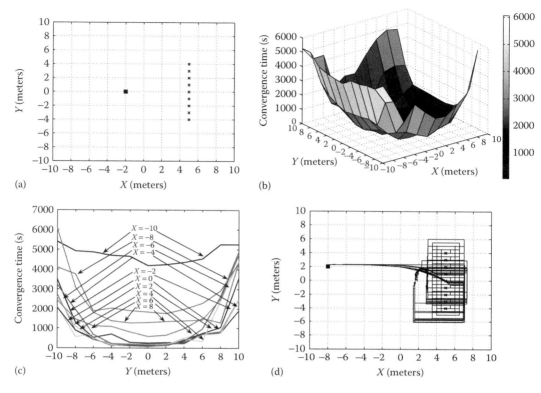

**FIGURE 34.9** Algorithm performance with variation in source position: (a) Initial placement of glowworms and the source location. (b), (c) Convergence time required for different source positions. In (c), $X = 6$ represents the $X$-coordinate of the source position in meters. (d) Delayed convergence when the glowworms are placed far from the source.

where $F(x_i(t))$ is the actual gas concentration, $\gamma(0, \sigma)(t)$ is zero mean Gaussian noise with a standard deviation of $\sigma$, and $C(x_i(t))$ is the concentration measured by the gas sensor. The glowworms were then placed in a line as shown in Figure 34.9a. With a source at $(-2, 0)$, the time for convergence was noted for different values of $\sigma$. Figure 34.10 shows the effect of noise on the algorithm. The algorithm is found to be robust even for practically high noise levels. This robustness can be attributed to the luciferin update mechanism of the GSO algorithm, which considers only the maximum concentration measured in the last $N_{mem}$ seconds.

## 34.4.5 Effect of Source Strength Variation

As explained in Section 34.2, the gas concentration profile is obtained by assuming the odor source to emit odor patches that are transported by wind and turbulence. By reducing the number of patches emitted by the source, the performance of the algorithm in dilute conditions is analyzed. A reduction in source strength results in a lesser probability of encountering a patch by a glowworm.

The algorithm is seen to converge successfully even under dilute conditions (Figure 34.11a). In Figure 34.11b, the glowworm trajectory for a source strength $N = 2$ patches per second can be seen. Initially, a glowworm close to the $X$-axis acquires the highest luciferin value and attracts other glowworms toward it.

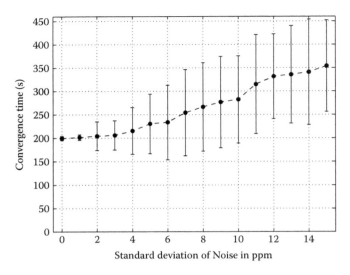

**FIGURE 34.10**   Algorithm performance in the presence of sensor noise.

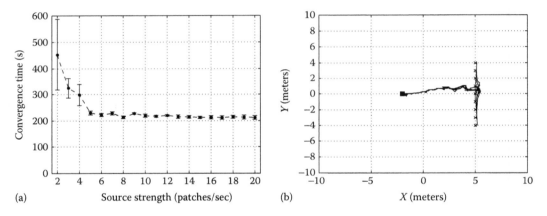

**FIGURE 34.11**   (a) Algorithm performance with variation in source strength. (b) Glowworm trajectory with source strength, $N = 2$.

This glowworm then moves upwind as it is in the anemotactic mode. As the frequency of arrival of patches is less, the upwind motion of the glowworm is delayed. This results in delayed convergence.

## 34.4.6 Effect of $N_{mem}$

The luciferin level of the glowworm is equal to the maximum concentration value encountered in the past $N_{mem}$ seconds. The choice of $N_{mem}$ is seen to have a considerable effect on the algorithm performance as shown in Figure 34.12. The value of $N_{mem}$ needs to be chosen such that it is large enough to remember the maximum concentration, which was measured close to the glowworm's current position, but small enough to ensure that the maximum concentration was not measured far from the current location of the glowworm. The maximum amplitude of the concentration within the patch decreases away from the source [17], but the concentration measured at any instant can be misleading. This is the reason for the delayed convergence for low $N_{mem}$ values.

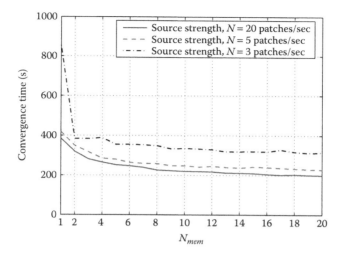

**FIGURE 34.12** Algorithm performance with variation in $N_{mem}$.

### 34.4.7 Effect of Variation in Number of Glowworms

In robot swarms, a single agent often has limited capabilities and hence its highly desirable that more than one robot gets to the source location to carry out a task after locating the odor source, say defusing mines, fixing gas leaks, and so on. Here, we assume that at least seven glowworms need to get to the source location for the task to be completed successfully. The glowworms are initially placed at random locations to the right of the $Y$-axis.

In Figure 34.13b, 15 randomly placed glowworms are used to locate the odor source. The algorithm was terminated in 249 seconds. As the glowworms are randomly placed, none of them are directly behind the source location. Hence, a simple anemotactic motion will not take them to the source. Here, chemotaxis takes all the glowworms close to the centerline of the plume as the probability to find a concentration higher than that at the centerline rapidly decreases with the lateral distance from the centerline [41]. An anemotaxic motion from this point will take the glowworms to the source.

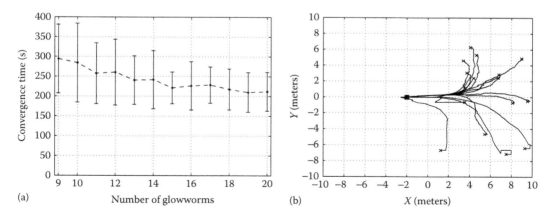

**FIGURE 34.13** Variation of swarm size: (a) Algorithm performance with variation in swarm size. (b) Trajectory for 15 glowworms used to locate source.

The performance of the algorithm with variation in swarm size is shown in Figure 34.13a. The plot clearly indicates the decrease in mean convergence time with larger swarm sizes. This improvement in performance with swarm size is an essential feature in swarm algorithms.

## 34.5 Obstacle Avoidance Behavior

Earlier sections had considered robots to be point masses, that is, a robot did not perceive other robots as obstacles and was assumed to move through each other. We have shown the algorithm to work well with such an assumption. Although this approximation helps us to assess the performance of the algorithm under idealistic conditions, it raises concerns about the algorithm's performance when implemented on real robots. To evaluate the algorithm's performance when the above-mentioned assumption is relaxed, we studied the performance of the algorithm after introducing obstacle avoidance between glowworms.

The obstacle avoidance technique introduced utilizes gyroscopic forces. The use of gyroscopic forces is known to prevent grid lock situations, encountered in multi-agent systems using potential function-based approaches [42]. In the gyroscopic approach, whenever a glowworm's path comes within the obstacle avoidance radius $d$ of another glowworm, it takes a step in the perpendicular direction. This may be explained using Figure 34.14, in which we assume $B$ to be a glowworm with a lower luciferin value than $A$. The glowworm $B$, which is in chemotactic mode, takes a step toward $A$ according to the GSO algorithm. As taking this step takes $B$ closer to $A$ than the avoidance distance $d$, obstacle avoidance forces it to take a step in the direction perpendicular to the one suggested by the algorithm. In the next instant, the suggested path is again toward $A$, and obstacle avoidance forces it to take the perpendicular path. Continuing this way $B$ can be seen to circle $A$. This circling behavior is seen to add an explorative component to the algorithm, this explorative behavior is later utilized for source localization without anemotaxis.

After the introduction of obstacle avoidance, the convergence is seen to be delayed considerably. This delay in convergence is caused by a hindrance to anemotaxis. The glowworm with the highest luciferin value attracts all other glowworms toward it. Thus, a cluster is formed around this glowworm. As the glowworm with the highest luciferin has no neighbors, it enters the anemotactic mode and tries to move upwind. The presence of glowworms around it prevents this motion. In Figure 34.15, the glowworm $L$ has the highest luciferin value and hence enters the anemotactic mode and tries to move upwind, but as that would take it too close to $A$ and $B$, obstacle avoidance forces it to move in the perpendicular direction, thus hindering anemotaxis and delaying convergence.

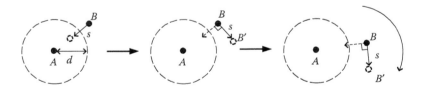

**FIGURE 34.14**   Obstacle avoidance in glowworms.

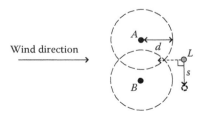

**FIGURE 34.15**   Anemotactic motion hindered by obstacle avoidance.

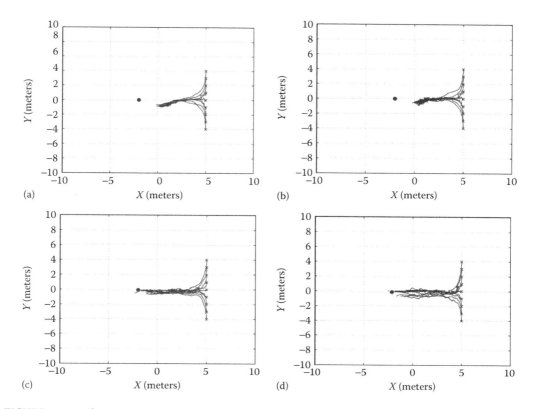

**FIGURE 34.16** Glowworm trajectories after 250 seconds: (a) $N_l = 1$ (b) $N_l = 3$ (c) $N_l = 5$ (d) $N_l = 7$.

This problem can be solved by introducing obstacle avoidance maneuvers at a larger distance, say $N_l$ times $d$, the normal obstacle avoidance distance, for the glowworm in the chemotactic mode. In Figure 34.14, if $A$ is attracting $B$, obstacle avoidance maneuvers can be introduced in $B$ at a distance of $N_l$ times $d$ from $A$. In case $A$ is not the glowworm attracting $B$, but one merely in its path, the obstacle avoidance distance will be $d$. Higher values of $N_l$ result in glowworms being spread out, thus minimizing anemotactic hindrance.

Figure 34.16 shows the variation of glowworm trajectory with $N_l$, with a source at $(-2, 0)$. All simulations assume an obstacle avoidance distance, $d = 0.1$ meter. For lower $N_l$ values as in Figure 34.16a and b, the glowworms bunch up together, hindering the anemotactic motion. In Figure 34.16c and d, due to high $N_l$ the glowworm trajectories are observed to be spread out. Hence, anemotaxis is unaffected, resulting in a faster upwind progress. Convergence is observed to occur in about 250 seconds. Figure 34.17 shows the improvement in convergence time observed by increasing $N_l$. With sufficiently large $N_l$, the convergence time is close to that of non-obstacle avoidance simulations.

## 34.6 Source Localization without Anemotaxis

### 34.6.1 Under Constant Wind Conditions

As explained in Section 34.3, anemotaxis is the explorative part of the algorithm. While in the anemotactic mode the glowworm takes a step in the upwind direction in the hope of getting to the source. As anemotaxis demands the knowledge of wind direction, the glowworms are required to have a wind sensor. This Section explains the possibility of utilizing the obstacle avoidance behavior to localize an odor source without anemotaxis.

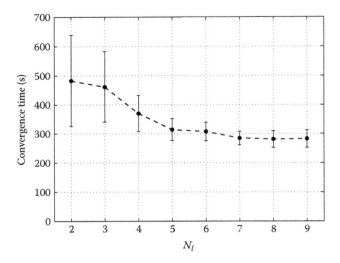

**FIGURE 34.17**  Convergence time variation with $N_l$ (with anemotaxis).

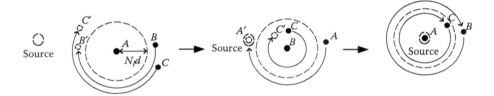

**FIGURE 34.18**  Glowworm convergence without anemotaxis.

Due to the absence of anemotaxis, the glowworm with the highest luciferin value remains stationary. When other glowworms, which are attracted toward this glowworm, get closer than a distance $N_l d$, they start circling it. In Figure 34.18, $A$ is the glowworm with the highest luciferin value at first. It stays stationary until it finds another glowworm with a higher luciferin value. Glowworms $B$ and $C$ are attracted by $A$. They revolve around $A$ due to gyroscopic obstacle avoidance. While revolving, the glowworms $B$ and $C$ get closer to the source than $A$. While at locations close to the source there is a high probability of one of them encountering a patch and thus acquiring a higher luciferin value. In Figure 34.18, $B$ is shown to encounter a patch while it is at location $B'$. Now $B$ has a higher luciferin value than $A$ and $C$. Hence, $A$ and $C$ are attracted by it. As they are close to $B$, gyroscopic obstacle avoidance makes them revolve around $B$. While revolving, $A$ reaches the source and thus acquires a higher luciferin value. Now $B$ and $C$ start revolving around $A$. In Figure 34.19a, the gyroscopic forces are seen to take the glowworms to the source without anemotaxis. However, after they get to the source, the glowworms keep circling the source. The simulation assumes an obstacle avoidance distance $d = 0.1$ meters and $N_l = 15$.

The parameter $N_l$, which controls the dispersion in the glowworm swarm, has a significant influence on the convergence time. Lower $N_l$ values result in glowworms bunching up close together, which makes exploration of concentration variations in the environment difficult, thus delaying convergence. As can be seen from Figure 34.19b, the convergence time decreases significantly with an increase in $N_l$. This is due to the improved exploratory behavior introduced by better dispersion of glowworms. The simulations assumed grid point glowworm placement and source placement as shown in Figure 34.9a. The obstacle avoidance distance $d$ was assumed to be 0.1 meter.

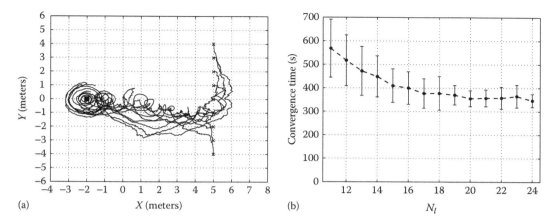

(a)    *X* (meters)    (b)    $N_l$

**FIGURE 34.19** Source localization without anemotaxis: (a) Glowworm trajectory under constant wind conditions after 500 seconds. (b) Convergence time variation with $N_l$ (without anemotaxis).

## 34.6.2 Under No Wind Conditions

As the above-mentioned algorithm does not demand wind information, it can be used to localize an odor source even in the absence of wind, which is considered to be the most difficult problem to solve [6]. Figure 34.20b shows the glowworm trajectory obtained by applying the algorithm on a concentration profile, generated using the model described in Section 34.2, assuming wind speed to be zero. Here, the glowworm that has the highest luciferin value and attracts other glowworms toward it. Once close, obstacle avoidance forces them to revolve around the glowworm with the highest luciferin value. While revolving, one of these glowworms might encounter a patch and acquire a higher luciferin value. This process, as explained earlier, will take the glowworms to the source. Once a glowworm reaches the source, it remains stationary, while others revolve around it. As, even after reaching the source, the glowworms continue to explore the environment, the algorithm may be used to solve a moving source problem too. However, this problem is left for future work.

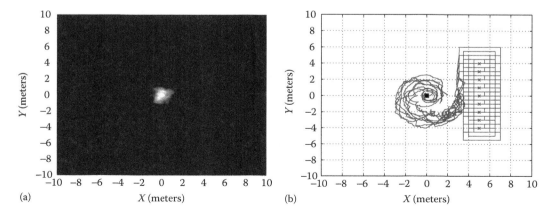

(a)    *X* (meters)    (b)    *X* (meters)

**FIGURE 34.20** Source localization under no wind conditions: (a) Odor concentration profile in the absence of wind. (b) Glowworm trajectory under no wind conditions after 1200 seconds.

## 34.7 Simulations Using Experimental Data

To further evaluate the performance of the algorithm, we have used concentration measurements obtained from a turbulent dye mixing experiment [37]. A small modification needs to be made to the GSO algorithm with anemotaxis for successful localization. The earlier rule by which an upwind step is taken only while measuring nonzero concentration is removed as it leads to inefficient anemotaxis. As the environment is dilute, and odor concentration above the threshold level is detected only for small intervals, thus the earlier rule for anemotaxis is seen to be insufficient. Hence, the explorative component of the GSO algorithm was improved by making the glowworm take upwind steps whenever it has no neighbors and has nonzero luciferin.

Figure 34.21 shows snapshots of a simulation using experimental data in which the GSO algorithm with anemotaxis is used. The glowworms are seen to be able to locate the source in about 350 seconds. The parameter values used for the simulation are shown in Table 34.2. The light patches in the figure show the presence of dye and the intensity of these patches represent the dye concentration. When converted

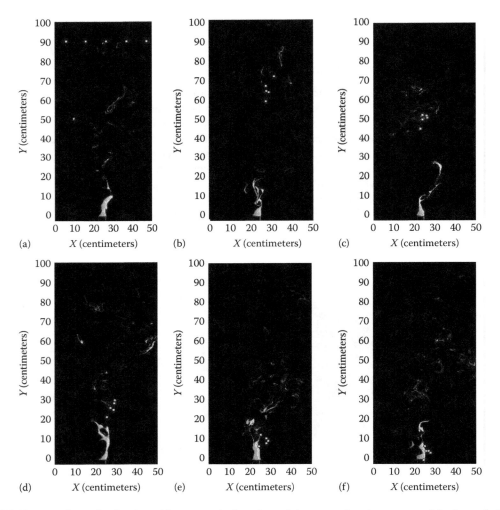

**FIGURE 34.21** Source localization with anemotaxis. Snapshots of glowworms locating a source of dye in a turbulent environment. The simulation assumes an $N_l = 5$: (a) time = 0 second, (b) time = 100 seconds, (c) time = 200 seconds, (d) time = 300 seconds, (e) time = 350 seconds, and (f) time = 400 seconds.

**TABLE 34.2**  Parameter Values Assumed for Simulations
Using Experimental Data

| Parameter | Value |
|---|---|
| Number of agents | 5 |
| Gas sensor threshold | 25 |
| Step size | 0.5 centimeters |
| $\beta$ | 0.25 |
| $n_t$ | 3 |
| Spiraling width | 10 |
| Communication range ($r_s$) | 35 centimeters |
| $N_{mem}$ | 35 |
| Field size | $100 \times 50$ centimeters |

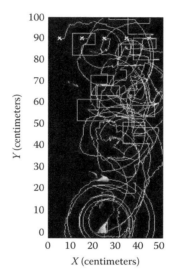

**FIGURE 34.22**  Glowworm trajectory while locating dye source without using wind information after 1400 seconds.
The simulation assumes an $N_l = 25$.

to gray-scale values, pure white is represented by 256. The dark regions indicate the absence of dye and
have a gray-scale value of around 20. We have assumed a dye concentration threshold equivalent to a
gray-scale value of 25.

Figure 34.22 shows the glowworm trajectory for the GSO algorithm without anemotaxis. As explained
in the earlier section, the glowworms use their gyroscopic maneuvers to move toward the source and thus
get to the source location. These simulation results support our claim about the capability of the GSO
algorithm in locating dilute odor sources in turbulent environments.

## 34.8 Experiments Using Gas Sensors

An odor source localization experiment using a set of sensors was carried out to evaluate the validity of
the proposed approaches in a realistic environment. An ethanol source, shown in Figure 34.23a, was used
for all of these experiments. An air pump was used to bubble air through ethanol so as to increase the
strength of the source. A fan was used to simulate the presence of wind. A set of four Figaro TGS-822 gas

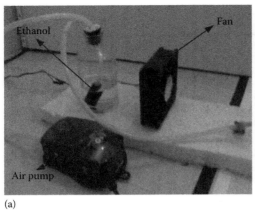
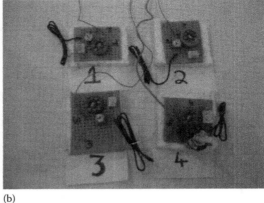

(a)                                                    (b)

**FIGURE 34.23**   (a) The ethanol source used in the experiments and the fan used to create a unidirectional air flow. (b) The set of sensors used for the experiments.

sensors, shown in Figure 34.23b, were used to measure ethanol concentrations. Experiments were carried out in a field of size 1.5 meter × 1.5 meter.

Experiments were carried out in the presence of a unidirectional wind and under no wind conditions. The sensors were initially placed at random locations within the field. The GSO algorithm was applied to the gas sensor output values. The communication range for the GSO algorithm was assumed to be large to ensure global communication. A step size of 5 centimeter was used. All gas sensors were calibrated so as to obtain a zero concentration output voltage of 2 Volt. Gas sensors usually have a cap with a flame-proof stainless steel gauze. This cap was removed for the experiments so as to improve the response time of the sensor [43].

The GSO algorithm provided new coordinates for the gas sensors based on the output of the gas sensors. The sensors were then manually moved to the location as specified by the GSO algorithm. Then, the gas sensor output at the new location is noted and the GSO algorithm evaluates the next location. This is repeated until the sensors converge to the odor source location. Obstacle avoidance maneuvers as described in Section 34.5 are also used. Figure 34.24a shows the sensor trajectory in the presence of a unidirectional wind. The wind here is directed from the source to the field along the negative $y$-direction. As can be observed from the figure, the sensors converged close to the source location in 18 steps.

Figure 34.24b shows the sensor trajectory in the absence of wind. It can be seen that the initial position of the sensors is the same as in the previous experiment. The difference in sensor trajectories in the above-mentioned two cases is due to the absence of anemotactic motion in the no wind experiment and the variation in the concentration profile due to the presence of wind. Here convergence was observed in 17 steps. These experimental results demonstrate the suitability of our approach for solving odor source localization problems under real-world conditions.

## 34.9 A Comparative Discussion with Earlier Approaches

Jatmiko et al. [16] had modified the PSO algorithm to solve the odor source localization problem. They utilized a filament-based model for the odor source, which is well suited for smoke-like profiles, in which small-scale eddies shred an odor plume into filaments [44]. However, the filament-based model has not been shown capable of modeling sources in which large-scale eddies are formed, resulting in random, disconnected patches [17]. They utilize the instantaneous gas concentration measurements to locate the source, which can be misleading, thus resulting in delayed convergence. Instead, we defined the luciferin

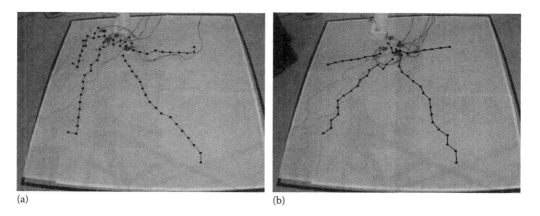

**FIGURE 34.24**  Source localization using gas sensors: (a) Sensor trajectory in the presence of a unidirectional wind. (b) Sensor trajectory under no wind conditions.

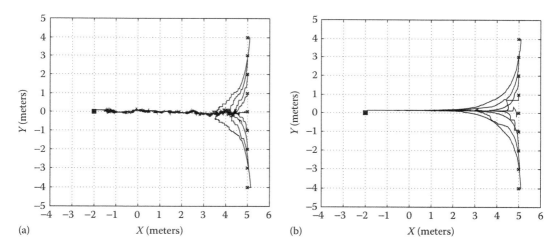

**FIGURE 34.25**  (a) Instantaneous concentration navigation. ($N_{mem}$ = 1, Convergence time = 405 seconds). (b) Navigation using maximum concentration measured in the past 20 seconds. ($N_{mem}$ = 20, Convergence time = 196 seconds).

value to be the maximum gas concentration value encountered in the last $N_{mem}$ seconds. As the maximum gas concentration encountered reduces as one moves away from the source [17], the difference in luciferin value can be relied upon for source localization. The improvement in convergence time, by utilizing maximum concentration instead of instantaneous measurements, can be clearly seen from Figure 34.25. The significant difference in convergence time (405 seconds and 196 seconds) and the distorted trajectory is a consequence of the reliance on instantaneous concentration measurements. The agents in our algorithm adapt to the environment by switching modes. This mode-switching behavior is seen to choose the most appropriate of the three modes. Other algorithms are seen to use the wind information only to aid chemotactic motion [16]. Moreover, our method has been able to solve the odor source localization problem in the absence of wind, which only a few algorithms, such as the recently proposed infotaxis algorithm [6] have been shown capable of solving.

## 34.10 Conclusion

The paper presents an approach to localize an odor source in a natural setting. The method utilizes an intelligent integration of the chemotactic and anemotactic approaches. To generate gas concentration profiles similar to ones in a natural setting, a patch-based model was developed. The model was shown to generate concentration profiles that resembled the complex, chaotically evolving structure observed in turbulent plumes.

The GSO algorithm that has been used earlier for multimodal function optimization was chosen as the basic algorithm for the chemotactic mode. To suit the problem, three behavioral modes were introduced in the glowworms. A behavioral mode was chosen by the glowworm according to the concentration measured and the presence of a neighbor. Moreover, as the maximum concentration in a patch decreases with an increase in distance from the source, the luciferin value of a glowworm is made equal to the maximum concentration measured by it in the past $N_{mem}$ seconds. The difference in luciferin level was utilized for chemotactic navigation.

The algorithm was applied to the concentration profiles generated by the patch-based model. The glowworms were able to locate the odor source by communicating with other glowworms and by utilizing the wind information. By switching modes appropriately, the glowworms were able to utilize the advantages of chemotaxis and anemotaxis behavior. In the absence of detectable odor patches, the spiraling mode was used to search the field.

The algorithm performance variations with different parameter values were studied next. The communication range was seen to have an effect on algorithm performance when it is small enough to isolate glowworms, thus forcing them into anemotaxis or spiraling mode. The variation of source location is seen to affect convergence time when all the glowworms are outside the plume. Under such conditions, the spiraling mode is utilized to capture the plume and get to the source. The algorithm is seen to be quite robust to sensor noise. Even high values of sensor noise are not seen to have a significant effect on algorithm performance. The algorithm has also shown to work in dilute conditions, although the convergence is delayed due to reduced frequency of arrival of patches, which affects anemotaxis. Lower values of $N_{mem}$ are seen to mislead glowworms, resulting in delayed convergence. The glowworms are misled by the fact that a lower $N_{mem}$ might result in a glowworm, which is far from the source, having a higher luciferin value than one closer to the source, due to instantaneous concentration variations.

Obstacle avoidance behavior was next introduced between glowworms. The obstacle avoidance resulted in a hindrance of anemotaxis, thus delaying convergence. To counter this, obstacle avoidance maneuvers were introduced at a larger distance for a glowworm in the chemotactic mode and this was shown to improve performance. To locate the source, in the absence of wind information, a novel approach utilizing obstacle avoidance was introduced. This approach was shown to locate odor source in the absence of wind.

The suitability of the proposed approaches is tested on data obtained from a turbulent dye mixing experiment. The algorithm is seen to be capable of locating the odor source in the presence and in the absence of wind information. The ethanol source localization experiments using gas sensors introduce the practical limitations of sensor delay, noise, and so on. The algorithm is seen to be successful in locating the odor source in spite of such limitations. This further demonstrates the suitability of the proposed algorithm in a practical scenario. Thus, on the whole, the algorithm can be considered a promising candidate for a robotic odor source localization algorithm that works well in a natural setting.

## References

1. Zarzhitsky, D., Spears, D.F., and Spears, W.M., Swarms for chemical plume tracing, *Proceedings of IEEE Swarm Intelligence Symposium,* Pasadena, CA, pp. 249–256, 2005.
2. Russell, R.A., Laying and sensing odor markings as a strategy for assisting mobile robot navigation tasks, *IEEE Robotics and Automation Magazine,* 2, 3–9, 1995.

3. Casbeer, D.W., Beard, R.W., McLain, T.W., Li, S.M., and Mehra, R.K., Forest fire monitoring with multiple small UAVs, *Proceedings of American Control Conference,* Portland, OR, pp. 3530–3535, 2005.

4. Fronczek, J.W. and Prasad, N.R., Bio-inspired sensor swarms to detect leaks in pressurized systems, *Proceedings of IEEE International Conference on Systems, Man and Cybernetics,* 2, 1967–1972, 2005.

5. Farrell, J., Li, W., Pang, S., and Arrieta, R., Chemical plume tracing experimental results with a REMUS AUV, *Proceedings of the Ocean Marine Technology and Ocean Science Conference,* San Diego, CA, pp. 962–968, 2003.

6. Vergassola, M., Villermaux, E., and Shraiman, B.I., "Infotaxis" as a strategy for searching without gradients, *Nature,* 445, 406–409, 2007.

7. Berg, H.C., Bacterial behaviour, *Nature,* 254, 389–392, 1975.

8. Kennedy, J.S. and Marsh, D., Pheromone-regulated anemotaxis in flying moths, *Science,* 184, 999–1001, 1974.

9. Zarzhitsky, D., Spears, D.F., Spears, W.M., and Thayer, D.R., A fluid dynamics approach to multi-robot chemical plume tracing, *Proceedings of the Third International Joint Conference on Autonomous Agents and Multi-Agent Systems,* New York, pp. 1476–1477, 2004.

10. Ishida, H., Kagawa, Y., Nakamoto, T., and Moriizumi, T., Odor-source localization in the clean room by an autonomous mobile sensing system, *Sensors and Actuators: B,* 33, 115–121, 1996.

11. Wandel, M., Lilienthal, A., Duckett, T., Weimar, U., and Zell, A., Gas distribution in unventilated indoor environments inspected by a mobile robot, *Proceedings of the IEEE International Conference on Advanced Robotics (ICAR),* pp. 507–512, 2003.

12. Cui, X., Hardin, C.T., Ragade, R.K., and Elmaghraby, A.S., A swarm-based fuzzy logic control mobile sensor network for hazardous contaminants localization, *Proceedings of the IEEE International Conference on Mobile Ad-hoc and Sensor Systems (MASS),* pp. 194–203, 2004.

13. Cao, Y.U., Fukunaga, A., Kahng, A., and Meng, F., Cooperative mobile robotics: Antecedents and directions, *Proceedings of the IEEE/RSJ International Conference on Intelligent Robots and Systems. (IROS 1995),* pp. 226–234, 1995.

14. Lytridis, C., Virk, G.S., and Kadar, E.E., Search performance of a multi-robot team in odour source localisation, in *Climbing and Walking Robots,* Tokhi, M.O., Virk, G.S., and Hossain, M.A. (Eds.), pp. 809–816, Springer, Berlin, Germany, 2006.

15. Hayes, A.T., Martinoli, A., and Goodman, R.M., Swarm robotic odor localization, *Proceedings of the IEEE/RSJ International Conference on Intelligent Robots and Systems (IROS 2001),* 2, 1073–1078, 2001.

16. Jatmiko, W., Fukuda, T., and Sekiyama, K., A PSO-based mobile robot for odor source localization in dynamic advection-diffusion with obstacles environment: Theory, simulation and measurement, *IEEE Computational Intelligence Magazine,* 2, 37–51, 2007.

17. Balkovsky, E. and Shraiman, B.I., Olfactory search at high Reynolds number, *Proceedings of the National Academy of Sciences,* 99, 12589–12593, 2002.

18. Krishnanand, K.N., and Ghose, D., Glowworm swarm based optimization algorithm for multimodal functions with collective robotics applications, *Multiagent and Grid Systems,* 2, 209–222, 2006.

19. Krishnanand, K.N. and Ghose, D., Theoretical foundations for rendezvous of glowworm-inspired agent swarms at multiple locations, *Robotics and Autonomous Systems,* 10, 549–569, 2008.

20. Liao, W.-H., Kao, Y., and Li, Y.-S., A sensor deployment approach using Glowworm Swarm Optimization Algorithm in wireless sensor networks, *Expert Systems with Applications,* 38, 12180–12188, 2011.

21. Dutta, R., Ganguli, R., and Mani, V., Swarm intelligence algorithms for integrated optimization of piezoelectric actuator and sensor placement and feedback gains, *Smart Material Structures,* 38, 105018, 2011.

22. Chetty, S. and Adewumi, A.O., Comparison study of swarm intelligence techniques for the annual crop planning problem, *IEEE Transactions on Evolutionary Computation*, 18, 258–268, 2014.

23. Mannar, S. and Omkar, S.N., Space suit puncture repair using a wireless sensor network of micro-robots optimized by Glowworm Swarm Optimization, *Journal of Micro-Nano Mechatronics*, 6, 47–58, 2011.

24. Khan, K. and Sahai, A., A glowworm optimization method for the design of web services, *International Journal of Intelligent Systems and Applications*, 10, 89–102, 2012.

25. Huang, Z. and Zhou, Y., Using glowworm swarm optimization algorithm for clustering analysis, *Journal of Convergence Information Technology*, 6, 78–85, 2011.

26. Wang, L., Zhao, L., and Yan, H., Application of GSO for load allocation between hydropower units and its model analysis based on multi-objective, *Journal of Computers*, 7, 1135–1141, 2012.

27. Manimaran, P. and Selladurai, V., Glowworm swarm optimisation algorithm for nonlinear fixed charge transportation problem in a single stage supply chain network, *International Journal of Logistics Economics and Globalisation*, 6, 42–55, 2014.

28. Yu, Z. and Yang, X., Glowworm Swarm Optimization algorithm for whole-wet orders scheduling in single machine, *The Scientific World Journal*, 2, 37–51, 2013.

29. Zhou, Y., Zhou, G., and Zhang, J., A hybrid glowworm swarm optimization algorithm to solve constrained multimodal functions optimization, *Optimization*, 64, 1057–1080, 2013.

30. Tang, Z. and Zhou, Y., A Glowworm Swarm Optimization algorithm for uninhabited combat air vehicle path planning, *Journal of Intelligent Systems*, 24, 69–83, 2015.

31. Yang, Y., Zhou, Y., and Gong, Q., Hybrid artificial glowworm swarm optimization algorithm for solving system of nonlinear equations, *Journal of Computational Information Systems*, 6, 3431–3438, 2010.

32. García-Segura, T., Yepes, V., Martí, J.V., and Alcalá, J., Optimization of concrete I-beams using a new hybrid glowworm swarm algorithm, *Latin American Journal of Solids and Structures*, 11, 1190–1205, 2014.

33. Zhou, Y., Zhou, G., and Zhang, J., A hybrid glowworm swarm optimization algorithm for constrained engineering design problems, *Applied Mathematics and Information Sciences*, 7, 379–388, 2013.

34. Gong, Q.Q., Zhou, Y.Q., and Yang, Y., Artificial glowworm swarm optimization algorithm for solving 0-1 knapsack problem, *Advanced Materials Research*, 143, 166–171, 2011.

35. Luo, Q.F. and Zhang, J.L., Hybrid artificial Glowworm Swarm Optimization Algorithm for solving constrained engineering problem, *Advanced Materials Research*, 204, 823–827, 2011.

36. Cui, H., Feng, J., Guo, J., and Wang, T., A novel single multiplicative neuron model trained by an improved glowworm swarm optimization algorithm for time series prediction, *Knowledge-Based Systems*, 88, 195–209, 2015.

37. Villermaux, E. and Duplat, J., Mixing as an aggregation process., *Physical Review Letters*, 91, 184501, 2003.

38. Siggia, E.D. and Shraiman, B.I., Scalar turbulence, *Nature*, 405, 639–646, 2000.

39. Krishnanand, K.N. and Ghose, D., Detection of multiple source locations using a glowworm metaphor with applications to collective robotics, *Proceedings of IEEE Swarm Intelligence Symposium*, pp. 84–91, Pasadena, CA, 2005.

40. Krishnanand, K.N. and Ghose, D., Glowworm-inspired swarms with adaptive local-decision domains for multimodal function optimization, *Proceedings of IEEE Swarm Intelligence Symposium*, Indianapolis, IN, 2006.

41. Webster, D.R., Rahman, S., and Dasi, L.P., On the usefulness of bilateral comparison to tracking turbulent chemical odor plumes, *Limnology and Oceanography*, 46, 1048–1053, 2001.

42. Chang, D.E., Shadden, S., Marsden, J., and Olfati-Saber, R., Collision avoidance for multiple agent systems, *Proceedings of 42nd IEEE Conference on Decision and Control,* 1, 539–543, 2003.
43. Martinez, D., Rochel, O., and Hugues, E., A biomimetic robot for tracking specific odors in turbulent plumes, *Autonomous Robots,* 20, 185–195, 2006.
44. Marques, L., Nunes, U., and de Almeida, A.T., Particle swarm-based olfactory guided search., *Autonomous Robotics,* 20, 277–287, 2006.

# 35

# Digital Reputation for Virtual Communities

Roberto Battiti

Anurag Garg

## 35.1 Introduction

A virtual community can be defined as a group of people sharing a common interest or goal who interact over a virtual medium, most commonly the Internet. Virtual communities are characterized by an absence of face-to-face interaction between participants that makes the task of measuring the trustworthiness of other participants harder than in non-virtual communities. This is due to the anonymity that the Internet provides, coupled with the loss of audiovisual cues that help in the establishment of trust. As a result, digital reputation management systems are an invaluable tool for measuring trust in virtual communities.

Trust is an important component of all human interactions whether they take place online or not. There is an implicit assumption of trust in each interaction that we participate in that involves some investment on our part. The Merriam-Webster dictionary defines trust as *assured reliance on the character, ability, strength, or truth of someone or something*. Even more pertinent to virtual communities is an alternative definition that states that trust is a *dependence on something future or contingent; reliance on future payment for property (as merchandise) delivered*. These definitions illustrate that the basis of trust is an expectation of future payment or reward, and that the transaction partner will behave in an honest fashion and fulfill their obligations.

The processes behind the creation of trust can be classified in four broad categories:

1. *Personality-based trust:* A person's innate disposition to trust others.
2. *Cognitive trust:* Through recognition of the characteristics of the transaction partner.

3. *Institution-based trust:* A buyer's perception that effective (third-party) institutional mechanisms are in place to facilitate transaction success [1].
4. *Transaction-based trust:* That relies on a participant's past behavior to assess their trustworthiness.

Personality-based trust does not have a significant use in virtual communities where decisions on trust are made algorithmically. Most cognitive factors that form trust in real-life interactions such as the manner of a person, their body language, and the intonations of their speech are also absent in virtual communities. However, there are alternative forms of cognition that can be used instead. These are almost invariably based on the virtual identity of the community member. In most *purely* online contexts—as opposed to contexts in which the online identity is linked to a offline identity such as with online banking—there are virtually no costs to creating a new virtual identity. Obtaining an email address on any of the free web-based email services such as Yahoo, Hotmail, and Gmail is trivial, and with each one comes a new virtual identity. Although these services are increasingly adopting strategies such as requiring image captchas to counter automated registration spam programs, it is not difficult to create say a dozen accounts in 10 minutes and as the eBay scam case [2] showed, that is all a malicious user needs to surmount simple feedback systems and commit online fraud.

Enforcing trust in virtual communities is hard not only because of the difficulties in recognizing the trustworthiness of participants but also because of the lack of adequate monitoring and appropriate sanctions for dishonest behavior. This lack of *institution-based trust* is because there are not enough trusted third parties with the power to punish to ensure honesty of all the players [3]. This problem is exacerbated in decentralized virtual communities and electronic marketplaces. An organization like eBay with a centralized infrastructure can act as the trusted third party to at least store feedback information in a reliable fashion even though the information itself may not be reliable. The problem becomes much harder with decentralized systems such as peer-to-peer (P2P) networks in which the absence of a centralized authority is a defining feature of the system.

Hence, it appears that the best strategy for creating trust in virtual communities is through *transaction-based trust* where feedback on participants' past behavior is collected and aggregated to decide their trustworthiness. A transaction-based trust strategy is far from perfect and suffers from many of the same shortcomings as the other strategies. For instance, the lack of verifiable identities [4] can make a transaction-based system vulnerable to manipulation. However, as many recent proposals have shown [5–10] such strategies are not contingent upon a trusted third party, and the community as a whole provides the institutional basis for trust. Hence, if the problem of identity can be solved, transaction-based systems are capable of providing the solution for virtual communities.

Transaction-based trust creation strategies are also commonly known as reputation-based trust management systems or just *reputation systems*. Some authors use the term reputation systems narrowly to include only those systems that monitor past transactions of participants to compute their trustworthiness. This information is then shared with other participants who decide whether or not to interact with the target participant on its basis. We use the term in a more general sense to include recommendations systems that recommend items as well as systems that perform distributed authentication by inferring trustworthiness. All these systems have one feature in common. The collection and aggregation of feedback to rate objects or people.

Reputation systems research lies at the intersection of several disciplines including evolutionary biology [11–13], economics (game theory and mechanism design) [14–17], sociology (social network theory [18–20]), and computer science (e-commerce, P2P systems, cryptography, etc.). In this chapter, we survey the approaches taken by researchers from these disciplines to provide the context for digital reputation schemes. We begin by listing the various applications of digital reputation systems. This is followed by a discussion of what motivates the participants of a virtual community to cooperate with each other. We then look at what are the requirements of a good reputation system. This is followed by a more detailed analysis of the components of reputation systems. We classify different types of feedback and their role in constructing reputations. We discuss second-order reputation and the motivations for providing feedback. Reputation modifying factors such as transaction context are looked at next followed by a discussion on how reputations can be interpreted. We conclude by looking at several specific digital reputation systems that have been proposed.

# 35.2 Applications of Reputation Management

The main uses of digital reputation management systems in virtual communities are:

1. Incentivizing cooperation
2. Identifying and excluding malicious entities
3. Authenticating users in a distributed manner
4. Providing recommendations

Resnick [21] further defines three requirements for a reputation system: (1) to help people decide whom to trust, (2) to encourage trustworthy behavior and appropriate effort, and (3) to deter participation by those who are unskilled or dishonest. To this, we can add the requirements that a reputation system (4) must preserve anonymity associating a peer's reputation with an opaque identifier and (5) have minimal overhead in terms of computation, storage, and infrastructure.

The participation of an individual in a virtual community strongly depends on whether, and how much benefit they expect to derive from their participation. If an individual feels that they will not gain anything from joining the community, they are unlikely to participate. Hence, a good distributed system must be designed such that it incentivizes cooperation by making it profitable for users to participate and withhold services from users that do not contribute their resources to the system.

An equally serious challenge to distributed systems and virtual communities comes from users who act in a malicious fashion with the intention of disrupting the system. Examples of such malicious behavior include users who pollute a file-sharing system with mislabeled or corrupted content, nodes that disrupt a P2P routing system to take control of the network, nodes of a mobile ad hoc network (MANET) that misroute packets from other nodes [10] and even spammers [22] who are maliciously attacking the email community. Therefore, a central objective of all digital reputation schemes is to identify such users and punish them or exclude them from the community. This exclusion is usually achieved by allowing members to distinguish between trustworthy and untrustworthy members. An untrustworthy member will not be chosen for future interactions. In contrast with incentivizing cooperation that motivates truth from participants, this is based on punishing falsehoods.

Another use for digital reputation systems is for distributed authentication. Distributed authentication does not rely on strict hierarchies of trust such as those using certification authorities and a centralized public key infrastructure that underpin conventional authentication. Such networks capture trust relationships between entities. Trust is then propagated in the network so that the trust relationship between any two entities in the network can be inferred. PGP [23], GnuPGP and OpenPGP-compatible [24] systems all use webs of trust for authentication. Closely related are virtual social networks including the Friend of a Friend system, LinkedIn, Tribe.net, Orkut, and Friendster that use some or the other form of trust propagation.

Recommendation systems are another application that relies on similar principles as digital reputation systems. Instead of computing the trustworthiness of a participant, recommendation systems compute the recommendation score of objects based on the collective feedback from other users. These systems are in widespread commercial use and examples include the Amazon recommendation system and the International Movie Database (IMDB) movie recommendations. In recommendation systems, objects instead of members of a virtual community are rated. The members may then be rated on the quality of feedback they provide just like in reputation systems.

When the number of objects to be rated is usually large compared with the user-base, the collected data can be very sparse. Collaborative-filtering based recommendation systems use the collective feedback to try to compute the similarity in the ratings made by any two users and using this similarity to weigh the rating of one user when predicting the corresponding choice for the other user. There are many ways in which this similarity can be computed. One method uses the Pearson's correlation coefficient [25]:

$$w_{u,i} = \frac{\sum_j (v_{u,j} - \bar{v}_u)(v_{i,j} - \bar{v}_i)}{\sqrt{\sum_j (v_{u,j} - \bar{v}_u)^2 \sum_j (v_{i,j} - \bar{v}_i)^2}} \tag{35.1}$$

where $w_{u,i}$ computes the similarity between users $u$ and $i$ based on their votes for all objects $j$ that both have voted for and $\bar{v}_u$ and $\bar{v}_i$ denote respectively the average ratings given by users $u$ and $i$.

Another method is to use vector similarity [26]:

$$w_{u,i} = \sum_j \frac{v_{u,j}}{\sqrt{\sum_{k \in I_u} v_{u,k}^2}} \frac{v_{i,j}}{\sqrt{\sum_{k \in I_i} v_{i,k}^2}} \tag{35.2}$$

where $v_{u,j}$ is the rating given by user $u$ for object $j$ as before, and $I_u$ is the set of objects for which user $u$ has a rating. Other model-based methods exist that use Bayesian network models or clustering models to group users together and use group membership to predict what a user may vote for a given object. Clustering models have also been shown to be useful in reputation systems for eliminating spurious ratings [27] made by malicious users in an electronic marketplace.

## 35.3 Motivating Cooperation

Motivating cooperation among participants has been a subject of research since long before the first virtual communities were formed. As long as there are shared resources, there will be users who are tempted to use more than their fair share for their own benefit even if it is at the expense of the community as a whole. This conflict between individual interests and the common good is exemplified in the "tragedy of the commons," a term coined by Hardin [28] who used the overgrazing of the English "commons" (property that was shared by the peasants of a village) as an example. In reputation systems research, such selfish users are termed as *free-riders* or *free-loaders*.

An example of free-riding can be found in MANETs. MANETs function on the premise that nodes will forward packets for each other even though forwarding a packet consumes power. However, if there are free-riders that inject their own packets in the network but do not forward any packets from other nodes, they can exploit the cooperative nature of other users.

Another example of free-riding in virtual communities is found in early file-sharing systems where an individual is allowed to download files without being required to contribute any files for uploading. In their measurement study of Napster and Gnutella, Sariou et al. [29] reported that this resulted in significant fractions of the population indulging in client-like behavior. They reported that the vast majority of modem users were free-riders who did not share any files and only downloaded files offered by others. However, equally interesting is the fact that many users with high bandwidth connections indulged in server-like behavior and offered their own files for sharing but did not download any files at all. This unselfish behavior is the flip-side of free-riding where users in a virtual community indulge in *altruism*.

Altruism has been studied by evolutionary biologists [11–13] in humans and other primates. These authors have sought to explain altruistic behavior by arguing that it signals "evolutionary fitness." By behaving generously an individual signals that they have "plenty to spare" and is thus a good mating choice. In the context of virtual communities, altruistic behavior is motivated in part by a desire to signal that interacting with the individual is likely to be beneficial in other contexts as well.

Another strand of research comes from economists who have studied cooperation by setting up nonzero sum games and determining the equilibria that result both through analytic game theory and through simulation. In game theory, it is usually assumed that players are "rational" or "selfish," that is, they are only interested in maximizing the benefit they obtain with no regard to the overall welfare of the community. This is distinct from an "irrational" player whose utility function depends on something more than just their own benefit and whose behavior cannot be predicted. An example of irrational players are "malicious" players who actively wish to harm other players or the game as a whole even if it means reducing their own personal benefit.

The "game" that is typically used for modeling the problem of cooperation is the Prisoner's Dilemma [30,31]. The classic prisoner's dilemma concerns two suspects who are questioned separately by the police and given a chance to testify against the other. If only one prisoner betrays the other, the betrayer goes free whereas the loyal prisoner gets a long sentence. If both betray each other, they get a medium sentence, and if both stay silent, they get a small sentence. Hence, if both prisoner's are "selfish" and have no regard for the other prisoner, we see the best strategy for a prisoner is always to betray the second prisoner, regardless of what the other prisoner chooses. If the second prisoner confesses, the first prisoner must confess too, otherwise the first prisoner will get a long sentence. If the second prisoner does not confess, the first prisoner can get off free by confessing as opposed to getting a short sentence by not confessing.

If the game is played only once, the solution is obvious. "Always Defect" is the dominant strategy.[*] Axelrod [14] studied an interesting extension to the classic problem that he called the iterated prisoner's dilemma. Here, members of a community play against each other repeatedly and retain memory of their past interactions. Hence, they have an opportunity to punish players for their past defections. Axelrod set up an experiment with various strategies submitted by fellow researchers playing against each other. He discovered that "greedy" strategies tended to do very poorly in the long run and were outperformed by more "altruistic" strategies, as judged by pure self-interest. This was true as long as cheating was not tolerated indefinitely. By analyzing the top-scoring strategies, Axelrod stated several conditions necessary for a strategy to be successful. A successful strategy should be (1) *Nice*: It will not defect before an opponent does, (2) *Retaliating*: It will always retaliate when cheated. A blindly optimistic strategy like "Always Cooperate" does not do well, (3) *Forgiving*: To stop endless cycles of revenge and counter-revenge, a strategy will start cooperating if an opponent stops cheating, (4) *Nonenvious*: A strategy will not try to outscore an opponent. Axelrod found that the best deterministic strategy was "tit-for-tat" in which a player behaved with its partner in the same way as the partner had behaved in the previous round.

The prisoner's dilemma is a game with a specific reward function that encourages altruistic behavior and discourages selfish behavior. Let $T$ be the temptation to defect, $R$ be the reward for mutual cooperation, $P$ be the punishment for mutual defection, and $S$ be the Sucker's punishment for cooperating, whereas the other defects. Then, the following inequality must hold

$$T > R > P > S \tag{35.3}$$

In the iterated game, yet another inequality must hold

$$T + S < 2R \tag{35.4}$$

If this is not the case, then two players will gain more in the long run if one cooperates and the other defects alternately rather than when both cooperate. Hence, there will be no incentive for mutual cooperation. Recall that the objective of a rational player is to maximize their individual score and not score more than the opponent.

Other virtual communities may have different cost and reward functions. The game-theoretic approach is to devise strategies to maximize individual utility in a fixed game. However, if a community is designed so that its interests as a whole are not aligned with those of an individual, the system is will be "gamed" by the rational users leading to the ultimate failure of the virtual community. Mechanism design [15] deals with the design of systems such that players' selfish behavior results in the desired system-wide goals. This is achieved by constructing cost and reward functions that encourage "correct" behavior.

---

[*] A dominant strategy always give better payoff than another strategy regardless of what other players are doing.

Another example of incentivizing cooperation can be found in collaborative content distribution mechanisms [32–34] such as BitTorrent [35], in which peers cooperate to download a file from a server by downloading different chunks in parallel and then exchanging them with each other to reconstruct the entire file. In fact, BitTorrent implements a "tit-for-tat" strategy to prevent free-riding. Basic reputation schemes have also been implemented in second generation file-sharing systems such as Kazaa that measure the participation level of a peer based on the number of files the peer has uploaded and downloaded. Peers with a higher participation level are given preference when downloading a file from the same source.

## 35.4 Design Requirements for a Reputation System

Although mechanism design may help achieve overall system goals through proper incentivization, it is ineffective against deliberately malicious (or irrational) participants. To solve this problem, a number of reputation systems have emerged. These operate by allowing a user to rate the transactions they have had with other users. This feedback is collected and aggregated to form a reputation value that denotes the trustworthiness of a user. This reputation information is made available to requesting users who then make their decisions on whether or not to interact with a given user based on its reputation. Several competing schemes for aggregating this feedback have been proposed [5–9,36,37].

The architectural and implementation details of the aggregation mechanism depend on the underlying network on which the virtual community is based. When the community is built on top of a traditional client-server network, a trusted third party exists, which can be relied on to collect and aggregate opinions to form a global view. eBay feedback, Amazon customer review, and the Slashdot distributed moderation systems are all examples in which feedback from users is stored in a centralized trust database. The aggregation is performed in this centralized database, and all users have access to the global reputations thus computed. When the community is built on top of a P2P network, the challenges of managing feedback become much harder. There is no centralized, trusted, reliable, always-on database, and the collection, storage, aggregation and dispersal of trust information must be done in a distributed way. Relying on third parties for storage and dissemination also makes the system vulnerable to tampering and falsification of trust information in storage and transit. Moreover, the system must also provide redundancy because users may drop out of the network at any time.

Hence, there are two separate but interrelated challenges that must be overcome by any distributed trust-management system. The first is the choice of an appropriate trust metric that accurately reflects the trustworthiness of users and is resistant to tampering and other attacks. This was recognized by Aberer and Despotovic as "the semantic question: which is the model that allows us to assess trust [of an agent based on that agents behavior and the global behavior standards]" [5]. The second is designing a system architecture that is robust against the above-mentioned challenges or the "data management" problem.

At this point, it is useful to ask why a reputation system would work in a virtual community and what may cause it to fail. The reasons for potential success are as follows: (1) The costs of providing and distributing reputations are negligible or zero. (2) The infrastructure for decentralized aggregation and dissemination already exists in the form Distributed Hash Table (DHT)-based* routing systems [38,39]. (3) It is easy to build redundancy in the reputation system. In addition the potential failings are as follows: (1) Users may lie about the feedback they provide. (2) Users may not bother to give feedback. (3) Untrustworthy users may mask their behavior or retaliate against negative feedback by sending negative feedback about other users. (4) Users may try to game the system by building up their reputation by acting honestly over several small transactions followed by cheating in a large transaction in a process known as *milking*. (5) Users may form malicious groups that give false positive ratings to each other to

---

* In systems using distributed hash tables or DHTs, objects are associated with a key (usually produced by hashing the object ID such as filename), and each node in the system is responsible for all objects associated with a key that falls in a certain range.

boost each others reputations. (6) Users may reenter the system with a new identity to conceal their past behavior. Hence, a good reputation system must try to overcome these potential failings.

## 35.5 Analysis of Reputation System Components

### 35.5.1 Direct versus Indirect Evidence

Two distinct types of evidence are usually combined by reputation systems to form the reputation of a user. These are (1) *direct evidence*, which consists of a user's first-hand experiences of another user's behavior and (2) *indirect evidence*, which is second-hand information that is received from other nodes about their own experiences. If only first-hand information were used to decide the reputation of another peer, the sparseness of feedback would be a problem because in large virtual communities, the interaction matrix is usually very sparse. A user would not be able to make a trust judgment on another user with whom they have never interacted before. Hence, users must rely on indirect evidence to compute the reputation of users that are new to them.

In a centralized system such as eBay all indirect evidence is collected at the central trust database and is made available to other users. In a decentralized system, indirect evidence can be shared in a number of ways. The interested user may ask for indirect evidence from its neighbors or other users it trusts as in [6]. The indirect evidence may also be propagated to all users in the network using a recursive mechanism like [7]. *Designated Agents*[*] may also be chosen from within the community to store indirect evidence that can then be furnished to requesting users [5,8,9]. In the latter schemes, designated agents responsible for specific users in the system are chosen using distributed hash tables. Multiple such agents are chosen to ensure redundancy in case an agent leaves the network or tries to falsify this information.

### 35.5.2 Second-Order Reputation

Using second-hand information in a decentralized system leads to another problem. How can a user know that the provider of the second-hand information is telling the truth? We can think of an individual's reputation for providing accurate reputation information on other as their second-order reputation information. Second-order reputation is often termed as *credibility* as it measures the truthfulness of an individual as a provider of information.

The problem was first addressed by Aberer and Despotovic [5] who were among the first to use a decentralized reputation storage system. They use the trustworthiness of an agent (first-order reputation) to decide whether to take its feedback into account. The feedback from agents who are deemed trustworthy are included, and that from untrustworthy agents is excluded. Kamvar et al. [7] also use the same strategy of using first-order reputation as second-order reputation as well. However, although it is reasonable to assume that an individual who cheats in the main market cannot be relied upon for accurate feedback, the reverse is not necessarily true. An individual can act honestly in the main market and enjoy a high reputation and at the same time provide false feedback to lower the reputation of others in the community who are after all his/her competitors. An example in which such a strategy would be advantageous is a hotel rating system where a hotel may provide very good service and thus have a high reputation but at the same time may give false feedback about its competitors to lower their reputation. Hence, its credibility is very different from its reputation.

The solution is to recognize credibility as different from first-order reputation and compute it separately. Credibility can be computed in several different ways. Credibility values can be expressed explicitly and solicited separately from reputation values. However, this leads to a problem of endless recursion. To ensure that users do not lie about other users credibility, we need third-order reputation to measure an individual's reputation for providing accurate credibility information and so on.

---

[*] The term *Designated Agents* was coined by the authors and includes all systems in which one or more users are made responsible for storing and sharing another user's reputation information by the system.

Credibility can also be computed implicitly. This can be done in two ways. In the first method, inspired by collaborative filtering, the similarity of user $j$'s opinions to those of user $i$ are used to compute the credibility of user $j$ in the eyes of user $i$ ($C_{ij}$). This can be done using Pearson's correlation coefficients as used by Resnick et al. [25] or by using vector similarity like in Breese et al. [26]. PeerTrust [8] uses yet another similarity metric that resembles the Pearson coefficient. In this method, if user $i$'s opinions are often at variance with those of user $j$, $i$ will have a lower credibility value for $j$ and vice versa.

The second approach for implicit credibility computation measures the credibility of a user by taking into account the agreement between the feedback furnished by the user and the views of the community as a whole. The community average view can be represented by the reputation value computed from the feedback from all the reporting users. If a user gives wrong feedback about other users, that is his or her feedback is very different from the eventual reputation value for computed, its credibility rating is decreased and its subsequent reports have a reduced impact on the reputation of another user. This method of computing credibility has an advantage in that it is more scalable, and it can operate in decentralized systems in which a complete record of all opinions expressed by other individuals may not be available due to privacy concerns. However, there is a danger of "group-think" in such systems. There is strong encouragement for an individual to agree with the opinion of the group as a whole as disagreements are punished by lowering the credibility of the individual who disagrees.

### 35.5.3 Motivation for Providing Feedback

A closely related issue is how to motivate community members to provide feedback to others when providing feedback consumes their own resources. In some respects, this problem is the same as that of motivating cooperation between community members as discussed earlier. However, there are some important differences. In a designated agent system, designated agents must use their own resources to store, compute, and report reputation values for other members. Most literature assumes that agents will perform this task because they are "altruistic" or because they too will benefit from a designated agent system. However, in a large network, there is little incentive for a particular individual to expend resources to maintain the reputation system. How do we prevent such an individual from free-riding the reputation system itself? Moreover, from a game theoretic perspective, not reporting feedback may be advantageous to an agent in a competitive situation. By reporting feedback to other agents, it is sharing information with them which if kept to itself may have given it an advantage.

One strategy to encourage truthful reputation reports is to create a side-market for reputation where accurate reputation reports are rewarded with currency that can be traded for accurate reports from others [40]. Such a system needs to be structured in such a way that providing more feedback and honest feedback results in more credit. However, this solution suffers from the above-mentioned recursion problem as third-order market would need to be created and so on.

An alternative way to avoid the recursion problem is to incorporate the credibility of an individual in the reputation system as a separate variable as discussed the Section 35.5.2. If the credibility is computed through direct evidence only and is not shared with others as in [9], the recursion problem can be avoided.

### 35.5.4 Motivation Is Not a Problem: Dissenting Views

A number of authors do not agree that motivation (of cooperation and of sharing feedback) is a problem at all. Fehr and Gächter [13] develop a theory of "altruistic punishment." They designed an experiment that excluded all explanations for cooperation other than that of altruistic punishment. They designed a McCabe style investment game [41] where players could punish their partners if they wished. However, punishing was costly both to the punisher (1 point) and the punished (3 points). Punishment induced cooperation from potential noncooperators, thus increasing the benefit to the group as a whole even though it was costly to the punisher. For this reason, punishment was an altruistic act. They concluded that negative emotions are the cause of altruistic punishment and that there is a tendency in humans to punish those that deviate from social norms.

Similarly, Gintis et al. [12] argue that behaving in an altruistic fashion sends a signal to other participants that interacting with that individual is likely to prove beneficial. This argument cuts at the heart of the game theoretic notion of how rational agents operate. Rational agent behavior now becomes probabilistic in that an agent may act in an altruistic fashion in the hope of future reward instead of only interacting when an immediate benefit is expected. This interpretation also depends on whether the individuals in the system are humans or are automated agents that do not share the "altruistic" characteristics and behave in a strictly rational sense.

This has been used as evidence that altruism serves an agent's self-interest. It also explains why greedy strategies are outperformed by more altruistic strategies in Axelrod's experiment. Hence, there is some theoretical basis to the claim that a virtual community can be be self-correcting and will exclude bad participants.

## 35.5.5 Other Design Issues

A number of other design choices must be made in a reputation system.

*Reputation Context*: There has been some research on whether reputation is contextual. It is often assumed that reputation is heavily dependent on context. This point of view was aptly expressed by Mui et al. [42]

> Reputation is clearly a context-dependent quantity. For example, one's reputation as a computer scientist should have no influence on his or her reputation as cook.

However, in a contrary argument, Gintis et al. [12] suggest that compartmentalizing reputation too strictly can have a negative effect. Their contention is that altruistic behavior is motivated in part by a desire to signal that interacting with the individual in question will be beneficial in other contexts as well. They argue that the notion of reputation in a real world is far more fuzzy and incorporates generosity as well. A generous participant is more likely to be honest as well. This gives participants an incentive to behave in an altruistic fashion in addition to behaving in an honest fashion.

*Transaction value*: A reputation system must also be able to distinguish between small and large transactions. Buying a pencil online is not the same as buying a car. If all transactions are rated equally, an individual may exploit the system through a strategy called *milking* where they act honestly in a number of small transactions to build their reputation and then cheat in a large transaction without hurting their reputation too much. In PeerTrust [8], this problem is solved by incorporating a transaction context factor that weighs the feedback according to size, importance, and the recency of the transaction.

*Interpreting reputation*: Once the reputation of a user (or object in case of recommendation systems) has been computed, it can be used in several ways depending on the application context. In a file-sharing system [6], a peer may choose the peer with the highest reputation to download the file from. On Amazon, the recommendation system may prompt one to buy a book that one was not aware of before. GroupLens [25] helps decide which movie a user decides to watch.

Equally common are applications that demand a Boolean yes/no decision on "Should $i$ trust $j$?" There are several methods by which this translation from an arbitrary range of reputation values to a binary value can be achieved. These include (1) using a deterministic threshold (I will trust you if your reputation value exceeds 6 on a scale of 10), (2) relative ranking (I will trust you if your reputation is in the top 10% of community members), (3) probabilistic thresholds (the probability that I trust you is a monotonic function on the range of possible trust values), and (4) majority rounding (I will trust you if I trust a majority of people with reputation values close to your reputation).

Another approach [19,43] is to include information that allows a user to decide how much faith it should place in the reputation value. On eBay, this information is the number of feedbacks a user has received. A user with a high reputation and a large number of feedbacks is much more trustworthy than one with a low number of feedbacks.

*Benefits of high reputation*: Many reputation systems particularly those proposed for file-sharing applications [5–7] do not consider the consequences of a peer having a high reputation. In file-sharing systems, the most reputable peer in the network will be swamped with requests as all peers are going to want to download the resources from it. Hence, peers with high reputations are "punished" for their high reputations by having to expend more of their resources to serve others instead of being benefited from their reputation. In this scenario, the interests of the system are not aligned with that of individuals, and the individual peers have no motivation to act honestly and thus increase their reputation.

To motivate individuals to try and acquire high reputations, there needs to be a mechanism by which nodes that have a high reputation are rewarded. In a file-sharing application, this could be achieved through preferential access for higher reputation nodes to resources at other nodes.

*Positive versus negative feedback*: A reputation system may be based solely on either positive feedback, negative feedback, or a combination of both. The disadvantage of a negative feedback only system [5] is that each new entrant into the system has the highest possible reputation. Hence, a misbehaving individual may create a new identity and shed their bad reputations to start afresh. Using only positive feedback, on the other hand, makes it hard to distinguish between a new user and a dishonest user. If old users choose not to interact with any users without a minimum level of positive feedback, a new user may thus find itself frozen out of the group and interacting only with malicious users.

*Identities*: A reputation system that allows unlimited creation of new identities is vulnerable to manipulation [4]. Hence, there must be a cost of a new identities. However, at the same time, this cost must not be so large as to discourage newcomers from joining. Friedman and Resnick argue that in social situations, there is inevitably some form of initiation dues [44]. They also find that, although these dues are inefficient, especially when there are many newcomers to a community, no other strategy does substantially better in terms of group utility.

## 35.6 Some Reputation Systems

We now look at some reputation-management algorithms that have been proposed in recent years.

### 35.6.1 Complaints-Based Trust

One of the first reputation management algorithms for the P2P systems was proposed by Aberer and Despotovic in [5]. This system is based solely on negative feedback given by peers when they are not satisfied by the files received from another peer. The system works on the assumption that a low probability of cheating in a community makes it harder to hide malicious behavior.

Let $P$ denote the set of all peers in the network and $B$ be the behavioral data consisting of trust observations $t(q,p)$ made by a peer $q \in P$ when it interacts with a peer $p \in P$. We can assess the behavioral data of a specific peer $p$ based on the set

$$B(p) = \{t(p,q) \text{ or } t(q,p) \mid q \in P\} \qquad (35.5)$$

In this manner, the behavior of a peer takes into account not only all reports made *about* $p$ but also all reports made *by* $p$. In a decentralized system a peer $q$ does not have access to the global data $B(p)$ and $B$. Hence it relies on direct evidence as well as indirect evidence from a limited number of witnesses $r \in W_q \subset P$:

$$B_q(p) = \{t(q,p) \mid t(q,p) \in B\} \tag{35.6}$$

$$W_q(p) = \{t(r,p) \mid t(r,p) \in B \land r \in P\} \tag{35.7}$$

However, the witness $r$ itself may be malicious and give false evidence. Aberer and Despotovic assume that peers only lie to cover their own bad behavior. If peer $p$ is malicious and cheats peer $q$, $q$ will file a complaint against it. If $p$ also files a complaint against $q$ at the same time, it could be difficult to find out who is the malicious peer. However, if $p$ keeps acting maliciously it will become easy to detect it as there will be a lot of complaints filed from peer $r$ about a set of good peers and a lot of complaints filed from these good peers all about peer $p$. Based on this, the reputation of $p$ can be calculated as

$$T(p) = \left|\{c(p,q) \mid q \in P\}\right| \times \left|\{c(q,p) \mid q \in P\}\right| \tag{35.8}$$

Aberer and Despotovic proposed a decentralized storage system called *P-Grid* to store reputation information. Each peer $p$ can file a complaint about another peer $q$ at any time as follows:

$$insert(a_i, c(p,q), key(p)) \ and \ insert(a_j, c(p,q), key(q))$$

where $a_i$ and $a_j$ are two arbitrary agents. Insertions are made on the keys of both $p$ and $q$ as the system stores complaints both by and about a given peer.

Assuming that an agent is malicious with probability $\pi$ and that an error rate of $\varepsilon$ is tolerable, then a peer $p$ will need to receive $r$ replicas of the same data satisfying $\pi^r < \varepsilon$ to ensure that the error rate is not exceeded. If a simple majority rule is employed, the total number of queries for each trust verification will never exceed $2r + 1$.

A peer $p$ making $s$ queries will obtain a set

$$W = \{(cr_i(q), cf_i(q), f_i, a_i) \mid i = 1, \dots, w\}$$

where $w$ is the number of different witnesses found, $a_i$ is the identifier of the $i$-th witness, $f_i$ is the frequency with which witness $a_i$ is found and $s = \sum_{i=1}^{w} f_i$. $cr_i(q)$ and $cf_i(q)$ are the complaints about $q$ and filed by $q$, respectively, as reported by witness $a_i$. Different witnesses are found with different frequencies, and the ones which are found less frequently have probably been found less frequently also when complaints were filed. So it is necessary to normalize $cr_i(q)$ and $cf_i(q)$ using frequency $f_i$ in this way:

$$cr_i^{norm}(q) = cr_i(q)\left(1 - \left(\frac{s - f_i}{s}\right)^s\right) \tag{35.9}$$

$$cf_i^{norm}(q) = cf_i(q)\left(1 - \left(\frac{s - f_i}{s}\right)^s\right) \tag{35.10}$$

Each peer $p$ can keep statistics on the average number of complaints filed ($cf_p^{avg}$) and received ($cr_p^{avg}$) and can determine if a peer $q$ is trustworthy basing on the information returned from an agent $i$ using this algorithm:

**Algorithm (Trust Assessment Using Complaints)**
$decide(cr_i^{norm}(q), cf_i^{norm}(q)) =$
**if**

$$cr_i^{norm}(q)cf_i^{norm}(q) \leq \left(\frac{1}{2} + \frac{4}{\sqrt{cr_p^{avg}cf_p^{avg}}}\right)^2 cr_p^{avg} cf_p^{avg}$$

**then return** 1; **else return** -1.

This algorithm assumes that if the total number of complaints received exceeds the average number of complaints by a large amount, the agent must be malicious.

## 35.6.2 EigenTrust

Kamvar et al. [7] presented a distributed algorithm for the computation of the trust values of all peers in the network. Their algorithm is inspired by the PageRank algorithm used by Google and assumes that trust is transitive. A user weighs the trust ratings it receives from other users by the trust it places in the reporting users themselves. Global trust values are then computed in a distributed fashion by updating the trust vector at each peer using the trust vectors of neighboring peers. They show that trust values asymptotically approach the eigenvalue of the trust matrix, conditional on the presence of pretrusted users that are always trusted.

A peer $i$ may rate each transaction with peer $j$ as positive ($tr(i,j) = 1$) or negative ($tr(i,j) = -1$). The local trust value at $i$ for $j$ can then be computed by summing the ratings of individual transactions:

$$s_{ij} = \sum tr(i,j) \tag{35.11}$$

Local trust values are normalized as follows:

$$c_{ij} = \frac{\max(s_{ij}, 0)}{\sum_l \max(s_{il}, 0)} \tag{35.12}$$

to keep them between 0 and 1.

To aggregate normalized local trust values, trust values furnished by acquaintances are weighted by the trust a peer has in the acquaintances themselves:

$$t_{ik} = \sum_j c_{ij}c_{jk} \qquad \left(\text{note that } \textstyle\sum_j t_{ij} = 1\right) \tag{35.13}$$

where $t_{ik}$ is the trust a peer $i$ places in peer $k$ based on the opinion of its acquaintances.

If we define $\vec{t_i}$ to be the vector containing the values $t_{ik}$ and $C$ to be the matrix containing the values $c_{ij}$, we get $\vec{t_i} = C^T \vec{c_i}$. The preceding expression only takes into account opinions of a peer's acquaintances. To get a wider view, a peer may ask its friends' friends and so on:

$$\vec{t} = (C^T)^n \vec{c_i} \tag{35.14}$$

Kamvar et al. show that if $n$ is large, peer $i$ can have a complete view of the network and the trust vector $\vec{t_i}$ will converge to the same vector for every peer $i$ (the **left principal eigenvector of** $C$), conditional on $C$ being irreducible and aperiodic.

They further add three practical issues to this simple algorithm. If there are peers that can be trusted a priori, the algorithm can be modified to take advantage of this. They define $p_i$ as $\frac{1}{|P|}$ (where $P$ is the set containing the pre-trusted peers) if $i$ is a pretrusted peer, and 0 otherwise. In presence of malicious peers, using an initial trust vector of $\vec{p}$ instead of if replace $\vec{e}$ generally ensures faster convergence. If a peer $i$ has never had any interaction with other peers, instead of being left undefined, $c_{ij}$ can be defined as

$$c_{ij} = \begin{cases} \frac{\max(local_{ij}, 0)}{\sum_l \max(local_{il}, 0)} & \text{if } \sum_l \max(local_{il}, 0) \neq 0 \\ p_j & \text{otherwise} \end{cases} \tag{35.15}$$

To prevent malicious collectives from subverting the system, Kamvar et al. further modify the trust vector to

$$\vec{t}^{k+1} = (1-a)C^T \vec{t}^k + a\,\vec{p} \tag{35.16}$$

where $a$ is some constant less than 1. This ensures that at each iteration, some of the trust must be placed in the set of pretrusted peers, thus reducing the impact of the malicious collective.

Hence, given $A_i$ the set of peers that have downloaded files from peer $i$ and $B_i$, the set of peers from which peer $i$ has downloaded files, each peer $i$ executes the following algorithm:

**Algorithm (Distributed EigenTrust)**
Query all peers $j \in A_i$ for $c_{ji}t^{(0)} = c_{ji}p_j$;
**repeat**
$$t_i^{(k+1)} = (1 - a)\left(c_{1i}t_1^{(k)} + c_{2i}t_2^{(k)} + \cdots + c_{ni}t_n^{(k)}\right) + ap_i;$$
send $c_{ij}t_i^{(k+1)}$ to all peers $j \in B_i$;
wait for $c_{ji}t_j^{(k+1)}$ from all peers $j \in A_i$;
$\delta = \left| t^{(k+1)} - t^{(k)} \right|$;

**until** $\delta < \varepsilon$;

Each peer thus obtains the same global trust value matrix for all other peers in the network. Kamvar et al. further describe a DHT-based solution to anonymously store multiple copies of the trust value for a given peer at several *score managers*. This eliminates the problem caused by malicious peers reporting false trust values for themselves to other peers to subvert the system.

### 35.6.3 PeerTrust

In PeerTrust [8], Xiong and Liu define five factors used to compute the trustworthiness of a peer. These are (1) feedback obtained from other peers, (2) scope of feedback such as number of transactions, (3) credibility of feedback source, (4) transaction context factor to differentiate between mission-critical and noncritical transactions, and (5) community context factor for addressing community-related characteristics and vulnerabilities.

Given a recent time window, let $I(u, v)$ denote the total number of transactions between peer $u$ and $v$, $I(u)$ denote the total number of transactions of peer $u$, $p(u, i)$ denote peer $u$'s partner in its $i^{th}$ transaction, $S(u, i)$ denote the normalized amount of satisfaction peer $p(u, i)$ receives from $u$ in this transaction, $Cr(v)$ denote the credibility of peer $v$, $TF(u, i)$ denote the adaptive transaction context factor for peer $u$'s $i$th transaction, and $CF(u)$ denote $u$'s community context factor. Then, the trust value of peer $u$ denoted by T(u) is

$$T(u) = \alpha * \sum_{i=1}^{I(u)} S(u, i) * Cr(p(u, i) * TF(u, i) + \beta * CF(u) \tag{35.17}$$

where $\alpha$ and $\beta$ are weight factors.

In their experimental study, Xiong and Liu turn off the transaction context factor and the community context factor and use two credibility metrics. The first is based on the trust value of the reporting peer (similar to EigenTrust), whereas the second is based on the similarity between the reporting peer and the recipient of the trust information. They further propose using a PKI-based scheme and data replication to increase the security and reliability of their system.

### 35.6.4 ROCQ

Garg et al. [9,45] proposed a scheme that combines local opinion, credibility of the reporter, and the quality of feedback to compute the reputation of a peer in the system. In their scheme, direct evidence in the form of local opinion is reported to score managers, which are chosen using distributed hash tables.

The reputation $R_{mj}$ of user $j$ at score manager $m$ is

$$R_{mj} = \frac{\sum_i O_{ij}^{avg} \cdot C_{mi} \cdot Q_{ij}}{\sum_i C_{mi} \cdot Q_{ij}} \tag{35.18}$$

where $C_{mi}$ is the credibility of user $i$ according to user $m$, $O_{ij}^{avg}$ is $i$'s average opinion of $j$ and $Q_{ij}$ is the associated quality value reported by $i$.

The quality value of an opinion depends on the number of transactions on which the opinion is based and the consistency with which the transaction partner has acted. Thus, an opinion is of greater quality when the number of observations on which it is based is larger and when the interactions have been consistent (resulting in a smaller variance). When the number of observations is high but they do not agree with each other, the quality value is lower.

The credibility of a user is based on direct evidence only and is not shared with other users. This prevents the recursion problem of calculating the third-order reputation and so on. The credibility is based upon the agreement of the reported opinion with the group consensus as reflected in the reputation value and is updated after every report received. The precise formula for adjusting the credibility of user $i$ by user $m$ is

$$C_{mi}^{k+1} = \begin{cases} C_{mi}^k + \frac{(1-C_{mi}^k) \cdot Q_{ij}}{2} \cdot (1 - \frac{|R_{mj}-O_{ij}^{avg}|}{s_{mj}}) \\ \qquad \text{if } |R_{mj} - O_{ij}^{avg}| < s_{mj} \\ C_{mi}^k - \frac{C_{mi}^k \cdot Q_{ij}}{2} \cdot (1 - \frac{s_{mj}}{|R_{mj}-O_{ij}^{avg}|}) \\ \qquad \text{if } |R_{mj} - O_{ij}^{avg}| > s_{mj} \end{cases} \tag{35.19}$$

where $C_{mi}^k$ is the credibility of user $i$ after $k$ reports to user $m$, $O_{ij}^{avg}$ is the opinion being currently reported by user $i$, $Q_{ij}$ is the associated quality value, $R_{mj}$ is the aggregated reputation value that user $m$ computed for $j$ and $s_{mj}$ is the standard deviation of all the reported opinions about user $j$. In this way, if a reporting user is malicious, its credibility rating is gradually reduced as its opinion does not match that of the community as a whole.

The authors also propose combining both direct and indirect evidence to create reputation. They proposed a threshold number of interactions between two users below which users will rely on the global reputation of their prospective partner and above which they would rely on firsthand evidence only. This eliminates the problem of sparsity of data while at the same time allowing for reputation to be tailored according to personal experience.

### 35.6.5 Propagation of Trust and Distrust

A number of mathematical approaches to propagating trust [46,47] and distrust [48] have been proposed. In particular, Guha et al. gave a number of models of atomic (single-step) propagation. Let $B$ be a belief matrix whose $ij$th element signifies $i$'s belief in $j$. $B$ can be composed of either the trust matrix ($T_{ij}$ is $i$'s trust in $j$) or both the trust and the distrust ($D_{ij}$ is $i$'s distrust in $j$) matrices (say $B_{ij} = T_{ij} - D_{ij}$). Then, atomic propagation of trust takes place by (1) *direct propagation* (matrix operator:$^*$ $B$) : assumes that trust is *transitive* so if $i$ trusts $j$ and $j$ trusts $k$ then we can infer that $i$ trusts $k$, (2) *co-citation* ($B^T B$): if both $i$ and $j$ trust $k$ and if $i$ trusts $l$, then $j$ also trusts $l$, (3) *transpose trust* ($B^T$): assumes that trust is *reflexive* so that if $i$ trusts $j$ then trusting $j$ should imply trusting $i$, and (4) *trust coupling* ($BB^T$): if $i$ and $j$ trust $k$, then

---

$^*$ The matrix operator when applied to a belief matrix would yield a new matrix indicating inferred trust.

trusting $i$ should also imply trusting $j$. They then go on to propagate trust using a combined matrix that gives weights to the four propagation schemes:

$$C_{B,\alpha} = \alpha_1 B + \alpha_2 B^T B + \alpha_3 B^T + \alpha_4 BB^T \tag{35.20}$$

Thus, applying the atomic propagations a fixed number of times, new beliefs can be computed. In a limited set of experiments, they study the prediction performance of their algorithm on real data and find that the best performance comes with one-step propagation of distrust, whereas trust can be propagated repeatedly.

## 35.7 Conclusions and Future Work

The area of reputation systems remains fertile for future research. Initial research in this area has focused on applying lessons from diverse fields such as evolutionary biology and game theory to computer science. In particular, game-theoretic models such as the iterated prisoner's dilemma were adapted to virtual communities. However, this analysis is limited by not considering the presence of irrational (malicious) players in the community.

Simultaneously, several new reputation schemes have been proposed. These schemes typically propose a new trust model followed by experimental simulation. However, it is difficult to compare the schemes side-by-side as each scheme makes its own assumptions about the interaction model, modes of maliciousness, and levels of collusion among malicious peers, not to mention widely varying experimental parameters such as the number of peers in the system and the proportion of malicious peers.

More recently, there has been some work on analyzing systems in the presence of malicious peers. For instance, Mundinger and Le Boudec [49] analyze the robustness of their reputation system in the presence of liars and try to find the critical phase transition point in which liars start impacting the system.

As the interest in virtual communities, particularly self-organizing communities, grows, we are likely to see a lot more research on the various facets of this topic.

## References

1. Zucker, L., Production of trust: Institutional sources of economic structure 1840–1920, *Research in Organizational Behavior*, 8(1), 53, 1986.
2. Kirsner, S., Catch me if you can, *Fast Company*, http://www.fastcompany.com/magazine/73/kirsner.html, retrieved August 13, 2005.
3. Atif, Y., Building trust in e-commerce, *IEEE Internet Computing*, 6(1), 18–24, 2002.
4. Douceur, J., The Sybil attack, in *Proceedings of the International Workshop on Peer-to-Peer Systems*, 2002.
5. Aberer, K. and Despotovic, Z., Managing trust in a peer-2-peer information system, in *Proceedings of the Tenth International Conference on Information and Knowledge Management*, ACM, New York, 2001, p. 310.
6. Damiani, E., di Vimercati, S.D.C., Paraboschi, S., and Samarati, P., Managing and sharing servents' reputations in p2p systems, *IEEE Transactions on Data and Knowledge Engineering*, 15(4), 840, 2003.
7. Kamvar, S.D., Schlosser, M.T., and Garcia-Molina, H., The eigentrust algorithm for reputation management in P2P networks, in *ACM Proceedings of the International Conference on World Wide Web*, ACM, New York, 2003, p. 640.
8. Xiong, L. and Liu, L., PeerTrust: Supporting reputation-based trust in peer-to-peer communities, *IEEE Transactions on Data and Knowledge Engineering, Special Issue on Peer-to-Peer Based Data Management*, 16(7), 843, 2004.

9. Garg, A., Battiti, R., and Costanzi, G., Dynamic self-management of autonomic systems: The reputation, quality and credibility (RQC) scheme, in *International Workshop on Autonomic Communication*, 2004.

10. Buchegger, S. and Le Boudec, J.-Y., A robust reputation system for P2P and mobile ad-hoc networks, in *Proceedings of the Workshop on the Economics of Peer-to-Peer Systems*, 2004.

11. Trivers, R.L., The evolution of reciprocal altruism, *Quarterly Journal of Biology*, 46, 35, 1971.

12. Gintis, H., Smith, E.A., and Bowles, S., Costly signalling and cooperation, *Journal of Theoretical Biology*, 213, 103, 2001.

13. Fehr, E. and Gächter, S., Altruistic punishment in humans, *Nature*, 415(6868), 137, 2002.

14. Axelrod, R., *The Evolution of Cooperation*, Basic Books, New York, 1984.

15. Jackson, M., *Mechanism Theory*, Encyclopedia of Life Support Systems, Oxford, UK, 2000.

16. Akerlof, G.A., The market for "lemons": Quality uncertainty and the market mechanism, *The Quarterly Journal of Economics*, 84(3), 488, 1970.

17. Morselli, R., Katz, J., and Bhattacharjee, B., A game-theoretic framework for analyzing trust-inference protocols, in *Proceedings of the Workshop on the Economics of Peer-to-Peer Systems*, 2004.

18. Avery, C., Resnick, P., and Zeckhauser, R., The market for evaluations, *American Economic Review*, 89(3), 564, 1999.

19. Resnick, P., Zeckhauser, R., Swanson, J., and Lockwood, K., The value of reputation on eBay: A controlled experiment, in *Proceedings of ESA Conference*, 2002.

20. Kakade, S.M., Kearns, M., Ortiz, L.E., Pemantle, R., and Suri, S., Economic properties of social networks, in *Advances in Neural Information Processing Systems 17*, Saul, L.K., Weiss, Y., and Bottou, L., Eds., MIT Press, Cambridge, MA, 2005.

21. Resnick, P., Zeckhauser, R., Friedman, E., and Kuwabara, K., Reputation systems: Facilitating trust in Internet interactions, *CACM*, 43(12), 45, 2000.

22. Boykin, P.O. and Roychowdhury, V., Personal email networks: An effective anti-spam tool, *IEEE Computer*, 38(4), 61, 2005.

23. Zimmerman, P. and Philip, R., *The Official PGP User's Guide*, MIT Press, Cambridge, MA, 1995.

24. Callas, J., Donnerhacke, L., Finney, H., and Thayer, R., *RFC 2440—OpenPGP Message Format*, Internet Engineering Task Force, Fremont, CA, 1998.

25. Resnick, P., Iacovou, N., Suchak, M., Bergstorm, P., and Riedl, J., GroupLens: An open architecture for collaborative filtering of netnews, in *Proceedings of ACM Conference on Computer Supported Cooperative Work*, 1994, p. 175.

26. Breese, J., Heckerman, D., and Kadie, C., Empirical analysis of predictive algorithms for collaborative filtering, in *Proceedings of the Uncertainty in Artificial Intelligence Conference*, 1998, p. 43.

27. Dellarocas, C., Immunizing online reputation reporting systems against unfair ratings and discriminatory behavior, in *Proceedings of the Conference on Electronic Commerce*, 2000, p. 150.

28. Hardin, G., The tragedy of the commons, *Science*, 162, 1243, 1968.

29. Saroiu, S., Gummadi, P., and Gribble, S., *A Measurement Study of Peer-to-Peer File Sharing Systems*, 2002.

30. Granovetter, M., Economic action and social structure: The problem of embeddedness, *American Journal of Sociology*, 91, 481, 1985.

31. Fudenburg, D. and Tirole, J., *Game Theory*, MIT Press, Cambridge, MA, 1991.

32. Ahlswede, R., Cai, N., Li, S.-Y.R., and Yeung, R.W., Network information flow, *IEEE Transactions on Information Theory*, 46(4), 1204, 2000.

33. Biersack, E., Rodriguez, P., and Felber, P., Performance analysis of peer-to-peer networks for file distribution, in *Proceedings of the Workshop on Quality of Future Internet Services*, 2004.

34. Byers, J.W., Considine, J., Mitzenmacher, M., and Rost, S., Informed content delivery across adaptive overlay networks, *IEEE/ACM Transactions on Networking*, 12(5), 767, 2004.

35. Cohen, B., Incentives build robustness in BitTorrent, in *Workshop on Economics of Peer-to-Peer Systems*, 2003.

36. Zacharia, G., Moukas, A., and Maes, P., Collaborative reputation mechanisms in electronic marketplaces, in *Proceedings of the Hawaii International Conference on System Sciences*, 8, 1999, p. 8026.

37. Dellarocas, C., Mechanisms for coping with unfair ratings and discriminatory behavior in online reputation reporting systems, in *Proceedings of the International Conference on Information Systems*, 2000, p. 520.

38. Ratnasamy, S., Francis, P., Handley, M., Karp, R., and Shenker, S., A scalable content addressable network, in *Proceedings of the SIGCOMM Conference*, 2001, p. 161.

39. Stoica, I., Morris, R., Karger, D., Kaashoek, F., and Balakrishnan, H., Chord: A scalable peer-to-peer lookup service for internet applications, in *Proceedings of the SIGCOMM Conference*, 2001, p. 149.

40. Jurca, R. and Faltings, B., An incentive compatible reputation mechanism, in *Proceedings of the IEEE Conference on E-Commerce*, 2003.

41. Berg, J., Dickhaut, J., and McCabe, K., Trust, reciprocity and social history, *Games and Economic Behavior*, 10, 122, 1995.

42. Mui, L., Mohtashemi, M., and Halberstadt, A., Notions of reputation in multi-agents systems: A Review, in *Proceedings of the International Joint Conference on Autonomous Agents and Multiagent Systems*, ACM, New York, 2002, p. 280.

43. Sabel, M., Garg, A., and Battiti, R., WikiRep: Digital reputations in collaborative applications, in *Proceedings of AICA Congress*, 2005.

44. Friedman, E. and Resnick, P., The social cost of cheap pseudonyms, *Journal of Economics and Management Strategy*, 10(1), 173–199, 2001.

45. Garg, A., Battiti, R., and Cascella, R., Reputation management: Experiments on the robustness of ROCQ, in *Proceedings of the International Symposium on Autonomous Decentralized Systems*, IEEE, 2005, p. 725.

46. Richardson, M., Agarwal, R., and Domingos, P., Trust management for the semantic web, in *Proceedings of the International Semantic Web Conference*, Springer, Berlin, Germany, 2003, p. 351.

47. Golbeck, J., Parsia, B., and Hendler, J., Trust networks on the semantic web, in *Proceedings of Cooperative Intelligent Agents*, 2003.

48. Guha, R., Raghavan, P., Kumar, R., and Tomkins, A., Propagation of trust and distrust, in *Proceedings of WWW 2004*, 2004, p. 403.

49. Mundinger, J. and Le Boudec, J.Y., The impact of liars on reputation in social networks, in *Proceedings of Social Network Analysis: Advances and Empirical Applications Forum*, 2005.

# 36

# Approximation for Influence Maximization

Jing Yuan

Weili Wu

Wen Xu

## 36.1 Introduction

Online social networks (OSNs) (including sites such as FaceBook, LinkedIn, Orkut, and messengers such as Skype) are among the most popular sites and communication tools on the Internet. The users of these sites and tools form huge social networks. These social networks provide a powerful means for sharing, organizing, and finding contents and contacts. Businesses have been harnessing the behavior of their participants and the social structures among participants. OSNs are a very powerful tool for many reasons. By using social networking is not about traffic, it's about influence: our influence on others and leveraging the power of others' influence. Social networking, when we actually take the time to participate on a meaningful level, has tremendous power in helping us influence others. More than just broadcasting our ideas in general advertisement channel, social networking provides us an unique opportunity to influence people in a person-to-person environment.

Research shows that people tend to accept recommendations from people they know and trust, rather than from other advertisement channels such as TV and commercial website. This observation motivates intense research about popularity boosting in viral marketing area. There are many problems based on this observation and raised from various considerations and backgrounds, A typic one is the influence maximization (IM): Given a social network $G$ with a diffusion model $m$ and a positive integer $k$, find $k$ nodes to maximize the total number (or expected total number) of infected nodes.

To explain the IM problem, let us first consider the deterministic diffusion model. In this model, each node has two states, infected and noninfected. If a node is infected, then every neighbor (or out-neighbor in directed network) will get infected. Therefore, after certain time, an infected node $x$ would make all nodes, which are reachable from $x$, infected. The $k$ infected nodes chosen initially are called *seeds*. Thus, the IM problem is to find locations of $k$ seeds to maximize the total number of nodes reachable from seeds. This total number is denoted by $\sigma(A)$ where $A$ is the set of seeds. It is well known that the IM problem in deterministic diffusion model is NP-hard [1] and has a polynomial-time $(1 - e^{-1})$-approximation as $\sigma(A)$ is a monotone nondecreasing and submodular function, and such an approximation exists for the submodular maximization [2]. This approximation is best possible as the IM in deterministic model has no polynomial-time $(1 - e^{-1} + \varepsilon)$-approximation for any $\varepsilon > 0$ unless NP = P [3].

## 36.2 Bharathi–Kempe–Salek Conjecture

There are two important probabilistic diffusion models, the independent cascade (IC) model and the linear threshold (LT) model.

The IC model is a probabilistic diffusion process consisting of discrete steps. Given a directed graph $G = (V, E)$, each edge $(u, v)$ is assigned with a probability $p_{uv}$. It is similar to the IM model that each node has two states, infected and noninfected, and initially every node is noninfected. To start, choose a subset of nodes as seeds and activate the process. At each step, each infected node $u$ intends to infect its noninfected out-neighbors. An out-neighbor $v$ gets infected from $u$ with probability $p_{uv}$. There are two important rules:

- Any infected node $u$ has only one chance to influence a noninfected out-neighbor $v$. That is, if $u$ cannot make $v$ infected in one step, then $u$ cannot make $v$ infected in another step.
- If a noninfected node $v$ receives influence from more than one infected in-neighbors $u_1, \ldots, u_k$ at some step, then $v$ becomes infected with probability $1 - (1 - p_{u_1v}) \cdots (1 - p_{u_kv})$, that is, all events "$u_i$ makes $v$ infected" for $i = 1, \ldots, k$ are independent.

The diffusion process ends if no more new infected node is produced.

The LT model is another probabilistic diffusion process consisting of discrete steps. Consider a directed graph $G = (V, E)$. Each node has two states, infected and noninfected. Each edge $(u, v)$ is assigned with a weight $p_{uv}$ such that for every node $v$, $\sum_{u \in N^-(v)} p_{uv} \leq 1$ where $N^-(v)$ is the set of all in-neighbors of $v$. Initially, every node is noninfected and randomly chooses a threshold from $[0, 1]$ with uniform distribution. To start the process, choose a subset of seeds and activate the influence process. At each step, every noninfected node evaluates the total weight of infected in-neighbors. If this total weight is bigger than or equal to its threshold, then the noninfected node becomes infected; otherwise, it keeps noninfected. The process ends if no more new infected node is produced.

It is proved in References 4, 5 that the LT model is equivalent to a cascade model in which only the second rule is different from the IC model. This cascade model is called the mutually-exclusive cascade (MC) model because the second rule in the IC model is replaced by following:

- If a noninfected node $v$ receives influence from more than one infected in-neighbors $u_1, \ldots, u_k$ at some step, then $v$ becomes infected with probability $p_{u_1v} + \cdots + p_{u_kv}$, that is, all events "$u_i$ makes $v$ infected" for $i = 1, \ldots, k$ are mutually-exclusive.

Although the IC model and the LT model are so close, they are a little different in term of computational property.

If we set $p_{uv} = 1$ for every edge $(u, v)$, then the IC model becomes the deterministic model. Therefore, the IM problem with the IC model is still NP-hard. It is also not hard to show the NP-hardness of the IM problem with the LT model [4,5]. However, if we consider the IM problem in arborescence, then the computational complexity would be different with the IC model and the LT model.

Actually, Bharathi et al. [6] made a conjecture about the IM problem on arborescence. In the introduction of their paper, they state that "the general version of influence maximization is *NP*-complete, and we conjecture it is so even for arborescence directed into a root."

The solution of this conjecture was given by Wang et al. [7] and Lu et al. [8]. They showed that in arborescence, the IM problem is polynomial-time solvable with the LT model and NP-hard with the IC model. Why in the LT model, a polynomial-time solution can be found? This is due to a special property of the MC model.

**Theorem 36.1** *In the MC model, following operation would not change the expected number of infected nodes: At a nonseed node $v$ with in-neighbors $u_1, \ldots, u_k$, break node $v$ into $k$ nodes $v_1, \ldots, v_k$, replace each edge $(u_i, v)$ by an edge $(u_i, v_i)$ with $p_{u_iv_i} = p_{u_i,v}$ and replace edge $(v, w)$ by $k$ edges $(v_i, w)$ for $i = 1, \ldots, k$ with $p_{v_iw} = p_{v,w}$.*

This property enable a polynomial-time dynamic program working well for the IM problem with the MC model.

## 36.3 Kempe–Kleinburg–Tardos Conjecture

Let $\sigma_M(A)$ denote the expected number of infected nodes with the diffusion model $m$ and the seed set $A$. When we treat computing $\sigma_M(A)$ as a black box or oracle, the IM problem is still NP-hard with the IC model and the LT model. However, they would have a polynomail-time $(1 - e^{-1})$-approximation as the expected number of infected nodes is a submodular function with respect to the set of seeds and hence the IM problem is still a submodular maximization.

When Kempe et al. [4,5] studied a generalization of the LT model, they made an interesting conjecture on the submodularity of the expected number of infected nodes as a function of the seed set. This generalized threshold (GT) model is defined as follows.

The GT model is a probabilistic diffusion process consisting of discrete steps. Consider a directed graph $G = (V, E)$. Each node has two states, infected and noninfected. Each node $v$ is assigned a function $f_v$ mapping from subsets of $N^-(v)$ to $[0, 1]$ where $N^-(v)$ is the set of in-neighbors of $v$. Initially, every node $v$ is noninfected and randomly chooses a threshold $\theta_v$ from $[0, 1]$ with uniform distribution. To start the process, choose a subset of seeds and activate the influence process. At each step, every noninfected node evaluates the value of $f_v(U)$ where $U$ is the set of infected in-neighbors of $v$. If $f_v(U) \geq \theta_v$, then $v$ becomes infected; otherwise, $v$ keeps noninfected. The process ends if no more new infected node is produced.

Kempe et al. [4,5] conjectured that if for all $v \in V$, $f_v$ is submodular and monotone nondecreasing, so is $\sigma_{GT}(A)$ where $A$ is a seed set. This conjecture was proved by Mossel and Roch [9] in 2007.

## 36.4 Randomized Approximation

Actually, computing $\sigma_M(A)$ is #P-hard for $m = IC$ [10] and $LT$ [11]. Kempe et al. [4,5] suggested to compute approximation solution of the IM problem using Monte-Carlo method. Their algorithm produce $(1 - e^{-1} - \varepsilon)$-approximation solution with probability $1 - n^{-\ell}$ within very high time. Later, this algorithm is improved [10–12]. The first significant progress was made by Borg et al. [13] whose algorithm runs in time $O(k\ell^2(m + n) \log^2 n/\varepsilon^3)$ where $n$ is the number of nodes, and $m$ is the number of edges in input network $G$. Tang et al. [14] speed up the running time to $O((k + \ell)(m + n) \log n/\varepsilon^2)$. Tang et al. [15] gave the same running time with a Martingale approach.

There are many influence-based optimization problems in the literature, such as active friending [16], community expansion [17], rumor blocking [18], and profit maximization [19]. The IM problem is also studied in other diffusion models [20,21]. Could above-mentioned randomized algorithms be extended to those influence-based optimization problems and those diffusion models? This is a question that contains a lot of research issues.

## References

1. N. Chen. On the approximability of influence in social networks. *The 2008 Annual ACM SIAM Symposium on Discrete Algorithms*, 2008, pp. 1029–1037, ACM, New York.
2. G. Nemhauser, L. Wolsey, and M. Fisher. An analysis of the approximations for maximizing submodular set functions, *Mathematical Programming*, 14, 265–294, 1978.
3. U. Feige. A threshold of $\log n$ for approximation set cover, *Journal of the ACM*, 45(5), 634–652, 1998.
4. D. Kempe, J. Kleinberg, and E. Tardos. Maximizing the spread of infuence through a social network. *The 2003 International Conference on Knowledge Discovery and Data Mining*, 2003, pp. 137–146, ACM, New York.

5. D. Kempe, J. Kleinberg, and E. Tardos. Influential nodes in a diffusion model for social networks. *The 2005 International Colloquium on Automata, Languages and Programming*, 2005, pp. 1127–1138, Springer-Verlag, Berlin, Germany.

6. S. Bharathi, D. Kempe, and M. Salek. Competitive influence maximization in social networks. *WINE*, pp. 306–311, Springer-Verlag, Berlin, Germany.

7. A. Wang, W. Wu, and L. Cui. On Bharathi–Kempe–Salek conjecture on influence maximization in arborescence, *Journal of Combinatorial Optimization*, 31, 1678–1684, 2016.

8. Z. Lu, Z. Zhang, and W. Wu. Solution of Bharathi–Kempe–Salek conjecture on influence maximization in arborescence, *Journal of Combinatorial Optimization*, 33, 803–808, 2017.

9. E. Mossel and S. Roch. On the submodularity of influence in social networks, *STOC*, 2007, pp. 128–134, ACM, New York.

10. W. Chen, C. Wang, and Y. Wang. Scalable influence maximization for prevalent viral marketing in large-scale social networks. *The 2010 ACM SIGKDD Conference on Knowledge Discovery and Data Mining*, 2010.

11. W. Chen, Y. Yuan, and L. Zhang. Scalable influence maximization in social networks under the linear threshold model. *The 2010 International Conference on Data Mining*, 2010.

12. Y. Zhu, W. Wu, Y. Bi, L. Wu, Y. Jiang, and W. Xu. Better approximation algorithms for influence maximization in online social networks, *Journal of Combinatorial Optimization*, 30(1), 97–108, 2015.

13. C. Borgs, M. Brautbar, J. Chayes, and B. Lucier, Maximizing social influence in nearly optimal time, *SODA'14*, 2014, pp. 946–957, Society for Industrial and Applied Mathematics, Philadelphia, PA.

14. Y. Tang, X. Xiao, and Y. Shi. Influence maximization: Near-optimal time complexity meets practical efficiency, *SIGMOD*, 2014, pp. 75–86, ACM, New York.

15. Y. Tang, Y. Shi, and X. Xiao, Influence maximization in near-Linear time: A martingale approach, *SIGMOD*, 2015, pp. 1539–1554, ACM, New York.

16. D.-N. Yang, H.-J. Hung, W.-C. Lee, and W. Chen. Maximizing acceptance probability for active friending in online social networks, *KDD*, 2013, pp. 713–721, ACM, New York.

17. Y. Bi, W. Wu, Y. Zhu, L. Fan, and A. Wang. A nature-inspired influence propagation model for the community expansion problem, *Journal of Combinatorial Optimization*, 2013. doi:10.1007/s10878-013-9686-9.

18. L. Fan, Z. Lu, W. Wu, B. M. Thuraisingham, H. Ma, and Y. Bi. Least cost rumor blocking in social networks, *ICDCS*, 2013, pp. 540–549, IEEE, Washington, DC.

19. Y. Zhu, Z. Lu, Y. Bi, W. Wu, Y. Jiang, and D. Li. Influence and profit: Two sides of the coin, *ICDM*, 2013, pp. 1301–1306, IEEE, Washington, DC.

20. J. Li, Z. Cai, M. Yan, and Y. Li. Using crowdsourced data in location-based social networks to explore influence maximization, *INFOCOM*, 2016, pp. 1–9, IEEE, Washington, DC, USA.

21. G. Tong, W. Wu, S. Tang, and D.-Z. Du. Adaptive influence maximization in dynamic social networks. IEEE/ACM Transactions on Networking (TON), 25, 112–125, 2017.

# 37

# Approximation and Heuristics for Community Detection

Jing Yuan

Weili Wu

Sarat Chandra Varanasi

## 37.1 Introduction

Recent advances in Internet technologies and Web 2.0 applications lead to a diverse range of evolving social networks. Studies have shown that most of these networks exhibit strong modular nature or community structure. Tremendous effort has been devoted to identify communities of interest and study their behavior over time. There are rigorous methods for discovering important community structures at multiple topological and temporal scales. In the current chapter, we attempt to survey the principle methods for community detection and identify current and emerging trends as well as open problems within this dynamic field.

Social networks depict the interactions between individuals and are represented by a graph of interconnected nodes, in which nodes represent individuals and edges represent the connections between the individuals. Community detection, also known as graph clustering, has been extensively studied in the literature. The goal of community detection is to partition vertices in a complex graph into densely-connected components, that is, communities. Extracting community structure and leveraging them to predict the emergent, critical, and causal nature of social networks in a dynamic setting is of growing importance.

Researchers have proposed different definitions of subgroups with various levels of internal consistency between vertices. Fortunato [1] identifies three levels to define a community: local definitions, global definitions and definitions based on the similarity of vertices. In local definitions, a definition of communities consists of parts of the graph with few ties to the rest of the system. In this partition, communities are studied from their inner structure independently of the remaining part of the graph. In global definitions, a global criterion associated with the graph is used to compute communities. This global criterion is dependent on the algorithm implemented to locate communities, either a clustering criterion or a distance-based criterion may be used. In the vertex-similarity-based community definitions, communities are considered as groups of vertices similar to one another.

Although there is no widely accepted unique definition of the term community, the most commonly used definition is that of Yang et al. [2]: a community as a group of network nodes, within which the links connecting nodes are dense but between which they are sparse. This intuitive definition has been formalized in a number of competing ways, usually by way of a quality function, which measures the quality of a partition. A large number of papers have been published with the aim of seeking a good community partition. When it comes to community detection algorithms, it is natural for us to ask *What is the definition of a good partition?* The answer to this question yields an measure function that quantifies the goodness of a given division of the network [3–7]. There is an important line of research focus on seeking appropriate objective functions and solving the community detection problem from the perspective of optimization [4,8,9]. For example, the normalized cut of a group of vertices $S$ is defined by Meila et al. [8] as the sum of weights of the edges that connect $S$ to the rest of the graph $\bar{S}$, normalized by the total edge weight of $S$ and that of $\bar{S}$. Intuitively, groups with low normalized cut make for good communities, as they are well connected amongst themselves, whereas sparsely connected to the rest of the graph. The *conductance* of a division of the graph into a set of clusters is defined as the sum of the normalized cuts of each of the clusters [4,10]. *Modularity* has recently become quite popular as a way to measure the goodness of a clustering of a graph [9]. The intuition behind the definition of modularity is that a subgraph corresponding to each community is denser than the one in a random graph.

In Sections 37.2 and 37.3, we will discuss both traditional methods for community detection proposed in the literature as well as new trends in social network analysis, with pros and cons of the methodologies.

# 37.2 Traditional Methods

## 37.2.1 Graph Partitioning

Graph partitioning methods provide a powerful framework for fast and effective community detection. The main idea is to shrink or coarsen the input graph successively so as to obtain a small graph, partition this small graph, and then successively project this partition back up to the original graph, refining the partition at each step along the way. Two important graph partitioning methods include Graclus [10] and Multi-level Regularized Markov Clustering (MLR-MCL) [11]. These are highly scalable methods that have been used in studies of some of the biggest graph datasets [11,12].

The main components of a multilevel graph partitioning strategy are as follows: (1) Coarsening—The goal is to produce a smaller graph that is similar to the original graph. This step may be applied repeatedly to obtain a graph small enough to be quickly partitioned with high quality. A popular strategy is to first construct a *matching* on the graph, defined as a set of edges no two of which are incident on the same vertex. For each edge in the matching, the vertices at the ends of the edge are merged into a single node in the coarsened graph. Coarsening can be performed fast using simple randomized strategies. Karypis et al. present a heavy-edge coarsening heuristic for which the size of the partition of the coarse graph is within a small factor of the size of the final partition obtained after multilevel refinement [5]. (2) Initial partitioning—In this step, a partitioning of the coarsest graph is performed. As the graph at this stage is small enough, one may use strategies like spectral partitioning which are slow but known to give high quality partitions. (3) Uncoarsening—In this step, the finer connectivity structure of the graph revealed by the uncoarsening is used to refine the partition on the current graph, usually by performing local search. For example, *Metis* [5] uses a variation of the Kernighan–Lin algorithm [13] for refining during uncoarsening, and *Graclus* [10] uses weighted kernel k-means for the refining step. This step is repeated until we derive the original graph.

## 37.2.2 Hierarchical Clustering

Hierarchical clustering techniques aim at identifying groups of vertices with high similarity, and can be classified into two categories: (1) agglomerative algorithms, in which clusters are iteratively merged if their similarity is sufficiently high and (2) divisive algorithms, in which clusters iteratively split by removing edges connecting vertices with low similarity [14]. Commonly used similarity measures include cosine similarity, Jaccard index, and Hamming distance between rows of the adjacency matrix. Newman and Girvan propose a divisive community detection algorithm based on the idea of *edge betweenness* [9]. Its underlying principle calls for removing the edges that connect different communities. Several measures of edge betweenness are computed, in particular the so-called intermediate centrality, whereby edges are selected by estimating the level of edge importance based on these measures. As an illustration, intermediate centrality is defined as the number of shortest paths using the edge under consideration. The steps involved are as follows: (1) compute centrality for all edges, (2) remove edges with the greatest centrality (ties are broken arbitrarily), (3) recalculate centralities on the remaining graph, (4) iterate the process from step (2) until a sufficiently small number of communities are obtained. The advantages of such algorithms lie in their intuitive simplicity but as noted elsewhere [15], they often do not scale well to large networks. Simply computing the betweenness for all edges takes $O(|V||E|)$ time, and entire algorithm requires $O(|V|^3)$ time.

## 37.2.3 Network Modularity Optimization

Modularity optimization is one of the most widely used methods for community detection. Modularity has been introduced to measure the quality of community algorithms. Modularity-based methods are designed based on the idea that a subgraph which is a community is denser than the one in a random graph [9]. Therefore, we can evaluate a community by the edge density difference between the community and a random subgraph, and the most community-like partition can be found by maximizing such difference [16].

Modularity is a benefit function that measures the quality of a particular division of a network into communities. Modularity-based methods rely on modularity function Q to determine optimal number of clusters in the network. Newman et al. [9] proceed with the initial introduction and provide the following formula:

$$Q = \sum_i (e_{ii} - a_i^2) \tag{37.1}$$

where $e_{ij}$ denotes the number of edges having one end in group $i$ and the other end in group $j$. In addition, we have $a_i = \sum_j e_{ij}$ denote the number of edges having one end in group $i$. This modularity function Q measures the fraction of edges in the network that connect vertices of the same type (i.e., intracommunity edges) minus the expected value of the same quantity in a network with the same community divisions yet with random connections between vertices.

The modularity optimization methods detect communities by searching over possible divisions of a network for one or more that have particularly high modularity. As exhaustive search over all possible divisions is usually intractable, practical algorithms are based on approximate optimization methods such as greedy algorithms [17–19], simulated annealing [20,21], or spectral optimization [22,23], with different approaches offering different balances over the tradeoff of speed and accuracy. The algorithm introduced by Girvan et al. [9] and then improved in Reference 16 is based on modularity. The "glutton" type algorithm maximizes modularity by merging communities at each step to get the greatest marginal increase. Only those communities sharing one or more edges are allowed to merge at each step. This method is performed in linear time; however, community quality is less than that of other more costly methods. Further, modularity optimization often fails to detect clusters smaller than some scale, depending on the size of the network [24].

The main benefit of the Louvain algorithm [25] lies in its capacity to operate very quickly on extremely large weighted graphs. This property however does not guarantee an optimal graph partition; an adaptive modularity formula, derived from the initial formula, is used for weighted graphs. Initially, all vertices are placed in different communities. For each node $i$, the algorithm computes the gain in weighted modularity when placing $i$ in the community of its neighbor node $j$ and then chooses the community offering maximal gain. At the end of this first loop, the algorithm yields the first partitioning stage before repeating the same step considering formed communities as new nodes. The algorithm stops once additional increases in modularity are no longer possible. This method has very short processing time thus it has been used to process very large social networks with over 2.6 million customers extracted from phone companies [25].

### 37.2.4 Spectral Clustering

Spectral algorithms estimate the number of communities according to the eigenvalue distribution of the Laplacian matrix [10,26]. Classical data clustering techniques such as k-means algorithm can be applied to derive the final partition of the network [27]. The main idea behind spectral clustering is that the low-dimensional representation induced by the top eigenvectors exposes the cluster structure in the original graph. Major disadvantage of spectral algorithms lies in their computational complexity. Modern implementations for eigenvector computation use iterative algorithms such as the Lanczos algorithm. In practice, spectral clustering is hard to scale up to networks with more than tens of thousands of nodes without employing parallel algorithms.

Spectral methods that target weighted cuts form an important class of algorithms that can be used for community detection and are shown to be effective [26]. Recent advances in this domain have targeted large scale networks (e.g., local spectral clustering). Dhillon et al. showed that normalized cut can be optimized by weighted kernel k-means algorithm, which can cluster graphs at a comparable quality to spectral clustering with affordable computational cost [10].

### 37.2.5 Statistical Inference

Methods based on statistical inference attempt to fit a generative model (e.g., Bayesian inference) to the network data, which encodes the community structure. The overall advantage of this approach is its principled nature and inherent capacity to address issues of statistical significance. Most methods in the literature are based on the stochastic block model [28], mixed membership [29], degree-correction [30], and hierarchical structures [31]. Model selection aims at finding models which are both simple and effective at describing a system. The modular structure of a graph can be considered as a compressed description of the graph to approximate the whole information contained in its adjacency matrix. Model selection can be performed using principled approaches such as minimum description [32] and Bayesian model selection [33].

## 37.3 Emerging Trends

### 37.3.1 Community Detection in Heterogeneous Networks

Most conventional community detection algorithms assumes a homogeneous network in which nodes and edges are of the same type. However, many real-world systems are naturally described as heterogeneous multirelational networks which contain multiple types of nodes and edges [34]. For example, different relationships are defined based upon various communication methods in Reference 35. Such diversity presents both opportunities and challenges, as valuable information can be gained from recognizing the network heterogeneity yet it is not obvious how to handle nodes and edges of different types appropriately.

Cruz et al. integrate structural dimension and compositional dimension and compose an attributed graph to solve the community detection problem [36]. Aggarwal et al. propose to use *local succinctness property* to extract compressed descritpions of the underlying community representation of the social network with a min-hash approach [37]. Liu et al. present a new method for detecting communities based on optimizing the composite modularity, which is a new modularity proposed for evaluating partitions of a heterogeneous multirelational network into communities [38].

## 37.3.2 Overlapping Community Detection

It is well understood that users in a social network are naturally characterized by multiple community memberships. For example, a person usually has connections to several social groups such as family, friends, and colleagues; a researcher may be active in several areas. In online social networks, the number of communities an individual can belong to is essentially unlimited as a person can simultaneously associate with as many groups as she wishes. Recent research shows that the overlap is indeed a significant feature of many real-world social networks [39]. For this reason, there is growing interest in overlapping community detection algorithms that identify a set of clusters that are not necessarily disjoint.

Nonnegative matrix factorization (NMF) is used to assign soft membership of a vertex to a community. The method involves factoring the given feature matrix $V_{m \times n}$ into two matrices $W_{n \times k}$ and $H_{k \times m}$ where $w_{i,j} \in W$ indicates the dependence of vertex $v_i$ to community $j$. Here $n$ denotes the number of vertices and $k$ the number of communities. NMF is nonparametric and not scalable due to high running time complexity of $O(kn^2)$. Psorakis et al. propose the hybrid method called Bayesian NMF [40]. By using the information about the number of interactions between each pair of nodes for NMF, a parameter based inference on a generative model is applied for community detection. The generative model is based on Gaussian Mixture Model and Overlapping Stochastic Block Model (OSBM) model [41] using a multivariate binomial distribution on a community assignment vector. Wang et al. [42] combine disjoint detection methods with local optimization algorithms. A partition is first derived from existing disjoint community detection algorithms. Then, two variances associated with each community with respect to the inclusion and exclusion of each node are calculated. The variances determine the soft membership of every node to each community. Model-based Overlapping Seed ExpanSion (MOSES) [43] uses OSBM with a local optimization scheme in which the fitness function is defined by a set of observed conditional probabilities. MOSES uses greedy approach to expand communities from edges. Unlike OSBM, MOSES is nonparametric in the sense that we do not need to feed connection probabilities. The worst case time complexity is $O(mn^2)$, where $m$ denotes the number of edges and $n$ denotes the number of vertices.

## 37.3.3 Community Detection in Dynamic Networks

Most of the community detection algorithms discussed so far are designed with the implicit assumption that the underlying network is unchanged. In real world, however, the social networks evolve over the time. *How should community discovery algorithms be modified to take into account the dynamic and temporally evolving nature of social networks?* Asur et al. present an event-based approach for understanding the evolution of communities and their members over time [44]. They introduced a structured way to reason about how communities and individuals within such networks evolve over time and what are the critical events that characterize their behavior. They showed that a diffusion model can be efficiently composed from the events detected by the framework and can be used to effectively analyze real-life evolving networks in an incremental fashion.

Lin et al. [45] propose a dynamic community detection scheme *FacetNet* using probabilistic community membership models. The advantage of such probabilistic models is the ability to assign each individuals to multiple communities with a weight indicating the degree of membership for each community. Kim et al. [46] revisit the cost functions used in existing research and show that temporal smoothing at the clustering level can degrade the performance due to iterative adjustment. Their remedy is to push down the cost to each pair of nodes, leading to a temporal-smoothed version of pair-wise node distance function, and then density-based clustering is conducted on this new metric. This model can account for the arbitrary creation/dissolution as well as growing/shrinking of a community over time. Wang et al. [47,48] propose a novel algorithm, *Noise-tolerance community detection*, to discover dynamic community structure that is based on historical information and current information. An updated algorithm is introduced to help find the community structure snapshot at each time step. One evaluation method based on structure and connection degree is proposed to measure the community similarity. Based on this evaluation, the latent community evolution can be tracked and abnormal events can be detected.

## 37.3.4 Community Detection with Edge Content

Most community detection algorithms use the links between the nodes to determine the communites in the graph. Such methods are typically based purely on the linkage structure of the underlying social network. However, in many recent applications, edge content is available to provide better supervision to the community detection process [49,50]. Many natural representations of edges in social interactions such as shared images and videos, user tags and comments are naturally associated with content on the edges. Although some work has been done on utilizing node content for community detection, the presence of edge content presents unprecedented opportunities and flexibility for the community detection process. Recent research shows that such edge content can be leveraged to greatly improve the effectiveness of the community detection process in social networks. Pathak et al. present the *community-author-recipient-topic* model in email communication networks, emphasizing how topics and relationships jointly affect community structure. In this model, same users in different conversations can be assigned to different communities [49]. Qi et al. propose an algorithm for community detection with edge content [50]. They show that edge content provides unique insights into communities because it characterizes the nature of the interactions between individuals more effectively. Further, the information available only at the nodes may not be able to easily distinguish the different interactions of nodes that belong to multiple communities. The use of edge content enables richer insights which can be used for more effective community detection.

## 37.3.5 Terminal-Set-Enhanced Community Detection

According to Malliaros et al. [51], most of the existing objective functions for community detection are too complicated to obtain an approximation algorithm, and the state-of-art approaches cannot provide provable performance guarantees when partition the network into three or more communities in general settings. Tong et al. [52] propose that if we devote our attention to the initial min-cut intuition there is a very simple objective function, that is, the number of the cut-edges. With this objective function, the problem becomes the classic minimum cut problem which has been well studied by the computation theory community. However, the approach solely based on the minimum cut problem is problematic in the sense that the produced communities are not always desirable. Their approach is motivated by the observation that although most of the graph partitioning problems are NP-hard in general case, we can find effective approximations if one of the terminal sets of the optimal partition is known to us.

The authors investigate the community detection problem based on the concept of terminal set [52]. A terminal set is a group of users within which any two users belong to different communities. Although the community detection is hard in general, the terminal set can be very helpful in designing effective community detection algorithms. They first present a 2-approximation algorithm running in polynomial time for the original community detection problem. Then, to better support real applications they further consider the case when extra restrictions are imposed on feasible partitions. For customized community detection problems, they provide two randomized algorithms which are able to find the optimal partition with a high probability. Demonstrated by the experiments performed on benchmark networks the proposed algorithms are able to produce high-quality communities.

# 37.4 Discussion

The field of community detection in social networks is still emerging, with a number of theoretical and empirical problems, and covers a gamut of core research areas both within computer science and across disciplines. In the current chapter, we first examined the basic concepts for community detection. Then we took a look at the core methods developed to extract community structure from large networks, from traditional ideas to the state-of-the-art approaches. We also discussed emerging trends of research that is gaining traction in this field. In the following we briefly highlight some open problems that are likely to play a significant role as we move forward within this field. Worth note that this is by no means a comprehensive list of cross-cutting problems but a highlight of some of the key challenges and opportunities within the field.

*Open problems*: One possible research direction is to design scalable algorithms for big data applications. With the size and scale of evolving social networks, researchers are increasingly turning to scalable, parallel and distributed algorithms for community detection. From algorithmic perspective, multilevel algorithms relying on graph coarsening and refinement offer a potential. For example, given recent trend towards cloud computing, researchers are beginning to investigate algorithms for community detection on distributed big data processing platforms such as Hadoop. Another promising research direction is the visualization of communities and their evolution. Visualizing large complex networks and honing in on important topological characteristics is a grand challenge. It is due to the unaffordable computation cost incurred when we attempt to characterize the behavior of networks with billions of nodes. This important area, particularly in the context of community detection in social networks, has seen limited research thus far.

# References

1. S. Fortunato. Community detection in graphs. *Physics Reports*, 486(3):75–174, 2010.
2. B. Yang, D. Liu, and J. Liu. Discovering communities from social networks: Methodologies and applications. In *Handbook of Social Network Technologies and Applications*, pp. 331–346. Springer, Berlin, Germany, 2010.
3. G. W. Flake, S. Lawrence, and C. L. Giles. Efficient identification of web communities. In *Proceedings of the 6th ACM SIGKDD International Conference on Knowledge Discovery and Data Mining*, pp. 150–160. ACM, 2000.
4. R. Kannan, S. Vempala, and A. Vetta. On clusterings: Good, bad and spectral. *Journal of the ACM (JACM)*, 51(3):497–515, 2004.
5. G. Karypis and V. Kumar. A fast and high quality multilevel scheme for partitioning irregular graphs. *SIAM Journal on scientific Computing*, 20(1):359–392, 1998.
6. J. Leskovec, K. J. Lang, and M. Mahoney. Empirical comparison of algorithms for network community detection. In *Proceedings of the 19th International Conference on World wide web*, pp. 631–640. ACM, New York, 2010.
7. F. Radicchi, C. Castellano, F. Cecconi, V. Loreto, and D. Parisi. Defining and identifying communities in networks. *Proceedings of the National Academy of Sciences of the United States of America*, 101(9):2658–2663, 2004.
8. M. Meila and J. Shi, Random walks view of spectral segmentation. *Proceedings of the 2001 International Workshop on Artificial Intelligence and Statistics*, Professional Book Center, Denver, CO.
9. M. E. Newman and M. Girvan. Finding and evaluating community structure in networks. *Physical Review E*, 69(2):026113, 2004.
10. I. S. Dhillon, Y. Guan, and B. Kulis. Weighted graph cuts without eigenvectors a multilevel approach. *IEEE Transactions on Pattern Analysis and Machine Intelligence*, 29(11):1944–1957, 2007.

11. V. Satuluri and S. Parthasarathy. Scalable graph clustering using stochastic flows: Applications to community discovery. In *Proceedings of the 15th ACM SIGKDD International Conference on Knowledge Discovery and Data Mining*, pp. 737–746. ACM, New York, 2009.

12. J. Leskovec, K. J. Lang, A. Dasgupta, and M. W. Mahoney. Statistical properties of community structure in large social and information networks. In *Proceedings of the 17th International Conference on World Wide Web*, pp. 695–704. ACM, New York, 2008.

13. B. W. Kernighan and S. Lin. An efficient heuristic procedure for partitioning graphs. *Bell System Technical Journal*, 49(2):291–307, 1970.

14. K. Nordhausen. The elements of statistical learning: data mining, inference, and prediction, by trevor hastie, robert tibshirani, jerome friedman. *International Statistical Review*, 77(3):482–482, 2009.

15. D. Chakrabarti and C. Faloutsos. Graph mining: Laws, generators, and algorithms. *ACM Computing Surveys (CSUR)*, 38(1):2, 2006.

16. A. Clauset, M. E. Newman, and C. Moore. Finding community structure in very large networks. *Physical Review E*, 70(6):066111, 2004.

17. L. Danon, A. Díaz-Guilera, and A. Arenas. The effect of size heterogeneity on community identification in complex networks. *Journal of Statistical Mechanics: Theory and Experiment*, 2006(11):P11010, 2006.

18. A. Noack and R. Rotta. Multi-level algorithms for modularity clustering. In *International Symposium on Experimental Algorithms*, pp. 257–268. Springer, 2009.

19. K. Wakita and T. Tsurumi. Finding community structure in mega-scale social networks:[extended abstract]. In *Proceedings of the 16th International Conference on World Wide Web*, pp. 1275–1276. ACM, New York, 2007.

20. R. Guimera and L. A. N. Amaral. Functional cartography of complex metabolic networks. *Nature*, 433(7028):895–900, 2005.

21. C. P. Massen and J. P. Doye. Identifying communities within energy landscapes. *Physical Review E*, 71(4):046101, 2005.

22. M. E. Newman. Modularity and community structure in networks. *Proceedings of the National Academy of Sciences*, 103(23):8577–8582, 2006.

23. T. Richardson, P. J. Mucha, and M. A. Porter. Spectral tripartitioning of networks. *Physical Review E*, 80(3):036111, 2009.

24. S. Fortunato and M. Barthelemy. Resolution limit in community detection. *Proceedings of the National Academy of Sciences*, 104(1):36–41, 2007.

25. V. D. Blondel, J.-L. Guillaume, R. Lambiotte, and E. Lefebvre. Fast unfolding of communities in large networks. *Journal of Statistical Mechanics: Theory and Experiment*, 2008(10):P10008, 2008.

26. J. Shi and J. Malik. Normalized cuts and image segmentation. *IEEE Transactions on Pattern Analysis and Machine Intelligence*, 22(8):888–905, 2000.

27. U. Von Luxburg. A tutorial on spectral clustering. *Statistics and Computing*, 17(4):395–416, 2007.

28. J. Reichardt and D. R. White. Role models for complex networks. *The European Physical Journal B*, 60(2):217–224, 2007.

29. B. Ball, B. Karrer, and M. E. Newman. Efficient and principled method for detecting communities in networks. *Physical Review E*, 84(3):036103, 2011.

30. B. Karrer and M. E. Newman. Stochastic blockmodels and community structure in networks. *Physical Review E*, 83(1):016107, 2011.

31. T. P. Peixoto. Hierarchical block structures and high-resolution model selection in large networks. *Physical Review X*, 4(1):011047, 2014.

32. M. De Domenico, A. Lancichinetti, A. Arenas, and M. Rosvall. Identifying modular flows on multi-layer networks reveals highly overlapping organization in interconnected systems. *Physical Review X*, 5(1):011027, 2015.

33. X. Yan, C. Shalizi, J. E. Jensen, F. Krzakala, C. Moore, L. Zdeborová, P. Zhang, and Y. Zhu. Model selection for degree-corrected block models. *Journal of Statistical Mechanics: Theory and Experiment*, 2014(5):P05007, 2014.

34. Y. Sun and J. Han. Mining heterogeneous information networks: A structural analysis approach. *ACM SIGKDD Explorations Newsletter*, 14(2):20–28, 2013.

35. I. Guy, M. Jacovi, E. Shahar, N. Meshulam, V. Soroka, and S. Farrell. Harvesting with sonar: The value of aggregating social network information. In *Proceedings of the SIGCHI Conference on Human Factors in Computing Systems*, pp. 1017–1026. ACM, New York, 2008.

36. J. D. Cruz, C. Bothorel, and F. Poulet. Integrating heterogeneous information within a social network for detecting communities. In *Proceedings of the 2013 IEEE/ACM International Conference on Advances in Social Networks Analysis and Mining*, pp. 1453–1454. ACM, New York, 2013.

37. C. C. Aggarwal, Y. Xie, and S. Y. Philip. Towards community detection in locally heterogeneous networks. In *SDM*, pp. 391–402. SIAM, Philadelphia, PA, 2011.

38. X. Liu, W. Liu, T. Murata, and K. Wakita. A framework for community detection in heterogeneous multi-relational networks. *Advances in Complex Systems*, 17(06):1450018, 2014.

39. F. Reid, A. McDaid, and N. Hurley. Partitioning breaks communities. In *Mining Social Networks and Security Informatics*, pp. 79–105. Springer, Dordrecht, the Netherlands, 2013.

40. M. N. Schmidt, O. Winther, and L. K. Hansen. Bayesian non-negative matrix factorization. In *Independent Component Analysis and Signal Separation*, pp. 540–547. Springer, Berlin, Germany, 2009.

41. P. Latouche, E. Birmelé, and C. Ambroise. Overlapping stochastic block models with application to the french political blogosphere. *The Annals of Applied Statistics*, 5(1):309–336, 2011.

42. X. Wang, L. Jiao, and J. Wu. Adjusting from disjoint to overlapping community detection of complex networks. *Physica A: Statistical Mechanics and its Applications*, 388(24):5045–5056, 2009.

43. C. Lee, F. Reid, A. McDaid, and N. Hurley. Detecting highly overlapping community structure by greedy clique expansion. *arXiv preprint arXiv:1002.1827*, 2010.

44. S. Asur, S. Parthasarathy, and D. Ucar. An event-based framework for characterizing the evolutionary behavior of interaction graphs. *ACM Transactions on Knowledge Discovery from Data (TKDD)*, 3(4):16, 2009.

45. Y.-R. Lin, Y. Chi, S. Zhu, H. Sundaram, and B. L. Tseng. Facetnet: A framework for analyzing communities and their evolutions in dynamic networks. In *Proceedings of the 17th International Conference on World Wide Web*, pp. 685–694. ACM, New York, 2008.

46. M.-S. Kim and J. Han. A particle-and-density based evolutionary clustering method for dynamic networks. *Proceedings of the VLDB Endowment*, 2(1):622–633, 2009.

47. L. Wang, Y. Bi, W. Wu, B. Lian, and W. Xu. Neighborhood-based dynamic community detection with graph transform for 0-1 observed networks. In *International Computing and Combinatorics Conference*, pp. 821–830. Springer, Berlin, Germany, 2013.

48. L. Wang, J. Wang, Y. Bi, W. Wu, W. Xu, and B. Lian. Noise-tolerance community detection and evolution in dynamic social networks. *Journal of Combinatorial Optimization*, 28(3):600–612, 2014.

49. N. Pathak, C. DeLong, A. Banerjee, and K. Erickson. Social topic models for community extraction. In *The 2nd SNA-KDD Workshop*, volume 8, ACM, New York, 2008.

50. G.-J. Qi, C. C. Aggarwal, and T. Huang. Community detection with edge content in social media networks. In *2012 IEEE 28th International Conference on Data Engineering*, pp. 534–545. IEEE, 2012.

51. F. D. Malliaros and M. Vazirgiannis. Clustering and community detection in directed networks: A survey. *Physics Reports*, 533(4):95–142, 2013.

52. G. Tong, L. Cui, W. Wu, C. Liu, and D.-Z. Du. Terminal-set-enhanced community detection in social networks. In *IEEE INFOCOM 2016—The 35th Annual IEEE International Conference on Computer Communications*, pp. 1–9. IEEE, 2016.

# Index